# The Urban Book Series

**Editorial Board**

Margarita Angelidou, Aristotle University of Thessaloniki, Thessaloniki, Greece

Fatemeh Farnaz Arefian, The Bartlett Development Planning Unit, UCL, Silk Cities, London, UK

Michael Batty, Centre for Advanced Spatial Analysis, UCL, London, UK

Simin Davoudi, Planning & Landscape Department GURU, Newcastle University, Newcastle, UK

Geoffrey DeVerteuil, School of Planning and Geography, Cardiff University, Cardiff, UK

Jesús M. González Pérez, Department of Geography, University of the Balearic Islands, Palma (Mallorca), Spain

Daniel B. Hess, Department of Urban and Regional Planning, University at Buffalo, State University, Buffalo, NY, USA

Paul Jones, School of Architecture, Design and Planning, University of Sydney, Sydney, NSW, Australia

Andrew Karvonen, Division of Urban and Regional Studies, KTH Royal Institute of Technology, Stockholm, Stockholms Län, Sweden

Andrew Kirby, New College, Arizona State University, Phoenix, AZ, USA

Karl Kropf, Department of Planning, Headington Campus, Oxford Brookes University, Oxford, UK

Karen Lucas, Institute for Transport Studies, University of Leeds, Leeds, UK

Marco Maretto, DICATeA, Department of Civil and Environmental Engineering, University of Parma, Parma, Italy

Ali Modarres, Tacoma Urban Studies, University of Washington Tacoma, Tacoma, WA, USA

Fabian Neuhaus, Faculty of Environmental Design, University of Calgary, Calgary, AB, Canada

Steffen Nijhuis, Architecture and the Built Environment, Delft University of Technology, Delft, The Netherlands

Vitor Manuel Aráujo de Oliveira, Porto University, Porto, Portugal

Christopher Silver, College of Design, University of Florida, Gainesville, FL, USA

Giuseppe Strappa, Facoltà di Architettura, Sapienza University of Rome, Rome, Roma, Italy

Igor Vojnovic, Department of Geography, Michigan State University, East Lansing, MI, USA

Claudia Yamu, Department of Built Environment, Oslo Metropolitan University, Oslo, Norway

Qunshan Zhao, School of Social and Political Sciences, University of Glasgow, Glasgow, UK

The Urban Book Series is a resource for urban studies and geography research worldwide. It provides a unique and innovative resource for the latest developments in the field, nurturing a comprehensive and encompassing publication venue for urban studies, urban geography, planning and regional development.

The series publishes peer-reviewed volumes related to urbanization, sustainability, urban environments, sustainable urbanism, governance, globalization, urban and sustainable development, spatial and area studies, urban management, transport systems, urban infrastructure, urban dynamics, green cities and urban landscapes. It also invites research which documents urbanization processes and urban dynamics on a national, regional and local level, welcoming case studies, as well as comparative and applied research.

The series will appeal to urbanists, geographers, planners, engineers, architects, policy makers, and to all of those interested in a wide-ranging overview of contemporary urban studies and innovations in the field. It accepts monographs, edited volumes and textbooks.

**Indexed by Scopus.**

Fernando Carrión Mena · Paulina Cepeda Pico
Editors

# Urbicide

The Death of the City

*Editors*
Fernando Carrión Mena
Facultad Latinoamericana de Ciencias
Sociales
FLACSO Ecuador
Quito, Ecuador

Paulina Cepeda Pico
Facultad Latinoamericana de Ciencias
Sociales
FLACSO Ecuador
Quito, Ecuador

ISSN 2365-757X         ISSN 2365-7588  (electronic)
The Urban Book Series
ISBN 978-3-031-25303-4      ISBN 978-3-031-25304-1  (eBook)
https://doi.org/10.1007/978-3-031-25304-1

© The Editor(s) (if applicable) and The Author(s), under exclusive license to Springer Nature Switzerland AG 2023
This work is subject to copyright. All rights are solely and exclusively licensed by the Publisher, whether the whole or part of the material is concerned, specifically the rights of translation, reprinting, reuse of illustrations, recitation, broadcasting, reproduction on microfilms or in any other physical way, and transmission or information storage and retrieval, electronic adaptation, computer software, or by similar or dissimilar methodology now known or hereafter developed.
The use of general descriptive names, registered names, trademarks, service marks, etc. in this publication does not imply, even in the absence of a specific statement, that such names are exempt from the relevant protective laws and regulations and therefore free for general use.
The publisher, the authors, and the editors are safe to assume that the advice and information in this book are believed to be true and accurate at the date of publication. Neither the publisher nor the authors or the editors give a warranty, expressed or implied, with respect to the material contained herein or for any errors or omissions that may have been made. The publisher remains neutral with regard to jurisdictional claims in published maps and institutional affiliations.

This Springer imprint is published by the registered company Springer Nature Switzerland AG
The registered company address is: Gewerbestrasse 11, 6330 Cham, Switzerland

# Contents

**Part I Introduction**

1 Urbicide: An Unprecedented Methodological Entry in Urban Studies? .................................................... 3
  Fernando Carrión Mena and Paulina Cepeda Pico

**Part II Urbicide. The Death of the City**

2 Urbicide. The Liturgical Murder of the City .................... 25
  Fernando Carrión Mena

3 Death By Theory and The Power of Ideas: From Theories of Cities to "Smart" Cities ................................... 47
  Michael Cohen

4 Urbicide: Towards a Conceptualization ........................ 59
  Maria Mercedes Di Virgilio

5 Urban Order and Disorder. Genealogy of Urbicide .............. 77
  Eduardo Kingman Garcés and Susana Anda Basabe

6 Imaginaries and Archetypes on the Death of the City ............ 93
  Alfredo Santillán Cornejo

7 COVID-19 and the City: Reframing Our Understanding of Urbicide by Learning from the Pandemic .................... 107
  Roberto Falanga and João Ferrão

**Part III Annihilation: The End of the Public Space**

8 The Ideology of Public Space and the New Urban Hygienism: Tactical Urbanism in Times of Pandemic ....................... 127
  Manuel Delgado-Ruiz

9   The Transformation of Urban and Digital Spaces
    from a Democratic Perspective .................................. 145
    Arnau Monterde and Joan Subirats

10  Streets, Avenues, Highways ..................................... 157
    Pablo Fernández Christlieb

11  The Post-automobile City From Deterritorialization
    to the Proximity City: The Case of Madrid ...................... 171
    José María Ezquiaga and Javier Barros

12  Mobility as an Expression of the Urbicide: The Risks
    of Transport Modernization in Latin American Metropolises ...... 205
    Pablo Vega Centeno, Jérémy Robert, and Danae Román

**Part IV  Decline of the Built Environment**

13  The Urbanisation of Risk ....................................... 235
    Andrew Maskrey and Allan Lavell

14  Urbicide or Suicide? Shaping Environmental Risk in an Urban
    Growth Context: The Example of Quito City (Ecuador) ............ 263
    Jose M. Marrero, Hugo Yepes, Paco Salazar, and Sylvana Lara

15  Between Greens and Grays: Urbanization and Territorial
    Destruction in the Sabana de Bogotá ............................ 293
    Alice Beuf, Germán Quimbayo Ruiz, and Olaff Jasso García

16  Overregulation, Corruption, and Urbicide ....................... 315
    Vicente Ugalde

17  Obsolescence of the Built Environment .......................... 333
    Valeria Reinoso-Naranjo and Manuel Martín-Hernández

**Part V  Dissolution of Social Interaction**

18  The (Un)made City: Spatial Fragmentation, Social
    Inequalities and (De)compositions of Urban Life ................ 359
    Ramiro Segura

19  The City and the Abandonment of Public Space. Between
    Neoliberal and Citizen Urbanism ................................ 377
    Patricia Ramírez Kuri

20  A "New" Urban Colonialism? North–South Migration
    and Racially Structured Gentrification in Latin America ........ 395
    Juan Pablo Pinto Vaca

| | | |
|---|---|---|
| 21 | **Urban Frontiers in the Fracturing City: Heritage, Tourism and Immigration** .............................................. | 419 |
| | Jesús M. González-Pérez, Josefina Domínguez-Mujica, Margarita Novo-Malvárez, and Juan M. Parreño-Castellano | |
| 22 | **The Production of Emptied Places in the Borderlands of the Metropolitan Area of Buenos Aires** ....................... | 447 |
| | Andrea Catenazzi and Julieta Sragowicz | |

### Part VI  Degradation and Abandonment

| | | |
|---|---|---|
| 23 | **Reconstructing Cultural Paradigms. Experiences in East Europe: The Historical Memory of the Historical Centers in Lithuania** ................................................... | 467 |
| | Olimpia Niglio | |
| 24 | **Lose the Memory, Lose the History, Lose the City** ............... | 489 |
| | Salvador Urrieta García and Veronica Zalapa Castañeda | |
| 25 | **Revolt and Destruction. The Public and Monument Landscape in Latin American Cities** ....................................... | 511 |
| | Francisca Márquez | |
| 26 | **Trends of Urban and Territorial Reconfiguration in Metropolitan Buenos Aires** ................................. | 531 |
| | María Carla Rodríguez | |
| 27 | **Anatomy of an Urbicide. Social Housing in Santiago 1980–2006** ... | 555 |
| | Alfredo Rodríguez and Ana Sugranyes | |
| 28 | **Urbicide. A Look Through the Mirror** ......................... | 565 |
| | Inés del Pino Martínez | |

### Part VII  Destruction of Common Life: Violence

| | | |
|---|---|---|
| 29 | **The Besieged City: Geographies of Crime** ..................... | 585 |
| | Alfonso Valenzuela-Aguilera | |
| 30 | **Urbicide, Violence, and Destruction Against Cities by Criminal Organizations** ..................................... | 603 |
| | Arturo Alvarado | |
| 31 | **Discursive Understandings of the City and the Persistence of Gender Inequality** ......................................... | 637 |
| | Nora Libertun de Duren, Diane E. Davis, and Maria Lucia Morelli | |
| 32 | **Border Cities Between Life and Death: Ciudad Juárez and El Paso** ............................................. | 655 |
| | Mauricio Vera-Sánchez and Luis Alfonso Herrera-Robles | |

## Part VIII  Contraction of Public Management: Privatization

**33**  **The Metamorphosis of Infrastructure in Latin American Urbanization: From Insufficiency to Presence as Fictitious Capital** ................................... 673
Beatriz Rufino

**34**  **Public Policies (Or Their Absence) as Part of Urban Destruction** ............................................... 693
Marcelo Corti

**35**  **Metropolitanicide? *Urbs*, *Polis* and *Civitas* Revisited** .............. 707
Mariona Tomàs

**36**  **International Tourism, Urban Rehabilitation and the Destruction of Informal Income-Earning Opportunities** ..................................................... 725
Alan Middleton

**37**  **De-urbanization: From the Shock to the Revolution of a New Urban Logic** ..................................................... 751
Paulina Cepeda Pico

## Part IX  Urbicide: Cities Cases

**38**  **Grassroots Spaces Make London Exciting: The Relationship Between the *Civitas* and the *Urbs*** .............................. 779
Pablo Sendra

**39**  **Rio de Janeiro: The Trajectory of the Wonderful City, Violence, and Urban Disenchantment** .......................... 793
Mauro Osorio, Maria Helena Versiani, and Henrique Rabelo

**40**  **The Implosion of Memory. City and Drug Trafficking in Medellín and the Aburrá Valley** .............................. 813
Luis Fernando González Escobar

**41**  **Caracas. Urbicide and Precariousness of Urban Life at the Beginning of the Venezuelan Twenty-First Century. The Worst of Capitalism and Savage Populism** .................. 843
Alberto Lovera

**42**  **Santiago, the Non-city? Destruction, Creation, and Precariousness of Verticalized Space** ....................... 865
Loreto Rojas Symmes, Alejandro Cortés Salinas, and Daniel Moreno

**43**  **Neoliberal Urbicide in Barcelona. The Case of Ciutat Vella** ........ 891
Pedro Jiménez-Pacheco

## Part X  Epilogue

**44  Epilogue. Remake Us from Ruins, Collective Memories
     and Dreams** .................................................. 917
     Víctor Delgadillo

**Index** .......................................................... 939

# Editors and Contributors

## About the Editors

**Fernando Carrión Mena** is Research Professor in Flacso-Ecuador. His focus of study are the topics of: housing, urbanization process, city, historic centers, cultural heritage, violence, security and drug trafficking, borders, decentralization and sociology of football, among others. He created 8 thematic magazines (political science, security, city, historical centers, borders), wrote more than 1000 journalistic articles and 306 scholars, published 74 books (editor of 52 and author of 22) and edited 12 book collections (97 volumes). He produced 4 cinematographic documentaries. He has worked as a consultant for multilateral organizations and as a university professor. For his career, he won 9 awards, 6 decorations and 5 distinguished citizens distinctions. He was director of CIUDAD, FLACSO, Municipal Planning, as well as a former Councilor and former Advisor to the Municipality of Quito. e-mail: fcarrion@flacso.edu.ec; URL: https://works.bepress.com/fernando_carrion/

**Paulina Cepeda Pico** is Researcher at Flacso Ecuador. Architect and Master in Urban Studies with a research scholarship at Facultad Latinoamerica de Ciencias Sociales, Flacso-Ecuador. Senior architect for the development of heritage home restoration and rehabilitation projects, design and construction of architectural projects and urban planning consultant. In academics, her main lines of research are: housing policies, urban planning, urbanization process, governance, urban feminism, digital city, market and land policies. She has been an advisor to some urban actors. e-mail: pccepedafl@flacso.edu.ec; URL: https://flacsoandes.academia.edu/PaulinaCepeda

# Contributors

**Arturo Alvarado** El Colegio de México (Colmex), Mexico City, Mexico

**Susana Anda Basabe** Facultad Latinoamericana de Ciencias Sociales, Flacso Ecuador, Quito, Ecuador

**Javier Barros** Polytechnic University of Madrid, Madrid, Spain

**Alice Beuf** Geography Department, National University of Colombia, Bogotá, Colombia

**Fernando Carrión Mena** Facultad Latinoamericana de Ciencias Sociales, Flacso Ecuador, Quito, Ecuador

**Andrea Catenazzi** Universidad de Buenos Aires (UBA), Buenos Aires, Argentina

**Paulina Cepeda Pico** Facultad Latinoamericana de Ciencias Sociales, Flacso Ecuador, Quito, Ecuador

**Michael Cohen** Milano School of Policy, Management, and Environment, The New School Milano, New York, USA

**Alejandro Cortés Salinas** Universidad Alberto Hurtado, Santiago de Chile, Chile

**Marcelo Corti** National University of Córdoba, Córdoba, Argentina

**Diane E. Davis** Harvard University, Cambridge, MA, USA

**Víctor Delgadillo** National Autonomous University of Mexico (UNAM), Mexico City, Mexico

**Inés del Pino Martínez** Pontifical Catholic University of Ecuador (PUCE), Quito, Ecuador

**Manuel Delgado-Ruiz** Universitat de Barcelona, Barcelona, Spain

**Maria Mercedes Di Virgilio** Universidad de Buenos Aires (UBA), Buenos Aires, Argentina

**Josefina Domínguez-Mujica** University of Las Palmas de Gran Canaria, Las Palmas de Gran Canaria, Spain

**José María Ezquiaga** Polytechnic University of Madrid, Madrid, Spain

**Roberto Falanga** University of Lisbon, Lisbon, Portugal

**Pablo Fernández Christlieb** National Autonomous University of Mexico (UNAM), Mexico City, Mexico

**João Ferrão** University of Lisbon, Lisbon, Portugal

**Luis Fernando González Escobar** National University of Colombia, Medellín, Colombia

**Jesús M. González-Pérez** University of the Balearic Islands, Palma, Spain

**Luis Alfonso Herrera-Robles** Autonomous University of Ciudad Juarez (UACJ), Ciudad Juárez, México

**Olaff Jasso García** College of Geography, Faculty of Philosophy and Letters, National Autonomous University of Mexico (UNAM), Mexico City, Mexico

**Pedro Jiménez-Pacheco** University of Cuenca, Cuenca, Ecuador

**Eduardo Kingman Garcés** Facultad Latinoamericana de Ciencias Sociales, Flacso Ecuador, Quito, Ecuador

**Sylvana Lara** Metropolitan Directorate of Risk Management, Metropolitan District of Quito (DMQ), Quito, Ecuador

**Allan Lavell** Facultad Latinoamericana de Ciencias Sociales, Flacso Costa Rica and Risk Nexus Initiative (RNI), San José, Costa Rica

**Nora Libertun de Duren** Harvard University, Cambridge, MA, USA; Inter-American Development Bank (IDB), Washington D.C., USA

**Alberto Lovera** Central University of Venezuela (UCV), Caracas, Venezuela

**Francisca Márquez** Universidad Alberto Hurtado, Santiago de Chile, Chile

**Jose M. Marrero** Repensar, Quito, Ecuador;
Mayor's Advisor, Metropolitan District of Quito (DMQ), Quito, Ecuador

**Manuel Martín-Hernández** ESARQ of Guadalajara, Guadalajara, Mexico

**Andrew Maskrey** Coalition for Disaster Resilient Infrastructure (CDRI) and Risk Nexus Initiative (RNI), Málaga, Spain

**Alan Middleton** Birmingham City University, Birmingham, England

**Arnau Monterde** Open University of Catalonia, Barcelona, Spain

**Maria Lucia Morelli** Architecture and Urbanism (PAU), New York, USA

**Daniel Moreno** Pontifical Catholic University of Chile, Santiago de Chile, Chile

**Olimpia Niglio** Department of Civil Engineering and Architecture, University of Pavia, Pavia, Italy;
Faculty of Engineering and Design, Hosei University, Tokyo, Japan

**Margarita Novo-Malvárez** University of the Balearic Islands, Palma, Spain

**Mauro Osorio** Federal University of Rio de Janeiro (UFRJ), Rio de Janeiro, Brazil

**Juan M. Parreño-Castellano** University of Las Palmas de Gran Canaria, Las Palmas de Gran Canaria, Spain

**Juan Pablo Pinto Vaca** Autonomous Metropolitan University (UAM), Mexico City, Mexico

**Germán Quimbayo Ruiz** Geography Department, National University of Colombia, Bogotá, Colombia

**Henrique Rabelo** Federal University of Rio de Janeiro (UFRJ), Rio de Janeiro, Brazil

**Patricia Ramírez Kuri** Social Research Institute, National Autonomous University of Mexico (UNAM), Mexico City, Mexico

**Valeria Reinoso-Naranjo** Technical University of Ambato (UTA), Ambato, Ecuador

**Jérémy Robert** Rennes 2 University, Associated with the French Institute of Andean Studies, Lima, Peru

**Alfredo Rodríguez** SUR, Corporación de Estudios Sociales y Educación, Santiago de Chile, Chile

**María Carla Rodríguez** Universidad de Buenos Aires (UBA) and National Council for Scientific and Technical Research, Buenos Aires, Argentina

**Loreto Rojas Symmes** Universidad Alberto Hurtado, Santiago de Chile, Chile

**Danae Román** Pontifical Catholic University of Peru (PUCP), Lima, Peru

**Beatriz Rufino** University of São Paulo (USP), São Paulo, Brazil

**Paco Salazar** Inteligentarium, Quito, Ecuador

**Alfredo Santillán** Facultad Latinoamericana de Ciencias Sociales, Flacso Ecuador, Quito, Ecuador

**Ramiro Segura** National University of La Plata, La Plata, Argentina; National University of San Martín, San Martín, Argentina

**Pablo Sendra** The Bartlett School of Planning, University College London (UCL), London, UK

**Julieta Sragowicz** Universidad de Buenos Aires (UBA), Buenos Aires, Argentina

**Joan Subirats** Autonomous University of Barcelona, Barcelona, Spain

**Ana Sugranyes** SUR, Corporación de Estudios Sociales y Educación, Santiago de Chile, Chile

**Mariona Tomàs** Department of Political Science, University of Barcelona, Barcelona, Spain

**Vicente Ugalde** Center for Demographic, Urban and Environmental Studies, El Colegio de México (Colmex), Mexico City, Mexico

**Salvador Urrieta García** School of Engineering and Architecture, National Polytechnic Institute (IPN), Mexico City, Mexico

**Alfonso Valenzuela-Aguilera** Autonomous University of the State of Morelos (UAEM), Cuernavaca, Mexico

**Pablo Vega Centeno** Pontifical Catholic University of Peru (PUCP), Lima, Peru

**Mauricio Vera-Sánchez** National Open and Distance University (UNAD), Bogotá, Colombia

**Maria Helena Versiani** Federal University of Rio de Janeiro (UFRJ), Rio de Janeiro, Brazil

**Hugo Yepes** Mayor's Advisor, Metropolitan District of Quito, Quito, Ecuador

**Veronica Zalapa Castañeda** School of Engineering and Architecture, National Polytechnic Institute (IPN), Mexico City, Mexico

# Part I
# Introduction

# Chapter 1
# Urbicide: An Unprecedented Methodological Entry in Urban Studies?

**Fernando Carrión Mena and Paulina Cepeda Pico**

> *If the sidewalk ends up favoring the automobile over the pedestrian, the street dies and the end of the city begins.*
> *Jane Jacobs (paraphrasing).*

**Abstract** This chapter introduces the debate of Urbicide as a methodological and theoretical way to understand urban studies. Cities are one of many urban territories and are the most relevant, with a social and historic construction. The configuration process of cities has an origin and also an apparent end. But the urban policies and planning do not consider the process of destruction of cities and try producing utopian cities. This chapter opens the debate about the way that urbanization and urban development are conceived to cause the death of today's cities and principally poses as a methodological theoretical framework from three central inputs: denial of denial, socio-spatial adjustments and transformations, obsolescence and destruction of the city, which then opens up into six components of analysis (annihilation, deterioration, dissolution, degradation, destruction and contraction).

**Keywords** Urbicide · Urban studies · City theory · Methodology

## 1.1 Introduction

The twenty-first century has been cataloged by the United Nations as the century of cities, due to the amount of agglomerated population that inhabits them. Globally, 57% of inhabitants live in cities, that is, more than 4.2 thousand million people.They also concentrate the greatest economic activity (80% of world GDP)

F. Carrión Mena (✉) · P. Cepeda Pico
Facultad Latinoamericana de Ciencias Sociales, Flacso Ecuador, Quito, Ecuador
e-mail: fcarrion@flacso.edu.ec
URL: https://works.bepress.com/fernando_carrion/

P. Cepeda Pico
e-mail: pccepedafl@flacso.edu.ec

(World Bank[1] 2022); acquiring a strong protagonism even comparable to that of States and large global corporations, according to Sassen (2001).

Nevertheless, from this relevant weight that the city has, the debate of its crisis and decadence is proposed, until the limit of the topic of the death of the city, due to the aggressions that it suffers, as well as the deep and structural transformations that it goes through. These phenomena have arisen from certain general, spatial or sectorial components, as well as from its most intimate and specific structures.

Therefore, we have some exemplary cases that have occurred systematically throughout this twenty-first century. According to the Mayor of New York, Bill De Blasio, September 11, 2001, is *"the darkest day of our city"*. He said this because of the attack on the Twin Towers of The World Trade Center by an incursion of three commercial flights that left the area in ruins and took the lives of 2,996 people and left around 2,680 people injured. In 2005, Hurricane Katrina destroyed New Orleans leaving floods, thousands of deaths and a loss of more than 125 thousand million dollars. Between 2000 and 2020, the foundational historic center of Quito, as in other cities, reduced its population by an annual average of 2.5% by forced eviction, emptying the society, reducing its time of use and reducing space in 2015, according to UN-HABITAT, 24% of the global urban population lives in informal human settlements or suburbs, it means, out of the city.[2] In February 2022, Russia carried out a targeted tactical bombardment of the largest cities of Ukraine. This resulted in destruction, death and mass expulsion of citizens. In May 2022, UN Commissioner Michelle Bachelet, visited Mariupol, stated that *"a city that was once prosperous, now lies in ruins"*.

Undoubtedly, it is about references of the destruction and death of cities around the world, with the aggravating factor that they are neither unique nor isolated cases. Oppositely, it generates a global effect that influences other territories. But they are even more representative because they impact the urbes in different ways, which are the most significant artificial asset produced throughout the history of humanity.

War aggressions, negative urban processes, the modification of ecosystems and historical transformations have generated devastating impacts on certain sectors of cities, as well as on the totality of some of them. Indeed, the public space—which is the city—is undergoing a process of dissolution and agoraphobia,[3] due to the monopoly of certain functions, to the rupture of diversity (Jacobs 2011), to the violence that reduces social interaction (Carrión and Velasco 2018), to the logic of the market that expels population (Abramo 2012); to the damage of the ecosystem in which it is inscribed, producing fires, floods and countless disasters (Lavell 2000) and finally the wars that rage against cities because of their cultural, social, economic and political weight (Graham 2002).

---

[1] World Bank: https://www.bancomundial.org/es/topic/urbandevelopment/overview.

[2] Suburb is a compound word, where *sub* is a prefix meaning: low, below or inferiority; and *urb* refers to urbe, that is, less than city or, as defined by the DRAE: "population nucleus located on the outskirts of a city".

[3] Borja (2003) states that it is a "class" disease that reinforces a segregating and denying discourse of the city.

These destructive facts have been brought to the attention of the media, universities, political parties, citizens, social movements and academia. This has led to severe questioning from citizens in general. But these phenomena are also expressed socially through a collective complaint that grows in the citizens' claims and, above all, in the presence of certain vindicating and contesting collective projects of the city. As a consequence of this perspective and this conjuncture, there is a confrontation of two predominant and different models of the city: one, governed by the sense of the market and the other by the meaning of the utopia of *living well* or of the respect for the citizen condition, inscribed in the right to the city (Lefebvre 2020; Harvey 2013).

The processes of selective and massive destruction of cities take place without the impediment of local, national and international institutions in charge of ensuring their existence and evolution. Such is the case that they have remained undaunted by the impact of wars, urban violence, the emptying of spaces, the expulsion of activities and people and climate change, among others, that is, of governments that maintain complicit silences, as well as, in many cases, direct, consistent or inconsistent public policy actions that lead to the deterioration of the city, causing, as a whole, both urbicide (destructive process) and urbanicide (process of complicit silence), as Carrión argues in his chapter of this book.

This institutional quality questions their structural capacities to counteract the negative impacts that occur in the city, either because they have entered into the logic of the minimal state or because the classic and traditional paradigms of action do not contemplate this problem. This leads to a type of government where conflict is present. On the one hand, between those who advocate the weight of the private world (accumulation), and on the other, those who demand the relevance of the public (democratization), both in the exercise of government and in its own institutional structure.

These transformations have a temporal and historical configuration, since a long time ago, but also spatially, distributed throughout the planet and converted into an essential part of the content of the urbe. Therefore, they are evident now in deep structural processes of destruction, which lead to the following questions that roam the world: Why and how do these destructive phenomena occur in cities? Why do institutions do very little to neutralize them? What is the social and private response to these processes?

This handbook seeks a collective response to these questions, beginning with a call for attention to this destructive fact that occurs on a daily basis and unnoticed. The publication does not seek to present this phenomenon in a pessimistic and unalterable way; on the contrary, it is a warning that seeks to be—from a critical point of view—a contribution to the orthodox approaches of urbanism and urban development, since cities are the main scenarios of progress and also of conflict and destruction, assuming the condition of a contradictory political community, which simultaneously develops and could die.

With this handbook called "*Urbicide or the death of the city*", we want to contribute to the knowledge of cities, but provoking a methodological turn within its hegemonic vision. First, overcoming the understanding of the city as if it were an exclusively physical material object; and second, from the design of public policies born exclusively from *urbanism*[4] and *urban development*[5] aimed at monopolistic and market logic.

Therefore, the purpose is to open a critique to the concepts of *urban development* and *urbanism*, which have been pigeonholed in mercantilized and unilateral trends, by adding them a new double connotation. On the one hand, by means the logic of *creator or creative destruction* (Schumpeter 1994) through the historical perspective. In other words, a process of decay of the old city (urbicide) that leads to generate a new one, negating the previous one. And, on the other hand, from the structural perspective of the city, as a social and political community, where common sense, social interaction and respect for collective rights are systematically eroded. These two connotations are framed within an incoherent recognition of the transformation of the territorial configuration or scale (city, region, system).

In view of this, we seek to deepen our knowledge of the dynamics of the creative power of negation, in order to understand the causes of its destruction and, based on them, to create the conditions of life, of "y" in the city, not only from the perspective of innovation, but also of the change of meaning that should follow urban development. It is, in other words, to draw attention to how the existing city (memory, buildings, population) is destroyed and eroded under explicit liturgies: that is, of an order with recognizable logics, actors and forms. This methodological turn will allow to look at the city from what is lost and is destroyed, not only in a critical way, for the sake of supposed "superior" values coming from politics, culture or economy, which end up by functionalizing the own institutions behind these purposes; and not to promote strategies aimed at mitigating the destructive effects, in order to begin (re)building a real city and urban condition to live better.

To this, the first thing to be discussed is the meaning of *urban development*, which cannot be conceived under a quantitative notion of growth, for example, demographic (range-size), economic (real estate investment) or territorial (urban sprawl, compact-dispersed city, verticalization). Rather, it should be aimed at solving the unsatisfied needs of the population, reducing inequalities and promoting good living. The second point to discuss is *urbanism*, so that it ceases to be a discipline that promotes urban development from the exclusive perspective of territorial *production and consumption* (territorial ordering), dialectically adding it the focus on the social, that is, on social justice and *civility* (Carrión and Cepeda 2021).

A logic of *urban development* and *urbanism* that breaks the social, economic and environmental balance leads, for example, to the erosion of the past in the form of memory, accumulated acquis[6] or the production of oblivion. Also to the logic of consumer freedom that segments markets (fragmentation), and denies the principle

---

[4] Urbanism was born at the time of the industrial revolution and its object is the study and intervention in the urbs; that is, in the organization of the territory and the own services of the city.

[5] Urban development, also a tributary of the industrial revolution, is understood as a set of technical and administrative measures aimed at territorial planning, without ignoring the production of services (health, education) and infrastructures (mobility, energy), as well as the control of spatial and population expansion.

[6] Acquis: a set of moral, cultural or natural assets accumulated throughout history.

of urban unity. In addition to the expansive or densified urban growth that violates the environment, producing, in turn, phenomena that revert against the city itself, destroying or even killing it. Finally, it leads to the construction of stereotyped cities, governed by the market and masculinities, due to planning that denies social and functional diversity.

In synthesis, they should focus more on analyzing and assuming the contraction, destruction and involution experienced by urbs and, therefore, their inhabitants. This means focusing less on the production of buildings and infrastructures and more on the effects of exclusion, expulsion and quality of life; less on consumerism and costs of services and more on the impacts generated by pollution (lack of recycling, watersheds that supply water to cities and then receive sewage); less on conservation that freezes history and more on *productive heritage* that adds value to history (Carrión and Cepeda 2022a); less on the dynamics of urban production and more on the logics of reproduction, which contribute to the quality of life of the population.

In this sense, the concept of *urbicide*, which was born in that decade of the sixties, refers to violence against cities, whether due to military strategies,[7] environmental effects or mismanagement. This concept structures the narrative of the liturgical murder that urbes experience, when aggressions are committed with premeditation and malice aforethought, by identified actors. In short, it is structured with the intention of understanding the practices aimed at the production of oblivion, the erosion of the material foundations of the city and the destruction of common life.

In this way, the concept of *urbicide* begins to interpret urban processes from the social sciences, with phenomena that were already present a long time ago and that produce contemporary structural crises. This is the case of what it meant at the end of the nineteenth century in Paris with Haussmann or in Barcelona with Cerdá; but also in the USA, with the penetration of the automobile and in Latin America, with the process of *urbanization without a city* (slums, young towns, popular colonies) since the middle of the last century.

This set of processes seeks to be scientifically understood in order to counteract them. To this purpose, this handbook is an attempt to generate a process of systematization of different thematic inputs, arising from the confluence of different voices from Europe, USA and Latin America. In this sense, it is a collective, theoretical and methodological construction of *urbicide*, born from the convergence of common efforts.

---

[7] Atrocious cases such as those of Hiroshima and Nagasaki in World War II, of Guernica after Franco's fascism, or of US President Bush's preemptive wars. The ex-mayor of Belgrade, Bogdan Bogdanovic, pointed out that the wars in the Balkans were anti-urban, with the aim of undermining the cultural values concentrated in the cities.

## 1.2 The Urbicide: New Methodological Entry for Urban Studies

Assuming the concept of *urbicide* implies a change in the method of analysis and understanding and, moreover, of acting in the city. It is a contemporary paradigm that highlights the *destructiveness* that emerges in the production of the urbe and, therefore, questions the traditional way in which it is understood and managed. It seeks to explain what is lost, what is destroyed and, above all, the reasons why all this occurs, that is, the causes of its own destruction, in order to act upon them. And this is not a recent, punctual or isolated process, because it has history and is structural.

The process of urbanization is generated by the logic of agglomeration of population and activities in specific territories, while the city is directly related to the production of public space, services and its collective condition. In this, amalgam can be, strong tensions, conflicts and contradictions, where the city and its vital cycles are affected. For example, with the industrialization process that has occurred since the mid-nineteenth century in Europe or since the Second World War in Latin America, the acceleration of migration from the countryside to the city has produced vast areas that cannot be considered, strictly speaking, cities, but on the contrary, only human settlements, as defined by United Nations Habitat.[8]

In this historical process, the city, understood as the space of agglomeration and social interaction, goes through multiple stages that can even lead to its death, as in fact happens, because it is a living organism. So, for example, if it loses its internal and structural conditions, such as life in common, the death of the city sets in. But also, if its condition as a living organism is not assimilated, it will not be understood that it is a process in which death is present. That is, the explicit practices that certain specific actors have to lead to the death of the city are unraveled. In general, they are carried out under established or liturgical practices in accordance with an explicit order, leading to the destructive end of the essence of the city.

The urbicide, then, is a death or demise of a being that has life and that, therefore, can be susceptible to losing it. Moreover, because everything that lives has an intrinsic value inscribed in the meaning and principles of ontology (being and its properties), that is, in a condition of existence that is expressed according to its social, spatial and temporal expressions. Ontology, as a principle of the strata of being, is historically established through a close relationship between social and environmental aspects to conform an ecosystem, because the city does not exist independently of society (community) and of the place where it is produced (nature). In this way, the community (society) and the territory (place) are the result of an indissoluble historical process.

In addition, cities concentrate the largest number of people on the planet, which humanizes them. This implies overcoming the orthodox and hegemonic conception

---

[8] Human settlement is a place where population groups are established in precarious shelters. These places do not become neighborhoods or part of the city, because they lack services, infrastructure, housing and public space. They are informal or illegal, which makes public policies ignore their existence.

of the city, inscribed in the rationality of an inanimate physical object or inert material (urbs), composed of a territory (urban structure), buildings (urban functions) and a set of infrastructures (material base of the city). In this way, it is indispensable to incorporate the essential condition of citizenship (civitas), which is formed in the public space (urbs), in order to make the city what it really is: a *political community (polis)*.

The starting point for this statement comes from the fact that men and women live in groups or communities, as they are social and political animals, due to their inability to solve their problems in isolation or individually. But they do so not only to survive, as is the case with animals, but mainly *to live well*. Therefore, it is not just any community, it is a *political community (polis)*, because the city is not the simple aggregation of people or families, but rather the presence of citizens (civitas), with its rights and duties (sovereignty), governed by regulated institutional frameworks.

Hence, the political community is contrary to the individualism that, for example, advocates the market (consumer sovereignty) as well as coming from the contraction and incoherence of the public (minimal state) and from the limitation of the use of public space (disturbance of social interaction). On the contrary, the city, understood as a political community, shows the ways and forms in which urban society constructs the modalities of coexistence among citizens and their ties with the polis, as a territorial and political unit of representation. This is defined by Lefebvre (2020) from the right to the city, both in its use and in the participation of its actions.

The city is a social construction born from the societal nexus that comes from conjugal life (home), domestic space (house), village dynamics (community) and citizen formation (public space) under a specific purpose: to live well. In this manner, the city is configured as the common place or town hall, where the house, the village and the public space coincide at the same time to form the political community, with the purpose of producing "the happy and virtuous city proposed by Aristotle (Carrión and Cepeda 2021).

If the purpose of living well is not achieved collectively (city of the commons), an *urbicide* process is structured, with which, human rights as a whole are violated, particularly the rights of nature (environment) (Acosta and Martínez 2009), the rights of the city (Carrión and Cepeda 2021) and the rights to the city (Lefebvre 2020; Harvey 2013; Carrión and Dammert 2018). If this happens, the rights of citizenship as a group are violated, because their lives are associated, such that if one of them fails, everything collapses; then cities lose many of their virtues, conditions and vital characteristics.

If citizenship acquires the rights OF and TO the city, as is already happening per se,[9] this means, on one side, that the land and the environment receive an explicit social function, which means that they establish a relationship between society and the ecosystem in a univocal and balanced way, that is, of the city with nature and

---

[9] The right to the city has gained ground in Latin American societies, first from the institutional perspective, when it was enshrined in the constitutions of Brazil (1988), Ecuador (2008) and Mexico City (2018). But also internationally within The New Urban Agenda (2016) and the Habitat International Coalition (HIC).

vice versa, and on the other side that the urbe be capable of self-regulating, which supposes having its own resources, competences and policy: self-government. It is with autonomy; otherwise, it is impossible to satisfy the rights to/from the city.

Therefore, when this happens, that is, when the collective rights of the citizenry are violated, there must be a *political community* capable of interpreting urbicide as a crime, in order to correct the behaviors which are opposed to living well and to the common sense of the polis (political community). This means, in an unpostponable way, having to typify urbicide as a crime, included in the penal codes of our cities and countries, because it is not a common homicide but a murder. In the meantime, it is worth asking: Who is responsible for the deterioration, destruction, abandonment, erosion of the city? Who should assume the costs of all that this implies? What cities are emerging after these crises?

These phenomena have been little studied, although they have some important antecedents that should be highlighted. Francoise Choay (2009) states that since the beginning of the twentieth century, some European cities began a process of deconstruction, through certain factors related to urbanism, which provoked a sudden modernity with, for example, the new road axes in Paris (Haussmann) or the widening of Barcelona (Cerdá). Along the same lines, but in the United States of America, Jane Jacobs (1973) analyzes life and death in cities, based on the weight of the motor car, the predatory verticalization and the insecurity of certain urban places in New York City. Likewise, some analyses have tried to explain what is lost, due to the effects introduced by: real estate investment (Huxtable 2010), destruction by dispossession of housing and population expulsion (Berman 1987). These are analytical processes that study the erosion or negative impact on cities, but which seek an explicit theoretical interpretation, because they remain in the observation and description of the facts. But, in addition, they cannot find a common dialectic that intertwines them and leads them in the same direction, both analytically and resolutely.

Subsequently, since the second half of the last century, new analyses appear describing the processes of negation of the previous urbanization by the installation of a new one, provoking the destruction of the old. But it will be in other cases, such as those of the wars, that this new theoretical approach is explicitly consolidated. There appear authors such as Coward (2006, 2008) and Graham (2002) who interpret the violent destruction of the material and human in the urbs, taking the example of the Aleppo wars in Syria. Likewise, the phenomena of terrorism is analyzed, especially of the Twin Towers in New York-USA (Shaw 2000; Campbell et al. 2007), something similar but not explicit is the case of Sendero Luminoso in Lima, as well as the violence in Medellín-Colombia or Celaya-Mexico, as in other cities of the world, where there have been relatively similar phenomena.

In recent times, some authors, such as Carrión (2014a, b), have added to this trend the processes of gentrification, colonization of memory and de-urbanization. To these thematic inputs, it would be necessary to include the negative impacts coming from the natural phenomena of volcanic eruptions (Armero–Colombia), earthquakes (Santiago–Chile), climate change (New Orleans–USA) or even plagues (Spanish Flu) and pandemics (COVID-19) (Herscher 2006; Carrión and Cepeda 2022b), among others. With all these thematic entries, the multidimensionality of the phenomenon and of the concept of urbicide appears.

In this way, the characteristics of the death of the city have been present throughout history, albeit with different expressions. Perhaps, for this reason, at least three organizing inputs of the area can be found:

- The first, coming from the explicit phenomenon of the *negation of the negation* (Hegel 1977), as it occurred throughout history when a city dies to give birth to the appearance of a new one. This process expresses the successive character of development through a process of substitution or negation of the old. But it should also be noted that in many cases vestiges of the previous phase remain to form what could be called a *palimpsest*.[10] This happened, for example, with the cities of the Inca Empire, such as Cusco and Quito, when, during the Spanish conquest, they were (re)founded with other logics that denied the previous ones. It also happened with the end of the walled city, which gave way to the emerging modern city, with the rupture of the limits or walls, to open up to what would later be known as the urban region or metropolitan area. These processes are inscribed, following Borja, in that: *"Periodically, when change seems to accelerate and is perceptible in the expansive forms of urban development, the death of the city is decreed"*.
- The second, born from the *spatiotemporal adjustments and transformations* (Harvey 2013), where the generalized problems of the presence of motor transport stand out, which breaks the pedestrian and proximity logic of the city, causing the development of *suburbs* (that is, less than a city), of *dormitory cities* (expression of the non plurifunctional nature of the urbe) or of *car cities*, where large road axes for motor vehicles dominate over demands on a human scale (North American and European urban highways). These phenomena occurred at the end of the nineteenth century and beginning of the twentieth century, which continue to this day, provoking the destruction of social interaction and of the city's own dynamics.
- The third, originated by the *obsolescence or destruction of the essence of the city*, where, on the one hand, external phenomena stand out, such as wars that use them as scenarios or places to settle political, cultural and economic conflicts. But also, on the other hand, internal events, born of the modification of *ecosystems* (pollution, climate change), of the sense of community on which the production of real estate projects is based according to profitability, located in the peripheries or in urban centers (densification and gentrification); and of the impulse of the dynamics of consumer sovereignty that produces fragmentation and segmentation. It is about intrinsic processes to the city, which took place 40 years ago, originated in the decade of the eighties and inscribed in the globalization process, which tend to erode the original sense of city as civitas.[11] For they were alerted more than

---

[10] According to the DRAE, Palimpsest is: "Ancient manuscript that preserves traces of a previous writing that has been artificially erased".

[11] Civitas, from Latin, means city, that is, the organized community where citizenship is integrated and where it satisfies their rights.

half a century ago by Lefebvre (2020), when he stated: *"The city is, therefore, everything that is experienced, known, represented, built or destroyed as a city."*

With this background, the current peculiar destruction of the city, which has never been so evident, is analyzed. In this process, these three explanatory ways converge and they do it from the concept of *urbicide*, as an analytical category. The purpose is to interpret and explain the processes of annihilation, destruction, degradation and contraction that have been occurring within cities, from a multidimensional perspective. Additionally, to understand these processes within dynamics that occur deliberately and not by chance or by natural logic, because behind them there are social, economic, public and cultural forces that explicitly drive them, with clear logic and order.

Therefore, it is about different forms of *un-urbanization* or denial of urbanization, which tend to question the meaning of *urban agglomeration*, through the modality of *sub-urbanization*, which is nothing more than an expression of something inferior to the city, deduced from the prefix *sub*, which produce precarious human settlements in Latin America, and of high income in the United States of America and Europe (dormitory cities, which are not cities). What is clear is that under no point of view it is the production of a city that allows its inhabitants to live well.

As a counterpart, there are the compact zones within cities, which are forming large interstitial voids, increasingly extensive and complex, in the manner of what Marc Augé (2008) defined as the *non-places* and Carlos de Mattos (2010) as the *artifacts of globalization*, which managed to attract strong capital investments and produce important urban brands, but not collective processes of urban integration, denying, then, the sense of city.

Finally, during its evolution, the existing city is denied in order to give birth to a new one. This happens in certain spaces within a city, such as those native or foundational places, where local society tends to reject, stigmatize or abandon them, configuring what could be denominated as *urban patricide*, that is, a process of denial of the historical origin of the city or of a particular space. It is a way of negating the past, whether remote (origin) or recent (modernity). This is the case of historical centers, ports or beaches, from which cities initially emerged, not only do they turn their backs on them, abandon them, but also degrade or kill them in the collective imagination. This is a process where cities tend to deny their original act, physically or symbolically, producing deterioration, destruction and denial.

## 1.3 The Structure and Content of the Book

This publication has an expository logic that is structured in ten sections, in which each of them represents a thematic approach to the concept. It begins with a first theoretical–methodological approach to the process under several inputs and finishes an epilogue that seeks to systematize the process in order to open the debate to an inverse logic of how the city has been thought, but with an optimistic sense. In the intermediate sections, the different topics that articulate this methodology are developed in a direct way.

This means that the book should be read in its totality and under the proposed sequential order. With this, a complete and comprehensive conceptual construction of the phenomenon is structured, that is, a theoretical notion, anchored in multiple heterogeneous realities, built from an inductive point of view, from different themes (environmental, economic, social, institutional), as well as types of cities (European or Latin American), which coexist as a whole. It is, then, the result of a collective construction project, where the parts integrate into the whole, to produce this theoretical and methodological contribution for the understanding of cities, from contemporary perspectives and paradigms.

The book starts with an exhaustive reflection in theoretical and historical terms to subsequently identify the processes that have followed the construction of the category urbicide, in order to place it in the present time. It continues with the sections that show the different dimensions of urbicide, exposed in the different chapters, which follow the logical sequence of the exposition, to later describe, in its interior, the articles that compose it which emphasizes that the intention of the book is the collective construction of the concept under its theoretical and historical discussion, as well as its specificities occurring in reality.

In that perspective, the book is structured with the following chapters or dimensions of analysis: ii. positioning of the concept and its multidisciplinary and epistemological inputs; iii. annihilation of commonplace systems, it is, of public space; iv. deterioration of the material basis of the city; v. dissolution of social cohesion as gentrification and inequality; vi. degradation and abandonment due to social conditions and denial of history; vii. destruction of common life by violent phenomena (wars); viii. contraction of institutional frameworks such as deregulation and privatization; ix. significant cases of study. This is in addition to the introduction and epilogue.

## 1.3.1  Part 1. Introduction

The handbook "*Urbicide or the death of the city*", begins with a study that seeks to situate this thematic as a methodological theoretical framework from three central inputs: denial of denial, socio-spatial adjustments and transformations, obsolescence and destruction of the city; which then open up into six components of analysis (annihilation, deterioration, dissolution, degradation, destruction and contraction). This construction is framed within the current debates on the city, relating them both to the literature and to the critical positions they carry. Finally, it includes the presentation of the theme and the general characteristics of the publication.

## 1.3.2 Part 2. Urbicide. The Death of the City

The death of the city has a liturgical dynamic that is instrumented from both physical and symbolic perspectives, that is, from the material and the urban imaginaries, in an integrated, historical and structural manner. In this first part of the book, the basis of the paradigm of urbicide is constructed from six different but complementary approaches. It starts with a theoretical perspective assumed by Michael Cohen and Fernando Carrión, then with the historical entry elaborated by Eduardo Kingman and Susana Anda. Subsequently, it continues with the vision of the crisis developed by Mercedes Di Virgilio, then with the presentation of the theme of disaster (COVID-19) by Roberto Falanga and Joao Ferrao, and concludes with the vision of urban imaginaries by Alfredo Santillán.

The construction of ideas and urban theories structure the scenarios that, according to Michael Cohen, can generate the life or progressive death of the city and, within it, of the citizens. Thus, the way in which the city is assumed by public and private actors makes them responsible for the strong impact on urban production and/or destruction. Highly changing and vulnerable urban territories develop multiple moments of crisis that, according to Mercedes Di Virgilio, become a challenge for Latin American cities. There, the housing crisis, which is structured under the neoliberal thesis, as an effect of the constant urban growth, does not mean processes that allow a dignified condition to house citizenship. Likewise, Roberto Falanga and Joao Ferrao suggest that the impact of pandemics in cities must be assimilated under certain public powers, which generate ambivalent processes, both destructive and constructive, toward the management of disasters of these global characteristics.

In that sense, Fernando Carrión suggests that cities are crossed by multiple threats originating in the territorial, symbolic and social spheres, leading to dramatic effects that generate forces of social resistance. In this sense, cities are built and destroyed, according to Eduardo Kingman and Susana Anda, to contribute to the proposal of urbicide from the historical perspective, and with it analyze the breaking point between urban order and disorder, imprinting a distinctive seal on governance. So the processes of destruction that accompany the historical condition of the city have an important dimension in the urban imaginaries, according to Alfredo Santillán's contribution, which adds a *moral dilemma*.

## 1.3.3 Part 3. Annihilation: The End of the Public Space

If public space is the city, it is experiencing its annihilation and devastation as a result of the processes of privatization, violence and exclusion, in the form of agoraphobia (Borja and Muxí 2003). The reality of the agony of the street due to the monopolistic use of the motor vehicle is evident, as well as the transfer of physical–material activities to the digital world of technological platforms, which have a global, virtual and private condition. In this sense, the group of chapters that make up this section discuss several turning points, all related to public space.

According to Manuel Delgado, public space is managed as an ethical scenario of good citizenship, with a strong idealistic tendency of morality, leaving aside its conflictive essence. Then, certain logics of tactical urbanism appear as emergent, generating a conflict between urban culture and urbanistic culture, where the aesthetic and formal order the space, annihilating it, because they leave aside its real conditions to live in.

On the one hand, the speed of cities generates inequitable ways of using and moving through them. According to Pablo Vega Centeno, Jeremy Roberts and Danae Roman, poorly designed sustainable mobility initiatives affect the daily mobility of citizens, as well as the lack of resources, control and appropriate public policies, since most of them reproduce the mercantile trend, which reduces the options for an adequate mobility for certain vulnerable groups. On the other hand, the direct relationship between urban production and the exponential growth of large mobility infrastructures is part of the displacement of weaker activities to the peripheries and the colonization of new urban territories. According to José María Ezquiaga, these models of individual mobility generate consumerism and degradation.

This monopoly also has a territorial and temporal base that is built from the street, the avenue and the highway, according to Pablo Fernandez Christlieb, when it structures the social condition from these infrastructures that make it impossible to experience the city. Finally, according to Joan Subirats and Arnau Monterde, in the cities the transit of people, events and flows must be assumed in their global condition, for being immersed in a world of digital networks, so understanding this double condition or analogy allows us to analyze the trajectory that cities take when they enter into these logics of tension.

### 1.3.4 Part 4. Decline of the Built Environment

The relationship between the city and its environment builds an ecosystem that generates constant imbalances, especially when the city and its urbanization process deteriorate its surroundings, causing its own destruction. To understand this dynamic, five moments stand out, addressed in this part: the urban production of risk, described by Maskrey and Lavell; the construction of vulnerability by Marrero, Yepes, Salazar and Lara; the destruction of the natural territory by Beuf, Garcia and Quimbayo; the obsolescence of the urban material landscape by Reinoso and Martin; and the incoherent regulations to the physical, expansive and densified production by Ugalde.

Capitalist urban production is a generator of risk, according to Andrew Maskrey and Allan Lavell, which is produced, accumulated and made visible in the urban structure, being an endogenous historical process. Therefore, risk and vulnerability in urban environments are factors created by human and collective action, as shown by Marrero, Yepes, Salazar and Lara, producing extreme situations, which come to qualify them as *kamikaze (suicidal) cities*. This process of accelerated urbanization at the beginning of the cities is still developing with new and particular capitalist logics. Large residential projects driven by private dynamics and located in protected areas, according to Alice Beuf, Jasso Garcia and Quimbayo, produce a high level of environmental degradation that simultaneously deteriorates the city and the environment.

This urban spillage constantly mutates, producing the abandonment of territories where the relationship between urbanism and architecture, according to Reinoso and Martin, enters into conflict, producing obsolescence and expiration of material goods. In this and any other process, the regulatory function of the city government determines the course of the relationship between city (artificial) and nature (originary), leading to Ugalde's reflection in which he describes the permissive and incoherent effect of urban regulations that is responsible for the cyclical process of urban deterioration.

## 1.3.5 Part 5. Dissolution of Social Interaction

Cities behave like a factory that produces inequalities, preventing spatial and social integration. In this sense, the neoliberal logic of the city is determinant, as it dissolves into a constellation of discontinuous spaces. This social dissolution is evidenced by the loss of rights, the reproduction of inequalities, the segregation that expels the poor, the urban voids that generate social ruptures and the contemporary fractures caused by demographic and migration trends. In this sense, Patricia Ramírez Kuri analyzes public space and describes how the influence of urban neoliberalism destructs historical and cultural configuration of the cities and causes social damage. But the citizens create forms of articulation and vindicate rights.

According to Ramiro Segura, the city is considered an organic form within the different modalities of existing contemporary urbanization, in this sense the city *is made and unmade*, and the author proposes that, from a double dynamic, on the one hand, the social and spatial powers that unmake the city and, on the other hand, the ways of remaking the city, the (dis)composition of urban life is generated. New demographic trends, in terms of aging and migration, transform territories and deploy new vulnerabilities. In this sense, societies that are at risk (social, political, natural) have developed important urban exodus processes in the last century, but opposite to this there is a strong movement of societies with purchasing power to less developed and more economic territories.

For Juan Pablo Pinto, the trend of relocation of retired U.S. migrants to Latin American cities generates processes of segregation and destructive production of urban spaces, because of the influence of mechanisms of foreignization and colonialism that are not new. Meanwhile, according to Jesús González, Josefina Domínguez, Margarita Novo and Juan Parreño, the urban borders or limits that are configured within cities are the consequence of the segmentation of social classes that reproduce impoverishment and gentrification as delimiting forces. Finally, Andrea

Catenazzi and Julieta Sragowicz approach a classic urban problem: the need for access to urban land generates new and, at the same time, vacant spaces, both in terms of their public, symbolic and social condition.

## *1.3.6 Part 6. Degradation and Abandonment*

One of the major problems of today's cities is the insufficient importance given to their inhabitants. Citizenship has lost its leading role in this configuration, even though there is no city without citizenship. This group of texts highlights two ways in which citizenship is destroyed: time (memory) and space (habitat).

The symbolic (belonging) and symbiotic (integration) inside the urban material landscape means recognizing its value in time, but from its inverse perspective, because belonging and integration tend to be diluted, especially in sculptures, monuments and nomenclature, which end up devaluing historical facts and producing oblivion. Local culture, according to Olimpia Niglio, is the "humus" of the community's life and represents the values that are transmitted through generations, which are often erased by foreign interferences, putting at risk not only the material condition but also the abandonment of the tradition and identity of the cities. Therefore, losing memory and history would mean losing cities, as stated by Salvador Urrieta and Verónica Zalapa.

Socio-spatial transformations change local ways of life and since the end of the twentieth century have been depredatory, considering that some cities, due to certain factors, have been able to preserve their historical patrimony. Urban planning and policies are fundamental axes in preserving and revaluing local heritage, from the sense of maintaining memory and not abandoning the historical condition that cities have.

Analyzing and understanding this temporality would also mean rethinking these spaces from their new logics. According to Francisca Marquez, the condition of public space of historic centers places them at a crossroads between the past and the present, where social revolts have transformed their principal function. In this sense, the author argues that public space is the site of resistance and dispute, but also the space for overcoming the historical–colonial matrix of power, which liberates or decolonizes. The transformation of the foundational centralities produces systematic processes of destruction, but also of resistance, in such a way that the constant depopulation they suffer is accompanied by development strategies, such as the one related to popular commerce, according to Inés del Pino.

Finally, neoliberal urban production policies have generated, on the one hand, the production of *private housing in the city*. As Alfredo Rodriguez and Ana Sugranyes state, the apparent success of housing policies generated a new problem that is expressed in the "the ones with a roof". On the other hand, the fragmented city that contains walled and privileged territories seeks, according to María Carla Rodríguez, strategies of resistance from self-managed urbanism.

### 1.3.7 Part 7. Destruction of Common Life: Violence

The increasingly powerful presence of acts of violence is a cause and consequence of the negative urban transformation and the destruction of the common life of citizen coexistence. In this group, urban violence, illegal markets and social vulnerability, both gender and territorial, are analyzed. Citizen insecurity transgresses interpersonal relationships, public and domestic spaces, becoming an urban planning principle that, for example, induces the construction of gated urbanizations with high rates of social exclusion.

Wars destroy cities in their privileged places, such as spaces of urban social integration (public spaces), infrastructures (factories and mobility), symbolic expressions (monuments, libraries) and instances of connectivity (ports or transmission towers). From this perspective, terrorism strikes in the essential places of citizen coexistence, as do the illegal markets (drug trafficking) when disputing the territories of production, transit and consumption of drugs. In addition, the city becomes a system of spaces for aggression and violence against women (femicide) and vulnerable groups (racism, xenophobia), as well as the increase in social vulnerability that is built in our cities. Terrorism, war and illegal markets (drug trafficking) have an impact on coexistence places and put into dispute the territories, turning them into zones of conflict, conquest and continuous struggle.

Violence is a factor that manages to transform the city to the point of generating restricted and inaccessible zones, according to Alfonso Valenzuela, which makes that time (hour, day) and place become correlating factors to certain crimes. In other words, the geography of crime allows the understanding of the lack of protection of citizens and the economic contraction it generates. Urban violence can be seen as the result of the way criminal organizations affect, control, destroy and reconfigure urban life and its spaces. Arturo Alvarado's approach is based on the fact that the power strategies of these illicit markets expand to populations, governments and even urban infrastructures, producing the control of territories, as well as strategies of exploitation, occupation and destruction of space.

The accelerated transformation of society and its territories reveals urban violence, which has been invisibilized for a long time. Border cities, for example, as Mauricio Vera and Luis Alfonso Herrera explain that generate a process of gradual destruction with their peers where the geopolitical dimension of control produces desolate scenarios of destruction, while the dimension of the imaginary finds opportunity in *complementary asymmetry* (Carrión and Enríquez 2022). Finally, the struggles for equality and equity vindicate the role of excluded sectors in cities, in order to guarantee the right to use and enjoy the city for women, migrants and everyone.

The city, configured from a patriarchal logic becomes a system of spaces of aggression, exclusion and violence against women. According to Nora Libertun de Duren, Diane Davis and Maria Lucia Morelli, the daily experience of women in the city shows how its management can, directly or indirectly, prevent or allow violence against the most vulnerable citizens. These dynamics destroy the development of urban, common and collective life, restricting, expelling and violating citizenship.

## 1.3.8 Part 8. Contraction of Public Management: Privatization

City government is a fundamental element in the production of urbicide, insofar as it is permanently under two complex forces of destruction. On the one hand, the loss of the sense of representative self-government, both due to the centralist visions of national states, as well as the process of globalization with the presence of virtual service platforms. And, on the other hand, the processes of privatization or deregulation that leads to have cities that apparently function by themselves (automatons), with no other purpose than profit and the abandonment of the social sense. Without leaving out the mediocre and erroneous urban policies, lacking of collective city projects, in this chapter, there is an analysis of the powers that rule in the city (capital) but without leaving aside the role of local public management, which contracts in such a way that it abandons its social logic of government. There, the transformation of the production and management of urban infrastructures, related to valuation, described by Beatriz Rufino, stays consigned.

Neoliberal policies have transformed the way infrastructures are produced, financed and managed, entering into a global and private logic, based on a fictitious capital function (financialization). Certain interests and powers put the life of the city at risk but, according to Marcelo Cortí, despite the fact that this is not the basic condition of the city. The attacks and threats to which cities are exposed come from the decisions of public policies, which according to the author are either erroneous, due to ignorance or a nefarious conscience, under a State that facilitates capital, or that introduces a mistaken notion of subsidiarity and abuse of urban planning strategies and tactics.

Mariona Tomas introduces the lack of understanding of urban territory (social, territorial and political) from its metropolitan condition, which induces logics of failure, since this scale is not conceived institutionally and in management policies. In this sense, Paulina Cepeda proposes that contemporary logic has produced sudden disasters and transformations in social and spatial trends, assimilated as an *urban shock* and not as an *urban revolution*. This incapacity or abuse generates that the city collapses simultaneously from the processes of un-urbanization and mainly of carelessness. Finally, Alan Middleton argues that governments become tools for destruction, for example, when managing rehabilitation and modernization policies that generate an impact on informal sectors to the point of expelling them, thus displacing other negative externalities.

## 1.3.9 Part 9. Urbicide: Cities Cases

Finally, it is interesting to incorporate to the debate and discussion of the proposal of the category urbicide in some specific cases of cities, where this process is clearly observed. This is to indicate not only that there are processes of urbicide in certain areas of the cities or in certain central themes of them, but also processes that affect the cities as a whole. For example, there is the death of the city resulting from the change in the model of global accumulation, which makes uncertain the functioning of a city like Detroit. This city of the automobile is used to produce vehicles within itself, whereas now the decentralized logic of assembly is promoted.

The loss of capital status due to the emptying of the functions of the state apparatuses located in the city of Rio de Janeiro generated an unusual impact. According to Mauro Osorio, Maria Helena Versiani and Henrique Rabelo, the aspects of the city's trajectory that give it the seal of capital status and its economic development conditions have generated marked inequalities. The exclusion of the integration processes in the European Union (Brexit) generated negative effects in the city of London. In the opinion of Pablo Sendra, cultural diversity in the neighborhoods is typical of London, and he analyzes the relationship between urbs and civitas that allows replacing it. According to Luis Fernando González in Medellín, the decade of the 90's has been marked by terror and urban violence process accompanied by drug trafficking investment. The logic of the city of the *XXI century socialism* in Caracas shows us the hardships it lives in all orders, as Alberto Lovera states, and neither could miss Santiago, where it is possible to speak, according to Loreto Rojas, Alejandro Cortés and Daniel Moreno, of the absence of "city due to the high territorial division of its government (36 autonomous communes without an integrating government), which shows that there are "cities that are never constituted as such because of the fractionation of its government and territory. Finally, Pedro Jiménez analyzes the historical centrality of Barcelona and the process of urbicide argument on real estate financial destruction.

### 1.3.10  Part 10. Epilogue

To conclude, it is interesting to close the publication with an optimistic view of the future of the city. From the sense of Urbicide; it is intended to denounce the key elements of the death of the city to, like the phoenix, project its best development guidelines, in order to achieve the good living of the population; that is, to provide it with a common sense of citizenship, in order to live happily and in peace. In that sense, according to Victor Delgadillo, social groups imagine better futures for cities, based on social capacities to resist adversities, by recovering collective memory and identities. And Saskia Sassen states that this condition is the result of the destructive power of capital, where local governments guarantee this process, producing abandoned territories and even with inhabitants in dead lands.

All this directed towards the need to build new and more democratic ways of sustaining accumulation processes that promote social cohesion, with cultural and political objectives to end up with the identity, symbols and collective memory concentrated in the cities.

# References

Abramo P (2012) La ciudad com-fusa: mercado y producción de la estructura urbana en las grandes metrópolis latinoamericanas. EURE. Santiago 38(114):35–69

Augé M (2008) Los no-lugares. Espacios del anonimato. (ed. or., 1992). Ed. Gedisa. Barcelona

Acosta A, Martínez E (2009) Derechos de la naturaleza: El futuro es ahora. Editorial Abya-Yala

Berman M (1987) A critique of three methodologies for estimating the probability of containment failure due to steam explosions. Nucl Sci Eng 96(3):173–191

Borja J (2003) Ciudad y planificación: La urbanística para las ciudades de América Latina. Cuadernos de la CEPAL

Borja J, Muxí Z (2003) El espacio público: ciudad y ciudadanía

Campbell D, Graham S, Monk DB (2007) Introduction to urbicide: the killing of cities? Theor Event 10(2)

Carrión F (2014a) Urbicidio o la producción del olvido. Observatorio Cult 25:76–83

Carrión F (2014b) Urbicidio o la producción del olvido. In: Durán L, Kingman E, Lacarrieu M (eds) Habitar el Patrimonio, nuevos aportes al debate desde América Latina. Ed. IMP, Quito

Carrión F, Cepeda P (2021) La ciudad pospandemia: del urbanismo al "civitismo". Desacatos: Revista de Ciencias Sociales (65):66–85

Carrión F, Cepeda P (2022a) Ciudad pandémica glocal

Carrión F, Cepeda P (2022b) From heritage fetishism to productive inheritance. In: The Routledge handbook of urban studies in Latin America and the Caribbean: cities, urban processes, and policies

Carrión F, Dammert-Guardia M (2018) Derecho a la ciudad: una evocación de las transformaciones urbanas en América Latina

Carrión F, Enríquez F (2022) Latin America's global border system. Ed. Routeledge, New York

Carrión F, Velasco A (2018) Is there a typical urban violence? In: Bhan G, Srinivas S, Watson V (eds) The Routledge companion to planning in the global south. Routledge Companions, New York, pp 287–297

Coward M (2006) Against anthropocentrism: the destruction of the built environment as a distinct form of political violence. Rev Int Stud 32:419–437

Coward M (2008) Urbicide: the politics of urban destruction. Routledge

Choay F (2009) El reino de lo urbano y la muerte de la ciudad. Andamios 6(12):157–187

De Mattos C (2010) Globalización y metamorfosis urbana en América Latina. Ed. OLACCHI-IMQ, Quito

Graham S (2002) Bulldozers and bombs: the latest Palestinian-Israeli conflict as asymmetric urbicide. Antipode 34(4):642–649

Harvey D (2013) Ciudades rebeldes: del derecho de la ciudad a la revolución urbana. Ediciones akal

Hegel GWF (1977) Phenomenology of spirit (Trad. Miller AV). Oxford University Press, Oxford

Herscher A (2006) American urbicide. J Arch Educ 60(1):18–20

Huxtable A (2010) On architecture: collected reflections on a century of change. Bloomsbury Publishing USA

Jacobs J (1973) Muerte y vida de las grandes ciudades americanas. Ediciones Peninsula, Madrid, p 2

Jacobs J (2011) The death and life of great American cities—50th anniversary edition. Modern Library, New York

Lavell A (2000) Desastres urbanos: una visión global. Woodrow Wilson Center and ASIES Guatemala publicación, pp 17–28.

Lefebvre H (2020) El derecho a la ciudad. Capitán Swing Libros

Sassen S (2001) Global cities and global city-regions: a comparison. In: Global city-regions: trends, theory, policy, pp 78–95

Shaw M (2000) New wars of the city: 'urbicide' and 'genocide'. In: Cities, war, and terrorism towards an urban geopolitics

Schumpeter J (1994) Capitalism, socialism and democracy. Londres y Nueva York, Routledge, pp 81–85

**Fernando Carrión Mena** Research Professor in Flacso Ecuador. His focus of study are the topics of: housing, urbanization process, city, historic centers, cultural heritage, violence, security and drug trafficking, borders, decentralization and sociology of football, among others. He created 8 thematic magazines (political science, security, city, historical centers, borders), wrote more than 1000 journalistic articles and 306 scholars, published 74 books (editor of 52 and author of 22) and edited 12 book collections (97 volumes). He produced 4 cinematographic documentaries. He has worked as a consultant for multilateral organizations and as a university professor. For his career, he won 9 awards, 6 decorations and 5 distinguished citizens distinctions. He was director of CIUDAD, FLACSO, Municipal Planning, as well as a former Councilor and former Advisor to the Municipality of Quito. He has published more than 1000 journalistic articles, 310 academic articles, 27 books as author and 53 as editor.

**Paulina Cepeda Pico** Architect and Master of Research in Urban Studies with a research scholarship at Facultad Latinoamericana de Ciencias Sociales , Flacso Ecuador. Senior architect of developing heritage housing restoration and rehabilitation projects, design and construction of architect projects and making of urban consulting. Her main areas of research are: housing policies, urban planning, urbanization process, governance, feminism urbanism, digital city, market and land policies. She is currently part of the team of researchers at Flacso Ecuador.

# Part II
# Urbicide. The Death of the City

# Chapter 2
# Urbicide. The Liturgical Murder of the City

**Fernando Carrión Mena**

> ¿Has the city died? Globalization is what is killing it now. Before it was the metropolization that was developed with the Industrial Revolution. And before it was the baroque city, which extended outside the medieval enclosure. Periodically, when historical change seems to be accelerating and is perceptible in the expansive forms of urban development, the death of the city is decreed.
> Borja, J. 2003:23.

**Abstract** *Urbicide* is a new analytical category that allows us to understand urban processes from the opposite angle to the one that has been hegemonically understood until now: urban development. It is about the liturgical assassination of the city produced by identified actors that cause the death of a city, of certain parts of it, as well as of some strata of the urban population. They are destructive facts that come from the economy that fragments society and space, from the culture that denies the meaning of diversity, from wars that destroy the material and human bases, from the climate change that plagues its essence, from the politics that does not represent society, and from public management that denies the sense of common space or town hall. Four types of urbicide can be found: historical (law of denial), natural (climate change), anthropic (expulsion), and symbolic (toponymy).

**Keywords** Urbicide · Urban development · Urban demolition · History

## 2.1 Introduction

Historically, the city is born when the population becomes sedentary; when it begins to take shape with the cultivation of the land in fertile places and with good river and land connections; when the concentration of the economic surplus begins in a few hands and in certain specific spaces; when the city was not just *in* the countryside, but was *from* the countryside (Mundford 1938).

---

F. Carrión Mena (✉)
Facultad Latinoamericana de Ciencias Sociales, Flacso Ecuador, Quito, Ecuador
e-mail: fcarrion@flacso.edu.ec

© The Author(s), under exclusive license to Springer Nature Switzerland AG 2023
F. Carrión Mena and P. Cepeda Pico (eds.), *Urbicide*, The Urban Book Series,
https://doi.org/10.1007/978-3-031-25304-1_2

Archeological studies determine that these processes occurred around 10,000 years before the Christian era. From that time to the present, many cities were born and all changed, because they are a historical elaboration. They also express, with unparalleled richness, that the city is the most significant and important product built by humanity throughout history.

Currently, more than 4200 billion inhabitants live in cities, representing 57% of humanity and with a tendency to continue increasing (UN-HABITAT). To this, demographic phenomenon must be added the concentration of most of the world economy: more than 80% of the planet's Gross Domestic Product (World Bank[1]) is incubated within it. Additionally, it should be noted that politics is also fundamentally urban, not only because of the concentration of voters and candidates, but also because of the weight of existing problems, which leads to cities directing global processes. That is, the large urban agglomerations and the vast complexity of their territories make them the great centers of political power and command of world economies.

The process of urbanization of the planet has led to the city having an unparalleled role. If in the nineteenth century the empires were the great actors of the world and in the twentieth century the States, for this twenty-first century, it is clearly the cities. Never before have they achieved such a significant position and presence. The processes of decentralization of the state, the demands arising from the territories, as well as the presence of a strong global inter-urban institutional framework (UCLG, UIM)[2] have contributed to this. Hence, the cities have such a role that has made them one of the three most important strategic actors worldwide, together with the decomposing States and the large global corporations on the rise (Sassen 1998).

These characteristics of overcrowding and globalization of urbanization generate a series of problems as a counterpart, ranging from: urban inequalities that demand spatial justice, the presence of climate change that requires a modification of the current model of urban development and increases in citizen violence that modify social interaction. They are difficult phenomena to understand, because they contain a very large and diverse variety of pathological situations.

This extermination comes from wars, from the colonization of capital in the built space, from the increase in the ecological footprint that induces climate change, from mutations of the population due to aporofication (Carrión 2022), from gentrification processes, among others, typical of the city or that occur in it.

Unfortunately, this set of realities shows us that the city can not only wear out, but also produce internal dynamics of destruction. In the same way, if this reality is not acted upon or is ignored, the only thing that can happen will be its decline, and in fact, it is happening now.

Hence, it is essential and necessary to enter into the discussion of this process, which is contrary to what has always been in the field of cities, based on urban development. Even more so, this methodological input has obscured the reality of

---

[1] https://www.worldbank.org/en/topic/urbandevelopment/overview.

[2] International relations between nations are overcome, and interurban relations are established between cities.

the permanent destruction that is embodied within cities throughout history and the world.

If a comparison is made with medicine, which in general terms studies life and its pathological vicissitudes, some interesting conclusions can be drawn. Medicine is the discipline that identifies problems of the human body (diagnosis) to then carry out prevention or healing practices, in order to reduce the death of people. In the case of *urbicide,* we can say that it is similar: It must serve to identify problems with the aim of correcting them, in order to prevent the city from dying. This means, first identifying urban pathologies, to then heal them, to which must be added the principle of justice, due to the prevention it entails and the results it provokes. A vision of this type demands that the events that occurred be penalized so that they do not happen again, and when that happens, they can be understood in their real legal and historical dimensions.

The area or focal point of analysis is the city, but from a perspective that arises from an inverse methodology to the one that has traditionally been used. An attempt is made to understand it from a little-explored angle and in a completely different way than it has been done until now, to focus less on production and more on its destruction, on what is lost or contracted: that is producing a change of method: less from the value of what is created and more of what disappears and squanders; less about how the city is produced and more about how it is destroyed, less about what is gained and more about what is lost; less from memory and more from oblivion. This means an entry from the causes of these processes as well as the effects they cause.

All these processes of destruction can be understood and understood under a new analytical approach to the city, where the concept of urbicide tends to be central, because it locates the fact and explains it within the framework of why the city is lost. But it is also a look from the hope that shows that the negative must disappear so that city-making prevails. Ultimately, it is about understanding the structural causes of its self-destruction, in order to understand it better, in order to subsequently act and think less about pessimism and more about optimism; as well as to confront it, counteract it by generating alternative proposals.

It is a deep look at the city from an unconventional point of view, since it is about knowing and analyzing what the city loses due to the action carried out by its own *urbanites*, both in the sphere of its production and its consumption. This position is directed toward its improvement, showing the face that is not seen but is lived by many, so that it can be the basis of a city for the people. It is a vision that assumes that the city does not have a linear sequential evolution, nor that each of its parts can be understood in isolation. It is a comprehensive phenomenon with history.

The category *urbicide* or the death of the city contributes theoretically and methodologically to its knowledge, as well as to the design of urban policies to neutralize the perverse effects. It is a methodological theoretical *castling*, which integrates the negative forms of the city in order to contribute to its positive construction. Thus, the most current theories and analyzes on the progressive death of the city are incorporated to build its reversal.

For this purpose, this article has an expository logic with the following four sections: first, where the historical background of the urbicide concept is presented; second, in which an approximation to the urbicide concept is attempted, third, where the types of existing urbicides are described, and the last one, where some general conclusions are elaborated.

## 2.2 The Trajectory of the Formation of the Urbicide Concept: Background

It is important to follow up on how the concept of *urbicide* has been procedurally constituted over time. According to the existing bibliography, urbicide can be used to interpret past times as well as current ones.

The formation of the *urbicide* concept has followed a logic with different thematic, methodological entries and moments, which until now has not been clearly defined. The existing literature is very dispersed, fragmented and varied, so much so that it is necessary to systematize and integrate the contributions made since its inception. Despite this, some clearly identified entries can be found, coming from an itinerary through the fields of literature, social sciences, urban, environmental and violence.

To understand the origin of this concept, some significant background information can be reviewed, which begin at the turn of the nineteenth century to the twentieth century when the processes of deconstruction of the city begin, due to the presence of certain factors from modernity. There they are, for example, the weight of the motor car and the growth of citizen insecurity. In this perspective, the emblematic case of Paris is located, where there is a process of urban deconstruction, originated in the changes promoted by the proposal of Haussmann's road axis,[3] which led, according to Choay (2009), to the qualifications of post city or post-urban era, which would give way to the divorce of the civitas with the urbs, typical of the closure of a cycle. Simultaneously, but in New York, the impact of urban regeneration projects is being experienced, which today could be classified as part of *project urbanism* (Carrión 2022), interpreted in the classic text by Jane Jacobs (1973), *Death and Life of Great American Cities*, a fundamental reference in the interpretation of these processes contrary to the sense of what the city is. The two cases come from different realities, although they have a common factor: urban policy actions are decisive in inducing the processes of destruction of certain places in large cities, although in their interpretation the concept of *urbicide* is not explicitly present yet.

Subsequently, and at the beginning of the second part of the twentieth century, the first explicit reference to *urbicide* appears, but not within the interpretation of urban processes. In this case, the use of it will come from the space of literature, thanks to Michel Moorcock who adopts it in his novel called "Eric: Dead God's Homecoming" (1963).

---

[3] It was a proposal that destroyed 75% of the original urban fabric and produced a compartmentalization of the urban social fabric (Marchand 2002).

Later, in the 1970s, the concept of urbicide sought an explicit space within the Social Sciences, when it timidly entered the interpretation of the urban restructuring processes that occurred in some US cities. This is the case of Ada Louise Huxtabl (1972), who analyzed the destruction of certain places in New York, within the framework of modernity introduced by the aggressiveness of real estate investments, superimposing on previously existing constructions. Then, Marshall Berman (1987) did the same, with the description of the destruction of the South Bronx, also in New York, in the sense of housing dispossession, with the consequent expulsion of the population, under the phenomenon of gentrification, as described by Glass (1964) for the city of London.

A totally different entry to the previous ones comes from the effects produced by wars: that is, from a phenomenon relatively external to the city, which makes it a strategic place, as the scene of a conflict. This is the case of Sarajevo (1982–1995), where Marshall Berman (1987) uses it explicitly to describe the destruction and siege of the city, which caused the loss of 36% of its population. This conceptual entry served to interpret the destruction that occurred in other cities as a result of so many war actions: situation that occurred at the beginning of this century with authors such as Coward (2006), Graham (2002) who describe the violent destruction produced by wars in the material world of cities. From this moment on, *urbicide* is projected with a connotation of violence against the city, an important thing, but still restrictive or reductionist.

More recently, with the expansive processes of globalization and urbanization developed within the new pattern of accumulation, authors such as Carrión (2014a, b, c) return to the concept to interpret what is happening in the central and consolidated areas of Latin American cities. But, additionally, also from the end of the century, in the set of urban structures, to analyze the processes of gentrification, colonization of memory and de-urbanization in the region, which lead to the interpretation under these same guidelines of urbicide (Carrión and Cepeda 2021).

In the twenty-first century, there are strong negative impacts on cities, generated by destruction, erosion and massive deterioration, coming, in this case, from the effects caused in the relationship of the city with the ecosystem in which it is based. Here are, for example, in the forefront, the causes and effects of climate change, which produce floods, alluviums, hurricanes, as well as, on the other hand, fires and droughts that strongly affect life in cities.

There is also another thematic entry, which is clearly delineated from the eighties when the reform of the State, the logic of the market, globalization and the development of new communication technologies entered with force, producing significant processes of alienation, which they generate urban dynamics contrary to city life and urban well-being. With this, the components from which urbicide is outlined are more clearly delineated; that is, defining the entrances to the massive, selective, sectorial or partial death of the cities. With this, it is possible to define, analyze and process the deep causes that are behind this dynamic, as well as visualize them and, above all, seek to restrict them (Carrión 2014b).

From this brief journey through the history of urbicide, two types of original sources can be systematized: the first, from external causes to the city, such as wars or imbalances produced in nature. And the second is the result of intrinsic problems or the internal structure of the city. But there may also be a combination of the two, as occurred with the zoonotic pandemic of COVID-19, which produced one of the greatest deaths in history, as it was an urban disease that was imported and became a community disease. Behind all these phenomena, the issues of vulnerabilities that cities have are present, which makes it very difficult to counteract or mitigate their effects.

From this journey, it can be concluded that its conceptual formation is in the making and that the amplitude of the field is significant. But, without a doubt, a lot has been done, so much so that today it is possible to publish this book. It should also be noted that it has been very useful that it has been considered as part of the Social Sciences, to understand urban processes. However, it is necessary to work on its methodological theoretical condition, even more, to understand its effects within the current globalization process. And today, it is much more necessary because cities have never been destroyed so massively, selectively and under a great diversity of situations, all of which can be understood within the concept of *urbicide*: being its antidote: the right to the city (Harvey 2013).

## 2.3 What Do We Understand by Urbicide?

As a starting point to attempt a theoretical approach to the concept of urbicide, a very important difference and relation is established between *urbicide* and *urbanicide*, because some authors use them interchangeably, as if they were synonyms. However, it is important to make a distinction, for which we follow the arguments of Marcela Lagarde (2006) to differentiate femicide and feminicide.[4]

From this perspective, it can be affirmed that urbicide is the death of the city, while urbanicide is the death of the city without public, private or community institutions doing anything to prevent or remedy it. It may even be the case that the institutions themselves are the cause of urbicide, with which its effect is double. Urbanicide can come from urban policies, in some cases, originating from the imposition of explicit actions that cause urbicide or, failing that, from maintaining a position of negligence, complicity or silence. It is a proposal that explicitly integrates local, national and international institutions, by omission or by explicit action.

In other words, there is urbanization when there is no impediment or, even, when institutional subjects in charge of ensuring its development promote it directly. Thus, there are cases in which the institutional framework has not reacted to, for example, the overthrow of the Library of Alexandria, the bombing of Baghdad and the tourist invasion in Venice, the construction of large residential towers in downtown Santiago

---

[4] According to Lagarde (2006), femicide is the murder of a woman for being female. Femicide, on the other hand, occurs when the set of criminal acts against women occurs in a framework of silence, omission, negligence and inactivity of the authorities.

or the emptying of society in the Historic Center of Quito.[5] Moreover, in many cases, their own actions have been those that have deteriorated the accumulated wealth of heritage throughout history.

The term urbicide is a concept composed of two words that have Latin roots: *urbs*, which the synonym of city, and *cide*, which refers to death or massacre; that is, to the death of the city. But it is not a natural death or a homicide, it is rather a murder, because it is an aggravated homicide, committed with premeditation and treachery[6]: nor is it the end of the city, as many have predicted.

Urbicide does not refer to the death of all cities, nor to the end of the city as a complex reality; but, rather, to the murder of a city or certain essential components and dynamics of it. These are processes, not facts or specific milestones, which are generated recurrently in cities, so much so that certain territories manage to adapt or survive, while others inevitably die.

Urbicide is the liturgical and deliberate murder of the city, carried out with premeditation, order and explicit forms. It is produced through actions promoted by identified actors that devastate or destroy the *urbs*, that is, the systems of significant places of life in common: squares, monuments, libraries; infrastructures and services; the *civitas*, due to the negative impacts on society and citizenship with the loss of rights; the *polis*, because they annihilate the institutional frameworks of government through processes of privatization, deregulation and centralization.

In principle, it is expressed under a multidimensional logic that has, among others, the following components: (i) The dismantling and annihilation of the systems of common places, that is, of public spaces (violence, privatization, agoraphobia); (ii) the destruction and deterioration of the material base of the city such as infrastructures (war, risk, vulnerability); (iii) the dissolution of social cohesion (gentrification, inequality, fragmentation); (iv) the degradation and abandonment of urban spaces (gaps, degradation, inactive); (v) the annihilation of institutional frameworks and the design of a type of urban policy (deregulation, flexibility, privatization); and (vi) the breakdown of the environmental balance (climate change, pollution, expansion of the agricultural frontier).

The destruction process can be selective or massive, through events in one or several cities, in one or some infrastructures, in an area or in a group, in a social group or in a community. With urbicide, the processes of urban destruction are questioned, in the form of a *punctual necrosis*, that is, the death of an area (a neighborhood); an infrastructure (housing); or a service (health), specific to the city. As it can also be ubiquitous, to be located in different places. And additionally, it can occur under a *general necrosis*, resulting from the death of a city due to natural (Pompeii), anthropic (Detroit) or war (Sarajevo) effects.

So, urbicide makes it possible to understand, among others, the phenomena of unemployment in urban areas, loss of memory due to modernization, colonizing

---

[5] In 1990, the population of the Historic Center of Quito was 81,384 inhabitants, and 20 years later, it was reduced to 40,913 (Del Pino 2013).

[6] According to the RAE, it is understood by: Homicide: "death caused to one person by another." And Murder: "to kill someone with treachery, cruelty or for a reward."

gentrification and the negative impacts introduced by the neoliberal city, by organizing territories in the manner of a constellation of discontinuous spaces (Castells 1999).

Additionally, it is possible to verify the processes of devastation of the public (privatization), strengthening of the private (market), change of the domestic (culture at home), dissolution of the social (violence) and decomposition of the urban structure (project planning). This contributes to the analysis of urban problems, as well as to raise the breaking point in the theoretical approaches of urban studies and city management policies.

In this way, urbicide allows us to understand urban processes from a reverse way of how they have been studied and prefigured until now: urban development, design and planning. It is the production of a change in the research method (it does not replace it) and in the design of urban policies (it does not replace it either). It is about understanding the city through the origins of its own destruction, to confront them for its benefit. For example, in the current context, with the COVID-19 pandemic, the global city suffered a significant *urban shock*, which produced an important change in the paradigms of understanding the city, coming from social interaction (civitas), from the public space (urbs) and the formation of the political community (polis).

Urbicide should be involved in urban development, with considerable weight, because it tends to explain the reasons why the city erodes. Therefore, it is a criticism of urban development models that strengthen inequalities, fragmentation, and exclusion. As well as that they affect climate change, preventing social resilience, conceived less as satisfying the needs of the population and more as capital subject to valuation. That is why, for example, the anti-seismic capacity of buildings is seen as an expense, as is the acceptance of the fragility and vulnerability of the city.

It is about understanding the city through the causes of its own destruction (structure and essence), to show how the prevailing city model is producing this negative effect. Therefore, urbicide does not simply serve to document or notarize the phenomenon: nor just to understand and describe it. It is basically to act on the city, to correct its anomalies, in such a way that they do not happen again.

It is a critique of the dominant models of urban development, with the aim of finding the dynamics of a new urban reality, which allows it to restore the sense of a *good place to live happily*, and not to undermine the quality of life of its population, the well-being of its neighbors and the good life of its citizens; more urgent because now it is conceived from a perspective that does not allow the sense of city to be strengthened; because its condition as a mercantile and business object prevails.

It is about critically showing how in the prevailing city model one can find the hope of a new urban reality, which will return to the city the sense of a good place (utopia); in short to be the starting point for what Aristotle said so many centuries ago:

The city emerged because of the necessities of life, but now it exists to live well.

Urbicide is a destruction that does not operate as a synonym for the disappearance or the end of the city, as some authors believe would be happening. The figures of homicide (criminal act of killing a person), genocide (act of extermination of a social group) or femicide (execution of a woman for gender reasons) do not mean that they lead to the end of humanity. Likewise, urbicide is not the end of the city, but of certain parts of it or of those cities in particular that have been attacked in different ways.

It is not about its natural death either, because it occurs with premeditation, treachery and an explicit order. So much so that it is a deliberate and liturgical process, behind which there are public and private actors and citizens (urbanites). That is why it is not a homicide but a murder, which should even be classified as a crime, included within the penal codes. Even more so, if the right to the city is regulated in some national constitutions and has also been recognized by the New Urban Agenda, approved by 193 countries in 2016. And especially if there are already environmental, patrimonial crimes and paralysis of public services, among others, and if not, who is responsible for the deterioration, destruction, abandonment, erosion of the city? If they are not typified, they will continue to occur in a general and permanent way. It is necessary to hold the perpetrators of urbicide accountable, not only to reduce the phenomenon, but also to try to end it.

It is clear that the concept cannot be understood in a linear way; because it does not have an explicit sequence; because it can go forward or backward; nor because it has a particular historical condition. But it must be taken into account that cities, being historically produced, were born at a specific moment and developed over time, they can, at a particular moment, produce something superior or, simply, perish.

An additional quality must be added as well: Its death, under the figure of urbicide, is accompanied by the sense that the city not only dies but also kills, as Eduardo Kingman states, in his article in this book. An example of this quality occurs when the city violates the ecosystems in which it is part of, causing climate change, which, in turn, violates the city: that is forming a cycle of circular causation consisting of the city producing environmental changes, which tend to destroy the city itself, in the form of an explicit *boomerang*. The phenomenon of the COVID-19 pandemic also clearly illustrates this.

### 2.3.1 Toward a Typology of Urbicide

The murder of the city comes, as has been described, from diverse conditions and situations, although always with the same general logic. Therefore, the set of them must be ordered, in such a way that it can be understood in its dimension and specificity. For this purpose, a classification has been made with four particular types of urbicides or subentries that make it up, each one with its particular nuances.

The first comes from a historical perspective (historical urbicide), where each city, in its evolution, can find phases in which the birth of a city is inscribed in the death of the previous one, establishing an evolutionary continuum. The second, originated in an entry born in the relationship of the city, as an artificial product, with the environment in which it operates (natural urbicide). The third, coming from the human conditions that order the city (anthropic urbicide), are proper to the essence of the city or, failing that, from the outside, such as wars; and the fourth, originated in symbolic situations (symbolic urbicide).

## Historical Urbicide

In the evolutionary process of cities, the presence of multiple stages is observed, typical of the existence of different historical moments. Hence, *historical urbicide* has to do with the need to understand and diagnose the main problems that cities embody throughout their time continuum, that is, of its constant transformation.

From this perspective, there is a stage of death of a city that gives way to the birth of a new one, but within it, which testifies that *urbicide* is part of the life of cities and not its end. It also shows that the city is a historically determined product, where in one phase there is one city and in the next there may be another totally different one: in short a city that dies and another that is born simultaneously; that is, a process in which one city gives way to another, but within it. It is a dynamic that is inscribed in the *law of the negation of the negation*, which embodies the progressive character, in the sense that the old gives way to the new (Hegel 2004).

Thus, we have, historically, that the death of the classical city gave rise to the emergence of the modern city and in this the global city originated. As a testimony of this process are, among other cities, the cases of Rome. Paris, London and Barcelona, which are a perfect sample of this trajectory. The interesting thing is: In each of them, the successive sequence of cities can be read in the form of a *palimpsest*.

It is a dialectical process that, according to Engels (1951), can be explained within the so-called *Law of the negation of the negation*, where the superior degree eliminates or denies the previous one, being present the nascent and the expired, the new and the old simultaneously (palimpsest). In urban terms, this proposal is verifiable in the fact that every city implicitly carries its own negation. Unfortunately, this phenomenon in the urban area has been little studied, which has prevented a better understanding of its historical evolution, which is not linear at all: but rather the opposite: that the death of the city is a structural and permanent component of its life.

Thus, we have, for example, how in the *medieval city* the walls are demolished to deny or violate the ordering of the character of a compact city, inscribed in specific limits, for refuge and defense purposes. In the *city of the industrial age*, its urban structure was prepared for mass production, mechanized and concentrated, far from rural areas, where there are inputs and energy sources at low prices and in a dispersed manner. Against this, the anti-urban and anti-industrial criticism was born, for producing too large and polluting conglomerates, which destroy human life (Engels 1976).

And now, in these last forty years, with the *global city* (Sassen 1998), urbicide has become more evident from this historical perspective, as it accounts for the destructive events that hang over it, mainly due to the weight of the logic of the market, of technological innovation and of global competition.

From this moment, urbicide seeks a space within the Social Sciences, to interpret and explain the processes of massive and selective destruction that have been occurring within cities at a planetary level.

## Natural Urbicide

*Natural urbicide* is based on the fact that the city is an artificial product produced by humanity, governed by social laws, while the environment, in which it is inscribed, does so based on natural laws. But now, humanity lives in a new nature, born of mutual links that break the balance, which produce problems or complex pathologies, which can lead to death.

*Natural urbicide* is configured in the interrelation of the city with the environment, forming an *ecosystem* (Figueras 2013); that, when the balance of the pairs in the equation is broken, urbicide is present. Clearly, there are alterations in the logic of life in nature that have an impact on the city[7]: but also some transformations in the functioning of the city, which negatively modify the behavior of the environment.

More recently, with the so-called *climate change*, the city itself has become the cause (victimizer) and the effect (victim). This is because it is the space where some of the natural problems originate, but also in the place that receives the greatest impacts. Nowadays, cities are the main consumers of fossil fuels, as well as producers of greenhouse gas emissions, with which they become the origin and destination of urbicide.

This means that cities are not mere passive victims of these events, but an explanatory part of the problem with which they die and kill. For this reason, it is the area where its exits must be found, an issue that if it is not assumed it will be the cause of urbanicide. For this reason, the city operates from urbicide as well as from urbanicide (passivity originated in its ignorance). This is because the current urban development models break the balance of the ecosystem, although in a different way depending on where they are located, be it in the countries of the South or in those of the global North.

Vulnerabilities are largely the cause of social disasters, because cities not only generate imbalances within the ecosystem, but also are not prepared to face the *boomerang* that the logic of urban development prints. Hence, cities are not mere passive victims of these events, but part of the causal explanation of environmental problems, because the city not only dies, but also kills.

In other words, *urbicide* comes from the attacks of nature, caused by different reasons, and has a varied expression in cities, for example, through events such as: earthquakes (Port-au-Prince, Haiti, Osaka, Japan), volcanic eruptions (Vesuvius, Italy, Antigua-Guatemala), hurricanes (San Antonio, USA, Havana, Cuba), floods (Quebec, Canada, Piura, Peru), fires (Valparaíso, Chile, Guayaquil, Ecuador), climate change (New Orleans, USA, San Juan, Puerto Rico) and droughts (Cape Town, South Africa, Chihuahua, Mexico).

For this reason, the definitions of *resilient cities* or *climate adaptation* do not go to the cause of the problems, but only to the search for the *adaptation* of the city to supposedly natural phenomena, with which cities do not modify their ecological

---

[7] In Ecuador, according to its Constitution, nature is subject to rights, as specified in its Article 71 when it states that: "**Nature** or Pachamama, where life is reproduced and carried out, has the right to full respect for its existence and the maintenance and regeneration of its vital cycles, structure, functions and evolutionary processes".

footprint (environmental impact indicator) but are adapted, producing more problems in the form of *circular causation*.

There is a second expression of natural urbicide that has to do with the impacts from certain phenomena such as plagues (Bubonic, in Los Angeles, USA or Sydney, Australia), epidemics (Spanish, Fever, Kansas, USA, Berlin, Germany) and pandemics (COVID-19, in the world's largest cities[8]), among others. What was interesting about these phenomena were the hygienist tendencies that began to suffocate the negative impacts in cities, but unfortunately they are not maintained over time, because economic interests prevail over those of human and environmental value.

*Anthropic Urbicide*

*Anthropic urbicide* is born from the activities promoted by public or private institutions, as well as by citizens, which alter the essence and structure of the functioning of cities. In short, it originates from the introduced pathologies, which may be of economic, cultural or political origin, through two determinations: one, coming from its own structure (internal factors) and two, from outside of it (external factors).

From the interior, part of the essence of the city, the causes of its destruction or urbicide can be understood. In this, the following statement by Lefebvre is clear:

> A city is, therefore, everything that is experienced, known, represented, built or destroyed as a city.

In this perspective, there is a variety of situations, among which are those related to urban violence (Carrión and Velasco 2018), where the following stand out: *common violence*, which erodes citizen coexistence as occurs in Caracas-Venezuela or San Pedro Sula-Honduras; the *violence of illegal economies* (drug trafficking) in Medellin-Colombia, Naples-Italy or Culiacan in Mexico that impacts daily living; and *discriminatory violence* (gender, xenophobia) that affects public and domestic spaces, with the case of Ciudad Juarez in Mexico standing out. And we cannot fail to mention the cases of urban terrorism that annihilate social interaction and affect certain specific spaces in cities, such as in New York, USA, on September 11, 2001, or in Lima-Peru with the actions of Sendero Luminoso. Some particular aggressions can also be highlighted, for example, from the blowing up of a bridge (Toledo, Spain), the attack on Churches (two churches burned in Santiago, Chile) or vandalism against the signage or the nomenclature in the Historic Center of Quito, Ecuador, or in the city of San Luis Potosi, Mexico.

Additionally, they can come from pathologies originating from the weight of economic logic. Here, for example, the reform of the State leads to undermine the self-government of the city through the centralization of powers and privatization processes, which functionalize urban planning, conceiving it as part of the urban planning of projects (Special Zones of Urban Development). Similarly, the dynamics of the market take weight, especially, linked to real estate businesses above the satisfaction of the basic needs of the population (Santiago, Chile; London, Great Britain).

---

[8] This is the case of what has recently happened with COVID-19 in most cities on the planet (Carrión and Cepeda 2021).

The phenomenon of inequalities has been tremendously devastating, as the gentrification processes can testify; that is to say, of the population changes due to the expulsion of those with low economic resources, producing verticalization of certain places of exception, eroding memory and increasing the economic returns of capital. With gentrification, places of memory are colonized, the pre-existing physiognomy is expelled, the historically constituted ways of life are modified, and the previous daily life is killed, as has happened in London, England, and Santiago, Chile, among many other cities in the world: nor can we fail to mention the weight that a sector of the economy such as tourism has been acquiring especially because of the impacts it has brought; now is called touristification (Calle 2019). These processes radically change the daily meaning of the city, as occurs in cities such as Venice, Italy, Barcelona, Spain, or Cusco, Peru, with devastating effects, because they homogenize internal supply with external demand and they empty the population originally from the historic centers and imposes foreign currency on local markets; it is enough to see these effects in Havana in Cuba. The idea that tourism is an industry without chimneys is not sustainable, because it is highly polluting, so much so that it produces urban devastation.

Within the inequalities it is necessary to take into account the phenomenon of the unprecedented presence of the so-called closed neighborhoods in almost all cities (gated communities), their excuse is that they guarantee the citizen security of the area, which abounds in residential segregation, it erodes the sense of public space and produces ghettos (Caldeira 2017), as occurs in many cities, such as Buenos Aires, Argentina (Countries), São Paulo, Brazil, Bogota, Colombia, and Quito, Ecuador.

A large part of these inequalities are built on the basis of evictions and dispossession, producing the denial of universal access to the city and the functional disarticulation of the city. These are processes that today are generalized under the concept of *urban extractivism*, which has to do with the appropriation by real estate corporations of the speculatively produced income, the expulsion and displacement of the low-income population, environmental imbalances and institutional degradation (Vásquez 2017).

Another of the degrading factors has to do with the weight of the motor car, producing various direct effects in the city, which are: the monopolization of the road infrastructure, the reduction of public space (parks, sidewalks, equipment) and the redirection of urban planning toward demands of capital and not of those coming from the population. North American cities like Miami or Detroit are the best and most obvious example; but also, for other reasons, the cities of Bogotá, Mexico or Lima, where there has been a notable drop in speed, generating dysfunctional pathologies. And, as if that was not enough, it is the cause of air, land and water pollution in cities.

Also, in the line of *anthropic urbicide*, the actions of structures originating outside the cities are presented, such as wars and economic or political conditions that, without a doubt, severely and directly affect them, because it is not only the scenario.

In contemporary wars, the cases of Sarajevo, Bosnia, Aleppo, Syria, Gaza, Palestine, or Kyiv, Ukraine, among many others, can be verified. However, they all have a common distinctive element: These wars, unlike the previous ones, are fundamentally urban wars, where there is a high density and diversity of the population. This new condition makes wars more destructive in material terms and more painful in terms of loss of human life. Perhaps, Hiroshima and Nagasaki are the most extreme and terrifying examples of the double urbicidal effect: of the material and the human.

Due to the sense of tragedy caused by wars, some authors ended up linking urbicide with the crime of genocide: first because they cause massive deaths of urban dwellers and second because the destruction of the material base of a city directly affects the life of its population. In other words, the destruction of the material base of the city leads to the annihilation of the human being, because it is destined to satisfy the basic needs of the population. There are the blows suffered by the infrastructure (transport, drinking water), services (education, health), production (goods, information), public space (squares, parks) and the government (institutions, policies).

These destructive effects tried to be nuanced by the definitions of wars. For example, *preventive wars*, promoted by President George W. Bush, such as those that occurred in Afghanistan and Iraq, or the *necessary wars*, produced in Syria and Libya by President Obama (Nobel Peace Prize) or the so-called *Arab Spring* in Tunisia and Egypt. This set of wars are nothing more than euphemisms that end up justifying the impact on spaces for social integration, the destruction of the material base of the city and the massive loss of human lives.

In the case of the general economic effects, it can be exemplified with the crisis experienced in Detroit, USA, under the presence of a new model of accumulation of the automotive industry worldwide, which introduced the logic of assembly.[9] It can also be verified with the meaning of national policies, as happened in Cuba with the precariousness of Havana, because the revolution placed more emphasis on the countryside than on the city. The case of Rio de Janeiro in Brazil should also be highlighted, which lost its status as capital, due to the transfer of this representation to the city of Brasilia.

It is worth pointing out that many economic policies originating in national states directly affect the life of cities, for better and for worse. The city becomes the stage where the institutionality of the economic system is expressed by imposing itself over the meaning of the social system.

This institutional weakness not only calls into question its structural condition, but also becomes an additional element for the destruction of the city.[10] In other words, heritage destruction, institutional weakness and conceptual obsolescence, in the framework of globalization, configure a situation marked by the production of

---

[9] Detroit, a symbol of the industrial power of the USA and of the American dream, goes bankrupt and begins to run out of economic activity and population. The city was called "motor city because it was the mecca of the automobile industry in the USA and the world, a cutting edge in the development of that time. However, it began to die due to the change in the logic of production at a global level: it is inefficient to have all the links in the production cycle concentrated in a single space: Detroit. For this reason, assembly was imposed, which is nothing more than bringing the places of production closer to those of consumption, seeking to improve competitiveness worldwide.

[10] The hegemonic paradigms have been functional to these processes. For example, thanks to tourism, gentrification and conservation policies, among others.

oblivion, which could well be characterized as a global crisis of heritage; very similar to the one produced after the Second World War,[11] with the difference that it was located in Europe and the current one is deployed in a generalized way in the territory and ubiquitously[12]; which is possible because the globalization of heritage has occurred, thanks to the scientific–technological revolution in the field of communications, the declarations of World Heritage,[13] the weight of international cooperation and the homogenizing tourism that breaks borders.

*Symbolic Urbicide*

Within *symbolic urbicide*, some determinations can be found that tend to affect the life and structure of cities. They come from various orders and from different constructions, as can be illustrated with the meaning of toponymy, perceptions and urban imaginaries.

In this perspective, there are certain instructive cases around urban toponymy, converted into a powerful instrument of *symbolic urbicide*, in at least two forms: the names of cities and the nomenclature of public spaces. Regarding the former, at least two singular entries can be found: the first originating from colonial subjugation and the second from the internal conflicts of the countries.[14]

The mutation of the names of a city has a logic and an explicit meaning, which has been a regular practice throughout history. There are the names of certain cities that illustrate these processes. The modification of the name of a city tries to kill its past and mark a new reality, in the manner of a *historical urbicide*, that is, inscribed in the law of the negation of negation, where the metamorphosis of the name of a city redefines a new form of possession, of dominion, but also of belonging.[15]

In these cases, it represents a change of hegemony, expressed in the struggle for toponymy, where symbolic urbicide makes a lot of sense, because it gives the inhabitants of a city a sense of belonging. It is the feeling that instills the notion of homeland,[16] but it can also produce a displacement toward another denomination, which represents a different city.

---

[11] A very similar situation was experienced with the Second World War in Europe, which gave rise precisely to a very strong impulse of the theses of monumental restoration and reconstruction.

[12] Humanity has experienced three great patrimonial conjunctures: the first with the first modernity, the second with the World War and now with globalization.

[13] At the moment, there are 187 cities considered world heritage, which decides to form the Organization of World Heritage Cities (QCPM), to exchange experiences, disseminate knowledge, and generate technical assistance, among others (ovpm.org/main.asp). In addition, it should be noted that currently (2013) 981 sites are cataloged: 759 cultural, 193 natural and 29 mixed, in 160 countries around the world.

[14] The original names of the cities made reference to the places where they were located, to the vocations, to the religious, natural or social symbols.

[15] "On the map of Europe, we find the same names as in the Middle Ages, we admire the long duration of these urban constructions that are named Paris, Naples, London, Milan, but also Barcelona, Prague, Zurich..." (Choay 2006).

[16] According to the dictionary, Homeland is: "Place, city or country where you were born."

The cases of the names of colonial cities in Latin America are directly linked to the founding processes of Spanish cities. The foundation is a colonial act of construction of a new administrative base of the territories,[17] to reorganize their domination politically, socially, economically and militarily. In this case, it is done in previously existing cities, to which the original name is replaced by an additional one of Spanish origin, precisely to symbolically show the colonial domain.

In these cities, their foundation goes hand in hand with the imposition of a name, in many cases composed, where the name of the conquerors goes first and then the original. There are the examples of San Francisco de Quito or Santa Fe de Bogota. But there are also those of Santiago de Nueva Extremadura, Chile, (originally with the Inca name of Laja, which disappears with the foundation); then, Nuestra Señora de La Paz, Bolivia, that was also imposed. However, the names of Mexico City, Mexico, and Cusco, Peru, still survive, which kept the original, due to the weight of the Inca and Aztec culture in those locations.

In the case of the change of the names of cities, originated by internal conditions to the countries, perhaps the most interesting, although not unique, is Russia, Soviet Union, with the city of Saint Petersburg. This name was originally given in homage to Peter the Great, its founder, in the seventeenth century. With the outbreak of the First World War, the German sound of the name was eliminated, to be renamed Petrograd. Then, it acquired the nomination of Leningrad, ten years after the death of Vladimir Ilich Lenin, Leader of the Soviet revolution (1917), with all the ideological weight that this change implies[18] and with the elimination of its capital status. Later, in 1991, it recovered the initial name of Saint Petersburg, within the framework of Perestroika.

What is perceived in this process of transformation of the names of cities is that the denomination can deny a type of city and give birth to another; but it always goes hand in hand with other changes, for example with the colonial imposition of the foundation of cities (the Spanish Crown in Latin America), with the change of political regime (Franco in Spain, Lenin in the Soviet Union) or with the functional modification of the city capital (Río de Janeiro).

The change of the nomenclature in the public space is also very revealing. It transits from the first moment when it is marked by the customary logic, coming from the society that names according to perceptions, traditions, and features of the place or the daily life of the population. This is how we have the names of the streets of silver shops or butcher shops; also from the Sol or Grande squares; or from the neighborhoods of La Pradera or Bellavista.

But, subsequently, the commemorative determination penetrates, with which it is sought to implant and legitimize the official history (of the dominant), above the socially constructed one. In this case, a different order is imposed on society, thanks

---

[17] Currently, there is an inverse process, described as *colonial*, which is expressed, for example, in the overthrow or destruction of commemorative monuments. These are the cases of Christopher Columbus in almost all the cities of the continent (Buenos Aires, Mexico City, Quito), as well as of the founders of cities (Benalcázar in Cali or Jiménez de Quezada in Bogotá) and of certain generals such as Baquedano in Santiago.

[18] Volgograd, or city of the Volga, is the initial name, to later give way to Stalingrad (1925–1961) and, finally, with the fall of the Soviet regime, return to its original name.

to the fact that the city is the most important instance of socialization. With this, the names of women or indigenous people are made invisible or ignored, while the names of soldiers or conquerors are consecrated through authoritarian imposition. In other words, uniformity is established that denies the essence of the city: its heterogeneous structure, which is the basis for respect for difference.

Another interesting case to be analyzed within symbolic urbicide is the one that originates from what could be called the conflict over language. It is in the colonial logics of power, as well as in authoritarian visions. This is the case of the dictatorship of Francisco Franco in Spain (1936–1975). In his authoritarian policy, he sought to impose the hegemony of the Castilian language over regional languages, for example in the Basque Country, in Valencia or in Catalonia. For this purpose, he uses the nomenclature of the cities, forcing the use of the Spanish language in the public space, as well as exalting certain historical facts typical of his political sign: in other words denying local languages and imposing a questionable historical view.

In Spain, the names of the streets and squares that were castilianized have not been completely reversed. There are the examples of 51 streets in Valencia and 52 in Madrid that changed (also in Barcelona and Bilbao), to reconcile memory with history. In Barcelona, the names of the streets begin to be feminized and to recover their Catalan meaning.

The case of Xátiva-Valencia should be highlighted. In 1707, Felipe V, ordered "to have it ruined to extinguish its memory" through the burning of the city, the irrigation of salt in its supply fields and the change of its name to Colonia Nueva de San Felipe. But history resists: Felipe V, is in the Museu de l'Almodí in Xátiva in a painting placed upside down, as an act of repudiation. This fact has become the emblem of the city and of the Valencian progressive and nationalist movements precisely in order not to lose memory. The dictator Franco also wanted the name of the city to be Játiva, in Spanish, but its inhabitants keep it as Xátiva, in Valencian. Undoubtedly, the conflict over the words, the language and the tradition inscribed in the toponymy of the place comes from the need to maintain identity and sense of belonging (Mascarel 2019).

Urbicide additionally comes from symbolic constructions called urban imaginaries. In this area, within Latin America, the greatest imaginary that exists in the city is that of fear, converted into an urban principle that liquidates public space (agoraphobia), producing the so-called closed urbanization or urbanization of walls, which leads to the loss of public space and community life.[19] It also reduces time, space and life in society.

And another powerful imaginary is the one of *territorial stigmas* (Wacquant 2007), inscribed in urban segregation and fragmentation, which end up generating social

---

[19] The walls in the cities, in their diffuse peripheries or, even, in the compact city and in its central zones, respond to the same logic. It is not about protecting free citizens but about excluding overexploited or marginalized social sectors. The supposedly protected ones are locked up in their neighborhoods; they are generically called: "closed neighborhoods", "gated cities, "private urbanizations, etc. In some cases, the names are sufficiently explicit, such as "El Encierro (The Confinement)." This closed population renounces the city to defend its position of privilege with respect to the excluded sectors.

grievances, leading to the denial of the other (otherness), putting an end to alterity, a substantial base of the city. In this way, for example, what is called the *dual city* formally composed of exclusion is established: the formal city/informal city or legal city/illegal city, which ignores the existence of the illegal and informal, being, paradoxically, a considerable part of the Latin American city (Hardoy and Satterwaite 1987).

## 2.4 Conclusions

Urbicide is a daily reality in today's cities, originating in the liturgical murder of the city. A set of social and institutional forces participate in it with the decision to make it happen, for which they draw up explicit strategies. The economy seeks the efficiency of the city, in order for it to become a space for business and profit. Politics uses it for the purposes of domination and hegemony. Wars have it as a setting for settling exogenous situations. The changing nature lies in wait for it with an explicit schedule.

All this is possible thanks to the concept of urbanicide, which makes the process more regulated and liturgical, both to destroy the city and to ignore the processes that undermine its existential foundations. Behind these processes are explicitly a set of public, private and social institutions—local, national and international—that do nothing or do very little to stop or, at least, mitigate the damage that has occurred. This means that behind urbicide there is an explicit order and actors that promote it.

After the urbicide and urbanicide, there are explicit tensions or contradictions, which lead to shaping the death of the city. They can be located in the urban peripheries (sub-urbs) as well as in the consolidated areas (centralities): also in the internal logics of the functioning of cities, such as in the presence of property (patrimony), which gives the city a sense contrary to that of living happily or living well. Now, here are the main tensions or contradictions.

*Urbanization that does not produce a city*

Since the post-World War II period, urbanization in Latin America has accelerated, producing a phenomenon of concentration of the population in peripheral areas of cities, thanks to migration from the countryside to the city, which produced an *urbanization without a city* (urbicide). The cities were filled with *human settlements* that, as the name indicates, do not become part of the city, because they are simply deposits of people. It is about *sub-urbs*, that is, of spaces (urbs) that do not become cities (sub).

*Memory and history*

Remarkable transformation processes are experienced in the consolidated city, which lead to its degradation. There are located, for example, the initiatives of gentrification and touristification, as economic processes of valorization of capital, which tend to expel historically resident population and colonize memory. With this, the fight against memory is established from a double perspective: the deterioration of daily life and the destruction of the material and symbolic base of the city. In this case, the tension is inscribed between memory and history, expressed in the production of oblivion (urbicide), when in reality all times should be present simultaneously, configuring the *value of history* (Carrión 2022), with a sense of future or project. Urbicide leads to a memory without history, because it entails oblivion. Recovering memory means recovering the value of use and the value of history. More clearly: when conservation policies are established, that seek to return to the origin and not to the *productive estate*, the importance of adding time to the past is left aside (Carrión 2022).

*The city that dies and the one that is born*

The historical condition of a city is structured on the basis of sequential moments, where the superiors deny the previous ones (historical urbicide). Currently, the global city tends to deny the modern city, with which cities are no longer urbanized but are urban regions. Much of this has to do with the weight introduced by strong technological innovation, which Bauman (2002) warned us about in his book on liquid modernity. In this sense, there is a need for cities not to forget what happened historically so that the memory of its destruction serves as a containment dam so that it does not happen again.

*Heritage of humanity and the market*

Cities are historic and everything in them is historic. Therefore, it is the place with the highest concentration of heritage. A condition of this type shows the voracity of capital, especially in those places where its presence and density is more pronounced. A situation of this type causes a position in which heritage is appropriated by the market, causing a notable change: the city assumes the strategic condition of business (capital valuation) and not of good living (urbicide), with which the sense of the heritage of humanity yields to the inclement market, which devours everything.

Everything that happens after the urbicide has incalculable consequences. That is why it must be prevented. In this perspective, the thesis of the right to the city becomes an obvious antidote: first because it gives society the possibility of leaving for a different city, in terms of access to the goods, services and wealth that are produced in the city; second because it will be possible when the population represented in the city government can design the exits linked to the majorities; third, that you can think of a totally different city, based on meeting the needs of the population and not those of the market: good living; fourth, recover history so that the memory of its destruction serves to prevent it from happening again; and fifth a good citizen urbanism has to be installed.

# References

Bauman Z (2002) Modernidad líquida. Ed. Fondo de Cultura Económica, Ciudad de México
Berman M (1987) Among the ruins. New Internationalist
Borja J (2003) La ciudad conquistada. Editorial Alianza, Madrid
Caldeira T (2017) Ciudad de los muros. Ed. GEDISA, Barcelona
Calle M (2019) Turistificación de centros urbanos. Universidad Complutense de Madrid
Carrión F (2014a) Urbicidio o la producción del olvido. Observatorio cultural, Santiago
Carrión F (2014b) Urbicidio o la producción del olvido. In: Durán L, Kingman E, Lacarrieu M (eds) Habitar el Patrimonio, nuevos aportes al debate desde América Latina. Ed. IMP, Quito
Carrión F (2014c) La ciudad y su gobierno en América Latina. Ed. UNAM, Ciudad de México
Carrión F (2022) La itinerancia patrimonial: De la propiedad a la apropiación por aporoficación. El Caso de Quito
Carrión F, Cepeda P (2021) Ciudad Pandémica Glocal. Ed. FLACSO-Ecuador, Quito
Carrión F, Velasco A (2018) Is there a typical urban violence? In: Bhan G, Srinivas S, Watson W (2018) The Routledge companion to planning in the global south. Ed. Routledge, New York
Castells M (1999) La era de la información. Ed. Siglo XXI Editores, Barcelona
Coward M (2006) Against anthropocentrism: the destruction of the built environment as a distinct form of political violence. Rev Int Stud
Choay F (2009) El reino de lo urbano y la muerte de la ciudad. Andamios. 6, Ciudad de México
Del Pino I (2013) Impactos del turismo en sectores patrimoniales. ponencia presenta da en: La intervención urbana en centros tradicionales con enfoque social, Bogotá
Engels F (1951) Dialéctica de la naturaleza (trad. de Roces W). Editorial Grijalbo, México
Engels F (1976) La situación de la clase obrera en Inglaterra. Ed. Akal, Madrid
Figueras E (2013) La ciudad como ecosistema urbano. ETS Arquitectura (UPM), Madrid
Glass R (1964) Introduction: Aspects of Change. London Aspects of Change: Centre for Urban Studies. London, Mac-Gibbon and Kee
Graham S (2002) Bulldozers and bombs: the latest Palestinian-Israeli conflict as asymmetric urbicide. Antipode
Harvey D (2013) Ciudades rebeldes. Del derecho a la ciudad a la revolución urbana. Ed. AKAL, Madrid
Hardoy J, Satterwaite (1987) La ciudad legal y la ciudad illegal. Ed. Grupo Editor Latinoamricano, Buenos Aires
Hegel F (2004) Fenomenología del espíritu. Ed Fondo de Cultura Económica, Ciudad de México
Huxtable AL (1972) Will they ever finish Bruckner Boulevard? A primer on urbicide. Collier Books, New York
Jacobs J (1973) Muerte y vida de las grandes ciudades americanas. Ediciones Península, Madrid, p 2
Lagarde M (2006) Del femicidio al feminicidio. In: Desde el jardín de Freud. Universidad Nacional de Colombia, Bogotá
Mascarell P (ed) (2019) Memòria de la destrucció La crema de Xàtiva i altres urbicidis. Ed. Ayuntamiento de Xativa, Xativa
Marchand B (2002) Paris, histoire d'une ville, s-XIXe-XXe. Ed. Seuil, Paris
Moorcock M (1963) Eric. Regreso al hogar del Dios Muerto. In: Science Fantasy # 59. Nova Publishing
Mundford L (1961) La ciudad en la historia, sus orígenes, transformaciones y perspectivas. Ed. Pepitas de Calabaza

Mumford L (1938) The culture of cities. Secker and Warburg, Londres
Sassen S (1998) Ciudades en la economía global, enfoques teóricos y metodológicos. Revista EURE, Santiago
Sasse S (1999) La ciudad global, New York, Londres, Tokyo. Ed. Eudeba, Bunos Aires
Vásquez AM (ed) (2017) Extractivismo urbano. Debates para una construcción colectiva de las ciudades. Ed. Fundación Rosa Luxemburgo, Buenos Aires
Wacquant L (2007) Los condenados de la ciudad: gueto, peripherias, Estado. Buenos Aires, Siglo XXI


**Fernando Carrión Mena** Research Professor in Flacso Ecuador. His focus of study are the topics: housing, urbanization process, city, historic centers, cultural heritage, violence, security and drug trafficking, borders, decentralization and sociology of football, among others. He created 8 thematic magazines (political science, security, city, historical centers, borders), wrote more than 1000 journalistic articles and 306 scholars, published 74 books (editor of 52 and author of 22) and edited 12 book collections (97 volumes). He produced 4 cinematographic documentaries. He has worked as consultant for multilateral organizations and as a university professor. For his career, he won 9 awards, 6 decorations and 5 distinguished citizens distinctions. He was director of CIUDAD, FLACSO, Municipal Planning, as well as a former Councilor and former Advisor to the Municipality of Quito. He has published more than 310 academic articles, 27 books as author and 53 books as editor.

# Chapter 3
# Death By Theory and The Power of Ideas: From Theories of Cities to "Smart" Cities

**Michael Cohen**

**Abstract** This chapter focuses on the power of ideas and how the adoption of certain ideas and theories serves as the substantive basis for policies. Ideas affect urban behavior. By themselves they do not "kill" urban areas, but their impacts in many forms and processes contribute to the worsening of urban conditions. The chapter cites the impacts of theories and concepts such as theories of legality which lead to exclusion, or intra-urban inequality; terms such as marginal; distinctions such as formal versus informal; and conventional economic concepts such as income, wealth, and investment. The use of these concepts in an international context of United Nations Habitat conferences, the adoption of Sustainable Development Goals and the New Urban Agenda, and growing concern about climate change raise the importance of clearly understanding how "theory".

**Keywords** Cities · Urban development · Urban policy

## 3.1 Introduction

This chapter focuses on the power of ideas and how the adoption of certain ideas and theories serves as the substantive basis for policies. Ideas affect urban behavior. By themselves they do not "kill" urban areas, but their impacts in many forms and processes contribute to the worsening of urban conditions.

Theory is an important tool in helping us understand how the world works. Its greatest analytic power lies in explaining patterns of causation through demonstrating the relationships between concepts, showing why and how things happen within a world of great complexity. Its normative power comes from asserting how things should be and what social outcomes will be best for society and/or categories of its members. The difference between the analytic and normative voice is very important, because *confusing how things work with how things should work is fundamentally different.*

M. Cohen (✉)
Milano School of Policy, Management, and Environment, The New School Milano, New York, USA
e-mail: cohenm2@newschool.edu

If theory is a necessary component of positivist thinking, theory also potentially has enormous negative power in asserting, often with unjustified certainty, that specific patterns of causation are the only reasonable and compelling explanations for the origin and significance of phenomena. To say, simply, "no, it does not work that way", is an authoritatively sounding assertion of fact.

However, while such assertions are frequently not proven, they are nonetheless often accepted and disseminated by the great majority of people so that an assertion is transformed from a hypothesis to a statement of fact. This process is very profound and affects many spheres of human behavior, regardless of the extent to which their assertions are true or false. As a supporter of former US President Richard Nixon replied in 1973 after hearing about the Watergate break-in, "don't try to confuse me with the facts".

This article examines how some theories applied to urban areas have had dramatic and long-term negative impacts on the management of urban areas, despite that their assertions are not either correct or proven. The consequences of these impacts can be characterized as contributing to the undermining of the quality of life of urban residents and, in the extreme, to "the death of cities".

If this assertion itself seems extreme, it is worth considering the relationship between ideas and specifically theoretical assertions and the existing levels of urban poverty, intra-urban inequality, environmental deterioration, growing challenges to public health, and increasing levels of unmet demand for urban services by urban residents. This essay suggests that ideas have a powerful role in determining public and private attitudes and understanding of urban issues. Further, it argues that some ideas have contributed to policies which have resulted in growing levels of deprivation, conflict, violence and, in some cases, outmigration from specific urban areas. These phenomena should not be assumed to be "natural" or "costless" to the families concerned, but rather as a direct consequence of policies that have, in many cases, failed to understand and/or resolve these urgent problems.

These issues are particularly relevant to the current historical moment for several important reasons:

**First**, it is now more than five years since the global community met at the United Nations Habitat III Conference in Quito, Ecuador in October 2016, and adopted the New Urban Agenda, a 22-page list of 54 commitments made by national governments to support the sustainable development of urban areas (UN Habitat 2016). This document is a noteworthy example of the application of theory as the remedy of massive problems at the urban level in countries around the world, when the "smart cities" movement gathers in strength, actively supported by technology companies, and the ubiquitous urban impacts of the COVID-19 pandemic suggest that relationships between cause and effect can be identified, understood, and even managed.

**Secondly**, urban issues were included in the list of Sustainable Development Goals (SDGs) adopted in 2015 by national governments at the United Nations General Assembly. The SDGs include a statement of global objectives about cities, in SDG #11, and targets relating to various urban challenges (UN 2015). SDG #11 is notable, because it does not include the words poverty, inequality, or productivity, all of which must be central concerns for cities and towns. Those issues are included in other

SDGs, suggesting that, while they are important, the authors of the SDGs believed that it was possible to divide up objectives and not to insist, in the end of a long negotiating process, that the SDGs needed to reflect some level of integration. This issue will be discussed later in this article.

**Thirdly**, the issue of *which* theories are appropriate foundations for urban policies by national and local governments is very much in debate. This debate has centered around the assertion that "smart cities" approaches can provide the basis for more "scientific" and "systematic" analysis of urban problems and the identification of solutions to address them. The subject of smart cities is largely based on the notion that more complete data and information about cities can become the basis of policy and budgetary decisions at the urban level. This is well-captured in the recent book, *Urban Operating Systems: Producing the Computational City* (Luque-Ayala and Marvin 2020) which presents the different analytic elements of how cities can be understood as computational spaces in which specific objectives can be identified and pursued using quantitative data. There has been a plethora of books about the promise of the smart city, as well as some very compelling counter arguments such as those provided by Adam Greenfield in *Against the Smart City*, (Greenfield 2013) who asserts that smart city approaches tend to assume that urban areas are homogenous, are all the same, and thus do not recognize the importance of local differences whether ecological, cultural, spatial, material, or economic. He argues that smart city approaches are therefore reductionist and unable to identify what specific cities require in terms of policies. Greenfield's position fits well with three decades of urban studies going back to the 1980s which articulated the singular importance of "the local". Many scholars and observers have insisted on this local perspective (Lippard 2007; Kemmis 1995).

This debate has evolved over the last decade and now very much reflects the central role of digital platforms in organizing and managing data which are used for a variety of social purposes. A 2021 article on this issue is titled "Platform Urbanism and the Impact on Urban Transformation and Citizenship", thereby asserting the central technological role of distinct digital platforms to produce different outcomes in cities as diverse as Singapore, Seoul, Rio de Janeiro, or London (Hanakata and Bignami 2021).

A **fourth** reason for the importance of this issue is growing appreciation of the impact of cities on climate change as well as the effects of climate change on urban areas themselves. Rising temperatures around the world very much confirm that climate change is indeed a global phenomenon with growing impacts on all forms of life. Inferences from the scientific evidence about climate change can imply different policy directions for cities.

Finally, a **fifth** arena in which the role of theory is very important is the experience and understanding of the COVID-19 pandemic in urban areas around the world. Patterns of contagion were first understood in terms of the role of population density as a contributor to higher rates of contagion. Later analysts observed that the different definitions of density at varying levels of scale suggested much more complicated patterns of causation, raising significant concerns. While some argued that density was "green" and therefore "better", the need for "social distancing" to limit contagion and reduce the demand for health infrastructure implied that density was to be avoided.

All of these examples suggest that "death by theory", therefore, is not simply a provocative title, but rather poses a much broader question: Does reliance on theories and explanations of urban behavior and phenomena lead to a situation where we can no longer see urban differences and, therefore, we implicitly accept the fundamental assumptions of smart cities that urban behavior is similar across urban areas? And going further, does a blindness to seeing local differences actually lead to bad policies that harm cities and their residents?

## 3.2 Powerful and Dangerous Theories

Given this context, it is worth examining examples of how some theories have proven to have had very dramatic and decisive consequences for urban areas. These examples are presented in this essay as a cautionary note to those who believe that well-articulated policies and examples of urban analysis can immediately be assumed to have positive effects. Rather, these examples suggest that the consequences these theories legitimize can also become justifications and instruments for social exclusion, discrimination, stigmatization, and can institutionalize intragenerational misery and deprivation. While some argue that we need an "urban science" or a "science of cities", this article argues that the theories implied in such systematic approaches may actually have serious negative consequences for cities and their lower income groups.

The essay briefly presents the following types of theories.

### 3.2.1 Theories of Legality Which Become Theories of Exclusion

One of the most powerful legal instruments in cities in developing countries are building codes. These codes, often inherited from earlier British, French, Portuguese, or Spanish colonial rule, legitimize certain kinds of buildings, the materials with which they are built, and their spatial and physical dimensions. Such codes describe what is "a proper house", regardless of whether their costs are too high for 80% of the local urban population to pay. In some cases, these codes are more than a century old and bear little relevance to the current local economy, building costs, and the availability of urban land. Yet, these codes are vigorously applied by municipal authorities and force large numbers of urban residents to build housing which is "out of code" and therefore illegal. The bulldozer has become the logical consequence of building codes in some cities (Turner and Fichter 1972).

In Buenos Aires, the local water supply company decided that it could not connect "out of code" houses to the municipal water supply system. The result was that families living in "out of code" housing were forced to pay higher rates for a liter of drinkable water, which they had to buy from private water vendors, than those in "proper" houses connected to the public water supply system (Cohen et al. 2004). This situation has been well-documented in Ethiopia, India, Indonesia, Yemen, Honduras, and many other countries, where poor households living in "out of code" housing pay private water vendors a multiple of the legal price for a liter of water. The result, as demonstrated during the pandemic, millions of urban households could not afford "legal" water, and, among other consequences, were unable to "wash their hands" as the World Health Organization was advising everyone on the planet to do. Indeed, in 2020, the United Nations once again repeated that there were more than 2 billion people without adequate water supply.

In the 1980s—declared by the United Nations as the International Drinking Water and Sanitation Decade—more than 550 million people received potable water as a result of international assistance and national efforts. Yet at the end of the decade, more people did not have potable water than at the beginning of the decade.[1] Population growth and making it illegal for poor households to receive "legal water" contributed to this problem.

The water example suggests that "theories of legality" within urban areas can become "theories of exclusion", justifying why it is not "legally possible" to provide public services to homes which have been built with other materials. The human cost in health, longevity, labor productivity, and environmental sanitation is enormous.

## 3.2.2 Theories of Recognition: Formality and Informality

One of the major theoretical themes repeated in the economic and social literature is the dualistic distinction between the formal and informal sectors in employment and housing. This distinction was established in 1972 by the International Labor Organization's Mission to Kenya to examine informal employment (ILO 1972). This was later followed by studies in India (Calcutta), Indonesia (Jakarta), Nigeria (Lagos), Philippines, Brazil (Sao Paulo), Cote d'Ivoire (Abidjan), and Colombia (Bogota) and first published in the academic literature in 1973 by Keith Hart in his article on economic activities in Ghana (Hart 1973).

Forty-five years later, the ILO published another major report, in 2018, asserting that 60% of global employment and 90% of employment in developing countries was informal (ILO 2018). This stunning conclusion completely turns analyses of national economies on their heads by demonstrating that statistical evidence proves that the dualistic representation of national and urban economies has been completely

---

[1] UNDP, *International Drinking Water and Sanitation Decade, 1970–1980: Results*, (New York: UNDP, 1981).

incorrect. Moreover, these analyses have generated a prejudicial attitude toward informality by national and local governments who believed that informal economic activities represented a minority of total activity and was, by definition, "not modern" and thereby not worthy of official recognition or policy support.

Theories of informality have led, therefore, to a failure to recognize the reality of employment and income generation in most economies and this failure has undervalued the enormous contribution of informal employment and production to national and global economic performance.[2] If 90% of employment in developing countries is informal, and about 80% of GDP in most developing countries is generated in urban areas, one can conclude that both national and urban economies are heavily fueled by the informal sector.[3]

Instead of focusing on economic policies to support formal sector enterprises, for example with loans for equipment or working capital, international agencies and national governments should have been recognizing the role of informal enterprises and alleviating their constraints on productivity. They should now ask who are the majority of employers and firms in economies and respond by affirming their support for informal enterprises. In simple terms, why should the majority of economic enterprises in developing countries be discriminated against while only 10% of employment, in the formal sector, is considered the only legitimate source of income?

### 3.2.3 Theories of Description: The Myth of Marginality

Another example is the use of the term "marginal" to describe slum dwellers in Latin America. This theory of description, going back to the 1950s, was disproven in 1976 by the North American social scientist, Janice Perlman, in her award-winning book, *The Myth of Marginality*, which demonstrated that slum dwellers in Rio were not unemployed, were not victims of broken families, were not promiscuous, and were practicing churchgoers (Perlman 1976). This research convincingly demonstrated that the characteristics of the so-called marginal populations living in the slums of Rio de Janeiro were not "deviant" or worthy of derision. Rather, the favela dwellers were poor people who, despite their low incomes, worked hard to improve their incomes and the lives of their children. The study contrasted sharply with the incorrect use of the descriptive term "marginal" by policymakers in Brazil and throughout Latin America. A follow-up study by the author in 2010 showed that this misuse of the term marginal continued to be incorrect (Perlman 2010).

---

[2] For an excellent analysis of the process of recognition of informality, see Maria Carrizosa, *Working Homes: Space-use Intensity and Urban Informality in Bogota*, Publication # 28498120, Doctoral Dissertation, The New School, 2021.

[3] The assertion that 80% of GDP is generated in urban areas has evolved from the World Bank analysis in 1991 that 50% of GDP came from cities and towns. See Michael Cohen, *Urban Policy and Economic Development: An Agenda for the 1990s*, (Washington; World Bank, 1991).

## 3.2.4 Theories of Validation

The use of derogatory terms to describe the urban poor not only reflects deep and pervasive ignorance in policy circles in Latin America and the international community. It also challenges the methods by which most economists have studied, explained, and apparently misunderstood what was happening in developing economies. By failing to perceive the scale and magnitude of informal economic activities, these economists have deliberately ignored what must be contradictions in their own assessments of the scale and quantitative composition of Latin American economies. For example, how could they be accurate in projecting national economic performance if they did not have accurate estimates of 90% of economic activity? In what professional field can projections be made on the basis of 10% of the sample? This question raises serious questions about the integrity of these economic analyses, whether carried out at the national government level or in international agencies such as the IMF, World Bank, Inter-American Development Bank (IADB), or the Caja Andino de Fomento (CAF).

Going further, if macro-economists calculate that 70–80% of GDP is generated in urban areas, this suggests that the majority share of national GDP is itself informal and, consequently, has therefore been badly under-estimated for decades. In fact, it is a shocking reality that economic analysis is so biased against informal employment that economists are unable to see what is happening in cities around the world. As a critic of economists in the World Bank once noted, "always certain, occasionally correct".

## 3.2.5 Theories of Profiling

Theories of racial and ethnic profiling have become widespread justifications for police stopping and, in some cases, arresting people in cities around the world. The so-called stop and frisk policy of New York City Police under the administration of Mayor Rudolph Giuliani and then continued under Mayor Michael Bloomberg, proved to be a violation of the human rights of "people of color" in New York. Similar profiling occurs in other cities, such as Buenos Aires, where young dark-skinned men from the Andes, whether Argentina, Chile, Bolivia, or Peru are similarly discriminated against by white Argentines.

Theories of profiling have become increasingly dangerous over the last few decades, as the police murder of African Americans all over the USA demonstrates systematic police bias against people of color. The most well-known case of this profiling is the case of George Floyd who was murdered on May 25, 2020, by Minneapolis, Minnesota police. This case led to global protest marches and the conviction of Derrick Chauvin, the police officer who killed Floyd.

## 3.2.6 Theories of Intra-Urban Inequality

It is also very significant that, in the so-called post-COVID-19 environment, so many people are only now discovering the enduring power of intra-urban inequality in most urban areas. Structural racism and the role of informality have long been important features of so-called industrial societies. This is hardly a surprise. Indeed, the pandemic might be viewed as the symptom and not the cause of this realization. With millions of the urban poor deprived of adequate housing, clean water, sanitation, and health services during the pandemic, this lack of services has certainly contributed to the scale and severity of the pandemic itself, whether in rich or poor countries. A horrifying photo in spring 2020 of slum dwellers fighting for food in the Kibera neighborhood of Nairobi, the largest slum in Africa with more than a million residents, demonstrated the acute lack of food for the urban poor. This deprivation most certainly contributed to the contagion in the pandemic as malnourished people are likely to have weaker immune systems than better-fed populations.

A critical dimension of intra-urban inequality is the difference in incomes between rich and poor in all cities. These differences reflect longstanding generational differences in economic and educational opportunities between rich and poor. School attendance and the quality of education are highly correlated with neighborhood per capita and household incomes.

But they also reflect differences in the distribution of public expenditures by national and municipal governments. A study in the late 1990s showed that 11% of the population of Buenos Aires received 68% of municipal expenditures for infrastructure and public services (Cohen and Debowicz 2004). A conclusion of this study was that "if I know your zip code, I can predict the future of your children". Similar patterns exist in many cities, even if such stark contrasts are rarely studied by urban specialists. As one Argentine colleague told me in 1998, when the article was being reviewed, "Doesn't the same pattern exist in New York City?" The answer is most certainly yes. One of the zip codes in the South Bronx of New York City is the poorest in the entire USA and not surprisingly, the social indicators are far below averages within New York and certainly within the country as a whole (WHEDco 2021).

This unequal concentration of public expenditures per capita stands in stark contrast to the average financial bonus received by Wall Street employees which reached $257,000 a year in 2021 (DiNapoli 2022). This payment is in addition to annual salaries and eight times more than the income required for a household of four to remain above the poverty line in New York City.

It is against this backdrop that it is important to recognize that New York lost a million jobs in the first three months of the pandemic and that, more than two years later, many of these informal sector jobs have not yet come back. From 2010 to 2020, New York had experienced a massive increase in wages for labor at the lower end of the income distribution, but this was abruptly interrupted by the pandemic (Parrott 2022). This unequal effect of the pandemic is supposed to be addressed by new municipal government policies as outlined in the new report by Mayor Eric Adams, "Renew, Rebuild, and Reinvent: A Recovery Strategy for New York" (Adams 2022).

The issue remains whether this will actually address the question of intra-urban inequality.

### 3.2.7 Theories of Income and Theories of Wealth

Another interesting and understudied subject is the focus on income rather than wealth. As French economist Thomas Piketty suggested in 2014 in his important book on inequality, if private wealth for some groups increases faster than GDP, then inequality will increase (Piketty 2014). Yet most studies of well-being in the USA focus on income differences rather than wealth differences, thereby understating the important economic differences between races. This finding by Darrick Hamilton represents an important case in which theory, in this case theories of well-being, badly distort our perceptions of what we should be studying and monitoring through time (Hamilton 2021). As Hamilton emphasizes, measuring income and not wealth dramatically fails to assess individual and family possibilities to invest in education of their children, undertake expensive medical procedures, and build assets such as housing, among other important lifetime objectives.

### 3.2.8 Theories of Density

Another important theoretical problem relates to spatial concepts such as density. Up to March 2020 and the outbreak of the pandemic, most urban theorists were arguing that higher density was good; i.e., it was environmentally friendly, it reduced the time and pollution costs of commuting from home to work, it helped to create the cultural identity of neighborhoods, and in general helped to reduce the negative externalities of urban living. This view was found within many statements of the United Nations' New Urban Agenda document approved at the UN Habitat III Conference in Quito in October 2016 (UN Habitat 2016).

This perception changed in a few days after the outbreak of COVID-19. People all over the world realized that if they were wearing masks to protect themselves against contagion with the virus, they also should be avoiding high-density situations, such as public transport, going to movies or sports events, attending school, and many other activities.

Density had suddenly become a problem to be avoided at all costs. The rich were able to move out of dense urban environments and to ride out the storm of contagion by living in rural areas or near the beaches.

Even though the debate about density and contagion has been raging since the outbreak of the pandemic in March 2020, it is not at all resolved. Analysts have pointed out that many cases appeared in small, low-density towns, so the correlation between density and well-being was not so tight. It remains to be seen how this debate will be resolved, but also whether individual families will decide whether they are "safer" in the suburbs or less dense residential environments.

The example of outmigration from New York City to surrounding states as a result of high housing prices over the last decade suggests that when existing financial burdens have become too onerous to overcome, urban residents vote with their feet. When a two-bedroom apartment in Harlem costs $2400 a month in rent plus utilities, it is time to consider paying half that amount in a mortgage to buy a modest home in nearby suburban New Jersey. The quantitative significance of the New York housing market has turned out to be significant at the national level in the USA. Two economists from the University of California at Berkeley, Enrico Moretti, and Chiang Tai-Hsieh, have demonstrated that high housing costs in New York City and Silicon Valley in California together have a significant statistical effect on reducing the GDP of the USA (Hsieh and Moretti 2019).

This outmigration also happened in New York City during the COVID-19 pandemic when high levels of hospitalization and death frightened people from continuing to live in a high-density environment lacking adequate medical care. Seeing freezer trucks filled with dead COVID-19 victims lining New York's streets was enough to scare people to seek other places to live, especially after the City Government erected emergency tent hospitals in 2020 in Central Park. One analyst of these movements wrote of New York "hemorrhaging people" in 2020, while conceding that many of these people returned from the upscale Hamptons area of Long Island in 2021 when COVID-19 deaths had declined (Offenhartz 2022). Yet real estate prices in New Jersey have continued to increase, much as the United States has experienced "The Great Retirement", as millions of workers have quit their jobs, preferring life at home and avoiding their prior work environments.

These phenomena are not just the privileged behaviors of wealthy people in wealthy countries. In the 1960s, when Calcutta was widely considered to have the worst urban conditions in the world, largely due to its slums, its lack of water and sanitation, and to the millions of people sleeping in the streets, migration to Calcutta from the desperately poor rural areas of West Bengal in India completely stopped. Rural families had heard that life in Calcutta was worse than their plight in rural villages. How people perceive urban phenomena affects how they understand them.

## 3.3 Smart Cities and the Death of Localism and Local Knowledge

As Adam Greenfield wrote in 2013, the theories about smart cities represents one of the most serious historical impacts about cities, (Greenfield 2013) assuming that urban areas are basically alike. This assertion of homogeneity failed to acknowledge and recognize that the local attributes of individual urban areas are often their most significant characteristics, including for example their ecology, their location, their economic functions, their cultural histories and values, and their accomplishments, whether local orchestras, art museums, political organization, and/or their policies which may or may not be particularly successful in dealing with a specific problem. The "lure of the local" is an assertion of something special about an individual locality, suggesting that these special features attract visitors, encourages repetition in another place, or explains why its local population wants to live there in the first place.

A particularly important aspect of this argument is the role of local knowledge as a critical input affecting local decisions and long-term decisions about the future of the place. Without local knowledge, there is no explanation of why the place has its present characteristics, as well as the likelihood that the narrative of the place will be considered worthy of attention and safe keeping.

## 3.4 Toward a Conclusion

This brief essay has not intended to claim that all theory is bad. The functions of theory which help us to understand causation are necessary to enable us to avoid making the same mistake twice. But that said, it is nonetheless true that there are many examples where simplified explanations have had enormous impacts on how we see the world and particularly cities. Flawed or overly simplified theoretical explanations have significantly damaged our understanding of urban phenomena and distorted our normative judgments, particularly in terms of policy, regulations, and rules more generally. It is time to revisit our theoretical assumptions to see if they are still relevant or whether their explanatory power is no longer justified.

## References

Adams ME (2022) Renew, rebuild, reinvent: recovery strategy for New York. New York City Government, New York
Carrizosa M (2021) Working homes: space-use intensity and urban informality in Bogota. Doctoral dissertation, The New School Publication #28498120
Cohen M (1991) Urban policy and economic development: an agenda for the 1990s. World Bank, Washington[4]
Cohen M, Debowicz D (2004) The five cities of Buenos Aires. UNESCO, Paris
Cohen M, Brailowsky A, Chenot Camus B (eds) (2004) Citizenship and governability: the unexpected challenges of the water and sanitary concession in Buenos Aires. The New School and Aguas Argentinas, New York and Buenos Aires
Di Napoli T (2022) State comptroller of New York. Weekly News
Greenfield A (2013) Against the smart city. Verso, London
Hamilton D (2021) Examining the racial wealth gap in the United States. Testimony Before the Joint Economic Committee, US Congress, May 12

---

[4] The assertion that 80% of GDP is generated in urban areas has evolved from the World Bank analysis in 1991 that 50% of GDP came from cities and towns.

Hanakata NC, Bignami F (2021) Platform urbanization and the impact on urban transformation and citizenship. South Atlantic Q 120(4):763–776
Hart K (1973) Informal economic opportunities and urban employment in Ghana. J Modern Afr Stud 11(1):61–89
Hsieh C-T, Moretti E (2019) Housing constraints and spatial misallocation. Am Econ J Macro-Econ 11(12):1–39
International Labor Organization (1972) Employment, incomes, and equity: a strategy for increasing productive employment in Kenya. ILO, Geneva
International Labor Organization (1972–1980) Studies on informality 1972–1980. ILO, Geneva. The author provided data for the Abidjan paper in the early 1970s
International Labor Organization (2018) Women and men in the informal economy: a statistical picture, 3rd edn. ILO, Geneva
Lippard L (2007) The lure of the local: senses of place in a multi-centered society. The New Press, New York. Or Kemmis D (1995) The good city and the good life: renewing the sense of community. Houghton Mifflin, Boston/New York
Luque-Ayala A, Marvin S (2020) Urban operating systems: producing the computational city. MIT Press, Cambridge
Offenhartz J (2022) New York has regained three quarters of residents who fled during COVID, data suggests. Gothamist. February 21. Greenfield, op.cit.
Parrott J (2022) Full employment and rising wages: New York's twin challenges of emerging from the pandemic. The Center for New York City Affairs, New York
Perlman JE (1976) The myth of marginality. University of California Press, Berkeley
Perlman JE (2010) Favela: four decades of living on the edge in Rio de Janeiro. Oxford University Press, Oxford and New York
Piketty T (2014) Capital in the twenty-first century. Harvard University Press, Cambridge
Turner JFC, Fichter R (1972) Freedom to build: dweller control of the housing process. Macmillan, Houndsmill
United Nations (1981) International drinking water and sanitation decade, 1970–1980: results. UNDP, New York
United Nations (2015) Sustainable development goals. United Nations, New York
United Nations (2016) The new urban agenda. United Nations, New York
Women's Housing and Economic Development Corporation (WHEDco) (2021) WHEDco, New York

**Michael Cohen** is Director of the Doctoral Program in Public and Urban Policy at The New School and a PhD in Political Economy from the University of Chicago. Chief of Urban Development Division at the World Bank. He worked in 55 countries. He has written widely on development policy, urban development, Africa and Latin America.

# Chapter 4
# Urbicide: Towards a Conceptualization

**Maria Mercedes Di Virgilio**

**Abstract** According to the United Nations (2018), since 2007, we are living in a predominantly urban world. By 2030, more people in all world regions will live in urban rather than rural environments, even in Asia and Africa. In Latin America, more than 80% of the population lives in cities. This chapter presents a systematization of contributions from various authors from different latitudes of the world to conceptualize urbicide. It examines different definitions and case studies, from North to South but with Latin American reality as a background, presenting perspectives and discussions around the concept. We are witnessing an era of growing cities, but they do not seem to have the capacity to house people facilitating their daily lives, but rather urbicidal practices and processes promote the destruction of the city. Therefore, this chapter suggests these could be understood as "urbicidal cities" by presenting other features of urbicide and different ways in which urbicidal violence is expressed, examining three cases of urbicide experiences from the South: (i) the eradication of slums (the case of the City of Buenos Aires); (ii) abandoned neighborhoods with empty houses (the cases of Querétaro, Monterrey, and Ecatepec in Mexico); (iii) urban catastrophes (the case of Port-au-Prince in Haiti).

**Keywords** Urbicide · Urban environment · Latin America · Urban sociology

## 4.1 Introduction

This chapter presents a systematization of contributions from various authors from different parts of the world to conceptualize urbicide. It examines different definitions and case studies, from North to South but with Latin American reality as a background, presenting perspectives and discussions around the concept.

According to the United Nations (2018), since 2007, we have lived in a predominantly urban world. By 2030, more people in all world regions will live in urban rather than rural environments, even in Asia and Africa. In Latin America, more than 80%

---

M. M. Di Virgilio (✉)
Universidad de Buenos Aires (UBA), Buenos Aires, Argentina
e-mail: mercedes.divirgilio@gmail.com

of the population lives in cities. In addition, cities have achieved such prominence that, together with states and large global corporations, they have become the three most important actors at the planetary level (Carrión 2018). However, we are heading a contradictory path: an increasingly urban world in a scenario that announces the death of the cities. Never before was the city threatened in such a massive way, and under a great diversity of situations, it could be classified under the concept of *urbicide*. The city death does not seem to attack the city as a geographical space but as an "experiential space" (Löw 2013). Therefore, urbicide appears to relate more to the death of cities by and for the people. According to United Nations statistics (2020), approximately 1.6 billion people live in substandard housing, and 100 million are homeless. In Latin America, one of the poorest regions on the planet, one out of every three families—a total of 59 million people-lives in inadequate housing or housing built with precarious materials or lacking essential services. Nearly two million of the three million families formed each year in Latin American cities must settle in informal housing because of an insufficient supply of adequate and affordable housing (Bouillon 2012). The housing deficit has systematically increased since 1990. Without a profound change in this trend, the housing shortage will continue to be one of the most significant challenges in the region in the coming years. Even when urban areas are expanding as cities consume more land, the housing crisis continues to deepen. Far from being a symptom of poorer economies and countries, the housing crisis has become a significant challenge in urban centers, impacting people, and communities worldwide (Fields and Hodkinson 2018).

We are thus witnessing an era of growing cities, but they do not seem to have the capacity to house people facilitating their daily lives. Therefore, I suggest we could understand those as "urbicidal cities that, "bound by its instrumental role for neoliberal urban growth, results in the urbicide of preexisting socio-spatial forms of urban life" (Lesutis 2021: 1197).

Based on a bibliographic review, we identify cases that allow us to characterize in a general way the urbicidal processes and practices in contemporary contexts. The chapter is organized into four sections, apart from this introduction and the conclusions. It begins by presenting the origins of the concept of urbicide and its different meanings associated with the changes in urban policies and renewal processes. It continues by pointing out the features of urbicide violence and how it relates to the destruction of the cities. The third section presents the mechanisms involved in urbicidal practices and processes. Then, the chapter points out the urbicide experiences in Latin American cities, presenting the City of Buenos Aires (Argentina) case to illustrate the urbicide processes through the eradication of shantytowns. Mexico's experience points out the abandoned neighborhoods with empty houses and the Haitian suffering as an example of urban catastrophe. Finally, we present the main conclusions.

## 4.2 The Concept of Urbicide

In the literature, the concept of urbicide emerges associated with urban renewal processes in the postwar USA.[1] Marshall Berman (1996) coins the notion by evoking the process of creative destruction that occurred in the Bronx since the middle of the twentieth century. The physical and social destruction of the area began in the late 1950s with the construction of the major freeway in the district—the Cross Bronx Expressway—and spread gradually southward during the early 1960s. The structure of the monumental work devoured hundreds of houses and blocks. In 1987, The New Internationalist journal declared that, in the 1970s, alone in the South Bronx, home destruction displaced more than 300,000 people. The use of the concept associated with urban renewal processes continues until now. For example, Günay (2015) shows how the renewal schemes employed to resolve the urbanization problem are turned into the instruments of "urbicide as a political model of urban destruction in Istanbul. In other latitudes, other studies show similar consequences. For example, Hall (2016) argues that redevelopment practices implemented to face the "low-access food desert" in a poor black Miami neighborhood (Overtown) far to secure food security are undermining networks of social and economic interdependency in the existing foodscape. Thus, the urban renewal practices reproduce the spatial and racial urbicide through more overt forms of racism and spatial violence. As a result, the so-called food desert landscape mutates toward gentrifying foodie districts. Rodgers (2009: 965) uses the concept to refer to the elite-led process of urban disembedding in Managua (Nicaragua). In their struggle against "the utopian attempts of the Sandinista revolution to foster more egalitarian forms of social organization," they underwent the citys urban transformation sponsored either by unregulated private initiatives or directly by the elite-captured Nicaraguan State.

However, the use of the concept became widespread by analyzing neocolonial military strategies to destroy built urban environments to crush organized resistance (Coward 2004) and deploying warfare and militarized urban planning in neo-imperial wars (Graham 2003, 2011). The most treated cases in the bibliography are related to military conflagrations (such as Baghdad, Iraq; Aleppo, Syria). In addition, though, other forms of para-militarization linked to terrorism (Lima, Peru; New York, USA) or drug trafficking (Ciudad Juarez, Mexico; Medellin, Colombia) are associated with urbicide (Carrión 2018). In all these scenarios, destruction goes through the annihilation of the enemy by killing large numbers of civilians (Coward 2018).

Urban militarization followed by displacement could be at the base of urbicide practices—in the war contexts or out of. It is a condition of "urban cleaning" or "pacification" interventions that generate the appropriate context to develop mega-events such as the World Football Cup or Olympics Games. "As it prepares to host the upcoming mega-events, Rio (Brazil) has sought to reduce crime in the *favelas* to

---

[1] The federal urban renewal program adopted in 1949 included planning, regulation of private land use, and financial aid to stop urban deterioration. Urban renewal was oriented to slums, deteriorated and deteriorating areas that could be vacant or have predominantly residential, commercial, or industrial use (Johnstone 1958).

improve the citys public image and secure the areas proximate to event sites (Watts 2013a). The city government has instituted the Favela Pacification Program (FPP) [...] Rio's "shock of peace" (Smale 2011) strives to "pacify" both the drug trade and social unrest through the militarized occupation of the *favelas*, the long-awaited provision of social services (Bailly 2011), and the integration of local businesses and properties into the formal economy" (Fisher 2014: 3). Tanks, armored vehicles, rifles, and trained police dogs occupied these areas. The result was hundreds of people injured or killed. Others suffered displacement due to peacekeeping destruction (Romero 2012; Garcia-Navarro 2014).[2] Residents who stayed in the neighborhoods fear that the FPP actions will also lead to gentrification and, consequently, more displacement.

Since then, we can identify more meanings for the concept of urbicide in the literature. For instance, urbicide is also linked to ethnic or religious wartime reconstruction strategies and the effects of urban redevelopment practices. Based on the case of Syria's civil war in 2015, Di Napoli (2019: 255 ss) "demonstrates that urbicide should not be seen as a discrete destructive event or even series of events occurring during a period of direct hostilities." Instead, it ought to be interpreted expansively. Furthermore, it is an ongoing process in which "postwar reconstruction is not the solution to urban destruction but rather the continuation of such violence." The underlying recovery process logic may result in post-conflict authoritarian economic patronage practices and selective recognition of property assets—urbicidal in purpose and effect. Traditional models of post-conflict property restitution usually fail to properly consider the linkages between methods of destruction and possibilities of future conflict, permitting urbicidal reconstruction and the persistence of authoritarianism.

More recently, the concept has been associated with the spatial violence of poverty or austerity policies.[3] As another way of creative destruction, austerity promotes ruined geographies and blasted landscapes, inflicting an insidious spatial trauma (Shaw 2019). The spatial trauma seems to be racialized (also gendered and class-based), addressing variegated race, class, gender, and austerity interactions. Thus, austerity policies do not govern in a homogenous manner but instead operate on historically and geographically raced and classed terrains of domination (see Phinney 2020). Other urban policies—or the lack of them—could also promote the creative destruction of the city. It can be related to "the weight that the logic of the market has around tourism (Venice, Italy; Barcelona, Spain), gentrification (London, England; Santiago, Chile), gated communities (Buenos Aires, Argentina; São Paulo, Brazil), the weight that the automobile acquires (Miami, USA; Lima, Peru) and the undermining of the self-government of the city, through centralization and privatization processes that eliminate urban planning, conceiving it as project planning that stimulates real estate businesses" (Carrión 2018: 6).

---

[2] "In these communities, residents are served eviction orders, minimally compensated, and resettled in areas on the periphery of the city." https://crownschool.uchicago.edu/shock-peace-intersection-between-social-welfare-and-crime-control-policy-rio-de-janeiro%E2%80%99s-favelas.

[3] "Austerity is a common name globally for neoliberal policies of public-service cut-backs and pro-market discipline" (Sparke 2017: 287).

Finally, Lesutis (2021: 1197) focuses on the urbicide "as constituted through natural resource extraction [...] [showing] how urbicidal extractivism is a part of a broader multi-scalar territorial reshaping of social relations of contemporary capitalist development characterized by multiple forms of socio-spatial destruction."

In this way, the social sciences have adopted the concept of urbicide to describe the different forms of violence applied to a city—or city areas such as the favelas in Rio de Janeiro—as a whole. The common point in these experiences is the destruction and weakening of the record that gives memory, culture and identity, and material support to cities and their inhabitants (Aguirre and Baez Gil 2021).

## 4.3 The Features of Urbicide Violence

Urbicidal practices and processes promote the destruction of the city. Destruction is always "widespread and deliberate" (Coward 2013: 201), reducing the city to rubble and making it practically impossible to inhabit and/or reuse any material. They represent a unique and paradigmatic form of material violence that aims to destroy the constitutive elements of urbanity (Stenberg 2010; Coward 2007). Coward's understanding of the killing of urbanity has three critical dimensions. Firstly, urbicide is seen as the killing or slaying of the citys structure, form, and experience. Second, the killing of urbanity is related to a necessary destruction process of the modern city. Multiple remodeling forces—the car, the architect, the urban planner, and the capital—are responsible for killing the modern city. "In some senses, then, this urbicide is neither avoidable nor, ultimately, condemned. Indeed, without this killing of the urban, the city would stagnate and become moribund. However, this brings us to the third facet of this conception of urbicide." Such destruction attempts a particular ethical and political impulse toward maintaining or revivifying an urban ideal characterized by diversity, modernity, and functional and aesthetic integration of the urban environment. Also, it is associated with "the democratic utility of the city to emancipatory politics (Davidson and Iveson 2015: 648). Ultimately, urbicide refers "to the loss of an experience specific to the city through the disruption, destruction or remodeling of the material infrastructure on which that experience is predicated" (Coward 2007: no page).

In all urbicidal practices and processes, we can recognize the symptoms of political violence instead of the signs of physical coercion. We talk about political violence because, in urbicide practices and processes, violence seems to be subordinated to achieving political goals. Sometimes as in the war context, political goals are the objectives of some national states with colonizing or domination desires. In a civil or ethnic-military conflict, political goals are the (social, racial, religious, etc.) regimen that part of the society—in general, the most powerful—aims to impose on the other. Finally, in policy implementation (e.g., urban renewal, reconstruction processes, austerity policies, etc.), the purpose of homogenizing lifestyles and the urban experience plow through poor built-up areas with impunity.

Through physical and political violence, urbicidal practices and processes promote the symbolic violence associated with destroying memory, history, and the collective effort. Urbicidal practices and processes colonize places of memory, expelling the historically constituted physiognomy and life forms to kill the previous daily life (Carrión 2018). Urbicide, then, as a fundamentally political issue, represents the violent exclusion of the possibility and the condition of being with others.

All the different forms of violence are involved in the urbicidal practices and processes. The aim of all these forms of violence acting together is the systematic destruction and weakening of the support of the public and the inhabitants as a condition for the possibility of urbanity.

Paraphrasing Brenner and Schmid (2015), it is argued that when the built environments can no longer manage such forms and levels of violence, they are creatively destroyed through urbicidal practices and processes. Then a new exclusive formation of the urban takes place, supported by other social hierarchies.

## 4.4 The Mechanisms Involved in Urbicidal Practices and Processes

In scenarios scared by combined forms of violence and their clear material manifestations, empirical evidence of urbicide cases shows some mechanisms involved in the destruction processes. We could associate urbicide with two essential tools that operate at the base of such practices: dispossession and spoliation.

Dispossession is materialized when people lose their house using different coercive public and private and material and symbolic means, becoming deprived of place, home, citizenship, and modes of belonging (Butler and Athanasiou 2013; Alexandria and Janoschka 2018). The bibliography identifies different means of dispossession. The housing bubble bursting has left thousands of families without homes and indebted for life. (i) Mortgage debts: mortgage financing arrangements permitted lending standards to become laxer. When the growth in household debt proves to be unsustainable, household defaults, foreclosures, in some cases, the loan balance exceeds the house value, and fire sales are undesirable consequences. (ii) Land grabs around the globe (Hadjimichalis 2014). Borras et al. (2012) show that contemporary land grabbing is associated with the power to control land and other associated resources, such as water, to derive benefit from such control. Land grabbing is also linked to the extraction of resources (Wolford 2010). (iii) Displacement in renewal or gentrifying neighborhoods (Janoschka and Sequera 2016; Shin and Kim 2016), especially when the primary buffers against gentrification-induced displacement of the poor (public housing and rent regulation) are being dismantled by policymakers. Finally, we could consider (iv) the social and material enclosure of commons and how they operate across scales and sites (Hodkinson and Essen 2015).

Urban spoliation "refers to the precariousness of collective consumption services that, together with access to land, are shown to be socially necessary for the urban reproduction of workers. […] spoliation can only be understood as a historical product that feeds on a collective feeling of exclusion and produces a perception that material or cultural good is lacking, when socially necessary." It is associated with a "process of denaturalization of violence that permeates the banality of [everyday life]" (Kowarick 1996: 737).

With these two concepts in mind, urbicide could be considered an expression of a spoliative process that dispossesses a city and its inhabitants of their rights to urbanity.

In the context of global capitalism, the State, through its policies, seems to accompany the processes of spoliative progress in globalized cities (Brenner and Theodore 2002; Harvey 2007). It happens through dismantling policies and institutions of the Welfare State and the construction and consolidation of new institutions aimed at facilitating all types of private instruments in production, financing, and access to housing for business purposes. The change of the State's regulatory functions and normative frameworks, the privatization process, and the subsidized transfer of resources—such as, for example, urban land or public debt—are enabling mechanisms that facilitate these processes. The tendency has been to focus on forms of de-regulation to support real estate firms in generating new supply levels (Brill and Raco 2021). However, even in this structural continuity, the instruments that operationalize the processes have varied under different circumstances, establishing local scars in institutional destruction/construction dynamics.

Within the frame of urbicidal practices and processes, urban dispossession and spoliation are often associated with colonial practices: methods of domination, which involve the subjugation of a social group, a state, or other forms of collective life and living organization to another, including political and economic control over a dependent territory (Hidalgo 2016). Moreover, they engage in the form of conquest that is expected to benefit a state or a country economically and strategically (Stanford Encyclopedia of Philosophy 2006).

## 4.5 Urbicide Experiences from the South

Most of the papers that give an account of cases, processes, or urban practices focus their analysis on traumatic experiences that have or took place in cities of the global north (for example, Sarajevo, Vukovar, etc.) or the Middle East (for instance, Aleppo or Jenin, among others). However, in Latin America, these experiences are repeated. In this framework, this section aims to recover cases of Latin American cities that allow us to think about other features of urbicide and different ways in which urbicidal violence is expressed.

## 4.5.1 The Eradication of Slums

Since the end of the 1970s, the City of Buenos Aires began to undergo important transformations in the context of the last civic-military dictatorship. Some authors (Oszlak 1991: 15) suggest that urban, infrastructure, and housing policies implemented during the dictatorship government (1976–1983) have had "serious consequences on the spatial distribution and living conditions of low-income sectors of the Metropolitan Area of Buenos Aires"—including the City of Buenos Aires.[4] Urban policies during the dictatorship in the city were the initial kick that unleashed the expansion of gentrification processes in many of its neighborhoods and continues even today. Even though these policies and the state stakeholders had different and often contradictory rationalities, they laid the foundations for a significant transformation. In fact, during the dictatorship period, urban policies often acted as a bastion for decisive intervention and state investment (Menazzi 2012). The interventions serving as the spearhead for the transformation of the city were many: (i) those that operated on the housing market—the implementation of the Building Code (1977) and the rental market Act 21342/76, (ii) land-use planning instruments—Code of Urban Planning (1977), (iii) those of infrastructure—construction of the highway network and the recovery of "green spaces" and, finally, (iv) eviction policies applicable to undesirable situations (Martel and González Redondo 2013)[5]—eradiation of *villas* (shantytowns) and the militarization of the public space.[6]

Although all the actions of the last civic-military dictatorship had dire consequences in the restoration of the free housing market and on the (in)access of low-income groups to housing in Buenos Aires, the process of eradicating slums was especially perverse. As an urbicide practice, it was based on State violence militarizing the right to the city among these populations and combining material, political and symbolic forms of violence.

In 1980, the former head of the *Comisión Municipal de la Vivienda* (Municipal Housing Commission), Dr. Del Cioppo, stated that: "*living in Buenos Aires is not for everyone but for those who deserve it, for those who accept the guidelines of a pleasant and efficient community life. We must have a better city for the best people*" (Competition, 1980 cited in Oszlak 1991: 78). Under this ideological framework, the military government carried out the most violent slum eradication policy in the citys history, while freeing up rental prices to restore the free market.[7] The coercive eradication of slums was also accompanied by an intense advertising political campaign

---

[4] The Metropolitan Area of Buenos Aires includes the common urban area composed of the City of Buenos Aires (central city) and 32 surrounding municipalities totally or partially corresponding to the urban agglomerate.

[5] They are the actions aimed at expelling "the squatters" who are not legitimate subjects of the right to the city (Martel and González Redondo 2013).

[6] During this period, political activity was suspended, the anti-subversive struggle was intensified, and popular demonstrations were repressed (Oszlak 1991).

[7] Moreover, the operation was initially focused on the settlements of the northern part of the city: the shantytowns located in Belgrano, Retiro, and Colegiales neighborhoods. However, in southern districts, slums were also violently eradicated.

aimed at stigmatizing and criminalizing *villeros* (Oszlak 1991). The dictatorship destroyed the neighborhood improvements population had achieved over years of struggle with great effort and scrapped the organizational fabric. Of the 224,335 inhabitants of the citys slums in 1976, only 40,533 remained in 1980 (Oszlak 1991). The total population of shantytowns in the area decreased from 109,601 inhabitants in 1976 to 14,578 in 1980. In some cases, they were relocated to areas practically unoccupied and reduced to grasslands and rubble. In others, these spaces were occupied by parks and new public spaces (Arqueros Mejica 2013).

Consistent with the period of economic restructuring, territorial interventions sought to promote a change in the citys productive profile, restricting industrial uses to some city areas and encouraging residential, service, and business uses.

> To promote city southern area urban development [where the industrial uses are located], the Municipality amended the Urban Planning Code. It is intended to facilitate the construction of housing and retail businesses without limits on the ground floor. Therefore, building applications [...] shall be exempt from payment of drainage and construction rights and shall not pay the tax for land contribution for five years (La Prensa 30/06/1978).[8]

Sectoral policies aligned with the economic restructuring process focus on market de-industrialization and liberalization. The measures' consequences in the city landscape became evident soon: de-industrialization of the central areas, displacement of numerous industrial activities to the suburbs, and of tertiary activities concentration in the city and clean industries. Thus, the industrial sector lost its ability to absorb labor and left productive facilities (Oszlak 1991). Moreover, these initiatives took place within an institutional framework in which the State abandoned its role as a redistributive agent and did not defend urban services as an object of social consumption (Herzer 1992).[9]

Dictatorial policies strongly impacted the physiognomy and dynamics of Buenos Aires' central neighborhoods and downtown area. The changes in the Urban Planning Code implemented during the civic-military dictatorship (1976–1983) and its subsequent amendments in 1989 generated conditions for the renewal of these areas in the following decades (Guevara 2010).

On the one hand, the changes in the Urban Planning Code enabled the transfer of the lands surrounding the old port of Buenos Aires to private hands. The old Puerto Madero[10] had been abandoned after a few years of its creation and left unused. After its inauguration (in 1889), it ceased to be a port for export ships to become a river port. The port activity was dramatically decreased as infrastructures became obsolete and the process of import substitution was exhausted. Warehouses and cargo cranes

---

[8] The paragraph highlights the positive aspects of this measure—elimination of open-air garbage dumps, construction of recreational parks, and widening and resurfacing of avenues—while avoiding the negative consequences of these processes.

[9] The eradication of slums was the most prominent and brutal measure that expressed the change in the role of the State.

[10] Puerto Madero did not meet the technical conditions to house large vessels and, about ten years after completion, due to the increase in the size of ships—by the end of the 1910s—it became obsolete.

were neglected, along with large unoccupied vacant lots.[11] Based on the regulations in force, in 1989, the area was declared unusable from the port point of view.[12] On the other hand, de-industrialization and rules changes also affected the neighborhoods in the South-San Telmo, La Boca, Barracas, etc.—and the north of the city—Palermo. The process of de-industrialization pushed the shutting of workshops, factories, and warehouses, making a large amount of land and buildings suitable for reusing or demolishing.

### 4.5.2 Abandoned Neighborhoods with Empty Houses

During the 1980s, Latin American countries began implementing a massive housing construction policy articulated with market initiatives and concentrated construction industry sectors. The forerunner of such initiatives in the region was Chile, where the Ministry of Housing and Urbanism, with the Chilean Chamber of Construction, produced social housing for 12% of the country's population, some 600,000 families, between 1980 and 2006. The production model—spread in other Latin American countries—was based on a financing mechanism in which public and private actors converged and combined savings, subsidy, and credit (Rodríguez and Sugranyes 2011). In this frame, the State provided the housing subsidy, and the beneficiary families saved and took out a housing loan. The construction companies were responsible for producing low-cost housing with a guaranteed financial profit. Although the State would hold tenders for the presentation of projects to provide land and housing, the construction company would define the location of social housing, usually offering the lowest possible cost land located in the periphery, which could generate future profitability in the surrounding plots. Consequently, the model, which blurs the role of the State and abandons to the market the criteria for urban development, produced housing without equipment, far from productive centers, and with scarce recreational areas.

The Chilean model was implemented in Mexico in the last decade of the twentieth century. In this case, the State articulated the shift of its regulatory role in the design, planning, and implementation of housing development programs with the liberalization of ejido lands to the urban market (Salazar 2014). Thus, the de-regulation of housing policy made it possible for private companies to oversee land production, infrastructure production, and the construction and marketing of housing. Mexican State's role was reduced to granting mortgage loans to "rightful claimants,"—which between 2001 and 2011 totaled 7 million loans for new housing. The titling transactions for this housing were supported by granting thousands of mortgages. The

---

[11] There were numerous proposals to reactivate it and transform its functions (in 1925, 1940, 1960, 1969, 1971, 1981, and 1985). However, none of them came to fruition.

[12] "The declaration of lack of necessity was recorded in the transfer of land and other state assets to the private sector, under Act No. 23,697 of Economic Emergency of 1989. Following trends from other cities in the world (Barcelona and London, for example), it was then decided that the area had to be urbanized to integrate it with the rest of the city—(Guevara 2013: 148).

policy shift was supported by the official discourse outlining that "the formal market should solve the continuous and growing housing demand through the construction of new housing, which also constituted a good stimulus to the economy" (Salazar 2014: 8).

The creation of low-income housing grew exponentially during the Fox administration and that of his successor, Felipe Calderón. Through credits from the National Workers' Housing Fund Institute (Infonavit), citizens could acquire a house in areas exclusively for this type of real estate. But unfortunately, the production of social housing financed by the State and in the hands of real estate developers produced housing complexes located in the peripheries, of small size and low construction quality, contributing to the existence of a large stock of substandard housing and, more recently, abandoned ones. Thus, as a trace of these social housing projects in the country, entire neighborhoods of houses already built have become almost deserted.

In the cities of Querétaro, Monterrey, and Ecatepec, we can find kilometers and kilometers of unused social housing, one next to the other, creating a desolate image. According to the National Council for the Evaluation of Social Development Policy, there are five million abandoned properties throughout the country. This way, neighborhoods with identical buildings house and millions of empty dwellings create long stretches of buildings without urban quality rules.[13] According to Oxfam Mexico's Research Coordinator, Milena Dovalí: "They promised people that factories would arrive and there they would find jobs. But unfortunately, that never happened, so the infrastructure never arrived, and they were left in very unsafe and underdeveloped areas, forcing many people to leave" (Cited in Transecto, 05/02/2020).

### 4.5.3 Urban Catastrophes

Twelve years after the catastrophe, there still seems to be no agreement on the number of deaths caused by the earthquake in Haiti in January 2010. Some sources put the death toll at 190,000, and others count it to 300,000 (DesRoches et al. 2011). Undoubtedly, the number of fatalities is shocking. Even more surprising, however, is the fact that all sources agree that these casualties could have been avoided. The 2010 earthquake seems to have synthesized a long-standing social process that fatally impacted the daily life of Haitian society in general and the inhabitants of Port-au-Prince in particular. The consequences of the disaster have been due more to the communitys vulnerability than the magnitude of the natural phenomenon.[14] Year after year, about fifty earthquakes of a magnitude similar to the one recorded in Haiti

---

[13] The picture of abandoned cities by different failed public policy initiatives is repeated in different latitudes: for example, Madrid Metropolitan Area (Spain) (Cañizares and Rodríguez-Domémech 2020) or Lanzhou region, Tieling in Liaoning, and Changxing in Zhejiang estate (China) (Duhamel and Trápaga Delfín 2015).

[14] The earthquake reached 7.3 on the Richter scale. The most powerful earthquake to hit the country in 200 years (GOH 2010).

in 2010 occur in different latitudes. However, they rarely produce the destruction and devastation recorded in this country and its capital city.

The earthquake struck west of Port-au-Prince: 65% of the buildings in the metropolitan area collapsed or had a high degree of destruction, affecting more than two million people who were left homeless. According to official data (GOH 2010), over 600,000 people had left the affected areas to seek shelter elsewhere in the country. In October 2012, almost three years after the catastrophe, 496 refugee camps were open, and 358,000 people lived in displacement (Feldmann 2013). Even today, the destruction of infrastructure (beyond housing) remains chilling: More than 1300 educational institutions and more than 50 hospitals and health centers have collapsed or are unusable. The country's main port cannot be used. The Presidential Palace, Parliament, law courts, and most ministerial and public administration buildings have been destroyed. The destruction of physical assets, including housing units, schools, hospitals, buildings, roads, bridges, ports, and airports, was estimated to be USD 4.3 billion. In addition to the collapse of urban infrastructure, there was the environmental collapse: The earthquake has put further pressure on the environment and natural resources, increasing the Haitian people's extreme vulnerability (GOH 2010: 7).[15]

What factors explain such figures? On the one hand, the existence of very unfavorable living conditions. At the time of the earthquake in Haiti, more than half of the population was poor (54% in 2009), and 80% survived on less than two dollars a day. In the metropolitan neighborhoods that were most affected by the earthquake, such as Carrefour Feuilles, Bel Air, or Cité Soleil,[16] hunger, overcrowding, filth, and material shortages still plague today. No drinking water, sewage, waste disposal, electricity is extremely limited (many houses have no electricity), street lighting is non-existent, most streets are made of gravel, and health services are highly precarious.[17] Illiteracy in the country reaches 50%, and the infant mortality rate is 70 per 1000 live births. Thus, the intense destruction seems to have been due more to the very precarious living conditions, urban density, and poor housing conditions than to the earthquake's intensity (Durán Vargas 2010).

On the other, the figures can be explained by the weak institutional capacity of public agencies and the national State itself. This weakness is associated with centuries of dispossession at the hands of imperial forces and, more recently, with political conflicts and economic crises fomented by transnational interests. It is worth

---

[15] "The main source of energy for Haitians is wood (it is cooked with charcoal), and thus the population consumes the trees, and the country's forests are almost non-existent. A key fact that explains another of the frequent natural disasters Haiti suffers, flooding" (La Capital, 03/01/2015).

[16] Cité Soleil is the poorest neighborhood in the country and perhaps in the entire region. An area that extends from the airport to the bay of the capital, on the Caribbean Sea. "Citi Soleil is the most dangerous neighborhood in Haiti; the situation here is very fragile. During the day, it is considered a yellow zone where you must patrol very carefully, and at night it becomes a red zone, very dangerous, where different criminal bands and gangs operate" (Interview with the press officer of the Argentine contingent of Blue Helmets, La Capital, 03/01/2015).

[17] At the time of the earthquake, less than 10% of the population has access to potable tap water, and less than one-third has access to electricity, even intermittently (DesRoches et al. 2011).

recalling that Haiti was the first country to pay an indemnity of 50 million gold francs[18] to the French Empire as compensation for gaining its independence in 1804. A little more than a century later, the yoke of imperialism would be felt again: Haitian society suffered at the hands of the armed forces of the USA, almost two decades of occupation (1915–1934) under the pretext of safeguarding the interests of US companies based in the country. To these milestones must be added the repeated military dictatorships that caused atrocious living conditions that remain over time and weakened public and private institutionality (Feldmann 2013; Gómez Gil 2010). Finally, multilateral actions and agreements to help the Caribbean country overcome the tragedy do not seem to have been entirely adequate. Gómez Gil (2010) points out that the interventions did not necessarily contribute to rebuilding and strengthening the country's institutional structures. Nor did they manage to avoid contradictions and uncoordinated actions among the major donors, especially of a regional and institutional nature. Finally, not all the organizations that acted in Haitian territory seem to have proven experience in implementing emergency aid mechanisms.

At the scene of the catastrophe, a multiplicity of urbicide practices was deployed that managed to consummate the violent destruction of urban materiality (Coward 2006; Graham 2002) inseparable from the violence of poverty: more than 80% of the population survives in the informal economy (*changas*, street vending, shoe shining, exchange of any merchandise or directly, begging).

Multiple forms of urbicide are present in the streets of Port-au-Prince: the *necrosis* on the postcards of the collapsed houses, the leaky walls, the lack of roads, and/or the *destruction of infrastructures* or public services (health, education, etc.). They seem to foreshadow the total death of a city. The worst of deaths, one executed for decades but never ends (finally) to kill it: the *agony*.

"In Port-au-Prince, everything but everything seems eaten away by the sun, the weather, and oblivion. Garbage is piled up on the streets […]. The lack of infrastructure and streets in conditions to move around (with ditches on the sides and very few sidewalks) makes it take more than fifteen minutes to travel 500 m by car, without exaggeration. The multitude of people on the street is doing nothing. They are there. They leave their homes to make a living" (La Capital, 01/03/2015).

## 4.6 Conclusions

In this chapter, I asked what urbicide is and what features have urbicidal practices and processes. Also, I wondered how violence is articulated with them and how dispossession and spoliation operate in these tragic urban experiences. Social sciences have adopted the concept of urbicide to describe the different forms of violence applied to the city and urbanity.

These practices focus on the destruction and weakening of the record that gives memory, culture and identity, and material support to cities and their inhabitants.

---

[18] Equivalent to the French state budget on those dates.

The concept of urbicide emerged associated with urban renewal processes in the postwar USA. It currently applies to analyze neocolonial military strategies to destroy built urban environments, urban para-militarization linked to terrorism and drug trafficking, urban militarization followed by displacement, ethnic or religious wartime reconstruction strategies, etc. More recently, the concept has also been associated with the spatial scars of austerity policies and natural resource extraction.

In these urbicidal practices and processes, we can recognize the symptoms of different types of violence: physical, political, and symbolic. In the destruction processes, dispossession and spoliation raze the cities and the right to urbanity. Often, they take place associated with colonial practices.

Focusing on cases from Latin American cities, I approached less examined urbicidal experiences and their impacts on urban life and city spaces. In all cases, the consequences are vulnerable groups' economic, social, and political destruction or marginalization. A common feature stands out in the review: urbicidal violence destroys cities and, above all, the right to the city.

# References

Aguirre Moreno A, Baez Gil EY (2021) Urbicidio: Sobre la violencia contemporánea contra las ciudades. Agora 40(1). ISSN 0211-6642

Alexandri G, Janoschka M (2018) Who loses and who wins in a housing crisis? Lessons from Spain and Greece for a nuanced understanding of dispossession. Hous Policy Debate 28(1):117–134

Arqueros Mejica MS (2013) Procesos de producción social del hábitat y políticas públicas en las villas de la Ciudad de Buenos Aires. El caso de Barrio INTA. Master dissertation, Universidad Torcuato Di Tella, Buenos Aires

Berman M (1996) Falling towers: city life after urbicide. In: Crow D (ed) Geography and identity: exploring and living geopolitics of identity. Maisonneuve, Washington, pp 172–192

Borras SM Jr, Kay C, Gómez S, Wilkinson J (2012) Land grabbing and global capitalist accumulation: key features in Latin America. Can J Dev Stud/Revue canadienne d'études du développement 33(4):402–416

Bouillon CP (2012) Un espacio para el desarrollo: Los mercados de vivienda en América Latina y el Caribe. Inter-American Development Bank, New York

Brenner N, Schmid C (2015) Towards a new epistemology of the urban? City 19(2–3):151–182

Brenner N, Theodore N (2002) Cities and the geographies of "actually existing neoliberalism." Antipode 34(3):349–379

Brill F, Raco M (2021) Putting the crisis to work: the real estate sector and London's housing crisis. Polit Geogr 89:102433

Butler J, Athanasiou A (2013) Dispossession: the performative in the political. Wiley

Cañizares MC, Rodríguez-Domémech MÁ (2020) "Ciudades fantasma" en el entorno del Área Metropolitana de Madrid (España). Un análisis de la Región de Castilla-La Mancha. EURE (Santiago) 46(139):209–231

Carrión Mena F, Cepeda P (2021) Urbicidio: la muerte de la ciudad. https://www.cidur.org/urbicidio-la-muerte-de-la-ciudad/. Accessed: 11 July 2022

Carrión (2018) Urbicide or the city's liturgical death. Oculum Ensaios 15(1):5–13

Coward M (2004) Urbicide in Bosnia. In: Graham S (ed) Cities, war and terrorism: towards an urban geopolitics. Blackwell, Oxford, pp 154–171

Coward M (2007) 'Urbicide' reconsidered. Theor Event 10(2)
Coward M (2018) Against network thinking: a critique of pathological sovereignty. Eur J Int Rel 24(2):440–463
Davidson M, Iveson K (2015) Beyond city limits: a conceptual and political defense of 'the city' as an anchoring concept for critical urban theory. City 19(5):646–664
DesRoches R, Comerio M, Eberhard M, Mooney W, Rix GJ (2011) Overview of the 2010 Haiti earthquake. Earthquake Spectra 27(1_suppl1):1–21
DiNapoli EK (2019) Urbicide and property under Assad: examining reconstruction and neoliberal authoritarianism in a "Postwar" Syria. Columbia Human Rights Law Rev 51(1):253–312
Duhamel F, Delfín YT (2015) Cuestión de política económica: ciudades fantasmas en China y México. In: Trápaga Delfín Y (Coord) América Latina y el Caribe y China Recursos naturales y medio ambiente 2015. Unión de Universidades de América Latina y el Caribe, México, pp 141–158
Durán Vargas LR (2010) Terremoto en Haití: las causas persistentes de un desastre que no ha terminado. Nueva Sociedad 226:13–20
Feldmann AE (2013) The 'phantom state' of Haiti. Forced Migr Rev FMR 43:32–34. University of Oxford. https://www.fmreview.org/es/estadosfragiles/feldmann
Fields DJ, Hodkinson SN (2018) Housing policy in crisis: an international perspective. Hous Policy Debate 28(1):1–5
Fisher B (2014) Shock of peace: intersection between social welfare and crime control policy in Rio de Janeiro's Favelas. ADVOCATES' FORUM. School of Social Service Administration, The University of Chicago, pp 9–19. https://crownschool.uchicago.edu/shock-peace-intersection-between-social-welfare-and-crime-control-policy-rio-de-janeiro%E2%80%99s-favelas
Gil CG (2010) Un análisis multifocal del terremoto de Haití. Papeles de relaciones ecosociales y cambio global 110:145–157
Government of the Republic of Haiti (GOH) (2010) Action plan for national recovery and development of Haiti, Port-au-Prince
Graham S (2003) Lessons in urbicide. New Left Rev 19:73–77
Graham S (2011) Cities under siege: the new military urbanism. Verso Books
Guevara T (2010) Políticas habitacionales y procesos de producción del hábitat en la Ciudad de Buenos Aires. El caso de La Boca. Master dissertation, Facultad de Ciencias Sociales, Universidad de Buenos Aires
Guevara T (2013) ¿La ciudad para quién? Transformaciones territoriales, políticas urbanas y procesos de producción del hábitat en la Ciudad de Buenos Aires (1996–2011). Doctoral disstertation, Facultad de Ciencias Sociales, Universidad de Buenos Aires
Günay Z (2015) Renewal agenda in Istanbul: urbanisation vs. urbicide. ICONARP Int J Arch Plann 3(1):95–108
Hadjimichalis C (2014) Crisis and land dispossession in Greece as part of the global 'land fever.' City 18(4–5):502–508
Hall W (2016) (Un)making the food desert: food, race and redevelopment in Miami's Overtown community. ProQuest ETD Collection for FIU. AAI10743588. https://digitalcommons.fiu.edu/dissertations/AAI10743588
Harvey D (2007) Espacios del capital. Hacia una geografía crítica. Akal, Madrid
Herzer HM (1992) Ajuste, medio ambiente y urbanización. A propósito de la ciudad de Buenos Aires. In: Zamora R, Lungo M, Maricato M (eds) Hábitat y Cambio Social. FUNDASAL, San Salvador, El Salvador
Hodkinson S, Essen C (2015) Grounding accumulation by dispossession in everyday life: the unjust geographies of urban regeneration under the private finance initiative. Int J Law Built Environ 7(1):72–91
Johnstone Q (1958) The federal urban renewal program. Univ Chicago Law Rev 25(2):301–354
Janoschka M, Sequera J (2016) Gentrification in Latin America: addressing the politics and geographies of displacement. Urban Geogr 37(8):1175–1194

Kowarick L (1996) Expoliación urbana, luchas sociales y ciudadanía: retazos de nuestra historia reciente. Estudios Sociológicos 14(42):729–743

Lesutis G (2021) Planetary urbanization and the "right against the urbicidal city." Urban Geogr 42(8):1195–1213

Löw M (2013) The city as experiential space: the production of shared meaning. Int J Urban Reg Res 37(3):894–908

Martell D, González Redondo C (2013( La expulsión de lo indeseable. La Unidad de Control del Espacio Público y la racionalidad política de la gestión del espacio urbano bajo el macrismo. Urdergraduate dissertation. Licenciatura en Ciencias de la Comunicación Social. Facultad de Ciencias Sociales, Universidad de Buenos Aires

Menazzi L (2012) Políticas y proyectos para Buenos Aires. Reconfiguraciones en los modos de hacer ciudad a partir de la cuestión del Mercado Nacional de Hacienda (1976–2003). Doctoral dissertation. Facultad de Ciencias Sociales, Universidad de Buenos Aires

Naciones Unidas, Consejo Económico y Social (2018) Ciudades sostenibles, movilidad humana y migración internacional. Informe del Secretario General, Comisión de Población y desarrollo. 51er período de sesiones, 9 a 13 de abril

Oszlak O (1991) Merecer la ciudad. Los pobres y el derecho al espacio urbano. CEDES-Humanitas, Buenos Aires

Phinney S (2020) Rethinking geographies of race and austerity urbanism. Geogr Compass 14(3):e12480

Rodgers D (2009) Slum wars of the 21st century: gangs, mano dura and the new urban geography of conflict in Central America. Dev Chang 40(5):949–976

Rodríguez A, Sugranyes A (2011) Vivienda Privada de Ciudad. Revista de Ingeniería 32:100–107, jun/dic

Salazar C (undated) Suelo y vivienda en la construcción de la ciudad neoliberal. El Colegio de México

Shaw IGR (2019) Worlding austerity: the spatial violence of poverty. Environ Plann D Soc Space 37(6):971–989

Shin HB, Kim SH (2016) The developmental State, speculative urbanisation and the politics of displacement in gentrifying Seoul. Urban Stud 53(3):540–559

Sparke M (2017) Austerity and the embodiment of neoliberalism as ill-health: towards a theory of biological sub-citizenship. Soc Sci Med 187:287–295

Stanford Encyclopedia of Philosophy (2006). https://plato.stanford.edu/about.html

Stenberg SH (2010) Urbicide: the politics of urban destruction by Martin Coward. Glob DiscourSe 1(2):193–195

United Nations (2020) First-ever United Nations Resolution on Homelessness. https://www.un.org/development/desa/dspd/2020/03/resolution-homelessness/ Accessed: 7 July 2022

Wolford W (2010) Contemporary land grabs in Latin America. Presentation at the FAO Committee on Food Security 36th Session, October, Rome

## Press Sources

La Capital, 3th January 2015. https://www.lacapital.com.ar/edicion-impresa/la-capital-puerto-principe-como-se-vive-haiti-un-pais-atrapado-la-miseria-n640945.html. Accessed: 10 July 2022

La Prensa, 30th June 1978. Archive footage

Transecto, 5th Feb. 2020. https://transecto.com/2020/02/urbanismo-fantasma-desafio-de-la-nueva-era-urbana/. Accessed: 11 July 2022

**Maria Mercedes Di Virgilio** is sociologist and Ph.D in Social Science (Universidad de Buenos Aires. Senior CONICET Researcher at, Instituto de Investigaciones Gino Germani, Universidad de Buenos Aires. Full Associate Professor in Research Methodology at the School of Social Sciences at the same university. Her research focuses on the residential mobility processes of lower and middle- class families, residential segregation, housing, urban policies, and their relations with spacial mobility.

# Chapter 5
# Urban Order and Disorder. Genealogy of Urbicide

**Eduardo Kingman Garcés and Susana Anda Basabe**

**Abstract** The aim of this chapter is to broaden the notion of "urbicide" by relating it to notions developed within conceptual fields other than those of urbanistics, such as "biopolitics", "governmentality", "anthropocene", and "climate change". The text is inscribed within a perspective of comparative historical analysis between Europe and the Andes (and more than historical genealogy, since it connects the present and the past within the same problematic axis). Its central concern is to relate "urbicide" not only to the destruction of the city but also to what the city destroys in terms of ways of life and the quality of life, of species and of landscapes. At the same time, our concern is to analyze how this problem could be managed. It is possible that notions such as governmentality could lead to more democratic and inclusive forms of governance of both cities and populations.

**Keywords** Genealogy · Historical research · Demolition · Town planning

## 5.1 Introduction

This chapter aims to enrich the notion of *urbicide*, relating it to the effects suffered by the city just as much as to those caused by the city, or in other terms that brings us closer to the dominating conceptual frameworks in the discussions about the Anthropocene and climate change phenomenon: both processes concern the fate of the city as well as that of the planet.[1] Every city is built on the rubble of other cities and/or their resources. Many European cities were not only destroyed but their fields were turned into wastelands. In his travels through America, Humboldt registered some of these ruins and its uses made by the conquerors:

---

[1] Cities are currently responsible for the production of 70% of $CO_2$, while by 2060 the number of buildings in the world will double in relation to the present (Chumpitaz Requena 2021).

E. Kingman Garcés · S. Anda Basabe (✉)
Facultad Latinoamericana de Ciencias Sociales, Flacso Ecuador, Quito, Ecuador
e-mail: gsanda@flacso.edu.ec

E. Kingman Garcés
e-mail: ekingman@flacso.edu.ec

© The Author(s), under exclusive license to Springer Nature Switzerland AG 2023
F. Carrión Mena and P. Cepeda Pico (eds.), *Urbicide*, The Urban Book Series,
https://doi.org/10.1007/978-3-031-25304-1_5

Despite the admiration that the spaniard conquerors showed for the roads and aqueducts of the Peruvians, not only did they not take the trouble to preserve them, but they deliberately destroyed them in order to use their artistically carved stones for new monuments[2] (Humboldt 1876: 550).

Cities are both constructive and destructive. They generate order as well as disorder in its interior and surroundings. In classical terms, we speak of powers, both affirmative and negative, and of agencies and complex systems. The city and its urbanism cannot be understood outside of this dialectical relationship in the deployment of its historical and natural forms. To what extent have scholars of the city incorporated this type of problem into their agendas? And to what extent has this incorporation occurred from politics rather than from the police?

In the chapter, we try to define what urbicide implies in social terms, leaving the much more complex discussion of urbicide and climate change for another moment. Our perspective is historical or more precisely genealogical. Our interest is to understand the starting point of the game of oppositions between urban order and disorder, urbanization and urbicide, to illustrate secondly, the transformations that the city generates in terms of governance.

## 5.2  The Urban Context

Unlike what happened until relatively a few decades ago, when capitalism did not reach everywhere or did so more in extension than in depth, without completely affecting many of the rural areas, mountains, and forests, today we are witnessing the functioning of an urban and conurban continuum capable of including all territories, including remote areas such as those of the Amazon, the "*punas*" or the high "*páramos*". There are also an infinite number of energy flows interacting with each other without necessarily taking spatial forms. Mongin (2005), for example, discusses the widespread urban, but we could also describe the urban as fluids or as agencies and relations located in different strata or nodes, in the sense of a new kind of palimpsest. We refer, among other things, to the connections generated by the informal circulation of resources (including those related to narcotraffic and organized crime), as well as to the increasingly complex relations of the city with geology and the ecosystem, the circulation of algorithms or viruses, such as the case of the actual pandemic. It is not a matter of aggregations but rather of systems and "machinic assemblages". If we follow Nietzsche, we could refer to powers, both affirmative and negative: This helps us to understand the notion of urbicide, not as something that is imposed from the outside, but as forces that breakthrough from the inside, crossing different temporalities.

---

[2] Author's translation. Original Spanish passage:
"A pesar de la admiración que los conquistadores mostraban por las vías y acueductos de los peruanos, no sólo no se dieron el trabajo de conservarlos, sino que los destruyeron deliberadamente a fin de utilizar en sus nuevos monumentos sus piedras talladas artísticamente" (Humboldt 1876: 261).

The city is covered with traces and imprints. Simultaneously, what we refer to as the past, continues to act on the present as survivals. In this sense, urbicide is consubstantial with the birth of cities and urban development, in the same way that death is consubstantial with life. It is expressed in the form of major catastrophes, but it is not limited to these. In the nineteenth century, European cities were perceived as great references of civilization and progress, but at the same time they had generated not only the destruction of the natural environment but also colonization and pauperism. Therefore, cities in Africa, Asia, and Latin America are now perceived as "cities of misery" while for others they are new spaces opened for opportunity, investment, and the idea of "progress".

Cities enabled long-distance trade and the emergence of modern States. Yet, the development of trade, manufacture and later on, industry, led to the transformation, frequently violent, of the old forms of production and its lifeworlds. It consisted in a dialectic that was both constructive and destructive at the same time, of which there are not necessarily any records and that was generated step by step, in a manner that is often imperceptible. Some of Marx's texts show the destructive effects of primitive accumulation or industrial development in cities like London:

> Every unprejudiced observer sees that the greater the centralisation of the means of production, the greater is the corresponding heaping together of the labourers, within a given space; that therefore the swifter capitalistic accumulation, the more miserable are the dwellings of the working-people. "Improvements" of towns, accompanying the increase of wealth, by the demolition of badly built quarters, the erection of palaces for banks, warehouses, etc., the widening of streets for business traffic, for the carriages of luxury, and for the introduction of tramways, etc., drive away the poor into even worse and more crowded hiding places (Marx 2007: 722).

Not only London or Paris but also in Latin America the cities were the point of deployment of civilizing projects directed to the indigenous and black population and to the poorest.[3] National and republican projects in Latin America took shape in the cities. These were the great referents for a new type of economy and a new type of aesthetics based on ornamentation and hygienism.

Cities and its inhabitants have passed through great transformations as a result, among other things, from the "epochal changes". The history of cities has not, in fact, followed an evolutionary lineal path. On the contrary, it has gone through great moments of change, caused by natural disasters, wars, plagues, famines, but also by social events, such as those of the commune of Paris, *la rebelión de los barrios in Quito* in the context of the independence movements (Minchom 1994) or by the simple urbanístic action generated by Cerdá, Haussmman, Vicuña Mackenna, Pereira Passos, and Torcuato de Alvear. All of this has led to radical modifications of their internal configurations, making them unrecognizable from one generation to the next (Almandoz 2006).

The history of architecture has made enough records of how significant buildings and spaces of an era were demolished to allow expansions, or to modernize uses,

---

[3] A current example is the one concerning the construction of "Millennium Cities" in the territories of indigenous communities located in the Ecuadorian Amazon (Wilson and Bayón 2017).

activate flows, or simply to change landmarks and historical references. At the same time, the natural environment has also been impacted by urban expansion, as have the different species that inhabit it. As Bruno Latour points out, although without referring directly to cities:

> all human activities turn out to be transformed, in part, into geological forms; everything that we used to call bedrock is beginning to be humanized – or, in any case, to bear traces of a tempestuously remodeled humanity! It is no longer a question of landscapes, of the occupation of land, or of local impact. From now on, the comparison is made on the scale of terrestrial phenomena (Latour 2017: 114–115).

As Latour himself shows, the amount of energy produced by human civilization is equivalent to the energy of volcanoes or tsunamis. It is possible that we are witnessing a sort of contradiction: on the one hand, the urban, as well as modernity or the idea of progress, as Western paradigms, have been generalized reaching the whole planet, but on the other hand, the city has lost many of its virtues as a living space, as a possibility of utopia and as a *polis*. In general terms, the city as a device of economic, political, and social order has left behind many of the characteristics to which we were familiar to, or that we imagined, related to everyday life, or to be more precise: It has acquired a new reality and a new complexity, within larger networks, spatial, territorial and terrestrial, but also biological and technological.

### 5.2.1 City Government, Governance, and Urbicide

When we discuss the city, we are interested in understanding more than just its spatial configuration. It is important to know how its government is organized.[4] A city depends on an order that makes possible supplies, the construction of roads, public hygiene, and the provision of labor. In principle, this order could be the effect of urban planning and demography, the natural result of the economy, or it could be imposed from above. In general, everything is resolved historically, within multi-located fields of forces.

One of the issues of concern in Occident since the mid-eighteenth and nineteenth centuries was the governance of large concentrations of population, with all that this implies in terms of resources, movements, needs, and interests. Before the emergence of modern society, cities developed in a context that was predominantly agrarian, involving relatively small concentrations related to trades and commerce.

The cities of the ancient regime, on both sides of the Atlantic, simultaneously strove to build the public image of the sovereign (Burke 1992) and at the same time protected economies and societies that were dominantly territorial. On the one side was the city, on the other were the localities in which power was exerted in a personalized way by local aristocracies.

---

[4] Many of the devices necessary for them can be operational during one period and cease to be so in another. Walls, for instance, responded to defensive and commercial needs, marking a difference between "an inside" and "an outside."

The aim of early modernity was to organize the management of entire populations related to relatively stable urban activities such as industries and services. We refer to the transition of societies organized on the basis of a bureaucratic order to properly defined urban societies. This was a much more complex exertion of power, since it was directed at an unknown, somewhat autonomous population, which required the development of much more sophisticated systems of intervention in administrative terms, but also in terms of bio-politics and political economy. This was directly related to the needs of accumulation, but also to the requirements of widening and deepening power in fields such as sexuality, health, education, and entertainment. The problem does not now lie in:

> that of fixing and demarcating the territory, but of allowing circulations to take place, of controlling them, sifting the good and the bad, ensuring that things are always in movement, constantly moving around, continually going from one point to another, but in such a way that the inherent dangers of this circulation are cancelled out. No longer the safety (sûreté) of the prince and his territory, but the security (sécurité) of the population and, consequently, of those who govern it (Foucault 2009: 93).

Each of these interventions was accompanied by changes in the organization of spaces. The eradication of the old market plazas and the type of culture that thrived in them and their replacement by gardened spaces is an example in this sense. But also the production of places and non-places destined to put into practice what would eventually be called "mass culture".

If cities had served as a base for sovereign power and acted as the basis for an aesthetics of representation, a new type of concern was now developing related to the government of populations, with all that this implied in urbanistic terms. It was a social concern that took urban forms. From the last third of the eighteenth century, but especially during the nineteenth century and even later in the twentieth century, in Latin America, the State and its devices were oriented towards integrating strips of life that had previously escaped its control and influence. The concentration camp, as well as the panopticon, are forms of power resulting from architectural practices, just like other spaces such as hospitals, psychiatric centers, housing programs, and schools. This is an initial moment of the control of bodies or what Agamben calls the *Bare Life*. A second moment, which is properly biopolitical, is related to the governance of populations (Foucault 2009; Agamben 2004).

In every city, there is a need for governance, but we must distinguish the way in which this is resolved in the old cities, whether corporative or based on the "power of the prince", from what happens later with the development of mercantilism and capitalism, with the administration of population flows as well as parallel flows of commodities, information, waste, viruses, and pests. Ordering implies to differentiate, classify, design, and to take control of spaces that while favored by agglomeration and by the concentration of resources, lead to conflicts of a new type, such as those generated by the presence of waves of displaced, unemployed, or permanently unoccupied people due to deforestation or by the growth of violence.

This dynamic of urbanization and urban modernization began to take hold from the second half of the nineteenth century in Europe and later in Latin America, leading to new forms of relationships between individuals and classes, based on political economy rather than on forms of mutual dependence and reciprocity. The conversion of labor force into commodities was a fundamental condition for the formation of societies of separation and discipline, but not necessarily the only one, since it was necessary to colonize the whole of individuals' lives, a process that has taken on new and unusual forms in recent times with the accelerated changes in technologies and biotechnologies. The city has been a favorable place for the development of urbanistic devices, technologies, and knowledge oriented to the governance of populations, in a modern sense. All this has involved a long process of destruction of old forms of social and cultural relations and the dismantling of traditional spaces of socialization, related, among other things, to religiosity and street commerce, but it has also given rise to the creation of new forms and indeed to the survival of old forms in the middle of new ones.

If we speak of urbicide, we must not lose sight of the creative as well as destructive character of these devices; their concern for disciplining in factories, the formation, and control of childhood, the healing and subjection of the insane, as well as the transformation and destruction of nature, civilization, and extermination of other cultures. The major problem, in any case, was and is the administration of flows, which is the reason for the existence of political economy, statistics, probability calculus, and more recently metaverses and informatics. It was, according to Debord (1994), a colonization of social life, including spare time. But it was also about devices and systems of government, capable of controlling both urban conglomerates and their different points of escape (in Deleuze's terms, their forms of territory and deterritorialization). Much of that control has now become invisible and at some points subtle, while remaining profoundly violent. The city, in order to exist, needs to destroy and to self-destruct. This destruction extends beyond its surroundings, in the elimination of forests, the extermination of terrestrial species, including human groups, as well as the pollution of oceans and rivers.[5]

### 5.2.2 Ruins, City and Modernity

There is not a single path in the development of modernity in Occident, and this is because it is a rather complex constellation of economic, social and political forces, biopolitics being one of them, but not the only one. Walter Benjamin (1999) showed how the deployment of the world of commodities in nineteenth-century Paris brought changes in the organization and functioning of the city, in the structures of sensibility and in the forms of public relations. These were, as we can now see, complementary or rather parallel to the development of biopolitics. It was the two-faced side of an urban model in which the ideals of great avenues, gardened spaces, specialties in

---

[5] In Peru as in other Andean countries, for example, it is more and more evident the transformation of Amazonian territories produced by unsustainable agriculture and a fallow crisis caused by the increased use of agrochemicals which generates widespread destruction of forests, including deforestation, and soil impoverishment (Bedoya et al. 2017).

trade, the theater, and the opera, had as a permanent counterpart the deterioration of the working-class neighborhoods, the underworlds of the sewers and the ports. On the one hand, there was an early development of consumption and consumption spaces, but on the other hand, there was a proliferation of motley, dark, labyrinthine spaces such as those described by Dickens, Poe, or Sevenson.

Benjamin's analysis focuses on the city, on the passages, the great avenues, the *Flâneur*, while a more recent perspective examines the place that the city occupies within a network of planetary flows. When we now speak of urbanization, we are referring both to population concentrations and to networks and flows of commodities, people, equipment, and resources at a global level, which is something that goes beyond cities, and at the same time includes them. These networks are now largely self-generated, virtual, and somewhat autonomous. The functioning of biopolitics, in the same way as economy, communication, administration, and distance control, depends as much on spaces with concentrated populations as on networks and flows, aggregations, and micro- as well as macro-policies. This whole scenario has led to new ways of organizing spaces and new forms of relating to them.

Not only the State and the institutions intervene in the lives of the populations; there are a series of dynamics that are generated from within the devices that in many cases enter into contradiction with the State itself. We refer to the demands of war industries, pharmaceutical industries, entertainment industries, financial capital, or the drug trafficking industry. But also to the control of plagues and diseases, security, the civilization of customs, the government of women and infants, the management of anomie; as well as, the organization of economy from *below*, such as the efforts of the population to rebuild social bonds or cope with plagues through networks of care. It is about needs that do not necessarily come from the State, but that originate in cities and out of them, with the passage from manufacturing and artisan forms to industrial and post-industrial ones, demographic growth, migratory movement, the expansion of markets and the "democratization of consumption", as something that starts in Occident and spreads throughout the rest of the world. These aspects, that are related to urbanization, as well as the formation of concentrations, networks and population flows; needed and need to be organized, governed, managed, not only directly but at a distance with resources that start both from the State and from machinic assemblages (Deleuze 2017).

The city, in itself, constitutes a powerful device that makes this operation necessary. In terms of governance, the city does not only operate from a space of concentration or territorialization, but also from deterritorialized networks, many of which are—as we have already pointed out-virtual. The city is constantly being made and unmade. It has, in fact, a much greater mobility than the State: This forces it to be much more creative and active and not only destructive.

Now we know, in addition, that life and the control of life must be assumed not only in relation to humans, and that a city, in spite of being an urban phenomenon or being defined in urbanistic terms, constitutes a form, more or less violent, of intervention in the natural and geographic environment. This means that we are talking about appropriation of life in a broader and more complex sense than the one conceived by Foucault. In fact, the destruction of forests and the extermination of biological species were parallel to the dynamics of subordination of human conglomerates, and not only follows, but precedes the policy of population administration in the world, although until now we had not been aware of it.

Walter Benjamin was one of the first to show the relationship between the city, the unfolding of the world of commodities, and the generation of ruins, overflows, and catastrophes. These ruins are not necessarily physical, they can be social: They are the expression of a mechanical order that at the same time as it grows, it collapses. The modern city, with its passages, arcades, buildings, fashions, universal expositions, and spectacles, is at the same time, according to Benjamin, a generator of misery, overcrowding, and waste.[6] Urban renewal and urbicide would be, in this sense, two sides of the same coin.

But there is something else in all this: changes in materiality, mobility, and consumption lead to transformations in the structures of sensibility. If we follow Benjamin, the most important critique to modern social functioning that occurs in cities, should not be oriented so much to their economy, although this is undoubtedly important, but rather to the loss of experience and with it the loss of meaning. What the city provokes (if we can still call the city itself as a "body without organs") is a sort of "narcotic historicism", an "addiction to the masks" of which Benjamin himself refers to. The problem with this reverie is that it does not allow us to assume the conditions to which we have been dragged into: the fate of the city, but also the fate of the planet.

If the city is presented as the best reference of progress, we must not lose sight of the fact that "The concept of progress must be grounded in the idea of catastrophe. That things are "status quo" *is* the catastrophe. It is not an ever-present possibility but what in each case is given" (Benjamin 1999: 473). In this way, Walter Benjamin has detached himself from the illusion (or dream) of progress. Progress and decadence are for Benjamin "two faces of the same thing".

### 5.2.3 City and Biopolitics in the Andes

The colonial city was initially perceived as a fortress city and an outpost of conquest. Many of the indigenous ruins served as a base for Spanish buildings, but this did not mean that the indigenous exchange networks necessarily ceased to function, as evidenced in Quito, La Paz, and Cuzco (Musset 2011). The Hispanic commercial activity itself relied on the network of tambos and roads left by the Incas (Poloni-Simard 2006). Although the colonial pattern of spatial organization aimed at separating the "republic of Indians" from that of Spaniards, this did not operate in practice:

---

[6] Architecture itself is, as Toyo Ito (2000) shows, an architecture made to be quickly discarded.

## 5 Urban Order and Disorder. Genealogy of Urbicide

In Lima, Viceroy Toledo had reduced the Indians to the suburb known as *El Cercado*, in 1570, but many continued to live in various districts of the city, mixed with the non-Indian population (Chocano 2000: 28).[7]

The idea of a good government, as it was assumed in Hispanic America by the Bourbon reforms, included urban planning policies that included population policies in the context of a predominantly agrarian world. The proposals focused on the generation of a certain disciplinary order in the context of non-disciplinary societies, as well as on the strengthening of cities prior to urbanization. This included the relocation of cemeteries, slaughterhouses, hospitals, and the control of food supplies. As the historian Gabriel Ramón writes in relation to Lima:

> In the words of the viceroy Marqués de Avilés, Lima, capital of the Peruvian viceroyalty, was the center from which Enlightenment derived and the model for arranging the rest of the provinces of the kingdom. With the continent as context, this titanic task had in each city a privileged scenario. Not only did it aim to refine the mechanisms of surveillance or urban policing, such as the subdivision of the city into barracks and neighborhoods and the introduction of new authorities, but it also attempted to use the physical urban elements as pedagogical instruments. The reorganization of the city under Cartesian parameters and the construction of sumptuous official buildings in neoclassical style were concomitant with the social reorganization of the city walls and had to convey a univocal message to its inhabitants (Ramón 1999: 96).[8]

Both the foundation of cities and their refoundation were based on the need to make them organized spaces. In Ecuador, the refoundation of Riobamba after the earthquake of 1797 was based on the model of separation, due to the fear of riots and indigenous uprisings (Coronel 2003). Many years later, the Ecuadorian president Gabriel García Moreno, took advantage of the destruction of Ibarra caused by the earthquake of 1868, to design a city model, conceived in terms not only of urban planning and policing but also of moral control. These were experimental actions of population administration and modernity under strata-based patterns.

Andean cities, far from developing during the first phase of the Republic, began to decline due, in part, to the deterioration of long-distance exchange. With the Republic we witnessed, in the Andean area, a double process in terms of territorial conformation: the constitution of a State centrality located in urban contexts and, on

---

[7] Author's translation. Original Spanish passage:
"En Lima, el virrey Toledo habìa reducido a los indios al arrabal llamado el cercado, en 1570, pero muchos continuaron viviendo en diversos barrios de la ciudad propiamente dicha, mezclados con la poblaciòn no india" (Chocano 2000: 28).

[8] Author's translation. Original Spanish passage:
"En palabras del virrey, Marqués de Avilés, Lima, capital del virreinato peruano, era el centro de donde se deriva la ilustración y el modelo que arregla las restantes provincias del reyno. Con el continente como contexto, esta titànica tarea tuvo en cada ciudad un escenario privilegiado. No solo se apunta hacia un refinamiento de los mecanismos de vigilancia o policía urbana, como por ejemplo, la subdivisión de la ciudad en cuarteles y barrios y la introducción de nuevas autoridades, sino que se intentó utilizar los elementos físicos urbanos como instrumentos pedagógicos. La reorganización de la ciudad bajo parámetros cartesianos y la construcción de suntuosos edificios oficiales de estilo neoclásico eran concomitantes con la reorganización social de intramuros y debían transmitir un mensaje unívoco a sus habitantes" (Ramón 1999: 96).

the other hand, the generation of necessarily decentralized forms of administration of populations by delegation (Guerrero 2010). It was a question of State constitution attempts generated from a center as well as from the plantations, rural parishes, or the hacienda system.

It was only towards the last third of the nineteenth century that cities began to gain weight in the Andes. These were cities that in turn gave rise to proposals for renovation and progress that allowed the reproduction of an oligarchic order. In social terms, modern urban devices were combined with forms of an ancient regime, oriented towards moral and racial control (Coronel 2003).

The control of indigenous and peasant migration, home visits, sanitization of public spaces and markets, towards the first decades of the twentieth century, were early forms of administration of populations anchored in the cities (Kingman Garcés 2006). This was not something that occurred only in the Andes. In the case of Rio of Janeiro, hygienist and urban improvement policies led to the overthrow of the cortijos and the locations where small popular enterprises operated (Popinigis 2007).

In historical terms, it is necessary to distinguish the personalized management of the *proper Indians, peones concierto*, colonists, and tributaries, from the administration of "loose populations", transformed into day laborers, porters, or small *trajinantes* who were limited to precarious labor inside the cities. It was about the management of the city as a whole, as well as the development of a police and an aesthetics, marking the differences between those who have a part and those who have no part, in Ranciere's sense. It is difficult to establish a clear separation between the forms of private administration of populations and public forms during the nineteenth century and until late in the twentieth century (Poole 2009; Guerrero 2010). In small towns in the interior of the Andes, local power remained in the hands of the hacienda owners for a long period of time. There was undoubtedly an urban dynamic—generated by regional markets—that pointed towards another field of forces, but society as a whole was in one way or another interwoven with agrarian economy and a form of daily relations marked by servitude and racism.

With early modernity, the city became an object of concern for different fields of knowledge related to State planning. The cities demanded the organization of mechanisms such as the police, the school system, hospital services, customs, and planning and statistical centers, while agrarian areas were organized on the basis of local micro-powers that only with time and as part of urbanization gained access to modern devices.

Sanitation of port cities and the control of epidemics responded to the requirements of a fledgling global economy. All this was accompanied with a concern about racial degeneration factors, European immigration to the detriment of Asians, the separation and control of spaces and the increase in birth rates. This assumed the organization of an urban order, in urbanistic, hygienic, and security terms. Epidemics and the measures generated to combat them became central in debates about urbanization, without, however, developing practical actions to improve the health of the vast majority of the population in the long term.

## 5.2.4 Urbicide in the Andes

One of the problems that would appear in the Andes with early modernity up to the late twentieth century was how to administrate a population that was losing its links with the hacienda and migrating to the cities, becoming urban, but not necessarily considered to be citizens. This involved a dynamic generated by the incorporation of the indigenous and peasant population in the dynamics of the modernization of the State and the economy. As José María Arguedas (1983) points out in relation to Peru:

> The opening of the highways broke the isolation that the barbaric geography had imposed on Peru. The penetration of the powerful and multiple modern factors that, inevitably, drive to the development or the rupture of excessively antiquated social structures, have made explosion, in part, in the still viceregal organization of the society of the Andean region. The Indians have invaded the cities escaping from the frozen villages or haciendas, frozen in the sense that no possibility of ascent existed and still does not exist in those haciendas and villages: whoever is born Indian must die Indian. On the other hand, the communities with more or less enough land found themselves, almost suddenly, with the opening of the roads, with a prodigious increase in their economy[9] (Arguedas 2012: 83).

Such waves of population from the Andes had become necessary for the expansion of trade and for the functioning of cities with increasingly complex networks. Given the nature of Andean societies, large parts of this population were incorporated in a precarious manner as servants and urban peonage, as well as in informal commerce. Even though the "legitimate citizens" continued to perceive them as "non-citizens", they were gradually opening their way both on the margins and in the heart of the cities. While there was a formation of labor sectors, necessary for the functioning of the cities, these were assumed in terms of disorder, pollution, and urbicide (extrapolating the times).

Cities such as Lima, Bogotá, Guayaquil and Quito contributed, in long and medium terms, to break many of the ties with the villages and small provincial cities as well as to the formation of semi-autonomous popular neighborhoods, plebeian cultures, and counterpublics. By occupying public spaces, historical areas and poor neighborhoods, this population was placing in question the hierarchical order traced by the "literate city" and "aristocratic modernity".

While urbanization had expanded by the first decades of the twentieth century, not all sectors that were inscribed within it followed a capitalist logic. There was

---

[9] Author's translation. Original Spanish passage:
"La apertura de las carreteras rompió el aislamiento que la bárbara geografía había impuesto al Perú. La penetración de los poderosos y múltiples factores modernos que, inevitablemente, impulsan el desarrollo o la ruptura de estructuras sociales excesivamente antiquadas, han hecho explosionar, en parte, la todavía virreinal organización de la sociedad de la región andina. Los indios han invadido las ciudades huyendo de las congeladas aldeas o haciendas, congeladas en el sentido de que no existían ni existen aún en esas haciendas y aldeas ninguna posibilidad de ascenso: quien nace indio debe morir indio. Por otra parte, las comunidades con tierras más o menos suficientes se encontraron, casi de pronto, por la apertura de las vías de comunicación, con un incremento prodigioso de su economía" (Arguedas 2012: 83).

undoubtedly a specificity inherent to the Andes, which did not correspond exactly to what had occurred in Europe. We are referring to a particular form of urban society closely imbricated with the agrarian space as well as with a mobility of populations, *trajines,* and merchandise that is at the same time capitalist and non-capitalist. The colonial and racist bases of this process mark some differences with what happened, long before, in Europe.[10]

As we can see, *the Andean* maintains various dimensions: *the Andean* as networks of kinship and reciprocity, as well as its relations with nature that differ from those of Western cultures (Degregori 2013: 324); *the Andean* as particular ways of being and doing that, although having a rural origin, continued (and in part continues) to be reproduced in urban contexts; *the Andean* as an urban–rural mechanism that included both a modern dynamic related to migration, the formation of poor neighborhoods, the incorporation in the labor world and urban consumption, as well as social and cultural patterns coming from the hacienda and the communal ways of being and doing: the cities of the Andes as cities of peasants, but not only of peasants, since they included mestizo and mixed indigenous populations, as well as Afro-Latin descendants and, more recently, displaced populations from other countries (Kingman Garcés and Bretón 2016). In this sense, on one hand, a large part of the peasant population was (and still is) related to commerce and street trades, with independent and semi-independent occupations established between the city and rural areas, in the form of "*trajines callejeros*" (Kingman Garces and Muratorio 2014). However, another part of the population formed part of the servitude and urban peonage or was gradually being incorporated into labor sectors. Therefore, the urban, in this case, did not necessarily coincide with industrialization or with a given model of industrialization, but neither with urbanization as it had occurred in Europe.

## 5.2.5 The Organization of Space and the Governance of Populations

The functioning of an order, in an economic, political and social sense, has depended on cities and imagined urban-based communities, as well as on the formation of communication and exchange networks to which the entire planet has been integrated. In order to function, States require the development of a certain administrative centrality and at the same time a decentralizing capacity, capable of relating them to what operates outside of their dynamics. In the city there are the ministries, as well as the main civil, military, and lettered authorities, and it is from there that a series of interested actions are deployed to integrate (and subordinate) other areas, even the most remote ones, but none of this is given more than as a possibility, because there are other forces at play that operate under other logics and even outside of all logics. It is possible to name it as urbanization, but it is a new form of urbanization.

---

[10] However, this does not mean to disregard what happened to the Irish laborers in London or to the immense mass of unemployed and underemployed in Paris or Berlin in the nineteenth century.

It is possible that we may be permeated by political, technological, and biological devices of unknown reaches, many of which have achieved a life of their own. These devices have put in crisis our old ways of understanding the organization of space and, as part of that, the place that cities and localities occupy in the dynamics of current transformations, related to the production of images and virtual worlds, but also in the destruction of species and climate change, as well as the colonization of life and its imaginaries. It is something that goes beyond what humanities conceive, and within them urbanism and urban sociology, as well as the partition established between what is human and non-human, biological and non-biological, sensible and non-sensible, robotic and organic, and even beyond the current ways of understanding domination and power. If we want to continue the discussion of urbicide, we can only understand it within this dynamic. If we want to continue the discussion of urbicide, we can only understand it within this dynamic. As part of radical transformations, partly future and partly present, capable of generating bewilderment and fear as well as fascination.

Michael Foucault showed the relationship between biopolitical devices and the organization of space, introducing, in this field, a genealogical perspective. It is possible that beyond Foucault it may be necessary to think about the way in which these devices function in the midst of transformations in the biosphere or in the non-places of media order. For Foucault, there is a difference between the administration of leprosy, based on the separation and expulsion from the cities, from the administration of the plague, oriented, on the contrary, towards the inclusion of plague victims within devices of control organized internally in neighborhoods, blocks and dwelling houses. "It is not a question of driving out individuals but rather of establishing and fixing them, of giving them their own place, of assigning places and of defining presences and subdivided presences. Not rejection but inclusion" (Foucault 2003: 46). Foucault was referring to specific forms of intervention in the lives of populations in urban contexts, oriented towards their government rather than to their expulsion. Deleuze (1995) perceived the technological, political, and social reaches of this phenomenon and did so in terms of a society of control rather than a disciplinary society.

It is worth discussing to what extent the structuring of space under urban and conurban forms conditions and is conditioned by modern forms of life administration. This is directly related to the development of capitalism, but also to the need to broaden and deepen power over sexuality, emotions, health and disease, and education, as well as overflows of natural and technological resources, consumer goods, information, and population segments. By power in its broadest sense, we refer to something related to individuals, but also to social conglomerates in need of supervision, evaluation, normativization, as well as to certain levels of permissibility and informality, making use of closed spaces but also of open spaces and those of circulation. If we consider the way in which the world-economy operates, or the way in which information is produced and circulates, we are not really discussing processes produced from one unique center, but rather about the incorporation, under different conditions, times, and forms, of a disparity of phenomenons, which are economical as well as social and environmental. For instance, let us think of the integration within the same production chains of cutting-edge technology resources with rudimentary technologies or of populations subjected to super-exploitation and even slavery in the midst of the euphoria of progress, consumption, and modernity. One might say that we are living a process of integration within a unique system, but under privileged conditions, or, on the contrary, under subordinate and subaltern conditions, both in terms of what enhances life and what destroys it.

By examining the current pandemic, we are confronted with new ways of structuring spaces and flows, both related to biological factors and to issues of control and security. But at the same time as the pandemic remained, we witnessed the multiplication of unemployment, the trafficking of populations, the multiplication of precariousness and of what Butler (2004) refers to as *precarious life*. The pandemic has put into question the previous devices of disease management, generating new challenges in biopolitical terms. The pandemic constitutes a new phenomenon, with its own specificities, on a scale never seen before, which has demanded the development of new forms of health organization and administration, but above all new ways of understanding governmentality. We now know that a pandemic cannot be resolved exclusively on the basis of hospital practices, since it depends on the management of flows, many of which are invisible, and on relations with economy and politics. It is a global phenomenon with local effects, difficult to solve if issues such as equity and redistribution of resources at the global level are not addressed. Although it is true that the pandemic is a biological phenomenon, it is not separated from politics and political uses. We know that its management has implications that go beyond the strictly medical field, and that are related to biopower, but also to the possibility of contributing to the construction of new proposals for coexistence, which are much more fair and less invasive with the planet.

The dynamics of urbanization not only raised an economic problem, but also a political, social, environmental, and human one. Even globalization could be assumed in terms of urbanization, only that we are dealing with a generalized urban or integrated urban scenario, like a spatial continuum in movement (which includes virtually, as well as non-places and non-localizable energy flows). We refer, among other things, to the gigantic waves of migration from one border to another or from one continent to a different one, which has been occurring for several decades, but also to the various forms of aggregation of small- and medium-sized populations, as well as the traditionally agricultural and jungle areas, within the urban framework.

Haussmann's strategy to modernize the city developed in terms of urban planning as well as security and urbicide. More recent ways of adjusting spaces have given rise to "cities of separation" and "cities of walls", as well as "cities of misery". The formation of large population conglomerates has made necessary the development of centralized forms and at the same time sufficiently flexible and decentralized forms to be able to control the flows and counterflows, but much of what has happened has escaped from any control, as in the case of drug trafficking. What we are currently witnessing is the formation of large machinic orders composed of places and non-places, in permanent change, construction, and destruction. It is possible that the significance of all of this will go considerably further than what we currently understand as a city.

# References

Agamben G (2004) The open: man and animal. Stanford University Press, Stanford

Almandoz A (2006) Urbanismo europeo en Caracas: 1870–1940. Equinoccio, Universidad Simón Bolívar-Fundación para la cultura urbana, Caracas

Arguedas JM (2012) El indigenismo en el Perú, en Obras Completas, vol XII. Editorial Horizonte, Lima, pp 75–88

Bedoya Garland E, Aramburú CE, Burneo Z (2017) Una agricultura insostenible y la crisis del barbecho: el caso de los agricultores del valle de los ríos Apurímac y Ene, VRAE. Anthropologica 38(XXXV):211–240

Benjamin W (1999) The arcades project. The Belknap Press of Harvard University Press, Cambridge, Massachusetts

Burke P (1992) The fabrication of Louis XIV. Yale University Press, New Haven, Connecticut

Butler J (2004) Precarious life. The powers of mourning and violence. Verso, London

Chocano M (2000) La América colonial (1492–1763): Cultura y vida cotidiana. Editorial Sìntesis, Madrid

Chumpitaz Requena F (2021) Las ciudades, el cambio climático y la energía incorporada. Revista Limaq (8):13–28. https://revistas.ulima.edu.pe/index.php/Limaq/article/view/5549/5248

Coronel R (2003) Orden local y orden público: El Municipio de Riobamba en la transición de la Colonia a la República (1790–1850). Procesos, Revista Ecuatoriana de Historia (19):11–22. https://revistas.uasb.edu.ec/index.php/procesos/article/view/2034

Debord G (1994) The society of the spectacle. Zone Books, New York

Degregori CI (2013) Del mito del Inkarri al mito del progreso. Migración y cambios culturales. Instituto de estudios peruanos (IEP), Lima

Deleuze G (1995) Negotiations 1972–1990. Columbia University Press, New York

Deleuze G (2017) Derrames II. Aparatos del Estado y axiomática capitalista. Cactus, Buenos Aires

Foucault M (2003) Abnormal. Lectures at the College de France (1974–1975). Verso, London

Foucault M (2009) Security, territory, population: lectures at the Collège de France (1977–78). Palgrave Macmillan, London

Guerrero A (2010) Administración de poblaciones, ventriloquía y transescritura. Análisis históricos, estudios teóricos. Instituto de Estudios Peruanos (IEP) FLACSO-Ecuador, Lima-Quito

Humboldt A. von (1876) Cuadros de la Naturaleza. Imprenta y Librería de Gaspar, Madrid

Ito T (2000) Escritos. Colegio Oficial de Aparejadores y Arquitectos Técnicos, Murcia

Kingman Garcés E (2006) La ciudad y los otros. Higienismo, Ornato y policía. Quito, 1860–1940. FLACSO, Quito

Kingman Garcés E, Bretón V (2016) Las fronteras arbitrarias y difusas entre lo urbano-moderno y lo rural-tradicional en los Andes. J Latin Am Caribbean Anthropol 22(2):235–253. https://doi.org/10.1111/jlca.12216

Kingman Garcés E, Muratorio B (2014) Los trajines callejeros. Memoria y vida cotidiana: Quito, siglos XIX-XX. FLACSO-Ecuador, Instituto Metropolitano de Patrimonio, Fundación Museos de la Ciudad, Quito

Latour B (2017) Facing Gaia: eight lectures on the new climatic regime. Polity Press, Cambridge

Marx K (2007) Economic and philosophic manuscripts of 1844. Dover. Publications, Mineola, NY

Minchom M (1994) The people of Quito, 1690–1810: change and unrest in the underclass. Westview Press, Boulder

Mongin O (2005) La condition urbaine: la ville à l'heure de la mondialisation. Seuil, Paris

Musset A (2011) Ciudades nómadas del Nuevo Mundo. Fondo de Cultura Económica, México D.F.

Poloni-Simard J (2006) El mosaico indígena. Abya-Yala-IFEA, Quito

Poole D (2009) Justicia y comunidad en los márgenes del estado peruano. In: Historia subalterna y cultura en América latina y los Andes. Editado por Pablo Sandoval. Instituto de Estudios Peruano, Lima, pp 599–638

Popinigis F (2007) Proletários de casaca. Tabalhadores do comércio carioca (1850–1911). Unicamp, Rio de Janeiro

Ramón G (1999) Urbe y orden. Evidencias del reformismo borbónico en el tejido limeño. In: El Perú en el siglo XVIII: La Era Borbónica, editado por Scarlett O'Phelan Godoy. Instituto Riva-Agüero, Pontificia Universidad Católica del Perú, Lima, pp 295–324

Rancière J (2000) Le partage du sensible. Esthétique et politique. La Fabrique Éditions, París

Wilson J, Bayón M (2017) Potemkin revolution: Utopian jungle cities of 21st century socialism. Antipode 1–22

**Eduardo Kingman Garcés** is historian and anthropologist, interested in introducing a conceptual perspective in his work and developing a creative relationship with fieldwork and the archive. His field of work is social and cultural history in urban contexts, as well as the debate on heritage, security and urban identities. His studies analyze the constitution of urban social sectors and disputes over memory and space in the city. Due to his solid knowledge in this area, he has been invited to teach classes and participate in seminars and conferences in different countries. Profesor Emeritus at Flacso Ecuador.

**Susana Anda Basabe** is an anthropologist and research associate at Flacso, Ecuador. Her lines of research are Anthropology, Anthropology of food, ethnography, political economy, political ecology, indigenous communities, Amazonia, Andes, forms of production, deforestation, colonization, symbolism, social economics and social change.

# Chapter 6
# Imaginaries and Archetypes on the Death of the City

**Alfredo Santillán Cornejo**

> *If the apocalypse is only a fantasy, the society that invents it to scare itself is very real and each end of the world is the reflection of its time (Musset* 2022*, p. 15).*

**Abstract** This essay draws on the historical persistence in Western culture of literary and cinematographic images about the destruction of big cities to present its reflection on urbicide. The constant repetition and renewal of these images in each age evoke the existence of an imaginary that associates urban destruction to superhuman forces, whether divine or coming from nature, following the imprint of the biblical Apocalypse. Following this reflection, the essay discusses how urbicide entails an archetypal meaning as punishment to the city in its double dimension, as physical and moral order. This archetypal basis allows us to extrapolate the reflection to the crisis of inhabiting in today's cities. Lurked by the logic of commoditization of urban space inherent to post-industrial capitalism, contemporary metropolises encourage the detachment and disinterest of people with the spaces they inhabit. Faced with this situation, the essay concludes that the dilemmas of urbicide can be approached from the perspective of philosophical anthropology, which focuses on the existential sense of inhabiting as reference that guides decisions about preserving or transforming the urban environment.

**Keywords** Urbicide · Catastrophe · Apocalyptic illustrations · Archetypes · Inhabiting

## 6.1 Introduction

The destruction of great cities has become a persistent image over time: from the cities razed to the ground by the wrath of God in the biblical stories of the Old Testament to the apocalyptic scenes offered by contemporary cinema and science fiction. As Musset (2022) shows, the conversion of monumental buildings into rubble

---

A. Santillán Cornejo (✉)
Facultad Latinoamericana de Ciencias Sociales, Flacso Ecuador, Quito, Ecuador
e-mail: asantillan@flacso.edu.ec

© The Author(s), under exclusive license to Springer Nature Switzerland AG 2023
F. Carrión Mena and P. Cepeda Pico (eds.), *Urbicide*, The Urban Book Series,
https://doi.org/10.1007/978-3-031-25304-1_6

mobilizes a collective imagery about the fate of civilization in the hands of colossal forces capable of generating such devastation: earthquakes or giant waves, asteroids from space, nuclear bombs, alien attacks, among many other agents. These images have a long-lasting historical persistence and can be considered as "stagings" of urbicide, that is, as representations of the killing of the city. Within the field of reflection on urban imaginaries, representations are a window to access the structures of signification that constitute the social labor of imagining. In his proposal for myth analysis known as "myth critique" Durand (2012) recognizes the historicity of imaginaries in the sense that each era is characterized precisely by a particular type of dominant imaginaries. However, the persistence and constant recreation of the representations of urbicide show us a deeper layer of cognition, in terms of Jung (1970) the presence of an archetype, an archaic matrix in which eschatological fantasies of the end of humanity are presented.

But what sense do these apocalyptic images make in the midst of the triumph of planetary urbanization? To think of the destruction of cities at a time when more than half of the world's population lives in them is an unprecedented event, typical of this millennium. Even more so during the global pandemic of COVID-19, which has shown us an image of the post-apocalyptic city very different from the one imagined in the movies. The intact but uninhabited infrastructure shows us a new urban life that develops on an "augmented reality" in which connectivity and telematic relations are consolidated as a daily mode of urbanity (Cortez and Finquelievich 2021). In this context, the current proliferation of images of urban dystopias constitutes a window to glimpse the fate of the urban environments we inhabit, not because they may or may not come close to concrete experiences or scientific predictions, but because they serve as a symptomatic manifestation of the discomforts of planetary urbanization.

This imaginary dimension of urbicide seeks to open a dialogue with the more consolidated reflection on this category. Although there are some very important ideas about the symbolic death of the city through toponymy or the loss of memory of many heritage policies (Carrión 2014), the major referent of urbicide is the physical destruction of the urban environment, either in the most direct sense as shown by the reflections of Aguirre Moreno and Báez Gil (2020) on the cities annihilated by military actions, or in its broader sense linked to the transformations that generate destructive impacts on the territories and the ways of inhabiting them as follows from the tradition of Jacobs (2011) which has generated a fundamental line of reflection in force to this day. In both cases, processes of urbicide are really dramatic, not at all spectacular as in the field of cinematographic imagery. However, we already have a thorough reflection on both and therefore the purpose of this essay is to strengthen a view of their significance, in other words, to look for clues in the imagined urbicide to understand the scope of urbicide actually practiced.

If, as Gravano says "there is no city without an image of a city" (2019, p. 263), we can expand this idea by considering that the repertoires of physical destruction and killing of cities go hand in hand with the ways of constructing images of destruction, whether prior to or accompanying in parallel the physical processes, or even after them. However, we do not currently have an inventory of images that would allow us to analyze how different urban processes are represented through visual narratives.

Thus, we are committed to an archetypal reading of urbicide, searching for meanings of its inscription in the archaic deposit of psyche and culture interweaving. We rely on a series of ideas of various authors who have coined reflections on the cultural component of urban life, and we project their scope as conceptual resources to capture the meaning of the material destruction of cities.

This essay's argument is structured in two parts. The first is dedicated to digging into the archetype of urbicide, having the existential dilemma of urban life as a self-imposed moral order as reference. The second takes up reflections within philosophical anthropology on human habitation as a conceptual resource to interpret the imagery on destruction of today's cities. It proposes a comparison between the metaphysical model and the anthropic model in which morality is found as a common denominator in the legitimization of urbicide.

## 6.2 God's Wrath Against Cities: Punishment as a Foundational Urbicidal Archetype

Although the term "city" is misleading as throughout History, it has encompassed very dissimilar human settlements, from small agricultural villages in ancient times to contemporary megalopolises, it maintains a substantial basic principle as the creation of an artificial environment for human life, a produced habitat, "unnatural" though always dependent on nature, that is intended to stand the test of time. Mumford's great work (2013) accentuates the triad *city, power, and civilization* in the description of urbanization's history from its origins, to show the interrelationships between emerging urban functions with modes of social organization, and their transformations over time. Following his line of thought, religious, and economic functions are fundamental to the emergence of cities, but both functions can be better understood as the material support of transformations in human sociability, that is, as part of the formation of the bonds that hold society together. To which we can add a bidirectional relationship, not only as a result of economic, religious, political, etc., spheres, but also as an agent that influences these spheres. Capel recognizes this agency by defining the city as "the best human invention" (2005, p. 2), in the sense that it is a creation that has substantially changed the future of its own creators.

However, it is important to be cautious of any eulogy of the city that easily falls into a civilizing prejudice that overestimates urban culture and technological development above the ways of life of the village or nomadism and its logic of mobile appropriation of the territory. If several authors are critical of the anti-urban ideology that tends to hold the city responsible for broader social problems (Capel 2002), it is also necessary to take a critical look at the common idea that considers the city as the center of cultural creation, as if this human activity were less intense in less urbanized environments. The notion of urban as a civilizing ideology tends to hierarchize the territory, ignoring the articulations and dependencies between city, countryside, and wilderness, to use Tuan's terms (1990), to differentiate artificial environment, domesticated nature, and uninhabited nature. It is important to recognize both types of prejudices because beneath them some common meanings about urbicide are implanted: as punishment or revenge when the defects of urban life are accentuated or as tragedy and catastrophe when its virtues are maximized. We start, therefore, by situating urbicide as the destruction of a particular human creation, the *topos* destined to make a way of life possible.

The most archaic sense of this destructive act places its agent in the divine. Although for different cosmologies the sacred origin of the human being is a basic principle -divine creation as *image and likeness of God* in Christianity, for example-, it is not as clear whether and in what manner creations of this creation are also subsidiary to this divine quality. Thus, an ambivalent field opens up around the possibility of thinking about this invention, which in turn reinvented the notion of human, as a manifestation, albeit indirect, of divinity. The Bible is an essential source for unraveling the tensions between the city, the sacred, and the human. In general terms, the conflict revolves around human self-determination and the challenge of establishing a moral order to guide collective life in the artificial habitat (city). As a varied bibliography shows (Capel 2002; Musset 2022) Jerusalem will become in the myth the ideal type of the *holy city*, which seeks to recreate on Earth the virtuous life oriented toward the sacred, while Babylon (and other cities) embodies the archetype of a radically self-determined moral order, beyond supra-human commitments. It is precisely these attempts at self-determination that are punished with urbicide in the Christian tradition, as we shall see further on.

In the Genesis account, the origin of the city appears with the second exile of humanity, the expulsion of Cain after the murder of his brother Abel, which implies the first mythical homicide, under the figure of fratricide. The story tells us that in his exile Cain founds the first city under the name of his son Enoch (Contemporary English Version of The Bible, n. d., Gen 4:1–17). But it is worth noting that he did not stay to live there, but continued his wandering life founding several cities and from his wide descendants arose several clans specialized in metal work, commerce, and music. For this reason, several interpretations see in Cain nothing less than a father of human civilization, since the technology and knowledge in the development of metallurgy, the extension and complexity of commercial exchange, and artistic development have substantially changed social organization (García, n. d.). Following this interpretative route, the myth of Cain and Abel has an archetypal background. Abel, related to animals, represents subsistence based on activities that require high mobility in the territory, hunting, later converted into grazing; while Cain represents agriculture, an activity that in contrast requires rootedness. In this symbolic sense, Cain kills Abel, sedentarism submits nomadism to its logic of rootedness. Spengler (1966) emphasized the importance of settlement as a milestone in human history insofar as human beings, when they domesticated plants, metaphorically became plants, in the sense that they chose to take root in the territory.

The mythical account has broad overlaps with archaeological work in which data abound about the role of human settlements in the transition between the Paleolithic and Neolithic periods, moment in which the first proto-cities emerged between 6000 and 8000 BC in the Euphrates Valley. However, Mumford (2013) argues that the urban functions thought to be characteristic of cities already existed in Neolithic villages. His hypothesis of the origin of the city points out that it was the result of the union between Paleolithic and Neolithic institutions. The success of the Neolithic village in the provision of food and population growth implied a new political function of protection that revived the importance of weapons, typical of the Paleolithic, but now inserted in a broader social organization:

> If one dare to call this a marriage of the two cultures, they were probably at first equal partners, but the relationship became increasingly one-sided, as the weapons and coercive habits of the aggressive minority were re-enforced by the patient capacity for work that the stone-grinding neolithic peoples showed. As often happens, the rejected component of the earlier culture (hunting) became the new dominant in the agricultural community, but it was now made to do duty for the governance of a superior kind of settlement. Weapons served now not just to kill animals but to threaten and command men (Mumford 1961, p. 39).

For the time being, we will not dwell on the economic and technological changes that made cities possible, but rather return to the archetype of Cain. Exile implies abandonment to one's own human capacities as the only resource for survival, and the success of this trajectory during several millennia, with the achievements in knowledge and mastery over seeds, plowing, irrigation, and soil fertilization, will turn villages into material evidence of the human triumph over their survival and autonomy. But the wound of exile remains open, leaving the duality between seeking a reunion with the sacred or separating from it definitively. Following Mumford (2013) the Neolithic village had already developed ritual functions offering its own spaces for ceremonial encounters, pilgrimage, and burial of the dead. This last element has been vital in the symbolic qualification of the territory because it allowed a spatial support to generational continuity in time, the material expression that successive generations are connected by sharing the same space.

If the characteristic that gives rise to the city is symbiosis between primitive institutions, development of religious functions, expressed in temples and sanctuaries as great architectural works, can be understood as a continuity of the conflict for self-determination. Deepening of the hierarchies of the new social organization implies the development of a more vertical law and morality in such a way that liberation from the law of God with exile leads to submission to human law. Thus, the sense of divinity of the human takes a new direction in which the rulers of the most important ancient cities achieved recognition as divinities. The archetype of the God-King is present in the Sumerian story of Gilgamesh, ruler of Uruk, considered by many as the first city properly speaking; later something similar happened to the city of Memphis, re-founded as the capital of Ancient Egypt by Nermer, considered the first pharaoh.

However, in contrast to the epic narratives that exalt the quasi-divine power of rulers, based on available archaeological evidence Ancient History presents the wars and successive plundering that destroyed these first cities to the point of rendering them unrecognizable today. And although these destructions are evidently anthropic, it is significant how deeply the biblical account, which establishes a supernatural will in the annihilation of the cities, has permeated Western culture. Thus, it is interesting that, in the long term, the image of divine power acting on the city has moved from its consolidation as the center of a civilization to its destruction.

But although the wrath of God is turned against cities and not against villages, not all of them are protagonists of this destiny. The "holy city" or "city of God" as St. Augustine called the archetype of Jerusalem, the city oriented by the fulfillment of sacred precepts exists as a moral world that has encouraged the imagery of many utopias as Mumford (2013) also presents. In contrast, Babel, Sodom, Gomorrah, and Babylon are killed by ignoring the sacred and implanting human order par excellence. Free will in these cases is punishable with urbicide. Although Babylon is mentioned in several biblical passages, even with the name of the "Great Whore," the most detailed description of its destruction appears in the Book of Revelation:

> Her sufferings will frighten them, and they will stand at a distance and say, "Pity that great and powerful city! Pity Babylon! In a single hour her judgment has come." Every merchant on earth will mourn, because there is no one to buy their goods. There won't be anyone to buy their gold, silver, jewels, pearls, fine linen, purple cloth, silk, scarlet cloth, sweet-smelling wood, fancy carvings of ivory and wood, as well as things made of bronze, iron, or marble. No one will buy their cinnamon, spices, incense, myrrh, frankincense, wine, olive oil, fine flour, wheat, cattle, sheep, horses, chariots, slaves, and other humans. Babylon, the things your heart desired have all escaped from you. Every luxury and all your glory will be lost forever. You will never get them back. The merchants had become rich because of her. But when they saw her sufferings, they were terrified. They stood at a distance, crying and mourning. Then they shouted, "Pity the great city of Babylon! She dressed in fine linen and wore purple and scarlet cloth. She had jewelry made of gold and precious stones and pearls. Yet in a single hour her riches disappeared" (Contemporary English Version of The Bible, Rev. 18: 10–17).

We can consider this story as a referential description of the commercial and territorial dynamics of ancient times, since we see the concentration of wealth in a small portion of territory in which there is an exuberance of objects extracted from distant places that have passed through various land and sea routes to reach the great market. In short, this description evokes the capacity of the city as a centripetal force to spatially assemble social organization. Moreover, we can take this account and imagine human actors that only appear as shadows before detailed objects and then the city appears as a place of multiple encounters, of intensity of stimuli, of heterogeneity of languages and faces. This sociability is what distinguishes it from the village. But in the biblical account, the description of material exuberance carries an implicit questioning of moral defects such as ambition, opulence, vanity, and pride.

Very much in consonance with the story of the destruction of Babylon, the materialist reading of the origin of the city broadly emphasizes the way in which the achievements in the production of surplus gave support to an economy that left behind the goal of subsistence and was erected on the accumulation of such surplus and its exchange, generating a social stratification of inequality that in turn derived in the political model of the city-state. Gravano (2018) uses the term "urban surplus" to characterize more precisely the substantial relationship between the urban and the concentration of material and symbolic surplus from the first cities to the existing ones. Thus, the urban surplus that initially resulted from the success of self-determination becomes the seed of the urban conflict that summons its destruction. Internally, social segregation, inequality, and the accumulation of power become a front of disputes. Outwardly, the more the magnificence of the city is flaunted, the more walls and armies become necessary to protect it and the more attractive it becomes as an object of conquest not only for the wealth it accumulates, but above all for the symbolic value of destroying and seeing the "great city" fall as an emblem of the decline of a civilization.

This journey leads us to suggest that at the symbolic level, the most archaic meanings of urbicide are constituted around moral sanction of cities through punishment and revenge. But despite destruction, a return to a pre-city moment does not appear on the horizon, much less to nomadism. Another urban project will emerge on its ashes, perhaps better than the murdered one, but equally as the materialization of the contradictions and ambiguities of its creators. The history of urban transformations can be understood as the attempts to redesign this invention to correct its imperfections.

## 6.3 The Imaginary of the Late-Modern Urbicide and the Crisis of Inhabiting

Musset calls this powerful imagery on the destruction of cities "Babylon syndrome" (2022) and finds a variety of similes to the biblical story in the literary and cinematographic production of science fiction, where he shows us that since the late nineteenth-century cultural production has strongly taken up the topic of dystopia where megalopolises of the near future come to collapse. The images of the emblematic cities of the present projected into the future and destroyed by asteroids, meteorites, gigantic waves, alien invasions, viruses and other laboratory experiments, etc., stand out. But a new element in the plot of catastrophic situations is social and political upheaval and chaos. The revolutions or social revolts in general that are simmering in the shadows of urban life and that eventually explode violently, are recurring images of contemporary urbicide.

The substantial difference between these images and the biblical images is that the foundation of the destructive force comes from the very bowels of the city and not from outside. Something similar occurs with the situations of viruses, genetic experiments, and underground scientific laboratories, as activities arising from within the cities, although they have analogies with biotic agents such as the biblical plagues. In any case, the central point of contemporary imagery is that they refer directly or indirectly to the anthropic action behind urbicide, whether due to catastrophes resulting from climate change, technological progress and/or social inequality. It is true that the primary meanings of punishment and revenge remain in force, but with different agents: punishment of nature for environmental damage, or revenge of oppressed groups against unjust order. But beyond these analogies, what is worth highlighting in this new production of images of urbicide is the possibility of human beings to remove God and recover for themselves the destiny of their own creations.

The very notion of "urbicide" has become valuable precisely because of the continuous wars of the last quarter of the previous century in which cities in the Balkans, and in the Middle East so far this century, have become war targets (Aguirre Moreno and Perea Tinajero 2020). These catastrophic scenarios can be an opportunity for a deep review exercise of the scope of current urbanization, like the destruction / reconstruction of European cities after World War II on which the model of heritage intervention was forged, as explained by Carrión (2014). For this essay, we rely on the case of standardized housing compounds, as a proposal for the reconstruction of housing infrastructure in the second post-war period, which motivated Heidegger's (1951) philosophical reflection on dwelling. He sustains that the human development of the capacity to build, in the sense of great engineering works, is undeniable, but he questions whether this capacity to build is accompanied by the capacity to inhabit. As he explains:

> However hard and bitter, however hampering and threatening the lack of houses remains, the real plight of dwelling does not lie merely in a lack of houses. The real plight of dwelling is indeed older than the world wars with their destruction, older also than the increase of the earth's population and the condition of the industrial workers. The real dwelling plight lies in this, that mortals ever search anew for the nature of dwelling, that they must ever learn to dwell (Heidegger 1951, p. 8).

In the philological analysis of the German term *baun*, which would be the root of the word *bauen* translated as "to build," the author identifies multiple meanings that go beyond the action of erecting something, but rather this term is associated with "caring for," "sheltering," "looking after." Furthermore, the term itself includes the meaning of "dwelling," which implies that it is never a matter of "just building" and then inhabiting, but rather that building already involves a way of dwelling. In short, the underlying theme is that in building-dwelling there is a certain tension between building and caring for, that is, intervening and modifying a space and at the same time taking care to preserve it. Even more complex, this idea which can be defined as a philosophical anthropology of dwelling, introduces a metaphysical dimension, since it is not limited to the harmonious coexistence between humans and nature, but this interrelation is one of the pivots, the earthly axis of existence "with mortals." The other axis is the supraterrestrial world, heaven, the abode of "the divine." Thus, he explains that true dwelling implies what he calls the *fourfold*: "In saving the Earth, in receiving the sky, in awaiting the divinities, in initiating mortals, dwelling occurs as the fourfold preservation of the fourfold" (Heidegger 1951, p. 4).

These ideas have become predecessor referents of the spatial turn perspective in social sciences. Since the nineties, it has widely shed light on the multiple angles of the notion of individuals and places "mutually shaping" each other, which allows us to think urbicide from an existential dimension. Before following this route, let us conclude the reference to Heidegger's essay by highlighting this connection:

> When we speak of man and space, it sounds as though man stood on one side, space on the other. Yet space is not something that faces man. It is neither an external object nor an inner experience. It is not that there are men, and over and above them space; for when I say "a man", and in saying this word think of a being who exists in a human manner-that is, who dwells-then by the name "man" I already name the stay within the fourfold among things (Heidegger 1951, p. 6).

The common image of urbicide powerfully emphasizes on how built materiality turns into ruins, the destruction of what was built, but in terms of inhabiting, it is worth asking whether or not what has been destroyed had achieved this existential purpose or, in other words, does one seek to preserve what is built, even if it functions with its back turned to dwelling as if they were separate processes? Management of architectural heritage is a perfect example. As shown by authors such as Carrión (2018) and Kingman (2004) heritage conservation tends to fall into the fetishism of protecting certain properties with historic value, but separated from the social relations that built them and those that keep them alive. And this example helps us avoid a naive criticism of urbicide that condemns all forms of destruction and exalts any action to preserve constructions. What is built, the cities, must always remain intact, and become eternal in spite of their creators' ephemeral nature?

These questions serve us to reflect on the documented forms of urbicide for which there is available research. The works of Aguirre and Báez (2020) and Aguirre and Perea (2020) discuss the use of the category of urbicide for studies on urban transformations, whose objective is the substitution of one materiality for another. They argue that, although these are dramatic changes that seek greater profitability and render the city restrictive for many populations, strictly speaking the city does not die, at least not as when bombings render unusable all materiality reduced to rubble. The difference can be interpreted as the attack on a *topos* that hosts a specific human existence in order to implement another, as opposed to a destruction that makes any human occupation impossible.

However, from the point of view of inhabiting, the most widespread form of urbicide is precisely that provoked by the implementation of the new capitalist economy. The distinctive trait of post-Fordist capitalism is that it is more mobile in the territory and more flexible in labor relations, and as proposed by Sennett (2004), its effect on cities is that it calls into question the very possibility of building a link with the environment and creating a habitat of existential value. His reflection is based on showing the changes of flexible capitalism in the world of labor, where he highlights the precariousness and the absence of a link that offers meaning to the labor relationship beyond the task performed under the motto "nothing in the long term." Indifference to the workplace arises precisely because it makes no sense to invest psychic energy in the bond if the job ends when the task is finished. If the rise of the metropolis was directly related to the Fordist model of productive capitalism, changes in planetary urbanization follow the logic of post-industrial capitalism. Thus, the author points out:

> My argument is precisely that flexible capitalism has the same effects in the city as in the workplace. In the same way that flexible production produces more superficial and short-term relations at work, this capitalism creates a regime of superficial and unbound relations in the city (…) The dialectic of flexibility and indifference appears in three forms: the first is expressed in the physical attachment to the city, the second by the standardization of the urban environment; the third by the relations between family and urban work (Sennett 2004, p. 17).[1]

This weak connection leads to geographic instability that encourages mobility and leads to a decline in rootedness. While human spatiality is precisely composed

---

[1] Translation of the original work is ours.

of the dialectic between pause and movement, rootedness and displacement, fluidity of today's capitalist economy puts mobility before a resurgence of nomadic logic, although with characteristics quite different from those of the Paleolithic. Current reflections on inhabiting, such as those by Giglia (2012), show us that domestication of space requires time, it is not an immediate process that occurs when occupying a territory. It is possible to "reside without inhabiting" and in this situation the qualities of the "anthropological place" in terms of Augé (2001) of being a relational space, of identity and memory, are relegated either to past times, in the nostalgia for the place that once served as home and that is no longer inhabited, or to uncertain future times, as the place that one hopes one day to inhabit.

Yet it is necessary to show the other dimension of the fluidity of capital and labor in global economy, since the displacements they generate do not operate in an empty space, but, on the contrary, investments are implanted in inhabited territories causing massive *expulsions*, to use the term in Sassen's work (2015) that best describes the extractive logic of current capitalism. It is precisely the advance of these processes of forced rupture of habitation that the literature on urbicide has questioned as a loss of the public, of the right to the city, and of environmental quality while the city as a commodity gains exchange value to the detriment of use value.

However, in all these cases what seems to us less discussed is urbicide as the materialization of an idea, that is, not as a specific act (demolition, bombing, etc.) tied to a cause (gentrification, war), but rather to understand it as a process, in which the image of the need for intervention has been on the rise, and has gained ground on the symbolic field. The act of urbanization is not possible without prior triumph of the idea that visualizes the need for the pre-existing to disappear, to create space for something new to exist with the promise of a certain future. A good example is how urban renewal processes rely on the image of "decayed space" in which the image of physical deterioration is often accompanied by moral discredit of its inhabitants. In abandoned, destroyed, dirty, vandalized spaces, only proscribed actions such as drug use, criminal acts, destitution, prostitution, etc., can take place. This association between space and morality has been effective in the public debate on the diagnosis of areas considered problematic in the city and the drafting of solutions, which has ended up legitimizing investments that promise to save the site. As Sarlo (2009) points out, the public city is degraded as a space, dangerous, dirty, noisy, etc., because the private city emerges as an ideal with its promise of calm, silence, asepsis, and security: the gated community for living, the shopping mall for consuming, and the private car for transportation.

Still, it is possible to develop a critical view that transcends the ideological scope. If we follow the archetypal model, the association between place and immorality can be interpreted as the contemporary version of the model of moral punishment on the city, but in this case, judgment occurs in reverse, since poverty is sanctioned in the material sphere. Media exposure of these spaces as dens of "bad living" debases and stigmatizes their inhabitants, degrades them morally, thus the physical destruction of the place becomes the remedy to sanitize bad behavior. Based on the assumption that space is social production, following Lefebvre's (2013) classic postulate, then we can consider observing the "social production of degradation" behind deterioration of the spaces assessed for intervention. Therefore, we are able to understand the rupture of the link that originally sustained a certain habitat in which urban designs converged with the life projects of its inhabitants.

These reflections lead us to a revision of urbicide that does not originate from the premise of maintaining the existing city at all costs, for this would imply the idealization of the dominant forces that have generated its production process. It is necessary to think of the present city as the materialization of triumphant ideas that have managed to impose themselves over other alternatives. That is to say that the city, which in principle we would seek to preserve, is the product of the correlation between the forces of various agents, where the capacity to determine it is not equal. Even the very tension between preserving or transforming a given urban environment is inscribed in this permanent dispute. This is where the idea of inhabiting can be useful as it enables us to focus attention on whether proposed conservation or transformation is close to the existential ideal of inhabiting. Public space, as the main collective good, is one of the key topics of debate in the face of the advancing privatization of cities. But public life has always been exclusive, despite the spatial designs to make it possible, precisely because it is the symbolic components that guide the ways of using spaces and making them meaningful. Spatial design in itself is not a sufficient condition for the intensification of public life, but it undoubtedly favors or restricts it. There can be spaces for public life and yet be devoid of it. Without a positive value assigned to the idea of sharing a space between unequal and different individuals (Duhau and Giglia 2008), without practical implementation of this coexistence, the ideal of public life is not possible.

If, as Jameson (2005: 123) said "it is easier to imagine the end of the world than the end of capitalism," the terrain of imagination becomes a substantial arena to dispute the urban future: can we thereby create an imaginary of urbicide that is restorative of human habitation? If capitalism destroys what little was previously common, its transformation may also require a process of destruction, a process of urbicide to build materiality with existential dwelling in mind. Can this potential imaginary break with the violence of destruction instituted by the myth of Babylon? Undoubtedly this implies a challenge to the imagination. The urban future full of flying cars characterized by verticality (Musset 2018) needs to be nourished less by the planning imagination that has characterized the history of urban utopias, many of them failed, as shown by Mumford (2013) and be inspired more by human creativity to weave a meaningful life on a certain space. That is to say, the planning ideal can turn to look in detail at how people effectively agency space, even when they fulfill their designs. As de Certeau (1996) shows, it is the different ways of walking and the agencies they entail that make the sidewalk effectively a sidewalk rather than just its layout. We can imagine this micro-labor of city making, like the action of millions of ants, constituting an imaginary of urbicide "from below" as opposed to the imaginary of great cataclysms driven by some great force coming "from above."

# References

Aguirre Moreno A, Báez Gil EY (2020) Urbicidio: Sobre la violencia contemporánea contra las ciudades. Agora: Papeles de Filosofía 40(1):87–110. https://doi.org/10.15304/ag.40.1.6603

Aguirre Moreno A, Perea Tinajero G (2020) Urbicidio: Violencia bélica contra las urbes. Bajo Palabra 24:319–336. https://doi.org/10.15366/bp.2020.24.016

Augé M (2001) Los no lugares. Espacios del anonimato. Una antropología de la sobremodernidad, Gedisa

Capel H (2002) Gritos amargos sobre la ciudad. Perspectivas Urbanas 1:17

Capel H (2005) La ciudad es el mejor invento humano [Bifurcaciones. Revista de Estudios Culturales Urbanos]. http://www.bifurcaciones.cl/2005/06/entrevista-horacio-capel/

Carrión F (2014) Urbicidio o la producción del olvido. Revista Observatorio Cultural 19:28–42

Carrión F (2018) Urbicidio o la muerte litúrgica de la ciudad. Oculum Ensaios 15(1):5–12. https://doi.org/10.24220/2318-0919v15n1a4103

Contemporary English Version of The Bible (n.d.) Retrieved 22 Sept 2022, from https://www.biblegateway.com/

Cortez S, Finquelievich S (2021) Ciudad aumentada y pandemia. El habitar en el Orden Digital. Cuaderno Urbano. Espacio, Cultura, Sociedad 31(31):203–227. https://doi.org/10.30972/crn.31315784

De Certeau M (1996) La invención de lo cotidiano. Artes de hacer.: Vol I. Instituto Tecnológico y de Estudios Superiores de Occidente: Universidad Iberoamericana

Duhau E, Giglia Á (2008) Las reglas del desorden: Habitar la metrópoli. Sigo XXI-UAM

Durand G (2012) La mitocrítica paso a paso. Acta Sociológica 57:105–118

García EA. s. f. Caín, fundador de la civilización humana

Giglia Á (2012) El habitar y la cultura. Perspectivas teóricas y de investigación. Antrhopos-UAM

Gravano A (2018) Hacia una arqueología de lo urbano. Urbania Revista Latinoamericana de Arqueología e Historia de las Ciudades 7:13–20. https://doi.org/10.5281/zenodo.2539725

Gravano A (2019) Cauciones epistemológicas en el trabajo sobre imaginarios urbanos. In: Vera P, Gravano A, Aliaga F (eds) Ciudades indescifrables: Imaginarios y representaciones sociales de lo urbano. UNICEN-USTA, pp 257–274

Heidegger M (1951) Building dwelling thinking. http://faculty.arch.utah.edu/miller/4270heidegger.pdf

Jacobs J (2011) Muerte y vida de las grandes ciudades (Segunda edición). Capitan Swing Libros

Jameson F (2005) Arqueologías del futuro: El deseo llamado utopía y otras aproximaciones de ciencia ficción. Akal, Madrid, p 2009

Jung CG (1970) Arquetipos e inconsciente colectivo. Paidós

Kingman E (2004) Patrimonio, políticas de la memoria e institucionalización de la cultura. Iconos: Revista de Ciencias Sociales 20:26–34

Lefebvre H (2013) La producción del espacio. Capitán Swing, Madrid

Mumford L (1961) The city in history. Its origins, its transformations, and its prospects. Harcourt Brace Jovanovich, Inc, New York

Mumford L (2013) Historia de las utopías. Pepitas de calabaza ediciones

Musset A (2018) Star Wars. Un ensayo urbano galáctico. Bifurcaciones

Musset A (2022) El síndrome Babilonia: Geoficciones del fin del mundo (Primera Edición). Bifurcaciones

Sarlo B (2009) La ciudad vista. Mercancías y cultura urbana. Siglo Veintiuno Editores

Sassen S (2015) Expulsiones. Brutalidad y complejidad en la economía global. Katz Editores

Sennett R (2004) El capitalismo y la ciudad. In: Ramos Á (ed) Lo urbano en 20 autores contemporáneos. ETSAB/UPC, pp 213–220

Spengler O (1966) La decadencia de Occidente. Bosquejo de una morfología de la Historia Universal: Vol II. ESPASA–CALPE, S. A. http://disenso.info/wp-content/uploads/2013/06/La-Decadencia-de-Occidente-O.-Spengler.pdf

Tuan YF (1990) Topophilia: A study of environmental perceptions, attitudes and values. Columbia University Press. New York

**Alfredo Santillán Cornejo** Professor-Researcher at Flacso Ecuador. Sociologist, Master in Anthropology and Ph.D. in Social Studies from Externado University of Colombia. His lines of research are urban anthropology, socio-spatial segregation and urban imaginaries. He is the author of the book "La construcción imaginaria del Sur de Quito" (2019) in addition to several academic articles, compilations and chapters in books.

# Chapter 7
# COVID-19 and the City: Reframing Our Understanding of Urbicide by Learning from the Pandemic

**Roberto Falanga and João Ferrão**

**Abstract** The magnitude and reach of the COVID-19 pandemic have brought multiple effects into our lives. Cities have been one of the most important stages of such changes in both material and symbolic dimensions. Against this backdrop, this chapter analyses the main effects of this pandemic through the lens of the concept of urbicide with a focus on Europe. Accordingly, the chapter discusses five sub-dimensions: the material reconfiguration of environmental balance and commonplace system, as well as the symbolic reconfiguration of social cohesion, consumer sovereignty and democratic institutions. The discussion points out both destructive and constructive sides of this pandemic, which ultimately demands the reframing of our understanding of the urbicide. To do so, we identify the main pandemic effects, briefly classified as revealing, accelerating and cluttering, along with the solutions in place to mitigate, adapt and/or transform local governance, in association with five main domains: environment, planning, society, economy and democracy. Based on the acknowledgement that pandemic's effects have not been linear, our main argument builds on the need to reframe our conceptual lens to improve the interpretive potentiality of the urbicide, thus incorporating a regenerative ethos about ongoing societal and urban transformations.

**Keywords** Governance · Pandemic · Urban regeneration · Social cohesion

## 7.1 Introduction

The concept of urbicide indicates either the murder of a city or the destruction of specific urban elements (Carrión 2018). In the latter case, drivers of destruction can bring dangerous impacts on multiple urban "layers". A first layer refers to the material dimension of the city, defined as the "urbs". A second layer holds the social character

---

R. Falanga (✉) · J. Ferrão
University of Lisbon, Lisbon, Portugal
e-mail: roberto.falanga@ics.ulisboa.pt

J. Ferrão
e-mail: joao.ferrao@ics.ulisboa.pt

of the city, thus the citizenry living in a specific urban settlement understood as the "civitas". A third layer is what can be referred to as the "polis", thus the urban political life of the city. Threats of destruction can intercept the three layers and tell of the overwhelming effects that modernist rationalism first, and neoliberalism later, have had over cities by (re)shaping their forms and the socio-political interactions within. As Carrión (2018) put it, threats to cities and citizens unfold through multiple forces that carry dramatic effects over systems of belonging and urban heritage. In fact, citizens can be both physically and symbolically eradicated from their places.

The concept of urbicide helps understand the magnitude and reach of different forces driving the destruction of the physical, social and political city. Therefore, the urbicide can unfold through two main sets of events and dynamics. The first set refers to catastrophic natural events and wars, as well as anthropogenic actions (e.g. climate breakdown and pandemics) and the negative consequences of modern urban planning. Such events and dynamics share the material destruction and the devastation of commonplaces and infrastructure as common elements. The second set of events concerns threats to social interactions due to social segmentation and polarisation. Increasing inequalities and the dissolution of social cohesion are paired by growing individualisation and the predominance of consumerist logic. In this set, the urbicide equally impacts on political life through processes of social annihilation from political institutions and the progressive loss of democratic ethos.

By bearing this in mind, in this chapter we examine the extent to which the application of this concept accounts for the main impacts of the COVID-19 pandemic in city life. To what extent does the concept of urbicide contribute to understanding the impacts of this pandemic in urban environments, and how does such understanding contribute to advance this concept for future debates? We believe that a systematic examination of the COVID-19 impacts through the lens of the urbicide can help answer these questions and stimulate further discussion on the concept of urbicide in light of the current transformations in our cities. Accordingly, the chapter draws primary inspiration from the conceptualisation of the urbicide by Carrión (2014, 2018) and Polanco (2021) and briefly systematises in the first section the main units of analysis adopted by those authors to discuss the material and symbolic destruction of cities during the COVID-19 pandemic. The second section reviews some of the main literature produced in the last couple of years on urban problems and responses to the health crisis through the lens of the concept of urbicide. By delving into some major impacts of this pandemic and some of the solutions that have been put in place by public powers, we privileged scholarly contributions in the field of social sciences with a focus on Europe. The third and last section advances knowledge on the concept of urbicide by relating the insights presented in the previous section to the inputs from a previous work of ours concerning the COVID-19 pandemic and the governance strategies to mitigate, adapt and transform the city before systemic global risks (Ferrão et al. 2021). Our purpose is to cast light on both destructive and constructive sides of the current pandemic crisis, thus contributing to integrate the concept of urbicide into a broader vision of urban transformations.

## 7.2 Examining the COVID-19 Impacts Through the Lens of Urbicide

In this section, we examine the urbicide's material and symbolic destruction as analytical dimensions that allow disentangling some major impacts brought about by the COVID-19 pandemic. At the outset of our analysis, we recall that epidemics and pandemics are not new phenomena to world history. In the last few years, the severe acute respiratory syndrome (SARS) affected 26 countries in 2003, with a fatality rate of 9.6% (WHO 2003). Between 2003 and 2015, the avian influenza A (H5N1 virus) caused 440 fatalities across 16 countries (WHO 2013). In 2009, as the H1N1 influenza (swine flu) hit a considerable number of young people, governments tried to put in place preventive plans. With the first cluster of pneumonia cases in Wuhan, in the Hubei Province of China, documented at the end of 2019, the COVID-19 pandemic was officially declared by the World Health Organization on the 11th of March 2020 (WHO 2020). While its outbreak found almost all countries unprepared to manage the impacts of this extreme event, international organisations orchestrated a set of policy recommendations and public restrictions that were incorporated all over the world in a relatively short time.

The multi-level governance strategy has given a significant contribution to spreading information and early warning signals, although the application of a common agenda in extremely diverse political environments has not always been effective. According to Carrión (2018), not only healthy countries performed better in supporting affected people thanks to greater economic resources, as a big difference was made by whether national health systems were predominantly public or privately owned. Structural inequalities at the global level were further shown by imposed limitations to international mobility and, since 2021, the unequal distribution of vaccines. Likewise, as different rules were applied from country to country, borders' control added fuel to the debate on the dominance of private actors in vaccination and, more broadly, the management of the health crisis. A great difference was also made by different urban planning traditions, with direct implication on the supply of services and goods, as well as on the access to quality public spaces and soft mobility systems.

Figure 7.1 proposes a synthesis of the main units of analysis adopted by main scholars to describe the concept of urbicide. Acknowledging their characterisation of material and symbolic destruction, this synthesis is functional to develop our original contribution on the COVID-19 impacts through the lens of urbicide.

Based on the synthesis shown in Fig. 7.1, our account for the material destruction (7.2.1) enumerates two main sub-dimensions that aim to open a reflection on the destructive as well as the constructive sides of the current pandemic crisis. Accordingly, we discuss the reconfiguration of environmental balance (7.2.1.1) and commonplace system (7.2.1.2). As for the symbolic destruction (7.2.2), we equally aim to combine a view on the destructive and constructive sides by discussing three main sub-dimensions that account for the reconfiguration of social cohesion (7.2.2.1),

# City's destruction

## Material destruction (URBS)

| Catastrophes | Anthropogenic risks | Urban production | Social segmentation and polarisation |
|---|---|---|---|
| Natural catastrophes floods, earthquakes, tsunamis, … | Biodiversity loss: greenhouse gas emission, air and noise pollution, traffic jam, … | Modernisation: social obsolescence/ heritage destruction; decline and abandonment of central urban areas; real estate speculation and suburbanisation | Functional and social segregation (modern rationalist planning) |
| Technological catastrophes | Climate changes | Neoliberal city: urban regeneration/gentrification; mass tourism impacts/eviction; megaprojects, privatisation; public spaces/enclaves | Inequalities, socio-spatial fragmentation, forced mobility (land market and real estate) |
| Wars | Pandemics (zoonoses) | | "New" social boundaries (migration, info exclusion, …) |
| | | | Violence and unsafety |

| Destruction of infrastructure as the material base of the city | Breakdown of environmental balance | Devastation of the common place system | Dissolution of social cohesion |

## Symbolic destruction

| Social interactions (CIVITAS) — Individualisation of social relationships | Institutions (POLIS) — Loss of urban democratisation |
|---|---|
| Predominance of the consumerism logic | Globalisation |
| Deterritorialisation of social interactions (ICT) | Deregulation and flexibility |
| | Burocratisation, corruption |
| | Privatisation of urban life |

| Privileging of consumer sovereignty to life of population | Annihilation of institutional frameworks |

**Fig. 7.1** Urbicide: a synthesis of the concept. *Source* Authors' own work based on Carrión (2014, 2018) and Polanco (2021)

consumer sovereignty (7.2.2.2) and democratic institutions (7.2.2.3). Both material and symbolic dimensions are discussed by considering key outputs from our review of the recent literature on the COVID-19 pandemic, with a focus on Europe.

## *7.2.1 On the Material Destruction*

The material destruction of the city refers to the aggravation of physical conditions of the built environment, the "urbs". From our understanding of the urbicide's conceptualisation, we identify three main sub-dimensions that account for the city's destruction. First, a city's infrastructure can be overwhelmed by catastrophic natural events, as well as by wars. Considering that this is not strictly the case of the current pandemic, here we focus on the other two sub-dimensions: the reconfiguration of environmental balance and commonplace. The former sub-dimension focuses on the (new) role played by the (urban) environment during this pandemic, whereas the latter concerns the difference made by urban planning in the face of the pandemic.

### 7.2.1.1 The Reconfiguration of the Environmental Balance

In our review of the literature, we found two main emerging trends engendering the impacts of the COVID-19 on the (urban) environment, referred to as the breakdown of the environmental balance in the concept of urbicide (Carrión 2014, 2018; Polanco 2021). The first trend regards the resurgence of nature in the public debate, which has been leveraged by an emerging attention to nature-based solutions in cities. Green areas and non-human species in the urban environment have been seen as an opportunity to reflect on the urban biodiversity and the relations between the built and the natural environment, which further triggered the debate on the physical and psychological quality of human life. The second emerging trend tells of the growing movement from the city to the countryside. During the pandemic, a considerable number of people were inclined to decamp in uncontaminated sites and rediscover nature. Both trends also show, however, the significant weight of social inequalities in securing equal access to all citizens.

About the first trend, nature has played a significant role in the public debate during the COVID-19 pandemic. In 2020, OECD scenarios indicated 60–80% decline in the tourism sector, with cities being the most affected by the lack of visitors in contrast to rural and natural areas (OECD 2020b). In addition, the therapeutic and preventive functions of nature have helped shift the public debate on the need for new investment in nature-based solutions, including urban flooding and heat, as well as energy-efficient buildings (Moglia et al. 2021). New ideas on the creation of green areas have waved the debate into the need to decrease noise levels and improve air quality in cities (Sharifi and Khavarian-Garmsir 2020). In the first stages of the pandemic, Acuto and colleagues (2020) found a high reduction in global emissions and a decreased concentration of $CO$, $NO$, $NO_2$ and Ozone, which are likely to have negative impacts on the health of urban populations. The resurgence of the debate on the positive effects of human exposure to greenery has been paired by the practice of open-air activities, such as sport and pet walking. Likewise, the improvement of the urban environment buffered the negative effect of shutdown/lockdown conditions on mental health by mitigating anxiety and stress caused by the pandemic (Haase 2021). The experience of reduced urban mobility further contributed to the debate on human well-being in association with the call for greater investment on mental health (Basu et al. 2021; Rumpler et al. 2020).

As regards the second trend, the ascent of nature-based solutions in cities and the impacts on mental health of prolonged shutdowns and lockdowns in the last two years have prompted new people's flows within and outside the cities. Nathan and Overman (2020) shed light on the "big city exodus", as inner-city office spaces have often been substituted by at-home office work. Coppola (2021) adds to this that long-distance mobility for leisure and business declined while mobility within urban regions and daily commuting increased through decamping movements, a specific behavioural strategy of this pandemic. The reduction of urban concentration in search for the uncontaminated nature foregrounded, however, the significant structural social inequalities of our cities. While higher-income and highly educated people either decamped to secondary residences or visited family and friends in the countryside, lower-income and educated people have rarely had the opportunity to get out of the city, thus continuing to experience higher risks of contagion within urban spaces.

### 7.2.1.2 The Reconfiguration of the Commonplace System

The reconfiguration of the commonplace, which Carrión (2014, 2018) and Polanco (2021) refer to as its destruction through the lens of the urbicide, points out two main trends during this pandemic. The first trend entails the success of some European cities in facing the spread of the coronavirus due to their relatively large size and compactness. Both characteristics are deemed to ease public access to healthcare and have encouraged public actors to think on new investments for more walkable and bikeable cities. A second trend was derived from the call for more diversity within cities, as strongly advocated by scholars defending the "15-min model city" that puts new emphasis on resources' provision and proximity accessibility in contrast to mobility-based accessibility.

As the airborne COVID-19 rapidly went global due to the advanced system of international connections, this pandemic has illuminated the multi-scalar relation between accessibility and density in the transmission of viral disease (Coppola 2021). First-hit places were characterised by great degrees of connectivity of local economies, including hubs for tourism and business (Florida et al. 2020). The risk of infection was considered higher for trips and public transport, thus pushing users to either stop travelling or opting for other modes of transportation. Cities favouring access to workplaces through soft mobility systems have had positive impacts on the quality of life. Likewise, from a comprehensive review of literature, Mouratidis (2021) found that city size and compactness have helped counteract the contagion by improving walkability and bikeability. In Europe, large and compact cities have provided easier access to healthcare infrastructure, and this outcome calls out the key role of urban planning in public health. In contrast, often-degraded suburbs have come to the public fore due to inadequate conditions of living quality, which calls out their potential for regenerative urban actions and the improvement of green infrastructure.

Regarding the second trend, scholars advocate greater diversity as a strategy that helps produce better, more just and more liveable cities in contrast to the car-dependent legacy of modernist urban planning. Hananel and colleagues (2022) recently found supporting evidence by analysing Ultra-Orthodox communities living in mixed neighbourhoods in the USA, who happened to show low degrees of infection when compared to gated communities because of the role that social capital played in protecting community health and rising resilience (see also Moglia et al. 2021). The hollowing out of specialised neighbourhoods also confirmed the fragility of clustering strategies based on real estate valorisation and fiscal revenues (Ramani and Bloom 2021). While old and new monofunctional downtowns and edge cities have been the most susceptible to the economic fallout, "super-neighbourhoods" and "15-min cities" have gained a centre stage in the public debate. According to Moreno and colleagues (2021), the "15-min city" unfolds urban diversity through the provision of residential, commercial and entertainment components to improve the quality of human interaction and their participation through the planning processes. The author recently acknowledged the role of density, proximity, diversity and digitalisation in defining this model, which presents an opportunity to rethink resource allocation on a citywide scale. By bringing into focus the concept of self-sufficiency, the "15-min city" aims to strengthen local service supply in place of amenities accessed elsewhere in the city. Nevertheless, as Pozoukidou and Chatziyiannaki (2021) put it, proximity requires redistributing functions based on geographical, economic and social principles, which require power transference to the lower level of the governmental system.

### 7.2.2 *On the Symbolic Destruction*

The symbolic destruction of the city refers to the worsening of social interactions in the "civitas" and its political implications in the "polis". In our review of the literature on the COVID-19 pandemic, we identified three main sub-dimensions that account for the ways in which this pandemic has affected the polis. First, the reconfiguration of social cohesion and its multiple impacts on human life that strongly rely on the persistence of structural inequalities. Second, the reconfiguration of the economic and labour spheres through the shift from in the presence to online activities based on the development of new technologies that bring new advantages and disadvantages. Third, the reconfiguration of the role played by local and national democratic institutions in the regulation of the urban sphere.

### 7.2.2.1 The Reconfiguration of Social Cohesion

Two main emerging trends were found as regards the reconfiguration of the social cohesion, referred to as dissolution in the concept of urbicide (Carrión 2014, 2018; Polanco 2021). The first trend concerns vulnerable groups, such as poor and at-risk of poverty people, minority ethnic groups and the LGBTQ + community. Besides material impairments, these groups have often suffered from discriminating behaviours with practical implications in their daily life. As De Rosa and Mannarini (2021) put it, the "othering" processes activated by the social representation of outgroup members have been based on the fear of contagion and have changed the meaning of proximity and closeness, with new prescriptive rules requiring a "defamiliarisation" of emotional linkages. The second trend regards the aggravation of inadequate housing conditions for poor people, especially migrants and, in some cases, students. The latter often found themselves in cohabited homes with serious difficulties in keeping on with blended education programmes.

As regards the first trend, the unequal distribution of communities in urban environments and the concentration of poor people in specific areas has brought harsh impacts on some vulnerable categories (OECD 2020a). Discriminating behaviours and practical impairments have especially affected lower-income and educated people, often suffering from racialised practices and living in poor housing conditions. These groups also turned into an accelerator of infection due to limited access to public transport, healthcare, healthy food and recreational opportunities (Almagro and Orane-Hutchinson 2020; Florida et al. 2020; Jay et al. 2020). In the USA, Ruprecht and colleagues (2021) found that Black, Latin and Native American populations, along with sexual and gender minorities, have been more likely to contract and suffer from COVID-19. Increased vulnerability is explained by underpaid jobs in "essential industries", which have required face-to-face contact, as well as by institutional discrimination and racist zoning laws, which have made physical distancing more challenging in some urban areas. Some scholars further notice that minority groups and the LGBTQ + community have been overrepresented in the prison population, thus suffering from having been locked with few visits from family and children for a long time, with limited access to healthcare (Harriot 2021). A study from the USA reported less secure access to financial, medical and educational resources, with a high percentage of job loss in LGBTQ + families (Gil et al. 2021).

The second trend concerns social injustices and the transformation of "homes" into multi-tasking contexts due to the intensification of forced coexistence in private space (De Rosa et al. 2021). In some extreme cases, homes have been a stage of domestic violence, with a dramatic escalation of abuses in caseloads and serious difficulties to ensure adequate response from public institutions. Housing has shown the sharp contrast between wealthier people's "self-sufficient" homes and inadequate living conditions (e.g. inadequate access to air and sunshine), difficult cohabitations (e.g. high number of occupants) and plumbing facilities (Ahmad et al. 2020). Several harms have been experienced by poor migrants living in overcrowded small apartments of often informal housing markets with associated difficulties in paying rents (Vilenica et al. 2020). Co-housing practices among workers and students have also shown how even a substantial portion of care workers has continued to be subject to "marginal" housing conditions (Buckle et al. 2020). In addition, students have lost out in a range of areas, such as accommodation contracts and academic services, with controversial outcomes from blended teaching worldwide. In Italy, Amerio and colleagues (2020) found a strong association between poor housing and depressive symptoms, due to small apartments, poor quality views and scarce indoor qualities, with implications on working performance. Likewise, other scholars have pointed out the need for greater support to young people, who are among the most exposed groups to mental health issues (Sinha et al. 2020).

### 7.2.2.2 The Reconfiguration of Consumer Sovereignty

Two main emerging trends were retrieved from the literature about the rampant role of the digital sphere during the pandemic, an aspect referred to as the primacy of consumer sovereignty to the life of the population by Carrión (2014, 2018) and Polanco (2021). On the one hand, the exponential growth of digital platforms for e-commerce and online delivery is triggering new opportunities and risks, especially about personal data protection. On the other hand, the reconfiguration of the labour market highlights dramatic transformations for small entrepreneurs and aggravating inequalities between skilled and low-skilled workers, with critical situations for vulnerable groups, especially women.

During this pandemic, the impacts on urban economies have been particularly harsh in some sectors, with a decline in the international tourism economy. OECD (2020b) pinpointed that tourism, retail sales of products and services, and small manufacturing were the most negatively affected sectors, as the digital shift was especially hard for small producers, sellers and consumers. Against this backdrop, the acceleration of the digitalisation of economic and labour activities has been compounded by the exponential growth of e-commerce with new business-to-consumer and business-to-business transactions. In fact, the role of digital platforms for interoperable data systems has enabled an unprecedented amount of information sharing and advanced new challenges for data protection in the digital sphere (Steen and Brandsen 2020). Fraudulent and deceptive online practices have spread along with fake news, sale of unsafe hand sanitisers, surgical face masks or disinfectants and price-gouging practices to profit from the surge in demand (WTO 2020).

The second trend regards the transformation of homes into "electronic cottages", which has allowed the continuation of some jobs (Doling and Arundel 2022), while others have dramatically fallen in the labour market. The opportunity of remote working calls for a global re-appreciation of the value and reward of skilled and low skilled jobs. White-collar workers have been more likely to be able to work from home, to have access to a personal car for transportation and to live in uncrowded homes (Florida et al. 2020). In contrast, being employed as "essential workers" or "frontline workers" is a condition that mostly involves minorities with high-touch jobs at the bottom of the wage scale, which has exposed them to higher risks of job loss due to automation and digital transition and infection to the coronavirus (Afsahi et al. 2020). The acceleration of online food delivery, for example, has been barely accompanied by an adjustment of job conditions as well as protection measures for health and safety. Evidence further shows that the impacts of the economic crisis have been more negative for women as they are likely to have lower income and fewer savings, being often single parents and more likely to be informal and/or unpaid care and domestic workers (Mouratidis 2021).

### 7.2.2.3 The Reconfiguration of Democratic Institutions

Literature review pointed out two main trends emerging about the relationship between citizens and democratic institutions that give account of the annihilation of the institutional frameworks in the urbicide (Carrión 2014, 2018; Polanco 2021). During the COVID-19 pandemic, local governments played a key role in the face of the ambivalent perception of the central state. Secondly, central states have produced public discourses and made decisions in the management of the health crisis within a highly complex field of forces.

As regards the first trend, the continuous state of emergency has nurtured public cynicism towards public institutions (Afsahi et al. 2020). Countries with an existing crisis of democracy have been at the stage of a growing destabilisation of the social order, whereas those countries that provided long-term pandemic plans have shown a more effective management of the coronavirus. This pandemic has illuminated the ambivalent perception of the central state, as either reinvigorated in the management of the health crisis or overwhelmed by that. In contrast, while local governments have often become more dependent on the state than it was before the crisis, pre-existing networks and partnerships between communities and external agencies have contributed to provide more effective responses (Leach 2021). Local authorities have more frequently gained centre stage in the public debate for putting in place solidarity-induced measures and encouraged self-aid networks via proximity democracy (Smith et al. 2021). Inhabitants have been a powerful resource to counteract the negative impacts of the pandemic at the community level through initiatives against lockdown-induced loneliness and lack of essential goods and services (Falanga 2020). The strengthening of elective ties for self-organised mutual aid has been indicative of the power of community action, as well as of the pervasiveness of structural inequalities (Smith et al. 2021). In fact, as noted by Steen and Brandsen (2020), the opportunity to break through procedural restrictions during the pandemic opened up options for new experiments often powered by people affected by the economic paralysis (Hall et al. 2021).

Connected to the trend above, the political discourse has developed at the crossroads of international, national and local forces by eventually playing a role in shaping final decisions. The media have given stage to scientists, who have been essential in understanding, tracking and responding to the virus. However, in some cases, the exposition of virologists and the integration of scientific knowledge into decision-making has generated confusion, apprehension and scepticism among citizens (see also: Cho and Gower 2006; Neuman, Just and Crigler 1992). As pointed out by Seebohm (2021), confusing and contradictory messages from governments and media have worsened trust towards decision-makers, which has exacerbated feelings of isolation and emotional distress. Furthermore, whereas this was perceived as a time of "command and control", delays in information release, rumours, and fake news have nurtured fear and disorientation, as well as critical behaviours and public outcries. On a more critical stance, Standring and Davies (2020) contended that the public discourse has often underlined existing social pathologies provoked by neoliberalism. The authors (ibidem) point to the emergence of new forms of "necro-socialism" through the adoption of a "socialist" language on the priority of health over finance through small acts of kindness out of a robust engagement with reverting policies on structural inequalities.

## 7.3 Reframing Our Understanding of Urbicide Based on the Impacts of the COVID-19 Pandemic

The COVID-19 pandemic has driven multiple changes, some of them temporary, some others more persistent. Ferrão and colleagues (2021) have discussed the magnitude and reach of the pandemic effects by identifying three main categories: (i) *revealing* weaknesses (e.g. vulnerability of the most exposed socio-professional, ethnic and age groups, the limited human and financial resources of public services and social organisations, the financial fragility of many companies), (ii) *accelerating* trends that had not reached a relevant expression until the disruption of the pandemic (e.g. remote working or home delivery) and (iii) *cluttering* phenomena, as well as *disrupting* actions towards positive changes, including human–planet relations. According to the authors (ibidem), the three categories of effects intercept different, at occasion interwoven, governance strategies put in place by public authorities against systemic global risks: (i) *mitigation*, through immediate responses in the face of the emergency; (ii) *adaptation*, through short-medium preventive measures aimed to improve urban resilience; and (iii) *transformation*, through medium-long term actions addressing the causes of risks, thus aiming to advance structural changes and transformative agendas.

Examination of the COVID-19 impacts through the lens of the urbicide allows acknowledging the magnitude and reach of the material and symbolic destruction in the short and middle terms and envisage trends to be either countered or enhanced based on a transformative agenda. Accordingly, Fig. 7.2 shows the subdimensions and associated trends discussed above as clustered within broader domains, namely environment, planning, society, economy and democracy.

In each domain, as synthesised in Table 7.1 and discussed in the following subparagraphs, we identify the main pandemic effects, briefly classified as revealing, accelerating and cluttering; as well as the solutions in place to mitigate, adapt and/or transform local governance.

## City's transformation

### Material transformation (URBS)

| Environment | Planning |
|---|---|
| Pandemic effects are **cluttering** our understanding of human/non-human relations. **Adaptation** strategies have mainly addressed short-term solutions and international commitment is needed for a transformative agenda. | Pandemic effects are **accelerating** our awareness about the pitfalls of modernist planning. **Mitigation strategies** have mainly sought for temporary initiatives (e.g., tactical urbanism) and only few cities have shown commitment to longer term changes. |

### Symbolic transformation

| Social interactions (CIVITAS) | | Institutions (POLIS) |
|---|---|---|
| Society | Economy | Democracy |
| Pandemic effects are **revealing** the fragility of some social groups, such as ethnic minorities, migrants, LGBTQ+ community, young people and women. **Mitigation strategies** have mainly relied on the articulation between public authorities and social groups, and little evidence is shared on action against structural inequalities. | Pandemic effects are **accelerating** existing trends in the growth of e-economy and delivery platforms, with risks for small sellers and data protection. **Adaptation** and **transformation** solutions are being put in place by improving digital devices and skills, although little has been done so far to enhance digital access and literacy. | Pandemic effects are **accelerating** existing trends related to citizens' disaffection towards the state, while enhancing trust in local authorities. **Mitigation** strategies have mostly built on initiatives of proximity democracy, while little evidence exists on longer-term strategies to recover citizen distrust at the national level. |

**Fig. 7.2** COVID-19 pandemic through the lens of urbicide. *Source* Authors' own work

# 7 COVID-19 and the City: Reframing Our Understanding of Urbicide ...

**Table 7.1** Pandemic effects and solutions in place in the identified domains

| Domains | Pandemic effects | Solutions |
| --- | --- | --- |
| Environment | Pandemic effects are cluttering our understanding of human/non-human relations | Adaptation strategies have mainly addressed short-term solutions while international commitment is needed to foster a transformative agenda |
| Planning | Pandemic effects are accelerating our awareness about the pitfalls of modernist planning as well as the lack of planning | Mitigation strategies have mainly sought to promote temporary initiatives (e.g. tactical urbanism) and only few cities have shown commitment to longer-term changes |
| Society | Pandemic effects are revealing the fragility of some social groups, such as poor and at-risk of poverty people, ethnic minorities, migrants, LGBTQ + community, young people and women | Mitigation strategies have mainly relied on the articulation between public authorities and social groups, while little evidence is provided on a more robust commitment to revert structural inequalities |
| Economy | Pandemic effects are accelerating existing trends in the exponential growth of e-economy and delivery platforms, which impair small sellers and put in risk personal data | Adaptation and transformation solutions are being put in place by improving digital devices and skills, although little has been done so far to enhance digital access and literacy |
| Democracy | Pandemic effects are accelerating existing trends related to spreading disaffection towards the central state, which contrasts to higher appreciation of local authorities | Mitigation strategies have mostly built on initiatives of proximity democracy, while little evidence exists on longer-term strategies to recover citizen distrust at the national level |

## 7.3.1 Environment

In the environment domain, the pandemic effects are mostly cluttering our understanding of the relation between human and non-human species in the city, within the frame of the ongoing climate breakdown. Temporary changes during the pandemic aimed to take advantage from the perceived suspension of urban life. However, inequalities emerged as to the access to quality green spaces within the city, and the possibility to commute from city to countryside. Solutions have mainly regarded short-term and on–off initiatives, which call for greater commitment of local, central and international agencies towards effective, fair and inclusive solutions.

## 7.3.2 Planning

In planning, the pandemic effects are both accelerating changes and cluttering our understanding of the world. On the one hand, the modernist legacy has accelerated our understanding of the need for more diversity in the city. This debate has been supported by effective solutions carried out in diverse, large and compact cities, which have provided good results in health access and sustainable mobility. On the other hand, these same effects push the debate on urban planning towards the centrality of the neighbourhood as a key unit of analysis. Like in the previous domain, adaptation strategies have been mainly short-term and one-off, although some big cities, such as Paris and Barcelona, are investing in medium-long term changes.

### 7.3.3 Society

The dissolution of social cohesion calls on impacts of the COVID-19 pandemic on urban and domestic lives. In both, social interactions have been dramatically changed by the health measures adopted by public authorities to contrast the virus contagion. Isolation at home and physical distance in public spaces have been particularly hard for groups that suffered from material (e.g. job loss) and symbolic (e.g. discrimination) harms. Poor migrants as well as some students have suffered from inadequate housing conditions, which increased risks of infection in overcrowded cohabitations, paired by problems at school and/or in mental health. Mitigation strategies against such revealing effects often relied on the action of social groups and the third sector, with immediate aid packages coming from public powers, however out of orchestrated strategies of long-term interventions.

### 7.3.4 Economy

In the economy field, major changes have come from the development of new technologies and the digital sphere. The acceleration of these phenomena has built on needs to provide alternative instruments to workers and effective ways of communication. In the labour market, the home has become a multi-operational centre for some, while others have continued to work in contact with the public, thus exposing themselves to higher risks of contagion. In parallel, spreading service and goods' delivery workers corroborated significant divides in the protection of skilled and unskilled workers. Responses in this field have relied on medium-long strategies as regards the digital shift of public and private agencies, with investment for skill improvement and updated devices.

### 7.3.5 Democracy

Finally, pandemic impacts on democracy have been rather ambivalent. While central governments have played a major role in managing the health crisis, some states have shown unpreparedness and have sent contradictory public messages. Others have been praised for effective actions, with local authorities often coming to the fore as effective proxies with social groups, community networks and the third sector. Yet, dependency on central funding may have constrained degrees of autonomy in this field. Emerging polarisation in the political spectrum and raising mobilisations are likely to change the democratic scenario in the days ahead. Against such accelerating effects, there is no evidence of proper adaptation strategies put in place by states.

## 7.4 Concluding Remarks

In this chapter, we aimed to systematise knowledge about the impacts of the COVID-19 pandemic through the lens of the urbicide. This original contribution is mostly based on a selected review of scientific literature in social sciences with a focus on Europe. While relying on situated knowledge, the chapter allowed examining two interconnected questions: to what extent does the concept of urbicide contribute to understanding the impacts of this pandemic in urban environments, and how does such understanding contribute to advance this concept for future debates about the city?

Some of the impacts of the COVID-19 pandemic may be linked to the material and symbolic destruction of the city, as postulated by the concept of urbicide. Like in similar extreme events, this pandemic has suspended the normal functioning of cities, worsened social vulnerabilities and disparities, and created problems to the urbs, the civitas and the polis. Nevertheless, impacts have not been linear, as the magnitude and reach of this pandemic have elicited a highly complex set of issues that broaden the analytical scope of the urbicide. An in-depth understanding of this concept has helped navigate through the systemic and, at occasion, contradictory urban issues that neither entirely nor comprehensively attach to it. Our effort has been to explain some of the most striking phenomena of this pandemic by problematising the concept of urbicide itself. In fact, it is through the lens of the urbicide that the asymmetric nature of cities can be more clearly seen in the articulation between urbs, civitas and polis. This conceptual lens allowed a crosscutting interpretation of the ways in which the impacts of the pandemic, as well as those of similar extreme events, can relate to the material and symbolic dimensions of the city. Did the pandemic exacerbate some of the harms of modern rationalism and neoliberal globalisation? Did it create new conditions, albeit transitory, for new transformative trends in our cities, thus opening room for new ways of living our common spaces?

One of the key lessons learnt from this literature review is that the pandemic is not only a source of some urbicide-related issues, such as the aggravation of structural inequalities generating discrimination and conflict, but also a new potentiality for urban transformations. And this lesson brings us to reflect on whether and how it is possible to advance knowledge on current urban transformations through the lens of this concept. This chapter points at one possible pathway: the improvement of the interpretive potentiality of the urbicide based on the acknowledgement of some conceptual and practical limits.

In our opinion, the concept's destructive ethos underpins an idea of cities stepping back from existing or ideal conditions. However, destruction and construction, permanence and transformation, death and life gain new meanings according to different interpretive angles. In other words, this chapter can stimulate the debate on the dynamic character of the urbicide and broaden its scope as to the ongoing societal and urban transformations. From a focus on absence caused by destruction, the urbicide can speak the language of new desired (per)formed presences.

While planetary processes of urbanisation undermined the collective production and use of cities, thus bringing to the surface some urbicide-related issues, we believe that our approach to this concept can bring new insights at the edge of city and urbanisation. If we aim to push forward a transformative and progressive urban agenda, the urbicide should incorporate a regenerative ethos, contrasting the present regressive forces against the urbs, the civitas and the polis.

## References

Acuto M, Larcom S, Keil R, Ghojeh M, Lindsay T, Camponeschi C, Parnell S (2020) Seeing COVID-19 through an urban lens. Nat Sustain 3:977–978

Afsahi A, Beausoleil E, Dean R, Ercan SA, Gagnon JP (2020) Democracy in a global emergency: five lessons from the COVID-19 pandemic. Democracy Theor 7(2), v–xix

Ahmed Z, Ahmed O, Aibao Z, Hanbin S, Siyu L, Ahmad A (2020) Epidemic of COVID-19 in China and associated psychological problems. Asian J Psychiatry 51:102092

Almagro M, Orane-Hutchinson A (2020) The determinants of the differential exposure to COVID-19 in New York City and their evolution over time, SSRN

Amerio A, Brambilla A, Morganti A, Aguglia A, Bianchi D, Santi F, Costantini L, Odone A, Costanza A, Signorelli C, Serafini G, Amore M, Capolongo S (2020) COVID-19 lockdown: housing built environment's effects on mental health. Int J Environ Res Public Health 17:5973

Basu A, Kim HH, Basaldua R, Choi KW, Charron L, Kelsall N, Hernandez-Diaz S, Wyszynski DF, Koenen KC (2021) A cross-national study of factors associated with women's perinatal mental health and wellbeing during the COVID-19 pandemic. Plos One 16(4):e0249780

Buckle C, Gurran N, Phibbs P, Harris P, Lea T, Shrivastava R (2020) Marginal housing during COVID-19, Australian Housing and Urban Research Institute Limited, Melbourne

Carrión F (2014) Urbicidio o la producción del olvido. Observatorio Cultural 25:76–83

Carrión F (2018) Urbicidio o la muerte litúrgica de la ciudad. Oculum Ens 15(1):5–12

Coppola A (2021) Looking through (and beyond) a truly total territorial fact. Notes on the pandemics and the city, Territorio, pp 7–13

Cho SH, Gower KK (2006) Framing effect on the public's response to crisis: human interest frame and crisis type influencing responsibility and blame. Public Relations Rev 32(4):420–422

de Rosa AS, Mannarini T (2021) COVID-19 as an "invisible other" and socio-spatial distancing within a one-metre individual bubble. Urban Des Int 26:370–390

de Rosa AS, Mannarini T, Gil de Montes L, Holman A, Lauri MA, Negura L, Giacomozzi IA, da Silva Bousfield AB, Justo AM, de Alba M, Seidmann S, Permanadeli R, Sitto K, Lubinga E (2021) Sense making processes and social representations of COVID-19 in multi-voiced public discourse: illustrative examples of institutional and media communication in ten countries. Community Psychol Global Perspect 7 (1):13–53

Doling J, Arundel R (2022) The home as workplace: a challenge for housing research. Housing Theor Soc 39(1)

Falanga R (2020) Citizen participation during the COVID-19 pandemic. Friedrich Ebert Stiftung

Ferrão J, Falanga R, Liz C (2022) COVID 19 – Preparar as cidades para riscos globais. In Ribeiro L, Noronha J, Rodrigues J, Oliveira R (Eds) Metrópole e pandemia: presente e futuro, Rio de Janeiro: Letra Capital, 102–135

Florida R, Rodríguez-Pose A, Storper M (2020) Cities in a post-COVID world. Papers in Evolutionary Economic Geography, Utrecht University

Gil RM, Freeman TL, Mathew T, Kullar R, Fekete T, Ovalle A, Nguyen D, Kottkamp A, Poon J, Marcelin JR, Swartz TH (2021) Lesbian, Gay, Bisexual, Transgender, and Queer (LGBTQ+) communities and the coronavirus disease 2019 pandemic: a call to break the cycle of structural barriers. J Infectous Diseases 224:1810–1820

Haase D (2021) COVID-19 pandemic observations as a trigger to reflect on urban forestry in European cities under climate change: introducing nature-society-based solutions. Urban Forestry Urban Greening 64:127304

Hall J (2021) The perfect storm? Emerging from the crisis stronger, through sharing what we have. In: Smith G, Hughes T, Adams L, Obijiakupp L (eds) Democracy in a pandemic: participation in response to crisis. University of Westminster Press, pp 17–24

Hananel R, Fishman R, Malovicki-Yaffe N (2022) Urban diversity and epidemic resilience: the case of the COVID-19. Cities 122:103526

Harriot (2021) No justice without us: respecting lived experience of the criminal justice system. In: Smith G, Hughes T, Adams L, Obijiakupp C (eds) Democracy in a pandemic: participation in response to crisis. University of Westminster Press, pp 75–78

Jay J, Bor J, Nsoesie EO, Lipson SK, Jones DK, Galea S, Raifman J (2020) Neighbourhood income and physical distancing during the COVID-19 pandemic in the United States. Nat Human Behav 4:1294–1302

Leach M (2021) Mutual aid and self-organisation: what we can learn from the rise of DIY responses to the pandemic. In: Smith G, Hughes T, Adams L, Obijiakupp C (eds) Democracy in a pandemic: participation in response to crisis. University of Westminster Press, pp 123–138

Moglia M, Frantzeskaki N, Newton P, Pineda-Pinto M, Witheridge J, Cook S, Glackin S (2021) Accelerating a green recovery of cities: lessons from a scoping review and a proposal for mission-oriented recovery towards post-pandemic urban resilience. Dev Built Environ 7:100052

Moreno C, Allam Z, Chabaud D, Gall C, Pratlong F (2021) Introducing the "15-minute city": sustainability, resilience and place identity in future post-pandemic cities. Smart Cities 4:93–111

Mouratidis K (2021) How COVID-19 reshaped quality of life in cities: a synthesis and implications for urban planning. Land Use Policy 111:105772

Nathan M, Overman H (2020) Will coronavirus cause a big city exodus? Environ Plann b: Urban Anal City Sci 47(9):1537–1542

Neuman WR, Just MR, Crigler AN (1992) Common knowledge: news and the construction of political meaning. The University of Chicago Press

OECD (2020a) Cities policy responses

OECD (2020b) COVID-19 and global value chains: policy options to build more resilient production networks

Polanco VD (2021) La muerte simbólica y material de la ciudad: una aproximación sobre el urbanicidio. Revistarquis 10(1):14–22

Pozoukidou G, Chatziyiannaki Z (2021) 15-minute city: decomposing the new urban planning Eutopia. Sustainability 13:928

Ramani A, Bloom N (2021) The donut effect: how COVID-19 shapes real estate. Stanford Institute for Economic Policy Research

Rumpler R, Venkataraman S, Göransson P (2020) An observation of the impact of COVID-19 recommendation measures monitored through urban noise levels in central Stockholm, Sweden. Sustain Cities Soc 63:102469

Ruprecht MM, Wang X, Johnson AK, Xu J, Felt D, Ihenacho S, Stonehouse P, Curry CW, DeBroux C, Costa D, Phillips G (2021) Evidence of social and structural COVID-19 disparities by sexual orientation, gender identity, and race/ethnicity in an urban environment. J Urban Health 98:27–40

Seebohm L (2021) Learning how to listen in a pandemic. In: Smith G, Hughes T, Adams L, Obijiakupp C (eds) Democracy in a pandemic: participation in response to crisis. University of Westminster Press, pp 69–74

Sharifia A, Khavarian-Garmsird AR (2020) The COVID-19 pandemic: impacts on cities and major lessons for urban planning, design, and management. Sci Total Environ 749:142391

Sinha M, Kumar M, Zeitz L, Collins PY, Kumar S, Fisher S, Foote N, Sartorius N, Herrman H, Atwoli L (2020) Towards mental health friendly cities during and after COVID-19. Cities & Health

Smith G, Hughes T, Adams L, Obijiakupp C (eds) (2021) Democracy in a pandemic: participation in response to crisis. University of Westminster Press

Standring A, Davies J (2020) From crisis to catastrophe: the death and viral legacies of austere neoliberalism in Europe? Dialogues Human Geogr 10(2):146–149

Steen T, Brandsen T (2020) Coproduction during and after the COVID-19 pandemic: will it last? Public Adm Rev 80(5):851–855

Vilenica A, Mcelroy E, Lancione M, Fernandez Arrigoitia M, García Lamarca M, Ferreri M (2020) COVID-19 and housing struggles: the (re)makings of austerity, disaster capitalism, and the no return to normal. Radical Housing J 2(1):09–28

WHO (2003) Consensus document on the epidemiology of severe acute respiratory syndrome (SARS). WHO/CDS/CSR\GAR\2003.11

WHO (2013) Pandemic influenza risk management. WHO Interim Guidance

WHO (2020) WHO director-general's opening remarks at the media briefing on COVID-19, 11 Mar 2020

WHO (2020) E-commerce, trade and the COVID-19 pandemic


**Roberto Falanga** Sociologist, Assistant Research Professor at the Institute of Social Sciences of the University of Lisbon, expert in participatory and deliberative democracy, urban policymaking, and urban regeneration. He has published extensively and is a member of national and international research projects on these topics. He has been a consultant to local, national, and international public organizations and evaluator of innovative practices of participatory policymaking.

**João Ferrão** Geographer, retired Researcher Professor at the Institute of Social Sciences of the University of Lisbon, expert in urban studies, territorial development policies, and planning. He was a lecturer in the Department of Geography of the Faculty of Arts of Lisbon, president of the Portuguese Association for Regional Development, expert for the OECD, the European Commission and the Portuguese Government, Secretary of State for Spatial Planning and Cities, and Pro-Rector of the University of Lisbon.

# Part III
# Annihilation: The End of the Public Space

# Chapter 8
# The Ideology of Public Space and the New Urban Hygienism: Tactical Urbanism in Times of Pandemic

Manuel Delgado-Ruiz

**Abstract** This essay interprets the implementation of "tactical urbanism" as the latest step in the urban technocratic project aimed at destroying or at least subduing all actual urban life, thus as a form of urbicide. It presents the case of Barcelona, where in the spring of 2020 the "new municipalism" city council developed several interventions based on tactical urbanism, aiming at guaranteeing a prophylactic environment against the spread of COVID19. Though presented as temporary, these transformations ended up being permanent, combining with other urban policies such as the "superblocks" (*supermanzanas*), justified as countering climate change. The essay argues that these policies aimed at refurbishing outdoor urban spaces reiterate the hygienist vocation of early nineteenth-century urbanism, born as a "science" precisely in the Catalan capital. The growing influence of tactical urbanism is analyzed in the framework of a left-wing municipal government that attempts to develop a new orientation in city governance and that employs as a crucial discourse a rhetoric of public space as an ethical arena for good citizenship. Though this urbanism is exhibited as environmentally friendly, it is inscribed in a long tradition of policies aimed at sanitizing cities by removing their natural tendency at being spaces for conflict.

**Keywords** Public space · Environmental hygienism · Social conflict · Urban policy

M. Delgado-Ruiz (✉)
University of Barcelona, Barcelona, Spain
e-mail: manueldelgado@ub.edu

## 8.1 New Citizenist[1] Municipalism

In the last decades of the last century, led by its Mayr Pasqual Maragall, Barcelona presented itself and was recognized worldwide as a model of progressive urban governance and organization. A set of urban initiatives led by architects modified the morphology of the city through groundbreaking urban design. Later, following her appointment to the municipal government in 2015, Mayr Ada Colau, respecting Maragall's vision of Barcelona, promoted transformations that once again positioned the Catalan capital as a benchmark in terms of creative adaptation to the climate crisis. Under Maragall, design focused on hard squares, large cultural containers, and avant-garde urban furniture, but the contributions that now placed Barcelona as a star city were the *superilles*—the superblocks—and the radical application of tactical urbanism. Both in the first "model Barcelona" (Borja 2010; Delgado 2017) and in the current "post-model Barcelona" (Montaner 2015), the common argument used to justify emblematic urban operations was (and is) the idea of a moral rebound—an argument based on values about good city planning and, by extension, planning of and for good citizenship.

Ada Colau won the municipal elections in June 2015 as the lead candidate for Barcelona en Comú (Barcelona in Common) and sought to express, politically, the spirit of the 15M Movement—the movement "of the indignant" (Pereira-Zazo, and Torres 2019)—who had occupied the squares of major Spanish cities in the spring four years earlier. The Comuns (the Commons) was a new political group related to Podemos, composed of militants from different social movements (Ada herself had been a squatter activist) that formed a government with the ecosocialist party Iniciativa per Catalunya. The members promoted a new form of local government that, until then, had only been applied in small towns by independent candidacies but which now became prevalent in almost all large- and medium-sized Spanish cities. They did so in the name of democratic principles that were considered to have been absent in the administration of public affairs, but which they aspired to incorporate into municipal management: transparency, citizen participation, equality, social redistribution, and dignity.

The new style of local management was soon projected as "new municipalism" (Romero and Boix 2015; Subirats 2006). It appeared on the Spanish and, later, the international political scene as a means to apply the principles of radical democracy to everyday management issues and fight from the institutions the neoliberal commodification of the territory and the consequences of its crises in the form of new and greater social demand. Its ideology corresponded to what has come to be

---

[1] Citizenism and citizenist are the terms that have been adopted to translate *citoyennisme* and *citoyenniste*, used to describe the ideology of ethical and aesthetic reform of capitalism typical of most new social movements. They first appeared in the pamphlet "L'impasse citoyenniste. Contribution à une critique du citoyennisme," which began to circulate on the internet in 2001 signed with the pseudonym Alain C., behind which was the geographer and sociologist Jean-Pierre Garnier. http://www.notbored.org/citizenism.html#_edn2. Accessed 11 August 2022.

called citizenism (Delgado 2016)—an updating of classical republicanism and left-wing liberalism that takes the political perspective of realizing cultural modernity projects and understands democracy not as a form of government but as a way of life and a moral imperative. Citizenism does not call for the dismantling of the capitalist system but instead seeks ethical and aesthetic reform to humanize it and temper its excesses.

Ada Colau and the Comunes were not the only candidates to promote new municipalism in the Spanish state. Other candidates who presented themselves as being "pro change" reached Mayral positions in important cities, such as Madrid, Valencia, Zaragoza, and Cádiz. The case of Barcelona was singular not only because, along with Cádiz, it was the only city that reelected the government in the subsequent elections in 2019, but also because it was accompanied by important doctrinal production and attracted international publicity, which led to new municipalism coming to be called international municipalism due to its ascendancy in the municipal policies of Europe, Latin America, and the USA (Chamock 2018). Thus, Barcelona is seen today as representing a new form of local governance paradigm, both in terms of theorization and its formal plans and in terms actions, which are compatible with the general idea of equipping cities with cutting-edge digital technology—so-called smart cities, of which Barcelona is claimed to be an excellent example (Charnock et al. 2021). New municipalism has revealed itself as an instrument for realizing universal ideals, such as the development of human capital, multiculturalism, quality of life, technological humanism, cosmopolitanism, citizen participation, and the "common good." Above all, its most admirable projects have been hailed as elements of a zealous crusade against automobilization and in favor of improved health conditions for urban citizens.

In practice, the emergence of new municipalism in the administration of large cities such as Barcelona has been limited, regarding economic redistribution and the acquisition of rights, to a downgraded, relaxed version of social democratic or even Christian social projects, which has positioned capitalist appropriation as a "social" and, increasingly, an "ecological" move (Mansilla 2017). This is consistent with the fact that the massive mercantile depredation of cities is being carried out in the name of the human good and the salvation of the planet. This constitutes a recycling of neoliberalism (Janoschka and Mota 2021) as a form of what we might call "virtuous capitalism," which employs new rhetorical repertoires to justify, claiming highly moral motives, the execution of plans that, although presented as urban development, often end up as mere real estate projects.

## 8.2 The Idealism of Public Space as a Moral Scenario

It is interesting that both citizenism as an ideology and the new municipalism in its government implementation conceive an idealistic vision of public space and, by extension, whole cities. Citizenism imagines public space as an ideal forum for consensus and reconciliation presided over by citizens. These hypothetical citizens have no antagonisms and foster truces between social segments with incompatible interests who agree to forget their differences and inhabit spheres in which the old social classes are conflated in the interests of shared goals and peaceful coexistence. In fact, the most persistent and central element of the discourses regarding both the model Barcelona and the post-model Barcelona seems to be the constant invocation of values in this chimerical public space.

The public space of which dogmatic citizenists speak, with its cleanliness and guarantee of ethical civic norms, is much more than a stage for social life, starring total or relative strangers. Ideologists, politicians, and technicians of new municipalism base the topographical concept of public space, adopted since the 80s of the last century,[2] on a notion taken from political philosophy (Arendt 1958; Habermas 1962)—that of using it to designate a bourgeois public sphere and deploy it for the benefit of civil society. This new use of the concept of public space locates in public places—spaces of public ownership and free concurrence—the proscenium on which abstract democratic principles are enacted as practical ways of being together. This constitutes a formula to ethically achieve the great transformations shaping the capitalist spatial turn—the conversion of urban spaces into major sources of production and the accumulation of surplus value.

Public space was thus introduced into official discourses on the city as an eidetic space, conceived and organized not according to what it was but to what it should be, and the discourses suppressed the access to or emergence of anything that contradicted or disagreed with its theoretical perfection. In other words, public space is a hypothetical ideal space that attempts to superimpose itself on real streets. It is this idealism of public space that, in Barcelona, is central to all discursive strategies for social pacification (Borja and Muxí 2003; Borja et al. 2004). Its task is to guarantee—not least in Barcelona—the fulfillment of neoliberal urban agendas that soften its ravages, making it a priority to discipline life in the streets.

This means that the notion of public space, in terms of both urban governance and the technocracy of urban citizenship, does not have a mere descriptive function, but serves as a pure ideology (Delgado 2015). Its task is not, as it pretends, to describe accessible voids between built volumes but to morally elevate territories as quality spaces—presumably accessible regions in which appropriate uses, desirable meanings, and good fluidity of displacements is ensured. It is a matter of configuring garrison spaces—in both the military and culinary senses—to guarantee safe, predictable environments for urban reform operations, making them appealing for speculation, tourism, and institutional purposes in terms of legitimacy. To this end, the urban exteriors should be meeting places for orderly multitudes of free and equal beings who will use them to enjoy friendly coexistence, free of conflict, and should become paradises from which anyone who does not espouse middle-class values is expelled or barred. This is the *ville en rose*, over which new municipalism in Barcelona would like to reign sweetly.

---

[2] The Google Books Ngram Viewer tool reveals that the concept of *public space* in books was almost nonexistent before 1980 but has enjoyed uninterrupted, exponentially increasing use since then.

## 8.3 Tactical Urbanism: From Tactics to Strategy

New liberalism corresponded to new urban globalization and to new urbanism. Smith (2002) stated that this was what he perceived to be happening at the planetary level at the end of the 1990s. New municipalism and its philosophical fantasy of public space are among the favorite instruments of new ways of global urbanism. The new urbanism of the new municipalism holds that capitalist accumulation in cities must be sustainable, inclusive, participatory, intercultural, and thus beneficial for the good of humanity (Anguelovski et al. 2018). In this framework, we can see the urgency of generating public spaces inspired by market ecologism that makes common people bear the responsibility for and price of the environmental disasters they are victims of. The green reconversion of urban capitalism has had to dynamically transform transit areas in cities to facilitate revolutionary "active" mobility—on foot or by bicycle—based on clean energies. This is the latest chapter in the long history of attempts to deconflict and depoliticize the streets as a so-called public space. It is tactical urbanism.

Tactical urbanism adopts various doctrinal and practical starting points to justify its proposals for intervening in urban mobility spaces. It claims to be the heir of Situationist International and its proposal to subvert urban order by generating occasions to upset bourgeois normality. It also claims a tempered reading of Henri Lefebvre, from which it adopts the proposal of urban models or forms developed outside the urban technocracy: an "experimental utopia" (Lefebvre 1961)—later, a "concrete utopia" (Lefebvre 1968)—made up of practices and proposals and attentive to implications and consequences. Tactical urbanism is postulated as executing the dream of Jacobs (1961: 29–142) for a recovery of "sidewalk society" based on cooperation and creativity, as well as for the management of cities linked to places and conducive to calm, non-invasive mobility.

de Certeau (1980) appears as a major reference with his concept of tactics. Tactics are the tricks that allow the subjugated to take advantage of junctures and occasions that allow memory to accumulate potentially useful experiences. Since this memory is pragmatic and serves the search for and exploitation of circumstances, it must itself be mobile and adaptive. Tactics differ from strategies, which are conceived of at (and applied from) positions of power by those who hoard privileges and organize controls. Tactics are ways of doing through which users reinterpret and reuse the spaces designed from above by specialists to make *other things* of them and do *other things* with them. For Certeau, tactics are a resource of the weak to counteract the control strategies of the powerful.

The antagonistic vocation of this type of urbanism was appropriately called "guerrilla" insofar as it began by challenging the authoritarianism of institutional urbanism and the despotism of the automobile. This type of urbanism was intended to be applied on a small scale as the fruit of neighborhoods' self-initiated projects. This was made possible by new forms of alternative financing and organization by multidisciplinary and ephemeral collectives, sometimes associated with transformative social movements. Isolated actions, such as Open Streets in Seattle, are often considered precursors of a movement that would later spread throughout the USA and Canada from 1965 onwards. Another precedent would be Bonnie Shreck's artistic works in Californian urban spaces in the early 1970s—the period during which bicycle lanes were initiated in Bogota. The "urban recipes" of architect Santiago Cirugeda in the second half of the 1990s are also important precedents for tactical urbanism in Spain (Cirugeda and De Nieves 2006).

Already widespread in the 2000s, tactical urbanism took the form of small, ephemeral, spontaneous interventions in North American cities such as New York, Dallas, and San Francisco. They involved modifying the local environments through the low-cost, playful use of rudimentary materials for folding picnic furniture, paint-to-zone sidewalks and driveways, plastic and concrete bollards, wooden planters, and site-made materials. This urbanism soon generated new types of street furniture with their own names: Berlin cushions (plastic obstacles on the road to force vehicles to slow down) or New Jersey fences (pieces of concrete painted yellow). Another action consisted of creating parklets and pocket parks as spaces for traffic or car parking.

Drawing inspiration from these experiences, tactical urbanism emerged as a movement after a meeting of the Next Generation of New Urbanist Group in 2010 (Lydon and Garcia 2015; McGuirk 2014) and underwent increasing formalization. Many of the interventions already undertaken by local administrations were initially carried out in the collaborative spirit in which they were born—from below. The tactical municipal projects were conducted in the manner of experiments whose concretization depended on consensus approval and the usufruct of the users themselves. This implies that each action of this type was (in theory) approved, revised, or even rejected by its recipients. It was the experience of these modified spaces (their subjection to daily appropriations) and the opinions of the communities concerned that subjected each action to gradual adaptation to the social needs of citizens in their environments, especially those related to the sociability of proximity, that they were intended to promote, alongside the objective of generating social capital.

In a few years, social urbanism became the reference model for municipal institutions, especially in the hands of governments aligned with new municipalism, which took over the administration of many medium-sized cities and, increasingly, large capital cities. In reality, urban planning procedures and the accompanying discourses were certainly a shrewd way for urban governments to use "poor" urbanism as a means to continue neoliberal urban development policies during the recession that followed the 2008 economic crisis. They made planning processes more flexible and acceptable (even desirable) to citizens by presenting them as ways to combat air and noise pollution, enhance sustainable mobility, and stimulate peaceful coexistence.

In practice, tactical urban planning and ecological measures ended up constituting the ingredients that made cities worthy of the appellation "creative." This is one of the attributes that urban promotion policies and competitions between cities currently employ as a benefit with which to justify what, in practice, are marketing techniques, as well as a source of prestige for political institutions in the eyes of their own and other countries' citizens. The results of interventions are presented as regenerating the urban fabric as a component of intellectual dynamism (if necessary, even rupturism) that corresponds to the growing dematerialization of the sources of economic growth, increasingly embellished with various avant-garde and ideological accompaniments. Tactical urbanism embodies combinations of innovation, imagination, art, and goodness that make it what Mould (2014) defined as "the latest political vernacular of neoliberalism."

Tactical urbanism aimed to promote the innovative recycling of urban space by encouraging the use of urban voids or the reuse of abandoned or decrepit land that could not be built upon, which resulted from the real estate crisis. The tactics ended up becoming a strategy through which municipal administrations could modify urban spaces in quick, cheap, and reversible ways, enabling them to test different solutions without mortgaging the urban spaces, waiting for projects to generate satisfactory results, or obtaining guarantees of consensual functional success. Entire cities were conceived of as open-air laboratories in which to test strategies without investing too many public resources (Cardullo et al. 2018; O'Callaghan and Lawton 2016). Despite this culturally anti-establishment patina, which allowed its achievements to be exhibited in contemporary art museums, the success of tactical urbanism served to depoliticize urban issues by varnishing initiatives with aesthetics and moralism and staging a social authenticity (Franco 2018) that won considerable political and media praise for its effectiveness.

It is no wonder that tactical urbanism is today an increasingly reputable resource for "city specialists," widely publicized through affirmative and propositional discourses. Its aesthetic is that of "do-it-yourself," a bricolage style that results in anecdotal, decorative, or peripheral actions and operations without any disruptive implications for neoliberal urbanism, harmlessly controlling the regulatory framework that governs territorial planning. Tactical urbanism embodies a political and economic program based on a diminished role for the public and an increased role for market forces. It does not propose planning that truly corrects deficiencies or excesses, such as those associated with social polarization, sociospatial segregation, or housing shortages. It does not even do so—contrary to its claims—with respect to environmental pollution. All these issues are at best alleviated but more often camouflaged or displaced. Tactical urbanism, with its low-cost initiatives, its informal air, and the moralizing ethos that surrounds and legitimizes it, far from its initial subversive pretensions, is today one of the favorite commercial brands for promoting business cities, including Barcelona.

## 8.4 Emergency Urban Planning in Times of Pandemic

Tactical urban planning began to be implemented in Barcelona in April 2020 as a necessary measure to deal with the worldwide state of emergency caused by the COVID-19 pandemic. Under the mobility restrictions that affected the entire population, the Barcelona City Council surprisingly activated the "revitalization of public space," presenting it as a response to the health crisis. This was an emergency urban planning action—a temporary adaptation to the catastrophic coronavirus epidemic. Overnight, colorful geometric shapes were painted on chamfers and pedestrian spaces; roads, bollards, planters, and concrete blocks were distributed as benches; obstacles were placed to inhibit vehicular traffic; deliberately rustic street furniture was installed; and the bike lanes were extended. As soon as the lockdown measures ended, many public parking spaces came to be occupied by terraces of catering establishments to compensate—it was said—for the losses caused by their closure during the lockdown.

This was repeatedly presented as the resilient resourcefulness with which Barcelona faced and overcame the catastrophe (Bermejo et al. 2022), with the aim of ensuring physical distancing between people and a healthy atmosphere to combat the disease.[3] These tactical urban planning measures, presented almost as medical prescriptions, occurred in other cities around the world: Milan, Auckland, Lima, Berlin, Bogota, Vienna, and New York.[4] In all cases, the modifications to urban exteriors were presented by the authorities as planned pilot responses to the health emergency, intended to facilitate the safe movement of street users, and the actions taken in this direction were based on a review of tactical urban planning as a tool for maintaining and promoting physical and mental health during the pandemic.

In Barcelona, the authorities and the media emphasized that the COVID-19 lockdown, especially the first phases of its relaxation and the gradual return of people to the streets, gave rise to an unusual image of streets. Suddenly, citizens could enjoy their immediate surroundings free of cars and communicate with their neighbors, who were no longer strangers, as they had been previously. Also, it was possible to enjoy moving around on foot or using clean transport, with fewer and shorter trips, in an environment free of atmospheric and acoustic pollution, reveling in the pure air and birdsong.[5] This was an idyllic setting that provided an image of what a city could be like when it was suddenly free of air pollution, as if the health crisis had a silver lining. Important lessons could be learned, and people had the opportunity to imagine an alternative renewal of urban life.

This approach was explicit. In fact, the state of emergency caused by the COVID-19 pandemic coincided with the process of preparing Barcelona's new Urban Master Plan, intended from the outset to ensure that Barcelona's entire conurbation improved the quality of life of its population by adopting an energy transition. In 2022, a document was issued a document (Servei de redacció del Pla Director 2022) in which they agreed on the importance of turning the experience of confinement into an opportunity for reflection and learning. The report proposed that urban planning should consider the crisis situation and make the desired city model a reality. In doing so, it stressed health-related aspects, not only in terms of combatting pandemics, but

---

[3] The health reasons for the tactical urbanism interventions in Barcelona appeared frequently in the local press as "emergency" measures required by the health crisis. See, for example, how the transformations were presented by Catalan public television, TV3: https://www.ccma.cat/324/lurbanisme-tactic-adapt-transform-the-city-durgency-per-pandemic/news/3056906/. Accessed 17 May 2022.

[4] Cases with consistent theoretical developments were Paris (Denis and Garnier 2022), Toronto (Hassen 2022), Cape Town (Jobanputra and Jennings 2021), and Warsaw (Majewska et al. 2022).

[5] During the lockdown, "with less traffic the noise also went down. And the sum of the two factors allowed the birds to descend from the tops of the trees to streets, banks and fountains" (Mouzo et al. 2020: 4).

as a preventive measure by generating healthier environments. Those responsible for the urban future of the city understood that the fight against the pandemic and the fight against global warming could be couched in terms of health. It could not be otherwise since the catastrophic spread of the virus occurred only weeks after Barcelona declared a state of climate emergency in January 2020. It was prepared to act urgently and forcefully against greenhouse gas emissions, and it seemed that the pandemic had providentially emerged to show us a way to cure the ills of carbon.

Moreover, it was "scientifically" argued that general confinement and the suspension of everyday life had paradoxical positive effects on health. Thus, at the height of the desertion of the streets, in the spring of 2020, nitrogen dioxide ($NO_2$) concentrations in Barcelona fell by an average of 50%, and daily noise levels were reduced by five decibels. These data were reported in a study by Koch et al. (2022), who estimated that if the confinement measures had lasted for a full year and $NO_2$ concentrations had been maintained at that level, 5% of myocardial infarctions, 6% of strokes, and 11% of diagnoses of depression would have been prevented in the Catalan capital. The authors pointed out the urgency of transforming cities by considering the experience of the coronavirus epidemic and rethinking public policies to remodel urban spaces along the lines of what Barcelona had done.[6]

The desire to generate empirical evidence of the value of tactical urban planning as a vehicle to prevent the spread of coronavirus was translated into research aimed at demonstrating the opportunity provided by the health emergency in Barcelona. In the middle of the lockdown (May 2020), street research was conducted to collect data on target citizens' receptions of urban adaptations to the health emergency (Martí and Espíndola 2020). The conclusions could not have been more positive, and based on their results, the authors stated that the COVID-19 crisis provided positive lessons about redesigning streets to make them greener, quieter, more inclusive, and healthier, and about reconciling our urban habits with nature. Another field study attested to the lessons learned from the Barcelona City Council's antiviral measures in terms of mobility, focusing on their optimal effects on logistics infrastructure and the delivery of goods (Castillo et al. 2022). Finally, again producing evidence of the beneficial effects of the pandemic, simulations of changes in a public bicycle network during total or partial lockdown were based on tactical urbanism (Bustamante et al. 2022).

The implementation of "new urbanism" in Barcelona cost almost 11 million euros of public funding—10% of what was required to combat the pandemic. The items included in this expenditure had titles such as "Works for the execution of the actions of the public space and mobility in the framework of COVID-19"; "Works related to the expansion of the public space for the reopening of the terraces, on the occasion of the relaxation of the restrictions established with the declaration of the state of alarm"; and "Supply of benches for the actions of public space and mobility in the framework of COVID-19." Regarding these charges, the Tribunal de Cuentas (Court of Auditors), the body in Spain responsible for auditing the public sector, investigated in addition to the irregular awarding of the works, precisely the fact that

---

[6] For a chronicle of this process, in its different stages, please see the work of Fernández (2014) on urban planning attempts to "disinfect" what is now the Raval neighborhood in Barcelona.

they did not comply with their stated nature as emergency actions to deal with the health situation.[7] Barcelona City Council defended itself against these allegations by claiming that the works "were intended to guarantee public health in the movement of travelers through the city, ensuring, in turn, efficient, healthy, and sustainable movement."

Tactical urbanism was presented in Barcelona not as an innovation, but as a next step in its tradition of creative and rupturing urbanism. The city had known precursors to the popular urbanism of resistance that was presumed to be at the root of the new measures implemented by the City Council. The self-managed but institutionally supported complex of Can Batlló has been mentioned as an early expression of tactical urbanism (Fontes 2022), but a more evident type of urbanist resistance was undertaken by the residents of the Ribera neighborhood in Forat de la Vergonya (Hole of Shame) in the early 2000s, with its successful combatting of the municipality's gentrification plans (Cordero 2016). Tactical urbanism eventually joined with another urbanist trend aimed at improving the vitality of the city—in this case, through a specifically Barcelona model of intervention that served as inspiration for similar reforms in other capitals. These were the *superilles*—updated superblocks conceived of by urban rationalists—which comprised units of up to eight blocks, fully pedestrianized or with reduced traffic, landscaped, and intended for leisure, local commerce, and social gatherings. The construction of superblocks in different sectors of the city—Poblenou, Eixample, and Sant Antoni—was inspired by a simple but strong objective: to clear streets of motorized traffic and provide them with street furniture that would turn them into squares (Rueda 2019; Zografos et al. 2020). Plans for the superblocks, initiated in 2016, earned unanimous international praise and won a special mention regarding the European Prize for Urban Public Space.

The combination of tactical urbanism and superblocks placed Barcelona's new urbanism on a plane that went beyond the fight against coronavirus®. It was no longer about preventing the spread of a deadly virus but about avoiding the global catastrophe that was about to be caused by mass automobilization and the environmental filth it generated. A study (Landrigan et al. 2018) showed that environmental pollution caused more deaths worldwide than COVID-19. In two and a half years, the COVID-19 epidemic claimed six million lives, but air pollution causes nine million premature deaths annually. Beyond the health emergency that justified its adoption in Barcelona and other cities, tactical urbanism, along with superblocks, became measures presented as facilitating drastic changes in mobility systems in favor of more sustainable and environmentally friendly modes of transport. Thus, Barcelona fulfilled its commitment as a member of the Climate Leadership Group of Cities (C40) by reducing the emissions responsible for global warming. It also adopted the guidelines issued by the UN (the Paris Agreement, Agenda 2030, and the New Urban Agenda) and by the European Union (the European Green Pact, Next Generation, and the New Leipzig Charter).

---

[7] Accessed 24 May 2022. https://elpais.com/espana/catalunya/2021-11-23/el-tribunal-de-cuentas-objeta-la-emergencia-del-urbanismo-tactico-realizado-por-colau-en-pandemia.html.

This shows that the policies of pacifying urban spaces around the world go hand-in-hand with campaigns to replace the use of gasoline for transportation. Hence, the Catalan capital is reinforcing measures to discourage the use of combustion engine vehicles, hindering their use as much as possible and favoring clean and versatile means of transport. The objective is to reduce car use by 24% in two years and to achieve 80% of intra-urban travel by foot, public transport, or non-polluting vehicles. Institutional publicity and proclamations in major international forums have ignored or forgotten to answer the question of whether these actions are contributing as they should to making cities more rational, egalitarian, and inclusive.

The results of implementing tactical urbanism are being questioned, above all, because of the uselessness of initiatives in reducing traffic congestion and air pollution levels. Barcelona closed 2021 with levels of environmental pollution three times higher than those set by the World Health Organization. Mobility problems have increased, some caused by the traffic jams generated by restrictions on the circulation of motor vehicles, and others from unregulated and chaotic "sustainable" traffic circulation that is contributing to new forms of accidents. Car noise has been suppressed, but the neighbors of the *superilles* now suffer the noise from terraces of bars and the young people who gather there in the early hours of the morning. The colorful decorations on the ground mimic a kindergarten aesthetic, are confusing to use, and in many cases fail to capture the interest of passersby. The privileged areas with tactical urban planning initiatives and supermarkets have been constructed at the expense of adjacent roads, causing an increase in traffic density and therefore in pollution. No one doubts that the improvement in the respective environments will increase the price per square meter of built-up land and lead to gentrification. A sustainable city will have to be more expensive in a society where most of the built-up land is privately owned.

However, in Barcelona, tactical urbanism was unexpectedly imposed by the government of Ada Colau and Barcelona en Comú, to the detriment of the plebiscite under which this model of the transformation of city mobility systems was born. There was nothing horizontal about how the authorities applied measures that were hypothetically democratic and communitarian, but that were not previously presented as projects and were not subject to any public discussion, even in the interests of political plurality. Tactical urbanism has been defined as "a mega tool for citizen participation" (Paz Serra in Cussent 2022), but the affected neighborhoods were not even warned that their environments would be radically transformed, nor that the proceedings, however much they were presented as provisional, were irreversible, irrevocable, and here to stay.

However, an interesting feature of the transformations was that the arguments used to justify them were based on studies claiming that the new measures would prevent up to 667 premature deaths in Barcelona per year. As a result of air pollution, the incidence of obesity and diabetes would be reduced, cases of depression caused by loneliness and isolation would decline, and Barcelonans would live an average of 200 days longer (Mueller et al. 2020). Prior to the COVID-19 epidemic in 2017, the Barcelona City Council evaluated the *superilles* constructed in various city neighborhoods. The campaign was entitled *Salut als carrers* (Palencia et al. 2020), which, in English, has two different meanings: "health in the streets," highlighting a sense of place, and "health to the streets," introducing streets as entities. In other words, streets are spaces full of users who enjoy physical, mental, and social well-being, and the streets themselves are cured of their ills.

The focus of the praise for this new urbanism was on streets as both healed and healing—a type of urban spa in which to enjoy peaceful leisure and quiet encounters—a world set apart and protected from real urban life and its disturbances.

## 8.5 Against Conflict Conceived as a Plague

The most distinctive aspect of the discourses endorsed in Barcelona—as in other cities—was the implementation of tactical urbanism to stop the spread of the COVID-19 pandemic, and the systematically used organic metaphors that reconceived the city as a living being, whose health could be improved by interventions that would cure its ills. This somatic reading applied diagnoses and urbanistic treatments to the suffering of the city as a physical body, but it was equally concerned with salvation and redeeming the sins of modern civilization, considering that some were guilty of unleashing a planetary cataclysm that poisoned the air and could only be fought by emptying the streets, first referring to family units, and then warning about the toxic nature of public life.

Based on ecosystemic theories, the fight against coronavirus rested on hygienist arguments that had been used to justify urban sponging policies since the mid-nineteenth century (Palero and Ávila 2020). Hygienism was a medically inspired doctrine concerned with public health that investigated and proposed actions against the endemic and epidemic pathologies that were depleting industrialized cities. The objective was to promote what today we would call the "quality of life" of the population. An emblematic example of this was the redesign of the banks of the Thames River in London, in the UK, in response to cholera epidemics (Johnson 2007). Such interventions were a kind of biological urbanism based on medical topographies and statistics about the incidence of certain infections attributable to the poor state of public and private facilities. The scrutiny, today in force as "spatial epidemiology" (Pina et al. 2010), never failed to warn of the harmful role of certain behaviors of the popular classes that constituted a plague. It was a matter of sanitizing not only the habitats in poor neighborhoods, but also their inhabitants, who had to be educated to eradicate their bad habits—their sanitary habits and the behavioral ones that led to alcoholism, promiscuity, prostitution, or the abandonment of family obligations.

Barcelona is another clear example of this. The history of urban planning in the Catalan capital has encompassed waves of morphological and infrastructural reforms that, since the nineteenth century and as a function of liberalization projects in a modern city, were conceived and applied simultaneously in the service of environmental health and social neutralization (Alcaide González 1999; Capel and Tatjer 1991), the latter with what today we would call "education in values" (i.e., moral lessons that induced good practices in what we call "urbanity," in the sense of a restrained and gentle way of living in the city).

This was the spirit of the great urban reforms of the nineteenth century, starting with the massive emptying of urban centers, the model for which was that of Baron Haussman in Paris, and continuing with models of urban expansion based on reticulation accompanied by the rejuvenation of old neighborhoods. This last case involved the demolition of the city walls and the expansion of Barcelona through the Eixample project, which was based on medical opinions about the urgency of opening up the urban fabric to ventilate it and to control the epidemics that periodically decimated the population. When Ildefons Cerdà, who invented urban planning as a science, analyzed the walled city of Barcelona, he identified three basic problems to be solved: mobility, density, and morbidity (Cerdà 1863). His proposal aimed to create a zoned city, socially cohesive and facilitating movement, but above all, healthier. He did this in the service of a bourgeoisie that was already thinking in territorial terms of its new hegemonic role and which, above all, was concerned about the periodic social upheavals that endangered it. The justification for this was the urgent need to apply a therapy that would prevent the diseases that plagued Barcelona, and it was no coincidence that it was a doctor, Pere Felip Monlau, who extended to the whole city the theories of Michel Lévy about the "atmospheric cube" and the ration of breathable air that each house (and now each fragment of the city) needed (Muñoz 2010).

The Baixeras Plan of 1889, the Jaussely Plan of 1903, and the Macià Plan in the 1930s continued these lines of urban planning in the name of public health. They were suspended at the outbreak of the civil war, although it could be said that the devastation caused by the Italian air force's bombing did not stop it from being brutally executed. After Franco's victory in 1939, urban prophylaxis resumed and was continued from 1977 onward by democratic city councils. The latter were responsible for resuming the work of removing the old parts of the city, and they were convinced that advanced urban planning and quality architecture could solve the chronic social problems that Barcelona suffered and continues to suffer.[8]

The last stage of this process was the one that engendered, theorized, and imposed the new Barcelona municipalism led by Ada Colau and Barcelona en Comú, based on tactical urbanism and superblocks, for which the circumstances imposed by the COVID-19 epidemic provided the great backbone. Now, the malignant microbes that the first hygienism tried to eradicate in action were replaced by a lethal virus on the loose. However, the changes were not, in fact, part of an anti-pandemic prophylaxis, despite being presented as such. On the contrary, the exceptional circumstances that forced the lockdown of cities were the excuse and impetus for updating capitalist urbanism on a global scale (Graziano 2021). This continues to reinforce the arguments that posited urbanism a surgical instrument to remove the urban tumors and impurities that endanger cities and, by extension, the entire globe. In the nineteenth century, environmental risks were associated with cholera, tuberculosis, and yellow fever; recently, it was coronavirus that motivated urban health measures; and now,

---

[8] A chronicle of this process, in its different stages, is presented in the work of Fernández (2014) on urban attempts to "disinfect" what is now the neighborhood of Raval in Barcelona.

the medicinal action has to be even more drastic, since the new miasma to be fought is the most harmful that humanity has ever known—atmospheric pollution and its accompanying global warming.

In recent times, the manipulations of urban space aimed at hiding new and old forms of misery, and the informalities that hinder neighborhoods and urban centers, have been described as new hygienism. It concerns the efforts—urban, normative, and, if these are not enough, policing—to make streets "quality public spaces" and, thus, a proscenium for exhibiting good bourgeois manners with no fuss or passion. Hence, the neo-hygienism of campaigns to clean up irregular commerce in the street (Espinosa Zepeda 2016), or what is known as "preventive urbanism," aims to contribute to security against criminals, vagrants, and loiterers (for the case of Barcelona, see Porretta 2010; Thomasz 2010).

Part of this new hygienism is the admired urbanism undertaken in the name of alternative municipalism in cities such as Barcelona. There, tactical urbanism was imposed as a means to counteract COVID-19, hiding the fact that its destiny was to be implemented, together with the superblocks, as a definitive measure in the fight against climate change. However, this new hygienism has the same mission as its predecessors: not only to ward off, as it claims, the pandemics of coronavirus and environmental pollution, but to keep conflict, conceived as a plague, at bay and prevent it from threatening the good health of the bourgeois city. Its objective is to calm life in the streets and turn them into orderly, predictable spaces, to deactivate threats to the peaceful government of cities, and to present appropriate scenarios so that the dominated (the urbanized) can and want to collaborate avidly with those who dominate them (the urbanizers).

## 8.6 Conclusion

The preceding text engages in theoretical inquiry regarding tactical urbanism, interpreted as the latest episode in a struggle between urban culture and urbanist culture (i.e., between the ways of living in urbanized spaces, where conflict is a persistent element, and the formal structuring of urban territorialities according to criteria that are presumed to be aseptic and neutral). Taking advantage of the opportunity provided by measures to fight the COVID-19 pandemic, the "new urbanism" of the no less new municipalism insists, updating its rhetorical justification, on the old objectives of the "science of the city." These have been, and are, those that determine the meaning of cities through devices that seek to give coherence to highly complex spatial ensembles, calming and making clearly intelligible the forms of spatiality that are difficult to control but that characterize urbanism as a way of life.

We can speak of tactical urbanism as a new manifestation of institutional urbanism—that is, as a will to destroy or deactivate the urban. The urban is what Henri Lefebvre (2000: 47–53) called urban society as a meeting of differences and an endless succession of simultaneities, dislocations, and confluences—the work of the citizens rather than imposed on them. It was this reality that Lefebvre denounced as attempts at intervention by "experts" in the service of the state and the capitalist order. Regarding that reality, the author of *Le Droit à la ville* never tired of repeating that the technical specialists know nothing or almost nothing (to the extent that it is a blind field for them).

Special attention has been paid to the case of Barcelona, not because it is singular in using the COVID-19 pandemic to justify changes in the organization of the city, nor because it has established tactical urbanism in the name of fighting against the ecological disaster that awaits us, but because it has added its own construction—the superblocks—to the urban hygienism of the twenty-first century. The distinction of urban policies in the Catalan capital is that they are an example of change processes accompanied by a moral-enrichment justification that wraps up the capitalist appropriation of cities in abstract universal values against which it is impossible to rebel, obvious and urgent though they are. This is why Barcelona continues to represent a paradigm of new forms of "good urbanism," exemplary for its virtuous intent, first by positioning its initiatives as promoters of the principles of citizenship, civic-mindedness, and participation, then by presenting itself as a planetary referent in the environmental battle against climate change, and finally by functioning as a tool in the service of the great struggles of humanity—feminism, anti-racism, the LGBTI movement, and decolonialism—from whose programs all the militant and questioning dimensions of the capitalist order have been eradicated.

The progressive government of Barcelona continues to pursue the line, already established in the 1980s, of adding to its avant-garde image in urban planning matters by presenting itself as a promoter of urban change based on ethical values. It has done this (and continues to do it) by placing in a central discursive role the notion of public space as a territory for reconciling differences, deactivating conflict, and eliminating antagonism at the confluence of those scenes of life and struggle that were once the streets. The agoraphobic logic of official urbanism consists, as Jacobs (1961: 30) wrote, of applying to cities an order inspired by a set of recipes openly designed not to improve cities but to murder them.

## References

Alcaide González R (1999) La introducción y el desarrollo del higienismo en España durante el siglo XIX. Precursores, continuadores y marco legal de un proyecto científico y social. Scripta Nova 50. http://www.ub.edu/geocrit/sn-50.htm. Accessed 23 Jul 2022

Anguelovski I, Cole H, Connolly J et al (2018) Do green neighbourhoods promote urban health justice? Lancet Public Health 3(6). https://doi.org/10.1016/S2468-2667(18)30096-3

Arendt H (1958) Human condition. University of Chicago Press, Chicago

Bermejo AD, Aja AH, Fernández AS (2022) Resiliencia urbana: discurso e institucionalización de un concepto. Ciudades 25:1–18. https://doi.org/10.24197/ciudades.25.2022.1-18

Borja J (2010) Luces y sombras del urbanismo de Barcelona. Universitat Politècnica de Catalunya, Barcelona

Borja J, Muxí Z (2003) El Espacio público: ciudad y ciudadanía. Diputació de Barcelona/Electa, Barcelona

Borja J, Muxí Z, Ribas C, Subirats J, Barnada J, Busquet J (2004) Public space development in Barcelona. Some examples. In: Marshall T (ed) Transforming Barcelona. The renewal of a European Metropolis. Routledge, London, New York, pp 161–172

Bustamante X, Federo R, Fernández-i-Marin X (2022) Riding the wave: predicting the use of the bike-sharing system in Barcelona before and during COVID-19. Sustain Cities Soc 83:103929

Capel H, Tatjer M (1991) Reformas sociales, servicios asistenciales e higienismo en la Barcelona de fines del siglo XIX (1876–1900). Ciudad y Territorio 3: 233–248

Cardullo P, Kitchin R, Di Feliciantonio C (2018) Living labs and vacancy in the neoliberal city. Cities 73:44–50. https://doi.org/10.1016/j.cities.2017.10.008

Castillo C, Viu-Roig Marta M, Alvarez-Palau EJ (2022) COVID-19 lockdown as an opportunity to rethink urban freight distribution: lessons from the Barcelona metropolitan area. Transp Res Interdisc Perspect 14. https://doi.org/10.1016/j.trip.2022.100605

Cerdà I (1863) Necesidades de la circulación y de los vecinos de las calles con respecto a la vía pública urbana, y manera de satisfacerlas. Madrid, sn

Charnock G (2018) Barcelona en Comú: urban democracy and 'the common good'. Socialist register 54

Charnock G, March H, Ribera-Fumaz R (2021) From smart to rebel city? Worlding, provincializing and the Barcelona Model. Urban Stud 58(3): 581–600. https://doi.org/10.1177/0042098019872119

Cirugeda S, De Nieves J (2006) Ciudad prótesis. Visions 5: 36–41

Cussen I (2021/2022) Urbanismo táctico: Qué es y dónde encontrarlo. Revista BCN/NY 12/13. https://www.ub.edu/revista-bcn-ny2022/2021/12/13/urbanismo-tactico-que-es-y-donde-encontrarlo/. Accessed 20 July 2022

de Certeau M (1980) L'invention du quotidien. 1. Arts de faire. Gallimard, París

Delgado M (2015) L'espace publique comme ideologie. Los Libros de la Catarata, Madrid

Delgado M (2016) Ciudadanismo. La reforma ética y estética del capitalismo. Los Libros de la Catarata, Madrid

Delgado M (2017) La ciudad mentirosa. Fraude y miseria del modelo Barcelona. Los Libros de la Catarata, Madrid

Denis J, Garnier N (2022) Une expérimentation urbaine en temps de pandémie: les coronapistes à Paris. Centre de Sociologie et d'Innovation. https://halshs.archives-ouvertes.fr/hal-03609658/ 163-174. Accessed 23 May 2022

Espinosa Zepeda H (2016) Neo-higienismo y gentrificación en el discurso de la movilidad urbana: Desalojo de "tiangueros" en Guadalajara, México. In: Aricó G, Mansilla JA, Stanchieri ML (eds) Barrios corsarios: memoria histórica, luchas urbanas y cambio social en los márgenes de la ciudad neoliberal. Virus, Barcelona, pp 173–196

Fernández M (2014) Matar al Chino. Entre la revolución urbanística y el asedio urbano en el barrio del Raval de Barcelona. Virus, Barcelona

Fontes A (2022) O processo de autogestão de Can Batlló, em Barcelona: uma feição radical do urbanismo tático. Oculum Ensaios 19. https://doi.org/10.24220/2318-0919v19e2022a4967

Franco D (2018) Tactical urbanism as the staging of social authenticity. In: Tate L, Shannon B (eds) Planning for authentiCITIES. Routledge, London, pp 177–194

Graziano T (2021) Smart technologies, back-to-the-village rhetoric, and tactical urbanism: post-COVID planning scenarios in Italy. Int J E-Plann Res 10(2). https://doi.org/10.4018/IJEPR.2021040101.oa7

Habermas J (2011) The structural transformation of the public sphere: an inquiry into a category of bourgeois society. Translated by Burger T, Lawrence F (eds) Polity Press, Cambridge, 1962. Strukturwandel der Öffentlichkeit. Untersuchungen zu einer Kategorie der bürgerlichen Gesellschaft. Suhrkamp Verlag, Berlín

Hernández Cordero A (2016) El Forat de la Vergonya: entre la ciudad planificada y la ciudad habitada. Hábitat y Sociedad 9: 11–27. http://hdl.handle.net/11441/56660. Accessed 24 May 2022

Jacobs J (1961) The death and life of great American cities. Vintage, New York

Janoschka M, Mota F (2021) New municipalism in action or urban neoliberalisation reloaded? An analysis of governance change, stability and path dependence in Madrid (2015–2019). Urban Stud 58(13): 2814–2830. https://doi.org/10.1177/0042098020925345

Jobanputra R, Jennings G (2021) Learning from covid-19 tactical urbanism: challenges and opportunities for 'infrastructure-lite' in sub-saharan African cities. In: Southern African transport conference 2021. http://hdl.handle.net/2263/82386. Accessed 21 July 2022

Johnson S (2007) The ghost map. The story of London's most terrifying epidemic and how it changed science, cities, and the modern world. Riverhead, Londres

Koch S, Khomenko S, Cirach M et al (2022) Environmental pollution. Impacts of changes in environmental exposures and health behaviours due to the COVID-19 pandemic on cardiovascular and mental health: a comparison of Barcelona, Vienna, and Stockholm, Environ Pollution 304. https://doi.org/10.1016/j.envpol.2022.119124

Landrigan PJ, Fuller R, Acosta NJ et al (2018) The Lancet Commission on pollution and health. The Lancet 391(10119): 462–512. https://doi.org/10.1016/S0140-6736(17)32345-0

Lefebvre H (1961) Utopie expérimentale: Pour un nouvel urbanisme. Rev Fr Sociol 2(3):191–198

Lefebvre H (1968) Le Droit à la Ville. Anthopos, París

Lefebvre H (2000) [1972]. Espace et politique : le droit à la ville II. Anthropos, Paris

Lydon M, García A (2015) Tactical urbanism: short-term action for long-term change. Island Press, Washington, DC

Majewska A et al (2022) Pandemic resilient cities: possibilities of repairing Polish towns and cities during COVID-19 pandemic. Land Use Policy 113. https://doi.org/10.1016/j.landusepol.2021.105904

Marti M, Espindola L (2020) Opportunity in the Time of COVID-19. J Public Space 5(3):23–30. https://www.journalpublicspace.org/index.php/jps/article/view/1373. Accessed 22 July 2022

Mansilla J (2017) Los nuevos municipalismos y el fin de la historia. Quaderns-e de l'Institut Català d'Antropologia 22(1): 1–4. https://raco.cat/index.php/QuadernseICA/article/view/329851. Accessed 11 Aug 2022

McGuirk J (2014) Radical cities. Across Latin America in search of a new architecture. Verso, Londres

Montaner JM (2015) El post-modelo Barcelona, El País, 5 Setembre. https://elpais.com/ccaa/2015/09/05/catalunya/1441471404_116226.html. Accessed 28 May 2022

Mould O (2014) Tactical urbanism: the new vernacular of the creative city. Geogr Compass 8(8):529–539. https://doi.org/10.1111/gec3.12146C

Mouzo J, Rodríguez M, Ríos P et al (2020) El año de la pandemia, El País, 30 December. https://elpais.com/espana/catalunya/2020-12-30/el-ano-de-la-pandemia.html. Accessed 11 Aug 2022

Muñoz F (2010) Contra la densidad. La ciudad higiénica y el urbanismo de los ensanches. In: López Guallar M (ed) Cerdà y Barcelona La primera metròpoli, 1853–1897. Ajuntament de Barcelona, Barcelona, pp 27–34

Mueller N, Rojas-Rueda D, Khreis H et al (2020) Changing the urban design of cities for health: the superblock model. Environ Int 134. https://doi.org/10.1016/j.envint.2019.105132

O'Callaghan C, Lawton P (2016) Temporary solutions? Vacant space policy and strategies for re-use in Dublin. Irish Geogr 48(1). https://doi.org/10.2014/igj.v48i1.526

Palencia L, León-Gómez BB, Bartoll X et al (2020) Study protocol for the evaluation of the health effects of superblocks in Barcelona: the "Salut Als Carrers" (Health in the streets). Int J Environ Res Public Health 17(8). https://doi.org/10.3390/ijerph17082956

Palero JS, Avila M (2020) Covid-19. La vigencia del higienismo decimonónico en tiempos de cuarentena. Cuaderno urbano 29: 9–26. http://www.scielo.org.ar/scielo.php?script=sci_arttext&pid=S1853-36552020000200001. Accessed 23 July 2022

Pereira-Zazo O, Torres S (ed) (2019) Spain after the Indignados/15M movement. Springer, New York

Pina MF, Alves SF, Ribeiro AI et al (2010) Epidemiología espacial: nuevos enfoques para viejas preguntas. Universitas Odontológica 29(63): 47–65. https://www.javeriana.edu.co/universitasodontologica. Accessed 22 May 2022

Porretta D (2010) Barcelona, ¿ciudad del miedo?: urbanismo " preventivo" y control del espacio público. DC. Revista de crítica arquitectónica 19–20: 183–192. https://upcommons.upc.edu/bitstream/handle/2099/10607/09_PORRETTA.pdf. Accessed 22 May 2022

Romero J, Boix A (2015) Democracia desde abajo. Nueva agenda para el gobierno local. Universitat de València, València

Rueda S (2019) Superblocks for the design of new cities and renovation of existing ones: Barcelona's case. Integrating human health into urban and transport planning. Springer, Cham, pp 135–153

Servei de redacció del Pla Director (2022) El PDU, la COVID-19 i la ciutat saludable. Àrea de desenvolupament de polítiques urbanístiques de l'AMB, Barcelona

Smith N (2002) New globalism, new urbanism: gentrification as global urban strategy. Antipode 34(3): 427–450. https://doi.org/10.1111/1467-8330.00249

Subirats J (2006) El poder de lo próximo. Las virtudes del municipalismo. Madrid. Los Libros de la Catarata

Thomasz AG (2010) Debajo de la alfombra de los barrios del sur: derecho a la ciudad o nuevas formas de higienismo. Intersecciones En Antropología 11(1):15–27

Zografos C, Klause KA, Connolly J et al (2020) The everyday politics of urban transformational adaptation: struggles for authority and the Barcelona superblock project. Cities 99. https://doi.org/10.1016/j.cities.2020.102613

**Manuel Delgado-Ruiz** Professor of social anthropology at the *Universitat de Barcelona*. Member of the *Observatori d'Antropologia del Conflicte Urbà, OACU*. He has worked on the social consequences of urban transformations and the social appropriation of public space. He is the author of the books La cité de la diversité (1996), El animal público (Anagrama Award for Essay 1999), Sociedades movedizas (2007), La ciudad mentirosa (2008), L'espace publique comme idéologie (2015) y Ciudadanismo (2016).

## Chapter 9
# The Transformation of Urban and Digital Spaces from a Democratic Perspective

**Arnau Monterde and Joan Subirats**

**Abstract** Large cities are in the midst of a process of transformation, having to face major challenges of global dimensions, such as the climate and energy crisis, the high levels of growing inequality, and the impacts of the global financial economy. Digital networks are in a similar situation with increasing risks: the growing privatisation of infrastructures and concentration of power of large technology companies, their energy and climate impacts, and new forms of violence and discrimination in social networks. Indeed, the urban and digital spaces share many characteristics, but most importantly, the common challenge of their democratic governance. Based on an analogy between the urban and the digital, this paper will discuss issues including their spatiality, ecological impact, policies and forms of governance, and social uses. From the dialogue between these dimensions, some possibilities are explored for thinking about the challenges that lie ahead of both spaces. Finally, taking into account that the urban and digital spaces fully converge in big cities, some challenges are pointed out on how to face the democratic debate and the forms of democratic governance of the urban and digital space within cities.

**Keywords** Digital space · Urban space · Governance · Democracy · Digital networks

## 9.1 The Global City and Digital Networks

Most large cities, towns and metropolises are now global cities in which people, knowledge, and financial flows circulate globally. Cities represent urban materiality in that meeting point that is public space and that responds to logics of functioning that transcend their tenuous borders. Indeed, these are increasingly becoming points

A. Monterde (✉)
Open University of Catalonia, Barcelona, Spain
e-mail: amonterde@uoc.edu

J. Subirats
Autonomous University of Barcelona, Barcelona, Spain
e-mail: joan.subirats@uab.es

of connection for the great global networks of information, circulation, and power. Large cities and their financial centres concentrate on a fundamental part of the organisation of a power that also operates on a global scale. Regarding the citizenry, although able to operate on a global scale, they also operate under more restricted and unequal way conditions. Yet, what they do find in cities is the space from which they connect to the world; a space in which they situate themselves and from which they can act. Besides, the impacts of the current model of urban development transcend the city itself and have a global reach: the energy consumption of cities, the management of the waste they generate, and the pollution of air or water—neither of which know administrative boundaries.

The city is global (Sassen 1999, 2008) and this globality has been accelerated in a muddled manner with the deployment of digital networks, connectivity and digitalisation of all the infrastructures and services that operate in the city. Not only have cities become large central nodes in this global digital network, but today, cities also operate at a digital level. These organise their services, collect and measure data, support telecommunications infrastructures, consume energy to maintain the digital system, and gather the new centres of innovation and development of the large global tech companies, embodying the most important productive space in the world. And the digital, far removed from this essentially immaterial idea, also concentrates a large part of the conflicts that will mark this century: the energy impact and material resources necessary to produce the current model of technological development, the accelerated privatisation and governance of this large and emerging global space, or all the political problems associated with the control, surveillance, and exploitation of data. These set of elements are not easy to govern from traditional democratic logic. And yet, while not free from threat, digital networks continue to have the capacity to act as spaces for free communication, cooperation, exchange, and the generation of emerging forms from which to think about society and democracy, in the same way, that cities have been and continue to be.

In a context of multiple and overlapping crises such as the ecological, energy, and climate crises, the cyclical economic crises that continue to generate a large increase in inequalities, the crisis of representative politics and the rise of the far right and populism (Gerbaudo 2018), and geopolitical conflicts in new scenarios of war, without forgetting the impact of COVID-19, the city continues to be the main scenario in which these multiple disruptions and conflicts converge. And this time it does so in an ultra-connected way through digital networks on a global scale. These networks are also subject to and affected by these multiple crises and are increasingly creating new digital public spaces, new spaces of sociability, spaces for the construction of information, spaces with new models of centralised and deregulated governance, new spaces of cooperation, and new spaces of conflict.

This chapter proposes an approach that we believe is more than necessary: to think of the city and the urban space in its global dimension in an analogous way to how we envision the world of digital networks and the digital space that is today's Internet. This analogy should allow us to understand both spaces through different layers of shared analysis, incorporating the existing tensions in terms of power relations, the forms of production and management of resources, their models of governance and social organisation, and its emerging processes.[1] The trajectories of cities and digital networks at the global level, with very different time scales of development, can tentatively help to identify similar processes, such as the privatisation of public spaces, the management and ownership of resources and infrastructures, models of democratic governance, regulation of shared spaces, their social function, and the challenges posed by highly uncertain future scenarios. This approach can serve as a starting point to address some of the threats posed by prospective planning towards the construction of more sustainable and more democratic global spaces in the aforementioned context of multiple overlapping crises.

## 9.2 Space, Power, and Democracy

The city has historically been conceived as a democratic space, or so this was the case in the first Greek polis such as Athens (Sinclair 1991). This is, as spaces of encounter, deliberation and construction of shared meanings with respect to the public. Even so, and accentuated by its exponential growth, the city and urban space have also represented a large part of the accumulation of inequality, with the concentration of large factories and their workers in exploitative conditions during the eighteenth and nineteenth centuries in Europe, or with the development without democratic planning of the large urban conurbations of the Global South. It is within big cities that this exploitation takes place, where socio-economic differences are expressed through the organisation of labour, property, and the development of new financial, economic, and social dynamics.

In this sense, the city is the space in which power genuinely operates and resides in the twenty-first century. Cities are the places where the large financial centres, the government of the dispersed offshore production centres, and the large infrastructures are located. Simultaneously, the city is the place where part of the speculative value economy that feeds these financial flows is articulated: stock exchanges, housing and its speculative assets, tourism, and the new urban services; all mediated by large digital platforms such as Uber, Airbnb, and Glovo, a new reality which has come to be called platform capitalism (Srnicek 2017). However, although we see more and more dynamics operating on a global scale and affecting many cities in a similar way, there are many differences in how cities are organised and governed in the present times, especially if analysed from the perspective of the Global South and in terms of inequality, accentuated by the concentration of power in the digital space controlled by the USA and China.

Historically, cities have had the capacity for self-government based on flexible and diverse political systems that were capable of internalising democratic dynamics

---

[1] The concept of emergence comes from network theory and complex systems, and explains the emergence of collective behaviours and systemic relationships that cannot be explained by the behaviour of the parts of the system separately (Boccaletti et al. 2006). As the scale of analysis increases, one can observe these systemic properties that illustrate emergent behaviours through forms of self-organisation.

in their governance models, allowing high levels of innovation, as in the case of the city of Barcelona in the eighteenth century (Martinez 2009) or the Greek polis. In addition, cities have also had to mediate cross-border trade, migration, or environmental flows, which has given them a certain self-regulatory capacity to deal with the impacts of these flows, while at the same time being forced to address problems that transcended their administrative competencies at the local level. With the rise of financial capitalism and the digital economy in recent decades, flows have globalised and new actors, services, and processes have emerged, impacting the city on multiple levels: large platforms for the management of tourist flats such as Airbnb, for mobility such as Uber, for home food delivery such as Glovo, to mention just a few. These services, with their main headquarters in the USA, are affecting the rental and tourist markets in the world's main cities, the local economy of mobility services, and the logistics and distribution systems that involve new forms of job insecurity. Added to this are the increases in air pollution on a global scale caused by an unsustainable mobility model, very high energy consumption, and the urban impact of climate change.

All this has generated a crisis in cities' capacity for self-government and their difficulties in influencing, intervening, and responding to problems that transcend them and operate at scales over which they have practically no power for influence or effective governance. This raises the fundamental question of how to address conflicts resulting from dynamics that transcend the limits of urban space but have a high impact within the urban space itself. Certainly, the answer to this question involves the exploration of new models of governance and the construction of new scales of political intervention in cities (Berber 2013), as well as a reformulation of their relationship with the state.

Global digital networks have a much more incipient trajectory in terms of the regulatory tradition of the public sphere, with barely 30 years of history anchored to the birth of the Internet. In its origins, the Internet was born under principles of cooperation and knowledge sharing (Castells 2004). The Internet was conceived under a premise that points to its decentralised nature so that it cannot be controlled or intervened in its entirety, and it is born from open and shared protocols, which makes its global interconnection possible, giving it a priori an undeniably democratic nature. But understanding the Internet today implies understanding that the global flows of information, code, and knowledge also constitute new power relations and that the materiality of this power lies in the capacity to sustain the infrastructures and codes that enable the interconnection of millions of people on a global scale. These include the large servers and data processing centres (physical computers that support the Internet) that are in specific material locations, subject to national legislation within the countries in which they are located, together with the cables and satellites that connect the servers internationally, and the energy consumption associated with these servers. It also includes the large social networks and their proprietary code and the massive personal data collected on them, or what has been called the new oil, as personal and user-generated data is one of the main sources of value of social networks.

The power we refer to is held by a few globally operating companies that make up the largest percentage of the companies listed on the National Association of Securities Dealers Automated Quotation (NASDAQ).[2] As the Internet is such a new space, transnational and national regulation has always lagged far behind the impact of rapid technological development. Moreover, the power of networks and networked communication (Castells 2013) also lies in the developments of the digital interfaces that today mediate social life as a whole, the code that makes up the set of applications that the Internet is today in private hands, and the data generated by the people who use such networks. As we have already mentioned, the billions of data generated daily that circulate today on the networks constitute one of the principal sources of value for the large companies operating on the Internet. And this is where the paradox of how a decentralised and initially democratic space has ended up becoming a space of control, or surveillance (Zuboff 2019), of shaping society, which, added to the dynamics of deregulation, usually ends up leading to and being associated with enormous social inequalities. Even so, digital networks, and the Internet as a decentralised space, continue to be a space where processes of communication and cooperation take place on a global scale, where new models of ownership and economy are developed, and where free communication still exists, albeit not without threats. The Internet today is also a public space with a particular chaotic nature, where there is conflict, where there is collaboration, and where there is also innovation and creation.

Analysing the urban space and the digital space, through some of their political, economic, social or environmental dynamics, such as their forms of government, conflicts and power relations, the set of social practices that take place in these spaces or their ecological impact, should be able to shed light on both spaces to tackle some issues that are becoming crucial in the twenty-first century. The future of cities and digital networks depends primarily on dealing with the question of democracy. In other words, to question and construct models of governance that incorporate emerging forms of democracy (Subirats 2011) is able to respond to multiple crises, making it possible to confront them so that we can defend and protect these common spaces. Analogously, thinking of these spaces opens up possibilities for understanding their respective and interrelated processes of globalisation and privatisation and their development dynamics. It also raises questions about the infrastructures and resources that sustain them, as well as the possibilities of handling the climate crisis that both spaces are forced to face. This perspective of joint analysis of the urban and digital space makes it possible to evaluate models of management of the material and the immaterial and their respective economic and social values. In short, engaging with the democratic governance of the city and digital networks makes it possible to study possibilities for the construction of democratic processes of intervention and defence of these public and common spaces. Due to the specificity of their networked structure, the exploration of new forms of governance on higher-level scales and their networks enable the emergence of new models.

---

[2] It is the second-largest automated and electronic stock exchange and securities market in the USA, the first being the New York Stock Exchange.

## 9.3 Territory, Ecology, Politics, and Social Processes in Public and Digital Space

Urban and digital spaces can be characterised based on particular elements of analysis. From a geographical or territorial perspective, the urban space, although its borders are sometimes blurred, has well-defined geographical and political jurisdictional limits. As a result, its local and territorial scale is conditioned by metropolitan dynamics, i.e. the urban space is built around the whole city. Besides, when they connect, cities can also generate a distributed network of cities. And despite they will always remain subject to state-level legal and political orders, they will, nonetheless, have a certain margin in the direct and indirect competencies of local political administration. This grants them multiple properties or capacities in terms of autonomy and political action, such as territorial proximity and local governance, the management of their resources, or the development of their services. Mobility in the urban space will always be conditioned by the means of transport available for any movement within that space, which is a relevant spatial, temporal, and economic limitation.

The digital space is practically unlimited and ultra-connected, with hardly any barriers to access beyond the material conditions of connectivity, i.e. having a device and access to the network. Moreover, its delimitation is reduced to the capacities and knowledge to access the different connection points, giving practically unlimited access to any open place in the network of networks. However, morphology also oscillates between forms of open networks, which, at the same time, are subject to high levels of centralisation. Although, as predetermined in its origins and thanks to its open protocols, its nature has a distributed form, there are large nodes in this network that play a central role in web traffic, data circulation and accumulation, processing capacity, and, in short, the machinery that supports this centralisation. Nowadays, these large nodes are owned by the large technology companies that support a large part of the Internet infrastructure (the so-called GAFAM[3]: Google, Amazon, Facebook, Apple, and Microsoft), which is therefore in private hands. To put it differently, since the infrastructures that sustain the system as a whole are still concentrated in the hands of a few global companies, the network structure and open network in the circulation of information do not necessarily imply either high levels of power distribution or distributed dynamics to promote the logic of organisation of this power.

With public attempts to generate Internet infrastructure still being very limited, the privatisation of the digital space is the norm. This privatisation implies that its governance is solely in the hands of private agents with particular interests, generally oriented towards the maximisation of economic profit. At the layer of the code, protocols and standards have been built to enable globally networked communication. Indeed, this is possible thanks to major agreements and the creation of legal frameworks and open and shared ownership (Kelty 2008).

---

[3] The most important companies in the world in terms of their NASDAQ listing. In China, the most prominent companies are Tencent and Alibaba Group.

Just as public services and public infrastructures were built and developed in urban space to guarantee fundamental rights to its inhabitants (such as schools, hospitals, roads, green spaces, public transport, and so on), so would the digital space require similar public investment to ensure that it regains its public and democratic dimension. Property regimes in urban space differ from those in digital space. Large-scale privatisation processes have been observed in large cities, but public interventions that guarantee quality of life are still essential to make these spaces habitable: air quality, mobility, access to basic services, etc.

The analysis of processes of privatisation of urban space and the responses from the public sphere can be an example to confront new forms of public policies on digital space. There are experiences of forms of ownership based on common spaces of communal management and ownership: housing cooperatives, forest areas, green spaces, etc., which operate with a logic that is relatively different from privatisation, such as the communal management of shared or public resources. These forms of ownership frequently occur in the digital space, especially in the case of intangible goods such as computer code. There are thousands of code projects (languages, programmes, applications, protocols, etc.) that have communal ownership models and build their governance models within their communities to decide the rules for managing this common intangible resource, also described as digital commons (Fuchs 2021). The digital space is a great asset of experiences and practices of projects that have communal ownership of code and knowledge, projects that then make the whole of society work through their infinite applications. This is why the open knowledge and free software movement, known as Free-Libre Open Source Software[4] (*FLOSS*), has had such an impact in recent decades (Kelty 2008).

If one looks at forms of governance and self-governance, like the distributed nature of digital networks, especially those developed during the origins of the Internet, networks of cities can also have high levels of decentralisation and autonomy in their forms of governance. While large cities operate through legal codes and have a long tradition of regulation and urban planning, the digital space operates through computer codes, which determine its development and are subsequently regulated by legal codes that, due to the speed of development of the digital age, tend to lag far behind their deployment. The relationship between forms of government, democracy, and the speed of regulation in accelerated processes of development will be highly conditioned by the regulatory capacities of the digital space, but also those that affect the urban space, which is strongly subject to the dynamics of digital transformation. Therefore, all of this requires agile operational systems of governance and regulation capable of responding to the rapid and profound digital transformation of society while at the same time generating scalable responses to public-community projects.

Concerning politics and power in the city and digital networks, the world's large megalopolises, which have grown in an uncontrolled manner, with barely any planning or regulation, and which are sustained in conditions of high precariousness, poverty, and profound environmental impact (Sassen 2008), could be compared

---

[4] Refers to all computer code that is freely accessible, i.e. that can be freely accessed or read, copied, modified, and distributed.

to the role of the large technology companies. These act as digital megalopolises, with low levels of regulation and without guaranteeing fundamental rights, high energy consumption and environmental impact, and low levels of democratic quality. The spatial geography of (urban and digital) networks will be a central element for analysing governance dynamics and processes, in addition to their democratic horizons.

This opens up questions about whether the morphology of networks can contribute to a greater distribution and/or centralisation of power. Still, the network structure itself is not in question, as it characterises the urban and digital reality. Even more so at a time when networks are not only human and digital but also ecological, where the interconnectedness of living systems and the effects on planetary ecosystems affect the whole of society.

Regarding the ecological question, urban and digital spaces are sustained by infrastructures that organise them and require multiple energy resources with their respective economic, political, and above all, climatic impacts. Because of their different time trajectories, the ecology of the urban space differs from digital space. More than 2000 years of building cities by mankind give urban space characteristics derived from the accumulated capacity to rehearse, transform, change, learn, and rebuild itself in the medium and long term. Cities today are the result of the development of multiple services and infrastructures that support them (water channelling systems, electrical systems, waste management systems, transport systems, communication systems, etc.). This has made it possible to accumulate experience in the management and development of these systems, but these same long-standing inertias can generate difficulties in providing rapid responses to problems and challenges that emerge and change life and the urban fabric. Even so, the city has its biophysical limits as limiters of its capacity for growth, and these are much less limited in the digital sphere, where there is still a profound lack of knowledge about the real impact of its functioning in terms of energy and emissions, with profound climatic effects. Tackling the ecological limits of urban and digital space requires thinking not only about its material limits and the energy impacts of unsustainable models of energy dependence but also about developing new spaces of material sustainability that put life at the centre (Herrero et al. 2019). In essence, that city and technology hybridise with the ecosystems with which they must coexist to guarantee present and future life.

Finally, and as an expression of the richness of urban and digital space, we must refer to the emerging forms of sociability, the re-appropriation of public space, and social cooperation, whereas urban space has historically been a space for the reproduction of inequality, it is also the space where protest, social conflict, and the constitution of ties that have allowed and continue to allow us to intervene and transform the city itself are produced. In the digital space, there is inequality of access, power, and communication capacity, but there are also emerging forms of organisation and communication in networks that point to new patterns of organisation and mass identity (Monterde et al. 2015), generating advanced tactics of mobilisation, collective action, social movements, and, in short, citizens connected in networks in the face of new urban conflicts.

The question that arises is whether open and/or public digital spaces can operate in a self-regulated manner under principles of a public nature when they are generally private. The dynamics of social networks under the ownership of large technological companies point towards a predetermination of the digital space and its social dynamics that are conditioned by decisions that have to do with the design of the interaction interfaces and of the digital space itself (Lovink 2019). In this way, the private nature of the spaces of digital sociability imposes a model based on capturing and maximising attention that allows value to be given to this social medium. Thinking about democratic designs implies going beyond simplistic approaches that only attend to the deregulation of public space as an opportunity (Franks 2021). The city and the urban space are forerunners in innovation trials based on design and its impact on its social dynamics, yet, their slow development and the capacity for adaptation of the people who live in cities have allowed democratic responses to be generated to assimilable changing contexts. On the other hand, the absence of the public dimension of the new spaces of sociability mediated by social networks and the privatisation of the design of interaction and collaboration dynamics highlight some of the main challenges of the digital era.

## 9.4 Towards a Democratic Urban Digital Space

Numerous elements distinguish and differentiate both the digital and the urban space. Analysing both spaces from their different layers allows us to understand the current models of governance by which they are governed, and in both cases, it is the forms of governance that end up conditioning their development, thus resulting in the existence of multiple models of city and network. This approach can also open up diversification of paths on how to orient processes to face the democratic challenges of both spaces. In addition, both spheres share similar issues as they face global problems of a very similar nature, such as resources and energy dependency, service-centred production models, the impacts of the financial economy, and connectivity through complex networks.

On the other hand, both spaces intersect, with digital technologies being another layer of the urban space, and this is a fundamental element in internalising the analysis of their interrelationships as another layer to be considered. Cities sustain a large part of global connectivity; they are organised and governed through networks and increasingly incorporate highly complex IT systems (sensors, data collection systems, development of infrastructure for connectivity, etc.) for their internal functioning, political management, or the construction of new service production models.

Technologies have put a face on and made visible new scales of power, at the same time as they operate in the urban space through new services based on platform capitalism with the deployment of services on logistics and private transport, tourist accommodation, and courier services, which, under the umbrella of collaboration, end up having an impact on the economy and on the form of the city itself and put a new local face on an economy that is increasingly globalised and at the same time centralised in very few hands.

The concentration of large global and private powers is surely one of the main threats to both digital networks and cities, where the former end up governing more or less indirectly, with practically no democratic control. Imagining and building new forms of democratic governance of cities and networks implies addressing this issue as a fundamental element. Forms of government in network structures open up the possibility of imagining and building new models of democratic governance based on participation and community management of common resources (Monterde 2019), as they incorporate the capacity to intervene in public spaces and society as a whole for the citizens as a whole.

Urban space, like digital space, faces the challenge of confronting its models of democratic governance, so that they become spaces of freedom and hope (Harvey 2000), guaranteeing fundamental rights and recovering the capacity of their inhabitants to intervene in their forms of management. Governing public and common spaces requires thinking about new instruments that allow us to face the challenges of the twenty-first century. Democratic innovation is undoubtedly a tool to consider to build these governance models based on free access to knowledge, open and innovative ways of community building, multiple forms of collaboration, the defence of common spaces able to guarantee life from an ecological and eco-social perspective, and the possibility of engaging in the shared management of these spaces. Because democratising urban and digital spaces implies democratising all the layers that compose them, from the resources and infrastructures to the services and social processes that take place in them.

The democratic challenges of urban and digital space share how both can be inspired to attain a more democratic society. Rethinking ownership models, decentralising and distributing power, and exploring new intermediate while sustainable scales of governance to respond to territorial and physical limits are some of the present challenges. As a result, generating spaces to foster democratic innovation for the emergence of new networked democratic systems for cities that are being built and transformed is a good start.

# References

Barber BR (2013) If mayors ruled the world why they should and how they already do. In: If mayors ruled the world. Yale University Press, pp 3–24
Boccaletti S, Latora V, Moreno Y, Chavez M, Hwang DU (2006) Complex networks: structure and dynamics. Phys Rep 424(4–5):175–308
Castells M (2004) La era de la información (Vol. I: La Sociedad Red) [The information age]. Siglo XXI, México DF, México
Castells M (2013) Communication power. Oxford University Press, Chicago
Franks MA (2021) Beyond the public square: imagining digital democracy. Yale LJF 131:427
Fuchs C (2021) The digital commons and the digital public sphere: how to advance digital democracy today. In: Westminster Pap Commun Culture 16(1)

Gerbaudo P (2018) Social media and populism: an elective affinity? Media Cult Soc 40(5):745–753
Harvey D (2000) Spaces of hope, vol 7. University of California Press
Herrero Y, Pascual M, Reyes MG (2019) La vida en el centro: voces y relatos ecofeministas. Libros en acción
Kelty CM (2008) Two bits: the cultural significance of free software. Duke University Press
Lovink G (2019) Sad by design. Pluto Press
Martínez G (2009) Barcelona rebelde: guía histórica de una ciudad, vol 18037, Debate
Monterde A (2019) De la emergencia municipalista a la ciudad democrática. En Ciudades democraticas. La Revuelta Municipalista en el Ciclo post-15M. Icaria Editorial, pp 25–53
Monterde A, Calleja-López A, Aguilera M, Barandiaran XE, Postill J (2015) Multitudinous identities: a qualitative and network analysis of the 15M collective identity. Inf Commun Soc 18(8):930–950
Sassen S (1999) Ciudad Global, La. Eudeba, Buenos Aires, p 50
Sassen S (2008) Territory, authority, rights. In: Territory, authority, rights. Princeton University Press
Sinclair RK (1991) Democracy and participation in Athens. Cambridge University Press
Srnicek N (2017) Platform capitalism. Wiley
Subirats J (2011) Otra sociedad ¿otra política?: De «no nos representan» a la democracia de lo común. Icaria Asaco, Barcelona
Zuboff S (2019) The age of surveillance capitalism: the fight for a human future at the new frontier of power: Barack Obama's books of 2019. Profile books

**Arnau Monterde** holds a Ph.D. in Information and Knowledge Society by the Open University of Catalonia. He is involved in several programs and projects about digital rights, technological sovereignty, open-source and the democratization of network society. He is one of the cofounders of the decidim.org project. He is the director of the Digital and Democratic Innovation Center in Barcelona City. He was the coordinator of the tecnopolitica.net project at the Internet Interdisciplinary Institute (IN3-UOC).

**Joan Subirats** is professor of Political Science at the Autonomous University of Barcelona, specialist in governance, public management and public policy analysis. He has also worked on issues of social exclusion, problems of democratic innovation and civil society.

# Chapter 10
# Streets, Avenues, Highways

**Pablo Fernández Christlieb**

> *Gracchus was city born and bred, and the urbs was his habitat. It was part of him and he was part of it, and he nursed and consummated contempt for far horizons and green valleys and babbling brooks. He had learned to walk and run and fight in the twisting alleys and dirty gutters of Rome. He had scrambled like a goat in his childhood over the roofs of the endless tenement houses. The smell of charcoal fires, which pervaded the city, was the sweetest perfume he knew. This was one area in his life where cynicism never conquered. To go through the narrow market streets, with their rows of pushcarts and stalls, where the merchandise of the whole world was displayed and sold, was always a new adventure for him. Half city knew him by sight. It was "Ho, Gracchus!" here, and "Ho, Gracchus!" there, with no ceremony or bother about it, and the vendors and cobblers and beggars and loafers and draymen and masons and carpenters liked him because he was one of them. And he, in turn, loved their world, the world of gloom where the towering tenements almost met over the dirty alleys and had to be propped apart with timbers, the world of the streets, the noisy, dirty, wretched streets of the world's greatest city.*
> HOWARD FAST,
> ***Spartacus***

**Abstract** Streets are made of everyday comes and goes since ever of its inhabitants, which are called neighbors; therefore, what prevails is memory and the assembly of what is similar. Avenues, in turn, are made of currentness, and of promenades and parades of its inhabitants, which are called citizens; there meets together what is different. And highways are made of oblivion, of speed, and of indifference, and is inhabited by automobiles. While streets and avenues come from the origins of city, highways do not, and they represent the destruction of cities: then begins urbicide, understood as the impossibility of experiencing cities.

---
Translated by José Alejandro Arrangoiz Arechavala.

---
P. Fernández Christlieb (✉)
National Autonomous University of Mexico (UNAM), Mexico City, Mexico
e-mail: pablof@unam.mx

**Keywords** Social history · Cultural identity · Cities

## 10.1 The Street

There are two types of streets: those where the houses are built first so that the street emerges in the space between them, and those where the streets are built first so that houses come to fill the gaps in between—that is, the streets emerging out of spontaneity and the streets emerging out of planning.

When houses are built first, people will typically take a plot of land that suits their needs and settle in as best they can. If there is a tree that is useful or too big, or if there is a large boulder sitting on the way, the house will simply make its way around it. And if a house sits on the path of high wind gusts, it will simply tilt away to avoid them. And then, when the whole settlement becomes established, the houses will end up sitting where they please, forming together a haphazard arrangement. And between one house and the next, almost by accident, there will emerge little passages, alleys, passageways, paths, and trails, some wider and some narrower, weaving their way between the walls, with the people following them twisting and turning along to the stark rhythm of the street. Many old towns share that feature, and people describe them as quaint. But many new towns share that feature as well, and people refer to them as shanty towns, squatter areas, *favelas*, Hoovervilles, *villas miseria*, or slums. All of them were built with whatever people found at hand, first with stones or bricks (fired bricks or mudbricks), or with clay or slate tiles, and then with zinc sheets and some very thirsty gray bricks—but preferably not with wood, because it will rot or burn. And their inhabitants love them very much, with that deep age-old love for the land, because they carried each stone like one would carry an illusion, and because they built them to last forever, passing them down to their children and expecting they'll do the same with their children. And the streets they form, they feel like they are somehow more ancient. And that is exactly what medieval cities feel like.

By contrast, when the streets are built first, people try to first find a flat stretch of land (to avoid waste of time and potential obstacles) and then rely on nothing but a string and Egyptian geometry, as Serres said (1993), or even a plow, to build an orthogonal street grid. And it is only after the orthogonal plan has been laid down that the walls of the houses can be put up to form a straight line, one after the other. The most ancient cities like those of the Sumerians or the Indus Valley civilization were designed with this system, and so were Roman camps and medieval bastides, and yet they somehow feel as if they were modern, perhaps because they are more technological.

When it comes to discussing streets in the city, history certainly matters, because we are not interested in looking at what we have won (that is history for the winners) but rather what we have lost—we want to be able to recognize that which we need to recover. And already from this point, we can see that the potential for pleasant dreams and terrifying nightmares about cities has been laid out from the beginning. But regardless, there are both badly twisted streets and perfectly straight streets: in the former, you may suddenly run into your neighbor on any bend, while in the latter, you will rather see them at the corner store. In both cases, going to the city square will involve some form of turning, either going around a corner or meandering about until you get there. And in any case, the streets are meant for children to play—and when they grow up and feel the need to sigh with nostalgia about their childhood, they will surely reminisce about them.

The street is itself a domestic space in the sense that it provides a space for life to come to life. But at the same time, domestic might be too big a word for it, as it comes from the Latin *domus*, which refers to those big noble houses with patios, guest rooms, and basements whose inhabitants have no need for streets—and streets have no need for them either. *Insulae* would be a more appropriate word: those city blocks encompassing apartment buildings filled with people who would make use of the street as if it were their home and their house as if it were an extension of the street, unworriedly pulling out as many chairs as gossips and knitting tools simply to hang out there for a while. A street should therefore be crowded with multiple doors opening and closing non-stop with people coming in and out, and with staircases further inside with people going up and down; but especially down because street dwellers always forget something—the damn bread, an onion, or a spool of thread—and use it as an excuse to go out again at every turn. And those who are most experienced go out to properly fulfill their duty of occupying the corners. And those who are older and no longer have any duty come out to soak up some sun and recharge their bodies like batteries. And there are lots of windows as well (the first windows recorded in history were found in Crete) so people can look outside and be on the street without leaving their houses, like Murillo's two women at a window painting. And just like that, neighbors who do not know each other by name now know each other by sight for who knows how long. However, contrary to that trend, the residential buildings that began to be built in the 1950s—a well-intentioned initiative, as Le Corbusier has suggested—surrounded by Ebenezer Howard style landscaped areas, indeed have all kinds of conveniences around them, but they lack the street, and that was historically what streets were meant to offer.

Because, strictly speaking, transit is a misnomer for what we do on the streets—not even the passing delivery trucks do that. We rather come and go because the street already has everything we use for life. In that effective extension of the house, people will attach things to the doors and walls to literally set up shop, be it a tailor's shop, a jewelry shop, a home appliances shop, an automotive repair shop, or a computer repair shop. And quite often it is the whole family who works there, including cousins, children, nephews and nieces, apprentices, and odd-jobbers who will also come inside the house. And thus, ordinary routine days will not involve any kind of separation or working hours, family time, resting time, time to see who's passing by on the street, time for leisure, which is the time for thinking, reflecting, wasting time, singing, getting one's hopes up, crying quietly while we do our job, and doing our job while we cry and get our hopes up. And still today, the best cities will have the typical street of the blacksmiths, the street of the carpenters, the street of the saddlers, and the street of the car part thieves. And if you go there, you will see not only tools but also families and friends, quarrels, and laughter. We can tell when we are turning into a particular street because we can clearly hear distinct sounds; we can tell them apart and locate the sources, with sounds of power tools, hammers, typewriters, or radios broadcasting all kinds of content—and also dog sounds, especially those kinds of dogs that you can no longer tell their breed or pedigree. And the actual speed on the streets is the general walking speed, with even buses having to slow down somewhere between 4.5 and 14.5 km per hour—spanning from walking speed to the speed of bicycles, with the speed of roller skates, skateboards, and joggers somewhere in between. That is the range of speed that allows for conversation, that allows us to roughly work out the details of the little white lies we tell about why we are arriving at this time of day, and that allows us to catch sight of people's faces, their clothing, their gait or their gestures, or of geraniums, street signs, facades, cracks on the wall, and any irrelevant detail or triviality. It is also the speed at which one can peep cautiously into the windows that are lit up by dinner lights. Or as Studeny put it (1995, p. 17), it is the speed of the soul.

Streets are not just randomly scattered around town. They are interwoven into bundles that will make up neighborhoods. The thing about neighborhoods is that you cannot really tell where they end or begin. They might be demarcated by an avenue. But whatever the case may be, each neighborhood is just as good as a tightly squeezed city, offering everything from a town square, a set of ongoing discussions, a distinctive style, and a body of locals. Perhaps the only thing they are missing is autonomy, as neighborhoods are often not self-sufficient. They tend to rely on each other. For example, we might typically find our first love on our own street, but we will certainly have to go to the surrounding neighborhoods to find our second love story. Neighborhoods are densely packed samples of a city that are made up of streets, and they always have enough people to crowd the streets in shifts, as Jane Jacobs rightly pointed out (1961, p. 152), with those who go out early on the morning shift to drop off the kids at school, those who will do the mid-morning shift to carry out distributions and deliveries, those who will go out at lunchtime and crowd the small bistros and the areas around food stalls and carts, those who will go out for ice cream around mid-afternoon, and those who will be watching all day long from the corner watchtower to make sure everyone is doing their shifts as expected. And thus, all day long you will see passing bicycles, glances, handshakes, birds, and one or two cars.

And when you see those by late afternoon heading back to their neighborhoods and streets after a long day of tirelessly roaming around the city, you can tell by their faces that they are now relieved. You can tell they will now be warmly welcomed with open arms and with some hearty soup, and they will be called by their name—even if their neighbors do not really know their name—because they look as if they had remained completely unknown while being out in the city, out of their streets and neighborhoods (even if their colleagues do know their name). Upon returning, they will feel recognized by their own streets, as if the streets were waving back at them. Indeed, streets are always our point of departure, to go out for the day, to go out for our lives, and to go out for our history. But then, streets most certainly have a duty, which is to remain, to have been there since forever, to wait, or to remain the same and in the same way. Because that specific layout, those specific twists and turns, those fountains, those corners, the rocks that make up those walls and the lights that light up those facades, and even the materials and the air, they are all impregnated with local perceptions, sensations, memories, and even hopes and possible futures, just as if local residents were constituted by their street, because that is where they came from, and that is where they still belong. Memory is the true essence of the street. And thus, the street is vitally, and historically, the place where we must return, where we can go back to find our hopes and our future.

## 10.2 The Avenue

Originally, the ideogram used to indicate the existence of a city used to be a circle crossed by two perpendicular lines, dividing it into four quadrants. And that is indeed the initial layout of a city: both the cross and the circle constitute its avenues. That was the traditional layout of Roman cities: the *cardo* running north to south, the *decumannus* running east to west, and both encircled by the *pomerium*, which had a somewhat sacred connotation—perhaps because it marked the boundaries of the city, which is the only truly sacred thing we have, because therein lies our humanity, as we are not really sure of what it was that existed before that. Such was the layout of the four avenues running through the historical lake in Tenochtitlan, a city built by one of the very few truly primordial cultures—as claimed by Spengler (1922, p. 58. Vol. II) and Morris (1979, p. 13)—understood as the cultures that determine the fate of history.

Initially, the purpose of avenues may be to offer the possibility to get to the city center (where the city square and the market are located) without having to take so many twists and turns as you would by going through the streets, and to fit in larger vehicles like carts or trucks that can carry bulky packages, which is one reason why the number of avenues may immediately increase, emerging as diagonals crossing the circumference as if they were diameters or radiuses. But they can also be aimed at marking the boundaries between the existing neighborhoods or with new neighborhoods when cities grow beyond their original circumference, creating additional concentric circles and turning the initial layout into a kind of spiderweb (Goitia 1968, p. 117)—which would make for a beautiful ideogram as well. In any case, however, it is inherent in the nature of avenues to eventually stop, to end at some point. That is why they will typically lead right into a monument, some kind of door, a wall, the city square, an obelisk, or anything that is nice to see and can offer some rest to the eye and to the feet. And they are also typically sprinkled with parks and gardens, serving as restful places and even as ornaments.

However, with the change of width, the air will grow bigger and the atmosphere will widen, making the little house, the little windows, and the little heights that characterized the streets look small to the eye and to the overall feel of the avenue. That is why the scale of such monumentality must emerge with the main buildings surrounding it, but, at the same time, it has to be endowed with some kind of spatial musicality composed of symmetries, harmony among the various heights, rhythm, and smooth curves that must be accurately calculated, just as if we were trying to achieve a conjunction between method and grace, as Giedion claimed (1965, p. 124). That is why avenues, if they are indeed proper avenues—regardless of when they were built—will typically be pretty baroque (just as streets will typically be pretty medieval, regardless of time) because, as Pascal stated in his time, they have both geometry and finesse.

The Baroque—all things baroque—is theatrical by nature. Indeed, the reason for the existence of avenues is not transportation but spectacle, as their spaciousness and greatness are not intended for the circulation of vehicles but rather for staging processions, parades, carnivals, demonstrations, uprisings, and competitions—be it for Easter Week, on independence day, at a Shrovetide carnival, in 1871 or in 1968, in bull runs and marathons, and on so many other collective celebrations that are undeniably theatrical. That is why we do not see windows along avenues but rather balconies, which have historically served as viewing galleries.

But avenues are not just about the spectacle of the great popular festivals; they are also about the ordinary daily parading of people, because corners on avenues do not have small neighborhood shops but rather restaurants, bars, and cafes where people can sit and watch other people pass by—or watch the passers-by. That is why some avenues are called *paseos* (promenades), like Madrid's *Paseo de la Castellana*, Mexico City's *Paseo de la Reforma*, or Barcelona's *Paseo de Gracia*, for those who bring sidewalks to life, fitting in tiny round tables to put a cup of coffee or a beer on top, and perhaps an ashtray. But people typically occupy those tables not because they are thirsty, but rather because they are seeing someone, on a date or just to talk. And while they are waiting, they get busy watching the people walking on the sidewalks. We can thus see not only how things like Spain's "Vermouth hour" originated but, more generally, how the leisure hour originated, which has so significantly contributed to culture and civilization. That is the time for new ideas, the time for reflection, and even the time for books (like Hemingway's), for new theories, and for new ways of being in the city, which, as Richard Sennett claimed, is a wonderful byproduct of the introduction of boulevards in the nineteenth century (2018, p. 36). And thus we find that it doesn't really matter in the end if the person was ultimately stood up, so long as passers-by stick to their job of wandering around leisurely and window-shopping with the aim of being seen—so they must make sure their attire has been defined by the art of randomness, which requires meticulous preparations. That includes all those stereotypical city dwellers, like dandies, flâneurs, clochards, vagrants, Baudelaire, Simmel, and Benjamin, all of whom became so vital to cities that they might as well be sponsored by the state. And that also includes misanthropes and hermits, because we also go to avenues in search of one's fair share of anonymity.

In any case, avenues are the place to gain exposure in the city, or to put it in other words, they are the setting and the stage for showcasing our persona, our own sense of self, before society. And so avenue dwellers no longer recognize each other as neighbors but rather as citizens. And by the same token, we can already see that the experience of avenues is not directly associated with the experience of memory but with the experience of the present moment, the experience of currentness and fashion. But when avenues degenerate and become corrupt, they will no longer be about appearing or about appearance, but only about status, about such disdain as that expressed by everyone's air of superiority over others, and about consumerism.

There are two types of avenues: those that must be laid out because the city is expanding and those that will raze through the city just so they can be laid out. The former, arguably represented by the American main streets, perhaps those of Soria y Mata's Linear City, those that delineate the expansion of cities, or those of Pope Sixtus V's baroque Rome, will typically end in some monument-style finish, driving away any temptation to keep on going endlessly and also reducing their speed. And the latter, just like Baron Haussmann's redevelopment, will raze through whatever they find in their path—typically, old neighborhoods filled with "the usual people," which is what they would appear intent on eradicating in the first place. And one might think that they are trying to stretch the city outwards just to be able to fit in more of these avenues, which, of course, will lack any kind of terminal point at their ends so that they never stop and can extend far and wide into the beyond.

But make no mistake: the latter kind are still avenues. The difference is that they do not know what awaits them, as they will give rise to a critical, traumatic event for cities. While the time of the streets is defined by the comings and goings, the time of the avenues is defined by the sittings and passings-by. And in both, life will come into being, and the place will certainly remain in place. But when the raze-through kind of avenues suddenly lose whatever restraint they may have, they will suddenly force themselves out, shooting out like an uncontrolled champagne cork and overflowing into the open. Indeed, considering their width and breadth, their straight or perhaps only just slightly winding path, and their lack of a terminal point, something odd will happen to the perception of those who find themselves there in the middle of an avenue, something will happen to their sensation, to their leisurely way of walking, and to their own expectations and intentions: they will experience some kind of anxiety, a rush, or a hurry, a nearly unstoppable instinctive response similar to vertigo, because nothing will certainly stop that flow. And thus, given its particular features and circumstance, the avenue cannot help but exude, secrete, or produce a new kind of substance, unknown until that moment, which we now refer to as speed—indeed, speed only started to be a topic of conversation until the nineteenth century. And no horse nor fiacre, no race nor Ferrari, will ever be able to catch up with it. Speed is an object that will go so fast that it will detach from things, leaving them as they lag way behind it. And anything that will get in its way will appear to it as an encumbering obstacle, be it pedestrians, intersections, or things to look at, and if they do not move aside, speed will run them over. And, as is entirely logical, this anxiety must be duly materialized into something, and naturally, into something that must be impressively technological, namely, the automobile and the watch. The watch will help to ensure there is always a considerable level of anxiety, and the car will help to ensure there is always constant acceleration. And that, in turn, might give rise to the speedometer, which is an amalgamation of the two. That will already create the conditions for what is coming next: just as the avenue gave rise to the automobile, the automobile will give rise to the highway. Leisure time will remain a thing of the avenue, and now will come the new era of the optimization of time; otherwise, what use are watches and hurries. Indeed, if the notion of a carriageway still accounted for the existence of carriages, highways definitely arose to exclusively account for the existence of automobiles.

## 10.3 The Highway

There are two types of highways: those that will go out of the city and those that will come into the city. The ones that will go out are often avenues that got too carried away. And the ones that will come into the city might have originally been made of iron—curiously enough, railways do not owe their name to the locomotive but to the rails—referring to a way made of rails, in English as well as in Spanish and French. The first highways of this kind were railroad tracks; and later on, with the advent of automobiles—which are like little individual locomotives meant for selfish or egotistical use—they turned into paved roads. Judging by the way they were advertised, highways were meant to connect cities. But in fact, what they did was to connect with the avenues that were already coming out of a city, forming a perfectly continuous highway that will not even realize the existence of a city, simply running through it—or piercing through it as one would pierce a tender steak.

And the inhabitants of highways are not neighbors or citizens but automobiles (Glancey 2016, p. 57), which is why some aliens might have thought that automobiles were the true earthlings, and human beings might be nothing more than viruses or bacteria that have attached to them. And the essence of their existence is speed. That is why highways have no corners, nor anything that would typically be on them—like a grocery store or a cafe. They offer nothing to look at, so no one will be looking out the car window; all passengers will be looking straight ahead, because at a speed of 150 km per hour, to say the least, the only thing you can make out or spot—provided it is sitting ahead and not on the sides—are these huge bulks that make up industrial buildings or shopping malls. The English word *mall* shares no common origin with the word bulk, but it certainly should. So we have massive bulks offering no detail to the eye and oversized signs giving us directions, including specific data like the highway number, the next exit, or the next intersection. And the only scenery is provided by information, because, as Dejan Sudjic rightly pointed out, unlike avenues and streets, highways have no name. And if highways will offer nothing to see, don't expect much in terms of offering somewhere you feel you are arriving at—except for the airport, which will kindly offer a continuation of the highway but in a different form—because, when all is said and done, reaching one of those dormitory towns of the lower classes or the residential areas of the middle classes is just as good as reaching nowhere, or reaching a place where you could rightly use Gertrude Stein's beautiful phrase: "There is no there there" (as cited by Sudjic 2016, p. 211). In other words, you might reach *suburbia*, which has been advertised or promoted as a state-of-the-art neighborhood. But they cannot be neighborhoods because they have no streets, and they have no streets because they have no people, and the windows won't open—let alone the doors—with any outsider being perceived as a highly suspicious stranger. There will be no shops or cafés on the corners. And you will see no one going on foot, except maybe an odd worker who will very much look like an outsider.

If the street will bring that which is similar together and the avenue will bring that which is different together, the highway will only lump that which is indifferent together. The slowness of the streets will elicit the way of thinking of memory; the hustle and bustle of the avenues will elicit the way of thinking of currentness. The speed of the highways, however, will elicit the way of thinking of oblivion—specifically, that abyssal kind of oblivion that will make people believe that what we see there in between the apartment buildings sitting in the wasteland and in the carefully fenced private residential developments is a street, and that all that over there makes up a neighborhood, and that that thing over there is an avenue, and that thing over there makes up a city. And also, that other thing over there is a human being.

## 10.4 The Killing of a City

The street and the avenue share a common origin. The traumatic nature of the highway (striking too hard suddenly; nearly deadly) lies in the fact that, while streets and avenues emerged along with the city itself (and thus share its destiny, its fate, or in other words, that which must be fulfilled under conditions of necessity, that which must be realized, that which must be made real) highways certainly do not share that same origin. They rather emerged in history almost as if they were an accident, and maybe that is why highways love accidents. They are so utterly unnecessary that they end up killing the city, along with any nightingale living in it.

The popular tabloids will say that it has been a case of urbicide, which, in all fairness and in all obviousness, can be defined as the disappearance of life from a city. According to one definition, architecture refers to that which lies between the walls, and thus walls are not architecture. By the same token, among whatever will remain of cities, there will be buildings, and even buildings with a lot of movement around them, but there will no longer be life there, because that movement is no longer human, as it will exceed all human scale in terms of speed, weight, noise, distance, and reasons: it is a movement driven by machines, data, production, and money, and no longer by people, neighbors, or citizens. So, those are not even ruins—which would be so evocative, so historical, so human—because they still serve a purpose; however, that purpose does not emerge from within, from their very constitution, but it rather comes to them, it is imposed on them from the outside, from that foreign place that is swarming with designers, urban planners, speculators, financial accountants, and global technologies. And therefore, the urbicide is defined by the fact that, even if cities may still have inhabitants, they have lost one thing. One thing is no longer there, and that is perception: that which Kevin Lynch (1960, p. 112) claimed allows us to apprehend, discern, and appreciate the harmony between the avenues, the differences between the neighborhoods. And then there will be no more memory, which Rossi (1966, p. 191) would refer to as something like the consciousness of the city, as it allows us to place its destiny at its origin; it allows us to understand that there is a destiny in every origin. And there will be no more sensations or experiences, nor that which is often called fondness, that which makes us cling to its walls and to its corners, and it will only let us leave the city because we have promised to return. Nor will there be any anticipation, expectation, or expectancy. The inhabitants of a city from which life has been eradicated will certainly remain barbarians or tourists, even if they were born there.

Because it all ultimately boils down to two possibilities: (1) to stop or (2) not to stop: chasing after destiny or chasing after accident. Ultimately, the ability to stop remains the province of the handicraft, the province of that which is crafted by hand, as it will stop as soon as one has accomplished what one desired, or as soon as one gets tired, or as soon as one bumps into something that simply cannot be continued, or something that must not be continued—which would mean there is some kind of ethic—perhaps because one does not see the point or because it is no longer needed, or because it may lead to an adverse outcome, like growing in ugliness instead of growing in beauty. And the inability to stop, in turn, remains the province of technology, the province of that which is crafted by force, especially the forces of nature and the act of challenging their laws and their subsequent uncontrollability. In other words, that which is incapable of stopping is whatever keeps accelerating: free fall, capitalism, desire, highways, and vertigo. And then it ceases to be important what it is that we do, because there is ultimately no ethic involved and because one will do it so automatically that we could easily just go on day in and day out. But when it comes to that which is capable of stopping, every single thing we do will matter, because it takes hard work, because there are few of such things, and because it has its own unique quality: beautiful, useful, comfortable, or sacred. When it comes to that which is incapable of stopping, all that matters is whatever can be counted, whatever has a quantity—that is, anything to which one could always add one more without end and without purpose. And thus the perfect quantity will naturally be that of money, which can now be stored in the form of digits on computers, so it no longer even has to occupy any space, and we can always add one more digit without end. That is why today's cities, already deprived of life, will serve as money-producing devices: Vázquez (2016) claimed that there are many (ranging from leisure-driven multinational organizations to public administrations to real estate developers) who are colluding to turn cities into objects of consumption (p. 148), like those cities that have been turned into merchandise and have their own trademark (think of the BCN or the NYC brands), or those that have been turned into museums (think of Paris or Venice, which in reality are not so much museums as they are gift shops, certainly the most visited galleries in museums), or those that have been turned into global cities (think of Shanghai or Dubai). Or like Silicon Valley, or those huge shopping malls that have already been equipped with airports and everything, or like those isolated and exclusive residential areas that are like earthly paradises just like the original (that is, of course, equipped with their own surveillance tower), or like those gentrified neighborhoods that look like interactive theme parks, like bohemian Disneylands.

Technology clearly does not know anything about stopping. And lest they appear anachronistic, in this our post-hippie era well into the twenty-first century, both intellectuals and academics, and even critics, have found technology (especially digital technology) to be very innovative and genuinely creative, almost liberating and revolutionary—something akin to wishing to remain a leftist but refusing to become obsolete, although in reality, people are actually struggling to find ways to justify having a mobile phone. And so it happens that technology gets disguised as the avant-garde, as the embodiment of correctness and fun, and it is thus used to plan smart cities, using friendly technologies and self-driving cars and bicycles, having make-believe neighbors and make-believe citizens who trade on the London Stock Exchange from their farms, keeping their own selves charged with only pure energies and keeping their farms charged with only clean energies, making sure they send out their share of love of nature from there, as if the old city, with its little avenues and streets, must now be a thing for unconscionable backward-thinking folks—even when this brand of cybernetic bucolicism is intent on killing the city in a hygienic way. The least we can anticipate for the future of these techno-romantics is nothing but terminal boredom.

## 10.5 The Way Back to the Street

Barbarians were those who lacked the ability to behave in a city: boorish, apprehensive, and apathetic (*or blasé*, as Georg Simmel called them—1903, p. 239). And tourists must be quite half-witted, because they will buy anything that is sold to them, even that very trip that will turn them into tourists, but they will unwittingly come loaded with a certain kind of memory, that rare visceral wisdom that will make them fascinated by Italian villages, by Latin American towns, by the Arab medinas, and by medieval cities with their twisted streets (even if they end up destroying them as they admire them). They will be told that history does not pass by those places, but the truth is that history has not passed by those places because it has already settled there; it has remained there alive. Indeed, the dreamed image of a city, for some reason of collective unconsciousness, is the medieval city. Perhaps people imagine that they will see their children running around with hoops, the cries of their mothers calling them as they hang out their clothes, neighbors telling each other about their lives in the courtyards, and those timeless things like freshly baked bread and local wine. As if the genius of the city, its *genius loci*, resided in its streets.

But, at the end of the day, the reason why the collective consciousness keeps looking back to medieval cities is that there used to be a world there where community existed, a circumstance where everyone feels that they belong and are part of the place that they built and that they now inhabit and preserve, a circumstance also where the various groups involved (guilds, church members, merchants, priests, monks, bards, acrobats, scholars, students, craftworkers, and peasants) have managed to achieve harmony between them while preserving a fair level of liberties, of cooperative equality, of horizontal participation, of autonomy and trust, and of negotiation with the feudal lords. And as Lewis Mumford claimed, "a city that could boast that the majority of its members were free citizens, working side by side on a parity, without an underlayer of slaves, was a new fact in urban history" (1961, p. 271), because "for a brief while '*communitas*' triumphed over '*dominium*' (p. 252). Well, as it turns out, the medieval city was indeed a city of streets.

And it is not innovation but memory that pervades the streets. It sprouts from its corners and its doors, and so in this day and age—yes, still today—the community will be once again founded every time the neighbors go out there and come together to save a fountain or a memorial building, to protect the children's territory from cars, to prevent a monopolistic franchise from launching operations there, to prevent the demolition of an impractical building, to oppose the construction of a business tower, to reject any bans on aspirational endeavors (those forbidding people from sitting on the curb, from skateboarding in the opposite direction, from hanging their clothes out their windows, or from buying things from street vendors), or to request the landscaping of an area or the paving of a public square, to hold a street market, or to save a local business from closure (Mongin 2005, p. 321). Indeed, that is where people will start talking, arguing, organizing, getting to know each other, helping each other, getting along, enjoying each other's company, laughing, and ultimately realizing that that very day, just for one day, they achieved what the city dreamed of being from the outset—because it carries its fate in its own memory.

The neighbors thus turn medieval, and they might as well turn baroque, because if the occasion warrants it, if they have not yet been heard, if their demands need to be taken a step further, they will take their memory out to the avenues, to that place meant for display and spectacle where currentness seethes with life, to be seen and to protest—meaning to make manifest expression—the fact that the city belongs to us.

## References

Chueca Gotia F (1968) Breve historia del urbanismo. Alianza, Madrid
García Vázquez C (2016) Teorías e historia de la ciudad contemporánea. Gustavo Gili, Barcelona
Giedion S (1965) Espacio, tiempo y arquitectura (El futuro de una nueva tradición). Dossat, Barcelona, p 1982
Glancey J (2016) Cómo leer ciudades. Una guía de arquitectura urbana. H. Blume, 2017 (Trad.: José Miguel Gómez Acosta)
Jacobs J (1961) The death and life of great American cities. Random House, New York
Lynch K (1960) La imagen de la ciudad. Gustavo Gili, Barcelona, 1998 (Trad.: Enrique Luis Revol)
Mongin O (2005) La condición urbana. La ciudad a la hora de la mundialización. Paidós, Buenos Aires, 2006 (Trad.: Alcira Bixio)
Morris AEJ (1979) Historia de la forma urbana. Desde sus orígenes hasta la Revolución Industrial. Gustavo Gili, Barcelona, 2016 (Trad.: Reinald Bernet)
Mumford L (1961) The city in history. Its origins, its tranformations, and its prospects. Harcourt, Brace & World, New York
Rossi A (1966) La arquitectura de la ciudad. Gustavo Gili, Barcelona, 1971

Sennett R (2018) Construir y habitar. Ética para la ciudad. Anagrama, Barcelona, 2019 (Trad: Marco Aurelio Galmarini)
Serres M (1993) Los orígenes de la geometría. Siglo XXI, México, 1996 (Trad.: Ana María Palos)
Simmel G (1903) The metropolis and the mental life. In: Simmmel G (Donald Levine, ed—1971). On individuality and social forms. The University of Chicago Press, Chicago
Spengler O (1922) La decadencia de Occidente. Bosquejo de una morfología de la historia universal. Espasa-Calpe, Madrid, 1966, Tomo II
Studeny C (1995) L'invention de la vitesse. France, XVIIIeme-XXeme siècle. Gallimard, France
Sudjic D (2016) El lenguaje de las ciudades. Ariel, Barcelona, 2018 (Trad.: Ana Herrera)

**Pablo Fernández Christlieb** is M.A in Keele University (UK), Ph.D. in *El Colegio de Michoacán (México),* and has a postdoctoral stay in the École des Hautes Études en Sciences Sociales (France) in social psychology. Has written around twelve books, both in academic issues and in everyday culture, with constant reference to the city; and is a regular contributor to newspapers and magazines. He has been visiting researcher in Barcelona, Santiago de Chile and Paris.

# Chapter 11
# The Post-automobile City From Deterritorialization to the Proximity City: The Case of Madrid

**José María Ezquiaga and Javier Barros**

**Abstract** There is a broad consensus in the scientific community to understand that the impact of private car-based urban mobility models is unsustainable in terms of urban environmental quality, energy economy, and health. The case study presented in this text, the Madrid Centro Project, approaches this situation through the simplest but richest element of the system of urban public spaces: the street. As the conclusions show, there is a need to reorient the character of urban planning toward greater flexibility and technological innovation ensuring sustainability in its environmental, land use, and mobility components. The interrelation between buildings and streets opens rich reconfiguration options, as does the growing number of motorized and non-motorized modes of transport. The Central Madrid Project has provided an opportunity to experiment with a new concept through which the pre-existing urban tissue is reconfigured in "new urban cells". These constitute a land use- mobility organization in which the streets that enclose the urban cells receive through traffic and those inside provide a more domestic local core, to which only residents can get by car. This allows a much larger allocation of public space to pedestrians and urban vegetation.

**Keywords** Urban planning · Urban mobility · Public space

## 11.1 Introduction

The exponential growth of private car ownership is associated to the urban sprawl process and feeds back on the needs of individual mobility generated by the very low-density territorial patterns. There is a broad consensus in the scientific community to understand that the impact of private car-based urban mobility models is unsustainable considering the key variables in the contemporary concept of quality of life: urban environmental quality, energy economy, and health.

---

J. M. Ezquiaga (✉) · J. Barros
Polytechnic University of Madrid, Madrid, Spain
e-mail: jmezquiagadominguez@gmail.com

© The Author(s), under exclusive license to Springer Nature Switzerland AG 2023
F. Carrión Mena and P. Cepeda Pico (eds.), *Urbicide*, The Urban Book Series,
https://doi.org/10.1007/978-3-031-25304-1_11

Regarding the first aspect, urban mobility based on the private car is inefficient, as it requires a high energy consumption per passenger kilometer and a disproportionate amount of urban public space for circulation and parking space. This applies as well comparing to collective transport modes of greater capacity, as to other shared transport modes, with or without a driver. In health terms, the captive mobility of the car is associated to a sedentary lifestyle, which has in recent years fueled the dramatic growth of cardiovascular diseases and obesity, as well as the deterioration of air quality in the urban environment, associated with respiratory diseases, allergies, and cancer.

The case study presented in this text, the Madrid Centro Project, approaches this situation through the simplest but richest element of the system of urban public spaces: the street. In the central urban fabric of Mediterranean cities, streets are the reference for buildings, the support of mobility, and through their commercial plinth, the membrane of interaction between the spheres of the public and the private. For this reason, working on the street-building relation offers in Madrid the opportunity to tap on the potential of a fabric rich in possibilities of collective appropriation and provides a chance to reconfigure the urban block in morphological and functional terms.

## 11.2 Metropolis and Territorialization

The representation of the traditional city rests on the idea of limit, either the physical demarcation of the urban enclosure—materialized in gates, walls, or boulevards—or the most ideal separation between the ordered artificial universe and the world of what is rural and natural. The metaphor of the delimited city has historically informed the urban culture and the planning instruments, assuming the objective of harmoniously formalizing growth over the surrounding free territory. This notion was common to both the theorists of the *Grosstadt*—the great city that grows by continuities—Baumeister or Eberstadt of the early twentieth century, as to the visionaries of the ruralization of growth, such as Howard, or Arturo Soria. In both options the condition for the aggregation of new urban pieces, or for the creation of satellite settlements, implied the previous existence of a central city. The urban debate has focused for years on the choice between the supporters of planned extensive growth and the proponents of the limitation of such growth and decentralization.

The harmonious image of continuous growth, organized around an urban center, where the managerial functions reside, and delimited by ring roads and green belts has been for decades the dominant *icon* of the *planned city* that inspires the first land planning laws. The role of this canonical imagery is by no means negligible, as Gould (1995) pointed out about the coercive character of certain representations of scientific theories: *"there is nothing more unconscious, and therefore more influential through its subliminal effect, than a broad image traditionally used for a subject that, in theory, could be represented visually in a hundred different ways, some of them with remarkably disparate philosophical implications"* (Figs. 11.1 and 11.2).

# 11 The Post-automobile City From Deterritorialization to the Proximity ...

**Fig. 11.1** Stadt mit turm, a sketch for Fritz Lang's 1926 film Metropolis. *Source* Kettelhut, Erich (1926). Deustche Kinemathek

In the last decades of the twentieth century, the inability of this discourse to account for a complex urban reality, that is, ambiguous, heterogeneous, multiple, and paradoxical, became evident. A new urban approach based on the recognition of the existing city was a significant first step by incorporating a new sensitivity to the geographical qualities of the city site and the landscape as the basis of the project. Planning thus detached itself from its origins as a mere instrument of distribution of land uses and values to become the embodiment of a *city project*. The vindication of the differential, the concrete and the local appeared as a response, sometimes conscious, sometimes intuitive, to the processes of dissolution of urban identity and deterritorialization that characterized the modern metropolis. Faced with the growing homogenization of spaces and places, urban interventions demanded a new dimension as instruments highlighting differential qualities in cities and territories. However, the vision of the architectural project as a lever for the transformation of the city—often linked to urban requalification/gentrification operations guided by strict market criteria—has relevant limitations.

A recent variant of the culture of projects is associated with the appreciation of the iconic role of architecture in the new symbolic economy based on culture, leisure, tourism, and communication (Zukin 1991, 1995) and the growing weight of this approach in the current scenario of globalization of competition between cities to attract investments in a context of accelerated mobility of capital (Sassen 1991).

**Fig. 11.2** F. Ll. Wright Broadacre City

The image, materialized in architectural projects and emblematic infrastructures, has become a relevant battlefield in the game of urban marketing (Boyer 1996).

The urban project must be both an effective space for social innovation and a lever for urban transformation, avoiding the common situation in which a project becomes just an image of modernity or efficiency that does not deliver real change. Additionally, it can encourage, from a nostalgic perspective, the mirage that it is possible to recover from the strict intervention on the urban form the civic qualities of the traditional city. In this line, David Harvey criticized the premises underlying the emergence of the so-called New Urbanism (Calthorpe 1993; Katz 1993), asking whether *"isn't the idea that formalizing a spatial order is or can be the basis of a new aesthetic and moral order so perpetuated?"* (Harvey 1997:68) (Fig. 11.3).

But even more important is the difficulty in understanding the new geographical and social dimensions of the contemporary metropolis. The modern metropolis has discarded any notion of aprioristic limit, inaugurating the *era of deterritorialization*. The British sociologist Giddens (1990) has analyzed the intimate relationship between modernity and transformations in time and space. The emptying of time—made possible by the uniform dimension of time of the mechanical clock—is the necessary condition for the emptying of space. Modern societies are increasingly stressing the division between space and place, favoring relations between subjects that are spatially different and therefore unable to maintain face-to-face contacts.

**Fig. 11.3** Hight line Nueva York. The urban project as a creator of civic value alternative to the car culture. *Source* Photo by José María Ezquiaga (2019)

Decades earlier Melvin Webber pioneered the formulation of the spatial consequences of the growing development of relationship domains unrelated to specific locations. Current notions of virtual community or cyberspace have taken this idea to its ultimate consequences (Boyer 1996; Mitchell 1995, 1999). "*As appropriate as the language of land uses and densities may be for describing the static characteristics of a site, it is incapable of dealing explicitly and specifically with the dynamic models of location of human communication, which occur in space but transcend any given place*" (Webber 1964:84).

In the social sphere, the space/time split is a necessary condition of the extreme dynamism that characterizes modernity and provides the gears for the development of rationalized organizations. They can "*connect the local and the global in ways unimaginable in more traditional societies and in doing so routinely affect the lives of millions of people*" (Giddens 1990:20).

The spatial consequences of these transformations has generated a profound alteration of the urban scenario. The exponential growth of metropolitan mobility tends to lead to a previously unknown diffuse occupation of the territory. The most significant feature of this phenomenon is that the weakest activities—as in the traditional European city—or residence—as in the formation of the Anglo-Saxon suburb—do not relocate to the periphery, but functions and characteristic elements of centrality abandon traditional locations to colonize a new suburban territory (Rowe 1991; Indovina 1990, 2007).

Therefore, the classic relations of dependence between the central city and the outer settlements get distorted: the segregated and hierarchical metropolitan model often becomes a *polycentric* or *reticulated structure*. Activities that previously took place in a concentrated space now consume a greater extension of the territory. The new periphery blurs the last conceptual boundaries between the city and the countryside (Soja 1989).

The new organizational structure of the developed metropolises implies a deep crisis of the roots of the idea of urbanity. In the emerging *diffuse city*, the characteristic elements of the conformation of the city are still present, but the conditions of density, functional interaction, and spatial continuity are absent, which define urban spaces and on which the conventional urban instruments are based (Ezquiaga 1990; Boeri et al. 1993; Macchi Cassia 1995).

On the scale of urban space, this translates into the obsolescence of conventional civic expressions of the public: avenues, parks, squares, facilities, and infrastructures, and their replacement by private spheres capable of mobilizing and bringing together in a flexible way the various forms of collective life, particularly around consumption, entertainment, and sporting and cultural events. The experience of the evanescent vision of social facts: "*everything solid vanishes into thin air*", Berman (1982), prepares contemporary subjects to assimilate without difficulty or risk a space without quality, lacking significant density, and therefore suitable for ephemeral consumption (Auge 1998; Koolhaas et al. 2000) (Fig. 11.4).

**Fig. 11.4** Seoul Reclamation of the Cheonggyecheon Canal. Renaturalization and recovery of the geographical memory of the city buried under an urban highway. *Source* Photo by José María Ezquiaga (2019)

## 11.3 Contemporary Formations of Metropolis

The academy has often portrayed cities through an equivocal image as a coherent and unitary entity, since the use by the sociologists of the Chicago School, in the twenties, of the organic metaphor to explain the life cycle of cities, up to the conception of planning as an expression of the spatial vocation of a city as a subject. More recently, the strategic planning associated with the economic discourse has deepened the dissemination of this icon by presenting cities as economic subjects competing in a universal scenario.

However, the city is not so much an actor as a place occupied by many actors (Marcuse 2000, p.256). It is no longer possible to speak of a direct relationship between the forms of centrality and a specific geographical reference, as in the past with Historic Cores or nowadays with Financial Centers. The contemporary expression of centrality assumes a multiplicity of spatial configurations, both in geographical scale and in quality. The new information- and knowledge-based economy has a global dimension, enabled by electronic links that allow certain activities, notably the financial markets, to function as "*a unit in real time*" (Castells 2002 and 1997–1998, Sassen 1991). Since the recognition of the primacy of virtual networks, authors (Webber 1964:84, Mitchell 1995 and 1999, Boyer 1996, Ascher 2009) have asked themselves about the future of large urban agglomerations in the face of the emerging processes of deterritorialization (Teyssot 1988, Burdett and Sudjic 2007) (Fig. 11.5).

**Fig. 11.5** Hong Kong. *Source* Photo JM Ezquiaga

Although cities will continue to play the role of command posts (Le Corbusier 1945), the extraordinary capacity for generating wealth associated with new activities and their unequal distribution depending on their place in the globalized networks of individuals and companies, determine an extraordinary variety of situations in the location and structure of centrality. Thus, it is possible to speak of geographical or electronic centralities, which respond to physical nodes of concentration of directional functions, or have a territorial meta character, linked to electronically generated spaces, e.g., financial markets (Castells 1995). In any case, paradoxically, the optimization of the use of information technologies always demands a material infrastructural support and a geographical territory on which to deploy. Global cities constitute, from this point of view and regardless of any other quality, hyper concentrations of infrastructure and the area where the conflict between the market and the public sphere materializes.

Considering the organizational form of the territory of centrality, Sassen (1991, p.333) notes the permanence of the conventional urban center as a key expression of centrality. She also detects simultaneous trends towards the expansion of centrality over the metropolitan territory, forming a network of intense tertiary activity nodes, and towards the formation of "transterritorial" centralities organized on telematic networks of economic exchange (Hall and Pain 2006). It is also possible to speak of an "infraterritorial" centrality, by virtue of the folds of time and space over concrete geographical centralities.

Telematics appears as a necessary condition for the decentralization and spatial dispersion of the activities previously associated with the Urban Center, by neutralizing physical distances. However, other gravitational forces tend to maintain the cohesion and importance of urban centers as concentrations of infrastructure and nodes of technological innovation networks associated with knowledge and higher education.

In this context, cities constitute the control centers and nodal points of location of key markets and companies, as well as the breeding ground for innovation and the symbolic and architectural expression of new activities. To this function of cities as an infrastructural support of the economy, Castells (2002, p.36) adds that of constituting the field of social values. Indeed, the modern network economy lacks any reference outside the strict logic of competitiveness and market. The city constitutes, therefore, the scene of the conflict between the market and the public sphere, overarching and explaining the modern construction of social space and its architectural expressions.

In summary, the modern transformations of the metropolises—with significantly diverse modulations between Europe and America and between developed and emerging countries—are a process of overcoming spatial constraints that does not operate gradually but in successive leaps of organization and scale:

- The formation of metropolitan areas, overflowing the limits of the traditional continuous and compact city.
- The *polynuclear* city-region, which implies a new expansion of the scale of interaction and the overcoming of the simple relationship of dependence of the metropolitan core.

- The post-metropolitan territory, which initiates the *fractal* organization of a territory formed around major axes of supra-regional development.

The formation of the city-region required overcoming the significant deficiencies of the formative phase of the metropolitan areas, but also the generation of new and significative territorial imbalances. We cannot fail to mention the two most important: the growth of the urbanized area and the decline of the traditional central core. Indeed, a faster growth in urban land area than in population or metropolitan GDP is a common feature of developed countries. Because of this sustained pressure of urbanization, the rural area tends to disappear, except in the explicitly protected areas, and the residual lands tend to multiply in expectation of development located on the margins of the urbanized areas (Boeri et al. 1993).

The evolution of the metropolis does not stop at the phase of consolidation of the city-region, but we are on the threshold of a new qualitative leap toward the formation of a new territory that, according to Soja (1994 and 2000), we could call *post-metropolitan* and whose most noteworthy features in the case of Madrid we identify below. The geological metaphor of a space structured in strata is probably more suitable than conventional zoning (or segregation of uses) to represent the complex dimensions of metropolitan reality. The strata account for different crystallizations of the social construction of reality, capable of overlapping over the same geographical space and, more importantly, allowing time to become an additional dimension of space.

The qualitative leap in the diffusion of accessibility made possible by the construction of the great metropolitan roads constitutes, without a doubt, the most important spatial transformation in modern metropolises. At first, interurban highways functioned as linear attractors (*strips*) on which gravitated a succession of architectural objects whose iconography portrayed the unique character of the functions they housed: corporate and institutional headquarters. However, the new ring roads not only enable movements between the radial axes, reticulating the road system, but also play an even more relevant role as colonizing elements of a new territory generating a constellation of strategic points of maximum accessibility at their intersections, links, and connections with transport interchanges. These nodes constitute the new central areas of the suburban territory.

In the traditional city the "Centrality" implied "difference" with respect to the ordinary fabric, which played a background role, and implied, likewise, urban "identity", produced by the presence of a public space (square, street, avenue …) that assumed a meaning of "civic institution" in relation to buildings and activities. The *new* peripheral centralities (provisionally admitting the semantic paradox), on the contrary, stand out in their environment because of their *autistic* character, lack significant public spaces, and base their identity on the attractiveness of the activities they house (generally large leisure and commercial areas) rather than on the architectural expression of them. The result of this form of colonization of the territory is the multiplication of spaces that escape the control of the built form: voids between the fragments of disconnected residential plots or between urbanized enclaves, abandoned productive areas.

The result is the decline of the public sphere (prophesied by Jacobs or Sennett) and an extremely simplified and impoverished experience of space. Accessing the interior of the building is the only way to get an experience of the place. The access from the highway, the parking lot, and the entrance of the building itself do not reach the condition of places. The buildings are autistic, introspective volumes, even without the will of external figurative presence. Architectural innovation appears distorted in the pieces intended for leisure that substitute the traditional public space by expressive media qualities of the new symbolic economy of entertainment.

## 11.4 The Contemporary Transformations of the Madrid Metropolis

The contemporary transformations of the Madrid metropolis have overcome the spatial constraints of the city, not gradually but in successive leaps of organization and scale (Ezquiaga 1990, 1993, 1994 and 2000).

The first stage, which covers the decades of the sixties and seventies of the last century, is well known. Subway lines and the M-30 beltway give a structure to the radio-concentric urban tissue in the urban core, and the radial roads and railways act accordingly in the metropolitan periphery. The central city brings together the institutional, commercial, services, and tertiary activities, surrounded by a residential urban periphery distributed in socially segregated pieces according to a quite simple pattern: concentration of the highest incomes in the North and West and of the lowest in the South and East. Outside the municipal limits of Madrid gravitate several swathes of metropolitan satellite settlements formed by fast-growing and discontinuous dormitory cities, supported by the primitive network of radial roads and rural settlements. The new cities are born with hardly any basic infrastructures are fully tributary to central Madrid for employment and essential services.

The imbalances associated with this form of growth are evident. Fragmented and spatially discontinuous development favors a double social polarization: the North/South based on environmental quality patterns and the Center/Periphery antinomy determined by the inefficiency of the transport system. In this way, the imbalances in the distribution of employment, services, and environmental quality associated with the insufficiency of the mobility system aggravate the inequity of the model by severely limiting the quality of life of the peripheral and metropolitan population.

This context explains the bases on which the renewal of urban planning was conceived in the period of democratic transition, during the 80 s to the mid-90 s: emphasis on the correction of urban quality deficits of the built peripheries and regional policies of territorial rebalancing through the distribution of large facilities (especially Universities), promotion of the decentralization of tertiary activity, and creation of a genuine metropolitan transport infrastructure (from the creation of the Regional Transport Consortium in 1986, to the construction of the Metro-Sur in 2003).

The policies of rebalancing, conceived in the first instance in a voluntarist way, found an adequate harmony with the endogenous tendencies of transformation of the metropolitan area. At the end of the twentieth century there is a new leap that affects the scale of functional interdependence of the metropolitan settlements, from 35 to 90 km, that is, covering a large part of the territory of the autonomous region of Madrid and the main cities of nearby regions such as Guadalajara or Toledo. Now, the most significant thing about this stage is that it involves a substantial change in the organization of the territory: the jump to a complex polycentric structure.

Indeed, at this stage, three key trends for the future of Madrid overlap. First, the transformation of the "geometry of mobility" with the opening of the large metropolitan ring roads (M-40, M-45, and beginning of the M-50), improvement of the capacity of the radial highways, railway modernization and the southern Metro-Sur subway ring. Again, infrastructures go beyond the mere function of strengthening consolidated settlement patterns to become vectors of a new scale of colonization of the territory. To this is added the strengthening of the most important metropolitan settlements, not only in population, but also in the attraction of economic activities of greater qualification (trade, services, and transport logistics), due both to the saturation and extraordinary rise of land prices in the city of Madrid, as well as to the substantial improvement in the conditions of quality and accessibility of those. But above all, it is interesting that at this stage a process of substantial transformation of the form of growth begins: an ex-novo generation of "centralities" associated with the metropolitan accessibility nodes, and therefore lacking an urban reference of traditional centrality and low-density suburbanization first of high and medium income households and later of a segment of directional activities of greater added value (financial, R & D, directional tertiary), specialized services (private universities, hospitals…), leisure and retail (large retail parks, theme parks…).

## 11.5 The Qualitative Leap in the Diffusion of Accessibility: Linear Attractors and Nodes

The qualitative leap in the dissemination of accessibility constitutes, therefore, the most important spatial transformation of Madrid on the threshold of the XXI century. As we have pointed out, at first the radial highways were "linear attractors" on which gravitated a succession of architectural objects that expressed in their iconography the unique character of the functions they house: corporate and institutional headquarters. However, the new ring roads not only fulfill the function of facilitating the movements between the radial axes, reticulating the road system, but also play an even more relevant role in the formation of the colonizing nodes of a new territory. Although its layout is autistic to its immediate spatial environment, the new ring roads have generated a constellation of strategic points of maximum accessibility at their intersections, links, and connections with transport interchanges. These nodes constitute the new central areas of the suburban territory.

In this way, a new territory appears; "interstitial" with respect to traditional metropolitan settlements, but "central" from the logic of metropolitan accessibility. Suburban tertiary corridors specialized in business services and consumption appear: first on the axis of communication with the airport, later on the La Coruña road, and recently in the form of an archipelago of corporate campuses and commercial parks on the N-I road; the development of an arc of industrial activities in the South, supported on the campuses of the Carlos III and Rey Juan Carlos Universities; the logistical and technological orientation of the SE arc of Madrid and the Henares Corridor, with the support of the University of Alcalá; and the consolidation of the node of scientific and technological innovation in the North, around the new city of Tres Cantos and the Autonomous University.

Centrality has historically been a quality acquired throughout a temporal process of singularization, spatial and functional, of certain places in the city. Centrality implies difference with respect to the ordinary fabric, which plays the role of background, and implies, likewise, urban identity, supported by the presence of a public space (square, street, avenue ...) that assumed a meaning of civic institution in relation to buildings and activities (Rowe 1978). As we have pointed out, the new centralities, on the contrary, stand out in their environment by their centripetal character and dependence on the automobile, lack significant public spaces, and base their identity on the attractiveness of the activities they house (generally large leisure and commercial areas) rather than on the architectural expression of them.

The new commercial and leisure formats, often referred to as urbanized parks, go beyond the elementality of the first formulas for exploiting the points of maximum accessibility of metropolitan highways through large, isolated containers such as big-box retail. These based their success on quantity management: parking facilities that made it possible to turn the car into the shopping cart and multiply the breadth and depth of the assortment of available consumer goods (Rowe 1991). However, in the big-box retail, the emphasis on spatial rationalization does not go through architecture. Suffice it to perceive the contrast between the sophistication with which *merchandising* techniques organize the functional distribution and presentation of products according to a logic of induced itineraries of consumers and the poverty and simplicity of the architectural support, reduced to the strict functionality of the industrial space. With no more iconological concessions than the ubiquitous advertising totems, which supplant the frequent banality of architecture (Auge 1998).

The speed with which the process of suburbanization has occurred in Madrid has generated a leap between the first suburban artifacts (big-box retail, unique buildings of equipment and offices) and the new typology of the urbanized park. This is a difference with the evolution followed in North America, where shopping centers assumed the role of backbone of the low-density residential suburb, replacing the lack of true public spaces and community centers. In fact, the introduction of the typology of the open pedestrian mall was somehow an attempt to artificially reproduce the character of the street in the traditional European city. Its subsequent evolution as a closed gallery in height has constituted the new paradigm of commercial space and its dissemination to other contexts has influenced the modern configuration of museums, institutional and teaching buildings, and leisure centers (Project on The City 2001).

In the supermarket or the suburban mall, the dialog between the building and the metropolitan system occurs individually through the road infrastructures and services. Urbanized parks are urban pieces with a self-sufficient vocation, indifferent to their context because they respond to a spatial logic of a higher scale: the great metropolitan movements facilitated by the new highways between functionally specialized areas. Its origin and destination are, therefore, linked to the future of the automobile. Its morphology arises from the hybridization of the models of the garden city and the campus (Rowe 1991). From the first they adopt the extensive occupation of low height and the importance of free spaces to configure an environment. From the Anglo-Saxon campus they adopt the autonomy and openness of the constructions. In short, the parks are thematic sets of autonomous architectural pieces, organized around parking facilities and common services, and located at the points of maximum accessibility of the metropolitan arterial network.

The organization of suburban space dramatically accentuates the decline of the public sphere that Jacobs or Sennett prophesized and a simplification and impoverishment of the experience of space. Nowadays, as the nature of consumption evolves from serving needs to the imaginary in order to integrate into the culture of leisure, the nature of the anchor stores of commercial parks tends to change. This can explain why new leisure items such as multiplexes, restaurants, and ice rinks constitute the iconic elements of the new symbolic economy of entertainment, incorporating expressive qualities hitherto denied to the "MacDonalized" architecture of large shopping malls (Ritzer 1996).

## 11.6 The Post-metropolitan Territory: From Fragmentation to Fractality

The formation of the city-region meant the overcoming of significant deficiencies of the formative phase of the metropolitan area but also the generation of new and large territorial imbalances. The increase in the urbanized area and the decline of the traditional central urban core are the two main imbalances.

Indeed, the increase in land consumption in a significantly higher proportion than population growth and metropolitan GDP growth is a phenomenon common to European developed countries—which on average have increased their urbanized area by 20% in the last two decades—attaining dramatic conditions in Madrid. The Madrid metropolis has doubled the area affected by urbanization since the mid-seventies. As a result of this sustained pressure of urbanization, the rural area tends to disappear, except in areas explicitly protected as Natural Parks, and "residual land" located on the margins of urbanized areas tends to multiply in expectation of development. As the utopia of *green rings and wedges* has been forgotten, the municipal planning developed in the last years of the century has responded to the phenomenon by proposing the urbanization to full capacity of most of the metropolitan voids. This is a wrong option that will limit in the immediate future the opportunities for environmental adaptation and renaturalization of the new territory.

Metropolitan Madrid has experienced in recent decades a qualitative leap towards the formation of a territory that, according to Soja, we could call post-metropolitan. Its defining features would be the following:

- The distant ex-urbanization supported by the expansion of the metropolitan arterial network (M-50 beltway, new radial toll roads…). This process initially adopts a low-density nebulous configuration towards the Sierra de Guadarrama and an extensive archipelago in the region of La Sagra in Madrid and Toledo. Although during the last decade real estate activity has decreased in the city of Madrid (Barros and Ezquiaga 2022) due to the impact of the economic crisis, the dynamics to the dispersion of growth have continued at a regional scale, albeit at a slower pace. The emerging Toledo-Madrid-Guadalajara corridor is already the main urban node and the main territorial attractor of international investment in the Iberian Peninsula.
- The anti-distance effect of the new high-speed railway lines on cities located between 70 and 200 km from the capital: Toledo, Guadalajara, Segovia, Cuenca, and Valladolid. The experience of the recent development of Ciudad Real advances the patterns of this phenomenon, which will further integrate these cities with Madrid.
- The transformation of the organizational patterns of the new territory. The geographically fragmented and functionally specialized city-region becomes, over time, a fractal territory whose emerging conditions deserve a closer comment.

The revitalization of central Madrid. Paradoxically, the polynuclear conformation and the increase in scale of the metropolis allocates a new value to its "headquarters." Investors and public officials alike have understood that and have opened a spring of public and private projects whose effects are still uncertain, but which, by themselves, reveal a new strategic valuation of the central space. The extension of La Castellana to the North and the refurbishment of Paseo del Prado to the South suggest a reinvention of the backbone of the city and paradoxically a return to the model of dense, continuous, and intense city in urban events.

An example can help understand the scope of this idea. At first glance, the establishment of a corporate campus of a major international bank in the suburban environment of the metropolitan west can appear only as a simple piece, a paradigm of a mode of occupation of the territory based on closed units and alien to their geographical context. From a wider spatial and temporal scale, it appears as the first fragment of a more complex territorial piece integrated by the pre-existences of the most diverse nature of its environment, the new low-density residential growth areas, the economic activities, and housing resulting from the growth in extension of the first-generation metropolitan cities and the new infrastructures of connection to the metropolitan network. Each fragment increases the diversity and complexity of the whole, even if the physical units are monofunctional, typologically monotonous or socially homogeneous. The mistake lies in confusing diversity and fragmentation with *chaos* (in the sense given to this term in information theory). The metropolitan development of cities like Madrid shows that this is not always the case.

However, this mode of development is not desirable because of a quite different problem: the accelerated consumption of non-renewable resources, land, and energy, which it entails. Defining efficient strategies to increase territorial complexity while ensuring an efficient use of resources becomes a key issue in projects (Mostafavi 2010). This implies the need to go beyond the option to open all the metropolitan territory to any kind of use and with the highest possible density, which can be deemed as saturation planning. Saturation planning does not consider the eco-systemic capacity of the land, reduces geographical and biological diversity, severely challenges any possibility of a consistent green infrastructure, and giver the private car and its infrastructures a disproportionate role. Overall, saturation planning fosters functional and typological monotony and reduces the opportunities for adaptation to future changes in the regional organisation.

## 11.7 Put People First: The Proximity City

The analysis of the *Sample of Cities* prepared by UN-HABITAT shows an uninterrupted reduction in the proportion of urban space allocated to streets over the last hundred years, going from reaching 25% of the urbanized area at the beginning of the twentieth century to an extension of 21% today. About 48% of cities allocate less than 20% of urban land to streets, and pedestrian access to arterial roads has also been reduced, generating for most new inhabitants a captive demand for motorized mobility, covered either by public transport or private car. In cities such as Madrid or Barcelona, cars get close to 70% of the total street space. The city car is also largely responsible for the problems of urban congestion and the consequent waste in terms of time and energy consumption.

On the other hand, the trend towards a growing individualization of urban life is triggering profound changes in the ways in which citizens organize urban space and, above all, in the way they spend their time. In recent decades, this can be attributed at least in part to the deep renewal in the means, motives, places and schedules of displacements, communications, and exchanges associated with changes in ways of life, urban morphology and demand for public facilities and services. Faced with these problems and from a global perspective, it is necessary to promote urban mobility models aimed at reducing motorized travel in the short term—especially in work-home commuting—alternatively prioritizing pedestrian mobility and collective or shared means of transport that respect the environment. In the medium and long term, it will also be necessary to change the modes of urban organization, which today are based on functional segregation based on land use planning and land values.

The modern architectural culture has defended the concept of *the open city*, understood in the double sense of unlimited urban extension and liberation of architecture from the contextual constrictions of the traditional city. As we have seen, the reality of the new metropolises does not respond so much to these principles as to a pragmatic logic of spatial fragmentation in tune with the acceleration of the consumption cycle. This is clear in the transience, temporary, and superficial qualities of the built environment.

Quite different is the notion of open city, defended among others by Sennett (2018). From this perspective, the city is porous and incomplete, that is, susceptible to a permanent transformation and recreation based on the real needs of its inhabitants. It also suggests the need to promote reasonable densities that allow the mixture of uses (residential, endowment, and economic activities), variety and social cohesion, favoring pedestrian mobility as a preferred means of access to daily services.

The idea of *neighborhood unit* (Fig. 11.6) is associated in the Anglo-Saxon urban culture with the experience of the loss of homogeneity identity that had historically characterized the neighborhoods that make up the city and the need to offer an organic alternative from rational planning. The traditional European city was an aggregation of autonomous neighborhoods with their own identity and history and not from the spatial fragmentation or territorial disintegration that would later characterize the industrial and post-industrial city. Likewise, during the period of formation of the great North American cities, the neighborhoods assumed an identity based on the community of geographical or racial origin, language, religion ... and later class distinctions.

The idea of neighborhood unity, as Gallion (1950:277–290) points out, is neither a sociological phenomenon nor a particular theory in the sphere of the social sciences. It is simply an environment in which a mother knows that her child will not have to cross traffic streets on his way to school, a school that will be a friendly pedestrian distance from his home. The neighborhood unit reorganizes the distribution of the residential fabric from the primacy of pedestrian access to essential services and equipment, starting with the school, and public transport. With various parametric modulations, the idea is present in Clarence Perry's proposal for the Regional Plan of New York and its surroundings of 1929 as well as in the Chicago Plans of 1942 and Abercrombie for the Greater London Plan of 1944.

Clarence Stein's outline for the Radburn Plan shows the grouping of three neighborhood units served by a high school and two shopping centers has a maximum distance radius of one mile. Each of these units does organize around the primary school and local shops in a maximum radius of half a mile as the streets prevent passing traffic.

In order to have a satisfactory public–private domains interaction, the street needs, in addition to being a generative compositive element, to keep playing a significant role as an element of relationship both between people and between activities. A necessary condition to achieve a lively street is to have a residential structure and activities of a certain density. However, it is common to note that the simplification of the typologies of public spaces, together with the lack of a sufficient building density to guarantee a minimum threshold of activity, led to the abandonment of the street and its transformation into a deteriorated and unsafe space.

The report *Towards an Urban Renaissance* (1999) prepared by the Urban Task Force, chaired by Richard Rogers, updated the Neighborhood Unit model (Fig. 11.7)

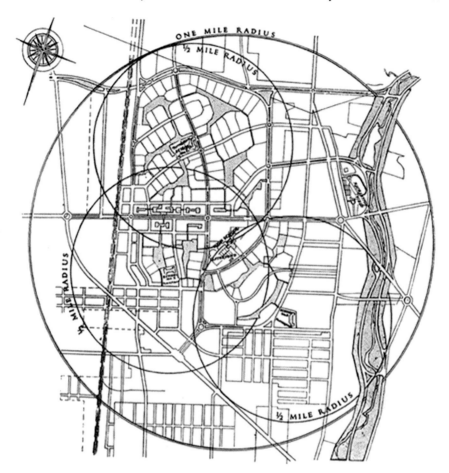

**Fig. 11.6** Radburn general plan showing articulated neighborhood units in half-mile radius environments. *Source* Clarence S. Stein 1950. Toward New Towns for America

incorporating intensity and mixed use as goals. The document considers that one of the main attractions of urban life is the proximity to the workplace, commerce, and essential services in the social, educational, and leisure sphere. This means understanding that a good urban design must favor that most people can live near such services thanks to the integration of uses at the scale of the neighborhood, the street, the block or in the vertical distribution of the buildings. The attached figure exemplifies how to integrate into neighborhood units a wide choice of services closes to residential areas without creating monofunctional areas of commerce, business, or housing. In short: the city of 15 min (Fig. 11.7).

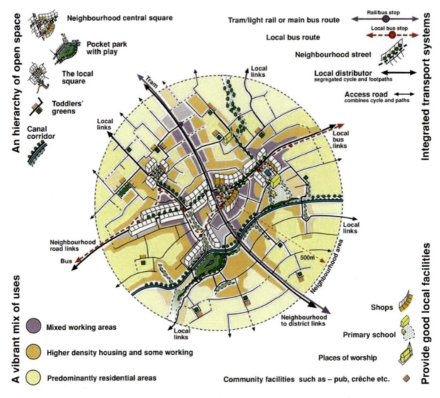

**Fig. 11.7** Key components of the mixed-use integrated neighborhood unit. *Source* Department of The Environment, Transports and The Regions (1999). Toward an Urban Renaissance. (Andrew Wright Ass)

## 11.8 Strategies for a New Sustainable Urban Mobility: The Proposal of the Madrid Centro Project

With the collaboration of the urban ecology expert Salvador Rueda and the sociologist Jesús Leal, we proposed in the Madrid Centro Project (Ezquiaga and Herreros & Perez Arroyo 2011) an alternative, radical but suitable for an incremental, low cost implementation. The proposal was based on the re-organization of the built space through a coordinated intervencion on mobility, public space and governance from the perspective of a *new urban* cell that redefines the basic unit of the central urban fabric. This allows prioritizing the liberation of public space for citizens and at the same time maintaining the flows required by the functions of urban centrality. The proposal was to specialize the use of street space according to the needs of circulation and the creation of pedestrian priority areas in which the increase in the vegetated surfaces delivered improved environmental quality overall quality of life of the residents. The simple segregation of transit traffic from resident-only traffic enables the creation

# 11 The Post-automobile City From Deterritorialization to the Proximity …

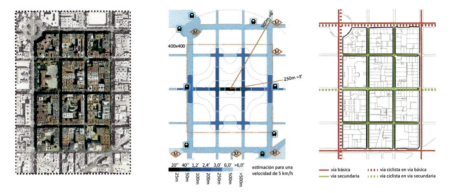

**Fig. 11.8** Proximity city and new urban cell implantation in the nineteenth-century expansion of Madrid. *Source* JM Ezquiaga, Juan Herreros, Salvador Pérez Arroyo. Madrid Centro Project (2011)

of environmental areas in which it is feasible to reverse the unequal distribution of street use between the car and the citizen. The structure of main streets guarantees access by public transport and car to the entire urban fabric, but the secondary mesh of streets of smaller section admits the limitation of access to residents and allows to create a complementary network in which pedestrian comfort, cycling accessibility, arborization, and economic and commercial activities become the main elements. On the other hand, the new urban cell, in line with the historical experience of the *neighborhood unit,* constitutes the coherent scope to reorganize citizen's access to proximity services and public facilities, a key instrument in correcting inequalities in access to public goods (Figs. 11.8 and 11.9).

In the center of Madrid, the private vehicle occupies between 65 and 70% of the streets and avenues directly or indirectly, while the car's share does not exceed 30% of total trips. The disproportion between space occupation and private vehicle travel is evident. The automobile continues to be the factor that generates the deepest urban dysfunctions and the largest obstacle to achieving the goal of a proximity city.

However, during the last decade there has been a hopeful change of trend in a context marked by economic crises and the growing social awareness of the relevance of climate change in urban policies. On the one hand, the volume of traffic in the municipality of Madrid has decreased by almost a third between 2005 and 2019 (the data for 2020 and 2021 are not comparable because of mobility restrictions linked to the COVID-19 pandemic), and this has occurred both in the center and in the peripheral areas. The recovery during the 2020–2021 period, and the data of other statistical series published by the municipality for the first months of 2022, suggest that the decreasing trend continues, and may even increase because of the use of teleworking (Fig. 11.10).

On the other hand, although the census of drivers registered by the General Directory for Transit for the Madrid Region has increased between 2012 and 2022 by 5%, this data reflects the inertia of the considerable number of drivers in mid-age groups. An analysis by age (the open data scheme does not allow a gender analysis) shows

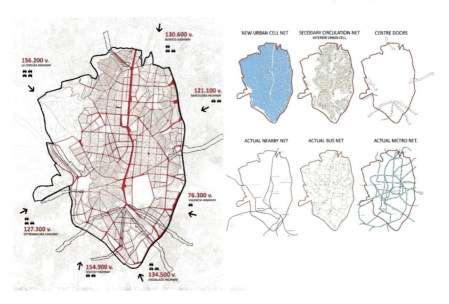

**Fig. 11.9** Traffic intensities and public transport infrastructures in Madrid. *Source* Madrid Centro Project op.cit

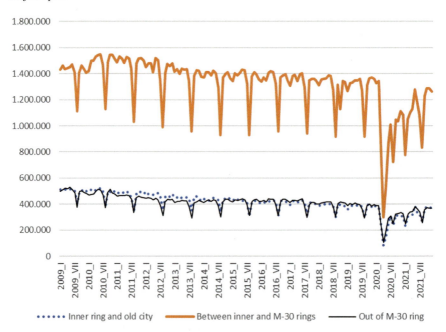

**Fig. 11.10** Evolution by months since January 2009 of the sum of the average daily intensities in the set of permanent capacity stations of the municipality of Madrid. The Madrid Center project operates between the old city and the M-30 ring. *Source* Authors according to data published in the Data Bank of the Madrid City Council (http://www2.munimadrid.es/CSE6/control/mostrarDatos)

that during that period the number of people with a permit in any of its categories under 45 years of age has decreased by 376,000 people. The context of economic crisis during that period and the difficulties experienced by the youngest in the labor market are conditioning elements, but not completely determinants, since they add to a change of trend in the attitudes of this segment of the population towards mobility and lifestyle.

Guaranteeing urban functionality and a new conception of public space means transferring part of the travels that today are made in private cars to the rest of the modes of transport. This redistribution implies a strengthening of public transport, which implies a greater ease of use of the urban fabric, especially for social groups that cannot access the car autonomously (the elderly, children or people with functional disabilities). In this way, the accessibility of the population to public goods, places of relationship and workplaces can improve, avoiding mobility issues resulting in a factor of exclusion (Fig. 11.11).

Effective mobility management will contribute to a better functionality of the central area. The role of the city of Madrid in the metropolitan area does not justify that travels from the periphery penalize the quality of life of those who inhabit the central districts due to the impact of the daily use of the car for commuting. The position of the central area as a privileged node of economic activity in the metropolitan area requires maintaining a rotation parking capacity inside, but also moving a part of this capacity to its perimeter, creating deterrent parking in the surroundings of the M40 ring road connected by high-frequency public transport passing through the central area. This should be applied in addition to specific policies at territorial level to decentralize economic activities and correct the imbalances between employment and residence in the metropolitan environment of Madrid as a whole.

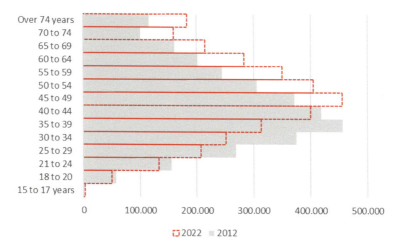

**Fig. 11.11** Evolution of the age structure of drivers (all categories of permits) in the Community of Madrid between 2012 and 2022. *Source* Authors according to data from the General Directorate of Traffic (https://sedeapl.dgt.gob.es/WEB_IEST_CONSULTA/subcategoria.faces)

Public transport is an inalienable asset of Madrid. For the implementation of urban cells, it is necessary to adapt the current bus network to the mesh of basic roads. The current distribution structure of surface public transport lines in the center of Madrid tends to adopt a cellular structure, finding that the number of lines that do not occupy the main roads is minimal.

Finally, the mobility ecosystem in a contemporary city has recently become much more varied and complex. From the perspective of mobility policies, it is necessary to assume this reality and promote a wide and varied offer of motorized and non-motorized modes of transport that rationalizes the indiscriminate access of vehicles to the center without putting essential accessibility in crisis. In the areas of lower density, there is an opportunity to implement on demand transport systems, and in the medium and long term electric autonomous vehicle systems of public service. As an alternative to the private car, these systems allow to modulate the size of the vehicles and adapt it to the requirements of urban fabrics in which the lower population density makes the implementation of large capacity collective systems economically unfeasible and can be articulated with high capacity systems that connect transport hubs distributed in the most remote peripheries with the heart of the city (Figs. 11.12 and 11.13).

**Fig. 11.12** Madrid Centro Project. Priority shared street residents. *Source* Madrid Centro Project op.cit

**Fig. 11.13** Madrid Centro Project. A new urban planning. System of urban cells applied to the whole of the central city. *Source* Madrid Centro Project op.cit

## 11.9 The Urban Cell as a Basis for the Articulation of Public Space

The understanding of public space as an organizational system, identity reference and mediating element between the city and its inhabitants implies:

- The relevance of increasing the size of the urban planning unit that can no longer be the block but a minimum grouping of them.
- The consideration that the quality of public space is the most solid foundation to operate the transformation and reconstitution of the city for its residents. A more efficient articulation with other convergent policies is possible if the essential support of public space is available
- The urban fabric conditions the open space and must evolve to address energy efficiency and climate change. The design of the interrelationship between street and building demands new dialogue based on contextual respect that values the rich heritage of architecture and the streets of the city.
- Working on the identity of the streets can offer the real opportunity to exploit the full potential of a city as rich in the proportions of available space as Madrid.
- The constituent elements of the urban scene are in Madrid disproportionately distributed, if not trivialized, in extensive portions of the urban tissue. The width of the sidewalks, the space ceded to cars, the presence and size of the trees, the lack or excess of public furniture, the lack of coordination of all of them, both weaknesses and levers of opportunity.
- Redefining the street opens the door to rethink the block, recover the possibility of acting on an obsolete urban fabric, which claims a second chance through which the urban scenario is innovative and confident in permanent evolution.
- Rethinking the city in large units that encompass a homogeneous set of blocks, allows to redefine the basic space for social interaction and for a new urban logic.
- Issues such as the energy balance and greenhouse gas emissions, construction of a new frame for social relations and coexistence with the network of public facilities, give this new city and its associated public space the role of main element for a new quality of life.
- The city and its urban fabric, on the one hand, the structuring axes and the large voids with facilities on the other, will constitute a new way of describing the Center at the same time as the support of all the transformation object of the Madrid Centro Project and as the place of implementation of a novel concept of Urban Cell (Fig. 11.14).

In the realm of land uses and activities, the new urban cell can bring a redefinition of the ground floor uses and in the block cores, as well as promote the renovation of entire blocks or groups of them that can implement the three-dimensional and programmatic hybridization beyond the current regulation of compatible uses that restrict important opportunities for innovation such as small offices, creative groups, "clusters" of startups, and unofficial educational uses that claim an ideal place in the center of the city mixed with residence and commerce.

In terms of mobility, the reduction of the use of private vehicles; bicycle extension and pedestrian movement; the rationalization of surface public transport, and the better accessibility of taxis to the interiors of the cell, will result in a more fluid, more efficient, more hierarchized, and more sustainable mobility. Providing non-surface parking to residents is a key objective as will be the possibility of renouncing to have their own vehicle. The interior of the cell has a restricted parking capacity,

# 11 The Post-automobile City From Deterritorialization to the Proximity ...

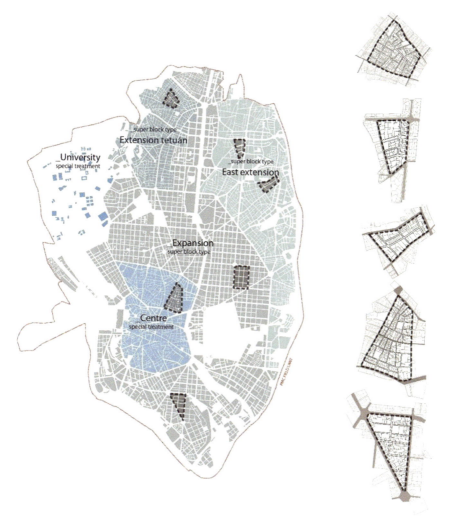

**Fig. 11.14** The urban cell as a flexible transformation unit to promote regulatory innovation, hybridization of uses and re-densification. *Source* Madrid Centro Project op.cit

while new typologies as automatic parking can become tools to encourage a more rational use of the private vehicle (Fig. 11.15).

Environmental quality, understood as a commitment to contemporary organoleptic sensibility, will be determined by the increase in urban green, traffic noise mitigation, reducing emissions on the narrowest streets or the increase in free spaces. Likewise, the urban cell also allows progress toward compliance with the European Air Quality Directive by reducing passing traffic (Figs. 11.16 and 11.17).

**Fig. 11.15** The new urban cell as a flexible transformation unit to promote regulatory innovation, hybridization of uses and re-densification. *Source* Madrid Centro Project op.cit

From the point of view of management, the implementation of the urban cell can be flexible, scalable, adaptable, and evolutionary: it is susceptible to a progressive, tentative, isolated implementation, by degrees of opportunity or ease from an initial phase of minimum cost limited to the regulation of one-way streets and maintenance of surface parking and traffic regulation with ephemeral obstacles. The reversibility

**Fig. 11.16** Madrid Centro Project. Indicators of use of the urban fabric before and after the implementation of the urban cell. *Source* Madrid Centro Project op.cit

**Fig. 11.17** Madrid Centro Project. A test of the urban cell concept in the fabric of the nineteenth century Madrid. *Source* Madrid Centro Project op.cit

of the actions acts as an incentive and allows tests with different transformative capacity.

Finally, the new urban cell becomes a useful and flexible transformation vector to promote regulatory innovation, hybridization of uses and a variety of densities. The implementation strategy of the new cell assumes that the traditional block is the basis for pedestrian displacement and the diversity of uses at the service of buildings, while the base for longer distance travel and public transport is better suited to a mesh of approximately 400 × 400 m. From this approach, the possibilities of specialization for the different areas that the new cellular organization promotes are:

- Location of urban services and non-residential uses on the basic streets at the perimeter of the urban cells, in accordance with the regulations in force on the transformation of uses.
- Controlled densification and diversity of offer consolidating the current activity nodes. Small activity nodes at the local level have a potential for densification strategies and an improved offer of services. This can apply to a Metro station, a zonal facility, a square or a market.
- Internal density reconfiguration and local community facilities. The public spaces in the cells can benefit from public use facilities located in the ground floors, in the residual spaces or in underused areas. These uses must be adapted to the objectives of social cohesion and empowerment of identity.
- Hybridization. Mixed typologies can constitute an important chapter of the innovation proposed for the Madrid Centre. The key emerging trends to overcome the rigidly zoned city are sustainability, recycling and rehabilitation, residential diversity, mix of work and housing to alleviate displacement, three-dimensionalization of the city. They all demand a firm commitment to the hybridization of uses.
- Typological innovation. Certain novelties hitherto impossible such as atomized housing, satellite rooms, shared work centers or other collective uses in residential buildings associated with leisure, study, work, physical health, could strengthen the idea of community, even becoming small production units (energy, services, products). There is a generation of clean productive activities that are related to new technologies and are compatible with residential programs that could make the center a welcoming environment for new uses. So, the center can become a "science park" or urban "technology park".
- Control of environment variables. The new urban cell allows a design of the variables of the street environment based on lower levels of emissions and noise. The reduction of car traffic allows to reduce the concentration of heat, and the possibility that these streets become authentic "ecological corridors" that cross the urban fabric

## 11.10 Conclusions

Changes in the forms of production, organization of consumption and mobility of capital, people, and goods, are profoundly affecting the character of cities. As a result, the contemporary expression of the urban condition assumes a multiplicity of spatial configurations, both in geographical scale and quality, in open rupture with traditional configurations, and demands new instruments and styles of urban planning.

The spatial consequences of these transformations have substantially altered the urban scenario. In developed countries such as Spain, the exponential growth of metropolitan mobility tends to lead to a previously unknown dispersed occupation of the territory. The most significant thing about this phenomenon is that the weakest activities do not relocate to the periphery, but that functions and characteristic elements of centrality leave urban locations to colonize a new suburban territory.

Therefore, there is a distortion in the classic relations of dependence between the central city and the outer settlements. These aggregated and hierarchical metropolitan models tend to transform into a more complex structure. Activities that previously took place in a concentrated space now consume a greater extension of the territory. The new periphery blurs the last conceptual boundaries between the city and the countryside. Although the impact of the economic crises, particularly the real estate crisis of 2008, has led to a slowdown in the consumption dynamics of the metropolitan territory, the underlying trends towards a dispersed occupation of this territory continue to determine the actual growth model.

The exponential growth of private car ownership is a necessary condition for the viability of the processes of territorial dispersion and low density and at the same time feeds back on the needs of individual mobility generated by the models of territorial occupation of very low density and its urbanization of economic activities and centralities. This has led to a substantial increase in mobility costs, difficulties in implementing an efficient public transport system, as well as increases in urbanization and in road building costs.

The impact of urban mobility based on the predominance of the private car is unsustainable in relation to two of the key variables in the contemporary concept of quality of life: urban environment quality and health. Regarding the first aspect, urban mobility based on the private car is inefficient because it demands a high energy consumption per kilometer of travel with respect to the modes of collective transport and additionally supposes a disproportionate occupation of urban public space, as well in road as in parking space (in cities like Madrid Barcelona around 70% of the total space of streets and squares), when compared to other modes of collective or shared transport. The city car is also largely responsible for the problems of urban congestion and the consequent waste in terms of time and energy consumption.

In relation to the second aspect: health. The territorial dispersion/captive mobility of the private car pair is associated with a sedentary lifestyle, one of the accredited causes of the dramatic growth of cardiovascular diseases, as well as the deterioration of air quality in the urban environment, associated with respiratory diseases and cancer, which alone demands regulatory and corrective public action.

From this perspective, there is consensus among experts on the need to face the challenges derived from globalization, climate change and social transformation, from a new urbanism, based on the transformation and recycling of the existing city. This translates into reorienting the character of urban planning toward a flexible instrument open to technological innovation, enabling an integrated approach with environmental, housing, mobility, and infrastructure strategies from a triple perspective:

- Environmental sustainability, so that the consumption of materials, water and energy resources does not exceed the capacity of ecosystems to replenish them.

This also means avoiding that the rate of emission of pollutants exceeds the capacity of air, water, and soil to absorb and process them.
- Sustainable land use, which contemplates a profound change of priorities: from a model of expansion and growth to saturation of the territory, to strategies for the transformation, rehabilitation and recycling of built fabrics, infrastructures, and existing activities; from mere land management to improving quality of life and health benefits for citizens. Under certain conditions, the densification of existing urban fabrics presents a reasonable alternative to neutralize the tendency to dispersion that makes the task of configuring a city with basic infrastructures and sustainable equipment more difficult.
- Sustainable urban mobility aimed at reducing captive car mobility in the short term, especially in commuting. As an alternative there should be a prioritization of collective or shared means of transport that respect the environment by planning their combined use, and an adoption of mid- and long-term models of urban organization based on a greater density and mixed uses, so reducing the need for motorized mobility.

The limitation and rationalization of the pre-eminence of the private car in urban space makes possible the recovery of the street as a civic space for pedestrian and cyclist mobility, economic activity, rest, and encounter. Working on the interrelationship between street and building also opens the possibility of re-thinking the functions, density, and volumetric configuration of the built spaces (Fig. 11.18).

It is also essential to design a new mobility based on a wide and varied offer of motorized and non-motorized modes of transport that rationalizes the indiscriminate access of vehicles to the center without putting essential accessibility in crisis. In the urban centers of most Spanish cities, the structure of main streets guarantees access by public transport and car to the entire urban fabric, but the secondary mesh of streets of smaller section admits the limitation of access to residents and would allow to create a complementary network in which pedestrians, cycling accessibility, arborization, and economic and commercial activities become the main elements.

Finally, the case study described, the Madrid Centro Project, proposes to rethink the organization of the built space from the perspective of what we have come to call "new urban cell". The discrimination of transit traffic from resident-only traffic allows the creation of environmental areas in which it is feasible to reverse the unequal distribution of street use between the car and the pedestrian. The structure of main streets guarantees access by public transport and car to the entire urban fabric, but the secondary mesh of streets with resident-only access allows to create a complementary network in which pedestrian comfort, cycling accessibility, arborization, and economic and commercial activities become the main elements. The new urban cell is also the coherent scope to reorganize citizen's access to local services and facilities, an important instrument for correcting geographical inequalities.

**Fig. 11.18** Hight Line New York. *Source* Photo JM Ezquiaga

## References

Ascher F (2001) Les nouveaux principes de l'urbanisme. L'Aube. Spanish Edition, Madrid, Alianza, 2004
Ascher F (2009) Organiser la ville hypermoderne. Grand Prix de l'Urbanisme 2009, Paris, Parenthèses
Auge M (1998) In: The non-places. Spaces of anonymity. An anthropology of over modernity. Barcelona, Gedisa
Barros Guerton J, Ezquiaga Domínguez JM (2022) Urban digital twins, morphology and open data: an initial analysis in Madrid. In: Annual conference proceedings of the XXVII international seminar on urban form. University of Strathclyde Publishing, Glasgow, pp 847–854. ISBN 9781914241161. https://doi.org/10.17868/strath.00080467
Berman M (1982) In: Everything solid vanishes into thin air. Madrid, Siglo XXI, 4th edn, 1991
Boeri S, Lanzani A, Marini E (1993) Il territorio che cambia. Ambienti paesaggi e immagini della regione milanese. Milano, Abitare Segesta
Boyer MC (1996) Cybercities. Princeton Architectural Press, New York
Burdett R, Sudjic D (eds) (2007) The endless city. London, Phaidon
Calthorpe P (1993) The next American metropolis: ecology, community and the American Dream. Princeton Architectural Press, New York
Castells M (1995) The informational city: information technologies, economic restructuring and the urban-regional process. Editorial Alianza, Madrid
Castells M (1997) The information age, vol I. The network society. Madrid, Editorial Alianza
Castells M (1998a) The information age. Economy, society and culture. Volume II. The power of identity. Madrid, Editorial Alianza

Castells M (1998b) In: The information age. Economy, society and culture. Volume III. End of Millennium. Madrid, Editorial Alianza

Castells M (2002) The city of the new economy. Pasajes n° 35, March, pp 34–37

Departament of the Environment, Transports and the Regions (1999) Towards an urban Renaissance. London, Department of the Environment, Transport, and the Regions

Dematteis G, Emanuel C (1992) La diffusione urbana: Interpretazione e valutazioni en DEMATTEIS, G. (ed) Il fenómeno urbano in italia; interpretation, pospettive, politiche, Franco Angeli, Milan

Ezquiaga JMª (1990) Las afueras. Transformations of the peripheral landscape Architecture n° 286–87, September-December, pp 72–87

Ezquiaga JMª (1993) Madrid, une dimensión de métropole. Cahiers de l'Iaurif n° 104–105, Aout, pp 73–80

Ezquiaga JMª (1994) The city of Madrid. A cohesive vision with a dynamic approach. De Architect September, pp 54–63

Ezquiaga JMª (1995) Horizontes post-metropolitanos, in a collective work: From the ancient city to the cosmopolis, pp 207–228. Cuadernos de la Fundación Botín n° 12 Observatorio de Análisis de Tendencias. Santander, 2008

Ezquiaga JMª (2000) The Madrid Region. In: Simmonds R, Hack G (eds) The global city regions. Their emerging forms. London, New York, Spon Press, pp 54–65

Ezquiaga JMª (2019) Photo of the cheonggyecheon canal

Florida R (2005) Cities and the creative class. Routledge, New York

Friedmann J (1993) Toward a non-euclidean mode of planning. J Am Plann Assoc 59(4):482–485

Gallion A (1950) In: The urban pattern. city planning and design. New York, D. Van Nostrand Co.

Genovese T, Eastley L, Snyder D (ed) (1998) Civitas/what is a city?. In: Harvard architecture review, n°10. New York

Giddens A (1990) The consequences of modernity. California, Stanford University Press, Stanford

Giddens A (2002) Sociology. Alliance, Madrid

Gould SJ (1995) Scales and cones: evolution limited by the use of canonical icons. In: Sacks O, Miller J, Gould SJ, Kevles DJ, Lewontin C (eds) vol 20. Historias de la Ciencia y el Olvido. Madrid, Siruela

Hall P (1998) Cities of tomorrow. Blackwell, Oxford

Hall P, Pain K (ed) (2006) The polycentric metropolis. London, Earthscan

Harvey D (1990) The condition of postmodernity. Blackwell, Third impression, Oxford

Harvey D (1997) The new urbanism and the communitarian trap. Harvard Design Magazine Winter/spring 1997:68–69

Healey P (2007) Urban complexity and spatial strategies. Routledge, New York

Indovina F (ed) (1990) La città diffusa. Venezia, DAEST-IUAV

Indovina F (ed) (2007) La ciudad de baja densidad. Barcelona, Diputación de Barcelona

Jane J (1961) The death and life of great American Cities. New York, Random House. Spanish Edition Capitan Swing Libros, 2011

Katz P (1993) The new urbanism: toward an architecture of community. McGraw Hill, New York

Kettelhut Erich (1926) Stadt mit turm, a sketch for Fritz Lang's 1926 film Metropolis

Koolhaas R, Boeri S, Kwinter S, Tazi N, Ulrich Obrist H (2000) Mutations. Act, Barcelona

Le Corbusier (1945) Manière de penser l'urbanisme. Boulogne-sur-Seine, Editor l´Architecture d´Aujourd´hui. Maneras de pensar el urbanismo, Buenos Aires, Infinito, 1976

Macchi Cassia C (1995) «I segni storici sul territorio, strumenti per il progetto della città diffusa». Rassegna di Architettura e Urbanistica, 86/87:7–22

Marcuse P, Van Kempe R (2000) Globalizing cities. a new spatial order. Blackwell, Oxford

Mitchell WJ (1995) In: City of bits. Cambridge (Mass.) MIT Press

Mitchell WJ (1999) *e-topia*. Cambridge (Mass.) MIT Press

Monclús FJ (ed) (1998) La ciudad dispersa. Barcelona, Centro de Cultura Contemporània de Barcelona, p 223

Mostafavi M, Doherty G (2010) Ecological urbanism. Baden, Lars Müller Publishers, Harvard GSD

Project on the city (2001) In: Harvard design school guide to shopping. Köln, Taschen
Ritzer G (1993) The McDonaldization of society. An investigation into the changing character of contemporary social life. Pine Forge Press
Ritzer G (1996) In: The Mac donalization of society. Barcelona
Rogers R, Gumuchdjian P (2000) Cities for a small planet. Gustavo Gili, Barcelona
Rowe P (1991) In: Making a middle landscape. Cambridge (Mass.), London, the MIT Press
Rowe C, Koetter F (1978) Collage city. The MIT Press, Cambridge
Sassen S (1991) The global city: New York, London, Tokyo. Princeton University Press, New York
Secchi B (1989) Un Progetto per L'Urbanistica. Einaudi, Torino
Sennett R (2018) Building and dwelling: ethics for the city. Allen Lane
Simmonds R (1993) The built form of the new Regional City. A radical view. In: Hayward R, Mc Glyun S (eds) Marking better places. Urban Design Now. Oxford, Butterworth
Soja, EW (1989) Postmodern geographies: the reassertion of space in critical social theory. Verso, London, New York
Soja EW (1994) Postmodern geographies. New York, Verso, Fourth impression, London
Soja EW (2000) Postmetropolis. In: Critical studies of cities and regions. Oxford, Blackwell
Sorkin M (1992) Variations on a theme park. The New American City and The End of the Public Space, New York, Hill, and Wang, Sixth printing
Stein CS (1950) Towards new towns for America. Cambridge, Mass. MIT Press
Teyssot G (ed) (1988) La città del mondo e il futuro delle metropoli. Esposizione Internazionale della XVII Triennale. Milan, Electa
Toy M (1994) The periphery. Architectural Design, London
Webber M (1964) The urban place and the non-place urban realm in Webber. In: Et alt M (eds) Explorations into urban structure Philadelphia. University of Pennsylvania Press (Spanish edition. Barcelona, Gustavo Gili)
Zukin S (1991) Landscapes of power. University of California Press, Berkeley and Los Angeles
Zukin S (1995) The cultures of cities. Blackwell, Oxford

**José María Ezquiaga** has a PhD in Architecture and Sociologist. Tenured Professor at the Polytechnic University of Madrid. President of the Spanish Association of Planning Practitioners. He has been Dean of the College of Architects of Madrid (2015–19) and held various responsibilities in Public Administrations: Director of Regional and Urban Planning of the Madrid City Council (1985–88), General Director of Urbanism (1988–91) and General Director of Planning Urban Planning and Coordination of the Region of Madrid (1991–95). He founded in 1995 Ezquiaga Arquitectura, Sociedad y Territorio EAST, and is its Principal, and having been distinguished with various professional awards and recognitions.

**Javier Barros** is an architect, he holds a title in Advanced Studies in Urbanism (DEA by the Institut d'Urbanisme de Paris, today part of the Ecole d'Urbanisme de Paris). His PHD is in progress in the Department of Urbanism and Territorial Planning of the Polytechnic University of Madrid, as he is currently developing his thesis about information systems and open data and its application to urban analysis. He has published a number of scientific papers about regional and urban matters. He has developed his activity mainly in the field of spatial planning, since 1998 at EAST, where he has been responsible for planning and consulting projects, both in Spain and in Europe and Latin America.

# Chapter 12
# Mobility as an Expression of the Urbicide: The Risks of Transport Modernization in Latin American Metropolises

**Pablo Vega Centeno, Jérémy Robert, and Danae Román**

**Abstract** This chapter reflects on the concept of Urbicide by analyzing experiences of inequality in daily mobility in metropolises throughout Latin America. It appears that despite allowing for improvements, contemporary transformations (particularly under the discourse of sustainable mobility) also have negative effects, either unwanted or not, that marginalize certain population groups. Three phenomena (or forms of Urbicide) are highlighted in this chapter: the deficiencies of large infrastructure projects in terms of costs, effectiveness, and coverage; the implications of the formalization of informal modes of transport; and finally, the forms of exclusion in the promotion of active mobility.

**Keywords** Mobility · Urbicide · Metropolis

## 12.1 Introduction

Lefebvre (1972) stated that "the right to the city" has become an urban aspiration, because of the need to improve the quality of life for all. One of the central themes of the discussion has focused on housing and its surroundings, to the extent that, it is the "place in the world" where the inhabitants locate and identify their urban experience. Although the right to the city is an indispensable and necessary approach, it does not exhaust the dimensions of urban life, particularly in a metropolis. In a large city, it is unlikely that the differing needs that guarantee human fulfillment will be located in dwellings or in close proximity to residential areas. On the contrary, it usually has

---

P. Vega Centeno (✉) · D. Román
Pontifical Catholic University of Peru (PUCP), Lima, Peru
e-mail: pvega@pucp.edu.pe

D. Román
e-mail: danae.roman@pucp.edu.pe

J. Robert
Rennes 2 University, Associated with the French Institute of Andean Studies, Lima, Peru
e-mail: jeremy.robert@cnrs.fr

several centralities in areas where economic, educational, or cultural activities are concentrated, but where residential spaces are limited.

Depending on where our home, work, family, friends, study centers, and cultural spaces are located, our daily lives will be constructed around a network of destinations. These destinations will mainly be accessed on foot or by bicycle if they are close to home, or motorized transport if they are more distant.

In the words of Duhau and Giglia (2008), living in a metropolis refers to the set of practices and representations that make the presence of subjects in space and their relationship with others possible (p. 24). Living in a big city, therefore, is an experience that considers the multiple places we occupy in everyday life as a set of movements linking them together.

In this chapter, we will use Harvey's (1989) mobility approach, which allows us to question the right to the city as a gateway to the opportunities and services it offers, and will also analyze existing modes of transportation within a city. In other words, this approach allows us to start from the conditions, experiences, and needs of the people. Then, in line with Vasconcellos (2010), we will distance ourselves from those views of mobility that only focus on displacement data.

Sustainable mobility is a concept that is often widely adopted by multilateral organizations, which emphasizes how to mitigate the adverse environmental effects of motorized transportation. From this perspective, instruments such as the New Urban Agenda of the United Nations (NAU) or the Sustainable Development Goals (SDG) bring together a set of agreements about what the cities of the future will aspire to. Among these, there is a commitment to significantly increase accessible and sustainable public transport infrastructure, such as promoting pedestrian and bicycle circulation (Habitat 2016). However, these commitments seem to focus more on improving road conditions and means of transportation than on the actual needs of the people.

Since the 2000s, and particularly since 2010, considerable improvements in public transport have been observed in many Latin American metropolises. In general, the typical urban transport models of the twentieth century—the Penny War and artisanal and informal services—have been replaced by reformed systems to greater or lesser extents (Errazuriz et al. 2017). However, although these transformations have been reflected in the improvement of mobility conditions for a significant proportion of the urban population, many challenges remain, particularly for the poorest citizens.

We can ask ourselves: If the public policies of large metropolises incorporate the commitments of instruments such as the NAU, without striving to reduce social inequality, what effects can these actions have on the mobility of the most vulnerable social sectors? Latin American metropolises are characterized by significant inequality among their inhabitants, expressed between the peripheral urbanizations that are occupied by low-income populations, the consolidated spaces that are generally located in central areas, and the expansion of certain areas of a city, such as the suburbs. This inequality is accentuated or aggravated by the fact that the poorest citizens often live in areas that are far removed from the areas that have the best job opportunities and specialized education and health services, meaning that they are the ones who have to travel the most, mainly to the central areas where these functions are often located. For this reason, we believe that despite the genuine improvements and a lot of good intentions, there is a severe risk of aggravating the conditions that exclude the poorest populations that live on the urban peripheries.

These innovative policies can contribute to the Urbicide, as they can be understood as actions that destroy places of significant meaning for everyday life (Carrión 2018), where constantly being on the move is a way of living in a city (Lindón 2014). We define Urbicide as the (re)creation of inequalities and forms of exclusion caused by the inherent contradictions of modernization and urban transformations in terms of mobility in the neoliberal context. The objective of the chapter is to highlight the adverse effects of the urban transformations that are currently underway, whether wanted or not, that marginalize and hide certain population groups.

In order to understand this, we need to examine the effects of these innovations on the structure of a city, how people live, and, in particular, if they mitigate or increase the social inequality gaps that tend to characterize Latin American metropolises. Therefore, we propose an examination of several Latin American metropolises, all of which are large, emerging cities that experienced explosive growth during the second half of the twentieth century. We will rely, in particular, on the ongoing research being developed within the framework of the MODURAL project, which focuses on sustainable mobility practices in Lima and Bogotá.[1] This will be complemented with a consideration of the literature relating to other metropolises around Latin America.

This chapter is organized into four parts. In the first section, the characteristics of the urban structure of both Lima and Bogotá will be examined, as well as the evolution of their transport policies. The second section explores the main innovations carried out in the framework of the reform of the transport systems, particularly the Bus Rapid Transit (BRT) system, and their effects on the living conditions of the poorest citizens. The third section will analyze the role currently played by "informal" transportation before the chapter concludes with a brief examination of the experiences of pedestrians and cyclists who should be protected by the new sustainable urban policies.

## *12.1.1 Daily Mobility in Latin American Metropolis*

### 12.1.1.1 Urban Structure and Travel Conditions

Latin American metropolises have common characteristics that can affect how their urban structures conform (Ciccolella 2012). In particular, the accelerated urban growth since the middle of the twentieth century has been mostly horizontal and outside official planning frameworks. This urban growth has also been accompanied

---

[1] The MODURAL program (2020–2023) "Sustainable mobility practices in Latin American metropolis: a comparative study of Bogotá and Lima" (https://modural.hypotheses.org/) is a program financed by the National Agency for Research of France (ANR). Université Rennes 2, the IFEA, the Pilot, Santo Tomás, Jorge Tadeo Lozano and National Universities in Colombia, and the Pontifical Catholic University of Peru participate.

by socio-spatial segregation, which, as its main feature, translates into relegating low-income sectors to increasingly remote peripheries. In parallel, these metropolises also present an over-concentration of activities in their central areas.

Therefore, despite certain variations and exceptions,[2] the typical Latin American metropolis follows a monocentric organization model (Borsdorf 2003) or a centralized polycentrism, where the main centralities are in proximity (Gonzales and Del Pozo 2012; Vega Centeno et al. 2019). A model like this requires the need for extended trips, where the quality of the daily commute is related to the social status of the commuters (Robert et al. 2022; Vasconcellos 2010).

However, despite the improvements made in public transport in recent decades, many inhabitants currently have to make long daily commutes in poor and often overcrowded conditions. The inequalities between those living in the center and the periphery are reinforced by the poor quality and lack of organization of public transport to and from the peripheries. Daily commutes for work and/or studies are often extensive. According to data from Moovit,[3] commutes on public transport in Bogotá last an average of 64 min, and more than 65% of passengers spend more than two hours on public transport daily. Similar figures are observed in Lima, where the average time is 59 min, and 60% of trips last more than two hours. This means that for people living on the peripheries, journeys take longer and are more costly on average, than for people living near the center.

These transport costs can represent between 4 and 13% of household expenses, according to Guzmán et al. (2019), which is a significant proportion. In Lima, household expenditure on public transport can be between 5 and 10%, although these figures can rise to more than 10% in Bogotá and Santiago, and can reach 30% in São Paulo. Nevertheless, despite these figures, subsidies for the operation of public transport are very infrequent or limited, except for Buenos Aires and Caracas, where they represent approximately 70% and 50% of the costs, respectively, according to 2014 data from the Observatory of CAF Mobility.[4] The systems are self-financing in Lima and Bogotá, and there is a 30% subsidy in Santiago.[5] It is also worth mentioning, however, that some cities (such as Bogotá) have targeted subsidies available for specific population groups (such as students) or depending on an individual's socioeconomic circumstances.

Because public transport is the only alternative that people living in the peripheries have, they also face other problems such as the lack of comfort and the recklessness of the drivers. People are often crowded together and are sometimes unable to fully fit into vehicles. Consequently, travel becomes uncomfortable and stressful, and the

---

[2] Some differences between metropolises are related to subcentralities, densities, rich condominiums in the periphery, and public transport coverage.

[3] https://moovitapp.com/insights/es-419/Moovit_Insights_%C3%8Dndice_de_Transporte_P%C3%BAblico-commute-time.

[4] The data can be accessed via this link: https://www.caf.com/es/conocimiento/datos/observatorio-de-movilidad-urbana/,.

[5] Despite this, the Transantiago fare is contested and its rise in 2019 was the reason for the social unrest seen in Santiago in October of the same year. The cost of tickets was also at the center of social conflicts in several cities in Brazil in 2013.

journeys can feel dehumanizing. Some even report feeling treated like merchandise or animals because of the confinement.

> (Public transport buses) carry many people in a metal box crushed like a can of sardines driven by people who are not afraid of anything and are willing to pass the mobility on you without the slightest fear" (Young user, August 14, 2020, Lima, Peru).

However, seeing as public transport is often the only way that many people can get to work, many choose to wake up earlier to avoid crowding and queues, in the hopes of having a seat or just not being late.

Traveling on public transport is also dangerous because of the risk of accidents. According to a CAF report (Daude et al. 2017), in 2004, the average rate of deaths in traffic accidents in Latin America was 17 deaths per 100,000 inhabitants, which is more than double the rate in both Europe (seven deaths) and North America (eight deaths). Approximately, 80% of fatalities in traffic collisions involve pedestrians or motorcyclists. Citizen insecurity and crime are other factors that also affect travel conditions. According to figures from CAF in 11 Latin American cities, feelings of insecurity while either waiting for or traveling on public transport represent almost 20% of the reasons for customer dissatisfaction (journey time comes later, with 18.3%). It should be noted that this survey did not consider households living in informal settlements.

Travel conditions are even more difficult for women as they can often be victims of sexual harassment. In Latin America, 16% of women state that they have been victims of some type of physical or sexual assault on public transport (this figure reaches 19% among those who have to commute daily on public transport) (Daude et al. 2017). In Lima, according to figures from the citizen observatory *Lima Cómo Vamos*, almost 30% of women reported having suffered harassment on public transport (Lima Cómo Vamos, 2019). In Buenos Aires or Santiago, more than 70% of women have felt unsafe on public transport, compared with slightly less than 60% of men (Allen et al. 2019). Harassment is identified as one of the main factors of insecurity, along with traveling alone, and can occur on all modes of public transportation, such as the Metro in Mexico City (Soto 2017). Many women adopt different protection tactics such as adopting certain postures, choosing their clothing, using only certain spaces in the vehicles, traveling in empty or sparsely populated vehicles, making frequent transfers, avoiding moving at certain times of the day, such as at night, traveling in groups, or using more expensive transportation modes (Allen et al. 2019; Muñoz 2013; Soto Villagrán 2017; Valenzuela Córdova 2020).

People with mobility issues also face many obstacles, such as a lack of adequate signage, infrastructure, or ramps to enter vehicles. This can translate into a perception not only of vulnerability but also of anguish and shame. Many are forced to get around during off-peak hours or to wait for empty vehicles, although many others simply choose not to go out anymore (Cohen 2006; Hidalgo et al. 2020; Prada 2021). The difficulties that characterize mobility disproportionately affect poorer citizens living in the peripheries, which translates into high figures of dissatisfaction. According to the CAF survey (Daude et al. 2017), the percentage of people showing dissatisfaction with public transport is 29.9% in Lima and 38.2% in Bogotá.

This overview of dissatisfaction figures shows a very complex and even contradictory situation. On the one hand, the trend is toward improvement, with a reduction in road accidents and the increase in infrastructure geared toward bicycle usage, for example, while on the other hand, problems such as congestion are worsening.

Starting from the notion of Urbicide in the analysis, our goal is not to measure the evolution of mobility conditions; rather, it is to show how the ongoing transformations, carried out in a certain way and with specific rationalities, inscribed in the experiences of citizens, can (re)produce inequality and exclusion.

### 12.1.2 Transport Policies, with Modernization and Sustainability in Sight

In the face of urban growth and the increase in travel demand, transportation systems have undergone significant changes in the last three decades (Rivas et al. 2019). Starting in the 1990s, these changes were implemented within the framework of an economic restructuring led by neoliberal policies. In the context of the economic crisis, the state promoted deregulation of the sector while simultaneously abandoning the operation of the systems under its responsibility. Lima is an extreme case, where a lot of leeway was left for an "informal" offer, leading to an explosion of small transportation vehicles, such as buses, combis,[6] taxis, and mototaxis (Bielich 2009).

The 2000s marked a return of the state control, with the promotion of private or public–private investment and the concession of mass transportation systems within the framework of transportation reforms. This period saw significant improvements in the public transport offer, symbolized by new infrastructures such as the BRT, subways, or urban cable cars (Vergel-Tovar 2021), as well as with the implementation of integrated systems. Bogotá is an emblematic case in this sense, with an almost exhaustive formalization of its public transport offer within the last ten years. Santiago also underwent significant reforms, with the implementation of Transantiago in 2007. However, unlike Bogotá, which underwent gradual reforms, Santiago brought in its reforms overnight (Briones 2009).[7] These transformations made it possible to reduce certain negative externalities such as road accidents and air pollution. Bogotá, for example, has made great progress in terms of road safety: There were 500 deaths in road traffic accidents in 2019, down from more than 1000 deaths in 2000.[8]

---

[6] Combi is a Peruvian term that refers to an informal minibus system operated by Toyota Coaster or other van models, which began with the Volkswagen Kombi at the 1980s.

[7] The implementation of these reforms allowed great improvements in the supply of public transport, although they present many difficulties and attract a lot of criticism. The case of the social protests in Santiago due to the increase in subway fares in 2019 is an example of this.

[8] See Anuario de siniestralidad vial de Bogotá 2019, Secretary of Mobility Bogotá, https://www.simur.gov.co/sites/simur.gov.co/files/2021-09-2021/Anuario/Anuario_Siniestralidad_2019.pdf.

However, the main question, as pertinently formulated by Figueroa (2005), is whether these investments have meant a real innovation or are more of an *aggiornamento* of the transport organization. On the one hand, Figueroa draws attention to the important public–private investments destined for urban highways with private tolls that occur in parallel with innovations in public transport. These urban highways are large viaducts, which cross the city for the benefit of anyone who has a car and can afford the tolls on a daily basis. The construction of raised toll roads in Mexico is an emblematic case, as is the "yellow line" project, which was inaugurated in Lima in 2018. Large-scale investments in projects like these do not constitute any transformation of the current urban structure or the transport system; on the contrary, they stimulate the acquisition of more private vehicles and accentuate the inequalities between those who have their own car and those who must rely on public transport. With the increase in motorization rates, congestion has become a major problem and one which justifies regulation policies for private motorized transport.

On the other hand, the aim is to rationalize the offer, replacing the "informal" offer with a formal, modern, and more integrated offer, with large-scale infrastructures such as those required by the BRT, the Metros, or routes concessioner to large transport operators (Errazuriz de Nevo et al. 2017). At the same time, these policies seek to respond to the concerns regarding sustainable development and environmental issues particularly raised by international agendas, which translate into the promotion of more active transportation modes, such as cycling (Montero 2017). Latin American metropolises broadly follow this pattern of transport supply transformation, albeit with significant differences in the progress of the reforms and the ambition of the policies (Fig. 12.1).

In Lima, the transport sector is characterized by the gradual implementation of an integrated transport system, unlike in Santiago or Bogotá, where Transantiago and SITP (Integrated Public Transport System), respectively, allow physical and financial integration. In São Paulo, the implementation of corridors, and then metropolitan trains, was never able to cover the entire urban agglomeration, which was the initial plan, and the extreme costs racked up by the project over the last ten years caused strong social discontent. The limitations of this transport reform led to an increase in

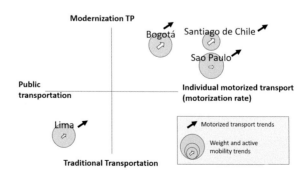

**Fig. 12.1** Overview and trends of change in urban mobility in some Latin American cities. *Source* Elaboration by the authors is based on the CAF Mobility Observatory (Vasconcellos and Mendonça 2016)

**Table 12.1** Individual motor vehicle fleet in selected Latin American metropolises

| City | Automobile fleet | | | Motorcycle fleet | | |
|---|---|---|---|---|---|---|
| | 2007 | 2014 | Evolution between 2007 and 2014 (%) | 2007 | 2014 | Evolution between 2007 and 2014 (%) |
| Bogotá | 958,281 | 1,770,895 | 84.8 | 116,197 | 667,926 | 474.8 |
| Lima | 453,198 | 1,191,113 | 162.8 | 27,000 | 36,372 | 34.7 |
| Santiago | 819,174 | 1,649,479 | 101.4 | 22,634 | 83,491 | 268.9 |
| São Paulo | 4,386,158 | 6,290,842 | 43.4 | 652,225 | 1,187,565 | 82.1 |

*Source* Authors' source based on CAF data (2007–2014)

car trips and especially motorcycle trips.[9] On the other hand, Lima is characterized by a robust informal transport network. Its large-scale infrastructures, implemented in the 2010s, comprise the metropolitan (BRT) and the Metro (electric train), which cover less than 7% of the trips (against more than 50% by informal buses and combis). Its motorization rate is still low, with 120 cars/1000 inhabitants against 200/1000 inhabitants in Bogotá, 250/1000 inhabitants in Santiago, and 300/1000 inhabitants in São Paulo, but its motorization rate is likely to increase. The stated objectives are the modernization of the transport network with new subway lines, the integration of modes, the rationalization of routes via concessions, and the formalization of informal operators, following either the Bogota or the Santiago model.

Despite the reforms, an increase in the number of individual vehicles, cars, and motorcycles is observed in all cases, albeit with marked differences depending on the city (see Table 12.1). The increase in motorcycles is particularly prevalent in Bogotá, followed by motorcycles in Santiago, and automobiles in Lima.

In relation to the active modes (cycling and walking), placed on a pedestal by international agendas, a first reading of the data shows they make up a relatively large proportion (e.g., more than 20% in Lima, Bogotá, and even in São Paulo, but less in Santiago). However, a closer look at the data shows the overrepresentation of walking compared with cycling. This is particularly the case in heavily populated sectors, where it is often the only option due to a lack of alternatives. Unlike Bogotá, which presents itself as a cycling city (Ríos et al. 2015), in Lima this is still a very rare mode of transportation.

For both cyclists and pedestrians, numbers increased as a result of the COVID-19 pandemic: from 3.7% in 2019 to 6.2% in 2021 in Lima, according to the citizen observatory Lima Cómo Vamos (2021), and from 6.6% in 2019 to more than 10% at the beginning of 2021 in Bogotá, according to the municipality. These changes reflect both new political priorities and changes in the habits of the citizens, but in the case of Lima, the changes are not necessarily related to the densely populated sectors found around the periphery.

---

[9] There are 56.7 motorcycles per 1000 inhabitants in São Paulo (according to data from the CAF Mobility Observatory in 2014), against 14.3 in Santiago and 3.7 in Lima. However, Bogotá has 73.6 motorcycles per 1000 inhabitants.

These overhauls of the transport systems seek to cover the mobility needs of the various social sectors and urban territories. One of the first contradictions to highlight is the concomitance of a double agenda with the promotion of sustainable mobility with active modes and public transport while continuously investing in the automobile, with large road networks, urban highways, and bypasses constantly being constructed (Moscoso et al. 2020; Stiglich 2021). The contradiction is exacerbated by looking at the beneficiaries of such policies, primarily the automobile, which is the preserve of the wealthier classes. However, less well known are the contradictions that characterize the policies and proposals for the modernization of urban transport that seek more sustainable mobility from both environmental and social viewpoints. For this reason, we aim to highlight these latest contradictions by researching the experiences of inequality in daily mobility as a manifestation of Urbicide.

## 12.2 Mobility Infrastructures: Changes and Permanence

One of the greatest transformations to have occurred in Latin American transportation has been the introduction of BRT as an efficient alternative to public transport, in relation to the prevailing informal system, and as a less onerous investment than the introduction or expansion of metropolitan trains or Metros. The BRT is an exclusively Latin American model that was first implemented in the city of Curitiba during the 1970s. The establishment of BRT systems in the first decades of the twenty-first century in cities such as Bogotá, Lima, and Santiago came to symbolize the political interest of the municipal governments, inspiring them to reform their transportation systems by eradicating informal transport from the city, supposedly the cause of the majority of traffic problems. The BRT also became the epitome of modernity and innovation that Latin American metropolises could show to the world, with the financial support of the IDB and the World Bank. It is also a transport policy that complies with the ODS and the NAU.

However, an important debate regarding the role of BRT in public transportation is currently ongoing. The debate centers on its relationship with the particular city, the cost of the fares, its effectiveness as a commuting alternative for the inhabitants, and finally, its coverage.

### 12.2.1 *The BRT in the Network of Public Spaces in the City*

There are several types of BRT systems found throughout Latin America. There are those of São Paulo or Santiago, where the investment in exclusive lanes is minimal, while in Bogotá two exclusive lanes per direction were implemented through some of the main avenues of the city (Rosas and Chías 2020). Likewise, coverage can vary: Bogotá, São Paulo, and Santiago's BRT systems have significant territorial coverage, while Lima has only one central line.

It is important to understand the relationship that this service coverage has with urban space and more directly with public spaces. In this regard, Salazar (2008) points out that experiences, such as those of São Paulo or Mexico City, are sectored approaches that are focused on the search for efficiency in the transport system. A different experience is the case of Bogotá, which exists within the framework of the application of the transportation-oriented development model (TOD) (Hobbs et al. 2021). In this case, there was an initial drive to include innovation in transportation within the framework of a policy of citizen public spaces, such as the one that occurred on Caracas Avenue, and the integration with pedestrian spaces in the central area. In the same way, some of the projects implemented in the periphery, "in conjunction with shopping centers", have led to the development of urban sub-centers (Beuf 2007). However, this led to serious difficulties because the planning process conceived mobility and public spaces as independent systems. However, in certain cases, the transformation of the space to enable the development of the TransMilenio bus system in Bogotá assumed the expulsion of traditional activities, breaking existing economic chains (Arteaga et al. 2017). The TOD model opens up potential opportunities for urban development, although it is vital that it also considers comprehensive urban policies (Quintero-Gonzales 2018) in order to avoid negative effects, such as the risks of financial speculation on the land and the expulsion of the poorer residents from these spaces (Padeiro et al. 2019).

The experience of Lima, which is limited to the authorization of a trunk line, is more in keeping with the first group of metropolises because of the approach adopted, since the trunk line was not related to the public spaces that the BRT crossed. One of the most traumatic cases of this was the rupture of the fabric of the traditional neighborhood of Barranco, where the building of the road separated the main residential area from the rest of the area's amenities (with the main square, shops, tourism, etc.) (Vega Centeno et al. 2011). The paradox was that the BRT, which was meant to boost the well-being of the residents, ended up rupturing the opportunities for inclusion in the city of popular neighborhoods. The same phenomenon can be observed with Lima's Metro (electric train), whose infrastructure has created an insurmountable barrier for the southern districts of the city.

Difficulties related to the implementation of BRT projects are also present with other types of infrastructure. For example, there were the shocking problems caused by the southern highway in the city of Santiago, where several events converged. On the one hand, priority was given to the construction of a highway that significantly reduced the travel time between Santiago and Rancagua (90 km south of Santiago) for private vehicle users. On the other hand, the replacement of informal passenger service lines with the Transantiago system implemented in 2007, rationalized territorial coverage, and removed many of the bus stops that the populous neighborhoods in the south depended on. This situation led to harmful effects on people's daily lives, their budgets, and their strategies for accessing the city center (Avellaneda and Lazo 2011). The redesign of the Transantiago routes with the introduction of the highway was carried out without any prior study of the characteristics of the social fabric of the surrounding area. The population that lived in the poor neighborhoods in the south of the city saw their access to bus stops reduced and left them highly vulnerable to robberies and assaults because they were forced to make their way through dangerous areas to reach a bus top.

Several neighborhoods had to organize collective systems to move to the nearest Metro station, which led to unnecessary expenses. In other cases, residents had to sacrifice job opportunities or even their access to health care due to the impossibility of reaching those destinations safely (Sagaris and Landon 2017).

In summary, the reforms of the transportation system articulate poorly with the public spaces of the social urban fabric, resulting in new instances of segregation and/or the exclusion of numerous popular neighborhoods.

## *12.2.2 Costs of the BRT for the Average Citizen*

The BRT, unlike other informal modes of transport, offers a transport system characterized by new, low-emission vehicles, drivers formally employed by a private company that holds the system's concession, and a route system planned by the municipality with exclusive corridors. However, in most cases, financing the system depends entirely on the fares paid by the commuters meaning that they are similar to, or even higher than those of informal public transportation.

Figure 12.2 shows that, with the exception of some cities such as Buenos Aires and in less proportion Caracas, most of the Latin American metropolis do not have strong subsidies policies, so the biggest part of the revenue for this transport systems comes from the collection of travel fares (Estupiñán et al. 2018). In other words, if innovation in public transport demands that a significantly larger proportion of familie's budgets be allocated to mobility, those on lower incomes will be adversely affected, restricting their possibilities of having a "right to the city." In fact, in many cities, the main users of the BRT systems are not the poorest sectors.

Fare integrations seek to mitigate these costs, which will allow travel across the entire network on a single ticket. Bogotá, for example, shows an advanced level of integration in the region. However, the validity window of its unlimited ticket—currently 110 min—can cause difficulties because this length of time is often insufficient for people on the periphery, particularly when considering that their trips are not always limited to a round trip to work but often need to stop multiple times along the way. This is the case, for example, for domestic workers, whose mobility conditions are among the most difficult because of distance, time, comfort, and cost (Fleischer and Marín 2019; Montoya-Robledo and Escovar-Alvarez 2020). The current intersectionality between gender and socioeconomic level highlights the limits of the reforms and, more generally, of an urban design that still sidelines the most vulnerable groups.

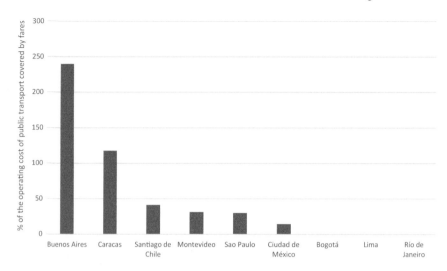

**Fig. 12.2** Percentage of the operating cost of public transport covered by fares in Latin American cities. *Source* Authors' source based on CAF data (2014)

### 12.2.3 The Effectiveness of the System

In addition to the cost, the experience of traveling on the BRT can be severely affected by the huge numbers of people using the system, which can lead to delays in transfers because of congestion at the stations and on the vehicles themselves. Rosas and Chías (2020) describe the enormous problems that the TransMilenio, the most emblematic of the BRT projects in Latin America, has been facing and seriously question the relevance of a single-modal system. The figures in the Origin–Destination Survey of 2019 show a decrease in total SITP trips between 2011 and 2019 (although TransMilenio trips did increase), accompanied by very high dissatisfaction levels, particularly for the TransMilenio.

Lima's sole trunk line also generates huge congestion problems, both because of user demand and the number of units that overlap in exclusive lanes, meaning that BRT has become a much less attractive option than when it began operations. While route times have been reduced, waiting times are longer, meaning that not much has changed overall. Additionally, its introduction has not meant the end of the informal modes of transportation, which continue to be used by the lower-income populations that live on the peripheries of the city.

Finally, a special mention should be reserved for the introduction of the tariff system of magnetic cards. This technologically advanced system is based on the assumption of purchasing power of users typical of countries with higher per capita income than those of Latin America. This can be seen in the experience of the only BRT line in Lima as well as its sole Metro trunk line, where two long lines form at the station entrance: one for loading the card with credit and the other to enter the station itself. A problem here is that most users cannot afford to load their cards with amounts much greater than the daily cost of their ticket, given their budget constraints, which means they have to load their cards daily, increasing their overall trip times.

### 12.2.4 Coverage of the BRT System in the Metropolis

The transport reforms, which began with the introduction of the BRT, have led to the rationalization of routes and bus stops in order to ensure greater fluidity within the system. However, these reforms have adversely affected its territorial coverage in that the system has lost capillarity, thereby limiting people's options for accessing the network or reaching their final destination from it (Borthagaray and Orfeuil 2013).

In the case of the Transantiago, which replaced the old informal transportation network, the effects have been overwhelmingly negative for the poorer populations living on the periphery. Due to the scrapping of various routes and bus stops, public transportation no longer has the same coverage, meaning that people often have to walk longer distances, make numerous transfers, and generally take longer to reach their final destinations, often at an increased cost. These changes disproportionately affect the women, who are mainly in charge of household and childcare and have to make a greater number of journeys, which causes more economic and time expenses (Jirón 2017).

On the peripheries of Bogotá, these reforms have led to a reduction in the frequency and number of public transport routes, which is why the offer has had to be supplemented with buses with decreased carrying capacity. In the case of Lima, it is clear that a single trunk line is unable to meet passenger demand, meaning that feeder buses, without exclusive lanes, have had to complement the service. This offer still only meets a fraction of the city's daily transport needs, meaning that informal transportation modes are still the only real option for many citizens, particularly for those on the periphery.

## 12.3 Informal Transport: Problem or Alternative for the Poorest?

### 12.3.1 The Stigmatization of Informal Transport

As mentioned above, one of the urban transport modernization project's central objectives is the eradication of so-called informal public transport, considered the source of a multitude of problems. The term "informal transport" includes artisanal transport systems, characterized by fragmented ownership of vehicles and not necessarily outside the law and openly illegal systems (Godard 2008). This type of transport can represent an important part of the offer in several Latin American cities. In the case of Lima, this category includes buses (27% of trips), combis (25%), motorcycle taxis (4.2%), and collective-taxis[10] (2.9%), which can account for 60% of daily journeys.[11] In Bogotá, informal transport currently only represents 3.3% of journeys according to the 2019 Origin–Destination Survey, but it was the dominant system until the end of the twentieth century.

According to the official discourse, informal transport is associated with a series of problems such as poor service quality, inefficient routes, reckless driving, and increased congestion, accidents, and crime levels. The fact that they do not pay taxes, have licenses, or submit to vehicle inspections is also a criticism. The position of the authorities is to seek the eradication of this offer by its criminalization, exacerbated by the media[12] (Sierra and Ortiz 2012; Urquiaga 2020).

In Lima, the informal modes of transport have a much lower satisfaction rating (18%) when compared with the more formal modes (BRT and the Metro with 41% and 62%, respectively). However, in Bogotá, the TransMilenio satisfaction rate is only 12%, while informal transport reaches 39% (Moscoso et al. 2020). Without denying these criticisms and the problems that characterize this offer, it can also be understood that these figures are a response to the lack of a formal public transport offer accessible to large sectors of the population.[13] In this sense, it is worth asking to what extent the eradication of informal transport within the modernization project can be understood as Urbicide and whether it results in increased inequalities.

### 12.3.2 An Alternative Offer

In Latin American metropolises, informal transportation has expanded hand in hand with the explosive growth of urban sprawl, as has occurred in Lima (Vega Centeno et al. 2011), and the increased deficiency of formal public transportation systems. Avellaneda (2008) underlined the relative "social efficiency" of informal transport, which covers most of the metropolitan territory, operates with high hourly amplitude, and provides its service more frequently, thereby facilitating access for the low-income population to the places, where the opportunities offered by the city are found.

---

[10] "Taxi-colectivos" are old or informal cars that carry several passengers on a defined route for a fixed price.

[11] WhereIsMyTransport. (2022). Understanding the Mobility Ground Truth in Lima. WhereIsMyTransport Ltd. https://landing.whereismytransport.com/lima-public-transport-insights-whitepaper.

[12] See, for example, the article in El Comercio, 10/04/2020, "The networks that control the avenues of Lima", https://especiales.elcomercio.pe/?q=especiales%2Fel-negocio-criminal-en-las-avenidas-de-lima-ecpm%2Findex.html.

[13] Similarly, the case of taxis working through online apps, such as Uber or Beats, is not strictly informal, but they reveal the precariousness of working conditions. Its use as an alternative to public transport should be considered.

In the case of Bogotá, although the buses no longer cover the entire metropolitan area, the so-called jeeps or vans run informally on routes neglected by the new system. The same configuration is observed in Lima, where the Taxi-Colectivo occupies the main and busiest avenues of the capital, competing with public transport on complementary corridors. This mode of transport offers faster service, is more comfortable (one per seat), and is more expensive than a bus, although cheaper than a taxi. However, these informal services are considered incompatible with the ongoing transport reform process and with the associated ideal of modernization. They are not officially authorized, and their formalization is currently the subject of controversy.[14]

From the point of view of the "informal" operators, they offer quality service and respond to the needs of the people. According to a bus driver in Lima, the authorities:

> (…) create a bad image of the group to discredit them, to harm them. This is why they do not formalize (...) there should be people who conduct surveys and who publish in the media to say that it is not what they think, highlighting the good points as well. But the press likes to be morbid (Mobile operator, March 2022, Lima, Peru).

The testimony of a mobile operator of a group of jeep drivers from El Lucero in Bogotá evidences the same logic:

> We are the informal transport; we transport children, the elderly, pregnant women, the disabled, so we transport too many people. We have a problem. We are not legal. We are informal. The operatives take our cars to the patios, having the papers in order, simply for carrying a sign that says San Francisco – Paradise. With that, we are already, as they say here, branded as informal transport (Mobile operator, March 2020, Bogotá, Colombia).

Some informal transport operators are very well-organized groups, with a directive, a car maintenance policy, and clearly defined routes, schedules, frequency, and fares. If the SITP takes 40 min to complete the route, then they will depart every seven minutes with up to eight passengers. They provide all kinds of services such as home delivery and taking patients to health facilities, helping by loading shopping bags, can also protect their passengers in the event of theft, and can even help the police by transporting wanted individuals to the police station. However, the authorities have not officially recognized them since the SITP was implemented, and they are at risk of having their cars confiscated by the police at any time. The role of small operators as agents of local solidarity, beyond commercial transport services, is highlighted in the cities such as Quito (Gamble and Puga 2019) and Lima (Sierra and Ortiz 2012). The reduction in public transport, and the fear of using public transport due to the possibility of contagion during the COVID-19 pandemic, led many to travel exclusively with known motorcycle taxi drivers with whom the same logic is applied in the face of insecurity.

Informal transport can represent an alternative, sometimes the only alternative, for people with mobility difficulties living on the periphery. For these people, who are generally on low incomes, these services complement public transport in terms of territorial coverage. They offer services that can allow them to cover the so-called

---

[14] The recently implemented formalization policy in 2022 authorizes the operation of these units throughout Peru, with the exception of Lima.

last mile, as well as carrying packages (which is often prohibited in other, more formal transportation). Its flexibility is adapted to the needs of the inhabitants, who often have to make multiple trips, as well as to the obstacles related to gradient, street narrowness, exposure to the weather (be heat, or rain), and feelings of insecurity when walking. On the slopes of the urbanized hills on the peripheries of Lima, motorcycle taxis are almost the only vehicles that can negotiate the winding, sloped streets.

On the other hand, informal transport represents a labor refuge for many low-income inhabitants in the face of economic crisis periods that have occurred since the 1990s. Buying or renting a taxi or motorcycle taxi has been one of the strategies to tackle the massive unemployment levels caused by the COVID-19 pandemic and/or by Venezuelan migrants looking for work in Peru. These activities contribute to local employment in the peripheries, albeit informal and precarious, where small agglomerations of commerce and services tend to develop. Despite its limitations, driving a taxi or motorcycle taxi is still an attractive way to generate income and, in some cases, is more attractive than working for a large transport company. One of the problems faced by transport modernization reforms concerns precisely the employment of small operators and their staff (drivers and conductors). In the case of Bogotá, some of these small operators managed to organize themselves and join the SITP, while others went bankrupt. As for the drivers, many continued to work in the provisional SITP (at least, in the bus lines that are in the process of integrating it), but some ended up rejecting job offers as formal SITP drivers, for various reasons, but mainly economic ones. SITP's inability to ensure its security in certain neighborhoods is also mentioned since small operators are more popular as they are more established in these areas.

Informal transport, therefore, offers certain advantages, both for users (and for the poorest, particular) and for operators and drivers, and, by extension, for the economy and dynamics of many marginalized urban territories. Therefore, if the modernization project is justified by deficiencies and problems associated with informal transport, it can also paradoxically accentuate greater inequalities if it ignores these realities.

## 12.4 Active Mobility: A Privilege for Some, a Burden for Others

As part of the modernization project, and following international agenda policies, the authorities in many Latin American metropolises have promoted active modes of transportation (i.e., cycling and walking) as a priority for sustainable mobility (Herce 2019; Obregón and Gómez 2021; Ríos et al. 2015; Sanz et al. 2018). These initiatives are justified by the objective of limiting greenhouse gas emissions and air pollution, combating vehicular congestion, and promoting a more active lifestyle in the long run. The COVID-19 crisis has also boosted the attractiveness of these modes as they limit the risks of contagion.

In recent years, initiatives in favor of active modes have multiplied. In 2022, Bogotá obtained the Sustainable Transport Award from the Institute for Transport and Development Policies (ITDP) for "its timely response to the need to decongest public transport and prevent the spread of COVID-19." The awards committee particularly valued the temporary bike paths, which Bogotá pioneered at the beginning of the pandemic, such as pedestrian spaces (Bogotá Abierta) and school mobility programs (Al Colegio en Bici and Ciempiés). Today Bogotá has almost 1000 km of cycle infrastructure, with more than 10% of journeys currently being made by bike. Bogotá also decreed the year 2021 to be the "year of the pedestrian." In comparison, Lima only has approximately 290 km of bike lanes, Santiago has about 350 km, and São Paulo has 700 km.

Despite the progress made in promoting active transport, the challenges remain great and the criticisms strong. It is worth mentioning, by way of illustration, the "blow"[15] given by the English newspaper "The Guardian" to Bogotá, branding it the "Cycling Capital of Death".[16] The situation is similar for pedestrians, with fearsome accident statistics. In Bogotá, a pedestrian is killed, on average every 37 h in a traffic collision.[17] Lima, for its part, was considered by the WHO to be the most dangerous city in the world for pedestrians in 2009, as they were the victims in nearly 80% of fatal accidents.[18] According to the *Lima Cómo Vamos* survey, only 6% of Lima pedestrians felt safe when crossing the road (Lima Cómo Vamos, 2018).

However, looking beyond these criticisms, we are interested here in highlighting the contradictions inherent in the promotion of active mobility and the way urban innovations under the seal of sustainability can have negative consequences, either direct or indirect, on certain territories or social groups. In this sense, we aim to reveal the difficulties faced by lower-income populations and the generation of new inequalities by prioritizing the practices of some users over others, as well as other forms of physical and symbolic violence.

### 12.4.1 *The Prioritization of the Automobile, the Forgetting of the Peripheries*

According to van Laake and Moscoso (2020):

---

[15] In reference to the article in Semana (10/27/2020), "The Guardian hits Claudia López hard after forceful accusation about Bogotá," [online], https://www.semana.com/bogota/articulo/the-guardian-da-duro-golpe-a-claudia-lopez-tras-contundente-senalamiento-sobre-bogota/202026/.

[16] The Guardian 25/10/2020; "'Cycling capital of death': Bogotá bikers battle violence on city's streets" https://www.theguardian.com/world/2020/oct/25/bogota-cycling-capital-of-death-colombia-violence.

[17] https://www.eltiempo.com/bogota/2021-sera-el-ano-del-peaton-en-bogota-583529.

[18] OMS, *Rapport de situation sur la sécurité routière dans le monde: il est temps d'agir*, Genève, 2009.

The current difficulties of active mobility in Latin America can be attributed to decades of abandonment by gliders. Although walking and cycling were once very common, the promotion of motorized transport was a crucial element in the modernization of the 20th century in Latin America.

The prioritization of the car in the city is the result of an "elite projection" process (Walker 2017) "[that] saw decision-makers facilitate the mobility practices of the well-off, assuming that what is good for those who can buy and maintain a car would be good for the entire urban public."

Cyclists and pedestrians are the most vulnerable individuals within a city because prioritization of the car in the design of the city generates multiple situations of vulnerability. It translates into obstacles such as tracks that do not have pedestrian crossings or pedestrian bridges that involve walking long distances before climbing stairs.[19] Faced with these difficulties, many often choose to cross in the middle of the track, avoiding the cars. The problem of road safety is usually treated by the authorities as simply a problem of the improper and dangerous behavior of pedestrians, who have been imposed regulations and fines (Dextre and Avellaneda 2014; Pérez López 2014; Santos 2014). The same is also being observed in relation to the lack of infrastructure for and the treatment of cyclists in Lima. It is worth mentioning that even though the law favors the pedestrian and the cyclist over the car, the opposite is often true. Not giving way, even at zebra crossings, speeding up when the traffic lights are amber, honking the horn, and verbal abuse are all common actions of car drivers, which pedestrians, cyclists, and even police officers are used to.

The priority given to vehicular traffic translates into the abandonment of investments for pedestrians and cyclists and is particularly evident in peripheral sectors. There is no shortage of examples of abandoned bike lanes that are now occupied by garbage, parked cars, or pop-up shops of various types. Sidewalks are also often neglected and have obstacles such as light poles, garbage cans, hollow drainage boxes, and poorly dimensioned infrastructure; sometimes, they are not even present (Román 2021). Blank walls and lack of public lighting are other elements that further increase the feelings of insecurity, particularly in cities, where harassment is commonplace (Páramo et al. 2021; Villagrán 2020). Investments in infrastructure, signage, and other facilities to encourage active mobility are often concentrated in certain areas of a city, such as commercial and tourist areas, parts of the historic centers within the framework of recovery projects, and middle–upper class residential areas. In the popular peripheries, the residents' priorities are assumed to be different, even though they are the ones who walk the most and are the traditional cyclists (Jirón and Mansilla 2013; Pérez López and Landin 2019). The bike lane map is one example among many where it only illustrates the concentration of the infrastructure in the center of the city (Map 12.1).

---

[19] The recent construction of a pedestrian bridge on the Costa Verde in Lima (in 2021) to cross a double lane, which needed an extension of almost 500 m, reflects how little the needs of the pedestrians were considered when constructing the projects. For more details on this

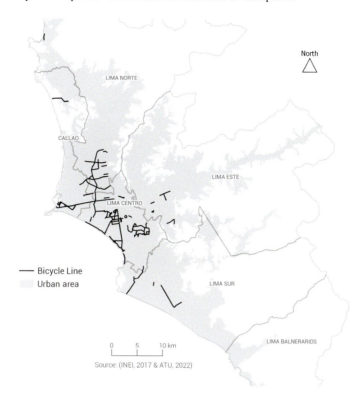

**Map 12.1** Bicycle lane network in Metropolitan Lima. *Source* Authors' source based on INEI 2017 & ATU 2022

**Fig. 12.3** Slope in Independencia (Lima) and in Altos de Cazuca (Bogotá). *Source* D. Román 2021; F. Demoraes, 2022

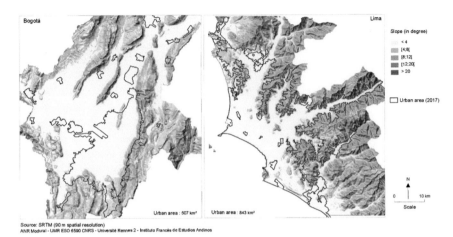

**Map 12.2** Relief and extension map of the urban areas of Bogotá and Lima. *Source* Authors' source based on SRTM

Paradoxically, the peripheral areas of the city are the areas that experience the greatest mobility difficulties, because of their geographical characteristics and the presence of steep slopes (Fig. 12.3 and Map 12.2). This characteristic of the territory, which is commonplace among Andean cities, led the city of Medellín to introduce urban cable cars—an idea that was quickly taken up by other metropolises (Vergel-Tovar 2021). However, these cable cars can only partially cover demand, which is why, for many, the slope of the terrain continues to be an obstacle, which is sometimes almost insurmountable. Using a bicycle becomes impractical and even walking is a challenge and is synonymous with dirt, fatigue, accidents, and difficulty with carrying objects, among other factors.

### 12.4.2 Unwanted Cyclists

If the lack of investment in infrastructure in certain sectors of the city is one of the more obvious sources of inequality, policies that promote bicycle use, which either make them invisible or even exclude certain groups, are another. This is the case when new obstacles are imposed on bicycle use, which disproportionately affects poorer people despite them having been cyclists long before cycling became fashionable (Pino 2017). These people, who rely on bicycles as a means of subsistence and transportation, include sellers of different types of products, including ice cream makers, pastry chefs, or locksmiths, or those who take their tools with them, such as electricians, cleaning workers, or gardeners.

---

case, see https://www.infobae.com/america/peru/2021/12/16/costa-verde-puente-peatonal-jorge-munoz-estas-son-las-criticas-que-recibio-por-el-puente-peatonal-de-458-metros-de-extension/.

A particularly vulnerable group of cyclists are the young migrants of Venezuelan origin, who work as delivery or couriers. Their experiences are more complex and difficult than the average cyclist, mainly due to the nature of their job, but also due to the control to which they are systematically subjected in public spaces (Román 2020). The number of Venezuelan deliveries has increased exponentially in Latin America in recent years and in particular since the pandemic. Most of them are migrants who use bicycles because they do not have the financial resources for a motorcycle. Their situation is precarious since the companies they work for do not consider them to be employees, but rather freelance workers, who are responsible for adapting to the provisions of each country, and also for any accidents that happen to them. Their jobs are physically exhausting and stressful because of the number of packages that they have to deliver, their deadlines, and the number of hours a day they have to work in order to earn a minimum wage. Their work also means they spend most of their time in public areas, where they also eat, rest, and socialize (Román 2020).

Migrant workers are easily identifiable by their suitcases and bicycles and the fact that they are mainly found in groups in public spaces. However, they are often subjected to surveillance and harassment by the authorities and security personnel. This is particularly true for the wealthier districts of Lima, where they are essentially prohibited from mobilizing and parking for a set of reasons: They do not have uniforms or security elements, they carry out propaganda for these companies, and they use these spaces for a long time and neighbors plaints (Román 2020). For example, ice cream vendors who travel on tricycles on the boardwalk in Miraflores (a wealthy district of Lima) are forbidden from using the bike lanes and are forbidden to stop (unless they are making a sale and must continue moving).

In the mentioned district, Miraflores has also gained notoriety in 2019 for being the first district to regulate active modes of transportation, including walking, cycling, and micro-mobility as scooters (Ordinance No. 518, 2019).[20] The provisions of the standard considerably complicate the work of delivery drivers, with the measures considered by analysts to be unnecessarily punitive, or even illegal.[21] The ordinance proposes, for example, to consider imposing a fine of 2520 soles on anyone "carrying a load that exceeds or limits the visibility of the maneuver", a measure that directly targets deliverymen. Along the same lines, the Miraflores authorities prohibited the parking of motorcycles and product delivery bicycles along any roads under their jurisdiction in May 2020.[22] In another wealthy district of Lima, La Molina, the Mayor proposed for his part to prevent foreigners from working in delivery, taxi, or motorcycle taxis in his jurisdiction.[23]

---

[20] The Miraflores district is one of the 43 that belongs to Metropolitan Lima. Each one has a municipality with a certain level of autonomy from the metropolitan authority.

[21] https://es.linkedin.com/pulse/prop%C3%B3sito-de-la-ordenanza-N-518-miraflores-servicio-en-allasi-ur%C3%ADa.

[22] https://www.miraflores.gob.pe/con-decreto-de-alcaldia-urgente-miraflores-regula-estacionamiento-de-motocicletas-dedicadas-al-reparto-de-productos/.

[23] https://larepublica.pe/sociedad/2022/02/21/alcalde-de-la-molina-propone-que-extranjeros-no-trabajen-en-delivery-taxi-O-mototaxi/.

These measures reflect the persistence of prejudices related to nationality, skin color, bicycle use, or socioeconomic level that aim to criminalize, discriminate against, and exclude various "undesirable" elements. When asking a Venezuelan delivery why they were invited to move along by authorities, he replied that:

> We couldn't be there because that's part of the mall and it's more than anything for *normal* people. Those who have delivery jobs have to look for another place to park. (Venezuelan Delivery, December 27, 2020, Lima, Peru, our emphasis from Román 2020)

Behind all this control, there is an ideological construct regarding which cyclists are "good" and "bad" users, as well the "correct" use of public space (Delgado 2014). So even though cycling is heavily promoted in discourses on sustainability and use of public space, in practice, there are clearly huge inequalities in how various groups of cyclists are treated. For this reason, these policies are clearly not inclusive, since those who depend on them to generate economic resources do not have the same opportunities when using them. It makes the inequalities more evident, particularly regarding the difficulties with daily mobility faced by the lower socioeconomic groups, especially migrants.

### 12.4.3 From Fear to Pedestrian Exclusions

Interventions in favor of walking can also, paradoxically, be a source of discrimination. A first category of intervention concerns the initiatives that are justified by security in public spaces. In Lima, as in many other Latin American metropolises, the manifestation of this increased "security" is the fencing-off of neighborhoods, condominiums, and other spaces. Faced with a climate of fear and increased feelings of insecurity, these barriers and defenses further reflect the desire of the residents to protect themselves from "strangers," who mainly comprise people of different skin color, different socioeconomic backgrounds, or simply people who are not their neighbors. Any "stranger" is therefore considered to be suspicious and a potential criminal (Alegre 2012; Vega Centeno et al. 2017). These fencing processes limit the right of mobility and the use of public spaces. The privatization of both the public spaces and security as a characteristic of a neoliberal city (Sequera 2015) translates into a fragmentation of urban space and an exacerbation of the dynamics of exclusion.

Another category of intervention that lends itself to the re-creation of inequalities is intervention projects in public spaces. These projects, which often include a pedestrian component, simultaneously translate into the imposition of new regulations regarding the "good" and "bad" use of public space. Sex workers, street vendors, or others who need to use public space for their subsistence are generally excluded and marginalized in this transformation process, which is happening across many Latin American metropolises. However, tourists or pedestrians who are "consumers" are welcome (Pérez López 2014). In these cases, these urban transformations can highlight and exacerbate both the discrimination and inequalities rooted in the societies, as well as showing how priority is always given to income generation over social welfare.

Finally, these projects facilitate gentrification processes. This may have been the case regarding many interventions in the historic centers of several Latin American metropolis. With recovery and pedestrian projects accompanied by the valorization of tourism and the prohibition of itinerant commerce, it translates into the expulsion of the popular classes, as has happened in Mexico City (Crossa 2018). These processes have also been observed with projects that favor cycling (Hoffmann 2016; Stehlin 2015) for certain groups.

## 12.5 Conclusion

By using the mobility approach, this proposal shows the different forms and processes in which situations of Urbicide can occur. This concept was understood as the destruction of the existing urban social fabric by the neoliberal and mercantile transformation of the city through policies and interventions regarding transport and mobility policies. This approach was first justified because mobility is one of the main objectives that guide modernization projects and contemporary transformations in Latin American metropolises. Secondly, it allows us to observe the attempts at inclusion, exchanges, meeting, and social diversity, as well as their collateral effects such as exclusion, discrimination, violence, fragmentation, or marginalization.

The notion of Urbicide has allowed us to go beyond descriptions of existing, structural difficulties (e.g., the costs and severe inconvenience of travel for lower-income populations in urban peripheries, harassment, etc.) to identify the contradictions inherent in the process of modernization and destructive transformations (opposite of the "creative destruction" of capitalism). In this sense, it demonstrated the manifestation of these transformation processes that affect daily life and reduce opportunities regarding how infrastructure, policies, and socioeconomic and urban changes can adversely affect the most vulnerable in society in different ways, such as physical barriers, and access to transportation (discrimination, cost, gender, etc.) as workers (delivery) or mobile operators (informal transport).

The case of the BRTs reflects this great crossroads: The public transport model that Latin America has exported to the world as a success has serious drawbacks that need to be addressed before it can establish itself as such. It also allows us to question whether informal modes of transport were really the cause of the transportation problems inherent in metropolises and particularly in the urban peripheries. Following the same logic, the discourse of sustainable mobility can express a face of Urbicide across metropolises. Many of these innovations increase the gap between the most accommodated sectors and the forever forgotten inhabitants.

An Urbicide approach allows us to observe the hidden dimensions and the concealment of the "good practices" that the official discourses of the international agendas disclose. Some solutions aim to improve the mobility conditions of some sectors of the population, but often have collateral effects such as excluding groups who are often already marginalized. They can also avail innovative transformations but can dredge up prejudices, stigmas, and inequalities that are rooted in the various societies and can end up recreating and exacerbating them. They then translate the processes of direct or indirect destruction either material or symbolic of various components of urban life. They seek to impose an idea of modernization and order, forgetting the essential role that this "disorder" that they seek to destroy actually plays in the daily lives of the majority of a city's inhabitants. This situation can be considered an unavoidable requirement for the gestation of a critical and emancipated citizenry (Sendra and Sennett 2021; Sennett 2001).

The chapter is not about romanticizing informality or popular initiative, nor about denying the need and benefits of urban transformations, instead it is about highlighting blind spots through which the reproduction of inequalities can be observed. It is a call for attention to urban policies and projects which prioritize the least visible in the city, being the majority of citizens in Latin America. This analysis then leads us to reflect on citizen participation in the production of the city and, therefore, on the right to the city. The protests against the increase in public transport fares that swept across Brazil in 2013 show the need to include the idea of justice in mobility transformations (Verlinghieri and Venturini 2018). More broadly, the protests in Santiago in 2019 that took the Paseo Bandera as a symbolic stage—imposed as an "icon of smart cities" to the citizens, but without a real democratic appropriation—were a sign of the rejection of this superficial, placebo, elitist, and exclusionary urbanism, the symbol of the neoliberal city (Jirón et al. 2021). These protests began precisely because of a rise in the cost of subway tickets and eventually led to a new Constitution. However, these movements may also reflect the same contradictions, such as the use of the bicycle as a new ideal, imposed by dominant groups while excluding others (Nikoleva and Nello-Deakin 2020; Stehlin 2019). The examples developed in this chapter show that those affected are also the most marginalized in the democratic process. On another line, the promotion of individual transportation modes—such as bicycles, but also motorcycles, and even new formal and informal individual services provided by Apps—needs to be studied carefully, to the extent that they can contribute to the weakening of the overall objective of proving a widespread and accessible transportation service for the peripheries.

The contradictions of the modernization process of mobility, and more generally the production of the city, reveal social fractures. The ongoing evolutions accelerated by the pandemic, particularly the digitization of mobility and work services and innovations for carbon–neutral transport, will therefore need careful and critical analysis to prevent them from reinforcing or translating into new social inequalities or new forms of exclusion.

## References

Alegre M (2012) Cómo se camina en Miraflores? Mallas peatonales y accesibilidad para personas con movilidad reducida en la zona central de Miraflores: impacto de las acciones realizadas y las necesidades por resolver. [Online] https://issuu.com/vilmouv/docs/informe-peatones-miraflores-final

Allen H, Cárdenas G, Pereyra LP, Sagaris L (2019) Ella se mueve segura. Un estudio sobre la seguridad personal de las mujeres y el transporte público en tres ciudades de América Latina. Books

## 12 Mobility as an Expression of the Urbicide: The Risks of Transport ...

Arteaga I, García D, Guzmán E, Mayorga J (2017) Los pasajes del BRT en Bogotá: los puentes peatonales del sistema "Transmilenio" como dispositivos que generan nueva urbanidad. Quaderns de Recerca en Urbanisme 7:142–167. [Online] https://upcommons.upc.edu/handle/2117/104822

Avellaneda P (2008) Movilidad, pobreza y exclusión social en la ciudad de Lima. Anales de Geografía de la Universidad Complutense no. 28

Avellaneda P, Lazo A (2011) Aproximación a la movilidad cotidiana en la periferia pobre de dos ciudades latinoamericanas. Los casos de Lima y Santiago de Chile. Revista Transporte y Territorio no. 4, Universidad de Buenos Aires, pp 47–58. www.rtt.filo.uba.ar/RTT00404047.pdf

Beuf A (2007) Ville durable et transport collectif: le TransMilenio à Bogotá. Annales De Géographie 657:533–547. https://doi.org/10.3917/ag.657.0533

Bielich C (2009) La guerra del centavo. Una mirada actual al transporte público en Lima Metropolitana. Instituto de Estudios Peruanos, Lima

Borsdorf A (2003) Cómo modelar el desarrollo y la dinámica de la ciudad latinoamericana. EURE (santiago) 29(86):37–49. https://doi.org/10.4067/S0250-71612003008600002

Borthagaray A, Orfeuil JO (ed) (2013) La fábrica del movimiento. Ciudad Autónoma de Buenos Aires, Café de las Ciudades, p 506

Briones I (2009) Transantiago: Un problema de información. Estudios Públicos (116)

Carrión F (2018) Urbicide, or the city's liturgical death. Oculum Ensaios 15(1):5–12. https://doi.org/10.24220/2318-0919v15n1a4103

Ciccolella P (2012) Revisitando la metrópolis latinoamericana más allá de la globalización. Revista Iberoamericana de Urbanismo no. 8

Cohen R (2006) Cidade, corpo e deficiência: percursos e discursos possíveis na experiência urbana. Universidade Federal do Rio de Janeiro, Rio de Janeiro, Programa de Estudos Interdisciplinares de Comunidades e Ecologia Social

Crossa V (2018) Luchando por un espacio en Ciudad de México. Comerciantes ambulantes y el espacio público urbano. México, El Colegio de México

Daude C, Fajardo G, Brassiolo P, Estrada R, Goytia C, Sanguinetti P, Álvarez F, Vargas J (2017) RED. 2017. Crecimiento urbano y acceso a oportunidades: un desafío para América Latina. CAF. https://scioteca.caf.com/handle/123456789/1090

Delgado M (2014) El espacio público como ideología, CATARATA, pp 120

Dextre JC, Avellaneda P (2014) Movilidad en Zonas Urbanas—Transitemos, Lima. https://transitemos.org/publicaciones-3/movilidad-en-zonas-urbanas/

Duhau E, Giglia A (2008) Las reglas del desorden: habitar la metrópoli. Siglo XXI

Errazuriz de Nevo M, Taddia A, Albero R, Pérez JE, Brennan P, Ortiz P (2017) Evolución de los sistemas de transporte público urbano en América Latina, Banco Interamericano de Desarrollo—BID

Estupiñán N, Scorcia H, Navas C, Zegras C, Rodríguez D, Vergel-Tovar E, Gakenheimer R, Azán Otero S, Vasconcellos E (2018) Transporte y Desarrollo en. América Latina 1(1):120

Figueroa O (2005) Transporte urbano y globalización: políticas y efectos en América Latina. EURE (santiago) 31(94):41–53. https://doi.org/10.4067/S0250-71612005009400003

Fleischer F, Marín K (2019) Atravesando la ciudad. La movilidad y experiencia subjetiva del espacio por las empleadas domésticas en Bogotá. EURE (Santiago) 45(135):27–47

Gamble J, Puga E (2019) Is informal transit land-oriented? investigating the links between informal transit and land-use planning in Quito, Ecuador, Working Paper WP19JG1. 2019. Lincoln Institute of Land Policy. https://www.lincolninst.edu/sites/default/files/pubfiles/gamble_wp19jg1_0.pdf

Godard X (2008) Transport artisanal, esquisse de bilan pour la mobilité durable, Conférence CODATU XIII, [Online] http://www.codatu.org/wp-content/uploads/Transport-artisanal-esquisse-de-bilan-pour-la-mobilit%C3%A9-durable-Xavier-GODARD.pdf

Gonzales E, Del Poz, JM (2012) Lima, una ciudad policéntrica. Un análisis a partir de la localización del empleo. Investigaciones Regionales 23:29–52. https://www.redalyc.org/articulo.oa?id=28924472002

Guzmán L, Oviedo D, Ardila AM (2019) La política de transporte urbano como herramienta para disminuir desigualdades sociales y mejorar la calidad de vida urbana en Latinoamérica, No. 2, CODS, Bogotá

UN-Habitat (2016) Urbanization and development: emerging futures. World Cities Report 3(4):4–51

Harvey D (1989) In: The urban experience. JHU Press

Herce M (2019) Sobre la movilidad en la ciudad: propuestas para recuperar un derecho ciudadano, vol 18. Reverté

Hidalgo D, Urbano C, Olivares C, Tinjacá N, Pérez JM, Pardo CF, Pedraza L (2020) Mapping universal access experiences for public transport in Latin America. Transp Res Rec 2674(12):79–90

Hobbs J, Baima C, Seabra R (ed) (2021) Desarrollo Orientado al Transporte: Cómo crear ciudades más compactas, conectadas y coordinadas. Monografía del BID No 841, BID, pp 332

Hoffmann M (2016) Bike Lanes are White Lanes: bicycle advocacy and urban planning, Lincoln, NE, University of Nebraska Press

Jirón P (2017) Planificación urbana y del transporte a partir de relaciones de interdependencia y movilidad del cuidado. In: ¿Quién Cuida en la Ciudad?, United Nations, pp 405–432

Jirón P, Mansilla P (2013) Atravesando la espesura de la ciudad: vida cotidiana y barreras de accesibilidad de los abitantes de la periferia urbana de Santiago de Chile. Revista De Geografía Norte Grande 56:53–74. https://doi.org/10.4067/S0718-34022013000300004

Jirón P, Imilán WA, Lange C, Mansilla P (2021) Placebo urban interventions: observing smart city narratives in Santiago de Chile. Urban Stud 58(3):601–620. https://doi.org/10.1177/0042098020943426

Lefebvre H (1972) Le Droit à la ville suivi de Espace et politique. Paris, Anthropos

Lima Cómo Vamos (2018) Encuesta sobre calidad de vida. Lima

Lima Cómo Vamos (2019) Encuesta sobre calidad de vida. Lima

Lima Cómo Vamos (2021) Informe urbano de percepción ciudadana en Lima y Callao. Lima

Lindón A (2014) El habitar la ciudad, las redes topológicas del urbanita y la figura del transeúnte. In: Sánchez D, Domínguez LA (Coords.) Identidad y espacio público. Ampliando ámbitos y prácticas, Gedisa, pp 55–76

Montero S (2017) Worlding Bogotá's Ciclovía: from urban experiment to international "Best Practice." Lat Am Perspect 44(2):111–131. https://doi.org/10.1177/0094582X16668310

Montoya-Robledo V, Escovar-Álvarez G (2020) Domestic workers' commutes in Bogotá: transportation, gender and social exclusion. Transp Res Part a: Policy and Practice 139:400–411

Moscoso M, van Laake T, Quiñones L, Pardo C, Hidalgo D (ed) (2020) Transporte urbano sostenible en América Latina: evaluaciones y recomendaciones para políticas de movilidad. Despacio, Bogotá, Colombia

Muñoz D (2013) Experiencias de viaje en Transantiago. La construcción cotidiana de un imaginario urbano hostil. Bifurcaciones 13

Nikoleva A, Nello-Deakin S (2020) Exploring velotopian urban imaginaries: where Le Corbusier meets constant? Mobilities 15(3):309–324. https://doi.org/10.1080/17450101.2019.1694300

Obregón KT, Gómez DMP (2021) El peatón como base de una movilidad urbana sostenible en Latinoamérica: una visión para construir ciudades del futuro. Boletín De Ciencias De La Tierra 50:33–38

Padeiro M, Louro A, Marques da Costa N (2019) Transit-oriented development and gentrification: a systematic review. Transp Rev 39(6):733–754. https://doi.org/10.1080/01441647.2019.1649316

Páramo P, Burbano A, Aguilar M, García-Anco E, Pari-Portillo E, Jiménez- Domínguez B, López-Aguilar R, Moyano-Díaz E, Viera E, Elgier A, Rosas G (2021) La experiencia del caminar en ciudades Latinoamericanas. Revista de Arquitectura (Bogotá) 23(1):20–33. https://doi.org/10.14718/

Pérez López R (2014) Movilidad cotidiana y accesibilidad: ser peatón en la ciudad de México. Cahiers Du CEMCA 1:3–21

Pérez López R, Landin JM (2019) Movilidad cotidiana, intermodalidad y uso de la bicicleta en dos áreas periféricas de la Zona Metropolitana del Valle de México. Cybergeo: Europ J Geograph

Pino R (2017) Ciclismo popular. Una práctica de desplazamiento y un campo de estudio por reconocer. Diseño y Sociedad 43:28–39. UAM

Prada LV (2021) Movilidad cotidiana de mujeres con discapacidad visual: un acercamiento a sus experiencias y estrategias para desplazarse en la ciudad de Bogotá. Tesis licenciatura, U.

Santo Thomas, Bogotá. Bitácora Urbano Territorial 29(3):59–68. https://doi.org/10.15446/bit ácora.v29n3.65979

Quintero-Gonzalez J (2018) Desarrollo Orientados al Transporte Sostenible (DOTS). Una prospectiva para Colombia

Ríos RA, Lleras N, Olivares C, Pardo CF (2015) Ciclo-inclusión en América Latina y el Caribe, Guía para impulsar el uso de la bicicleta, BID

Rivas ME, Suárez-Alemán A, Serebrisky T (2019) Políticas de transporte urbano en América Latina y el Caribe: dónde estamos, cómo llegamos aquí y hacia dónde vamos, Monografía del BID, pp 719

Robert J, Gouëset V, Demoraes F, Vega Centeno P, Pereyra O, Flechas AL, Lucas M, Moreno Luna C, Moreno MM, Pardo CF, Pinzón Rueda JA, Prieto G, Saenz Acosta, H. y Villar-Uribe JR (2022) Estructura urbana y condiciones de movilidad en las periferias populares de Lima y Bogotá: desafíos y método de análisis. Territorios 46. https://doi.org/10.12804/revistas.urosario.edu.co/territorios/a.9942

Román DL (2020) Movilidad en bicicleta y acceso desigual al espacio de trabajo. Caso comparativo de su uso como medio de transporte y como herramienta de reparto en el distrito de San Isidro. Tesis para obtener el grado de Licenciatura. Pontificia Universidad Católica del Perú. Lima, Perú

Román DL (2021) Movilidad en bicicleta en ciclovías de zonas distintas de Lima a través de la etnografía móvil. Ponto Urbe. Revista do núcleo de antropologia urbana da USP 28

Rosas J, Chías L (2020) Los BRT ¿nuevo paradigma de la movilidad mundial? Investigaciones Geográficas 103:1–14. https://doi.org/10.14350/rig.60045

Sagaris L, Landon P (2017) Autopistas, ciudadanía y democratización: la Costanera Norte y el Acceso Sur, Santiago de Chile (1997–2007). EURE (santiago) 43(128):127–151

Salazar C (2008) Los corredores confinados de transporte público en las metrópolis latinoamericanas: ¿una oportunidad para hacer ciudad ? En: Salazar C, Lezama JL (eds) Construir ciudad. Un análisis multidimensional para los corredores de transporte en la ciudad de México. México, El Colegio de México

Santos CD (2014) S. Corpo e mobilidade urbana: uma experiência pedestre na cidade de São Paulo (Doctoral dissertation, Universidade de São Paulo)

Sanz A, Kisters C, Montes M (2018) Sobre espejos y espejismos en el auge de la bicicleta. Transporte y Territorio 19:57–80

Sendra P, Sennett R (2021) Diseñar el desorden, Alianza Ensayo, pp 232

Sennett R (2001) Vida urbana e identidad personal: los usos del orden. Barcelona, Eds. Península

Sequera J (2015) Ciudad, espacio público y gubernamentalidad neoliberal/City, Public space and neoliberal governmentality. Urban [S.l.], n. 07 July 2015. ISSN 2174-3657. Disponible en: <http://polired.upm.es/index.php/urban/article/view/3082>

Sierra A, Ortiz D (2012) Las periferias, ¿territorios de incertidumbre? El caso de Pachacútec, Lima-Callao, Perú. Bulletin de l'Institut français d'études andines [En línea] 41(3). https://doi.org/10.4000/bifea.400

Soto P (2017) Diferencias de género en la movilidad urbana. Las experiencias de viaje de mujeres en el Metro de la Ciudad de México. Revista Transporte Y Territorio 16:127–146. https://doi.org/10.34096/rtt.i16.3606

Stehlin J (2015) Cycles of investment: bicycle infrastructure, gentrification, and the restructuring of the san francisco bay area. Environ Plan A 47(1):121–137. https://doi.org/10.1068/a130098p. [Crossref],[WebofScience®],[GoogleScholar]

Stehlin J (2019) Cyclescapes of the unequal city: bicycle infrastructure and uneven development. University of Minnesota Press, Minneapolis, p 328

Stiglich M (2021) Unplanning urban transport: Unsolicited urban highways in Lima. Environ Planning A: Econ Space 53(6). https://doi.org/10.1177/0308518X211007867

Urquiaga JV (2020) La criminalización de los 'colectiveros,' nota de blog. https://politeama.pe/2020/02/03/la-criminalizacion-de-los-colectiveros/

Valenzuela Córdova P (2020) Mujer y Transporte: La experiencia de movilidad urbana en Transporte Público de las mujeres que realizan labores reproductivas y productivas en el Gran Santiago. [Online] http://repositorio.uchile.cl/handle/2250/173852

Van Laake T, Moscoso M (2020) Movilidad activa: La promoción de la caminata y la bicicleta en América Latina. In: Moscoso M, van Laake T, Quiñones L, Pardo C, Hidalgo D (eds), Transporte urbano sostenible en América Latina: evaluaciones y recomendaciones para políticas de movilidad. Despacio, Bogotá, Colombia, pp 57–70

Vasconcellos EA (2010) Análisis de la movilidad urbana: Espacio, medio ambiente y equidad. CAF. https://scioteca.caf.com/handle/123456789/414

Vasconcellos EA, Mendonça A (2016) Observatorio de Movilidad Urbana: Informe 2015–2016 (resumen ejecutivo), CAF, pp 34. http://scioteca.caf.com/handle/123456789/981

Vega Centeno P (2017) La desigualdad invisible: el uso cotidiano de los espacios públicos en la Lima del siglo XXI. Territorios 36:23–46. https://doi.org/10.12804/revistas.urosario.edu.co/territorios/a.5097

Vega Centeno P, Dextre JC, Alegre M (2011) Inequidad y fragmentación: Movilidad y sistemas de transporte en Lima Metropolitana. En Lima_Santiago. Reestructuración y cambio metropolitan, RIL Editores, pp 289–328. http://arquitectura.pucp.edu.pe/investigacion/centro-de-documentacion/inequidad-y-fragmentacion-movilidad-y-sistemas-de-transporte-en-lima-metropolitana/

Vega Centeno P, Dammert Guardia M, Moschella P, Vilela M, Bensús V, Fernández de Córdova G, Pereyra O (2019) Las centralidades de Lima Metropolitana en el Siglo XXI. Una aproximación empírica (Vol. 1–1). Fondo Editorial PUCP. http://arquitectura.pucp.edu.pe/publicaciones/

Vergel-Tovar CE (2021) Sustainable transit and land use in Latin America and the Caribbean: a review of recent developments and research findings. In: Advances in transport policy and planning. Academic Press. https://doi.org/10.1016/bs.atpp.2021.05.001

Verlinghieri E, Venturini F (2018) Exploring the right to mobility through the 2013 mobilizations in Rio de Janeiro. J Transp Geogr 67:126–136. https://doi.org/10.1016/j.jtrangeo.2017.09.008

Villagrán PS (2020) Construcción de ciudades libres de violencia contra las mujeres. Una reflexión desde América Latina. Revista "Cuadernos Manuel Giménez Abad 7:17–26

Walker J (2017) In: The dangers of elite projection. [Online] https://humantransit.org/2017/07/the-dangers-of-elite-projection.html

**Pablo Vega Centeno** Principal Professor at the Architecture Department of the Pontifical Catholic University of Peru (PUCP) and researcher at the Center of Architecture and the City (CIAC-PUCP). Doctor in Architecture from the University Catholic of Louvain and Sociology graduate of PUCP. He specializes in urban studies, particularly mobility and public spaces, has published numerous articles in scientific journals and book chapters. Currently, he is head and coordinator of the Interdisciplinary Group on Cities and Urban Territories (INCITU-PUCP).

**Jérémy Robert** PhD in Geography from the University of Grenoble. Researcher at the Rennes 2 University, associated with the French Institute of Andean Studies in Lima, in the framework of the MODURAL project: "Sustainable mobility practices in Bogotá and Lima" (https://modural.hypotheses.org; 2020–2023). His research deals with the relationship between policies, urban services and inequalities in Latin American metropolis. He is also a member of the Interdisciplinary Research Group on Cities and Urban Territories (INCITU PUCP).

**Danae Román** Sociology graduate from the Pontifical Catholic University of Peru (PUCP) and researcher at the Interdisciplinary Research Group on Cities and Urban Territories (INCITU-PUCP). She was a teaching assistant at PUCP, and researcher in diverse projects, both academic and in the private sector, with vulnerable groups. She has participated in seminars and has published papers about urban affairs. Currently, she is part of the MODURAL project, that researches sustainable mobility practices in Latin America.

# Part IV
# Decline of the Built Environment

# Chapter 13
# The Urbanisation of Risk

**Andrew Maskrey and Allan Lavell**

**Abstract** Urban growth is normally associated with progress and human development. Today, over half the global population lives in urban centres. In high-income economies and middle-income regions such as Latin America, over 80% do. Urban development and growth, however, are often also accompanied by increased exposure to hazards, vulnerability, risk, creating conditions for disasters and even catastrophe. While many cities were founded and have expanded in hazard-exposed areas, the way they have grown has magnified risk over time. Speculative urban development has become a primary circuit for capital accumulation in all income and regional geographies. Today's cities are shaped by market forces, as opportunities for short-term gain override considerations of longer-term sustainability. Multi-dimensional poverty and lack of access to safe land and protective infrastructure have constructed every day, extensive and intensive risk for large contingents of urban dwellers. Environmental degradation of water courses, vegetation and groundwater reserves amplifies hazard. Urban planning is often little more than a veneer on this landscape of speculative development and destruction. Finally, global challenges such as anthropic climate change, the loss of biodiversity, depleting water resources and land degradation magnify risk but should be understood, not as external threats, but rather as endogenous attributes of the total urbanisation of society.

**Keywords** Disaster · Natural hazard · Climate change

---

A. Maskrey (✉)
Coalition for Disaster Resilient Infrastructure (CDRI) and Risk Nexus Initiative (RNI), Málaga, Spain
e-mail: andrew.maskrey@gmail.com

A. Lavell
Facultad Latinoamericana de Ciencias Sociales, Flacso Costa Rica and Risk Nexus Initiative (RNI), San José, Costa Rica

© The Author(s), under exclusive license to Springer Nature Switzerland AG 2023
F. Carrión Mena and P. Cepeda Pico (eds.), *Urbicide*, The Urban Book Series,
https://doi.org/10.1007/978-3-031-25304-1_13

**Terminology**

Except where otherwise cited, the following standardised definitions were adopted by the United Nations General Assembly in 2017 (UNGA 2016)

**Disaster**: A serious disruption of the functioning of a community or a society at any scale due to hazardous events interacting with conditions of exposure, vulnerability, and capacity, leading to one or more of the following: human, material, economic and environmental losses and impacts.

**Disaster risk**: The potential loss of life, injury, or destroyed or damaged assets which could occur to a system, society, or a community in a specific period of time determined probabilistically as a function of hazard, exposure, vulnerability and capacity. **Intensive disaster** risk: The risk of high-severity, mid- to low-frequency disasters, mainly associated with major hazards. **Extensive disaster** risk: The risk of low-severity, high-frequency hazardous events and disasters, mainly but not exclusively associated with highly localised hazards.

**Disaster risk reduction**: Disaster risk reduction is aimed at preventing new and reducing existing disaster risk and managing residual risk, all of which contribute to strengthening resilience and therefore to the achievement of sustainable development.

**Disaster risk management**: Disaster risk management is the application of disaster risk reduction policies and strategies to prevent new disaster risk, reduce existing disaster risk and manage residual risk, contributing to the strengthening of resilience and reduction of disaster losses.

**Exposure**: The situation of people, infrastructure, housing, production capacities and other tangible human assets located in hazard-prone areas. Measures of exposure can include the number of people or types of assets in an area. These can be combined with the specific vulnerability and capacity of the exposed elements to any particular hazard to estimate the quantitative risks associated with that hazard in the area of interest.

**Hazard**: A process, phenomenon or human activity that may cause loss of life, injury or other health impacts, property damage, social and economic disruption or environmental degradation. Hazards may be natural, anthropogenic or socio-natural in origin. **Natural hazards** are predominantly associated with natural processes and phenomena. **Anthropogenic hazards**, or human-induced hazards, are induced entirely or predominantly by human activities and choices. Several hazards are **socio-natural**, in that they are associated with a combination of natural and anthropogenic factors, including environmental degradation and climate change. Hazards may be single, sequential or combined in their origin and effects. Each hazard is characterised by its location, intensity or magnitude, frequency and probability. Biological hazards are also defined by their infectiousness or toxicity, or other characteristics of the pathogen such as dose-response, incubation period, case fatality rate and estimation of the pathogen for transmission.

**Resilience**: The ability of a system, community or society exposed to hazards to resist, absorb, accommodate, adapt to, transform and recover from the effects of a hazard in a timely and efficient manner, including through the preservation and restoration of its essential basic structures and functions through risk management.

**Vulnerability**: The conditions determined by physical, social, economic and environmental factors or processes which increase the susceptibility of an individual, a community, assets or systems to the impacts of hazards.

## 13.1 Introduction

Nearly forty years ago, *The Urbanisation of Capital* (Harvey 1985) highlighted how urban development had become a primary circuit for capital accumulation. Traditionally, economists viewed the production of urban space and infrastructure as a secondary circuit of capital accumulation that supported and underpinned activities such as industrial production, services and government. From that perspective, the production of urban space was required to facilitate the primary circuit of capital accumulation by providing housing for workers and infrastructure for the movement of raw materials and finished goods.

From different starting points, Harvey, together with other prominent urban researchers in the 1970s and 1980s (Lefebvre 1970; Castells 1972; Mingione 1981), challenged that view and proposed that urban development had emerged as a primary circuit of capital accumulation in itself. The production of urban space, for residential use, services, commerce, or leisure, had become an end, stimulated on the supply side by speculative capital investment (it was a particularly effective mechanism for money laundering), and on the demand side by a range of hard and soft strategies to manipulate and shape the use and consumption of urban space, including through the promotion of urban lifestyles, culture, amenities and landscapes. Today's cities in low and middle as well as high-income countries are fundamentally shaped and driven by market forces.

As a primary circuit of capital accumulation, urban development necessarily experiences periodic crisis of over-accumulation of capital, when an excessive and speculative supply of urban space loses contact with the limitations of real or imagined demand. For example, the sub-prime mortgage crisis of 2007–2008, exemplified a crisis of over-accumulation in the production of urban space that ultimately threatened global financial stability (Castells et al. 2012; Robinson 2014). More recently, the threat of collapse of major Chinese property developers show that the underlying risk has not been fully or adequately addressed.

The production of urban space, however, not only contains within itself the seeds of periodic crisis of capital over-accumulation. As we will show in the present chapter, it is also a driver of accelerating disaster and climate risk. The generation and accumulation of disaster and climate risk are endogenous to the logic of speculative urban development, to the extent that such risk now threatens social welfare, economic and urban sustainability.

As market forces and speculation shape cities and city regions, the economic benefits are privatised at the same time as the accumulated risks are socialised and transferred to other territories and social groups, both within and between cities. This process of risk accumulation, however, is increasingly global and systemic in character. Ultimately, contemporary urban societies are characterised by lifestyles and patterns of consumption, distribution and production that generate systemic global risks. From that perspective, the disaster and climate risks associated with drivers such as loss of biodiversity, overconsumption of soil and water resources, and the accumulation of greenhouse gases in the atmosphere should be viewed as endogenous to contemporary urbanisation rather than externalities to be protected against. As such, climate change itself cannot be regarded as an *environmental* problem or as an exogenous threat to cities. It is a humanly induced attribute of contemporary urban society, a *Frankenstein* that now menaces its own creator.

In this chapter, we will explore the evolution of urban disaster and other associated manifestations of risk at different levels. Starting with an understanding of the basic components of disaster risk, we will examine how urban risk is generated, accumulated and becomes crystallised in urban structures and systems. The evolution of risk will be examined from a historical perspective, including in epochs and regions where urbanisation was emergent and still not dominant in demographic and economic terms. As most major metropolitan and urban centres today have origins in the distant past, such an understanding is critical to make explicit the process of social construction of risk in cities and urban areas in general.

After examining the accumulation of risk in cities, we will then examine the urbanisation of risk as the dark side of the urbanisation of capital. We will trace how global disaster and climate risk is generated and accumulated systemically through the urban process to the point that it now increasingly threatens the sustainability and indeed survival of cities and urban society in general.

The conclusion of the chapter is that the seeds of *urbicide* are endogenous rather than exogenous to contemporary urban process. Urban society is germinating the seeds to its own destruction. Recognising the endogenous nature or risk, however, means that cities can, in principle, correct such processes and guarantee more secure and sustainable growth and development.

## 13.2 Disaster and Catastrophe in Cities

The concept of *urbicide* is closely related to the term *catastrophe* as used in disaster studies. *Catastrophe* refers to extreme circumstances that negatively affect human society. Its attributes include extreme loss and damage, both material and non-material, where even the structures and institutions created by society to react to emergency and disaster conditions are damaged or lost, or become inoperable (Quarantelli 1985, 2000).

Disasters and catastrophes can materialise at different scales, from the household or neighbourhood level, through events that affect whole cities, city regions and countries, to global manifestations of systemic risk such as the impacts of global warming and climate change and the recent COVID-19 pandemic (Lavell et al. 2020). Quantitative evidence indicates that disasters and catastrophes associated with a diversity of hazard events of a physical-biotic-natural, technological, and anthropogenic origin are not only recurrent but are growing in numbers and intensity as time passes (United Nations 2015a, b). While attention has conventionally been given to disaster risk in large metropolis quantitative evidence highlights rapid increases in risk accumulation in small and medium scale urban centres, repeating or even exceeding a process already followed in well-established metropolis (United Nations 2015a, b).

Historically, the destruction of Pompei by the eruption of Vesuvius in 79 A.D is one of the best know examples of a city experiencing catastrophe. The destruction of Callao in Peru in 1746 and Lisbon in 1753, associated with destructive earthquakes and tsunamis, were similarly considered major catastrophes. Importantly, they led to reflections by enlightenment philosophers such as Voltaire and Rousseau, that prefigured modern socially based, disaster studies (Garcia Acosta 2019).

Subsequently, and throughout the twentieth and twenty-first centuries, a myriad of large urban disasters has been documented. Earthquakes were the trigger for damage and destruction in San Francisco, USA (along with urban fire), in 1910, Skopje, Macedonia in 1963, Managua, Nicaragua in 1972, Guatemala City in 1976, Mexico City in 1985, Kobe, Japan in 1995, Bam, Iran in 2003, Port-o-Prince, Haiti in 2010 and Kathmandu, in Nepal in 2017, just to mention some of the most prevalent.

In terms of meteorological hazards, New Orleans, USA, was heavily damaged by Hurricane Katrina and its associated storm surge in 2005, New York, USA, by Super-Storm Sandy in 2010 and Houston, USA, by Hurricane Harvey in 2017. In some countries, new capital cities have been built in safer areas to reduce hurricane hazard exposure, for example, Belmopan in Belize built to substitute Belize City which was severely affected by Hurricane Hattie in 1961 due to its coastal location.

In terms of technological disasters, Halifax, Canada was severely impacted by the explosion of an ammunition ship in 1917 with the death of over 1000 persons. Entire neighbourhoods in Guadalajara, Mexico, were destroyed by explosions in sewerage ducts in 1992 due to infiltration of petroleum. Bhopal, India, 1984, and Chernobyl,

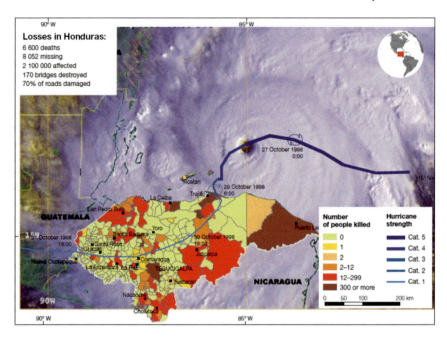

**Fig. 13.1** Mortality in Honduras associated with Hurricane Mitch, 1998 (United Nations 2011). *Sources* Image (NOAA 1998), Damage (COPECO 1998), Hurricane path (USGS 1998). Collage by UNISDR

Ukraine, 1986, are examples of nuclear and chemical technological disasters. Well-publicised fires in garment factories in Dhaka, Bangladesh in 2012 led to over 1000 deaths. Finally, smaller urban centres have been seriously damaged by wildfires in the USA, Canada and Australia over the last 3 years or by tornados as in the case of Mayfield, USA in 2021.

Generalised catastrophes are still relatively rare in major modern metropolis given that most physical hazards are spatially constrained. Even in cases such as New Orleans, USA or Tegucigalpa, Honduras, affected by Hurricanes Katrina and Mitch, respectively, severe damage was limited to quite specific areas (see Fig. 13.1). Total loss and damage are normally restricted to smaller cities where there may be generalised hazard exposure: for example, Yungay, Peru, was destroyed by an earthquake generated avalanche and rock fall in 1970 and Armero, Colombia, by a volcanic lahar in 1985. In both cases, over 20,000 deaths were registered, representing over 90% of the total population. In fact, risk generation and accumulation are now predominantly associated with rapidly growing small and intermediate urban centres, particularly in low- and medium-income countries, rather than well-established metropolitan areas. As such, *urbici*de increasingly resembles infanticide rather than death by old age.

However, other hazards, including biological vectors such as the SARS-CoV-2 virus, volcanic ash clouds and nuclear contamination, are not spatially constrained and affect extensive areas. Similarly, hazards associated with climate change, including sea level rise, and falling ground water tables, can affect entire cities in coastal and in arid or semi-arid areas, respectively.

## 13.3 The Basic Constituents of Urban Risk

### 13.3.1 Hazard

Hazard signifies a state of imminent or potential damage and loss and is finally materialised in real-world events or occurrences, such as earthquakes or hurricanes.[1]

Hazards are classified according to their origins and genesis as natural, socio-natural, technological, anthropogenic, social and biological.

Natural hazards refer to hydrometeorological, geological, geomorphological, oceanographic and other planet forming processes that are intrinsic to the natural world and where the influence of human intervention cannot be identified or measured. Seismic and volcanic activity, and tsunamis, for example, epitomise natural hazard. There is little humans can do to avoid or reduce their occurrence or intensity. All risk-reducing activity when faced with these must be through reduction of exposure or vulnerability (see Sects. 3.2 and 3.3).

Human intervention in the environment, in contrast, is intrinsic to the development or scale of so-called socio-natural hazards. Many small-scale but recurrent events, and some larger scale floods, droughts, landslides, and cases of land subsidence, for example, are socio-natural in their origins. There are few river basins in the world, whose hydrology has not been modified through the construction of dams, irrigation systems, embankments and other hydraulic infrastructure or through environmental change, such as deforestation or urbanisation. Surface water flooding in urban areas, for example, is normally associated with inadequate drainage infrastructure as is flooding on the sides of roads and highways, where adequate drainage has not been provided. The destruction of mangroves contributes to coastal erosion and magnifies storm surge; deforestation and the construction of infrastructure, such as roads, destabilises slopes in mountainous areas, increasing landslide hazard. Riverine flooding is magnified by the creation of large impermeable surfaces in urban areas that increase run-off in storms. Finally, attributes of human-induced global warming and climate change, such as sea level rise and extreme temperatures are clearly socio-natural in character. In fact, a defining feature of the Anthropocene is that due to human-induced climate change, all hydrometeorological hazards should now be considered socio-natural.

---

[1] See Terminology (above for the official UN definitions).

Technological and/or anthropogenic hazards include fires, explosions, contamination and industrial accidents that are entirely human induced. In large metropolitan areas, particularly those with heavy industry and infrastructure for refining petroleum products, technological hazard becomes woven into the urban fabric, as does fire hazard in dense and unregulated informal settlements.

Social hazards include threats associated with wars, violence and civil unrest, which while not a feature of urban society per se often find their expression in urban areas, particularly centres of political power or of strategic importance.

Biological hazards refer to vectors such as the SARS-CoV-2 virus that, being spatially fluid, are not exclusive to cities. However, such vectors are transmitted through the communication networks that link cities, regionally and globally, and through population density and concentration (Lavell et al. 2020). Thus, rapid global transmission of biological hazards can also be considered an attribute of urban society.

### 13.3.2 Exposure and the Resource-Hazard Continuum: The Beginnings of All Urban Risk

How cities are planned and managed, influences how they address the hazards to which they are exposed and hence the risks they face. However, risk in many urban areas is characterised by an *original sin* of cities being founded in hazard-exposed areas. In urban contexts, therefore, risk always begins with exposure to hazard, in other words the founding or location of a city in an area exposed to potentially damaging floods, hurricanes, tornados, tsunamis, landslides, volcanic activity or earthquakes. Many are in areas subject to multi hazard conditions.

Historically, the siting of cities, or their evolution from an initial trading post or settlement, mining community or fortification has always been influenced by a variety of locational factors, including access to essential resources such as water, food and energy, connectedness, and the existence of sheltered ports or defence from potential enemies. The location of cities has rarely been a random occurrence but reflects visible locational advantages. Consequently, many cities were founded on seaboards, along navigable rivers or in easily defendable locations, which however, were also hazard exposed.

Examples include many large cities, such as Mexico, Bogota, Quito, Lima and Santiago, Los Angeles, and San Francisco, in the Americas, Tokyo in Japan or Istanbul in Turkey located in areas of high seismic, flood and even hurricane hazard. Other cities, originally established as trading and transport hubs were in areas exposed to cyclonic wind and storm surge, such as New Orleans, USA or La Havana, Cuba. Cities, such as Naples, Italy were located near to active volcanos to exploit the rich volcanic soil for agriculture or, in the case of Dhaka, Bangladesh, in flood-prone deltas with nutrient-rich soil. The factors that made cities prosperous were nearly always influenced by location. To what extent those who founded the cities, or were responsible for their posterior expansion, were aware of the level of hazard can be debated. However, the frequent reconstruction of cities following damage and destruction in disasters implies that locational advantages normally take precedence over risk.

> **Box 13.1: The Chao Phraya River floods in Thailand**
>
> In 2011, the river Chao Phraya flooding in Thailand produced significant asset damage in the manufacturing industry around Bangkok, in the agriculture and fisheries sectors, as well as loss of life and injury. Shortages of critical industrial components were transmitted through global supply chains paralysing production in geographically discontinuous regions. The total loss of operating profit to Toyota and Honda alone was estimated at US$1.25 billion and US$1.4 billion, respectively (United Nations 2013). Flood risk in the basin had never been modelled, and the scale of the disaster took global businesses, the government and the insurance industry by surprise.
>
> However, despite the scale of these losses, very few companies decided to relocate to less hazardous areas of Thailand or to other countries. A survey conducted among Japanese businesses in Bangkok in 2012 showed that almost 80% had decided to stay in the same locations, compared to 16% that had already moved or were planning on moving to other locations in Thailand and 6% that planned to move overseas (Government of Japan 2012). For most businesses, the value creation opportunities provided by the location outweigh any contingent liabilities posed by future floods. Moreover, the rapid return on investment associated with many coastal or river locations in tourism and industry amply compensate for their owner's future disaster loss. Not so, however, for those dependent on these investments for their livelihoods and social welfare or for governments that need revenue from taxes.

In many contexts, indigenous knowledge and accumulated experience could have informed locational decisions to avoid or at least reduce hazard exposure. However, for the reasons expressed above this is generally not the case. Many major early cities simply ignored the geophysics and hydrology of their locations due to lack of information and experience by colonisers or conquerors or due to naturalist or fatalistic interpretations of disaster as *Acts of God*. Even today, with detailed geospatial information and knowledge available on most hazards, few existing cities or rapidly growing small and medium urban centres, use information on hazard exposure to inform land-use and locational decisions (see Box 4 with the case of Choloma, Honduras).

Over 90% of recorded cases of COVID-19 occurred in urban areas, highlighting a different relationship between hazard and exposure. Biological hazards have always had a devastating impact in urban areas, for example, the Black Death in fourteenth century Europe, Asia and North Africa or the Great Plague of London, which killed an estimated 20% of the population of London in the seventeenth century, the Spanish flu of 1918–21, and evolving cholera epidemics in Asia, Europe and the Americas in the nineteenth and twentieth centuries.

Unlike, geological or hydrometeorological hazards, biological hazards are spatially fluid. Thus, while an earthquake has clear spatial constraints, the SARS-CoV-2 virus rapidly flowed through transport networks from Wuhan, China, to wherever humans went locally, nationally and internationally. Cities and the relations between them in a globalised economy are the nexuses and pivots for viral contagion.

### 13.3.3 Vulnerability: Multi-dimensional Poverty and Resilience

Risk, however, is not only a function of hazard exposure but also of vulnerability, in this case understood as the susceptibility to damage of building structures, infrastructure networks and other elements of urban morphology, as well as to humans, and their livelihoods. A well-constructed seismic-resistant building will not collapse in even a powerful earthquake. A poorly constructed building will. Similarly, social groups experiencing multi-dimensional poverty are likely to experience greater economic vulnerability to disaster loss than groups with substantial reserves and covered by insurance. Vulnerability not only conditions the probability of loss in a disaster but also the capacity to absorb the loss and recover. Vulnerability is ameliorated by resilience, the inbuilt capacity and resources possessed by individuals, communities, countries, or businesses to recover post-disaster and to adapt for the future.

Vulnerability, understood from physical, economic, social, cultural and political perspectives (Wilches-Chaux 1993), has been a core concept used to interpret the social construction of risk and the differential socio-territorial risk in cities. Understanding vulnerability together with the social construction of hazard-exposure highlight that risk generation and accumulation is an endogenous characteristic or attribute of urban processes, as opposed to an externality that impacts on *normal* urban development. As we will highlight in the next section, disaster risk is never exogenous to urban process. Rather than an Act of God or of Nature it is a core attribute of urban processes. In many cities, particularly in low- and middle-income countries, layers of disaster risk are generated and accumulate on a foundation of quotidian risk associated with poverty, exclusion and inequality in access to land, infrastructure, services, health and security.

In cities, vulnerability is primarily a result of inequality in access to land, services and opportunities for productive employment and income, political participation, reciprocal recognition and mutual care, and is closely related to conditions of multi-dimensional poverty. Socio-territorial inequality is further configured by other drivers such as an absence of or ineffective urban planning and by corruption in land and property markets. These drivers increase the likelihood that those experiencing multi-dimensional poverty live in areas highly exposed to hazard events, due to lack of access to safe land for housing; in precarious and highly vulnerable buildings and are faced with a deficit (or non-existence) of protective infrastructure, such a drainage or access to emergency services (see Box 2 on Lima).

**Box 13.2: Inequality and urban disaster risk in Lima**

Lima is the capital and principal city of Peru with a population of around 10 million. With a Pacific Ocean sea-board location, it has been a primary destination for migrants from many areas of the country searching for improved life opportunities but also reflecting the marginalisation from development benefits of non-urban, rural territories in the country over several decades. Poverty, exclusion and marginalisation are dominant characteristics of a vast segment of Lima's population, including through informal work and income opportunities, lack of access to safe land due to urban land speculation processes and lack of access to urban infrastructure. The construction of social housing was never predominant in Lima but has all but disappeared over the last 30 years of neoliberal economic development. Consequently, over 90% of Lima's territorial expansion over the period 2000–2018 and beyond has taken place on the urban periphery in areas declared to be unfit for urban development by planning authorities and characterised by mudslides (huaicos) and landslides, flood and drought as well as severe earthquake hazard (multi hazard scenarios) (Fig. 13.2). The process of urban expansion reveals the multiple causal relations between a lack of access to the formal urban land market and safe land, compounded by land acquisition through urban land sharks, low socio-economic status, lack of access to basic potable water and sanitation, land degradation and urban risk construction and distribution. Figure 13.3 shows the spatial correlation between urban expansion and hazard-prone land and between this and socio-economic status, which is reflected in unregulated processes of urban housing and service provision expansion in such areas. Inequality is the basis for urban risk construction.

*Source* Chavez et al. (2022), in press.

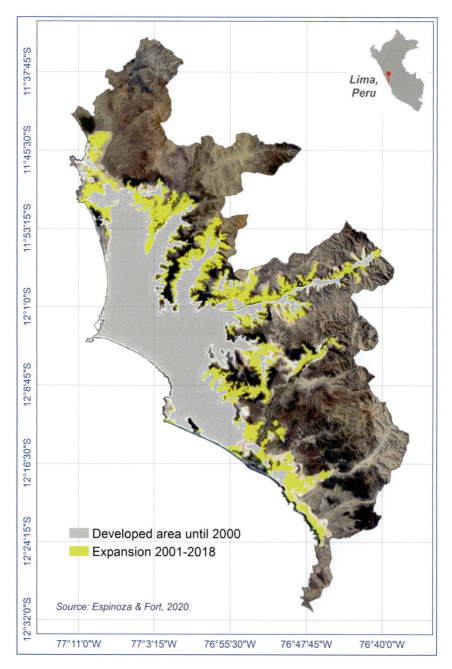

**Fig. 13.2** Urban expansion and hazard exposure in Lima, Peru. *Source* Chavez Eslava et al. (2022)

13 The Urbanisation of Risk

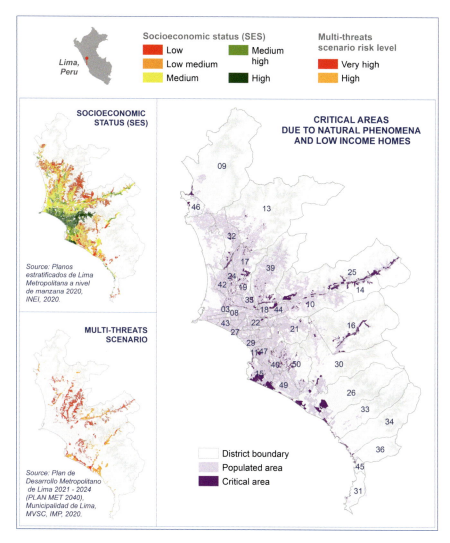

**Fig. 13.3** Spatial coincidence between natural hazard exposure and low-income homes in Lima. *Source* Chavez Eslava et al. (2022)

## 13.3.4 Disaster Risk: The Concatenation of Hazard, Exposure and Vulnerability Factors

Disaster risk is an expression of the probability of experiencing a loss when a hazard event occurs. Mathematically, it is an expression of the concatenation of a range of hazard, exposure and vulnerability factors. When any one of these factors is managed and addressed risk is reduced. A city located in an area that does not experience hazard, by definition, would not be at risk. Cities in hazard-prone areas, but whose urban morphology reduces or minimises exposure would also not be at risk; for example, cities in flood-prone areas, where buildings and infrastructure are located on higher ground. Cities exposed to hurricanes but with buildings engineered to withstand high winds would not be vulnerable and therefore would not be at risk.

As a probability, different risk layers or strata can be identified, according to the frequency of occurrence and the severity of loss.

**Every day or quotidian risk** refers to conditions such as poor health, crime, pollution, overcrowding and lack of social infrastructure that translates into low life expectancy, vulnerability to pathogens and poor health, which are intimately related to multi-dimensional poverty. As Box 3 highlights, life expectancy in areas in cities characterised by multi-dimensional poverty can be decades lower than in neighbouring high-income areas.

> **Box 13.3: Quotidian and extensive risk In Glasgow**
> In the wealthy town of Lenzie, East Dunbartonshire, Scotland, men live to an average age of 82. Just 12 km down the road, in Calton, Glasgow, they die at 54 on average. Male life expectancy in Calton is lower than in the Gaza Strip, where men can expect to live until the age of 71. Poverty, drugs and alcohol, crime and poor health are rampant. A full 44% of its inhabitants are on incapacity benefits and 37% live in workless households. The case of Calton epitomises how economic poverty, together with other poverty factors such as powerlessness, exclusion, low literacy and discrimination, translates into conditions of quotidian risk. The high levels of deprivation in Calton also coincide with extensive disaster risk. In addition to inequality and poverty, the inadequate design of the city's sewage systems and small urban watercourses mean that local flooding occurs on a regular basis. Diminishing floodplains along the Clyde River have further exacerbated flood hazard in the area (United Nations 2015a, b).

As Box 3 also shows, social groups characterised by quotidian risk often also experience **extensive risk**, which is associated with frequent, localised hazard events such as surface water flooding, storms or landslides. Extensive risk is a particular attribute of informal settlements, located on unstable slopes or in low-lying flood-prone areas and manifests as regular small-scale disasters. As Fig. 13.4 shows, most

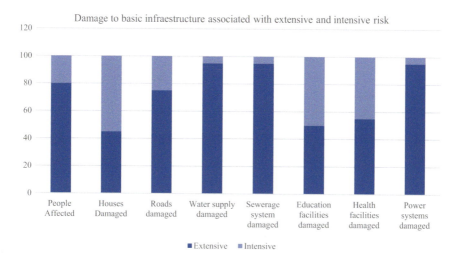

**Fig. 13.4** Intensive and extensive risk and basic infrastructure. *Source* Adapted from United Nations (2015a, b)

of the damage to housing and local infrastructure (water and sanitation, electricity, health and education facilities and roads) is associated with extensive risk.

Many cities report almost daily incidences of extensive risk associated with surface water flooding, landslides and subsidence, wind damage and other hazards. While extensive risk rarely produces a *shock* except at the neighbourhood or local level and does not cause major mortality, it represents a continuous erosion of already precarious living conditions and livelihoods, feeding back into and aggravating multi-dimensional poverty.

**Intensive risk** refers to infrequent but severe losses associated with major earthquakes, tsunamis, floods or hurricanes. Unlike extensive risk, intensive risk can lead to major structural damage and large-scale mortality and affect strategic urban infrastructure such as metropolitan scale power, transport and water systems. If exposure and vulnerability increase as cities expand then what begins as patterns of extensive risk can gradually accumulate until a layer of intensive risk is generated.

Given that infrastructure systems are interdependent, intensive risk may become **systemic risk** characterised by cascading non-linear impacts throughout the urban system. The impact of Super-storm Sandy on urban infrastructure in New York in 2012, where power failures lead to concatenated impacts in transport and water infrastructure illustrates how systemic risk becomes internalised in complex metropolitan areas (Maskrey et al. 2021).

## 13.4 The Social Construction of Urban Disaster Risk

The decision to locate a city in a given place and build it in a certain way responds, as was discussed above, to a range of social and economic imperatives, including locational advantages. However, while the founding of cities in hazard-exposed locations generates *risk by origin,* the way cities grow and evolve often generates and accumulates additional risks.

The concept that risk is socially constructed was introduced in modern disaster studies in the 1970s and 1980s (see O'Keefe et al. 1976; Hewitt 1983). In the 1990s, and through the work of the Network for Social Studies on Disaster Prevention in Latin America (LA RED) in particular, the concept gradually filtered into international policy frameworks, becoming central to the Sendai Framework for Disaster Risk Reduction: 2015–2030 (United Nations 2015b). Subsequently, it has been adopted as a method and concept for analysing the process of risk construction to inform more effective public policy (Oliver-Smith et al. 2016).

The first way in which risk is socially constructed is through increasing hazard exposure. As cities grow and expand from their original location, so does hazard exposure and hence risk. As cities which once had 100,000 people grow by orders of magnitude to 1 million or 10 million, so does the value of assets and the number of people at risk. Similarly, as countries urbanise, risk becomes concentrated in cities. While the growth of megacities has been well studied, the most rapidly growing cities in many countries are small and intermediate in scale represent an ongoing front of accumulating hazard exposure, vulnerability and risk (see Box 4).

**Box 13.4: The case of Choloma: rapid urban growth and short-term gain versus long term sustainability**

Choloma, along with Quebrada Seca and La Jutosa comprise the third largest urban concentration in Honduras today. Located in the Department of Cortes, this agglomeration forms part of the larger metropolitan area of San Pedro Sula and Puerto Cortez near to the Caribbean coast. Choloma municipality has grown from a predominantly rural population of some 36,000 in 1974 when Hurricane Fifi led to over 1500 deaths to over 300,000 today of which near to 60% is urban. Population density grew from 80 km$^2$ in 1974 to over 850 in 2018 and between 2001 and 2018 the number of housing units grew over 60%. Hurricanes are regular features and the area has been affected seriously by such events as Alma, 1966, Marco, 1969, Fifi, 1974, Kate 1999, Francelia 2017 and Eta and Iota 2020 with large-scale flooding of the Ulloa and Chamelecon rivers and its smaller tributaries—Choloma, Blanco and Cauce. The rapid urban growth of Choloma city and surrounds has been driven by the installation of garment and electronic industries promoted by the 1987 law on maquila, outsourcing industry. Principally employing female labour, this fact explains that 55 percent of the population in the area are women. However, urban growth has not been accompanied by adequate infrastructure development, environmental management, housing or urban services. The contamination of water sources and quotidian risk are prevalent. Vulnerability to disaster grows continually, converting high urban hazard exposure into urban disaster risk. Although improvements to early warning systems have occurred since Fifi and fewer deaths will occur in future events of that size, overall losses to assets and infrastructure can be expected to substantially increase. The short-term gains for the city from industrial activity and tax returns have not led to investments in risk management and reduction. Choloma, therefore, is a typical example of rapidly growing small and intermediate sized cities experiencing economic opportunities and boom, but which do not plan or assign sufficient resources to secure sustainability. The national disaster risk agency COPECO declared in 2018 that in Choloma: *"there are no legal instruments that allow declaring high-risk areas and developing a policy aimed at limiting their occupation, defining housing uses and declaring protection zones. The maintenance of the protection works is null, and void and the new ones obey the conjuncture and immediate need and not to an integral planning format"* (Fig. 13.5).

*Source* World Bank (2018).

**Fig. 13.5** Choloma 1985–2018. *Source* Google earth

Why and how certain cities in hazard-prone areas experience rapid growth can only be understood in terms of their role in national, regional and the global economies. Cities in hazard-exposed areas but with strong locational advantages grow in the context of a globalised development model that prioritises short-term economic gain at the cost of longer-term risk and unsustainability. Given that all cities ultimately participate in the global economy, risk patterns when cities in different income and regional geographies are compared, reflect the economic and territorial role each city plays to what extent the accumulation of disaster and climate risk is a necessary condition to maximise economic opportunity and the existence of capacities and resources to invest in resilience.

Urban development and expansion, therefore, do not generate or accumulate risk per se. It is the processes through which cities expand that socially construct disaster risk. Well-planned and well-managed cities can grow in ways that can reduce risk over time. However, in most contexts it is the urban process itself that generates new risks.

The underlying logic behind the urban process in most cities can be characterised in terms of the privatisation of gains and the socialisation of risks: a manifestation of moral hazard in which some gain without concern or facing consequences for risks transferred to others. This leads to a socio-territorial transfer of risk at different levels and between social groups and territories within cities and between cities in different regions. The trade-off between profiting from locational advantages and resources and the generation and transfer of risks is ultimately the foundation stone for the generation and accumulation of urban risk. Many locations that were rational and functional in their origins have over time been transformed into hazard-prone locations that now make infrastructure provision, for example, increasing difficult to manage, finance and protect.

For example, as Box 5 shows, cities built by colonial powers, often had little understanding of the characteristics of local hazard when importing structural systems and building styles from metropolitan centres. In Peru, for example, following the Spanish conquest vulnerability was systematically locked into urban morphology due a lack of understanding or contempt for vernacular building.

**Box 13.5: Vulnerable building in Lima in the colonial period**

Due to vulnerable colonial building, not suitable for a seismically active area, Lima was damaged by repeated earthquakes in the sixteenth and seventeenth century. It was not until the devastating earthquake and tsunami of 1746, that efforts were made to adapt the city to reduce seismic vulnerability. French mathematician Louis Godot was retained by Virrey Jose Antonio Manso de Velasco to propose a new urban morphology that was less vulnerable to earthquakes. Godot's meticulously calculated proposals to widen streets and reduce building height to one-story were seen as a threat to their prestige by Lima's aristocracy and ecclesiastical authorities and were turned down. However, a

# 13 The Urbanisation of Risk

> radical change in building technology was approved. All upper stories after 1746 were built using lightweight *quincha* walls (wood frame with bamboo and mud infill) which drastically reduced earthquake vulnerability. For that reason, many buildings from the second half of the eighteenth century and nineteenth century in Lima are still standing today (Walker 2009).

There are several processes, endogenous to the construction and consolidation of urban space that drive the social construction of risk. An early delimitation of such processes (Lavell 1998) included.

- Informalisation, exclusion, marginalisation and poverty.
- Socio-territorial segregation associated with factors such as social inequality, speculative market forces, corruption, non-existent or ineffective urban planning and mechanisms to provide infrastructure and services that reinforce segregation.
- Concentration, centralisation and density of the urban economy, population and political power.
- Environmental appropriation and degradation, including of land, water and forestry resources and urban hinterlands.
- Interdependence of infrastructure and other urban systems, systemic risk and lack of redundancy.
- Governance arrangements that exclude and co-opt, reinforcing socio-territorial segregation.

In the context of these six processes, hazard, exposure and vulnerability are socially constructed leading to increasing risk generation and accumulation. As dynamic and constantly changing entities, cities evolve. Urban risk is different as a concept to risk in cities given that urban processes per se are behind the former while more social or sector defined risk generating processes are behind the latter. Thus, a badly built health facility or school, for example, can exist in semi-rural or rural areas as well as in cities and be explained by the same causal factors such as cost, corruption or failed risk analysis. But risk due to urban socio-territorial segregation cannot be reproduced in rural areas, nor can that associated with urban density and concentration of critical functions and infrastructure (see Lavell 1998).

To illustrate the operation of the risk drivers listed above, we will examine in detail two of the mechanisms through which urban risk is generated and accumulated: increasing hazard exposure and socio-territorial inequality.

## 13.4.1 Increasing Hazard Exposure

As we have discussed, hazard exposure is a primary determinant of risk in cities. However, as urban functions change and the city grows, even locations with low hazard exposure can accumulate risk. Urban growth is often consistent with the occupation of new hazard-exposed areas and at the same time generates new hazards, compounding risk. These include technological hazards associated with gas and petroleum installations and industry, through to air and water pollution and the socio-natural hazards generated by urban expansion.

There are varied processes through which urban expansion generates or aggravates socio-natural hazards (Mitchell 1999). In the case of Dhaka, Bangladesh, (see Box 6), water bodies are often infilled for speculative residential or industrial development. However, this reduces the drainage capacity of the city and increases flood risk, particularly for low-lying low-income and informal settlements as well as earthquake risk due to soil liquefaction.

> **Box 13.6: Urban growth generating hazard in Dhaka, Bangladesh**
>
> The 1897 Assam earthquake (also known as the Great Indian Earthquake), one of the largest ever recorded in South Asia, caused extensive damage to the city's buildings and infrastructure. At that time, Dhaka's metropolitan population was less than 100,000. Now, it is estimated to be around 22 million. However, it is not only the 200-fold increase in exposed population that has led to Dhaka's current level of earthquake risk. The city has also been unable to address the processes that shape and accumulate that risk over time.
>
> Many areas surrounding central Dhaka are flood prone during the rainy season and until recently were occupied by natural water bodies and drains, vital to the regulation of floods. Land-use planning instruments such as the Dhaka Metropolitan Development Plan restrict development in many of these areas. Despite the Plan, these areas are still being rapidly urbanised through private- and public-sector projects.
>
> Large areas of Dhaka are highly susceptible to liquefaction during earthquakes, and many have been used as construction sites for buildings and infrastructure in recent decades. Destroying retention ponds and drains increases risks of seasonal flooding just as building in drained wetlands increases earthquake risk. During an earthquake, sands and silts can liquefy to the point where the soil no longer supports the weight of buildings and infrastructure, which may subsequently collapse or suffer heavy damage. Dhaka's wetlands, drained and filled with sand for housing development, are prime candidates for liquefaction. With little contemporary experience of earthquakes, Dhaka is vulnerable and ill-prepared. The older part of the city is home to densely populated, multi-storey, unreinforced brick buildings predisposed to heavy damage in a strong earthquake. And despite guidelines for earthquake-resistant construction, faulty design and poor-quality materials and workmanship mean that many modern reinforced concrete buildings are also vulnerable (Rahman 2010).

Similarly, the vertiginous growth of impermeable asphalted surfaces in cities, through the construction of roads, car parks and build areas, reduces infiltration and increases run-off during storms and flood events. Extreme run-off then overwhelms what are often inadequate draining systems, increasing flood risk particularly in low-lying informal settlements. Similarly, building on landfills and reclaimed coastal land, may increase seismic risk in earthquake prone areas.

Socio-natural hazard may also be magnified as cities expand into their hinterlands, generating new hazards through processes such as deforestation, over-exploitation of surface and ground water resources or destroying natural coastal protection such as mangroves. As Box 7 highlights, unregulated extraction of ground water can cause cities to sink, drastically increasing flood risk.

---

**Box 13.7: Jakarta is sinking**

Jakarta is rapidly sinking due primarily to an excessive level of groundwater extraction, given that in the colonial period few areas of the city were connected to a piped water network. The problem has been exacerbated by a failure to replenish groundwater at a sufficient rate. In 1995 industries, hotels and private consumers in Jakarta were drawing more than 300 million cubic metres of groundwater a year, which was about three-times the rate that the aquifers were being replenished.

Since then, the problem of non-replenishment has increased as massive buildings and concrete take over natural drainage sites, green areas and open spaces and human waste and rubbish clogs waterways. Freshwater floods now surge up from the ground during the rainy season and rainwater flows straight into the sea, making no contribution to groundwater. The increasing extraction levels also contribute to rising levels of groundwater salinisation caused by seawater intrusion, which damages the piped network through corrosion.

Groundwater extraction is one of the main causes of land subsidence in Jakarta, which has now reached critical levels. As more luxury hotels, apartments and shopping malls are constructed, the city gets heavier, which combined with groundwater extraction and the creation of vacuums in the aquifer, pushes the city downwards. Many such constructions are reported to grossly exceed their building floor coefficients, meaning they are too big and therefore too heavy for the land space on which they are built.

As Jakarta slowly sinks, flooding is becoming worse and more frequent. Perhaps even more catastrophic is the estimation that in less than 20 years' time, the sea will permanently flood the first two to four kilometres of the coastal area of Jakarta, rendering almost one third of Jakarta uninhabitable and causing the displacement of millions of people. Due to these reasons, the government is now undertaking a relocation of the country's capital to another site with World Bank support.

(Colbran 2009)

## 13.4.2 Socio-Territorial Inequality

Clearly, while urban expansion and the changing dynamics of land-use and the urban economy increase hazard exposure, the subsequent accumulation of risk is not evenly distributed in social or territorial terms. Risk accumulation is normally a reflection of socio-territorial inequality at different scales, both between cities in different regions and within cities.

In terms of socio-territorial inequality within urban areas, most cities in low- and middle-income countries have relied on informal settlement to house low-income social groups. Informal settlements are generally located in marginal land with low value. Numerous case studies over the last 40 years have highlighted how much of that land is in hazard-exposed areas within and around cities. These include low-lying areas prone to flooding and unstable slopes prone to landslides, just to mention two of the most frequent scenarios. Hazard exposure is then further aggravated by the lack of protective infrastructure. Much surface water flooding in informal settlements is associated with a lack of, or inadequate, drainage infrastructure. At the same time, housing may be poorly built and precarious, highly vulnerable to earthquake or wind. At the same time, the same settlements often lack access to emergency services or social infrastructure such as health facilities (see Box 2 on Lima).

Consequently, risk accumulates in informal settlements, with a sequenced layering of quotidian, extensive and intensive risk. Quotidian risk underpins extensive risk to frequently occurring, low intensity hazards such as surface water flooding, landslides and storms that in turn further aggravate multi-dimensional poverty. As informal settlements grow and multiply, they gradually internalise and accumulate intensive risk to infrequent, major hazards such as powerful earthquakes or tropical cyclones. Intensive risk, in other words, may materialise decades of accumulated everyday and extensive risk.

As informal settlements consolidate over time, building quality may increase and infrastructure will eventually be installed, including protective infrastructure to stabilise slopes or protect against flooding. In this way, extensive risk may be managed and reduced over time, while however, exposure to intensive risk continues to increase.

Manifestations of socio-territorial inequality in cities, however, are not restricted to informal settlements. In many cities in low-and middle-income countries, the traditional inner-city areas where well-to-do families once lived have gone through dramatic processes of change over the last century. As wealthy families moved to new and modern suburbs, their original city houses were often subdivided and turned into tenements for multiple low-income households, often combining residence with informal economic activities. The resulting process which combines rapid deterioration through over-use of the buildings structure with lack of maintenance due to low-rents and unclear legal responsibilities leads to an accumulation of structural vulnerability to earthquake, fire, wind and rain. In Lima and Callao, inner-city areas that were once structurally sound now concentrate earthquake risk (Maskrey and

Romero 1986). Similar scenarios play out in Havana Centro, Cairo or Mumbai as vulnerability accumulates over time.

Socio-territorial inequality also plays out between different income and regional geographies given that cities have widely different capacities to manage and reduce risks and to adapt to a hazard-exposed environment. In general terms, as cities enjoy economic success the standards or buildings and infrastructure improves over time, reducing vulnerability.

In high-income countries, local and regional governments may be able to invest in major protective infrastructure to reduce exposure and vulnerability. For example, the Thames Barrier in London was a multi-billion-pound investment to reduce the exposure of the city to a potentially lethal combination of spring tides and high-water levels on the Thames. Many Japanese cities have invested heavily in seawalls to protect themselves from tsunamis (though the 2011 East Japan earthquake and tsunami showed the limitation of that approach).

Investments in protective infrastructure of this kind, however, is generally unthinkable in low-income countries. Annual local government spending per inhabitant in high-income countries is often orders of magnitude greater than in low-income countries and reflects the capacity to invest in risk-reducing infrastructure, effective land-use planning and building standards and emergency services. Economic capacity is normally reflected in differential institutional quality and the technical capacity to plan and manage urban growth in a way that reduces rather than increases risk.

The differential capacity to invest in resilience, however, veils an ongoing process of risk transfer from high to low- and middle-income countries. Some cities in high-income countries, may have been able to strengthen resilience through transferring their risks to cities in low- and middle-income countries, for example, through outsourcing production to cities with low-labour costs and unregulated industrial facilities. The growth of the ready-made garment industry in countries such as Bangladesh or Honduras drives an accumulation of risk in these cities, whereas the capital accumulated through these activities, may be invested in strengthening resilience elsewhere (see Box 4, Choloma).

Another factor that contributes to socio-territorial inequality is ineffective or distorted systems for urban planning and building control. While most countries apparently have planning laws and norms in place to manage land use and development, these systems often represent little more than a thin veneer on the operation of the market forces that really shape urban growth and condition urban morphology. While urban and regional planning, in theory should lead to improved risk management and reduction and increasing urban sustainability and quality, changes in land-use and territorial occupation in most countries reflect not planning but rather the logic of speculative development, leading as we have seen not only to the urbanisation of capital but also to the urbanisation of risk.

Hazard levels are not adequately considered in speculative urban development, for commercial, industrial or residential purposes. Often the short-term profit that can be made through speculating with urban land values is so great that the risks associated with a 1-in-20-year flood event can be largely discounted. At the same time, there is little incentive for speculative developers, to consider hazard levels given that, once the development has been sold on, then so have the internalised risks (Johnson et al. 2012).

Another issue is that the institutional frameworks for disaster risk management and reduction that now exist in most countries and cities as a specialised sector have had little or no influence on the process of speculative urban development (Lavell and Maskrey 2014). Conceptually, disaster risk reduction remains rooted in a concept of *protecting development from disasters,* in other words treating risk as an externality and exogenous to urban process. Consequently, on an instrumental and operational level, the disaster risk reduction sector concentrates on disaster preparedness and emergency response, rather than on addressing risk generation processes. It has neither the political leverage nor technical capacities to intervene in the decisions that shape cities, such as land-use changes and strategic infrastructure projects. Ultimately, if risk is endogenous to urban process, then its management must be in the sectors and institutions involved in planning and managing the city and not in an exotic disaster risk reduction or management sector.

## 13.5 Urbanisation as a Driver of Systemic Risk

Through unmanaged risk accumulation, cities like Jakarta effectively start to consume themselves, as the costs or risk start to outweigh the original locational benefits of the city. The decision of the Government of Indonesia to move the capital to a new location reflects its inability to address the underlying drivers of urban risk: a clear recognition of *urbicide*. While Jakarta may represent an extreme case, similar processes are underway in many cities in low- and middle-income countries.

However, the urbanisation of risk has further ramifications: systemic causes and effects that now threaten the sustainability of the global urban society as a whole. The urbanisation of capital as described in the introduction to this chapter is fully internalised in the global economy. As such, the construction of urban space has become a principal circuit for capital accumulation at the global level.

In 1970, Henri Lefebvre put forward the hypothesis of the *total urbanisation of society,* in which the urban–rural dichotomy would lose meaning. Fifty years later, a totally urbanised global society, in all its territorial, economic, social, cultural and political dimensions, now shapes and moulds global demand for land, water, food, energy and other resources. Contemporary lifestyles are now predominantly urban in all countries of the world. Global demand for concrete and steel for building and infrastructure, for energy for transport, heating and cooling and for the resources required to produce food, clothing and consumables can therefore only be understood as endogenous attributes of a totally urbanised global society, that even Lefebvre might have had difficulty imagining.

Before the industrial revolution, cities were largely dependent on their hinterlands for natural resources, such as water, food and firewood (for energy). This placed limits not only on urban expansion but also on the global impact of urbanisation.

In contrast, today's totally urbanised global society depends on an over-exploitation of natural resources on a global scale. From that perspective, problems such as global warming (from carbon intensive energy use) the over-exploitation and exhaustion of water and marine resources, deforestation, land degradation and biodiversity loss (to produce food, in particular animal protein) are not autonomous *environmental* problems or even less *rural* problems. They are endogenous attributes of the urbanisation of capital and hence are global drivers of the urbanisation of risk.

As such, the challenge of adapting cities to climate change, cannot be reduced to that of strengthening the resilience of urban areas in coastal locations, exposed to sea level rise. It is rather that the contemporary urban process itself *systemically* generates risks that negatively feeds back into unsustainable cities. The logic of privatising benefits and socialising risk, that characterises the global economy finds one of its maximum expressions in the urban process. The degradation of marine resources through over-fishing and rising water temperatures, the exhaustion of groundwater reserves through over-exploitation or the deforestation of the Amazon basin for soya production, are therefore manifestations of systemic risk endogenous to and driven by a totally urbanised global society.

These risks are increasingly concatenated, with systemic, non-linear and cascading impacts on urban areas. Cities sink due to uncontrolled groundwater extraction at the same time as they are threatened with rising sea levels and increased river flooding due to the degradation of catchments. Extreme temperatures in urban areas increase the demand for energy for cooling from already deficient infrastructure which in turn lead to increasing carbon emissions.

This accumulation of systemic risk endogenous to a totally urbanised global society now affects the urban system, with non-linear and cascading ripple effects at different spatial and temporal scales that are difficult to model and predict (Maskrey et al. 2021). While many large metropolises today consider themselves to be *victims* of their original location in hazard-exposed areas and may have developed capacities to manage most extensive and some intensive events, they now face rapidly accelerating, concatenated risks for which they are not prepared.

At the same time, systemic risk has differential impacts on cities with different levels and capacities for resilience, in other words, there is a social and territorial transfer of risk to those cities and countries that are less able to invest in resilience. Increasing levels of biological hazards, together with consequences of climate change such as urban heating and sea level rise point to increasingly complex future risk scenarios, not only in major metropolitan areas but also in rapidly growing small and medium size urban centres. In the best of all possible worlds, these more recently growing cities should have the foresight to learn from history and to develop in a way that allows them to manage and reduces risk. Unfortunately, these cities are often those with the weakest capacities for urban planning and management and thus seem condemned to repeat the errors of history.

If the urbanisation of risk is understood as an attribute of the urbanisation of capital, then addressing the existential nature of systemic risk, whether human-induced climate change, soil degradation, loss of biodiversity and exhaustion of fresh water supplies, requires a radical rethink of our totally urbanised global society. If *urbicide* is understood as the endogenous generation and accumulation of risk in that society, then that risk now poses an existential threat to society itself. Before cities sleepwalk into their own future nightmares a radical transformation of urban society is required.

# References

Castells M (1972) La Question Urbaine. Francois Maspero, Paris
Castells M, Caraca J, Cardosa G (ed) (2012) Aftermath: the cultures of the economic crisis. Oxford University Press
Chavez Eslava A, Salas CB, Lavell A, Sandoval DM (2022) (in press) La desigualdad y la construcción social del riesgo de desastre urbano: de los desencadenantes de amenazas físicas a bióticas. El caso de Lima, Perú. In: Lavell A, Eslava AC (eds) Miradas sobre la desigualdad, el riesgo y la resiliencia en tres ciudades de América Latina: un acercamiento desde la construcción social y la coproducción del conocimiento. FLACSO Costa Rica. San Jose
Colbran N (2009) Will Jakarta be the next Atlantis? Excessive groundwater use resulting from a failing piped water network. Law Environ Dev J 5(1):18. Available at http://www.lead-journal.org/content/09018.pdf
COPECO (2018) Bases para la incorporación de la Gestión del Riesgo en la planificación del municipio de Choloma. Informe IV, Entorno Regional y Municipal, Escenarios de Riesgo Municipal, Análisis de Actores y Propuestas de Intervención (prepared by Gomez FR)
Garcia Acosta V (2019) Desastres Históricos y Secuelas Fecundas. Academia Mexicana de la Historia
Government of Japan (2012) Floods in Thailand that caused a significant impact on trade environment, etc. of neighbouring nations/regions, including Japan. Section 3
Harvey D (1985) The urbanization of capital. Basil Blackwell, Oxford
Hewitt K (1983) Interpretations of calamity from the viewpoint of human ecology. Geograp Rev 74(2)
Johnson C, Adelekan I, Bosher L, Jabeen H, Kataria S, Wijitbusaba A, Zerjav B (2012) Private sector investment decisions in building and construction: increasing, managing and transferring risks. Background Paper prepared for the 2013 Global Assessment Report on Disaster Risk Reduction. UNISDR, Geneva, Switzerland. http://www.preventionweb.net/gar
Lavell A (1998) Gestión de Riesgos Ambientales Urbanos. FLACSO-LA RED. Mimeo, San José
Lavell A (2004) Concepto y Práctica de la Gestión Local del Riesgo. CEPREDENAC-PNUD, Panamá
Lavell A, Maskrey A (2014) The future of disaster risk management, environmental hazards. 13(4)
Lavell A, Mansilla E, Maskrey A, Ramirez F (2020) The social construction of the COVID-19 pandemic: disaster, risk accumulation and public policy. LA RED & Risk Nexus Initiative
Lefebvre H (1970) La Révolution Urbaine. Editions Gallimard
Maskrey A, Romero G (1986) Urbanización y Vulnerabilidad Sísmica en Lima Metropolitana. Centro de Estudios y Prevención de Desastres (PREDES), Lima, Perú
Maskrey A, Jain G, Lavell A (2021) The social construction of systemic risk: towards an actionable framework for risk governance. UNDP, New York
Mingione E (1981) Social inequality and the city. Basil Blackwell, Oxford
Mitchell J (1999) (ed) Crucibles of hazards: mega cities and disasters in transition. United Nations University Press, Tokyo
O'Keefe P, Westgate K, Wisner B (1976) Taking the naturalness out of natural disaster. Nature 260
Oliver-Smith A, Alcantarra I, Burton I, Lavell A (2016) Forensic disaster research. International Council for Science (ICSU)

Quarantelli E (1985) What is disaster? The need for clarity in definition and concepts in research. Article #177. Disaster Research Center, University of Delaware
Quarantelli E (2000) Emergencies, disasters and catastrophes are different phenomena. Preliminary Document 304. Disaster Research Centre, Delaware
Rahman A (2010) Dhaka's peripheral development and vulnerability to earthquake liquefaction effects. Asian Disaster Manage News 16(1):6–8
Robinson (2014) Global capitalism and the crisis of humanity. Cambridge University Press
United Nations (2011) Revealing risk: redefining development. Global Assessment Report on Disaster Risk Reduction, UNISDR, Geneva, Switzerland
United Nations (2013) From shared risk to shared value: the business case for disaster risk reduction. Global Assessment Report on Disaster Risk Reduction, UNISDR, Geneva, Switzerland
United Nations (2015a) Making development sustainable: the future of disaster risk management. Global Assessment Report on Disaster Risk Reduction, UNISDR, Geneva, Switzerland
United Nations (2015b) Sendai framework for action on disaster risk reduction: 2015b–2030. Geneva, Switzerland
United Nations General Assembly (2016) Report of the open-ended intergovernmental expert working group on indicators and terminology relating to disaster risk reduction, vol A/71/644
Walker C (2009) Diálogos con el Perú: Ensayos de Historia. Fondo Editorial de San Marcos
Wilches-Chaux G (1993) La Vulnerabilidad Global. In: Maskrey A (ed) Los Desastres No Son Naturales. Tercer Mundo Editores, Bogota
World Bank (2018) Choloma, Honduras, FORIN research paper. World Bank (prepared by Lavell A, Brenes A)


**Andrew Maskrey** is an urban and regional planner, who worked since 1979 in Peru on urban development and disaster risk issues. In 1992 he was founder of LA RED and between 2007 and 2017 was coordinator of the UN Global Assessment Report on Disaster Risk Reduction. He currently heads the Risk Nexus Initiative.

**Allan Lavell** founding member of the Network for the Social Study of Disaster Prevention-LA RED; and independant consultant. Ph.D. in Geography, London School of Economics. Published over 150 items on disaster risk and given 148 international conferences in five continents. U.N. Sasakawa Laureate, 2015.

# Chapter 14
# Urbicide or Suicide? Shaping Environmental Risk in an Urban Growth Context: The Example of Quito City (Ecuador)

**Jose M. Marrero, Hugo Yepes, Paco Salazar, and Sylvana Lara**

**Abstract** The present article uses the literal definition of urbicide, that is violence or destruction of a city to evaluate the dimension of trigger effects of natural hazards in the process. Some natural hazards are directly influenced by people and their interventions, above all in urban environments where the man-made changes appear to be broadly much more substantial and wider reaching on a local scale. In this regard, the city of Quito has been studied in detail here to attempt to comprehend the environmental risks faced by mega cities and taking advantage of the recently developed tool of the Multipurpose and Multihazard Exposure Model (MMEM-DMQ). This tool allows us to quantify the degree of exposure and vulnerability of a city to floods, the natural hazard most commonly affected by human interventions on the environment. Our conclusions evidence that man-made interventions at all levels of human society have created extremely complex high-risk situations that are difficult to solve. In the case of Quito, the environmental risk is such that "urbicide" falls short as a description, with the authors preferring to call it a "kamikaze city" on account of the enormous risks that have been taken with no possible short-term expectation of solution.

**Keywords** Natural disaster · Man-made disaster · Urban planning · Urbicidio

J. M. Marrero (✉)
Repensar, Consulting, Quito, Ecuador
e-mail: josemarllin@gmail.com

J. M. Marrero · H. Yepes
Mayor's Advisor, Metropolitan District of Quito (DMQ), Quito, Ecuador

P. Salazar
Inteligentarium, Quito, Ecuador

S. Lara
Metropolitan Directorate of Risk Management, Metropolitan District of Quito (DMQ), Quito, Ecuador

© The Author(s), under exclusive license to Springer Nature Switzerland AG 2023
F. Carrión Mena and P. Cepeda Pico (eds.), *Urbicide*, The Urban Book Series,
https://doi.org/10.1007/978-3-031-25304-1_14

## 14.1 Introduction. The Urbicide Concept from the Point of View of the Environmental Risk

Do cities really die? If so, do they die outright or is the process slow and selective, sometimes imperceptible, taking one to two generations? Or is the annihilation violent and overnight? Or the result of changes in trends of what our understanding of a city is, and that are not accepted to? And who kills cities and how do they do it? These are some of the many questions that we can reflect upon here where we are invited collectively to reflect upon what we understand by "urbicide" from our different perspectives. And, although "urbicide" is open to many different interpretations (Campbell et al. 2007), in this chapter, we are using it in its most literal sense of violence focussed on the destruction of a city, to specifically analyse how natural hazards cause deterioration or destruction and how human beings contribute towards shaping the premises for "urbicide", as conspiring agents and victims of the destruction, at one and the same time.

Environmental risk is usually conceived of as deriving from three factors, that is, natural hazards, vulnerability and exposure (or value) (Fournier d'Albe 1979). The first, that is natural hazards, is invariant when we refer to geological processes such as earthquakes, volcanic eruptions and tsunamis. By invariant, we mean that these respond to their own cycles and origins that are not man made nor respond to or are influenced by the presence of human beings on the planet, rather converting them into passive elements that receive their impact (Hewitt 1983). However, when considering ocean-atmospheric phenomena, there is growing evidence of human activities influencing the same that would demand a broadening of the concept environmental risk (Rebotier 2012). But, in any case, when a natural hazard of great magnitude takes place, our capacity to contain it or manage it is nil. This leads us to understand that the other two factors, vulnerability and exposure, are what are going to help us to answer our questions.

Large natural hazards that have their origin in the core of the Earth cause whole cities to be wiped out. This can be defined as a "long-term suicide" rather than an urbicide. Nobody perpetrates the murder of the city. Its destruction can rather be considered to be the result of voluntary exposure, conscious or not, to the possible lethal and catastrophic effect of a large natural hazard event. The suicide begins from the moment the site is chosen for a human settlement and continues through its process of development towards consolidation. In the process, the population may lose sight of the natural hazard or not fully understand it, or perhaps, they may simply ignore the natural hazard until they are affected by its impact. The maximum exposure resulting and the enormous energy concentrated in events such as these will make the city highly vulnerable and fragile, with the clear probability that it may not be able to withstand the impact and die.

We have only to look at Pompeii to find a prime example of this situation. Pompeii was seriously affected by an earthquake (Cubellis et al. 2007) before being totally destroyed by the paroxysmal eruption of Vesuvius in 79 AD (Scandone et al. 2019). Its total destruction was the result of its high exposure. The final destruction was violent and almost instantaneous as the result of intensely hot pyroclastic flows that covered the 10 km between the city and the crater in scarcely a few minutes. It was, as we know, so violent and instantaneous indeed that the expressions of horror on the locals' faces were forever cast in the pyroclastic materials for all eternities to see. Two thousand years later, Vesuvius is still a highly active and dangerous volcano. There were significant eruptions again in 472 and 1631 AD (Guidoboni 2008) and in 1906 (Chester et al. 2015). But even in the face of all that evidence, the present towns of Pompei and Herculano are still within the 10 km radius of the volcano and not only them but also part of the present-day Naples (Pesaresi et al. 2008). The high-risk area (vulnerability to pyroclastic flows) is now populated by 550,000 people (Rolandi 2010) and could be defined as a "kamikaze" urban development since the towns have moved rapidly (in volcanic terms) in the direction of the threat, instead of fleeing in the opposite direction, thereby exposing themselves, on Vesuvius' slopes and in its ravines, to the next large to major eruption (Zuccaro and De Gregroio 2013).

Another example of a city affected by urbicide due to volcanic hazard is St. Pierre on the island of Martinique that was totally destroyed by pyroclastic flows from the Mount Pelée volcano in 1902, an event that caused the death of some 28,000 residents (Chrètien and Brousse 1989). Yet another is Armero in Colombia, where massive lahars triggered by the eruption of the Nevado de Ruiz volcano killed over 23,000 people in 1985 (Voight 1990). Armero is now nothing more than a cemetery, and in St. Pierre, known in the twentieth century as the "Paris in the Caribbean", there are at present 4000 inhabitants.

There have been cities that have died as a result of earthquakes too. Dotted around the world's seismic belts (Sherman and Zlogodukhova 2011), it is common to find toponyms that make reference to cities that were abandoned and re-established in other places after massive earthquakes. This is the case, for example, of Santiago de los Caballeros de Guatemala, the first capital of the General Captaincy of Guatemala through to 1776; only three years after, it was obliterated by an earthquake and abandoned for the new settlement of Nueva Guatemala de la Asunción, now the capital (Peraldo Huertas and Montero Pohly 1996). It is also the case of Riobamba Antigua in Ecuador, thus named to differentiate it from the present city of Riobamba that was moved 14 km eastward after the earthquake in 1797 or of Pelileo Nuevo, the city that replaced Pelileo Viejo after the earthquake of 1949, both in the Interandean Valley of Ecuador (Beauval et al. 2010). The locations of these sites, in close proximity to active geological faults or on soils with high ground motion amplification capacity, again indicate that exposure to hazard was what led to their destruction.

At the opposite end of the spectrum of destruction, we have the small, frequently repeating climatic events that do not destroy the city as such, but that gradually deteriorate the settlements that are close to the riverbeds, river slopes or ravines. In these cases, it is human intervention that participates in the destruction process breaking the environmental equilibrium. This chapter concentrates on this type of hazard in the city of Quito to show how repeated processes caused by incorrect human intervention may produce progressive urbicide or a long-term suicide. To this end, we have used the recent developed tool known as the Multipurpose and Multihazard Exposure Model of the Metropolitan District of Quito (MMEM-DMQ) described hereinafter (Yepes and Marrero 2021). The tool was used to produce a flood and debris flow risk assessment of Quito, given that both these hazards are strongly influenced by the significant modifications to the hydrographic network that were carried out in order to further develop the city, with drastic alterations of land use. These two factors have affected the dynamics of natural processes such as patterns of rainfall, the absorption of the water by the soil and the runoff into the hydrographic network altered as a result of the drain system built. The data collected as a result show that all the different levels of society have played a part in the environmental risk that has been shaped and that this is not a result, as is often proposed, of poorer and illegal settlements of marginal segments of the population.

## 14.2 Quito City

The city of Quito in the Metropolitan District of Quito (DMQ in Spanish) represents the most important urban continuum area. The city is, at present, home to a population of around 2 million inhabitants in a densely built-up area of 197.33 km$^2$ (Fig. 14.1). It is surrounded by other important centres in a discontinuous urban system that, nevertheless, are linked socio-economically with incoming and outgoing flows of people and merchandise on a daily basis, even beyond the administrative boundaries of the DMQ (Salazar et al. 2021).

The city extends some 50 km further SSW-NNE along the Interandean Valley. It was built on an active tectonic thrust fault with a deformation rate of between 1 and 4.5 mm/year (Beauval et al. 2018). The accumulated tectonic deformation along the fault for the late Quaternary Period (a geological period that covers hundreds of thousands of years) produced a 500 m elevation of the plateau of Quito over the trough of the Interandean Valley, producing what is known as a piggy-back basin (Yepes et al. 2016). Besides the evident activity and thus danger of the geological fault, this topographic elevation increases the possibilities of avalanches and erosion at the base of ravines and slopes.

The city is bordered to the West by the active volcanoes of Atacazo, Pichincha and Pululahua, together with the extinct Casitahua, whereas to the East, it is limited by non-volcanic hills that are the morphological evidence of the Quito Fault. There may be as much as 1200 m (shortest horizontal distance) between the lowest point of the Quito piggy-back basin and the heads of the micro-basins on the highest reaches of the western volcanic massifs (Fig. 14.1). This means that the potential for generation of debris flows is greater on the western edge of the city. Moreover, the materials emitted by the explosive activity of Pululahua and, in the past, Casitahua have filled, with extremely young pyroclastic deposits, the former beds of the ravines and rivers in the extreme North of the city being actually highly susceptible to erosion and deepening due to the effect of runoff rainwater.

14 Urbicide or Suicide? Shaping Environmental Risk in an Urban Growth ...

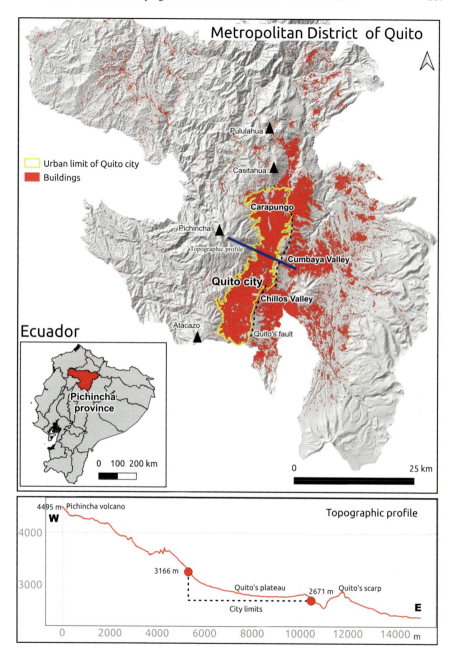

**Fig. 14.1** City of Quito is within the Metropolitan District of Quito, in the province of Pichincha, Ecuador. *Source* Own elaboration

## 14.2.1 Growth of the City of Quito

The area that is the site of the present city of Quito was an important hub of paths between the Amazon, the Andes and the Coastline. The first settlements date back to the Formative Period in the region of Ecuador (6000–2500 BP) as has been evidenced from the archaeological discoveries at Alangasi, Cotocollao and Chillogallo, among others. These remains not only indicate the first movements to build socio-political entities in the valley of Quito and its surroundings (Lozano Castro 1991, 25), but also a period of deepening of knowledge of environmental control with an agriculture system adapted to specific ecological floors. Moreover, the first relevant changes in the environment took place in the Formative Period, but they ended abruptly due to the Pululahua eruption (Torres 2017). It was in the Integration Period (500–1532 BP) when settlement in the area was recovered. Then, in the sixteenth century, the area was occupied by the Spaniards who were interested here, as in other cities in Latin America, in strategically exerting control over the conquered territory (Portes and Walton 1976).

In more recent times, Quito has undergone a few key moments that have shaped the present city. The first of these (1895–1910) was marked by profound social changes in the structure of agriculture and the urban reality as a result of the introduction of Capitalism and the growth of a more bourgeois society in the country (Carrión and Erazo Espinosa 2012). This did not produce an expansion of the urban area of Quito but rather an intensification of the city's problems, such as the overcrowding and social segregation existent in the polygon of the old colonial area that was still surrounded by estates and agricultural land. The first processes of urban expansion[1] occurred in later years, with the longitudinal growth as a result of regional integration, domestic industry and real estate processes, typical of capitalism. In 1942, the city was to formalise its expansion with its first modern regulatory plan (Odriozola 1945). This plan was to define the growth in the North and South through specific uses and originally obeying preliminary respect to the ravines and slopes where some parks and public areas were established, although later phases were shown a certain disrespect for these norms (Molina et al. 2017). The second important phase (1960–80) of capitalist modernisation, however, was a much more profound transformation of the North and South that gave rise to the present city and, by extension, to the DMQ (Salazar et al. 2021). This process also involved the inclusion of budding rural areas on the outskirts via the consolidation of a longitudinal transport network that allows mobility, economic integration and use of urban services.

It is important to highlight that in Quito, as in many other Latin American cities, people outside the formal labour market set up residence on the outskirts (Portes and Roberts 2008). Thus, a significant part of the "popular" neighbourhoods of Quito corresponds to this type of land use, where the owners paid for the land ownership, but never received the corresponding deeds. These processes intensified over the 2000s and through to the present, accounting for over half of the plots in Quito since the 60's

---

[1] The process was fostered and moved by the Cabildo itself, the municipal body, to solve the problem of overcrowding.

(Moscoso Rosero 2013). Due to informal processes occur outside the land planning policies, the territorial occupation was non-selective and, in some cases, made use of land in ravines or agricultural plots and even areas of environmental protection. The urban plan and processes of regularisation of property deeds are to partially resolved the formal problem of property ownership but did little to tackle the environmental risks run by these settlements (Larco Moscoso 2018). This allowed for occupation of areas potentially affected by primary lahars (Mothes 1992) or by flooding and secondary lahars on the slopes of the Pichincha (Barberi et al. 1992). This latter case is significant in that it affects an area where variety of the most important hospitals in the city are to be found and that had formal plans from the very beginning of its development.

The expansion of the city towards the Eastern Valleys began in the 80's and followed a real estate vision focussed on importing urban planning regulations. The process intensified in the 90's, as a result of the scarcity of available urban land. This process coincided with the expansion of the motorways crossing the city and the formalisation of the transfer of the airport to the eastern edge of the Interandean Valley from the Quito Plateau and private real estate developments that boomed with the change of currency to the dollar, motivating thus tremendous speculation on the land market (Salazar 2013; Carrión 2016; Durán et al. 2016). The process of development in the valleys produced a disperse city with low densities, built on the margins of the new access routes and the old rural paths. The result was settlements on slopes, developments on the edges of ravines and intense waterproofing of the soil.

### *14.2.2 The Multipurpose and Multihazard Exposure Model of the MDQ*

As of 2019, efforts have centred on developing the MultiPurpose and Multihazard Exposure Model of the Metropolitan District of Quito (MMEM-DMQ) at a plot and building scale (Yepes and Marrero 2021) in order to tackle different types of problems by offering support for informed decision-making. The methodology involved includes various different processes of data cleaning, constructing a geographic layer that represents the buildings in as updated a version as possible, where both the formal city tissue and informal are represented and specific algorithms allow the alphanumeric data to be transferred to the spatial elements. This process is continuous in design so is constantly open and moreover allows for permanent revision and incorporation of indicators.

From the technical perspective, the basic idea of the MMEM-DMQ consists in two files holding multiple indicators that allow for a full description of the various aspects of the two elements of interest (plots and buildings). The first file relating to the plot (land property) is made up of 58 indicators and the second associated to the building itself has 142. The information, when structured in this way, makes it easier to work and allows for efforts to focus on the design of specific queries (like filters of data selection) moulded to the operator or decision maker's needs.

The multipurpose nature of the MMEM-DMQ is the direct result of the need to implement more realistic and sustainable strategies of handling information in disaster risk reduction (UNDRR 2021). Data, we believe, not only should help in understanding the risk of disaster but should also incorporate other social aspects. The more people who use the MMEM-DMQ to different ends on a daily basis, the better the guarantees that the data contributed to the MMEM-DMQ are constantly updated over time. However, this use of the tool in no way detracts from its basic significance in disaster risk assessment and planning of mitigation strategies. The data sources used to build the most relevant indicators that are taken into consideration here are the following:

- The Property Register Data are exported (not available to the general public) (Dirección Metropolitana de Catastro 2020c). These data allow for population to be assessed from the number of houses and special services required such as electricity and water. Another value obtained is the reposition cost of the building in US dollars.[2] The description of the building is also used as a socio-economic indicator as it allows us to discriminate between low-, medium- and high-income households (DMQ 2019).
- The Geographic Building Layer is generated from five different sources in order to represent the entirety of buildings in the city and not only the city according to the property register:
  - The layer of blocks of buildings (Dirección Metropolitana de Catastro 2020a).
  - The layer of building units (Dirección Metropolitana de Catastro 2020b).
  - The layer of buildings from the Instituto Geográfico Militar (2012).
  - The layer of buildings from the Open Street Map (2019).
  - The layer of buildings/properties from the National Institute of Statistics and Census (INEC 2010b).
- Various different Digital Terrain Models (DTMs) or Digital Surface Models are giving priority to the ones available with greatest precision. From them, we obtain values such as height and slope of terrain of the exposed element.
  - A partial 1 m DTM of DMQ recalculated to 2 m, available for internal consultation (Dirección Metropolitana de Catastro 2020d).
  - A total 3 m DTM and a partial 10 m DTM of MDQ (Sigtierras 2010).
  - A total 30 m DTM of DMQ (NASA-JPL 2021).
  - A total 30 m DSM of DMQ (European Space Agency, Sinergise 2021).
- Database of geographic accidents/landforms (Dirección Metropolitana de Catastro 2021) to identify open and filled ravines.

---

[2] The reposition cost should not be confused with the building value. The reposition cost remains fixed over time and is calculated according to the characteristics of the building, whereas the latter decreases over time as a result of ageing processes and other related regulatory factors.

- The hydrographic network from the blending of the geographic layers from the Military Geographic Institute (IGM) and the DMQ (Secretaría de Territorio del DMQ 2018).

## 14.3 Shaping Environmental Risk in the City of Quito

Here, we have selected flood and debris flow risk assessment to attempt to illustrate how human activity shapes environmental risk in the city of Quito. Flood and debris flow hazards are usually geographically limited events with an annual or multi-annual frequency. The extreme climate events they are associated with, however, produce intense, high-severity risk and constitute medium–low frequency disasters (UNISDR 2015). In reality nevertheless, extreme climate events account for a substantial accumulative number of victims and destruction of property (Cáceres and Nuñez 2011).

### 14.3.1 Modification of the Hydrographic Network

Previous to the charted development of Quito, the hydrographic network was formed as the result of the tectonic compression that created the piggy-back basin (Hibsch et al. 1997; Beauval et al. 2014). The progressive upward growth of the East border of the Quito plateau as the result of a compressional reverse fault (at present, presenting as a 400–600 m scarp) progressively modified the slope of the terrain and filled the piggy-back basin with eruptive material and subsequent erosive processes of the western volcanic massifs dating from the Pleistocene through to the present. This constant elevation of the Quito plateau due to the fault activity obstructed the drainage to the East, thus favouring the formation of lagoons and swamps at the bottom of the basin.

In recent times, the hydrographic network has suffered further alteration as the result of human intervention as the city of Quito grew, a process that is still ongoing. In Fig. 14.2, the hydrographic network can be compared in extension with the present layout of the city of Quito. Thanks to important work *Secretaría de Territorio, Hábitat y Vivienda del DMQ* (2018), an approximation of the old hydrographic network was re-constructed at digital level, following the lines of the former ravines and areas covered by lagoons and swamps. The hydrographic network was altered as a result of three processes:

- The lagoons and swampy areas were dried. This required a constant process of water extraction to keep a balance between the depleted watershed (due to absence of surface water) and the compacting of the soil that has been left at present (Peñafiel 2009).

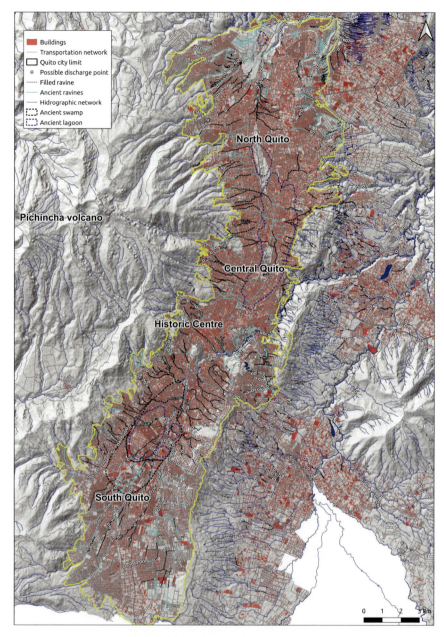

**Fig. 14.2** Geographic distribution of the city of Quito and the former hydrographic network and possible discharge points of sanitation network. *Source* Own elaboration

- The beds of many ravines were filled either by engineering compacting of soil, certified with public permits, or via informal and uncontrolled processes (Guéguen 1997).
- Channelling of riverbeds together with greywater pipes (mixture of rainwater and wastewater) via underwater tanks. However, these were deficient to the needs and overflowed due to excess population (Pourrut and Leiva 1989). To solve the problem, various systems of galleries were built as of 2004 to allow for deep drainage and to relieve the former tanks that had been built in the beds of the former ravines, but they are still not optimal solutions (Paredes Méndez 2017).

Another significant modification of the territory that affects the dynamics of the flood and debris flow hazard is due to the soil waterproofing. About 49.2% of the 197.33 km$^2$ of land available in Quito is totally waterproofed as the result of building roads, households, facilities and areas of artificial coverage such as synthetic tennis courts, among others. This limits the soil filtration capacity or even to reduce the speed of water runoff. Moreover, if we consider the degree of waterproofing indicated, the runoff coefficient in almost half the city is close to 1, meaning that the present amount of rainfall in the city gets to the ravines and rivers in at least double the amount it did before and with greater force in less time. This implies that the impact of rainwater before was much less intense, both in its capacity to flood as to cause erosion downstream.

This effect is directly related to the design of the sanitation network (EPMAPS 2012) that dumps grey water directly into the ravines and rivers. Figure 14.2 shows the estimated 435 possible discharge points (white/black circles) detected in the city of Quito from a spatial analysis of the polyline vectors that represent the pipelines. The analysis focussed on finding the ends of the pipelines in rivers and ravines. The water in these open-air discharge points is then collected by deep sewers that take the water to the rivers Machangara (South) and Monjas (North).

Quito is one of the few capitals in Latin America where the sanitation network does not separate rainwater from wastewater. As a result, three aspects have to be taken into account (Campaña et al. 2017): (1) an increase in the daily flow of watercourses where before the flow was sporadic; (2) as previously noted, that the waterproofing of the city has led to a vast increase in the flow received by the sanitation network and dumped in the rivers and ravines; and (3) the elevated water pollution levels.

Thus, the beds of the rivers and ravines are subjected to constant cross and longitudinal erosion which is favoured by weak basin-fill volcanic material, above all in rural parishes in the North. This process has increased the exposure of those buildings located close to the banks of water channels. The end result is a space that has been altered by human intervention that has almost destroyed the hydrographic network and has artificially carpeted the surface, thereby resulting in serious problems when attempting to control surface water.

## 14.3.2 Heavy Rains Near the City of Quito

The main trigger of flood and debris flow hazards in the area is the heavy rain that takes place in the rainy season near Quito, highly located in time and space (Peltre 1989a). This phenomenon is also combined with periods of intense drought and has been documented as far back as the XVII century (Domínguez-Castro et al. 2018).

The rain tends to be intense and short-lived, generating huge volumes of water and eventually mud and rocks that are channelled down the slopes, above all on the sides of the Pichincha volcano (Perrin et al. 2001). This environmental process has undergone changes due to climate change, with the rainy season now occurring later (Villacis Rivadeneira and Marrero de León 2017) and increased intensity of events of heavy rain (Serrano Vincenti et al. 2016) producing expected higher impact of the hazards in the coming years.

The database of the Municipal Emergency Operational Committee describing the emergency calls in 2011–2019 period was taken to pinpoint the areas most affected by high impact hydrometeorological events. A spatial intensity index was calculated with a 500 m bandwidth factor and a 0.006 transparency filter on a 30 m grid (Jaquet et al. 2008). The coordinates assigned to each event usually correspond to the crossroads of streets closest to where an emergency call for help was made (for damage to housing or facilities or flooding in the streets, for example), as a result not necessarily close to a ravine or river. In Fig. 14.3, it can be observed how most events are located on both sides of the central axes of the city, the areas where the hydrographic network was most seriously altered, above all in the central historic area of Quito where the ravines leading off the Pichincha volcano influence most. An example of how these processes play out in reality was offered on the 31 January 2022 in the El Tejado ravine (Fig. 14.3), an event that seriously damaged the neighbourhoods of La Comuna and La Gasca and that caused 28 deaths, 48 wounded and 12 missing persons (El Universo 2022; BBC News 2022).

## 14.3.3 Flood and Debris Flow Risk Assessment Based on MEMM-DMQ

The arbitrary spatial-physical vulnerability index (ASPVI) was created to carry out flood and debris flow risk assessment using a building base MMEM-DMQ query (see Appendix for further details). The ASPVI took three factors into account: (1) the spatial vulnerability factor, the distance (D) between the centroid[3] of the building and the nearest point located in the centre of the water course of the river or ravine; (2) the spatial vulnerability factor of the terrain slope (TS) where the building stands; and (3) the physical vulnerability factor, that is, the type of house frame (HF) conditioned by the house status (HS).

---

[3] The centroid is the geometric centre of a plane or the arithmetic mean position of all the points in the figure.

14 Urbicide or Suicide? Shaping Environmental Risk in an Urban Growth …

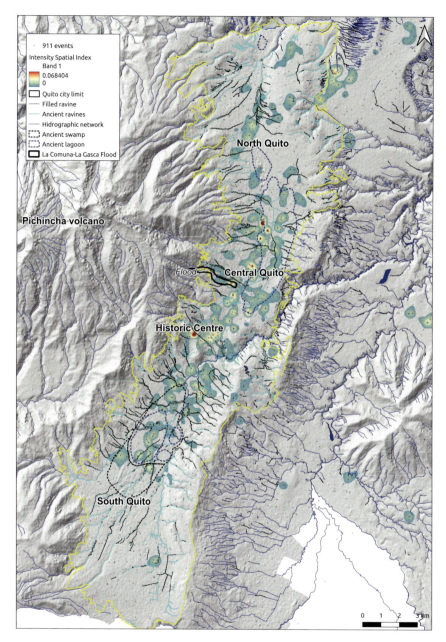

**Fig. 14.3** Distribution of hydrometeorological impact emergency events and their spatial intensity index plotted over the current and former hydrographic network. *Source* Own elaboration

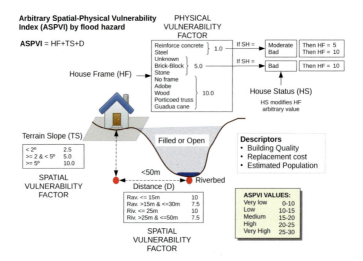

**Fig. 14.4** Indicators and arbitrary values used to build the ASPVI. The final values are expressed in arbitrary units from 0 to 30. *Source* Own elaboration

Figure 14.4 shows the categories and the arbitrary value assigned to each vulnerability factor to obtain the final ASPVI expressed as the sum of all these factors. The value of the physical vulnerability factor is established from the HF which may be increased if the house status (HS) is bad or moderate. The ASPVI gives values from 0 to 30 with the highest values attached to the greater conditions of vulnerability, with the worst risks assigned to buildings closest to the centre of the water course, with higher angled slopes and greater physical vulnerability. Besides, only the buildings whose centroid is less than 25 m from the centre of the water course of the ravines or 50 m from the centre of the rivers are included. All the rest are excluded.

The final result is shown in Fig. 14.5 where buildings on filled or open ravines are marked with black dashed lines along the edges. The ASPVI shows the highest values in the sloped areas on both sides of the Interandean Valley, especially where the terrain slope is most pronounced. There are also high values in some central areas in the extreme North.

If we analyse the distribution of values for each factor (Fig. 14.6), we can observe how there are low arbitrary values for the house frame (HF) due to the fact that most of the city is built with reinforced concrete (top-left). Besides, the influence of the HS is limited with greater repercussion in low-quality constructions (22.6%), whereas in medium- and high-quality buildings, its influence is negligible at respective values of 1.9 and 0.7%. If we attend the distribution of the quality of the constructions (top-right), we can see that the high values are predominant above all in the northern ravines, whereas the low-quality value constructions are concentrated in the South, on slopes and lower reaches. With respect to the distribution of the terrain slope (bottom-left), high and medium values predominate in line with the morphology of the Interandean Valley where the central axis offers more moderate slopes. Last

14 Urbicide or Suicide? Shaping Environmental Risk in an Urban Growth … 277

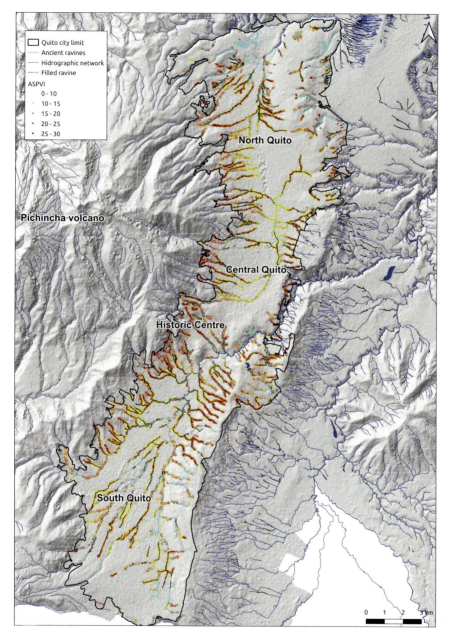

**Fig. 14.5** Results of the flood and debris flow risk assessment expressed as ASPVI values. *Source* Own elaboration

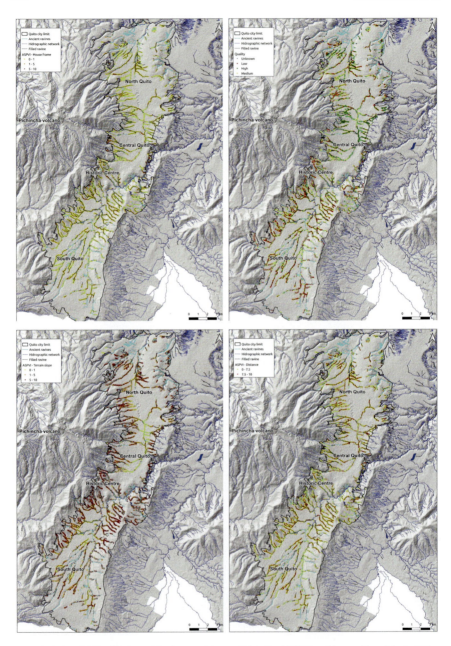

**Fig. 14.6** Spatial distribution of the factors that condition the ASPVI and the quality of the building as socio-economic indicators for the city of Quito. *Source* Own elaboration

but not least, the former hydrographic network of the city of Quito appears highly populated close to the centre of the watercourse or thereabouts (bottom right). There are only a few stretches of ravines or rivers where said proximity is not given.

Overall, in spatial terms, the scenario is highly complex as reflected in Table 14.1. We can see how 9% of the population and buildings reveal greater or lesser exposure to this hazard, in other words, the equivalent of some 200,000 people and over 4 billion dollars in reposition costs. The relative weighting of high-quality buildings (10%) is greater than the low- and medium-quality constructions (8–9%), thus indicating clearly that the problem is not restricted exclusively to low-income groups. Besides, the spatial concentration of the most significant socio-economic classes is focussed on the northern part of the city that makes the real impact also relevant, despite their being overall lesser in number. In general, the level of exposure is extremely high although the vulnerability values obtained via the ASPVI oscillate, with most buildings with low arbitrary values (10–15) or medium (15–20) and the highest group (20–25) third in importance. The most important relevant factor is how there is a predominance of buildings in filled ravines, where the ravine no longer exists visibly in the urban layout but not the danger deriving from the debris flows that continue upstream.

In a second interpretation of the data (Table 14.2), it is clear that the majority of building are closer to ravines than to rivers, since in the study area, there are only the River Machangara in the South and Monjas in the North. In any case, the number of buildings in very close proximity to the centre of the watercourse is extremely high. The data also show the predominant low arbitrary values for HF which, as previously stated, are due to the use of reinforced concrete and the consideration of good or very good HS given to most buildings, especially those of medium and high quality. With respect to the TS, the highest values are significant for all the levels of quality of buildings, since the problem is related to both the East and West slopes of the Quito basin where a large part of the city of Quito is built.

## 14.4 Discussion

The arbitrary spatial-physical vulnerability index (ASPVI) is simply calculated thanks to the characteristics of the MMEM-DMQ that affords a quick analysis using simple or complex criteria to assess exposure features. The ASPVI is useful to conduct a flood and debris flow risk analysis in high populated cities as Quito and to offer support for decision-making processes. It also permits a global risk assessment of the entire analysis area using more accurate social data, as opposite to those approaches based in remote sensing (Velez et al. 2022).

However, there are certain weaknesses, some of which are related to the deep-reaching modifications in the hydrographic network. In our approach, the spatial vulnerability factor D is related to the centre of the water course of ravines and rivers, but gravitational flows do not necessarily follow these courses when they enter urban areas. Obstacles encountered along the way may produce serious modifications in

Table 14.1 Distribution of the number of buildings, estimated population, reposition cost and quality of the building expressed in absolute and relative numbers

| ASPVI | Buil | % | Pob | % | Rep | % | Low | % | Med | % | Hig | % |
|---|---|---|---|---|---|---|---|---|---|---|---|---|
| City | 322,535 | | 2,059,278 | | 47,816,928,497.10 | | 32,241 | | 264,371 | | 10,738 | |
| *Filled ravines* | | | | | | | | | | | | |
| 0–10 | 0 | 0.00 | 0 | 0.00 | 0.00 | 0.00 | 0 | 0.00 | 0 | 0.00 | 0 | 0.00 |
| 10–15 | 6501 | 2.02 | 47,794 | 2.32 | 1,291,500,533.31 | 2.70 | 266 | 0.83 | 5831 | 2.21 | 305 | 2.84 |
| 15–20 | 8436 | 2.62 | 51,585 | 2.51 | 1,219,907,023.35 | 2.55 | 618 | 1.92 | 6888 | 2.61 | 312 | 2.91 |
| 20–25 | 4891 | 1.52 | 26,924 | 1.31 | 571,622,324.19 | 1.20 | 717 | 2.22 | 3465 | 1.31 | 150 | 1.40 |
| 25–30 | 607 | 0.19 | 2737 | 0.13 | 33,052,026.35 | 0.07 | 174 | 0.54 | 417 | 0.16 | 16 | 0.15 |
| Total | 20,435 | 6.34 | 129,040 | 6.27 | 3,116,081,907.20 | 6.52 | 1775 | 5.51 | 16,601 | 6.28 | 783 | 7.29 |
| *Open ravines* | | | | | | | | | | | | |
| 0–10 | 0 | 0.00 | 0 | 0.00 | 0.00 | 0.00 | 0 | 0.00 | 0 | 0.00 | 0 | 0.00 |
| 10–15 | 3014 | 0.93 | 20,863 | 1.01 | 625,217,197.03 | 1.31 | 171 | 0.53 | 2658 | 1.01 | 127 | 1.18 |
| 15–20 | 3814 | 1.18 | 24,593 | 1.19 | 534,215,806.39 | 1.12 | 295 | 0.91 | 3174 | 1.20 | 131 | 1.22 |
| 20–25 | 2361 | 0.73 | 11,972 | 0.58 | 236,968,850.93 | 0.50 | 377 | 1.17 | 1560 | 0.59 | 50 | 0.47 |
| 25–30 | 294 | 0.09 | 1435 | 0.07 | 20,045,579.85 | 0.04 | 51 | 0.16 | 237 | 0.09 | 6 | 0.06 |
| Total | 9483 | 2.94 | 58,863 | 2.86 | 1,416,447,434.19 | 2.96 | 894 | 2.77 | 7629 | 2.89 | 314 | 2.92 |
| Global | 29,918 | 9.28 | 187,903 | 9.12 | 4,532,529,341.40 | 9.48 | 2669 | 8.28 | 24,230 | 9.17 | 1097 | 10.22 |

**Table 14.2** Distribution of number of buildings, estimated population, reposition costs and quality of the housing expressed in absolute and relative numbers considering D, HF and TS factors

| ASPVI | Buil | % | Pob | % | Rep | % | Low | % | Med | % | Hig | % |
|---|---|---|---|---|---|---|---|---|---|---|---|---|
| City | 322,535 | | 2,059,278 | | 47,816,928,497.10 | | 32,241 | | 264,371 | | 10,738 | |
| *Distance (D)* | | | | | | | | | | | | |
| Rav. ≤ 15 | 11,720 | 3.63 | 73,835 | 3.59 | 1,846,856,459.54 | 3.86 | 1023 | 3.17 | 9294 | 3.52 | 447 | 4.16 |
| Rav. > 15 < 30 | 18,134 | 5.62 | 113,841 | 5.53 | 2,675,223,419.71 | 5.59 | 1643 | 5.10 | 14,884 | 5.63 | 650 | 6.05 |
| Riv. ≤ 25 | 18 | 0.01 | 55 | 0.00 | 1,922,886.98 | 0.00 | 1 | 0.00 | 17 | 0.01 | 0 | 0.00 |
| Rav. > 25 < 50 | 46 | 0.01 | 172 | 0.01 | 8,526,575.17 | 0.02 | 2 | 0.01 | 35 | 0.01 | 0 | 0.00 |
| Total | 29,918 | 9.28 | 187,903 | 9.12 | 4,532,529,341.40 | 9.48 | 2669 | 8.28 | 24,230 | 9.17 | 1097 | 10.22 |
| *House frame (HF)* | | | | | | | | | | | | |
| Resistant | 22,621 | 7.01 | 162,337 | 7.88 | 4,108,926,948.04 | 8.59 | 1036 | 3.21 | 20,570 | 7.78 | 1015 | 9.45 |
| Medium | 5782 | 1.79 | 18,581 | 0.90 | 309,837,101.02 | 0.65 | 1175 | 3.64 | 2642 | 1.00 | 43 | 0.40 |
| Weak | 1515 | 0.47 | 6985 | 0.34 | 113,765,292.34 | 0.24 | 458 | 1.42 | 1018 | 0.39 | 39 | 0.36 |
| Total | 29,918 | 9.28 | 187,903 | 9.12 | 4,532,529,341.40 | 9.48 | 2669 | 8.28 | 24,230 | 9.17 | 1097 | 10.22 |
| *Terrain slope (TS)* | | | | | | | | | | | | |
| < 2° | 5140 | 1.59 | 34,691 | 1.68 | 1,104,967,173.07 | 2.31 | 387 | 1.20 | 4,125 | 1.56 | 244 | 2.27 |
| ≥ 2 < 5° | 10,562 | 3.27 | 67,611 | 3.28 | 1,604,773,098.55 | 3.36 | 782 | 2.43 | 8,798 | 3.33 | 377 | 3.51 |
| > 5° | 14,216 | 4.41 | 85,601 | 4.16 | 1,822,789,069.78 | 3.81 | 1500 | 4.65 | 11,307 | 4.28 | 476 | 4.43 |
| Total | 29,918 | 9.28 | 187,903 | 9.12 | 4,532,529,341.40 | 9.48 | 2669 | 8.28 | 24,230 | 9.17 | 1097 | 10.22 |

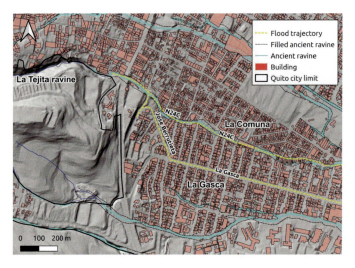

**Fig. 14.7** Main direction of the flow (yellow-dash line) was shifted at some point, affecting the street of La Gasca, further South. *Source* Own elaboration

the direction of gravitational flows, as happened in the case of the debris flow event on the 31st of January 2022 in the El Tejado ravine (El Universo 2022; BBC 2022). In Fig. 14.7, the alternative direction taken by the debris flow is detailed in the yellow-dash line (Calles José Berrutieta and La Gasca, in the South) as opposed to the course it would have followed if it had met no obstacles (N-24C La Comuna, to the North). This modification produced as the result of human occupation of the territory, meant that the buildings in La Gasca street did not figure in our analysis. So, in order to fully cover Flood and Debris Flow risk assessment, previous events must be studied or simulations should be carried out to see what possible flow courses can be expected in the modified urban environment.

An evaluation of the Flood and Debris Flow Hazard has been carried out in the city of Quito. However, there are important differences in the ravines located around the sides of the Pichincha (with higher transport capacity) than in the lower areas where the flows have lost their velocity (deposition area) or where the dynamics of the gravitational flow may be different, more characteristic of the overflowing of rivers. Using the MMEM-DMQ on the slopes of the Pichincha, we found that 13,889 buildings are exposed to risk, corresponding to an approximate population of 83,113 and with a reposition cost of 2,049,140,000 $US.

The effect of the HS on the HF factor may be greater than was calculated from the registry data used in the MMEM-DMQ. When the percentage of the buildings in poor/bad or moderate state on a block according to the registry is compared with the data given by the National Institute of Statistics (INEC 2010a), we can see that although use similar semantics, the institute registers a higher level of overall deterioration (Fig. 14.8). The reason for this difference could be due to two main factors, among others, that is (1) the Registry's tax implications, not shared by the

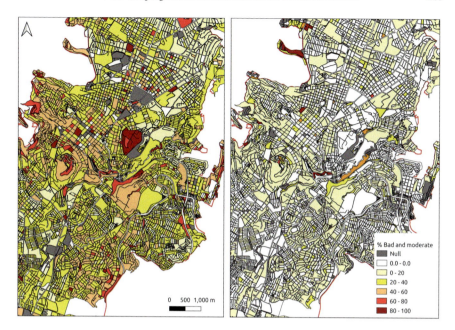

**Fig. 14.8** Differences between the percentages of building in bad or moderate state according to the property register (right) and the INEC (left) at street block level. *Source* Own elaboration

INEC and (2) in the case of the Register, the data are collected by a civil servant, whereas in the INEC it is the owner or the resident who gives their impression with respect to the state of the building. So, this overwhelming difference leads us to suspect that the importance of the HS may be greater when defining the final value of the physical vulnerability factor HF, although this has not been taken into account in the present analysis.

Other types of natural hazards take place in the city of Quito, such as landslides (Puente-Sotomayor et al. 2021), subsidence (Jaramillo Castelo et al. 2018), volcanic eruptions (Aguilera et al. 2004), earthquakes (Beauval et al. 2014) and wildfires (Estacio and Narváez 2012). In general, the city presents a high environmental risk level overall given the extreme vulnerability and exposure of many of the existing buildings. Besides, if we use a multihazard perspective, we can see that the high slope areas are more sensitive to landslides when there is heavy rain as many of these areas are precisely on the banks of rivers and edges of ravines where floods and debris flow hazards also occur (Peltre 1989b).

In this context, it is highly complex at the level of territorial planning to forbid any type of development. Moreover, little space would be left available if all the natural hazards that could affect the city of Quito be factored in, apart from the fact that these areas are already highly built-up. In reality, the only option available is to quantify the environmental risk and assume it, looking for ways to gradually reduce it and/or shifting it during re-construction processes. Territorial planning has various possible options open, of which we would like to highlight two:

- One is the present situation where the risk is almost unknown (potential impact in people and infrastructure) and where scarce preventive strategies exist or are effectively applied, thereby increasing physical and social vulnerability. This is due to the lack of fulfilment or respect for many of the control policies, most of which are associated with prohibition or excessive urban development limitation in a city such as this where there is an enormous demand for housing at all levels of society. In fact, if we consider the levels of exposure obtained here for only flood and debris flow hazards, re-location processes are not feasible on account of the high costs involved and because alternative places are not exempt from their own risks of suffering other types of natural hazard impacts.
- The other infinitely more desirable option would be to work on plans of risk management that go beyond response to emergency situations. What is needed are plans that allow for the territory to be managed in real time in order to generate early warning, supported by tools towards to mitigate and reduce physical and social vulnerability. This, although it would not totally rule out economic losses, would, however, reduce the number of casualties besides making people more aware of the place they live in.

## 14.5 Conclusion

The development of the city of Quito from the mid-twentieth century onwards obeyed a great many factors but was carried out with total disrespect for the environment. Although it is true that formal and informal land occupation development is related to socio-economic status, in terms of risk areas, different levels of society generally live side by side and the transformation they have exerted on the territory is similar from the perspective of environmental risks: that is, filling legally or illegally ravines, artificially covering surfaces or by simply channelling the riverbeds in other directions inside the city. So, it is a process of joint construction not exclusively related to low-income groups as is traditionally affirmed. A similar circumstance was already observed in Portoviejo after the earthquake on 16 April 2016 (Ye et al. 2016). In that case, the more complex and, in many cases, legalised buildings ($\geq 3$ stories) suffered greater damage in relative terms (Marrero et al. 2019).

If all natural hazards are factored into the picture, it must be said that all the citizens of Quito assume a high level of environmental risk, apart from the normal social hazards of a large city (violence, lack of safety, among others). Despite all legislation and urban planning, the city continues to grow, mostly outside the margins of the norms and occupying areas unaware of the natural hazards. At the same time, the area already built up presents serious problems that are not easy to manage using a multi-risk perspective.

To date, the city has proved unable to adopt a risk management plan based on real-time analysis of the environment nor to consider plans that go beyond emergency response. This lack of vision is most manifest in processes of reconstruction after a natural hazard impact where few lessons are learnt, and the emphasis is on wiping out the traces of the tragedies that occurred. This type of behaviour does not help to disaster risk reduction, so it breaks the historical memory and leaves the populations open to be affected by the same events, above all in the case of floods.

The city of Quito, therefore, in its evolution and growth is not so much typical of a process of urbicide as it is of a kamikaze city that has turned its back on the environment and now finds itself highly limited in its options to reduce the risks. The multihazard analysis shows just how complex the challenge is for a city that was basically built with all the premises of a long-term suicide.

## Appendix: Query Description to Calculate ASPVI

The data columns used to construct the ASPVI index are described here (see Fig. 14.4):

- Original column header names in MMEM-DMQ:
  - **HIDISQU**: Distance (D) from the building's centroid to the nearest centre of the ravine water course.
  - **HIDISRI**: Distance (D) from the building's centroid to the nearest centre of the river water course.
  - **TSLOPG**: Terrain slope (TS) grouped in intervals' values from 0 to 6.
  - **ARMA**: House frame (HF) in numerical values, from 0 to 9.
  - **ESTA**: House status (HS) in numerical values from 0 to 5.

- New column header names added to MMEM-DMQ:
  - **DIVAL**: Arbitrary value of D factor.
  - **TSVAL**: Arbitrary value of TS factor.
  - **HFVAL**: Arbitrary value of HF factor.
  - **ASPVI**: Arbitrary spatial-physical vulnerability index value.

When a new assessment is made, the MMEM-DMQ can be easily modified by adding new data columns.

The numerical categories present in TSLOP, ARMA and ESTA data columns are showed in Tables 14.3, 14.4 and 14.5.

The following query was designed using the Panda library (The pandas development team 2020) in Python programming language (van Rossum 2022). The query follows these steps:

(1) Select only those buildings located close to a river or ravine that meet the following distances (in metres):

- HIDISQU $>$ 0 and HIDISQU $\leq$ 50 or HIDISRI $>$ 0 and HIDISRI $\leq$ 50.

**Table 14.3** Terrain slope classified by groups

| Angle | Group | Description |
|---|---|---|
| < 2° | 1 | Very slightly leaned/sloped |
| ≥ 2° TS < 5° | 2 | Slightly leaned/sloped |
| ≥ 5° TS < 10° | 3 | Leaned/sloped |
| ≥ 10° TS < 15° | 4 | Heavily leaned/sloped |
| ≥ 15° TS < 30° | 5 | Moderately steep |
| ≥ 30° TS < 60° | 6 | Steep |
| ≥ 60° | 7 | Very steep |

**Table 14.4** House frame (HF) classified by groups

| Group | Description |
|---|---|
| 0 | Unknown |
| 1 | Reinforce concrete |
| 2 | No house frame |
| 3 | Brick block |
| 4 | Adobe |
| 5 | Steel |
| 6 | Wood |
| 7 | Porticoed truss |
| 8 | Stone |
| 9 | Guadua cane |

**Table 14.5** House status (HS) classified by groups

| Group | Description |
|---|---|
| 0 | Unknown |
| 1 | Good |
| 2 | Very good |
| 3 | No HS |
| 4 | Moderate |
| 5 | Bad |

(2) Add arbitrary value in D factor:

- if HIDISQU $\geq$ 0 and HIDISQU $\leq$ 15, then DIVAL = 10.0.
- if HIDISQU $>$ 15 and HIDISQU $\leq$ 30, then DIVAL = 7.5.
- if HIDISRI $\geq$ 0 and HIDISRI $\leq$ 25, then DIVAL = 10.0.
- if HIDISRI $>$ 25 and HIDISRI $\leq$ 50, then DIVAL = 7.5.

(3) Add arbitrary value in TS factor:

- if TSLOPG $\leq$ 1, then TSVAL = 2.5.

- if TSLOPG $==2$, then TSVAL $= 5.0$.
- if TSLOPG $>$ 2, then TSVAL $= 10.0$.

(4) Add arbitrary value in HF factor:

- if ARMA $==1$ or ARMA $==5$, then INDEB $= 1.0$. Then, check HS factor:
    - if ARMA $==1$ or ARMA $==5$ and ESTA $==5$, then HFVAL $= 10.0$.
    - if ARMA $==1$ or ARMA $==5$ and ESTA $==6$, then HFVAL $= 5.0$.
- if ARMA $==0$ or ARMA $==3$ or ARMA $==8$, then HFVAL $= 5.0$. Then, check HS factor:
    - if ARMA $==0$ or ARMA $==3$ or ARMA $==8$ and ESTA $==5$, then HFVAL $= 10.0$.
- if ARMA $==2$ or ARMA $==4$ or ARMA $==6$ or ARMA $==7$ or ARMA $==9$, then HFVAL $= 10.0$.

(5) Calculate ASPVI value:

- ASPVI = DIVAL + TSVAL + HFVAL.

# References

Aguilera E, Pareschi M, Rosi M, Zanchetta G (2004) Risk from lahars in the northern valleys of Cotopaxi Volcano (Ecuador). Nat Hazards 33(2):161–189. https://doi.org/10.1023/B:NHAZ.0000037037.03155.23

Barberi F, Ghigliotti M, Macedonio G, Orellana H, Pareschi M, Rosi M (1992) Volcanic hazard assessment of Guagua Pichincha (Ecuador) based on past behaviour and numerical models. J Volcanol Geoth Res 49(1):53–68. https://doi.org/10.1016/0377-0273(92)90004-W

BBC News (2022) Ecuador: deadly landslide after heaviest rainfall in years. https://www.bbc.com/news/world-latin-america-60211539

Beauval C, Marinière J, Yepes HA, Audin L, Nocquet JM, Alvarado A, Baize S, Aguilar J, Singaucho JC, Jomard H (2018) A new seismic hazard model for ecuador. Bullet Seismol Soc Am 108(3A):1443–1464. https://doi.org/10.1785/0120170259

Beauval C, Yepes H, Audin L, Alvarado A, Nocquet JM, Monelli D, Danciu L (2014) Probabilistic seismic hazard assessment in quito, estimates and uncertainties. Seismol Res Lett 85(6):1316–1327. https://doi.org/10.1785/0220140036

Beauval C, Yepes H, Bakun WH, Egred J, Alvarado A, Singaucho JC (2010) Locations and magnitudes of historical earthquakes in the Sierra of Ecuador (1587–1996). Geophys J Int 181(3):1613–1633. https://doi.org/10.1111/j.1365-246X.2010.04569.x

Cáceres L, Nuñez AM (2011) Segunda comunicación nacional sobre cambio climático. Reporte, Ministerio del Ambiente. Subsecretaría de Cambio Climático. https://unfccc.int/resource/docs/natc/ecunc2.pdf

Campaña A, Gualoto E, Chiluisa-Utreras V (2017) Evaluación físico-química y microbiológica de la calidad del agua de los ríos machángara y monjas de la red hídrica del distrito metropolitano de quito. Bionatura 2(2):305–310. https://doi.org/10.21931/RB/2017.02.02.6

Campbell D, Graham S, Monk DB (2007) Introduction to urbicide: the killing of cities? Theory Event 10(2). https://doi.org/10.1353/tae.2007.0055

Carrión A (2016) Megaprojects and the restructuring of urban governance: the case of the new quito international airport. Lat Am Perspect 43:252–265. https://doi.org/10.1177/0094582x15579900

Carrión F, Erazo Espinosa J (2012) La forma urbana de quito: una historia de centros y periferias. Bulletin De L'institut Français D'études Andines 41(3):503–522. https://doi.org/10.4000/bifea.361

Chester D, Duncan A, Kilburn C, Sangster H, Solana C (2015) Human responses to the 1906 eruption of Vesuvius, southern Italy. J Volcanol Geoth Res 296:1–18. https://doi.org/10.1016/j.jvolgeores.2015.03.004

Chrètien S, Brousse R (1989) Events preceding the great eruption of 8 May, 1902 at Mount Pelèe, Martinique. J Volcanol Geoth Res 38(1–2):67–75. https://doi.org/10.1016/0377-0273(89)900 30-9

Cubellis E, Luongo G, Marturano A (2007) Seismic hazard assessment at Mt. Vesuvius: maximum expected magnitude. J Volcanol Geoth Res 162(3):139–148. https://doi.org/10.1016/j.jvolgeores.2007.03.003

Dirección Metropolitana de Catastro (2010) Modelo Digital del Terreno a 1 metro de resolución. Distrito Metropolitano de Quito

Dirección Metropolitana de Catastro (2019) Norma técnica para la valoración de bienes inmuebles urbanos y rurales del Distrito Metropolitano de Quito. Reporte, Distrito Metropolitano de Quito

Dirección Metropolitana de Catastro (2020a) Datos de cartografía catastral. Distrito Metropolitano de Quito. Accedidos en julio de 2020a

Dirección Metropolitana de Catastro (2020b) Datos de Cartografía Catastral. Capa geográfica Bloque Constructivo. Distrito Metropolitano de Quito. Accedidos en julio de 2020b

Dirección Metropolitana de Catastro (2020c) Datos de Cartografía Catastral. Capa geográfica Unidad Constructiva. Distrito Metropolitano de Quito. Accedidos en julio de 2020c

Dirección Metropolitana de Catastro (2020d) Datos de Cartografía Catastral. Modelo Digital del Terreno a 1 metro de resolución. Distrito Metropolitano de Quito. Accedidos en Octubre de 2019

Dirección Metropolitana de Catastro (2021) Base de datos de accidentes geográficos. Distrito Metropolitano de Quito. Accedida en mayo de 2021

Domínguez-Castro F, García-Herrera R, Vicente-Serrano SM (2018) Wet and dry extremes in Quito (Ecuador) since the 17th century. Int J Climatol 38:2006–2014. https://doi.org/10.1002/joc.5312

Durán G, Martí M, Mérida J (2016) Crecimiento, segregación y mecanismos de desplazamiento en el periurbano de Quito. Íconos. Revista de Ciencias Sociales 156:123–146. https://doi.org/10.17141/iconos.56.2016.2150

El Universo (2022) El aluvión fundió a La Comuna y a La Gasca, dos barrios de Quito con realidades y orígenes distintos. https://www.eluniverso.com/noticias/ecuador/el-aluvion-fundio-a-la-comuna-y-a-la-gasca-dos-barrios-con-realidades-y-origenes-distintos-lluvia-desaparecidos-fallecidos-escombris-lodo-historia-barrios-de-quito-nota/

EPMAPS (2012) Datos cartográficos de la red de saneamiento de Quito. Empresa Pública Metropolitana de Agua Potable y Saneamiento de Quito

Estacio J, Narváez N (2012) Incendios forestales en el Distrito Metropolitano de Quito (DMQ): conocimiento e intervención pública del riesgo. Revista Letras Verdes 11:27–52. https://doi.org/10.17141/letrasverdes.11.2012.914

European Space Agency, Sinergise (2021) Copernicus global digital elevation model. Technical report, Distributed by OpenTopography.https://doi.org/10.5069/G9028PQB.Accessed: 2021-06-27

Fournier d'Albe EM (1979) Objectives of volcanic monitoring and prediction. J Geolog Soc 136(3):321–326.https://doi.org/10.1144/gsjgs.136.3.0321

Guéguen P (1997) Microzonificación de Quito, Ecuador. Reporte técnico, Instituto Geofísico. Escuela Politécnica Nacional. https://www.researchgate.net/publication/32970136 Microzonificacion de Quito Ecuador

Guidoboni E (2008) Vesuvius: a historical approach to the 1631 eruption "cold data" from the analysis of three contemporary treatises. J Volcanol Geoth Res 178(3):347–358. https://doi.org/10.1016/j.jvolgeores.2008.09.020

Hewitt K (1983) The idea of disaster in a technocratic age, vol 74. Allen I& Unwin Inc. Winchester, USA, pp 171

Hibsch C, Alvarado A, Yepes H, Perez VH, Sébrier M (1997) Holocene liquefaction and soft-sediment deformation in Quito (Ecuador): a paleoseismic history recorded in lacustrine sediments. J Geodyn 24:259–280. https://doi.org/10.1016/S0264-3707(97)00010-0

INEC (2010a) Base de datos del Censo de Población y Vivienda 2010a—a nivel de manzana. Instituto Nacional de Estadística y Censos. http://www.ecuadorencifras.gob.ec/base-de-datos-censo-de-poblacion-y-vivienda-2010a-a-nivel-de-manzana/

INEC (2010b) Cartografía digital. Instituto Nacional de Estadística y Censos. http://www.ecuadorencifras.gob.ec/cartografia-digital-2010b/

Instituto Geográfico Militar (2012) Base cartográfica nacional, escala 1:50000. http://www.geoportaligm.gob.ec/portal/index.php/cartografia-de-libre-acceso-escala-50k/

Jaquet O, Connor C, Connor L (2008) Probabilistic methodology for long-term assessment of volcanic hazards. Nucl Technol 163(1):180–189

Jaramillo Castelo CA, Cruz D'Howitt M, Padilla Almeida O, Toulkeridis T (2018) Comparative determination of the probability of landslide ocurrences and susceptibility in Central Quito, Ecuador. In: IEEE Xplore. 2018 international conference on eDemocracy and eGovernment (ICEDEG). 4–6 April 2018. https://doi.org/10.1109/icedeg.2018.8372341

Larco Moscoso MA (2018) Quito: 40 años de políticas de regularización de suelo, aportes al mapeo de enfoques sobre urbanización informal, el período de la Revolución Ciudadana (2009–2014) y los retos en el paradigma del Buen Vivir. Master Tesis, Facultad Latinoamericana de Ciencias Sociales, FLACSO Ecuador. Departamento de Asuntos Públicos. http://hdl.handle.net/10469/14654

Lozano Castro A (1991) Quito ciudad milenaria: forma y símbolo. Quito, Ecuador: Ediciones Abya-Yala. http://190.57.147.202:90/xmlui/handle/123456789/1879

Marrero J, Yepes H, Pastor J, Palacios PB, Erazo C, Ramón P, Estrella C (2019) Integrating and geolocating post earthquake building damage surveys: the 7.8 Mw Jama-Pedernales earthquake Ecuador. Spat Inf Res 27:317–328. https://doi.org/10.1007/s41324-018-0230-y

Molina E, Ercolani P, Ángeles G (2017) La planificación del espacio público de ocio como oferta para el residente y el visitante de la ciudad de Quito. Siembra 4:141–148. https://doi.org/10.29166/siembra.v4i1.508

Moscoso Rosero R (2013) Dinámicas socio-espaciales urbanas. Una exploración desde las lotizadores irregulares de Quito, negociantes de la pobreza. Cuadernos De Vivienda Y Urbanismo X Seminario de Investigación Urbana y Regional pp 1–10

Mothes PA (1992) Lahars of Cotopaxi volcano, Ecuador: hazard and risk evaluation In: Geohazards, McCall G, Laming D, Scott S (eds). Springer, pp 53–63. https://doi.org/10.1007/978-94-011-2310-5_7

NASA-JPL (2021) NASADEM merged DEM global 1 arc second V001. Distributed by OpenTopography. https://doi.org/10.5069/G93T9FD9. Accessed: 27 June 2021

Odriozola J (1945) El Plan Regulador o Director de la Ciudad de Quito 1942–1945. Reporte Técnico, Municipalidad de Quito. Open Street Map. 2020. Planet OSM. https://planet.openstreetmap.org/. Accessed: 01 Nov 2020

Paredes Méndez DF (2017) Hydraulic analysis of urban drainage systems with conventional solutions and sustainable technologies: Case study in quito, ecuador. J Water Manage Model 26(C440). https://doi.org/10.14796/JWMM.C440

Peñafiel LA (2009) Geología y análisis del recurso hídrico subterráneo de la subcuenca del sur de Quito. Tesis de grado, Facultad de Ingeniería en Geología y Petróleos. http://bibdigital.epn.edu.ec/handle/15000/1147

Peltre P (1989a) Quebradas y riesgos naturales en Quito, periodo 1900–1988. Estudios de Geografía 2:45–90

Peltre P (1989b) Riesgos naturales en Quito: lahares, aluviones y derrumbes del Pichincha y del Cotopaxi, Volume 2 of Estudios de Geografía. Corporación Editora Nacional. http://www.doc umentation.ird.fr/hor/fdi:31647

Peraldo Huertas G, Montero Pohly W (1996) La secuencia sísmica de agosto a octubre de 1717 en guatemala. efectos y respuestas sociales, In: Historia y desastres en América Latina, ed. García Acosta, V., 295–324

Red de Estudios Sociales en Prevención de Desastres en América Latina (La Red). Centro de Investigaciones y Estudios Superiores en Antropología Social (Ciesas). http://cidbimena.desast res.hn/pdf/spa/doc8262/doc8262.htm

Perrin JL, Bouvier C, Janeau JL, Menez G, Cruz F (2001) Rainfall/runoff processes in a small peri-urban catchment in the Andes mountains. The Rumihurcu Quebrada, Quito (Ecuador). Hydrological Processes 15(5):843–854. https://doi.org/10.1002/hyp.190

Pesaresi C, Marta M, Palagiano C, Scandone R (2008) The evaluation of "social risk" due to volcanic eruptions of Vesuvius. Nat Hazards 47(2):229–243. https://doi.org/10.1007/s11069-008-9214-x

Portes A, Roberts B (2008) La ciudad bajo el libre mercado. La urbanización en América Latinadurante los años del experimento neoliberal, Chapter Introducción. Universidad de Zacatecas, pp 13–59

Portes A, Walton J (1976) Urban Latin America. University of Texas Press, The Political Condition from Above to Below

Pourrut P, Leiva IS (1989) Las lluvias de quito: características generales, beneficios y problemáticas. Estudios de Geografía 2:33–44

Puente-Sotomayor F, Mustafa A, Teller J (2021) Landslide susceptibility mapping of urban areas: Logistic regression and sensitivity analysis applied to Quito. Ecuador. Geoenviron Disasters 8(19):1–26. https://doi.org/10.1186/s40677-021-00184-0

Rebotier J (2012) Vulnerability conditions and risk representations in Latin-America: framing the territorializing urban risk. Glob Environ Chang 22:391–398. https://doi.org/10.1016/j.gloenvcha. 2011.12.002

Rolandi G (2010) Volcanic hazard at Vesuvius: an analysis for the revision of the current emergency plan. J Volcanol Geoth Res 189(3):347–362. https://doi.org/10.1016/j.jvolgeores.2009.08.007

Salazar E, Henríquez C, Durán G, Qüense J, Puente-Sotomayor F (2021) How to define a new metropolitan area? the case of Quito, Ecuador, and contributions for urban planning. Land 10(413):1–23. https://doi.org/10.3390/land10040413

Salazar P (2013) Transformaciones inmobiliarias en el DMQ 1970 -2010. Instituto de la Ciudad. https://www.researchgate.net/publication/359722591

Scandone R, Giacomelli L, Rosi M (2019) Death, survival and damage during the 79 AD eruption of Vesuvius which destroyed Pompeii and Herculaneum. J-Reading J Res Didactics in Geography 2:5–30. https://doi.org/10.4458/2801-01

Secretaría de Territorio, Hábitat y Vivienda (2018) Quito sensible al agua. Diseño urbano sensible al agua. Distrito Metropolitano de Quito. https://territorio.maps.arcgis.com/apps/MapSeries/index. html?appid=1c26e5eeb5fa4a4d9ea7bc935cb9b747

Serrano Vincenti S, Carlos Ruiz J, Bersosa F (2016) Heavy rainfall and temperature proyections in a climate change scenario over Quito, Ecuador. La Granja: Revista de Ciencas de la Vida 25(1):16–32. https://doi.org/10.17163/lgr.n25.2017.02

Sherman SI, Zlogodukhova OG (2011) Seismic belts and zones of the earth: Formalization of notions, position in the lithosphere, and structural control. Geodyn Tectonophysics 2:1–34. https:// doi.org/10.5800/GT-2011-2-1-0031

SIGTIERRAS (2010) Modelo digital del terreno a 3 m resolución. Sistema Nacional de información de Tierras Rurales e Información Tecnológica. http://www.sigtierras.gob.ec/

The Pandas Development Team (2020) February. pandas-dev/pandas: Pandas. Zenodo. https://doi. org/10.5281/zenodo.3509134

Torres Jiménez KV (2017) Sistemas socioecológicos en la prehistoria del Valle de Quito: un estudio de escala temporal amplia. Tesis de graduación, Escuela de Antropología, Facultad de Ciencias Humanas, Pontífica Universidad Católica del Ecuador. http://repositorio.puce.edu.ec/handle/22000/13238

UNDRR (2021) UNDRR Strategic Framework 2022–2025. Technical Report, United Nations. https://www.undrr.org/publication/undrr-strategic-framework-2022-2025

UNISDR (2015) Global assessment report on disaster risk reduction 2015. Technical Report, United Nations. https://doi.org/10.18356/919076d9-en

Van Rossum G (2022) Python language reference. Release 3.10.4. Technical report, Python Software Foundation. https://fossies.org/linux/python-docs-pdf-a4/reference.pdf

Velez R, Calderon D, Carey L, Aime C, Hultquist C, Yetman G, Kruczkiewicz A, Gorokhovich Y, Chen RS (2022) Advancing data for street-level flood vulnerability: evaluation of variables extracted from google street view in quito, ecuador. IEEE Open J Comput Soc 3:51–61. https://doi.org/10.1109/OJCS.2022.3166887

Villacis Rivadeneira E, Marrero de León N (2017) Precipitaciones extremas en la ciudad de Quito, provincia de Pichincha–Ecuador. Ingeniería Hidráulica y ambiental XXXVIII(2):102–113

Voight B (1990) The 1985 Nevado del Ruiz volcano catastrophe: anatomy and retrospection. J Volcanol Geoth Res 42:151–188. https://doi.org/10.1016/0377-0273(90)90027-D

Ye L, Kanamori H, Avouac JP, Li KF, Cheung L, Lay T (2016) The 16 April 2016, Mw 7.8 (Ms 7.5) Ecuador earthquake: a quasi-repeat of the 1942 Ms 7.5 earthquake and partial re-rupture of the 1906 Ms 8.6 Colombia-Ecuador earthquake. Earth Planet Sci Lett 454:248–258. https://doi.org/10.1016/j.epsl.2016.09.006

Yepes H, Audin L, Alvarado A, Beauval C, Aguilar J, Font Y, Cotton F (2016) A new view for the geodynamics of Ecuador: implication in seismogenic sources definition and seismic hazard assessment. Tectonics 35:1249–1279. https://doi.org/10.1002/2015TC003941

Yepes H, Marrero J (2021) Construcción del modelo de exposición multipropósito y multiamenaza de edificaciones y población para el dmq. Informe técnico, Alcaldía del Municipio de Quito. Asesoría en riesgos

Zuccaro G, De Gregorio D (2013) Time and space dependency in impact damage evaluation of a sub-plinian eruption at mount vesuvius. Nat Hazards 68(3):1399–1423. https://doi.org/10.1007/s11069-013-0571-8

**Jose M. Marrero** Geographer researcher at Montserrat Volcano Observatory, Montserrat Island, Caribe. He is an expert in hazard and risk assessment, scientific communication in volcanic crises. He was an advisor in natural risk at the Mayor's Office of Metropolitan District of Quito.

**Hugo Yepes** is an advisor in natural risk at the Mayor's Office of Metropolitan District of Quito. He got the prize "Frank Press Public Service" of the Seismological Society of America—SSA on 2011.

**Paco Salazar** Inteligentarium, Quito, Pichincha, Ecuador. Expert Ingenier in urban economy. Former team member of Consejería Presidencial para el Hábitat y el Medio Ambiente de Ecuador.

**Sylvana Lara** is a functionary of Dirección Metropolitana de Gestión de Riesgos at Metropolitan District of Quito.

# Chapter 15
# Between Greens and Grays: Urbanization and Territorial Destruction in the Sabana de Bogotá

**Alice Beuf, Germán Quimbayo Ruiz, and Olaff Jasso García**

**Abstract** This text discusses the concept of urbicide, understanding it as a contradictory process in which the production of the modern city implies the destruction of a given territory's conditions, such as habitability and ecological sustainability. Such a destruction process occurs in multiple territories and territorialities and threatens human and non-human urban natures. For our discussion, we explore the case of the Sabana de Bogotá region in Colombia, where the city of Bogotá is located, and the multiple territorial destructions related to the region's metropolitanization. As a focus, we draw upon the development of the housing megaproject Ciudad Verde in the municipality of Soacha, which involved different types of territorial destructions reflected in environmental degradation and loss of cultural practices or heritage. We argue that territorial destructions are part and parcel of the Sabana de Bogotá region's urbanization.

**Keywords** Urban deterioration · Sustainable development · Degradation

## 15.1 Introduction

This text discusses the concept of urbicide, understanding it as a contradictory process in which the production of the modern city implies the destruction of a given territory's conditions, such as habitability and ecological sustainability. We explore urbicide as a process of territorial destruction through the transformation of the Sabana de Bogotá region in Colombia. The concept of urbicide often is defined as the process of symbolic and material destruction of the urban experience (Carrión Mena 2018). We

---

A. Beuf (✉)
Geography Department, National University of Colombia, Bogotá, Colombia
e-mail: aabeuf@unal.edu.co

G. Quimbayo Ruiz
Geography Department, National University of Colombia, Bogotá, Colombia

O. Jasso García
College of Geography, Faculty of Philosophy and Letters, National Autonomous University of Mexico, Mexico City, Mexico

© The Author(s), under exclusive license to Springer Nature Switzerland AG 2023
F. Carrión Mena and P. Cepeda Pico (eds.), *Urbicide*, The Urban Book Series,
https://doi.org/10.1007/978-3-031-25304-1_15

challenge this conceptualization, arguing that urbanization itself is the destructive process of multiple territories and territorialities.

In this text, we use a definition of territory (and territories) as a set of relationships between humans and between humans and non-humans. Far from conventional conceptualizations, territory is not merely the physical support of human activities in need of policies that aim to "master and plan" such geographical space. Instead, territory is a socio-ecological space that is produced by relationships through asymmetric power and inequalities within an urbanization. In addition, we understand the ecological dimension of urbanization as a comprehensive way to address other dimensions of the same urban reality. In doing so, there are different urban territories in time and space articulated in different ways in which the city, as a geo-historical entity, is sustained by the territory's ecology. Therefore, the urbicide destroys territories through uneven geographical relationships at the expense of territorial sustainability. We argue that contemporary urbanization, compared to prior ages, causes multidimensional and multi-scalar significant destructions over human and non-human existences and heritages.

Urbanization as a process of territorial destruction not only is limited to the historical entity known as the "city" nor the urban space (Choay 1969; Harvey 1993). We refer to urbanization as the extended (or planetary) urbanization (Angelo and Waschmuth 2015; Arboleda 2020), which includes all the territories and territorial processes (i.e., rural landscapes or suburbanization) involved in urban space making: extractive spaces, water supply reserves, cultivated fields and all cities' spaces of dependency. Moreover, territories' destruction dynamics and their analysis should not be restricted to build public spaces or architectural heritage, but encompass altogether the destruction of cities' ecological dimensions.

Having introduced our theoretical perspective, we shall analyze how urbanization in the Sabana de Bogotá region implies multiple destructions of urban and rural metropolitan territories. We show that urbanization is more than a process of "creative destruction", given that contemporary ecological destruction has reached its irreversibility point. Therefore, we increasingly need to reevaluate the destroyed territories of urbanization that have been invisibilized in the name of "progress" or development. Likewise, we attempt to assess the excessive anthropocentrism surrounding how urbicide is understood (Coward 2006), taking into account the intricate socio-ecological relationships that explain territorial transformations and destructions such as urbanization. It is not about ignoring the humanistic concerns in urbicide, but of including more explicitly the environment–society relations in territorial destruction analysis. The non-human sphere, that is the ecological dimension, is active and fundamental in processes of territorial transformation and allows us to understand territorial destruction and its role in the production of urbanization in late capitalism. Usually, in analyzing environmental (territorial) conflicts, the non-human is seen as an object of dispute or a stressful factor in the interactions between the environment and the society. However, ecosystems, animal and plant species, soil, rivers, geomorphological formations and air, among others, have an active role in environmental dynamics and controversies that show the mutual dependencies between social actors and non-human entities (Quimbayo Ruiz 2018; Stengers 2015).

In turn, these ecological concerns in urbicide connect with the tension between processes of territorial destruction and dispossession, as well as the types of resistance that are reflected in other forms of territorial socio-ecological coexistence (Lesutis, 2020). This also suggests that urbanization encompasses the production of multiple territorialities (Haesbaert 2013), which expose fundamental and urgent questions about spatial justice (Soja 2014) in times of climate emergency and planetary crisis. The establishment of sacrificial zones, such as for the extraction of building materials or landfills, the destruction of green areas, and the loss of possible access and space appropriation by the most vulnerable populations, is illustrative examples. Likewise, the development of an urbanization model that prioritizes real estate speculation and profit threatens the loss of peasant (*campesino*) and ancestral territorial settings and occupation or ones different from those imposed by capitalist urbanization. Furthermore, the destruction of territories exacerbates climate risks and vulnerabilities and accumulated socio-ecological inequalities alike.

In what follows, we introduce a geographical characterization of Sabana de Bogotá. Later, we specify the different types of territorial destruction identified in our case study. We conclude by offering a balance between the divergent understandings of the meanings of these destructions and the theoretical and conceptual contributions that we propose to discuss the concept of urbicide.

## 15.2 The Sabana De Bogotá: Massive Urbanization in a High Andean Plateau

The urbanization process of the Sabana de Bogotá can be explained through different territorial transformations, which arise from socio-ecological relationships at different space and time scales. The historical socio-spatial segregation in the metropolitan space is mutually intertwined with situations of environmental suffering and injustice. In what follows, we shall introduce the socio-ecological features and specificities of the Sabana de Bogotá region.

The Sabana is the region where the Colombian Capital District, Bogotá, is located besides the municipalities and towns of its metropolitan area.[1] The Sabana is a high Andean plateau ecosystem that is between 2600 and 3800 m above sea level in the central axis of the Colombian Eastern Cordillera (part of the Northern Andes). This particular physical geography explains why the region was declared a national strategic ecosystem in 1993 (Article 61 Law 99/1993). The Sabana is of a fluvial-lacustrine origin, setting a geoform that constitutes the surrounding hills and mountains of the city region, characterized by high levels of biodiversity, mainly consisting of sub-ecosystems such as high-Andean forests, rivers, wetlands and moors (*páramos*).

The region's hydrological system (the upper basin of the Bogotá River) has a total area of 4321 km$^2$ comprising the following sub-basins: the Teusacá, Tibitó, Negro,

---

[1] Around 20 municipalities depending on the scale and the definitions of the limits of the Sabana.

Frío, Chicú, Balsillas, Tunjuelo, Fucha and Soacha rivers. The region's climate is cold and dry on the plain, with average temperatures in the range of 12–17 °C. Precipitation averages are around 800 mm per year, decreasing to 500–600 mm per year toward the southwestern part of the Sabana. A colder humid climate prevails in the higher areas, with temperatures lower than 10 °C and annual rainfall above 1000 mm, while cold humid conditions prevail in the foothills, with average temperatures between 10 and 12 °C and annual rainfall between 800 and 1000 mm. Temperature and precipitation regimes in the Bogotá region are affected by the El Niño/La Niña Southern Oscillation (ENSO) (Carvajal and Navas 2016; IDEAM-FOPAE 2015).

In this unique and vulnerable ecosystem, the city of Bogotá was founded by the Spanish conquerors in the same place where an important settlement of the Muisca people existed. The initial site lay in the foothills of the eastern hills (Cerros Orientales) between two watercourses, the San Francisco River and the San Agustín River, and the city's expansion took place toward the Altiplano ecosystem. Yet, the region is not at all a high Andean plateau without topographic features. Instead, it is a complex interconnected network of wetlands, which saw its ecological functionality affected by its depletion and territorial transformations. After the colonial and independence periods, during the nineteenth century, the privatization of the indigenous reservations and the concentration of land as large cattle farms constitute the main territorial changes (Delgado Rozo 2010), besides the growth and expansion of urban spaces at the beginning of the twentieth century.

Urbanization exceeded its historical limits of a compact city in the middle of the twentieth century and expanded at a very fast pace from the 1950s until the 1990s, fueled by strong demographic growth (population growth in this period of demographic transition and the arrival—or expulsion—of waves of migrants from rural areas and small towns). In addition, in 1954, five neighboring municipalities were annexed and a continuous semi-circular urban sprawl formed, from Bogotá's eastern hills (*Cerros Orientales*) toward the edge of Bogotá River, within the current limits of the Capital District. In the last two decades, however, the dynamics of this process overflowed district boundaries and configured a complex functional metropolitan area, which gradually polarized an urban region that encompasses medium-sized cities at a distance time of up to 3 h from Bogotá (Thibert and Osorio 2017) (also see Table 15.1).

In our present analysis, we are primarily interested in the formation of the metropolitan area because it is entirely included in the Sabana de Bogotá, whereas the urban region is more extended. Since 1997, there has been an acceleration in the rate of expansion toward the 20 municipalities of the metropolitan area (see Table 15.1). These 20 municipalities show a rate of expansion and demographic growth systematically higher than Bogotá. In addition, due to the saturation of urban space within the Capital District, Bogotá grows at a densifying rates (demographic growth is higher than urban expansion), while in the 20 municipalities, on the contrary, the rate of urban expansion has been stronger than the demographic growth, meaning that urbanization has reached very diverse territories in the Altiplano. Spatial and urban planning historically contributed to limiting urban sprawl within the District Capital of Bogotá (along with other structural factors), but at the present time, it

**Table 15.1** Metropolitan growth and population figures

| Years | Geographical unit | Urban sprawl (ha) ||| Demographic growth ||| Population density (hab/ha) |
|---|---|---|---|---|---|---|---|---|
| | | Total (ha) | Absolute increase (ha) | Relative increase (%) | Total (inhabitants) | Absolute growth (hab) | Relative growth (%) | |
| 1997 | Bogotá DC | 31,334 | – | – | 5,956,995 | – | – | 190 |
| | 20 municipalities of the Sabana of Bogotá | 6530 | – | – | 722,052 | – | – | 111 |
| 2005 | Bogotá DC | 33,506 | +2172 | +6.9% | 6,778,691 | +821,696 | +13.7% | 202 |
| | 20 municipalities of the Sabana of Bogotá | 7853 | +1323 | +20.2% | 1,036,586 | +314,534 | +43.5% | 132 |
| 2010 | Bogotá DC | 35,667 | +2161 | +6.4% | 7,363,782 | +585,091 | +8.6% | 206 |
| | 20 municipalities of the Sabana of Bogotá | 20,995 | +13,142 | +167% | 1,410,799 | +374,213 | +36.1% | 67 |
| 2016 | Bogotá DC | 36,143 | +476 | +1.3% | 7,980,001 | +616,219 | +8.3% | 221 |
| | 20 municipalities of The Sabana of Bogotá | 27,309 | +6314 | +30% | 1,969,893 | +559,094 | +39.6% | 71 |

*Source* Own calculations from: IDOM (2018)

is completely unable to stop such processes in the metropolitan area. In particular, since 2005, there has been unbridled urban growth in these municipalities, due to the recomposition of socio-economic productive systems and the relocation to the metropolitan periphery of many economic activities such as warehouses, industries, and logistics platforms that heavily consume space.

Sabana de Bogotá's urbanization also has been possible due to the construction of major hydraulic infrastructures during the late 1970s and throughout the 1980s and early 1990s. In fact, the Andean plateau is fed only by the Bogotá River's waters; therefore, it was necessary to build two dam lakes connected together to the city's water supply network in order to bring water to Bogotá from the Chuza River in the Chingaza Páramo at 2990 m above sea level. This latter action corresponds to a water transfer from another basin (the Orinoco's) to the Bogotá-Magdalena watershed. These major hydraulic engineering works provide more than 80% of the water consumed in Bogotá. The second water supply system is connected to the highly polluted Bogotá River, which is channeled and distributed to a network of water treatment plants (Bolívar Molano and Montoya Garay, 2021). Altogether these infrastructures guarantee a water supply not only for the growing population of the city of Bogotá, but also for the other municipalities of the Sabana. The Bogotá Water Company (*Empresa de Acueducto de Bogotá*) sells a specific amount of water for use in these municipalities. This is a driver of the region's increasingly massive urbanization.

In our analysis, we emphasize the situation within the metropolitan area of the municipality of Soacha, which has its urban perimeter in conurbation with Bogotá. Soacha is the most populated municipality in the metropolitan area and its population growth has been one of the highest and most exponential in all of Colombia: 23,997 in 1973, 398,295 in 2005, and 645,205 inhabitants in 2018. This demographic dynamic is due to high population growth and the waves of migrants that have arrived in recent decades from different parts of the country, including refugees from the Colombian armed conflict. This population is attracted by the proximity to Bogotá and the availability of cheaper land and housing. Soacha presents the highest poverty rate in the whole metropolitan area and is connected to the southern marginalized and impoverished neighborhoods of the capital. This situation reflects that the north/south socio-spatial urban segregation is deepening at a metropolitan scale.

Most of Soacha's urban space corresponds to marginalized or informal settlements. Since 2010, a change of pace in the scale of formal urbanizations has been noticeable. The Ciudad Verde housing project, with more than 45,000 housing units finalized in 2020, is an example of such an urban transformation in the making of the marginalized and peripheral city. This project was part of the execution of the so-called "Housing Macro-projects" (*Macroproyectos de Vivienda*) or low-income housing macro-projects under the national administration of Álvaro Uribe's government (second term) in Soacha. With the adoption of Decree 4260 of 2007, the national government set the figure for developing the macro-projects of national social interest, to guarantee the rapid turnover of capital required by housing developers and construction companies because of the exhaustion of available land and a possible decrease in the profitability of social housing projects. The adoption of Decree 4260 facilitated the transformation of land use—from rural to urban—intended for the mass production of social housing, justified by the argument of the urgency of reducing the housing deficit.

For the formulation of these macro-projects, the owners of rural lands have organized themselves to lead the development initiative, setting up trusts and presenting their projects to the Ministry of the Environment, Housing and Territorial Development (after 2011 to the Ministry of Housing). Initially, the macro-projects could be directly approved by the ministry, without any coordination with the municipalities, even going around and not recognizing their municipal Land-Use Master Plans (Planes de Ordenamiento Territorial, henceforth POT), as actually happened in Soacha. Yet, some councilors from Bogotá's District challenged the figure set for developing the macro-projects in the courts, winning the case after the Constitutional Court ruled and declared Decree 4260 unconstitutional in 2010. The Court ruled that the Decree does not acknowledge municipalities' autonomy in their land-use regulations. However, the projects already started before the court decision is continued as planned, among them the Ciudad Verde Project in Soacha. Nowadays more than 200,000 inhabitants live there.

The urbanization model described above for the Sabana de Bogotá has caused territorial destruction at various socio-ecological levels. What has prevailed is a developmental vision in the name of urban "progress" and a quantitative approach to housing production. The Ciudad Verde project was based on guaranteeing the realization of influential private and speculative interests of such sectors as real estate, housing development and construction companies. It has not considered ecological considerations or the quality of urban life after a few years.

These kinds of housing macro-projects have become problematic because of the complex and diverse specificity of the Sabana de Bogotá ecosystem, which is already depleting its ecological integrity. Territory destruction has been pointed out and denounced for decades by various environmental, academic and activist sectors. All local sub-ecosystems and protected areas have been framed according to the scientific political concept of the Main Ecological Structure—*Estructura Ecológica Principal*—(henceforth, MES), proposed originally by the Dutch-Colombian ecologist Thomas van der Hammen (1998) for the Environmental Action Plan for the Bogotá River watershed. Several Colombian environmental scientists have since used it (Fig. 15.1).

The MES might portray the Bogotá regional ecosystem as a hybrid socio-natural entity (cf. Swyngedouw 1996). This urban-regional ecosystem has undergone constant land-use changes, and its original ecology of streams, wetlands and swamps has been continuously and severely modified during the prehispanic period (cf. van der Hammen 1992; Delgado Rozo 2010). Although the urban-regional ecosystem structure and ecological values already had been modified for a long period of time,

Main Ecological Structure Bogotá's river upper basin (Van der Hammen, 1998).

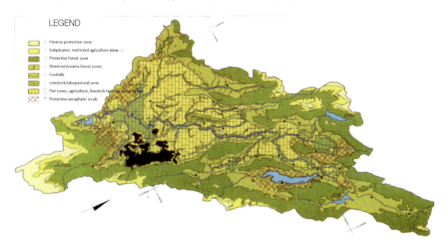

**Fig. 15.1** Main ecological structure proposal in the Sabana de Bogotá region by Thomas van der Hammen (1998). *Source* Own elaboration

they have been markedly simplified and depleted, especially since the human population of the city of Bogotá skyrocketed during the twentieth century and later with the acceleration of the urban sprawl over all the Sabana de Bogotá.

In parallel to the MES and ecological structures transformations, recent climate change projections for the city region predict an increase in average temperature close to 1 °C for the next 40 years (IDEAM 2015), which will increase rain patterns. The territory faces a significant climate risk and vulnerability, especially in issues such as water supply, food and settlements (IDEAM 2017), taking into account the complex and intricate regional climatic conditions and variability. Yet, despite the transformations associated with urbanization that the ecological structure has undergone, the region still holds important levels of biodiversity, including unique high Andean ecosystems such as wetlands and subxerophytic shrubs or native and endemic fauna and flora species that are threatened highly now (Quimbayo-Ruiz 2016).

## 15.3 Types of Territorial Destruction: Between Resilience and Points of no Return

For our understanding of urbicide as a process of territorial destruction, it is important to take into account that there are thresholds of both resilience and points of no return for ecological processes, as well as the existence of different territorialities (urban, rural) constituting the Sabana de Bogotá as a territory. This alludes to the configuration of an unsustainable urbanization model that underlies a strong metropolitan socio-spatial segregation, as well as the occurrence of multiple socio-ecological inequalities. In Fig. 15.2, we introduce a typology of territorial destruction based on empirical information and a review of different sources of information such as studies on social and population dynamics and ecological, hydrological and geophysical regional characterization, as well as oral sources and interviews. We have summarized all these sources of information in order to propose a synthesis of socio-ecological processes related to territorial destructions identified in the Sabana.

### *15.3.1 Deterioration of Hydrological Processes (Pollution and Flood Risk Management)*

The Sabana de Bogotá's urbanization process has involved a complex relationship with its hydrological and hydrogeological systems, which comprises a set of diverse ecosystems such as páramos, rivers, wetlands and geological features (soils) that make possible hydrological dynamics. The population in the Bogotá region exceeds eight million inhabitants who consume about 17 m$^3$/s of drinking water (Gaviria Melo, interview February 2022). Yet, hydrological issues are not only about the

# 15 Between Greens and Grays: Urbanization and Territorial Destruction …

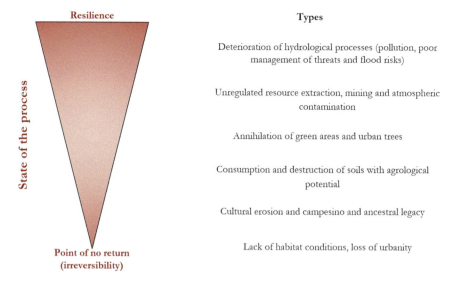

**Fig. 15.2** Types of territorial (territories) destruction and state of the urbicide process. *Source* Own elaboration

supply of drinking water that makes life possible in the region, but also the wastewater treatment and how water bodies such as rivers, streams and wetlands absorb pollution.

On the one hand, the urban space has sprawled at the expense of environmental quality, through infrastructure development like the aqueduct and sewage system (including dams and treatment plants by the Bogotá Water Company), adaptations of water rounds, canalizations and hardening of water bodies. Molano and Garay (2021) offer an interesting map of the water supply, purification and sanitation systems in the Sabana de Bogotá. They introduce an actor mapping of the region's water management governance (ibid.: pp. 489 and 497). Due to ecosystemic and climate-changing conditions, infrastructure shortcomings and challenges, Bogotá is considered one of the capitals with the greatest vulnerability in its aqueduct system worldwide (Gaviria Melo, interview February 2022) (Fig. 15.3).

This vulnerability occurs for two main reasons. First, the city of Bogotá's growth has led to processes of the occupation of territory, mainly through urban and industrial activities, that have caused the deterioration of the quality of water in the internal basin, especially due to the fact that the adequate treatment of water was not ensured. The polluted waters have been discharged mostly into the rivers of the basin until today (Gaviria Melo 2021). Secondly, the process of urbanization—or rather urbicide—has implied the need to supply water to Bogotá by bringing water from other basins, since the capitalist city raises the need to increase the availability of water for development projects in the region, with changes in the use of soils of high agrological quality to convert them into developable land and for industry.

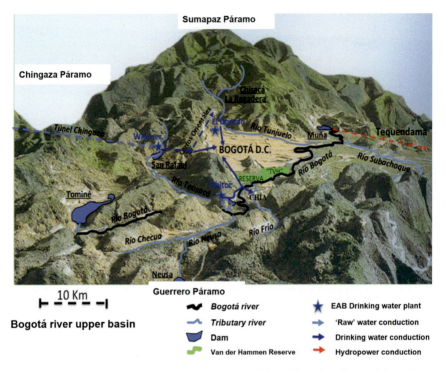

**Fig. 15.3** Drinking water supply system for Bogotá and the EAB region. *Source* Adapted from Gaviria and Sierra (CAR 2011) and modified by the authors. *Image courtesy* Sergio Gaviria Melo

The construction of this built environment (Harvey 1990) for the city's water supply, in turn, implies a contradictory process of urbanization understood as urbicide. The constant growth of the city implies carrying more and more water and therefore the creation of new infrastructure; however, the transfer between basins has serious effects on the territories of subtraction, for example, as it can affect the hydrological cycle of the region and cause a situation of scarcity (Esparza 2014). On the other hand, the lack of treatment for contaminated water within the city puts its inhabitants at risk, especially within the most impoverished populations. These two elements indicate that urbicide is a multi-territorial and multi-scalar process. We will illustrate how the latter affects the Ciudad Verde housing macro-project area.

Water sources that supply Bogotá and several municipalities from the metropolitan region are the Chingaza System, Tunjuelo System and Bogotá River system. The water supply for Soacha and Ciudad Verde is provided by the Bogotá Aqueduct Company (EAB, acronym in Spanish). Initially, the Ciudad Verde project contemplated the construction of 22,000 housing units, which ended up increasing to 42,000 and later to 49,000 units, which has generated a burden on the general urbanization process, including the aqueduct. The construction of primary aqueduct and water supply networks (e.g., connected to the Chingaza system) was left to urban developers and their companies. Amarilo, one of these companies, participated in the construction of the "Cazuca" water storage tank, later giving it to the EAB for its administration.

Regarding sewerage and wastewater treatment, the Canoas Wastewater Treatment Plant (PTAR, acronym in Spanish) was built in Soacha, to which wastewater is supplied by the Fucha–Tunjuelo interceptor and the Tunjuelo–Bajo interceptor. The PTAR receives 11,666 L per second. The Canoas PTAR receives wastewater from the Fucha, Tunjuelo and Tintal water currents and from the municipality of Soacha. The PTAR was built through a public–private partnership; the National Planning Department contracted the temporary public–private partnership Union APP Canoas, co-formed by the firms WSP, Castalia Advisors and Durán and Osorio Abogados, to set up the technical, legal, economic and financial structure of the PTAR. The Public–Private Association Project, the EAB, the Regional Autonomous Corporation of Cundinamarca-CAR, the Capital District and its Environmental Office (Secretariat) also participated in the project. The Argos company supplied the concrete for the construction of the plant.

In fact, the environmental quality of the Bogotá River and its basin (with all its rivers and water bodies) has taken dramatic overtones due to its level of pollution and degradation by urban and industrial activities in the region. Different courts of justice have ordered and ruled (in 2004, first case before the Administrative Court of Cundinamarca, and in 2014, the second case before the State Council) in favor of the rehabilitation of the river and its basin. These court rulings required that all new urban projects had to provide pretreatment before delivering water to the aqueduct. But, the ruling guidelines have been widely ignored in their mandates, despite some specific marginal actions. The long-term constraints of the institutional setting diminished ecosystem protection and management and remain trapped in endless legal disputes and high judicialization of environmental issues. Besides conflicts of interest, the lack of clarity in spatial planning rules has also created an uncertain scenario for urban-regional ecosystems, such as the upper Bogotá River basin, whose governance regimes overlap with the different district (local), regional or national levels.

On the other hand, regional flood threats and risks are reinforced through human interventions. For example, due to the natural disturbances of the ENSO phenomenon (Niño/Niña) exacerbated by the effects of anthropogenic climate change, the rainy periods are getting wetter, but the territory increasingly has been impermeabilized by the loss of the soil's green covers, hardening of land surfaces and Bogotá's rivers channelization and other urban interventions. This has made the territory more vulnerable and affected by floods, which have had already catastrophic effects such as those that occurred in the 2010–2011 period, when multiple low-income housing and infrastructure projects were affected. Despite some adaptation measures taken by government authorities and the state, the drivers of transformation and territorial destruction still are not taken into account, increasing the probability of recurrence of such disaster situations in the short and long terms.

## 15.3.2 Unregulated Resource Extraction, Mining and Atmospheric Contamination

The resource and mineral extraction activities have been implicated significantly in the Sabana de Bogotá's urbanization process. In particular, the southern parts of the region have been one of the most important places in the extraction of construction materials in the country, such as clay, sandstone, gravel and aggregates (Portilla and Laura 2021; Reina Rozo 2013). A clear example of this situation is present in Soacha, where Ciudad Verde was developed. According to the National Agency of Mines—ANM (2020), there are 61 permits for the exploitation of construction materials in Soacha, of which only 12 have an environmental management plan. Legal mining and resource extraction activities cover 2300 ha of the municipality, which represent 15% of the total municipal area (182 km$^2$). Furthermore, 70% of legal mining is in the hands of only seven companies. Nevertheless, there also exists an undetermined number of illegal enterprises with zero environmental protection measures. It is estimated that there are approximately 200 mining zones in the municipality and that the level of illegality is between 60 and 75%. In this way, the available area in Soacha for the development of mining projects covers nearly 30% of its territory (Marin 2014).

The main real estate companies involved in Ciudad Verde's development are Constructora Amarilo, Constructora Bolívar and Hogares Soacha; three companies are belonging to ProBogota, while the first two belong to the Colombian Chamber of Construction—Camacol—that is a national trade association. All these organizations are grouped together with the most important political and business groups in Colombia. The companies that supplied the construction materials to these real estate companies were mainly Cementos Argos, CEMEX, Holcim and Ladrillera Santa Fe, four companies that are also members of Camacol, which has had mines and extraction sites for materials such as clay, gravel, sand and aggregates around the Tunjuelo river basin, near the city of Bogotá and Soacha.

The situation described above implies major problems. On the one hand, the little or zero regulation of mining operations has allowed the expansion of this activity in places that in theory are under environmental protection. Yet, oral evidence from residents of Soacha indicates that the mining could be expanded until the Paramo de Sumapaz ecosystem if actions are not taken, resulting in damage to an ecosystem that is an essential part of the sustainability of Bogotá city. On the other hand, the illegality of the mines has encouraged many of the extraction zones to be located very near to living units, less than 200 m away, when according to the law they should be located at the very least 400 m. All these issues, besides the low technical standards of illegal mining and a lack of enforcement of environmental regulation, have caused air pollution derived from the extractive processes (Fig. 15.4), among other environmental problems and conflicts. Moreover, this situation has also caused human health problems in Soacha. According to the municipality's health office, the main cause of visits to the local hospital is related to acute respiratory infections (Marin, op. cit).

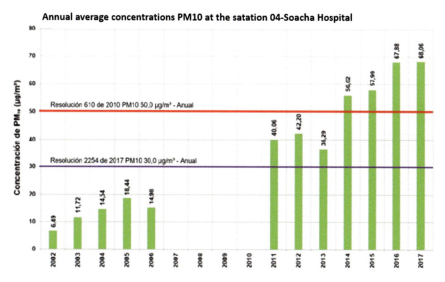

**Fig. 15.4** Annual average concentrations PM10 at the station 04—Soacha Hospital. Adapted. *Source* Corporación Ambiental Caminando el Territorio. [WWW Link. Retrieved: 14 August, 2022]

In this way, the extraction of building materials that make possible expanded urbanization in the Sabana de Bogotá is simultaneously related to a process of territorial destruction, loss of habitability and sustainability.

### 15.3.3 Annihilation of Green Areas and Urban Trees

At the metropolitan level, the urbanization model represents a reduction and constraining of green areas and depletion of urban trees, despite Sabana's ecological potential for such spaces and values. There is a significant number of green and protected areas in Bogotá and its surrounding region, but the population suffers from a lack of effective access to green spaces, especially in marginalized areas (Quimbayo-Ruiz 2016). According to our field observations, housing projects, such as Ciudad Verde in Soacha, have not guaranteed the full harmonization between landscape and local ecosystem, nor the provision and management of green areas and trees and a lack of effective access to the population that needs it most. This situation in Ciudad Verde reproduces a pattern already identified in the Bogotá region (Scopelliti et al. 2016), in which this urban development brought a deep transformation of former green and rural areas into a low-income housing citadel in a period of 13 years (see Figs. 15.5 and 15.6).

**Fig. 15.5** Ciudad Verde area, Soacha, 2009. *Source Photo* Alice Beuf

**Fig. 15.6** Ciudad Verde in 2021. *Source Photo* Alice Beuf, 2021

## 15.3.4 Consumption and Destruction of Soils with Agrological Potential

For years now, several scientists and experts on environmental issues in the Sabana de Bogotá region have been denouncing the loss of areas with agrological potential due to the urbanization process and related land-use changes like infrastructure building (Gaviria et al. 2004). Thanks to the knowledge of the geo-environmental history of this region (van der Hammen 1998), it is known that its Andisol soils, especially in plain areas, have exceptional characteristics that allow territorial sustainability. In fact, in the plain areas of the region, some soils have been classified as being of high agrological quality (Classes 2 and 3), which prohibits their use in industrial activities and urbanization, according to national regulations on rural spatial planning (Decree 3600, 2007). These are fertile agricultural soils used for food crops and livestock on the slopes and in flat areas and also by the flower industry.

According to Sergio Gaviria Melo (2021), excess water can be managed to sustain food security in the Sabana de Bogotá's fertile soils. In fact, there is prehispanic evidence of the indigenous peoples who inhabited (and their descendants who still inhabit) the region who were aware of this phenomenon and cultivated agroecosystems with hydro channels, taking advantage of periods of rain and floods as well as those of drought and frost.

Despite the evidence of the ecological functioning of the region, urban development and growth continue to be the priority without taking into account environmental considerations and climate variability, driving territorial destruction (urbicide). The region's agricultural capacities, in which eight million inhabitants live, have been compromised due to urbanization instead of guaranteeing food security and sovereignty. Quoting again Gaviria Melo (2021: 6), "Soil conservation has been recurrently controversial by the interests of builders and investors, in agreement with the district and municipal administrations in the revision of the land use of the different POT during the present century".

In addition, a technical characterization of the soils was carried out some time ago (IGAC 2012). The characterization's results supported the protection of the rich agricultural lands in many plain sectors of the region. The characterization reclassified the soils at a property scale (1:10,000) for their physical and chemical properties that determine capacity use, freeing 16,875 ha for non-agronomic uses (Gaviria Melo, op. cit). This classification change was in part due to excess moisture in the soil—a limiting factor for use—which was temporary. This also only occurs under extreme conditions such as those that occurred during the time the study was prepared (2011), when severe flooding from the most intense La Niña period recorded in the country's history affected the region.

With all the situations above mentioned, it can be assumed that the development of urban projects and low-income housing, as well as areas destined for logistics operations and services (known locally as *"zonas francas"*) related to land and air transport routes, are prioritized over agricultural capabilities.

## 15.3.5 Lack of Habitat Conditions, Loss of Urbanity

There is much more discussion about the dimensions of urbicide such as the "death of the city", but there is less attention to how to overcome the unsustainable city model that makes it possible. Therefore, it is important to focus the analysis on how to achieve a right to the city in the production of urban areas, especially those that are most marginalized. The Sabana de Bogotá and its urban environments have strong socio-spatial segregation, especially concentrated in the South of the region (Thibert and Osorio Ardila 2017). Such a segregation process comes hand in hand with a degradation in the conditions of habitat and housing, transportation and mobility, public space and environmental degradation, soil and air pollution and loss of effective access to green areas, among others. There have been a historical and unjust allocation and development of high-impact activities such as landfills, dumps or extractive activities, in low-income and marginalized areas. It is clear that the relationship between this process of urban loss and the threat of eroding other manifestations of "being a city" with the other types of territorial destruction is described in the previous sections.

Ciudad Verde's peripheral location and lack of accessibility, equipment, services and employment are some of the most important issues in the local urban life. For example, the construction of the main avenues and roads for accessing the area (Avenida de los Terreros and Avenida Ciudad de Cali) and proposed in the POT—at the expense of the promoter Amarillo—were not finished before the arrival of the inhabitants. This occurred due to land legalization problems and the location of the aqueduct facilities within the area. When the inhabitants arrived, the promoter Amarillo offered a bus system that connected the 2 km that separated Ciudad Verde from the Autopista Sur Highway, which is the main way of connection between Soacha and Bogotá. Today, these avenues and roads are finished, and there are public buses that provide the connection for a relatively high price. Moreover, a serious problem is traffic congestion because of the very low number of entry points to the housing macro-project.

The mobility issues are related to a broader transportation problem that exists throughout the metropolitan area and particularly in the southern part of Bogotá, which is very congested and far from the metropolitan center where most of the jobs are located. Accessibility conditions accentuate the problem of socio-spatial segregation, from which projects such as Ciudad Verde directly suffer and create a risk of producing impoverished settlements on the outskirts of the city. Next to the highway is the shopping center where the inhabitants of Ciudad Verde get their supplies; the shopping center planned within the project has not yet been built. Inside Ciudad Verde, inhabitants have transformed some areas into grocery stores, but the architecture does not permit flexibility in the uses of spaces. It is not possible to make the houses "productive" as people do in informal settlements (stores, workshops, apartment rentals).

Another problem, however, which is not specific to these housing projects but to social housing development in general in Colombia, is housing quality. Added to buildings that break architecturally with the rest of the rural fabric are the housing areas proposed in these projects that are too small to accommodate large families, as low-income families tend to be. The designs of the houses are very similar, with surfaces that vary between 41 and 58 m$^2$ in Ciudad Verde (Beuf and Garcia 2016).

Additionally, this low-income housing macro-project, like others nearby, is built in the Bogotá River's flooding zone, below the level of the Soacha River and the Tibanica stream. Despite the system of dams and containment works built to protect the houses, the inhabitants have to deal with important problems such as the backflow of sewage in the pipes. In Ciudad Verde, the main issue is not water supply but drainage which is problematic and has a strong impact on the quality of life of the inhabitants. Recent low-income housing macro-projects, due to their massive nature, also have to be understood in relation to their surroundings and in particular, according to Humberto Medellín (local environmental leader and guide), to the saturation of urban space and the loss of quality of public spaces, the lack of parks and green areas throughout the municipality of Soacha, as well as the degradation of its historic center.

## *15.3.6 Loss of Cultural Practices or Heritage*

In the metropolitan territory of the Sabana de Bogotá, there are not only urban territories. Metropolitan rurality encompasses different territorialities and populations such as peasant communities (*comunidades campesinas*), small agricultural entrepreneurs and industrial flower growers. It is a space characterized by the presence of economic and productive activities such as agriculture, livestock husbandry, rural tourism and cultural crafts, surrounded by the most important protected areas and urban-regional ecosystems (around 734 km$^2$). About 70% of rural land is made up of *paramo* ecosystems, 9% by high Andean forest and scrub, <2% by forest plantations, 16% by pastures and 3% by crops (Secretaría Distrital de Ambiente, s.f.). Even more critical in this context, there are manifestations and modes of rural, peasant and ancestral occupation—represented mainly in surviving groups of the Muisca indigenous people (Chaves Agudelo 2016; Sánchez-Castañeda 2020), which are affected by the dynamics of urbanization, peri-urbanization and suburbanization.

In the area of Usme, part of the Capital District in the southeast of the metropolitan area, the inhabitants organized themselves to propose an alternative peasant occupation model to the advance of urbanization (Quimbayo Ruiz et al. 2020). The district recognized the formation of a Rural–Urban Edge in that area, characterized by a diversity of overlapping urban and rural activities and territorialities (Bernal Mora 2020). This scenario of negotiation today is still in dispute and emerged since the declaration of an urban expansion zone in the south in the first Territorial Plan of Bogotá in 2000. This peasant (campesino) struggle is unique in the entire metropolitan area. In other areas like Ciudad Verde, the imposition of the logic of urbanization prevailed over pre-existing uses and territorialities.

According to the interviews carried out in our research, community ties in large low-income housing projects like Ciudad Verde which are very weak. Therefore, the modalities of access to housing are very different compared with housing in self-built informal neighborhoods. Under informality, it had been possible to build an entire community and consolidate collective ties alongside social struggles for the improvement and legal recognition of neighborhoods. In the case of social and low-income housing macro-projects, individualization is the rule: in the acquisition of housing, in daily solidarity practices and in resident committees. According to an interview with Humberto Medellín:

> The residential units [complejos residenciales] break the structures of the popular organization. The first one is individualized in his apartment and he is individualized as a whole. There is no solidarity. In the popular neighborhoods before, if the situation was not fixed, the inhabitants began to confront each other and already with councilors in the municipality, because they fought and managed to do that through those fights. Right now people go to work early, arrive at night at 9 to pick up their children, half feed them, and pick them up already sleepy, and asleep. Even that is breaking. That is one of the ugliest things there is. That sitting down to eat with the family is breaking. All of that is breaking. So there is a breakdown of the family nucleus, a breakdown of the social nucleus. (Interview carried out on February 3, 2022)

However, new forms of social organizations are emerging in these new urban settings, particularly led by young people. For example, the Environmental Movement Walking the Territory (Caminando el Territorio) and Suacha Youth Network (La Red Joven Suacha) both bring together young environmentalists. These movements and organizations develop strong and smart analyzes of the urbanization's environmental impacts and mining in Soacha, such as the study on atmospheric pollution in Soacha, whose results were presented previously. Likewise, young organizations work on guided tours with members of the community to promote the social and cultural appropriation of the territory, nurture the sense of place belonging, and recognition of environmental, social, and economic problems. In this case, the environmental perspective prevails and even nostalgia for natural landscapes is present, as it is described as follows:

> Suacha con U, of xerophytic valleys and wetlands, of mountains kissed by the mist goddess, governed by frailejones and eagles, where the waters and at some point the majestic Suacha River flowed freely and gave themselves kindly to those men and women that it is still time and we unearthed it, that it is still time and we see the marks of their fingers. (Environmental Movement Walking the Territory and Suacha Youth Network, Rural Press 2015 and internal documents, cited by Beuf 2019)

Yet, all the young people are not organized like in these organizations and movements. The feeling of hopelessness in these new places in the contemporary city is deep, and it is also difficult to organize within the current social and environmental struggles. The mobilizations take place rather spontaneously, under an explosive register, such as what happened in the last National Strike of April–May 2021, when young persons from the impoverished outskirts led the revolts. In Soacha, as in other places in the Bogotá metropolitan area and the country, the young people from the front lines (Primeras Líneas) blocked strategic roadways and the entrance to the city and to Bogotá from the south for several weeks. The lack of prospects for improving their lives through employment or education and, in general, the lack of meaning in the face of everyday life experiences enclosed in these urban areas of the peripheral city crystallize the motivations of young people. The mass production of standardized urban spaces disconnected from the natural environment goes hand in hand with the destruction of the hope of finding one's own place in the world.

## 15.4 Divergent Understandings of Territories Destruction

As one of the environmental leaders interviewed from Soacha, it seems that the goal of spatial (territorial) planning led by local and regional entities in the last decade was summed up as "getting the water to continue paving the Sabana". In this chapter, we show that this instrumental conception of the territory and its planning on a metropolitan scale systematically ignores the multidimensional and multi-scalar destructions that are inherent and constituent to contemporary capitalist urbanization.

The planetary crisis that we can no longer avoid forces us to re-examine our own understanding of the nature of the urbanization process. Such a process is not only the production of built spaces to offer housing solutions to people who need them. It is also a complex process of territorial transformation that unites spaces and places that are sometimes distant or which depend on each other. Articulated with extractive processes, urbanization is nourished by the destruction of multiple territories and their ecological, cultural and social dimensions, configuring an urbicide. Urbanization has involved irreversible damage and depletion of ecological processes, and it cannot be conceived as "creative destruction". These destructive interventions should be taken into account, systematically measured, made visible and recognized in order to rethink how we value urbanization, how we balance the construction of new neighborhoods and our relationship with the non-human. Recognizing the ecological dimension of urbanization as a source of urban life (and death) should be fundamental for planners, decision-makers and urban society as a whole, before authorizing or developing new urban projects.

Amid contemporary territorial destructions, some processes are coming to reach points of no return for the ecosystems. Not all destruction can or should be compensated either socially or ecologically speaking, and therefore, we advocate the term "territoral destruction", which may sound too radical for some lecturers. For example, in the case of the Sabana de Bogotá, the increasing resource extraction and mining throughout the Altiplano has caused filtration and loss of hydrological regulation, which is why rivers and currents are reducing their flow. These processes are usually analyzed as "environmental impacts", but they are destructive. Furthermore, these destroyed territories also belong to the non-humans whose lives depend on them; thus, the urbicide concept urgently needs to engage with this socio-ecological dimension.

Acknowledging territorial destruction without possible compensation should not lead to inaction but rather to greater creativity in the potential responses toward resolution in the current urban models. The relational character of the geographical space and of the territories could shed light to elaborate alternative paths to massive contemporary capitalist urbanization. Indeed, as the technological dimension of the territory imposes limits on urban expansion, it would be very appropriate to recall the notions of decentralized spatial and land-use planning models on a regional scale. This perspective is useful to understand that the configuration of the metropolitan area also depends on migratory, economic and social terms, on the situation of the other rural and urban territories of the country and beyond. Supporting better living conditions in rural territories, even very distant ones could alleviate the pressure on the ecosystems on which cities depend and nurture.

**Acknowledgements** We want to express our gratitude to the people who helped us to develop and write this text. Special mention goes to Humberto Medellín, environmental leader and inhabitant of Soacha. Humberto, thank you for sharing your knowledge and for your fight to protect your territory. We want to thank the huge contribution to this text provided by Dr. Sergio Gaviria Melo, member of the alliance in the defense of the Sabana de Bogotá. Without his knowledge of the ecological dimensions of this region, this text would not be the same. We also would like to thank to Paul Fryer for help us in improving our English language writing. Finally, we thank the Mexican society who, through the Mexican government, financed Olaff Jasso García's research stay in Colombia through the scholarship: "PROGRAMA DE INICIACIÓN A LA INVESTIGACIÓN CNBBBJ-UNAM-2021". Awarded by the Universidad Nacional Autónoma de México.

# References

Angelo H, Wachsmuth D (2015) Urbanizing urban political ecology: a critique of methodological Cityism. Int J Urban and Reg Res. https://doi.org/10.1111/1468-2427.12105

Arboleda M (2020) Planetary mine. territories of extraction under late capitalism. Verso, New York

Bernal Mora M (2020) Multiterritorialidades en los bordes urbano-rurales de Usme Bogotá-Colombia. Tesis de Maestría en Geografía. Universidad Nacional de Colombia. https://reposi torio.unal.edu.co/handle/unal/77775

Beuf A (2019) 624. Los significados del territorio. Ensayo interpretativo de los discursos sobre el territorio de movimientos sociales en Colombia. Scripta Nova. Revista Electrónica de Geografía y Ciencias Sociales, 23. https://doi.org/10.1344/sn2019.23.22452

Beuf A, García C (2016) La producción de vivienda social en Colombia: un modelo en tensión. In: Abramo P, Rodríguez Mancilla, M, Erazo J (eds) Procesos urbanos en acción. CLACSO/Abya-Yala, Quito

Carrión Mena F (2018) Urbicidio o la muerte litúrgica de la ciudad. Oculum Ensaios 15(1):5–12. https://doi.org/10.24220/2318-0919v15n1a4103

Carvajal JH, Navas O (2016) Bogotá "Savanna." In: Hermelin M (ed) Landscapes and landforms of Colombia. Springer, Switzerland, pp 115–126

Chaves Agudelo JM (2016) (Re) Meanings of nature in a neoliberal globalized modern world: three cases of Colombian indigenous ethnicities (Pijao, Cofán and Muisca-Chibcha). Doctoral Dissertation. Melbourne: University of Melbourne. School of Geography.

Choay F (1969) The Modern City: planning in the 19th Century. George Braziller, New York

Coward M (2006) Against anthropocentrism: the destruction of the built environment as a distinct form of political violence. Rev Int Stud 32:419–437. https://doi.org/10.1017/S026021050600 7091

Delgado Rozo JD (2010) *La Construcción Social del Paisaje de la Sabana de Bogotá*. Monografía para optar al título de Magíster en Historia. Universidad Nacional de Colombia. Facultad de Ciencias Humanas. Departamento de Historia, Bogotá

Esparza M (2014) La sequía y la escasez de agua en México. Situación actual y perspectivas futuras. *Secuencia,* (S.I.) 89:195. https://doi.org/10.18234/secuencia.v0i89.1231

Gaviria Melo S (2021) Algunos elementos para la reconstrucción geoambiental de la Sabana de Bogotá. Documento inédito

Gaviria S, Faivre P, Van der Hammen T (2004) Los suelos y la erosión. In: Aspectos geoambientales de la sabana de Bogotá. Publicación No. 27. Ingeominas, pp 135–168

Harvey D (1990) Los límites del capitalismo y la teoría marxista. Fondo de Cultura Económica. México, pp 466

Harvey D (1993) The nature of environment: dialectics of social and environmental change. In: Miliband R, Panitch L (eds) Real problems, false solutions. Socialist Register. The Merlin Press, London, pp. 1–51

Haesbaert R (2013) Del mito de la desterritorialización a la multiterritorialidad. Culturas y Representaciones Sociales 8(15):9–42

IDEAM -Instituto de Hidrología, Meteorología y Estudios Ambientales (2015) Nuevos escenarios de cambio climático para Colombia 2011–2100. IDEAM / PNUD / MADS/DNP/Cancillería, Bogotá

IDEAM-Instituto de Hidrología, Meteorología y Estudios Ambientales (2017) Análisis de vulnerabilidad y riesgo por cambio climático para los municipios de Colombia. IDEAM/PNUD/MADS/DNP/Cancillería, Bogotá

IDOM (2018) Estudio de crecimiento y evolución de la huella urbana para los municipios que conforman el área Bogotá Región. SDP/Alcaldía Mayor de Bogotá

Instituto Geográfico Agustín Codazzi IGAC (2012) Levantamiento detallado de suelos en las áreas planas de catorce municipios de la sabana de Bogotá. Escala 1:10.000. Bogotá

Instituto de Hidrología, Meteorología and Estudios Ambientales (IDEAM) and Fondo Para la Prevención y Atención de Emergencias de Bogotá (FOPAE). IDEAM-FOPAE 2005. Estudio de la caracterización climática de Bogotá y cuenca alta del río Tunjuelo. Bogotá.

"Investigación sobre la calidad del aire en el municipio de Suacha". 19 de febrero, 2019. Caminando el Territorio. [WWW Link. Retrieved: 14 August, 2022] https://caminandoelterritorioblog.wordpress.com/2019/02/19/investigacion-sobre-la-calidad-del-aire-en-el-municipio-de-suacha/

Lesutis G (2020) Planetary urbanization and the 'right against the urbicidal city'. Urban Geography. https://doi.org/10.1080/02723638.2020.1765632Publishedonline:20May

Marín A 22 de mayo, 2014. Soacha, una sola cantera, *El Espectador.* [Retrieved: 14 August, 2022]. https://www.elespectador.com/bogota/soacha-una-sola-cantera-article-494003/

Molano B, Alejandra V, Garay M, Williams J (2021) El sistema tecnológico ampliado hídrico del Área Metropolitana Funcional de Bogotá: un análisis desde la gobernanza del agua. Cuadernos de Geografía: Revista Colombiana de Geografía 30(2):481–503. https://doi.org/10.15446/rcdg.v30n2.8.93586

Portilla D, Laura T (2021) Análisis de la actividad minera con un enfoque territorial, consideraciones y acciones para el desarrollo minero, en el municipio de Soacha-Cundinamarca. Universidad de ciencias Aplicadas y Ambientales. Bogotá D.C., pp 113

Quimbayo-Ruiz GA (2016) Gestión Integral de la Biodiversidad en el Distrito Capital: aportes para una gobernanza urbana. Biodiversidad en la Práctica. Documentos de Trabajo del Instituto Alexander von Humboldt 1(1):44–76

Quimbayo Ruiz GA (2018) People and urban nature: the environmentalization of social movements in Bogotá. J Pol Ecol 25(1):525–547

Quimbayo Ruiz GA, Kotilainen J, Salo M (2020) Reterritorialization practices and strategies of *campesinos* in the urban frontier of Bogotá, Colombia. Land Use Policy, 99:105058. https://doi.org/10.1016/j.landusepol.2020.105058

Reina Rozo JD (2013) Metabolismo Social: Hacia la sustentabilidad de las transiciones socioecológicas urbanas. Tesis de investigación presentada como requisito parcial para optar al título de: Magister en Medio Ambiente y Desarrollo. Universidad Nacional de Colombia. Facultad de Ciencias Económicas. Instituto de Estudios Ambientales. Bogotá D.C, Colombia

Sánchez-Castañeda PA (2020) Memory in sacred places: the revitalization process of the Muisca community. Urban Planning 5(3):263–273

Scopelliti M, Carrus G, Adinolfi C, Suarez G, Colangelo G, Lafortezza R, Panno A, Sanesi G (2016) Staying in touch with nature and well-being in different income groups: the experience of urban parks in Bogotá. Landsc Urban Plan 148:139–148

Secretaría Distrital de Ambiente (s.f.). Bogotá es más campo que cemento. Retrived march 23, 2022, from: http://ambientebogota.gov.co/de/ruralidad-sda

Soja E (2014) En busca de la justicia espacial. Traficantes de Sueños, Madrid

Stengers I (2015) In catastrophic times: resisting the coming Barbarism. Open Humanities Press

Swyngedouw E (1996) The city as a hybrid: on nature, society and cyborg urbanization. Capital Nat Social 7:65–80

Thibert J, Ardila OG (2017) Segregación urbana y metropolítica en América latina: el caso de Bogotá. En: Alfonso O (ed) Bogotá en la encrucijada del desorden, estructuras socioespaciales y gobernabilidad metropolitana. Serie Economía Institucional Urbana. Número 13. Bogotá: Universidad Externado de Colombia, pp 463–497

Van der Hammen T (1992) El hombre prehistórico en la Sabana de Bogotá: datos para una prehistoria ecológica. En: van der Hammen, T. Historia, ecología y vegetación. FEN, COA, Banco Popular, Bogotá, pp 217–232

Van der Hammen T (1998) Plan ambiental de la cuenca alta del río Bogotá. Análisis y orientaciones para el ordenamiento territorial. Corporación Autónoma de Cundinamarca-CAR

**Alice Beuf** is an assistant Professor at the Geography Department, National University of Colombia, Bogotá since 2014. Doctor (Ph.D.) in Geography from the University Paris Ouest Nanterre, Master of Science in Geography of Development and Geography of Health and Philosophy from the University Paul Valéry Montpellier III. She has been a resident researcher (pensionnaire) at the French Institute of Andean Studies in Bogotá (2010–2014). Her lines of research are "Expansion of global capitalism, society and territories" and "Territories, territorialities and ordering".

**Germán A. Quimbayo Ruiz** Ecologist from the Pontificia Universidad Javeriana (Bogotá), Master in Geography from the University of the Andes (Colombia) and Doctor (Ph.D.) in Social and Cultural Encounters with subject (thesis) in Environmental Policy from the University of Eastern Finland, Finland. He is currently a Ph.D. researcher and member of the research groups Economy, Environment and Alternatives to Development (GEAAD) and GEOURBE (Studies on regional urban dynamics in Colombia), both affiliated with the National Universtiy of Colombia, Bogotá.

**Olaff Jasso García** Geographer from the Faculty of Philosophy and Letters of the UNAM. Interested in research and teaching in the field of geography and social sciences. He has taught classes as a teacher's assistant in the subject of political economy. His lines of research (thesis) are "Social construction of water scarcity in urban spaces" and "Society-nature relationships in capitalism".

# Chapter 16
# Overregulation, Corruption, and Urbicide

**Vicente Ugalde**

**Abstract** This chapter seeks to discuss how the law can affect urban development. To explain the difference between actual construction and the buildings registered in Mexico City, this chapter analyzes the current urban regulations of the latter. The underlying hypothesis is that overregulation plays a key role in explaining this gap. The increase in regulations, coupled with the accumulation of procedures and requirements related to obtaining permits in terms of construction and land use in Mexico City, creates more scope for rules to be broken, drives corruption, and serves as an incentive for the non-observance of building laws. The expansion of urban regulations affects activities that enable cities to function and grow, thereby contributing to urbicide.

**Keywords** Urban norms · Urban legislation · Corruption · Execution of works · City · Mexico

## 16.1 Introduction

The observant visitor coming to a city like Mexico for the first time will soon realize that in addition to the repeated protrusion of water tanks occupying the rooftops, in low-income neighborhoods, these cisterns are surrounded by vertical fragments of the steel rods used to reinforce buildings. The skyline of these extensive areas of the urban territory is not marked by the glamorous outline of modern skyscrapers, but

---

The research underpinning this contribution was undertaken for a project entitled What is governed? Comparing governance in Mexico and Paris: conflict resolution, governance failures, new partnerships and public policies. The author acknowledges the financial support from 299297 Fund Conacyt-EcosNord. He also thanks Elvia Palma for her useful help reviewing early draft of this text and Reviewers for their comments and suggestions. The manuscript has been translated from Spanish by Suzanne Stephens.

---

V. Ugalde (✉)
Center for Demographic, Urban and Environmental Studies, El Colegio de México (Colmex), Mexico City, Mexico
e-mail: vugalde@colmex.mx

by the constant presence of water tanks and steel reinforcement bars, often inclined and sometimes with their ends covered with inverted plastic bottles.

Although this dashes the hopes of passersby of enjoying views such as those afforded by San Cristóbal de las Casas or Antigua, for the inhabitants of these neighborhoods, it leaves open the possibility of a future building. But, what does this panorama mean in legal terms? What does it reflect in terms of compliance with urban regulations? Are these unfinished buildings an expression of illegality? Are they exclusive to irregular settlements? Does this occur in central neighborhoods and populations with higher incomes? If we stop at this last question, it is clear that this phenomenon is not restricted to low-income, peripheral neighborhoods.

According to official data, between 2011 and 2015, 147,553 new homes appeared in Mexico City, although only 7679 were registered with the department responsible for authorizing construction. In the borough whose population has the highest average income, Benito Juárez, the difference between the numbers of new and authorized housing is no less significant. During that period, out of 18,497 new homes, only 2186 construction procedures were registered.

To shed light on this difference, this chapter examines compliance with urban regulations. The underlying hypothesis is that part of this difference can be explained by non-compliance with regulations, which, in turn, is a consequence of the growth of urban regulations since the increase in regulated situations creates more opportunities for rules to be broken. This chapter is part of the line of research proposed by Azuela (2016) a few years ago, which involves discussing the effects of urban regulations observed in the organization and functioning of cities. For this author, considering legal rules, not in terms of their instrumental use for establishing prescriptions and sanctions, but rather of the effects they have on social reality in their instituting or performative function, suggests that these effects may be destructive or unwanted and, in some cases, may also be contributing to urbicide in regard to the shift to more fragmented cities (Carrión 2014). With the caveat that (Azuela 2016:13):

> the social effects of the rigorous enforcement of regulations may be worse than those of their non-compliance.

This chapter suggests elements to discuss how overregulation can affect the process whereby the informal and illegal, as a constituent part of the juridical order, become part of what is legal (Connelly 2012) and therefore affect urban development and the habitability of cities.

The first part reviews approaches in the academic literature to the association between overregulation and the incidence of practices regarded as typical of corruption. The aim is to explore the idea that the increase in regulations and corruption affects activities that enable cities to function and grow. This is followed by a review of the changes that regulations have undergone in Mexico City, both those concerning buildings and the procedures, whereby the government authorizes these activities. The point is to show how the accumulation of procedures and requirements related to obtaining authorizations in terms of construction and land use in Mexico City could constitute an incentive for the non-observance of legal rules.

Subsequently, based on information obtained from boroughs in that city, the chapter presents data on the difference between construction activity and the buildings registered in accordance with current urban law regulations, which is one of the consequences of overregulation and excess red tape. The chapter concludes with considerations on the role of overregulation as a factor in the destruction of cities and therefore conducive to urbicide.

## 16.2 Pernicious Effects of Overregulation

The increase in the regulation of urban development activities leads to situations that may not comply with the law. Although many of these regulations were designed to limit the voracity of the real estate sector, they affect ordinary citizens attempting to make home improvements. Thus, when a person attempts to add a bedroom, floor or wall to their home, they are required to follow lengthy procedures that demand numerous requirements and in addition are processed in different offices.

In a detailed analysis of French town planning regulations regarding land use, construction, and urban planning, Tribillon (2016: 135) notes that the endless normalization of urban planning behaviors and actions through the law prevents the direct access of actors to urban planning, which means that they must resort to legal professionals at all times. In the case of India, Perry (1998) finds that the law serves as a tool for those who are able to manipulate it, while those who live in illegal settlements rarely benefit from it. Based on what was observed for this study and findings presented previously (Ugalde 2016), we think that Mexico City has seen the creation of an ecosystem that encourages intermediation, not by legal professionals but by paperwork expeditors, as pointed out by these two authors. What has been observed in Mexico is also similar to what Tribillon has noted in France regarding the fact that the state ceased to be the urban planner and industrializer of the territory, instead emphasizing its role as the provider of a legal framework. In France, the state has also set itself up as the guardian of legality and maintained and expanded that role, due to the deepening and improvement of the legal framework for planning. In Mexico, what may well be happening is that the state has limited itself to massively producing rules of law, but unlike what has been observed in France (Tribillon 2016: 37), without guaranteeing their enforcement and perhaps only pretending to do so.

Given the pernicious effect of compliance with urban regulations that increase the cost of processes that give life to and make the city work, namely real estate construction of infrastructure, one should also add the effects of the endless increase in regulations fated to be disobeyed. This leads to the spread of the illegal city, the demographic growth of that illegal city, and as Fernandes and Varley (1998: 9) note to the fact that these residents continue to be obliged to reinvent social practices on a daily basis to respond to the exclusionary legal system with which they also coexist.

In the specific case of regulations concerning construction and urban development, it has been argued that although the legal framework is generally designed to organize real estate construction that meets the needs of the middle- and high-income classes or those inserted in the formal labor market, it impacts informal settlements (Santiago 1998; Fernandes and Rolnik 1998). In the case of Mexico, this overregulation may be operating as an incentive to perpetuate the condition of illegality, not only with respect to the way they acquired land or housing, but also to the fact that they built or improved their homes outside urban law.

The proliferation of rules, procedures, and documents that must be provided and requirements that must be met in urban regulation procedures is a factor that hinders the activities, whereby cities develop and operate. Although they are key to its organization and operation, the regulation of economic activities such as housing construction can increase their cost. In this case, the multiplication of rules, procedures, and documents to be provided and requirements to be fulfilled in urban regulation procedures creates scope for the exercise of discretionary actions by public servants and therefore the number of occasions on which acts of corruption may occur. These practices also contribute to increasing building costs. A decade ago, it was estimated that corrupt practices occurred in twenty percent of the encounters between citizens and officials, regardless of the legal and administrative traditions of the country involved (Transparency International 2011). It is a problem that cuts across countries and levels of development. In the case of corruption in land-use procedures, its scope and costs are a matter of concern in both Mexico and societies as disparate and distant as Malaysia (Malpezzi and Mayo 1997) and Iran (Tilaki et al. 2014).

Indeed, the spread of corruption in government land-use management is unstoppable and present in every region in the world. This is borne by the *Doing Business* report published by the World Bank on the practices and conditions that affect business activity,[1] as well as the specialized literature. While it is true that corruption has not always been the object of disapproval: it was thought that corruption could generate benefits in the case of complex and restrictive regulation, which inhibited economic growth (Leff 1964 cited by Chiodelli and Moroni 2015), and this has gradually been qualified. The latest Doing Business reports record that companies facing lawsuits related to bribery have been forced to wait 50% longer to obtain the urban planning authorization that gave rise to the act of corruption compared to those who do not engage in these practices (World Bank 2014).

The study of corruption related to land-use regulations has diversified in recent years, providing explanations of its modalities, generalization (Transparency International 2011; Chiodelli and Moroni 2015), the conditions that drive its emergence and expansion, such as social inequality (Barr and Serra 2010) and the lack of accountability and control (Chiodelli 2019), and emphasizing features of the specific context that are conducive to this type of practice, for example, in Italy (Chiodelli and Moroni 2015), Spain (Quesada et al. 2013), and Sub-Saharan Africa (Boamah et al. 2021).

Among the pernicious effects of corruption, Chiodelli and Moroni (2015) highlight the dysfunction or inefficiency of bureaucracies, the loss of trust in the latter

---

[1] The *Doing Business* document presents and evaluates indicators on business regulations in twelve areas of the life cycle of a business (setting up a business, managing building permits, obtaining electricity, registering property, obtaining credit, protecting minority investors, tax payment, cross-border trade, contract enforcement, and insolvency resolution) in 190 countries.

and governments and political systems and institutions in general, which affects budget allocation criteria and inhibits economic growth. This is not only because of what the Doing Business reports suggest, but because corruption distorts prices and modifies incentives for economic agents in sectors such as housing and public works, vital sectors for the city.

This chapter explores whether overregulation distorts the functioning of the economic sector of construction in general and housing construction in particular. As explained below, the length of the processes to obtain permits increases the production costs of real estate, creating opportunities for corrupt practices. By impacting the final price of goods, this delay operates as an incentive to flout regulations as well as contributing to the deterioration of the conditions of the rule of law.

A survey conducted by the World Justice Project agency revealed aspects of the perception of the rule of law in Mexico. The survey covered topics such as limits on government power, absence of corruption, open government, fundamental rights, order and security, civil and criminal justice, and regulatory compliance.[2] It is found, for example, that regulatory compliance is determined by situations such as the effective, efficient, timely, and corruption-free resolution of administrative procedures. It also found that regulatory non-compliance is related to whether the government respects property rights and avoids excessive delays (World Justice Project 2018). The survey provides indices by state, one of which refers to compliance with corruption-free procedures. In the latter, Mexico City obtained 0.22 (on a scale of 0–1, in which 0 assumes less or non-existent respect for the rule of law and 1 assumes that there is greater respect), which is below the national average (0.44) (World Justice Project 2018). This suggests that conditions for compliance with the regulation in the city are clearly affected by corruption.

The perception of the harm caused by corruption to compliance and therefore to the effectiveness of urban regulation is borne out by the National Survey of Government Quality and Impact (*ENCIG*) undertaken by the National Institute of Statistics and Geography (*INEGI*).[3] In response to questions related to the effectiveness, duration, and involvement of corruption in procedures such as land-use permits, demolition, construction, alignment, and official number permits, the 2019 survey shows that of nearly 40,000 respondents in Mexico City who engaged in a procedure related to property and had face-to-face contact with a public servant, 41% experienced an act of corruption. The survey defines an act of corruption as:

---

[2] More than 25,000 surveys were administered in which lawyers were included. Over 600 variables were analyzed, systematized into forty-two factors, and grouped into these seven issues (World Justice Project 2018).

[3] The ENCIG is conducted every two years. It is designed to collect information on the experiences, perceptions, and evaluation of the population aged 18 and over in cities with 100,000 inhabitants or more on the procedures and services provided by the levels of government. It also provides estimates on the prevalence of victims of acts of corruption and their effect on undertaking procedures, payments, requests for public services and other types of contact with the authorities. See INEGI (2021).

"a practice that occurs when a public servant or government employee abuses their functions to obtain personal benefits such as money, gifts, or favors from the citizen."[4]

In certain areas of the city located in the boroughs of *Benito Juárez, Coyoacán, Magdalena Contreras,* and *Tlalpan,* this figure is close to 70% (INEGI 2019).

This chapter does not argue that overregulation invariably leads to corruption. However, the records presented in the following section suggest seriously considering the links between this condition, the overregulation of certain activities, the perception of the way interactions between the government and the governed take place regarding building permits, and what this means for the functioning of a central activity in the development of cities, such as housing production.

## 16.3 An Intractable Challenge: More Issues, Fewer Rules

The proliferation of regulations is due to numerous factors and, at the same time, has generated concern and led to efforts to introduce order and halt the increase in the number of regulations. Although one consequence of the burgeoning of regulations has been the creation of procedures, conditions, and requirements for obtaining urban planning authorizations, efforts have also been made to standardize the process whereby officials in the one-stop windows of the boroughs or *alcaldías* (previously called *delegaciones*) undertake those procedures. This homogenization has mainly been achieved through successive administrative agreements.[5]

### 16.3.1 Building Regulations and the Regulation of Procedures

Following the earthquake that hit Mexico City in September 1985, the government reviewed the building regulation, and two years later, the one that had been in force since 1976 was replaced. The aim was to reduce the risk level, especially because of the ever-present threat of earthquakes, and to improve the urban image through the standardization of buildings.[6] Since then, building regulations have accumulated tasks and increased expectations about their ability to organize the participation

---

[4] This is the definition included in the ENCIG 2021 survey questionnaire, available at https://www.inegi.org.mx/contenidos/programas/encig/2021/doc/encig21_cuestionario.pdf. Accessed August 30, 2022.

[5] In Mexican law, an administrative agreement is an administrative act of a general nature (in the same way as a regulation, a decree, an official Mexican standard, a circular or a format, which is intended to establish specific obligations.) According to the Federal Law of Administrative Procedure, in the event that the conditions of competition are not met, agreements must be published in the *Official Gazette of the Federation* to produce legal effects.

[6] Decree publishing the Federal District building regulations, published in the Official Gazette of the Federation (DOF) on July 2, 1987.

of public and social agents in construction. The regulation published in 1987 was replaced in 1992 and again in 2004.

This regulation brings together the rules concerning building. It not only concerns rules that seek to foresee and organize actions strictly associated with real estate construction, but also to anticipate their impacts in that they are inserted in a specific area of the city. It includes aspects of urban planning, construction engineering, density, and structural safety, as well as legal and administrative instruments for the management and control of the activities it regulates. The inclusion of these aspects in the regulation translates into new requirements and procedures to obtain authorization for urban actions such as construction, mergers, divisions, extensions, and special permits; authorization for issues related to urban development[7]; and authorization for technical aspects of building, environmental protection, and risk management.[8] The regulation also organizes the intervention of civil servants in each procedure, defines the functions and responsibilities of those who contribute to their implementation, such as *project managers* and the *jointly responsible directors,* and organizes the verification, surveillance, and definition and imposition of sanctions, comprising the deterrent aspect of the regulation.

As noted earlier, the rise in the number of regulations governing urban actions in Mexico City is the result of the adaptation of those regulations to the emergence of issues requiring greater regulation. This increase has not been greater only because cautious efforts have been made to optimize regulations as a means of expressing the prescriptions, orientations, and definitions required to organize this social field.

Despite the growing number of issues included in the regulation, there have been attempts to optimize its text. As shown in Table 16.1, the amendments have been accompanied by an effort to prevent an increase in the number of articles, which has in fact decreased, albeit without reducing their length. A comparison of the 2004 regulation, with 257 articles, with earlier versions reveals a worrying increase in its length. The 1976 regulation had 394 articles, while the 1985 and 1992 versions both had 353. This reduction of more than 130 articles has not, however, led to a significant decrease in the length of the text because whereas the 1976 regulation had roughly 45,000, the current regulation has approximately 42,000, despite the administrative simplification process contemplated in the customer service model adopted by the government during that period.

The inclusion of issues affects the scope of regulations, as well as the diversity and complexity of their contents. Understanding them, therefore, requires specialized knowledge. This difficulty is not, as noted earlier, the only consequence. The incorporation of issues has led to more paperwork and requirements to obtain authorization.

---

[7] Land uses, densities, buildable levels, availability of public services, urban impacts, environmental impacts, public roads, aerial and underground conduits, nomenclature, alignment, and integration into the urban context and image.

[8] Soil mechanics, design, structural safety, plans, materials, and construction and demolition procedures, as well as those related to hygiene, environmental conditioning, sustainability, energy efficiency, communication, and safety in emergencies.

**Table 16.1** Construction regulation amendments

| Year | Official publication date | Length (articles) |
|------|---------------------------|-------------------|
| 1976 | 14-December-1976 | 394 |
| 1985 | 18-October-1985 | 21 |
| 1987 | 03-July-1987 | 353 |
| 1992 | 02-August-1993 | 353 |
| 2004 | 29-January-2004 (last amendment on 15-Dec.-2017) | 257 |

*Source* Compiled by the author based on the Official Gazette of the Federation in the dates indicated

To organize these procedures, actions have been undertaken to systematize the granting of building permits since 1989. Subsequently, in 1994, the Department of the Federal District (DDF) established One-Stop Windows. The aim of this material and organizational device was to respond to complaints regarding the lack of homogeneity in the operating procedure criteria. To this end, the new One-Stop Windows were authorized to receive, integrate, manage, and deliver documentation and to deal with procedures such as those related to water and drainage, public lighting, potholes, commercial establishments, advertisements, parking lots, billboards, land use, and buildings.[9] The Manual of Procedures and Services to the Public of the Public Administration Department of the Federal District[10] was subsequently issued in 1996. In 1997, another agreement incorporated clarifications regarding the functions that had been granted to the One-Stop Windows in 1994. It stated that One-Stop Windows would have to establish a file identification system and have a Government Ledger to register procedures, establish opening hours, and respect the requirements set forth in the Manual of Procedures and Services to the Public.[11]

In 2004, the organization and functions of One-Stop Windows were modified once again. It was decided to divide One-Stop Windows into four types: advice, reception, follow-up, information, and delivery. The list of procedures and services about which citizens must be informed and advised and on which they must receive, register, manage, and deliver documentation was expanded. In that agreement, the government announced the

> acknowledgment in administrative procedures and paperwork of the principle that good faith prevails in the actions of the vast majority of citizens […]. It is therefore announced

---

[9] An agreement creating One-Stop Windows for the reception and delivery of documents in the sixteen boroughs in the Department of the Federal District was published in the DOF on September 23, 1994, in the Benito Juárez Borough. It also authorized these windows to help individuals with the regulation "through information brochures, signage for each procedure and indication of the corresponding requirements and formats, with the support of the computerized system for management control" (DOF, May 23, 1994).

[10] Published through an agreement in the Official Gazette of the Federal District (GODF) on September 13, 1996.

[11] Agreement modifying and specifying the functions of the Borough One-Stop Windows, published in the DOF on November 25, 1997.

that public administration should not condition or hinder the legal activities they decide to undertake.[12]

This resulted in provisions that could help citizens. On the one hand, it was decided that in the case of construction declarations (which in certain cases replaced building permits), One-Stop Windows should only check the documents in the application, without examining their contents. On the other, it was decided that when a document was missing, the application would be returned to the interested party at the window itself with the corresponding stamp and signature and an indication of what was missing.[13] The city government subsequently published a Specific Operating Manual for Borough One-stop Windows.[14] This manual included the procedures and stages, a diagram, as well as the description of tasks assigned to both the operator and the person responsible for the One-Stop Window. A procedure applicable to all paperwork, specifically in relation to the Construction Declaration, was implemented. Since forms and legal bases were not included, it was more useful for civil servants than citizens.

In 2012, the city government published an agreement establishing single procedures. It included a list of procedures and services, the administrative unit responsible for each one, the objective, requirements, and forms, as well as the operating policies and regulations. It even mentioned the length of the procedure, expressed, in many cases, in minutes.[15] In the introduction, the agreement referred to the World Bank's 2009 Doing Business report, noting that 138 days were required to obtain construction permits in Mexico City, as opposed to other countries where the processes were simpler.

Despite these amendments (Table 16.2), processing times have not been reduced, and on the contrary, it seems that the pernicious effects of overregulation have mushroomed.

## 16.3.2 More Requirements, Lengthier Procedures

If, by way of an example, only the construction or expansion of a property in Mexico City is considered, building regulations include different procedures depending on the size of the work. In addition to proving ownership of the property, it is necessary to accompany the application with documents that usually involve a procedure. In the construction declaration procedure, applicants must submit documents related

---

[12] Agreement modifying and specifying the functions of the Borough One-stop Windows, published in the DOF on November 19, 2004.

[13] The agreement once again included aspects concerning the file identification system, including a progressive number, a code for each issue, the date, and as in the 1997 agreement, its registration in a Government Ledger. Agreement modifying and specifying the functions of the Borough One-Stop Windows, published in the DOF on November 19, 2004.

[14] GODF, October 8, 2004.

[15] Published in GODF on July 2, 2012.

**Table 16.2** Structuring and amendments to the functioning of the one-stop windows in Mexico City Boroughs

| Year | Name of instrument | Official publication |
|---|---|---|
| 1989 | Agreement creating the central management offices for building permits and documents | 29-September-89 |
| 1990 | Agreement delegating various functions to the General Director of Urban Reorganization and Ecological Protection | 06-August-90 |
| 1994 | Specific operating manual that regulates and ensures the transparency of the activities of the borough One-Stop Windows | 23-May-94 |
| 1994 | Agreement creating the sixteen boroughs of the Department of the Federal District, Borough One-Stop Windows for the reception and delivery of documents | 23-September-94 |
| 1996 | Manual of procedures and services to the public of the Public Administration Department of the DF | 13-September-96 |
| 1997 | Agreement modifying and specifying the functions of the Borough One-Stop Windows | 25-November-97 |
| 2004 | Agreement modifying and specifying the functions of the Borough One-stop Windows | 19-May-04 |
| 2004 | Specific operating manual of the Borough One-Stop Windows | 08-October-04 |
| 2012 | Agreement establishing Single Procedures for Processing Procedures and Services | 02-July-12 |

*Source* Compiled by the author based on the Official Gazette of the Federation published on the dates indicated

to land use, the alignment, and official number; obtaining each of these documents requires going to different government offices and performing a procedure. The same occurs with other documents required to prove both the feasibility of drinking water and sewerage services, paying the taxes provided for by law, and other procedures that vary depending on the type and size of the construction. Real estate construction also involves civil protection, road impact, environmental impact, and urban-environmental impact procedures, to mention just a few requirements which, in addition to Mexico City, exist in several Mexican municipalities with regulations concerning this issue. By way of an example, Fig. 16.1 shows the documents that must be submitted to begin the process required to start building.

In addition to the documents and, therefore, procedures that are included in Fig. 16.1, in cases where the property is located in areas with historical, artistic, or archaeological value, applications must be accompanied by a permit from the National Institute of Anthropology and History (INAH) and in some cases by approval from the National Institute of Fine Arts (INBA). Certain situations require presentation of a Favorable Opinion of the Urban Impact Study or the Urban Environmental Impact Study, which also constitute a procedure in itself. In the case of extensions, modifications, or repairs, a Demolition Permit must also be submitted. In the case of enlargements, the original plan of the building must be presented, which, depending on the date, is located in a specific archive in the borough office.

# 16 Overregulation, Corruption, and Urbicide

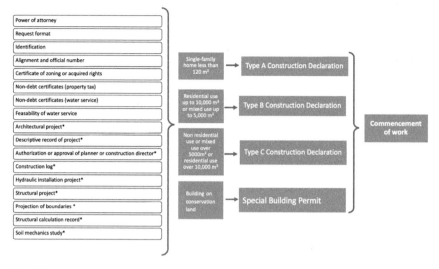

**Fig. 16.1** Documents required to begin construction procedures. *Source* Compiled by the author based on building regulations and the Agreement on Single Procedures for processing paperwork and services. *Not required in the case of the Type A Construction Declaration

The intended use of the property affects some of the procedures. This is the case of the Report on the Feasibility of Drinking Water and Sewerage Services issued by the agency in charge, the Mexico City Water System, as well as certain procedures performed by the City's Environment Secretariat regarding waste management and environmental impact.

Completion of the works requires another type of procedure that varies depending on the type of building but which, in general terms, are essential for validation by the authorities of the completion of the work on the property. This procedure makes it possible to register a property with the Public Property Registry and subsequently update the basis on which the value of the property and the property tax are estimated.

The proliferation of technical terms has turned construction regulations into texts that can only be understood by experts, not in construction but in the management of bureaucratic procedures. These are individual *expeditors* or offices that provide their services based on several types of knowledge (urban law, engineering, architecture, and urban planning) as well as their contacts with the public servants responsible for procedures. These contacts are a valuable asset since they pave the way for obtaining permits, which is virtually impossible for ordinary citizens or firms that do not use the services of these expeditors.

Despite efforts to organize the operation of the Single Windows and procedures, the behavior of bureaucracies creates an ecosystem in which obtaining permits is only possible for these expeditors. This intermediation benefits the boroughs because it perpetuates the distance between the applicant and the public servant, while at the same time delegating to the associated third party, a kind of para-bureaucratic organization, namely the expeditor, the task of defining the terms of the exchange is entailed by obtaining the permit. Thanks to the intermediation provided by the expeditor, the act of corruption is no longer based on the interaction between the public servant and the person applying for a permit. The mechanism whereby the transaction takes place, in which the applicant pays to save time to obtain documents, uses an intermediary who acts as a kind of safety valve that protects the public servant in the event that the transaction is exposed.

The increase in legal rules and corruption acts, such as the payment of bribes or commissions to civil servants through expeditors, impacts the activities regulated by these rules. One effect is the delay caused by the accumulation of paperwork and procedures, which entails considerable costs. The government agency responsible for regulatory improvement, the National Commission for Regulatory Improvement (formerly the Federal Commission for Regulatory Improvement), analyzed the impact, especially on the cost of housing, of excessive paperwork to obtain building permits, from the time they are requested until the delivery of the property in 51 municipalities across the country (Cofemer 2016: 3). The study found that an average of 19 procedures was necessary and included 186 requirements, which took 339 days. According to the National Chamber of the Housing Development and Promotion Industry, although the authorities admit that these procedures take from 450 to 500 days, they actually take 1675 days, in other words, just over four and a half years. According to Cofemer, this delay impacts opportunity costs and therefore the final price of housing. On the other hand, although the study does not include Mexico City, the increase in procedures in this city also leads to an increase in construction costs related to the delay in procedures, which some years ago was estimated at 138 days, a longer period than that required in other countries.[16]

In this chapter, we consider that another cost of this overregulation lies in the fact that it discourages ordinary citizens and small businessmen from respecting the rule of urban law. Unable to afford a paperwork expeditor, citizens choose to act without permits.

This system of granting permits encourages social actors to opt for illegality as regards town planning rules and leads to the underreporting of town planning activity.

The following section presents data that point, on the one hand, to an association between the complexity and slowness of obtaining permits in urban matters, and on the other, to the illegality of certain actions. We wish to highlight the fact that this incentive to build without authorization has consequences for the functioning of the city and its political organization. As mentioned earlier, it is difficult to register buildings that have been built or expanded and therefore update the tax base for property taxes, a key source of tax resources for city governments. At the same time, if the illegal city is not only explained by the means of access to urban land but also by the condition in which the buildings were built (Fernandes and Varley 1998; Calderon 2016), another consequence is associated with the fact that it encourages

---

[16] This is recognized by the Agreement issued by the government of the Federal District establishing Single Procedures for Processing Procedures and Services in 2012. Published in the Official Gazette of the Federal District on July 2, 2012.

a situation of illegality in regard to urban regulations. This has a major impact on cities in the region.

## 16.4 Mexico City: The Enigma of Underreporting

Comparing the number of homes registered in the city by the National Institute of Statistics and Geography (INEGI) with the number of construction declarations registered illustrates what we mean when we speak of illegality in the urban built environment. To perform this exercise, we compared data from population and housing censuses and information obtained from the Mexico City boroughs for the periods 2006–2010 and 2011–2015. We obtained records from seven boroughs (*Azcapotzalco, Benito Juárez, Cuajimalpa, Cuauhtémoc, Magdalena Contreras, Tláhuac* and *Tlalpan*) for an initial exercise. During the second exercise, we obtained data on 15 boroughs, in other words, all of them except *Milpa Alta*.

Figure 16.2 shows the difference between the number of new dwellings and the construction declarations registered. This difference could be explained by the fact that type B and C declarations can include several dozen dwellings.[17] On the other hand, the fact that the declarations of modifications or extensions fail to specify whether the number of dwellings they cover are only those added to an existing building or whether they refer to the cumulative number of dwellings does not make comparison easy. However, based on the data on the number of homes comprising each construction declaration in the borough of *Benito Juárez*, we have estimated the number of new dwellings not covered by the construction declarations registered.[18]

If we consider the accumulation of housing built regardless of what is expressed in the construction declarations, one could argue that cities like Mexico City are seeing the development of housing stock of dubious legality, either because property owners failed to request the registration of the declaration of the corresponding construction or because as we will see, despite having submitted this document, they did not continue the procedure until they obtained the Notice of Completion.

---

[17] A Construction Declaration is a statement made by the owner or possessor of a property that they are aware of their responsibility regarding the legal requirements to build, expand, repair, or modify a construction. Through this declaration, the owner or possessor assumes the obligation to comply technically and legally with the applicable legal provisions. The regulation foresees three modalities. A Type A Construction Declaration corresponds to the construction of no more than one single-family home of up to 120 m$^2$ built. A Type B declaration corresponds to non-residential or mixed uses of up to 5000 m$^2$ or up to 10,000 m$^2$ in the case of residential use. A Type C declaration corresponds to non-residential or mixed uses of more than 5000 m$^2$, or in the case of residential use, more than 10,000 m$^2$. See Building Regulations of the Federal District (GODF January 29, 2004).

[18] According to the INEGI, between 2005 and 2010, the number of dwellings in the borough of Benito Juárez increased by 18,914. Based on information on the number of dwellings covered by each of the declarations registered in that period, we have estimated that each of the 938 Construction Declarations covers an average of 18.2 dwellings. This means that 17,027 dwellings were registered, in other words, nearly 1900 fewer dwellings than those recorded by INEGI.

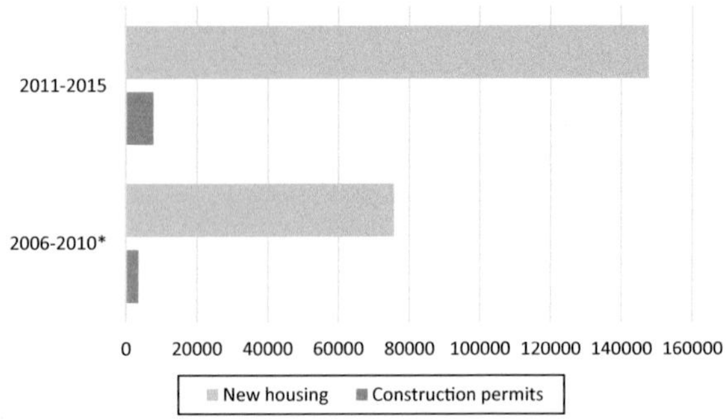

**Fig. 16.2** Comparison of construction declarations and new housing in Mexico City (2006–2010 and 2011–2015). *Source* Compiled by the author based on information obtained through INFOCDMX on Mexico City (CDMX) Boroughs for 2006–1010 and 2011–2015

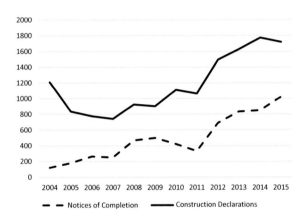

**Fig. 16.3** Construction declarations and notices of completion 2012–2015. *Source* Compiled by the author based on information obtained through INFOCDMX on Mexico City Boroughs)

Building regulations provide that once construction works have been completed, a Notice of Completion must be delivered to the borough office or the Department of Urban Development and Housing (SEDUVI). In the case of Type B and C declarations and if the construction was subjected to what is indicated in the Construction Declaration, in the Hydraulic Services Feasibility Report and the Urban Impact Study, when a citizen delivers the Notice of Completion, the authorities grant them an Authorization of Use and Occupation.[19] Figure 16.3 shows the difference between the Construction Declarations and Completion Notices registered in each municipality from 2012 to 2015. Although a declaration can be valid for three years and although it can have two extensions, it is possible to infer that the large number of constructions registered with Construction Declarations have not been *formally* completed or that if they were, the Notice of Completion process was not undertaken, and the Use and Occupancy Notice was not obtained. In other words, constructions whose declarations were registered do not obtain the Authorization of Use and Occupation.

---

[19] This procedure is provided for in Article 65 of the Building Regulations of the Federal District (GODF, January 29, 2004) but is not included in the Agreement that establishes Single Procedures for Processing Paperwork and Services (Official Gazette of the District Federal of July 2, 2012).

The gap between Construction Declarations and Notices of Completion suggests that the legality of a broad range of urban development actions remains pending in this city. In both cases, that of new dwellings built without a Construction Declaration or that of those that did not reach the Notice of Completion stage, a kind of non-compliance is taking place, thereby not only compromising the physical completion but also the completion of the property in the abstract legal sphere. Part of the city is illegal not in regard to achieving access to the land, but because its construction fails to adhere to the legal provisions concerning land use and construction.

From our perspective, the proliferation of procedures and requirements for obtaining building permits in Mexico City as a result of overregulation is a condition that leads to delays in completing paperwork and drives an increase in corrupt practices, thereby multiplying the costs of housing construction among other economic activities. Beyond the fact that excessive legal rules play a role in this economic distortion, this mindless use of regulation trivializes the rule of law and undermines its ability to guide urban development and the functioning of society in the city. Overregulation not only affects informal settlements as some authors have observed (Santiago 1998; Fernandes and Rolnik 1998) but also other sectors of the social body of cities, since they encourage them to engage in construction outside the regulations and permits, in other words, illegally.

## 16.5 Final Remarks

What has been observed in Mexico City suggests that as a system of restrictions and sanctions, urban regulation was conceived by legislators and decision-makers concerned with establishing checks and balancing for major players in the real estate sector, who overlooked ordinary citizens, particularly low-income ones. Faced with iron-clad, restrictive, and convoluted regulations, these citizens choose to build outside them.

This perpetuates an ambivalent idea regarding the role of legal regulation in social life. On the one hand, it builds up expectations about its ability to organize and improve social coexistence while on the other, it elicits accusations about its chronic inability to achieve these goals. Growing frustration over the ability of the rules of urban law to organize the activity whereby the city develops, coupled with the increasing difficulty of building within the narrow scope for maneuver related to its overregulation, serves as an incentive to sidestep urban law. Paradoxically, a citizen's attempts to act within the provisions of the law lead them to do so outside it. Urban actions such as building occur more frequently in the illegal sphere, thereby jeopardizing other goals of urban regulation such as organizing the production and reproduction of the city.

Despite being far removed from the concerns of academic production on the issue of regulation and corruption in the management of land use and construction, the notion of urbicide in the sense defined by Carrión (2014) is useful for the analysis proposed in this chapter. Since it refers to a process that can include the fragmentation and erosion of institutionality (Carrión 2014), this notion also refers to the inability to steer social behavior. Rather than referring to the political will to destroy the city, the revenge of the rural on the urban world, phenomena such as closed communities, privatization, the closure of local businesses, or the proliferation of skyscrapers (Mongin 2015), or embracing the catastrophic view of urban globalization others have associated with the notion (Nahoum-Grappe 2015), this chapter has attempted to highlight the way excessive expectations of the rules of urban law encourage its intensive use. Overregulation and the proliferation of paperwork and procedures may well be intended to limit the excesses of globalization and the ambitions of the elites noted by Logan and Molotch (2015) that make the city a growth machine with questionable or non-existent collective benefits. Nevertheless, in this chapter we wanted to underline the fact that the excessive use of legal rules multiplies the opportunities for non-compliance with the law and therefore the scope for acts of corruption while at the same time encouraging ordinary citizens to engage in illegal building. By affecting the ability of the law to effectively organize the development and operation of the city, overregulation tends to encourage the expansion of the illegal city, with all its attendant disadvantages.

## References

Azuela A (2016) Introducción. Una especie de neorrealismo jurídico. In: Azuela A (ed) La ciudad y sus reglas: sobre la huella del derecho en el orden urbano. Universidad Nacional Autónoma de México, México, pp 9–40

Barr A, Serra D (2010) Corruption and culture: an experimental analysis. J Public Econ 94(11–12):862–869. https://doi.org/10.1016/j.jpubeco.2010.07.006

Boamah EF, Watson V, Amoako C, Grooms W, Osei D, Kwadwo VO, Nyamekye AB, Adamu K, Appiah GK (2021) Planning corruption or corrupting planning? J Am Plann Assoc. https://doi.org/10.1080/01944363.2021.1987969

Calderon J (2016) La ciudad ilegal. Punto Cardenal Editores, Lima

Carrión F (2014) Urbicidio o la producción del olvido. Observatorio Cultural 25:76–83

Chiodelli F (2019) The illicit side of urban development: Corruption and organized crime in the field of urban planning. Urban Studies 56(8):1611–1627. https://doi.org/10.1177/0042098018768498

Chiodelli F, Moroni S (2015) Corruption in land-use issues: a crucial challenge for planning theory and practice. TPR 86(4):2015. https://doi.org/10.3828/tpr.2015.27

Cofemer (2016) Construcción de vivienda en México. Trámites para autorización, Secretaría de Economía, México

Connolly P (2012) La urbanización irregular y el orden urbano en la zona metropolitana de la Ciudad de México de 1990 a 2005. In: Salazar C (ed) Irregular. El Colegio de México, México, pp 379–425

Fernandes E, Varley A (1998) Illegal cities. Law and Urban change in developing countries. Zed Books Ltd., London

Fernandes E, Rolnik R (1998) Law and Urban change in Brazil. In: Fernandes E, Varley A (1998) Illegal Cities. Law and Urban change in developing countries. Zed Books Ltd., London, pp 140–156

INEGI (2019) Encuesta Nacional de Calidad e Impacto Gubernamental. Instituto Nacional de Estadística y Geografía, México. Disponible en: https://www.inegi.org.mx/programas/encig/2019/. Accessed Apr 2022

INEGI (2021) Encuesta Nacional de Calidad e Impacto Gubernamental. ENCIG Principales Resultados. Instituto Nacional de Estadística y Geografía, México. Pdf. Disponible en: https://www.inegi.org.mx/contenidos/programas/encig/2021/doc/encig2021_principales_resultados.pdf

Logan JR, Molotch H (2015) La Ciudad como máquina de crecimiento, X, Madrid, pp 157–210

Logan J, Molotch H (2015) La Ciudad Como Maquina de Crecimiento. In: El Mercado Contra la Ciudad: Sobre Globalización, Gentrificación y Políticas Urbana, Madrid, edited by Observatorio Metropolitano de Madrid, pp 157–210

Malpezzi S, Mayo SK (1997) Getting housing incentives right: a case study of the effects of regulation, taxes, and subsidies on housing supply in Malaysia. Land Econ 73(3):372–391

Mongin O (2015) Introduction. Tous Urbains 11:28–31

Nahoum-Grappe V (2015) L'Urbicide: Le meurtre du social. Tous Urbains 3(11):32–37

Perry A (1998) Law and Urban Change in an Indian City. In: Fernrandes, Varley (eds) Illegal cities: law and urban change in developing countries Zed Books. London, pp 89–104

Quesada MG, Jiménez-Sánchez F, Villoria M (2013) Building local integrity systems in southern Europe: the case of urban local corruption in Spain. Int Rev Adm Sci 79(4):618–637

Santiago AM (1998) Law and Urban challenge: illegal settlements in the Philippines. In: Fernandes, Varley A (eds) Illegal cities. law and Urban change in developing countries. Zed Books Ltd., London, pp 104–122

Tilaki MJM, Abdullah A, Bahauddin A, Marzbali MH (2014) An evaluation to identify the barriers to the feasibility of urban development plans: five decades of experience in urban planning in Iran. J Urban Environ Eng 8(1):38–47. http://www.jstor.org/stable/26203408

Transparency International (2011) 'Corruption in the Land Sector', (TI working paper 4), available at http://www.transparency.org (Accessed Mar 2014).

Tribillon J-F (2016) Le droit nuit gravement à l'urbanisme. Edition de la Villete, Paris

Ugalde V (2016) Del papel a la banqueta: testimonio del funcionamiento de la regulación urbano-ambiental. In: Azuela A (ed) La ciudad y sus reglas: sobre la huella del derecho en el orden urbano. Universidad Nacional Autónoma de México, México, pp 113–137

World Bank (2014) Doing business 2015. Going beyond efficiency. International bank for reconstruction and development/the world bank. Washington

World Justice Project (2018) Rule of law index. Available at https://worldjusticeproject.org/sites/default/files/documents/WJP-ROLI-2018-June-Online-Edition_0.pdf. Accessed 30 Aug 2022

**Vicente Ugalde** Research Professor at Centre for Demographic, Urban and Environmental Studies at *El Colegio de México* where he teaches Local Government. His primary research area is environmental and urban regulation in Mexico with a special focus on legal decision-making in control of regulation.

# Chapter 17
# Obsolescence of the Built Environment

**Valeria Reinoso-Naranjo and Manuel Martín-Hernández**

**Abstract** Architecture is the creation of a society that seeks to give itself a way of living. These forms build a material language that expresses both the individual and the collective, that is, a field of permanent communication of a space that speaks and transmits plurality, heterogeneity and diversity. Despite the fact that many architectures do not always communicate social values and are strictly related to elements of imposition and tyrannical forms of power, any architecture may be subject to its possible disappearance. This disappearance, in any case, represents the loss of all those social values that it is capable of communicating. The urbicide to which we want to refer here is caused by urban planning itself, through the permanent search for the so-called progress that is capable of exerting strong—and also subtle—mechanisms of destruction of architecture, such as seeking its expiration. Observing the destruction of the material base from the exercise of expiration or obsolescence also constitutes a link to look at the processes associated with the production and de-production of the city from other logics.

**Keywords** Architecture · Urban design · Historic cities · Suburbs

## 17.1 Introduction

Architecture is the creation of a society that seeks to give itself a way of living. These forms build a material language that expresses both the individual and the collective, that is, a field of permanent communication of a space that speaks and transmits plurality, heterogeneity and diversity.

Despite the fact that many architectures do not always communicate social values and are strictly related to elements of imposition and tyrannical forms of power, any

---

V. Reinoso-Naranjo (✉)
Technical University of Ambato (UTA), Ambato, Ecuador
e-mail: vc.reinoso@uta.edu.ec

M. Martín-Hernández
ESARQ of Guadalajara, Guadalajara, Mexico
e-mail: mmartin@esarq.edu.mx

© The Author(s), under exclusive license to Springer Nature Switzerland AG 2023
F. Carrión Mena and P. Cepeda Pico (eds.), *Urbicide*, The Urban Book Series,
https://doi.org/10.1007/978-3-031-25304-1_17

architecture may be subject to its possible disappearance. This disappearance, in any case, represents the loss of all those social values that it is capable of communicating. The urbicide to which we want to refer here is caused by urban planning itself, through the permanent search for the so-called progress that is capable of exerting strong—and also subtle—mechanisms of destruction of architecture, such as seeking its expiration.

Observing the destruction of the material base from the exercise of expiration or obsolescence also constitutes a link to look at the processes associated with the production and de-production of the city from other logics. For this, we propose to develop this text in three sections:

The first section, "city, power and obsolescence", opposes the "city for dwelling" to the "city of power" as spaces in dispute, a situation directly related to the crisis of urbanism and architecture used to deploy constant actions in the task of seek order and homogenization, counteracting attempts at otherness and that activates mechanisms of differentiation, substitution and expiration.

We entitled the second section "spatial condemnations", places whose result expresses both the physical and discursive exercise of obsolescence and where the deliberate provocation of a series of tensions and non-tensions is verified, that is, a certain logic of excesses and indifference, and places exposed to the violence of expiration and its times.

Finally, in the third section, "critiques of expiration", we reaffirm the importance of time for the city and architecture as a reflection of its authenticity and durability, establishing the trialogical relationship between modification, belonging and permanence, directed to a re-politicization of architecture and urbanism.

## 17.2 City, Power and Obsolescence

### 17.2.1 The City for Dwelling versus The City of Power

Since ancient times, the conformation of the *urbs* (the physical territory of the city) responds to the tireless ordering of the *civitas* (the citizens who inhabit it), where the predominance of the heterogeneity of society makes the *polis* (the political), display constant actions in the "task of ordering" society in its space. For authors such as Lefebvre (2013), this ordering task consists of fixing differentiated spaces with a homogeneous appearance destined for their control. This clearly contrasts with the ideals that have been pursued for the creation of the city for dwelling, capable of reflecting the experiences, the decisions of the social production of the city and the habitat throughout processes that go from the bottom up, imagining the city as a container of universal coexistence and unlimited differences, which grows and multiplies according to those who adhere to it under the laws of equality and belonging, erasing borders that favor heterogeneity, adaptation and transmutation (Cacciari 2012).

However, all those virtues that are directly related to what would be obtained by enjoying the right to the city come into contradiction before an abstract and atopic representation, which imposes hierarchies and spatial mandates from top to bottom, typical of the city that lives and recreates power through planning. For Rogério Haesbaert, "[t]he territory is *always* linked to power and to the control of social processes through the control of space" (2013, p. 13). This, which historically became the essence of the *polis*, brings us closer to understanding that the city (*urbs*), in its search for order, reflects the permanent process of differentiation and mutual exclusion between the city for dwelling and the city of power, which finds in urbanism the ideal instrument for this task, and this concludes in the imposition of a social and material vision of the world (Bourdieu 1999).

## 17.2.2  *Crisis of Urbanism*

The history of the modern city is closely linked to the history of modern urbanism as an instrument apparently created to guarantee order in a heterogeneous society. In this framework, the political dimension that urbanism acquires (understood as the science of the organization of spaces) would correspond to guaranteeing all its inhabitants equal access to all common goods and services, necessary to contribute to social, economic and political transformations, respecting the heterogeneous and the diverse (Borja 2003), an ideal that is finally exceeded by the clash with the norm and the difficulty of the administration of the territory. But the city, understood as a container of differences, is also a space of conflict and disputes, where the *polis* assumes a vocation for normative planning to purge conflictive relations based on immutable principles, fixed and unquestionable rules, creating political instruments and the permanent modification of the institutions or technocratic instances to establish said order and the materiality of political ideals (Choay 1970; Mongin 2006).

The historical absence of citizens, the *civitas*, throughout this process constitutes an essential part of the crisis of urbanism. As Richard Sennett states, "[…] the real and immediate experience of man, in all his possible freedom and diversity, is considered less important than the creation of a community free of conflict" (1975, p. 113). Thus, urban life, which is precisely a complex of conflicts, and the city, which should be the territory of that confrontation, are stripped of their own codes of social life and the capacities for their transmutation.

Now, Lefebvre in *El derecho a la ciudad* (1969) unmasks urbanism by exposing it as a biased ideological doctrine within the universality of its foundation and poorly legitimized because it lacks social consensus as its political origin is absent. It is a "pseudoscience" obedient to power and the predominance of the status quo, which avoids confrontation in the spatial dispute from the establishment of an order that appeals to the original conciliatory discourse of urbanism as a link that would guarantee, apparently, the cohesion of the political community and, therefore, "progress".

The truth is that the tasks of order to which modern urbanism responds refer to the framework of modernity and the ideas of modernization made up of three processes that François Ascher (2004) defines as: social individualization, which operates behind the spectrum of the political community; the rationalization or progressive substitution of tradition, which displaces the values of the community toward its homogenization; and social and aesthetic differentiation, which eliminates the heterogeneous experience and coexistence.

These purposes created to live in a modern society are nothing more than the permanent fear of the others, of disorder and diversity, which finds in the very logic of urbanization, capitalism and its urban practices the appropriate mechanisms to create an appearance of community based on the imposition of an abstract-instrumental order, crossed by a rationality focused on manifesting a harmonious appearance, organized and projected under specific parameters, but governed by dominant logics and imposed knowledge (Martínez 2014).

### 17.2.3 *Urban Planning, Urbicide and Obsolescence*

Bogdan Bogdanovic, former mayor of Belgrade and scholar of post-Yugoslav cultures, uses the word "urbicide" to refer to the war against cities in the Balkans (Serb-Croatian-Bosnian war in the former Yugoslavia, 1992–1995), where cities like Dubrovnik, Sarajevo or Mostar were intentionally attacked as symbols of multiculturalism. That seems to be the future of war: destruction of cities and death of civilians, and at the same time a "patrimonial terrorism" in search of identity amnesia (Boyer 2013).

The built environment represents the social action and the political act of inhabiting a community with values, customs and memory. The different destructions carried out on the material base of cities or architecture[1] show that the attack on a built fact, whether public or private, eliminates a shared spatiality that safeguards both individual and collective memory and the possibility of expressing otherness. As Martin Coward emphasizes, "Urbicide, which is a planned assault on the built environment; represents, therefore, an assault on the conditions of possibility of such heterogeneity" (2009, pp. 54–55).

The act of physical and symbolic violence involved in material destruction, as Coward refers again, becomes a disproportionate action with respect to the very objectives of the conflicts, but it becomes a constitutive part of a genocide and, in addition to the disintegration of family ties or "homeicide", a term coined by

---

[1] The destruction of the material base of the city has been a constant in history. From Carthago, to mention an important event from the mid-second century BCE, to Mariúpol (Ukraine) right now (March 2022), passing through the destruction of the great pre-Hispanic cities in Latin America, or Rotterdam or Warsaw at the hands of the Nazis, or Aleppo, in the endless Syrian conflict, motivated by countless reasons that are often incomprehensible, Walter Benjamin's maxim shown in the IX thesis of his Philosophy of History (1940) is inexorably fulfilled: "[…] a single catastrophe that tirelessly piles up ruin after ruin […]" (2008, p. 310).

Douglas Porteous and Sandra Smith that makes explicit "the massive destruction of the home" (Coward 2009, p. 10). Therefore, destroying, demolishing and dismantling an architecture or an urban space, totally or partially, in itself represents eliminating the experience of plurality and disintegrating a social fabric.

But, regardless of the warlike reasons behind the original use of the word "urbicide", there are other reasons cities are subject to annihilation. The primary logics for the formation of a city, that is, the agglomeration and accumulation of social groups in a specific space, linked to modern capitalism, make the production of the city a chaotic and complex framework on which it becomes necessary a certain order. By opposing the city of power to the city for dwelling—a permanent dispute—the role established for the practice of urban planning also becomes part of an exercise in urbicidal violence. According to Françoise Choay, the city, a place of exchange of goods, information and affections, is in crisis, and it is precisely because of urbanism, a pragmatic instrument, often utopian, regularizing, organizing, "a totalitarian instrument of conditioning" (2009, p. 163), capable of substituting "the city" for the imposition of "the urban".

Till (2009), referring to Zygmunt Bauman, points out that the search for order or, more precisely, the imposition of order, seamlessly intertwined social individualization, rationalization and social and aesthetic division with the practical ideas of the modern urbanism, that is, with the desire to adopt rules (since the treatises of Leon Battista Alberti in the fifteenth century), to regulate and control (the activity of Baron Haussmann in Paris in the mid-nineteenth century), or to rationalize the occupation of the territory (the invention of urban planning as a discipline by Ildefonso Cerdá to plan the future Barcelona). All this was done in pursuit of the utopian model of the ideal city to "create the city that does not exist", in which to live, as Victor Hugo said (Mongin 2006, p. 135). This is the "ideal" ideological base for the reproduction of the city and therefore of capitalism, subjecting society to what Neil Leach—based on Walter Benjamin–points out as "the shock of the modern" (2001, p. 72) or the "shock of the new" (1997, p. 3).

Thus, during the heyday of industrial society, plan models aimed to project urban complexes based on "progress and order", creating what does not exist (the new), from surgical and partial executions that established the primacy of circulation and the market, under the rules of order, regularity and control (Mongin 2006) and that always pursued a misunderstood or strategically orchestrated modernity.

Here, urbicide, as Coward (2009) points out, acquires qualities of spatial administration, which, focused on the production of the new, had serious implications for the twentieth century: *the tabula rasa*, a principle that characterized urbanism proclaimed by Le Corbusier or Ludwig Hilberseimer as a symptom of modernity in the interwar period that, although they could not carry it out, became almost the only recipe to solve urban chaos and disorder. The interventions carried out by Robert Moses after World War II in New York to build the Cross Bronx Expressway or the first World Trade Center (1972), with the aim of revitalizing lower Manhattan—as portrayed by Marshall Berman (Coward 2009)—were projects of "urban renewal" (actually, of planned destruction of large sectors of the city), executed under slogans such as "demolition for progress" or "reinventing the world" (Byles 2005, p. 183). The "passion to demolish" (Rosero 2017, p. 101), under the mythical progress, always hid dark purposes: the Cross Bronx Expressway involved the displacement of fifteen hundred families; the demolition of the neighborhoods for the creation of the WTC displaced an entire popular artistic culture. Even the famous demolition of the Pruitt-Igoe housing blocks in St Louis (Missouri) in 1972—which supposedly paved the way for architectural postmodernity—was motivated by racism and xenophobia that was hidden behind an urban renewal presented as a "cure for the disease" (Rosero 2017, p. 74).

The danger of this, as Jeff Byles unmasks in his book *Rubble*, is the adoption of a culture of destruction by the society, as an exercise of a naturalized and socially accepted violence, which turned "[…] architecture into the only art whose destruction society allows" (2005, p. 20), and as the ideal justification to displace or annul the "others" and thereby impose a new order that is justified by expiration.

### 17.2.4 Urban-Architectural Obsolescence

After the "Haussmannization of urbanism", the ease with which the culture of demolition was installed in the society not only progressively replaced the heterogeneous and diverse city for dwelling with the homogeneous city that gives priority to flows and standardization, rather, it showed the vulnerability and susceptibility that any built environment is exposed to a deliberate destruction to achieve it. What is behind this logic? Richard Sennet responds: "[…] the only constant of capitalism is instability" (2006a, p. 19) and that condition has caused a constant definition and redefinition according to the needs and requirements of capital rotation, from the innovation and the reduction of spatial barriers to accumulation (Harvey 1994). Of course, this contradicts the logic of living and the notions of stability and durability that architecture has historically sought, by placing time as the worst enemy of what is already built (Till 2009).

Here, there is no room for a city for dwelling; Paradoxically, capitalism needs impermanence to perpetuate itself, which brings us back to what Frederick Gutheim said in 1949: "[…] the moment a building is finished, its destruction begins" (Byles 2005, p. 23). It is a fact that, in the face of the dispossession of social values, as result of submission to the flows of consumption, the material space, within the framework of capitalism, is just one more object. This reification gives urban space and its material basis the condition of prescription, expiration or obsolescence.

The term "expire" refers to that which loses validity and/or effectiveness and happens when something is ruined and/or worn out. For Giancarlo De Carlo, architectural obsolescence, both physical and technical, occurs when the motivation for use "[…] or when the building's tissues are exhausted and have lost their regenerative capacity" (2005 p. 21) granted by the social component. These patterns can be identified as functional obsolescence, physical obsolescence and economic obsolescence.

The *functional obsolescence* of buildings and public spaces arises when these structures no longer fulfill the functions for which they were originally designed. Examples of this problem are the traditional houses of high-income families in historic centers that are abandoned by their occupants as result of changing fashions and aspirations for modern conveniences (Rojas et al 2004, p. 9).

However, functional obsolescence has little to do with architecture, since functionality is not an objective and eternal attribute, "[…] utility per se does not exist, nor does objective, universal and incontrovertible functionality; there is, on the contrary, the subjective and complex utility, the changing and fragmentary function" (Lampugnani 1999, p. 66). Architecture of all ages has always known this: "Old interiors are modern because they welcome and satisfy the most diverse, complex, changing and unexpected needs" (Lampugnani 1999, p. 89); hence, also the modernity of the historic, hybrid centers full of urban "intensity".

The other two obsolescences have to do with building deterioration due to lack of maintenance, as a result of natural disasters or urban activities (physical obsolescence), or due to lack of profitability, speculative operations or economic pressures typical of urban growth (economic obsolescence) (Rojas et al. 2004, p. 9). In all three cases, it is clear that obsolescence is not neutral or devoid of an exercise of power, since expiration would operate as something intentional and, at the same time, imposed or, as Lefebvre (2013) reveals, it is a strategic game that causes everything previously built to waver in the search for the new.

## *17.2.5 Tyranny of the New*

There is a phrase by Till, "[a]ll architecture is but waste in transit" (2009, p. 67), which gives an account of the Benjaminian catastrophist vision while it described the destructive character of progress that "[…] does not perceive anything lasting" (Benjamin 2010, p. 347). "Goodbye to the old, welcome to the new" (Rosero 2017, p. 101); the condemnation of architecture in the face of its possible obsolescence would respond to the transfer of power that capitalism needs to reproduce itself (Sennet 2006a), a process typical of monopolistic and oligopolistic competitions that Joseph A. Schumpeter already called in 1942 "creative destruction" (1996, p. 118).

The history of capitalism and modernity is the history of the perpetual reproduction of the new and its debris. For Rosalind Krauss, the ideology of the new is fully identified with avant-garde aesthetics; Nevertheless,

> […] its covert message is that of historicism. The new is made comfortable by becoming familiar, since it is seen as having gradually evolved from the forms of the past. Historicism acts on what is new and different to diminish novelty and mitigate difference (1996, p. 289).

Even the idea of conserving, rehabilitating or revitalizing, misunderstood, is made to consume and prescribe. This is, in short, the task of capital: to reduce cultural forms and habits to needs and desires, in such a way that, having reached a certain level of consumption, it commits to a reinvention that manages to change places, landscapes, symbols and even materials for the transitory or the ephemeral, achieving an architecture willing to fade to the rhythm of its consumption (Harvey 1997; Mongin 2006), in such a way that urban and material life resides in a "permanent state of becoming" (Bauman 2013, pp. 2–3).

The cycle of urbicide that contains consumption and expiration generates collateral effects in terms of the phenomenon of urban extinction; Ignacio González-Varas points out two of them:

> [...] in the first place, and obviously, by *consumption* - in terms of its extinction by abandonment -, but its death can also be caused by the opposite cause, that is, by *excess of success* or by *consumption* -in terms of its extinction due to *excessive use* (2014, p. 199).

This is the case of "thematic" cities designed, above all, for mass tourism, but which, however—and the case of Venice is the most paradigmatic but not the only one—"is no longer a suitable space to live and coexist and it becomes a space for consumption" (2014, p. 199). From the hand of this "consumption space" can come the destruction of the city by converting it into an image. We see it in many cities around the world, with an architecture concerned exclusively with itself—with the help of advanced representation and construction technologies—and busy decorating speculative investments that turn the city into a commercial brand.

In any case, these material and symbolic orderings in the city indicate that there is a submission of the collective rhythms to these experiences (Bourdieu 1977). For this reason, even if it is unnecessary and there are no problems of obsolescence to solve, the city or the portion of the city that does not give political returns (historic city), or does not offer economic attractions (urban centers), is subject to an urbicide of expiration. It is the case of Latin American cities and their primary centers where their material bases were dismantle to adapt to the languages of the new in at least two forms; with the adoption of imported and superficial aesthetics to create social differences as Eduardo Kingman (2008) refers, or in operations of material destruction to adapt cities to international architectural canons as Silvia Arango (2012) explains.

## 17.3 Spatial Condemnations

The force with which modernity and progress are established not only manages to settle mechanisms for constant transformation, but also manages to build criteria and canons of truth and aesthetic quality, in such a way that everything that does not conform to the control criteria established by the city of power can be declared non-existent and even obsolete (Santos 2010).

All this capacity granted to the city that lives and recreates power is capable of defining a series of itineraries aimed at the conversion of landscapes and architectures (Muñoz 2005), verifying itself in the urban space through concrete operations on the material support. Here, every thought related to segregation, standardization or consumption has the ability to become stronger and naturalized, becoming a condemnation, a spatiality that hides the faces of expiration.

### 17.3.1  Gentrification

The neologism "gentrification" was introduced by Ruth Glass in her analyzes of the city of London in the mid-sixties of the last century. Neil Smith assumed the concept, as he himself says, imbued with a certain rehabilitative utopia, but without the meanings of expulsion and deterritorialization that came to characterize that gentrification practically from the beginning, from the moment private capital intervened. Furthermore, at the same time, "[…] vibrant working-class communities were becoming culturally devitalized" (Smith 2005, p. 31). It is interesting to see here a certain "revanchism" on the part of capital against the liberal social policies of the second half of the twentieth century, translated into "[…] revenge against minorities, the working class, women, environmental legislation, gays and lesbians, immigrants, [including campaigns] against political correctness and multiculturalism" (Smith 2005, p. 43).

Instead of "gentrification", a concept that is somewhat restrictive, of dubious meaning and even stigmatizing, the geographer García (2001) has proposed "elitization". This other neologism is directly related to the three capitals that are combined in these processes of supposed urban requalification to which Giandomenico Amendola refers, so elitization is

> […] able to communicate social *status* well. The part of the city that has been made one's own, inhabiting it, expresses a complex combination of economic capital (purchasing power and heritage), cultural capital (skills, *expertise*, taste), social capital (interpersonal relations, group belonging and possibility of be accepted) (2000, pp. 124–125).

For what has to do with the effects of gentrification or elitization on historic urban areas, we should take look at the document of ICOMOS—the organization associated with UNESCO dedicated to the conservation, protection and enhancement of cultural heritage—called "The Valletta Principles", designed for the safeguarding and management of these areas. There it is stated that gentrification can cause

> […] the loss of a place's habitability and, ultimately, of its character, [so that] historic urban areas and towns are at risk of becoming a product of mass consumption, which can lead to the loss of its authenticity and heritage value (ICOMOS 2011, s/p).

There is an interesting theoretical debate about gentrification that has to do with the validity of its use as a concept applicable to the global South, given that the most widespread literature refers to cases almost always located in Anglo-Eurocentric locations. For example, Jennifer Robinson puts in crisis the possibility that any urban methodology or concept developed in the USA or Europe—as is the case with gentrification—is applicable in principle to any city in a "world of cities", since "[…] the danger is that a comparative urbanism equal to engaging with a wider range of cities invites a new round of imperialist appropriation of international urban experiences to service Western and other well-resourced centres of scholarship" (2011, p. 19).

However, López-Morales (2015) defends the theoretical usefulness of gentrification, as has happened in certain urban centers of cities in Mexico, Brazil or Chile, understood as an urban phenomenon not limited only to the mercantile interests of capital or tourism—which is its classic interpretation—but as a result of well-intentioned policies of the State or certain NGOs that seek to appropriate their symbolic values, even in a process of "aporofication", that is, the appropriation of a central place that has been previously abandoned by the elites and occupied by popular classes, as Carrión and Cepeda (2022) point out for the case of the city of Quito and other historical centers. Faced with these situations, López-Morales himself (2015) refers to a distinctive anti-gentrification movement typical of an active "insurgent citizenship" that is characteristic of Latin America.

## 17.3.2 The City of the Peripheries

### 17.3.2.1 Peripheries and Urban Centers

The destruction of the city at the hands of the peripheralization and the decomposition of urban centers had already been pointed out by Lewis Mumford in the classic *The City in History* of 1961:

> What began as an escape from the city by families, became a more general withdrawal that has produced, more than independent suburbs, a suburban belt that expands. […] Unfortunately, the sum of all these dispersions does not produce a new urban constellation. While they potentially provided the building blocks for a new kind of city with many centers, administered on a regional scale, their effect so far has been to corrode and undermine the old centers (Mumford 2012, pp. 838–839).

The multiplicity of urban centers, greater accessibility to services, avoiding long distances, less real estate pressure on the historic center could be advantages of peripheralization, but this is not the case; For Mumford, this situation "automatically" means the destruction of the city. According to Rojas et al., the negative consequences are numerous,

> Central areas abandoned by the most dynamic activities and higher income families are gradually occupied by productive activities or services of lower productivity or informality and by lower income families. […] The abandoned or deteriorating residences are occupied by low-income families, who sometimes rent space in these informally subdivided buildings or occupy the abandoned houses free of charge (2004, p. 3).

As we will see later on, the urban center can become an inner periphery. It is also in the peripheries where a series of conflicts are uncovered, the product "more of a political and cultural impotence than an expression of a society evolving towards greater sincerity and individuality" (Herzog 1989, p. 113). Before them, the classic instruments of urban planning are no longer valid—with their unstable balance between the public and the private, the full and the empty—but an architecture that is often banal. We believe that Edward Soja is correct in calling all these diverse peripheries "exopolis", giving it a sense of "simulacrum", that is, "an exact copy of a city that has never existed" (1996, p. 19), a real city but at the same time fictitious: a real estate endeavor whose characteristics have nothing to do with what we recognize as a "city", but are set by the market.

### 17.3.2.2 Fortress Communities

One of the most sinister effects of peripheralization is the privatization and depoliticization of cities due to (real or imagined) urban fear. In the last chapter of *The Postmodern City*, Giandomenico Amendola describes how in those places "[the] smiling community has been transformed into the purified and shielded community. A nightmare […]" (2000, p. 342). In this "privatopia"—a term of EvanMcKenzie—a strict norm regulates

> […] lifestyles and architectural typologies, behaviors, access and characteristics of the guests. […] It is a subtle form of ethnic cleansing and purification that simultaneously tends to create a socially and culturally homogeneous community and when you try to separate it and defend it from the hostile world, [so that the inhabitants of these peripheral suburbs] no longer know how to live with what is different to them (Amendola 2000, p. 344).

These closed residential peripheries are places of an "imaginary hyperreality" full of simulations, because, to begin with, the apparent community is not what we think it is, since all its inhabitants belong to the same economic group, in addition

> [t]he residential area simulates order, but it is the result of a disintegrated and disjointed city. The public space inside is at the same time very private and marks its difference with that of those who live outside the walls. […] It is a daydream that follows the rules of consumption and is promoted through advertising and credit (Levi et al., 2006, p. 168).

### 17.3.2.3 Central Peripheries

A few years ago, Manfredo Tafuri wondered how to reconcile the reality of historic centers with "[…] the breaking of any static balance, typical of modern life and culture and, consequently, of the modern city" (1964, p. 28). To this end, Tafuri proposed the following: "[to] not reduce the old city to a more or less convenient corollary of the new city […] but to rediscover a function of close complementarity that allows the continuity of fabric and functions even in the inevitable differentiation" (1964, p. 28). It is necessary to build that balance that should exist between the existing and new cities based on the "temporal stratification" that the city possesses through the history of its transformations. That balance is not understandable for the forces of speculation that would like to convert the historic centers into "periphery", "[…]

understanding by 'periphery' any part of the citizen body that is in a state of inferiority compared to a ' center' in any case active" (Tafuri 1964, p. 29).

The strengthening of the central city that can be achieved contributes to highlighting a series of lessons that we can learn from these historic centers, seen not as a timeless and frozen heritage, but as an open book where values of spatial continuity, homogeneity, compactness, in short, historical and urban values of high educational interest.

## 17.4 Criticism of Expiration

### 17.4.1 Date of Expiry

Do cities and architectures expire? The city, according to Claude Levy-Strauss, "[…] is both an object of nature and a subject of culture; it is individual and a group, it is lived and imagined: A human thing par excellence" (1988, p. 125). In addition, now according to Mumford, the city is not only a place of culture, is at the same time of extermination, the result of an endless dialectic between its production and its destruction:

> Thus, the most precious collective invention of civilization, the city, preceded only by language in the transmission of culture, became from the beginning the receptacle of noxious internal forces, oriented towards constant extermination and destruction (2012, p. 94).

The way to avoid that constant destruction to which Mumford refers would be in the recognition of the complex, objective and subjective values of the city that allow us to identify its "face", its personality, product—as Lampugnani recalls—of "countless" political, social, economic, cultural and, therefore, also architectural events. "Urban design, as a discipline, must measure itself -and itself be measured- against this face. Its objective is not abstraction"—the classic method of urban planning—"but a complex, subtle, and often submerged reality", a reality that consists, especially, "of human memories, experiences, dreams" (2021, p. 164).

That reality is built from the history of the city and its citizens. For this reason, the criticism of the most orthodox modern urban utopias is based, mainly, on the way in which the drawings of the most conspicuous architects of the time tried to expel time from the city. After the Second World War, the conservationist criteria of Italian origin, allied to the theories of the urban project, helps us whenever we extend their recommendations to the entire city. Let us see, for example, Saverio Muratori's already classic reflection on urban life and history:

> The key to the urban problem, the only point of agreement for an urban plan worthy of the name, is the understanding and conservation of the character of our city. Renewing this character but preserving it must be the only urban criteria; and if it is necessary to expand the city by creating new roads, squares and neighborhoods, we must avoid emptying or replacing the old with the new by destroying it (1950, p. 40).

The idea of history in Henri Focillon can help to decide the best action when developing an urban project on the existing city: "History […] can be considered as a superposition of presents that extend with great amplitude" (2010, p. 100). Then: "[…] what we call history is constituted precisely by the diversity and inequality of the currents […] in such a way that each fraction of the elapsed time is at the same time past, present and future" (1966, pp. 9–10). Therefore, there is no place for urban expiration, since all times, including those of the project or planning—paraphrasing Focillon—are included in the reality of the city.

## 17.4.2 Time in Architecture and City

As Till (2009) says, architecture needs to continue facing its temporary condition because it is unavoidable and also multiple, continuous and powerful. Architecture must be understood as a framework that maintains the enormous number of times that occupy it (linear, cyclical, subjective, social, long and short times). Essentially, the experience of architecture is temporary, from the punctual journey through its environments to its vital development. It is precisely, in the words of Juhani Pallasmaa, "[…] through the temporal strata of our built scenarios where we grasp the past and the fluidity of cultural time, [therefore] we are mentally incapable of living in chaos or in a timeless condition" (2016, pp. 117–118). Here lies the importance of urban history: this one from Pallasmaa is, without a doubt, a beautiful definition of the city, precisely because it treasures and represents time: "The city structures the capture and preserves time in the same way as literary or artistic works. Buildings and squares allow us to return to the past, to experience the slow healing time of history" (2014, p. 43).

That temporary coexistence leads us to one of the fundamental conceptual instruments for the urban project: the "modification". This implies not repeating the canons and principles, nor the previous methods and processes, which gave rise to the existing city. The respect for the real territory implies not leaving it frozen in a timeless present, but ensuring its transfer to the future, through its inevitable modification. What is durable is what is changing, adaptable and continually reworked. The time of city and architecture opens up to the future, keeping the past alive. After all, to exist, according to Henri Bergson, is a continuous change, because "[…] what does not change does not last" (1963, p. 441). Consequently, duration does not consist in the substitution of one instant in time for another, but in the incessant growth of the past. Thus, the identity of architecture and the city lies precisely in its ability to adapt to the passage of time.

"Modification" was the subject of one of the most famous editorials by Vittorio Gregotti in *Casabella* magazine. It depends directly on the consideration of the real to be modified, and this gives rise to another important concept: the "belonging" to places, an idea that "[…] becomes an authentic pedagogy of the project" (Gregotti 1993, p. 77), since the place (the relationships between morphology and typology, the environment, its history) is the foundation of architecture. For this reason, the idea of belonging tells us that in any urban intervention, in any inevitable modification of the real, a *tabula rasa* is not conceivable, nor is the transferability to one place of project experiences designed for other places or the internationalization of an abstraction, which was the quintessential modern aspiration.

This "belonging" must also be addressed from the idea of "permanence". Piero Sanpaolesi, one of the references author in the theory and practice of heritage intervention in Italy in the third quarter of the twentieth century, is very clear in his criticism: "The destruction of old quarters […] is the result of widespread ignorance and indifference, common even in educated circles, so far as the city and its architecture are concerned" (1972, p. 245), caused, especially, by economic reasons and unproven obsolescence. To avoid these alterations or destruction, Sampaolesi recommends taking into account the age of the city, its shape, its history, its traditions, the accumulation of works and records, as well as its beauty.

Therefore, the plan or the project should not freeze history, on the contrary: life in the city demands a creative and open attitude in the relations between the history of architecture and architecture itself. An attitude that should be based on modesty, common sense and sensitivity to those architectures and urban environments on which it intervenes, knowing, moreover, that the city is not infinitely malleable, since

> […] it is not infinitely available in the face of changes in the economy, institutions and politics. Not only because of the resistance that spatial inertia itself opposes, but also because to some extent it constitutes the trajectory along which these same changes can occur (Secchi 2015, pp. 28–29).

### 17.4.3 Re-politicization of the City. Toward Another Urbanism

#### 17.4.3.1 Post-politics and Re-politicization

Governance—"[…] an arrangement of Governing-beyond-the-State" (Swyngedouw 2007, s/p)—is the practical manifestation of the so-called "post-politics". According to Žižek (2006), post-politics is the search for compromises between interests— almost always economic—that leave the State out, in such a way that it is reduced to a simple agent in the equation, but far from that "space of litigation" that should characterize politics. Here, we defend a re-politicization of the urban process that must recognize, first, that traditional planning with modern roots is "[…] a violent act that erases at least part of what existed in order to erect something new and different, [and also that] such interventions hold a totalitarian moment, the temporary suspension of democracy" (Swyngedouw 2011, p. 61). However, if we want a democratic planning process, it must recover politics, it must build a "[…] space for the cultivation of dissent and disagreement, to claim the presumption of equality of each and every one of us, [because the result] is contingent, often unpredictable, enormously varied, risky, [...] deeply conflictive" (Swyngedouw 2011, p. 62), and also equitable and inclusive.

Therefore, one way to re-politicize the city would be to recover urbanism as a territory of conflict, in such a way that in the face of the orthodox ideology of the plan—imaginary, decontextualized, falsely coherent, arrogant and imposing-, typical of post-political governance, we would have "another urbanism".

### 17.4.3.2 Neo-urbanism

Like other authors, Anthony Vidler has insisted on the importance of the existing city as the foundation of urban intervention processes; therefore, he defines urban planning as "[...] the instrumental theory and practice of constructing the city as a memorial of itself" (1994, p. 179). At the same time, we would have to speak of a "neo-urbanism" that develops other methodologies, substituting the classic linearity "that linked the diagnosis, the identification of needs and the final elaboration of a plan" by a "heuristic, iterative, incremental and recurrent management" (Ascher 2004, p. 73), that is: elaboration and testing of hypotheses, partial realizations, long-term measurements, feedback processes, etc. To put this into practice, we should refer to two disciplinary developments that equally affect architecture and the city: the "urban project" and the "historic urban landscape".

The "urban project", the central theme of a congress held at the Harvard University Graduate School of Design in April 1956, was defined by its dean Jose Luis Sert (Sert et al. 2006) as a new work of synthesis, capable of bringing together all disciplinary approaches to the urban phenomenon, from architecture or engineering to urban planning or landscape architecture. Through the urban project, it seeks to intervene in cities based on multifunctional and hybrid proposals that avoid zoning, taking into account streets, squares, gardens, blocks or neighborhoods as instruments of the plan. It trying not only to put order in the city but also to architecturally control the elements (buildings and free spaces) that make it up, incorporating urban infrastructures and facilities into the same plane of reflection. The urban project is nothing more than a way of rediscovering architecture with urban planning, disciplines that should never have been divorced, although today, unfortunately, their practice "[...] has been largely relegated to commercial or anachronistic approaches" (Aureli 2007, p. 195), purely formalistic. It would therefore be necessary to recover the dimension of the intervention—which has nothing to do with scale or size—because it would not only review the very conception of urbanism, but also that of architecture.

On the other hand, the "historic urban landscape" emerged at the beginning of the twenty-first century as the fundamental concept for the management of cities— initially of historical value. In the so-called Vienna Memorandum, UNESCO (2005) included under that concept—and in a holistic way—values such as archaeological, architectural, historical, scientific, aesthetic, sociocultural or ecological, in such a way that all the elements that define the urban character of a city could be brought together. The interesting thing about the concept is that, after its assimilation, the old notions of "historic complex" or "historic center" have been superseded. In addition, it includes, as stated in the subsequent *Recommendations*, "[…] the general urban context and its geographical environment" (UNESCO 2011, section 8), that is, the entire city, and also "[…] the uses and social and cultural values, the economic processes and the intangible aspects of heritage in their relationship with diversity and identity" (UNESCO 2011, section 9). The intention of this new principle is to combine the protection of heritage (used here in the broadest sense of the term) and the dynamics and social, cultural, functional and economic development of the entire city—not just the historic or central one—where the community emerges as the main actor in these processes.

The complementarity between contemporary architecture, the city, and the values of the historic urban landscape are thus a good principle allied with the urban project that we referred to earlier. Urban planning and project methodologies originally applied to the protection and intervention in historic centers allows us to use them throughout the city.

### 17.4.3.3 Participation and Insurgent Urbanism

The crisis of urbanism to which we have referred previously is also that of the urban planner and the agents who make their decisions about the city. As Paul Davidoff and Thomas A. Reiner, *Advocacy Planning* theorists, said in the early 1960s:

> If an ultimate objective of planning is to widen choice and the opportunity to choose, then the planner has the obligation not to limit choice arbitrarily. […] This is crucial: we maintain that neither the planner's technical competence nor his wisdom entitles him to ascribe or dictate values to his immediate or ultimate clients (1962, p. 108).

Despite the fact that it is a text that is already sixty years old, it is surprising that the professional practice criticized there is still present in the supposedly democratic agendas of urban decision-making, since planners and architects still maintain their exclusive ability to forecast the future and prediction of the needs of citizens through the instruments of the plan or project. However—and this is an example—the unpredictable should not be taken out of urban reflection, as happens with informality, often caused, according to Ananya Roy, by planning itself or its intentional absence.

> Engagement with informality is in many ways quite difficult for planners. Informal spaces seem to be the exception to planning, lying outside its realm of control […]. To deal with informality therefore partly means confronting how the apparatus of planning produces the unplanned and unplannable (2005, pp. 155–156).

According to James Holston, the utopia of modern urban planning seeks to transform what exists outside of its historical and social reality. An idea of a plan without contradictions or conflicts is unreal because "[…] it assumes a rational domination of the future in which its total and totalizing plan dissolves any conflict between the imagined and the existing society in the imposed coherence of its order. [Nevertheless] it fails to consider the unintended and the unexpected as part of the model" (1995, p. 46). By introducing "the other", the unpredictable, the unregulated, the self-managed, we would be facing an "insurgent urbanism" capable of re-territorializing its processes through the alternative development of "[…] new kinds of practices and narratives about belonging to and participating in society" (Holston 1995, p. 53); a planning that knows how to learn, therefore, from lived experiences.

Today it is more evident than ever "[…] the need for the citizen to be a substantial part of the decision-making system on the organization of the territory, in which he lives or in which he is going to live" (Fariña Tojo 2015, p. 74). Assuming, therefore, participatory practices, these can be seen in two different ways. According to Leonie Sandercock, one has to do with the traditional negotiation and consensus processes typical of participatory urbanism, but where qualified technicians and managers lead the entire process.

> The flip side is less benign. It can frown at you and insult you. It does not normally wear a suit and is not interested in planning institutions, since they have traditionally excluded it. […] They are neighborhood and community movements, working from the bottom up (as opposed to from the top down), that teach […] that marginal communities find their voice, and not speak for themselves (1998, pp. 6–7).

The result has to do directly with what Richard Sennet—from Jane Jacobs—calls "open city", a democratic and participatory space built from a special attention to the edges ("passage territories"), the architectural-urban spatial stimulation ("incomplete form") and conflicts and agreements ("narratives of development") (Sennet 2006b).

The urban planners should therefore abandon their privileged situation, the "overhead arrogance", as Stefano Boeri would say, "incapable of supporting the multidimensional and dynamic nature of urban phenomena" (2010, p. 182), and start collaborating at a street level with other agents and with citizens willing to make decisions about the places where they are going to live. Giulio Carlo Argan, an art historian who became mayor of Rome in the second half of the 1970s, defined urban planning—probably the fruit of his experience—as "the science of managing urban values": "The task of urbanism is not to project the city of the future but to administer in the common interest a heritage of values, certainly economic […], but also historical, aesthetic, moral; collective and individual" (1984, p. 223). Thus, the urban planner should qualify as an "educator", promoter of an ethical sense that seems to be the appropriate attitude toward the present and future city during these coming years.

## *17.4.4 Architecture for Uncertainty*

Architectural culture—many times without being aware of it—carries with it an abuse of power and domination, which implies the substitution of ethical concerns for aesthetic ones; the disappearance of all ethical discourse at the hands of the image, in such a way that the attractiveness of its appearance makes us forget the ethical judgments of a certain political or architecturally masked ideology. In this way, through certain strategies of seduction, architecture has become—especially after the so-called postmodernism and its added practices, such as deconstruction—in a simple game of empty forms destined for consumption, absent from any discourse as it is not the appropriation of an artificially translated philosophical or scientific terminology, which masks economic and representational interests that are not openly confessed (Leach 2001).

Now it is necessary to think about other practices that have to do, directly, with times of uncertainty, of complexity, with the irremediable contingency for the project in architecture and the city. "Contingency" means that things may turn out differently than expected. According to Till, "[…] in the contingent world the exact end is uncertain and the choices made along the way are exposed to other forces, and in particular the hopes and intents of others" (2009, p. 59). That is why architects and urban planners should abandon "[…] their delusions of autonomy and engage with others in their massy, complex lives" (Till 2009, p. 61). An "architecture for uncertainty" means building for open functionality, sparingly, in a durable way and, of course, looking for beauty (Lampugnani 2021). Architecture in times of uncertainty is not concerned about extreme novelty, but rather about its inhabitants, making frugal and appropriate use of resources and about what has already been built.

About building on what has been already built, we must be aware of the relation that exists between a global idea of sustainability and the safeguarding of the *locus* of urban areas. According to Lampugnani,

> [...] we must use our existing architectures more thoughtfully. Our cities, villages, and rural lands are filled with underused or abandoned buildings. We can restore and revitalize them instead of building new ones next to them (2021, p. 58).

If it were necessary to build new buildings, these should be sustainable, durable, made of recyclable materials, energy savers, passive environmental control, etc. However, it happens that the market is interested in expiration and demolition, which benefits land speculation and the profit percentages of construction operations; but this, is ecologically irresponsible and economically unacceptable. Lampugnani describes the sustainable and contemporary quality of construction on what is built in this text: "A restored, renovated, and revitalized building prevents the construction of a new one -prevents the consumption of land, materials, and energy. [It] is to make parsimonious use of resources" (2021, p. 62).

The principle of action should be, as Anne Lacaton and Jean-Philippe Vassal—Pritzker Prize winners in 2021—remind us, "[…] not to tear down, nor undo, nor cut off what is alive, but, on the contrary, to reinforce, thus contributing to the balance of existing urban structures" (2017, p. 101). For this, the architecture should have a certain quality of "generic space"; this means being ready for its "alterability" (changes of all kinds), "extendability" (horizontal or vertical lengthening) or "polyvalence" (multiple use of spaces without architectural or structural modification) (Leupen 2006, p. 25).

In relation to extensibility, progressive housing (which has to do with the theory of architecture as a process) "[…] responds to social practice, to the way in which most people produce their housing, according to the dynamics of their resources, possibilities, needs and dreams" (Ortiz Flores 2007, p. 12). The architect has an important political responsibility here, since they should move from their traditional idea of "creators" or "artists" to adopt for a moment the idea of the architect as an artisan, "[...] as a translator of desires and needs; process mediator and facilitator; catalyst of situations or connector between subjects and interests" (Verdaguer 2011, s/p).

Architects are educated—by necessity—in generality and therefore can address complex problems while developing particular or specialized issues (Morin 2001). However, dealing with those problems—this is extremely important—they must also abandon their traditional position of authority (the myth of the "author"), perpetuated in self-propaganda and in the educational system. Now, they must take position themselves as anti-heroes, as co-authors who are part of a much more complex and broad process that includes other actors and, above all, those who are going to inhabit their architectures.

## 17.5 Conclusion

Urbicide, from the logic of expiration or obsolescence, opens a critical debate that combines several theoretical perspectives on the cycle of production and destruction of the city from a mechanism of materialization and at the same time dematerialization of architecture from a regulatory urbanism. Thus, urbicide converts the city, understood as a space to inhabit, into a space for consumption and expiration. In our opinion, to reach this urbicide, there are three tyrannies: (a) the functional, physical and economic obsolescence of architectures; (b) the autocracy of power, which wants to leave its mark on the city; and (c) the tyranny of the new, which forces constant renewal. These actions, which are not neutral or devoid of an intentionality that favors the rotation of capital over the urban space, are verified in at least three aspects:

- As a mechanism of physical order that guarantees capitalist accumulation through real estate profit, favoring flows, consumption and the ephemeral, over the experience of building places and architectures that express appropriation and identity.
- As an action present in the materialization of spatial homogenization and, therefore, in social and aesthetic differentiation and control, condemning spaces to uniformity and standardization as patterns of consumption typical of a tyranny that establishes rather a form of urbanization based on expiration and the new.
- As a discursive exercise present in the construction of exclusivities and supported by social imaginaries in favor of a dubious progress, capable of declaring states of inferiority and substituting the values of heterogeneity, plurality and community, that is, the annulment, from the material, from the value of time and spaces of authenticity.

The destructive sense of expiration condemns spaces to abandonment and, at the same time, to gentrification, excess consumption, privatization, overcrowding and indifference, where the sense of the city fades. For this reason and against expiration and its consequences—advocated both by traditional urban planning and by the architectural project in its search for novelty—we speak of a "re-politicization of the city". Through this re-politicization we defend the existence of "another urbanism" and, at the same time, of an "architecture of complexity", in such a way that, in the face of the orthodox ideology of the urban plan and the architectural project—which are usually being decontextualized, falsely coherent, arrogant and imposing documents—we would have an "insurgent urbanism" and an "architecture of uncertainty".

The insurgency in the practice of urbanism makes it possible to re-territorialize its processes through the alternative development of new types of practices and social narratives about the sense of belonging and participation for the construction of alternative futures. Meanwhile, the architecture of uncertainty means building for an open functionality, in a durable way, which does not care about extreme novelty but about its inhabitants, making a frugal and adequate use of resources and what has already been built. In one case as in another, there is an "ethical sense", which seems to be the appropriate attitude toward the present and future city during these coming years. According to this agenda, there would be no possible obsolescence or expiration.

# References

Amendola G (2000) La Ciudad Postmoderna. Madrid, Celeste
Arango S (2012) Ciudad y arquitectura. Seis generaciones que construyeron la América Latina moderna. Fondo de Cultura Económica, México D.F.
Argan GC (1984) Historia del arte como historia de la ciudad. Laia, Barcelona
Ascher F (2004) Los nuevos principios del urbanismo. Alianza, Madrid
Aureli PV (2007) Architecture after liberalism: towards the form of the European capital city. In: Aureli PV et al (eds) Brussels-A manifesto. Towards the Capital of Europe. The Berlage Institute and NAI, Rotterdam, pp 185–204
Bauman Z (2013) La cultura en el mundo de la modernidad líquida. Fondo de Cultura Económica, México
Benjamin W (2008) Sobre el concepto de historia. In: Obras completas, libro I/vol 2. Abada, Madrid, pp 305–318
Benjamin W (2010) Imágenes que piensan. In: Obras completas, libro IV/vol 1. Abada, Madrid, pp 249–390
Bergson H (1963) La evolución creadora. In: Obras escogidas. Aguilar, Madrid, pp 433–755
Boeri S (2010) Atlas eclécticos. In: Walker E (ed) Lo ordinario. Gustavo Gili, Barcelona, pp 177–204
Borja J (2003) La ciudad conquistada. Barcelona, Alianza
Bourdieu P (1977) Structures and the habitus. In: Outline of a theory of practices. Cambridge University Press, New York, pp 72–96
Bourdieu P (1999) La miseria del Mundo. Akal, Madrid
Boyer MC (2013) Collective memory under siege: the case of 'heritage terrorism'. In: Greig Crysler C, Cairns S, Heynen H (eds) The SAGE handbook of architectural theory. SAGE, Los Angeles, pp 325–339
Byles J (2005) Rubble. Unearthing the history of demolition. Harmony Books, New York
Cacciari M (2012) La Città. Pazzini Editore, Rimini
Carrión F, Cepeda P (2022) Historic centers in Latin America and the Caribbean: from heritage fetishism to productive inheritance. In: González-Pérez JM, Irazábal C, Lois-González RC (eds)

The Routledge handbook of urban studies in Latin America and the Caribbean. Routledge, New York, pp 202–220

Choay F (1970) El Urbanismo, utopías y realidades. Lumen, Barcelona

Choay F (2009) El reino de lo urbano y la muerte de la ciudad. Andamios 6(12):157–187

Coward M (2009) Urbicide. The politics of urban destruction. Routledge, Milton Park y Nueva York

Davidoff P, Reiner TA (1962) A choice theory of planning. J Am Inst Plann 28(2):103–115

De Carlo G (2005) Architecture's public. In: Blundell Jones P, Petrescu D, Till J (eds) Architecture and participation. Taylor & Francis, London and New York, pp 3–18

Fariña Tojo J (2015) Cambiar el modelo urbano. Ciudades 18(1):69–79

Focillon H (1966) El año mil. Alianza, Madrid

Focillon H (2010) La vida de las formas. UNAM, México

García Herrera LM (2001) Elitización: propuesta en español para el término Gentrificación. In: Biblio 3W, vol VI, no 332, diciembre. http://www.ub.edu/geocrit/b3w-332.htm. Accessed 22 Nov 2021

González-Varas Ibáñez I (2014) Las ruinas de la memoria. Siglo XXI, México

Gregotti V (1993) Desde el interior de la arquitectura. Península, Barcelona

Haesbaert R (2013) Del mito de la desterritorialización a la multiterritorialidad. Cultura y representaciones sociales 8(15):9–42

Harvey D (1997) Condición de la Postmodernidad. Amorrortu, Buenos Aires

Harvey D (1994) The social construction of space and time: a relational theory. Geograph Rev Jpn, Series B, 67(2):126–135

Herzog J (1989) La ciutat i el seu estat d'agregació. Quaderns 183:113–121

Holston J (1995) Spaces of insurgent citizenship. Plan Theor 13:35–56

ICOMOS (2011) Principios de La Valeta para la salvaguardia y gestión de las poblaciones y áreas urbanas históricas. https://www.icomos.org/charters/CIVVIH%20Principios%20de%20La%20Valeta.pdf. Accesed 10 March 2022

Kingman E (2008) La ciudad y los otros. Quito 1860–1940. Higienismo, Ornato y Policía. FLACSO y Universitat Rovira i Virgili, Quito

Krauss R (1996) La escultura en el campo expandido. La originalidad de la Vanguardia y otros mitos modernos. Alianza, Madrid, pp 289–303

Lacaton A, Jean-Philippe V (2017) Actitud. Gustavo Gili, Barcelona

Lampugnani VM (1999) Modernità e durata. Skira, Milán

Lampugnani VM (2021) A radical normal. Propositions for the architecture of the city. DOM, Berlin

Leach N (1997) Rethinking architecture. Routledge, New York

Leach N (2001) La an-estética de la arquitectura. Gustavo Gili, Barcelona

Lefebvre H (1969) El derecho a la ciudad. Península, Barcelona

Lefebvre H (2013) La producción del espacio. Capitán Swing, Madrid

Leupen B (2006) Frame and generic space. 010 Publishers, Rotterdam

Levy-Strauss C (1988) Tristes trópicos. Paidós, Barcelona and Buenos Aires

López Levi L, Saínz EM, Chumillas IR (2006) Fraccionamientos cerrados, mundos imaginarios. In: Lindón A, Aguilar MÁ, Hiernaux D (eds) Lugares e imaginarios en la metrópolis, coord. Anthropos y UAM Unidad Iztapalapa, Barcelona, pp 161–169

López-Morales E (2015) Gentrification in the global south. City 19(4):564–573

Martínez E (2014) Configuración urbana, hábitat y apropiacion del espacio. Scripta Nova 18(493):33. https://revistes.ub.edu/index.php/ScriptaNova/article/view/15022/18375. Accesed 6 March 2022

Mongin O (2006) La condición urbana. La ciudad a la hora de la mundialización. Paidós, Buenos Aires

Morin E (2001) La mente bien ordenada. Seix Barral, Barcelona

Mumford L (2012) La ciudad en la historia. Pepitas de calabaza, Logroño

Muñoz F (2005) Pasajes banales: bienvenidos a la sociedad del espectáculo. In: de Solá-Morales I, Costa X (eds) Metrópolis. Gustavo Gili, Barcelona, pp 78–93

Muratori S (1950) Vita e storia della città. Rassegna Critica Di Architettura 3(11–12):3–52
Ortiz Flores E (2007) Integración de un sistema de instrumentos de apoyo a la producción social de vivienda. Coalición Internacional para el Hábitat, México
Pallasmaa J (2014) La sensación de la ciudad. La ciudad en tanto percibida, recordada e imaginada. In: Gávez AH (ed) Habla ciudad. Arquine, México, pp 39–44
Pallasmaa J (2016) Habitar. Gustavo Gili, Barcelona
Robinson J (2011) Cities in a world of cities: the comparative gesture. Int J Urban Reg Res 35(1):1–23
Rojas E et al (2004) Volver al centro: la recuperación de áreas urbanas centrales. Banco Interamericano de Desarrollo, Washington D.C.
Rosero V (2017) Demolición: El agujero negro de la modernidad. Textos de arquitectura y diseño, Buenos Aires
Roy A (2005) Urban informality: toward an epistemology of planning. J Am Plann Assoc 71(2):147–158
Sandercock L (1998) Towards cosmopolis: planning for multicultural cities. John Wiley, Chichester
Sanpaolesi P (1972) The conservation and restoration of historic quarters and cities. In: Gazzola P et al (eds) Preserving and restoring monuments and historic buildings. UNESCO, Paris, pp 245–251
Santos BS (2010) Refundación del Estado en América Latina, Perspectivas desde una epistemología del Sur. Instituto Internacional de Derecho y Sociedad, Lima
Schumpeter JA (1996) Capitalismo, socialismo y democracia. Folio, Tomo I. Barcelona
Secchi B (2015) La ciudad de los ricos y la ciudad de los pobres. La Catarata, Madrid
Sennet R (1975) Vida urbana e identidad personal. Los usos del desorden. Edicions 62, Barcelona
Sennet R (2006a) La cultura del nuevo capitalismo. Anagrama, Barcelona
Sennet R (2006b) The open city. Urban Age. LSECities. https://urbanage.lsecities.net/essays/the-open-city. Accessed 7 March 2022
Sert JL et al (2006) 'Urban design': extracts from the 1956 first urban design conference at the GSD. Harv Des Mag 24:4–9
Smith N (2005) The new urban frontier. Routledge, London and New York
Soja E (1996) The third space: journeys to L.A. and other real-and-imagined places. Blackwell, Oxford
Swyngedouw E (2007) The post-political city. In: BAVO (ed) Urban politics now: re-imagining democracy in the Neoliberal City. NAI Publishers, Rotterdam. https://www.researchgate.net/publication/281318055_The_post-political_city. Accessed 13 Feb 2022
Swyngedouw E (2011) ¡La naturaleza no existe! La sostenibilidad como síntoma de una planificación despolitizada. Urban NS01:41–66
Tafuri M (1964) Il problema dei centri storici all 'interno della nuova dimensione cittadina. In: Aymonino C et al (eds) La citta' territorio. Leonardo da Vinci, Bari, pp 27–30
Till J (2009) Architecture depends. The MIT Press, Cambridge (MA) y Londres
UNESCO (2005) Memorándum de Viena sobre el Patrimonio Mundial y la Arquitectura Contemporánea. Gestión del Paisaje Histórico Urbano. https://whc.unesco.org/archive/2005/whc05-15ga-inf7e.pdf. Accessed 6 March 2022
UNESCO (2011) Recomendaciones Paisaje Urbano Histórico. https://whc.unesco.org/uploads/activities/documents/activity-638-100.pdf. Accessed 6 March 2022
Verdaguer C (2011) Marco conceptual: La participación ciudadana como instrumento de sostenibilidad urbana. http://www.diba.cat/c/document_library/get_file?uuid=22bff79a-64a2-4b22-9f11-05d1435d547a&groupId=527890. Accessed 6 March 2022
Vidler A (1994) The architectural uncanny. The MIT Press, Cambridge, MA
Žižek S (2006) The lesson of Rancière. In: Rancière J (ed) The politics of the aesthetics. Continuum, London, pp 69–79

**Valeria Reinoso-Naranjo** PhD in City, Territory and Sustainability at University of Guadalajara, Master in Urban Studies—Flacso Ecuador 2013–2015 and Architect by the Central University of Ecuador and International Polytechnic Exchange Program of Milan 2009–2010. She has been a teacher at the Faculty of Architecture and Urbanism, Central University of Ecuador and is currently a teacher at the Technical University of Ambato-Ecuador in the Faculty of Architecture and Design.

**Manuel Martin-Hernández** PhD in Architecture and Retired University Professor of the University of Las Palmas de Gran Canaria (Spain) of which he is honorary professor and in which he was director of its School of Architecture and the Department of Art, City and Territory. He has been a visiting professor at CUAAD (University of Guadalajara, Mexico), with which he continues to collaborate in the development of master's degrees and doctorates. He is a professor of the Master's Degree in Architectural Design of the FAU of the Central University of Ecuador and coordinator of a diploma in Theory, Criticism and History of Architecture at ESARQ (Guadalajara).

# Part V
# Dissolution of Social Interaction

# Chapter 18
# The (Un)made City: Spatial Fragmentation, Social Inequalities and (De)compositions of Urban Life

**Ramiro Segura**

**Abstract** This chapter reflects on the (un)made city and the dynamics of (de)composition of urban life. In recent decades, various voices have questioned the persistence of the modern idea of "the city" as a self-evident and unquestionable assumption of urban studies. Regardless of the positions and reasons suggested in a broad debate that involves the contemporary social processes of space production as well as the theoretical and epistemological perspectives from which these processes are analyzed, there is an underlying shared conviction that it is necessary to re-imagine and re-map the urban. In this sense, here urbicide refers to an open and complex urban process, which is neither teleological nor linear. Rather than a given reality (or an inevitable future), the idea of urbicide constitutes a working hypothesis through which to explore the socio-spatial dynamics in contemporary urbanism, where the life and death of the urban are intertwined; that is, the city is made and unmade. Analyzing these processes implies a double movement: acknowledging the powerful social and spatial processes that "unmade" the city and, at the same time, showing (based on an ethnographic case) the ways in which its inhabitants are actively involved in "remaking" the city, in composing the urban.

**Keywords** Town planning · Urban space · Social process · Social inequality · Urban sociology

## 18.1 Introduction

This chapter reflects on the (un)made city and the dynamics of (de)composition of urban life. In recent decades, various voices have questioned the persistence of the modern idea of "the city"—a large, dense, self-contained and bonded nodal socio-spatial unit—as a self-evident and unquestionable assumption of urban studies. From the theories of the social production of space and by considering some classic

R. Segura (✉)
National University of La Plata, La Plata, Argentina
e-mail: segura.ramiro@gmail.com

National University of San Martín, San Martín, Argentina

proposals made by Lefebvre (2013), Brenner (2017) has recently pointed out that research should move from the city to the urbanization process, understood as a contradictory dynamic of implosion and explosion that produces new forms of urbanized landscape. For Brenner, these processes generate a diversity of "agglomeration" phenomena, being the city only one of the forms assumed by contemporary urbanization. Precisely, many of the *other forms* that destabilize the image of the city as an organic, coherent and hierarchical unit have been explored and described since the 1990s by the Los Angeles School (Dear and Flusty 1998) in terms of *privatopias* (McKenzie 1994) and *postmetropolis* (Soja 2008), among other neologisms that sought to capture urban transformations.

On the other hand, postcolonial theories have made a call to "open" the geography of urban studies theory, traditionally restricted to the urban experience of a few paradigmatic cities of the Global North (Robinson 2002; Roy 2013). This claim is not reduced, however, to expanding the empirical variability of the urban by adding "interesting cases" from the Global South, but rather to rethinking the historical difference between asymmetrically interconnected urban processes which require new ways of theorizing the urban (Roy 2016), a phenomenon that is not limited to the restricted geography of the theory of the Global North (Robinson 2011). Likewise, in a critical dialog with both perspectives, due to the assumed deconstructive emphasis from postcolonial theories and the supposed re-centering of the cities of the Global North in the case of theories of space production, Schindler (2017) has recently developed the guidelines for a southern urbanism that would both avoid the stereotyped labeling questioned by postcolonial perspectives and provide the key to our understanding of contemporary urbanization in a situated manner.

In short, regardless of the positions and reasons suggested in a broad debate that involves the contemporary social processes of space production as well as the theoretical and epistemological perspectives from which these processes are analyzed, there is an underlying shared conviction that it is necessary to re-imagine and re-map the urban.

The question of what we can call (by paraphrasing what Jane Jacobs already stated about "inurban urbanizations" in 1961, Jacobs 2011: 33), the life and death of the urban become particularly relevant in the current scenario of contemporary urbanism. In this sense, here *urbicide* refers to an open and complex urban process, which is neither teleological nor linear. Rather than a manifest and given reality (or an inevitable future), the idea of urbicide constitutes a working hypothesis through which to explore the socio-spatial dynamics in contemporary urbanism, where the life and death of the urban are intertwined; that is, the city is made and unmade. Specifically, the processes of (de)composition of urban life will be analyzed in depth in this chapter. This implies a double movement: acknowledging the powerful social and spatial processes that "unmade" the city and, at the same time, showing (based on an ethnographic case) the ways in which its inhabitants are actively involved in "remaking" the city, in composing the urban.

The starting point for these reflections is a collective investigation into the effects of the COVID-19 pandemic on daily life on the outskirts of intermediate cities in

Argentina,[1] which in recent decades have shown a trend of urban growth toward "extended metropolitan morphologies" (Prévot-Schapira and Velut 2016). In particular, the focus here will be on some socio-spatial dynamics in the western periphery of La Plata, Argentina, not because the processes occurring there are regarded as paradigmatic, prototypical or generalizable, but because they are strategic to show the dynamics of (de)compositions of urban life, i.e., death *and* life of the urban, which are likely to unfold differently and take on other forms in other places.

Since its foundation as a planned city in the late nineteenth century, La Plata has been the capital of the province of Buenos Aires, the main province of Argentina. The city of La Plata is located 56 km southeast of Buenos Aires city and, together with the municipalities of Berisso and Ensenada, it integrates the conglomerate of Gran La Plata with a population of 893,844 inhabitants for the year 2020. The original design was a square of 40 blocks (five kilometers) per side, clearly delimited by a 100-m-wide ring road, whose function was to separate the urban area from the rural one. Inside the square, grid arrangement predominates, with avenues every six blocks. At the intersection of the avenues, there are equidistant green spaces (parks and squares). Two main diagonals and six secondary diagonals seek to provide agility to driving within the square and connect the city center with the periphery. A monumental axis running along 51st and 53rd Avenues symmetrically divides the foundational plan. On the axis, there are the main public buildings: the House of Government, the Legislature, *Teatro Argentino*, the City Hall and the Cathedral. This axis is perpendicular to Río de La Plata. In addition to distinguishing public from private spaces, it symbolically connects the port with the pampas.

However, its historical course can be condensed in the displacement from the "ideal city" of the nineteenth century to the "broken square" of the early twenty-first century (Segura 2019). Indeed, contemporary urban dynamics combine the trend toward densification and verticalization of the planned foundational layout together with the expansion of low-density peripheries, spatially fragmented, unequal in economic and residential terms, and culturally heterogeneous.

On the western outskirts of La Plata, there are spatial proximity and sharp contrasts between farms of horticultural production (*quintas*) mainly run by a Bolivian population, "settlements" resulting from the informal land occupation by heterogeneous popular sectors (*asentamientos*), recent middle-class urbanizations largely composed of young professional families and upper-class gated communities. These contrasts—reinforced by urban voids, road infrastructure, walls, fences and various security devices—immediately refer to the paradigmatic images of urban fragmentation that theoretically breaks down or *decomposes* the city into segments that are supposedly insular and relatively autonomous.

Nevertheless, our ethnographic fieldwork showed a multiplicity of practices that, against the asynchronous image of nearby and separate islands, the inhabitants daily

---

[1] "Flows, borders, and foci. The geographical imagination in six urban peripheries of Argentina during the COVID19 pandemic and post-pandemic", project directed by Ramiro Segura and financed by the National Agency for Scientific, Technological and Productive Innovation Promotion within the framework of the PISAC-COVID-19 Call "The Argentine society in the post-pandemic".

deploy *to compose* an unequal and, at the same time, shared social world. Then, it is not a question of ignoring the strong contrasts and multiple inequalities that cross the peripheral space, but of accounting for the practices deployed to compose (Descola 2016) the place that is inhabited and establish forms of conviviality (Gilroy 2004) that unfold in its interstices. They are compositions and convivialities seeking to recompose an urban and social fabric that the dominant forms of space production (Roy 2013) do not stop breaking down and fragmenting.

The French anthropologist Philippe Descola's first and truncated fieldwork experience (Descola 2016) in Chiapas, Mexico, in 1973, gave rise to my reflections on the (de)compositions in this contemporary peri-urban scenario. In his research, Descola was interested in the interethnic relationships between the Lacandones and the Tzeltales, as well as the differential relations that both groups established with the rainforest. To achieve this aim, he worked in a town in the Lacandon jungle where some Tzeltales, who had fled from the highlands, their native habitat, expelled by the landowners, settled. The Tzeltales were forced to rebuild, in an environment completely different from the highlands, a world analogous to it, and this turned out to be very difficult for ecological reasons. Without the landmarks of the highlands, Descola explains, the Tzeltales.

> obsessively tried to reconstruct the physical and symbolic landscape that they had lost, as well as the social logic that was associated with it (…). I witnessed the permanent struggle waged by its inhabitants against an environment that was unfamiliar to them, and the strategies they put in place to try to tame it (Descola 2016: 29).

Although distant and different from the Lacandon jungle analyzed by Descola, the vertiginous transformation of the environment in the west of La Plata, which involves growing tensions between residential urban expansion and rural land uses, the negotiated spatial proximity between social classes, and the reciprocal glances between old settlers and recent residents, among other processes, constitute a scenario in which the inhabitants are actively involved in trying to compose a world in decomposition. This work of composition unfolds both in the place they inhabit and regarding the relations of conviviality with those sharing the place. While the concept of "fragmentation" tends to place emphasis on the distance, separation and reciprocal isolation between groups and social classes in the city (Segura 2020), here "conviviality" is considered to be an analytical tool (Nowicka and Vertovec 2014) that focuses on the processes of daily cohabitation across the borders of class, ethnicity and race (Gilroy 2004), paying attention to the contexts of interaction, negotiation and conflict of social positions and cultural identifications (Mecila 2017) as well as the struggles for the creation of habitable and shared worlds (Heil 2020).

Hence, urbicide is not a pre-established and inevitable destiny of urban life, since inhabitants are not passive objects of a process that they do not control but rather active agents who, in contexts of socio-spatial fragmentation, compose their new habitat—not without negotiations, conflicts and paradoxical effects—and establish modes of conviviality, that is, processes of cohabitation and interaction between different and unequal people with whom they share a place.

The chapter is organized in five sections. Firstly, it critically reflects on the concept of fragmentation from a socio-anthropological perspective attentive to interactions, mobilities and borders in urbanization processes. Secondly, it describes the practices of place composition. Thirdly, it focuses on the production of conviviality in the western outskirts of La Plata. Fourthly, it analytically addresses the limits of these works of composition of place and production of conviviality from the installation of a peripheral public space. Finally, it draws a conclusion on the processes of (de)composition of urban life.

## 18.2 A Socio-anthropological Critique of the Concept of Fragmentation

*Fragmentation* is a relatively new concept that seeks to specify the relationships between urbanization and inequality in contemporary metropolises. It is also probably one of the most powerful *metaphors* to represent the social and urban transformations of recent decades. In this sense, beyond strictly academic uses, "fragmentation" and related categories such as "dualization" and "polarization" have a negative semantic charge and have been mobilized to criticize (and even denounce) the destruction of a previous social and urban fabric, all of which brings the concept of fragmentation closer to the idea of urbicide. Briefly reconstructing the history of this concept, its uses in Latin American urban studies and its heuristic limits will allow a better understanding of the dynamics of (de)composition of urban life.

The most remote antecedent of the concept is found in *The Fragmented Metropolis: Los Angeles, 1850–1930* by Fogelson (1967), a book where it was pointed out that the fragmented conurbation of Los Angeles was the archetype, for better or worse, of contemporary urbanization. And its generalization probably coincides with the publication of the article "Post-modern Urbanism" by Dear and Flusty (1998), where they argued about the existence of the Los Angeles School, a school of urban theory not only located in the city of Los Angeles, but which regarded the urban planning of Los Angeles as the key to reading the predominant trends in world urban planning (Segura 2021a).

The exceptional status of a dispersed, discontinuous and polycentric metropolitan area such as Los Angeles in twentieth-century North American urbanism began to change when various cities, within the framework of the restructuring of neoliberal capitalism unfolded from the 1970s and 1980s, started to show characteristics similar to those that until then seemed exclusive to the city of Los Angeles: diverse and heterogeneous communities instead of homogeneous communities, consolidation of post-Fordism in the urban economy and the emergence of "new urban forms" that the researchers of the Los Angeles School cataloged with neologisms such as *exopolis, edge-cities, privatopias, technopoles* and *cybercities*, among others. Fogelson's prophecy seemed to be coming true, and perhaps the fragmented metropolitan area of Los Angeles heralded the future of cities around the world.

In the book *From Chicago to L.A.: Making Sense of Urban Theory* (Dear and Flusty 2001), the movement from Chicago to Los Angeles summarizes the course of North American urban theory throughout the twentieth century. The main idea is that cities have been transformed from ordered and hierarchical geographical landscapes that revolve around a nucleus (like those described in an ecological key by the traditional Chicago School) toward expanded urban agglomerations without a center, which inexorably envelop everything along their path. Another key idea is that urban future lies largely in Los Angeles' past and present. Thus, Los Angeles was no exception to the rule of contemporary urbanization processes.

Also in Latin America, during the processes of economic opening and globalization at the end of the twentieth century, large cities were reconfigured in structural, functional and territorial terms (De Mattos 2010; Ciccolella 2011): decline of productive functions and restructuring toward a service economy; passage from a compact metropolitan space with defined borders and limits, toward a metropolitan growth with diffuse borders and a polycentric structure; private suburbanization of the elites; increase in precarious habitat, both in the center and in the urban periphery; proliferation of "new urban objects" linked to consumption, such as shopping malls, hypermarkets, entertainment venues, international hotels, restaurants, theme parks and private developments, among other transformations.

Given this scenario, a debate was unfolded around the fragmentation of Latin American cities (Bayón and Saraví 2013). While authors such as Janoschka (2002), Borsdorf (2003) and Prévot-Schapira (2001) highlighted the consolidation of a new model of the city, authors such as Caldeira (2007) and Duhau and Giglia (2008) emphasized the continuity of the structure of both the social division of urban space and the patterns of segregation, setting the trend toward fragmentation in the previous model. However, apart from the degree of (dis)continuity in this process, the literature agrees on highlighting a change in scale in the processes of socio-spatial differentiation. Whereas a process of social mixture greater than that present in the traditional center-periphery pattern is observed on a large scale, the pattern of segregation is reinforced at the micro-level through walls and fences, barriers with which the islands of wealth and exclusivity are separated and secured against poverty. *Insularization* is the prevailing trend in this model, which underlines the consolidation of different types of islands (residential, productive, commercial and precarious) on a background inherited from concentric circles and previous urbanization axes.

Research on fragmentation has branched off in two main directions: on the one hand, the study of discontinuities in the urban fabric because of new urbanization processes; on the other, the analysis of both social inequality and the material and symbolic limits present in this process. However, in recent years, critical approaches have emerged that have pointed out its limitations and blind spots, without necessarily discarding the notion of fragmentation. Saraví's work (2015) in Mexico City suggests that fragmentation refers not only to disconnection but also to forms of connectivity and union. It is, then, not only about isolation and separation, but also about the way in which daily interactions take place, relationships that are generally based on differentiated and unequal integrations that usually reinforce social hierarchies.

In this sense, ethnographies such as that carried out by Elguezabal (2018) in "the towers" of Buenos Aires (buildings for the upper classes with private security, common spaces and various amenities) show that, contrary to what the notion of "enclave" entails, there is no coincidence between material and social borders. In the everyday life of the towers, these borders are blurred, labile and porous and, therefore, must be continuously marked, pointed out and reinforced in an "enclave work" that strives to separate the inside from the outside of the towers.

On the other hand, Jirón and Mansilla (2014), from the mobility paradigm, highlighted that the usual uses of "fragmentation" overlook three dimensions: its temporality (representing it as a synchronous event); the interactions between the fragments (regarding them as isolated); and the inhabitants' space–time experience (absent in many studies). Inequalities should not only be considered in terms of "fixed enclaves" but also in relation to "mobile gradients," that is, to the differential possibilities of moving and accessing the benefits and opportunities present in the urban environment. Therefore, the proliferation of barriers, the establishment of discontinuities and the production of distances that are typical of "fragmentary urbanism" undoubtedly have significant impacts (and differentials according to class, gender, ethnicity and age, among other dimensions) upon access to activities, people and places.

However, through mobility practices (and not without effort), the inhabitants seek to overcome these barriers and, as the authors beautifully write, "mend" a fragmented urban fabric. By placing the focus on the subjects' space–time practices and their meanings, the study of mobility makes it possible to build an "intermediate plane" (Magnani 2002) to analyze the city: neither the panoramic map of the city nor the individual fixed at a point in the city, but rather people "crossing the thicket of the city" (Jirón and Mansilla 2013), taking tours, making circuits together, encountering (and, sometimes, overcoming) obstacles, establishing relationships and (re)producing differences (Segura 2018).

In short, recent socio-anthropological research in fragmented urban spaces has highlighted the relevance of interactions, mobilities and borders in the production of (in)accessibilities, inclusions/exclusions and social asymmetries, as well as in the construction of forms of coexistence and cohabitation between different and unequal groups. Ultimately, it is about exploring the conditions and ways in which social actors produce places and create ways of living together in urban spaces characterized by being spatially fragmented, socially unequal and culturally heterogeneous. To this end, the concept of composition (Descola 2012) allows us to know the identification operations (establishment of similarities and differences between oneself and others) and relationships (establishment of ties) that people display to shape the world they inhabit and build links with the people with whom they share the place, even within the framework of processes of socio-spatial fragmentation.

## 18.3 Place Composition

*The west* (*el oeste*), as is often colloquially referred to, corresponds to an axis of recent expansion of the urban fabric on land with a rural vocation, historically dedicated to the production and supply of food for the city. In recent decades, there has been a significant multiplication of land uses as well as the spread of various residential types (popular settlements, middle-class urbanizations and gated communities) that overlap and juxtapose a landscape of former small rural towns, dominated by vegetable farms, large storage sheds, a refrigerator and state infrastructure (a public hospital, a prison and other prison service institutions) and crossed by large road infrastructures connecting the outskirts with both the city and other locations in the Metropolitan Region of Buenos Aires.

The inhabitants of the western periphery, regardless of where they live, tend to value the west as "a quiet place." The notion of *tranquility* articulates with landscapes, objects and people in an affective economy of a place in which the presence of country houses, large vacant areas and extended residential occupation contributes to a geographical imagination that values a certain distance and separation from the city (Segura et al. 2022). If in the study by Araujo and Cortado (2020) in the west of Rio de Janeiro, tranquility appears associated with the absence of certain types of violence, in the west of La Plata, tranquility unfolds from the distance with respect to the urban and its dizzying pace, greater contact with nature and a distance that implies larger and more spacious places (and requires extended daily mobility).

At the same time, the increase in material enclosure devices and different types of controls over access to neighborhoods and homes are a sign that this tranquility requires specific support to maintain it. "Tranquility" refers to a mode of establishing relations between desired and undesired presences, in a place crossed by inequalities and conflicts, which combines spatial proximity and social contrast, tensions and negotiations between urban expansion and rural uses, as well as between old settlers and recent residents.

The experience of the inhabitants of the middle-class urbanization self-styled "El Gigante del Oeste" (The Giant of the West) shows, in a privileged fashion, this dynamic of (de)compositions. "The neighborhood is surrounded by empty lots. Near it there is another neighborhood called El Centinela (The Sentinel), which is also a large neighborhood, but a piece of land without any construction separates us," describes Verónica,[2] a 40-year-old lawyer who lives with her partner and their two young children. Indeed, a narrow and long paved road that begins in one of the great avenues running through the western area constitutes the only access to the neighborhood. This road also functions as a boundary with El Centinela, a pre-existing neighborhood of popular sectors. The other sides of the neighborhood are occupied by horticultural production fields, animal farms and, a little further away, a popular settlement, giving the complex a heterogeneous and contrasting aspect of land use and social sectors.

---

[2] The interviewees' names have been changed to preserve anonymity.

"We didn't move to the Giant of the West, we invented the Giant of the West," Dominga (54 years old, a school principal) told me one afternoon at her home. It is, in fact, a neighborhood arising from the organization among 432 beneficiary families of a state real-estate loan for the first home granted between 2012 and 2015, who (even counting on the loan) found it impossible to access an unregulated and dollarized formal urban land market.[3] Faced with this situation, the families became involved in a collectivization process to acquire cheap rural land relatively close to the city and also to demand, from the Municipality of La Plata, a modification of the land use code that would enable the construction of the neighborhood and finally provide it with services and basic infrastructure to make it suitable for urbanization: building roads and installing public lighting, water, electricity and gas services.

Once these objectives were achieved at the end of 2015, the residents worked on the design of the neighborhood: they divided up a 22-block plot of land, leaving space for a public square and for a future community center, and then distributed the lots among the families, where later each one began to build their homes.

The relationship between associative forms of participation and city production is a persistent problem in Latin American urban studies. However, as Ventura (2021) has shown from the study of other collective experiences of beneficiaries of the same credit in La Plata, the salient feature of the production of neighborhoods such as El Gigante del Oeste is the leading role of young middle-class families in this process. Although it is not a process of self-construction of the kind that the popular sectors have carried out in most parts of the cities of the Global South in recent decades, it shares the logic of "peripheral urbanization" (Caldeira 2022). According to this author, this type of urbanization has the following characteristics: a distinctive form of agency in which the inhabitants are agents of urbanization; a transversal link with the official logics in search of solving problems of ownership and regularization of land tenure; new forms of political action that mobilize citizen demands and expectations; and the creation of highly heterogeneous and unequal cities.

In this sense, the tasks of "domestication of space" (Giglia 2012), an experience generally unambiguously associated with the popular sectors that first "occupy" a place, then "inhabit" it and progressively "build" a house and a neighborhood, was constitutive of the process of "production of locality" (Appadurai 2015) in El Gigante del Oeste. This process was not exhausted in the efforts to search for and access the land to build the house and collectively design the neighborhood, but rather extended over time, involving various facets of daily life. The residents continued to organize,

---

[3] The Argentine Credit Program of the Bicentennial for Single Family Housing (El Programa de Crédito Argentino del Bicentenario para la Vivienda Única Familiar; ProCreAr) was an innovative public policy, aimed primarily at the middle sectors that, despite the sustained improvement in economic and social indicators since 2003, did not have access to the private mortgage market. During the implementation of the program, there was a decoupling between the economic and territorial dimensions of the program, observing mismatches between the success of the economic reactivation policy through construction and the multiplicity of problems in urban policies at the local level. Regarding this last aspect, access to urban land was the main limiting factor in the implementation of the program, deepening the trend toward the production of extended, diffuse, and low-density urban space that entails high costs (social, economic, and environmental) in the medium and long term (Segura and Cosacov 2019).

with a representative for each of the 22 blocks that constitute the neighborhood and with the establishment of commissions devoted to solving different common problems like road maintenance, afforestation, security, transportation and sewerage, as well as to develop cultural and recreational activities, such as film cycles, Children's Day celebrations and national commemorations such as the May Revolution or Memorial Day, among others. In short, it is a neighborhood "invented" by its inhabitants, as Dominga maintained.

As Benson and Jackson (2013) showed for middle-class neighborhoods in London, there is no doubt that in these "place-making" practices, place of residence, sense of belonging and class position are related. The inhabitants of El Gigante del Oeste do not only reside there, but "make a middle-class place" through interventions in the neighborhood space aimed at generating "improvements" in the built space such as street maintenance and lighting, public space, the afforestation of the square, among other activities, as well as fostering a sense of belonging to the place where they reside through a set of daily practices such as organizing shared activities, shopping in the neighborhood and renovating houses, that is, territorializing it as a form of belonging.

At the same time, the composition of place is not exhausted in the production of a middle-class neighborhood but is also linked to a deeper anguish of living in the periphery. It is, in effect, a work of (middle-class) place composition in a peripheral space previously unknown to the families who migrated there, which lacks many of the typically urban facilities, occupies a subordinate place in the urban hierarchy and has a negative semantic load in the imaginary of the city.

By reflecting on the experience of displaced highland Tzetales in the jungle lowlands, Descola ventured a suggestive hypothesis: "One of the means [of fighting against an unfamiliar environment] was to enclose the town permanently in a bubble of "civilized" noise to keep the disturbing otherness of the jungle and its occupants at a distance" (Descola 2016: 29). The "civilized" noise, then, is a means to silence the surrounding jungle. In an analogous way, the place-making practices of a middle-class neighborhood by families who recently migrated to the western periphery of La Plata to access their own home in an environment combining the double proximity of rural production and the popular sectors seek to produce urbanity. Through the organization of meetings, activities and celebrations, they give vitality to the local public space, one of the ways through which to compose a place to live in the most comfortable way possible.

## 18.4 Production of Conviviality

The compositional operations deployed by the inhabitants are not limited to the built space of the neighborhood, nor are they limited to the relationships between its residents. On the contrary, these are operations that, to continue with Descola's (2012) repertoire of analytical categories, identify and seek to establish ties with the environment and with those who live there. The fast construction of a neighborhood like El Gigante del Oeste in an area largely dominated by agricultural production and popular neighborhoods does not only impact on the experience of the middle-class inhabitants who moved to the new neighborhood. We know that displacement affects the displaced and also modifies the experience of those who were previously in the place (Gupta and Ferguson 2008). Indeed, the growing installation of middle- and upper-class neighborhoods in peripheral areas historically reserved for popular sectors breaks the spatial and temporal logic of dominant urban expansion during much of the twentieth century. This is what Durham (2000) called "the periphery as a process" in the case of Brazil. It was a dynamic of urban expansion that involved the slow and laborious process of land occupation and urbanization of residential space, as well as successive waves of new settlers who tended to locate in the most disadvantaged and least urbanized peripheral areas. The recent urbanizations of the middle and upper classes, in addition to appropriating land that was potentially for the extension of popular urbanization and/or land destined for rural production, destabilize the spatial distribution of the rich center and the poor periphery, and also the temporal logic that, according to Elias and Scotson (2000), used to assume the figuration of those established there and newcomers. Although the inhabitants of El Gigante del Oeste have recently arrived, they are effective and fast in the production of a "middle-class neighborhood" due to the conjunction of social networks, cultural capital, familiarity with red tape and the undoubted social selectivity of state bureaucracies (Ventura 2021). Thus, while for these inhabitants the nearby neighborhoods—paradigmatically, El Centinela—were a source of danger and insecurity, for the pre-existing inhabitants of the areas surrounding El Gigante del Oeste, the transformation generated by the installation of this neighborhood was a cause for suspicion and conflict. In particular, the relatively quick access to services and urban infrastructure by its residents for which they waited a long time and many still did not have in their own neighborhoods.

Anthropological research has highlighted that, in scenarios like the one described here, actors deploy multiple practices to ward off physical closeness and emphasize moral distance (Carman 2015). These practices do not only imply distancing, but also interaction. In effect, seeking to compose the fragmented space, plagued by contrasts, which they inhabit, the residents of El Gigante del Oeste deployed a "policy of good neighbors," as Martín (44 years old, public employee) called it in an interview. "Our neighborhood," continued Martín, "has a policy toward the surrounding neighborhoods. When there was a conflict situation with the El Centinela, a commission was set up, we got together, we chatted, we did things together." In short, it is about identifying the actors and the predominant conflicts in the immediate environment and generating devices to strengthen ties with them to reduce social distance and establish forms of conviviality with others.

Insecurity is a persistent concern for the inhabitants of El Gigante del Oeste and involves fears related to both housing security and personal safety when entering and leaving the neighborhood through the only paved road that connects the neighborhood with the rest of the city. As many of its inhabitants recognize, the possible robberies and the generalized fears were initially located in the adjoining neighborhood and its inhabitants. However, despite the persistence of prejudice among many of the inhabitants of El Gigante del Oeste and the growing fortification of homes and the installation of neighborhood alarms in search of protection (Segura 2021b), over time the neighborhood's security commission encouraged initiatives in conjunction with the residents of El Centinela. Regarding this issue, Dominga recalled that it was a rational strategy by a group of neighbors who said, "it can't be, we have to go look for them, let's get together, let's chat, think together about the issue of security." In addition, Luciana (30 years old, university student) recounted: "some time ago there were some robberies in the neighborhood, two or three in a row, so the neighbors mobilized to meet with the municipal delegate and have a meeting with the police, but the other neighborhoods were invited to participate." Thus, Martín acknowledged that although "the people from the neighborhood thought that the robberies were committed by people who were from there [El Centinela]," the joint work showed that in El Centinela "there are people who want to take care of their neighborhood more than we do ours."

The ties they established with the surrounding inhabitants were not limited to specific meetings on specific issues related to a restricted security agenda, but instead involved other dimensions to build common forms of coexistence. In this sense, an "entrepreneur fair" was organized. This fair takes place both in the central square of El Gigante del Oeste and in the square of El Centinela, and is open to residents of both neighborhoods who would like to offer products. Likewise, as Martín related, "the people from the El Centinela began to come here to buy some things and we go to the greengrocer's there." All of which, in Dominga's assessment, "totally changed the relationship. The conflict between this neighborhood and the El Centinela disappeared." Furthermore, these ties were also established with the Bolivian families from the surrounding farms, who generally view with concern the advance of "the neighborhoods" over "the farms" (Musante 2021). As Luciana stated, "there was an attempt to accompany those around us. Many of us began to consume the bag of vegetables from the small local producers, and girls from nearby farmland began to come to the neighborhood with the few things they had to sell."

Opening bridges of dialog, building meeting spaces and establishing commercial exchanges are ways of intervening in a fragmented plot that the installation of El Gigante del Oeste deepened. This *work of composition* of a fragmented socio-spatial plot reminds us of the forms of collaboration between very different people that Simone (2015) described and analyzed in contemporary African cities. He linked these emerging and changing forms of social collaboration "with the proliferation of constraints to ensure subsistence and maneuver within the city" (2015: 136). Likewise, he located the heterogeneity of urban opportunities and the possibilities for shaping alternative forms of subsistence in "the fragmented urban space" defined as "neighborhoods with highly divergent characteristics and the relationships between them" (2015: 141). Rather than defining informality as a compensation for the lack of urbanization, or fragmentation as the absence of connections, Simone invites us to think about the approximations, articulations and interdependencies between people that cross categories of class, ethnicity and social position, in order to elaborate ways of using the city and collaborating with each other. "Faint signs, flashes of creativity" that, even in their fragility, account for "efforts to create viable forms of urban life" (Simone 2015: 143–144).

Precisely at this point, the work of place composition and the emerging forms of collaboration between different people and groups are related to conviviality. We cannot think of conviviality as a reality taken for granted. As can be deduced from the case that we have been analyzing, the processes of producing, negotiating and sharing a common world require, instead, people's active involvement. Likewise, the establishment of forms of conviviality such as those described here does not imply the dissolution of differences and asymmetries, but rather provides tools to live and share the space with others. And this also has its paradoxical effects and limits.

## 18.5 Two Squares and the Limits of Conviviality

Reflecting on the production process of the neighborhood, Luciana recalled that "one of the projects was to build a square on some vacant land that separates El Centinela, the closest neighborhood, from El Gigante del Oeste. This is something that we hope for us neighbors, because then it would also strengthen the bond between the neighborhoods, and we would not be so separated. It would be good to remove the limits a little."

That square has never been built and the boundaries between both neighborhoods have not been erased. However, through various forms of collaboration, i.e., joint work on security issues, fairs in both neighborhoods and reciprocal buying and selling of goods and services daily relationships were fostered, which made it possible to reduce the social distance initially expressed as stereotypes and reciprocal mistrust. The residents' active involvement to compose negotiated ways of living together between different and unequal people, contrary to the normative and utopian visions that the term often carries, does not dissolve the differences and inequalities through which cohabitation, interaction and interdependence processes unfold, which is here referred to as conviviality. People cohabit with others, relationships are negotiated, spaces are shared, all of which does not necessarily erode limits, differences and asymmetries, even when people like Luciana would like this to happen (at least "a little").

The construction of another square in the area, resulting from a new association and collaboration between residents of El Gigante del Oeste and El Centinela, will allow us to advance in our reflection on the composition of places and the limits of conviviality. On one of my visits to the area in a sunny July afternoon during the COVID-19 pandemic, we toured the neighborhood with Sandra (42 years old, a teacher) and Dominga. During the walk, they proudly showed me the latest achievement of the collective organization: a bus stop inaugurated a few days before in one of neighborhood boundaries, product of a persistent collective demand by the neighbors so that one of the city's public transport lines would pass through the neighborhood. In addition to the obvious convenience of having public transport service in the neighborhood, the claim was based mainly on security reasons: to avoid traveling

500 m on foot to the nearest avenue along the only access road to the neighborhood, where several of the thefts suffered by the inhabitants of El Gigante del Oeste have been concentrated. Also, having managed to obtain an urban public transport line to change its route and deviate from the avenue to "enter the neighborhood"—as my interlocutors called it—once again shows the inhabitants' management skills and, as occurred with access to other urban services, it differentiates this middle-class neighborhood from others in the area, whose inhabitants must travel to the avenue to take public transport.

Very close to the bus stop, I saw the opening of a new street connecting El Gigante del Oeste with El Centinela and a square with some children playing and a few adults accompanying them. It is a large rectangular green space, delimited by used car tires painted white, with fixed children's games inside. When asking about this new square, my interlocutors reported that a few months before, in that same space, there had been an attempt to take over land by people from other areas of the city, who were finally evicted. Although Sandra and Dominga did not directly participate in the events taking place that day, from their account it can be deduced that the land seizure was in its infancy: people were "lotting" the land and distributing it among the families who participated in the seizure, although they had not yet begun to erect boxes in those lots.

Following a quick complaint filed by a neighbor from the area, these families were evicted. Not only did the justice system and the police participate in this eviction (agents formally in charge of this type of action), but also some residents of El Centinela and El Gigante del Oeste neighborhoods, who rebuked the families who were participating in the occupation and supported the authorities' work. The square built precisely in the place threatened by the seizure expresses a work of alliances and collaborations between neighbors, the judiciary and the police, to which we must add the Municipality of La Plata, later opening the street, enabling the arrival of the bus and collaborating with the fixed games that were installed in the square.

The constitutive ambivalences of urban public space have been highlighted by various authors: tensions between material form and social practice; dilemmas between encounter, recreation and dialog and, at the same time, demonstration, conflict and demand. The neoliberal urban policies of recent decades have deepened these tensions. On the one hand, Gorelik (2008: 44) critically referred to the growing romance of public space as an "idealized place where we deposit all the virtues of the city so as not to have to face the difficult commitment of putting them into practice in the reality of our cities." He argues against the projects of production, recovery and/or enhancement of public spaces by culturalist urban policies that do not intervene or reverse the dominant trends of space production that systematically deny the public, or resolve the urgent social problems (housing, work, violence, among others) that cross a fragmented urban space, depositing their (false) expectations in the productivity of public space.

On the other hand, Carman (2011) has even pointed out that, in many of these urban policies, the calls for "culture," "nature," and "public space" function as an alibi to deepen the asymmetries and the exclusion of actors and non-tolerated practices. It is no longer just a question of depositing false illusions in the capacity of urban public space to reverse the trends toward a fragmented and unequal socio-spatial fabric, but of the paradoxical appeal to the defense of public space to deepen that trend.

Both western squares, namely the one designed as a space for integration between two neighborhoods that Luciana and other neighbors dreamed of, which has never materialized, and the one built in the same space where the land grab was evicted with the active collaboration between inhabitants of both neighborhoods, condense the extremes of this ambivalence of public space. What occurred with these squares also shows the limits of conviviality and the paradoxical effects of place and relationship compositions in which social actors are involved. The public space (the square) shared by the residents of both neighborhoods results from the collaboration between the justice system, the police, the municipality and residents of both neighborhoods as a common rejection of land seizure in the area. A public space tense between integrating neighborhoods and avoiding land grabs by sectors that have nowhere to live.

Events like this remind us that urbanization (and eventual urbicide) is not a dual and polar process between state and mercantile practices that undo the city, on the one hand, and the inhabitants' practices that remake it, on the other. It is not about equalizing the different agents who, with differential power and resources, participate in the process of social production of space. Without ignoring the productivity of the practices of place composition and conviviality that the inhabitants deploy (paraphrasing Marx) in conditions that they have not chosen, it is about pointing out that in these processes differences are also established, limits are set, and even exclusions, such as the one giving rise to the square between the two neighborhoods, are produced.

## 18.6 Concluding Remarks

Based on a situated ethnographic experience, in this chapter the dynamics of (de)composition of urban life have been examined. The analyzed situation and the specific findings are not paradigmatic, nor can they be generalized. In their own specificity, the processes studied here remind us that people are always involved in the composition of the places they inhabit. By referring to the "locality production processes," the Indian anthropologist Arjun Appadurai pointed out that "it is necessary to recognize that stories produce geographies, and not the other way around," and specified that "we must move away from the notion that there is some kind of spatial landscape in which time writes its history. Instead, it is the agents, the institutions, the actors, and the historical powers that make the geography" (2015: 95).

Faced with an urbanization process tending toward socio-spatial fragmentation and increasing inequalities, the ways in which the inhabitants deploy practices and are actively involved in the composition of places and the production of conviviality in heterogeneous and unequal contexts have been described here. The urban is undone and remade in creative ways through compositions and convivialities that are often fragile and quite contradictory, but in any case, unavoidable in the process of inhabiting and establishing links with the built environment in which different and uneven people and groups cohabit. The analysis of the ways in which different people get involved to produce, regulate and/or share a common space by paying attention to the differences and inequalities that are identified, negotiated and not necessarily diluted in this process, allows us to describe the contemporary urbanization processes in a situated way. Beyond their fragility, limits and contradictions, these practices of composition and conviviality appear as a creative alternative to urbicide as the dominant trend. There is not only destruction of the city and the urban, but there are also situated practices of urbanity production in a scenario that changes vertiginously due to the convergence of multiple agents and processes that shape urban space.

In conclusion, this chapter has aspired to explore a way of approaching these processes that engage in critical dialog with—or, at least, maintains a methodological alert regarding—the idea of urbicide as an inescapable destiny. Neither the nostalgia typical of eurocentric views for urban spaces that never existed in most of the world, nor the resignation to the supposed inevitability of an overwhelming process of destruction of the urban has been put forward here. Instead, the situated analysis of processes of (de)composition of urban life with its creative and destructive dimensions of social relations and urban places has been proposed. The (un)made city regarded, in sum, as an open and conflictive process in constant transformation, which involves negotiation, dispute, projection and agreements between actors and sectors that cohabit in a shared and unequal world.

# References

Appadurai A (2015) El futuro como hecho cultural. Ensayos sobre la condición global. Fondo de Cultura Económico, Buenos Aires
Araujo M, Cortado TJ (2020) A Zona Oeste do Rio de Janeiro, fronteira dos estudos urbanos? Dilemas-Revista De Estudos De Conflito e Controle Social 13(1):7–30
Bayón MC, Saraví G (2013) The cultural dimensions of urban fragmentation: segregation, sociability, and inequality in Mexico City. Lat Am Perspect 40(189):35–52
Benson M, Jackson E (2013) Place-making and place maintenance: performativity, place and belonging among the middle classes. Sociology 47(4):793–809
Bordsdorf A (2003) Cómo modelar el desarrollo y la dinámica de la ciudad latinoamericana. Revista EURE 86:37–49
Brenner N (2017) La era de la urbanización. In: Sevilla Buitrago A (ed) Neil Brenner. Teoría urbana crítica y políticas de escala. Icaria, Barcelona
Caldeira T (2007) Ciudad de Muros. Crimen, segregación y Ciudadanía en São Paulo. Gedisa, Barcelona
Caldeira T (2022) Peripheral urbanization: autoconstruction, transversal logics, and politics in cities of the global south. In: Reis N, Likas M (eds) Beyond the Megacity. New dimensions of peripheral urbanization in Latin America. Toronto University Press, Toronto
Carman M (2011) Las trampas de la naturaleza. Medio ambiente y segregación en Buenos Aires. FCE, Buenos Aires

Carman M (2015) Cercanías espaciales y distancias morales en el Gran Buenos Aires. In: Kessler G (Dir.) El Gran Buenos Aires. EDHASA, Buenos Aires

Ciccolella P (2011) Metrópolis latinoamericanas: más allá de la globalización. OLACCHI, Quito

De Mattos C (2010) Globalización y metamorfosis metropolitana en América Latina. De la ciudad a lo urbano generalizado. Revista de Geografía Norte Grande 47:81–104

Dear M, Flusty S (1998) Post-modern Urbanism. Ann Assoc Am Geogr 88(1):50–72

Dear M, Flusty S (2001) From Chicago to L.A.: making sense of urban theory. SAGE, Los Ángeles

Descola P (2012) Más allá de naturaleza y cultura. Amorrortu, Buenos Aires

Descola P (2016) La composición de los mundos. Capital Intelectual, Buenos Aires

Duhau E, Giglia A (2008) Las reglas del desorden. Habitar la metrópoli. Siglo XXI, México

Durham E (2000) Viewing society from periphery. Brazilian Rev Social Sci 1:7–24

Elguezabal E (2018) Fronteras urbanas. Los mundos sociales de las torres de Buenos Aires. Café de las Ciudades, Buenos Aires

Elias N, Scotson J (2000) Os Estabelecidos e os Outsiders. Jorge Zahar Editor, Río de Janeiro

Fogelson R (1967) The fragmented metropolis: Los Ángeles, 1850–1930. Harvard University Press, Cambrigde

Giglia A (2012) El habitar y la cultura. Perspectivas teóricas y de investigación. Anthropos Editorial, Barcelona

Gilroy P (2004) After empire. Melancholia or convivial cultures. Routledge, London/New York

Gorelik A (2008) El Romance Del Espacio Público. Alteridades 18(36):33–45

Gupta A, Ferguson J (2008) Más allá de la "cultura". Espacio, identidad y las políticas de la diferencia. Antípoda 7:233–256

Heil T (2020) Comparing conviviality. Living with difference in Casamance and Catalonia. Palgrave, Basingstoke

Jacobs J (2011) Muerte y vida de las grandes ciudades. Capitán Swing, Madrid

Janoschka M (2002) El nuevo modelo de la ciudad latinoamericana. Fragmentación y Privatización. Revista EURE 28(85):11–20

Jirón P, Mansilla P (2013) Atravesando la espesura de la ciudad: vida cotidiana y barreras de accesibilidad en los habitantes de la periferia urbana de Santiago de Chile. Revista De Geografía Norte Grande 56:53–74

Jirón P, Mansilla P (2014) Las consecuencias del urbanismo fragmentador en la vida cotidiana de habitantes de la ciudad de Santiago. Revista EURE 40(121):5–28

Lefebvre H (2013) La producción del espacio. Capitán Swing, Madrid

Magnani J (2002) De perto e de dentro: notas para uma etnografía urbana. Revista Brasileira De Ciencias Sociais 17(49):11–29

McKenzie E (1994) Privatopia: homeowners associations and the rise of residential private government. Yale University Press, New Haven

Musante F (2021) Habitar desde el periurbano: entre quintas y barrios. Algunas reflexiones sobre la localidad de Abasto del periurbano platense. XIV Jornadas de Sociología. Universidad de Buenos Aires

Mecila (2017) Conviviality in unequal societies: perspectives from Latin America. Thematic scope and research programme. Mecila Working Paper Series 1:1–34

Nowicka M, Vertovec S (2014) Comparing convivialities: dreams and realities of living-with difference. Eur J Cult Stud 17(4):341–356

Prévot-Schapira MF (2001) Fragmentación espacial y social: conceptos y realidades. Perfiles Latinoamericanos 19:33–56

Prévot-Schapira MF, Velut S (2016) El sistema urbano y la metropolización. In: Kessler G (Comp.) La sociedad argentina hoy. Radiografía de una nueva estructura. Siglo XXI/Fundación OSDE, Buenos Aires

Robinson J (2002) Global and world cities: a view from off the map. Int J Urban Reg Res 26(3):531–554

Robinson J (2011) Cities in a world of cities: the comparative gesture. Int J Urban Reg Res 35(1):1–23

Roy A (2013) Las metrópolis del siglo XXI. Nuevas geografías de la teoría. Andamios 10(22):149–182

Roy A (2016) Who's afraid of postcolonial theory? Int J Urban Reg Res 40(1):200–209

Saraví G (2015) Juventudes fragmentadas. Socialización, clase y cultura en la construcción de la desigualdad. FLACSO/CIESAS, México

Schindler S (2017) Towards a paradigm of southern urbanism. City 21(1):47–64

Segura R (2018) La ciudad de los senderos que se bifurcan (y se entrelazan): centralidades conflictivas y circuitos segregados en una ciudad intermedia de la Argentina. Universitas Humanística 85:155–181

Segura R (2019) La Plata, Argentina. In: Orum T (Dir.) The Wiley-Blackwell encyclopedia of urban and regional studies

Segura R (2020) In search of conviviality in Latin American cities. An essay from urban anthropology. In: Scarato L, Baldraia F, Manzi M (eds) Convivial constellations in Latin American. From colonial to contemporary times. Routledge, New York/London

Segura R (2021a) Las ciudades y las teorías. Estudios sociales urbanos. UNSAM Edita, San Martín

Segura R (2021b) Protective arrangements across class: understanding social segregation in La Plata, Argentina. Int J Urban Reg Res (IJURR) 45(6):1064–1072

Segura R, Cosacov N (2019) Políticas públicas de vivienda: impactos y limitaciones del Programa ProCreAr. Ciencia, Tecnología y Política 2:1–12

Segura R, Musante F, Pinedo J, Ventura V (2022) Entrar, quedarse y salir. Formas de habitar la periferia durante la pandemia. Bitácora Urbano-Territorial 32(3):253–266

Simone A (2015) Reconfigurando las ciudades africanas. Íconos. Revista De Ciencias Sociales 51:131–156

Soja E (2008) Postmetrópolis. Estudios críticos sobre las ciudades y las regiones. Traficantes de sueños, Madrid

Ventura V (2021) Las clases medias y los desafíos de la participación: procesos de ciudadanización en la producción de ciudad (La Plata, Argentina. 2013–2015). Hábitat y Sociedad 14:223–241

**Ramiro Segura** is an anthropologists (UNLP), Holds a PhD in Social Sciences (UNGS-IDES) and a postdoc at the Freie Universität (FU) Berlin. He is Researcher at the National Council of Scientific and Technical Research (CONICET, Argentina) and Professor at the National University of La Plata (UNLP) and National University of San Martin (IDAES/UNSAM). He is author of the books Vivir afuera. Antropología de la experiencia urbana (2015) and Las ciudades y las teorías. Estudios sociales urbanos (2021).

# Chapter 19
# The City and the Abandonment of Public Space. Between Neoliberal and Citizen Urbanism

**Patricia Ramírez Kuri**

> ...*a basic distinction: one thing is the built environment and another is how people live in it.*
> ...***the way people want to live should be expressed in the way cities are built...***
> Sennett, R. (2019)

**Abstract** This text reflects on the city and the influence of urban neoliberalism in the destruction of historical and cultural elements causing social damage. The focus is related to public space because it is the active scene of citizen expression, where the social-cultural, environmental and political effects of this model of development converge. In this context, the crisis of the city imagined as a social and symbolic space where links can be created between different people, forms of urban articulation and creative solutions to conflicts over citizen rights, is emphasized. In the debate on this crisis, which is visible in public space, the concept of "urbicide" is useful, because alludes to the deprivation, ruin, or abandonment of ideas and of social and spatial foundations that make the city a reference for urban identity and cultural heritage accessible to the different social groups that they use and live. It is interesting to reflect: in what sense do we speak of urbicide? How do we understand and distinguish that condition? In an actual city, what is left of what is common for the reconstruction of the public?

**Keywords** Public space · Town planning · Neoliberalism

---

Study carried out within the framework of the research project "From domestic space to public space in times of pandemic. Social inequalities, urban violence and conflicts over rights centralities, peripheries and local and global borders (2022–2024)" under my coordination, PAPIIT IN306822-DGAPA-UNAM.

P. Ramírez Kuri (✉)
Social Research Institute, National Autonomous University of Mexico (UNAM), Mexico City, Mexico
e-mail: patricia.ramirez@sociales.unam.mx

© The Author(s), under exclusive license to Springer Nature Switzerland AG 2023
F. Carrión Mena and P. Cepeda Pico (eds.), *Urbicide*, The Urban Book Series, https://doi.org/10.1007/978-3-031-25304-1_19

## 19.1 Urbicide. Destruction and Reinvention of the Real and Imaginary City

> ... experience in a city...is full of contradictions..
> Sennett, R. (2019).

The ideal city seems to be dying in the face of the destruction of elements that have made it possible to desire and experience it as a reference for identity, as a place shared between different people, where plural forms of citizenship are constructed. Although the word "city" has not had a single meaning before or now, it has been conceived as "a living heritage with its own identity defined by the social and urban fabric, it is par excellence the place where shared values are learned," so "inhabiting the city is both an imaginary act and a political act" (Barré 1999: 382). Nowadays, the idea of a city open to cultural diversity, to political plurality, to spatial justice, to the care of people, nature, common goods, historical and environmental heritage, is in tension with the real city. This real city is experiencing the fragmentation of public space, the degradation of nature, the privatization of collective resources, the deepening of inequalities, and the displacement of people and social groups in disadvantaged conditions and poverty toward the peripheries.

Historically, cities have suffered damage for different reasons, ranging from aggressive actions for the appropriation and control of space and social resources, because of traffic on borders and boundaries, through forms of urban development that alter the social fabric; to war and political-cultural conflicts, socio-environmental disasters such as earthquakes and tremors, health crises such as epidemics and pandemics, attacks and forms of violence, which can cause the death of hundreds, thousands, and even millions of people. These urban realities that destroy hopes fracture identities and solidarity based on places that disappear with the inhabitants that gave them meaning; they produce suffering that accompanies diasporic migrations and violate human rights. Cities in ruins and the ruins of cities are testimony to public dramas that cause the loss of references of belonging, while these scenarios open up the possibilities of physical and symbolic reconstruction of the place, the collective, the memory, and the historical and cultural heritage.

Historical crises in cities have had heterogeneous, contradictory, creative, and destructive effects on communities, people, cultures, social, and environmental resources that leave visible and hidden traces on the built environment and on nature, in the places we use and inhabit. Faced with the social, material, and environmental effects that jeopardize the survival of civilization, cities try to reconstruct themselves with the impetus of different social and urban actors, resist oblivion, revalue the traces of memory embedded in devastated landscapes that underlie fragments and coexist with the built city and inhabited on the vestiges of the pre-existing one. The ruins of cities, with the voices and words of those who inhabit them in very different social contexts, tell us about ideas, dreams, desires and social relations of power and lack of power, of strategies of government, of forms of domination and colonization, as well as of human misery. They set up testimonies of what happened and bring us closer to understanding how it happened. The cases of Berlin (1939–1945), London (1940–1941), Stalingrad (1942–1943), Hiroshima and Nagashaki (1945), Saigon (1965), New York (2001), Damascus (2011), Beirut (2006) and Kiev (2022) among many others reveal how some cities that in the twentieth century faced attacks, bombings and military combat that have devastated an important part of their physical, social and environmental space, while breaking subjectivities and affecting generations of people. In Latin America, among others, the city of Santiago de Chile (1972), Guatemala (1954), Port-au-Prince (2010), Bogotá (1985).

Without diminishing the magnitude of the damage to these cities, it is interesting to reflect on the damage caused to the city and on the damage that the city causes to urban society and nature due to the logics of urban space production that subject and segregate social groups, fragment public spaces, exhaust and privatize common goods and environmental resources. These are processes of "creative destruction" (Harvey 2007) that threaten historical memory, weaken social ties and collective identities. In the critique of the form of predatory development that has driven the economic order of flexible capitalism in cities, regions and countries over the past three decades, Sassen (2015) states that they face an "… escalation of the destruction of the biosphere around the globe, the resurgence of extreme forms of poverty and brutalization where we thought that had been eliminated or were in the process of disappearing." Faced with this situation, he explains that what happens in cities are forms of "expulsion from a living space… for those who are in the lowest part or in the poor center… -while-, for those who are above it apparently meant getting rid of the responsibilities of being a member of society through self-separation, the extreme concentration of the wealth available in a society and the total lack of inclination to redistribute that wealth" (Sassen 2015: 23–26).

In this line of thought, the concept of "urbicide" is useful for thinking about and understanding urban processes and phenomena that damage the city and erode social life in circumstances of neoliberal globalization. The word "urbicide" appeared in science fiction literature in the 1960s when it was used by the English writer Moorcock (1963) and was transferred to the social sciences and urban movements in the following two decades. It is used in the critical discourse and in denouncing the actions of urban restructuring plans and programs in districts of New York City that revealed the reasoned elimination of some identity traits and elements. Berman (1987) talks about the ruins of the city and the victims of destruction, of expulsion. Since the second half of the twentieth century, they have appeared as nameless crimes caused by what he names as urbicides that eliminate, disintegrate, devour and displace hundreds or thousands of people. When referring to the period of destruction and loss of homes in the Bronx in the seventies due to violent actions, this author points out that in order to understand what was happening, he began by discovering and unraveling the networks that link seemingly disconnected groups, but which are involved in the process that led to the expulsion of people and communities leaving defeated victims.

Recovering this discussion, Carrión (2014) explains that talking about urbicide does not refer to the death or end of cities, but rather to "violence against the city for urban reasons," eliminating a particular city or elements that are inherent to it by military, economic, cultural or political interventions. From this approach, which considers that urbicide is substantially related to the urban economy, he proposes that cities are scenarios of social production of ruins, destruction and oblivion due to urban violence, climate change, the logic of urbanization and innovation that prioritize economic and market criteria. This situation, affirms this author, "leads to the erosion of institutionality and self-government (polis) through privatization or corruption, as well as the deterioration of the material base of a city (urbs), for the sake of a supposed urban development based on the logic of the neoliberal city" (Ibid. 2014: 127–128). The transformations inspired by this logic have had an impact on ways of life and on spatial divisions that show urban realities where new physical, social and cultural frontiers appear.

The notion of urbicide, from this perspective, alludes to the decline of the city understood as *urbis, civites* and *polis*. It is useful to understand what is happening in the urban experience of the twenty-first century and the way in which certain processes, phenomena and actions break the social fabric, damage cultural and natural heritage, violate aspirations of childhood and youth, and damage the living conditions of generations. In the context of the health crisis of recent years caused by COVID-19, which has disrupted social and urban life in the world, capital cities and metropolises capture the effects of illness and death in very different societies, dramatically evidencing the contradiction between the city imagined as a living space, that is open, diverse and inclusive, and the city lived through poverty, uncertainty, risk, violence and fear.

The experience of loss caused by the pandemic is intertwined with the crisis of institutional, employment, environmental, economic and habitability legitimacy, with forms of violence and crime that sow fear in daily life, with war conflicts and migratory exoduses, among others. These are manifested in contemporary cities, expanding into regions and communities that inhabit them, unfolding realities that seem to exhaust their attributes. This experience converges problematically in public space, where the ruins and shortcomings of the city are exhibited, as well as the fragmentation that distinguishes it as a common place for different members of urban society, as a space for politics, communication, encounter and participation in matters of general interest. For this reason, real public space allows us to think about what is absent and must be created in order to reconstruct the city, based on an inclusive and consistent idea of the urban environment as a living space.

## 19.2 The Neoliberal City and the Abandonment of the Public

> The planners' obligation is to serve the community rather than impose an alien set of values...the rough edges between what has been lived and what has been built are not resolved by the simple display of ethical rectitude by the planner...it can provoke anger in some...
>
> Sennett, R. (2019).

Although cities are explained in a relational way, as fulfilling strategic functions in social, regional, national and international processes, the globalization that occurred since the 1980s has imposed a neoliberal current of thought with a model of urban development and government converges that reduces the public spending, commodifies services and leads to the weakening of collective rights (Subirats and Martí-Costa 2014). In the capital cities of Latin America, these urban processes associated with interventions and disputes over social space have been taking place for almost half a century, as a result of capital investments in the built environment. These occur in circumstances of neoliberal globalization and the restructuring of capitalism, promoting a new economic order that has manifested itself since the eighties of the twentieth century and during the first decades of the twenty-first century, producing profound transformations in the urban landscape, in public space and in social life. One of the lines of action of these processes is the creation of urban real estate markets with headquarters in global cities (Sassen 2001). This trend intensifies in times of capital overaccumulation, which is driven toward productive uses through real estate and financial investment (Harvey 2015). Cities concentrate the social, spatial and environmental effects of these processes, perhaps unprecedented in the last century, which transform structure, form and functions through real estate investments in strategic locations, mechanisms for the privatization of public goods, as well as commercialization and financialization of the urban economy.

Large urban projects are one of the most visible expressions of these processes, which occur through direct capital interventions in the built space of cities for construction. These macro-interventions have an influence by emphasizing forms of segregation on a micro-geographical scale and moving local communities to peripheral places where the lack of infrastructure and services is greatest. Two emblematic cases showing the relationship between the State and civil society are emblematic. On the one hand, the case of Brasilia in 2014 due to the construction of the Mané Garrincha stadium, then considered the most expensive in the world, made with part of the budget for services such as health and education and which displaced hundreds of families. On the other hand, the case of the Mitikah project (2022) in Mexico City is emblematic of the implementation in a historic town of a megaproject that brings together a group of high-rise buildings at the cost of displacing original members of the community, causing property damage and weakening the social fabric.

In this line of reflection, talking about a neoliberal city alludes to these processes that name the problematic configuration of social space, resulting from profound changes in the very conception of the urban, expressed in the absence of a city project that takes into account the different dimensions of social life, the ways of inhabiting, collective and nature rights. This reasoning acts in favor of capital through business and strategic urban planning, supported by government strategies as well as urban policies, instruments and actions that favor the domination of private over public in the production of social space. The latter also moves away from the democratic planning that permeates the social and urban fabric, and that coexist and overlaps with different, heterogeneous and subaltern ways of inhabiting, working, consuming, such as popular urbanization, housing cooperatives, alternative and supportive economies, such as bartering and other forms of non-monetary exchange of goods and services.

In the discussion of the category of neoliberal city, Hidalgo and Janoschka recover, on the one hand, the argument that it is the capitalist city in the current phase of accumulation, which makes it possible to recognize the changes that have occurred and the government that drives them.[1] On the other hand, they claim the idea that it is a city where business and speculation predominate, without a clear social counterweight that challenges the mercantilist approach to decision-making.[2] They state, in the discussion about urban neoliberalism, that it is a process "geographically variable and unequal, with multiple scales, and interconnected." It is manifested through a diversity of selective urban policies, which are adjusted to socio-territorial and political urban contexts in different cities, for example, through property markets and urban land, which are key elements in speculative dynamics (Hidalgo and Janoschka 2014: 12–16).[3]

Following this discussion, talking about the neoliberal city is not limited only to enclaves where large financial, real estate, commercial and closed housing projects materialize in cutting-edge architecture, multifunctional high-rise buildings, with new technologies that favor the development of tertiary activities, and introduce significant changes in land use and in the urban landscape. It refers above all to the discordant social and symbolic relationship between these macro-urban projects with the local environment where they are implemented and with the entire inhabited space of the city, as well as to the urban policies, instruments and actions that make them possible by favoring the flow of large local and global capital invested in the environment and the private appropriation of collective resources. The socio-spatial consequences of neoliberal urbanism in public, domestic and private spaces are expressed in the privatization of collective resources—such as water and air, and in the social segregation evident in the displacement of social groups outside the places they inhabit and in the damage to nature. Faced with this situation, disputes over the city are growing, socio-environmental and political-cultural conflicts as forms of resistance by communities and social organizations in defense of collective rights.[4]

In Mexico City, this model promotes the centrality of the market and impacts the local and metropolitan dimension, producing new and greater spatial inequalities, enclaves of wealth and poverty, real and symbolic boundaries between strategic

---

[1] Ornelas (2000).

[2] Rodríguez and Rodríguez (2009: 7).

[3] Along the lines of Brenner and Theodore (2002).

[4] Ramírez (2021).

places for investment, separating places and social groups that have been left out of benefits of this form of urban development that weakens urban rights. Neoliberal urban planning is implemented with very different socio-territorial effects in the towns and municipalities that comprise it. These are expressed in the access of citizens to the system of social and urban resources; in the emergence of new gender inequalities that are added to those existing in the socio-economic structure; and in the growing tension between the aspirations and struggles of social organizations and the limitations to achieve better living conditions and of habitability in matters such as employment, housing, education, public space, health, care, security, territorial, social and environmental justice.

These problematic issues are manifested in public space as representations of profound urban inequalities and as demands for collective rights in a capital city such as Mexico, where 7.6% of the population is located in the highest strata, while more than half (58.8%) are social groups in different conditions of poverty. Between one strata and the other, there are 35% of the inhabitants, made up of middle classes (20.8%) and popular middle classes (14.3%) people who are not poor but who are on the verge of poverty (EVALÚA 2021).

The most visible representation of the neoliberal city in its hegemonic form is urban macro-projects. These are representations of power resulting from large investments of local and global real estate, financial and commercial capital in urban land, favoring the private appropriation of collective resources such as water and building space. These large enclaves show different types, scales and designs, both in terms of infrastructure and services and in terms of multifunctional corporate, housing, recreational and commercial complexes. They are implemented in urban space with peculiarities according to the context of the city where they are built defined by economic criteria and interests; they transform the form, structure and functions of the places where they are developed, deploying in urban space, the hegemony of private over public. The way of building and the risks involved were visible in the earthquake of September 2017, the urban policies that favored the real estate were developed with omissions and irregularities that did not comply with building regulations; the granting of permits and authorization of works in some housing buildings resulted in collapsed; thus the impact of this physical and social phenomenon, evinced forms of corruption that were far from being eradicated. Due to the earthquakes, around three hundred and twenty people lost their lives and one and a half million lost their homes, revealing the absence, until then, of an articulated institutional policy for local and regional reconstruction, all of which challenged the newly elected government in the capital city and at the federal level as of 2018. Currently, this reconstruction process in Mexico City has advanced by 57.5% in the rehabilitation and reconstruction of adequate housing for those who suffered total or partial loss.[5]

---

[5] See the website for the reconstruction of the Commission for the Reconstruction of the Government of Mexico City.

## 19.3 Urban Macro-Projects and the Power of Private Over Public

Neoliberalism, as a doctrine assumed by institutions and the State, and developed as a political project of class domination, favors power elites as opposed to policies of social and economic redistributive and welfare resources because they would jeopardize the accumulation of capital. Anti-collectivism and the supremacy of the private sector are justified by the argument that the public sector is ineffective, prone to corruption, to particularist agreements and not very transparent, which is why a continuous process of privatization is promoted, supported by technical evidence of efficiency through structural reforms that point to a "new society, marked by a systematic prejudice against the public" distributing public goods as merchandise and not as rights (Escalante 2015: 199–202). This logic considers a strong State that favors the centrality of the market as an expression of freedom and as a key device for efficiently resolving economic obstacles by generating information on consumption and production, and on competition, prices and the use of resources (Ibid. 2015). The social effects of this development model that besieges common goods are expressed in the urban, economic, political and environmental crises that arise in the places where people live and that converge in the public spaces of the city.

In this context of neoliberal globalization involving changes in the relationship between the State, society and the economy, Mexico City has experienced an unprecedented real estate boom evident since the 1980s of the twentieth century. The shift that occurred then in the direction of urban planning and policies favored changes and actions toward privatization, market liberalization and deregulation.[6] With this logic, the idea of a world capital, open to trade and consumption, prevailed and ranked Mexico City among global cities. The incorporation of new urban policies facilitated this boom whose representation on a large scale were urban megaprojects as powerful symbols of the development model imposed on the city, new social, urban and cultural realities that emphasize inequality and segregation. Explaining the meaning of these megaprojects, Negrete (2017: 108) points out that there are great works made and exhibited in the urban space that link capital and power, transforming the landscape with an impact on the "… social and environmental order, on the prevailing forms of life." This impact on the one part is expressed in the subsequent emergence of a set of physical and social elements, including commercial, residential, educational, corporate, cultural, leisure and mobility features, which expand over time in connection with detonating buildings. The effectiveness of this cluster of buildings, which this author calls the "assembly of megaprojects," is defined by the vocation of capital accumulation (Ibid. 2017: 111).

On the other hand, megaprojects result in the social production of self-segregated and segregated local environments, demaging to the environment, and fragmenting urban structure. The latter affects generations of inhabitants by imposing a model

---

[6] See Pradilla (2018) and Carlos De Mattos (2007).

of segregation that corresponds to "fortified enclaves… privatized, closed and monitored for residence, consumption, recreation and work…" which, with the argument of fear of violence and crime, motivate the abandonment of everyday and popular public space, such as streets that are delegated to the use of the popular classes, the urban poor, the marginalized and those who have no place to live (Caldeira 2007: 257). These spaces draw physical, social and symbolic boundaries that separate residents and users from other different groups because of their class status, origin and economic stratum, which are expressed in cultural practices linked to forms of consumption, purchasing power, tastes and preferences, bodily and social behaviors that they are enrolled in different ways of life.

The spaces of neoliberalization are characterized by important institutional changes in urban policy, by original forms of inter-institutional coordination, by the creation of new institutions on a regional scale that promote intergovernmental ties and the commercialization of the city. These spaces range from the creation of business networks led by public–private agreements and partnerships, to original models of local economic development policy that promote collaboration between private companies (Hidalgo and Janoshcka 2014: 9). In these places, the social and cultural practices of citizens respond to the codes and living conditions of groups with high- and very high-income levels, to new forms of regulation, security and private control of public functions, users, users and consumers. This is the case—among others—of the most exclusive shopping centers, linked to local and global consumption, of closed residential subdivisions, or of high-rise buildings that mix luxury housing, office and retail spaces, sports, entertainment and health services spaces within the complex. In contrast, particularly city streets, in central and peripheral localities, are reduced, with notable exceptions, to places of pedestrian crossing, mass use, sales and unpaid work, insecurity, risk and fear of violence for groups and social classes that are predominantly middle class and urban poor. Motorized mobility, through the prevalence of private cars in dispute with public transport, predominates in urban structures and prevails in the urban experience of public space.

## 19.4 Public Space: Closure and Abandonment

*The city is flawed because of its diversity, because of its inequalities, because of its tensions...*
*Sennett, R. (2019)*

In Mexico City, the development of the Santa Fe urban and corporate macro-project sets the tone both for the development and reproduction of these walled environments and for the major transformations of the capital inspired by the dominant neoliberal ideas and policies at the turn of the twentieth century to the twenty-first century, as well as for notable change in the meaning of the public, which makes greater emphasis on the private meaning, as a place, as a political sphere and as a state sphere. This change is expressed in the fact that the public domain, as a common good, is degraded as a space for encounter, communication and relationship, open and accessible to different people and social groups, and moves to closed, semi-public, selective places, guarded in person and through video surveillance systems. The Santa Fe project highlights multifunctional interventions that are designed with a mix of corporate, housing and service uses for upper-middle and high-income users, for economically prosperous social classes, with purchasing power and high purchasing and consumption capacity (Ramírez 2021).

Public space is transformed toward greater fragmentation based on a logic of that audiences and social groups segmented according to tastes, class, interests and purchasing power. In this line and following Caldeira, the street as a "space of public life has been annihilated and with this the possibility of the coexistence of diversity and difference, whereas the type of space that is created promotes not equality - as intended-, but only more explicit inequality" (Caldeira 2007: 376). The Santa Fe complex is representative of this logic that produces a particular form of urban enclave that is distinguished by rigid land uses, small areas for "mixed uses," favoring dependence on cars (Moreno 2011).

It should be emphasized that in the transition from the twentieth to the twenty-first century, the real estate market is more intensely attracting global investors who consider Mexico City a strategic location for capital investments in Latin America (Panreiter 2011). In Mexico City, large urban projects are not new, nor are private stocks, public–private agreements, and speculative real estate and financial investments in urban, semi-urban or rural land for more than a century. But the macro-urban projects that emerged in the last three decades name interventions of a monumental scale, promoted by considerable investments of global and regional financial capital in urban land, in places identified as strategic, where the initial real estate potential generates large and even excessive amounts of capital gains. The Santa Fe complex started in the second half of the 1980s (1987). It is the display of an assembly of mega-projects in ten residential neighborhoods, configuring an enclave of fortified sites developed in different stages, which continued until recent years with the opening of La Mexicana Park (2017), corporate projects and vertical housing units projected around this. It is an enclave of enclaves, surrounded by native peoples and by a group of neighboring popular colonies that emerged mainly during the second half of the twentieth century and with which there is no urban articulation or social integration.[7] These localities are geographically close; while they are socially and economically distant from the macro-project, they are predominantly inhabited by popular classes and urban poor people. These traits underline the obvious and latent conflict making the project since its inception under a problematic urban development model that distinguishes the neoliberal city.

The real estate boom linked to financial and commercial investments, which began at the end of the twentieth century, particularly the implementation of macro-projects, continued in the first two decades of the twenty-first century, with the development of grand buildings in strategic locations in the capital, mostly in central municipalities. On the one hand, what is happening on Paseo de la Reforma Avenue stands

---

[7] The towns surrounding the corporate complex are San Mateo Tlaltenango, San Bartolo Ameyalco and Santa Fe de los Altos.

out schematically, where the construction of the building called Torre Mayor (1999–2003) whose producers and users, as mentioned above, are neither Mexican nor do they have a clear or defined national identity, has resulted in a transnational consortium with control and coordination functions of operations in different countries on a cross-border scale (Parnreiter 2011: 17). This tower is the forerunner of the subsequent emergence of an ensemble of skyscrapers that currently compete for height and transform the urban landscape by deploying a coupled set of projects along the Reforma Avenue, which since the nineteenth century has been a leading public place in the capital, connecting the Historic Center with the Bosque de Chapultepec and currently extending to the roads that lead to the Santa Fe corporate complex.[8]

Among these, the Reforma Tower (2008–2016) stands out, with an unprecedented height of 246 m until 2021, deploying a monumental vertical complex with a multiplicity of corporate and service functions, chosen as the best skyscraper in the world two years after its opening.[9] At the same time, the BBVA Tower (2008–2016) emerged, and almost during the same period, the Diana Tower (2013–2016); the Reforma Latino Tower (2012–2015); and the Chapultepec Uno Tower emerged as well (2014–2019). These include Fibra Uno (FUNO), the largest real estate investment trust in Latin America, founded in 2011. This expansion of buildings emerged in the same period with the participation of the real estate company Pulso Inmobiliario, the New York Life tower (2009–2012); the Mapfre tower (2011–2013); and the Impera Reforma tower (2016, still under construction). This process is aimed at mixing corporate functions with the creation of this Reforma-Centro Histórico corridor, one of the most exclusive residential spaces in the country that currently hosts the development of fourteen projects that are under construction.[10]

On the other hand, in the logic of urban neoliberalism, promoting competition for height and innovation based on author's architecture and design, and of private appropriation of the public and of the city's building space, conceived in 2008 the construction of the urban project represented by the Mitikah Tower, the tallest in a group of buildings outside the central nucleus but still in the space within the inner city.[11] Located in the old town of Xoco, with 267.3 m from the ground to the

---

[8] Through Av. Reform-Lomas, Constituyentes and the Mexico-Toluca highway.

[9] This complex, which includes a restaurant, a shopping center and entertainment areas, was managed by the construction company of the Capital Vertical Grupo Inmobiliario building, while "LBR y Arquitectos" was responsible for the planning. In 2018, it was chosen as the best skyscraper in the world by the International Highrise Award. It was included in the list of the "50 most influential skyscrapers in the world in the last 50 years" by the Council on Tall Buildings and Urban Habitat. See Infobae (2022) which are the three tallest buildings in Mexico City.

[10] These works amount to an investment of 10 billion pesos and are being built on the stretch from Hidalgo Avenue to the Fuente de Petróleos, and their construction will result in an investment of 600 million pesos as a result of mitigation measures, which will be provided to a private trust whose main destination will be mobility projects and improvement of public space in Reforma and surrounding areas. See Zamarrón (2022). The new real estate boom in Reforma will provide 600 million pesos to Mexico City. Forbes Mexico.

[11] Located in the Benito Juárez area on the border with the Coyoacán area. The authorization included the construction of 1 million 28 thousand 71.96 m$^2$; however, the volume almost doubled, despite insisting complaints from affected neighbors (Proceso Magazine: 2019).

sky and sixty-five levels, it is considered the tallest in Latin America and becomes the tallest building in the capital after the Reforma Tower. From the beginning, this project generated dissatisfaction among residents, who organized themselves by expressing their rejection of a construction that would cause mobility problems, water shortages and pollution. The lack of information from developers and authorities to the community emphasized the discomfort that increased when excavation work caused cracks in the seventeenth-century temple of San Sebastián and fractures in the walls of neighboring homes (Cruz 2012).[12]

The Mitikah case (2022) is emblematic of conflicts over urban and human rights that have not been resolved in a just and creative way. This is an ensemble of projects that incorporate nine buildings: six towers, the highest of 65 levels and which bears the name of the mega-project; three with 35 levels each; another with 23 levels and another with more than 10 levels. A shopping center of 5 levels, an 11-level hospital and the Bancomer Center are added, both with five floors each (Seduvi, 07/2019). The Mitikah Tower is representative of the building complex, consisting of six hundred and sixty-seven luxury apartments, mostly sold.[13] The construction of the real estate complex, published as the largest in Latin America, was authorized in 2008, located on two buildings in the town of Xoco, in the Benito Juárez area in Mexico City. The authorization included the construction of 1 million 28 thousand 71.96 m$^2$; however, it has been reported that the volume almost doubled, despite persistent complaints from affected neighbors. Urban developers did not face any obstacles from the capital or federal authorities (Proceso Magazine: 2019).

After fourteen years of construction, developed by FUNO, one of the largest real estate companies in México, and an estimated cost of 22 thousand and 500 million pesos, the first two phases have come to an end.[14] With the opening of the huge shopping center, **with** 120 thousand meters of leasing area, five levels and 280 commercial spaces,[15] this powerful enclave is exhibited, which breaks the local social and urban fabric, taking over the emblematic Real Mayorazgo street, hence, showing the lack of planning, spatial justice, environmental damage and urban ethics, resulting in environmental damage which stands before an aggrieved and segregated community.[16] Mitikah in Mexico City is considered the most recent

---

[12] Listed as a historic monument by the National Institute of Anthropology and History.

[13] There are 14 types of apartments with dimensions ranging from 68 to 314 m$^2$, with an approximate price of 70,000 pesos per m$^2$, meaning that the smallest apartment is worth approximately 4.7 million pesos, while the largest one costs between 22 and 28 million pesos. In addition, it has a profitable office area of 64 thousand 649 m$^2$, distributed in 25 commercial spaces; its main tenants are: Loreal, Sanofi, WeWork and Total Play. Within the complex, there are different amenities such as cinema, swimming pool, toy library, children's games, event room and guest rooms, games room, sauna, spa, steam room and a gym.

[14] ¿Quién es dueño de Mitikah?, la polémica obra de la CDMX. La Silla Rota, 24/09/2022.

[15] Mitikah, El cartel inmobiliario y los sismos, Alejandro de la Garza, Sin Embargo, 24 de septiembre, 2022.

[16] The company Fibra Uno bought the construction from Ideurban and Prudential in 2015 for 185 million dollars. See Emilio Gómez, Mitikah: Neoliberal Emblem, The Mayan Day, opinion, September 24, 2022.

example of irregularities and arbitrariness, of temporary suspensions of the work, of resistance and neighborhood mobilizations, of complaints and conflicts with the community and with the organization of neighbors.[17] This experience of symbolic and real violence caused by the power of the private over the public, despite the fact that the residents of the town of Xoco resisted to prevent the privatization of Real de Mayorazgo Street, as proposed in phase II of the Mitikah Tower real estate project, and they reported the felling of more than eighty species of trees on that street, which caused a fine imposed on the real estate company by the Ministry of the Environment of Mexico City (Contreras: 2021).[18]

Today, most of these high-rise multifunctional enclaves—housing, offices, entertainment, health, technology—and the fifty large shopping centers anchored by large department stores are located in the west, center and south of Mexico City. Taken together, these globally linked mega-works represent highly profitable urban interventions, which have diversified the type, design and housing, corporate and commercial offerings according to the consumer profile.[19] In the expansion of real estate and financial capital investment in urban land, in the first two decades of the twenty-first century, the high concentration of business headquarters stands out on the one hand. By the end of 2021, they accounted for 58.6% of the top five hundred companies, equivalent to 293, and 64.6% of the commercial sales of these companies.[20] This, despite the fact that the configuration of this corporate geography shows a decrease of 10.9% compared to the number of companies existing in 2006, which, as explained by Parnreiter (2011), shows the relationship between the division of office space, that of foreign-owned business headquarters and that of producer services.[21] In the first decade of the twenty-first century, the enormous increase, especially in the west by the Santa Fe complex, but also in the south and even in the Historic Center-Reforma corridor, of "the supply of high-quality office properties is accompanied by a change in the spatial structure of the market," forming a Central District of Business.

In this formation, it is notable that, during the second quarter of 2022, Mexico City reported a total of 7.4 million m$^2$ of class A and A + office space, which means an increase of 246% in a decade, from 3 million m$^2$ in 2008 to 7.4 million m$^2$ in 2022. The main office corridors in Mexico City are: Santa Fe with 1.4 million m$^2$, Polanco

---

[17] The developer of the FIBRA UNO project is analyzing the second phase of construction of the project, so far they have land use authorization for the entire comprehensive project, but private licenses for this stage have not yet been granted. See Noguez (2022). After 14 years of construction, Mitikah Shopping Center will open at the end of the year. Forbes Mexico.

[18] A fine of 40 million pesos was imposed, but the lack of clarity in payment persists until the time of writing this article. Residents fear that an attempt will be made to continue with the work of the overpass that would privatize the street and the fine could be resolved by creating mitigation works in the area.

[19] The real estate boom in the period 2013–2018 is expressed in the support and authorization of 292 real estate projects, most of which are considered to have high impact, which represent housing developments, offices, centers and shopping malls, to a lesser extent hotels and hospitals (Cruz 2018). Most of it is concentrated in the municipalities of Álvaro Obregón, Miguel Hidalgo, Benito Juárez, Cuauhtémoc and Cuajimalpa.

[20] See Expansion (2022) "The Naked 500".

[21] Panreiter (2011), cites the existence of 329 companies in 2006.

with 1.3 million m$^2$, Insurgentes with 1.1 million m$^2$ and Reforma with 942,000 m$^2$. Currently, 26 new office buildings equivalent to 685,631 m$^2$ are under construction in the Insurgentes, Reforma and Polanco corridors, which will be the fastest growing: Insurgentes will concentrate 33% (226 thousand m$^2$), Reforma 29% (198 thousand m$^2$) and Polanco 18% (123 thousand m$^2$). In these strategic and central corridors, the capital that flows through real estate and financial development drives a series of exclusive corporate, commercial and residential projects with a high impact on the urban environment in which they are implemented.

The predominant strategic urbanization, presented in a non-exhaustive way, was carried out on the basis of an innovative, bold and effective discourse, aimed at responding to the aspirations and ways of life of middle and upper classes and social groups in convergence with legal and institutional devices, in order to implement the centrality of the market, the privatization of collective resources, common goods and public services. Constitutional articles were amended, new laws and regulatory instruments were introduced, new policies and programs were designed that facilitated selective and differentiated, lucrative interventions and actions for local, regional and global financial, commercial and real estate investment. The idea of hegemonically consolidating the global city of services, by making it a place of consumption which positions it on a global scale, largely guides the development of these major urban interventions, the result of public–private agreements. In this process, the sprawl and contradiction of territorial planning regulations and programs, the discretionary use of code and the irregularities in procedures became the rule. This situation is expressed in the significant proportion of buildings that exceed the number of floors and height allowed by regulations.[22]

Large physical and social formations tend to increasingly dominate the urban landscape, silently and visibly causing bodies to behave with reverence, respect and distance, acting as central mechanisms of "the symbolic of power and of the totally real effects of symbolic power" (Bourdieu 1999: 120–122). In this relationship between the body and the city, the street on a human scale is diminished not only as a computing element and an articulating element of pedestrian trajectories, but also as a public space for placemaking, with different people and social groups. In a city with metropolitan dimensions such as Mexico City, these conditions are intertwined on the one hand, distancing public space from its integrating and articulating attributes, reducing it to a place of passage and circulation, of contingency, risk and violence. On the other hand, making visible both the concentration of power and wealth in a minority and discriminating conditions of poverty in the majority. The resurgence of urban movements and citizen resistance voiced in public space transformed it into a communal place to the expression of conflict and advocate for urban rights.

The crisis of public space and domain is perhaps expressed above all in the fragmentation experienced by different people and social groups as a common place, as a political space, as a space for communication and for the construction of democratic forms of participation in the urban experience. Public spaces and green areas,

---

[22] The highest proportion of violations of the law occurred in two central municipalities in the capital: Benito Juárez and Cuauhtémoc (PAOT 2017–2018).

indispensable in a city with a good quality of life and options for coexistence, are insufficient; they are mostly deteriorated and are also an expression of Mexico City's marked socio-territorial inequality. Improving and expanding these spaces, based on territorial justice, is a key priority, along with the protection of the valuable historical and cultural heritage that characterizes the city. The tensions and disputes, as well as social and gender inequality that converge in public space, reveal the political and urban contradictions of the city, the polarization between social groups and the fragmentation of public life. This situation, which is expressed with greater emphasis in localities on the outskirts of the capital, is contrary to the idea of the city as an urban, civic and political space, a common reference, open and accessible to the whole society.

## 19.5 Final Note, ¿Toward the Open City?

In the neoliberal city, and facing the abandonment of public space, we ask ourselves what is left of what is common for the reconstruction of the city? Under the current circumstances of privatization of the public realm and the weakening of the collective space, it is relevant to discussed what remains of "the common" and how to govern it, in order to articulate a firm, effective political proposal contrary to the current period of active dispossession typical of neoliberal capitalism, from the point of view of urban movements and not just of them.[23] Does the idea of the city remain as a form of sociability and as a possible space for the construction of utopian dreams of the social order? The profound transformation of urban society and of the living space of the city hinges on the daily experience of what flexible capitalism economic order means, supported by urban ideas, policies and actions promoted by different social, political and economic actors, local and global, that make possible its materialization in very different cities and societies.

By reflecting on the destructive effects of neoliberalization processes on people, on the social fabric, on urban life, on culture and on nature, it is possible to debate and understand more and in greater depth what is happening and how its taking place. Researching problematic realities not yet considered, finding and deciphering the hidden or inconspicuous networks of apparently unrelated actors that nonetheless are involved in processes that result in the expulsion of individuals and communities, leaving victims broken, would lead to greater social justice. This situation confronts society with the challenge of reconstructing the common space as a public good, creating forms of collective organization, social ties that open spaces of resistance and participation to influence decisions that affect and damage the living conditions of all people. On the other hand, this line of discussion raises the need for a paradigm shift in the policies that support urban interventions, as well as the recovery of the common references that exist and that can provide elements for planning and co-creating cities in a democratic way based on social and collaborative urban planning.

---

[23] Di Masso et al. (2017).

The city lived as a complex, heterogeneous and unequal social space is transformed through individual and collective actions driven by different and even conflicting ideas about urban life and life in common between different groups in society. These ideas influence government policies and actions, ways of living, organizational and participatory forms in matters of general interest that together shape public, domestic and private experience. Sociable relations between culturally diverse social groups and classes, with different needs and interests, even irreconcilable in specific space–time contexts define urban experience and introduce it to the inescapable debate about the city and public space, about realities we don't see, about violence in urban areas and violence against women and girls. It seems like a utopian dream to realize the demand for a socially and spatially just city based on a paradigm shift from neoliberal urban planning to citizen urbanism, which recognizes the social function of land, human rights, urban rights, nature, cultural diversity and different sexualities. The events of urbicide experienced by twenty-first century cities lead to the demand for the right to the city and the rights of the city that converge in public space, and the scene of the battles that must be fought to rebuild a city with peace and justice.

## References

Berman M (1987) Among the ruins. New internationalist. Recuperado de: https://newint.org/features/1987/12/05/among

Bourdieu P (1999) "Los efectos del lugar", en P. Bourdieu (Dir.), La miseria del mundo, Buenos Aires, Fondo de Cultura Económica [1a. ed. en francés, París, Éditions du Seuil, 1993]

Caldeira T (2007) Ciudad de muros. Gedisa, Barcelona

Carrión F (2014) Urbicidio o la producción del olvido. En Habitar el Patrimonio, nuevos aportes al debate desde América Latina, editado por Lucía Durán, Eduardo Kingman y Mónica Lacarrieu. Instituto Metropolitano de Patrimonio/Flacso/Universidad de Buenos Aires, Ecuador, pp 116–129

Di Masso A, Berroeta H, Vidal T (2017) El espacio público en conflicto: Coordenadas conceptuales y tensiones ideológicas. Revista de Pensamiento e Investigación Social, Athenea Digital, vol 17, no 3, pp 53–92

Escalante F (2015) Historia mínima el Neoliberalismo. El Colegio de México, Ciudad de México

Harvey D (2007) Breve historia del neoliberalismo. Akal, Madrid

Harvey D (2015) Breve historia del neoliberalismo. Ediciones Akal, Madrid

Hidalgo R, Janoschka M (2014) La ciudad neoliberal. Gentrificación y exclusión en Santiago de Chile, Buenos Aires, Ciudad de México y Madrid, Santiago de Chile, Pontificia Universidad Católica de Chile (Serie Geolibros, núm. 19)

Moreno M (2011) "Terciarización económica y la creación de clusters: el megaproyecto de Santa Fe en la Ciudad de México", en Alejandro Mercado y María Moreno (coords.), La Ciudad de México y sus clusters, México, Juan Pablos/UAM Cuajimalpa, pp 143–188

Parnreiter C (2011) Formación de la ciudad global, economía inmobiliaria y transnacionalización de espacios urbanos: El caso de Ciudad de México. EURE (santiago) 37(111):5–24. https://doi.org/10.4067/S0250-71612011000200001

Pérez Negrete M (2017) "Los megaproyectos en la Ciudad de México", en Ana María Portal, Ciudad global, procesos locales: conflictos urbanos y estrategias socioculturales en la construcción del sentido de pertenencia y del territorio en la Ciudad de México, Ciudad de México, UAM-Iztapalapa/Juan Pablos

Ramírez P (2021) La ciudad neoliberal en Santa Fe. El sentido privado del espacio público, en Espacios públicos y ciudadanías en conflicto en la Ciudad de México, coordinado por Patricia Ramírez. Instituto de Investigaciones Sociales UNAM/Juan Pablos, México, pp 391–426

Sassen S (2001) The global city: New York. London. Tokio. Princeton University Press, Princeton

Sassen S (2015) Expulsiones. Brutalidad y complejidad en la economía global, Buenos Aires, Katz

Sennett R (2019) Construir y habitar. Ética para la ciudad. Ed. Anagrama, Barcelona

Subirats J, Martí-Costa M (2014) Ciudades, vulnerabilidades y crisis en España. Sevilla, Fundación Centro Estudios Andaluces

## *Hemerographic Sources*

Becerra D (2019) La Torre Reforma Latino. Smart Building, webste: https://smartbuilding.mx/la-torre-reforma-latino/

Carrasco J (2019) Mitikah, un megaproyecto plagado de irregularidades. Proceso, No 2234

CNN (2019) 19 de septiembre, la fecha fatídica que dejó huella entre los mexicanos. CNN en españo. Website: https://cnnespanol.cnn.com/2019/09/19/cientos-de-muertos-miles-de-damnificados-y-millones-de-dolares-en-perdidas-asi-fue-el-terremoto-del-19s-en-mexico/

Desinformémonos DE (2022) El Mundial de la exclusión en Brasil - Desinformémonos. Website: https://desinformemonos.org/el-mundial-de-la-exclusion-en-brasil/&gt

Ed. Mex. (2013) Torre Reforma Latino. Edificios de México, website: https://www.edemx.com/site/torre-reforma-latino-2/

Ed. Mex. (2016) Torre diana. Edificios de México, website: https://www.edemx.com/site/torre-diana/

El país (2022) El fuerte terremoto de Chile causa al menos 300 muertos. El País, website: https://elpais.com/internacional/2010/02/28/actualidad/1267311601_850215.html

Gamboa J (2017) "Historia detrás del desarrollo inmobiliario en México". Real Estate Market and Lifestyle, website: https://realestatemarket.com.mx/mercado-inmobiliario/turismo/21404-historia-detras-del-desarrollo-inmobiliario-en-mexico

Infobae (2022) Cuáles son los tres edificios más altos de la Ciudad de México. Infobae México, website: https://www.infobae.com/america/mexico/2022/04/10/cuales-son-los-tres-edificios-mas-altos-de-la-ciudad-de-mexico/

Noguez R (2022) Tras 14 años de construcción, centro comercial Mitikah abrirá a fin de año. Forbes México, website: https://www.forbes.com.mx/tras-14-anios-de-construccion-centro-comercial-mitikah-abrira-a-fin-de-anio-alistan-segunda-fase/

Noticias A (2017) "Se inaugura el parque "La Mexicana" en Santa Fe, con una inversión de 2 mil mdp". Aristegui Noticias web site. https://aristeguinoticias.com/2411/mexico/se-inaugura-el-parque-la-mexicana-en-santa-fe-con-una-inversion-de-2-mdp/

Reyna A (2019) "La Torre BBVA es reconocida como 'Edificio del año 2019' por Edificios de México". BBVA website: https://www.bbva.com/es/mx/la-torre-bbva-es-reconocida-como-edificio-del-ano-2019-por-edificios-de-mexico/

Univisión. Atlanta, Londres o Tokio: estas ciudades quedaron arrasadas por la guerra en los últimos 200 años. Univisión noticias website: https://www.univision.com/noticias/mundo/ciudades-destruidas-por-la-guerra-fotos

Vázquez, Ricardo. "Paseo de la Reforma nuevos proyectos de clase mundial". Real Estate. Market & lifestyle, website: https://realestatemarket.com.mx/articulos/mercado-inmobiliario/vivienda/12172-paseo-de-la-reforma-nuevos-proyectos-de-clase-mundial

Wikipedia (2022) Terremoto de Chile de 2010. Wikipedia, la enciclopedia libre website: https://es.wikipedia.org/wiki/Terremoto_de_Chile_de_2010

Zamarrón I (2022) El nuevo boom inmobiliario en Reforma dejará 600 mdp a la CDMX. Forbes México, website: https://www.forbes.com.mx/el-nuevo-boom-inmobiliario-en-reforma-dejara-600-mdp-a-la-cdmx

## Documents

Brenner N, Theodore N (2002) Cities and the Geographies of "Actually Existing Neoliberalism". Antipode 34(3):349–379. https://doi.org/10.1111/1467-8330.00246

Cobos EP (2018) Cambios neoliberales contradicciones y futuro incierto de las metrópolis latinoamericanas. Cadernos Metrópole 20(43):649–672. https://10.org/10.1590/2236-9996.2018-4302

Cruz Flores A (2018) Invadieron la ciudad 19.4 millones de m2 de concreto y acero en esta administración, La Jornada, 3 de diciembre: https://www.jornada.com.mx/2018/12/03/capital/026n1cap

De Mattos CA (2007) Globalización, negocios inmobiliarios y transformación urbana. Nueva sociedad, 212:82

EVALÚA (2021) Resultados de la medición de la pobreza 2016–2020. México: Consejo de Evaluación del Desarrollo Social de la Ciudad de México (Evalúa). Disponible en: https://www.evalua.cdmx.gob.mx/principales-atribuciones/medicion-de-la-pobreza-y-desigualdad

LaSalle JL (2022) Panorama de Mercado de oficinas clase A. Ciudad de México 2Q 2022. JLL online

Moorcock M (1963) "Dead God's Homecoming", in Science Fantasy #59, Nova Publishing

Ornelas J (2000) La ciudad bajo el neoliberalismo. Papeles de población 6(23):45–69

Pradilla E (2018) Cambios neoliberales, contradicciones y futuro incierto de las metrópolis latinoamericanas. Cadernos Metrópole 20:649–672

Rodríguez A, Rodríguez P (2009) Santiago, una ciudad neoliberal. Quito: Organización latinoamericana y del caribe de centros históricos (OLACCHI), pp 1–26

**Patricia Ramírez Kuri** Sociologist dedicated to the study of the city, to the logics that produce it and that transform the urban social space. She is a researcher at the Institute of Social Research, National Autonomous University of Mexico. She has been a fellow of the Rockefeller Foundation in the Urban Culture Program, Autonomous Metropolitan University and did a postdoctoral stay at the School of Government and Public Policies of the Autonomous University of Barcelona.

# Chapter 20
# A "New" Urban Colonialism? North–South Migration and Racially Structured Gentrification in Latin America

**Juan Pablo Pinto Vaca**

**Abstract** Through the analysis of the relocation of U.S. immigrants to Latin American cities, this chapter problematizes the necessary correspondence between migration and spatial segregation, usually raised by studies exploring the relationships between urban space, alterities, and inequalities. This chapter demonstrates that U.S. migration and the urban changes generated around it are not new phenomena in the region. Unfolded in the midst of uneven geographical development, this migration is understood as a symptomatic expression of the externalization of the U.S. border. I argue that such externalization has lasted as a result of the "destructive production" of physical, symbolic, and interactional urban spaces. This chapter not only reveals the mechanisms of foreignization of the natives and the deployment of racially structured urban gentrification, but most importantly, it demonstrates the renewed persistence of urban colonialism in Latin America.

**Keywords** Urban space · Immigration · Colonialism · Cities · Latin America

## 20.1 Introduction

The twentieth and twenty-first centuries witnessed the emergence of a form of migratory governmentality centered on a securitarian approach. This form of population management has been implemented through various strategies. For instance, straightening the global regime of migratory control (Düvell 2003), the making illegal of certain forms of mobility through legal mechanisms to incorporate them, in a subordinated form, into the labor market (De Génova 2002), the globalization of punitive and surveillance border public policies (Varela 2015), and the implementation of

---

Juan Pablo Pinto Vaca Ph.D. in Cultural Studies from the Autonomous Metropolitan University (Mexico). Thanks to Dr. Karla Encalada for the translation of this article and for her invaluable comments.

---

J. P. Pinto Vaca (✉)
Autonomous Metropolitan University (UAM), Mexico City, Mexico
e-mail: juanppintov@gmail.com

a "cost–benefit" language—which produces a distinction between mobilities with "high added value" and those that are undesirable (Sassen 2006).

This form of governmentality has deeply affected immigrants from the "global South." They are frequently conceived as undesirable and radical otherness. Numerous, epidemiological, military, and watery metaphors have been deployed around them. To be specific, they are typically represented as "invaders," "threats," "waves," "dangers," "problems," and intrusive agents that disturb the—always illusory—harmony of the community.

From an urban perspective, some scholars have repeatedly demonstrated how the production of immigrants as undesirable otherness constitutes a spatially produced process. This literature argues urban space is not so much a replica of social divisions as a constitutive and constituent mechanism for the production of hierarchical differences among groups (Bourgois [2003] 2010; Wacquant [1994] 2001, [2006] 2013; Santillán 2019). Consequently, the "others" are either placed in a separate urban habitat or expelled from certain locations (Monkkonen 2012; Boy and Perelman 2017). They are perceived as the incarnation of whatever is conceived as ominous (Kingman 2006; Carman et al. 2013). These scholars have also demonstrated how historically established groups in an urban space tend to close ranks to new groups (Elias and Scotson [1965] 2016; Caldeira [2000] 2007). As a consequence, internal and international immigrants have to experience and navigate numerous practices of stigmatization, marginalization, and segregation. In essence, these scholars argue that there is a necessary correspondence between migration and spatial segregation, produced as a result of differences in power existing and acting either autonomously or synergistically.[1]

In this chapter, I aim to problematize this academic *doxa*, namely the idea of an assumed correspondence between migration and spatial segregation. Through the study of migration and the relocation of United States citizens among Latin American cities, I propose the following questions: What happens to spatial segregation and the stigmatization of immigrants when mobility occurs from the "North" to the "South"? How do xenophilia and xenophobia work when the traditional direction of migration flow is reversed? How are urban frontiers produced in this scenario? What types of transformations take place in urban spaces when "high added value" migrations settle in Latin American cities?

Even though Latin America is more of an "emigration region" than an "immigration area" (Stefoni 2018, 9); at the end of the twentieth century and the beginning of the twenty-first century, "North–South" migratory displacements to Latin American countries, such as, Mexico, Ecuador, Guatemala, and Colombia, among others, have gained straight, visibility, and importance.

At a global level, so-called North–South migration remains statistically marginal. This type of migration, motivated by labor, academic, and recreational purposes, barely represents 5% of the aggregate number of "North–South" migrants worldwide (OIM 2020). Nonetheless, there has been an increase in the mobility of North

---

[1] Some studies have partially questioned the correspondence between segregation and migration. For more information, see for example, Segura (2012).

American retirees to postcolonial locations in recent decades. Multiple processes explain the increment in retirement migration. Among the most significant ones are the deterioration of living standards of white middle classes in the United States and the privatization of the pension system.

In this context, hundreds of thousands of senior citizens, affected by the structural forces that caused the decline of the "white American dream," develop diverse strategies to navigate their decline. For instance, the decision to move and live in "less wealthy" countries, where they are capable of stabilizing and improving their present-day standards.

The North–South retirement migration constitutes a novel phenomenon, defined in diverse ways—residential migration, residential tourism, amenity migration, geronto-immigration, lifestyle migration (Hayes 2013). From a critical perspective, considering that human mobility is an unequally distributed resource (Sheller and Urry 2018), North American retirement migration has also been described as "privileged migration" (Croucher 2009).

Little research has examined North–South migration from an urban perspective. Despite this, Latin Americanists investigating this phenomenon argue this type of transnational relocation "entails a neocolonial attitude on the part of the migrants, interested in creating and controlling space artificially" (Janoschka 2011, 94). For this reason, these scholars insist on the necessity to explore "the economic impact, the racial imaginaries, the unequal distribution of power, as well as the rights, and privileges that Northern migrants claim in their new destination country" (Hayes 2013, 3).

To contribute to the study of this phenomenon, I propose two complementary arguments. On the one hand, I will examine the complex interactions among urban space, mobility, race, class, nationality, and power, as a strategy to question the idea of the existence of a necessary correspondence between migration and spatial segregation. On the other hand, my research furthers the understanding of the debates on urbicide, through the analysis of how "new" forms of production of urban spaces in Latin America destroy the existing symbolic, physical, and relational dimensions of Latin American cities. In conclusion, this study will try to demonstrate the renewed persistence of a "new" form of urban colonialism, which is not so new in Latin America.

My first argument is that "North–South" U.S. migration constitutes a symptomatic expression of an—little-explored—externalization of the U.S. border.

I formulated the concept of border externalization from a critical and imaginative position. This concept is useful because it allows me to engage with two different approaches. On the one hand, the first proposition focuses on the importance of the idea of border in regulating the geographic, political, racial, economic, migratory, and urban imagination of U.S. society. Under this view, in 1893, Frederick Jackson Turner described border as a mobile entity tending to move to apparently empty spaces. As he states: "The U.S. people have gotten their temperament as a result of their constant expansion […] It would be a bad prophet who states that the expansive character of North American life has already ceased" (Turner 1987, 207). From his perspective, the externalization of the U.S. border constitutes the historic foundation of its civilizatory and evolutive model, the very essence of the North American nation. On the other hand, the second proposition has critically reformulated the concept of border externalization to demonstrate how U.S. securitarian logics are implemented as mechanisms that neutralize and maintain, as far as possible, what is seen as migratory "threats" (Varela 2015). My perspective, as I will show, is different.

I assert that the U.S. border, a global symbol of a set of inequalities and systemic advantages, is refracted in Latin American cities where North Americans have relocated. My argument is that the externalization of the U.S. border has operated through spatial, economic, and symbolic mechanisms and techniques. I will demonstrate how this form of border externalization is inseparable from processes of dispossession, which have been produced in the United States from the middle of the twentieth century to the present. I will also show how this border externalization has worked and how it has been constructed on multiple spatial scales, from a geopolitical arena to everyday urban relations.

I also argue that such externalization represents a process through which several relations have been produced between geographically dispersed populations and spaces. I will demonstrate how these relations are constituted by an aporetic logic of spatial proximity and social distance, of contact and separation, of articulation and differentiation. And I will also show how this logic permeates numerous fields of social and urban life.

My second proposition is that U.S. border externalization implies the deployment of "destructive production" dynamics (Gordillo 2018, 108) that affect urban spaces in which North Americans have relocated. I argue that these dynamics work through three urban dimensions that are related to each other.

The first dimension is related to the destructive production of physical and material spaces, which leads to a gradual process of racially structured gentrification in the city. This process generates a dynamic of spatial elitization and urban reconstruction. These dynamics are articulated around a migrant population that possesses a "high added value" and causes the expulsion of the local population to the peripheries of the city.

The second dimension, which refers to the productive destruction of symbolic spaces, discusses Occidentalist imaginary geographies and discourses tending to exoticize and inscribe Latin American cities and populations within specific temporal, social, economic, and racial coordinates. Not only United States institutions and lifestyle global promoters, but also North American expatriates have encouraged and deepened the development of this "urban Orientalism." I demonstrate how these geographic imaginaries produce spatial fetishes and mask, with a romantic crust, new forms of local and global inequalities.

The third dimension is related to the destructive production of urban social relations. This dimension reveals the existence of asymmetric social links caused by the quotidian interactions that take place among unequal social groups. When U.S. expats move to so-called less developed countries, they tend to become local elites, even though in their local societies they experience important processes of dispossession. This is possible not only due to the expats' social characteristics—income, nationality, migratory status, racial capital—but also, and overall, this is related to further structural asymmetries between the United States and Latin America. For this reason, this dimension puts emphasis on a form of urbicide produced when the "conditions of sociability that define a spatial and particular node" are destroyed or transformed (Gordillo 2018, 110).

I developed my arguments and problematizations on the basis of research I did between the years of 2019 and 2022. I studied U.S. immigration and its spatialization in San Miguel de Allende (Guanajuato-México) from a genealogical, discursive, and ethnographic perspective.

San Miguel is small heritage city located in the South of the Mexican central highlands, which has 174 thousand inhabitants. I conceive of the city of San Miguel as a paradigmatic study case. This is because "North–South" migration, racially structured gentrification, and the destructive production of space—usually studied either as a contemporary phenomenon or as a mechanical effect of neoliberal capitalism—all have a profound historical density in this city. Further down, I will discuss some essential aspects that characterize the history of San Miguel.

## 20.2 From "Ghost City" to U.S. Paradise

Progress, as Benjamin ([1940] 2018) highlights, leaves a trail of rubble and ruins in its wake. Among them are abandoned cities, desert towns, and limbos full of murmurs. However, in some occasions, progress—as a modern theology founded on material and symbolic processes of disinvestment and investment—can reverse the process of desertification, transform its debris into revered ruins, and build a paradisiacal city for U.S. retired citizens from these ruins. Even if this causes the expulsion of the native population as well as the destruction of particular forms of producing space. From 1940 to the present, this long-standing process has centrally defined the history of San Miguel de Allende.

In Mexico as well as in other Latin American countries, the decade of 1940 witnessed not only the creation of urban industrial poles but also the population densification of the main urban macrocephalias. However, San Miguel de Allende became the other side of these structural tendencies after experiencing a profound "urban crisis." From the perspective of local elites, this crisis threatened to transform the city into a ghost town. The census data of the time shows that this city, in fact, expelled its population from 1910 to 1930; it ranged from 10,547 to 8716 inhabitants (INEGI 2011). Nevertheless, as the census figures show, it is clear that the city was far from becoming a "ghost town," as elites imagined.[2]

The idea of the San Miguel's "urban crisis" was rather a political and ideological operation. This operation made readable the quotidian and structural transformations that San Miguel de Allende's elites experienced, as a result of the articulation of

---

[2] The stories about San Miguel's desertification are unmanageable; however, this "urban mythology" has positioned it as a regime of truth.

endogenous and exogenous processes[3] that gradually reconfigured social, ethnic, and territorial frontiers and, at the same time, supported local social hierarchies during the 1930s and 1940s. The possibility of San Miguel becoming a "ghost town"—a *horror vacui* or fear of empty spaces—was the way in which elite families codified and projected their concrete experiences while living in a dystopian scenario. Even though it is possible to question these projections, the chain of equivalences between "crisis" and "desertification" demonstrated symbolic efficacy. As a result of this efficacy, these elite families were able to design, at least, two strategies to achieve the city's "revitalization" through a comprehensive demographic, racial, urban, and economic reengineering.

The first strategy consisted of the promotion of a slow, sustained, and still ongoing process of urban museification and patrimonialization. As an alternative to descriptions that portrayed San Miguel as a destroyed and decadent city, the idea of an artistic, patrimonial, and cosmopolitan city emerged. This new city would lead its way without stepping on its rubble. This is because architectural and monumental remains from the colonial and the independent periods—when San Miguel was one of its protagonists—became totems and fetishes of the city's renaissance.

Consistent with the first federal legislation on patrimonial protection, approved during the Mexican post-revolutionary period, San Miguel became the subject of numerous interventions. These interventions promoted the conservation and restoration of numerous houses, streets, and temples. The aim was to bring back a colonial atmosphere to the city. Moreover, several projects were also implemented in relation to social services, health, and the "beautification" of the population. Another objective of the city's renaissance was to transform it into a sanctuary of leisure and recreation. This aim was achieved through the organization of events that were consistent with the taste of the literate and economically powerful elites—who did not abandon the city, despite the "urban crisis."

This patrimonial fever achieved optimistic results in 1939, when the Mexican state recognized the city as a "Typical Town." This effervescence was strengthened during the twentieth and twenty-first centuries, when San Miguel was declared a "Historic Monuments Zone" in 1982, a "Magic Town" in 2002, and a "Cultural Heritage of Humanity" in 2008. Throughout these decades, this process of "accumulation of symbolic collective capital" (Harvey [2001] 2014b) not only tilted the local economy toward the service sector (tourism), but also attracted vast amounts of economic and population flows.

From the beginning, the patrimonialization process in San Miguel was linked to a second strategy. Even though this strategy was developed during the 1940s, it still projects its long shadow into the present. I refer to the process of attracting North American immigrants, a foreign population perceived as desirable in economic,

---

[3] This "crisis" emerged as an effect of the 1910 Mexican Revolution, that is, as a result of the fragmentation of the accumulation of the territorial structure, the transformation of the existing hierarchical links among social classes, the destabilization of the position of the elite families in the social structure, and the implemented provisions aiming to restructure the relations between the church and the state. In addition to this, the impact of the Great Depression in the production and exportation of grains in Mexico made it possible to promote the idea of the city's decline.

urban, and racial terms—that would make it sustainable the process of "revitalization" of the city.

To persuade and attract men and women from diverse nationalities and "races" (especially from the United States) to live in San Miguel, this city was gradually transformed into an artistic cosmopolis. Different fine arts schools, like the Escuela Universitaria de Bellas Artes (1938) and the Instituto Allende (1948), were created during this period. Prominent Mexican artists, like Rufino Tamayo and Diego Rivera, became teachers in these schools. In addition, famous individuals were invited so that they could promote the new "modern" academic city among influential national and international artistic circles. In this context, I argue that San Miguel was an urban laboratory where numerous actors tried to materialize the idea of "Universópolis." An idea that Vasconcelos ([1925] 2003, 19)—who was the most prominent theorist of the miscegenation ideology and the creator of the "cosmic race" theory—envisioned as being realized next to the banks of the Amazon River.

As a result of the hegemony of a miscegenation ideology, the arrival of the first U.S. citizens to San Miguel was influenced by a selective incorporation and exclusion of foreigners. The state xenophobia and xenophilia—regulated through the idea of the existence of attributes as well as cultural and biological racial differences—made it possible for foreign populations of distinct nationalities to be conceived as either positive or harmful, in a context determined by the attempt to consolidate the Mexican Mestizo state.

During 1930s and 1940s, these selective rationalities were implemented through numerous mechanisms, such as: the coercion and expulsion of the Chinese population (Augustine-Adams 2015); the siege and inadmissibility decree promulgated against the Jewish population (Yankelevich 2015); and the naturalization privileges granted to Republican exiled Spanish, as they were perceived as a population that was able to assimilate the mestizophilic and hispanist projects (Gleizer 2015, 155). It was in this context that, despite historical conflicts between Mexico and the United States, the arrival of the first U.S. migrants took place in San Miguel. This arrival was endorsed by the principal ideologues of the Mestizo state.[4]

In the case of the initial arrival of North Americans to the city, it was not without conflicts[5]; however, in general terms, it was positive. This favorable reception was related to the fact that U.S. migrants were perceived as a revitalizing population in

---

[4] According to Vasconcelos: "The fifth race does not exclude but includes all forms of life; for this reason, excluding the Yanquis as well as excluding any other human being would imply the anticipation of self-mutilation [...] If we are not interested in excluding even races which could be considered inferior, it would be much less wise to exclude from our project a race which is full of drive and has social values" (Vasconcelos [1925] 2003, 20). It is important to consider that, at the time, the U.S. was seen as a symbol of progress because, as Vasconcelos puts it, Latin America owes "a great part of the construction of its railways, bridges, and companies to North Americans" (Vasconcelos [1925] 2003, 20).

[5] Some U.S. migrants were accused of encouraging immorality and communism. On a few occasions, they were deported. These exceptional cases were analyzed by Rudolph (2017) and Pinley (2017).

terms of achieving racial perfectibility, urban modernization, productive transformation, and the dynamization of the economy. From then to the present, U.S. migrants were co-produced as a highly desirable otherness.

At the beginning of the 1940s, it was registered the arrival of a few dozen North Americans to San Miguel. Since this historical juncture, the city has been part of the migratory circuits from the "North" to the "South" of the American continent.

Some data show that there are currently about 3500 U.S. citizens living in San Miguel (INEGI 2020). Other studies show that the number of expats living in the city oscillates between 8000 and 11,000 (Flores and Guerra 2016, 188). It is a type of migration mainly composed of females and qualified older adults.[6] The most critical reasons for their relocation are to improve their quality of life (in health, housing, food, mobility, and care) and to access leisure, shows, art workshops, tourism, and other "amenities" the city offers (Sloane and Zimmerman 2019).

In recent years, the massive arrival of U.S. retirees to San Miguel has been accompanied by a series of awards that the city has received from several global lifestyle promoters. For instance, San Miguel was designated as "the best destination to visit" (2013), "the best city in Mexico, Central and South America" (2016), "the best city in the world" (2017, 2018), "the best small city" (2018), and "the American capital of culture" (2019).

What is striking, in this case, is that the "accumulation of symbolic collective capital" and the attraction of a type of migration with "high added value" are two strategies that, even though established in the 1940s, have renewed persistence to the present. These strategies have survived even an "urban requiem."

According to Rajchenberg and Heau-Lambert (2008), since the emergence of the colonial regime to the formation of the nation-state, during the nineteenth and twentieth centuries, important processes of "symbolic desertification" have affected the history of Latin America. These processes justified the conquest and occupation of so-called tierras baldías (wastelands) by new settlers. Their aim was to externalize the civilizational border to spaces viewed as "anomic." However, the symbolic and political production of "empty spaces" is not merely a matter of the past. It can be found at the heart of some contemporary urban processes. The persistency of certain ideas about "unoccupied," "obsolete," and "degraded" urban spaces tends to justify the implementation of "regenerative," "rescue," and/or "gentrification" projects. The implementation of these projects frequently results in population cleansing or in its replacement. The history of San Miguel synthesizes and links these processes.

Below, I will describe some structural dynamics produced in the United States. I will illustrate how, in different periods, these dynamics made it possible U.S. immigrants—initially composed of war veterans and later by retirees—to be relocated to San Miguel. In the following section, I will analyze how the political practices of the "North–South" migration have produced the emergence of a novel form of U.S. border externalization, deployed not only in San Miguel but also in other Latin

---

[6] According to Sloane and Zimmerman (2019), 78% of the U.S. migrants who live in San Miguel are more than 65 years old, 58% are women, and 56% have gone to graduate school.

American cities. I will try to overcome methodological "localism" and "nationalism"; to this end, I will study San Miguel not as a self-contained space but as a space that shows how changing power relations are produced and worked outside the city.

## 20.3 U.S. Migration and Border Externalization

Several scholars[7] investigate "North–South" migration to Latin American cities, especially in relation to U.S. *baby boomers'* mobility, as a novel and contemporary phenomenon. However, the analysis of the historical process of San Miguel reveals the limits of an academic arrogance that tends to exaggerate the singularity of our times.

Migration from the United States to San Miguel has been recorded since the 40 s of the last century. Since then, the city has been positioned as a constitutive and a constituent space of the U.S. imperial rise. In this section, subdivided into two parts, I will examine the structural processes occurring in the United States that have motivated the relocation of North Americans to San Miguel. More importantly, I will also indicate how this process of relocation has produced a novel form of U.S. border externalization.

### 20.3.1 The U.S. Migration to San Miguel During the First Half of the Twentieth Century

Two decades after World War I (1914), some years after the Pearl Harbor Attack (1941), and nearly at the end of World War II (1945), not only the U.S. society, but also some state and military officials, publicly expressed their concern about the possible mental disorders and potential psychotic outbreaks that some ex-combatants might exhibit upon their return to the country.

As Mettler (2005, 16) argues, these concerns were not divorced from an economic scenario; to be specific, the fact that the last bastions of the U.S. *crash* of 1929 were still in force aroused deep social anxieties. During this period, it was believed that the United States would not be able to reintegrate into the labor market neither more than fifteen million war veterans who had been demobilized from various U.S. fronts, nor more than ten million civilians who worked for war industries (Mettler 2005, 16). War veterans were viewed as a potential social, political, labor, and economic problem that needed to be managed. Therefore, under the framework of the *New Deal*, in 1944, the U.S. state implemented the GI Bill program. This program allowed the U.S. state to design public policy instruments that facilitate good governance and the reinsertion into civilian life of millions of ex-combatants.

---

[7] For more information, see Croucher (2009, 2010), Hayes (2013, 2017, 2020), Janoschka (2011, 2013), Korpela (2010).

The reinsertion strategy adopted by the U.S. state puts emphasis on education. It principally consisted of providing subsidies and quotas to accredited institutions that offered diverse educational and training programs—from agriculture to fine arts—so that, through this process, U.S. war soldiers could become citizens again.

Even though the GI Bill program was initially established among formative education centers in the United States, the demands of war veterans (scattered around the world), the existing institutional incapacity, the overcrowding of some educational institutions, and the ex-combatants' decision not to settle in the United States redefined the plan's political orientations, as well as its operative components and spatial dimensions.

The spatial solution adopted consisted of accrediting academic institutions around the world to make possible their articulation to the GI Bill program. According to Pinley (2015), the subsidies and agreements signed between the United States and other nations allowed ex-combatants to study in more than 500 university centers, located in 58 different countries. Of this number, 85 accredited institutions were located in Latin America and 25 in Mexico. San Miguel, first with the Escuela Universitaria de Bellas Artes (1938) and then with the Instituto Allende (1948), would actively be part of the U.S. GI Bill program.

A "domestic problem" in the United States became a global problem. As a consequence, San Miguel de Allende was transformed into a new "zone of global contact," or a space where "different cultures meet, crush, and confront each other, usually mediated by highly unequal relations of domination and subordination" (Pratt 2010, 32).

From the perspective of the U.S. state, the implementation of the GI Bill program in San Miguel and other "underdeveloped" cities offered an effective "spatial solution" (Harvey [1981] 2014a 303) to a set of social problems that characterized the postwar context. This program produced a novel form of border externalization that consisted of relocating U.S. citizens, who were perceived as problematic or as surplus lives in other countries. Nevertheless, because of structural inequalities between Mexico and the United States, this residual *demos*, composed of impoverished ex-combatants, allowed them to become a revitalizing and privileged population only by crossing the U.S.-Mexican border. In other words, for impoverished U.S. ex-combatants, San Miguel represented a promise of well-being outside the United States.

In the context of the emergence of the United States as an imperial power, the GI Bill program was an innovative technique of global governmentality targeting an impoverished population. This form of governmentality produced specific routes, spaces, logics, and modes of circulation of people, narratives, and capitals. These routes, spaces, and logics have worked as the sediments that sustain the new "North–South" forms of migration. In fact, after the first GI Bill impulse—which stimulated the arrival of dozens of ex-combatants and other North American travelers to San Miguel—different transnational and transgenerational migratory networks were established. For thousands of U.S. migrants, who arrived in the city during the following decades, these networks functioned as their support.

At the end of the twentieth century and the beginning of the twenty-first century, in a radically different U.S. context—determined by the primacy of casino capitalism, the gradual disappearance of the "semi" welfare state, and the privatization of the pension system—the routes, networks, and spaces of circulation described above became activated again by segments of populations that had experienced the effects of the U.S. pension system. A brief description of the life story of a U.S. couple will allow me to better illustrate how this process took place.

## 20.3.2 The U.S. Migration to San Miguel De Allende During the Twenty-First Century

Leslie and Chuck[8] remain a U.S. couple who are over 65 years old. They relocated to San Miguel in 2019 and live there with 2000 dollars that—together—they receive as their retirement pension. Almost all their lives, they lived in New York City. Leslie steadily worked as a clerk in a real estate company, and Chuck, with a more flexible work trajectory, worked as a journalist, a bartender, and as an appraiser in the business of buying and selling art pieces.

"We were close to being *homeless*," says Leslie. At the beginning of the twentieth century, they requested a loan to purchase an apartment in New York City. However, given the increase in credit interest, their monthly installments became unpayable. As a result, they received constant notifications stating that if they did not pay off their debt, they would be evicted. As a strategy, they moved to Vermont, where they could live indefinitely on the condition that they would have to work even in their old age. In Vermont, the cost of living for a family of two was about 4000 dollars per month. In the United States, they had to work because their pensions only covered half of their expenses. In contrast, in San Miguel, they absolutely do not have to work.

There were other reasons that also motivated Leslie and Chuck's transnational relocation. For instance, the rise to the presidency of Donald Trump had an impact on Leslie and Chuck as well. In fact, this circumstance became the origin myth of their decision-making. According to them, it was impossible to understand how millions of people voted for someone who said that "all Mexicans are rapists." For this reason, when Trump was elected, they sold everything, took their car, and drove thousands of miles from Vermont to San Miguel de Allende.

However, the most important motivation that explains their relocation was the high cost of medical services in the United States. This is crucial because Leslie doesn't merely suffer from epilepsy, but additionally has had three back surgeries; each surgery cost her 120 thousand dollars. In the case of the two first surgeries, they paid the insurance deductible, yet they were not capable of paying Leslie's last surgery deductible. For this reason, they left the United States permanently. Even though their insurance company could sue them and take them to court, this is not an enormous concern for them. Leslie and Chuck are convinced they will not return

---

[8] These life stories are the result of multiple interviews I did with Chuck and Leslie between April and November of 2019, in San Miguel de Allende. The names have been modified.

to the United States. They recognize that private medical services in Mexico are unquestionably cheaper than paying 30 percent for an insurance deductible in the United States. They were also affiliated to the, now extinct, Mexican Seguro Popular (popular insurance) state program. With a look of relief and a smile on her face, Leslie comments that living in Mexico makes her feel free.

Leslie and Chuck's cases are similar to the experiences of hundreds of other expats who have decided to relocate to San Miguel or other Latin American cities. Their life trajectories illustrate and express, as a symptom, the processes of dispossession that pauperized white middle-class retirees currently experience in the United States. Additionally, these trajectories not only illustrate how a form of systemic expulsion occurs in one of the wealthiest nations in the world, but also show the mechanisms by which "Souths" proliferate in the "North"—or how the "North" becomes a form of "global South"—that is, the processes that make possible the emergence of a form of "North" in the "global South."

In the face of social and individual decline, Leslie and Chuck's personal strategies are similar to those of other expatriates, that is, adopting a "spatial solution" to "ensure 'low cost' lifestyles" (Hayes 2013, 6). In other words, their *"New Deal"* is now, similar to other cases, "a self-fulfillment individual project achieved through "North–South" mobility" (Korpela 2010; Hayes 2020).

It is equally critical to take into account that, during the last four decades, U.S. retirees have witnessed the dismantling of several institutions that had previously guaranteed the possibility for them to be included in a good retirement plan (Sassen 2015). Social security and pension provision plans were some of the most affected systems during this period (Blackburn 2010). These processes of dispossession, as San Miguel's case shows, have produced a specific form of historical geography that either linked populations and spaces that were disseminated or intensified the preceding spatial and population relations.

San Miguel de Allende has been the scene of these changes. The profound transformation of the demographic profile of U.S. people living in the city gradually occurred during the twenty-first century. In recent years, migratory flows to San Miguel have included neither students nor war veterans. Instead, these flows are composed of aging populations and individuals retired from working life. These thousands of pensioners and retirees have been seen as socially productively obsolete by their origin society.

Nevertheless, by relocating transnationally to a city in a "less developed" country, the declining U.S. white middle class, to which Leslie and Chuck belong, is often upwardly mobile. To achieve this form of well-being and materialize a new "lifestyle," they merely had to cross, in a "North–South" direction, the "magic threshold" that embodies the Mexican-U.S. border. The very same threshold that, for thousands of migrants from the "global South," has become some kind of mass grave.

In the disadvantageous scenario, the postwar period posed in the mid-twentieth century and in the context of the growing deterioration of the living conditions of U.S. retirees, San Miguel became like that Turnerian border, "always in retreat" and capable of breaking "custom ties," of "offering new experiences," of representing a "new field of opportunities," and of creating "new institutions and activities" (Turner

1987, 207). To achieve all this, it is only necessary to cross or expand the U.S.-Mexican border. In this way, U.S. austerity measures will be transformed into a set of pleasures, good experiences, and more economic services, inaccessible to former war veterans and retirees in the United States.

San Miguel has become a kind of promised land where it is possible to materialize, even in contexts of pauperization, the "white American dream," but in Mexico. In this sense, this city—as well as other Latin American cities—can be conceived as a figurative projection of the idea of the "Old American West." That is, as an urban fractal which expresses the externalization of the U.S. border.

## 20.4 The Destructive Production of the City

The externalization of the U.S. border (expressed in the "North–South" migration of U.S. citizens) was initially produced through bilateral agreements. However, over time, it gained relative autonomy from them. To a large extent, this occurred because so-called border externalization implied the deployment of "destructive production" processes in the city. Put differently, the externalization of the border is not a project floating in the air; in contrast, the processes of inequality and perdurability that define it have been produced, organized, and reproduced in a physical, imagined, and relational urban space. Further down, I will analyze how these process took place.

### 20.4.1 Destructive Production of Physical Space: Racial Gentrification in the City

Since the second half of the eighties, literature on urban inequalities has highlighted the presence of two novel phenomena in Latin America. Despite their similarities, these phenomena have been examined from various perspectives—that is, not only through the analysis of the articulation of their economic structures, but also through the study of class differences that define them. First of all, some scholars emphasized the existence of a new segregation pattern, founded on urban fragmentation. That is, a type of micro-scale segregation that functions through the existence of spatial proximity relationships among heterogeneous groups that are, at the same time, separated by frontierization, fortress practices, and distinction mechanisms (Janoschka 2002; Borsdorf 2003; Sabatini 2006; Ward 2012). Secondly, scholars have also registered several dynamics of gentrification or spatial elitization. That is, a process of substitution of users and land uses in certain urban areas. This process is the result of both the dynamics of capital reinvestment, in areas considered degraded, and the transformation of consumption patterns and lifestyles among the gentrifying classes (Atkinson and Bridge 2004; Díaz 2015).

Urban fragmentation and gentrification are studied as contemporary phenomena in Latin America. However, in San Miguel de Allende, these processes of destructive production of urban spaces not only contain a profound historical density but also make evident the existence of a racial dimension that is usually untaken into account in most studies.

At the end of the 1930s and the beginning of the 1940s, integral urban transformations took place in San Miguel. For instance, the city did not have the necessary residential infrastructure to receive the hundreds of U.S. citizens who progressively arrived in the city. As a result, a profound spatial rearrangement was implemented. This process became visible not only in the construction of new distinction sites, like hotels, galleries, and restaurants, to satisfy the foreigners' demands, but also in the evident transformation of private residencies into lodging places for foreigners.

These land use changes—from family housing uses to commercial uses—and land user changes—from Mexican residents to U.S. residents—predominantly affected the city's urban center and radically reorganized the social and economic life of the city.

As noted above, during the 1940s, San Miguel sought to become an artistic cosmopolis. In particular, a gradual structural change began to take shape in the city with the arrival of a "creative class" (Florida 2010). This class was composed of a vast community of U.S. artists (painters, photographers, writers) around whom the urban and economic "revitalization" and the reverse of the city's degradation presumably took place. Put differently, eighty years ago, San Miguel was already the scenario of a dynamic that is now widespread on a global scale—namely, the creation of cultural enclaves as a strategy to produce an "upward filtering" of the city, its infrastructure, and population. The relocation of a "creative" and transnational population to San Miguel was not only an engine of change in the city's socio-spatial structure and an element tilting the economy toward the service sector, but also a strategy to valorize the city's depreciated lands to generate a potential land rent. Repopulation and land investment in San Miguel represented two sides of the same coin.

A woman from San Miguel, born in 1934, explains this process: "North Americans began to arrive here between 1946 and 1947. Then, there was a little more movement in the city. A change began to take place. We no longer depended on the owners of the rancherías. [...] However, when North Americans arrived, San Miguel changed. It became a tourist attraction. At first, there were few; then, more and more. They began buying houses and becoming residents. And this is how a total change took place" (interview cited in Rudolph 2017, 46).

In the process of material and symbolic transformation of the city, U.S. migration functioned as a collective agent, which caused the elitist revalorization of the city's depreciated lands. San Miguel's lands were transformed, factually, into a U.S. competition in which some "rich Mexicans" equally participated, although to a lesser extent. The unregulated competition to gain access to urban lands caused a kind of spatial eugenics—namely, the gradual—although incomplete—whitening of space and the formation of a racially structured gentrification.

The displacement of Mexican people, direct or indirect, was the other face of gentrification and "regeneration." The testimony of a former San Miguel inhabitant illustrates this process: "I could no longer pay rent because when the Americans began to arrive, rents began to increase a lot" (interview cited in Rudolph 2017, 27). Through the mechanism of buying and selling of undervalued properties and the impossibility of affording rents and services, urban spaces were gradually "purged" in economic, racial, and demographic terms.

As I said formerly, in the twenty-first century, U.S. migration to San Miguel is no longer composed of students or war veterans. Instead, those who have arrived in the city are, mostly but not exclusively, retirees who, predominantly, belong to a white middle class in decline.

The "creative class" that formed the initial U.S. migration to San Miguel cleared the path for the arrival of a transnational "leisure class" (Veblen 2004). This class is characterized by the consumption, possession, and enjoyment, more or less ostentatious, of the spaces, services, and experiences that this city offers.

The U.S. retirees who currently live in San Miguel do not only settle in San Miguel's central area but also practically throughout the city, even in rural areas and ejido lands. This has fueled a speculative real estate logic through which properties are acquired in Mexican pesos and resold in dollars to foreigners, mainly U.S. citizens. For instance, in 2017, the San Miguel Real Estate Association recorded that 65% of the buyers of the 362 properties sold downtown—at an average price of 400,000 dollars—were foreigners (El Universal 2017).

Currently, in Colonia Centro, the central neighborhood, residential houses can cost between one million and three hundred thousand dollars. For instance, in Colonia Guadalupe, residences can be quoted at 400,000 dollars. In Colonia San Antonio, houses can reach 850,000 dollars. In Colonia San Rafael, stigmatized because of insecurity in the neighborhood, houses can be quoted for more than 300,000 dollars. Even in rural and communal areas like Atotonilco, prices can reach 850,000 dollars.[9]

These speculative dynamics in urban, peri-urban, and rural areas (e.g., increase in the cost of living, real estate harassment, and the excessive increase in rents, services, and taxes, among others) have caused the local population to be displaced and expelled, directly or indirectly, from their own city. According to some testimonies, "Before, whoever conquered the plaza, conquered the town. We lost the plaza" (author interview, August 2019) or "People sell their houses and their lands to the gringos to later become their servants" (author interview, October 2019). These testimonies illustrate the expansion and strengthening of an elitist and continuous process of whitening the space. A process that began in the twentieth century and was consolidated in the twenty-first century.

The contemporary processes of urban fragmentation and spatial gentrification that currently define San Miguel de Allende are a rhizomatic and intensified version of the historic racially structured gentrification of the city. Put differently, it is a process of "original accumulation" with a racial sign, a destructive production of the city that has not stopped happening.

---

[9] This information is part of real estate advertisements published in the *Atención* (Attention) newspaper between May and November of 2019. *Atención* is a newspaper of the U.S. residents' community who live in San Miguel.

## 20.4.2 Destructive Production of Symbolic Space: Occidentalist Imaginary Geographies and Urban Orientalism

With the gradual arrival of U.S. migrants in San Miguel, a novel imaginary geography was produced for the city and the country—that is, an Occidentalist form of inscribing the space and population into certain temporal, economic, social, and racial coordinates. This form of Occidentalism presupposes the destructive production of symbolic and urban spaces and the dissemination of an "urban Orientalism."

These "works of representation" (Hall 2013, 459)—materialized in several reports, autobiographies, personal chronicles, novels, travel guides, poems, photographs, and essays—have been carried out for more than eighty years by U.S. institutions and expatriates. These discourses are produced to circulate, be consumed, and be possessed in the United States. Yet they also operate as "acting imaginaries" that regulate social relations within the city (Hiernaux 2007, 18).

A recurring theme in these geographical imaginaries is the idea that the U.S.-Mexico border constitutes a magical threshold that takes one to a fetishized and exotic place. Heath Bowman and Stirling Dickinson,[10] two young U.S. citizens, formulated this idea. These young men, "escaping the winter blizzards and the distractions of civilization," entered Mexico and said: "How strange that by the simple procedure of crossing the Rio Grande, by exchanging the name Laredo for Nuevo Laredo, we should suddenly be subjected to such a new set of experiences. It is the quickest transformation you will find anywhere in the world" (Bowman and Dickinson 1935, 3). Almost eighty years later, essayist Morris Berman, a U.S. expatriate who has settled in Guanajuato since 2006, suggested a similar argument: "Despite their [Mexicans] efforts to emulate the United States, there is something archaic, primitive, and timeless about Mexico. This is exactly what I wanted for my life since it is characteristic of traditional societies. Crossing the border was like diving into a mirror: instantly, everything becomes its opposite" (Berman 2012, 199–200).

From certain U.S. perspectives, the border has operated as a geography capable of instituting a cognitive order that distinguishes the familiar space of "civilization" (United States) from a "primitive" space (Mexico and San Miguel), both considered as opposite but complementary.

In addition, U.S. citizens' displacement south of the border has equally been described as a "retrospective journey" (Sontag 2007, 309)—that is, a temporary journey into the past, where reminiscences of the "archaic" would still be alive. The border, within the United States' imaginary geography, is viewed as a spatialized temporality that operates as a chronopolitics that produces the "negation of coetaneity" (Fabian [1983] 2014). Because it is considered "eternal," the "traditional society" to which U.S. citizens travel—in this case, San Miguel and Mexico—is viewed as outside the course of history or is inscribed as the permanent infancy of the "West."

---

[10] Dickinson was one of the most important promoters of U.S. migration to San Miguel during the 1940s.

Moreover, according to imaginary U.S. geography, San Miguel appears as a promising land predisposed to be occupied. In this way, San Miguel constitutes an "Old American West" but located in Mexico. In fact, the representation of San Miguel as an Edenic location for U.S. immigrants—belonging to a white impoverished middle class—is the core element of the established spatial symbolization of the city. This is symptomatically illustrated in the report *"GI Paradise. Veterans go to Mexico to study art, live cheaply, and have a good time,"* published in 1948 by Life Magazine:

"To GI students in U.S. colleges, crowding into Quonset huts and scrimping on their $65-a-month government subsistence, the Escuela Universitaria de Bellas Artes in Mexico would be paradise. The Escuela is a fine-arts school, accredited under the GI Bill of Rights, to which 50 U.S. veterans and their wives have come to study painting, ceramics, murals, sculpture and languages. They find it very pleasant in the quiet little town of San Miguel de Allende, up in the mountains north of Mexico City. The air is crisp, the flowers are bright, the sun is warm, apartments are $10 a month, servants are $8 a month, good rum or brandy is 65¢ a quart, cigarets are 10 ¢ a package" (Life 1948, 57).

Crossing the border toward Latin America, Mexico, or San Miguel enables U.S. citizens to capitalize on a form of "aristocratization" (Lomnitz 2011, 157). In particular, because it allows them to "live like royalty in Mexico" (Franz and Havens 2006, 486).

However, crossing the border and relocating to San Miguel do not merely constitute a geographical descent profitable in economic terms, but also a descent that is capitalized in artistic and creative terms. This is because the U.S. geographic imaginaries constantly show how proximity to the social and racial Mexican landscape injects new vitality into the artistic and identity projects of U.S. migrants who see themselves as dissidents from their nation.

This ethnophagic drive was formulated in a photographic book on San Miguel de Allende, produced by Allan Kahn and James Norman. Citing the words of Charles Flandrau, it is stated that "Mexico is a prolonged romance. For even the brutal realities—of which there are many—are the realities of an intensely pictorial people among surroundings that, to the Northern eye, are never quite commonplace" (Kahn and Norman 1963, 27). In this sense, according to U.S. geographic imaginaries, crossing the border to Mexico and San Miguel has historically been not so much a matter of distance, but of difference (Staszak 2012, 181).

Lastly, another component of the U.S. imaginary geography is the inscription of what is Mexican as "servitude" and what is North American as a "lifestyle." This inscription appears in countless statements. For instance, "forty more Americans here [in San Miguel] means forty more maid jobs" (cited in Croucher 2009, 481); "servants [in San Miguel] are $8 a month" (Life 1948, 57); "In the mornings, the Mexican maids arrive/At the empty houses/Where the shadows are cold/And there aren´t any echoes/Of the often absent owners [...]" (Bellavance 2018, 18); or "Mexican servants have often had little schooling; many cannot read or write. Work and money are not always really important to them" (Norman 1972, 77). From this U.S. perspective, what is "Mexican" and "Mexicans" is constantly inscribed in the niche of what is servile.

The division between "lifestyle" and "servitude" does not represent a reflection of a differentiated economic status—also produced through the conquest of San Miguel's urban space—but a symbolic mechanism through which racial imaginaries about what is Mexican or Mexicans have been produced—during the twentieth and twenty-first centuries—on both sides of the border. Through being reproduced time and time again—in different ways and in diverse historical contexts—these statements transform into "the natural" or "nature" what is certainly the result of an "uneven geographical development" (Harvey 2007).

Through different strategies like fetishization, racialization, allochronism (temporary distance from an "other"), and spectacularization, the U.S. imaginary geographies about Mexico and San Miguel express a slow process of destructive production of the symbolic and imaginary dimensions of urban space. These mechanisms of space fetishization, urban Orientalism, and the ontologization of social inequalities are linked not only to the subalternization of local forms of giving meaning to a space that, only a few years ago, was considered familiar, but also to the concealment of local/global inequalities and "brutal realities" with a romantic aura.

### 20.4.3 Destructive Production of the Relational Space: The "Foreignization" of the Native

The condition of being a foreigner is frequently associated with someone who comes from outside, from overseas—that is, the outsider or the immigrant. However, the historical, social, economic, and spatial configuration of San Miguel has produced a different idea of what being a foreigner means.

The privileged position of expatriates related to uneven geographical development, the oscillation of the local economy toward tourist activities, the dominant urban entrepreneurship in the city, the territorialization of transnational dynamics, and the political will of successive local governments to attract capitals and high-income white populations, all have caused many San Miguel citizens, established and living in the city, to be seen and to see themselves as intruders and foreigners within their own city and society. This particular situation—which can be described as a process of "foreignization of the natives"—combines a set of structural processes of global inequality while making visible the renewed persistence of Mexicans' racism and classism.

The foreignization of the natives implies a complex network of experiences of alterity that local people from San Miguel have had to face. As a citizen of San Miguel explains, "instead of us being the locals, the 'colados' (intruders) are us, the ones from here" (interview cited in Flores and Guerra 2016). Multiple modes of exclusion, lack of protection, expulsion, and dispossession, compound this process of foreignization. A process that also makes visible a form of urbicide that emerges when certain "sociability relations that gave life [to a city] have dissolved," that is, when the "conditions of sociability that define a particular spatial node" are transformed and destroyed (Gordillo 2018, 110).

These experiences of alterity and processes of foreignization of the natives have been historically manifested in different ways at a local level. The most classic example is San Miguel's native families' direct or indirect residential expulsion from the urban city center and its surrounding neighborhoods. A citizen from San Miguel summarizes this process: "why is it that in our own country, on our own land, we are displaced anyway?" (author interview, November 2019). This form of spatial violence also enhances a process of symbolic dispossession. This process implies not only the destruction of affective, social, and familiar dimensions that symbolize physical spaces but also the promotion of new meanings, created by new users or by people who have greater purchasing power. Consequently, vast segments of the local population feel alienated within a space that was once considered their home.

Additionally, native foreignization represents an integral process that results from the relocation of a set of popular practices, such as religious and other celebrations, processions, and symbolic and material exchanges. These cultural practices were extirpated and removed from the symbolic heart of the city—the urban hyper-center—to the city's peripheries. Some of these practices have been "sanitized" and "adecentadas" (tidied up) to be a delight to the eyes of international community and triple-A tourists.

Natives' foreign status is also reflected in the fact that their lives are more exposed and vulnerable than those of the U.S. citizens who reside in the city. Locals' lives are considered superfluous; that is, their deaths do not have the same public, political, media, and economic impact as those of the U.S. migrants. If violent deaths in the city were investigated, the border that defines who is secure and who lacks protection would become clearer.

To be specific, Guanajuato remained the most violent Mexican state in 2019, registering a rate of 61 homicides per 100,000 inhabitants. San Miguel's municipality exceeded the Mexican state homicide average because it registered a rate of 64 homicides per 100,000 inhabitants (INEGI 2020). Of the total homicides registered in this municipality in the last thirty years, only nine have been foreigners.

The Mexicans and San Miguel's people are not considered to have lives that lack value, but rather lives that do not have the same value as others. Ximena, a middle-aged resident from San Miguel, explains this problem as follows: "They [the U.S. residents] are extremely supported by the Mexican government. They [the government] pay more attention to them in relation to security. I have noted this because my husband's family still lives downtown. They live in Guadiana [neighborhood] surrounded by foreigners. The police routinely visit them several times, so they are better protected. Maybe they pay more attention to them than to us, who live here, because they are foreigners" (author interview, August 2019). Ximena's story allows us to illustrate how a border of belonging and exclusion is drawn, a border in terms of (in) security and (lack of) protection. Additionally, it clearly shows who is the "other" in the city.

Violence has physically and symbolically rearranged the distance between U.S. citizens and Mexicans in the city. At a local level, "crime speech" (Caldeira 2007, 53)—in which Mexicans appear as authors, accomplices, accessories, and victims—makes fear circulate, legitimizes the construction of physical borders, and stimulates a model of segregation in which residential fortification and fencing proliferate (Caldeira 2007, 28).

Lastly, another mechanism of foreignization of the local population—and an example of magic realism with Kafkaesque airs—was the implementation of a municipal project that, precisely, transformed the local residents of San Miguel into tourists in their own city. As a municipal official who, in 2019, presented this program in front of dozens of inhabitants of the stigmatized San Rafael neighborhood stated: "Would you like feeling like a tourist in San Miguel de Allende? Would you like to be taken for a walk? If so, you barely need to register. They [municipal officers] will pick you up, and they will bring you back. Just imagine! Seeing San Miguel without paying for the local bus, sitting down and enjoying yourselves."

The political and institutional practice of transforming the local population into tourists constitutes the consummation of a progressive dynamic, namely the museification of the city. Simultaneously, this practice also makes visible the production of "an impossibility of using, of inhabiting, of experiencing" (Agamben 2017, 110), a possibility reserved only for those with substantial incomes. This process of foreignization reveals the existence of an exclusionary-inclusive dialectic through which locals are engaged with the community via their production as temporary visitors who do not possess the capacity to intervene. Put differently, they are deprived of any possibility of desecration.

In their relevance or apparent banality, the foreignization of the natives reveals the process of externalization of the U.S. border at a micro-scale level. This urban border also makes evident the mutation of the conditions of sociability that gave life to the city and the socio-spatial configuration in which "becoming-other" is "becoming-native."

## 20.5 Conclusions

In this chapter, I have described a novel form of externalization of the U.S. border that unfolded in the twentieth and twenty-first centuries. This form of externalization has not only been symptomatically manifested through the U.S. "North–South" migration, but it has also been produced through molar and molecular power structures. This externalization was possible with the support of the central government and the implementation of interstate policies like the GI Bill. It has also operated through diverse mechanisms and techniques. For instance, the construction of a racially structured type of urban gentrification of the city, the production of Occidentalist geographical imaginaries, the fabrication of transnational lifestyles targeting impoverished U.S. populations, and the deployment of various everyday border practices. All of them have made possible the "foreignization of the natives."

The study of the border externalization and the transnational relocation of "North–South" migration—with the inequalities and systemic privileges that they both produce—made it possible to discuss the idea of the assumed existence of a necessary correspondence between migration and segregation that some scholars take for granted. In fact, the paradigmatic case of San Miguel illustrates that the intersection of diverse forms of power, such as race, class, mobility capability, and nationality, implemented within historically specific contexts—that produced the "desertification" and "revitalization" of the city—and influenced by the existence of an uneven form of geographic development (between the United States and Mexico), has caused U.S. migration to Latin American cities to turn over the traditional relationship in which "newcomers" are the marginalized ones.

To conclude, it is necessary to address that the destructive production of physical, symbolic, and relational spaces is a constitutive and constituent process of the externalization of the U.S. border, the foreignization of the natives, and the asymmetrical restructuring of urban relations. This process demonstrates the renewed persistence of urban colonialism in Latin America, which functions through a simultaneously productive and destructive logic. It is precisely through this logic that the perpetuation of current material, spatial, symbolic, and day-to-day relations of domination and subordination is guaranteed.

## References

Agamben G (2017) Profanaciones. Adriana Hidalgo Editora, Barcelona
Atkinson R, Bridge G (ed) (2004) Gentrification in a global context: the new urban colonialism. Routledge
Augustine-Adams K (2015) Hacer a México: la nacionalidad, los chinos y el censo de población de 1930. In: Yankelevich P (ed) Inmigración y racismo: contribución a la historia de los extranjeros en México. COLMEX, México, pp 155–194
Bellavance L (2018) Assets. In: Cerda M (ed) México hoy. Editorial La Zonámbula, México, pp 18–19
Benjamin W (2018) Iluminaciones. Taurus, Bogotá
Berman M (2012) Las raíces del fracaso americano. Sexto Piso, Barcelona
Blackburn R (2010) El futuro del sistema de pensiones. Crisis financiera y Estado de Bienestar. Akal, Madrid
Borsdorf A (2003) Cómo modelar el desarrollo y la dinámica de la ciudad latinoamericana. Revista EURE 86:37–49
Bourgois P [2003] (2010) En busca de respeto. Vendiendo crack en Harlem. Siglo XXI Editores, Buenos Aires
Bowman H, Dickinson S (1935) Mexican odyssey. Willet, Clark & Company, Chicago & New York
Boy M, Perelman M (ed) (2017) Fronteras en la ciudad. (re)producción de desigualdades y conflictos urbanos. Teseo, Buenos Aires
Caldeira T [2000] (2007) Ciudad de muros. Gedisa, Barcelona
Carman M, Vieira da Cunha N, Segura R (ed) (2013) Segregación y diferencia en la ciudad. FLACSO Ecuador, CLACSO & Ministerio de Desarrollo Urbano y Vivienda de Ecuador, Quito
Croucher S (2009) Migrants of privilege: the political transnationalism of Americans in Mexico. Identities: Glob Stud Culture Power 16:463–491

Croucher S (2010) The other side of the fence. American migrants in Mexico. University of Texas Press, Austin
De Genova N (2002) Migrant "Illegality" and deportability in everyday life. Annu Rev Anthropol 31(2002):419–447
Díaz I (2015) Introducción. Perspectivas del estudio de la gentrificación en América Latina. In: Delgadillo V, Díaz I, Salinas L (eds) Perspectivas del estudio de la gentrificación en México y América Latina. UNAM, México
Düvell F (2003) The globalisation of migration control. OpenDemocracy. http://bit.ly/2bDM10T
El Universal (2017) Oferta Inmobiliaria en San Miguel de Allende. México
Elias N, Scotson J [1965] (2016) Establecidos y marginados. Una investigación sociológica sobre problemas comunitarios. Fondo de Cultura Económica, México
Fabian J [1983] (2014) Time and the other: how anthropology makes its object. Columbia University Press, New York
Flores M, Guerra P (2016) Entre lo local y lo foráneo: Gentrificación y discriminación en San Miguel de Allende, Guanajuato. Revista Legislativa de Estudios Sociales y de Opinión Pública 9(18):183–206
Florida R (2010) La clase creativa. La transformación de la cultura del trabajo y el ocio en el siglo XXI. Paidós, Barcelona
Franz C, Havens L (2006) The people´s guide to Mexico. Avalon Travel, USA
Gleizer D (2015) Los límites de la nación. Naturalización y exclusión en el México posrevolucionario. In: Gleizer D, Caballero PL (eds) Nación y alteridad. Mestizos, indígenas y extranjeros en el proceso de formación nacional. Universidad Autónoma Metropolitana.& Ediciones Educación y Cultura, México
Gordillo G (2018) Los escombros del progreso. Ciudades perdidas, estaciones abandonadas y deforestación sojera en el norte argentino. Siglo XXI Editores, Buenos Aires
Hall S (2013) El trabajo de la representación. In: Restrepo E, Walsh C, Vich V (eds) Stuart Hall. Sin garantías. Trayectorias y problemáticas en estudios culturales. Corporación Editora Nacional, IEP, Instituto Pensar, UASB, Quito, pp 459–496
Harvey D (2007) Notes towards a theory of uneven geographical development. In: Harvey D (ed) Spaces of global capitalism. Verso, UK/USA, pp 69–116
Harvey D [1981] (2014a) La solución espacial: Hegel, Von Thünen y Marx. In: Harvey D (ed) Los espacios del capital. Hacia una geografía crítica. Akal, Madrid
Harvey D [2001] (2014b) El arte de la renta: la globalización y la mercantilización de la cultura. In: Harvey D (ed) Los espacios del capital. Hacia una geografía crítica. Akal, Madrid
Hayes M (2013) Una nueva migración económica: el arbitraje geográfico de los jubilados estadounidenses hacia los países andinos. Andina Migrante 15:2–13
Hayes M (2017) North-South migrations and the asymmetric expulsions of late capitalism: global inequality, arbitrage, and new dynamics of North-South transnationalism. Migrat Stud 5:116–135. https://doi.org/10.1093/migration/mnw030
Hayes M (2020) Gringolandia. Migración norte-sur y desigualdad global. Abya-Yala, Quito
Hiernaux D (2007) Los imaginarios urbanos: de la teoría y los aterrizajes en los estudios urbanos. Revista EURE 99:17–30
INEGI-Instituto Nacional de Estadística y Geografía (2011, 2020) General census of population and housing of Mexico
INEGI-Instituto Nacional de Estadística y Geografía (2020) Deaths by Homicide in San Miguel de Allende 1990–2019
Janoschka M (2002) El nuevo modelo de la ciudad latinoamericana: fragmentación y privatización. Revista EURE 85:11–29
Janoschka M (2011) Imaginarios del turismo residencial en Costa Rica. Negociaciones de pertenencia y apropiación simbólica de espacios y lugares: una relación conflictiva. In: Mazón T, Huete R, Mantecón A (eds) Construir una nueva vida. Los espacios del turismo y la migración residencial. Editorial Milrazones, Barcelona

Janoschka M (2013) Nuevas geografías migratorias en América Latina: prácticas de ciudadanía en un destino de turismo residencial. Scripta Nova, vol 439

Kahn A, Norman J (1963) Mexican hill town. Fisher-Edwards Book Publishers, California

Kingman E (2006) La ciudad y los otros. Quito 1860–1940: higienismo, ornato y policía. FLACSO Ecuador & Universitat Rovira i Virgili, Quito

Korpela M (2010) A postcolonial imagination? Westerners searching for authenticity in India. J Ethn Migr Stud 36:1299–1315

Life (1948) GI Paradise. Veterans go to Mexico to study art, live cheaply and have a good time. January 5, pp 56–59

Lomnitz C (2011) Los orígenes de nuestra supuesta homogeneidad: breve arqueología de la unidad nacional en México. In: Grimson A (ed) Antropología ahora. Debates sobre la alteridad. Siglo XXI Editores, Buenos Aires

Mettler S (2005) Soldiers to citizens. The G.I. bill and the making of the greatest generation. Oxford University Press

Monkkonen P (2012) La segregación residencial en el México urbano: niveles y patrones. Revista EURE 38(114):125–146

Norman J (1972) Terry´s guide to Mexico. Doubleday & Company Inc., New York

OIM's Global Migration Data Analysis Centre (2020) Global migration trends. Factsheet

Pinley L (2015) The GI bill abroad: a postwar experiment in international relations. Dipl Hist 1–24. https://doi.org/10.1093/dh/dhu074

Pinley L (2017) San Miguel de Allende: Mexicans, foreigners, and the making of a world heritage site. University of Nebraska Press, Lincoln

Pratt ML (2010) Ojos Imperiales. Literatura de viajes y transculturación. Fondo de Cultura Económica, México

Rajchenberg E, Heau-Lambert C (2008) Para una sociología histórica de los espacios periféricos de la nación en América Latina. Antípoda 7:175–196

Rudolph K (2017) Voces de San Miguel. Una historia oral. Personal Edition, San Miguel de Allende

Sabatini F (2006) La segregación social del espacio en las ciudades de América Latina. BID

Santillán A (2019) La construcción imaginaria del Sur de Quito. FLACSO, Quito

Sassen S (2006) La formación de las migraciones internacionales: implicaciones políticas. Revista Internacional De Filosofía Política 27:19–39

Sassen S (2015) Expulsiones. Brutalidad y complejidad en la economía global. Katz Editores, Buenos Aires

Segura R (2012) Elementos para una crítica de la noción de segregación residencial socio-económica: desigualdades, desplazamientos e interacciones en la periferia de La Plata. Quid 16(2):106–132

Sheller M, Urry J [2006] (2018) Movilizando el nuevo paradigma de las movilidades. Quid 16, N 10—Dic 2018–Mayo 2018, pp 333–355

Sloane P, Zimmerman S (2019) Report of study of international retirement migration from North America to historic colonial cities in Latin America. Report for Participants

Sontag S (2007) Cuestión de énfasis. Alfaguara, Colombia

Staszak JF (2012) La construcción del imaginario occidental del "allá" y la fabricación de las "exótica". In: Lindon A, Hiernaux D (eds) Geografías de lo imaginario. Anthropos & Universidad Autónoma Metropolitana Iztapalapa, México

Stefoni C (2018) Panorama de la migración internacional en América del Sur. CEPAL

Turner FJ (1987) El significado de la frontera en la historia americana. Secuencia 7:187–207

Varela A (2015) La "securitización" de la gubernamentalidad migratoria mediante la "externalización" de las fronteras estadounidenses a Mesoamérica. Contemporánea, vol 4. https://revistas.inah.gob.mx/index.php/contemporanea/article/view/6270/7104

Vasconcelos J [1925] (2003) La Raza Cósmica. Misión de la raza iberoamericana. Paginadura Ediciones, Chile

Veblen T (2004) Teoría de la clase ociosa. Alianza Editorial, España

Wacquant L [1994] (2001) Parias urbanos. Marginalidad en la ciudad a comienzos de milenio. Manantial, Buenos Aires

Wacquant L [2006] (2013) Los condenados de la ciudad. Gueto, periferias y estado. Siglo XXI Editores, Buenos Aires

Ward P (2012) Segregación residencial: la importancia de las escalas y de los procesos informales del mercado. Quid 16(2):72–105

Yankelevich P (2015) Judeofobia y revolución en México. In: Yankelevich P (ed) Inmigración y racismo: contribución a la historia de los extranjeros en México. COLMEX, México

**Juan Pablo Pinto Vaca** Ph.D. in Cultural Studies from the Autonomous Metropolitan University (Mexico), Master in Anthropology and Specialist in Collective Memories and Human Rights. He has worked as a professor and researcher at different universities in Ecuador and in Mexico. In recent years, his research has focused on the urban inequalities, racism, migratory studies, and studies on violence in urban and border contexts.

# Chapter 21
# Urban Frontiers in the Fracturing City: Heritage, Tourism and Immigration

Jesús M. González-Pérez, Josefina Domínguez-Mujica, Margarita Novo-Malvárez, and Juan M. Parreño-Castellano

**Abstract** Two of the main problems of the current urbanization process are job insecurity and the increase in social inequalities in the city, which, among other factors, are linked to impoverishment and greater vulnerability, segregation and social polarization. Ultimately, these factors are mainly responsible for the configuration of a new spatial order, in which urban boundaries are consolidated. These are of various kinds and are recognizable throughout the history of the city, but those of an intra-urban nature stand out, which, in almost all cases, are delimited on the basis of the concept of social class. In this chapter, we seek to contribute new reflections on the construction of urban borders in the dual city, from three clearly interrelated perspectives: the territorial (heritage as a gentrifying agent), the sectoral (tourism and urban inequality) and the social (migrants and segregation). We conclude by highlighting the transversal role that tourism plays in the drawing of urban boundaries and its capacity to generate processes of inequality. Thus, the city is torn between the intensity of the expansion of two urban blocs, that of gentrification and that of impoverishment, and these two forces are delimited by insurmountable barriers.

**Keywords** Social inequality · Poverty · Urban history · Sightseeing · Immigration · Borders · Town planning

---

J. M. González-Pérez (✉) · M. Novo-Malvárez
University of the Balearic Islands, Palma, Spain
e-mail: jesus.gonzalez@uib.es

M. Novo-Malvárez
e-mail: m.novo@uib.es

J. Domínguez-Mujica · J. M. Parreño-Castellano
University of Las Palmas de Gran Canaria, Las Palmas de Gran Canaria, Spain
e-mail: josefina.dominguezmujica@ulpgc.es

J. M. Parreño-Castellano
e-mail: juan.parreno@ulpgc.es

## 21.1 Introduction

Post-Fordist capitalism has modified the qualities of the city, favoring the urban concentration of capital, work and culture through the transformation of its economic base, with the vertical disintegration of the production process, an intensive use of information and the impact of new services turning the urban space into an object of tourism development. This concentration of power, wealth, knowledge and leisure overlaps and contributes to the dynamics of social exclusion, differentiation and marginalization, i.e., to the social division disposed by the economic income, age or the ethnic and geographic origin of the population, allowing the identification of situations of fracture in the urban fabric.

From a spatial perspective, the images of renovated neighborhoods whose valorization promotes business interests contrast with those affected by situations of abandonment and impoverishment. This can be perceived in all types of urban spaces, but it is especially noticeable in those neighborhoods and enclaves with strong gentrification pressure (especially when we are in the compact city, such as historic centers) or tourism (mainly in sun and beach resorts).

The multiplicity of the actors intervening and the constant renewal in search of benefits modify the urban landscape generating more complex dynamics and altering the specialization of the old neighborhoods. Thus, the urban centers containing the most outstanding heritage values are affected by the dynamics of tourism that, while reifying the cultural legacy, promote the renovation of buildings for tourist use. The intervention of new agents, especially investment funds and the so-called collaborative economy, yields dynamics of expulsion of residents that increase the urban social fracture.

In this context, the objective of our work is to contribute new reflections on the construction of urban borders and the dual city, from three clearly interrelated perspectives: the territorial (heritage as a gentrifying agent), the sectoral (tourism and urban inequality) and the social (migrants and segregation). For this purpose, after placing the work in a theoretical framework on the dual tourist city, we analyze these three processes and urban agents individually. From a methodological point of view, each section combines the predominant theoretical analysis with case studies, individualized in boxes.

## 21.2 The Dual Tourist City: Urban Patches of Gentrification and Impoverishment

The scientific literature studies the dual tourist city mainly from the strong inequalities that occur in sun and beach destinations. However, these imbalances are equally visible within the consolidated city, mainly due to the expansion of the gentrifying stain and the associated segregation processes.

Tourist urbanization tends to respond to an urban practice that is singular, functional and structurally differentiated from the conventional city. The tourist functionalization of space and time (Agarwal 2012) has had the most relevant consequence of creating urban and regional structures with singular characteristics (Donaire et al. 1995; Antón 1998). Early on, Ash and Turner (1976) had already described the less developed riverside and island spaces that were integrated into international economic circuits through tourism development as "pleasure peripheries" of western metropolises. Three decades ago, Mullins (1991) wrote that tourist cities represent a unique form of urbanization because, among other things, they have been built solely for consumption. More recently, Judd (2003) has pointed out that tourist enclaves have become ubiquitous features of cities. For this author, these hubs facilitate the authoritarian control of urban space, modifying consumption and replacing and suppressing local culture with "Disney environments." Therefore, the city and tourism are two realities, among many others, in which the global and local dimensions converge and intertwine. Each city is a node variably connected to global networks of structures and flows of capital, population, goods and information.

The moment at which the process of touristization begins is important when it comes to constructing a specific urban-tourist typology. Although there are important regional and local particularities, the resulting territorial structure has common features in most sun and beach enclaves. The most prominent is the dual city growth model (Marcuse 1989; Mollenkpof and Castells 1992). That is to say, the city is divided in two: on the one hand, the beach strip modified by hotel towers and residential condominiums and, on the other hand, the tourist backwater composed of new residential developments and the genetic city. The latter is intensely transformed, where the remains of old economic activities (fishing, port, commercial…) remain, but are progressively displaced by commerce and services linked to tourism.

In some areas, these cities are creating real urban backbones and built-up continuums that are increasingly distant from the traditional concepts of the city. These "tourist conurbations" are spaces that have become increasingly complex and have expanded their urban functions (Costal del Sol in Spain, Florida in the United States, Australia's Gold Coast, etc.) and where their interpretation is not only due to tourist causes (Knafou 2006). In those more recent developments, the tourist monofunctionality is absolute, with no mix of uses and dominated by the presence of hotel resorts located on the seafront, with direct access to the beach. The hotel chains are practically the same, the all-inclusive model is the only lodging possibility, the architectural typologies of the hotels and their extensive recreational areas are almost identical and the predominance of bunkerized architecture is a valuable reference of these territories (González-Pérez 2012). A McDonaldization of the hotel industry turns them into authentic non-places of sun and beach tourism. The concept of McDonaldization was introduced by George Ritzer (1993). The McDonaldization of society is what happens when society, its institutions and its organizations demonstrate the same characteristics (efficiency, calculability, predictability and control) that are found in the fast food restaurants of McDonalds. In the field of tourism, the culture and true character of a destination are rationalized into an idealized, safe and easy-to-consume vacation package. The thematization of spaces and the absence of local identities are characteristic elements. Territories that lack past and place. However, beyond the urban backbones and the continuous construction of tourist resorts, hidden behind these cities of fantasy, there are extensive urban areas that have been born or have grown under the impulse of tourist activity. In some cases, we face authentic cities of misery.

In Mexico, this dual system has been corroborated in cities with older tourism development (Acapulco, Cancun, Los Cabos) and even in those with more recent take-off (Puerto Peñasco, Rosarito Beach) (Enriquez 2008). In Acapulco, one of the oldest sun and beach destinations in the Americas, but currently suffering from the problems of maturity, the location of each new hotel determined the conformation of the urban layout and human settlements (Valenzuela and Coll-Hurtado 2010). The city has grown by aggregation of tourist spaces, and each urbanized territory can be identified with a stage of tourism. This explains why the tourist city is currently divided into three zones: Acapulco Tradicional, Acapulco Dorado and Acapulco Diamante. Something similar occurs in Varadero (Cuba). The stages of urbanization, tourist development and Urban Development and Land Use Plan of the Varadero Tourist Enclave have zoned and structured the Hicacos Peninsula into tourist sectors, each associated with a stage of tourist urbanization: Old Varadero (Oasis section, Kawama section, Historic Varadero section-1883-, International section, Las Américas section) and New Varadero (Chapelín-Los Taínos section, Hicacos Point section) (González-Pérez et al. 2014).

The result of this urban development process linked to tourism is the construction of an unequal, polarized and segregated city, where together with intensely gentrified spaces we find severely impoverished territories. We are faced with socially fractured and urbanistically fragmented cities. In this model, defensive urban planning is fundamental. Without it, we cannot understand the construction of the urban and tourist territory. The planned walls, which are erected in the residential areas (gated communities) and in the hotel zones, separate the cities of fantasy from the cities of misery.

In the 1970s–1990s, the predominant urban processes and debate in most urban centers in First World countries, mainly those classified as historic, were dominated by degradation (exclusion, vulnerability…) and loss of functionality. Since 2000, the revitalization of these urban centers, in many cases thanks to the support of ambitious urban rehabilitation plans and major speculative investments, has led the issues to evolve into new debates: the impacts produced by urban tourism and gentrification (hotels, holiday housing, cruise ships…). Thus, in a few decades, we have gone from degraded spaces to eliticized spaces. For decades, the degraded neighborhoods of the urban centers or the old cities within the walls were the gateway for many labor immigrants, while the rehabilitated ones became an attractive settlement space for inhabitants of higher economic level, with a succession of processes of shantytownization and gentrification taking place in a reduced space. However, the expansion of the gentrification stain, after an important commitment to urban rehabilitation and a repositioning of historic centers as a priority area for urban tourism, is "expelling" those groups of lower social classes, many of them foreigners from impoverished countries.

Tourist gentrification advances territorially and spatially at the same time that societies and cities increase their inequality and segregation (González-Pérez 2019). Gentrification becomes an urban strategy that takes over from liberal urban policy (Smith 2002). There is a growing tension between the increasing role of tourism in the production of urban space and the increase in social inequality that implies the deterioration of the financial situation of particularly vulnerable neighborhoods and social groups. Any city in almost any country in the world is familiar with this process. Moreover, in the most touristic nations, gentrification is a variable independent of their level of development. This may be seen in some Caribbean islands (González-Pérez 2015, 2017).

Gentrification is a movement of return to the city, but not so much of people as of capital. Gentrification has spread both spatially and by sector and has great capacity for mutation (Smith 2006; Clark et al. 2007). Spatially, it expands, as an urban stain, from the historical centers to any neighborhood susceptible of tourist-real estate exploitation or with some type of attraction for tourist consumption. Sectorally, tourism produces gentrification (Smith 1996; Judd and Fainstein 1999; Wilson and Tallon 2011; González-Pérez 2019). Post-crisis urban restructuring found a fundamental ally in tourism and the rent gap as a necessary mechanism for gentrification. Unlike other types of urban processes, gentrification is visible. It is easily perceived by citizens to the point that it is possible to draw urban boundaries of gentrification with a simple field trip or city tour. And, more than other urban processes, the ordinary citizen suffers from it (Sánchez-Aguilera and González-Pérez, 2021).

In short, one of the main problems of current urbanization processes is wider social inequality in cities, linked, among other factors, to situations of impoverishment and greater vulnerability, segregation and social polarization (Lennert 2010; Harvey 2012; Vale 2014; Koutrolikou 2016; Dorling and Ballas 2018). These factors are primarily responsible for new spatial configurations. From the space-class dialectic, gentrification and impoverishment draw urban borders within our cities and intensify new processes of unequal geographic development, urban borders generated by the tension between the expansion of space that is being gentrified and that which is being impoverished.

**Box 21.1** The city of fantasy and the residual city in Bávaro-Punta Cana (Dominican Republic)

The Bávaro-Punta Cana tourist enclave is one of the most important in the Caribbean, in terms of number of beds and tourist arrivals. Since 1990, the population growth rate in this municipality has exceeded 50,000 per decade. This intense urban growth responds to a disjointed system in the form of a discontinuous, low-density urban sprawl. This is a clear example of a dual tourist city. Urban segregation is the only means for the development of this tourist model. It entails an extensive pre-coastal urbanization in the spaces of social reproduction. The dual city and the favelization of pre-coastal urban spaces, characterized by inequality and social fragmentation, expand under the logic of segregation. Exclusion is the predominant mode of social and spatial organization.

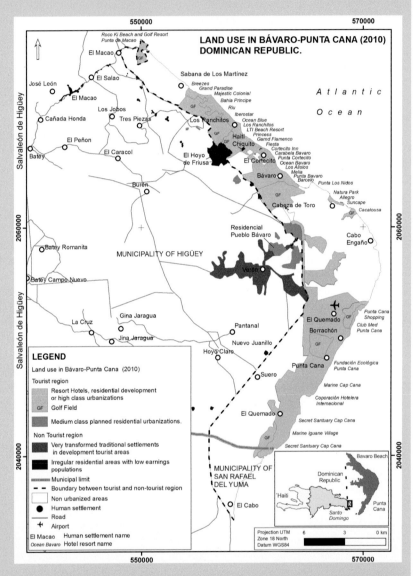

The Tourist Boulevard of the East is the main highway that delimits and encircles the whole tourist territory in Bavaro. However, what is more, this new boulevard separates, like a wall, the tourist land from the hinterland. Within the tourist land, many traditional nuclei that have now been totally transformed are preserved, such as El Cortecito; or small irregular settlements have sprung up, such as Haití Chiquito; but it is in the hinterland where the main spaces of social reproduction are to be found. Although it is highly variable depending on the area, these spread out along a close to 8 km radius from the Boulevard of the East to the West.

Beyond this first dividing line there stretches an extraordinarily complex urban space with a population of nearly 73,000 inhabitants. This manpower is made up of a significant number of interior immigrants, from all over the country (essentially from the provinces of Santo Domingo, El Seibo, San Pedro de Macoris and La Romana), and Haitians. The main town is Verón (a squatter settlement with 80 homes by 1993 and between 6000 and 8000 inhabitants in 2005); Macao is one of the few remaining traditional nuclei to remain; and El Hoyo de Friusa is probably the most dynamic. This diffused system, with blurred boundaries and complex spatial organization, is completed by planned precarious settlements (Nuevo Juanillo, Mata Mosquitos) and a long list of unplanned, irregular ones (El Manantial, Samy, Kosovo, Barrio Nuevo, Cristinita, etc.).

In Bávaro, the tourist space surface is 8620 ha in 2010, over twice the size of the hinterland (3727 ha). Each new room has influenced the growth of urbanization in the hinterland depending on the period considered. It is more significant in 1984–1990, during the period of spread of non-tourist urban sprawl: 16,914 m$^2$ of growth in the hinterland for new room.

*Source* González-Pérez et al. (2016)

## 21.3 Urban Agents and Processes

### *21.3.1 The Territorial. Heritage as a Gentrifying Agent*

Since the celebration of the IV International Congress of Modern Architecture in 1933, which culminated in the formulation of the Athens Charter (1934), heritage conservation has been assumed as part of the planning of cities (Oviedo 2014). This precept will have continuity years later with the approval of the Washington (1987), which warns of the need to "consider historic urban areas in territorial planning, due to the importance of historic values shaped by material and spiritual elements that determine the image of the city" (Washington, 1987, 2). The Krakow (2000) reinforced this idea, considering heritage as a determining factor in the configuration of the city, warning that "its conservation should be an integral part of the planning and management processes of a community, because of its contribution to sustainable, qualitative, economic and social development" (Krakow Charter, 2000,

3). With the arrival of the twenty-first century, we notice how the conservation and uses of heritage have been affected by urban policies oriented to the promotion of tourism. In practice, the gentrification and revitalization strategies that affect many of our cities have been applied in neighborhoods that, in general, have a rich and well-preserved monumental heritage and, consequently, have become places coveted by the wealthier classes and tourists (Romero and Lara 2015).

In view of this, we can affirm that heritage has become a differentiating element and main attraction in many cities, a circumstance that translates into an increase in demand, although it is true that it can also constitute a dangerous gentrifying agent, when restoration interventions end up breaking the traditional dynamism of neighborhoods and expelling or discriminating against the local population (Oviedo 2014). From our point of view, this is not how a historic center should be conceived today. The goal should be to achieve a balanced coexistence with services within everyone's reach, including the enjoyment of heritage, not only for tourists and the wealthy classes, but for all citizens. Therefore, the future challenges for our cities require post-gentrification and reconversion urban strategies, based on the authenticity of everyday life, prestige, status, but also on the uniqueness conferred by their heritage.

The diversity of positive and negative impacts generated by tourism in historic centers and the implications on heritage has been addressed by different authors at the European level (Ingallina 1994; Troitiño 1998; De la Calle and García 2016) given the multiple examples of cities with great heritage wealth (Madrid, Barcelona, Venice, Amsterdam or Lisbon) that have been affected by urban transformation processes. In Spain, de la Calle (2002, 2008) and Troitiño (1998, 2003) analyzed tourist dynamics and the influence they have on the heritage of historic centers, also recognizing the importance of cultural assets in the territorial and strategic planning of our cities. In Latin America, this issue has also been widely addressed by different authors (Cordero and Meneses 2015; Delgadillo 2015; Hiernaux and González 2016; Roldán 2017; González-Pérez et al. 2022; Carrión 2005, 2014) in relation to enclaves where the presence of a rich monumental heritage and a strong influx of tourists coincide, which, on occasions, has put them on the verge of collapse. These authors confirm the idea that tourism has taken a very relevant role in the rehabilitation of the central spaces of heritage character of the most monumental cities or of some of their most emblematic neighborhoods, as perceived in San Francisco, New Orleans, Quito, Lima or Bogotá, and in neighborhoods such as Getsemaní in Cartagena de Indias, Colonia Condesa in Mexico D.F or Jardim in São Paulo, among others, which have seen in tourist activity their main development mechanism (Roldán 2017, 80). However, in some of these, the increase of tourism has negatively impacted a loss of public space function.

Carrión (2014) and David Navarrete (2017) also refer to the heritage affectation in Latin American cities where these dynamics first appeared (Quito, Guanajuato, Cartagena de Indias…), since they note the usual process of renovation of historic buildings that take on new life to serve new needs and that, in return, change their meaning, are decontextualized or lose their identity. On many occasions, they have been the object of large-scale interventions to adapt them to a new use. Carrión also states that one of the greatest dangers of the application of these policies is the implementation of monumentalist measures that privilege the physical heritage, but detract from its authenticity and distort its true essence (Carrión 2007), as has happened in some American cities, especially those that have been supported by local heritage designations or through UNESCO.

A representative example of buildings that have transformed their original function to serve tourism can be found in boutique hotels. The concept first emerged in America (New York in 1980) and, from there, the model was exported to historic European cities. As a whole, these establishments are characterized by luxury, architectural uniqueness and the richness of the properties that serve as their infrastructure, usually located in a unique heritage environment. Their main clients are people of high cultural level and high economic capacity who seek personalized attention in a quality environment. Almost all of these hotels are included in a four- or five-star category. For the most part, they occupy old monumental buildings that adapt their old factories to the new hotel use and whose heritage recovery is being faced only as a business, adapted to the consumption and enjoyment of the visitor. This hotel typology allows us to clearly visualize the relationship between gentrification by tourism and/or touristification and the architectural mutation of heritage. A link that is not accidental, since the recovery of monumental heritage has been a strategic objective of urban policy and planning to boost tourism.

This type of hotel is usually located in neighborhoods that have undergone major heritage intervention, and where we also find gentrified establishments that have emerged as a result of the synergy caused by them. In some cases, such as in Barcelona, these new businesses coexist with premises closed due to price increases and with fashionable restaurants, nightclubs and cultural and consumer spaces. Hernández (2015) has made a classification that characterizes the commercial offer of the gentrified neighborhoods of this city that is exportable to others, while warning that both boutique hotels and the establishments located in their surroundings refer us to a new urban process derived from gentrification known as boutiqueization, which is directly related to the implementation of heritage rehabilitation policies.

Boutiqueization in Latin American cities has been the subject of study by Fernando Carrión, who defines it as the process of transformation in land use that leads to the emergence of new businesses that have arisen to cover prestigious activities that are more profitable than residential ones (Carrión 2014). For his part, David Navarrete warns that the excessive proliferation of this type of establishment refers us to the excessive share that tourism occupies in urban activities (Navarrete 2017). In Spain, this phenomenon has not yet been sufficiently studied from the academic field and those authors who have approached this subject have focused their analysis almost exclusively on the case of Barcelona (Romero and Lara 2015).

However, processes such as those described above are setting the pace of the times in the heritage areas of Spanish cities, which is contributing to a new social fracture. In contrast to the aging and decadent spaces of the urban heart of the past, tourist areas are being consolidated, in which the heritage is being artificialized, while at the same time losing their neighborhood identity.

**Box 21.2** Expansion and problems of boutique hotels in Palma (Mallorca, Spain)

The tourist centrality achieved by the historic city of Palma since the beginning of the twenty-first century and the expansive strategies of tourism-hotel promotion that have contemplated actions to revitalize the heritage have led to the appearance of *boutique* hotels in the most emblematic neighborhoods of the historic center, although there is a higher concentration in the *Sa Llotja-Born* statistical area and sustained growth in *La Calatrava* and *La Gerreria* (Novo-Malvárez 2019a). Practically, all of these hotels are located in old heritage buildings (palaces, stately homes, convents…) that have a degree of protection as catalogued assets or assets of cultural interest. Nowadays, they constitute an accommodation option with a high demand compared to the traditional high category hotels belonging to the big Majorcan chains. Luxury, personalized attention and the richness of the properties located in a unique heritage environment are the strong points of these establishments whose main clients are foreigners with a high purchasing power and cultural profile, in some cases exceeding 90% of the total (Es Convent de la Missió, Es Princep Hotel, Petit Palace Icon Rosetó or Hotel Brondo). Among their guests, the most common profile is that of couples between forty-five and sixty years of age, who stay an average of three days and whose priorities, in order of importance, are gastronomy, shopping and cultural practices (Novo-Malvárez 2019a).

# 21 Urban Frontiers in the Fracturing City: Heritage, Tourism and Immigration

Hotel *boutique* Brondo and bar of the Hotel boutique Es Convent de la Missió.

The appearance of *boutique* hotels in Palma is circumscribed to the turn of the century, although most of the existing ones have been inaugurated since 2015. According to statistics from the Govern de les Illes Balears (Allotjaments Turístics Mallorca), in 2022 there are thirty-five, thirty-eight if we include the Hotel Princep with a higher number of beds than usual (136), and the Nou Balears and Hotel Hostal Cuba, located in the vicinity of the historic center. All of them have in common their belonging to a four-star, four-star superior, five-star or five-star superior category (Can Bordoy), and two of them (Ca N'a Alexandre and Palacio Can Marqués) are classified in the Inland Tourism subgroup. This category, together with the care devoted to design issues and the standardization of the interventions, gives them a certain similarity, although each one has its own individualities related mainly to the building's past and the scope of the renovations carried out. The actions carried out to convert the old buildings into boutique hotels have been of great magnitude in all cases, altering structural, expressive and meaningful aspects that, on occasions, have threatened their heritage value. Among the existing hotels, there are two that stand out for their uniqueness, having taken advantage of the structure of two religious buildings: the hotel Es Convent de la Missió (2004) and Petit Palace Icon Rosetó (2018), whose former convent function is a differentiating element and represents added value (Novo-Malvárez 2019b).

The increase in the number of boutique hotels in Palma in recent years has created a bubble that the City Council has tried to control through the approval in July 2017 of a municipal moratorium (Novo-Malvárez 2019a). This restrictive measure, which is part of the revision of the General Urban Development Plan and was agreed with the Federation of Neighborhood Associations and the Association of City Hotels, came into force in order to control the situation and

> put limits to the avalanche of new opening applications, whose number in this same year tripled the number of establishments open so far. Since then, twenty-one have been inaugurated, the last three in 2021 occupying buildings declared BIC and with a five-star category (Concepció, Nivia Born Boutique Hotel and Palma Riad). Likewise, the multiplication of hotels of this type entailed the boutiqueization of their closest areas and the intensification of the use of some emblematic spaces of the city (Plaza Mayor, Plaza de Cort, Plaza de Santa Eulàlia, Plaza del Olivar, Colom street, Oms street, La Rambla or Paseo del Born) that were traditionally enjoyed by the people of Palma and that, due to this expansion, became massified spaces of consumption and sightseeing tour (Novo-Malvárez 2019a).

### *21.3.2 The Sectorial. Tourism and the Unequal City*

Tourism is one of the main agents of production and reproduction of global capitalism and, as such, it provokes drastic territorial transformations. Since it became an object of mass consumption starting in the 1960s, tourist activity has created its own cities or transformed pre-existing ones, while its effects have been globalizing with the generation of successive peripheries linked to international tourism (Gormsen 1997; Navarro-Jurado et al. 2015). Depending on the level of development of the tourist activity, the space is organized according to the needs of the tourist system itself, generating new roles, interactions between agents and territorial structures at different scales. The limited presence of the administration as a regulatory and planning agent can give rise to socioeconomic and territorial structures that are not very sustainable or not at all sustainable.

In this way, the tourist development of part of the European third periphery can only be understood, besides by the intervention of foreign investors and by the inadequacy of the planning practices of the public administrations. The example of Maspalomas Costa Canaria in the south of Gran Canaria is representative. In this area, there was a confluence of spatial segregation, scarce presence of the residential function, which will be added from the eighties onwards, and lack of internal articulation (Parreño-Castellano 2001) as a result of the planning deficiencies of the public administration in the territory. The under-presence of the residential function in the different planning impulses in this destination causes it to have to be developed partly in remote spaces, while the anti-urban concept of the destination itself will be a major obstacle in the coming years for its restructuring (Parreño-Castellano and Hernández-Calvento 2020).

The territorial extension of international tourism with the production of the fourth periphery has reproduced this scheme of action since the eighties, but now, in a globalized market in which converge the interests of large construction and lodging companies in the process of internationalization (and, therefore, of the international funds and financial entities that support them) and of undeveloped countries that offer favorable conditions for attracting investors. In this scenario, a monofunctional and segregated space built around tourist activity is generated once again, but now structured through the addition of large lodging establishments that are exploited with a business model based on the all-inclusive. These resorts, located in the territory according only to the logic of tourist production, that is, the accumulation of capital through the tourist business, are erected ignoring the territorial impacts, the migrations that could be triggered and the urban development of the surroundings. The insufficient corporate social commitment of the investing companies and the lack of financial and technical resources of the governments cause the residential needs of the working population to go unmet. As a result, informal and precarious settlements are growing around the destinations. The example of recent urban development on the islands of Sal and Boavista in Cape Verde is paradigmatic, with the aggravating factor introduced by the insular fragmentation of space (Cáceres Morales and Martínez Quintana 2019; Parreño-Castellano and Moreno-Medina 2021).

**Box 21.3** Informal growth in Espargos and tourism development (Sal, Cape Verde)

View of Espargos and transfer of tourism workers to Alto Santo Cruz.

Sal, with an area of just over 216 km$^2$, is the main tourist enclave of Cape Verde. The island, of low altitude and with scarce availability of water, had very little population until the eighties of the last century, despite, since the forties, having the only international airport in the country.

With the approval of the 1990 Constitution and the implementation of a liberalized economic model, Cape Verde's economic policy was based on attracting international investment attracted by favorable conditions of low taxation. In this context, tourism activity became a strategic sector, making possible the sale at low prices or the temporary transfer of land for the construction of tourist resorts.

The island, due to its territorial resources for sun and beach tourism, soon began to attract foreign investors interested in the construction of large lodging establishments that would later be put into operation through agreements with British, Portuguese and Spanish lodging chains. As a result, in 2016, the island received nearly 300,000 international tourists who stayed in hotels of 4 or more stars under the all-inclusive formula around the large beaches of Santa Maria and Murdeira Bay, in the south and west of Sal

The tourist-construction development of Sal necessarily implied the arrival of workers from outside the island, so at the same time the number of Cape Verdeans from the rest of the country and foreigners, mainly Europeans and Senegalese, attracted by the job opportunities offered by construction, tourism and the supply of goods and services to hotels, began to grow. As a result, the island, which in 1980 had 5,826 inhabitants, reached an estimated population of 40,000 in 2020.

The economic development of Sal and other areas of the country made it possible for Cape Verde's gross domestic product to grow to the point that the World Bank classified it as a middle-income country in 2007. However, this macroeconomic improvement was accompanied by an intense, disorderly and spontaneous growth of its main urban centers. Espargos, the island's capital, is the most representative case. Around the traditional nucleus (Preguiça), the city has grown rapidly since the 1990s, without the development of an adequate urbanization model and a sufficient housing and service provision policy. The lack of housing led to an increase in prices, which pushed a large part of the working population into low-quality self-construction, substandard housing or overcrowding. The result has been the creation of a suburbanized, self-built and partially slum-like space in which part of the population has no access to basic infrastructures such as sanitation, waste collection, water, electricity, public lighting, roads or road access. All this is an expression of the generation of an impoverished social fabric, unstructured and unsuited to urban life, caused by the tourist growth of the island (UN-Habitat, 2013).

The spontaneous and informal neighborhoods of shantytowns and self-built housing in Alto Santo Cruz, Alto São João, Alto de Tanque or Terra Boa are some of the marginal settlements generated by Sal's tourism model, enclaves inhabited by tourist workers and where there are worrying crime rates and serious environmental problems. The fact that the workers are moved by their employers to their homes from their places of employment is a sign of the precarious housing conditions in which this population lives.

At the same time that international tourist consumption generates coastal tourist resorts around which precarious and impoverished urban settlements emerge or grow in the fourth tourist periphery, digital capitalism has allowed tourist consumption of the city or of residential areas that were not planned for this purpose to increase in the last decade to unsustainable limits (Cocola-Gant et al. 2020). This spread of tourism to urban spaces, attracted by urban life itself or its cultural resources, is supported by new accommodation formulas, such as short-term tourist rentals and new hotel business models, or the development of other consumption modalities such as cruise tourism or residential tourism and are an expression of the new accumulation mechanisms of digital capitalism.

Touristification is the process by which tourist activity transforms cities to turn them into objects of mass tourism consumption and implies a change in the use of the city's real estate and public space, and three different situations can be recognized. First, an increase in activities directly related to tourism, with a proliferation of lodging establishments and other complementary offerings, both with traditional forms of management and with new ones within the framework of the digital economy. Secondly, an increase in other uses related to urban life that adapt to the massive presence of tourists. The most relevant example is the transformation of retail trade. Finally, a progressive increase in the occupation and economic use of public space by tourist activities.

The touristification of the city has been increasing as society has advanced in the interconnection and dissemination of information and tourism consumption has become a central element in the model of leisure and social relations. However, it is the economic strategies of governments and tourism agents that are the direct causes. For example, in the Spanish context and in relation to public intervention, the 2008 crisis and the measures taken for recovery increased the presence of this activity in many cities as it became one of the main sectors of the economic strategy. Similarly, the intervention of municipal institutions, with planning practices such as the pedestrianization of public space and the subsequent granting of licenses for use, has been a determining element.

The final result of touristification is the replacement of the resident by the tourist and of residential use by other uses related to the tourist production and consumption model. In this sense, touristification, especially when linked to overtourism, implies the death of the city as a space that allows the production and reproduction of the local population and involves the emergence of an urban-looking space at the service of tourist production and consumption (Pinkster and Boterman 2017), as has been detected in many cities in the first three tourist peripheries.

Social substitution occurs through the activation of gentrification processes through which the population of an area or neighborhood is displaced and replaced by another with greater purchasing power (Lees et al. 2016). This is caused by the increase in the sale and rental prices of real estate, and motivated by the growing attraction that some areas have in the housing market. However, gentrification should not be interpreted only as a process of revaluation of real estate in the city, but should be reformulated taking into account the strategies of the agents involved in the supply and the relationships it maintains with the life cycle of the neighborhoods (growth, abandonment or decline and revitalization) (Smith 1979; Domínguez et al. 2020). The factors that may determine the interest of tourist activity in space usually coincide with that which originates medium and high standing residential demand, so that touristification and gentrification appear in the same areas. Moreover, both processes are mutually reinforcing (Maitland and Newman 2008) in terms of the expulsion of less favored social groups.

In recent years, with the development of digital real estate marketing formulas, the internationalization of housing markets and the consolidation of certain international economic spaces, some transnational residential mobilities such as lifestyle migrants, digital nomads, remote workers, super-rich and wealthy people seeking to obtain visas, among others, are favoring the growth of gentrification processes in tourist cities. We are, therefore, facing international processes of gentrification or transnational gentrification that are especially important in areas and cities that are destinations for international tourist flows (Sales 2019).

Touristification also entails the tourist gentrification of the city. The gentrifying capacity of tourism has been analyzed by different authors (Gotham 2005; Lee 2016) and, in economic terms, is based on the better cost–benefit ratio that allows tourist activity to pay higher real estate rents, causing a progressive change of use, leading to a substitution of residents for tourists (Logan and Molotch 2007). This process, which had already been occurring in recent decades in some cities, is further intensified in the current context of digital and collaborative economy, to the extent that formulas of accommodation commercialization appear (holiday housing or housing for tourist use) that are capable of replacing residential use with very little capital intervention (Cocola-Gant and Gago 2019; Schäfer and Braun 2016; Wachsmuth and Weisler 2018). Both due to the growth of regulated lodging establishments or vacation housing and due to the general increase in prices, residents, especially those with lower purchasing power, are subject to direct displacement (Marcuse 1985). In a complementary manner, the increase in prices and the decrease in the supply of housing mean that progressively larger segments of the population cannot afford to buy or rent housing in areas undergoing a process of touristification, which can be interpreted as a process of exclusionary dislocation.

Finally, tourist gentrification is also produced by displacement pressure or displacement due to a change in the community environment or in the neighborhood itself. This may involve different situations such as discomfort due to the presence of tourists in residential buildings, feelings of insecurity, loss of supportive and trusting community environments; transformation in the nature and quality of private services; growth in the number of consumer service establishments for tourists and the disappearance of others linked to the resident population; reduction of public services for residents; transformation and occupation of public space by tourists and economic activities aimed at tourists, saturation of public transport or generalization of other forms of mobility aimed at tourists; disturbances in daily life, increased difficulties for residents to use public space, disappearance of elements of identity, decline in social life among residents, etc. In synthesis, all these situations produce an impairment of the right to the use of community and public space by the resident in conditions of tranquility and perceived security and a greater limitation for the development of community life.

In short, touristification, accompanied by the different processes of gentrification mentioned above, must be interpreted as a new process of appropriation of the city, especially present in the first three tourist peripheries, caused by the development of the recent dynamics of post-Fordist accumulation. It is, therefore, a more refined and flexible instrument of transformation of the city and the territory to the productive needs of tourism and, as in the case of the creation of the new global peripheries, it generates the loss of the city for the local population and the impoverishment of a part of its residents.

### *21.3.3 Urban Borders and Immigration*

The diversity of geographic origins has been a constant in urban history, as cities have been preferred areas of attraction for populations of a very varied nature (Vaughan and Arbaci 2011). The urban kaleidoscope shows these signs of plurality of origins through processes of socio-territorial differentiation or segregation, both of immigrants and of ethnic communities. This is a fact which is intrinsically related to the spatial expression of hierarchy and socioeconomic inequalities, which appeal to the division of social classes, turning the study of segregation into a hot topic in geographic science (Musterd 2020).

Most of the world's cities recognize the presence of ethnic minorities in underserved areas with poorer housing conditions and lower-income populations. "As the incomes of immigrants are, on average, lower compared with natives, their neighborhood choice is restricted by various constraints such as their lower purchasing power on the housing market and clustered location of affordable housing" (Tammaru et al. 2020, 452). It is also possible to identify some well-endowed neighborhoods, where high-income ethnic groups, who run or manage transnational organizations and companies, reside. In both cases, ethnic/geographic origin is a decisive factor in the intra-urban structure and reflects the imbalances inherent in the logic of capital.

### 21.3.3.1 Studies of Social Differentiation of Urban Space: Fordism, the City and Residential Segregation

Concern for the spatial distribution of social groups in cities dates back to the beginning of the twentieth century, increasing after World War II, when authors such as Shevky and Bell (1955) introduced ethnic status and foreign origin as factors of social differentiation in their studies. These works were joined by the first maps showing the location of ethnic minorities, especially in North American counties and cities, as evidenced by the map attributed to Samuel Fitzsimmons (1956).

As international immigration flows were activated, the model was validated in Western Europe and later in Southern Europe (Malheiros 2002; Domínguez-Mujica et al. 2010). In his work "The Urban Mosaic," Timms (1975) argued that the variation in the social characteristics of populations in different parts of the city may be summarized in terms of social rank, style of life preferences and ethnicity. Specifically, the labor and economic status of immigrants (Berry and Kasarda 1977); their cultural characteristics (Parreño-Castellano and Domínguez-Mujica 2008); and the economic, cognitive, social and political resources of their households were identified as factors of segregation. As a result, those with lower incomes and greater legal fragility ended up living in obsolescent historic neighborhoods (Peach 1975). On the other hand, the strengthening of the housing market in suburban areas favored the concentration of foreigners with high purchasing power in them. Finally, a fourth line of analysis focused on assessing the impact of the market and of housing and welfare state policies (King and Mieszkowski 1973).

### 21.3.3.2 The Consolidation of Socio-urban Inequalities and Ethno-spatial Segregation in Times of Globalization

At the turn of the century, Walks (2001, 407) expressed the importance of the urban transformations favored by globalization with these words "the internationalisation of the world economy, fuelled by the growth of trade, the development of information technology and the increasing mobility of capital and labour, has spurred the fragmentation, dispersal, and reorganisation of productive activities." These trends in the productive system, in which the mobility of capital and labor emerges, were accompanied by an intra-urban reorganization that increased the levels of inequality and social polarization in the conformation of what Saskia Sassen called the dual or divided city (1994, 2001).

Consequently, studies of residential differentiation in relation to the population of foreign origin multiplied, extending to the entire globe, while different models were established according to regions with a differentiated migratory history, which have in common the influence exerted on them by the globalization process (Van Kempen 2007). The thesis of the divided city harmonizes with the principles of a multidimensional logic in which imbalances increase and social cohesion dissolves through processes of gentrification and inequality, because the inequalities of income and wealth and the social polarization and residential segregation do not disappear in the late-capitalist metropolises.

A good example is the impact of the international financial crisis that began in the United States in 2007, as a result of the bank's policy of issuing real estate bonds that offered high profits and low risk through deregulation mechanisms. To keep the capital flow of these bonds constant, mortgage loans were granted in bulk, the so-called subprime mortgages, but when investors demanded payment of these bonds, interest rates rose and the banks were unable to respond to this demand. Consequently, a liquidity crisis, a credit crisis and an employment crisis ensued (Domínguez-Mujica 2021), multiplying the processes of housing dispossession. These affected lower-income segments of the population, including immigrants and ethnic communities.

At the same time, within a few years, the market revived and new opportunities arose for those "financialized" players seeking to profit from falling property prices. Many private equity funds became Real Estate Investment Trusts (REITs) and listed real estate companies that absorbed rental housing portfolios 2.0 (Aalbers et al. 2021). Through their selective real estate acquisition policy, they contributed to the transformation of the urban fabric, favoring the gentrification of certain neighborhoods from which migrants were displaced (Lisbon, Paris, Cape Town, Amsterdam, Athens, New York, Las Palmas de Gran Canaria, etc.)

In the case of the latter city, where tourism has contributed, since the mid-twentieth century, to productive diversification and urban development, the neighborhood where the supply of real estate for tourist use is concentrated, Isleta-Puerto-Guanarteme, has been the most affected by gentrification processes. Thus, since the middle of the second decade of the twenty-first century, it has seen a decrease in the number of low-skilled labor immigrants, mostly from Latin America and Africa, as opposed to an increase in lifestyle immigrants, digital nomads and other migrants with greater purchasing power, mostly from Western Europe. Recent investments in the modernization of accommodation and the acquisition of rental housing for new tourist uses by private capital funds have contributed to this. The consequent revaluation of real estate and the increase in rental prices have favored the expulsion of residents with lower purchasing power, especially the non-EU foreign population, which has been affected, in many cases, by eviction processes.

In other sociopolitical contexts (Mardin in Turkey, Manchester in New Hampshire and Halle/Saale in Saxony-Anhalt), Çaglar and Glick Schiller (2018) have confirmed these processes of dispossession and migrant displacement, highlighting that people who have migrated either within or across borders also face another cycle of displacement and insecurity due to the urban regeneration brought about by neoliberal measures.

**Box 21.4** The arrival of EU-immigrants versus the displacement of African and Latin American immigrants in the tourism district 3 (Las Palmas de Gran Canaria-Spain)

In Las Palmas de Gran Canaria, the main city of the Canary Islands (Spain), "the economic crisis that began in 2009 brought an abrupt halt to real estate renovation and refurbishment of the neighborhood of Isleta-Puerto-Guanarteme, its most important tourist district. The crisis had a special impact on the purchasing capacity of the lower and middle social classes, especially non-EU migrants, who were unable to pay mortgages and rents" (Domínguez-Mujica et al. 2020, 248).

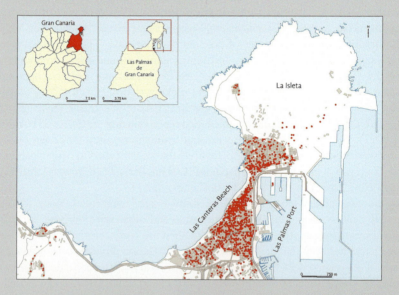

Location of foreclosures and evictions for non-paid rents District 3 Isleta-Puerto-Guanarteme.

Since 2014, a new and intense process of tourism gentrification in this area can be witnessed. With the beginning of the economic recovery, the consequent expansion of the local demand as well as the growth of the holiday rentals caused an increase in the arrival of new foreigners from European developed countries, such as Italians and Germans. Consequently, the neighborhood became a consolidated destination for international lifestyle migrants, including the traditional pensioners, and new ones, such as the digital nomads.

> However, the tourism reactivation kept displacing some immigrants with lower-income level (labor immigrants), as demonstrates the increase in foreclosures and evictions affecting these population groups. Therefore, the district 3 Isleta-Puerto-Guanarteme has been the most affected by eviction processes in the entire city.
> *Source* J. Domínguez Mujica y J.M. Parreño Castellano

### 21.3.3.3 The Effects of Ethno-Spatial Segregation During the Pandemic

Although it is still difficult to assess the impact of the pandemic on urban structures, research mostly conducted in the United States has shown that it has deepened intra-urban inequalities as the ethnic minorities have been disproportionately impacted because of social factors and residential segregation (Yearby and Mohapatra 2020).

Thus, in the Bronx, in New York City, in the first wave of the pandemic the highest rate of COVID-19 diagnoses and deaths among New York City's boroughs were observed, (Ross et al. 2020). This is due to the fact that immigrants in the Bronx—more than half a million—are disproportionately represented in the essential workforce at risk for exposure to contagion, including physicians, nurses, nursing aides, home health aides, subway and bus drivers, grocery clerks and others. A cause that is also argued by Clark et al. (2020) referring to Texas, when they point out that the economic situation of immigrant communities requires continuation of work and because the types of jobs most worked by immigrants often require face-to-face interactions. In addition, the fact that many of these immigrants are living in close quarters with multiple generations sharing bedrooms and bathrooms is also significant, making it more difficult to isolate those who may have been infected.

A situation similar to this has been described by Schanke-Mahl and Sommer (2020), in their study on New York, Seattle, Phoenix and Miami, and by Gil et al. (2021), in Chilean cities. For the first, the pandemic has hit hardest in racially and economically segregated areas with limited economic opportunity, large immigrant communities, overcrowded housing, air pollution and insufficient access to health care. And for the second, the pandemic has exacerbated pre-existing social inequalities, affecting immigrants more than non-immigrants in Chilean informal settlements. In the United States as a whole, the study developed by Islam et al. (2021) concluded that, in the early stages of the pandemic, COVID-19 adversely affected the socially vulnerable and race/ethnicity communities in the United States. In Spain, Madrid and Barcelona are another clear example of this differentiated incidence, since the greatest vulnerability was recorded in the urban peripheries built between 1960 and 1970, where the immigrant population (originally internal immigrants and, recently, foreigners) resides (Instituto Geográfico Nacional 2021: 73).

In addition to these facts, which allow us to deepen our understanding of the concept of the fractured city from the perspective of migrant morbidity, there is also the question of the urban restructuring that may occur in post-pandemic times, since there are no apparent signs of modification in the forces of capital and markets, and changes in the residential preferences of the population will continue to be conditioned by their socioeconomic status, predicting a long life for the borders that delimit the areas in which immigrants and ethnic communities reside.

## 21.4 Conclusions

Throughout history, it is possible to recognize the existence of "urban borders" within cities, a fact that alludes to a process of fragmentation, if not fracture, which can be seen in all the great stages of humanity and in the different civilizations. Boundaries (social, economic, cultural, ethnic…) are recognizable between cities in relation to each other, but also within them.

In recent times, in what we call the post-industrial city, new barriers and boundaries have been built, if possible, within what is called the dual city, sponsored by the dynamics and new agents acting within the framework of the post-Fordist economy and the global financial world. In this chapter, we have attempted to provide new analyses of these urban frontiers in the dual city, based on the dynamics produced by three important actors: heritage, which has been acting as a gentrifying agent; tourism, when it favors processes of socio-urban inequality; and immigration, which is inextricably linked to the consolidation of segregated cities. In all three cases, we defend the hypothesis that tourism is the transversal axis on which the fractured city is sustained and, therefore, favors the strengthening of the frontiers that are recognized within it.

Two types of territory are particularly vulnerable to this process. On the one hand, there is urban fragmentation between the spaces of production (first line of hotel resorts) and social reproduction (residence of workers employed in economic activities linked to leisure), in both the sun and beach tourist enclaves. On the other hand, socio-urban inequalities can be observed in the interior of the consolidated city, fundamentally based on the expansion of revaluation spots and, conversely, on impoverishment, depending on the dynamics of gentrification. Both in the first and in the second case, these are processes of unequal geographic development.

As far as heritage is concerned, its commodification is a sign of today's city. Its revaluation has led it to become the brand image of many cities, which allows it to act as a major tourist attraction. As a result, it produces speculative practices when its value is enhanced or its recovery is promoted. As a factor of gentrification, it can eliminate the traditional dynamism of neighborhoods and expel or discriminate against the local population. In too many cases, in many historic centers, heritage success is a social failure.

For its part, tourism development can be an urbanizing factor in a double sense. On the one hand, with touristification, cities are transformed to the point of becoming the object of mass tourism consumption and, therefore, a mechanism for capital accumulation in the current context of digital capitalism. The change in the use of real estate and public space is a verifiable evidence. At the same time, touristification expels the resident population or undermines their right to the city, as it favors gentrification (displacement of residents in favor of more affluent classes) and tourist gentrification (displacement of residents in favor of tourists), besides being a factor of impoverishment of residents in a general sense and an increase in inequalities. On the other hand, the growth of the tourist offer, due to the internationalization practices of large lodging chains, causes the appearance of informal settlements for the residence of tourist workers, mostly immigrants, in the hinterlands of coastal resorts. This tourist model, typical of today's developing countries, destructures the network of settlements and causes the appearance of urban centers that do not meet the conditions of habitability typical of the city.

Finally, social changes are one of the best indicators for interpreting the city's transformations and evaluating its internal dynamics. Among them, ethnic or geographic origin stands out, which has been, since the consolidation of industrial capitalism, a good reflection of the imbalances inherent in the logic of accumulation. The COVID-19 crisis seems not to have modified the forces of capital and markets. The distribution of residents by social class, intimately related to the economic level of the population, will continue to mark intra-urban inequalities. Public interventionism does not seem to be able to bend the interests of capital nor to break the iron urban boundaries that delimit the eliticized areas from the impoverished ones, the neighborhoods inhabited by high social classes or located in the speculative market (tourist, financial…), from those that are the residence of labor immigrants and ethnic communities. And the intermediate spaces, inhabited by middle classes, lose prominence.

All this leads us to conclude that, increasingly, the city is torn between the intensity of the expansion of two urban stains, that of gentrification, on the one hand, and that of impoverishment, on the other, and how these two forces are delimited by insurmountable barriers.

**Acknowledgements** Grant "Housing and international mobility in cities of the Balearic Islands: the emergence of new forms of urban inequality" (RTI2018-093296-B-C22) funded by MCIN/AEI/ and by "ERDF A way of making Europe." Grant "The post-COVID-19 territorial balance in the Canary Islands. New strategies for new times" funded by the CANARY ISLANDS GOVERNMENT (Smart Specialization Strategy of the Canary Islands RIS-3) ProID2021010005. Grant "Cities in Transition. Urban Fragmentation and New Socio-spatial Patterns of Inequality in the Post-pandemic Context. The Case of the Urban Area of Palma (Mallorca)" (PID2021-122410OB-C31) funded by MCIN/AEI/ 10.13039/501100011033 and by "ERDF A way of making Europe".

# References

Aalbers M, Hochstenbach C, Bosma J, Fernandez R (2021) The death and life of private landlordism: how financialized homeownership gave birth to the buy-to-let market. Hous Theory Soc 38(5):541–563

Agarwal S (2012) Relational spatiality and resort restructuring. Ann Tour Res 39(1):134–154. https://doi.org/10.1016/j.annals.2011.05.007

Antón S (1998) La urbanización turística. De la conquista del viaje a la reestructuración de la ciudad turística. Doc Anàl Geogr 32:17–43

Ash J, Turner L (1976) The golden hordes. International tourism and the pleasure periphery. Saint Martin's Press, New York

Berry BJL, Kasarda JD (1977) Contemporary urban ecology. Macmillan, New York

Cáceres Morales EM, Martínez Quintana L (2019) Turismo en Cabo Verde: de la dicotomía a la integración. Pasos 17(3):489-507

Çaglar A, Glick Schiller N (2018) Migrants and city-making: dispossession, displacement, and urban regeneration. Duke University Press

Carrión F (2005) El centro histórico como objeto de deseo. En: Carrión F, Hanley L (eds) Regeneración y revitalización urbana en las Américas: hacia un Estado establecido. Flacso, sede Ecuador, Quito, pp 35–57

Carrión F (2007) El financiamiento de la centralidad urbana: el inicio de un debate necesario. En: Carrión F (ed) Financiamientos de los centros históricos de América Latina y el Caribe, FLACSO, Quito, Ecuador, Lincoln Institute of Land Policy, pp 9–24

Carrión F (2014) Urbicidio o la producción del olvido. Observatorio Cult 19:28–42

Carta de Cracovia (2000) Principios para la Conservación y Restauración del Patrimonio Construido. UNESCO. Cracovia, Polonia. Recuperado de. https://ipce.mecd.gob.es/dam/jcr:b3b6503d-cf75-4cb0-adaf-226740ebd654/2000-carta-cracovia.pdf

Carta de Washington (1987) Carta Internacional para la Conservación de Ciudades Históricas. VIII Asamblea General del ICOMOS. Washington D.C., Estados Unidos. Recuperado de. https://www.icomos.org/charters/towns_sp.pdf

Clark E, Fredricks K, Woc-Colburn L, Bottazzi ME, Weatherhead J (2020) Disproportionate impact of the COVID-19 pandemic on immigrant communities in the United States. PLoS Negl Trop Dis 14(7):e0008484

Clark E, Johnson K, Lundholm E, Malmberg G (2007) Island gentrification and space wars. World Islands: Phys Hum Approach 14:483–512

Cocola-Gant A, Gago A (2019) Airbnb, buy-to-let investment and tourism-driven displacement. Environ Plann A: Econ Space, pp 1–18

Cocola-Gant A, Gago A, Jover J (2020) Tourism, gentrification and neighbourhood change: an analytical framework. Reflections from Southern European cities. In: Oskam J (ed) The overtourism debate. NIMBY, Nuisance, Commodification. Emerald, Bingley, pp 121–135

Cordero J, Meneses C (2015) La resignificación del patrimonio cultural en el centro histórico de Guanajuato. La apropiación material y simbólica por los habitantes y gentrificadores. Pragma, espacio y comunicación visual, 13:17–29

De la Calle M (2002) La ciudad histórica como destino turístico. Editoral Ariel, Barcelona

De la Calle M (2008) El turismo en las políticas urbanas: Aproximación a la situación de las ciudades españolas. En: Ivards JA, Vera JF (eds) Espacios Turísticos. Mercantilización, paisaje e identidad. Universidad de Alicante, Instituto Universitario de Investigaciones Turísticas: Alicante, pp 507–529

De la Calle M, García M (2016) Políticas locales de turismo en ciudades históricas españolas. Génesis, evolución y situación actual, PASOS. Revista de Turismo y Patrimonio Cultural 14(3):691–704

Delgadillo V (2015) Patrimonio urbano, turismo y gentrificación. En: Delgadillo V, Díaz I, Salinas L (eds) Perspectivas del Estudio de la Gentrificación en México y América Latina. UNAM, Ciudad de México, pp 113–132

Del Romero L, Lara L (2015) De barrio-problema a barrio de moda: Gentrificación comercial en Russa-fa, el "Soho" valenciano. Anales de Geografía 35(1):187–212

Domínguez-Mujica J (2021) The Urban mirror of the socioeconomic transformations in Spain. Urban Sci 5(1):13

Domínguez-Mujica J, Parreño Castellano JM, Díaz Hernández R (2010) Inmigración y ciudad en España: integración versus segregación socio-territoriales. Scripta nova XIV(331):50

Domínguez-Mujica J, Parreño-Castellano JM, Moreno-Medina C (2020) Vacation rentals, tourism, and international migration: Gentrification in Las Palmas de Gran Canaria (Spain) from a diachronic perspective. In: Handbook of research on the Impacts, challenges, and policy responses to overtourism. IGI Global, pp 237–260

Donaire JA, Fraguell RM, Mundet L (1995) La nueva configuración espacial del turismo en la Costa Brava". ¿España, un país turísticamente avanzado? Instituto de Estudios Turísticos, Madrid, pp 273–287

Dorling D, Ballas D (2018) Spatial divisions of poverty and wealth. In: Ridge T, Wright S (eds) Understanding poverty, wealth and inequality: policies and prospects. Bristol University Press, Bristol, pp 103–134

Enríquez JA (2008) Las nuevas ciudades para el turismo. Caso Puerto Peñasco, Sonora, México. Scripta Nova, Revista Electrónica de Geografía y Ciencias Sociales XII(270). https://raco.cat/index.php/ScriptaNova/article/view/115764

Fitzsimmons (1956) Distribution of Negro population by county 1950. USA Library of Congress. https://www.loc.gov/item/2013593062/

Gil D, Domínguez P, Undurraga EA, Valenzuela E (2021) Employment loss in informal settlements during the Covid-19 pandemic: evidence from Chile. J Urban Health 98(5):622–634

González-Pérez JM (ed) (2015) Ciudades en transición. Procesos urbanos y políticas de rehabilitación en contextos diferenciados: Centro Histórico de La Habana y Ciudad Colonial de Santo Domingo. Universitat de les Illes Balears, Palma

González-Pérez JM (2017) A new colonisation of a Caribbean city. Urban regeneration policies as a strategy for tourism development and gentrification in Santo Domingo's Colonial City. In: Gravari-Barbas M, Guinand S (eds) Tourism and gentrification in contemporary metropolises. International perspectives, Routledge, Abingdon, pp 25–51

González-Pérez JM (2019) The dispute over tourist cities. Tourism gentrification in the historic Centre of Palma (Majorca, Spain). Tourism Geographies 22(1):171–191. https://doi.org/10.1080/14616688.2019.1586986

González-Pérez JM (ed) (2012) La nueva fiebre del oro. Las otras ciudades del turismo en el Caribe. Universitat de les Illes Balears, Palma

González-Pérez JM, Salinas E, Navarro E, Artigues AA, Remond R, Yrigoy I, Echarri M, Arias Y (2014) The city of Varadero (Cuba) and the Urban construction of a tourist enclave. Urban Aff Rev 50(2):206–243. https://doi.org/10.1177/1078087413485218

González-Pérez JM, Remond R, Rullan O, Vives S (2016) Urban growth and dual tourist city in the Caribbean. Urbanization in the Hinterlands of the tourist destinations of Varadero (Cuba) and Bávaro-Punta Cana (Dominican Republic). Habitat Int 58:59–74. https://doi.org/10.1016/j.habitatint.2016.09.007

González-Pérez JM, Lois-González RC, Irazábal C (ed) (2022) The Routledge Handbook of Urban Studies in Latin America and The Caribbean Cities, Urban Processes, and Policies. Routledge, Taylor and Francis, New York. https://doi.org/10.4324/9781003132622

Gormsen E (1997) The impact of tourism on coastal areas. GeoJournal 1(42):39–54

Gotham KF (2005) Tourism gentrification: the case of New Orleans' Vieux Carre (French Quarter). Urban Stud 42:1099–1121

Harvey D (2012) The urban roots of financial crises: reclaiming the city for anti-capitalist struggle. Socialist Reg 48:1–35

Hernández A (2015) En transformación...Gentrificación en el Casc Antic de Barcelona (tesis doctoral). Universidad Autónoma de Barcelona: Barcelona. http://www.tdx.cat/bitstream/handle/10803/310607/ahc1de1.pdf?sequence=1

Hiernaux-Nicolas D, González-Gómez C (2016) Patrimonio y turismo en centros históricos de ciudades medias. ¿Imaginarios encontrados? URBS. Revista de Estudios Urbanos y Ciencias Sociales 5(2):111–125

Ingallina P (1994) Urbanisme et gestion des flux touristiques. L'exemple de Florence. Tourisme Et Culture, Cahier Espaces 37:208–215

Instituto Geográfico Nacional (2021) La pandemia COVID-19 en España. Primera ola: de los primeros casos a finales de junio de 2020. https://www.ign.es/web/resources/acercaDe/libDigPub/Monografia-Covid.pdf

Islam SJ, Nayak A, Hu Y, Mehta A, Dieppa K, Almuwaqqat Z, Quyyumi AA (2021) Temporal trends in the association of social vulnerability and race/ethnicity with county-level COVID-19 incidence and outcomes in the USA: an ecological analysis. BMJ Open 11(7):e048086

Judd DR (2003) El turismo urbano y la geografía de la ciudad. Revista Eure XXIX(87):51–62

Judd DR, Fainstein SS (1999) The tourist city. Yale University Press, New Haven

King AT, Mieszkowski P (1973) Racial discrimination, segregation, and the price of housing. J Polit Econ 81(3):590–606

Knafou R (2006) El turismo, factor de cambio territorial: evolución de los lugares, actores y prácticas a lo largo del tiempo (del s. XVIII al s. XXI). In: Lacosta A (ed) Turismo y cambio territorial: ¿eclosión, aceleración, desbordamiento? Prensas Universitarias de Zaragoza, Zaragoza, pp 19–30

Koutrolikou P (2016) Governmentalities of Urban crises in inner-city Athens Greece. Antipode 48(1):172–192. https://doi.org/10.1111/anti.12163

Lee D (2016) How Airbnb short-term rentals exacerbate Los Angeles's affordable housing crisis: analysis and policy recommendations. Harvard Law Policy Rev 10:229–253

Lees L, Shin HB, López-Morales E (2016) Planetary gentrification. Wiley, Cambridge, Malden

Lennert M et al (2010) FOCI Future orientations for cities. HAL. https://hal.archives-ouvertes.fr/hal-00734406

Logan JR, Molotch HL (2007) Urban fortunes: The political economy of place. 2nd ed. University of California Press, Berkeley and Los Angeles

Maitland R, Newman P (2008) Visitor-host relationships: conviviality between visitors and host communities. In: Hayllar B, Griffin T, Edwards D (eds) City spaces–tourist places: Urban tourism precincts. Elsevier, New York and London, pp 223–242

Malheiros J (2002) Ethni-cities: residential patterns in the Northern European and Mediterranean metropolises–implications for policy design. Int J Popul Geogr 8(2):107–134

Marcuse P (1985) Gentrification, abandonment, and displacement: connections, causes, and policy response in New York City. J Urban Contemp Law 28:195–240

Marcuse P (1989) Dual city: a muddy metaphor for a quartered city. Int J Urban Reg Res 13(4):697–708

Mollenkopf JH, Castells M (eds) (1992) Dual city: restructuring New York. Russell Sage, New York

Mullins P (1991) Tourism urbanization. Int J Urban Reg Res 15(3):326–342

Musterd S (ed) (2020) Handbook of urban segregation. Edward Elgar Publishing

Navarrete D (2017) Turismo gentrificador en ciudades patrimoniales. Exclusión y transformaciones urbano-arquitectónicas del patrimonio en Guanajuato. Revista Invi 32(89):61–83

Navarro-Jurado E, Thiel-Ellul D, Romero-Padilla Y (2015) Periferias del placer: cuando turismo se convierte en desarrollismo inmobiliario-turístico. Boletín de la Asociación de Geógrafos Españoles (67):275–302

Novo-Malvárez M (2019a) Los nuevos usos del patrimonio: la expansión de hoteles *boutique* en Palma (Mallorca), Estoa. Revista De La Facultad De Arquitectura y Urbanismo De La Universidad De Cuenca (ecuador) 8:83–95

Novo-Malvárez M (2019b) La nueva centralidad turística de Palma (Mallorca). La transformación de conventos en hoteles, Revista MEC-EDUPAZ. Patrimonio: Economía Cultural y Educación para la paz 15:27–63

Oviedo MS (2014) Centro histórico de Quito: Cambios en la configuración residencial y usos de suelo urbano asociados al turismo (tesis de maestría). Instituto de Estudios Urbanos y Territoriales Pontificia Universidad Católica de Chile, Santiago, Chile. Recuperado de. http://estudiosurbanos.uc.cl/images/tesis/2014/MHM_Soledad_Oviedo.pdf396208

Parreño-Castellano JM (2001) El proceso de urbanización del espacio turístico. En: Hernández Luis JA, Parreño Castellano JM (eds) Evolución e implicaciones del turismo en Maspalomas Costa Canaria. Ayuntamiento de San Bartolomé de Tirajana, Las Palmas de Gran Canaria

Parreño-Castellano JM, Domínguez Mujica J (2008) Extranjería y diferenciación residencial en Canarias: la perspectiva del microanálisis espacial. Investigaciones Geográficas 45:163–199

Parreño-Castellano JM, Hernández-Calvento L (2020) Maspalomas Costa Canaria: el difícil equilibrio entre el desarrollo turístico y la conservación del litoral. En: Parreño Castellano JM, Moreno Medina C (eds) Geografías urbanas de Gran Canaria y Fuerteventura. Mercurio, Las Palmas de G.C., pp 75–87

Parreño-Castellano JM, Moreno-Medina C, Do Nascimento JM (2021) Diasporic links and tourism development in Cape Verde. The Case of Praia. In: International residential mobilities. Springer, Cham, pp 191–214

Peach C (ed) (1975) Urban social segregation. Longman, London; New York

Pinkster FM, Boterman WR (2017) When the spell is broken: gentrification, urban tourism and privileged discontent in the Amsterdam canal district. Cult Geographies 24(3):457–472

Ritzer G (1993) The McDonaldization of society. SAGE, Los Angeles

Roldán OA (2017) Gentrificación en centros históricos: una discusión conceptual. Devenir 4(7):69–82

Ross J, Diaz CM, Starrels JL (2020) The disproportionate burden of COVID-19 for immigrants in the Bronx New York. JAMA Intern Med 180(8):1043–1044

Sánchez-Aguilera D, González-Pérez JM (2021) Geographies of gentrification in Barcelona. Tourism as a Driver of Social Change. In: International residential mobilities. Springer, Cham, pp 243–268

Sales J (2019) Especialització turística, gentrificació i dinàmiques residencials en un entorn urbà madur, el cas de Barcelona. PhD Dissertation, Universitat Autònoma de Barcelona.

Sassen S (1994) The urban complex in a world. Int Soc Sci J 139:43–62

Sassen S (2001) Cities in the global economy. In: Handbook of urban studies, pp 256–272

Schäfer P, Braun N (2016) Misuse through short-term rentals on the Berlin housing market. Int J Hous Markets Anal 9(2):287–311

Schnake-Mahl AS, Sommers BD (2020). Places and the pandemic—barriers and opportunities to address geographic inequity. In: JAMA health forum, vol 1, no 9, pp e201135–e201135. American Medical Association, Sept 2020

Shevky E, Bell W (1955) Social area analysis, theory illustrative application and computational procedure. Stanford University Press

Smith N (1979) Toward a theory of gentrification. A back to the city movement by capital, not people. J Am Plann Assoc 45(4):538–548

Smith N (1996) The new urban frontier. Gentrification and the revanchist city. Abingdon, Routledge

Smith N (2002) New globalism, new urbanism: gentrification as global urban strategy. Antipode 34(3):427–450. https://doi.org/10.1111/1467-8330.00249

Smith N (2006) Gentrification generalized: from local anomaly to urban ´Regeneration´ as global urban strategy. In: Downey G (ed) Frontiers of capital: ethnographic reflections on the new economy. Duke University Press, London, pp 191–208

Tammaru T, Marcin´Czak S, Aunap R, van Ham M, Janssen H (2020) Relationship between income inequality and residential segregation of socioeconomic groups. Reg Stud 54(4):450–461

Timms D (1975) The urban mosaic: towards a theory of residential differentiation (No. 2). CUP Archive

Troitiño MA (1998) Turismo y desarrollo sostenible en ciudades históricas. Ería 47:221–228

Troitiño MA (2003) La protección, recuperación y revitalización funcional de los centros históricos. Colección Del Mediterráneo Económico 3:131–160

Vale M (2014) Economic crisis and the Southern European regions: towards alternative territorial development policies. In: Salom J, Farinós J (eds) Identity and territorial character, Universitat de València, Valencia, pp 37–48

Valenzuela E, Coll-Hurtado A (2010) La construcción y evolución del espacio turístico de Acapulco (México). Anales De Geografía 30(1):163–190

Van Kempen R (2007) Divided cities in the 21st century: challenging the importance of globalisation. J Hous Built Environ 22(1):13–31

Vaughan L, Arbaci S (2011) The challenges of understanding urban segregation. Built Environ 37(2):128–138

Wachsmuth D, Weisler A (2018) Airbnb and the rent gap: gentrification through the sharing economy. Environ Plann A 50(6):1147–1170

Walks RA (2001) The social ecology of the post-Fordist/global city? Economic restructuring and socio-spatial polarisation in the Toronto urban region. Urban Stud 38(3):407–447

Wilson J, Tallon A (2011) Geographies of gentrification and tourism. In: Wilson J (ed) The routledge handbook of tourism geographies. Routledge, Abingdon, pp 103–112

Yearby R, Mohapatra S (2020) Law, structural racism, and the COVID-19 pandemic. J Law Biosci 7(1):lsaa036


**Jesús M. González-Pérez** is a full Professor of Geography at the University of the Balearic Islands (Spain). He is the President of the Spanish Geographical Association. He has stayed at about twenty universities in a visiting scholar capacity. In Stanford University, he was a Visiting Scholar (2015) and a Tinker Visiting Professor (2016). He has published nearly 200 papers in numerous high impact journals and prestigious publishing houses, and has participated in over 30 research projects. His main research lines are urban inequality and fragmentation, social segregation and immigration, and tourist gentrification.

**Josefina Domínguez-Mujica** is a full Professor of Human Geography at the University of Las Palmas de Gran Canaria (Spain). She is Chairperson of the Commission of the International Geographical Union on 'Global Change and Human Mobility' and she co-coordinated the Expert Group on 'Immigration' appointed by the Canarian Parliament (2008–2011).

**Margarita Novo-Malvárez** is Senior Lecturer of History of Art at the University of Balearic Islands (Spain). Her main area of research is linked to heritage management and its ties to tourism, a topic covered in most of her publications. This research profile has been complemented by several international research stays.

**Juan M. Parreño-Castellano** is a Full Professor of Human Geography at the University of Las Palmas de Gran Canaria (Spain). He is director in the Department of Geography in this university and chairperson of the Group of Urban Geography belonging to the Spanish Association of Geography (AGE).

# Chapter 22
# The Production of Emptied Places in the Borderlands of the Metropolitan Area of Buenos Aires

**Andrea Catenazzi and Julieta Sragowicz**

**Abstract** The Metropolitan Area of Buenos Aires concentrates almost one-third of the population of Argentina. A broad scope of social, urban, and habitational public policies was found insufficient to guide low-density dynamics of growth, without the basic urban services and with significant processes of self-management for many decades now. Within this context, renewing the ways of questioning the processes of urbanization is central for the academic and political urban agenda. This article proposes rethinking the metropolitan borderlands in their own terms, based on the notion of urbicide, and, with this approach, it identifies a series of territories which recent transformations refer to various modes of violence on the city. This work distinguishes three ways of production of emptied places: the prefigurations of urban and environmental land management in the inhabited banks of polluted rivers, the networks of illegality in informal settlements and the privatization of public spaces in gated communities. By approaching the metropolitan borderlands from the notion of urbicide, a revision and rewriting of urbanization processes is developed from the perspective of the displaced ones.

**Keywords** Border territory · Urbicide · Production of emptied places · Public politics · Urban policy

## 22.1 Introduction

The search for a plot of land where to build a dwelling on, the hours spent on trains and buses to get to work, are fragments of a metropolitan experience which thousands of men and women repeat daily. These are not new experiences nor diffuse problems, conversely, these are classic problems identified by urban planning and housing policies. But it is precisely in that passage from the metropolitan experience

---

A. Catenazzi (✉) · J. Sragowicz
Universidad de Buenos Aires (UBA), Buenos Aires, Argentina
e-mail: acatenaz@campus.ungs.edu.ar

J. Sragowicz
e-mail: julieta.sragowicz@gmail.com

to a sectoral policy that the management of what is metropolitan acquires the label of *technical problem*, and in that operation, the construction of a city crossed by processes of increasingly more visible, and at the same time less questioned, urban inequality is depoliticized.

It deals with a way of urbanization which structures the daily life of a good part of its inhabitants. In Argentina, a country where over 90% of the population is urban and the Metropolitan Area of Buenos Aires concentrates almost one-third of the population of the country, renewing the ways of questioning the processes of urbanization is central for the academic and political agenda. The notion of urbicide challenges us in this sense. Fernando Carrión and Paulina Cepeda propose to use this notion as a new theoretical-methodological approach to analyze the city in a converse way as it has been studied and prefigured so far, not only from the logics of production but also from its logics of intended destruction.

In Latin America, there is a huge production on Urban Studies with significant conceptual, methodological, and empirical contributions which is added to an important accumulated experience in territorial work, a production of knowledge and practices which must be repositioned, related, and debated. Within this framework, the purpose is to contribute to the widening of this debate from the characterization of diverse modes of production of emptied places in the borderlands of the Metropolitan Areas of Buenos Aires, understood as logics of destruction of a determined kind of city.

Why is it useful to analyze metropolitan borderlands from the notion of urbicide? What does the operation of approaching them from a reverse way which restitutes that what has been destroyed, add? The notion of urbicide restores the relation between violence against the city and the construction of discourses and representations as regards the otherness in the border territories of the Metropolitan Area of Buenos Aires. By examining the processes of constitution of the city from the borderlands, we refer to processes which are at the same time, territorial, social, and political. Rethinking about metropolitan borderlands in their own terms, from their potentiality to enlighten forms and processes that have not been visible, is a commitment to provide intelligibility to those *nameless and expressionless* spaces typical of the metropolitan (Novick 2017). As regards those border territories, three ways of production of emptied places are observed: the prefigurations of the urban and environmental land management in the inhabited banks of polluted rivers, the networks of illegality in informal settlements, and the privatization of public space in gated communities. Those are fragile territories where different actors define strategies to shorten, increase, or modify distances to others, places where certain urban practices have been interrupted (neighborhood solidarity, social organizations, meetings) and where an intentional emptying has been achieved in public spaces which no longer represent *the common thing*.

In summary, the notion of urbicide leads us to analyze those metropolitan borderlands from the long processes of urban sedimentation and not only from the great gestures of private initiative. It is from there that this work aims to propose conceptual and methodological guidelines to design instruments of an urban policy that contributes to a more egalitarian, diverse, and democratic city. The questions that guide this work restitute how the city was produced before and stop at reconstructing the political and social fabric in those territories which went through a material and symbolical emptying. To do that, in the first place, metropolitan borderlands are characterized to recognize those territories which recent transformations refer to diverse modes of violence on the city. Then, the modes of production and destruction of those territories are analyzed in three situations of borderlands, restoring how they have been studied before and examining the contribution of the notion of urbicide to a renewed comprehension of these same processes. At last, in the final reflections, we inquire into the common aspects and specificities of each mode of production of emptied places and share some interrogations with the intention of renewing and widening the explanation of the processes of metropolitan urbanization.

## 22.2 Metropolitan Borderland Territories, *Nameless and Expressionless Spaces*

During the second half of the twentieth century, the Metropolitan Area of Buenos Aires was subject to a broad scope of social, urban, and habitational public policies which turned out to be insufficient to guide low-density dynamics of growth without basic urban services and with important self-management processes. The expansion was organized around the city of Buenos Aires and around the traces of the main railway lines which penetrated the rural space triggering a strong process of land valorization. Until the 1970s, that dynamic was progressively consolidated, but, since then, it has not been able to articulate with the restrictive logics of the real estate market and current urban regulations. Thus, the metropolitan conformation was the result of different layers of long-term land plotting that coexisted with *barrios parque* ('park neighborhoods')[1] and weekend villas, together with the consolidated neighborhoods near railway stations and with a series of progressive occupations which began with very precarious houses and gradually became the neighborhoods of the workers. During the last decades, slums have set in the interstices of consolidated areas and informal settlements have been established in ever farther lands from downtowns or which are flood-prone—hence the relevance of riverbanks—with little aptitude for urbanization, typical of border territories. Families arrive after having been displaced by land values and the location of new real estate products such as gated communities. Besides, a good share of those territories characterized by the environmental vulnerability of their locations are not covered by urban infrastructure networks nor by basic social facilities.

The notion of urbicide as a theoretical and methodological device leads us to return to those territories and comprehend the dynamic of urbanization through the causes of their own destruction. Thus, this strategy of approach helps to understand what is lost, but, at the same time, it contains a collective political power, because

---

[1] Translator's note: *Barrio Parque* is a low-density residential suburb with predominance of private green spaces.

it allows focusing on that which must be kept and built (Carrión Mena 2014). Even though the notion of urbicide originates in the field of literature, its insertion in urban studies occurs through two different methodological inputs. On the one hand, by means of the relationship with the devastating effects produced by wars on cities, and, on the other hand, explicitly linked to the impacts of the re-functionalization of cities, especially in those places where popular sectors inhabit, such as happened in New York (Bronx) or in Chicago. Carrión Mena and Cepeda (2021) propose to recover the use of this concept to explore its analytical richness in relation to some of the phenomena of neoliberal urbanism which cross the cities of Latin America. The authors point out that urbicide refers to practices aimed at the production of oblivion. Therefore, it seeks to destroy the historical memory of the citizenship which operates as a mechanism of social cohesion and collective identity (civitas), so as to subdue those peoples to the logics of societies which are supposedly more developed. At the same time, urbicide linked mainly to urban economy leads to the erosion of institutions and self-government (polis) by means of privatizations or corruption, as well as the deterioration of the material basis of a city (urbs), for the sake of a supposed urban development inscribed within the logics of the neoliberal city.

Thus, through this analytical matrix, they have been reviewed and acknowledged some recent transformations as ways of exercising violence on certain city components. From the very specificities of the border territories in the Metropolitan Area of Buenos Aires, we distinguish three modes of production of places that have been emptied of certain components of the existing city: the prefigurations of urban and environmental land management in the inhabited banks of polluted rivers, the networks of illegality in informal settlements, and the privatization of public space in gated communities.

## 22.3 The Prefigurations of Urban and Environmental Land Management

One of the ways of exercising violence on the city finds justification on certain ideas and practices of urban and environmental land management. Some categories such as those of urban order and disorder, formal and informal city, cross and organize territorial debate and policies, and come into play during the elaboration and implementation of most of management instruments. Particularly, these categories justify the prefiguration of green areas as the only option for the inhabited banks of rivers that are crossed by urbanization. A first approach leads to characterize the actions of eviction and eradication as unforeseen or contradictory consequences of a *good environmental cause*. Nevertheless, the notion of urbicide questions this same process in another way. What is destroyed when an area landscaped as a park is projected on those inhabited riverbanks? What is lost and what is forgotten in the name of urban and environmental land management? This approach makes it visible a particular way of violence exercised through the eradication of informal settlements on the riverbanks in a situation of environmental risk.

The environmental and climate crisis explained by the pattern of production and consumption based on the pressure over natural resources bring environmental problems along with very unequal effects on affected territories and population. In the last decades, various researches have analyzed the urban and environmental conflicts that take place in the inhabited banks of polluted rivers in Latin America. Avalanches, alluviums, droughts, fires, and floods are some of the events and manifestations of a continuous social process which impact on the daily life conditions of a society. In the Metropolitan Area of Buenos Aires, and as from the beginning of this millennium, diverse groups of people affected by problems of water access and exposure to recurrent floods have developed multiple strategies of collective action to achieve accessibility to this resource and also to exercise influence on modes of risk management (Merlinsky and Tobías 2021). An emblematic case, and a broadly studied one, is that of the Mendoza Case and the National Supreme Court of Justice's decision for the comprehensive sanitation of the Matanza—Riachuelo River Basin. On the one hand, we find those studies focused on the environmental suffering experienced by the inhabitants adjoining the river, due to industrial pollution (Auyero and Swistun 2008). Other studies inquire into the effects and social productivity of the judicialization of such conflicts (Merlinsky 2013). Other researches focus on the resettlements, especially analyzing the repertoire of collective actions of the affected ones (Najman and Fainstein 2019) and the controversies, tensions, and disputes during the implementation (López Olaciregui 2019; Scharager 2017, 2019). In this line, Carman (2011) points out that the symbolical operation of dehumanizing the inhabitants smoothes the way for exercising public violence. The research done in two slums located in the City of Buenos Aires on the riverbank of the River Plate makes us question what representation of the territory, of the otherness, is built to legitimate violence.

Sragowicz (2021) delves into the labile nature of the borders of the State and State administration, by observing how the actors of the State apparatus located *at street level* (Lipsky 1980) or *in the trenches in the territories* (Olejarczyk and Demoy 2017) are in charge of the usual execution of public policies and the power granted by the direct control of operative routines and activities, which powerfully influences the direction and the form that policies take daily. The research work analyzes the judicialization of the demand of environmental sanitation in the Matanza Riachuelo River Basin and outlines that the only option considered for the cleaning and liberation of the riverbanks was the displacement of the families who inhabited that place and, in turn, that those massive and involuntary displacements were not considered as a problematic question when designing the urban and environmental land management. The author argues that the insertion of the initial objectives in the public policy on the territory and the new framework of actors and interactions that it brought about, complicated the understanding for the liberation of the riverbanks by part of said State actors in charge of its implementation, which enabled, by the addition of diverse forms of representing the displacements in the territory and the otherness, transformations in the ways of addressing them. The trenches laborers found in this location a space for the exercise of cheating, for building relations with the different social actors and for redirecting politics, setting a dialogue between *mud* and institutionalization, managing the interstice and the porosity of the State by an active exercise of the translation role.

Stopping on what is destroyed from the notion of urbicide, invites us to reflect on how that territory was occupied, who the actors were, the logics and practices which configured this city border and, particularly, to unravel what was destroyed with the proposal of urban and environmental land management with the *good objective* of recomposing the habitational conditions of families in a situation of environmental risk.

One of the borders of the city of Buenos Aires is the Riachuelo River, which landscape was linked to large factories and port structures. Since the last century, it has been a privileged place for the location of industries which dumped their waste into the river. Thus, the process of deindustrialization started in the decade of the 1970s left large, abandoned plots of land. During this period, low-income population occupied vacant interstices progressively (Perelman and Fernández Rey 2014). The repeated announcement of sanitation works, the unfinished policies of cleaning and dredging and the rehearsals of public bureaus of interjurisdictional coordination for the management of the basin, are part of the historical heritage of the interventions aimed at the Riachuelo River and configure an antecedent for the process of judicialization started in 2004. In a context of growing environmental conflict across the country, the problem of the Riachuelo River entered the public agenda forcefully.

The Comprehensive Environmental Sanitation Plan derived from the judicialization of the conflict, proposed a series of actions aimed at the management, prevention, and control for the recovery and preservation of the Matanza Riachuelo River Basin. The lines of action ranged from the elaboration of indicators and monitoring of industrial activity, the development of water and sewage infrastructure to policies for the urbanization of slums and informal settlements along with cleaning actions of banks and the recollection of garbage accumulated on the riversides to build a park. Nonetheless, and in relation to this last guideline, it was when the works for the building of the park began, in the meeting with the framework of territorial actors and in the interactions with the officials of the executive and judicial power, that the social production of an existing city made itself present, so far invisible. In that moment, the action of uninhabiting became important.

When reconstructing this process of urban and environmental land management, it is possible to notice that a concept of *territory to free and park* predominated from the side of the State, as an object which had to be dominated. The idea of control, conquest, and land management with a strong presence of the State apparatus (linked to the monopoly of force and the weight of the Law) were recurrent in the court rulings. The predominant use of directives aimed at *eradicating, cleaning, free, remove*, among others, denoted a comprehension of the territory as a mere scenario for the implementation of a sanitation policy which objective was to mitigate environmental risk without considering the conflict of the displacement of population. In fact, there predominated a restricted vision of the complexity of the process and the reasons for the occupation of those border territories. Consequently, the displacement of population was reduced to simple tasks of logistics to remove obstacles and the organization of removals (Sragowicz 2021).

The population living on the banks, sometimes deemed as obstacles, sometimes as usurpers, but very few times as subjects of rights, undoubtedly patient, polluted, and polluters, were synthetized in the other, the otherness. Carman (2015) points out that the policies of displacement—including those which are so far the most arbitrary that do not even guarantee resettlement—legitimate themselves on the supreme good of these inhabitants' lives and the supposed interruption of the suffering of their bodies; bodies that must be separated from their present habitat to be rehabilitated. The neighborhood—as a space of life for the neighborhood organizations and the collectivity organized in the Body of Delegates affected by the liberation of the banks of the Riachuelo River, did not call nor understood the displacement as a benefit nor as a solution to their sufferings, but rather as a violent irruption in their daily lives, a breaking of their histories, of their bonds and ties.

The process of reoccupation of the border territory, in terms of a parked public space, as a result of the action of urban and environmental land management, meant, at the same time, an act of expulsion, an obligation to move, a destruction of the way of life and a rupture of the existing social fabric which characterized one of the modes of production of emptied spaces. Nonetheless, this same process activated new ways of social organization and of resistances aimed at influencing on public action, while producing a resignification of environmental demand. Thus, the attempts of the organized collective body of the affected ones to give focus to the housing demand found a way of expression in the environmental issue. The urbicide appears to account for the necessity to vindicate the right to the city and to produce a citizens' urbanism—Carrion (2014) states—which allows us to restitute these collective actions, the processes of reorganization, and the politicizing effect of some of the actions in the urban and environmental land management (Catullo 2006; Scharager 2019).

## 22.4 The Networks of Illegality and the Forms of Circulation of Violence

The networks of illegality and the forms of circulation of violence in the settlements on metropolitan borders constitute another mode of production of emptied places because they destroy the forms of social organization, while delimiting sector that can be crossed and other ones which circulation is forbidden. In the Metropolitan Area of Buenos Aires, land seizures are a strategy of land occupation developed by several generations of social organizations as part of a political and social framework which, in the last years, has been by increasing situations of insecurity and violence. A first approach leads to a characterization of the networks of illegality in informal settlements as delimited sources of insecurity. Nevertheless, the notion of urbicide questions this same process in another way. What is destroyed when the networks of illegality find conditions for their expansion in contexts of high vulnerability and rights deprivation? What is lost and what is forgotten about the self-managing process of the habitat? This approach inquires into the ways of disorganization of the self-managing processes of the habitat and the circulation of violence.

A good deal of researches about land seizures have tried to comprehend the housing strategies behind urban informality, a classic theme of Latin American cities (Hardoy and Satterthwaite 1987, Clichevsky et al. 1990, Cravino 2006). These works outline the capacity of social organizations and the forms of interlocution with the different edges of a complex and heterogeneous State. The question of violence in land seizures was not a main topic in the pioneering works about this subject. In fact, it was in the last decades that issues related to violence emerged in the urban agenda linked to State violence in cases of eviction from occupied plots of land.

More recently, studies have been oriented to case analysis to understand the specificities of this phenomenon. Trufo (2017) analyzes the land seizures in the Metropolitan Area of Buenos Aires and points out that institutional violence applied directly or indirectly by police forces is intertwined with the violence produced by groups which conform networks of illegality—with different modes of connivance and participation of the segments of the State—robbery situations and internal evictions which occupants undergo, and disputes over questions which are many times related to coexistence and end up in very violent resolutions. Other works highlight the fundamental role played by the actors constituted around the process of occupation and the emergence of the figure of the informal plotter is stopped (Dombroski 2020a, b, 2022).

Approaching this process in a reverse way, such as the notion of urbicide suggests, invites us to return to the land seizures in the metropolitan border territories and review which the forms of organization that upheld this habitational strategy were and, particularly, to unravel how the capacity of action of that form of social organization was destroyed through the circulation of violence and the increasing presence of networks of illegality in informal settlements.

The Metropolitan Area of Buenos Aires expanded through diverse forms of self-management during long processes of urbanization. A vast experience in social and organized production of habitat crosses the forms of habitation of low-income social sectors. Since the late 1960s, different forms of self-managing construction with technical assistance have been systematized, based on this collective experience, and propelled by nongovernmental organizations. The participation in housing construction and the design of seed and evolutionary housing with rationalized technologies and limited maintenance were the key components of this initiatives of assisted self-management. Since the 1980s, the strategy of land seizures has emerged with the expectation of achieving urban regularization. The settlements originated in land seizures, with a very few exceptions, have located in zones of urban border in the intersection of several municipalities or in the resulting interstices of large works of regional infrastructure, in degraded soils with diggings or floodable land. In this border territories, settlements not only kept the surrounding urban fabric, but also tried to comply with current regulations.

Nevertheless, this organizational and social mobilization experience around the topic of access to land and housing was displaced by the multiplication of unemployment and subemployment along with the rise in poverty and indigence. Thus, while housing conditions were a key axis in social organization during the import substitution model (up to the middle 1970s), labor became the organizing axis during the neoliberal model (Cerruti and Grimson 2004).

In the 2001–2002 political and institutional crisis, Argentina engaged in new ways of collective action as a product of resistance to neoliberal policies. These became visible in the unemployed mobilizations (picketers), the formation of neighborhood assemblies, the recovery of broken factories and service companies by its employees and the multiplication of cultural collectives (Svampa 2008). In a context of high social mobilization, public policies promoted the conformation of new and the consolidation of old social collectives. The fundamental axes of struggle were the social economy as a way out the unemployment situation and the social production of habitat as a way out of the habitat emergency (Catenazzi and Reese 2016). The creation of the Forum of Organizations for Land, Infrastructure and Housing in the Province of Buenos Aires in 2004, based on the articulation of more than sixty technical and grassroots organizations, recovered that wide set of scarcely visible experiences. Years later, the forum was a leading actor in the collective process developed around the formulation and sanction of the Fair Access to Habitat Act (*Ley de Acceso Justo al Hábitat*). The historical accumulation of this social collective which promotes improvements in the access to the habitat, made it possible to move from a logic of expectation associated to a habitat demand-proposal, in terms of concrete results (housing, infrastructure or facilities), toward a political demand-proposal of change in the conditions of reproduction of urban inequality through the collective construction of a new fair access to the habitat act (Barousse 2021).

In this context, Dombroski (2022) rebuilds the process of urbanization in the Northwest of the Metropolitan Area of Buenos Aires through different urban projects in dispute of the same plot of land. On the one hand, the initial project of the neighborhood organization which in the land seizure looked to comply with the current regulations to strengthen a strategy of land occupation aimed at urban regularization. On the other hand, the project of informal land plotters who tried to go on with subdivisions in the interstices of the original land plotting. This completion in the interior of the neighborhood was made from a gradual land plotting where the lotter ensured the maximum use of the property by reducing the streets to narrow interior corridors. In this process, the networks of illegality linked to drug dealing began to permeate and confront social organization. They set in the last plots, using the intricate network of corridors which ensured a selective mobility in the territory for them, operating openly against grassroots political organizations, neighborhood organization, and public authorities. The original neighborhood organization began to disappear, and this situation set the population—fundamentally, women and diversities in their different ages—in conditions of a higher insecurity and vulnerability.

The notion of urbicide invites us to reflect about the destruction of the city, but not in global nor definitive terms, but about a particular city or certain essential components of it. The networks of illegality in the informal settlements have transformed and weakened a complex political and social framework, an organizational experience deployed around the access to the habitat. This process of circulation of violence contributes to the internal violence and to the delegitimization of the process of organization and characterizes another one of the modes of production of emptied spaces.

## 22.5 The Privatization of Public Space and the Gated Communities

Gated communities occupied vacant land plots which had traditionally been urbanized by popular sectors. The improvement of the conditions of accessibility and the existence of vacant land made gated communities the best option for high-income sectors as their permanent residence. It is a real estate product which tried to motorize and capitalize a movement toward the suburbanization of those social groups. However, this suburbanization process of the elites (Torres 2001) was not made on an empty territory. To the contrary, popular settlements among walls and security checkpoints were the other face of this process. The late appearance of gated communities on a territory urbanized by popular sectors increased the visibility of internal frontiers, an urban condition which not only reinforced the fragmentation, but also increased the conflict over the use of the same territory (Catenazzi and Quintar 2009).

A first entry leads to characterizing gated communities as gestures of private initiative. Nonetheless, the notion of urbicide questions this very process in another way. What is destroyed when gated communities locate in popular peripheries? What is lost and what is forgotten about the use of public space? Restriction on uses, gentrification, and homogenization of public space characterize another mode of producing emptied spaces. If urban is a synonym of density and diversity, the delimitation of frontiers and the privatization of public space question the very notion of city because these components turn from being the valued attributes of the city to that what is wanted to run away and be isolated from.

The metropolitan organization, in line with what various authors denominated *multifragmented city* (Lombardo 2005), began to show high- and low-income sectors located in increasingly limited geographical spaces. This distribution required a materialization of limits by perimeter walls which became a common urban landscape in border territories. The diminishing of physical distance implies a greater visibility of the frontiers because one of the social sectors increase its efforts to separate from the others to keep distance (Catenazzi 2007). In front of the material and symbolical distance in the city, different actors have historically defined their strategies to modify them.

Revising the privatization of public space from the notion of urbicide, invites to go back over the different layers of the process of metropolitan urbanization and, particularly, unravel what was destroyed with the construction of internal frontiers and the subsequent privatization of public space in the metropolitan border territories.

The metropolitan area of Buenos Aires expresses the features of a self-managing city from the beginning of the last century in tension with the growths made as enclaves linked to an economic model characterized by social inequality. As from the decade of 1990s, the metropolitan periphery was subject to an intense territorial reconfiguration propelled by improvements of access highways to the city of Buenos Aires along with large magnitude real estate investments to attract the localization of gated communities and large consumption and entertainment facilities. Gated communities could occupy floodable areas by incorporating labor, capital, and specialized technologies (Pírez Ríos 2008), which allow them to qualify the urban land of the developments by granting them the urban infrastructure and raising the level of soil over the flooding level of the land. The paradoxical articulation between the environmental discourse and the public–private partnership (between real estate developers and municipal administration) as a strategy of legitimation of these developments alerted different environmental collectives about some contradictory uses of the sustainable development paradigm.

Gate communities have been in company of a new economic actor, the urban developer, who combines the management of the real estate agent—who organizes the production of urban land and its commercialization—with the mortgage credit management and the services of the building company. Likewise, the urban developer guaranteed the articulation of an economic demand which was capable of affording these new residential products. Most of the municipal governments favored the location of gated communities in vacant lands, in some cases and in an explicit fashion, to avoid land seizures by popular settlements. Thus, gated communities displaced popular sectors to more distant, less accessible places, and in greater environmental risk conditions.

The insertion of gated communities within the urban fabric marked abrupt cuts because they were located in the middle of economic subdivisions, slums, old urban centers and separate physically by means of security devices (closed walls, surveillance posts) which have altered not only important parts of the peripherical urban landscape (in the sense of creating and consolidating situations of enclave) but have also originated a significant number of focalized social urban conflicts (Torres 2001). The coexistence of different social sectors which are very close increased, in many cases, fantasies about *invasion* and the experience of insecurity by middle- and middle-high-income sectors (Filc 2002).

Gated communities represent the archetype of self-segregation insofar as they have their perimeter enclosure as a basic condition; they need to close in on themselves because, precisely, the difference they offer within their limits is that what turns them eligible. A distinctive feature is that they have been defined, generally, by their ability to create an illusion of *microcity* which can be gone around freely, when it comes up to enclosed spaces and controlled entry. These urbanizations live from the contrast implied by the fact that the open city goes on its own way, while they, in a secluded way, guarantee well-being, security, and contact with nature. At the same time, the low-density neighborhood fabric which configures the most part of the metropolitan territory, when located among gated communities, acquires the paradoxical configuration of secluded fragmented neighborhoods. The urban insertion of these gated communities breaks the grid plan and open urban fabric which was characteristic of the formation of our cities. In this way, parts of the cities get enclosed and public space privatized, consolidating a more homogeneous environment, while other parts of the cities are crossed by multiple internal frontiers.

## 22.6 Final Reflections

These reflections have an open character, they are rather considerations of an exploratory and stimulating process to approach with new lenses about already studied phenomena. The analysis focused on the border territories of the Metropolitan Area of Buenos Aires which concentrates almost one-third of the Argentinean population and which is characterized by a low-density growing dynamic with scarce basic urban services and with significant processes of self-management along with a strong process of land valorization. This unequal configuration is the result of different layers intertwined. Long-term subdivisions of land coexist with *barrios parque*, weekend villas and gated communities, consolidated neighborhoods next to railway stations and the slums which located in the interstices of consolidated areas and the informal settlements located in ever more distant lands from downtowns or which are floodable with few aptitudes for urbanization, an own feature of border territories.

This article tried to renew the modes of interrogating these processes of urbanization, re-think metropolitan borderlands in their own terms, as from the notion of urbicide. A series of territories were identified which recent transformations refer to diverse modes of violence on some of the city components to characterize the modalities of production of emptied spaces understood as intended logics of destruction of the city. In principle, three modes of production of emptied spaces have been recognized: the prefigurations of urban and environmental land management in the inhabited banks of polluted rivers, the networks of illegality in informal settlements and the privatization of public spaces in gated communities. The questions which guided us allowed examining how the city was produced before and stopping at the reconstruction of the political and social framework in those territories which went through a material and symbolic emptying.

These reflections restitute the centrality of thinking urbanism, revisiting ideas and practices of urban and environmental planning and housing policy as from the processes of urban sedimentation, as it was necessary to historicize the territory to review, again, the discourses and representations of the framework of actors and their relation with their capacity of transforming the city, crossed by increasingly visible and, at the same time, less questioned urban inequality processes. In that key, the dialogue between the notion of urbicide and the long processes of urbanization allowed unravelling the diverse nominations of violence, the representations and interactions among the different actors and, therefore, reconstructing the political and social framework which went through a material and symbolic emptying. We have tried to bring agreements and representations into play during the process of urbanization, not exclusively from force relations, but also considering the repertoire of arguments and the rules employed by actors to make their demands count through acts of resistance. Thus, we stopped at putting emphasis on how and to what extent each actor defines and redefines conflict, at centering in the analysis of repertoires of arguments and strategies oriented to questioning other ones and placing as something common and general that what is being affected.

When going back to the notion of urbicide to recover the usage of this concept in relation to some of the phenomena of neoliberal urbanism which Latin American cities go through, such as Carrión Mena (2014) proposes, and, certainly, in an exploratory way, we consider that it is possible to associate the modes of production of emptied places characterized in the borderlands of the Metropolitan Area of Buenos Aires with actions which the author had differentiated in the exercise of violence on the city, those which devastate with the significative places of life in common (urbs); those which end with identity (civitas) and those actions which subordinate public policies and institutions to the interests of the market, losing government and representation capacities (polis). In the process of displacement derived from eradications, some ideas and practices of urban and environmental justified actions which destroyed the significative places of life in common (urbs). The weakening of neighborhood organizations in the informal settlements derived from the circulation of violence associated to the networks of illegality have put in question the identity and memory of collective action of social organizations (civitas). In gated communities, the market privatizes public space making it lose the possibilities of representation of the common (polis).

The different modes of production of emptied places refer to and have, as an effect, the dissolution of social interaction by means of the breaking of the social fabric because of the process of displacements, of the restriction of circulation or of the privatization of public space. In all cases, it is possible to recognize forms of connivence between the State and the market, the absences and presences of the State to guarantee the consolidation of these forms of production of empty places.

About the prefigurations of urban and environmental land management, the notion of urbicide allowed the recognition of a particular mode of violence exercised by means of the eradication of riverside settlements in a situation of environmental risk. The order for the landscaping of riversides within the actions undertook toward the urban and environmental land management, cast a light on the consequent destruction of the significative places of life in common of those who inhabited and deployed their daily life there. The work showed the tension and divergence in the significations and interpretations of the problems and solutions for the sanitation of a water body and its surroundings. The territory as a scenario of environmental land management, an object of control over which there are actions of eradication, cleaning, and remotion to conduct for its own preservation and enjoyment, was, at the same time, the neighborhood, the history of self-management, daily and community life. The solution for the environmental sufferings of those families was not necessarily guaranteed by simple logistic tasks of organization of the removals. To the contrary, the eradication was a violent irruption in their daily lives, an act of expulsion, an obligation to displacement, a rupture of their histories, of their ties, reactivating an organizational memory and a process of union around the dispute over the future of the policy of environmental land management and the housing policy.

About the networks of illegality and the forms of circulation of violence in informal settlements in the metropolitan borderlands, the notion of urbicide allowed to recognize them as another mode of the production of emptied places, because they destroyed the forms of social organization, while at the same time they delimited sectors which could be crossed and other ones which circulation was forbidden. As Carrion (2010) states, the crime geographies which slowly take control of parts of the city, impose a significant reduction of the use of it, restricting circulation spaces. We add that this restriction fundamentally affects determined genders. This produces devastating effects in social coexistence and daily life, so much that solidarity conditions are reduced, and multiple modalities of self-justice area expanded, from acquiring guns, learning self-defense, lynching people, and becoming a client of the buoyant industry of private security. But also, because every unknown one turns into a potential aggressor and because public space is considered a space out of control. Approaching the study of the networks of illegality in the informal settlements of the metropolitan border territories from the notion of urbicide cast a light on the forms of disorganization of the processes of self-management of the habitat and the circulation of violence, on the effect in the dissolution of social cohesion, on a central actor as community organizations in the interlocution with the State in contexts of high vulnerability and deprivation of rights. Thus, the capacity of action of that form of social organization was destroyed with the circulation of violence and the growing presence of networks of illegality in informal settlements. In turn, it allowed casting a light on other forms of the exercise of violence in those territories, not from the explicit violence of the State through the policies of eviction of land seizures, but also from the networks of illegalities which show themselves as other modes of territorial control superposed to the State control.

About the privatization of public space and the gate communities, it is also possible to recognize processes of urbicide from the increasing weight that real estate promotional capital has within urban economy, forming a true enclave which breaks with the logic of public space, of the homogeneous provision of services and the expulsion of low-income population; strengthening urban segregation, eroding social capital, and weakening city government. The materialization of a social homogenization fantasy through enclosure and conformation of the *microcity*, has, as a counter-face, not only the dissolution of the diverse, but also, the distancing of the other, in an operation which represents him or her as a menace to one's own security.

Throughout this work, we discovered the potence of approaching the metropolitan borderlands from the notion of urbicide and, with this approach, reposition the relation between the forms of violence on the city and the construction of discourses and representations as regards the otherness in these territories. The analytical operation of returning to these processes from said concept led us to a review, re-reading, and re-writing of the processes of urbanization from the perspective of the displaced. So, we tried to recover an underground and alternative memory of the resistances.

Someway, it is an exercise of reconstruction of a social memory which validates the knowledge which rises from the historical experience of popular sectors. In our case, reviewing those processes as modes of production of emptied places, enabled us to focus on the absent, a reading against the grain of the historical experience to, in Walter Benjamin's words, light in the past the spark of hope. When reviewing what is forgotten, when spinning from the resistances, we proposed ourselves to contribute to the inquiry of the present and provide intelligibility to those nameless and expressionless spaces of the very metropolitan. Reading and re-reading the different layers of the metropolitan configuration of the notion of urbicide with the focus set on what is destroyed has also a democratizing potence, because it enables nuances and cast light over that what is damaged, is subdued. What is at stake for whom suffer this exercise of violence? In contexts of growth of the political right-wing and hate speech which refresh the destruction of the different, reviewing practices and discourses paying attention to what is violated, rather than to what is presented as beneficial, overcoming, or inevitable, is a necessary political action.

## References

Barousse A (2021) Actores colectivos y política urbana en contextos de desigualdad. El caso del FOTIVBA (Foro de Organizaciones de Tierra, Infraestructura y Vivienda de la Provincia de Buenos Aires) y la Ley de Acceso Justo al Hábitat. (Tesis para optar por el título de Magíster en Estudios Urbanos) [Collective actors and urban policy in contexts of inequality. The Case of FOTIVBA (Forum of organizations for land, infrastructure and housing of the province of Buenos Aires) and the Fair Access to the Habitat Act. (Thesis to opt for the degree of Master in Urban Studies)]. Universidad Nacional de General Sarmiento, Buenos Aires

Auyero J, Swistun DA (2008) Inflamable: Estudio del sufrimiento ambiental. [Inflammable: study of the environmental suffering]. Paidós, Buenos Aires

Carman M (2011) Las trampas de la naturaleza. Medio ambiente y segregación en Buenos Aires. [Nature's traps. environment and segregation in Buenos Aires]. Fondo de Cultura Económica, Buenos Aires

Carman M (2015) Una mirada sobre cuerpos sufrientes: las relocalizaciones de villas ribereñas en Buenos Aires. [A look at suffering bodies: the relocations of riverside villages in Buenos Aires]. Antropología Social y Cultural del Uruguay 13:65–74

Carrión Mena F (2010) El laberinto de las centralidades históricas en América Latina. [The labyrinth of historical centralities in Latin America]. Quito: Ediciones Ministerio de Cultura

Carrión Mena F (2014) Urbicidio o la producción del olvido. [Urbicide or the production of oblivion]. Observatorio Cultural 25:76–83

Carrión Mena F, Cepeda P (2021) La ciudad pospandemia: del urbanismo al "civitismo". [The post-pandemic city: from urbanism to "Civitism"] Desacatos 65. Revista de Antropología Social, pp 66–85

Catenazzi A (2007) Fronteras y ciudad. En: Ciudades Fragmentadas. Fronteras Internas en el Caribe. [Borders and city. In: Fragmented cities. Internal Borders in the Caribbean.] Dilla Haroldo y Villalona Maribel (coordinators). Santo Domingo, pp 205–216

Catenazzi A, Aida Q (2009) El retorno de lo político a la cuestión urbana. [The return of the political to the urban question]. Buenos Aires: Prometeo—Universidad Nacional de General Sarmiento.

Catenazzi A, Eduardo R (2016) Argentina. A 20 años de Hábitat II, las asignaturas pendientes. En: Hábitat en deuda. Veinte años de políticas urbanas en América Latina. [20 years after Habitat II,

Pending Issues. In: Habitat in Debt. Twenty years of Urban Policies in Latin America]. Cohen M, Carrizosa M, Gutman M (eds) Edit. Café de las Ciudades y The New School

Catullo MR (2006) Ciudades relocalizadas: una mirada desde la antropología social. [Relocated cities: a look from social anthropology]. Editorial Biblos.

Cerrutti M, Grimson A (2004) Buenos Aires, neoliberalismo y después. Cambios socioeconómicos y respuestas populares. [Buenos Aires, neoliberalism and after. Socioeconomic changes and popular responses.] Cuadernos del IDES, Buenos Aires, vol 5, pp 3–63

Clichevsky N, Schapira MFP, Schneier G (1990) Loteos populares, sector inmobiliario y gestión local en Buenos Aires: El caso del Municipio de Moreno. [Popular land plottings, real estate sector and local administration in Buenos Aires: The case of the municipality of Moreno.] Buenos Aires: Centro de Estudios Urbanos y Regionales vol 29

Cravino MC (2006) Las villas de la ciudad. Mercado e informalidad urbana. [The slums of the City. Market and urban informality] Buenos Aires: Universidad Nacional de General Sarmiento

Dombroski LJ (2020a) Los territorios de asentamientos en el borde metropolitano de Buenos Aires, desde 1980 a la actualidad. [The territories of informal settlements in the metropolitan area of Buenos Aires, from 1980 to the present.] Revista Urbano. Bío Bío. Universidad del Bío Bío. 23(41):84–101

Dombroski LJ (2020b) Las formas de los barrios informales. Casos del Área Metropolitana de Buenos Aires. [The shapes of informal neighborhoods. cases of the metropolitan area of Buenos Aires.] 2006–2016. Revista Quaderns de Recerca en Urbanisme QRU 10:100–123

Dombroski LJ (2022) Tomas de tierras promovidas, loteos informales y proyectos de viviendas de interés social, en barrios del noroeste del Gran Buenos Aires. [Promoted land seizures, informal subdivisions and social housing projects, in neighborhoods in the Northwest of Greater Buenos Aires]. A&P Continuidad 9(16). https://doi.org/10.35305/23626097v9i16.372

Filc J (2002) Territorios Itinerarios Fronteras. La cuestión cultural en el Área Metropolitana de Buenos Aires 1990–2000. [Territories itineraries frontiers. The cultural question in the metropolitan area of Buenos Aires 1990–2000.]. Universidad Nacional de General Sarmiento. Ediciones Al Margen, Buenos Aires

Hardoy JE, Satterwhite D (1987) La ciudad legal y la ciudad ilegal. [The legal city and the illegal city]. GEL, Buenos Aires

Lipsky M (1980) Street-level bureaucracy: dilemmas of the individual in public services. Russell Sage Foundation, Nueva York

Lombardo J (2005) El espacio urbano global en la sociedad latinoamericana del siglo XXI. El caso de seis municipios en la Región Metropolitana de Buenos Aires. [The global urban space in the Latin American Society of the 21st Century. The case of six municipalities in the metropolitan region of Buenos Aires.]. Revista Diseño y Sociedad Universidad Autónoma Metropolitana 18:14–25

López Olaciregui I (2019) Tensiones en la implementación de un programa de relocalización. La perspectiva de los actores involucrados en el caso de la Villa 26. (Tesis para optar por el grado académico de Magíster en Diseño y Gestión de Programas Sociales). [Tensions in the implementation of a relocation program. The perspective of the actors involved in the case of villa 26. (Thesis to opt for the academic degree of Master in Design and Management of Social Programs)]. Facultad Latinoamericana de Ciencias Sociales, Buenos Aires

Merlinsky MC (2013) Política, derechos y justicia ambiental: el conflicto del Riachuelo. [Politics, rights and environmental justice: the Riachuelo conflict.] Fondo de Cultura Económica, Buenos Aires

Merlinsky MC, Tobías M (2021) Conflictos por el agua en las cuencas de los ríos Matanza-Riachuelo y Reconquista. Claves para pensar la justicia hídrica a escala metropolitana. [Conflicts over water in the Matanza-Riachuelo and Reconquista river basins. Keys to think about water justice on a metropolitan scale.]. Punto sur 5:24–40

Najman M, Fainstein C (2019) Lo nuevo con sabor a viejo. Relocalizaciones de asentamientos de la ribera del riachuelo al complejo Padre Mugica en la Ciudad de Buenos Aires. En ¿Cómo pensamos las desigualdades, pobrezas y exclusiones sociales en América Latina? luchas, resistencias y actores emergentes. [The new with the flavor of the old. Relocation of settlements from the riverbanks of Riachuelo River To The Padre Mugica complex in the city of Buenos Aires. In: How do we think about inequalities, poverty and social exclusion in Latin America? struggles, resistances and emerging actors.]. Comp. Pallarés Custodio, Lorena, Itatí Alicia y Vigna Ana. Buenos Aires: Tesseo, pp 131–154

Novick A (2017) Configuraciones metropolitanas: palabras, problemas e instrumentos. En Producción de vivienda y desarrollo urbano sustentable. [Metropolitan configurations: words, problems, and instruments. In: Housing production and sustainable urban development.]. Fidel C, Romero G (ed) CCC-UNQui, Buenos Aires, pp 177–198

Olejarczyk R, Demoy B (2017) Habitar la trinchera: potencia y política en el Trabajo Social. [Inhabiting the trench: power and politics in social work.] Ts. Territorios-Revista de Trabajo Social 1:13–28

Perelman P, Fernández Rey L (2014) Análisis sobre el proceso de relocalización de los pobladores de las villas ubicadas en el camino de sirga de la cuenca baja del Matanza Riachuelo. [Analysis of the process of relocation of the inhabitants of the villages located on the towpath of the lower basin of the Matanza-Riachuelo river.]. UCES, Buenos Aires

Pírez P, Ríos D (2008) Urbanizaciones cerradas en áreas inundables del municipio de Tigre: ¿producción de espacio urbano de alta calidad ambiental? [Gated communities in flood-prone areas of the municipality of Tigre: production of urban space with high environmental quality?] EURE. Revista Latinoamericana de Estudios Urbanos Regionales 34(101):99–119

Scharager A (2017) Cuando la justicia toca la puerta: relocalizaciones y política en una villa de Buenos Aires. (Tesis para optar por el título de Magíster en Antropología Social). [When justice knocks on the door: relocations and politics in a slum in Buenos Aires. (Thesis to opt for the degree of Master in Social Anthropology)] Universidad Nacional de San Martín, Buenos Aires

Scharager A (2019) Judicialización, política y conflicto social. Resistencias y controversias en un proceso de relocalización de villas en Buenos Aires (2008–2018). (Tesis para optar por los títulos de Doctor en Ciencias Sociales y Doctor en Geografía). [Judicialization, politics and social conflict. Resistances and controversies in a process of relocation of slums in Buenos Aires (2008–2018). (Thesis to opt for the degrees of Doctor in Social Sciences and Doctor in Geography)] Universidad de Buenos Aires y Université François Rabelais de Tours, Buenos Aires—Tours

Sragowicz J (2021) Tramas, márgenes y controversias: una aproximación al proceso de territorialización del Camino de Sirga en la Cuenca Matanza - Riachuelo en la Ciudad de Buenos Aires (2008–2015). (Tesis para optar por el título de Magíster en Estudios Urbanos). [Plots, margins, and controversies: an approach to the territorialization process of the towpath in the Matanza–Riachuelo Basin in the City of Buenos Aires (2008–2015). (Thesis to opt for the degree of Master in Urban Studies).]. Universidad Nacional de General Sarmiento, Buenos Aires

Svampa M (2008) Argentina: una cartografía de las resistencias (2003–2008). [Argentina: a cartography of resistance (2003–2008).]. Observatorio Social de América Latina N°24. Consejo Latinoamericano de Ciencias Sociales, Buenos Aires

Torres H (2001) Cambios socioterritoriales en Buenos Aires durante la década de 1990. [Socio-territorial changes in Buenos Aires during the 1990s.] EURE. Revista Latinoamericana de Estudios Urbanos Regionales 27(80):33–56

Tufró M, Brescia F, Lefevre CP (2017) Aguantamos contra el Estado, perdemos contra las bandas. Reflexiones sobre la circulación de violencias en tomas de tierras y asentamientos de la RMBA. [We hold out against the State, we lose against the gangs. Reflections on the circulation of violence in land seizures and informal settlements in the metropolitan region of Buenos Aires.] Quid 16. Revista del Área de Estudios Urbanos 7:146–167

**Andrea Catenazzi** Architect and Specialist in Planning and Social Policies Management (Universidad de Buenos Aires). Geography and Urban Development Ph.D. (University of Sorbonne Nouvelle – Paris 3). Her research issue is the relationship between metropolitan inequalities and instruments of urban policy.

**Julieta Sragowicz** Political Scientist (*Universidad de Buenos Aires*). Urban Studies Master (National University of General Sarmiento). Experience in local and national government administration. Social activist in the popular political field.

# Part VI
# Degradation and Abandonment

# Chapter 23
# Reconstructing Cultural Paradigms. Experiences in East Europe: The Historical Memory of the Historical Centers in Lithuania

**Olimpia Niglio**

**Abstract** The history of nations is the result of natural and anthropogenic stratifications that have induced changes to the territories and the organization of societies often with dramatic consequences that have questioned local cultural instances. Meanwhile, these local cultural values represent a fundamental "humus" for the life of the communities and the knowledge and transmission of these values to future generations are important actions for the continuity and development of the territories in respect of local needs. Meanwhile, these values are the main targets to be erased when external expansionist interference intervenes on the territories. Once these expansions took place through conquests and wars; today these wars are also fought electronically with often much more devastating consequences. The city, together with "cives", is the mirror of these transformations and its existence not only depends on the citizens but on the ways in which they themselves are able to transmit their cultural heritage to the future. This chapter intends to analyze some experiences that have strongly characterized the history of Lithuania immediately after the Second World War and how cities have been sacrificed in their essential and fundamental values. Although the forms of colonization have often been devastating, the death of the cities was followed by a period of cultural regeneration that is interesting to analyze to understand the meaning of local values for the development of territories.

**Keywords** Lithuania · Vilnius · Politics · Colonization · Culture · Paradigms · Community · Education

## 23.1 Introduction

> […] Crossing time is not just a matter of aesthetic judgment or taking positions with respect to other attitudes that change rapidly over time, but it is a fundamental question that concerns how we intend to build the future.

---

O. Niglio (✉)
Department of Civil Engineering and Architecture, University of Pavia, Pavia, Italy
e-mail: olimpia.niglio@gmail.com

Faculty of Engineering and Design, Hosei University, Tokyo, Japan

© The Author(s), under exclusive license to Springer Nature Switzerland AG 2023
F. Carrión Mena and P. Cepeda Pico (eds.), *Urbicide*, The Urban Book Series,
https://doi.org/10.1007/978-3-031-25304-1_23

*Bernardo Secchi*, 2008 (Andriani ed. 2010).

In Europe, from the second half of the nineteenth century, there was a profound transformation of cities. All this had been determined by the growing phenomenon of industrialization that had put in the foreground the urgent problem of urban reorganization, especially in the main inhabited centers. Around the middle of the nineteenth century in Manchester, the English city most involved in the process of industrialization, the situation was dramatic. This is how the philosopher Friedrich Engels (1820–1895) describes the state of the city:

> [...] the streets, even the best, are narrow and winding, the houses dirty, old and falling [...]. Individual rows of houses or groups of houses rise here and there, like small villages, on the new clay soil, on which not even grass grows; the streets are neither paved nor served by sewers but are home to numerous colonies of pigs closed in small enclosures or courtyards or wandering without restriction to the neighborhood. To the left and right of the river, a number of covered passages lead from the main street to the numerous courtyards, entering which you come across a revolting dirt [...] (Marcus 2015, 98)

The design and political choice for this process of reorganization of the cities were to adapt the old urban structures with new buildings and new roads that, except for very rare exceptions, were still characterized by structures of medieval origin, with narrow and winding streets, with the ancient walls that in many cases had represented a real barrier to the development of the nineteenth-century city. Numerous projects began with the aim of modernizing but also rehabilitating part of the main European population centers. This urban policy manifested itself with the realization of works of gutting the neighborhoods considered unhealthy, the consequence of which was the construction of large new streets, and squares as well as the construction of new buildings both residential and institutional.

These urban planning choices had encouraged the demolition of several historic districts on which adequate cognitive investigations had not been carried out but simply destroyed because they were considered lacking services and functions adequate to modern housing and work needs. This program covered most European cities, both western and eastern, with the aim of rehabilitating many neighborhoods no longer adapted to the needs of the contemporary.

Obviously, the main reasons were mainly political and economic. The theme of hygiene and functionality was the most adopted pretext also to justify many demolitions as well as the demolition of entire historic districts of Paris (18,531,869) with the plans of Eugéne Hausmann (1809–1891) in France, the total demolition of the valuable medieval walls of Vienna with the consequent construction of the Ringstrasse (18,591,872) in Austria, the city walls of Cologne (1862) in Germany and again the gutting of a part of the historic center of Barcelona (1859) with the project by Ildefonso Cerdá in Spain and a few years later al-so of the historic center of Stockholm (1866) designed by the urban planner Klas Albert Lindhagen in Sweden (Niglio 2009).

In Italy too, the interventions undertaken in Italy with the demolition of the medieval walls of the city of Florence (1865–1875) on a project by Giuseppe Poggi (1811–1901) were no less invasive. It was a fate that also touched many other Italian cities including Milan and Bologna. Thus, the most historically consolidated neighborhoods were partly destroyed to make room for the new buildings and their urban layout; otherwise, the monumental buildings were saved on which policies of isolation from the urban context were applied. These interventions of isolation of the monuments were carried out through works aimed mainly at giving life to large green spaces or squares around the monument (Calabi, 2004).

Unlike in the eastern cities, the situation was less dramatic because the industrial transformations had been less important with the positive consequence that the populations had not moved from the countryside to the cities. In fact, the historic centers of eastern European cities had been preserved quite intact until the second world war and the great debates that had opened in the West on the future of historic cities examined very little the urban realities of the eastern continent. There is no doubt that, at the end of the nineteenth century, throughout Europe, an important theme was that of the "aesthetic representation" of the city and the resolution of this theme had seen different methodologies applied.

Meanwhile, in the first decades of the twentieth century, the first important debates began to develop on the meaning of historic cities and their future in relation to the needs of a contemporaneity increasingly distant from the social and economic logics that, over the centuries, had generated these urban contexts.

But the historic center of a city is nothing more than a biographical book made with various materials: from stone (buildings), the earth (the natural landscape with its trees, flowers, etc.), the air (the cultural context and its relations with the environment) and where the community lives and transforms everything thanks to the correct use of these materials. The historic center is therefore a book that tells its story from birth to today, and its existence is closely linked to the knowledge of its biography and the correct use of its materials. But when the memory of this biography and the correct use of its materials are lost, the historic city dies and the root that generates the correct development of contemporaneity is lost forever.

The historic city is like a musical harmony where everything is in relation and collaborates to create a perfect sound between different forms and cultures. The contemporary city without the historical city is like a musical disharmony, where nothing interacts but everything prevails with its own noise without considering the dialogue between "stone, earth and air" and therefore between the different cultures that make up the cultural biography of a city (Niglio 2020). That is why it is very important to reflect on the dialogue between the historic city and the contemporary city and how contemporaneity itself enters the biographical history of the city. In fact, it is very complex to understand the reasons for contemporary cities in the world without knowing the evolution of their history.

## 23.2 Sacredness of the City and Modernization

Howard Moody, American writer, in 1962, wrote that a city is about to die when it cares only for its material values; when this city shows only understanding for the movement of machines and not for that of men; when the community is competent in technology but devotes little time to the knowledge of the moral life; when human values are absent from the core of decisions affecting city planning and governance (Moody 1962, 154).

The contemporary city offers us this sad reality, an expression of economic-industrial evolution, not always interconnected with the social and cultural one, the latter denounced in 1849 also by the English sociologist John Ruskin (1819–1900) about the insensitivity and estrangement that he had found in the community prisoner of the mechanistic and functional logic, especially of the new urban settlements, built as a result of the industrialization of the territories.

Thus, the problem of the "historic center" was born as one of the most obvious contradictions of the twentieth century, as it was first isolated, then abandoned while currently, it is the subject of discussion in important national and international debates in order to recover and safeguard it. All this, especially during the pandemic that hit the whole world at the beginning of 2020, when local communities rediscovered the authentic values of the ancient city, and its human dimension, thus rediscovering a close relationship with nature, of which the community is an integral part, but in recent decades had forgotten it.

Historical cities owe their aesthetic and monumental beauty to the values of sacredness that are present in the places. Thus, the sacred city received its imprint from the buildings intended for worship that became its converging point, reflecting that Aristotelian vision that gave the city itself the character of circularity in which each element is both a point of arrival and a point of departure. For example, in the Greco-Roman city we can identify celestial references: the *decumanus*, which connects the east with the west, represents the path of the sun, the *cardus*, on the other hand, which unites the north with the south, represents a cosmic axis. The point of intersection of the two directions represents the center of the space in which the city is divided, which becomes an image of the harmonic principles, taken from nature, theorized by Vitruvius, and taken up in the Renaissance treatises of Leon Battista Alberti, Filarete and in the projects of Leonardo da Vinci. We talk about the ideal city, the city of the logos, a city where geometry and technique harmonize on stable principles that make the history of the city eternal. This is the city that the Italian philosopher Rosario Assunto (1915–1994) defines as the city of Anfione, that is, the one built respecting the harmonic rules of music and therefore the rules of Zeus (Assunto 1983).

Opposed to these rules is the modern city, the Metropolis, dominated by functionalist science that has seized the images of the Logos transforming them into functional spaces, without any harmony. This is the city that Rosario Assunto defines as the city of Prometheus, rebellious to Zeus, and therefore to the laws of nature, which destroys the ancient cities, he seizes it in the name of that capitalism that in the twentieth century has transformed the history of humanity. In this way, we are witnessing the reversal of the relationship between contemplative life and active life. What we observe today, especially in the great metropolises of the world is the victory of Prometheus who against the laws of nature has created ur-ban contexts that have produced only the death of the concept of the city.

In fact, while in the cities the communities relate and dialogue, in the metropolis isolation prevails within a great confusion.

Meanwhile, capitalist culture has solved the problem of the historic city by placing on it the embalming label of the historic center which has been flanked by the most recent denominations such as: residential, commercial, executive, industrial centers, etc. ... Indeed, today the historical part of the city has assumed above all a tourist role and therefore of consumption; differently, this urban layout represents instead the poetic and sacred heart of the existence of the same community.

These topics have also been extensively treated by Roberto Pane (1897–1987), an Italian, historian of architecture, who analyzes the irreparable transformations of European historic centers in the years after the Second World War, states that it is evident moral laziness that has not allowed to realize projects respectful of the historical roots of the places. The fever of modernization has actually involved the whole world in an absolutely questionable path and of which today many pays for the loss of their cultural references (Pane 1967).

Roberto Pane analyzes these themes also by observing the cities of Eastern Europe strongly compromised by the damage of the Second World War and then by the Soviet colonization until the end of 1989.

Meanwhile, it was precisely this "moral laziness" that led the communities to no longer study history through which to trace good references for the construction of the contemporary city. Here, then, we are witnessing the death of the historic city to give space to the contemporary city that does not guarantee any cultural content and does nothing but attacks what remains of the historic city.

There is no doubt that contemporary ephemeral fashion has renounced dialogue with history and arbitrarily follows its path, without any respect for the environment, traditions, and people. The city evolves without rules and the im-prints of history are often trampled by arbitrary functional but only temporary needs. Everything is constantly changing.

All this not only produced the death of the historic city but also led to the loss of local identities, of the *stabilitas* loci of which Christian Norberg-Schulz (1926–2000) speaks, and therefore of those cultural values that are the basis of the spatial configuration of the city. In fact, it is difficult to imagine building the future by renouncing the rich baggage of cultural experiences handed down by history.

Even the German architect Walter Gropius, in 1953, in an article published in the Italian magazine "Casabella" stated that it is essential to find a point of tangency between history and contemporaneity, between the ancient city and modern city. However, this point of tangency is only possible when the architect knows how to implement his creative and artistic talents to build an organic architecture, rich in relationships and with the right proportions (Gropius 1953). So also, some architects of the Modern Movement encouraged that cultural continuity without which, even today, it is not possible to imagine the development of a sustainable city.

This is also reiterated in *The Washington Charter: Charter on the Conservation of Historic Towns and Urban Areas* (1987) which reaffirms the importance of cultural policies in urban planning. Above all, the Charter confirms the importance of historical continuity in the contemporary city because when this continuity is erased, then the city dies. In fact, a city that does not look at its cultural origins and its identity values is not aesthetically conceivable and in turn, a city without a cultural landscape is not aesthetically thinkable (Assunto 1983).

## 23.3 The Cultural Landscape of the City

The cultural landscape is the result of human actions that generate the transformation of the territories. When these actions take place without respecting the laws of nature then we are close to the creation of "non-places", where the community has no meaning other than the arrogance of functionalism. But this is not at odds with what history has handed down to us.

Man's interest in the landscape and the rule of nature finds ancient roots, in fact, in all cultures and in different historical periods man has always tried to establish a dialogue with the landscape, that is, that set of natural resources to which important aesthetic values are also recognized.

Since ancient times, from East to West, artistic and conceptual elaborations on the landscape have been fundamental to knowing and appreciating the dialogue between man and nature.

Questo dialogo si è concretizzato attraverso documenti descrittivi e rappresentazioni artistiche che hanno cercato di dare un'interpretazione completa dei contenuti materiali e immateriali e quindi della natura scientifica del paesaggio. Con riferimento all'Occidente fin dall'epoca greca, la pratica della *sophia*, cioè la conoscenza del paesaggio, è stata fondamentale per raccogliere informazioni e descrivere la diversità culturale dei territori. Proprio questa conoscenza è stata alla base della definizione del paesaggio culturale urbano nell'ambito della pianificazione delle città (Turri 2003; Venturi 2002).

Meanwhile, only starting from the new Millennium, first with the European Convention on Landscape (2000) and then with the UNESCO Conventions of 2003 and 2005, the concept of "Cultural Landscape" has been more established, which identifies a specific and unrepeatable identity of the places, the result of the interaction between the individual good and the context, architecture, and the environment, art and society. It is called Cultural Landscape, as man has organized and shaped the space creating a fusion between nature and culture. In this way, the Cultural Landscape identifies spiritual values as distinctive features of a territory, characterized by a stratigraphic richness of settlements and cultures. In this way, it is possible to speak of "landscape heritage" because of the constant evolution of social and cultural processes that have determined the characterization of a territory, as a collective living space in continuity with historical traditions.

Here the Cultural Landscape becomes a "living" and "active" element, an expression of the social and cultural actions of a constantly evolving territory. In fact, the landscape helps us to read the relativity and plurality of values that can be attributed to cultural heritage, and their variability in relation to different historical moments, and social and cultural contexts. Finally, the identification of the landscape as heritage is the result of a process of assigning values that cannot be defined in absolute form, but only in relation to the specificity of each place and each time (Venturi 2002).

From all this, it is not difficult to understand that history is a fundamental reference that has always contributed to identifying that intangible dimension of the cultural heritage of the landscape and hence the affirmation of identities, creativity, and diversity that have allowed to outline an innovative approach able to dialogue elements of the environmental context with factors of the historical and cultural context. For this reason, it is very much a city that has its own clear connotation when it is possible to clearly recognize the "Cultural Landscape" of this city and therefore it's intrinsic and extrinsic values that conform to both its material and intangible heritage. When it is not possible to recognize this "landscape" then the city has assumed a configuration that is no longer identifiable and therefore is close to its own end (Norberg-Schulz 1979).

Precisely on these principles have worked many of the governments of Eastern Europe that after the dissolution of the USSR, starting from 1990, began a path of social and political regeneration and therefore of reappropriation of their cultural landscape with interesting feedback. The following paragraphs describe the case of Lithuania with important projects for the reconstruction and regeneration of cultural identities. With the war in Europe breaking out on February 24, 2022, now this theme returns to be very topical with reference to the reconstruction projects that will have to be implemented in all the cities of Ukraine.

## 23.4 Vilnius, Capital of Lithuania

In 1971 Jonas Glemja, a Lithuanian restorer architect stated that the preservation of monuments was a methodology applied and theoretically confirmed already to the beginning of the twentieth century (Glemja 1971).

More than forty years after this interesting contribution to the state of monuments in Lithuania, new pages of history can be written today. But before dwelling on the news concerning this country that since 1991 has gained its independence from the Soviet state and has become part of the international community, let's analyze, albeit briefly, the main interventions carried out. The case of Lithuania is certainly an interesting example to understand the meaning of history in the configuration of the "Cultural Landscape" of cities.

In Lithuania, the first law addressing the problem of the protection of monuments is only from 1919 and is issued by the Provisional Government of Lithuania, born after the independence of the country conquered following the defeat of Russia in the First World War. But in 1940 the country was re-annexed to the Soviet state and in this same year law for the protection of monuments was enacted by the Soviet Lithuanian Government. In those years its capital, Vilnius, belonged to the Polish government which, unlike the Soviet one, was more attentive to promoting initiatives aimed at the conservation and enhancement of cultural heritage, even if the political conditions clearly poured the greatest resources on other socially and economically more urgent problems. In fact, it was evident that in the Vilnius Region in those years, unlike the rest of the country that was under the Soviet regime, restoration work was carried out on the main national monuments: especially in the historic center of Vilnius and nearby Trakai. With the Nazi occupation between 1941 and 1943 the historic center of Vilnius, characterized by an interesting Jewish quarter, is divided into two ghettos called the "small" and the "big" and then destroyed by persecution against the Jews.

The Polish writer Czeslaw Milosz (1911–2004) describes the Jewish quarter as a labyrinth of medieval streets, with houses connected by arches and sidewalks two or three meters wide but all bumpy. Currently very little of this situation survives also because the houses were mostly made of wood and the Nazi persecutions against the Jews had destroyed everything. Most of the neighborhood reduced to piles of ash and stones after the war was the subject of reconstruction with new buildings, and large open spaces, and therefore, if not a few cases, the original urban and typological connotation was lost.

In 1945, with the end of the Second World War, Vilnius has proclaimed the capital of the Lithuanian Soviet Republic and therefore most of the Poles present returned to Poland, while what little remained of the Jewish quarter was repopulated by Lithuanians. Even the few surviving Jews in Vilnius no longer had their own ghetto, and even today only a few stone plaques recall a history erased forever.

In the meantime, since 1950 in Lithuania, the indications given by some international documents for the protection of monuments and in particular the Athens Charter of 1931 have spread. For the first time, an Atelier was established consisting of a group of researchers who, according to precise scientific methods, dealt with interventions on monuments. This Atelier was established at the behest of the Lithuanian Government precisely to define criteria for conservative intervention. An Institute was then created within which the group of researchers was integrated, and which still operates on conservation issues.

In 1967, again by the Lithuanian Soviet Government, a new national law was issued for the protection of monuments and works of art. But what emerges is above all the attention paid mainly to the theme of the search for compatible functions for the reuse of ancient monuments, many of which are used for public functions such as government buildings, town halls, schools, museums, etc.

In general, there are four typological categories of reference:

(a) military monuments (castles, towers, city walls, etc. …);
(b) public buildings (government offices, town halls, churches of the different religious orders but with reference to Catholic ones);
(c) private houses and historic buildings;
(d) historic centers (in which interventions prevail mainly of reconstruction).

Certainly, it is the military architecture on which the greatest attention is poured, both for the value as well as for the meaning that these monuments have always had in the memory of the Lithuanian people.

In addition to the properly typological ones, it is then possible to identify in three other categories the principles and methods of intervention on monuments:

1. monuments that are preserved as received and on which only maintenance interventions of the state of affairs are carried out (e.g., simple protection of the walls);
2. monuments that have undergone partial reconstructions in order to understand their original styles and forms are now lost;
3. monuments that have undergone a total reconstruction, with the reconstruction of the plants, facades, and roofs, and that have been used for new functions, more responsive to the needs of modern life.

A significant example of the type of intervention that is aimed at preserving the monument as received without making changes is the case of Medininkai Castle, in the Vilnius region and about 30 km from the center of the capital (south-east side). The castle was built between the end of the thirteenth and the early sixteenth century but was completely abandoned in the following century. It is a large defensive quadrilateral that is spread over an area of about 185 hectares of extension. This castle was restored in 1951 to a design by the architect S. Lasaritskas and the direction of the architect E. Pourlis. Since these are mainly defensive walls, the restoration mainly concerned partial recomposition of the destroyed parts and the construction of wall protections with canopies that are completely reversible. The reconstructed parts are clearly legible thanks to the use of red bricks in contrast with the original gray stone.

An example of restoration like Medininkai Castle (Fig. 23.1) is that of the fortress of Kaunas, the second center of Lithuania and located about 100 km west of the capital. The city is crossed by two rivers, the largest in Lithuania: the Neris and the Numenas. These two rivers meet precisely in this city and at the point where they converge forming a peninsula on which stands the historic center of the city of Kaunas and its castle. Few remains of the latter, only a tower and a few stretches of walls partly preserved with roofs with a completely reversible wooden structure (AA. VV. 1971).

**Fig. 23.1** Kaunas, Medininkai Castle. *Source* Olimpia Niglio (2015)

The castle today is inserted within a very large and suggestive green area and from which you can see the beautiful fifteenth-century church of S. Giorgio, a gothic structure in red brick that has been in a state of total abandonment for over 50 years but whose external stylistic forms testify to its ancient splendor. The whole thing was completely looted.

A completely different case compared to the case studies analyzed so far but closer to the second category of intervention is the example of the castle of Trakai, today a small urban center about thirty kilometers west of the center of the capital, but once a military garrison and seat of the Karaite community of which today remain some wooden houses whose main elevations are characterized by three windows that tradition wants to be destined one to God, one for themselves and one to Grand Duke Vytautas, son of Duke Kestutis, founder of the castle.

The castle is spread over an island about 500 m from the coast in the middle of three small lakes Galvé, Totoriskiai and Bernardinai.

The castle built between the end of the fourteenth and the early fifteenth century is completely abandoned around the seventeenth century (Fig. 23.2). It was built as a military outpost against the Crusader invasions. The first interventions were carried out only starting in 1905. Some restoration work is carried out by some Polish scholars as the Vilnius Region was under the government of Poland, but the most important works were carried out only after 1950.

In fact, it is precisely from 1956 that the reconstruction works of most of the castle began on a project by the architects B. Krouminis and S. Raoudonvaris whose pear ended in 1962, the year in which the castle was reopened to the public.

The military structure is characterized by a large central keep separated from the body of the castle by a moat that encloses a courtyard with galleries. Today the castle

**Fig. 23.2** Trakai castle in the early XX century. *Source* Trakai Museum

**Fig. 23.3** Trakai castle. *Source* Mariusz Kluzniak (2010)

houses a museum where in addition to furniture, ceramic objects, and clothes of local popular culture it is also possible to observe a section dedicated to the reconstruction of the castle, with models, drawings, and photographs (Fig. 23.3).

The reconstruction of the castle is clearly linked to the principles emanating from the stylistic culture of the late nineteenth century and therefore to that French of Eugène Emmanuel Viollet-le-Duc (1814–1879) but finds justification in that principle that Roberto Pane (1897–1987) rightly defines as "psychological instance" and that in a devastated and oppressed country like Lithuania can only find value precisely the reconstruction of a deleted historical memory.

As shown by some of the relief drawings and watercolors made by landscapers at the end of the nineteenth century, there were very few remains of the original structure, if not the perimeter walls and part of the corner tower. Nothing remained of the roofs, of the horizontal structures that are made of wood had been the first to be destroyed.

Respecting the use of materials typical of local construction techniques (where the use of wood prevails for the abundant coniferous forests) all the structures are made of wood again and those stylistic forms are restored, perhaps original but in any case, also derived from other castles of Lithuania better preserved. The use of red bricks used to reconstruct all the missing parts makes it now clearly legible, even to an inattentive observer, how much of the truly original is preserved and what has been made again.

The studies also conducted by the Atelier, established at the behest of the Lithuanian Government, had highlighted the need to rebuild a national identity by restoring the two most interesting periods that had characterized the nation: the Renaissance for its military and religious architecture and then the nineteenth century for the palaces.

Even the very harsh climatic conditions for long periods of the year made it necessary, to be able to take advantage of these historic spaces, to proceed with reconstruction works and above all coverage. Certainly, the religious buildings in these interventions have always occupied a place of primary importance.

In general, we can say that the restorations carried out in Lithuania until the mid-seventies of the twentieth century mainly followed two directions. The first is essentially aimed at reconstruction and repair work, especially of the exteriors of buildings, therefore close to the stylistic vein; the second instead aimed only at partial repair works but mainly conservation.

In fact, this last operational direction has made it possible to carry out historical, archaeological, materials, and construction techniques on the monuments and therefore has developed more of that so-called "scientific restoration" strand to which the Atelier itself and the subsequent Institute have increasingly approached and recognized. But we are in the 60s when especially in Italy the theories of scientific restoration were now past history, and the theories of Cesare Brandi (1906–1988) and of the so-called "critical" restoration found development, which lies precisely in the full historical awareness of the distinction between past and present, in the critical detachment that allows defining the ancient bringing it back to its real and historical dimension (Brandi 2000).

The first of the Lithuanian monuments to have been the subject of a careful and "scientific" restoration was the Vilnius Cathedral which, although founded at the end of the fourteenth century, does not preserve anything of its original Gothic forms since it has undergone strong transformations up to the current nineteenth-century neoclassical appearance. Important studies were conducted in the 30s of the twentieth century following the floods due to the overflow of the Neris river that also involved the church.

But among the examples of perfectly preserved Gothic-style churches that certainly should be remembered is the church of St. Anne whose beauty is said to have struck Napoleon Bonaparte so much that he wanted to transport it to Paris. It is a work dating back to the late Gothic period and the façade seems to have been finished only in 1582, at a time when In Vilnius the Baroque culture was taking ample space, as shown by many exteriors and interiors of churches. The thin red brick ribs support the vaults of the interior less interesting when compared with the façade. Next to this then stands the Cistercian church with an adjacent monastery now the location of the Vilnius Academy of Arts. The interior rich in medieval decorations is rather devastated due to the Nazi raids but is currently the subject of a restoration of all pictorial surfaces.

Meanwhile, since the 50s of the twentieth century, the restoration of the numerous national monuments in the city of Vilnius has also been the subject of more stylistic than scientific interventions. As well as the case of the tower of the Upper Castle of the city, a large stronghold surrounded by a defensive palisade of the tenth century. The only section still standing today, which is also the symbol of the city, is the Gediminas Tower (1271–1341) founder of the city of Vilnius (Figs. 23.4 and 23.5).

It is an interesting octagonal construction in red brick but whose construction of a belvedere built at the end of the nineteenth century well placed has preserved its original structure, already largely tampered with even by reconstruction works of

**Fig. 23.4** Vilnius, Gediminas Tower (1838) before the restoration. *Source* M. Butkovskis

the early nineteenth century, as shown by an image by Mironas Butkovskis of 1838 (AA.VV. 1995).

In the meantime, many reconstruction and restoration works are carried out in the historic center of Vilnius of baroque churches such as that of St. Casimir where important internal and external interventions are still underway, the church of St. Catherine, the church of St. John of Vilnius University as well as the few Gothic-style houses remaining among the most historic districts of the city (Glemja, Jaloveckas 1984).

The historic center of Vilnius, in 1977, with a law of the Lithuanian State for the Protection of Monuments and Historical and Cultural Heritage is subject to a permanent protection constraint, but already a previous law of 1969 identified the value of the historic center of the capital, subsequently then reconfirmed in 1990 in a provisional law for the identification of monumental assets to be safeguarded.

**Fig. 23.5** Vilnius, Gediminas Tower after the restoration. *Source* Olimpia Niglio (2015)

Following the national independence gained in 1991 and studies carried out through international organizations such as ICOMOS (of which we recall the interesting report drawn up, in January 1994, by Prof. Panu Kaila, president of Icomos Finland, following his visit to the Lithuanian capital) in 1994 the Historic Center of Vilnius is included in the UNESCO World Heritage List, to the progressive number 427, with its 1487 buildings for a total area of 1,497,000 m$^2$ (AA.VV. 1994).

The signing in the UNESCO List of the historic center of Vilnius is also confirmed within the extraordinary session of the Office of the World Heritage Committee, held in Kyoto on 28 and 29 November 1998.

In the same year the Lithuanian government decided to initiate an important intervention within the capital. It is a work of total reconstruction of a monument that was destroyed at the end of the eighteenth century at the behest of the tsarist authorities, whose methods and criteria of intervention place this work in the third category previously described, that is total remaking of the plants, facades, roofs and search for new functions more responsive to contemporary needs (Venclova 2019).

This is the Palace of the Grand Dukes, a lower castle (Zemutinés pilis) wanted by Sigismund Augustus, King of Poland and Grand Duke of Lithuania who established his ducal court in Vilnius, making the city an important cultural center. The palace, located at the foot of the hill where stands the upper castle and the tower of Gediminas and behind the Cathedral, was completely abandoned at the end of the eighteenth century and from here demolished. Few and insignificant traces of the outer perimeter walls remain, so insignificant that even archaeologists have found very few traces to reconstruct its original appearance. A drawing made at the end of the XVIII century helps us to know the style of the castle (Fig. 23.6).

Meanwhile, in 1998 the Lithuanian government decided to make a copy of the old castle in the same place where it stood, inviting the architects to formulate hypotheses of reconstruction also using some prints of the XVIII from which the original stylistic

# 23 Reconstructing Cultural Paradigms. Experiences in East Europe: The …

**Fig. 23.6** Vilnius. The ruins of the Palace of the Grand Dukes, drawn between 1785 and 1786. *Source* governmental archive, Vilnius City

and typological forms can be deduced. At the deconsecrated church of St. Michael, museum of Architecture of the city of Vilnius, you can see some of the models made to formulate hypotheses of reconstruction of the castle.

Observing at the construction site it is rather difficult to define this as a restoration work but the reasons for its reconstruction are certainly to be found in the will of the people to revive the culture and history of the Lithuanian nation. In this regard, what Roberto Pane wrote in 1959 after visiting Warsaw devastated by Nazi destruction is entirely current. On this specific theme Roberto Pane affirms.

> […] the reconstruction of the Polish capital, as it was before the Nazi ferocity decided to erase its face, so that there would no longer be any image of the past that could speak of culture and national history (...), finds its full justification as a denial of those same infamous reasons for which the destruction had been meticulously perpetrated" (Pane 1987, p.137).

In fact, the devastating destruction wrought on Lithuanian monuments first by the tsarist, then the Soviet, and again by the Nazi government also find in the total re-construction certainly a strong nationalist motivation, as well as a clear sociocultural, will be aimed at restoring the memory of the past. Thus, as in the case of the total reconstruction of the Castle of Sigismund Augustus, everything is reproduced according to the possible original styles: stone jambs of windows and doors, frames, shelves, decorations, and after more than a century of good or bad restorations, Viollet le Duc has once again proved right. But in this case, rather than giving value to a method, reason and strength are given to the cultural needs of a nation that only tries to reconstruct a broken historical-cultural identity (Fig. 23.7).

This is what has happened and continues to happen in many of the realities not only of the countries of the East (Warsaw, Dresden, Leipzig, etc. …) but also of western Europe. Let us think in this regard of the current example of the reconstruction still in progress of the city of Jeper in Flanders or of the still operational construction sites of

**Fig. 23.7** Vilnius. The reconstructions of the Palace of the Grand Dukes. *Source* Olimpia Niglio, 2018

the castles in the Loire Valley in France or of the most important construction site of the church of Notre Dame in Paris (Niglio 2004, pp. 27–30). This is just to mention a few examples but also to take the opportunity to underline how each nation has its own precise concept of heritage and conservation and being the monuments bearers of values, these can change over time and this variable must be kept in mind and recognized in respect of the plurality and cultural needs of the individual European communities and beyond. In this perspective, respect for the tools and methods of conservation that each nation develops and implements in relation to the different territorial, cultural and temporal situations.

Today the reconstruction of Sigismund Augustus Castle in Vilnius represents a significant example of the reconstruction of the nation's cultural identity and in fact, this place was a symbol of the rebirth of the city and celebrated during the numerous cultural events that took place in 2009 when the Lithuanian capital was named European Capital of Culture 2009.

However, since 1996 numerous political and cultural initiatives have been undertaken to define Vilnius, European Capital of Culture 2009, also participating in events organized by other European capitals. During these years, important architectural and urban renewal projects have also been carried out with the construction of new public spaces and buildings for culture. An example of good practice of how a city destroyed by war and colonization has found so much energy to rise again and become a very interesting case study to understand that at the base of rebirth there is always a cultural project and that without the support of culture it is not possible to imagine the future.

## 23.5 The Labyrinth of Cultural Identity

The Lithuanian experience opens new perspectives and observations on the theme of the reconstruction of cities, as well as territories, which in different parts of the world have been the object of destruction both of material values (monuments, residences, schools, public and private institutions, etc.) and of intangible contents (traditions, religions, knowledge, scientific development, etc.).

The history of humanity, through migratory movements, wars, and devastation, tells the constant regeneration of local identities and how these roots do not die but survive and then be reborn because the history of a people cannot be erased. A city dies when these identity roots of the communities are eradicated and destroyed and, on these actions, we must now reflect to appreciate also the Lithuanian experience that helps us to understand that nothing is lost even when we think that everything is now destroyed forever.

The identity of a community consists of multiple elements that are not limited to those that appear on the official registers such as the date of birth, the place of birth, the residence. Unlike this formal information, much more important identity principles follow such as belonging to a religious tradition, a nationality, an ethnic and linguistic group, a family, a community. All these memberships obviously do not all have the same importance, but none is totally insignificant. Each of these identities defines a person's "soul genes" and never the same combination is found in other people. Here identity is a unique, unrepeatable, and irreplaceable information (Maalouf, 1998). These individual identities conform the reality that surrounds us, generate the transformations of the territories to adapt these to the specific needs of the communities but in many cases the conviction, that there are superior identities compared to others, generates processes of cultural colonization and mistreatment of the territories.

In truth, these actions do not concern isolated cases; the world is full of wounded communities that suffer persecution destined to erase the identities of the communities. In fact, it is not known how it is possible to affirm the legitimacy of one's identity especially when it is offended by other identities.

All this opens up a labyrinthine scenario that helps us to observe reality from different points of view and to seek, in the multiple dimensions of society, an answer to the meaning of an often-ambiguous identity. In fact, the very word identity is an image of the legitimate existence of the person but at the same time, it is also an instrument of war and destruction.

This happened in Lithuania as it continues to develop in many countries of the world. All the changes that have taken place in recent years for pandemic or war reasons are always linked to very complex identity dossiers that find their roots in the pages of history. Often the victims are always the same and above all the scenarios that lie ahead are always the same, namely the death of the communities and therefore the death of the cities.

This ambiguity of the same identity now requires greater participation and public discussion capable of fostering the exchange of information and different opinions. In fact, in this labyrinthine social reality, it is necessary to implement constructive actions in which public discussion can play an important role in reducing cultural divergences represented above all by a lack of education and knowledge. The theme, therefore, has a cultural root that necessarily invests the ability of the individual to know how to relate to social multiplicity. When this relational capacity is questioned and people are invited to enter a perimeter, so that they can conform to a univocal behavior, then all those evolutionary processes that favor dialogue with diversity are canceled. Precisely this perimeter bond with the passage of time becomes a lethal weapon both for the individual and for the community to which he belongs (Sen 2004). These observations allow us to understand the inapplicability of universal values and conformity to principles that everyone must pursue regardless of their respective cultures. In fact, it is precisely this universal bond that prevents us from knowing each other and therefore from relating to diversity.

So, in analyzing the processes that have led to often irreversible transformations of our territories it is essential to refer to our respective cultures and from these understand the reasons for these changes both positive and negative, as well as the death of cities. In fact, our cultural roots have a strong influence on our behavior, our actions, and opinions.

Even the quality of life we enjoy cannot but be influenced by our cultural system. The same principle applies to our sense of identity and our perception of affiliation to the groups of which we consider ourselves to be an integral part. Our cultural identities are extremely important, but they must never prevail over the construction of dialogue with other cultures. All this finds important diplomatic motivations even within a social system characterized by different cultures within the same country or region. Meanwhile, it is precisely this lack of willingness to listen to diversity that is now connected with the war that marked the history of the European continent on February 24, 2022, when Russia began the invasion of the territories of Ukraine for very complex reasons but difficult to understand in a millennium that is projected to confirm the development and value of different cultures and its communities. All reasons are important to analyze deeply the history of Lithuania's periods of great transformation.

The experience in Lithuania and the current experience in Ukraine allow us to highlight the value of cultural freedom that constitutes that creative energy capable of generating and revitalizing our territories through actions aimed at social, economic, and political sustainability. But when this cultural freedom is put at risk, this means that some identities do not allow other identities and therefore other communities to pursue a traditional and freely determined lifestyle. The social repression of certain lifestyles is common in many countries of the world, but these actions only facilitate conflicts and therefore the death of communities.

Allowing individuals to live as they wish, and encouraging them to do so, can be a great stimulus for cultural diversity and the development of territories. In fact, only the freedom to pursue different lifestyles can make a society more culturally diverse, more economically prosperous, and more politically stable. The importance of cultural diversity will directly depend on the value of cultural freedom and this diversity will play a fundamental and positive role in the regeneration of urban spaces and territorial development. In fact, a culturally diverse society will bring only excellent benefits to others by providing them with a wide range of experiences useful for a fruitful and constructive evolutionary process (Sen 2000).

All this must allow humanity to reflect on the values of its identities in order to understand that the world is both spectacularly rich and desperately poor. Contemporary life is characterized by unprecedented opulence, and the control over resources, knowledge, and technology that we now take for granted is something that would have been difficult for our ancestors to even imagine. But the world in which we live is also a world of appalling poverty and terrifying deprivation ac-companied by desperate social contradictions that continue to create wounds in so many communities.

For this reason, Lithuania's experience is an interesting example of how the cultural will of a people can generate fundamental principles of rebirth to activate programs of social, economic, and political renewal of nations.

## 23.6 Conclusions

The application of new cultural paradigms has allowed Lithuania to start a path of cultural, social, economic, and political regeneration of great interest. In this context it is also useful to quote the words of D. Paul Schafer, Canadian educator, and UNESCO advisor, that affirms:

> There are many different ways to perceive and define development.

> However, most would agree that development is concerned with human needs and their fulfillment in the final analysis. Since people have a variety of needs that must be satisfied if they are to function effectively in society and survive—to breathe, bond, eat, love, create, procreate, recreate, work, and the like—this give rise to a complex set of social, economic, scientific, artistic, educational, recreational, spiritual, technological, political and environmental requirements. How these requirements are dealt in specific situations and particular parts of the world is what development in general—and the development of culture and cultures in particular—are all about. (Schafer 1994)

Therefore, starting from these cultural assumptions, the Lithuanian experience helps us to reflect on the difficult and complex compromises that communities, governments, and both national and international organizations face every day to implement development models that are capable of respecting cultural diversity and finding in these diversities the motivations for development.

So it is easy to understand how development models are closely connected to the cultural specificities of communities and that we cannot imagine sustainable models without considering these peculiar characteristics of individual territories.

In fact, as also promulgated by the 2030 Agenda and therefore by the Sustainable Development Goals issued in 2015 by the United Nations, the development of communities is guaranteed when programs are designed on the real needs of people and their needs: education, food, employment, health, security and much more. Otherwise, when everything is designed without taking into account the local cultural specificities and therefore the "genes of the respective identities" then only conflicts and revolutions arise.

The reasons well expressed by the Lithuanian experience, and the current Ukraine crisis underline that no development and no peace is possible without a cultural project that is capable of addressing social, educational, commercial, technological but also aesthetic, and spiritual issues. This shows that the economy is not the factor of development, but one of the many architects whose results will be positive if properly based on cultural principles. This principle is underpinned by the outcome of the economies of developing countries. When these countries think they are progressing by copying other realities and forgetting their cultural roots, then we are not witnessing progress but only a regression facilitated by colonization which in turn produces internal conflicts.

So, a social system dies, and therefore the city dies, as the principles that govern its development are external to the needs and requirements of people.

It is now necessary to work on the components of different cultures, on the enhancement of local cultures, and above all to enhance the uniqueness, creativity, excellence, integrity, and diversity of everyone, because all this favors the growth of the main financial resource represented by human capital without which it is not possible to achieve any goal. For this reason, the Lithuanian experience actively involves us in a local regeneration program thanks to which to activate programs of humanization and education to heritage and culture, fundamental for a sustainable future and based on the centrality of human heritage (Niglio 2016, 2021).

## References

AA.VV (1971) Lietuvos pilys. Mintis Publisher, Vilnius
AA.VV (1994) Report. Convention Concening the protection of the world cultural and natural heritage. World heritage committee, Phuket, Thailand, Dec 12–17, pp 44–55
AA. VV (1995) Vilnius Pilis, The Vilnius Castle in old Photografhs. National Museum Press, Vilnius
Andriani C (2010) Il patrimonio e l'abitare. Donzelli Editore, Roma
Assunto R (1983) La città di Anfione e la città di Prometeo. Idee e poetiche della città. Jaca Book, Milano
Brandi C (2000) Teoria del restauro. Einaudi, Torino
Calabi D (2004) Storia dell'urbanistica europea. Mondadori, Milano
Glemja J (1971) Conservation et restauration des moniments en Lituanie, in *Monumentum*. VII: 54–58
Glemja J, Jaloveckas R (1984) The renewal end restoration of part of the old town of Vilnius. Monumentum 27(1):71–75
Gropius W (1953) Un nuovo capitolo della mia vita. Casabella 199:19–22
Maalouf A (1998) Les identites meurtrieres. Éditions Gresset & Fasquelle, Paris
Marcus S (2015) Engels, Manchester, and the working class. Routledge, London
Moody H (1962) The city: Necropolis or New Jerusalem? C&C. 17:154–156

Niglio O (2004) Conservazione e valorizzazione del patrimonio architettonico ed ambientale in Belgio. Progetto Restauro 30(X):27–30

Niglio O (2009) Consigli edilizi, deputazioni e commissioni d'ornato nell'Italia del primo Ottocento. In: Iacobone D, L'ospedale Serbelloni a Gorgonzola (Milano). Gangemi Editore, Roma, pp 95–104

Niglio O (2016) Il Patrimonio Umano prima ancora del Patrimonio dell'Umanità, in *Cities of Memory*. Int J Cult Heritage Risk 1(1):47–52

Niglio O (2020) La biografía cultural de la ciudad histórica en la edad contemporanea. EdA Esempi di Architettura 2:1–9. http://www.esempidiarchitettura.it/sito/journal_pdf/PDF%202020/21.%20EDA_2020_2_NIGLIO%20SORIENTE.pdf

Niglio O (2021) Towards a humanist education: understanding cultural heritage to redesign the future. Acad Lett. https://doi.org/10.20935/AL3223

Norberg-Schulz C (1979) Genius loci: towards a phenomenology of architecture. Rizzoli, New York

Pane R (1967) *Attualità dell'ambiente antico*, Firenze: La nuova Italia

Pane R (1987) Città antiche edilizia nuova. In: Civita M (ed) Attualità e dialettica del restauro. Educazione all'arte, teoria della conservazione e del restauro dei monumenti. Solfanelli Editore, Chieti

Schafer DP (1994) The challenge of cultural development. World Culture Project, Markham

Sen A (2000) Lo sviluppo è libertà. Perché non c'è crescita senza democrazia. Mondadori, Milano

Sen A (2004) La democrazia degli altri. Mondadori, Milano

Turri E (2003) Il paesaggio degli uomini. La natura, la cultura, la storia. Zanichelli, Bologna

Venclova T (2019) Lietuvos istorija visiems (Lithuanian History for Everyone), vol 1 and 2. R. Paknio Leidykla, Vilnius

Venturi Ferriolo M (2002) Etiche del paesaggio. Il progetto del mondo umano. Editori Riuniti, Roma

**Olimpia Niglio** Ph.D., is a Professor in Architectural Restoration at the University of Pavia (Italy) and a permanent visiting professor at Hosei University in Tokyo where she worked until December 2021. Until 2019, she was professor at Kyoto University Graduate School of Human and Environmental Studies. She is the director of EdA Esempi di Architettura International Research Center.

# Chapter 24
# Lose the Memory, Lose the History, Lose the City

**Salvador Urrieta García and Veronica Zalapa Castañeda**

**Abstract** The built city remains in crisis and faces an eventual "disappearance" due to anthropic causes, in a process of urbanization that deconstructs the historical city, is the case of Mexico City. Some factors that combine and give rise to a phenomenon that transmutes the urban universe, these are: The social, cultural, economic, spatial, and built heritage transformations, product of the modernities that put cultural heritage in doubt. An "incontinent" urbanization that reveals itself not only in socio-spatial degradation, but also in a dematerialization and a deterritorialization that dislocate the inhabitants of their geography. The historical city represented by polycentrism and socio-spatial diversity is geographically dislocated, digital space comes into play and the urban heritage becomes disjointed from urban planning. The future of the urban heritage falls into a dichotomy around its benefit, in a dynamic that has tourism as its guiding axis-social benefit or mercantile benefit, that is, predatory tourism or responsible tourism. Living and dynamic urban conservation opens a possibility of mitigating urban mutations, through local urban projects, which nuance the future of neighborhoods, supported by citizen action.

**Keywords** Urbicide · Urban heritage · Cities · History · Local project · Incontinent urbanization · Cultural tourism

## 24.1 Introduction

Since 2009 the United Nations reported that most of the inhabitants of our planet live in the urban environment, within it, numerous cities have faced various problems and inequalities that have been accumulating in recent decades translating into a crisis.

---

S. Urrieta García (✉) · V. Zalapa Castañeda
School of Engineering and Architecture, National Polytechnic Institute (IPN), Mexico City, Mexico
e-mail: surrieta@ipn.mx

S. Urrieta García
National System of Researchers (CONACYT), Mexico City, Mexico

© The Author(s), under exclusive license to Springer Nature Switzerland AG 2023
F. Carrión Mena and P. Cepeda Pico (eds.), *Urbicide*, The Urban Book Series,
https://doi.org/10.1007/978-3-031-25304-1_24

The above can be considered a socio-spatial crisis that had already been manifested by Rene Shoonbrodt since 1987, the newly imposed process of globalization, the author spoke of the "destruction of the cities and the countryside," and a critique of the neo-liberal model for the accumulation of capital, and because it tends to hoard the innovation, the research, the discoveries, to adapt it to the environment of consumption, concentrate the organs of production, also control the demand and, therefore, control the mode of life of the inhabitants; their morality, says Shoonbrodt becomes a collective morality (Schoonbrodt 1987, p. 9).

The economic crisis of the 80s, derived from the oil crisis of the 70s, was revealed in the urban universe at an international level, which is why it is analyzed collectively, thus, in 1989 a book on Latin American cities in the crisis ("Las ciudades latinoamericanas en la crisis") was published, we are located in a significantly urbanized region of the world: Latin America. It is a collective work coordinated by Schteingart (1989), where different topics are addressed, such as the negative features of development models (supported by neoliberal policies), which are contrasting with the dynamics of each Latin American country. This text points out how the crisis affects urban life in different areas such as housing, public services, health, education, among others.

The planetary diffusion of this phenomenon manifests itself only a few years later in the journalistic sector; "Le Monde Diplomatique" (1991) points out a kind of urban pathology, in a special issue of this newspaper entitled "Way of seeing: the city everywhere and everywhere in crisis".

Among other authors who give us their vision of this urban crisis we have Mike Davis and Oriol Bohigas. In 2006 Mike Davis published his well-known work called "Planet of Slums", which tells us about a really pessimistic urban future, but at the same time realistic because it talks about world statistics that pose "a growing inequality between cities of different sizes and different economic specializations, within each of them" (Davis 2006, p. 12). It is about a demographic growth that does not go hand in hand with an economic growth, or "a superurbanization whose engine is the reproduction of poverty and not the creation of jobs" (Davis, Op. Cit., p. 20), is the neo-liberal model as a reproductive machine of "favelas, bidonvilles, or lost cities" (Davis, Op. Cit., p. 21) notes that in place of crystal cities that rise towards the sky, a good part of the urban world of the twenty-first century lives so sordid in pollution, excrement, and decay (Davis, Op. Cit., p. 22).

Trying to point out some of the reasons for these uncontrolled growths, it is appropriate to refer to the architect Oriol Bohigas who in his work proposes an extremely suggestive title for our purpose, "Against urban incontinence: Moral reconsideration of architecture and the city (Contra la incontinencia urbana: Reconsideración moral de la arquitectura y de la ciudad)" (Bohingas 2004), in his analysis (in a European context, but applicable to other geographies) relates architecture and the city, makes a critique of architecture that occurs simultaneously as part of a social reality, pointing out: "This architecture has lost attributes that were its own" and that of course is revealed in the urban landscape. Meaning that, the production of the contemporary urban space, tends on the one hand, to magnify the architecture of the spectacle of a "Star system" and on the other hand, to stigmatize the architecture minor, taking it as "low quality", due to the overproduction of a modest home in the cities and by "the pre-eminence of commercial and financial factors" with which it is produced. The latter is serious, given that as Bohingas (2004) himself mentions in his critique, this modest architecture gives consistency and congruence to the traditional city.

However, Bohingas (2004) takes a position contrary to the "conservation of the extreme", opens a way of analysis regarding the "historical heritage" and advocates the search for the continuity of the city, which we interpret as the articulation of the historical space with the new ways of inhabiting the city.

We maintain that historical cities have lost socio-spatial qualities that refer us to the issue of the urban crisis, that is, the loss of urban heritage, the loss of the ancient city; this loss implies losing memory and history, then. Does the crisis of the historic city put us on the threshold of an urbicide?

Certainly, the urban crisis is not new as already pointed out, and the urban problem is very diverse, it has to do with different countries and cities, according to their geographical location, political situation, social, economic, financial, and cultural conditions. Of course, there are rich and poor, solid, and vulnerable cities, all products of their history and contemporary realities, all this reflects an unequal world. At this point, how to approach the loss of the city?

Among the scholars of the city who have written about the death of the city, we have Françoise Choay, who in 1994 on an exhibition on the "The city, art and architecture in Europe" wrote an article called "the kingdom of the urban and the death of the city", referring to the European context, Choay seeks to answer one question:

> Is it not time, then, to admit without remorse the disappearance of the traditional city and also to ask ourselves about what has replaced it, in short, about the nature of urbanization and about the non-city in which the fate of advanced societies seems to have become? (Choay 2006, p. 167)

The reading (between the lines) of this author, who talks about the construction of "the new Babel", of "a confusion of scales, which entangles the urban scene and makes indiscernible the difference of the interests at stake and the actors who confront each other here" (Choay 2006, p. 196).

Almost twenty years later Choay takes up the subject in three texts produced with a difference of more than forty years (1969–2011). The first "Espacements" (space-ments) addresses the role of spatial planning in different time periods: the Middle Ages as a space of contact, Classical (Renaissance and Enlightenment) as a space of spectacle, the nineteenth and twentieth centuries as a space of circulation and the current period as a space of connection; to remember the role of territory as a basis of human societies. The second text "Le De re aedificatoria et l'institutionnalisation de la société: Patrimoine: quel enjeu de société?[1] L'évolution du concept de patrimoine"; as a key element of a globalized society, which seeks to benefit from heritage in the life of historical cities. The other text alluded to is "La terre qui meurt" (Choay 2011) (The land that dies), as an indication of the development of the built heritage in the new urban territories, with new technical and scientific knowledge and new social practices.

---

[1] ¿Patrimonio que compromiso de la sociedad? Translations are our own.

We have taken these references from Françoise Choay to ask ourselves about the future of our cities, of the built universe, particularly about the so-called urban heritage, in the face of a globalized, commodified, and individualized world that requires another conception and practice of the territory. It is a social and historical heritage in jeopardy, with the risk of disappearing.

To address this issue of an uncertain urban future and particularly of the historical city, we propose, first, to present a conceptual scheme that aims to synthetically explain the phenomenon of the loss of the historical city or of a possible preservation of the patrimonial wealth. To expand the analysis, five subtopics are analyzed:

- The transformations of the historical city, starting from a fact, cities are transformed over time, in this sense we take as a reference case the City of Mexico that has historically been mutating in its social and spatial components.
- The arrival of "intense" urbanization that is phagocytizing rural and urban land, focusing on the latter, including its inherited spaces, that is, urban heritage. At the same time, we observe hypermodernity, with it cyberspace; a deterritorialization and a dematerialization of urban space that disturbs the ways of daily life in the city depending on various local factors.
- The disconnection of urban planning with urban conservation, breaks with the principles of the latter and prescribes with this, an uncertain future for the inherited city.
- According to the general trends in the historical sites, a disjunctive is presented in terms of the use of heritage spaces, in terms of use value or exchange value, in terms of tourism versus daily life and social benefit or tourist benefit.
- In connection with the previous points, the issue of the treatment of heritage space based on conservation projects that are usually absent or have an inappropriate logic is addressed.

## 24.2 Lose or Preserve the Historic City

The loss of the historic city supported by its urban heritage has been the subject of the work of Françoise Choay (among other authors) who underlines, among other things, the loss in the competence to build (Choay 1992, pp. 187–198), which refers us to the trivialization of cities, without nostalgic tints this author shows under a humanistic perspective that this architectural production of the past with its historical and scientific values have proven the ability of the human race to create a spatial wealth, which affects the well-being of the inhabitants and from this fact the relevance of conserving that human production. It is about looking at the experience of cities, including their mistakes, but above all their values, for an opportunity for the future of the built city. That is, to work on the planning of the city supported by the politics of memory.

On these premises we have built a conceptual schema to visualize with some optimism, a horizon on the urban future (Fig. 24.1).

# 24 Lose the Memory, Lose the History, Lose the City

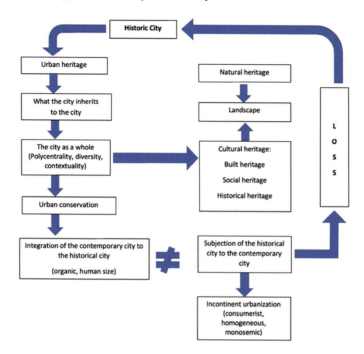

**Fig. 24.1** Conceptual basis from the theories of F. Choay and G. Giovannoni. *Source* Own elaboration

Thus, the scheme begins by pointing to the historical city, the one that carries a cluster of narratives and facts materialized in the inherited spaces. These spaces have been called ancient and/or historical sets and in general terms as architectural urban heritage.

This heritage can be very diverse due to the spaces that accumulate or juxtapose in such a way that, over time, socio-spatial legacies are generated that are successively bequeathed in a transgenerational way, hence we see urban heritage as the city that inherits the city.

The city as it grows contains differentiated spaces, meaning, historical sets and other pericentral and peripheral spaces that, without being equal, maintain socio-spatial relationships, thus, historical sets cannot be conceived without their neighboring and peripheral contexts, that is, the city must be seen as a territory that functions as a whole, although differentiated and with obvious inequalities. In this totality, different urban centers that create an urban diversity manifested in urban and architectural contexts are present.

From this historical city, the natural and cultural heritage emerges; from the first we can point out, in the case of Mexico City, a landscape already lost, by losing its lakes and rivers; from the second we consider the social, historical, and built heritages; they are cultural heritages that one aspires to recover or preserve.

Since the theory and practice of urban conservation began at the beginning of the twentieth century, the idea postulated by Giovannoni (1931) was to incorporate modern urbanism to the traditional city, in order to maintain the values of the organic city as its human scale, neighborhood contact; adding technical innovations linked to urbanism and other disciplines.

What has happened has been the opposite phenomenon, meaning that, the ancient and historical complexes have been subjected to the neoliberal dynamics of an urbanization that does not stop growing, that is predatory of space and society, consumerist, supposedly homogeneous and monosemic, since the value that prevails and means the city is economic, where exchange value prevails over use value.

From the patrimonial perspective, this type of wild urbanization tends to the loss of the natural heritage that is fractured vis-à-vis of the cultural heritage and all this is revealed in the loss of memory, the loss of history and finally of the historical city, with its landscapes and local cultures.

To delve into the loss of the historical city or an eventual recovery of cultural heritages we present the following elements.

## 24.3 The Transformations of the Historical City

It is a fact that, as time goes by, cities change, although in the best of cases the urban tissue seems to remain the same. The truth is that the generations of inhabitants are inexorably changing and with it also the ways of life and spatial contexts.

We have seen that in the history of cities, framed in urban revolutions, the first (in general terms) corresponds to a long period that Marcel Hénaff calls "the monument city" (2008), a long period that goes from the Neolithic period when the ancient cities began to form, both in the east and in the west. The above, up to the arrival of the industrial revolution that brings another urban revolution that the same Hénaff called "the city machine", which, among other things, for its technical innovations will lead to the expansion of the old cities, but also the emergence of new cities that reveal the hand of modernity. In a second period that ends until the end of the twentieth century when another urban revolution will appear, accompanied by the so-called "new technologies" of information and communication (ICT's).[2]

In these first two periods, mainly the aforementioned cultural heritage would be generated. This is the story of the cities (Mumford 2012).

Nowadays the issue of cultural heritage is a planetary issue, which has involved the official interest of many countries and the creation of international organizations such as UNESCO. And it is basically during the twentieth century, particularly in its second half, that the theoretical and methodological foundations are laid, as well as numerous interventions for cultural conservation.

---

[2] See: To deepen in the relationship of modernity and urban revolution see: François Ascher, Las tres revoluciones modernas ("The three modern revolutions") (2001, pp. 9–53).

With this logic, a movement called "New Urbanism" was born around 1980, which advocated an urban conservation policy in a letter called "Principles of the new urbanism" ("Principios del nuevo urbanismo") (1993) aimed at revaluing the neighborhoods and their plural uses, prioritizing pedestrianization, collective transport.[3] It is worth highlighting the desire to value the local context with its climate and an architecture that preserves its traditional forms, materials, and construction systems.

As can be seen, there is a whole movement that leans towards the principles of urban conservation instituted internationally; in this way there is a list of historical cities that, in principle, are protected by national and international organizations and legislations. Above all, those who signed the "Convention concerning the protection of the world, cultural and natural heritage" (UNESCO 1972) and especially those that make up the UNESCO list of "World Heritage Cities", which are more than two hundred around the world.

It should be noted that the degree of conservation of these cities is a function of their cultural, technical, and scientific resources. Among the cities that includes as world heritage cities is Mexico City.

As it has already been documented, Mexico City has undergone many transformations in what was the ancient pre-industrial city, both social and spatial changes, product of the passage of time and different generations that adopted and adapted the inherited spaces according to the lifestyles of the moment and new customs; the city tended to systematically reproduce on the same surface.

In order to illustrate these spatial transformations, we will point out some of those changes and destruction, which explain and contextualizes the inherited city, as well as lost spaces and monuments. At the end of the twentieth century, this historic city deserved to be declared a monument zone.

The transformations took place at two important moments, in different senses and linked to modernity: the first occurs at the end of the eighteenth century, with the arrival to New Spain of the Bourbon imperial family that brings another ideology regarded to illustration's current and form of administration for the viceregal capital and that has a significant socio-spatial impact on the urban context.[4] The second moment of transformation occurs with the establishment of the Mexican Republic, which will imply the beginning of the end of an urban way of life and the disappearance of religious hegemony signified by its monuments as well as an expansion of the Mexican capital city that grows exponentially in the twentieth century.

Around 1778 the Viceroy Antonio María de Bucareli built and inaugurated "Paseo Nuevo" (which later became "Paseo de Bucareli") that marked a social and spatial change in the mobility of the city. But above all it is necessary to emphasize the

---

[3] See: Perspectives for a new urbanism ("Perspectivas para un nuevo urbanismo") (Krier 2013, pp. 111–143).

[4] Regina Hernández Franyuti offers a broad outlook on this socio-spatial transformation in order to the ideas on the "new city" pursued through Ignacio Castera's proposals (1794) and later in Independent Mexico, with the ideas of Ortiz de Ayala (1932) (Hernández 1994), Ideología, proyectos y urbanización en la ciudad de México, 1760–1850 (Hernández, pp. 116–160).

changes that took place at the time of Viceroy Vicente Güemes called de Revillagigedo from 1789. Thus, with regard to urban services, after the failed attempts to illuminate the city at night, in 1790 Bucareli inaugurated the public lighting with "12128 glass lanterns with tin leaf lamps, supported by crow's feet" accompanied by a regulation (Castellanos 1998, p. 22), for a city of about 80 thousand inhabitants in 1753 (Santiró 2004, p. 36),[5] also "forbids littering in the streets that he also paved, built sewers and increased the ditches and built gateways and jetties, with which the flow of people and goods was controlled" (Jiménez Vaca 2017, p. 133).

The second moment in the transformation of the city marked by the "Ley Lerdo of 1856", on the alienation of the assets of the church, which would change the religious landscape with towers and domes that represented that religious life of the city.[6]

For Guillermo Tovar y de Teresa, a prestigious specialist in New Spanish art, points out that: "although it is dangerous to make calculations, if we consider that by the 1770s there where about 80 churches, parishes, temples holders, hospital chapels, convents, and schools in the city… in Mexico City there should be more than 350 altarpieces" (Tovar de Teresa 1992, pp. 12–13). This author tells us that, "given the lakeside condition of the city, it had to eat itself to grow…. Each year of the many of the 17th and 18th centuries, a church was demolished to build another one in the same site… The sets that were kept before the neoclassical and liberal destruction of the 19th century, the monastery cities, were harmonious by motley, rich samples - veins forever lost - that if they had existed, they would have been occasion of astonishment for the public and of endless sight for the historian of art worried by the 17th and 18th centuries" (Tovar de Teresa 1992, pp. 7–9).

The heritage inherited by the viceregal city and subsequently lost, was not only physical but also socio-cultural, that is, it also deserves to take into account the New Spanish society and in particular the fraternities in Mexico City, which had a primary role in the city, in everyday social practices and on holidays.

Although the economic power of the powerful fraternities had already been affected since the end of the nineteenth century, this involvement deepened in the first decade of the twentieth century, the dissolution of the different fraternities (of Indians, Spanish, guilds, African-Americans and mulatto), which meant the end of a very ancient form of social organization, which, according to Alicia Relying:

> Forged relationships between the established powers (civil, church, etc) and the civil society; the economic mechanisms of circulation and accumulation of capital; the history of the religious forms and rituals, expressions of everyday life, as well as the formation of the first systems of social assistance and mutual assistance in Mexico. (Bazarte Martínez 1989, p. 15)

---

[5] La población de la Ciudad de México, en 1777, Ernst Sánchez Santiró, Secuencia, 60, septiembre-diciembre, 31–56, Instituto Mora, 2004. There is a different data on the of Mexico's city inhabitants in 1811 found in the document referred of Sánchez Santiró vs Ma. Dolores Morales (En: Ma. Dolores Morales La distribucion de la propiedad en la Ciudad de México, 1813–1848, Ensayos Urbanos. UAM 2011, p. 67).

[6] The religious communities of both sexes comprised: fraternities, confraternities, congregations, brotherhoods, parishes, town halls and schools (Bazarte Martínez 1989, pp. 129–137). Dolores Morales refers in regard to the confiscation and its influence in the structure of the property and the building permit changes. In Ensayos urbanos, Morales op. cit. pp. 80–231.

These social groups had a freedom of management with respect to the church and also "contributed to the conservation and aggrandizement of the artistic heritage" (Bazarte Martínez 1989, pp. 191–193). As Mexico City has been described, it was transformed in its physical and social forms, with what has been called, modernity.

So far we have referred to the city that according to Mexican laws would belong to the historical order, that is, until the end of the nineteenth century (Diario Oficial de la Federación [DOF] 1972), but what about the transformations of that "other city that made history during the twentieth century?

As Adolfo Gilly (1990) said "The first modernity arrived in Mexico as a catastrophe, it was the destruction of the pre-Hispanic world, the destruction of the old and the coerced and forced implantation of the new. Urban modernization came a short time later" (Perló Cohen 1990, p. 13).

In this regard, we cannot omit the post-revolutionary and modern city that emerged between the 30s and 80s of the last centuries, of which only reminiscences remain. That is five decades of a city with a new architecture and an urban infrastructure that became "obsolete", according to the socio-spatial dynamics that "modernity and postmodernity" would demand.

During the period from 1930 to 1980 the city not only grew exponentially (geographically and demographically), but also had important transformations in various areas. As happened in Europe in the nineteenth century with the arrival of modernity, which came on wheels. Thus, the increasingly intensive use of cars changed the landscape of Mexico City, both on the rolling surface, and in all those places susceptible to parking cars.

During this period, the morphology of the city underwent important transformations. In the architectural field, some styles emerged, namely: Academic (Neoclassical, Republican and Porfirian) and post-revolutionary (Nationalist, Art-deco, Functionalist, Emotional, and International) (De Anda Alanís 2019) and others referred to as eclectic and Californian.

With regard to streets and avenues of the old city today the Historic Center, new roads were opened (Pino Suarez, Leandro, San Juan de Letrán, Izazaga, Leandro Valle, November 20, May 5, etc.), for this, buildings had to be demolished and the old urban fabric modified.

Another aspect that needs to be pointed out regarding the transformations of the city is the toponymy, which changes with the arrival of modernity. Thus, by referring only to the ancient city, along with the buildings and socio-spatial practices, the names of the streets that at the time identified the city with its inhabitants left, the names of the new heroes replaced the names that occupied those roads according to some religious buildings. This change of urban nomenclature will be intense throughout the twentieth century and will extend, for example, on fast roads where the names of old streets will be replaced by "functional" and anodyne names such as axis 1 south, axis 3 north, among many others. We wonder, does modernity imply the loss of memory and the city?

Beyond the Historic Center, in the early 70s there was also a transformation in the road structure of the city with the creation of the "road axes". It should also be noted the disappearance, modification, or reduction of gazebos.

In relation to facilities that had an important social life, we will mention an entire cinema park that housed large theaters, which disappeared or became, at best, "multiplexes" (set of small theaters), but no longer fulfilled the social role of before, especially in many neighborhoods of the city, which enjoyed this local entertainment. The neighborhoods were the object of social practices, which made the use of local space something more organic, which brought neighborhood contact closer.

As if it were necessary to mark a date in the transformation of the city, on September 19, 1985, an earthquake came to change the picture of the city, especially in the central city, which survives with the sign of the dispute.

It is necessary to recognize that urban spaces have their temporalities in terms of the creation and use of urban spaces, however, these temporalities do not necessarily have the need to destroy to build, although this may be valid in some cases, for the collective benefit that it may entail.

## 24.4 Incontinent Urbanization and Urban Heritage

Since the different types of towns were created in Mexico City in the second half of the nineteenth century, the city showed a heterogeneity in them, due to the socio-economic differences of the population. This heterogeneity continues as the city grows; in the post-revolutionary period, particularly from the 40s, the exponential growth of the city begins, the first housing complex, called "German President", emerges, this prelude to the housing units of the mass city, a city prone to losing all kinds of scale, especially the human.

Urban growth has to do with two substantive elements that make up the city: the urban fabric, and the social conglomerate (civitas) that gives reason to be to the physical space, thus also establishing a social heritage.

It is necessary to consider the entire city as heritage, and so, urban heritage over time is what the city inherits to the city, including the narratives that the city produces, a universe of stories that give meaning to the city. Thus, the historical city is produced. In this way, the history of the places can also be considered, within which the neighborhoods are counted, their open spaces and monuments, main or contextual.

In Mexico City, in addition to the growth of the city that had been taking place for decades in an intense way, it presents a socio-spatial metamorphosis; with the advent of neoliberalism and globalization, the aim is to homogenize the city and its citizens, especially from the consumption driven by the large chains of stores that inhibit the local trade, with this type of consumption the neighborhoods are blurred (many had already been disturbed by the tertiary sector), their daily use and their social cohesion, with the above, also begin to dilute places and trivialize the monuments that signify to the urban space by providing it with symbology and memory. We are facing an urban mutation of heritage.

In the incessant growth of cities there is a rupture between the city and nature, this brings to mind what Berque (1987, p. 16) tells us: modernity gradually created an abyss between culture and nature. A rupture of the "Homo urbanus" (Thierry Paquot dixit) with its environment, it generates a kind of shipwreck of the human being in that sea of asphalt that we call a city, but which in reality has become an urban magma, where the time-use of individuals is accelerated or distorted, especially with the use of electronic devices that become the prosthesis of the "hypermodern" human being.

With these devices the time is measured according to the duration of the messages issued or received, the rhythm of life is subject to the rhythm of the phone, to its sounds or alerts. The conversation is possible with several groups in a short span of time, you can also have the "gift of ubiquity" on the planet; all this relativizes time. At the same time, we enter the dematerialization of places, previously created by people, places are no longer necessary to communicate face to face. A romantic date takes place in the same space as a business date, an online purchase, a theater performance, or a football match, that is, in cyberspace.

This dematerialization of the city had already been visualized by Melvin Weber since the 60s, as stated by Choay (2006, pp. 199–217). According to this author, Webber's position is based "on his relationship with technology ... as a global social fact linked by a spiral of feedback to all the physical and mental dimensions of our societies" (Choay 2006, p. 202) and with these dimensional transformations, Webber (year) estimates the obsolescence of the city and advocates "a new kind of urban society of vast scale ... increasingly independent of the city" (Choay 2006, p. 207).

Within Webber's ideas, he also advocates the "erosion of localism" along with the "de-spatialization" of the city, the latter leads us to review the value of the local in our cities. But also, the values of the human being with the environment that surrounds him, meaning, his local geography. In this regard Berque links ontology to geography to affirm that:

> The place is necessary to the being for it to exist, it is not a simple place or simply chartable: this place supposes an indivisible relationship between two parts: the topos and the chora. The first would be a physical envelope, limited and totally identifiable; the second, which corresponds to the dark mass of astrophysicists, would be of the order of the symbol, even more of the feeling of things. (Bouteille 2005)

For Berque "the human being is a geographical being" (Berque 1987, p. 10), hence claiming that the inhabited world is fully the abode of the human being, which establishes a relationship to the ecological, technical, and symbolic of humanity in the land (Berque 1987, p. 17). So the inhabited world represents every human environment, but it is first and foremost, the geography of the, there are things where our existence begins (íbid.).

The work of Berque refers us to the thought of Alberto Magnaghi, who presents a territorial approach with the perspective of sustainability that seeks to "heal the human environment", to put the value characteristic of each place, this with the idea of a local self-sustaining development (coming from below), thus, the territory could "experiment with new forms of communities, economies of solidarity and public space" (Magnaghi 2000, p. 7), which would go beyond a simple localism. All this would have, a very different vision from Webber's.

Magnaghi, takes a position contrary to the figure of the metropolis, because he says, denies and destroys the city and maintains that:

> Since the second half of the 20$^{th}$ century, the metropolis differs from the historic city in all its forms, it has an absolute contempt for the constitutive rules of the identity of places, to which replaces an abstract, artificial rule, different from the relationship with the territory, to which, with different modalities, the city always conformed. (Magnaghi 2000, p. 14)

This author tells us about the morphogenesis of the contemporary metropolis in terms of hypertrophy and "topophagy", given that in the long term the metropolis lives and grows destroying the reproductive capacities of its own environment, but also engulfing the resources of more remote territories. Is this another side of urbicide?

The deterritorialization seen by Magnaghi disturbs the relationship of human settlements and their environment and leads to a territorial amnesia before the indifference of the places in which to live, which results in the rupture with the memory and the history of the places. Faced with this vision, it is opportune to remember that history and memory are at the base of the conservation of the historical city and its places. The incontinent growth of the city dislocates us in time and space and with this we will really know that it is being well inherited to future generations.

## 24.5 The Disconnection of Planning with Urban Conservation

Urban planning constitutes one of the substantive elements for safeguarding and enhancing cultural heritage; at the international level, since the 70s, the importance of urban planning plans has been established, where the principles of integrated conservation, already expressed by Giovannoni (1931), precursor of conservation and author of the term urban heritage, are taken up. In fact, the basic principle is the integration of the new urban building to the old city, with all that this implies according to the character of each historical ensemble (function, size, formal value, land use, technical state for its conservation and degree of animation) (Ostrowski 1976, p. 99).

The growth of the city, the appearance of new centers of gravity or new centralities, make it necessary that the plans for the rearrangement of the city take into account the values of the historical sets or inherited fabrics.

To the extent that the practice of urban conservation becomes an institutionalized international phenomenon, multiple associations arise that defend the social and spatial values of the inherited, even producing from recommendations laws and norms that seek to consolidate this conservation of cultural heritage. But this "heritage legislation" does not always have an empathy with local urban legislation, or they may even have conflicting or outright irreconcilable positions. The different historical cities of the Western world have kept different urban policies, logics and temporalities, according to their resources.

With regard to the relationship between urban planning and heritage conservation, it arrives belatedly in Mexico City with the 1979 Federal District Urban Development Plan that spoke of "recovering and protecting the sites and buildings of the urban cultural heritage" (Garza Villareal 1987, p. 404). In 1980 the Historic Center was declared a "Monument Zone" and in 1987 (the year in which the Historic Center and Xochimilco were named a World Heritage Site), a last version of the General Program of Urban Development of the Federal District (PGDUDF its acronym in Spanish) was published, based on a "public consultation" carried out in 1986. Here eight urban sectors were established corresponding to the Urban Centers and differentiating the Historic Center that corresponded to the Metropolitan Center. The version of the PGDUDF 1987–88 in its instrumentation provided for two types of special programs the ZEDEC (Zona Especial de Desarrollo Controlado) in certain spaces of the city and the Program of Improvement of the Historic Center translated into a "Revitalization Program" conducted by the "Subdepartment of Heritage Sites" (Departamento del Distrito Federal [DDF] 1989).

In general terms, urban planning obeyed the centralist logic of the state for the benefit of the ruling classes. The process of expansion or metropolization led to a degradation, both in the quality of the physical space of the central area of Mexico City and in the quality of life of its inhabitants. This area lost some functions, particularly that of housing for the benefit of the nearby periphery or inner city and distant or metropolitan, expelling population. At this time, François Tomas maintained that "the expulsions and the suppression of certain economic activities from the central city, did not obey a natural and inevitable evolution of the metropolis but the capitalist will to eliminate all obstacles and to allow a high profitability function" (Tomas 1990, p. 13).

At the end of the 90s, a democratizing process took place in Mexico City and the citizens were finally able to choose their authorities, this allowed to approach the planning of the historical complexes from another perspective, the new head of government stated that the Historical Center of Mexico City represented a priority for his administration. Before this event, the protection of the built heritage was given more at the level of listed buildings than the whole. The new administration proposed the "Program for the Integral Development of the Historic Center of Mexico City", which had priority areas and was structured based on four strategic lines of action, namely: (a) Rescue of centrality, (b) Housing regeneration, (c) Economic development, (d) Social development.

The integrality was postulated as the main objective and inescapable requirement in the historical complex, so it became part of the Planning Law of the Federal District (OGDF 2000b). In this administration the partial programs of the "Historic Center and the neighborhood of La Merced" came out.

During the twenty-first century efforts have been made by the City Government to order the planning of the Historic Center, through the agencies that the city has counted on for these purposes such as the Historic Center Trust and the Historic Center Authority, the latter has sought to coordinate the resources and interests of the public administration through Management Plans. Also, important actions have been undertaken by the Historic Center Trust to give it greater functionality and better habitability to it, however, the degradation suffered by its monuments has not been resolved, or also, the repopulation of the Historic Center, where many spaces are uninhabited or used as cellars. The social problems of the vulnerable population are still to be solved and economic interests prevail as hegemonic in the use of land and in the general functioning of this patrimonial space.

About the urban heritage-polycentrality relationship in Mexico City, the planning task expands beyond the Historic Center or "hypercentre", at least in the 175 Heritage Conservation Areas (ACP) that the city has, formulated by the Secretariat of Urban Development and Housing (SEDUVI, for its acronym in Spanish).

If already taking the guardianship of the Historic Center generates enormous difficulties and challenges for its preservation, even more so with its Management Plan and its Partial Plans[7] as instruments of regulation. This administrative entity does not have sufficient and efficient instruments for the management or of the ACP. However, the Law on the Safeguarding of the Architectural Urban Heritage of the Federal District of 2000 (OGDF 2000a) it extends its "protection" to other areas of the city (neighborhoods, colonies, complexes, monumental open spaces, and urban monuments), which already implies a recognition of the heritage diversity, including the "centers". It is necessary to have rules and regulations that are the vehicle of understanding and operability between the guardianship authorities of the urban heritage and the citizens who live the space on a daily basis.

Regarding this category of centers, the relationship of polycentrality and urban planning arises, which is taken up by Coulomb (2021) to point out the importance of centrality depending on the elements of urban dynamics present in spaces with this unobjectionable category. In these spaces, the role of public space is fundamental, according to Coulomb "they are a centrality par excellence, due to their multifunctional character and collective appropriation… the public space is centrality because it fully assumes the attribute of heterogeneity, both of uses, functions and users" (Coulomb 2021, p. 12).

Contrary to the benefits entailed by the centralities shown in heritage spaces (centers, villages, and traditional neighborhoods), their deterioration implies the loss of space, of encounter between generations, between genders and social classes. "The new shopping "centers" would be assuming, for certain sectors, a falsely socializing role: the socialization of the ghetto" (Coulomb 2021, p. 8).

For this author, the preservation of heritage must be seen with a general vision, not sector-based because they move away from urban problems, hence the importance of public space, which can federate the different historical and social centers of the city, including central and metropolitan peripheries. That is to say, a good management of the various centers can make the substantive public spaces of the city more accessible and articulated. It is not only about the return to the central parts of the city for an eventual compaction, but as proposed by Coulomb:

---

[7] The Historic Center of Mexico City, in addition to its World Heritage status, has some instruments such as tax "incentives" or the "transfer of potentiality" program. In addition to the Law on the Safeguarding of the Architectural Urban Heritage of the Federal District, which certainly has a broader spatial spectrum, but lacks sufficient rules that make the law efficient.

The return to the existing city would imply an important change in the forms of production of the city, and in particular of housing production. (Coulomb 2021, p. 5)

The return to the consolidated city implies a good collaboration between urban planning and the existing urban fabric, particularly with the ancient or historical fabric that may be dispersed in the contemporary city, as is the case of Mexico City. From the foregoing, we underline the importance of the ancient public space in the composition of the city as a whole, assimilating the old.

## 24.6 The Logics for Urban Conservation (Local Daily Life or Tourism)

The diversity of cities is part of the urban wealth, related to the function of these, for example, the cities of industrial, administrative, commercial, as well as those who have accumulated various riches as its built heritage and historical significance that contribute to make the city, a space full of narratives, constituting the "substance of the historical cities". They are expressions of the inhabitants and the built space, evoked to constitute a socio-spatial reference in each geographical territory. Thus, this historical space faces dynamics of socio-spatial change.

As its own history shows, urban heritage demands the need for economic support both for its preservation and for its management, which leads us to investigate where do the resources that are destined for such purposes come from?

The privilege of having a heritage has a cost, so sites and monuments are likely to be valued, not only culturally and socially, but also economically and politically. Some historic cities make use of this cultural resource as a source of economic support for its preservation or for profit.

For more than five decades, the creation of bodies, guidelines, instruments, oriented to the protection and conservation of heritage has been manifested. This task arises from the interest of those groups and institutions that protect the heritage, underlining the value that this cultural and natural wealth supposes for the local inhabitants. In this sense, the document called the Quito Norms (ICOMOS 2005), can be mentioned, it is one of the texts in which a clear position is expressed regarding the recognition of the social benefit that the economic value of heritage should have.

In this always diverse relationship between the economic, cultural, and natural, represented by heritage, the figure of tourism arises as a vehicle of financing and economic benefit, so we ask ourselves: who benefits from tourism?, tourism for what and for whom?, is tourism the way of development, progress and economic growth of historical cities?

In the world of the conservation of historical sites (which has become a planetary interest), many cities finance their interest in the cultural industry and the resources represented by tourism activity. In some cases, that is the main interest about the heritage, which explains the labels about some ancient, historical, or simply picturesque ensembles. It is for this reason that the tourism management of the built heritage is revealed as a cultural industry.

As part of the questions that we raised earlier about the benefit of preserving heritage, two areas seem to arise: one, the benefit for local inhabitants; and two: for all actors who are involved with tourism practices in certain heritage sites.

In the second case, the matter seems to derive in two directions, the first towards wild and predatory tourism, which ignores the rights and interests of local inhabitants; and a second is that which presents itself as a "reasoned, responsible, respectful" tourism of sites and monuments. It is towards the latter that we refer to in order to grant or not the benefit of the doubt as to its social will.

Among the instruments that promote responsible tourism, there is the International Council on Monuments and Sites (ICOMOS) that operates internationally.

As a result of the ideas and intentions of the members of this institution sponsored by UNESCO, there have been some letters that point to instruments that cater to this economic circumstance, such as the International Charter on Cultural Tourism (tourism management in heritage places significant) (ICOMOS 1999), in her, the estate is under the perspective of their value in both economic benefit to the local level, that is to say, through tourism tempered by the sustainability. There is talk of a bidirectional tourism—economy benefit referred to as the sustainable tourism industry (ICOMOS, Op. Cit., p. 2), which can provide the inhabitants who act as hosts, in addition, important means and motivations to care for and maintain their heritage and living traditions (ICOMOS, Op. Cit., p. 2).

## 24.7 Tourism for What and for Whom?

The previous question will have a variability of answers, in both approaches, contexts and temporalities. On the one hand, it is undeniable the presence of interests from personal (interests of the technicians who carry out the project proposals) coupled with the interests of the administration in turn, who seek to add to electoral political ends, which reduces the benefit that the local community could eventually have (Prats 2003).

This leads us to talk about the commodification to which the patrimony is subjected. We propose two paths to such a panorama, related to tourism; the first, "responsible or sustainable tourism"; the second, "predatory" tourism.

According to Carlos Hiriart, tourism can be both the best friend and the worst enemy of economic development […] irrefutable premise (Hiriart Pardo 2012, p. 18). The position of this author, specialized in tourism and heritage, converges with that expressed by ICOMOS.

Responsible or sustainable tourism obeys a logic that seems to point to the local scale, demanding the social benefit that entails a favorable management and development for the conservation of heritage and local development. That is, it seeks the consolidation of an ethical direction of tourism (Hiriart, Op. Cit.), which agrees with the institutional discourse.

On the contrary, the tourism predator is associated to an exploitation of the heritage on a plane which puts the value in use versus value in exchange, much of the soil as well as of their qualities, what Choay (1992) refers to as perverse effects, in which is immersed the dynamic ejection of the native inhabitants with the input of real estate, which leads, in some cases, the blurring of identity social as well as the venue itself, flanked by a series of trivialization.

Faced with such effects, the action, and the relative effectiveness of the intervention of existing international organizations is questioned, around the heritage issue and the consequences that a recognition of a site as part of the World Heritage List, a UNESCO instrument, can have for the space and its inhabitants.

On the one hand, it activates a certain awareness around the value, protection and conservation of the heritage, but on the other, awakens a sort of fashion for the heritage, recognition and legitimacy, by attracting attention to places somewhat unknown in which, as a good tourist attraction, sees parade both domestic and international tourists, a fact that adds to the factors that contribute to the commodification of heritage under the banner of the support granted by the UNESCO (Delgadillo 2014).

Faced with what is being said from the institutional field of heritage, Delgadillo (2016) takes a different position from ICOMOS. When studying housing in historical centers, he puts patrimonial wealth in front of social poverty (Delgadillo, Op. Cit.) and points out that in the case of Mexico City, urban conservation is selective and exclusionary, so it would end up privileging turistification and somehow gentrification (Delgadillo, Op. Cit.).

However, not all the territories and their societies allow them to be involved in such a homogenizer process, due to the resistance of the inhabitants, combined with a proper management of the policies that regulate the exploitation of heritage, tourism-related, which is conducive to such homogenization so accentuated that tend to be made invisible to the peculiarities of each one of the sites that are confronted with this phenomenon, making them similar in such a way that both the tourists as well as multinationals feel at home there (Choay 1992, p. 207).

Then, another opinion arises that suggests the future to us in terms of a dichotomy, that is, social heritage or touristification. Among the intellectuals who imbricate tourism with cultural heritage is Llorenç Prats, who debates the adoption and indistinct use of terms such as "sustainability", "development", among others, introduced in a quasi-fetishistic way in political and social discourses related to the heritage-tourism phenomenon (Prats 2003).

Prats, points out that heritage responds to a social construct that appears in the beginnings of modernity, as a kind of secular religion, which serves to sacralize discourses around identity (Prats 2006, p. 72). It is there that temporalities and interests intervene, according to the historical, political, social, cultural and ideological moment, the heritage is constituted and instituted.

These two sides of heritage lead us to ask ourselves: what can be expected from tourism and its effects?

Prats (2003) coincides with Choay (1992) in stating the existence of what he refers to as perverse effects and practices that also have counterproductive results in tourism activity that would harm the local economy. So, there is a kind of tendency to the generalization of tourism accompanied by forced visitors, an excessive patrimonial schooling that far from implanting interest, will generate a certain antipathy in the attendees, as well as the tourism of elderly and seasonal visitors (Prats 2003).

The above factors can lead to a patrimonial distortion in the local inhabitants, who may eventually fall into practices of identity theatricalization aimed at the economic purposes represented by the tourist introduction in their communities, in a kind of scenic invention that gives tourists what they want to see according to their expectations. On the other hand, it is possible to implement an adequate tourism and heritage management that is a participant in the local benefit.

So, who benefits from such tourism practices? Strictly speaking, it should be the locals who receive the greatest benefit, both economically and in their quality of life, taking care of the effects that are aggressive for their daily life, their culture, identity, and their economy. Since heritage is a fundamental element of cities, it contributes to the narrative of the world's cities, so it is necessary to prioritize its uses, and experimentation.

It is necessary to rehabilitate the housing function, instead of supporting real estate speculation that seeks to make urban land more profitable (Coulomb 2021).

## 24.8 The Lack of Local Projects

The concern and desire to safeguard the building heritage and its population, led since the 60s to the authorities of the City of Bologna to generate an urban project that marked a milestone in the history of urban conservation. This urban project has continued for decades but adapting to new contexts "making new nuances of the use of the city", critically requalifying the city and the territory and developing a living and active preservation. Emerges a conscious around the limits of the built environment's physical development regarding the availability of the built territory; and the urban expansion tent is questioned. Starting from idea that the monuments are not an obstacle for the development, there is adopted new politics of built space "retransformation" (as an only space available resource). Therefore, based in the conservation criteria (*restauro*, restructuration and reconstruction, it is critically requalified the territory and the city); and the social meaning is highlighted as well as the collective use of the ancient monuments through the time and they are defined such a "historical content", that explains and bases its reuse and restorative recovery (cfr. Scannavini 1998, p. 333).

The case of Bologna is inspiring to think about a local urban policy, which considers the memory and history of the places.

Local urban projects seem to offer an alternative for cities that have been constituted as an urban palimpsest, as is the case of Mexico City. The traces of other generations in the city are social and spatial contributions of a city that survives in the midst of the various crises of these times, oblivion contributes to the disappearance of the city, by losing the values that memory fixes in time and space.

One of the experiences we have been able to study (Urrieta García 2018, pp. 173–193), is the case of the district of la Medeleine-Champs de Mars in the town of Nantes, in western France, has allowed us to verify the importance of the urban projects that seek to reconcile the traditional city, with its values of social and spatial modern life, even with the idea of digital innovation. As a project, it was an urban recomposition in the continuity of pre-existing urban forms, to reassess the urbanism by which the neighborhood has been governed and to promote a contemporary architecture that dialogues with the historical and contemporary components (Urrieta García 2018, p. 174). The result of this project is in the current quality of life of its inhabitants.

In Latin America, the wide cultural richness of its historical sites, its urban dynamics, the economic interests at stake (such as the rent of the land), make urban heritage tend to dilute, this is the case of Mexico City. This loss occurs especially in those well-located places or neighborhoods, with a "minor" or contextual architecture and neighbors who do not have an economic solvency and who do not want or fail to organize to defend their cultural heritages, which are disappearing over time.

The action of preserving the urban heritage occurs in a differentiated way from city to city, from set to set and from neighborhood to neighborhood. This has its logic because in the case of historical centers, even with their similarities, all historical centers are different. In this sense, the contrast becomes more evident between the Latin American and European historical centers. At the international level, European references (due to their interest and long experience) are basic and illustrative; but as Paolo Ormido de Azevedo, who analyzes the Historic Center of Bahia in Brazil, points out: "European solutions do not apply to our reality. We have to look for solutions right here, depending on our specificities" (Ormindo de Azevedo 2009, p. 109).

Given the experiences in the conservation treatment of historical, ancient or traditional spaces, it is evident that those places where a rehabilitation project was generated, the socio-spatial heritage is maintained through memory, history and its forms that enrich the whole city. Hence, the current demand for these spaces in terms of neighborhoods or areas of the contemporary city. It is not only their location that makes these spaces attractive, it is the environment and the degree of habitability they offer.

## 24.9 Conclusions

In one way or another, socio-spatial transformations have occurred in historical cities and with them local ways of life have been changing, however, since the second half of the twentieth century these mutations have been more severe and even more predatory. The difference between one and another city has been the local exercise of urban planning that has been given to the historical sites by the heritage guardianship bodies, coupled with the support of citizens.

Against an possible "urbicide" that would imply the loss of memory and history that entails the loss of the historical city in its monumental and contextual or "minor" dimensions, it is necessary to propose an urban vision that is not necessarily pessimistic. One possibility to mitigate the degradation and loss of heritage is to use the city's memory policy. It is for this reason that we propose the creation of local urban projects, of urban spaces considered as heritage; projects where neighbors are present, as fundamental actors in the preservation of their place of daily life. The idea is to propose a strategy, where the urban project is not the last planning action. The intention of the local project, based on citizen action, allows managing the scales and temporalities of the projects, making the policies, strategies, and actions of the preservation of urban heritage flexible. In this way, the idea of a possible urbicide could be diluted.

# References

Ascher F (2001) Les nouveaux principes de l'urbanisme. La fin des villes n'est pas à l'ordre du jour, 625th edn. Éditions de l'Aube

Bazarte Martínez A (1989) Las cofradías de españoles en la ciudad de México: 1526-1860, 1st edn. Universidad Autónoma Metropolitana. División de Ciencias Sociales y Humanidades (Azcapotzalco)

Berque A (1987) Écoumène. Introduction à l'étude des milieux humains, 005177-02 edn. Alpha; Belin

Bohingas O (2004) Contra la incontinencia urbana: reconsideración moral de la arquitectura y la ciudad, 1st edn. Electa-espacio público

Bouteille C (2005, June 24) Ecoumène. Introduction à l'étude des milieux humains (Augustin Berque) (Review of the book Écoumène. Introduction à l'étude des milieux humains, by Berque A). Les Cafés Géographiques. http://cafe-geo.net/wp-content/uploads/ecoumene.pdf

Carranza Castellanos E (1998) Crónica del alumbrado de la ciudad de México, 1st edn. Instituto Politécnico Nacional

Choay F (1992) L'allégorie du patrimoine, 1st edn. Du Seuil

Choay F (2006) Pour une anthropologie de l'espace. Seuil

Choay F (2011) La terre qui meurt. Fayart

Coulomb R (2021) Patrimonio cultural, centralidad urbana y "modelo de ciudad". Gremium 8(NE 3):15–28. Retrieved 1 Apr 2022, from https://gremium.editorialrestauro.com.mx/index.php/gremium/article/view/165/167

Davis M (2006) Le pire des mondes possibles: De l'explosion urbaine au bidonville global (trans: Mailhos J). Editions La Découverte

De Anda Alanís E (2019) Historia de la arquitectura mexicana, 4th edn. Gustavo Gili

Delgadillo V (2014, May 5–10) La política del espacio público y del patrimonio urbano en la Ciudad de México: discurso progresista, negocios inmobiliarios y buen comportamiento social (Paper). XIII Coloquio Internacional de Geocrítica El control del espacio y los espacios de control, España, Barcelona, Universitat de Barcelona. chrome-extension://efaidnbmnnnibpcajpcglclefindmkaj/http://www.ub.edu/geocrit/coloquio2014/Victor%20Delgadillo.pdf

Delgadillo V (2016) Patrimonio urbano de la Ciudad de México. La herencia en disputa. Universidad Autonoma de la Ciudad de México

Departamento del Distrito Federal (1989) Programa de revitalización del Centro Histórico de la ciudad de México. DDF

Diario Oficial de la Federación (1972) LEY Federal sobre Monumentos y Zonas Arqueológicos, Artísticos e Históricos (Secretaría de Educación Pública. 06 May 1972). SEGOB, p 16. https://www.dof.gob.mx/index_113.php?year=1972&month=05&day=06

Garza Villareal G (ed) (1987) Atlas de la ciudad de México. Departamento del Distrito Federal, El Colegio de México, México

Giovannoni G (1931) Vecchie città ed edilizia nuova, 1st edn. CittàStudi

Giovannoni G (1998) L'Urbanisme face aux villes anciennes. Du Seuil

Hénaff M (2008) Ville qui vient. L'Herne

Hernández R (ed) (1994) La ciudad de México en la primera mitad del siglo XIX, Instituto José María Luis Mora

Hiriart Pardo C (2012) Panorama mundial del turismo cultural. In: Bruno Aceves H (ed) Patrimonio cultural y turismo. Cuadernos #18. Turismo cultural. Gobierno Federal, CONACULTA, pp 13–32. https://patrimonioculturalyturismo.cultura.gob.mx/publicaciones/

ICOMOS (1999) International cultural tourism charter, managing tourism at places of heritage significance. Adopted by ICOMOS at the 12th General Assembly in Mexico, Oct 1999. [PDF]. https://www.icomos.org/charters/tourism_e.pdf

ICOMOS (2005, Jan) Normas de Quito (1967). UNESCO-ICOMOS, Documentation centre. Retrieved 1 May 2022, from https://www.icomos.org/charters/quito.htm

Jiménez Vaca A (2017) Las acequias en la cuenca de México. Ediciones Navarra

Krier L (2013) La arquitectura de la comunidad, la modernidad tradicional y la ecología del urbanismo (The architecture of community) (trans: Pérez-Porro I). Reverté

Le Monde diplomatique (1991, Oct 1) La ville partout, et partout en crise. In: Ramonet I, Decornoy J, De Brie C (eds). Retrieved 1 Apr 2022, from https://www.monde-diplomatique.fr/mav/13/

Magnaghi A (2000) Le projet local. Mardagá

Morales MD (2011) Ensayos urbanos. UAM

Mumford L (2012) La ciudad en la historia, sus orígenes, transformaciones y perspectivas (The city in the history. Its origins, its transformations and its prospects) (trans: Revol E, 1st edn). Pepitas de Calabaza (original work published 1989)

OGDF (2000a) Gaceta Oficial del Distrito Federal (Décima época; 04 Apr 2000; 64). Organo de Gobierno del Distrito Federal. https://data.consejeria.cdmx.gob.mx/index.php/gaceta

OGDF (2000b) Gaceta Oficial del Distrito Federal (Décima época; 27 Jan 2000; No. 16). Organo de Gobierno del Distrito Federal. https://data.consejeria.cdmx.gob.mx/index.php/gaceta

Ormindo de Azevedo P (2009) El centro histórico de Bahía revisitado. Andamios 6(12):95–113. Retrieved 1 Apr 2022, from http://www.scielo.org.mx/scielo.php?script=sci_arttext&pid=S1870-00632009000300005

Ostrowski W (1976) Les ensembles historiques et l'urbanisme. Centre de recherche d'urbanisme

Perló Cohen M (ed) (1990) La modernización de las ciudades en México, 1st edn. [PDF]. Universidad Nacional Autónoma de México. http://ru.iis.sociales.unam.mx/handle/IIS/4924

Prats L (2003) Patrimonio + turismo = ¿desarrollo? PASOS, Revista de Turismo y Patrimonio Cultural 1(2):127–136. https://doi.org/10.25145/j.pasos.2003.01.012

Prats L (2006) La mercantilización del patrimonio: Entre la economía turística y las representaciones identitarias. Revista PH, Boletín del Instituto Andaluz del Patrimonio Histórico 14(58):72–80. Retrieved 1 Apr 2022, from https://doi.org/10.33349/2006.58.2176

Sánchez Santiró E (2004) La población de la Ciudad de México en 1777. Secuencia 0(60):31–56. Retrieved 1 Apr 2022, from https://doi.org/10.18234/secuencia.v0i60.880

Scannavini R (1998) Trent'anni di tutela e di restauri a Bologna. Costa ed

Schoonbrodt R (1987) Essai Sur la destruction des villes et des campagnes. Pierre Mardaga

Schteingart M (ed) (1989) Las ciudades latinoamericanas en la crisis. Problemas y desafíos. Trillas

Tomas F (1990) El centro de la ciudad de México: crisis y revaloración (trans: Silva H). Trace: Travaux et recherches dans les Ameriques du Centre (17):11–19

Tovar de Teresa G (1992) La Ciudad de los Palacios: Crónica de un patrimonio perdido. Vuelta.

UNESCO (1972) Convención sobre la protección del patrimonio mundial, cultural y natural (La Conferencia General de la Organización de las Naciones Unidas para la Educación, la Ciencia y la Cultura, en su 17a, reunión celebrada en París del 17 de octubre al 21 de noviembre de 1972). https://whc.unesco.org›archive›convention-es

Urrieta García S (2018) Espacio público: de la memoria urbana al proyecto local, 1st edn. Instituto Politécnico Nacional

**Salvador Urrieta García** Architect Engineer (National Polytechnic Institute—Mexico), specialized in urban heritage and public space, Masters and Ph.D. of Urbanism (University of Paris VIII and French Institute of Urbanism). Professor at the IPN's School of Engineering and Architecture since 1982. Member of the National System of Researchers (CONACYT).

**Veronica Zalapa Castañeda** Ph.D. Student, M.Sc. Architecture and Urbanism, Architect-Engineer and Lecturer (ESIA-IPN Mexico). Enroled in research projects on public space and urban heritage linked to the Public Space and Urban Project Workshop. Academic mobility and academic link to The University of Edinburgh and Heriot-Watt University (UK).

# Chapter 25
# Revolt and Destruction. The Public and Monument Landscape in Latin American Cities

**Francisca Márquez**

**Abstract** Thinking about the violence and destruction of the city and its historic centers after social uprisings requires placing oneself at the crossroads between the historical construction of these public spaces and the practice of *vita activa* (Arendt in La condición humana. Paidós Ibérica, Madrid 2005). In other words, making us think about and discuss the *urban condition* (Mongin in La Condición Urbana. La ciudad a la hora de la mundialización. Serie Espacios del Saber N° 58. Editorial Paidós, Buenos Aires 2006). The premise of this article states that the destruction, the rubble, as well as the bodies in motion during revolts are a material expression of the social bond that binds us and unties us from the memory of past time; but this is also, the material expression of struggle, dispute, and will that hides in our present-day cities. It is at this crossroads, between the past and the present, where new paths for urban transformation projects are built. Given that since the revolts the historic center has, for the most part, been undergoing a harsh transformation from its original vocation. The reactualization of its public calling demands to be rethought and redesigned in light of what has happened here. Hence the importance of reading historical and cultural cues, the actions, and expressions of the destruction and de-monumentalization of public spaces in our Latin American cities. A first working hypothesis developed throughout this chapter puts forward that, even though destruction and debris may dominate in protest, they also merge as expressive signs of subjectivities and desires for *vita activa*. These deep impressions contained in the revolts lead to rethinking the nature and quality of public spaces as places of dispute, resistance, and debate. In this same way, a second working hypothesis shows that revolt processes and being present are processes that may reverse agoraphobia (Agoraphobia: a concept that refers to the fear and anxiety of being in open places, the fear of crowds or being alone in public spaces.) and are progressive in an opposing direction to urbicide, paving a way for new public spaces and historical centers where decolonial processes (Decolonial processes: the Latin American epistemic,

---

This chapter takes up the results of the investigation Fondecyt 1180352, "Ruinas Urbanas, réplicas de memorias en ciudades latinoamericanas. Bogotá, Quito y Santiago". IR Francisca Márquez.

F. Márquez (✉)
Universidad Alberto Hurtado, Santiago de Chile, Chile
e-mail: fmarquezb@gmail.com

theoretical and methodological proposals to understand the relations of power and dominance in space-time, as well as to overcome the historical-colonial matrix of power and the liberation of subaltern subjects from that matrix.) and the celebration of the *urban condition* may prevail. In terms of the ethnographic work, the protests that took place in the historic centers of Latin American cities were analyzed: acts of demonumentalization; performances and barricades; and finally, the fate of the insurgent monuments. The chapter has two central conclusions. Firstly, *the common space of the different*. After the revolt and the landscapes of disorder that are left behind, all that remains is to protect and secure those common spaces, spaces of the *vita activa* of "us" and of the "others", from identification principles and also distinction and difference principles. Because public space is not necessarily the place where the inhabitants of a city exercise their equality, but perhaps where they exercise their enormous differences. Secondly, public squares that safeguard the urban condition. Overwhelmed by urban aesthetics, the city becomes a blackboard whose complex text will have to be unraveled and questioned in order for it to be interpreted and (re)thought. The form, shattered, twisted, fragile and displaced from its original site, then appears as a new way of structuring the social and its intelligibility within the framework of overwhelmed democracies. Transforming the minefields of the "patrimonial heritage" in our cities requires being open to the possibilities of intercultural coexistence that, instead of neutralizing our affective or moral disagreements and our ontological, ideological, and epistemological conflicts, allows us to explore everything that is in dispute or friction. In short, in the words of the anthropologist Claudia (Briones in Antropología Contemporánea. Intersecciones, encuentros y reflexiones desde el Sur Sur. Ed. Universidad Católica de Temuco, Temuco, pp 83–103 2020), lose the fear of living in friction to find ways of being together while being different.

**Keywords** Protest · Heritage · Historic city · Monuments · Public space

## 25.1 Introduction

Latin American cities are born broken; since the management and foundational design of conquest, everything stems from an order built from death, erasure, segregation, and the expulsion of their first inhabitants. The Latin American city is a product of overlapping plots and the remnants imposed by what has preceded them. However, despite this imposition on conquered territory, the image and shape of the city is irretrievably immersed in historicity woven by the practices of urban life. This is what we have been able to observe throughout urban history and more specifically in the recent revolts that have occupied and "assaulted" the historic and monument centers of our cities. In them, the dispute for the right to the city—as a collective right and action by organized subjects (Carrión and Dammert 2019)—is imposed and superimposed on the historical form of public space and its historic centers.[1]

Thinking about the violence and destruction of the city and its historic centers after social uprisings requires placing oneself at the crossroads between the historical construction of these public spaces and the practice of *vita activa* (Arendt 2005). In other words, making us think about and discuss the *urban condition* (Mongin 2006). The premise of this article states that the destruction, the rubble, as well as the bodies in motion during revolts are a material expression of the social bond that binds us and unties us from the memory of past time; but this is also, the material expression of struggle, dispute, and will that hide in our present-day cities. It is at this crossroads, between the past and the present, where new paths for urban transformation projects are built. Given that since the revolts the historic center has, for the most part, been undergoing a harsh transformation from its original vocation. The reactualization of its public calling demands being rethought and redesigned in light of what has happened there. Hence, the importance of reading historical and cultural cues, the actions, and expressions of the destruction and de-monumentalization of public spaces in our Latin American cities.

A first working hypothesis developed throughout this chapter puts forward that, even though destruction and debris may dominate in protest, they also merge as expressive signs of subjectivities and desires for *vita activa*. These deep impressions contained in the revolts lead to rethinking the nature and quality of public spaces as places of dispute, resistance, and debate.

In this same way, a second working hypothesis shows that revolt processes and being present are processes that may reverse agoraphobia and are progressive in an opposing direction to urbicide, paving a way for new public spaces and historical centers where decolonial processes and the celebration of the *urban condition* may prevail.

Together with the protests, revolts, and popular insurrections, the historic center—as a progressively uninhabited public space—thus acquires an imprint transforming it into a tool for political expression in the city, opening ways to activate memory to build a more inclusive society project. The protest reflects and reuses the historic center, its main squares, buildings, and monuments, but the question on how the new certainty in the destruction of the old prevails. After the uprising, what is now at stake is how to rethink the historic center from its centrality, that is, to encounter and represent an *imagined community* as a public space in the city (Anderson 2000). Overcoming heritage and monumentalism paradigms—patriarchal, military, and colonial—relies on the ability of urban policies to meet these demands.

---

[1] Anibal Quijano estimated that 65 million native people were exterminated in America in a period of less than 50 years: "Coloniality and Modernity/Rationality", Cultural Studies, 2007, 21.2, 168–178. An extermination that is estimated to be the equivalent to 95% of the native population existing before the conquest.

Like many qualitative studies, we begin with the formulation of one or more assumptions—working hypotheses—on possible answers or solutions to the problems considered. They are, by the way, assumptions based on empirical evidence that serve as reference points for further research.

Clearly the insurrection, multitudinous and sometimes violent expressions have left their mark on the materiality of these historic centers that form a protest landscape. Defending its infrastructure, boarding up and barricading free transit, the closure of local businesses, as well as the burning and destruction of heritage buildings and the interventions to its monuments and statues, make us fear not only for the destabilization of the forms of representation, but also the market forces and public services toward safer neighborhoods. During the social revolution, businesses, banks, pharmacies, and multinational companies that were in the city center were destroyed and many of them had to be protected, relocated, or simply abandoned. The protest challenges the historic center's priorities and drives out the symbols of the capital projecting the fight for a new awareness of the city. The functions inherent to urban centrality (goods, services, and information) are clearly under threat; and in many sectors, fear and desire for safer policies and police control prevail. The risk of whitewashing or safe citizen policies lurks and struggles under its principles of order and regulation prior to mobilization and protest on its return to this historic center.

Within the context of these mobilizations—under a State in crisis and the weight of neoliberal policies in terms of urban development decisions—urban planning and the founding center face not only the question of social integration, but also an understanding of the remnants left by the protests in the public space. There are many tensions and doubts: How to resolve and guarantee the (dis)encounter in the common space? How to ensure the diversity of symbolic representation in urban landmarks? How to guarantee the much-needed political processes and debates to all polis and the realization of the urban condition?

Without ignoring depopulation processes, increases in the price of land, segregation and the gentrification that affect historic centers today, in this chapter we will look at the transformation of the urban and political condition of these founding centers since the recent protests. In fact, the thesis that is developed is that simultaneously with these depopulation processes and the weakening of historic centers, there has been increased occupation of these places for citizen demonstrations and insurrections in recent decades. As a result of the protests and mass expressions that have taken place there in recent decades, the historic centers have been visibly destroyed. However, paradoxically they have regained their status as spaces of public domain where collective will is expressed by society representing its rights and duties (citizenship) and these places have become diverse meeting and representational spaces (Carrión 2019).

In summary, the thesis that is developed reveals that what is being demanded with the ongoing social revolts that have taken place in our Latin American cities is the right to the city in terms of a collective right. The massive and sometimes violent occupations of public space are the best expression of the desire for appropriation and revitalization of public space as a space for *vita activa*, and in Arendt's words, where work, labor, and creation (action) have a place in the design of a just city. Hence the importance of carefully observing these destructive processes of cities and their historic centers, both from a material perspective and from the urban imaginary that the protests disperse. Destruction as we have seen expels, but at the same time,

**Fig. 25.1** Santiago, Chile 2019. *Source* Alvaro Hoppe (2019)

persistently attracts demonstrators and citizens eager for new social and political horizons (Fig. 25.1).

## 25.2 Unrest, Revolt, and Rights

Reading on revolution and social explosion in Latin American cities at the beginning of the twenty-first century entails analyzing its link with neoliberalism, acute inequality, and urban segregation. An erosion of the fantasy of a fairer and egalitarian society is imposed everywhere in Latin America. Inequality indicators, which have been dragging on since the mid-twentieth century, confirm that economic growth alone is not enough. Increasingly, within Latin American societies the perception of the unforgiving living conditions within the lowest socio economic and middle-class sectors is shared. The daily awareness of a *hard life* is imposed on the illusions and promises of the imaginary of quality of life and social mobility (Martucelli 2020, 87–109). Expectations that are increasingly strained by working conditions, inequality, and mistreatment. In fact, the overwhelming, discontent, and the excess of a social and economic system defines the experience of society in Latin America (Araujo 2019).

At the Panorama Social de América Latina (Social Panorama of Latin America) 2019, ECLAC warned that "not only have various advances in social matters slowed or stagnated in a non-dynamic economic context, but there are also important signs of regression. After a five-year period of slow growth, the structural deficiencies of the region have become more evident, and their solution is part of the demands of widening social groups in particular of the new generations. These demands include the rejection of the persistence of the culture of privilege in its multiple dimensions, those linked to the concentration of wealth, segmented access to quality public and cultural services, and the lack of recognition of the dignity of individuals and communities. This is what many actors express under the demand to end the abuses."

Poverty reduction and extreme poverty rates in the region between 2000 and 2014 resulted in the improvement of living conditions and an increase in social expectations and demands for a significant proportion of the population, but at the same time, they were not able to meet the expectations and eliminate a cycle of vulnerabilities. ECLAC confirmed that the middle-income strata continue to face a number of vulnerabilities such as low educational levels, low quality of labor insertion, low coverage, and insufficient benefits in the pension system.

Likewise, despite this sustained reduction in poverty, inequality continues to be a historical and structural feature in Latin American and Caribbean societies. An inequality based on a highly heterogeneous and little diversified productive matrix and in a culture of privilege that is a historical constitutive feature of Latin American societies. In them, inequalities of socioeconomic origin intersect with gender, territorial, ethnic, racial, and generational inequalities.[2]

This inequality and erosion of a common social imaginary finds its clearest expression in segregated and latticed cities. Violent and unsafe cities, where the confrontation between social groups has intensified since the last decades of the twentieth century. Disillusionment with the state apparatus and its institutions, with access to opportunities to improve the quality of life, with the ability of families and neighborhoods to contain sociability increasingly translates into a series of criminal life patterns based on consumption. This explains why, beyond national differences, individuals feel alone, unprotected, and compelled to their own efforts in Latin America (UNDP 1998).

During the social revolution this growing unrest was at the center of the revolt, raising the question about and demand for *equal rights*, human and non-human, women's, dissident, indigenous peoples' rights, the right to affection, the right to nature, among many other rights. The unrest connected with the human rights revolution and established citizen demand for the right to have rights (Martuccelli 2020). However, in this search for one's own rights, the conviction for other knowledge also arises, knowledge subjected to centuries of "epistemicide"[3] by a culture of domination; rage and fury towards "cognitive imperialism"[4] is echoed in every cry, performance/act and the graffiti marking the city's walls (Fig. 25.2).

## 25.3 Protests in Historical Centers

Revolts at monument bases, in the historic squares and centers of our cities is nothing new in Latin America. To name a few from the twentieth century, Mexico City on

---

[2] Income distribution inequality shown in household surveys has continued its downward trend, but at a slower pace than observed in the past decade. Between 2014 and 2018, the Gini income index per person fell in six countries. In some, however, the reduction of this indicator has not prevented the absolute income gap from widening between groups at the extreme ends of the distribution. According to the most recent household surveys available, the Gini index reached an of average

**Fig. 25.2** Bogotá, Colombia 2019. *Source* Jefferson Ortiz Cala–Adepto Captura (2019)

August 27, 1968, stands out where the students seized the Plinth demanding the release of their imprisoned comrades. Also, the counterdemonstration on August 28 of that same year, organized at that very plinth but by the government with the surprising disobedience of its public officials who shouted, "We are the sheep of the administration!" In Buenos Aires, Argentina, the Madres de Plaza de Mayo (Mothers of Plaza de Mayo) who every Thursday since 1977 have held the "circle" around the Pirámide de Mayo in the city in protestdemanding news of their disappeared relatives. In Plaza San Martín Lima, Peru, the massive demonstration in November 2020 protested against the presidential vacancy of Martin Vizcarra and the swearing-in of Manuel Merino as the new head of State. Examples are numerous and confirm that the historic and foundational center, despite the evident discredit of our democracies and government, continues to summon the people in terms of its symbolic power.

This happened again in October 2019 with the social revolution in Quito, Ecuador, a prelude to mobilizations that took place in parallel in cities in Chile and then Colombia. In Ecuador, its protagonist was the indigenous movement, voiced by the Confederation of Indigenous Nationalities of Ecuador (CONAIE); in Chile, it was the student movement following years of protest and demands for free and quality education; in Colombia it was sectors grouped into the so-called National Strike Committee called for the "National Strike #21N". In all three countries, the trigger

---

0.4651 in Latin America. The lowest values, under 0.400 were recorded in Argentina, El Salvador, and Uruguay, while in Brazil and Colombia they exceeded 0.520.

[3] **Epistemicide**: a concept that refers to the systematic destruction of the system of knowledge and knowledge of an ethnic group, especially to assimilate these to a European and colonial worldview. A term coined by Boaventura de Sousa.

[4] Cognitive Imperialism: a concept coined by Boaventura De Sousa Santos in The End of the Cognitive Empire. It alludes to the domain of Eurocentric thought versus an epistemology that guarantees global cognitive justice.

was the rise in the cost of public transport. However, the roots are much deeper and more complex. In all cases, despair with the model of inequality and abuse is the consistent reason (Durán 2021; Martucelli 2020).

In these three cities, just as in so many other cases, the center stage of the protest was the historic center of these capitals. In the case of Quito, the location was El Arbolito Park, an old stadium that is part of the nostalgic heritage image and constitutes a kind of border between the monumental colonial city and the modern city. In Santiago, the protest lasted for five long months in Plaza Italia, now re-christened Plaza Dignidad by the people, thus summing up in one word the collective voice at the trigger point that divides the city in two: those with and those without. In Bogota, the protest was violently felt in the historic center, especially in Plaza de Bolivar and its surrounding avenues which were was the nerve center of the large urban demonstrations.

In these three cities, public transport and its infrastructure was brutally damaged. However, how should we comprehend these historic centers, the monumentality, patrimonial architecture and the city squares that have been so destroyed and affected in these revolutions? To respond to this question and provide answers to our working hypotheses, two protest devices were investigated: The destructive violence and demonumentalization. As we will see from the description and interpretation of these two devices deployed in the historic centers, the protest landscape covered everything, not only revealing the depth of social unrest, but also the refoundational and politically creative character of the acts of urban destruction (Fig. 25.3).

**Fig. 25.3** Quito, Ecuador 2019. *Source* Eduardo Sánchez de León Herencia (2019)

## 25.4 Barricades, Fires and Acts

From the first day of the protests, violence and destruction quickly transformed the historic and heritage center into a battlefield; *ground zero* as it was called in Santiago. In Quito, the daily use of public spaces was disrupted for about two weeks by the might of the bodies, the sounds from historical slogan chants and the spontaneous artifacts of the struggle: barricades made out of rocks and cobblestones extracted from the park itself, metal and cardboard shields, tires, fires and pots. All under the scenario of excessive force from the repressive state power (Durán 2021). Days and weeks later the center continued to be boarded-up and secured with barbed wire by the authorities to defend the heritage buildings.

In Santiago, something similar happened days later. On the night of October 18, 2019, the city's metro stations, and large supermarket chains began to be burned down. The public space was filled with gritty metallic noises like war drums filling the air and nightscape. "We are at war!" announced President Sebastián Piñera. In response to this provocation, bodies crowded onto the streets on mass banging pots and metal spoons. Barricades, fires and looting, just as in Quito, stripped the city of its glow and shine in a few days. From that night, the riots, dust and ash did not cease. Twisted fences, looted pharmacies and supermarkets, burned banks and metro stations and fallen monuments left the city demolished.

A few days later on November 21 large mobilizations erupted in Bogota and throughout the country. Clashes in Plaza de Bolívar, between *encapuchados* (hooded protesters) and the ESMAD of the National Police Force resulted in homemade bomb attacks on the Mayor's Office of Bogotá, the Palace of Justice and the Congress of Colombia. The demonstrators, in never seen before acts, called for a day of c*acerolazos* (banging pots in protest) in all the country's main cities, and Plaza de Bolívar was the main point of concentration.

The police forces used their riot gear, weapons and tear gas as a way to suppress the social protests in all the cities, resulting in hundreds of wounded and dead.

Notwithstanding, along with the violence inflicted on the bodies of the demonstrators and urban architecture, solidarity practices sprung up in the cities by groups of students, women and neighbors created peace areas, improvised meeting centers, soup kitchens, first aid and help strategies for the injured demonstrators (Durán 2021). The so-called *young people of the first line* with their improvised shields and elaborate fighting strategies who created a certain heroism on the battlefield were given special prominence. From the different frontlines, acts of spontaneous solidarity were being organized despite the evident violence from the police forces inflicted on the protesters. The destruction of the city and its monumentality must in part be understood as part of this confrontation, which left behind dozens of dead and hundreds injured.

As in any performance, in manifestations in the public space the bodies speak of memories, traditions and claims for one's place in history. In each dance and movement of the bodies demands for cultural agency are established (Taylor 2015). A playful dramaturgy in a collective and liminal scenography; collectives are each

**Fig. 25.4** Bogotá, Colombia 2019. *Source* Jefferson Ortiz Cala–Adepto Captura (2019)

of the movements, collectives are the rhythms, collectives are the cries and the acts of representation. Knowledge embodied, alive and always inescapable and for that reason, politically heterotopic. Practices of the body indomitable to tear gas bombs and repression, where the only thing that seems to matter is to destabilize the social frameworks of memory (Halbwachs 1970).

Cities and their historical centers—both as a metaphor and a mirror—provide accounts through the depth of the disorder of revolt and unrest, as well as the irreversibility of social demands. As in any social revolt, it is these moments of protest and violence where the masses on the street oppose their own order against the State apparatuses and the city's daily use. Moments that interrupt and divert the temporary flow of the city, and then—with some delay—a certain air of normality is recovered. But the truth is that, after the revolution, the city is not same anymore, its landscape and chromatic are mutated. In Quito, for example, in the face of disorder and destruction, the same actors who had participated in the mobilization returned to the park to collect their debris. Hundreds of citizens joined together in this shared and peaceful feeling to paint and repair the enormous damage to the walls, sidewalks and cobblestones in the historic center (Durán 2021). In the terms of Ranciére (2010), all possible world transformations take place in the historic center. And these traces are still present in the materiality of the streets and buildings, but above all in the memory and record of those who were there (Fig. 25.4).

## 25.5 De-monumentalizing the Public

In cities and their public spaces, symbolism overflows. This is often built on the ruins of previous buildings and monuments. In the terms of Adrian Gorelick; lessons from the statues would be nothing without the political and cultural effect of this spatial organization. Although as Castoriadis points out (1983), often the form overcomes the rigid link to a precise meaning and can lead to imaginaries and unsuspected actions. This has happened historically. In the centuries of republics, not only have the meanings cracked, but also the illusion of social consensus around that public space and its monumentality (Pizarro 2021).

In Latin America most of the monuments erected in cities in the nineteenth century were inspired by the neoclassical and civilizational spirit from European political and sculptural traditions. The historic centers, built with profits from modernization, the avenues, public buildings, squares and monuments constitute the privileged scenography for the rituals of power. However, this representation of the public of the first national and republican monumentality was built with obliviousness to the indigenous, black and Creole populations, and under the principle of silencing those differences. The metaphorical and allegorical resource of this monumentality and urban aesthetics always refers to that silencing, even when it uses powerful visual, spatial and temporal resources to impose its authority on the public space of the city (Masotta 2021).

However, in the twenty-first century, the monumental and elitist character of such urban devices has been dramatically confronted by social protest and multitudinous demonstrations by the historically silenced. Until that moment, the monumentality of heroic deeds seemed to fulfill the function of reifying the past into the promise of eternity. Patrimonial monuments were trapped in their own Pater-Patria-Patron etymology (Masotta 2021; Vignolo 2021). Indeed, together with the destruction of the historic centers of the cities, recent decades allow us to verify that the iconoclastic inventory of the monuments fallen historically in Latin America is vast and seems to continue growing. In recent decades, sculptures, busts and heritage buildings have been subjected to a certain ritual anthropophagy[5] by angry masses of citizens (Metraux 2011). Among this list of toppled and demolished monuments are statues of Christopher Columbus; the beheading and offering of the head of the aviator Dagoberto Godoy (believed to be Pedro de Valdivia the conqueror) the statue of Caupolicán, in the city of Temuco; the demolished monuments of the entrepreneur José Menéndez the exterminator of Fuegians that was deposited at the foot of the statue of the Patagonian Indian in the Plaza de Armas in Punta Arenas and the iconic statue of Sebastián de Belalcázar de Cali by indigenous communities, among many others.[6]

---

[5] Ritual Anthropophagy: Métraux conceived the anthropophagic ritual as an active cultural process, which cannot be restricted to the aesthetic sphere alone or limited to the Latin American experience. The myth lives in the heart of the modern.

[6] Although the debate proposed here mainly addresses events that occurred in Latin America, we cannot ignore the series of events that have occurred historically on other continents. After the death of George Floyd (2020) by a white policeman in the United States, a series of other statues were torn down: the statues of Christopher Columbus in Boston, Miami, Baltimore, Richmond, Minneapolis, among other cities; in Bristol, UK the statue of the seventeenth-century slave trader, Edward Colston, and many more. This should include the removal of statues by authorities, as a form of official de-monumentalization to avoid public backlash; or by simply making uncomfortable

These are some of the iconoclastic acts that respond to and oppose iconoclasm "from above" (Gamboni 2014) through their anthropophagic rituals (Metraux 2011). It should be noted that the power of the demolished and re-dressed monument seems to have little to do with the value of antiquity or historicity contained in any block of stone, marble or bronze. All of them are part of the spoils of war of the revolts and insurrections of the twenty-first century that are toppled from the public space, the reminders that in the name of progress and civilization, establish the colonial, patriarchal, racist and classist powers. The dismemberment, impaling and decapitation, like acts of war that are, after 500 years of Hispanic and national-state colonialism, offered back to us as a symbolic shift in the pagan reappropriation of the metonymic act in part for all (Allende 2019).

The monument's enduring time, which seemed to be its main promise, is then its greatest weakness. The deconstruction and resistance to the monuments topple the society that has erected them placing doubt on their eternity (Masotta 2021).

Added to these iconoclastic acts are sprayed on stencils, graffiti and protest posters thus building an archive—a protest blackboard on the cities' walls (Góngora 2021; Molina 2021; Márquez et al. 2021). The walls, as well as the fallen and re-constructed statues cry out and denounce not only a defiant attitude and position, but also a way of knowing and being in the world that asks us to re-think and un-learn some of the principles and devices imposed by hegemonic thought. Protests and revolutions condemn the recognition of these Others as subjects of knowledge and citizens in social, political and intellectual construction processes (Krotz 1993; Restrepo 2012). No longer conquered, enslaved and excluded beings, but as active opponents of a colonialist project that nullifies reciprocity and relationality. In these writings on the walls, burned heritage, headless sculptures, generals dragged from their horses and the mutilated eyes and dismembered bodies of presidents, what is being denounced is the annihilating power and exclusive cognitive categories and classifying white Western heterosexuals. The de-monumentalizing processes are then offered up as processes that break with and disrupt the so pleasant monocultural order and safe patrimonial estates (Taylor 2021: 50). De-monumentalizing choreographies that have become a privileged space to dispute the heroic narratives of the nation.

To paraphrase Freedberg (2007) we should question what is it about this monumentality and in these statues that people want to destroy, damage or mutilate? For Freeberg, all iconoclasm poses the issue of "embodiment," or more specifically, the thought or perceived reality that is "embodied" in these images. Iconoclastic acts of destruction, elimination or mutilation suggest that they are identified as real physical aggressions to a living prototype (the image). The first iconoclastic instance is rejection (De Nordenflycht 2001). There is fear not only of the representation of the body, but of the threat that it could be living or come back to life. However, creative and destructive acts are not two sides of the same coin but are the same: to create is to destroy and to destroy is to create. In this sense, one reacts critically and controversially to the reality of all iconoclasms, but often to change them because to create

---

symbols disappear, as was the case with the José Martí bust in the right-wing municipality in Santiago de Chile.

you have to want to destroy *something*—including the past—and to destroy means to create something new.

This "something new" is, in fact, one of the most relevant aspects of the questions at hand. In unequal and segregated cities, destroying the public space and the historic center, to the point of making it similar your own neighborhood, your own slum or favela, may certainly be a democratizing act of the public space by creating an aesthetic and an urban form similar to one's own. However, according to Latour (2018), we should also think about and admit that iconoclasm not only destroys an image, an icon, a representation that produces rejoicing or indignation, it can also simply generate doubt, restlessness and uncertainty, leaving the future horizon hanging in the balance.

But what should we do with these disgraced statues and monuments? Perhaps repair them, clean them and put them back in their place. The truth is that even when this restorative act takes place again and again, the persistence of the de-monumentalizing and iconoclastic act does not seem to end. The restoration of the statue of Columbus in Buenos Aires or the police protection of the statue of General Baquedano in Santiago, in contrast to the iconoclastic interventions of the anti-racist demonstrations of that year in other countries, seem to suggest that the dispute is still open-ended, says anthropologist Masotta (2021). And it is—he argues—that all monumental iconoclasm, when it cancels on itself, can confuse the symbol with the symbolized thing.

## 25.6 Uprising Monuments

> Cómo hacer para que los pueblos se expongan a sí mismos?
> Didi-Huberman, 2014

If some statues have been removed from the historic centers of cities, temporarily from the epistemic cages and the interpretive chains that link them to an official reading, perhaps they can be protagonists and allies of the social and political transformations that take place there (Vignolo 2021). This is the case with counter-monuments, uncomfortable or negative monuments, which remind us that a nation's history includes subaltern pain; and that history, even when less violent, engulfs memories. On the contrary, history may always be revisited, subverted and updated not only to remove the horrors from the pedestals but also to build its counter-monuments and honor the victims of that power.

Some examples are seen in the destruction and burning of the monument of the Spaniard Francisco de Aguirre who decimated the Diaguita people, this space was later replaced by a monument to Milanka, a Diaguita woman in the city of La Serena, Chile. The sculpture of Isabel la Catolica in La Paz, Bolivia, was dressed in a skirt, a yellow shawl, an *aguayo (woven cloth used to carry babies)* and hat like an altiplano chola (indigenous person) by activists from the Colectivo Mujeres Creando. The monument of the Inca Cahuide resisting the Spanish at the Battle of Saqsaywaman in Maca, Peru. The bust of Comanche, the great Afro-descendant leader of the El Cartucho slum in Bogota, Colombia. The Afro wigs that the Afro-Caribbean Nelson Fory placed on the statues of protagonists such as Christopher Columbus, Pedro de Heredia or Sebastián de Belalcazar in Cartagena, Cali and Bogota. Something similar happens with the "cross-re-dressing" of the General Baquedano statue and three wooden sculptures of Mapuche, Diaguita and Selk´nam peoples turning their backs on General Baquedano in Santiago de Chile.

As Doris Salcedo points out in terms of counter-monuments, these commemorative acts fly freely, no case is made for a political leader, government, or conforming to an ideology. It is the agency of the *multitude* that makes way for fragmentary and heterogeneous ways of community living.

Certainly, in these counter-monument movements, as far as the materiality is changing, the works take on a fleeting and ephemeral transient value. Where traditional monuments were destined to endure and perpetuate over time overcoming their own physical death under the conservationist principle, the counter-monument has an intransitive character by mutating from one state to another which may even lead to its own evanescence. In this way, memory is not housed in the object and its materiality, but in the community on whom its transmission now depends.

The counter-monument reverses the classical relationship between builder and observer, between creator and audience. If traditional monuments are fundamentally designed by a principal (State, Church) to teach new generations, the counter-monument delegates the task of elaboration and commemoration to the collective. It is the public intervention and observer that legitimize the existence of the work producing an active and communal relationship with the historicity of the monument.

Far from the spirit of traditional monuments whose commemorative nature ends, washes away and petrifies the past, depriving a motherland history of conflicts and updating, one of the characteristics of the counter-monument is being antiheroic, and therefore, to constitute non-submissive spaces against twentieth century totalitarianism and colonialism. This implies new ways for memory and commemoration that far from seeking to chronicle time and providing single answers, look to experience the ambivalence and uncertainty of memory by always actively merging them. The counter-monument seeks to account for absences, hybrid identities and disputed memories, its cause is always unresolved and dynamic rather than stable and monolithic (Bustamante and Márquez 2022) (Fig. 25.5).

## 25.7 Notes for a Democratic Public Space (Or How to Avoid Urbicide)

It was proposed at the start of this chapter that within the mobilization context, which have taken place in our Latin American cities, urban planning faces the challenge of understanding the feelings and opportunities that the protest installs in the public space. We certainly cannot deny the paradox that these uprisings leave behind. Along

**Fig. 25.5** Santiago, Chile 2021. *Source* Alvaro Hoppe (2021)

with the violence that has been part of the revolts in the city and the great devastation of infrastructure, architecture and monumentality, it has also been observed that the historic center reclaims its prominence as a place of political expression. In this sense, we point out that the protests open the possibility to reverse the prevailing agoraphobia in our cities. However, in order for public spaces to secure their prominence as places of expression and encounter they should: safeguard the diversity of symbolic representation in urban landmarks, and above all guarantee political processes and the realization of the urban condition for encounter in the common space. In these conclusions we seek to respond to our unease and the two working hypotheses that were raised at the beginning of the research.

### 25.7.1 The Common Space of the Different

A first working hypothesis that is developed throughout this chapter states that, although destruction and debris dominate in protest, they also merge with expressive gestures of *vita activa* subjectivities and desires. We share the thesis by Didi-Huberman (2017) when we ask ourselves what the reasons are for the uprising, when he points out that behind every revolt lies the strength of memories and the strength of desires, when they are inflamed. It should also be mentioned, that according to Butler (2010), no uprising is worthy of the name without "a certain radical inner experience" in which desires are fed by one's own buried memories. It is hard to imagine revolts with such violence and anger. Anger and such deep discontent that when it exploded exposed generations of the long brewing and enduring inequality.

Hence, it should not surprise[7] us that it is young people, women, indigenous people, Afro-descendants and sexual dissidents that planned the assault on the present with rebellious performances in the street spaces. In the words of Didi-Huberman (2017), an assault that thus reverses the heaviness that nails them to the ground.

In this question about public spaces, we should not forget that many of those who march, organize, resist the tear gas bombs and disobey the state of emergency carry the banners denouncing the abuse and calling for the emancipation of the "coloniality of power" understood as the deep asymmetries that characterize the structural aspects of our Latin American societies (Quijano 2000). It is this depth of feeling within the revolts, which forces us to rethink the character and quality of public space. It is clear that the force and violence seen during these de-monumentalizing processes in the public square does not mean the citizens really know who the figures being damaged are or why they should be toppled. What is clear is that we are facing gestures of anger, pain and deep political meaning. Gestures that chastise the res public and its works located in the public and monumental space where the civilizational ethos of the project of nation unfolds. Insurrectionist gestures that confront the poetics of exclusionary power (Geertz 1999) to dismantle the apparent consensus and regimes of representation (Durán 2021; Márquez 2020a, b, 2021). Gestures that remind and teach us that history can always be revisited, subverted and updated. It is there, in the space of the public, that the protesters oppose the government agenda to their own order. Apparently ephemeral interruption of the temporal flow, but where progressively changes in the landscape, in the decibel, in the perceptible and in the thinkable are visible (Rancière 2010).

Whether or not the historic centers and their monuments are restored, moved or made to disappear, the dispute that iconoclastic and creative practices provoke in the monumental-homeland-narrative always questions the definition of what deserves to be remembered and protected. Cleaning the statues, monuments and streets will not erase the depth of feelings and demands for a more diverse and just society. The question is then, how to rethink monuments and public spaces based on this diversity of gestures and voices. How and who defines what is worth being protected and preserved? How do we make a living memory of these demolished and absent monuments?

For some the acts of de-monumentalization (destruction, mutilation and toppling) are a problem limited to public order, and therefore they put all their efforts into mitigating their material effects (De Nordenflycht 2001). For others, the underlying problem lies in how the iconoclasm and counter-monumentality that stems from here contributes to accounting for these damaged subjectivities and therefore the decolonization and democratization of our cultures. And with it, the redefinition of common spaces in cities. The debate asks the question not only for the presence

---

[7] In the case of Chile, it has been pointed out that the rebellion must include the "excessiveness" of a neoliberal model that was built on the principle that every subject is responsible for their own destiny (and capital); that the principles of generalized competition are those that govern the day to day; and that the promise of integration is always via consumption and credit (Araujo 2019: 19).

that fits diverse heritages, but also in terms of the forms of participation for the expressions, decisions and occupations of that public space.

After the revolt where the unrest landscape is left behind, what remains is to protect and secure these *vita activa* common spaces of the "us/we" and of the "others", from identification principles and also distinction and difference principles. Because public space is not necessarily the place where the inhabitants of a city exercise their equality, but perhaps, the place where they exercise their great differences. In an orderly or chaotic way. Chaos as the principle of order (Del Sol 2022). Therefore, the city and its public spaces as the revolution teaches us—will be made and nourished by these moments of expression and rebellion, always in a complex conversation as befits the *vita activa*. The democratic and contemporary construction of the civitas in these terms demands the articulation (and not the elimination) *of the landscape of protest in the urban landscape*, so that the desires for expression and conviviality of those who inhabit and come to express themselves and make their political demands visible may live together in a more diverse and just city.

But whatever the answers to these challenges are, the planning and design of our public spaces the mirror of the society we want—can no longer be only sanctioned by experts and officials sitting in a governmental or real estate office. Learning to read, listen and safeguard these insurrectionary and iconoclastic gestures as blackboards and creative manifestos of the society we want is an urgency to guarantee the (dis)encounter in the common space.

### 25.7.2 *Squares for the Protection of the Urban Condition*

A second working hypothesis suggests that the revolt and being present processes may allow to reverse the agoraphobia and move in the opposite direction to urbicide, while opening a way forward to new public spaces, where the decolonial processes and celebration of the *condition of the urban* would be imposed. If at any time in Latin American history cities and their historic centers were thought of in an orderly manner, the untamed and creative nature of the revolt arrived to break down and question these imaginaries.

"Be Present!" may be a war cry, but it is also an act of solidarity, a testimony, or just a positive cry when walking and talking with others. "Be Present!" sometimes expresses a political movement, being together, even if the "other" has been disappeared or hides their face. Being present in the public space is always a call for mutual recognition to talk about injustice (Taylor 2021). Hence the relevance of safeguarding these gestures to Be Present! in each of our cities.

The philosopher Humberto Giannini pointed out that the cultural history of peoples begins, if it ever begins, with movements that are concrete, literary, pictorial and musical. First is the experience, the aesthetic experience, as a solid and expressive practice. Then comes philosophy, Giannini said, as a later grace, as a reflection that is placed behind this experience (1992: 36). If we apply this premise to the revolution, we may say that experience and creative expression is first and foremost.

Therefore, to paraphrase Rodolfo Walsh, we must never forget that *the walls are the printing presses of the peoples* and sheltered there is a language of many layers and coats, leaving the door open to multiple and contradictory readings. Politics and public order come later.

Richard (2007) asks how to read the *conflict of speech* between, on the one hand, the knowledge qualified from power and, on the other, the unarmed and angry knowledge of subalternity. Indeed, integrating *other knowledge* into the dialogue and understanding of the dis-order and dis-concertation of our cultures is part of the dialogue to be built. In times of revolt and pandemic, these other knowledges have their own modulations, on the streets, burning monuments, looting supermarkets, marching indefatigably, shouting, singing and also creating. In all these embodied gestures there is *other knowledge* that demands sovereignty, that is, the right to a place in the construction of culture and decision-making. But re-thinking public spaces means above all crossing the limits, binding and merging into and with this knowledge in motion. Because in these expressions "cultural agency" is not individual, but collective; it is neither institutional nor structured, but rhizomatic and flexible. The administration of difference and critical otherness are woven hand in hand with the historical production of "the others of the nation" (Segato 2007). Hence, the redefinition of history and urban life today inevitably has to deal with the spatial, temporal and symbolic redefinition of the social relations that are woven here, between the center and subalternity. Hence the urgency to ask ourselves how do we read these worlds from distant horizons so that we dare to lose our fear of them?

Overflowing with urban aesthetics, the city becomes a blackboard whose complex text must be translated and questioned in order to interpret and (re)think it. In the context of democracies, the twisted, fragile and displaced form rises as a new way of social structure. In some way, the alterations, interventions, and overthrows constitute a call to rethink and revisit new ways of commemorating in the urban space (Mora et al. 2020). To guarantee the political processes and the realization of the urban condition is in this sense offering promises and conditions that may be installed in the public space as a space of diverse and shared emotions. In this sense, one square—the main square—is not enough but many squares, so that the expressions can be diverse, local and from the territories that are inhabited. Wide and dignified spaces filled with demonstrations, rituals, markets, fairs, parties and protests. This was urban history in Mesoamerica and South America before the arrival of the Spaniards. The cities had large squares or ceremonial centers for organizing and shaping the urban and political culture developed there. As in the Inca cities, where the *kanchas (squares)* were political, administrative and religious centers, modeling a conception of urban life where everyday culture interacted with the market space, ceremonies and political work. Thinking about our Latin American cities after the revolts and unrest, requires recognizing that within the disorder of the bodies and souls that inhabit them, resides and will reside the poetic force of that noise that the *kancha* installs in our modern urbanity.

In short, to transform the "patrimonial-homeland" minefields in our cities will require opening up to the possibilities of intercultural coexistence that, instead of neutralizing our affective or moral disagreements and our ontological, ideological and epistemological conflicts, may let us explore everything that is in dispute or friction. In short, in the terms of the anthropologist Briones (2020), losing the fear of living in friction to find ways of being different together.

## References

Allende M (2019) Monumentos y gestos anticoloniales. En: Palabra Pública, U. de Chile. https://palabrapublica.uchile.cl/2019/11/12/la-parte-por-el-todo-monumentos-y-gestos-anticoloniales/
Anderson B (2000) Comunidades Imaginadas: Reflexiones sobre el origen y la difusión del nacionalismo. F.C.E., México
Araujo K (2019) Hilos tensados. Para leer el octubre chileno. Idea–USACH, Santiago
Arendt H (2005) La condición humana. Paidós Ibérica, Madrid
Bustamante J, Márquez F (2022) Contramonumento - Countermonument. In: De Nordenflycht J (ed) Diccionario del Patrimonio. Editorial Bifurcaciones, P. Universidad Católica de Chile, Santiago
Butler J (2010) Cuerpos que importan. Sobre los límites materiales y discursivos del sexo. Paidós, Buenos Aires
Briones C (2020) Interculturalidad y patrimonialización: La invisibilización de escenificaciones del ser juntos siendo otros. En: Diaz G (ed) Antropología Contemporánea. Intersecciones, encuentros y reflexiones desde el Sur Sur. Ed. Universidad Católica de Temuco, Temuco, pp 83–103
Carrión F, Dammert M (2019) Derecho a la ciudad. Una evocación de las transformaciones urbanas en Latinoamérica. Clacso, Flacso, Ifea, Buenos Aires
Del Sol G (2022) "El espacio común", El Mercurio. Suplemento Artes y Letras. Debate. Cuerpo E2. 16.01.2022:2
De Nordenflycht J (2001) Iconoclasia, patrimonio y arte en el espacio público. En: El Arte en el Espacio Público. Temas de la Academia. Academia Nacional de Bellas Artes – ANBA, Buenos Aires
Didi-Huberman G (2017) Sublevaciones. Untref, Jeu de Paume, Buenos Aires
Durán L (2021) Estallido Social. Espacios y Monumentos Insurrectos de octubre en Ecuador. Corpus. Archivos virtuales de la alteridad Americana 11(1)
Freedberg D (2007) Iconoclasia. Historia y psicología de la violencia contra las imágenes. Sans Soleil Ediciones, Vitoria Gasteiz
Gamboni D (2014) La Destrucción Del Arte. Iconoclasia y Vandalismo Desde La Revolución Francesa. Cátedra, Madrid
Geertz C (1999) Negara: El estado-teatro en el Bali del siglo XIX. Paidós, Barcelona
Góngora A (2021) El busto de Comanche. O de cómo entró un habitante de calle al Museo Nacional de Colombia. Corpus. Archivos virtuales de la alteridad Americana 11(1). http://journals.openedition.org/corpusarchivos/4505
Halbwachs M (1970) La morphologie sociale. Collin, Paris
Krotz E (1993) La producción de la antropología en el Sur: características, perspectivas, interrogantes. Alteridades 3(6):5–11
Latour B (2018) Sobre el culto moderno de los dioses factiches seguido de Iconoclash. Dedalus, Buenos Aires
Márquez F (2020a) Anthropology and Chile's Estallido social. Am Anthropol 122(2):666–687. ISSN 0002-7294, by the American Anthropological Association
Márquez F (2020b) Por una Antropología de los escombros. Revista 180(45):1–13. https://doi.org/10.32995/rev180.num-45
Márquez F (2021) «Monumentos en Latinoamérica: Entre la épica patria y la insurrección» Corpus. Archivos virtuales de la alteridad Americana 11(1). http://journals.openedition.org/corpusarchivos/4505

Márquez F, Colimil M, Jara D, Landeros V, Martínez C (2020) Paisaje de la Protesta en Plaza Dignidad de Santiago, Chile. Revista Chilena de Antropología (42):112–145. ISSN: 0719-1472

Martucelli D (2020) El estallido social en clave latinoamericana. Lom Ediciones, Santiago

Masotta C (2021) Las falsas promesas de los monumentos. Corpus. Archivos virtuales de la alteridad Americana 11(1). http://journals.openedition.org/corpusarchivos/4505

Metraux A (2011) [1928] Antropofagia y Cultura. El cuenco de plata. Cuadernos de plata, Buenos Aires

Molina R (2021) Hablan los muros. Lom Ediciones, Santiago

Mongin O (2006) La Condición Urbana. La ciudad a la hora de la mundialización. Serie Espacios del Saber N° 58. Editorial Paidós, Buenos Aires

Mora P et al (2020) Monumentos incómodos. En: Artishok. Revista de Arte Contemporáneo, 05.07.2020

Pizarro C (2021) La colonia siguió viviendo en la república. Corpus. Archivos virtuales de la alteridad Americana 11(1). http://journals.openedition.org/corpusarchivos/4505

Quijano A (2000) Colonialidad del poder, eurocentrismo y América Latina. In: Lander E (comp) La colonialidad del saber: eurocentrismo y ciencias sociales. Perspectivas latinoamericanas. CLACSO, Buenos Aires

Ranciére J (2010) La noche de los proletarios: archivos del sueño obrero. Tinta Limón, Buenos Aires

Restrepo E (2012) Antropologías disidentes. En: Cuadernos de Antropología Social, N° 35, pp 55–69. © FFyL, UBA

Richard N (2007) Figuras de la memoria: Arte y pensamiento crítico. Siglo XXI Editores, Buenos Aires

Segato RL (2007) Las Nación y sus Otros. raza, etnicidad y diversidad religiosa en tiempos de políticas de la identidad. Prometeo Libros, Buenos Aires

Taylor D (2015) El archivo y el repertorio. La memoria cultural performática en las Américas. Ed. Universidad Alberto Hurtado: Colección Antropología, Santiago

Taylor D (2021) ¡Presente! La política de la presencia. Ediciones Alberto Hurtado, Santiago

UNDP (1998) Informe de Desarrollo Humano. Santiago de Chile

Vignolo P (2021) "¿Sueñan las estatuas con ovejas de bronce?", Corpus. Archivos virtuales de la alteridad Americana, vol 11, N°1. http://journals.openedition.org/corpusarchivos/4505

**Francisca Márquez** Anthropologist from the University of Chile, Master's in Development and Doctorate in Sociology from the Catholic University of Louvain, Belgium. Currently working as an academic in the Department of Anthropology at the Alberto Hurtado University, Chile. Until 2014, she was Dean of the Faculty of Social Sciences at Alberto Hurtado University, Chile. She was National President of the College of Anthropologists of Chile and Vice President at the National Foundation for Overcoming Poverty. She has directed several research projects supported by the Science and Technology Fund in Chile (ANID) and published books and articles on urban identities, imaginaries, heritage and inequality in Latin American cities.

# Chapter 26
# Trends of Urban and Territorial Reconfiguration in Metropolitan Buenos Aires

**María Carla Rodríguez**

> *And I learnt that the memory has to be reinvented.*
> *Only then is it capable of lasting and going through time.*
> *Liliana Bodoc. Los días del Fuego*

**Abstract** The chapter explores the urban and territorial reconfiguration trends of the metropolitan Buenos Aires in pandemic times. It identifies the effects of the symbolic and symbiotic operations inherent to the metabolism of capital in its neoliberal phase considering: the signature urban policies, the treatment of public spaces, and different modalities of the popular habitat. In the face of the urbicide scenario—the death of an exploded and fragmented city in divergent and walled up territories—it presents a transduction exercise, which gives visibility to material and symbolic practices of a self-managed urbanism where the dividing wall becomes a transitional space that connects, combining the horizon and the right to beauty.

**Keywords** Urban transformation · Neoliberalism · Urbicide

## 26.1 Introduction: "Cities of Walls" in Pandemic Times

In 2020, most of the world's population dwelled in cities that, as Henry Lefebvre predicted in the early 1970's (1972), have become "urban developments". In Latin America, the new human settlements, promoted by the State, the market or, to a great extent, self-built, are located in outskirts that are increasingly further away and geographically segmented. Historical centers depopulate and become tourist areas, turned into objects of consumption for wealthy groups. Telecommunications have radically changed the way we experience space and time. Borders are reconfigured and, as Peter Marcuse stated in the mid 1990's (1995), gaps, barriers and walls of

---

M. C. Rodríguez (✉)
Universidad de Buenos Aires (UBA) and National Council for Scientific and Technical Research, Buenos Aires, Argentina
e-mail: trebol1968@gmail.com

different textures—tangible and intangible—increase between those who can and cannot afford material and symbolic goods.

The covid-19 pandemic quickened the pace of transformations in an unprecedented way, strengthening the capitalism of digital platforms: the expansion of home office, remote education, the consumption of information and communication technologies, and the viral spread of all kinds of delivery services for those who have a home and can afford to stay there and pay. Meanwhile, millions of people have been forced to be on the front line of essential services to earn their living or because they lack proper housing. During 2020, the public spaces in the cities depopulated, and, through the screens, we watched clear skies, clear-looking waters, animals boldly wandering about the concrete jungles in different parts of the planet. We also witnessed massive human outflows from major cities, such as New York, London, and Lima, some of them looking for better quality of life, and others looking for survival. The central areas of the cities stressed their "zombie" pattern: neither living nor dead, drained from the functions that have been abruptly taken from them and decentralized in other locations.

Sennet (2004) points out two distinct qualities that mark the configuration of sociability and the spatiality in the current stage of the financialized capitalism: flexibility and indifference. Just as flexible production leads to shallow and short-term relationships, which increase labor inequalities and an environment that is the opposite of democracy in work places, capitalism has also created a regime of shallow relationships with the city, underscored by the standardization of the urban setting. Work and consumption spaces have become neutral, and they fight against the local history of cities and the common stories of their inhabitants. Even though, on the one hand, at the turn of the twenty-first century, examples of decolonial, feminist, youth, ethnic identities, and environmental, etc., resistance emerged non-stop, on the other hand, their political coordination has been highly difficult and uncertain.

To sum up, the city that integrated and brought together the different ones, favored the social heterogeneity, convergence, gatherings, and urbanity as the mediation of conflict, is now facing a metropolitan and globalizing dynamic in which the supremacy of profit divides, scatters, pulls apart, privatizes, decentralizes, and creates new urban and territorial hierarchies. (Delgadillo 2021).

The "creative destruction" of the capital rationale (Polanyi 2007) underlies this dynamic that metabolizes matter, culture, and subjectivity. It partly recycles them; it partly shreds, destroys, and/or deserts them, depending on the reproduction and growth of profit. Cities, in particular metropolis, acting as the secondary circuit of the capital accumulation process (Harvey 2007), have been a privileged object of these dynamics (Berman 1989; Caldeira 2007).

Carrión (2018a) defines this multidimensional and complex process as "urbicide",[1] referring to the massive and selective destruction, carried out with overt premeditation and planning in the cities, and which encompasses their different

---

[1] It revisits the conceptualization suggested by Marshall Berman (1989) to account for the urban impact of the capitalist modernization in his hometown neighborhood, the Bronx, in the city of New York.

dimensions: as agora (squares, libraries, monuments), urbs (infrastructure and services), civitas (citizenship and vicinity bonds), and polis (administration capacity). This murder of the city, considered as a whole, stems from varying situations, which Carrión organizes in three types: natural, anthropogenic, and symbolic.

This conceptualization, in turn, exposes the sensitive nature of "facts"—as some pearls that change their color and shine in different hands, as said by writer Ursula Le Guin in *The Left Hand of Darkness (1968)*—articulating productively the narratives that question a key aspect of the complexity inherent to the experience of urban life: getting together with others, starting off the recognition of the unfathomable abyss of differences. On the contrary, the equalizing naturalization of capital, and its consumerist illusion of mandatory happiness, builds the planned production of oblivion, which sweeps away the material basis, the institutional frameworks and the citizen experiences.

In Latin America, the biggest imaginary in the cities is fear (Silva 2004), turned into an urban principle that eliminates the public space (agoraphobia), produces the gated, or walled, community (Caldeira 2007) and leads to the loss of community lifestyles. Carrión (2018b) adds that the weight of urban violence and crime contributes to the legitimization of different punitive policies of public safety. A "habit of fear" is naturalized, bred in the outskirts and pervading all the social sectors (Rodríguez 2014). However, during the covid-19 shutdown, crimes on the street and burglaries decreased, and at the same time, domestic violence went up, especially against women and children (Delgadillo 2021).

Even though toward the end of the 1990's, the public space seemed to be residual, in the first decades of the twenty-first century, and even in the midst of the pandemic, some countertrends were seen. The street was once again taken in different ways: by graffiti, wall paintings, installations, or big urban parades of different nature. Popular vindications of massive protests, such as the ones spearheading the constitutional process in Chile after the fuse sparked by the student outrage of the "penguins", the strike started in 2020 in Colombia or the great parades for the legal abortion in Argentina, events that drive the masses whose representation nobody can assume or channel politically, together with the emergence of the worrying novelty of right-wing traditional and conservative forces taking to the streets and the formation of militias, with the clear, but not exclusive, example of Brazil.

In this chapter, I will explore these symbolic and symbiotic operations and trends of the urbicide metabolism of the capital in my hometown, the City of Buenos Aires. From a qualitative perspective, I will analyze the territory reconfiguration trends at different scales, locations, and typologies of the habitat of the metropolitan region of Buenos Aires.[2] It is in part an autobiographical memory exercise, pierced by the implications of my citizen and activist experience as related to the processes

---

[2] The definition of Metropolitan Region of Buenos Aires (MRBA) is considered to be the most adequate description to capture the territory reconfiguration processes analyzed here. It is a continuous geographical area which includes a population of approximately 14 million inhabitants, according to the latest census, and it combines national, provincial and municipal governments.

considered.[3] Lastly, the analysis brings forward a transduction operation[4]—the epistemological device suggested by Henry Lefebvre—which postulates the emergence of the "right to beauty" as a political and theoretical response to the contemporary urbicide, as a discourse and material practice projected from the self-managed resistance urbanism anchored in the praxis deployed from everyday life, which politicizes anew the symbolic, territorial, and incarnate dispute for the right to the city, weakened in its institutional, juridified and technified reinterpretations.

## 26.2 Buenos Aires, Neoliberal City: Territory and Symbolic Reconfigurations in Dispute

I will start the journey through the central city of the MRBA, Buenos Aires, which is undergoing an urban renewal process with different turnarounds, whose continuity may be recognized since the early 1990's and its foundational conditions, the coup d'état of 1976. By and large, the regional transformations on a metropolitan scale in these decades included the broadening of the central area (started with the Puerto Madero operation), the regional urbanization of the riverside (with milestones such as La Boca, Puerto Madero itself and Costanera Sur in the City of Buenos Aires, and the costal urban developments in the districts of Avellaneda and Quilmes toward the South) and the northern riverside, started in the district of Tigre with the Nordelta project. The construction of highways, feeder roads, and major works of the metropolitan mobility system with regional and macro-regional importance accompanied the changes. The progressive realization of these operations led to the permanent dynamization of the process of valuation and the sustained increase in the price of land (Herzer 2008).

The alliance between culture and tourism represents another focal point in the transformation of the City of Buenos Aires. In a progressive and sustained manner, some actions were added, such as the scheduling of cultural events in the city, traditional tourism exhibitions, and the growth of the city's cultural tourism agenda in the offers of travel agents. The successive administrations boosted several neighborhoods of the city in terms of tourism. The flagship cases were the more traditional

---

[3] In particular, related to the development of self-managed cooperativism in central areas of Buenos Aires city by the Federation of Self-Managed Cooperatives MOI, Movimiento de Ocupantes e Inquilinos. Methodologically, I combine the review and analysis of records and field notes of these activist experiences with the results of a longitudinal research that I have directed in the context of the Area de Estudios Urbanos del Instituto de Investigaciones Gino Germani de la Universidad de Buenos Aires referring to the territorial transformations of the metropolitan area of Buenos Aires and the role of housing policies implemented under different designs and locations in that territory (within the framework of UBACYT UBA programming from 2002 to the present).

[4] Traduction, from information related to reality—everyday life—as well as an issue stated by this reality, elaborates and makes up a theoretical object, a possible object. It entails a balance between the concept framework used and empirical observation. It presents thoroughness in the invention and knowledge of utopia. (Lefebvre 1972).

neighborhoods of La Boca and San Telmo in the South, and Palermo and Recoleta in the North. The idea of "current" culture was promoted, which is evident in vital functions attributed to the urban entity: "culture is breathed in", "tango is lived", "Buenos Aires lives in all the corners of the city".

Simultaneously, a scenario of many houses without people and many people without houses has exacerbated. The right-wing political force that took office in the city after the demise of the local progressivism in 2008, and has stayed in power up to now,[5] has systematically upheld the decision to weaken or destroy institutional and legality frameworks promoted and useful for popular sectors, which had been created after the return to democracy in 1983 and, in particular, with the political autonomy of the city, expressed in the constitution of 1996. It was enacted in the context of a process of participation and gathering in the local public arena between progressive political forces and a broad range of urban social movements and organizations of civil society.[6]

The strategy of local neoliberal governance (Theodore et al. 2009) included the violation of certain laws, such as article No. 31 of the Constitution of the City[7] the Act 148 and the Act 341, which establish guidelines and instruments that have a strong public and decommercializing element to approach the different situations of the urban public habitat: slums, tenancy, hostels, squatting, overcrowding, unoccupied buildings, and building deterioration (Rodríguez 2009). The flipside to this is that, according to a survey conducted in 2019, 9.2% of *porteño* housing for residential use were empty, twice as high as the average in other Latin American cities, and which is concentrated in the neighborhoods with the highest m2 value (Puerto Madero, Retiro, Recoleta, and Palermo) (IVC 2019).

With the continued increase of the value of land and the lack of alternatives provided by public policies for access to habitat, the "slums" were the fastest growing, self-produced form of habitat, and it has become visible in the urban fabric of the city in the last decades. In 2008, the PRO wanted to eradicate them, but as of 2015, it has redefined an "integration" policy that is an example of the neoliberal governance operation, guided by the rationale of creative destruction at an institutional and territorial level. This dynamic determined the four big interventions in the slums, which were chosen according to their importance for the objectives of the overall urban

---

[5] In spite of the national electoral failure in 2019, the PRO kept its local stronghold, to a great extent based on its urban policy.

[6] A progressive experience that, when it was expanded at a national level, as the Alliance (Alianza), collapsed during the 2001 crisis.

[7] ARTICLE 31. The City acknowledges the right to decent housing and a proper habitat. To that end: (1) It gradually solves the housing, infrastructure, and service deficit, prioritizing the people from sectors in critical poverty and with special needs of scarce resources. (2) It sponsors the incorporation of empty housing, promotes self-managed plans, the social and urban integration of marginalized inhabitants, the restoration of precarious houses and the land-title and cadastral regularization, with the criterion of permanent establishment. (3) It regulates the facilities that provide temporary lodging, making sure to leave out those that conceal leasing (Constitution of the Autonomous City of Buenos Aires).

policy of the PRO for the southern area of CABA as of 2015: the historical neighborhood 31, Rodrigo Bueno (both on the riverside of the central area), the Playón de Chacarita (Chacarita Lot), and slum 20 (Villa 20), in Lugano, a historical district of popular houses and housing funded by the State, that the PRO set out to "seduce" in their quest for popular hegemony).[8] Those interventions intended to create positive effects on the renewal and valorization operations in the consolidated urban settings. (Rodriguez 2018).

Planning by projects, slowly but surely, created and reinforced social and spatial segregation effects, both at an urban scale and a micro, neighborhood one. New retaining walls, barriers, and screens appeared in the mediated slums, which worsened the segregation of these popular neighborhoods from their urban settings. Each project intervention led to new sets of works on the borders, intended to relocate part of the inhabitants. Highways or avenues fenced and limited their expansion and took the self-built blocks out of sight, like the Manchester of young Engels. Facades were painted in pretty and quaint colors; lightning was placed; some public spaces were improved, and new works of strong symbolism were built, emulating the urban model of Medellin.[9]

At the same time, new borders were reconfigured in its inner fabric. Improvements conceal the deterioration hidden in the heart of the housing structures where constructions have become steadily higher... Energy, water, access to Internet, lighting, and ventilation of houses are conspicuous by their absence, but the price of the $m^2$ of the room for rent—true and thriving market whose capital is not controlled by anybody—always gets higher and is expressed in dollars. This type of rent accounts for up to 60% of the income of those who can afford it, determined by the position of the monopolistic leasing on a public mostly made up by female heads of household, migrants, and youth, always under the threat of the possible "express eviction" that is not controlled by anybody either (CESBA 2018). Likewise, in some housing structures, the authoritarian methods of despondency are currently being used again; they are typified by Oszlack (1991) with practices such as breaking drinking water pipes or leaving mountains of debris that get filled with vermin, so as to make everyday life more complicated for those holding on, who cannot or do not want to leave the stockade of debris surrounding their habitat. Finally, in some occupations taking place in 2021, people are evicted, and everything is burnt (with the social, psychological, and environmental effects that this selective violence breeds).

Repairs were dismissed, and the organizations linked to the community lifestyles, and in the name of civil law and contractualism between individuals—ignoring the nuance of the original public property of the land intervened—the circuit of the informal commodification was encouraged: a "motley crew" mixing capital coming from drug and other trafficking to families that have an additional room to improve

---

[8] In the primary elections (PASO) of 2021, Juntos por el Cambio won in the city with 63% of votes, but it lost in the commune 8 and in the slums of the city.

[9] The social urbanism of the Medellin model entails (1) a production practice of governable spaces (2) that it creates spaces for citizens' control and standardization, and (3) the productive adaptation of the city (Quinchía Roldán 2013).

their income. The storyline of the drug dealer relies on the poverty setting and has territorial force, under this urban policy that is virtuous for their operation flow. Thus, the historical, social, organizational, and cultural characteristics of populations are subordinated, selectively promoting the implicit acknowledgement and the whitewashing of the informal commercial dynamic together with the pastoral ode to micro-entrepreneurship (Rodriguez 2018).

In the same years, also in a systematic and planned manner, the local government choked in the budget the housing operations generated in popular organizations, such as Act 341/964 for the Housing Self-management, a program that allowed the purchase of 118 lots in consolidated neighborhoods between 2002 and 2007, and the construction of 1261 houses in 46 projects currently inhabited in centric areas of the City of Buenos Aires, with an average surface of 71 m2, through the self-management of public resources by cooperatives and other social organizations. These processes, focused on individuals' development, are rooted in the stage of the return of democracy, and they led to the resilient installation of territorial milestones of a self-managing urbanism that is still active in the city and has a national outreach (Rodriguez and Zapata 2020).

However, urban neoliberalism operates at symbolic and material levels, in the invisibility and delegitimation of the popular and decommodificating alternatives for the production of habitat. A paradigmatic example was the active boycott to the building restoration of the former Padelai, in the heart of the historic center, an action carried out by its occupying dwellers together with the professional teams of the Architecture School of the University of Buenos Aires and housing NGOs at the end of the 1980's. At present, one of the buildings of the property was used for the Neighborhood Management and Participation Center. However, little is recalled about the long history and the fight for the use of that public venue, nor about the original content of the proposal passed by the City Legislature in 1990, which included the construction by the cooperatives of 118 houses and an extensive program to provide equipment for the neighborhood and stores, whose income should be reused in popular housing in the neighborhood of San Telmo. Additionally, the way the city government got rid of these regulations, implemented a variety of strategies to break up, selectively co-opt and displace the target population of this project, and how it cleared its judicial standing (the property was registered in 1991 under the cooperative's name as co-owner with the city government, which owns 25% of the total surface) have been pushed into oblivion.

The former Padelai is the foundational experience of the urban paradigm supported by Act 341/00, in whose executed projects it currently lives, through the concrete restoration of the urban heritage with a substantial participation of its inhabitants to exercise their right to the city. It is an urban restoration model opposed to the one promoted by the PRO administrations, which goes beyond the local scale. In this sense, in 2009, the Urban Development Minister back then, pointed out that their renewal policies "are part of a transformation process", which "goes beyond the achievements of a specific administration and should be taken as a State policy". It is about the "assimilation of the South to the North", formulated in 1998 by the then City Mayor De la Rúa[10], with expressive symbols in the relocation of the City Hall to the neighborhood of Parque Patricios and the decentralization of different ministries, among them Education and Social Development, which were relocated to the slums 31 in Retiro and 15 in Lugano, as icons of the social and urban integration achieved.

Other individuals made invisible by this urban policy were the dwellers of hostels, tenement houses, and squatters, the housing forms of the scattered urban poverty. The dispute dealt with in the void of urban centrality had led, in the midst of the return of democracy, during the 1980's, to massive processes of occupation of hundreds of public and private buildings, around 250,000 inhabitants, and surrounded by an institutional context (partisan youths, Human Rights agencies, organizations emerging from the buildings themselves, university groups), who advocated for many of the contents that were later materialized in the constitution of the autonomous city of 1996 (Rodriguez 2009). All along the twenty-first century, that memory has tended to be eradicated and exorcized. Nowadays, a rerun of the Argentine TV series Okupas can be seen on Netflix. Made in the early 2000's, its plot romanticizes marginality and draws a veil over the survival efforts of thousands of families of squatters who actually try to escape it.

Moreover, temporary lodgings "concealing leasing" (article 31 of the Constitution) have been overlooked; on the contrary, they have been selectively subsidized by means of different mechanisms of the local housing emergency policy, currently through modern "vouchers" for recipients, which go to certain hotel landlords operating in the formal urban fabric, as well as, increasingly, rental inside the slums, as a local expression of the growing financiarization of informal real state markets in Latin American cities.

Whenever it was possible, the powers that be resorted to some illegal practices associated with direct repression, such as the creation of "units for the protection of public spaces" aiming at ousting homeless people and, sometimes, pushing them toward the slums, which faced the novelty of having people sleeping in their narrow corridors. The urban and housing policy thus strengthened the exclusionary nature of the city, naturalizing the political decision, actively transformed and reinforced as common sense that "you can live where you can afford it".

Additionally, it is important to highlight that the national administration in the period 2005–2015 was not alien to these trends. A clear example, involving federal fiscal property in the South, was the land of the then called "Barrio Huracán" (Huracán Neighborhood), bordering the slum 21, in the neighborhood of Barracas, which was already the most densely populated one in the city. The land comprises 19 hectares and was managed by the National Administration Agency for State Property (ONABE). After the eviction of 1500 people who had built precarious houses there, in 2008, one hectare was given to the University of Buenos Aires, another hectare to the Argentine Industrial Union, and the rest was allocated to a great public housing operation for middle-income sectors, now called "Estación Buenos Aires", which

---

[10] Fernando de la Rua, Argentine President between 2000 and 2001, and head of the Alliance. The 2001 crisis was also the first major crisis of progressivism.

**Fig. 26.1** Map of AMBA (Area Metropolitana de Buenos Aires) and main cases referenced. *Source* Own elaboration

involved building 2476 houses and about 900 parking places. In early 2021, two thirds of them had not been assigned yet, that is to say, they were empty. The urban operation, funded with resources of the ANSES within the context of the PROCREAR program, occupied the area that should have been used for the swelling of the slum 21–24, where the densification, degradation of dwellings, and the informal room-renting market had reached unprecedented levels in those same years, together with the lack of basic water, sanitation, and infrastructure services (Fig. 26.1).

To sum up, the sustained creation of new quality public spaces led to losing sight of the intensification of spatial segregation and the growing barriers against wide social sectors to inhabit this renewed city. Constitutional rights have been distorted and violated, and collective memories of popular struggles have been weakened. The flexible planning and the market culturalism, persistently activated by means of the urban, touristic, and cultural policy along the decades, make up an economic and ideological aspect that aimed strongly at the production of hegemony, blurring the perceptions and classifications of the benefits and damages unequally produced and distributed.

## 26.3 Old and New Trends in the Urban Segregation During the Pandemic Present

*Sometimes you just want to have a glass of water to drink, you can do it anywhere in Buenos Aires, except in slum 31 and the rest of the slums in the city.*
One neighbor who cannot stay home.

In the pandemic, the urban renovation policy promoted in the City of Buenos Aires continued the previous trends, supported by the political force of the local executive power with a majority. In this stage, there was an attempt to revitalize those areas particularly affected during the pandemic (downtown) by means of tax exemptions, public lands continued to be given away (Costa Salguero, Parque de la Ciudad), 16 new urban consortiums were approved, as well as the feasibility of a major urban project on the riverside, boosted by the developer IRSA, by means of its rezoning (former Sport City of Boca Juniors). With a rhetoric that stresses the possibility of generating public spaces for all citizens, the disposal of public lands and the passing of specific legislation for developers of big lots are advanced, achieving the convergence of two seemingly opposing processes.

The executive power in CABA also worked on a plan to renovate the downtown area, which seeks to give a new function to the zone by recovering the residential use of the land[11], as a response to the sharp decline of activities and traffic, given the close of offices and stores. The granting of financial and tax benefits is proposed "(…) the intention is to turn the downtown area into a residential neighborhood with a mixture of uses, public spaces, and vicinity stores. The spirit of the plan is the creation of a "15-min city" or "proximity city", where citizens' commuting times are reduced when it comes to labor or leisure activities, following the Colombian-Parisian urban model[12].

Meanwhile, the housing policy continues to be underfinanced, and investment for 2021 was 25% lower than in 2016, in values updated according to inflation (ACIJ 2020). The pandemic arrived in the context of a hegemonic market rationale, where the extended and growing process of formal and informal tenancy underscored the condition of residential alienation[13]. Although evictions were suspended by decree (No. 320/2020) for a year in March 2020, rent prices continued to go up (67.2% during this year), and properties were taken off the market. Tenants in slums saw their pre-existing vulnerable conditions become more exacerbated: as the economic impact of the quarantine was greater in the informal labor market, those households that were not able to pay the rent faced a greater risk of eviction. (ACIJ 2020).

"Come and see the desperation at not having water and the fear of catching the virus", said Ramona Medina, a neighbor and leader in the slum 31, at the end of a video that went viral on May 3rd, 2020. A few days later she died of coronavirus. Ramona lived with her family (seven people), overcrowded in house 79, block 35 of slum the 31, and she was insulin-dependent. Many of the members of her family also

---

[11] LA NACION 11-08-2021.

[12] From the need to decarbonize cities, given the environmental crisis of the climate change, a planning that resumes the proximity city as a solution is proposed, driven by the Mayor of Paris, Anne Hidalgo, reelected in 2020 for a second, six-year term, and Carlos Moreno, a mathematician, an expert in urbanism, scholar and entrepreneur (Mayorga 2021).

[13] As Madden and Marcuse (2016) point out, the idea of "feeling at home" is opposed to residential alienation. The concept of alienation, applied to the residential field, helps understand the experience of the struggles for housing and shed light on the connections between the housing crisis and personal crises. The residential alienation happens when a capitalist class captures the production process of dwelling and exploits it for its own benefit (Madden and Marcuse 2016, p. 59, own translation).

caught the virus. During her funeral, the new house was allocated. The organization where she belonged, La Poderosa, stated:

> "Gritting our teeth, hitting the keyboard, containing our rage and spitting tears, we have to write this shit, to tell you all the things that Ramona said in the past, all the things we were tired of yelling for two months, all the things they didn't want to hear…". "Ramona did not die! Ramona was killed…!"
> 
> La Poderosa. Public statement. May 2020
> 
> On https://www.instagram.com/tv/CASnNDugc6B/?utm_source=ig_embed

They exposed that the relocation of her family was postponed during four years, that the request to be included in the risk group was ignored, that the health posts did not have the necessary basic supplies, that the first cases in the slum were not isolated in time, that the neighborhood lacked water during 12 days in the midst of the pandemic, and that reality was camouflaged with ghost programs that were outside of the territory.

Ramona's commitment was echoed in the sanitary brigade that was named after her, in the University of La Plata, and it replicates that line in their epidemiological work in the popular neighborhoods in the periphery of the capital city of the province of Buenos Aires. In 2021, in one of the narrow streets of the slum, Ramona's face appears impressive on a wall, painted a few meters from the place where her house used to be, and now only a pile of debris remains. Just as Ramona's house, hundreds of houses were demolished after the relocation. The Government of the City of Buenos Aires left the debris in situ, up for grabs for anyone who wants to take it.

"Honestly, there isn't much urbanization. The block where Ramona lived is totally filled with debris, destroyed, with rats, full of stuff and garbage", said Pamela Andrade, a leader in the House of Women and Gender Variance Groups, also renamed "Ramona Medina", a space where Ramona assisted in a health post. "In this block, there are families living that have to coexist with the concern to not get covid or any other illness created by these conditions". This way, the self-built housing structure continues with irregularities and marked by inaccessibility to basic services. During all of 2020, with no salaries or rest, the community cooks, as well as gender and health workers, among others, kept the neighborhoods together, sharply impoverished, such as the slum 31. The same situation of vulnerability made visible by Ramona Medina's death is faced by so many other community workers that, even though they are the first one to leave to provide care, are the last ones in the lines of the State.[14]

Living in poor conditions on the street was another hot topic in pandemic times. Homeless people were particularly vulnerable: apart from the risk of catching coronavirus, many places that helped them with food closed, and the possibility of having little jobs disappeared. It is a risk group, as 38% have existing health conditions and over 10% are 60 years old or older, the age with the highest mortality rate for covid-19. Data from the latest popular census carried out in 2019 estimated their number in

---

[14] http://anccom.sociales.uba.ar/2021/05/17/un-ano-sin-ramona-Y-la-villa-31-sigue-sin-agua-potable/.

7251 people (ACIJ 2020). This vulnerability became tragic, with criminal, violent, and fascist reactions, such as the case of a woman who was intentionally set on fire and burnt alive in July 2020, under the 25 de Mayo Highway, in the centric neighborhood of Constitución. After setting her on fire together with her belongings, the murderer-neighbor walked away undisturbed, as registered by surveillance cameras nearby. The video became viral and in many centric neighborhoods, such as San Cristóbal, Balvanera, and Monserrat; the mattresses and belongings of the homeless were intentionally set on fire. Although the context of health, housing, and socioeconomic crisis made the situation worse, the budget for 2021, as compared with the one in 2020, does not increase the amounts for that population. (ACIJ 2020).

## 26.4 Guernica and Nordelta: Divergent Peripheries of the MRBA

*Sow pesos and reap dollars*
Real Estate advertisement of Propiedades SA. 2021
*The crisis is huge because two different worlds are being created: a ghetto locked up in the slums and another one in the private neighborhoods. These worlds are increasingly further apart, not only in terms of income. They do not share codes. They have very different dialects and aesthetics. The government and the powerful cannot or do not want to bridge the gap between those two worlds.*
Rodrigo Zarazaga. Jesuit priest[15]

Buenos Aires is an example of the way the peripheries of Latin American cities have tended to a growing socioeconomic polarization and fragmentation, intensified at a small scale. Since the 1990's, high-income sectors were involved in a scattered, low-density urban expansion that was juxtaposed with the historical precarious popular urbanization, self-built from irregular divisions into plots and informal settlements on degraded and flood-prone lands, so far undisputed (Pirez 1994). Gated communities, country clubs, and sailing clubs were deployed ignoring the negative impact on the local and regional environmental conditions (Pintos and Narodowski 2015).

At present, there are about 600 gated neighborhoods in the metropolitan area of Buenos Aires that occupy approximately 50,000 hectares (two and a half times the surface of the Autonomous City of Buenos Aires, but a third of its population, about 300,000 inhabitants). (Venturini et al. 2020). Data collected between 2006 and 2016 show that 46% of the surface added to the conglomerate of the Greater Gran Buenos Aires belonged to these gated neighborhoods, as opposed to the 14% of the informal settlements (Lanfranchi et al. 2018); 28% of this surface belong to 168 neighborhoods developed in the expansion strips of the second and third belts

---

[15] La Nación, 1-9-2019.

between 2001 and 2020, and 55% of them were settled on wetlands and decapitated soil (Apaolaza and Venturini 2021).

In the middle of the pandemic, this expansion model by means of gated neighborhoods unfolded together with the construction of the Presidente Perón Highway, which shapes the third beltway ring of the City of Buenos Aires[16], and continues the Camino del Buen Ayre up to Provincial Route 2. This road project changes radically the accessibility conditions and the land value trends in districts on the SW of the MRBA (for example, Ezeiza, Esteban Echeverría, San Vicente, and Presidente Perón), where gated neighborhoods projects have multiplied.

The opposing side of the gated communities is the massive phenomenon of land occupation, which, over time, is consolidated as self-built settlements. This process started at the same time and with more visibility than building squatting in the central city, in the early 1980's, as a consequence of the growing impoverishment and the exclusionary policies carried out by the civic-military dictatorship. From then onward, with its cycles and fluctuations, land occupation has not stopped. Between 2001 and 2015, 300 plot occupations were recorded, which were later consolidated as neighborhoods, involving approximately 400,000 inhabitants. Most of them, 24 out of 25 attempts, fail and end up in evictions.

The complement of these divergent spatial forms is a deeply unequal division and ownership of the economic, social, and environmental benefits and costs, with strong income transfers to the private sector and the generation of scattered, low-density, privatized and segregated peripheries (Apaolaza and Venturini 2021, Ciccolella 2014). According to Smith, a key component of gentrification is the construction of the border myth, with the idea of "taking the city to the countryside", to enjoy the urban comforts in a setting infused with an idyllic rural context and an alleged environmental high quality (Pintos and Narodowski 2015).

Far from the fictional storytelling, the peripheries of the MABA are settings of contradiction and disputes between the concentrated real estate capital and the popular sectors that fight for the access to land and housing. The conflict that arose in July 2020, in the pandemic, in the district Presidente Perón, between the gated neighborhood San Cirano and the occupation of land known as Barrio Guernica Unido, is a clear example of it, and evidence that behind this tension there are also different city models at stake (Venturini et al. 2020).

In the town of Guernica (Pesidente Perón district), on the South of the Metropolitan Area of Buenos Aires (MABA), 98 hectares were occupied between July and October 2020 by approximately 2500 poor families in need of access to land and housing. The owners of the land, a firm ironically called *El Bellaco* (Rogue) S.A., focus on real estate enterprises, and they plan to build the gated community complex of the San Cirano Rugby Club there, and they have not been able to reliably demonstrate ownership of the land with the corresponding documentation. However, the intricate process of negotiation and factual pressure involving the province and local governments and the Justice, ended up in a huge operation with 4000 police officers and

---

[16] After General Paz and the Camino de Cintura, and before the Provincial Route 6.

the violent eviction of the resisting families, the burning of shelters made with sticks and plastic, and scenes of rage and panic.[17]

Thus, it was evidence of a State that, through Justice and a repressive instrument, operates as guardian of the speculative private ownership of the land, which strives to preserve that privilege to take ownership of the social surplus at all costs. This is opposed to the popular self-managed housing. During the period that the retrieval of the Guernica lands lasted[18], the neighbors also recovered popular organization schemes and memories of resistance, assigning delegates by block, zone, and neighborhood. In addition to sustaining the immediate everyday life, accompanied by organizations, students, and professionals of the public university (in this case, from the Schools of Geography and Architecture of the UBA), they started an emerging urbanization process through the division into plots, laying out streets and squares, or access to services such as electricity. The made surveys onsite and designed a unified neighborhood project, in agreement with the assemblies of delegates and organizations. A product of urban quality was prepared in a participatory manner, which considers the integration of the new neighborhood to the existing urban fabric, within the provisions of Act 14,449 for the Fair Access to Housing.

It was not enough… legislation, as is the case with Act 14,449 for the Fair Access to Housing, is harmless and insufficient. Its implementation is very limited, and it depends on the criteria of the towns, which, in this instance, show their agreement with real estate interests and a lack of critical sense in terms of the effects of the city model that they call for.

Therefore, one year after the swift eviction operation[19], only now does the urban proposal promoted by the province take shape, involving 860 land plots for the families surveyed that, to a great extent, have scattered and live crowded where they can… On the other side, the *Country & Club San Cirano,* projected over 361 hectares with 1600 plots of 950 $m^2$ each divided into five neighborhoods. Among their amenities, the project includes three artificial lagoons for water sports.

This urban model was launched by Nordelta (district of Tigre) in the 2000's, and it has all the elements of the neoliberal urban planning. As Janoschka (2006) points out, the change of hands of 1600 hectares was made as state payment to two private companies that were very important in the construction of social housing at that time. As a result of the regular flooding and the heavy investment needed for its conditioning, that surface bordering on the urbanized limit of the district of Tigre was unused and empty. In order to comply with regulations (Act 8,912/77) and urbanize it, it was necessary to lift the embankment four meters in all, which required considerable investment to dig great surfaces for artificial lakes and significant blindness regarding degradation and adverse effects—flooding—both for the people from the

---

[17] (Source) https://www.dw.com/es/argentina-violento-desalojo-en-predio-de-barrio-de-guernica/a-55442116.

[18] The symbolic dispute was expressed between "usurp" (private), "occupy" (political actors, a more expanded term), and "retrieve" (social organizations with a rights perspective).

[19] Led by the Security Minister of the Province, who has a long history of intelligence activities regarding picketer organizations, and who has political aspirations to become the governor of the province, showed up at the place dressed as Rambo and carrying war weapons.

adjacent popular neighborhoods and at a regional scale (lower basin of the Luján River) (Pintos and Narodowski 2015).

Initially, the project, with a modernizing drive, proposed building a new urban center with a social mixture. It was designed in collaboration with the Societé des Villes Nouvelles, the state-owned French association responsible for the development of the new cities in the Parisian outskirts (Paris is always present). The master plan was approved in 1991 by means of a decree of the province governor, but with a substantial modification, stating as the maximum occupation limit possible the proposed occupation density. This maneuver opened the door to the interest of real estate investment. The company Nordelta SA was created, and Eduardo Constantini made an appearance, becoming the president of the society, with a majority stake. The project transformed drastically its concept and took the shape of a setting with about twenty gated neighborhoods that operate independently, with a separate security service. Each neighborhood was assigned, through selective marketing, its own identity, segmenting groups and profiles of high and upper-high income. Each neighborhood in particular is coordinated in smaller units by means of physical barriers and their own security checkpoints, the layout of streets and the lot division that currently accommodates approximately 40,000 people.

Nordelta is consistently present in major newspapers. As of late, in 2020 during the pandemic crisis, it has been so because of the proliferation of capybaras, which the popular imagination has translated on social media as processes of taking and recovering their habitat. These small animals are, lucky for them, treated by the public opinion more kindly than homeless families.

Nordelta residents are used to living together with these animals, but the conflict arose because their number has grown in the last months, and they have been involved in different attacks to pets, traffic accidents, and damage to gardens. This overpopulation happened because in the last years, there has been an important destruction of areas that were not intervened, and they were deforested to build, and this pushed capybaras to venture into the areas with houses to look for new spaces and food. Among the proposals to address the conflict, one of them was breaking the relationship between neighbors and animals, and fencing houses to avoid their access. "Then, we would obtain a reduction of the population by their own will" (one of the residents, who is a biologist). An eradication plan that reminds us of some eviction protocols "with rights" suggested for human populations.

In fact, the audio message of one of the residents had become viral, where she complained about the presence of some neighbors who were drinking mate on the riverside: "They don't seem to be bad people, but they come from neighborhoods that are not very good visually". "I want to rest visually, because I have moral and aesthetic values".

The capybara debate gave voice one more time to those who claim for a Wetlands Act to stop the advance of the neoliberal capitalist urbanization over these ecosystems, vital as fresh water reservoirs, regulators of flooding and home to great biodiversity. In the Paraná delta, the second most important river in Latin America after the Amazon River, wetlands are jeopardized by the real estate business, in addition to the fires provoked to get lands for cattle or agriculture. This way, the discussion about the environmental impact of the advance of big urbanizations came to the forefront, as well as the other discussion regarding wealthy people who self-isolate in exclusive areas, disrespectful of the environment: people and nature are objectified.

## 26.5 Between Urbicide and the Right to Beauty: Walls that are Murals and Landmarks of Self-managed Urbanism

*It is not to stay home,*
*That we make a home...*
Juan Gelman

On the margins and urban interstices, resistance and resignifications arise in the face of these urbanicidal trends. Re-ownership of the walls that become murals and landmarks of a self-managed urbanism built from small circles, among the symbolic material debris of our cities. Fragments that rewrite the right to the city and come from the need to regain history, to banish the oblivion of destruction and build the citizen memory, typical of the city's self-government.

In the city, there is a myriad of interventions on walls, facades, and even blinds. In historical neighborhoods, in gentrified neighborhoods and in popular neighborhoods. Urban art in its diversity, and mural art in particular, is taking over the streets; there are mural art meetings and congresses, and murals that were painted just months, weeks, or days ago, that converse with architecture and the sense of belonging of the communities that contain them. An interesting example, to invite to walk about the neighborhood and fight insecurity, fifteen artists[20] intervened the blinds in the centric commercial area commonly known as "Once". The blinds of the stores tell the story of the neighborhood, since Plaza Miserere was a slaughterhouse until the present, reviving highlights such as "La Balsa", written in the legendary bar La Perla, birthplace of local rock 'n roll. The intervened blinds are displayed once stores close, and the restlessness of the streets gives way to the possibility of a poetic excursion. In the neighborhood of La Boca, a current muralist, Segatori, contributed with the longest mural in the world (2002 m$^2$) to pay homage to the great muralist of the first half of the twentieth century, Benito Quinquela Martín, who captured in his murals the everyday lives of the port city and its migrant workers, thus giving continuity to the historical narrative of the working-class neighborhood. Mural artists also seek to "make the invisible, visible" and raise their voice in public about vital topics, such as climate change, as can be seen in the mural along Juan B. Justo Ave., near the revamped Palermo Hollywood. Sometimes, they go along neighbors' participatory decisions, as is the case of the mural tribute to the Argentine musician Gustavo Ceratti, leader of Soda Stereo, connecting the neighborhoods of Agronomía and

---

[20] Some of the participating artists were Alfredo Segatori, Nora Basilio, Ignacio Pomilio, and Darío Parvis.

**Fig. 26.2** Mural in the neighborhood of La Boca. *Source* Own elaboration

Villa Devoto, and whose theme was chosen by popular voting. Murals and the wider spectrum of urban art reclaim stories, biographies, and symbols, and they outline another urban geography, more or less ephemeral and diverse, reliving the sense of public space. Self-managed, driven by the public sector—because its contribution is increasingly valued in cities—or also by companies, which deem it a communication tool. In the pandemic, routines have been more strongly linked to the places where we live, and that seems to have encouraged the connection with the activities in the neighborhoods of urban artists (Fig. 26.2).

The fight for the realization of the right to the city has also created milestones of a self-managed habitat (Jeifetz 2012), which revealed the significant differences when addressing the covid pandemic, as could be acknowledged in the cooperative associations of the MOI Federation. The participatory research[21] made it possible to learn the following characteristics:

(a) Having a proper and beautiful habitat as a trench for resistance.

Where residential complexes were built, with their community equipment and catered for with infrastructure and services that materialize the right to the city (such as the cooperative La Fábrica in the neighborhood of Barracas, or the cooperative El Molino in Constitución), people and families have been able to adopt the isolation,

---

[21] Carried within the framework of UBACyT Project 20020190100082BA. *Urban territories in transformation: public policies of habitat and infrastructure and dynamics of living in the AMBA (2010–2022)*—directed by María Carla Rodríguez. The analysis of the emergency conditions of this particular type of organization and resistance to neoliberalism, which differs from other predominant responses configured from the condition of subalternity, is expanded in Rodríguez (2020).

hygiene, and care measures because the materiality is there (housing, its size, the infrastructure). This was also verified where there are Temporary Housing Programs self-managed by the members of the cooperative societies. It is important to point out that the families of the cooperative are mainly part of the same social sector, made up by informal workers, with unstable jobs of self-employed who, given the absence of appropriate public policies, become the customers who rent rooms in the slums and bad hostels, or else have devotedly created the self-built popular neighborhoods themselves (Rodríguez and Zapata 2020).

(b) Having common "savings" and its community use in the emergency.

Creative and efficient methods were also observed, in which resources previously created as permanent and everyday practices developed by the self-managed cooperative organizations of the entire country were used (with their raffles, their fairs, their "saving installments", their "ice cream sales", their fixed accounts, surplus carefully put aside in the context of the planning of works done). The collective capacity to generate and manage some community and work capital has functioned as "safety savings" available in the crisis, which is very different from the situation of isolated families completely indebted. Cooperatives were able to determine the use of those common resources to address the most pressing needs of some of their partners. This capacity for action, which is not infinite, and would need the State and public policies to gather support and volume, also shows powerfully the virtuous aspects of the potential of the popular economy with a self-management perspective.

(c) Coexistence and care in self-managed cooperation.

There is also a set of virtuous immaterial effects of the existence of the community and organizational bond under the principles of self-managed cooperation.

The management capacity acquired has been translated into small and meaningful actions, for example, people who can do digital procedures to help those who have a harder time, support to manage the delayed subsidies established by the national administration to face the emergency, collective claims to service companies, and a connection that is even more subtle to share takes place between the families themselves. Each cooperative has displayed some capacity for action toward their associates, as well the network of inter-consultation mechanism within the framework of the MOI and with others.

Living together for years, implementing collective organization practices, strengthened another capacity, which became relevant in the context of the emergency: carefully looking at and listening to children, women, the most vulnerable, to crying and disruptions, setting up an alert collective agency, supported within the framework and the guidelines of the federation to stop domestic violence. This way, the network of the cooperative has also been discussing collectively their coexistence with small, big conflicts, and cultural risks regarding topics such as respecting the shutdown, common payments, and the challenges of everyday proximity commuting.

This local community capacity, in turn, is boosted by the organizational construction network, with its specific devices and environments, especially kindergartens, a space that is watchful of children' education in the current context, their food needs, the relationship with the neighborhood, the dialogue and support to the most vulnerable people and families, a physical nutrition, but also subjective, cultural, and spiritual. At the same time, popular artists of the cooperatives, who were confined, found time and space to deepen their creative production, popular painters and educators embarked on new expressive projects and share their works on social media (Figs. 26.3 and 26.4).

**Fig. 26.3** Cooperativa La Fabrica. MOI Federation. *Source* Nestor Jeifetz

**Fig. 26.4** Cooperativa La Fabrica. MOI Federation. *Source* Nestor Jeifetz

## 26.6 Colophon. In the Face of Urbicide, the Right to Beauty

Among the debris of this civilization crisis, I would like to give visibility to the emergence of a resilient and proactive self-managed urbanism, focused on the recreation of what is common, the neighborhood interaction and the reordering of the public space; all of them are aspects opposed to the hegemonic discourse flexibility/indifference, repeated in all the urbicide processes of creative destruction inherent to the neoliberal urban rationale, and the obsession of the urban project.

Its main characteristics are:

- The reasoning of participation, both individual and collective, focused on people's development, opposed to the rationale of profits.
- The production of habitat as a fixed asset, opposed to the production of commodities.
- The transfer of resources to popular organizations to carry out productive processes of habitat.
- The practice of individual-collective construction of rights against the assistance mentality and the establishment of public–private monopolies to manage poverty.
- Comprehensive vision of habitat as opposed to a sectoral vision that is "a roof".
- Historical approach, oriented toward practices and processes that nurture senses of life compared to an ahistorical idea.
- Right to the city understood as the universalization of access to the urban centrality (full access to all the flows, networks, services, and structures of opportunity typical of the urban life) as opposed to the exclusionary social-spatial rationale (you can live where you can afford it).
- City architecture as a party that interacts with urban neighborhood collective settings as opposed to the object architecture in the materialization of habitat.
- Democratization of knowledge, gathering of wisdom and access to "design": beauty as a right.
- Dispute and transformation of the state, regulatory, and programmatical institutionality in keeping with previous criteria.

The development of unique and collective subjectivization methods related to these practices starts off the acknowledgement of other forms of sensitivity, an intersubjective relationship with the other, creativity, production; of an existential uniqueness that seeks to agree with a desire, an enjoyment for living, a will, all that aimed at making changes and/or openings in the pervasive subjectivity system. That is why, at the symbolic and cultural levels, the self-managed collective practices for the production of habitat tend to put a strain on and pierce the binary categories that make up organization aspects of the hegemonic common sense or objective knowledge: feminine/masculine; normality/madness; public/private; productive work/reproductive work, formality/informality.

History is regained in its role as a shared and intergenerational experience that gives identity and sense to the subjectivization process. That is why the circle, as in dances of old, is a permanent figure of the layout of bodies in the spaces of self-management organization. Mediator of the recreation of social space, it redefines the transitional continuous from the intimate to the public, providing it with a democratizing political nature, where we see each other's faces, and the collective and personal energies are interwoven and channeled. This is expressed in the programmatical and project definitions, that is to say, by means of the treatment of the different residing spaces and the treatment of their transitions. The transitional spaces (the gradient of unveiled spaces of private use, streets, and common yards, settings such as empty plants, covered, and uncovered equipment open for neighbor use.) arise as an alternative against the walls of neoliberal urbanism and its dichotomous violence.

From there, the transduction operation, following the epistemological model suggested by Lefebvre in the 1970's, which propounds the emergence of the "right to beauty" as a theoretical and political response in the face of the contemporary urbicide, as a discursive and material practice projected from a self-managed urbanism, from everyday life, which updates and repoliticizes the dispute—symbolic, territorial, and incarnated—for the right to the city.

The urban architecture programs, whose material expression is born in the interaction with users, materialized through the control of its inhabitant producers, owned and recreated on a daily basis all along the housing cycle, who channel smoothly the complexity, allowing shades and diversity, in growing scales and complexities, is what we call a "beautiful habitat".

Therefore, beauty means the spatial expression that is appropriate to shelter and allow the development of the different needs of inhabitants' everyday life, in a spectrum that goes from the intimate environment to the different hues of private, community, neighborhood and the urban and public coordination. The intersubjective and negotiated treatment and recognition of transitional spaces, always open to the interaction and negotiation at these scales of greater complexity, is Ariadne's thread, the alchemy to transform the walls and regain the city and citizenship.

# References

ACI (2020) Derecho a la vivienda en ciudad de Buenos Aires. Baja presupuestaria para el déficit habitacional. On https://acij.org.ar/informe-derecho-a-la-vivienda-en-caba-baja-presupuestaria-para-el-deficit-habitacional/

Apaolaza R, Venturini JP (2021) Cambios de usos del suelo en la periferia del área metropolitana de Buenos Aires. Aportes para una teoría de la rent gap periurbana. Geograficando 17(1):e087. https://doi.org/10.24215/2346898Xe087

Berman M (1989) Todo lo sólido se desvanece en el aire. La experiencia de la Modernidad. SXXI Editores, Buenos Aires

Caldeira T (2007) Ciudad de muros. Gedisa, Barcelona

Carrión F (2018a) Urbicidio o la muerte litúrgica de la ciudad. Occulum Ensaios (online) 15(1):05–12. University of Campinas, Brazil. https://doi.org/10.24220/2318-0919v15n1a4103

Carrión F (2018b) Urbicidio o la producción del olvido. Observatorio Cultural 19(1):28–42

CESBA (2018) La ciudad de Buenos Aires inquilinizada. Un análisis acerca del mercado formal e informal de alquiler como estrategia de acceso a la vivienda en CABA. Consejo Economico de la Ciudad de Buenos Aires

Ciccolella P (2014) Metrópolis latinoamericanas, más allá de la globalización. Café de las Ciudades-OLACCHI, Buenos Aires

Delgadillo V (2021) La muerte simbólica y material de la ciudad: una aproximación sobre el urbanicidio. RevistArquis (online). https://doi.org/10.15517/RA.V10I1.45258

Harvey D (2007) Breve historia del Neoliberalismo. Editorial Akal, Madrid

Herzer H (2008) Con el corazón mirando al sur. Transformaciones en el sur de la ciudad de Buenos Aires. Espacio editorial, Buenos Aires

IVC (2019) Report on the occupation condition of housing in CABA. Instituto de Vivienda de la Ciudad. Gobierno de la Ciudad de Buenos Aires

Janoschka M (2006) El modelo de ciudad latinoamericana. Privatización y fragmentación del espacio urbano de Buenos Aires: el caso Nordelta. Revista BA deriva 04(1):80–117

Jeifetz N (2012) Reflexionando sobre la autogestión del hábitat. Desde una mirada de las contradicciones. Contribution to the 1st module of the 2nd cycle of the Escuela de SELVIHP. Quito. In www.moi.org

Lanfranchi G, Cordara C, Duarte J, Gimenez Hutton T, Rodriguez S, Ferlicca F (2018) ¿Cómo crecen las ciudades argentinas? Estudio de la expansión urbana de los 33 grandes aglomerados. CIPPEC, Buenos Aires. Retrieved from http://www.cippec.org

Lefebvre H (1972) La revolución urbana. Península, Barcelona

Madden D, Marcuse P (2016) In defense of housing. Verso, New York

Marcuse P (1995) Not chaos, but walls. In: Watson S, Gibson K (eds) Postmodernism and the partitioned City. Blackwell, London

Mayorga M (2021) París: la Ciudad de los 15 Minutos. In: Ruiz-Apilánez B, Solís E (eds) A pie o en bici. Perspectivas y experiencias en torno a la movilidad activa. Ediciones de la Universidad de Castilla-La Mancha. https://doi.org/10.18239/atenea_2021.25.17

Oszlack O (1991) Merecer la ciudad. Cedes-Humanitas, Buenos Aires

Pintos P, Narodowski S (2015) La privatopía sacrílega. Efectos del urbanismo privado en la cuenca baja del río Luján. Imago Mundi, Buenos Aires

Pírez P (1994) Buenos Aires metropolitana. Política y gestión de la ciudad. Centro Editor de América Latina, Buenos Aires

Polanyi K (2007) La gran transformación. Crítica del liberalismo económico. Quipu editorial, Madrid

Quinchía Roldan S (2013) Discurso y producción de ciudad: un acercamiento al modelo de urbanismo social en Medellín, Colombia. Cuadernos De Vivienda y Urbanismo De La Universidad Javeriana 6(11):122–139

Rodríguez C (2009) Autogestión, políticas del hábitat y transformación social. Espacio editorial, Buenos Aires

Rodríguez P (2014) Santiago de Chile y el hábito del temor a la periferia: la precarización de lo urbano en una ciudad neoliberal (1973–2010). Thesis for the Doctorate in Social Sciences. UBA

Rodriguez C (2018) Políticas de hábitat, villas y ciudad. Tendencias actuales y futuros posibles. Occulum Ensaios (online) 15(3):495–517. https://doi.org/10.24220/2318-0919v15n3a4179

Rodríguez C (2020) Desafiando la alienación residencial. Producción autogestionaria del hábitat y comunes urbanos en ciudad de Buenos Aires. ACME. An Int J Critic Geograph 19(3):647–664

Rodríguez C, Zapata C (2020) Organizaciones sociales y autogestión del hábitat en contextos urbanos neoliberales. ICONOS Revista De Ciencias Sociales 67(29):195–216. https://doi.org/10.17141/iconos.67.2020.3964

Sennet R (2004) El capitalismo y la ciudad. In: Martín Ramos A (Comp) Lo urbano en 20 autores contemporáneas. ETSAB

Silva A (2004) Imaginarios Urbanos: hacia el desarrollo de un urbanismo desde los ciudadanos. Convenio Andrés Bello; National University of Colombia, Bogotá

Theodore N, Jamie P, Brenner J (2009) Urbanismo neoliberal: la ciudad y el imperio de los mercados. Temas Sociales, Ediciones SUR 66(1):1–11

Venturini JP, Franchicchia A, Apaolaza R (2020) La ciudad que se pierde: autopistas, countries y desalojos…*Revista Ignorantes(s/n)* In https://rededitorial.com.ar/revistaignorantes/especial-guernica


**María Carla Rodríguez** Sociologist (UBA) and Ph.D. in Social Sciences (UBA). Full Professor of Urban Theory in the School of Social Sciences at the University of Buenos Aires. Researcher at CONICET and the Research Institute Gino Germani of the UBA. Activist for the right to habitat at the MOI (Squatters and Tenants Movement of Argentina) since 1991, the SELVIHP (Latin American Secretariat of Popular Housing and Habitat) since 1992, and the HIC (Habitat International Coalition) since 2000.

# Chapter 27
# Anatomy of an Urbicide. Social Housing in Santiago 1980–2006

**Alfredo Rodríguez and Ana Sugranyes**

> Mit einem Haus kannst du die Leute erschlagen wie mit einem Beil. (Heinrich Zille, Berlin, ca.1895 ) [You can kill people with a house just as with an axe (Heinrich Zille, Berlín, ca. 1895).]

**Abstract** The article describes the case of an urbicide, the social and urban effects of the production of low standard social housing in the city of Santiago between 1980 and 2006. This process had a paradoxical result: on the one hand, many units were built in large housing complexes, reducing the quantitative housing deficit; on the other hand, the qualitative deficit increased, leaving a footprint of precariousness in the city, which persists. The text presents elements of the origin, take-off, rise and fall of the housing subsidy, the public financing instrument delegated to the private sector for the location and construction of housing for low-income families. The complexity of the city shows that housing production cannot be reduced to a macroeconomic strategy, a growth machine or a financial capital instrument.

**Keywords** Urbicide · Social housing · Urban development

## 27.1 Introduction

Building a lot of houses does not always solve an urban problem. To illustrate this point, in this book on *urbicide* we review the massive construction of social housing in a city, Santiago de Chile, during a given period: from 1980 to 2006.[1]

This housing policy only contemplated financing the construction of housing developments for low-income families. A policy that disregarded, i.e. the structure

---

[1] Period marked by the recommendations of Harberger (MINVU 1978) and closed by the statements of Poblete (El Mercurio 2006), as pointed out in the text.

---

Translation: Clemen Talvy.

---

A. Rodríguez (✉) · A. Sugranyes
SUR, Corporación de Estudios Sociales y Educación, Santiago de Chile, Chile
e-mail: arsur@sitiosur.cl

of the city, the specificity of each place, the architecture of the housing complexes, let alone the livelihood of the families who were going to live there. The political interest was limited to the abstraction of the number and the amount of the granted subsidies.

While this financial process built fragments of non-city, the authorities, international organisations and their consultants were spreading the word about the success of Chile's housing policy throughout Latin America. Dazzled either by the successful figures of the new housing units, the development of economic and financial activities associated with construction, or by the reduction in the number of families living in land invasions, the authorities did not take into consideration the material products or their urban impact; only figures and statistics.

As critical observers of this process and its subsequent effects, we were shocked to see that as more housing was built, new urban and social problems appeared. So much so that, by the end of the period analysed (2006), the main housing problem in Santiago was not that of the families "sin casa",[2] as it had traditionally been. A new housing problem appeared with families "con techo"[3] (Rodríguez and Sugranyes 2005), i.e. families with inadequate new housing, in neighbourhoods without services, with impossible to pay mortgage debts, and deprived of the city.

This public housing policy has left destruction footprints on the city that persist up to now and are difficult to erase. This is the issue that we study in Santiago since late nineties up to date, and which Raquel Rolnik picks up in *Urban Warfare* (2019) by emphasising the greed of real estate capital in what she calls the "laboratory of Chile".

## 27.2 Housing Policy 1980/2006

### 27.2.1 Origin

The origin of this housing policy is precisely traceable. In 1978, in the context of a strong process of privatisation and deregulation promoted in Chile by the civil military dictatorship, Arnold Harberger, then director of the School of Economics at the University of Chicago, gave advice to the officials in charge of the Ministry of Housing and Urban Planning (MINVU 1978). He laid the foundations for the urban and housing policies of that time, which are still in force today.

The message he established was that cities grew in a "natural way" and that by not recognising this natural way, urban planning destroyed the dynamism of cities.[4]

---

[2] Translator's Note: people living in precarious housing conditions.

[3] TN: people living in subsidized housing.

[4] As the División Técnica de Estudio y Fomento Habitacional points out in the presentation of the conversation with A. C. Harberger (MINVU 1978): "Perhaps the most significant contribution of these conversations was the concept that there is a "natural" way of occupying space, which corresponds to the behaviour of the largest part of the most dynamic population of the city; a natural

Consequently, the MINVU's role in guiding urban development was not to impose an exogenous order, but to capture market signals, which were to form the basis for its action. In this sense, the optimal use of urban land was given by that activity which ensured the highest economic profit, which implies that the horizontal growth of cities is "a natural economic phenomenon" and that it is "foolish for governments to try to put an end to it" (MINVU 1978: 1). Harberger's influence is in the statements of the National Urban Development Policy issued the following year, in 1979. It mentions, on the one hand, that a flexible planning system would be applied, removing urban limits, with the minimum of state intervention, supported by technical norms and generic procedures; on the other hand, that the state would encourage and support the creation of an open housing market, which would be the responsibility of the private sector (MINVU 1981).

### 27.2.2 Take-Off

From then on, MINVU transferred housing production to the private sector and established a financing system that linked state contribution, housing subsidy, family savings and a bank loan. The material product of this financing policy was, and still is–forty years later–the so-called "social housing", as a commodity, whose price is established in the abstraction of the *Unidades de Fomento* (UF).[5] The top value of social housing was initially 240 UF; it then increased to 400 UF, 600 UF and is now over 1200 UF.[6]

This housing production finance system, launched in the 1980s in the context of the privatisation of public utilities and social security funds, became a central point of the financialisation of the economy. At the same time, Santiago's urban land was being reorganised with the eviction of informal settlements located in areas of expansion of middle and high-income sectors, and their inhabitants taken to the poorer outskirts. Municipal territories were being subdivided, creating homogeneous governance areas of high, medium and low incomes. Santiago was ready to implement Harberger's advices.

The first step taken by the private sector in the framework of this state initiative was the purchase of large tracts of low-priced land in what were then the outskirts of Santiago. With these private-land banks, the building companies, supported by the public housing subsidies, defined the location of social housing and the fragmentation of the city.

---

way that often does not agree with the traditional ideas of urban planning applied until today in our country".

[5] UF: a unit of account that is regularly adjusted, calculated and authorised by the Central Bank of Chile for money lending operations in national currency carried out by banking and savings companies. https://si3.bcentral.cl/estadisticas/Principal1/metodologias/EC/IND_DIA/ficha_tecnica_UF.pdf.

[6] To date, equivalent to € 40,000.

### 27.2.3 Boom

The boom of the system took place in the 1990s. The return to democracy in 1990 did not change the direction of the housing policies nor their financing instruments; rather, it accelerated quantitative production. During the 1990s, a notable decrease in the quantitative deficit was achieved, reaching the production of one dwelling for every thousand inhabitants, equivalent to the rate of post-war housing reconstruction in Germany. The number and population of informal settlements (land invasions), was reduced to about two per cent of the total number of households in the country. The international consultants began to disseminate this success story throughout Latin America, coinciding with the prescriptions of *Housing: Enabling Markets to Work* (World Bank 1993).

In the mid-1990s, academic articles began to appear criticising the urban effects of the massive construction of large housing estates–ghettos–on the periphery of the city, far from services.

At the end of the 1990s, a sense of dissatisfaction started to grow among the residents of the new social housing developments with their living conditions, due to the small size of the houses or the lack of facilities. In Santiago, a heavy rainfall damaged 40,000 new homes, showing that, in its eagerness to reduce the quantitative deficit, the MINVU had lowered the quality of the technical specifications of housing materials to reduce construction costs. It was a sign of exhaustion due to the progressive lack of interest of building companies, which were discovering more profitable niches real estate market. The MINVU was still walking the path of successful housing subsidy. Added to this were the effects of the Asian economic crisis, which led to an increase in poverty and made it difficult to pay back bank loans.

The living conditions of hundreds of thousands of poor families, inhabiting in concentrations of poor quality housing in deficient urban spaces, led to a situation that Skewes (2005) analysed by pointing out that "the solution to housing problems was to reduce construction costs to a minimum. (…) It is worth questioning whether political peace was indeed achieved at the cost of civil violence. This would mean that the city 's poor, with their insecurity, would be paying the social costs of the housing subsidy policy". (p. 122).

### 27.2.4 Decline

Thus, the 2000s began with a weakened social housing policy. After twenty-five years, land invasions began again in Santiago, such as the *Toma de Peñalolén*. Housing debtors organised themselves around the National Association of Mortgage Debtors of Chile (ANDHACHILE), demanding to pay their debts in accordance with their incomes. Between 2005 and 2006, the government cancelled the mortgage debts of some 180,000 families,[7] and transferred another 200,000 to private banks.[8] It initiated new housing subsidy programmes, such as the so-called *"Programa de vivienda social dinámica sin deuda"* (dynamic social housing programme without debt), which provided subsidised housing with low pre-savings and no subsequent payments. One of the arguments for eliminating credit was the high cost of charging low dividends that were appropriate to the income of poor families. The model was running out of steam.

The new government in 2006 acknowledged the urban disaster. That year saw the start of the "Quiero mi Barrio" ("I Love my Neighbourhood") programme, which consisted, as Minister Patricia Poblete pointed out, of "recovering these large settlements built in the 1980s (…) as in manageable and habitable neighbourhoods". "We don't want future governments to have to undo what we have done wrong in housing, just as we are recovering neighbourhoods because they were not thought of before".[9]

## 27.3 The Footprint

It is relevant to ask why it took so long to reach this decision.

The housing subsidy policy, distilled in Chile's laboratory and sold around the world, marked the urbicide. One of the biggest obstacles that prevented innovation and the proposal of alternatives was the social housing production model imprisoned in a captive market, with fully satisfied protagonists. The basis of understanding between the state as the funder and a few building companies that had the capacity to mass-produce was perfect. More than half of the 220,000 social housing units built in Santiago during that period were in the hands of six companies. The business was simple: the MINVU assigned the voucher or subsidy title to the applicant families, supported them in obtaining mortgage loans (directly or through banks), financed the building companies (transferring the funds from the individual subsidies) and, at the end of the year, the state returned 31% Value Added Tax (VAT) of the costs of building materials. However, the state not only protected the companies, but also the banks that granted the credit: the MINVU financed insurance on the loans and assumed responsibility for the auction of the property in the event of the debtor's insolvency. With these early securitisation measures, Chile avoided the effect of the subprime crisis.

---

[7] Between 2005 and 2006, out of 262,755 SERVIU (Regional Housing and Urbanisation Service) debtors in the country, 181,538, or 70% of the total, had their debts cancelled. Cámara de Diputados (2009) (Chamber of Deputies) report, 31/07/2009.

[8] Credits guaranteed by the state and transferred to private banks, corresponding to the Special Programme for Workers (PET), called off in 2006. As of July 2008, there were 204,011 debtors, and between 33 and 40% were defaulters. Cámara de Diputado (2009) (Chamber of Deputies) report, 31/07/2009.

[9] Interview with Minister of Housing and Urban Development Patricia Poblete in *El Mercurio* (2006).

It was a housing model that implied no risks. There was no competition: there were very few companies specialised in the field capable of winning large annual quotas for the construction of social housing complexes. In this captive market, there was no reason to innovate: the technology of social housing did not change in twenty-five years; the companies in charge of the construction of these low standard dwellings did not need the contributions, ideas and tests developed by different NGOs, professional organisations and universities.

The idea of improvement was not part of the social housing agenda. Why change? The different political spheres evaluated positively the massive and sustained production of hundreds of thousands of dwellings in the country. Since 1990, governments and the opposition praised the performance of housing ministers that generated votes for the elections. By the end of the 1990s, the first signs of the model's exhaustion emerged.

## 27.4 Outcome

### 27.4.1 Large Fragments of No-City

The location of social housing in Santiago led to the layout of large urban extensions covered with housing complexes, isolated from each other. The location and design of the housing developments was left completely in the hands of the building companies, as a senior MINVU official explained to us in the early 2000s: "Who better than the companies know the city and where to locate the new developments?" This criterion led the MINVU even to hand over the design of secondary urban roads to the companies, resulting in an urban patchwork of disconnected developments. It is worth noting that the housing developments were only groups of houses, without services. The theory was that there was no need for public intervention, given that these new concentrations of population would attract, by the laws of the market, the necessary services and facilities that would settle nearby. This did not happen.

The criteria for the design of the housing developments, subordinated to the interests of the private developers, resulted in a monotonous repetition of dwellings and unused urban residual spaces. Reviewing projects of the time, one can see that the buildings are distributed on a sort of no man's land, almost like the work of an inkpad repeated on the map, and the buildings are like a loaf of bread that is cut off when it reaches the street, without a façade. As a MINVU architect explained to us at the time: "The companies work with very tight construction costs, and we cannot therefore demand that they have windows facing the street".

## 27.4.2 Dissatisfied Dwellers

In a survey carried out in 2002, 64.5% of the dwellers of the social housing stock in Santiago stated that they wanted to "move out of their homes" (Rodríguez et al. 2005: 222). The reasons behind their intention were of a social nature: situations related to coexistence among neighbours; perceptions of insecurity, crime and drugs. Others referred to the small size of the house and its isolation from the urban fabric. However, despite their dissatisfaction, they could not leave the place: since they were mortgage beneficiaries and debtors, they could not apply for another house and had to pay their debt.

Several studies on the level of satisfaction of inhabitants in the social housing stock (see Arriagada and Sepúlveda 2002) have insisted on the disenchantment of users who had dreamed of their own home, a feeling that appeared between six months and two years after moving into the housing development. The intention to leave their home and the perception of affection–or disaffection–for the development demonstrate the importance of people's feelings towards the place and the urban environment, an aspect fully ignored by the housing policy makers of the period.

## 27.4.3 Difficult Coexistence

In comparative terms with other Latin American cities, Santiago had a low level of criminal violence in the period mentioned, and even so, the perception of violence was proportionally very high.[10] As Tudela (2003) explains, in terms of citizen security, there is no direct relationship between actual violence and its perception. For coexistence in the city, the fact that the population perceives insecurity is as serious as the crimes themselves. Talking (at the time) with dwellers of social housing complexes, especially those located in large homogeneous concentrations, the first recurring issue of daily concern was violence: "Living here is like being in jail". "We keep the children locked up at home".

Several newspapers began to show the conflictive urban and social landscape created by the housing developments. One of the first reports entitled "Santiago ocupado" (Santiago under Siege), identified eighty settlements in 2009 in which dwellers felt abandoned by the state (Qué Pasa Magazine 2009). According to the publication, neither the officials of the various public services, nor the employees of the water, electricity and telephone companies dared to get into these neighbourhoods.

---

[10] As explained by several authors compiled in *Conversaciones públicas para ciudades más seguras* (Acero et al. 2000) and *Seguridad ciudadana, ¿espejismo o realidad?*(Carrión 2002), comparing violence issues among the main Latin American cities, they found that violent crime rates in Santiago were among the lowest in the region. However, disorders due to the perception of violence are much higher than in other cities. "Santiago ocupado", Qué Pasa Magazine (2009) (Santiago, July 2009). https://bit.ly/3p7sDgL.

### 27.4.4 Difficult Recovery

Starting from 2006, began several programmes for the recovery of housing developments. We will not review them here in detail, but will just mention a few. For example, "Quiero Mi Barrio" (I Love My Neighbourhood), which continues to date, has allowed the recovery of public spaces in traditional neighbourhoods and former working-class areas, but has proved to be limited in solving problems in large social housing complexes. Other programmes, such as the so-called "Second Chance" programme, showed the complexity of urban regeneration and housing de-densification, because just demolishing buildings is not sufficient.

In the meantime, the social housing standards and design are improved, i.e. larger spaces in each unit, community services and equipment. However, in practise, it has proved very difficult to reverse the dependence on a trajectory of more than thirty years. Thus, large concentrations of poor families in isolated and precarious peripheries have continued to grow. This is because in a subsidiary state like Chile, despite the constitutional process, housing production still is considered "as a macroeconomic strategy, an engine of economic development and an important tool of financial capital" (Ortiz 2003).

## 27.5 Today

Solving the inherited problems is difficult because the city's situation has become more complex. The social and economic effects of the social outbreak and the pandemic have further highlighted the imbalances in housing and urban policies. The constituent process continues, and its results are to be seen in the long term.

The housing issue in Santiago for the lower and middle-income sectors is worsening. In the formal market, there has been an insane increase in housing and land prices, and a proliferation of a private rental offer of 20 square metre nano-homes in buildings of 30 to 40 floors. In the informal market, with the rise of precarious work and the large migratory flow to Santiago, housing has become a matter of survival in the face of abusive rental prices. Land invasions have multiplied in recent years, as they did in the middle of the last century.

People and families living in poverty are more than these counted in the official figures (Duran and Kremerman 2021) and their housing conditions are worsening in social housing complexes, in traditional neighbourhoods of progressive development, in new tenements, in subdivided houses and sublet rooms, and in land invasions.

Empirical observation leads us to fear that more than a third of the population suffers currently from some form of housing precariousness (MIDESO 2018).

Faced with this enormous challenge, the current government's housing policy review proposes to resume a massive production of housing through a range of different programmatic offers. The proposal considers, i.e. the comprehensive and careful renovation of deteriorated neighbourhoods; the reconversion of areas whose uses are in obsolescence; the radical remodelling of urban areas; the settlement of land invasions for their recognition as city neighbourhoods; and the sum of urban acupuncture interventions with the effect of recovering entire neighbourhoods.

Will these measures be enough to prevent a new urbicide? At least this time there is a proposal of diverse programmatic offers, and there are footprints left on the city that remind us of the mistake of only considering housing as a commodity.

## References

Acero H, Bruneau S, Burgos J, Galilea S, Lahosa JM, Laub C, Maeko H, Ochoa O, Orrego C, Rodríguez A, Vanderschueren F, Vézinal C (2000) Conversaciones públicas para ciudades más seguras. Ediciones SUR, Colección Estudios Sociales, Santiago de Chile. http://www.sitiosur.cl/r.php?id=28

Arriagada C, Sepúlveda D (2002) Satisfacción residencial en viviendas básicas Serviu: la perspectiva de capital social. Colección Monografías y Ensayos. Serie VII Política Habitacional y Planificación. Ministerio de Vivienda y Urbanismo, División Técnica de Estudio y Fomento Habitacional, Santiago de Chile. https://www.researchgate.net/publication/277955983_C_Ministerio_de_Vivienda_y_Urbanismo_2002

Cámara de Diputados C (2009) Comisión Especial sobre Deuda Histórica. In: Informe de la Comisión Especial relativa a las denominadas deudas históricas. Cámara de Diputados, Santiago. https://www.bcn.cl/obtienearchivo?id=repositorio/10221/22405/1/Referencia%20BCN%20-%20Deuda%20Hist%C3%B3rica%20-%20Estado%20del%20arte.pdf

Carrión F (ed) (2002) Seguridad ciudadana, ¿espejismo o realidad? FLACSO Ecuador—OPS/OMS, Quito

Durán G, Kremerman M (2021) La pobreza de "modelo" chileno: la insuficiencia de los ingresos del trabajo y pensiones. Fundación Sol. https://fundacionsol.cl/blog/estudios-2/post/la-pobreza-del-modelo-chileno-2021-6791

El Mercurio (2006) Entrevista a la ministra de Vivienda y Urbanismo, Patricia Poblete

Ministerio de Desarrollo Social (MIDESO), Chile (2018) Casen 2017. Síntesis de resultados de viviendas y entorno. http://observatorio.ministeriodesarrollosocial.gob.cl/encuesta-casen-2017

Ministerio de Vivienda y Urbanismo (MINVU), Chile (1978) Problemas de viviendas y planeamientos de ciudades. Conversación con Arnold C. Harberger. Monografías y ensayos, Serie 1. no. 103: Arquitectura y urbanismo. Ministerio de Vivienda y Urbanismo, División Técnica de Estudio y Fomento Habitacional, Santiago de Chile

Ministerio de Vivienda y Urbanismo (MINVU), Chile (1981) Política Nacional de Desarrollo Urbano 1979. Ministerio de Vivienda y Urbanismo, División de Desarrollo Urbano, Santiago de Chile

Ortiz E (2003) La producción social del hábitat: ¿opción marginal o estrategia transformadora? Unpublished. Online version at https://hic-al.org/wp-content/uploads/2019/01/Texto_EOF.pdf

Qué Pasa Magazine (2009) Santiago ocupado. https://bit.ly/3AaOieg

Rodríguez A, Sugranyes A (eds) (2005) Los con techo. Un desafío para la política habitacional. Ediciones SUR, Santiago de Chile. http://www.sitiosur.cl/r.php?id=81

Rodríguez A, Sugranyes A, Tironi M (2005) Resultados de una encuesta. Annex 1 of Los con techo. Un desafío para la política habitacional. Ediciones SUR, Santiago de Chile, pp 221–227. http://www.sitiosur.cl/r.php?id=81

Rolnik R (2019) Urban warfare: housing under the empire of finance. Verso Books

Skewes JC (2005) De invasor a deudor: el éxodo desde los campamentos a las viviendas sociales en Chile. In: Rodríguez A, Sugranyes A (eds) Los con techo. Un desafío para la política de vivienda social. Ediciones SUR, Santiago de Chile, pp 101–122. http://www.sitiosur.cl/r.php?id=81

Tudela P (2003) Espacio urbano e implementación de programas de prevención del crimen, la violencia y la inseguridad en el Gran Santiago a través de Sistemas de Información Geográfico-Delictual. Ministerio del Interior, Departamento de Información y Estudios, División de Seguridad Ciudadana, Santiago de Chile

World Bank (1993) Housing, enabling markets to work. Washington DC, A World Bank Policy Paper. https://doi.org/10.1596/0-8213-2434-9


**Alfredo Rodríguez** Architect and Master in urban planning. Director of SUR Social Studies and Education.

**Ana Sugranyes** Architect and PhD. President of Housing and Land Rights Network (HIC-HLRN).

# Chapter 28
# Urbicide. A Look Through the Mirror

**Inés del Pino Martínez**

**Abstract** The paper deals with the term *urbicide* (death of the city) according to the parameters outlined by Fernando Carrión, which are natural, anthropic, and symbolic. In the first case, two earthquakes that occurred in 1797 and 2006 in Ecuador, in different temporalities, show that the physical death of the city is temporary, since with the recovery of its materiality, history is restored and a renewed identity is put into effect. Anthropic and symbolic death are analyzed in the historical center of Quito, whose long-standing history is currently undergoing a social crisis that is expressed in systematic depopulation and repeated actions for its destruction, however, the popular commercial vocation of this space, the complexity and volume of this activity became evident during the global health emergency between 2020 and 2022 to show that the popular market, the street vending and floating population are the few activities that remain in this space and keep the historical center alive. A finding in this analysis was to identify that with a political decision and collective resistance to the death of this part of the city, it is possible to change the sign of the *urbicide*.

**Keywords** Urbicide · History · City · Quito

## 28.1 Introduction

In the city and architecture field, the idea of capitalist city is emerging as a new way of life in which technology permeates all strata of society, from the economy to everyday life and the domestic. By way of visible examples in Latin America, the economy is expressed in the city by giving cost to urban land, which is bought and sold like any other merchandise, subject to valuation and capital gains, the cost of land influences the social and spatial distribution of urban sectorization. This transformation was carried out in some cases on the pre-existing city, which led to the demolition of old neighborhoods. An example in the region was the opening of

---

I. del Pino Martínez (✉)
Pontifical Catholic University of Ecuador (PUCE), Quito, Ecuador
e-mail: idelpinom@puce.edu.ec

© The Author(s), under exclusive license to Springer Nature Switzerland AG 2023
F. Carrión Mena and P. Cepeda Pico (eds.), *Urbicide*, The Urban Book Series,
https://doi.org/10.1007/978-3-031-25304-1_28

*Carrera Décima* in Bogota, an urban project implemented between 1945 and 1960, a period in which, through a modernizing action, a significant part of the colonial and republican city was lost to make way for the opening of an avenue that forever changed the urban experience of that place (Niño, 2010). In this case, the *urbicide* (death of the city) occurred for anthropic reasons in the name of development and progress, a radical transformation that in four years gave space to a six-lane road and a parterre in the middle, tree lined, to give priority to cars and public transport, high-rise buildings, and thus changed the land value and symbols of the public space, giving public space of pedestrians and producing the oblivion of the daily walk on foot.

In other cities, instead of destroying the consolidated city to overlay new projects, the demolition of which generated expenses, vertical collective housing or residential complexes were created outside the built city, in line with the principles of International Congress of Modern Architecture (CIAM, for its initials in French), and architectural projects along the lines of BAUHAUS, planned and built with new technology and with economy, function, form, and profitability in mind, but without considering social diversity. Derived from this urban order, gated urbanizations emerge that privatize public space with the argument of security or seen on the other side of the coin: the fear of the *other*, in imaginary or real terms. This way of constructing the city transforms and kills public space, the interaction between neighbors, and the development of collective urban processes that create identity.

To differentiate the *urbicide* from the natural transformation of the city, it is worth to point out that the twentieth century was radical with the urban transformation related to previous periods; the support of the technology and urban engineering, real state profitability, planning from a specialized elite and not from the interest of citizens were elements that contributed to the transition to a city of individuals and not of collective dialog. In other aspects, the promotion of new housing sectors was accompanied by a discourse that values and spreads a language with terms that allude to health, hygiene, open air, healthy climate, green spaces, and leisure time as synonymous with transformation and structural changes of a new way of life in the city. Modern architecture also created a discourse that is modeled on the basis of economic interests and investment; the term *comfort* is an aspiration induced by propaganda, behind which are technological instruments that facilitate daily life, give the sensation of freedom, speed, efficiency, and innovation; among others, it can be the incorporation of elevators for vertical circulation, air conditioning or heating, electrical energy, electrical and electronic appliances that represent innovation; the enjoyment of vacations makes leisure a business. In short, the idea of the future is thinking that the past is degraded and obsolete, it does not serve for the requirements of modernity, and a set of terms is created that allude to a new way of life, the modern style. These changes, although they maintain distance with the historical city, are part of a natural transformation, of a cycle that dies to create new life from the seed of the old.

On the other hand, another sign of progress has been the obsolescence of the objects in the short term, the convenience of discarding what no longer serves instead of fixing it, the replacement of a part, or the creativity of inventing a mechanism that gives new life to an object that no longer exists, the new order discarding things not only because they do not work, but because a new equipment appeared on the market that replaces the previous one with technological and design improvements, with an attractive cost in favor of a market. That is, technology, function, and esthetics are incorporated into modern city and architecture, establishing a new order, the modern one. The excessive increase in urban waste without recycling alternatives contributes in the medium or long-term to *urbicides* that alter the natural balance.

Visions of the future such as the *generic city* visualize in some way the *urbicide* or death of the city as a certain possibility, they are not synonymous but complementary, the generic city builds a discourse that starts from the lack of identity and history, that is, according to Koolhass: "the generic city is what remains after large sectors of urban life moved to cyberspace… the generic city is sedated and is usually perceived from a sedentary position" (Koolhaas, 2011, pág. 15), that is, it arises: "from all that is left of what used to be the city", in this sense it could be provisionally affirmed that the *urbicide* could be one of the inputs to reach the generic city, at least in theoretical terms. In the process, the city would have to be emptied of identity and culture, and for this, one of the mechanisms is the degradation of historical cities, the strengthening of massification processes through technology, social isolation, and with it the elimination of collective initiatives. What is left if [the city's] identity is taken away? The generic? Asked Rem Koolhas (Koolhaas 2011, p. 6).

The degraded is that which has no value or is not used for a human purpose. That which has no useful results is lost and abandoned and ends in death (Lynch 2005, p. 15). Degradation and loss are identified by Kevin Lynch as the shadows or dark sides of change when it comes to the destruction of cities. It could be added that the abandonment of the historical city is not only the departure of people, but has several consequences: it produces material degradation, but at the same time it is the silent expression of the loss of meaning of a place, or that a life cycle has concluded, a condition that contributes to the transformation and death of the place. With this reference, it is worth pointing out that every city is transformed over time; each society and culture sets the measure and rhythm to the changes, whether physical or of meaning; every transformation emerges from the traces of the preceding city, its history and memory; transformation means change in the form of the city; therefore, it is a concept related to the natural life cycle of cities; while *urbicide* is the death of the city because of a premeditated destruction, a murder, in the words of Fernando Carrión:

> *Urbicide* is about understanding the urban processes in a different way; that is, generating a change of method: less how the city is produced and more how it is destroyed, less from the memory and more from oblivion. It is about understanding the city through the causes of its own destruction; that is, to show critically how from the prevailing model of the city we can find the hope of a new urban reality, which gives back to the city the sense of a good place (utopia). *Urbicide* is a concept composed of two words: 'urbs', which is the city, and 'cidio', which means death; that is, the death of the city. But this is not a natural death or a homicide, it is rather a murder. The *Urbicide* is the liturgical murder of the city, carried out with premeditation, order, and explicit form, which comes from actions that raze the systems of significant places of common life: squares, monuments, libraries (agora); they destroy the material base of a city: infrastructure, services (urbs); they

exterminate society and citizenship (civitas); as well as annihilate institutional frameworks of government: privatization, deregulation, centralization (polis). This murder of the city comes from various situations that can be ordered from three particular types: natural, anthropical, and symbolic (Carrión 2018).

In this scenario, the question arises how to reflect on the Latin American city? A vast continent with a complex history of cultural pre-existences that tell of several deaths and resurgences of societies settled in multicultural territories, complex topographies, and natural risks. With the Spanish conquest and the foundation of cities, the dispersed pre-Hispanic settlement pattern was eradicated, to move on to life in human conglomerates organized in a grid organized in a checkboard pattern, this change forced the indigenous population to live in *police* or in *city*, without considering that pre-Hispanic communities had a dispersed settlement pattern with a shared center of power, different ways of life, and deep cultural differences between them, although they were close to each other (Marin 2005). In this example, the term *urbicide* is not applicable, which by definition starts from the *urb*, but one could talk about the death of *the space with belong to*, a term that in free translation from Quechua means collective space, representing a local settlement pattern that allowed governance in multicultural societies that created rotating systems of exchange and community activities of collective benefit (Del Pino Martínez 2017, p. 26). The loss produced by war and a pandemic due to unknown diseases was the end of *the space with belong to*, that way of life in which the character of collective, local identity, and dialog and agreements predominate to make governance possible.

The *urbicide* is applicable to colonial and modern cities since the birth of the *urb* or the city in occidental terms, *the space with bewlong to* has been in the background for centuries; however, it is present in social genetics and emerges in crisis conditions. Is *urbicide* possible in cities that are intercultural and where there is a dense historical heritage rooted in all strata of the population, sometimes unconsciously? How do cultures today interact in everyday life and how do they manage the differences to be able to stay together? A provisional answer is that there are multicultural territories that, in the daily practice of centuries of orality and visuality, have produced changes, mutations, and differentiations between social groups. In some cases, survival conditions force the circulation between cultures without losing one's own identity, a sort of negotiation or force field that permanently reconfigures itself and evades the definitive death of the culture.

In the 1990s, Néstor García Canclini studied the cultural legacies of modernity and postmodernity, and culture, as a form of social reproduction inside capitalism in Latin America. One of the conclusions was to find that the continent is the result of "sedimentation, juxtaposition, and cross-linking of indigenous traditions, with the colonial tradition in popular sectors" (Canclini 2013, p. 6), which originates the concept of *hybrid cultures*. The term has been debated and sometimes questioned, but it allows us to understand the particular ways in which each culture adopts and adapts to live in the city. Another reflection allows us to understand the ways of experimenting social relations in interculturality when it speaks of *other forms of being in the world*, that is, a parallel world, practiced above all by popular culture in cities, in an effort to maintain its survival, with strategies of adaptation to modernity and postmodernity. According to this conception, *urbicide* is unlikely in practice because interculturality is a form of social interrelation that, on the one hand, keeps the identity and, on the other hand, transforms it in time. It uses the grammar and the same resources of modernity to interpret and transgress the sense of politics and the market, avoiding conflict, or in other situations, circulating in parallel with the imposed urban order. In this case, popular culture evades *urbicide* through actions of collective resistance. In this context, culture is an element that sustains life and keeps away the intentional end of the programmed death of the city. The forms of death have their complexities; one of them is the acceptance of the end, which in some cases would mean defeat.

## 28.2 Is Physical Loss of the City Possible?

The physical death of the city would seem to be the end; difficult to accept and assimilate, even more so from the concept of place where it is noted that *everything has a place in our lives*, this is present even when we mentally move from one place to another, that is, the notion of place has a cultural and historical dimension, and is related to the territory (Casey 2002, p. Prefacio). In this sense, the dissolution of places brings an existential crisis, associated with time and space in which it occurs. An experience in which the known space disappears in a matter of minutes. In countries with seismic risk, this experience is associated with the earthquake and has in the past and present meant *the loss of place* for a while. As a recent event, the earthquake of April 2016, off the coasts of the provinces of Manabi and Esmeraldas in Ecuador, made visible the geological vitality in the Andes and the partial loss of cities.

The 2016 earthquake occurred on a Saturday afternoon when families were outside their homes, on the beach, or in shopping centers, when suddenly they saw how large concrete structures collapsed in seconds, transforming the streets into a pile of rubble, people trapped or lifeless, a moment of collective crisis, that is, the city went into crisis. However, after 6 years, the destruction can be evaluated in the sense that the loss triggered in some cases temporary abandonment, the presence of empty properties was the occasion to rebuild or renovate; to buy or sell; in an attitude that could be called *resilient*. The example allows us to explain that material loss can be recovered or renewed; people abandon the city for a moment, but return, this is a gesture of belonging to the place, but also a will to rebuild a material good that has economic value, represents the work of a part of their life, and a long-term investment: the house is lost, but not the land. This collective attitude allows us to recognize an unconscious experience of the cycles of nature. Nevertheless, there is a duality between the violence of the material change and the psychological experience lived during the natural event that marks the life of the people, leaving a trace in the memory of the people and transforming their way of being: fear to living at high altitude, avoiding living between large windows for fear of the damage and rumble of glasses, an immaterial fact, the noise, was internalized in some people and remained in the memory of the crisis.

The effect of earthquakes or eruptions is a didactic case to understand the violent destruction and the loss of the place for an instant, because it does not give time to think what to do, the natural phenomenon is radical, it is the great auditor of the constructions in general, it has no regard with interventions of dubious invoice that hides what did not accomplish the norm, or that demands the urgent revision of the norm to avoid future losses. The earthquake, like eruptions and other natural phenomena, leaves a trace in the collective unconscious that becomes part of an identity associated with such circumstances, even for generations, that is, with the idea of death being activated in memory, reactions to sound, images, and vital experiences. To speak of death is to remember life in a look in the rearview mirror.

The physical death of the city due to earthquakes is an experience difficult to avoid due to the geological nature of the territory and has had different repercussions over time. However, in the present and past, physical death is not total as long as the imaginary and ruins exist in the devasted site. The destruction of ancient Riobamba, of colonial origin, by a grade VIII MSK earthquake in February 1797 evidence how the concept of *urbicide* is applicable in this context. The earthquake affected the coast and central highlands of Ecuador; Riobamba was the most affected populated center. The event occurred on Good Friday, at a time when a large part of the population was inside the churches of the city. With the earthquake, the roof tiles were first blown off, and when the structure of the roofs was weakened, several walls lost stability. As people ran, roof tiles and walls fell, producing the loss of human lives, estimated at twelve thousand; very few were saved. The natural phenomenon was aggravated by the landslide of a hill at the foot of the city called *Cusqui* which buried it and caused the damming of the river. With the dimension of this event, Riobamba, an stately urban center that competed with Quito was abandoned for being uninhabitable; the authorities decided its relocation to another site, twenty kilometers to the north, the transfer was not only of the toponym but also a symbolic gesture of belonging and collective identity to the city of Riobamba, which is reaffirmed by carrying in procession the stones of the façade of the cathedral and its placement in the new building.

Around 1800, an urban plan with a radial grid layout was proposed, like the ideal cities of the Italian Renaissance, an innovative design; however, the population opposed the materialization of this project due to the fact that the shape of the layout did not correspond to the social and spatial hierarchy of the ancient city. In this sense, everyone wanted to maintain their spatial location in relation to the square of the old city. According to these ideas of the population, the new Riobamba has a checkerboard design with wide streets to avoid future deaths due to falling tiles and walls. The toponym of the destroyed city was reproduced in the new location with the name Riobamba, leaving the name of *old Riobamba* to the one that was buried. Over time, the abandoned site was occupied by the indigenous community of *Sicalpa*, which was settled a few kilometers from the destroyed city, thus the old Riobamba became the *new Sicalpa*, leaving the original settlement with the name of *old Sicalpa*. For more than two centuries everything remained under the ground;

however, the tales and stories that were heard in the new Riobamba, especially in the wakes, re-entered the memory of the old city and stories of indigenous farmers of *new Sicalpa*, who became rich from night to morning due to alleged findings of treasures, coins, gold, and precious metals from the old city. Two hundred years later, a project to enhance the value of the old Riobamba found the city frozen: finely laid stone floors, the sewage system intact; the foundations were in place, it was possible to rebuild the history of the city and the sense of the chronicles about its importance in relation to Quito, this finding was as valuable as the wealth of the urban imaginaries and tales built collectively over time to raise the memory of the ancient city, thus avoiding its symbolic death (Del Pino 1986). The story leads to think that, although there is abandonment and material loss, there are traces that when studied the city reborn in a vision of mirrors, sometimes distorted and sometimes in an inverted image, sometimes imagined and in others in its tangible materiality; those who relive the tales and remember it correspond to the eighth generation, the power of orality transmits an experience that is told as a reality, and in part, fantasy. To answer the question, the physical death of the city is not possible in cities with its own identity, consolidated, with long-standing historical antecedents. Given the abandonment, archeology, plan, and memory of the population for several generations remains.

## 28.3 The Symbolic and Anthropical Death

Looking at death through the mirror of life, one could say that the symbolic value of a city, an architectural object, or a landscape is a collective construction that in time gives character to a place. Christian Norberg-Schulz spoke of *genius*, a Roman term, to name the guardian spirit that gives life to people and places and is present during their lives (Norberg-Schulz 1979). It is deduced from this that every place has a spirit that rests in certain spaces and cultures settled in a territory. In countries with high cultural diversity, this reflection is evident in the geography, in the toponyms of the towns, in surnames of the people. To illustrate this idea, the province of Imbabura is identified with a high and cold zone characterized by lagoons and the greenery of the countryside, and at the same time, a few kilometers away there is a warm zone of dry forest; cultural diversity, gastronomy, markets full of color are part of that first look of the tourist. The province must be known with time to identify its character, culture and history; more in depth, one discovers a premodern world that survives in parallel with urban modernity, associated with the local, that is, the understanding of sacred geography that rests on two tutelary volcanoes with powers and knowledge, they dialog with each other, with the lagoons, rivers and minor mountains of the region, they command respect and also benefits for the farmer: the invocation to *Taita* Imbabura and *Mama* Cotacachi, accompany the treatments to heal health affectations through knowledge on local herbal medicine, and are part of the farmer's spirituality. The kind geography creates roots and identity with the place, the cultural diversity has its own way of being in this space, often encloses conflicts and inequities, and the difficulty of reaching agreements, however, the identity with the specific place smoothes out the differences, creates roots and provides a sense of belonging.

Turning the mirror, we can deduce that the spirit remains in place as long as conditions of balance are maintained, conflict can mean its degradation and loss. The death of the city or part of it depends on its social vitality, on the sense of belonging, on the permanent strengthening of local culture. This antecedent leads us to think about the historical centers of some Latin American and Ecuadorian cities; one of them is Quito, which was declared as Cultural Heritage of Humanity along with Galapagos Islands as a Natural Heritage site by United Nations Educational, Scientific and Cultural Organization (UNESCO) and the Ecuadorian government in 1978, and thus became part of the list of cultural and natural world heritage sites. The arguments for the declaration were, on the one hand, the natural landscape in which the city is located, that is, a unique place. The general director of the organization expressed his experience upon returning to Quito: "I am thrilled to find myself once again in this city made of stones and clouds…". The second argument was for culture, which is expressed in the quantity and quality of colonial art guarded in the convents and monumental churches, whose relevance is known in the world and accounts for the creativity of its artisans and artists. The third argument was due to its historical character: its leadership in the most relevant episodes of history, that is, history and identity. Jorge Salvador Lara's speech on the Ecuadorian side added as a local value, multiculturality, and social diversity: "center of blood fusion, important field for the contact of dissimilar cultures" (Ministerio de Relaciones Exteriores. Ecuador, 1979).

The situations identified as causes of *urbicide* are natural disasters, anthropic, and symbolic causes. In this context, after the declaration of World Heritage, the historical center of Quito was in a process of deterioration, for physical recovery there were two projects of architectural restoration led by international cooperation; Spain, Belgium, France, among others, and the Ecuadorian government: the convent of San Francisco and the convent of Santo Domingo, which were in the worst condition caused by the systematic damages produced during four earthquakes between the eighteenth and nineteenth centuries: 1755, 1797, 1859, and 1868. In the other religious buildings, architectural and structural restoration works were carried out, as well as medium-size rehabilitation in other works by the Quito Salvage Fund (FONSAL, for its initials in Spanish) between 1989 and 2010, after which the responsibility passed to the Metropolitan Institute of Heritage (IMP, for its initials in Spanish), an entity that is part of the municipal agency. Parallel with the creation of FONSAL, a mixed economy company called Company for the Development of the Historical Center of Quito was created, which took the expropriation and purchase of housing, execution of a series of works in heritage buildings; the Junta de Andalucía carried out the rehabilitation of several houses that were expropriated by the municipality and were in very poor condition, located on Rocafuerte Street.

The declaration was timely because years later, when the 1987 earthquake occurred, it was possible to have advice for rehabilitation, international cooperation contributed with technical assistance, studies, consultancy to other affected cities, and urgent works were carried out. Subsequently, the Workshop Schools Program for Ibero-America of Spanish Agency for International Cooperation (AECI, for its initials in Spanish) contributed to the training of construction technicians through two Workshop Schools in Quito. Seminars and development projects of structural restoration of religious buildings were carried out. Furthermore, the municipality also worked on the publication of the Municipality of the Metropolitan District of Quito (MDMQ, for its initials in Spanish) land management regulations in 2005. This document contains the plan for the use and occupation of the land in the center, as well as the architectural and urban planning ordinances that regulate the incorporation of new housing, land use, heights, the incorporation of mechanical installations such as elevators and escalators, among others. For its part, the Pon a Punto tu Casa program (agreement between MDMQ and Junta de Andalucía), gave the opportunity to carry out maintenance works in private housing through small loans, prior inspection of the property, since, by law, neither the government nor the municipality can invest in the interior of private property; these initiatives changed the image of the center between the years 2000 and 2014 (El Telégrafo 2014).

After 44 years, and with the background described succinctly, it is worth asking what happened? It is clear that the results are not consistent with the efforts to preserve this space and strengthen the lives of its inhabitants, and instead of maintaining its population, it has lost it. It is estimated that from 90,000 people who lived in 1990, in 2022 it will reach approximately 30,000, most of them living in poverty. Significant abandonment accompanied by degradation inside the buildings could announce a process of *urbicide* or slow death of this space of the city for anthropic and symbolic reasons. Among the reasons, along with depopulation, the increase in the value of rents and purchase, prices compete with the areas of the city with the highest added value, which disappoints those interested in belonging to this space of the city; it seems that prices are inflated, and, on the other hand, people value more the new than the rehabilitated, they do not appreciate the privilege of having access to all services: communication, mobility, commerce. Neither are they sensitive to history nor are they clear about the benefits of future investment.

Is it possible to stop the degradation and abandonment? What to do to dodge the death of the center? And for the moment, who sustains the vitality of the center? Reviewing the press until 2014, reveals that loans for the realization of housing repair and maintenance projects were in progress; however, the outflow of population between the years 1990 and 2010 was 20,000 people, among other reasons due to the increase in rents, the eviction caused by the realization of the works was an excuse to change tenants with the expectation of a renewal with people of another social and economic level. Artisans of all specialties and people who had lived for several years paying low rents left; they knew the history of the sector and of the neighbors. On the other hand, convent restoration attracted a group of artists, artisans, and restoration workshops that also left at the end of the projects. The eviction of the inhabitants of La Ronda Street was radical, forcing the closure of the brothels on this street, and prostitution spread throughout the squares and streets of the entire center; the two that remained in the backyards of the houses had to leave due to the pressure of the new inhabitants of the street, with the argument that *the street now is a tourist attraction.* Since 1990, the departure of schools, private, and public institutions that generated an economic movement and kept social activity until the early hours of the night also left. Failed projects also contributed to the deterioration of the public space because people who do not feel that they own the space designed by those who did not think about the vocation of the place or of the people who inhabit it.

The misunderstanding of the vocation of a place correlated with the city itself is the series of at least four failed projects of urban rehabilitation in the same space, one of them on the filling of the Jerusalen ravine, on the southern edge of the first-order area of the historical center of Quito. Perhaps, a few reflections on the ravine can serve as background: the ravine is a topographic feature that has characterized the Quito plateau since its remote origins, it forms a system and has its own logic in relation to the geology of the place. The ravines in the history of this city have naturally divided sectors or neighborhoods. Water flows at its bottom, that is, it has direction and is part of a flow system if we consider its tributaries. The systematic filling carried out by the modernizing need of urban expansion in the twentieth century resulted in the union of urban sectors to facilitate transit in a north–south direction also thinking of the neighborhood integration, however, because of history is known that there is no empathy on either side of the ravine, on the contrary, with the filling the differences and conflicts between neighborhoods that did not have a similar growth experience were accentuated. In short, the meaning of the ravine was not considered at the time of generating a public space project that aspired to be compatible with people and the natural environment. What public works contribute is the facility of vehicular flow, but not social integration. In 1922 the boulevard over the Jerusalen ravine built as a civic space for the independence centennial parades did not transcend. At first, stately houses appeared around Victoria Square, but later *chicherías*, brothels, and canteens returned on the southern edge. As a hypothesis, it can be thought that one of the reasons could be linked to the place vocation, that vocation is expressed in a flow line: vehicles through the subsoil or bottom of the ravine and people on the surface. The viaduct is a flow line that maintains the east–west direction of the old river. The banks of the old ravine are incompatible in social, urban, and architectural terms, and its population will not exert a great deal of effort to meet, so the result of the investment of public works does not generate the appropriation of the population of either bank.

In another order, the weight of pre-existences and the vocation of the place appear in the vision of the mirrors in an oblique, deformed, blurred, or backlit way (Borges 2017, p. 79). Their examination allows us to recognize that they emerge in everyday life and in circumstances of crisis. Quito was born in a place of cultural pre-existences related to interregional exchange, involving interculturality, recognition of the other, family, and affinity bonds for the realization of alliances between traffickers; human activity is inserted in a sacred geography of which we have few traces, a nonhomogeneous and nonmercantile place, since there was no currency. The traffickers did not leave traces of buildings, temples, or large pyramids, in contrast, it was an *empty* space in which periodically, in symbolic cycles, several communities met to exchange objects, animals, and even people, called *catu* and then *tiánguez*, which has the toponym *Quito*. This space corresponds to the area of first order of the historical center and was the object of desire of the Incas and Spaniards. At the end of the dispute and wars, the site was taken by the latter, who founded the town of San Francisco de Quito. Not all agreements were fulfilled, but the conquest did not completely eradicate the previous social and spatial structure (Minchon 2007, p. 36). The local toponym *Quito* kept the original name to denominate the Audiencia and the newly founded town; it is preceded by a name from the calendar of the Spanish Catholic saints, which is *San Francisco*, as it happens in other cities, Christianizing places.

In the nineteenth and twentieth centuries, commercialization practices were not compatible with urban sanitation policies, such as street commerce and indigenous markets in the squares. Since then, merchants have been repressed, persecuted, fined, and their products confiscated. This is the daily history of a group of people who have sold on the street for more than two hundred years, in which the activity resists to abandon this common space, whose vocation and long-term reference has been the exchange. If this is so, it is worth asking: What are the material and immaterial traces that determine the vocation of a place? How to interpret and reorient with sensitivity the potential of this form of exchange, communication, and relationship between merchants and consumers? How to understand and give new life to a form of commercialization different from the occidental way of selling and buying? What to do to reconstruct its meaning based on the place vocation and the value of the local, to prevent the *urbicide* from taking place?

If everything takes place in our lives, the slow loss of culture and the identification with common places: such as squares, streets, atriums, corners, also worries because it leaves no trace but objects and spaces without history. The indiscriminate expulsion of the inhabitants of La Ronda Street also displaced the heritage actors, this is a group of people who keep in their memory the references of urban life, there was a blindness about their knowledge and experience, and they did not have the recognition of local or heritage institutions. In this sense, the departure was indiscriminate, people were dispersed to other sectors. Their displacement also distances social knowledge from the historic area, which is waiting for someone who finds it outside its context to give an account of what gave meaning to collective life and to understand the present. This memory is useful when proposing architectural and urban projects with a social approach, with sensitivity to the place, incorporating heritage stakeholders in the process and at the same time reinventing forms, uses, and meanings that highlight the spirit of the place or its vocation. In this case, the expulsion had to be selective to avoid oblivion and loss of collective memory.

At this moment, the floating population, street vendors, public and private employees, and a few inhabitants are the ones who sustain and give life to the center. The idea is uncomfortable for those who have tourist or real estate businesses; however, the tourist does not choose as an experience the subway ride or to know the modern and contemporary zone, but the experience of a city with a strong local identity, different from their own, whose history is perceived superficially through the tour and the density of public space, the urban landscape, the type of people, colors, odors, sounds of the city, among others.

The *urbicide* in Quito is slow, with facts that are not natural, but the product of inadequate urban policies in practice: expulsion of poor people, depopulation, degradation of the architecture in its interior, and a popular type of activity, associated with the market and street vending, which constitutes a collective way to face the death of the city and avoid this from happening. Some mechanisms of survival in the sanitary crisis allowed us to understand forms of communication that are rooted in ways of being of popular culture that give an understanding of other *ways of being in the world*. The impact of the exit of the popular market would be the *urbicide* of the center, moved by anthropic circumstances.

The activity of the San Roque market and wholesale market during the sanitary emergency of 2020 and 2021 shook Quito and the ghost of the *urbicide* arrived to position itself in them, transforming their operation, at least momentarily. It is known that moments of crisis produce turbulences that evidence contradictory situations, the interests of social groups become evident in a confrontation to keep their status, there are those who attack and others who defend the change, while the death of the city is silent, and in these cases accelerates. The fight against *urbicide* is a long-term resistance with the purpose of counteracting and weakening its effects, in this sense, the pandemic made visible the complexity of the food supply in the city and is an example of what one should understand. How the popular type of commercialization in the historical center of Quito defends itself to not lose its spaces and survival. In the context of the pandemic, the poverty in the city in general and the weakness of the institutions became visible but also, the responsibility of hundreds of farmers who did not stop producing in their chacras to maintain the cities supplied; on their own initiative and despite the circulation restrictions issued by the health authority and provisions, which instead of facilitating accessibility, impeded it in the name of the health regulations, causing shortages in major cities and trans regional: Quito and Guayaquil, causing economic losses and perishable products. The crisis situation of the city with shortages evidenced several particularities, interests, needs, and contradictory situations of that moment, in this sense, it is worth explaining the case of Quito.

Although urban planners know that Quito is supplied with food from distant points since agricultural land is scarce in the province, with the pandemic this situation became tangible in daily experience. As mentioned above, agricultural production did not stop and overcome several transportation problems to reach urban markets, including the closure of storage centers due to fumigation and the closure of access roads to the city. The shortage was not caused by lack of food but by the prohibition of trucks entering the city, especially if drivers were not vaccinated. However, not all the transport systems had the same treatment: vehicles that transported products for export to airports and trucks that supplied supermarkets had credentials for free circulation, while transport that goes to municipal markets, wholesale markets, and fairs did not have the same luck in reaching the places of sale (Fig. 28.1).

Quito has a traditional market called San Roque, located on the periphery of the historical center. It could be inferred that it is a space resulting from the displacement of the old markets located in the squares of the city; and a second one, called wholesale market, modern, located outside the center, intended for wholesale. Both are regulated by the municipality, however, the San Roque market, without being wholesale market, plays this role for several reasons, among them: it is a symbolic space recognized by the community, it inherited the tradition of the pre-Hispanic market, *catu* or

**Fig. 28.1** Religious complexes and colonial squares in relation to the San Roque market. *Source* Own elaboration based on information of Municipio del Distrito Metropolitano de Quito, 2010

*tiánguez* mentioned by several authors (Marin 2005). Its location is on one of the most important roads, in a place that quickly connects the north with the south.

The San Roque Fair is held in the early hours of Saturdays, a moment of great activity in which trucks arrive from all over the country to unload the sale in the market, to intermediate merchants who distribute in the markets of the neighborhoods and rotating fairs of the city. The living conditions of these merchants are not the best, they unload the merchandise to other transports in the street, often sleep in the same vehicle and continue to other destinations, generally buy on the way to the producer, and sell in the final destinations of Quito or Guayaquil. On the other hand, this neighborhood has some buildings enabled for refrigerators and warehouses, and rooms for rent for a group of temporary merchants, the housing conditions are of very low constructive quality, and services that are generally are communal use, most of this population is indigenous, some bring to the city their small loads of products to sell at the fair and supply themselves with products that they do not have in the countryside. During the pandemic and without a vehicle, they kept the products that were sold in nearby towns, exchanged their harvest with other neighbors or sold to intermediaries at very low prices. In other cases, communities closed the accesses to their farms for fear of contagion and destined the agricultural production to local consumption due to the difficulties of transportation, they avoided accessing the city where they were required to carry circulation and sales permits, vaccination certificates, and due to the bad treatment on the market and on the street (FIAN

**Fig. 28.2** Interior of the San Roque market. April 2022. *Source* Galo Benítez

Ecuador, Instituto de Estudios Ecuatorianos, Observatorio del Cambio rural, Tierra y vida, FIAN Internacional, 2020) (Fig. 28.2).

The sale of fresh products in San Roque is closely related to street food commerce in the historical center of Quito and is made up of family members of those who have their stalls inside the market and temporary traders. The pandemic was an opportunity to eliminate this type of sales around the San Roque market. Due to a case of COVID produced inside the market, the premises were fumigated, all traders were expelled, there was violence, removal of merchandise and resistance from the merchants; as a result of the repression some decided not to come back, a collective reaction that saw an immediate alternative: the dispersion of merchants all over the city with the occupation of hallways, garages, and corners where stalls appeared, particularly of fruits and legumes. Other traders formed alliances with van drivers to sell their product outside the market, and a third group contacted their *reliable buyers* to carry groceries to their homes, that is, the street trader exposed himself to police repression to deliver his products.

This practice was observed in almost all the neighborhoods located in the periphery of the historical center. Once the contagion had subsided, the traders gradually returned to their stalls, however, the street commerce in the center and around the San Roque market has decreased significantly. As a result of the repressions and discrimination, it would seem that an immediate solution was to abandon the countryside and emigrate outside the country as soon as international flights became available, particularly of young families of farmers in the province of Cotopaxi, where there are agricultural centers, farms, flower companies, suggesting that the closure of job opportunities in the city is directly related to the abandonment of the countryside and that in the future it will be seen in the lack of labor and the shortage of supplies in the city.

The Quito wholesale market is located in the south of the city, outside of the historical center, and is controlled by the municipality. Its function is to receive merchandise from containers and sell wholesale products called *abastos*, however, with the pandemic the trucks did not enter the city and the absence of buyers became evident; to this was added the application of a toll for entry to this market that conditioned the entry of suppliers, with the purpose of compensating the losses of the owners of the premises inside, this bothered the suppliers. In this scenario, merchants were forced to sell at retail, organize routes using social networks and electronic media to place the product in the hands of their regular customers, that is, the movements between seller and buyer were reversed. In this relationship, loyalty was a very important aspect in moving commerce. However, over time, the return to the market tends to slowly recover the economic losses.

An aspect that the pandemic revealed in the markets is that simultaneous wholesale and retail are necessary to sustain a business, whether in normal times or in crisis moments such as the one that produced the pandemic. Another aspect derived from unexpected situation was adaptation capacity and flexibility of the merchant to reach the customer through virtual media, and, above all, something that is beyond technology id the trust and reciprocity between people, maintained for years as an added value to the business. In this sense, it is clear that these are *other ways of being in the world*, in a raw world, full of uncertainty and vulnerability, which is the hidden face of the city.

In this scenario, how do we avoid the death of historic space? How do we escape the physical and symbolic crisis of the city? The immediate question would be how to change the sign of deterioration and abandonment? The solution is not immediate, nor could a bet on a project of great visual impact in depressed areas, as in Medellín, without a previous analysis, although it is a fact that a good urban and architectural design contributes to improve the quality of public space and contributes positively in the mood of the inhabitants as illustrated by the proposals of Jan Gehl or Gordon Cullen (Gehl 2014); however, the project can not be epidermal but considered about the needs of those who live in the place, who will deal in the future with the maintenance of the renovated spaces. Legislation that respects the rights of neighborhood residents, rather than punitive or over-controlling provisions, also contributes.

The *other ways of being in the world* in interaction with digital technology are of interest to understand popular culture; virtual communication has great power, its versatility is not seen, but felt, the efficiency and speed of transmission is a sign of the times, it has a differentiated rhythm in relation to the controlling institutional devices. With the closing of the markets due to lack of supplies in municipal spaces, the dispersion of the market in all sectors of the city evidenced the dimension and complexity of this activity, and the type of people who perform it, in addition to the creativity in the solutions generated to reach the buyer. For the merchant, this experience turned *the world upside down*: the merchant approached the client, that is, in the space of trade networks, flows were imposed on the nodes; direct communication between differentiated social groups became evident, greater solidarity and flexibility in the measure of the family requirements. Young people created routes to distribute merchandise, and a network was created, integrated by groups with age differences. The routes through the rugged topography of Quito, dodging the municipal authority led to also imagine the topography of communication that jumped from the market to the neighborhoods, to the streets, plunged through tunnels and shortcuts to reach the customer and the contacts of retailers in the city. In this sense, technology took its side and also showed its positive side in the *other ways of being in the world*, in resistance to *urbicide*.

With this example, it is worth to dimension the communication power, it has been said that is the fourth power in the twentieth century, in the face of the weakness of institutions, the slowness of the processes of adaptation to new collective circumstances, the weight of the legislation that often focuses on punishment rather than on problem solutions. This is not new, it has been said in several forums, but in the case of market traders it does not depend on their education level but on their creativity in dealing with problems and avoiding obstacles, their adaptation to new circumstances, and their enormous capacity to improvise with few resources. This cultural attitude, together with solidarity, is part of the subaltern identities that are not present today, for example, when the country went to dollarization it was thought that indigenous traders would have problems in calculating the *vuelto* or change, and the conversion between currencies, it was possible to verify that nothing happened, which allowed one to remember that this group had a previous experience from years ago with the sale of handicrafts in the squares around the world, in border places where currency conversion is usual. This experience allows us to be optimistic, the historical center is degraded, but it has not reached the *urbicide*. Those who sustain the life of this space of the city are the market traders, the street commerce, the people who come to work, those who still live in the center, the poor and elderly people who live in the center or its surroundings and go out very early to occupy the benches of the squares to receive the morning sun.

## 28.4 Final Reflection

The reflections and examples outlined in this article demonstrate that *urbicide* does not occur from moment to moment, but over time, and is generally anthropic in nature. To understand the scope of the term *death of the city*, it seems that the idea pointed out to several authors that *city is made by people* is of interest. The *urbicide*, then, is a collective process that has at least two antagonistic forces, one, from the community that defends the survival of the life of the city and its citizens, and, on the other hand, from the power, are premeditated actions to extinguish the history and the memory of a space of the city or the entire city. Mechanisms and tactics that come into play to end the life of the city and its citizens are possible through the *dilution of mirrors* that Borges rightly glimpsed in the recollection of his memory and are inspiring for the case at hand. Through the look in multiple mirrors, one could find the labyrinths for the extinction of the most outstanding human and collective work: that is the city, that is, after going through its history and memory, let it go.

However, accepting the death of the city and its citizens would be to accept a loss and oblivion, a collective suicide, an image that, at least for the moment, is not in sight. The health crisis of 2020–2022 was a proof, an external agent that no one imagined, brought to light the conclusion that citizens bet on life, not death, even in cities or sectors of cities that are degraded since the materiality of architecture during the crisis acquired the quality of shelter. In the process mechanisms of survival are invented that are generally repressed by the local authority, however, the collective resistance faced and overcame all barriers.

On the other hand, it would be necessary to differentiate between the *urbicide* and the natural transformation of the city; in both cases, the common denominator would be the time and speed of change; the natural transformation of the city is a palimpsest, the new is born taking as a reference the past, while in the *urbicide* the change or transformation is economic and of utility for the market, it is mechanical, and does not recognize the social actors that make the city, because the social heterogeneity is uncomfortable. The *urbicide* invents a new city, it strives to reinvent everything, including the history, however, it is a project without society, and it would have to be invented. Hence the urgent need for the population to lead processes of appropriation of historic spaces, to seek their rehabilitation; reestablish open spaces and exercise the right to the city; to exert leadership in the role of the city as a living space; to recognize its material and symbolic values; to promote priority projects that in Quito were initiated with positive experiences such as stimulating the rehabilitation of housing for heterogeneous groups and age differences; returning activities that left the center or creating new uses with multiplying effects in the sector, such as education and recreation; incorporating services necessary for the contemporary way of life, adding meaning to this space of the city. It is worth rethinking and evaluating initiatives carried out in the past to generate programs that are positioned above the political cycles and the support of international cooperation and initiatives that have been lost in the last ten years.

The *urbicide* or the death of the city is a slow process that invites one to face death in a collective way, through its mirror that is life, with a critical attitude and positive character, in a subversive and optimistic line to propose a new cycle for the historic city from the seed of the values of the past and that contribute to new *ways of being in the world* with emphasis in social diversity, integrating the memory of the inhabitants that still remain in the center, instead of their death.

# References

Borges JL (2017) La dilución de los espejos. Revista Universidad Complutense de Madrid. https://revistas.ucm.es/index.php/ESIM/article/download/58233/52399

Carrión F, 1 de agosto de (2018) Urbicide, or the city´s litugical death. Oculum ensaios 15(1):5–12. Obtenido de. https://periodicos.puc-campinas.edu.br/seer/index.php/oculum/article/download/4103/2575/13967

Casey E (2002) Representing place. University of Minessota, Minessota

Del Pino I (1986) Proyecto integral de rehabilitación de la antigua Riobamba. Inédito, Quito

Del Pino Martínez I (2017) Espacio urbano en la historia de Quito: Territorio, traza y espacios ciudadanos. Recuperado el 25 de 05 de 2022, de. http://bdigital.unal.edu.co/57661/1/TESIS%20ARTES%20Ines%20del%20Pino%20210617pq.pdf

El Telégrafo 27 de abril de (2014) Las casas patrimoniales perviven entre la restauración y el descuido. Obtenido de. https://www.eltelegrafo.com.ec/noticias/quito/1/las-casas-patrimoniales-perviven-entre-la-restauracion-y-el-descuido

FIAN Ecuador, Instituto de Estudios Ecuatorianos, Observatorio del Cambio rural, Tierra y vida, FIAN Internacional (2020) De quién nos alimentamos. La pandemia y los derechos campesinos del Ecuador. Informe, Quito

García Canclini N 1 de julio de (2013) Culturas híbridas. Estrategias para entrar y salir de la modernidad. Obtenido de ACADEMIA. https://www.academia.edu/20586759/Culturas_hibridas_Estrategias_para_entrar_y_salir_de_la_modernidad

Gehl J (2014) Ciudades para la gente. Ediciones Infinito, Argentina

Koolhaas R (2011) La ciudad genérica. GG mínima, España

Lynch K (2005) Echar a perder. GG, Barcelona

Marin LD (2005) Algunas reflexiones sobre el Ecuador prehispánico y la ciudad inca de Quito, Primera. Junta de Andalucía, MDMQ, PUCE, Sevilla

Minchon M (2007) El pueblo de Quito, Primera. FONSAL, Quito

Ministerio de Relaciones Exteriores. Ecuador (1979) Quito, Patrimonio Cultural de la Humanidad. Ministerio de Educación y Cultura, Quito

Niño CR (2010) La carrera de la modernidad. Construcción de la carrera Décima, Bogotá (1945–1960). Alcaldía Mayor de Bogotá, Bogotá

Norberg-Schulz C (1979) Genius-Loci. Paesaggio, ambiente, architettura. Electa Editrice, Italia

**Inés del Pino Martínez** Ph.D. of Art and Architecture and Master's in city government, mentioned in Urban Centrality and Historic Areas, FLACSO-Ecuador. Architect: Central University of Ecuador. Principal Professor: Pontifical Catholic University of Ecuador. Coordinator of the network of Ibero-American Historical Heritage (PHI-Ecuador).

# Part VII
# Destruction of Common Life: Violence

# Chapter 29
# The Besieged City: Geographies of Crime

**Alfonso Valenzuela-Aguilera**

**Abstract** This chapter addresses the spatial dimension of insecurity and decodes how territory neutralizes, encourages or inhibits the commission of a crime. For this purpose, we identify spatial parameters that characterize urban space. Although certain sites may be associated with a perception of fear under certain circumstances, a combination of both physical and psychosocial elements make one location a more favorable setting for the commission of a criminal act than another. As the city is a complex entity, a direct reading of the environment may prove to be intricate. Therefore, we propose a topology that highlights key aspects in order to understand the spatial perception of security in a territory. A correlation of both quantitative and qualitative information provides a topology that identifies the type of space associated with different types of crimes.

**Keywords** Crime · Territory · Town planning · Real estate markets

## 29.1 Introduction

The spatial configuration of cities has been transformed by insecurity. Essentially, violence has restricted the use of urban spaces by threatening citizens and hindering its use based on the logic of survival. Accessibility in both in time and space is being limited, with risk factors increasing depending on the time, day and place. Consequently, these sites acquire certain stigmas and territories take on novel configurations in accordance with new functions within the criminal economy. A collateral effect of the reterritorialization of crime is the growing rate of population that perceives the city as an unsafe place to live (Haesbaert 2004). This indicator has, among other things, a direct impact on social coexistence, solidarity mechanisms, and interpersonal relationships and on the use of public space. Moreover, affectation of the constitutive functions of the urban environment contributes to the expansion of the concept known as *urbicide*, considering how organized crime takes control

---

A. Valenzuela-Aguilera (✉)
Autonomous University of the State of Morelos (UAEM), Cuernavaca, Mexico
e-mail: aval@uaem.mx

© The Author(s), under exclusive license to Springer Nature Switzerland AG 2023
F. Carrión Mena and P. Cepeda Pico (eds.), *Urbicide*, The Urban Book Series,
https://doi.org/10.1007/978-3-031-25304-1_29

of strategic points throughout the city as well as the most meaningful places in the social life of its inhabitants.

This chapter addresses the spatial dimension of insecurity, decoding the way in which territory neutralizes, encourages or inhibits the commission of crime. For this purpose, we identify spatial parameters that characterize urban space. Although certain places may be associated with a perception of fear under certain circumstances, the combination of both physical and psychosocial elements makes one location a more favorable setting for the commission of a criminal act than another (Brantingham and Brantingham 1981; Wortley and Townsley 2008; Weisburd et al. 2016). As the city is a complex entity, a direct reading of the environment may prove to be complex.

Therefore, we propose a topology that highlights key aspects to be able to understand the spatial perception of security in territories. A correlation of both quantitative and qualitative information provides a topology that identifies the type of space associated with different types of crimes. (Weisburd et al. 2012).

A spatial analysis of crime in the city is based upon the assumption that a territory is socially produced and has the power to shape our behavior (Brantingham and Brantingham 1981). Following this rationale, contemporary society produces violent territories that can lead to defensive behavior in citizens. But how accurate is this spatial determinism? There are prominent theories on the importance of environment in the incidence of crime and these have derived into actionable strategies, including: the defensible or defensive space (Newman 1972); the search for predictive capacities through the use of geo-economic indicators that can predict crime through spatial risk modeling (Digital Terrain Modeling); the inclusion of spatial factors in crime analysis to promote safe environment design (Crime Prevention through Environmental Design or CPTED); as well as critical views that question the objects and generators of fear (Mawby 1977; Smith 1987; Gilling 1997). In addition, we cannot overlook the fact that crime and violence determinants have cultural, economic and social roots, and so public action also plays a fundamental role in the expansion or containment of crime by intervening in the environment where these events take place, either by repressing or socializing their resolution.

Over the last several decades, public security policies have focused on identifying areas with the highest crime incidence (known as hotspots), yet have devoted scarce attention to analyzing the way in which crimes occur in space (Sypion-Dutkowska and Leitner 2017). Undoubtedly, the development of criminal mapping techniques has been indispensable in locating the areas of greatest risk, but it is equally important to be able to identify the patterns that are articulated in territories through criminal activities. Locating crime in space could be one of several elements used in a system that involves different moments, spaces, mechanisms and interconnections among formal, informal and illegal actors (Valenzuela and Monroy 2014).

This section is based on a series of empirical analyzes of specific places in the city carried out in recent years (Valenzuela 2016, 2019, 2020) and focuses on the study of the location of crimes rather than on the motivations of the offenders or the social causes behind the crime. Because criminal behavior is directly related to the nature of the immediate environment in which it occurs, we examine the instrumental role of the environment in activating certain behaviors and guiding the course of events. Based on environmental criminology studies (Brantingham and Brantingham 1981; Skogan 1988; Herbert and Hyde 1985), which maintain that the distribution of crime in time and space is not random but rather depends on situational factors, we arrive at a definition of a spatial typology of crime. The concept of topology arises in a mathematical context that characterizes the geometric invariants that prevail even when undergoing continuous transformations. If we focus on the topological space, we can classify its multiple attributes, including the most distinctive ones, such as connectivity, continuity, compactness, convergence and proximity.

## 29.2 Territory and Perception

The perception of place involves complex cognitive mechanisms for identifying spatial components that have meaning, convey a specific connotation and include mental processes where memory and imagination intervene. Another stance suggests that perception depends on internal representations that respond to available information and mental models that each person acquires through previous conscious or unconscious references (Bartlett 1932). If we analyze the case of spaces where the local imaginary is linked to violent events, we find that associations that derive from previous experiences will result in similar reactions to these and will probably generate annoyance, aversion or fear. Within this analytical framework, territory is defined as a space of power where criminal acts take place (Bourdieu 2004; Haesbaert 2004; Lopes de Souza 1995). Therefore, territoriality necessarily involves a social tension involving the defense and appropriation of physical space in its different connotations (Pastalán 1970; Sommer and Becker 1969; Altman 1975). Hence, the personal hierarchy that an individual assigns to a territory will depend on the importance of the role that the individual plays in the daily life of that territory. It is in this context that Altman (1975) proposes three modalities:

(1) *Primary territories*, which are essential in the life of an individual, and over which the individual has almost absolute control in the personal and private space, such as a person's home;
(2) *Secondary territories*, which have a shared and semi-public character, such as social clubs, local bars and places with limited interaction;
(3) *Public territories* that are temporarily open to occupation, such as restaurants, parks, public transportation, beaches and work spaces where users subscribe to certain unwritten social norms.

A fourth modality involves the territories of transgression, located in violent cities and structured through a complex configuration of relationships of dominance within different social groups. On the one hand, territoriality fulfills practical functions such as the possibility of developing cognitive maps that allow types of behavior to be decoded and anticipated in a given space. It is possible to assign probable relationships between specific places and associated practices through this type of representation, allowing us to structure and program our daily activities within a margin of reliability. On the other hand, if we consider an urban or metropolitan scale, the structure of these maps may be determined both by the socioeconomic level and by the dominion that certain groups profess over spaces, as in the case of a violent city. This dominion is exerted through control mechanisms that allow individuals or groups to influence the behavior of others, in the sense proposed by Foucault (1977). Moreover, the exercise of power does not necessarily require the use of force, but is based, rather, on the establishment of these dominant relationships.

In this same way, criminal groups signal their hegemony over a territory by delimiting the dominated spaces, an action that is subject to possible negotiations with the formal authorities in charge of these territories (mayors, governors, trustees or security forces at different levels). These negotiations include agreements in which territorial control is distributed according to complex *translegal* configurations (Valenzuela, 2016), in such a way that certain areas can remain under the dominion of criminal groups in exchange for keeping other areas of the city protected from crime.

A key aspect of territoriality has been the intrinsic exchange value of the potential use of a given space (Altman 1975). Land acquires greater value in accordance based on an increase in demand and then become a control priority for criminal groups. Initially, interest in a space may be based on its geographical location, that is: proximity to markets, connections to marketing routes, articulation with centers of production, processing and distribution (in the case of drugs), mimesis or isolation (in the case of safe houses or shelters for drug traffickers); or because they are located in areas of high impunity in the face of generalized and institutionalized corruption. Likewise, when a destabilizing action occurs in the hierarchy of dominion, such as the capture of an important drug lord, a series of readjustments will take place within the organization that have a direct impact on the territory, whether it is the atomization of the different factions of the organization into independent cells, or a regrouping into new coalitions that may or may not include the original group. In this way, criminal organizations reconfigure themselves to secure both places and transfer routes that include cities, highways, seaports, borders and airports. This spatial arrangement is susceptible to other changes based on these events, as they also lead to intervention by armed forces or the Navy in an attempt to contain the advancement of criminal organizations in the territory, or simply as a restructuring of dominance over the space. Hence, territoriality and dominance are strongly linked and can maintain a stable structure over extended periods of time (Sundstrom and Altman 1974).

Another important aspect linked to the environment is the sense of identity produced by sharing a territory. Proximity often encourages the creation of social ties (Edney 1976) and that is why in towns infiltrated by organized crime, local drug dealers can live with their neighbors without the latter necessarily sharing in or approving of the activities of the former. Similarly, in some territories dominated by local gangs, it is common for residents who are not affiliated with these organizations to be able to move around the neighborhood, as long as the dominant group recognizes them. It is also interesting to note differences in the perception of territoriality based on gender, as highlighted by the findings of Mercer and Benjamin (1980), who noted just how social functions of territoriality vary in accordance with gender: men use public spaces as neutral areas where they can withdraw from contact with friends and associates, while women reserve similar spaces for socialization. This is relevant given that women use the local space consistently and more intensely, which may be a reason why they perceive greater vulnerability related to experiencing crime in their daily lives.

Territorialities also has a temporal component that allows individuals to settle in a certain place for a certain time and begin to gain possession rights (Becker and Mayo 1971) that can later become territorial rights. Once again we are faced with the articulation of power systems, with the territory defended firstly as a biological necessity (Ardrey 1966), and secondly as a way to exercise power over a physical-social environment (Edney 1972). That is why the appropriation of territory is a central concern for the recovery of degraded neighborhoods dominated by organized crime. We know that communities with good internal cohesion have better capacities when it comes to identifying and resisting the infiltration of groups from outside the neighborhood. Neighborhood-watch organizations can lead to initiatives to monitor the territory, react to threats and reduce vulnerability, thereby creating a feeling of relative control over the environment.

According to the semiological approach of Eco (1989: 295), architectural forms do not necessarily denote the specific functions of an object or space. Rather, these require a reference system; in other words, a code. If spaces are not linked to a shared code, they are not understandable or usable by the user. That is why our approach is based on the fact that before a violent city can exist, spaces of transgression and risk are configured through new codes in such a way that corresponding symbolic connotations will be more useful than the functional denotations of the space. Consequently, the semantic codes that define urban and spatial elements (such as the typological genealogies of hospitals, houses, schools, parks, etc.) are contrary to new configurations linked to the culture of violence such as residential areas, safe houses, prisons, etc.

Taking this into consideration, urban planning can be considered a spatial code that communicates messages, functions and meanings and persuades users to use the space in a certain way, in accordance with values linked to a particular way of life. This is how an exclusive gated community conveys a message of impenetrability through its surrounding high walls and controlled access guarded by armed security personnel. Similarly, an elevated toll highway that passes over marginalized places to reach the international airport sends out a message that abysmal socioeconomic differences are being circumvented through physical-spatial solutions. These territorial solutions not only circumvent undesirable areas of the city, but also activate, through the device of exclusion, the mechanisms that keep specific sectors of the population disconnected, while also bringing individuals with the economic capacity to pay tolls closer so they can use the spaces that this infrastructure interconnects.

In line with this reasoning, highways and communication routes interrupted by checkpoints (whether they be military, police, community or organized crime), send a message to both the antagonistic groups, and to the general public, of the former's relative control over the territorial domain. Along the same lines are the highly mediatic messages of bodies of rival drug traffickers hanging from emblematic vehicular bridges in the city, the disposal of human remains along highways and executions in broad daylight in public places or malls (Reguillo 2021). It is in these ways that we witness the instrumentalization of urban space for the mediatic effects of power, seeing how effective it is for the transmission of messages directed at a rival (government and/or criminal group) or, for influencing the collective imaginary with a general sense of fear. A comparable case is when messages began circulating through social networks in the later part of 2011 in the state of Morelos. The messages claimed that criminal groups were recommending that the population at large refrain from circulating in the city because scores were going to be settled between rival organized crime groups. The effect of this kind of message is magnified when state governments or the different public security corporations neither deny nor react to this type of threat, leaving citizens convinced of their own defenselessness and wondering who actually has control of the territory.

Another example of the complexity of territorial relations took place in the favelas of Rio de Janeiro, where marginalized settlements that were dominated by local drug traffickers in the 1990s were occupied by elite military troops known as the Battalions of Special Police Operations (BOPE). The troops carried out operations using military strategies because the sites are considered high risk. After several months of occupation, community police to restore the rule of law replaced the battalion. However, after a few months, elements from these same groups of police––both in office and others expelled from the force for their links to criminals—created their own translegal groups known as militias, which were in charge of selling security to residents in exchange for a fixed rent. These last groups have been the most damaging within the scheme of security because they are very familiar with both the institutional machinery of the police force and the groups of organized crime, and so they move in a translegal maneuvering that combines the formal, the informal and the illegal, thereby taking criminal violence to another level of complexity (Valenzuela and Monroy 2014).

## 29.3 Spaces of Transgression

The study of a city has notable references in the field of territoriality that include the works of Whyte (1956), Sommer (1969) and Lynch (1960). Lynch proposed the reading of urban spaces through the classification of perceptual elements that identify an order underlying the urban construct. While critics of his model argue that cultural, emotional, evaluative and experiential elements were left out in his approach, in this paper we recover some of these approaches but place them in the field of cognitive and perceptual processes of crime, exemplifying them with case studies and validating the critical apparatus through new interpretation codes of the city.

Based on this approach, we have identified five fundamental types that are key to reading the urban fabric, and that we believe will allow new paradigms of understanding of the configurations of a violent city and spaces of transgression, defined as borders, territories, routes, plazas and hubs. It must be mentioned that the five elements used to visualize spaces of transgression in the city are not exclusive to a specific place or time; on the contrary, they merge to form hybrids of the different elements. However, by identifying them separately, each component can be analyzed to throw light on particular aspects of their spatial configuration in a Mexican context and how they interconnect in a global criminal economy. For this purpose, we describe each element in its relation to the city as a whole so that we can then analyze their interactions as a whole, to identify configurations and structures as follows.

## *29.3.1 Borders*

These are territorial divisions or boundaries marked by physical, perceptual and symbolic discontinuities. In the context of violence, they warn that a material or imaginary line dividing two or more territories, dominated and disputed by different groups, has been crossed. They can be made up of walls, borders, railroad tracks, highways or avenues that serve as divisions, unions or sutures. They form a tacit barrier or allow continuity to be maintained under new behavioral conditions. In violent cities, borders are places of conflict since their definition is imposed by force, chance or negotiation.

A border is the conventional space that divides or marks the limits of a territory. It originated as part of an international convention to delimit the boundaries between one country and another, thus establishing sovereignty. However, in the current analysis, we consider that this division is instrumental in establishing the dominance of a certain criminal group over a certain territory, even when there are also demarcations such as maritime, river, lakes and air borders. Evidently, the establishment of a border is linked to the concept of routes, since it intercepts communication channels that are vital for the functioning of illicit markets. These borders are created by geographical accidents such as mountains, valleys, rivers and oceans, but also through physical infrastructures such as the border between the United States and Mexico where, in addition to a perimeter wall between the two countries, there is a series of surveillance devices both at crossings and along the border itself.

Although the concept of a border can be approached from different angles—cultural, imaginary, geographical, etc.—in this analysis we view it from the perspective of a territorial border linked to the economy of crime. This way, we can identify the constitutive elements that make a border a structural element for the operation of organized crime. The idea of a border serves both to unite members of a demarcation and to differentiate and separate them from what is beyond this limitation. The strip around the border is where exchanges are recorded—violent ones in this case- —between members of criminal groups and another or others, so that transgressed territories can be the cause of reprisals, revenge and compensation.

Within the criminogenic context, borders are elements that change constantly, especially when criminal groups experience hierarchical reconfigurations between them, or even within their own organizations. Likewise, they can become more or less porous over time, as is the case of binational borders, which, despite the existence of sophisticated control systems, do not prevent the subsistence of a billion-dollar market, as is the case of Mexico and the United States. It is necessary to traverse maritime, fluvial, or aerial borders to enter the US, and yet drug trafficking industries cross them constantly and continuously, thanks to a structure of corruption that traverses different entry controls. In the dispute between criminal groups for territorial control, there is not necessarily a formally defined record or geographical map, and yet territories are defended head-on against threats of occupation or transgression by antagonistic groups (including state forces). In this sense, there may be areas of influence where a cartel or criminal group maintains geographic control over a region and negotiates authorization for another criminal group to move illicit merchandise through its domains.

From the perspective of human geography, borders are defined either as absolute spaces, or as socially constructed spaces that are fluid, ever changing and complex. The border stands as a barrier that divides the territory and becomes a contact zone, generally conflictive, that can find a dynamic balance at different times, depending on arrangements between different criminal groups or even with the authorities in question at different levels. "Drug trafficking maps" are published periodically in different media, informing on the demarcations of influence of criminal groups, which may or may not coincide with state geopolitical demarcations. One can even speak of an extended border where there is hegemony over entire regions of the country, which sometimes cover drug trafficking routes, as in the case of criminal corridors in the Gulf of Mexico or between Acapulco and Mexico City.

Around the world, borders have become a porous concept as mergers between multinational criminal groups (for example, the Neapolitan *Ndrangheta* and the Mexican Zetas) are integrated into the criminal economy, with borders extending across seas and continents, creating multinational criminal regions. Traditionally, territories are considered to be "open to expansion and conquest" (Zusman 2006: 179), so they must be protected and monitored. Depending on whether it is a question of recognized territories or spaces of transgression, there will be checkpoints, access points, coast guards, aerial radars, customs, booths or checkpoints. However, in all cases, access and circulation control is carried out with authorizations, safe-conducts or provisional permits. Territories maintain these borders as functional spaces for larger-scale processes, and yet they send coded messages to outsiders about the impenetrability of a neighborhood (in the case of gated communities), the dominance of a criminal group, or the prevalence of enclosed spaces where the state still maintains relative territorial control.

## *29.3.2 Territories*

Territories are sectors of the city that preserve a distinctive character for their inhabitants over other demarcations. These polygons can be identified by both the crime incidence rate and by the perception of insecurity, which divide them into high, moderate, or low risk areas, and sometimes depends on the intensity of surveillance to which they are subject to. In some cases, territories can be neighborhoods or colonies associated with social stigmas or they can be characterized by the following features: particular architectural typologies, perimeter walls that confine areas, uniform grids, housing size typology, or other physical aspects such as topography, location, or widespread abandonment of the area.

The *Real Academia Española* defines territory as "[...] the geographical space limited by borders where a population is established and provides the necessary physical conditions for the existence of the State, which holds sovereignty." This includes land, air, maritime and inland water space, although this geographical definition does not include the human, symbolic and identity dimensions associated with the territory, which allow individuals or groups to project their conceptions of the world (Giménez 1999). In the crime economy context, territory has a precise and instrumental function in market dominance, as it allows the control of merchandise production, processing and distribution, so that criminal groups value and capitalize on the territory to claim it as their own by establishing their own borders.

Territories make it possible to articulate the operations of criminal organizations, which create production circuits, operating networks and special configurations that make the crime industry more efficient. Territories can vary in scale, circling streets, passing through neighborhoods and colonies, and reaching sectors, cities and regions that do not necessarily coincide with traditional spatial classifications. Instead, in their territorial definition, one scale can encompass others or overlap, so that a cartel can dominate a space in a particular way, while at the same time forming part of a federation or association of criminal groups that maintain a collaboration agreement within certain parameters with those who carry out activities for them.

The occupation of territories of transgression involves a certain degree of legitimacy, which is acquired by being a native of the place or by making symbolic gestures toward the population that may include reducing daily crimes against people and their resources, distributing aid in cases of disasters or contingencies, and even the construction of sports, educational or health facilities in the community. All this as part of symbolic and material exchanges with the population, where hegemony and control of the territory is transmitted either openly or in a veiled manner through expressions of violence directed at anyone who does not respect the hierarchy that is established in the place.

Territory became a defining resource in Mexico during the presidency of Felipe Calderón Hinojosa (2006–2012) who began a bloody war on drugs and whose key objectives included recovering territories. He stated that it was time that the government "[...] take back the control that the State should have never lost and should never lose over its territory." Halfway through his six-year term, nonetheless, he emphatically affirmed: "We cannot lose territories; there are states in which authority has been violated. The military and police operations that we have launched evidently do not make criminal activity disappear, but they do allow the State to strengthen, recover or fully assume the rule of the State over its own territory" (Zepeda Patterson 2009).

Given that criminal economies revolve around illicit markets, the valorization of territory plays a central role in the drug trafficking industry. Locations have strategic value in their interconnection capacities with other territories, as production spaces and as facilities with transformation and distribution potential, and so they represent a vital element for the operation of criminal activities. Furthermore, territory allows the real estate market to be used for transactions that involve money laundering, which is instrumental for reinvestment within the same criminal activities, thus allowing for the expansion of markets and growth as a multinational industry by expanding its territories.

Territories of transgression are those spaces in dispute, which corporations, institutions and individuals can claim as their own through expropriation, leasing and sales, and where criminal groups can use raiding (unauthorized use of a property), occupation (permanence in the transgressed territory) or appropriation (legitimation of territorial rights) to claim their rights. These organizations can break into, occupy and appropriate areas considered valuable for their own purposes. Faced with these territorial occupations orchestrated by criminal organizations or created as a hybrid between these groups and rulers, politicians, police and military, there are always expressions of resistance against the transgression of spaces, such as those of self-defense groups, community guards and even paramilitary groups, who call into question the dominion not only of criminal groups but of the State itself.

### 29.3.3 Routes

Routes are road or pedestrian connectors that communicate, articulate and organize urban spaces and emerge as highly vulnerable places for the commission of criminal acts, drug transferring and patrolling of territories. These connectors make up a road network that articulates the territory by interlacing the different areas of the city through neighborhoods of variable levels of danger and belonging to various demarcations, and where clear visibility and direction are important factors for the perception of safety. People observe the city as they move through it while organizing and connecting other environmental elements. A route includes roads, highways or roads that allow travel from one point to another, even when they have a connotation that indicates the direction taken for said purpose. In the organized crime context, roads are linked to transfer routes of illicit merchandise, which can be by land, sea or air.

Routes can also be represented as linear patterns of movement that are followed by commuters. Routes can follow roads or trails or go across territories where there are no established trails, such as open fields. According to Montello (2005), routes denote behavioral patterns while roads denote physical and linear entities. The former have an origin and destination, and can be articulated in a transportation network, intermodal if need be, with the transported product being integrated into established routes, or alternatively, using different roads so that the product transfer goes unnoticed. Routes are essential for the distribution and transfer of illicit shipments, but they are also instrumental in articulating the development of criminal activities. On the one hand, they are the means by which members of criminal organizations move to carry out orders and coercive actions throughout the territories, while on the other, they represent the possibility of having mobile and articulated points of sale along arteries, highways or at intersections, thus making it possible to boost the market by streamlining capital flows.

Routes use communication channels but differ from them because they have a defined meaning and direction. Another dimension can be found in collective underground transportation routes (metro), where people are transported without making eye contact with others or noticing the surrounding environment, and so they offer no interference, making it an ideal system for drug dealing activities, both inside the train as well as metro station entrances. When we consider the mobilization of illicit merchandise through transfer routes, we can refer to the geopolitics of capitalism, where continuity in the circulation of capital is very important so that an accumulation does not lead to a devaluation of the product. This means drug markets must seek expansion in order to reduce the turnover time of capital. For this purpose, routes are supported by road, rail and airport infrastructures that reduce transportation costs and allow the expansion of markets. Sometimes approaching the operation scale of large multinational corporations, drug cartels have managed to expand across the Atlantic to reach European, African and Asian countries via sea and air routes. This exponential growth has been possible with financial mechanisms and instruments that have leveraged the production of illicit substances.

Transportation infrastructures have the function of allowing and encouraging both the production and consumption of narcotic drugs, whether for domestic use or for export, in addition to maintaining reduced costs that benefit production and the market in general for extended periods of time. In this way, processing areas, such as laboratories or packinghouses, are configured as transfer spaces before entering collection points and distribution routes in the national and transnational territory, forming a complex network throughout the territory.

When we understand the economic functions that are linked to specific places, we can predict the intensity that will be used to guard them, as well as the type of defensive tactics that will be used to protect them. On the basis of this economic geography we can analyze the municipalities with the highest crime rates and map transfer routes and important collection, transfer or distribution points. In this context, the areas that stand out the most are those that connect drug production sites with illicit trade routes and other transfer points such as airports, runways, seaports or border access points. As concerns air terminals, the veiled role that facilities as important as international airports come to play is remarkable. Mexico City has been the scene of lethal confrontations, including with federal police agents who protect the facilities but who are also linked to organized crime. Seaports also have important connections through routes with drug markets (see Pérez 2014), so it is not surprising that coastal cities such as Acapulco, Cancún, Veracruz or Los Cabos head the list of cities with violent criminal activity. Likewise, the strategic location of border cities makes them particularly susceptible to prevailing violence, as is the case of Ciudad Juárez, Tijuana and Matamoros.

## 29.3.4 Plazas

Drug trafficking plazas are the most valued and defended points in territories and lead to frequent bloody confrontations to control them. The control of a plaza requires initial, continuous and systematic surveillance of a territory to identify possible competitors or threats that affect the execution of organized crime operations. Plazas may be found in the heart of a territory or the end of a route or avenue. In the context of a city of crime, this typology corresponds to the neuralgic points through which organized crime is articulated and structured in space. This element can be encompassed at varying scales so that a specific space, such as an urban park, will be found within a municipality, within a metropolitan district, within a federal entity and even as part of an international network. Plazas are nodes of confluence in the territory where activities are intensified, incorporating intersections, public transport terminals, or intermodal transferring. They can also materialize as public squares or parks, as corners, interstitial spaces and even alleys.

The transgression of the territory by criminal organizations is exemplified by what is known as the *Culiacanazo* of October 2019, an operation to arrest Ovidio Guzmán López, son of El Chapo in Culiacán. The army was involved in a combat against members of the Sinaloa Cartel, who took over the city with high-powered weapons to force the capo's release. This also resulted in 8 persons dead, 16 wounded and 49 inmates who escaped from the Aguaruto prison. This operation evidenced the tactical and logistical power of this criminal group, which not only achieved its immediate objective, but also completely paralyzed the city and made clear its ability to respond and to control the territory.

Plazas can be categorized based on models of the regional economy; a relevant spatial unit can be divided into several spatial units with the same type of relevance, which can then be grouped into one or several categories of lower hierarchy. Thus, a plaza is a nuclear spatial unit with a higher hierarchy that groups together other articulated nuclei (hinterland) and that generally forms part of a larger-scale production and distribution network. In this way, certain plazas will hold predominance over the different categories such as production, processing, distribution, land and air transfer, money laundering, creating asymmetric spatial dependencies within the functions of each bastion. The operation of a plaza is characterized by concentrating activities and being connected to transportation infrastructures, as well as by its interconnection with other networks of economic interaction within the criminal industry, which allows the movement of people, merchandise and products through these networks. Furthermore, the market for illicit products functions as an integrated economic system that is defined by the effectiveness of its interconnections with other points of articulation that are beyond its borders, corresponding to its decision-making capacities, displacements, as well as with the distribution of goods and services.

There are transaction costs in the establishment of business relationships, which can be minimized by making interactions routine. However, the first inspections, negotiations and hiring also imply an additional cost. Likewise, the distances implicit in the transactions involve a great expense, since increasing the distance increases the time, thus increasing the value of the product. When there is a sustained demand, the price increase of the product can be covered without jeopardizing the production process, as is the case with the production of cocaine, which costs $700 USD per kilo to produce in Colombia, goes up to $30,000 USD wholesale when sold in New York City, while in the retail market it can be sold for up to $120,000 USD (Aguilar 2013).

In addition, the different territorial components associated with the plazas can be defined based on their functional specialization, such as the production of raw material (opium, coca, marijuana); processing, which can be carried out in clandestine laboratories located in various countries; and distribution, which involves different modalities such as proximity to borders, seaports or landing strips. Plazas have different features that correspond to the production phase of a drug in particular, being that certain territories will have better conditions for the agricultural production of raw materials, and better locations with respect to potential markets and complicity with officials and authorities, which will allow them to ensure an economy of scale with the possibility of expansion.

A plaza comprises a specific territorial area where a criminal group maintains hegemony and monopoly of certain activities within the criminal economy (production, processing or marketing) in close collaboration with governments, police or military with whom they share a percentage of the profits for this activity. However, this demarcation is not permanent; it serves as a platform for the transfer of merchandise and also allows for the collection of a *derecho de piso*, a fee demanded from both legal merchants and from other criminal groups for the use of shops and commercial lots.

## 29.3.5 Hubs

Hubs are strategic and multifunctional sites where activities, means of transportation and communication routes converge. They can also form around public spaces used for commercial purposes or that have a symbolic meaning, around public facilities, or even as temporary installations. Hubs mark points in the territory that citizens identify by their physical features, such as an outstanding element in the landscape, or historical sites. There are many opportunities for moving drugs at these places of confluence, as well as for street theft, as there are often crowds of people circulating continuously.

Hubs condense activities and are strategic points of confluence, intersection or concentration of people and uses and can centralize the social life of a neighborhood. They are nuclear in nature and form as points of concurrence along the different routes or paths, in addition to articulating with other hubs to form a structure, network or system. The concentration of functions is the mechanism that originates cities, based on the dynamics deriving from the confluence of activities and the volume of production that attracts different production elements. The site itself can be a point of accumulation of fixed social capital (infrastructure, transport and processing means), and it can also generate a particular efficiency when these activities take root in the population.

Given the nature of the drug market, a concentration of activities results primarily from the need to concentrate production, processing and, to a certain extent, distribution in certain points of the territory. While the areas of the market may overlap depending on the different products, criminal groups may specialize in a particular activity (extortion, kidnapping, drug trafficking, protection, human trafficking, etc.), or combine several of them, allowing them to diversify risk in the event that actions of containment or eradication of a certain product or activity are initiated. Popular hubs are shopping centers or places offering multifunctional services because they have parking lots and ATMs, increasing the risk of incidences and where the time of day is an important factor in terms of prevention. However, they can also be apartments or office towers, facilities of metropolitan importance, supply centers, truck terminals or self-services identifiable not so much by their uniqueness but by their specialization.

A paradigmatic case took place in the state prison of Topo Chico in Nuevo León, which became the center of operations for the Zetas Cartel. It served as a refuge from persecution by drug traffickers, a source of income through the extortion of inmates and forced labor in workshops, and a reserve army when required by organized crime (Aguayo and Dayán 2018). This prison combined all the spatial elements of crime: it was a hub for drug traffickers, it was immersed in one of the largest plazas in the city of Monterrey, it was directly connected to the road that leads to the borders with McAllen and Nuevo Laredo, and it controlled most of the surrounding territories.

With other sources of crimes, such as the theft of auto parts, there are sites with a greater incidence of this type of activity that articulate at collection, distribution and retail sale points. These sites eventually become integrated into the formal and informal economy within an institutional framework that makes it possible to inhibit or stimulate the illegal market. The case of a center for the sale of stolen auto parts in the district of Iztapalapa in Mexico City is also paradigmatic. Because of its size, it is known as "the Ford" and it offers a wide range of stock and even offers order placement services.

The importance of hubs is evident in the study of spatiality because of their impact on communities. There have been notable urban interventions, such as the Library-Parks in the city of Medellín, the *Favela-Bairro* improvement program in Rio de Janeiro, and the strategy for the consolidation and recovery of public space in Bogotá. At the same time, important cultural initiatives have emerged, such as the Arts and Crafts Factories in Mexico City, the *Central Única de Favelas*, or the Afro Reggae Cultural Group in Rio de Janeiro, which have managed to involve the inhabitants through cultural centers with a such a high rate of success that even local organized crime factions have threatened them precisely because of their success in keeping the young population from popular settlements away from criminal activities.

## 29.4 Conclusions: Characterizing the Spatial Structure

An ultra-contemporary war field is continuous, flat, simultaneous, ubiquitous, systemic and productive, and affects sea, air, land, space and cyberspace. (González 2014: 11).

The significance of the field concept in the theater of war operations is illustrative in the context of conflicts produced within the drug trafficking industry. The elements that make up the spatial structure of crime acquire various meanings according to their functional interrelationships, and so the articulation of these units allows for the activation of a criminal economy. This is how routes connect with plazas, where the concentration of activities produces merchandise, later crossing territories and borders. Even while the five elements are empirical categories that help to concentrate information for a better understanding of the dynamics of spaces of transgression, while also highlighting parallels with different cities, there is a common structure that is revealed beyond their specific characterizations.

Organized crime has a spatial dimension that is essential to understanding the nature of the criminal economy. These spaces are as diverse as they are complex, but there is an interdependence between them and their spatial configuration that is key to understanding the geography of crime. Territorial control has been instrumental for the domination of specific markets, as well as for the operation of other criminal activities such as extortion, kidnapping, dispossession, human trafficking and the recruitment of new elements for criminal organizations. Territorial hegemony allows organized crime to consolidate its presence in territories and metropolitan areas, securing a criminal fence all around to turn them into strategic resources that are instrumental for eventual expansion within the regional illicit market. That is why territorial control becomes such an effective expression of power, allowing organizations to establish a collective imaginary in which the vulnerability of the population prevails in the context of crime marked by impunity. Paradigmatic examples in Mexico are highway blockades by drug cartels, which paralyzed 31 avenues in Monterrey in 2010, Jalisco in 2012, Michoacán in 2018, Tabasco in 2019 and most recently, in the previously mentioned case when the Sinaloa Cartel almost militarily occupied the city of Culiacán to release one of its leaders.

The structure of a territory is complexly organized (Salingaros 2005), with its elements interacting at many levels and with countless connections, and concentrations of activities interconnecting to form a network. In an extended metropolitan area, drug production can be concentrated at a certain point and connected to other points of concentration by means of routes that cross territories within a plaza controlled by a criminal organization, or by means of safe-conduct issued by associate cartels to cross state, regional and international borders.

The geography of crime integrates a complexity of formal/informal/illegal dimensions (Valenzuela and Monroy 2014). A legally recognized space integrated into the formal economy is also the locus of illicit or translegal processes with an organized and articulated complexity (Weaver 1948). Hubs may be specialized in one or more forms of criminal activity (production, processing, distribution, recruitment, extortion, kidnapping, etc.) and may be connected to other hubs that carry out complementary activities, forming a trajectory that can become a transfer route.

The diversification of criminal activities among criminal groups is becoming increasingly widespread and may acquire greater preponderance depending on market conditions and links with the authorities and/or law enforcement agencies. Criminal networks maintain overlapping connections that are projected in territories, establishing different circuit routes depending on the type of narcotic, criminal activity or scale of the illicit market in question. Hierarchies among criminal organizations require negotiations between rival groups so they can become articulated in a global and transterritorial market.

It is important to underline that the market of crime is articulated by land, air and sea, creating routes through different channels and generating profits that will often be invested in the financial or real estate market, thus closing the cycle of the geography of crime. Cities will also often react to the insecurity and violence with changes in daily routines, with the fortification of the environment, with residential isolation or the permanent policing of urban spaces, in an ambiguous context where translegality provides the de facto form of resolution when conflicts arise, leaving inhabitants unprotected and contributing to the dissolution of the city.

## References

Aguayo S, Dayán J (2018) El Yugo Zeta: Norte de Coahuila, 2010–2011. El Colegio de México
Aguilar R (5 Nov 2013) La cocaína: producción? y precio, El Economista, from https://www.eleconomista.com.mx/opinion/La-cocaina-produccion-y-precio-20131105-0013.html
Altman I (1975) The environment and social behavior. Brookes/Cole
Ardrey R (1966) The territorial Imperative. Atheneum
Bartlett FC (1932) Remembering: a study in experimental and social psychology. Cambridge University Press, Cambridge
Becker FD, Mayo C (1971) Delineating personal space and territoriality. Environ Behav (3):375–381
Bourdieu P (2004) Los tres estados del capital. In: Bourdieu P (ed) Campo del poder y reproducción social. Elementos para un análisis de la dinámica de clases. Ferreira, pp 195–200
Brantingham PJ, Brantingham PJ (1981) Environmental criminology. Sage
Eco U (1989) La estructura ausente. Introducción a la Semiótica. Lumen
Edney JJ (1972) Property, possession and permanence: a field study in human territoriality. J Appl Soc Psychol 3:275–282
Edney JJ (1976) Human territories: comment on functional properties. Environ Behav 8:283–295
Foucault M (1977) Microfísica del poder. La Piqueta
Gilling D (1997) Crime prevention. UCL Press
Giménez G (1999) Territorio, cultura e identidades. La región socio-cultural, Estudios sobre las Culturas Contemporáneas 25 Época II 5(9):25–57
González Rodríguez S (2014) Campo de Guerra. Anagrama
Haesbaert R (2004) O mito da Desterritorializaçao. Do "fim dos territórios" à multiterritorialidade. Bertrand
Herbert DT, Hyde SW (1985) Environmental criminology: testing some area hypotheses. Trans Inst Brit Geogr 10:259–274

Lopes de Souza M (1995) O territorio: sobre espaço e poder, autonomia e desenvolvimento. In: De Castro I, Da Costa Gómez P, Lobato Correa R (eds) Geografia: conceitos e temas. Bertrand, pp 77–116

Lynch K (1960) The image of the city. MIT Press

Mawby RI (1977) Defensible space: a theoretical and empirical appraisal. Urban Stud 14:169–179

Mercer GW, Benjamin ML (1980) Spatial behavior of university undergraduates in doublé-occupancy residence romos: an inventory of effects. J Appl Soc Psychol 10:32–44

Montello DR (2005) Navigation. In: Shah P, Miyake A (eds) Handbook of visuospatial thinking. Cambridge University Press, Cambridge

Newman O (1972) Defensible space: crime prevention through environmental design. Macmillan

Pastalan LA (1970) Privacy as an expression of human territoriality. In: Pastalan LA, Carson DH (eds) Spatial behavior for older people. University of Michigan Press

Pérez AL (2014) Mares de Cocaína. Las rutas náuticas del narcotráfico. Grijalbo

Reguillo R (2021) Necromáquina. Cuando morir no es suficiente. ITESO/Ned Ediciones

Salingaros NA (2005) Principles of urban structure (Design science planning). Techne Press

Smith SJ (1987) Design against crime? Beyond the rhetoric of residential crime prevention. Prop Manag 5(2):146–150

Skogan W (1988) Disorder, crime and community decline. In: Hope T, Shaw M (eds) Communities and crime reduction. HMSO, pp 48–61

Sommer R, Becker FD (1969) Territorial defense and the good neighbor. J Personality Soc Psychol 11:85–92

Sommer R (1969) Personal space. Prentice Hall

Sundstrom E, Altman I (1974) Field study of dominance and territorial behavior. J Personality Soc Psychol 30:115–125

Sypion-Dutkowska N, Leitner M (2017) Land use influencing the spatial distribution of urban crime: a case study of Szczecin, Poland. Int J Geo-Information 6(74):1–23

Valenzuela Aguilera A (2016) El Estado paralegal: México dentro de la estrategia de seguridad hemisférica. J Lat Am Geogr 15(3):5–22

Valenzuela Aguilera A, Monroy Ortiz R (2014) Formal/Informal/Ilegal: Los Tres Circuitos de la Economía Espacial en América Latina. J Lat Am Geogr 13(1):117–135

Weaver W (1948) Science and complexity. Am Sci 36:536–544

Weisburd D et al (2016) Place matters: criminology for the twenty-first century. Cambridge University Press, Cambridge

Weisburd D, Groff ER, Yang SM (2012) The criminology of place: street segments and our understanding of the crime problem. Oxford University Press, Oxford

Whyte WF (1956) The organization man. Doubleday

Wortley R, Townsley M (eds) (2008) Environmental criminology and crime analysis, Crime Science Series. Routledge

Zepeda Patterson J (27 Feb 2009) Entrevista a Felipe Calderón. El Universal

Zusman P (2006) Geografía histórica y frontera. In: Hiernaux D, Lindón A (eds) Tratado de Geografía Humana. Anthropos-UAM

**Alfonso Valenzuela-Aguilera** Professor of Urban Planning at Autonomous University of the State of Morelos, Mexico. A Fulbright and Guggenheim scholar, his work intersects with urban and regional planning policies and practice. He has held visiting professorships and chairs in several universities in Europe and North America, including the John Bousfield Distinguished Visitor in Planning by the University of Toronto in 2020. He recently authored The Financialization of Latin American Real Estate Markets: New Frontiers (Routledge, 2022).

# Chapter 30
# Urbicide, Violence, and Destruction Against Cities by Criminal Organizations

**Arturo Alvarado**

**Abstract** The article proposes a critical examination of the concept of *Urbicide* and its applications to the phenomena of urban violence produced or associated with the actions of criminal organizations, which include drug trafficking organizations and other criminal enterprises engaged in many illicit activities, such as smuggling, piracy, money laundering; human, arms, corruption, extortion, homicide, traffic of goods and properties in cities, among other transgressions. In all these activities, they capture and destroy cities or parts of them to obtain illicit profits because most of their activities take place within them, and on several occasions, the object of their control is the city itself. The concept of urbicide has been widely used and modified in such a way that many attributes have been added to it and are applied to numerous examples where its meaning has become unprecise, ambiguous. The author revises the literature on the topic and makes an effort to specify, classify, and measure what its call urban violence. It is a conceptual exercise of clarification, delimitation, and measurement that shows its usefulness in the analysis of the multiple different and sometimes fragmented expressions of urban violence, as well as its limits. His proposal refers to activities they capture and destroy cities or parts of them to obtain illicit profits because most of their activities take place within them, and on several occasions, the object of their control is the city itself. In particular, the way in which criminal organizations affect urban life in general; they occupy, control, destroy, and reconfigure certain urban spaces in order to exploit their inhabitants, its resources and take advantage of the urban environment for their illicit businesses.

**Keywords** Violence · Urban space · Organized crime · Catastrophe · Cities · Urbicide

---

I appreciate the assistance of Alejandro Ugarte, Oscar Ávila, Candy Hurtado & Karen Franco in the partial compilation of the references and data. I also thank the anonymous reviewers

A. Alvarado (✉)
El Colegio de México (Colmex), Mexico City, Mexico
e-mail: alvarado@colmex.mx

## 30.1 Introduction

The purpose of this work is to make a critical examination of the concept of *Urbicide* and its applications to the phenomena of urban violence produced or associated with the actions of criminal organizations, which some authors call cartels, and which include drug trafficking organizations and other criminal enterprises engaged in many illicit activities, such as smuggling, piracy, money laundering; human, arms, and drug trafficking, corruption, extortion, homicide, traffic of goods and properties, among other transgressions. In all these activities, they capture and destroy cities or parts of them to obtain illicit profits because most of their activities take place within them, and on several occasions, the object of their control is the city itself.

Even though the topic is not new in urban literature, we propose to show one side of the phenomenon associated with different and sometimes fragmented expressions of violence against cities and their inhabitants, as well as to present an essay that specifies, classifies, and measures what we call urban violence. I refer to the way in which criminal organizations affect urban life in general, they occupy, control, destroy, and reconfigure certain urban spaces in order to exploit their inhabitants, its resources and take advantage of the urban environment for their illicit businesses. This leads us to present the violence problems in the city from a different angle and also to propose some correspondences with other forms of urbicide, such as conflicts between criminal organizations, confrontations with the coercive forces of the states, wars, destruction plotted by terrorist organizations. All of them exert destructive behaviors on cities. Criminal organizations often use and occupy urban territories to acquire and control illicit markets, to distribute merchandise, or to control trade routes. Among their strategies for acquiring market and political power, they not only seek to gain (monopolistic) control over the distribution and consumption of illicit goods and services (sometimes also over urban production), but also over the populations where they are based, as well as over governments and urban public infrastructure (occasionally also private). This includes markets, shopping malls, the construction industry, housing, urban entertainment, as well as customs, airports, and jails. Also, the goal to control and subjugate public personnel, like police, administrative officers and public authorities. Their products can be illicit drugs; they also engage in other predatory activities such as extortion, kidnapping, human trafficking, arms trafficking, smuggling, piracy, and money laundering; they also often capture and regulate markets for the distribution of goods and services, such as the distribution of water, cable television or local utilities, tourism, commerce, restaurants and bars, the entertainment industry, or intervene in the real estate market and even alter urban policy. Their actions produce a combination of dread, fear, insecurity that also invades and configures public and private spaces, as well as creates spaces of apparent security. The magnitude of their destruction is associated with economic power, firepower, with the existence of competitors, with government surveillance and strategies to attack them, contain them, let them operate or collude with them.

Conflicts between criminal cartels can be a mechanism that generates destructive practices with bloody manifestations in cities.[1] The control of territories usually combines coercive and clientelistic strategies, but all of them destroy not only the physical aspects of the built environment, but also the trust, cohesion, urban social networks, and in general damage the urban fabric and networks. Their territorial (spatial-urban) strategy can be varied; sometimes they are concentrated in certain informal markets (for example, informal markets such as Tepito, or merchandise distribution facilities such as the central supply center in Mexico City); in other cases, in neighborhoods under the control of their trafficker's groups (such as the PCC in various locations in Sao Paulo, or various organizations in Rio de Janeiro, in Medellin, Colombia, or in border and gulf coast cities in Mexico). In contrast, in other occasions, they come to occupy and destroy entire cities, as in Allende, Coahuila or as happened in the last decade with the predation-systematic extortion of cities in the gulf of Mexico, in Tampico, Tamaulipas, where practically all businesses and services in the urban space were attacked-predated and many destroyed, or the owners were forced to close or abandon their businesses (Alvarado 2022; Zavaleta et al. 2018b). Also, the conflict with governments—the state—can generate occupation strategies and devastation of sectors or entire cities (to build their security areas, as in Culiacán, Sinaloa).

Reactions from affected populations to aggressions from both illegal actors and the government tend to configure spaces of protection, struggle, dissent, or consent of those orders. They don't react homogeneously as occurs in different popular neighborhoods of many Latin American cities, but also in middle and upper class neighborhoods, where they "enclose" themselves in their neighborhoods and condominiums, producing spatial configurations such as those narrated by Caldeira (2000). At certain times, neighbors and communities have created vigilant organizations.

---

[1] Examples of this can be seen in the different attacks committed by members of criminal organizations on public and private offices, such as small businesses (*OXXO* convenience stores in Mexico, banks, gas stations, stores, as well as police offices or posts and blockades on streets, highways, and urban circuits. These are sometimes committed as a reaction to police and military operations to arrest criminals, sometimes to block their access, others to prevent them from carrying out an operation and arresting a gang leader. A case that was widely publicized was the attempted apprehension of Ovidio Guzmán López, son of Chapo and head of the Sinaloa criminal organization, on October 17, 2019. The military attempt triggered a reaction that showed how the city of Culiacán was controlled by the Sinaloa organization, with blockades, attacks against police and military posts, which forced the authorities to withdraw (see among other references, newscast Televisa: https://www.youtube.com/watch?v=bnzdL0drR7U). They are also planned to instill fear in the population, provoke chaos and reactions against the government. Some include destruction of public infrastructure. Among the examples that illustrate these phenomena are the attacks organized by the PCC against the authorities and with the purpose of controlling important portions of the city, in the years 2006 and 2012. Other examples can be found in Rio de Janeiro, as well as in Guatemala, El Salvador, in Central America. Sometimes an authority is the target of an attack, to disarticulate the urban security policy, as represented by the attack against Omar Garcia Harfuch, Secretary of Public Security of Mexico City on Friday, June 25, 2020 (Diego Caso, El Financiero, June 26, 2020). Some acts are planned to defend or capture territories, others to attack authorities, and some others are counterattacks to public operations.

In large metropolises such as Mexico City, there are some areas where neighbors have created vigilant mechanisms; in other cases, they have enclosed entire neighborhoods where they hire private security personnel to protect themselves. Another collective phenomenon produced by insecurity has been the fencing of houses, parking lots, or businesses with iron fences to prevent access to possible crooks. This phenomenon has produced large avenues, commercial areas, houses, residential and commercial units with a spatial and esthetic configuration of fences in large areas of the metropolis. Illicit activities produced by organized criminal actors with resources destroy and alter the configuration of urban social space and produce certain spatial forms of organization, of territorial control.

All this leads us to ask ourselves: do these behaviors constitute urbicide practices? Whether against a city as a whole or in most cases as a strategy of partial occupation they ended up altering urban space.

Police and military strategies also occasionally produce territorial control and destruction of urban space even though some practices may be considered lawful. This occurs with operations against criminal organizations in cities or even against their inhabitants. Many police strategies to control populations, combined with territorial control, also have a destructive purpose in imposing another political and urban order; they may have the formal purpose of urban reconstruction with displacement, but in reality, they maintain a damage. Examples of this have occurred when authorities occupy territories to formally displace guerrilla organizations such as was the case in Medellín (Operation Orión in 2002 in Comuna 13 of Medellín, Colombia; or operation Todos Somos Juárez in Mexico in 2010–12, or in small towns in Michoacán, which are in continuous dispute between community self-defense organizations and other criminal gangs) but whose effective actions damage the ability of entire populations to free themselves and build a free and shared urban space, or change the actor that had illicit political-coercive control of a territory for another that is also illicit, as was the case in Medellin. Another example is the militias in Rio de Janeiro (see Paes Manso 2020, which manifests the imposition of a criminal and political order in the favelas controlled by these groups) or by another autocratic power, to benefit certain urban political-economic groups. In many cases, the means and strategy of intervention becomes a destructive end, as has been the occupation of university facilities by the army and the police, producing systematic destruction of the built environment and guiding principle (*leitmotif*) of the universities (as has occurred with military occupations in the National university (UNAM) and the Polytechnic (IPN) in Mexico in July 1968 and in 1999 with the Federal Police intervention against the university strike in February 2000).

Furthermore, in some cases, urban policies that are supposedly aimed at combating these types of organizations, as well as protecting the population and the urban environment, sometimes alter, isolate, segregate, limit access, and use of the city in the interest of improvement, urban development or the protection of an exclusive order and security, which only benefit some privileged groups against the lack of protection of large urban groups. In this sense, one of the issues critically debated by experts is the use of "security design" in cities. In fact, in Latin American cities, their inhabitants have implemented a sort of "people's defensive architecture" resulting from the actions of homes, communities, and businesses.

In all the above cases, the proposed interpretation does not coincide or share all the definitions (and propositions) used in literature regarding urbicide, which is why we now present a critical discussion of the concept, its annotation and denotaation, that is, the meaning and correlates-examples. In addition, we propose certain limits. Then, we will present a way of classifying, specifying, and measuring some forms of organized criminal violence in cities, in order to try to understand how these behaviors produce general damage to the urbs. They are connected to high murder rates, femicides, disappearances, massacres, and other forms of extreme violence such as gang violence, aggressive policing, and militarization; they generate distortions in political representation; citizen participation, political representation, the *polis*, while at the same time they reinforce segregation, polarization and create new precarious spaces. It is pertinent to mention that this violence does not always destroy all spaces but alters and reconfigures those that exist.

## 30.2 Urbicide and Violence in General

It is recognized that the term "Urbicide" was coined by Michael Moorcock in 1963 and originally meant "violence against the city" (Moorcock 1963; Carrión 2014). In addition, since the 1960s, various authors wrote several ideas about the destruction and death of the city. In 1961, Jacobs denounced what she considered the self-destruction of diversity in cities as active communities, produced by the postwar American urbanization trends, the decline and the regeneration of cities.[2] In subsequent years, the concept of destruction as death continued as a way of criticizing urban development and planning.

The term urbicide has recently been used more broadly, which includes not only this critique of urbanistic trends, but also the so-called conscious, premeditated behaviors, and practices of destruction of the *urbe* as an object of attack, whether of the physical environment created, its memory or its diversity. This would include material, subjective and symbolic dimensions. But, its application has led to include much broader themes and examples of destruction, not only of the built environment, but of the very meaning of the city and its main physical-spatial (built environment), community and symbolic components, as it has come to be equated with war, genocide, or the application of certain public policies. In many cases, it is not determined or defined who the actors are, nor is it precisely defined who the victims are (*sometimes the city is the target, the victim, sometimes the instrument*).

---

[2] Despite the fact that Jacobs' idea in The Dead of the American City (1961) is linked to the impacts generated by the refunctionalization of cities (As Carrión mentioned later in 2014:43), that is, public policies and the decline of American neighborhoods and cities, we cannot conclude that all the new urban renewal trends and public policies oriented by what several authors label as neoliberalism, are part of his idea.

Most contemporary authors consider it a behavior, an explicit conduct (to a certain extent), whose purpose is to destroy the *urbe* by resorting to violence. Other times it is a systematic and continuous behavior, while in certain situations, it is a passive behavior. Destruction can also result from the inaction of urban actors, governments, or communities; other moments it results from the unintended consequences of a strategy, such as the regeneration of urban areas or the implementation of large projects that are sub-optimal for the urban community. Some authors include attempted destruction. It is also a collective action, for which it is necessary to understand its aggregate, collective effect, an example of these would be the actions of vigilante groups, which increase randomly in urban communities; there are also the enclosures of businesses, neighborhoods, territorial portions that are isolated with the purpose of making them safe, unprotecting those who are outside the fence. In all cases, it means the (non-creative) destruction of the object *city* because it is a *city*. Several authors make parallels with wars and armed, warlike confrontations, including attacks, "hybrid" combats (when it means to destroy in order not to allow the passage or the benefit of the enemies). In other examples, they include gentrification processes, and in still others, the application-sustainability of urban policies in *theories* or *ideologies* considered *neoliberal*. In this sense, the concept today has numerous imprecise, contrasting meanings, its limits (properties, dennotation) and the examples used are very diverse, so it is necessary to make an exercise of precision before using it to examine the violent and destructive behaviors of criminal organizations in and against cities.

The concept is not clear either in its connotation (meaning, significance, intent) or its denotation (correlates, examples). Examining and establishing precise limits to the properties and empirical application of the concept will avoid further broadening the meaning, reducing its ambiguity. The following exercise proposes to clarify the meaning and the different applications of the concept. That is, to clarify the significance as well as the extension with examples registered in recent literature. Through a systematic bibliographic search, we selected a set of texts published by experts on this subject; we compiled the different contents and the examples used.

## 30.3 Clarification and Conceptual Limits

Table 30.1 includes definitions and examples from almost 20 authors, some of whom have several texts related to the topic. The table is organized by themes, beginning with general definitions of physical, symbolic, and even historical space; then by urban public policies (which generated the first period of debates); followed by themes such as wars, to which are added exclusion, destruction of heterogeneity and plurality, until integrating genocide, ethnic cleansing, and, also, themes of abandonment and ecological destruction. The examples are scarce and in many cases are confused with situations of national, state, and ethnic violence.

As shown in Table (30.1), the concept has an original connotation (meaning) in at least three dimensions or properties, but over the years, a wide range of properties have been added to it, many of which are not shared by all authors; neither are there clear examples that are directly and solely associated with cities (the more repeated example is wars). Moreover, the examples proposed are very diverse, ranging from historical issues of destruction of monuments and urban equipment, up to entire cities. The core of the concept is certainly a violent process of destruction (or destruction by violent means), in some ways purposive, conscious, programmed; but, in other cases, it is the consequence of other projects, which may or may not have as their purpose the destruction of the city as such, but a transformation of it, even by non-violent means. It is not a process of creative destruction, that is, to recreate the city, but to change its meaning-content-and forms of occupation of space (in a violent way). Nor is it a destruction of the entire city, but rather of its main components. In many cases, general violence is confused with national-community, ethnic violence and is integrated with urban violence. Many violent processes and acts are neither proper nor unique to the city, nor do they occur in the city nor are they directed against it, even though they end up damaging it.

The last part of the table includes several dimensions and ideas proposed by the author in order to critically reflect not only what is urbicide, but also what is urban violence, another vague category that confuses terms, examples, situations. This is not the purpose of this paper. To indicate that it is necessary to take into consideration all the elements from the different political, environmental or criminal spheres, but to point out how these also have an impact on urban destruction. Even some examples illustrate that the categories are not neutral and do not mean the same thing for all populations and at all times.

Some authors use encompassing concepts; others build them with correlates. The application of the concept has expanded to various urban issues, such as the destructive effects beyond wars against cities, adding effects such as disasters; another on rebellions, uprisings, or even what some consider the abandonment of the city (Herrera 2007), the oblivion that some associate with the transformation of cities (particularly in the global north), a product of the decadence of urban-industrial forms, of the decadence of old cities, both in areas that were dynamic centers or in working class suburbs, as occurred in cities of the industrial belt of the North American Northeast since the end of the 1960s. For example, in many cities in the industrial corridors of cities, rivers, and riverbanks of the North American northwest, Massachusetts, New York, Michigan or Illinois in the USA (it should be mentioned that this is very different in many post-industrial and post-Fordist European cities, some of which have had creative reconversion processes), in other European cities that have not achieved their transformation and urban renewal or in some Mexican cities of the Gulf (where large industrial and agricultural enclaves emerged and are being dismantled as an industrial, labor, and social-urban order) as well as in several border cities that were born and decayed when the manufacturing industry and the oil and agricultural enclaves were transformed. This is the result of the disappearance of an industrialization-urbanization model.

Literature concurs on a set of elements that constitute the concept, the experts agree that they are aggressions and premeditated actions, with an explicit order and form, but they do not share the rest of the dimensions. One example is the idea of death of the city or death of the urban; not everybody shares the idea because it is neither death of a city (that has occurred in history because of several causes), but the destruction of their constructed social space; there is destruction of certain spaces and portions of it, by violent processes, but other forms have no purpose and urban decay in general does not have that irresistible violent-destructive tendency.

I agree with Sara Fregonese that if the concept of urbicide: *"is to be employed as a theoretical and methodological tool to investigate urban contested spatialities across multiple settings, then it needs to be contextualised with a sensitivity towards the specific ways in which the material fabric of the city played a part in shaping the multiple experiences of those who were caught up in conflict"* (Fregonese 2009:310). The category is neither unidirectional nor univocal. There is no single theoretical actor, no single and specific purpose; sometimes it is combined and confused with several. The city ends up being the site where the damage occurs, but it is not always directed against it.

Consequently, I do not agree with the broad definition and the attributes and destructive logic proposed by many authors. It is true that many actions are premeditated, but in many cases, devastation is the consequence of unplanned actions and inactions or the undesired result of non-violent changes.

In any case, I share the idea that it is a violence against the city, against its identity; sometimes against the population, the citizenship, for urban economic, cultural, or political, warlike. I would add obliteration by military, police, and criminal reasons; sometimes they alter significant places such as squares, monuments, infrastructures. But, they do not always privatize *"nor subordinate public policies and institutions to the interests of the market or central power"* as proposed by Carrión (2014:35).

There is a set of nuclear elements that constitute the concept, plus the aggregation of various attributes, and their application in historical situations leads to the question of how to avoid the concept having multiple meanings. Other examples are the consequences of disasters, earthquakes in cities (Mexico, Philippines; other cities Asia, Arab countries or the Caribbean), or accidents, when subsequent actions mean abandonment, negligence and forgetfulness or transformation to obtain benefits for a group (for example, the intervention of North American federal agencies (particularly FEMA) after Hurricane Katrina in Louisiana in 2005, with negligent actions that maintain large portion of its population in dismal conditions and led to a deep housing crisis of populations without resources and a rearrangement of the real estate market in favor of other groups two years later). Another case is the explosion in the port of Beirut, Lebanon, on August 4, 2020 (which left more than 127 dead, more than 700 thousand wounded; and the destruction of thousands of homes, businesses and thousands of millions of dollars in damages. Two years after, there is no active reconstruction even though there was an international mobilization to help). And what about the public actions after or in response to the riots of Los Angeles, California in 1965 (Watts) and 1992 (R. King Case) or Baltimore, Maryland (2008)? Is the destruction directly associated against the city? Or the protest is directecd against an oppressive order that intends to maintain an oppressive order? Furthermore, how to include other topics such as population displacement, which in Central American countries and in Mexico have had remarkable and devastating proportions in cities?

and how to solve the issues of climate change, globalization? all of them produce urbicide?

It would also be necessary to be more precise on the effects of gentrification projects and processes, because not all renovation produces (non-creative) destruction. Not all these processes are urbicides; it is necessary to establish limits and exceptions, maybe only in those that are not supported by democratic conditions, or directly attack the democratic polity and includes violent methods.

## 30.4 Urbicide and Criminal Organization Actions

Because of their own history and attributes, cities are spaces where different types of transgressive, illicit, and violent behaviors take place; some of them are produced by external factors and others are endogenous.

Drug trafficking and in general the several local and transnational criminal organizations that trade drugs affect cities; they are present in all cities of the planet, in different forms, such as bands, gangs, mafias, cartels, guerrillas, militias, criminal organizations, paramilitary, parapolice and even as corruption networks. Sometimes, the coercive forces of the state create their own illicit organizations (mafias, syndicates, private monopolies of public coercion) or associate with others, such as the militias of Rio de Janeiro (Paes Manso, 2020) and capture portions of the city to extract illicit rent. They have economic, social, political expressions and territorial presence. In some situations, they only use the city's infrastructure for their benefit. In others, they colonize, transform, and destroy regular economies and public and private spaces.

The combination of activities of these criminal organizations makes their impacts multiply among populations, neighborhoods, economic, and political sectors (for example, the sale of illicit merchandise, protection-extortion schemes, influence peddling and political pressures). On several occasions, they have been a threat or danger to the life of cities. In recent decades, particularly, in the producing, distributing, and consumption of illicit drugs (such as poppy seeds, marijuana, and cocaine in Colombia and Mexico), the forms of criminal organization for the production and commercialization of these drugs have grown disproportionately (to the magnitude of transnational corporations), along with the competition to control these territories. At the same time, they have been increasingly incurring in the cities that are means of distribution and places of consumption.

The result of these processes are disproportionately high levels of homicidal violence, as well as other expressions of armed confrontations and capture of urban territories (including other associated crimes, like corruption, extortion, kidnapping, control of informal and illicit retail business, restaurants and bars and traffic with properties). There are other very violent cities in different regions of the world, but the content of that violence appears to have different causes, even if it also combines drug trafficking and other associated activities.

Presence of criminal organizations, their particular modes of operation as well as its consequences in city life, is diverse. One way of looking at it is the violence that manifests itself urbanely (homicidal violence, extortion, kidnapping, crime, transgressions to order and sociability and to political representation and participation; occupation, destruction, or trafficking of land). The records we present show that Latin American cities have been affected by these forms of homicidal and criminal violence (Tables 30.1 and ff.).

## 30.5 Urban Violence and Urbicide, Approaches, Proximities, Limits

To continue, we present two types of expressions of violence associated with urban destruction. The first has to do with what we call urban violence. The second has to do exclusively with the practices of criminal organizations in urban spaces.

To introduce the topic of urban violence, we start by defining it from several facts and examples (denotation): interpersonal violence in cities, homicides, femicides, partial or complete destruction of built spaces-build environment, urban equipment. The destruction of the spatial configuration of cities produced by illicit and armed actors, and the (violent) destruction by collective actions that is not resolved with restorative public measures.

Let's start with the idea that urban violence is part of urbicide, especially because, as Carrión notes, *"there is a considerable increase of homicides in cities"* (Carrión 2014:37), although the trend is not general for all cities. On the contrary, in many cities of the region, homicidal violence has tended to decrease consistently (See Concha et. al. Ibid), but this does not mean that the problems and causes of this violence disappear; on the contrary, in some situations, a low homicide rate may indicate the dominance of a criminal organization, its hegemony in a city (as could be the cases of Culiacán, Sinaloa, reviewed; or that of Sao Paulo under a kind of hegemonic dominance over the world of crime by the PCC. See (Dias 2009; Feltrán 2010 and Willis 2015). Moreover, it is necessary to specify not only its magnitude but also its trends, its directionality and the ways in which it affects urban life, as we will show below. Several reports and studies on urban violence have shown us that in the region we have a large number of cities with high levels of homicidal violence (Table 30.2).

Six of the largest and most dynamic cities in the region have homicide death rates above global and regional averages. Only two were bellow regional and global average of homicides. To add to our knowledge, InSigth Crime (2020) compiled data from 17 capital cities in the region, resulting in an average of 21.3 murders per

**Table 30.1** Concept of urbicide, Connotation and Denotation

| Urbicide as a concept (theory, epistemic, ontological, and other assumptions) | | | |
|---|---|---|---|
| Operational definitions: Connotation and Denotation | *Connotation*, defined as the set of properties that characterize the cities that are object-victims of urbicide (Bunge 2000:58) | *Denotation*, the set of all objects, real or not, to which the concept can be applied (Bunge 2000:60–61). ...[...] can be an infinite set[...] or a finite set, but limited, as in the case [...]"country"[....] or, for our case, city (following Bunge 2000:61) "Denotation of is equivalent to the set of objects that satisfy the condition C(x), or that have the property C'. (idem) This could add not only the city as an object, but urban parts, spaces or processes | Comments: For our case, the city At least three dimensions (properties) and numerous correlates |
| *Authors, definitions, examples, comments* | | | |
| Carrión (2018) defines it as: Violence against the city for urban reasons | *Connotation*: Military; economic, cultural, or political actions that: "(i) destroy with identity; (2) privatize, concentrate or subordinate public policies and institutions to the interests of the market or central power; (3) raze the system of significant places of common life, such as squares and monuments." | Denotation, examples | Includes disappearance of identity Adds privatization |
| Jacobs (1961) Death of the city | Destruction of the urban community by renewal policies of the city | Cities in the United States of North America, suburbs. Effect of urban renewal | Refers to public action, to public policy and also to processes |

(continued)

**Table 30.1** (continued)

| Urbicide as a concept (theory, epistemic, ontological, and other assumptions) | | | |
|---|---|---|---|
| Carrión (2014): Murder, annihilation, premeditated aggressions of cities for urban reasons<br>Destruction | Intentional, premeditated murder for urban reasons<br>"Liturgical murder of the urbs when aggressions and actions are produced with premeditation, order and explicit form." (Carrión 2014:80). "It is the murder or violence against the city for urban reasons." (Carrión 2018:5 et seq.) | Examples: wars; public policies; urban constructions: the library of Alexandria | Adds various meanings and general behaviors beyond the cities |
| Destruction, annihilation, razing the meanings of built spaces (material urban goods) (Carrión 2018:5 et seq.) | They raze places with meaning, squares, libraries, infrastructures "… they annihilate the institutional frameworks of government: privatization, deregulation, centralization…" | | |
| Death and destruction (Moreno and Gil 2021:90–91; Jacobs 1961) | Destruction of cities, attacking the material basis of the city, which destroys space, body, habitation through two practices: war and gentrification; spatial construction manifested in architecture and its identity, cultural, and historical components | The author specifies that not all gentrification is violent because not all urban renewal is urbicide, even if it operates in the city | It equates wars to urban wars<br>Gentrification as an urbicide requires specification<br>Include all forms of war violence?<br>What about police violence? |

(continued)

**Table 30.1** (continued)

| Urbicide as a concept (theory, epistemic, ontological, and other assumptions) | | | |
|---|---|---|---|
| Moreno and Gil (2021) | Urbicide: "comprises the city as (i) victim; (ii) there is always widespread or total (?)destruction inflicted; (iii) the place of destruction is dehumanized before being destroyed, as a space of criminals, enemies, and quasi-humans; (iv) destruction is an exercise to achieve the reconfiguration and spatial control of the city, and v) destruction is always premeditated, intentional, and planned." (Moreno and Gil 2021:107) | It does not include concrete examples | It proposes that the victim is the city, either as an object or as an actor Includes premeditation. Generalized damage, not partial Dehumanization (idea linked to the crime of genocide). And the reconfiguration of space |
| Fregonese (2009): Urban geographies in dispute | Urban destruction is also a way of dispute, of confrontation | War in Beirut (1982 siege?) | Try to specify its category and avoid multiple meanings |
| *Death by urban policies or wars* | | | |
| Bernard-Moulin and Macarena (2018) | Purposive and disproportionate destruction of the urban environment by conflicts that become forms of illegitimate warfare Reduce geographic complexity Collateral damage ("Wanton Damage"); | Destruction in Guatemala due to illegitimate war, with disproportionate acts of destruction, out of standards (Arendt 1979 cited in Bernard-Moulin and Gonzalez 2018, p. 5). Bombings, explosions, villages intentionally razed to reduce geographical complexity in southern Quiché (Aquilué 2021: 14 cited in Bernard-Moulin and Gonzalez 2018, p. 12) | Includes urban conflicts of magnitude comparable to wars Includes ethnic destruction but does not clarify whether it is an attack on a community or a city. Indirectly designates the aggressor Could ethnic violence be added? |

(continued)

**Table 30.1** (continued)

Urbicide as a concept (theory, epistemic, ontological, and other assumptions)

*Urbicide as war*

| | | | |
|---|---|---|---|
| Fregonese (2009) | Urbicide as a tactic of rival militias. Hostility against universally accepted civilian way of life. Attacks by rural dwellers who consider the city corrupt and indolent (Bogdanovic cited in Fregonese 2009:310) | Beirut, Lebanon; Vukovar, Sarajevo; Mostar, Dubrovnik | It manifests conflicts and city-countryside differences. Could militia and paramilitary practices be added in L.A.? |
| Graham (2022) | Urbicide as bulldozer | Result of deep antagonisms between ultra-right military and political elites against the demographic nature and urban growth of the Palestinian population. Weapon that distorts the military, geopolitical, and demographic balance between Palestinians and Israelis (Graham 2022:645) | War between national towns and communities Again, it includes ethnic conflict could it include terrorist tactics and thus their practice in cities? City is more and instrument than the target of destruction |
| Campbell and Monk (2007) pp 15:16 | (1) War with urban impacts because it produces a particular spatiality and (2) as "Theater" and (3) propositional violence where urbanity is the object of violence | | Cities are theaters—media—of major, national, international violence. Not necessarily the target |

(continued)

**Table 30.1** (continued)

| Urbicide as a concept (theory, epistemic, ontological, and other assumptions) | | | |
|---|---|---|---|
| | War: urban environments destroyed for the purpose of war or for tactics during wars | Guernica in 1937; Balkan wars. 2ª. World War II: Tokyo, Nagasaki, Hiroshima; Berlin; Warsaw. Ukraine. Islamic State's aggressions against cities and peoples | Indirect and direct warfare. Certain sites are selected as a specific military target against a nation or state or its people |
| Abujidi (2014) | Bombing and mass destruction of cities; concept that encompasses all forms of urban destruction (Abujidi 2014:23); includes cultural "cleansing" genocide and identicide Sometimes adds the geographical environment with the urban (problematic) Classifies the literature on urbicide into three streams: (1) it is anti-urban violence (Berman, 18 7th; Simmons (2001) cited in Abujidi 2014: (2); it is a politics of exclusion anti-heterogeneity (Coward 2001b cited in Abujidi 2014) (3) (includes) a war of or against terror (Graham 2003; Gregory 2003 cited in Abujidi 2014) | Mostar, Sarajevo in Bosnia and Herzegovina; Jenin camp and Nablus in Palestine; Fallujah, Iraq; Lebanese villas Destruction against Churches (mosques), libraries, archives and symbolic buildings (Abujidi 2014:16 electronic) | Includes war and religion Eliminating heterogeneity (religious, ethnic? Terrorism Identicide Very wide concept |
| Abujidi (2014) | *Domicide*, involves "razing" the opinions and perspectives of its victims (otherness) by *annihilation*; it is the devastation of the life of spaces, ecologies, infrastructure, landscapes | There are no concrete examples | Again, annihilating spatial-urban identity, but communal? Could it include megaprojects? |

(continued)

**Table 30.1** (continued)

| | | | |
|---|---|---|---|
| Urbicide as a concept (theory, epistemic, ontological, and other assumptions) | | | |
| Abujidi (2014) (and Hanafi 2006:93) | Spacio-cide (Hanafi 2006, in Abujidi): two levels of destruction, micro- and macro-territorial; not only against cities but also landscapes and built environment. Hanafi includes 3 principles: colonization, separation, and state of exception (Hanafi 2013: 190 y ss) | Targeting the place where Palestinians live. Occupied Palestinian territories, not only destroyed by bombs, but by destruction Attacks property | Spacio-cide oppose to genocide, from Terrorism? Micro- and macro-destruction Could include ideas of colonization, segmentation, state of exception |
| Urbicide and abandonment, oblivion | | | |
| Herrera Robles (2007) | Abandonment: A product-result of decay and other factors that are subsequently associated with criminal and public violence | Ciudad Juárez; post-Fordist border cities | Urbicide would be the result of abandonment |
| *"Violent Destruction"* | | | |
| Carrión (2014) | Patrimony destruction and "oblivion." Selective and massive act of destruction of the built environment, such as patrimony | Possible examples: *constructions that destroy, abandon, and "erase" urban patrimony on purpose (downtown Ciudad Juárez; downtown Sao Paulo)*, | Does it include urban "patrimony" as a specific attribute of violence? *Include "historic centers"?* |

(continued)

**Table 30.1** (continued)

Urbicide as a concept (theory, epistemic, ontological, and other assumptions)

*Exclusion*

| | | | |
|---|---|---|---|
| Environmental destruction *(urbiecocide)* | Destruction of monuments | Islamic State and contenders: Aleppo; Tunis, Mosul *Destruction of monuments* | Add ecocide in cities & Environmental Damage *What about collective destruction of monuments of civil wars, racial, or colonial segregation?* |
| | Destruction of a city or certain spaces and portions of it, by violent processes | | |
| Coward (2008) | Urban assassination of its specific qualitative existence and its heterogeneity. It is the territorialization of antagonism. Destruction as acts against American capitalism-imperialism; Terrorism as an attempt to cities and shared spaces; against heterogeneity. They attempt against (1) the buildings as a condition of heterogeneity; (2) the nature of that heterogeneity; (3) they produce effects on the built environment and on the identity | | Includes action against capitalism by terrorists |

(continued)

**Table 30.1** (continued)

| Urbicide as a concept (theory, epistemic, ontological, and other assumptions) | | |
|---|---|---|
| Perea Tinajero (2021):205 | Violence that disarticulates inhabitants and built environment (streets)<br>War against urban environments. Includes urban guerrilla and other "hybrid" wars (with little clarity). (Perea 2021:206), Military and police responses<br>Means disarticulating elements of the built space | No examples, but could it include Allende Coahuila? Border cities? | What is hybridism? Does it mean new war strategies or just destructive ones?<br>Adds urban guerrilla and police action (when does it disarticulate the space created?)<br>Does it include struggles of criminal organizations to control, occupy, and conquer territories? |
| *Urbicide as Genocide* | | | |
| Shaw (2004), pp 141 and ff | The concept is not only applicable to the killing of an ethnic group, but against their cities, their urban culture (p.7 electronic) | Cambodia's genocide was an urbicide<br>The destruction of urban culture | Adds genocide to the notion |
| Mendieta (2007) | Urbicide is genocide, a process of total war. Civilians and cities are military targets of tyrannical governments | | Equates urbicide to genocide against populations by (military) tyrants |
| *Violent disturbances to the "Urban order"* | | | |
| Ramadan (2009): 153 and ff | Murder of the city. In the 1960s, it referred to developments that damaged cities, but nowadays, Bosnian architects apply it to urban destruction in general (see Coward 2004:157) | Bosnia | From developments and urban regeneration to damage in cities because of war<br>Dominant-authoritarian reconstructions |

(continued)

**Table 30.1** (continued)

| Urbicide as a concept (theory, epistemic, ontological, and other assumptions) |
|---|
| Other possible areas of application of the concept: disasters, accidents, violent disturbances to urban order; violence due to drug trafficking and political violence |

| Other areas | Accidents and disasters. They are not necessarily intentional, but have precedent social practices (of carelessness-forgetfulness) and have noxious consequences | Explosion at Beirut Port in Lebanon on August 20, 2020; nuclear, industrial and environmental accidents | Urbicide as passive destruction of cities, by negligence |
|---|---|---|---|
| | Riots, rebellions, revolts, attacks, or counterattacks violent social-urban expressions and manifestations | PCC attacks in São Paulo (2006) and (2012) El Salvador, Guatemala, Rio de Janeiro. Reactions in Culiacan, 2019, in Monterrey, 2011, in Acapulco, in Laredo, Silao-Salamanca, Mexico, 2022 Public and urban responses to the Rebellions in Los Angeles, Watts-1965, 1968, 1992 | Destruction of zones and regions of cities as a result of "popular" rebellions, revolts and in other cases as a planned action of criminal organizations. Explicit purpose of destroying the policy, with collateral damage How can urban protest be explained by symbols of colonial or violent pasts? ¿How to explain violent demonstrations of feminists destroying urban objects? |
| | State's actors' behavior that destroys or attempt to destroy or inflict violence against urban actors | Police and military invasion and occupation of cities, urban facilities in Mexico (vs People's assembly-APPO- in Oaxaca; in Unam; military intervention programs in Medellín and Ciudad Juarez) | Militarism and illegal and illegitimate policing as expressions of urban destruction by autocracies |

(continued)

**Table 30.1** (continued)

Urbicide as a concept (theory, epistemic, ontological, and other assumptions)

| | | | |
|---|---|---|---|
| | | Conquest, occupation, and destruction of Tenochtitlán and construction of colonial Mexico-city on top of the ruins of the ancient ceremonial center | Urbicide as conquest. Colonization |
| | Invasions that destroy patrimony (tourism) | Venice and other "theme" cities (Paris, Florence, downtown Rome). Capture of idyllic cities by capital interests | Predatory tourism. Environmental impacts could be added |

*Urbicide as political violence*

| | | | |
|---|---|---|---|
| Coward (2008): 419 and ff | Destruction of the built environment as a specific form of political violence; not cultural confrontation; not necessarily the identity of a group but of the substratum, the constructed way of existence. Attempts against the built environment rather than against the population (contrary to drug trafficking), to the life-habitation spaces (Coward 2008:426) It coexists with other types of violence such as genocide or repression supported-instituted by the State Destruction of heterogeneity to build an antagonistic homogeneity, creation of a separation zone by demolition, as a border separating enclaves (Coward 2008:434) | No examples | All types of war violence? What about police violence? Delimits idea of destruction but continues the link with genocide Creation of special zones Narco-guerrillas could be part of this? |

(continued)

## 30 Urbicide, Violence, and Destruction Against Cities by Criminal …

**Table 30.1** (continued)

Urbicide as a concept (theory, epistemic, ontological, and other assumptions)

| | | | |
|---|---|---|---|
| Urban violence (proposal by the author of this text) | Component or dimension that accompanies the core of the concept. Homicidal violence, femicides, juvenicides in cities and by the fact of occurring in cities; urban gender violence (harassment, aggression, persecution, rape & murder of women in transportation and in urban public spaces-urban areas) | Homicides (on urban rates); injuries; femicides; juvenicides. Urban insecurity as a manifestation and production of "defensive" and segregating behaviors. Sexual and homicidal violence against women in public spaces. Assassinations of politicians, authorities | Partial and complementary dimension to the idea of urban destruction of its populations. Its impact on urban life. It is necessary to add a gender perspective to urban violence and urbicide. Political assassinations of authorities and police officers in urban contexts because of being urban authorities? |
| Alvarado (2019) | Attacks by criminal organizations; coercive nature of these interventions; predatory behavior; reactive and defensive violence in certain spaces | Confrontation between drug gangs in Ciudad Juarez between 2006 and 2010. (Arratia 2017) Maras, Salvador; CJNG in Michoacán and border cities; Zetas in northeast: Allende, Coahuila, Matamoros, N. Laredo; C. Juárez, Mexico; Gangs in Ecuador. Transnational criminal organizations in different urban territories | Narcowars? Illicit and violent political struggles to control cities |
| Gender, religious and ideological violence | The concept lacks gender connotation. It is not a central component, but urbicide must include gender perspective | Violence against women and other free expression of different gender orientations in cities | This generates defensive behaviors from women; insecurity also generates divisive consequences and some violence in demonstrations |

*Source* Author's elaboration based on the referred texts. Author translations and emphasis made by the author.
*Note* Due to translation from Spanish to English, some of the references quoted do not correspond to the literal-original-text

**Table 30.2** Homicide trends in the largest metropolis of Latin America

| Metropolitan | Population | Cases | Rate |
|---|---|---|---|
| São Paulo (2018) | 22,000,000 | *3727* | 9.5 |
| Ciudad de México (2018) | 21,900,000 | 1400 | 15.5 |
| (Gran) Buenos Aires | 15,100,000 | 904 | 5.2 |
| Río de Janeiro | 13,005,430 | 1987 | 29.7 |
| Lima (2019) | 10,555,000 | 745 | 8.48 |
| *Average* | *10,501,359* | *1197* | *16.92* |
| Santiago | 7,400,000 |  | 5.1 |
| Bogotá (2018) | 7,181,469 | 1052 | 12.7 |
| Guadalajara (2015) | 5,179,874 | 370 | 24.7 |
| Monterrey (2015) | 5,061,732 | 220 | 19.7 |
| Brasilia (2018) | 4,200,000 | 978 | 32.4 |
| Medellín (2018) | 3,931,447 | 583 | 23.1 |

Rates per hundred thousand inhabitants. *Source* Author's elaboration based on the referred texts Inegi, Ibge, Dane, Igarapé Institute, Insight Crime, Instituto de Pesquisa Econômica Aplicada–ipea 2020, y OCJ. Estimated cases are in italics

100,000 inhabitants. These trends are above the global murder rate of around 6.2 and the regional rate of just over 12 points, estimated by UNODC for 2018.[3]

As the table shows, the homicide levels allow us to characterize the vast majority of the region's metropolises as extremely dangerous and with destructive tendencies due to interpersonal, group, and organized violence.

To understand better the type of cities, we have made a classification according to a measure grouped in five categories.

This table (30.3) provides us with a metric that shows the differential magnitude of violence, and allows us to conjecture that it may be useful to classify and measure the degree of destruction generated by different actors and actions and add it as a measurable dimension in the notion of urbicide. However, it will be necessary to discern whether these violent deaths are not only products of structural factors such as inequality, lack of an effective justice apparatus, or lack of development or poverty (Concha Muños and Santos 2020). It is also linked to the presence of violent actors who carry out their practices within urban territories and in which we observe their main effects.

The vast majority of cities in the Latin American region have high levels of violence, and to understand its concrete manifestations, we suggest to consider the magnitude, the scales and spaces where it operates and where three aspects are

---

[3] There are other records of urban homicidal violence, such as those of the Citizen Council for Security, Justice and Peace, as well as InSigth Crime and the studies of the United Nations organization against Drugs and Organized Crime (UNODC). We use some of them because not all of their records are consistent or complete, and in some cases, they equate violence in the cities to national figures.

**Table 30.3** Ranking of cities by groups of homicidal violence rates

| Cities | Average rates (per each 100,000 inhabitants) |
|---|---|
| Average rate in the Americas | 15 |
| Group 1. Cities with extremely high violence (10 times the global rate) | 69.4 |
| 17 Cities | |
| Group 2. Cities with very high violence | 40.3 |
| 31 Cities | |
| Group 3. Cities with high violence | 31.1 |
| 6 Cities | |
| Group 4. Medium violence | 10.1 |
| 0 Cities | |
| Group 5. Moderate violence | 5.81 |
| Cities with low average rate 5.81 5 cities | |

*Source* Citizens Council on Security, Justice and Peace 2019; UNODC; InSigth Crime & El Tiempo
There are no cities with moderate violence. The list includes 4 cities in South Africa

combined: (1) The presence and form of criminal organizations, which operate to control markets, territories, and populations; many of them are intertwined with the urban form; (2) The urban public security policy (when it exists), its forms of territorial deployment, in particular with the police and relations with citizens; (3) The different responses of their communities (which can be of consent, subjection, passive or active resistance). By combining these three categories, we can find territories where one authority dominates in a legitimate and legal way, and other extremes of spaces under the control of illicit and destructive violent actors (because we have associated violence to illicit activities as destructive behaviors). We are also going to find hybrid situations were there seems to be under dual governance, and many others of territories in dispute. In them, we find spaces of high, medium, or low criminality as well as high to low capacity of inhabitants and authorities to confront and reduce violence. For example, there are some spaces of high violence and tight criminal control, with low resistance of communities (as many cases of precarious neighborhoods in cities of the *global south*, but also in the suburbs of Marseille or Paris, because this does not only happen in Latin America, but in all cities of the planet); in contrast, there are places of low criminality and high resilience of inhabitants (for example, some upper class neighborhoods or organized working classes communities) with parallel presence of democratic authorities.

What is it that produces urban violence? What we propose here is that there is a set structural causes, such as inequality, segregation or the ineffectiveness or inexistence of a justice system. At meso-level, we can find and measure the criminal actors, with their destructive behaviors. Together with them, we identify police behavior, where in some places, it is an effective deterrence and protective actors, but in some cases produces violent and predatory work. At this level, we can also identify the effect of public policies, such as ineffective security, health, housing, transportation, or commerce policies. And at the micro-level, we can identify a set of many actions and conditions of urban everyday life. These actions tend to reinforce urban decay and destructive trends of entire neighborhoods or cities (This could be part of an agenda for future works).

## 30.6 Expressions of Urban Violence Associated with Practices of Criminal Organizations

In relation to organized criminal activities, and not only drug trafficking, we have some evidence of its destructive processes that capture urban spaces for reorganizing them into illicit activities under their control. These are processes of destruction, distortion of cities (in economic, social, and political-civic aspects). But, massive urban destruction does not always occur (with the exemptions marked in the literature on Table 30.1).

Most attacks on cities by criminal organizations are not always consciously planned or premeditated; rather, they arise as reactions to attacks between criminal organizations or by state coercive forces. These is the interpretation we offer to the numerous blockades and concerted collective actions in cities such as Monterrey, N.L. (in 2012 and 2015), or Guadalajara in Mexico (in 2022 and before, called narcoblockades), or in Culiacán Sinaloa in 2019. There are more examples of other cities partially or entirely controlled by them in the urban corridors of the gulf, Veracruz, Tabasco, Tamaulipas (Zavaleta 2018b), and in border cities between Mexico and the United States of *North America*, E.U.A. (Tijuana, Ciudad Juarez, Nuevo Laredo, Reynosa, Matamoros, among others), which are subject to a triple dominance, criminal, political (national), and security policies of the USA. Similar phenomena occur in Rio de Janeiro in Brazil, in Medellin, Colombia, El Salvador, cities in Honduras and the capital of Guatemala or Ecuador, such as Cuenca, Guayaquil, or Quito (sometimes the expansion of activities of criminal actors first affects their cities, but ends up affecting the stability of a country, as recently happened in Ecuador). The list can be extended much further. In these cities, many neighborhoods are under the control of illicit organizations and are affected by their territorial presence, by their armed control of the territory and of the population. But, there are cases of strategic planning to control and produce disruptions and terror among the population, as we find in the interventions-attacks of PCC in Sao Paulo, particularly after 2006. Also, recent violent irruptions in cities in Mexico (Tijuana and border cities by CJNG), in late August 2022, seems to be a promotion plan and a combat between transnational organizations based in Mexico against the national government (not the local authorities). In these case, cities are the scenario, not the direct and unique target for the attack.

In the northern territories of Mexico, illicit groups have destroyed and turned certain cities into territories of death, in their jails and barracks, as happened in Allende, Coahuila, and in other places on the Mexican northern border, such as Matamoros, but also in large portions of Tampico. Can we qualify these activities as urbicide? This is violence against the city or in an more specific sense the death of the city (Moorcok 1963). It is the deliberate material destruction of partial urban areas and in one case the appropriation of an entire city. The criminal intervention in Allende, Coahuila on March 18, 2011, by the *Zetas* (Coord 2016: 12–14), and in other cities in northeastern Mexico, are deliberate acts of destruction of the urban environment, as criminal purposes.

Some of these border cities have created their own economy of services, transportation, merchandise, control of the media, the police and local authorities. New actors have emerged from these peculiar dynamics, such as the smuggler, the middleman, the human trafficker, the *serial narco killer (sicario),* the Falcon, the collective *commando* (see Carrión and Gottsbacher 2020-Handbook).[4] There have been situations in which the illicit group culminates its struggle by integrating itself as part of the local political elite (such as Mayors in several municipalities in Michoacán and other places). Hence, an interpretation of this capture considered a form of criminal governance has emerged in several contemporary studies (Lessing 2021). Beside this interpretation, what we are interested in emphasizing is that this violent destruction also generates other urban social and political-economic orders or is combined with the dominant ones.

There are some cities where the wars between drug traffickers and the government army have produced the destruction of the urban environment, such as in C. Juarez, accompanied by the invasion of gangs that fight each other to control portions of the city and then the military occupation that subjugates everyone, which is followed by its abandonment. It was also used as a policy of renovation of the urban territory in the program: *Todos Somos Juarez of 2010*. These policies were highly criticized by residents.[5] Also, the occupation-destruction as in Allende, Coahuila by the Zetas organization in the first decade of the century (*vid supra*).

To provide recent examples of urban violence in other regions of the world, we can quote several authors who refer to the ethno-national conflicts in Eastern Europe in the 1990s (Vukovar by the *Yugoslav people's army* and Serbian

---

[4] The presence and extension of the activities of criminal organizations generates subcultures around drug trafficking, the figure of the drug trafficker (hypermasculine mortal power), the smuggler; and urban and social aesthetics that have very violent contents and subordinated feminine roles (with exceptions).

[5] The so-called "Estrategia Todos Somos Juarez" … was a federal program in partnership with the municipal government of Ciudad Juarez, as well as with members of civil organizations in 2010. They installed a *consultative table* with all these actors to implement federal and local programs of public security, urban renewal, health, and education. Was an ambitious renovation plan, which aimed to respond to the intense violence generated by the fierce competition of this border territory between transnational criminal organizations. It has precedents in other programs in Colombia, and its results have only been partially evaluated, but it is precisely a program that pretends to solve the destruction of the city by the Narco-guerrillas. For a better understanding see: Meyer, 2010. Castillo and Ochoa, 2012. Arratia, 2017. There are other *experiments* in conflict resolution and violence reduction programs implemented by the governments of Medellin, Colombia (see Alvarado, 2014, as well as in Ecuador (Brotherton and Gude, 2021).

paramilitaries in 1991), which indifferently destroyed the cities, its buildings, schools, factories, hospitals, housing; the destruction of Sarajevo between 1992 and 1995 and the *ethnic* cleansing actions. In another place, in Harare, Zimbabwe, the government implemented the Murambatsvina operation (Clear the Filth). In May 2005, 700,000 people were displaced from their settlements, and their houses and infrastructure were destroyed (Misachi 2017). This has also been the case during the war against the *Islamic State* with the destruction of Aleppo and other localities.

Although the topic of gender is not the subject of discussion in this paper, it is necessary to consider the urgent necessity to integrate gender-based violence (particularly against women) into the concept of urbicide. It is a form of violence in urban public spaces.[6]

The organizational forms and spatial deployment of illicit organizations and their strategies are linked to the urban form. This issue should be explored in greater depth in the future, as well as the link with the activities and criminal markets of human trafficking, arms trafficking, extortion, kidnapping, money laundering, urban land trafficking, and corruption-collusion with authorities.

It should also be noted that these planned proactive behaviors do not always culminate in urban destruction. In some situations, communities have organized themselves and have managed to resist or repel these attacks, forming self-defense organizations. This has occurred in the state of Michoacán (in Cherán, Michoacán, by organizing its own community defense) and in the Mexican Bajío region, where cities are spaces of confrontation, defense, and government's recovery efforts. Likewise, there are examples of organized urban communities that have managed to build and maintain security strategies for their neighborhoods, either with the community itself or in partnership with legitimate governments.

What we also wish to manifest with these examples is the plasticity and innovation of the inhabitants in defending their ways of life and their urban spaces, trying to rebuild and keep their localities alive. In this sense, destructive actions also generate resistance and the formation of communities, groups, neighborhood-community associations, self-defense groups, and various forms of vigilantism. In certain situations, these forms of resistance manage to maintain, recover, or recreate their cities with community authorities. But, this shows that the destruction of cities is not inevitable and with a single direction.

---

[6] The deep and systemic violence against women in public spaces has been noted by UNWOMEN studies (see: https://mexico.unwomen.org/es/digiteca/publicaciones/2017/03/diagnostico-ciudades-seguras). And the dominance of Muslim-Islamic male power coalitions in several countries of the world has been the object of violence against women and social responses against them, as has happened with the protests in Iran since September 13, 2022, as a result of the death of Mahsa Amini, a young woman who was arrested and was in custody of the Iranian Islamic religious police (guidance patrols) was detained and beaten for breaking the law that requires women to cover their hair and virtually the entire body in public spaces. It is a manifestation of a violent urban order by a political coalition with coercive police power being exercised against women. See: https://www.bbc.com/mundo/noticias-internacional-62994373

**Table 30.4** Inventory of dimensions that may be associated either with the destruction, urban decay, and poor governance of cities, or, in contrast, with the resilience and creation of strong urban communities

| Antecedents/possible causes | Positive and Negative Consequences |
|---|---|
| | Resulting forms: impact on governance |
| 1. Violence (Homicides, illegal activities, violence in public spaces) | Impunity, high victimization; low or null governance. A high homicide rate can be an indication of processes of massive destruction; a low rate can mean an improvement, but also the capture of the city by an organization that imposes an order. A *criminal pax* |
| 2. Organized violence (criminal organizations, paramilitary, parapolice, terrorist agents) National and local expressions | Organized violence: local criminal governance networks or clusters Expansion of transnational criminal enterprises and *cartelization* of local gangs as well as portions of cities; expansion into regional-urban structures and networks |
| | Construction or consolidation of criminal power as market power and urban political power (not only for illicit matters). Capture and use of cities as sites of destruction, watchtowers, and places for crimes against humanity |
| 3. Government actions (public) Public policies on security. Reconstruction and urban renovation Support or conflict with communities | Punitive actions such as excessive imprisonment These policies generate unintended consequences that sometimes produce other damages. And they produce some of the following consequences: Limited territorial control (and corruption) Deterioration of democratic and governmental rule of law; deficient public policies Limited territorial sovereignty |
| | Control and criminal dominance in social organizations and neighborhoods |
| | Capture and co-opted reconfiguration of the state (Garay and Salcedo, 2008) |
| | Association, symbiosis between authorities and criminal entrepreneurs Building safe cities and effective communities in the fight against organized violence (In contrast and as a positive effect) Proactive reconstructive actions, articulated with communities and other authorities, which can generate new urban orders |

(continued)

**Table 30.4** (continued)

| Antecedents/possible causes | Positive and Negative Consequences |
|---|---|
| 3.1. Police work | Confrontation with criminal power as market and political power that can result in: |
| Macro- and micro-policing strategies | Limited, co-opted, bought police power: criminal sovereignty<br>Military interventions in the fight against crime<br>o<br>Legitimate construction of safe communities |
| 4. Citizens and neighborhood's Organizations | Three alternatives: Resistance, building collective efficacy, resilience and democracy |
| Regional Organization and Mobilization | Respect for law and order or impunity<br>Passive or complete acceptance of criminal power: No legal and justice order |

Source Author's elaboration based on Lessing 2021, and Adorno and Alvarado 2022

Finally, we present a possible classification of factors that can be associated not only with urbicide, but also with the problems of deterioration, destruction, and poor governance of our cities (Table 30.4).

## 30.7 Ecuador's Western Cities, an Example of Recent Regional Crime Growth

Besides looking at major trends of violence in metropolis, we need to include what is happening in medium-sized and port or border cities in countries that are located on drug trafficking routes in South America (particularly cocaine). They are increasingly affected by the presence of transnational criminal organizations, and this has altered the structure of security services, law enforcement and prisons, with some evident consequences in the increase of urban violence and others latent conflicts in terms of the territorial presence of crime. The expansion of activities of transnational organizations, particularly the so-called Mexicans (such as the Sinaloa and Cártel Jalisco Nueva Generación, but also Colombians, although all are transnational), has transformed the transit territories of illicit drug trafficking, as in the case of Guayaquil, in the province of Guayas, as well as in Manabí and its port (Mantas) and in Esmeraldas, bordering Colombia. These criminal enterprises increased violent competition between gangs associated with them. They were originally located only in that portion of the country. The city authorities had managed to contain the violence through police strategies of arrest and mass incarceration, but this has led to the prisons now being the territory of dispute and confrontation, with consequences for security in various cities. These processes date back to the end of the twentieth century, but in recent years, they have had a notorious expansion, even during the pandemic.

Even though this criminal violence has not reached the point of producing city-wide blockades and violent confrontations, the evidence shows that they have being affected much more than what the government, the press, and experts report. To begin with, there are numerous (previously) local and now regional organizations: Los Choneros, Los Pipos (New Generation), Los Tiguerones, Los Chone-Killers, Los Lagartos, Los Lobos, Los Gánster negros, Los Ñetas, Los Latin Kings-a transnational gang; they are associated with drug trafficking and homicidal violence (taken from Redacción La Hora, September 30, 2021). Many gangs have local origins, while others take the form of transnational gangs and others of the so-called Colombian *Bacrim*. Gang confrontations and homicides have been concentrated in the city's prisons and in some territories linked to the main gangs that have operated in this territory in western Ecuador. In past years, the effect was concentrated in two points, on the one hand in the prisons, and on the other in the streets and neighborhoods where these gangs operate. The number of contending gangs has produces various phenomena of splitting, rupture, and confrontation between them (questions can be raised and need more attention in future research: is it an internal cartelization phenomenon? Or is it caused by the arrival and entry of transnational gangs?).

Thus, Ecuador and particularly its mid-sized cities on the west coast have been the scene of the growth of this criminal world since the 1990s, but which in recent months has produced part of the prison crises and the extraordinary increase in homicides occurring in several areas of these cities. Homicides in Ecuador doubled between 2019 and 2022, from 1,187 cases in 2019 to 2496 in 2021, and 2116 so far in 2022. A similar increase occurred in Guayaquil.[7]

To confront homicidal violence in prisons (which has even led to massacres. See: Paredes 2020), Lenin Moreno declared a penitentiary crisis in the country in 2019 and began a strategy of transferring several prison gang members to other prisons, which led to the expansion of their presence in other provinces; an increase in their network and territorial influence. This also forced the intervention of the army (S.I. October 1, 2021; Olan 2021). This was a spillover effect to other prisons and other cities.

According to investigations carried out by Cesar Ulloa-Tapia, as well as reports from journalists' associations and the press, various cities beyond western Ecuador are the scene of confrontations between criminal organizations over drug trafficking. These spaces show how the prisons were the basis for re-building the internal structure of the gangs, but above all, for creating several external criminal networks. The evidence is obvious in the prisons, but it indicates the existence of a deep and growing criminal economy of local and transnational trafficking. In this sense, Ecuadorian urban territories are another example of how cities located in transnational trafficking corridors are affected despite the government's policies to negotiate, control, or combat them. Recall that within the Ecuadorian security strategies, there was a gang pacification plan, and current trends suggest that these mechanisms have been

---

[7] For the case of Guayaquil, the government reported 279 homicides in 10 months of 2020 and 613 so far in 2022. Data from El Universo, news, 2022. Also, Ministry of Government, DINASED and Ecuador Police; Insight Crime, 2020; INEC Integral.

exhausted, and perhaps we are at a breaking point of the transformation of these local structures into trans-territorial networks, with greater coverage and capacity for predation. They also manifest that the phenomenon of criminal presence and damage in the cities does not occur only in the metropolises that we have been describing in the text, but also in medium-sized urbs and in several of their scales. Until now, the most violent expressions have been concentrated in certain urban spaces, but there are already signs of greater impact in commercial areas and, in general, in the perception of urban security. It is likely to have other important effects on the governance of ports and customs, on security and eventually on urban governance.

## 30.8 Urban Violence, How to Avoid It

Cities systematically face and are the nest of multiple transgressions, because this behavior is one of its core components. These factors can be produced and are associated to situations where cities tend to deterioration, decline, partial, or total destruction; partly by premeditated violent actions, partly by unintended public and collective actions; sometimes as a result of planned premeditated actions (but other times not adverted), by blunder or omission.

As we have tried to show in the text, the concept of urbicide has been widely used and modified in such a way that many attributes have been added to it and are applied to numerous examples where its meaning has become unprecise, inexact, ambiguous. We have also shown that the extension of the concept's properties has led to the inclusion of borderline situations that are applied to phenomena of different nature. In this paper, we made a conceptual exercise of clarification, delimitation, and measurement that shows its usefulness in the analysis of the multiple different and sometimes fragmented expressions of urban violence (including its mechanisms, its actors, their strategies). Likewise, on the one hand, we specified the concept and narrowed down the correlates to which it can be applied, while on the other hand, we carried out an exercise to explore how we could integrate issues of drug trafficking, criminal violence, other forms of homicidal violence, gender violence, and other transgressions, as components of the system. While the exercise is useful in exploring the possible ways to apply the concept to urban violence, we also utilize several examples which demonstrates that broadening properties and examples will produce an unnecessary stretching of the meaning of the term. I believe that this exploratory exercise will allow us to advance in the efforts to maintain core properties of violence against the city, not only physically but also in a symbolic, collective, communitarian, and democratic way. Still, we need to improve our rigor and produce better classifications and measures. We need more examples that ratify our ideas, or, on the contrary other cases that will forces to reject our conjectures.

# References

Alvarado A (2022) Documentos de trabajo de campo y entrevistas en Tamaulipas en el año 2017. Conferencia—Recepción premio medalla al Mérito Universidad Veracruzana. Xalapa, Ver. 7 de septiembre de 2022

Alvarado A (2019) Organizaciones criminales en América Latina: una discusión conceptual y un marco comparativo para su reinterpretación. Revista Brasileira De Sociologia 7(17):11–32

Alvarado A (ed) (2014) Violencia juvenil y acceso a la justicia en América Latina. México, El Colegio de México. 2 tomos

Abujidi N (2014) Urbicide in Palestine: spaces of oppression and resilience. Routledge, London

Adorno S, Arturo A (2022) Crime and Governance in large metropolis de Latin America. Mexico City and Sao Paulo. Dilemas, UFRJ. Número especial 4:79–115

Aguayo S (ed) (2016) En el desamparo. Los Zetas, el Estado, la sociedad y las víctimas de San Fernando, Tamaulipas (2010), y Allende, Coahuila (2011). México, El Colegio de México

Moreno AA, Gil EYB (2021) Urbicidio: sobre la violencia contemporánea contra las ciudades. Agora: papeles de Filosofía 40(1):87–110. https://doi.org/10.15304/ag.40.1.6603

Arratia E (2017) Todos Somos Juárez. Competition in State-Making y la guerra contra el narcotráfico (2006–2012). Todos Somos Juárez. Competition in state-making and the war on drugs (2006-12)". En: Revista Española de Ciencia Política. Núm. 43. Marzo 2017, pp 83–111

Arendt H (1979) The origins of totalitarianism. Harcourt, London

Bernard-Moulin A, Gonzalez M (2018) Urbicide in the Guatemalan Civil War

Brotherton D, Gude R (2021) Social control and the gang: lessons from the legalization of street gangs in Ecuador. En: Critical criminology, vol 29. pp 931–955. https://doi.org/10.1007/s10612-020-09505-5.

Bunge M (2000) La investigación científica: su estrategia y su filosofía, 1a edn. Siglo XXI, Ciudad de México, Ciudad de Buenos Aires

Caldeira TP (2000) Cidade de Muros: Crime, Segregação e Cidadania em São Paulo. São Paulo: Editora 34/Edusp. pp 399

Campbell D, Graham S, Monk DB (2007) Introduction to Urbicide: the killing of cities?" Theory and Event 10(2). https://doi.org/10.1353/tae.2007.0055

Castillo N, Alberto O (2012) La política pública del combate al narcotráfico en Medellín, Colombia, y Ciudad Juárez, México En: Chihuahua hoy… México. documento en línea: https://elibros.uacj.mx/omp/index.php/publicaciones/catalog/download/62/57/478-1?inline=1

Cardoso G (2021) Entrevista a Coronel Mario Pazmiño, Exdirector de Inteligencia. Antena Uno. Recuperada de: https://www.antenauno.com/Detail/Article/6127/New/Mario-Pazmino-Exdirector-de-Inteligencia

Carrión F (2014) Urbicidio o La Producción Del Olvido. Observatorio Cultural de Chile 28–43

Carrión F (2018) Urbicide, or the city's Liturgical Death/Urbicidio o la Muerte Litúrgica de la Ciudad. Oculum Ensaios 15(1):5–13

Carrión F, Gottsbacher M (2020) Border violence in Latin America. An expression of complimentary asymmetries. In: Rivera L, Xóchitl B (eds) The Oxford handbook of the sociology of Latin America. Oxford, Oxford University Press

Caso D (2020) El Financiero, 26 de junio de 2020: https://www.elfinanciero.com.mx/cdmx/atentan-contra-omar-garcia-harfuch-secretario-de-seguridad-publica-de-la-ciudad-de-mexico/

Concha-Eastman A, Muños E, Santos R (2020) "Homicides in Latin America and the Caribbean. In: Rivera L, Xóchitl B (eds) The Oxford handbook of the sociology of Latin America. Oxford, Oxford University Press

Coward M (2006) Against anthropocentrism: the destruction of the built environment as a distinct form of political violence. Rev Int Stud 32(3):419–437

Coward M (2008) Urbicide: the politics of urban destruction. Routledge, London

Dias CCN (2009) Ocupando as brechas do direito formal: O PCC como instância alternativa de resolução de conflitos. Dilemas 2(4):83–105

El Comercio (noviembre 24, 2021) Recuperado de: https://www.elcomercio.com/actualidad/seguridad/bandas-pelea-contro-trafico-drogas-ecuador.html

El Universo (Agosto 5, 2022) Así están las cifras de la inseguridad en el primer semestre del 2022 comparadas con los tres años previos. Recuperado de: https://www.eluniverso.com/noticias/seguridad/asi-estan-las-cifras-de-la-inseguridad-en-el-primer-semestre-del-2022-comparadas-con-los-tres-anos-previos-nota/

Feltran GDS (2010) Crime e castigo na cidade: Os repertórios da justiça e a questão do homicídio nas periferias de São Paulo. Caderno CRH 23(58):59–73.https://doi.org/10.1590/S0103-49792010000100005

Fregonese S (2009) The Urbicide of Beirut? geopolitics and the built environment in the lebanese civil war (1975–1976). Polit Geogr 28(5):309–318. https://doi.org/10.1016/j.polgeo.2009.07.005

Garay J, Salcedo E (2008) La Captura y reconfiguración Cooptada del Estado en Colombia. Bogotá, Fundación Avina, Col

Gobierno de Ecuador (2019a) https://www.ministeriodegobierno.gob.ec/dinased-evidencia-una-resolucion-del-60-1-en-muertes-violentas/#:~:text=Hasta%20el%2024%20de%20diciembre,entre%20un%20a%C3%B1o%20y%20otro

Gobierno de Ecuador (2019b) https://www.policia.gob.ec/wp-content/uploads/downloads/2019b/09/rendicion-cuentas-2018.pdf

Graham S (2022) Teoría y práctica del urbicidio. New Left Rev 39–54

Hanafi S (2013) Explaining spacio-cide in the Palestinian territory: colonization, separation, and state of exception. Curr Sociol 61(2):190–205. https://doi.org/10.1177/0011392112456505

Herrera LA (2007) El desgobierno de la ciudad y la política de abandono: miradas desde la frontera norte de México. Universidad Autónoma de Ciudad Juárez. Cd. Juárez, México

Jacobs J (1961) The death and life of Great American Cities. Random House, New York

Instituto Nacional de Estadística y Geografía. INEGI

Insight Crime (2019) https://insightcrime.org/

InSigth Crime Organization (Octubre 5, 2021) Los Choneros. *InSigth Crime Organization*. Recuperado de: https://es.insightcrime.org/noticias-crimen-organizado-ecuador/los-choneros/

InSigth Crime Organization (2020) (https://es.insightcrime.org/noticias/analisis/balance-insight-crime-homicidios-2020/)

InSigth Crime Organization (2020) (https://insightcrime.org/wp-content/uploads/2020/01/112019_Cifras_Seguridad.pdf)

InSigth Crime Organization (Junio 1, 2021) Los Choneros. *InSigth Crime Organization*. Recuperado de: https://insightcrime.org/ecuador-organized-crime-news/los-lagartos/

La Hora (Septiembre 30, 2021) Seis bandas se disputan el control de las cárceles. *La Hora*. Recuperado de: https://www.lahora.com.ec/pais/carceles-control-bandas-narcotrafico-ecuador-muertes/

Lessing B (2021) Conceptualizing criminal governance. Perspectives on Politics 19(3):854–873. https://doi.org/10.1017/S1537592720001243

Mendieta E (2007) The literature of urbicide: Friedrich, Nossack, Sebald, and Vonnegut. Theory and Event 10(2). https://doi.org/10.1353/tae.2007.0068

Meyer M TNI (2010) ¿Todos somos Juárez? S.p.i. https://www.tni.org/es/art%C3%ADculo/todos-somos-juarez.

Misachi J (2017) "What Is Urbicide?" World Atlas. Recuperado (https://www.worldatlas.com/articles/what-is-urbicide.html#:~:text=Urbicide%20is%20a%20term%20that,of%20rapid%20globalization%20and%20urbanization).

Moorcock M (1963) The stealer of souls. Neville Spearman Limited, London

Noticieros T (2019) 17 de octubre de 2019. Así fueron las balaceras en Culiacán. Detienen a Hijo del "Chapo" Guzmán. Recuperado de: https://www.youtube.com/watch?v=bnzdL0drR7U

Paes Manso B (2020) A República das Milícias. Sao Paulo, Todavia

Perea Tinajero, GP (2021) Urbicidio y destrucción material de la ciudad contemporánea: formas del ejercicio de la violencia. Puebla

Paes B, Dias C (2018) A guerra: A Ascensão do PCC e o mundo do crime no Brasil. Sao Paulo

Paredes N (2021) Septiembre 30, 2021. Ecuador: 4 claves que explican qué hay detrás de la masacre carcelaria que dejó al menos 119 muertos, la peor de la historia del país . BBC News Mundo. Recuperado de: https://www.bbc.com/mundo/noticias-america-latina-58748756

Ramadan A (2009) Destroying Nahr El-Bared: Sovereignty and urbicide in the space of exception. Polit Geogr 28(3):153–163. https://doi.org/10.1016/j.polgeo.2009.02.004

Redacción Plan V. Febrero 24, 2021. Ecuador: radiografía del crimen organizado y sus actores. Plan V. Recuperado de: https://www.planv.com.ec/historias/sociedad/ecuador-radiografia-del-crimen-organizado-y-sus-actores

Octubre SI (2021) 1, 2021. Los choneros, Los lagartos, Los lobos, las bandas rivales que provocaron la última matanza en una cárcel de Ecuador *ABC International,* pp 1 Recuperado de. https://www.abc.es/internacional/abci-choneros-lagartos-lobos-bandas-rivales-provocaron-ultima-matanza-carcel-ecuador-202110011326_noticia.html

Swwisinfo.ch. (2021). https://www.swissinfo.ch/spa/ecuador-crimen-organizado_el-crimen-organizado-alza-su-voz-en-un-ecuador-perplejo/46948776

Shaw M (2004) New wars of the city: relationships of 'Urbicide' and 'Genocide'". en Cities, war, and terrorism. Wiley & Sons, Ltd, pp 141–53

Toledo-Leyv C (2022) Agosto 15, 2022. El narcotráfico y el miedo se apoderan de las calles de Ecuador. Deutsche Welle (DW), pp1. Recuperado de. https://www.dw.com/es/el-narcotr%C3%A1fico-y-el-miedo-se-apoderan-de-las-calles-de-ecuador/a-62816441

United Nations Office of Drugs and Crime. UNODC

Willis GD (2015) the killing consensus: police, organized crime, and the regulation of life and death in urban Brazil. University of California Press, Berkeley

Zavaleta Betancourt JA, Alvarado A (ed) (2018) Violencia, seguridad ciudadana y victimización en México. Ciudad de México, Colofón/Universidad Autónoma de Ciudad Juárez. ISBN 978-607-520-310-2 UACJ, 978-607–8622-22-1 Colofón.-

Zavaleta Betancourt JA, Alvarado A (2018b) Interregnos subnacionales. La implementación de la reforma de justicia penal en México. El caso de la región Golfo-sureste. Ciudad de México, Colofón/Universidad Autónoma de Ciudad Juárez. ISBN 978-607-520-309-6 UACJ, 978-607-8622-23-8 Colofón.-

**Arturo Alvarado** Phd in Social Science with a specialty in Sociology from *El Colegio de México*. Since 1986 he has been a professor-researcher at the Center for Sociological Studies of *El Colegio de México*, and was director of the CES from 2012 to March 2018. Member of the National System of Researchers (level III). Specialist in justice, human rights, security and urban violence prevention. Consultant for international programs of the United Nations and the World Bank.

# Chapter 31
# Discursive Understandings of the City and the Persistence of Gender Inequality

Nora Libertun de Duren, Diane E. Davis, and Maria Lucia Morelli

**Abstract** This chapter explores some of the urban conditions that can impact violence against women levels the Latin American and Caribbean cities. After delving more deeply into the connection between urban violence in general and the dynamics of violence against women, it focuses on how the latter unfolds in the different urban spaces, distinguishing between the private and the public realms of the urban space. The chapter also offers a review of some of the most common ways, in which urban policies have tackled gender violence in cities and offer some guidelines for how to develop urban policies that support gender equality. It highlights how there is a tendency to support policies that address the outcomes rather than the causes of gender inequality, whether in the household or the city. These policies tend to revolve around seeing women as victims and segregating women and men, without addressing the problem of abusive men and violence in society. It calls for urban planning practitioners and policymakers to internalize the notion that preventing violence against women is the first step in producing equitable cities.

**Keywords** Inequality · Gender · Violence · Cities

## 31.1 Introduction

In the contemporary era, cities have become known as sites of growing class, social, and spatial inequities. Such conditions have fueled conflicts over "the right to the

N. Libertun de Duren (✉) · D. E. Davis
Harvard University, Cambridge, MA, USA
e-mail: nlibertun@gsd.harvard.edu

D. E. Davis
e-mail: ddavis@gsd.harvard.edu

N. Libertun de Duren
Inter-American-Development-Bank, Washington D.C., USA

M. L. Morelli
Architecture and Urbanism (PAU), New York, USA
e-mail: morellim@mit.edu

city," which could be understood as the guarantee of equal access to justice, prosperity, and citizenship, including rights to move freely, rights to housing, rights to work, and even rights to live without fear or harm, with the latter called into question by random or targeted violence that is increasingly likely to predominate in cities (Lefebvre 1968; Caldeira 2000; Harvey 2003). These concerns have captured the attention of urbanists, who tend to trace most contemporary urban problems to the political economy of urbanization and the failures of governing authorities to enact policies that reduce socio-spatial exclusion, guarantee access to affordable housing, invest in transport infrastructure, provide security, create public spaces for all, and make it easier to find work (Sassen 2005; Castells 2008). In cities of the global south, how authorities enable or constrain informality also has a bearing on rights to the city, as does the question of security—particularly as rates of violence continue to rise because of organized or illicit criminal behavior, some of which finds its roots in the distinction between the formal and informal city, itself a reflection of inequality (Davis 2014).

Among the growing number of scholars who examine the ways that urban social and spatial inequality sets the stage for these urban concerns, whether violence related or not, surprisingly, few have focused on the distinct experiences of women. To the extent that certain populations are called out as egregiously disadvantaged by the recent trends in urbanization, the first line of categorization is usually class or income, with a secondary focus more likely to be racial, ethnic, migrant, or immigrant identity as well as how these experiences intersect with class position to determine where and how residents experience the city. Yet these general categories are not routinely disaggregated to examine women and how they live in the city, even though there is growing academic interest to deconstruct the complexities of gender identity in the contemporary era (Young 1994; Chant 2020). Perhaps the principal exception here is research focused on domestic violence in cities, which documents the ways that women are at risk in their own homes or vis-à-vis their partners (Garcia-Moreno et al. 2015). Even so, most of the research on domestic violence tends to ignore the larger urban context, in which these harms occur, thus relegating studies of women's exposure to violence to studies of the family, the psychology of masculinity, the social psychology of motherhood, and other interpersonal dynamics within the domestic realm.

Given the growing rates of violence in cities, combined with the fact that women are particularly vulnerable both within and outside the home, it is time to revisit and question the analytical frameworks urban scholars use to study, plan, and service cities. Why has there not been more concerted attention to women's experiences in the city? How might we better understand whether women living in cities are faced with additional barriers or constraints than men or other so-called identity groups? What aspects of urban social and spatial life have the most direct bearing on women? What responsibility does urban planners and policymakers have in generating or even reinforcing the silences of women in the city? And what, if anything, have they gotten right concerning planning or creating cities that work for women? These questions are important not just because femicides have started to attract considerable attention in cities of the global south but in recognition of the fact that the zones of danger for women are continually expanding from inside to outside the house. The focus on women in cities is also important because women are more likely to stand on the frontlines of a series of binaries that have come to characterize the urbanization process, particularly in cities of the global south (Falú 2014). Specifically, women—and in particular, low-income women—bridge the dichotomy between us and exchange-value through their work in and outside the home; their activities often straddle the formal and the informal sector, especially when it comes in the form of domestic work for other families; their work contributes to production and consumption dynamics that unfold at both the scale of the city and the household; and they are more likely to be denied access to the conditions that provide for stability in a world of precarity, including legal land titles and homeownership. As such, when urbanization patterns undermine or curtail women's capacities to safely inhabit and/or move through these multiple sectors, the overall social and economic functioning of the city may be destabilized or destroyed, with any negative impact happening at the expense of women and their households who will inevitably bear the personal costs. In this sense, a city that is not safe for women is a city that fails to guarantee Lefebvre's notion of the universal right to the city (Marcuse 2014).

Our aim in this chapter is to bring a focus on women into the study of cities but to do so with specialized attention to some of the mechanisms that connect the discourses and practices of urbanization to the pervasive occurrence of violence against women, particularly in many of the cities of Latin America. In the Latin American and Caribbean region, where more than 80% of all residents live in cities, urban policies should be at the forefront of the prevention of violence against women (Libertun de Duren et al. 2018). That is, planners need to acknowledge and incorporate in their practice that the way cities are designed, built, and managed plays a significant role in preventing—or enabling—violence against women. In what follows, we explore some of the urban dimensions that can impact violence levels in the region. After delving more deeply into the connection between urban violence in general and the dynamics of violence against women, we focus on how violence against women unfolds in the different urban spaces, distinguishing between the private and the public realms of the urban space. Next, we review some of the most common ways in which urban policies have tackled gender violence in cities. We conclude by reflecting on the direct and indirect ways that cities enable violence against women and offer some guidelines for how to develop urban policies that support gender equality. With the latter, we aim to support constructive actions to reduce the fears and realities of violence that women experience in cities, and by so doing to push back against the gendered forms of "urbicide" in ways that might engage the readers and contributors to this collective volume.

## 31.2 The Connection Between Urban Violence and Violence Against Women

There is already mounting evidence of the many ways in which urban violence impacts women's lives. It is estimated that one in three women living in the Latin America and Caribbean (LAC) region has been the victim of violence at least once in their lifetime (UN Women 2021a, b), a statistic that ranks the region as the one with the highest rates of femicide in the world (CEPAL 2021). Moreover, after the government-imposed lockdowns to prevent the spread of COVID-19, the incidence of violence against women drastically increased (UN Women 2020). For example, in Colombia, the percentage of women who reported living in households with frequent conflict was almost three times higher about a year since the COVID-19 quarantine began (UN Women 2021a, b). While most of these acts of violence happen in the domestic space and by intimate partners, it is important to underscore that violence in public spaces is also omnipresent in the Region (Sen et al. 2018). While it is true that sexual harassment and other forms of sexual violence against women and girls in public spaces are a global phenomenon, harassment in LAC cities appears to be higher than in other regions of the world (UN Women 2017a, b). The combination of fragile institutions, slow or lack of enforcement of the rule of law, with an inherited long tradition of biases against women, worsens the nature and intensity of harassment against women.

It is important to highlight that social violence, in all its forms, is a central concern for the Latin Caribbean region, which stands as the most violent in the world. In 2017, the Region's documented homicide rate was 24 per 100,000 inhabitants, which equals 33.5% of the world's homicides, a staggering amount considering that less than 9% of the world's population lives in the Region (UNODC 2019). The geography of crime in LAC is highly uneven and densely concentrated in cities, (Muggah and Aguirre 2018a, b), which may not be coincidental given the persistent and deep levels of social inequalities at the urban level. Many LAC cities exhibit extremely elevated levels of inequality, also making it the region with the highest income gap between high- and low-income households. Levels of inequality are higher in large cities than in small cities and more pronounced at the urban than the national level. In this context, it comes as no surprise that robberies are frequent in urban areas. It is worth noting that violence in robberies is more common in Latin American and Caribbean cities than in other parts of the world. In 2014, about 15% of robberies led to homicides, compared to the world average of 9.1 and 5% in the United States (Jaitman 2019). The pervasiveness of violence points out both an entrenched level of social unrest and inadequate channels to give voice to it. Likewise, as institutions fail to function and provide services to citizens, crime and violence are associated with both informal and illegal economies that have boomed across the region (Davis 2015).

At the same time, we need to comprehend that there is a continuum of violence in LAC cities. Each of these different acts of violence reflects the inability of institutions and social arrangements to operate for the common good, as well as of a societal structure in which multidimensional poverty is deeply entrenched. Against this background, violence against women is a lens through which we reveal how structural power inequalities are exercised and reproduced in the urban space. By violence against women, we understand all acts of physical and psychological violence in which women become victims solely because of their being identified as females. In that way, we depart from the dichotomous view of this kind of violence in which women are victims and men are perpetrators of violence. Rather, we root violence against women within the gendered norms, social practices, and active cultures that perpetuate and reproduce an unequal distribution of power between men and women (Parkes 2015).

Violence against women cannot be understood—let alone stopped—without acknowledging that the divisions between public and private spaces are artificial constructions and that failing to connect the two undermines effective actions to prevent it, both now and in the future. From the perspective of the city, it is important to underscore that there is a continuum of violence between the urban and the domestic, as *"rare are the cases where structural violence in cities does not boil over, at some point, into physical violence between individuals. Equally rare are the moments when domestic violence exists in an urban space free from other types of violence"* (Salahub 2019). Thus, the porous and fluid reality of the city goes against a rigid and dualistic conception of violence, in which women are identified as fearful victims, and the spaces they inhabit are either public or private (Pain 2001). A clear example of this is how fear of violence in the public space undermines women's ability to work and earn income. There is evidence women are likely to reject job opportunities if they demand commutes they perceive to be dangerous. And as women's earnings decrease, their vulnerability to gender violence inside the home increases, all the time that economic dependence is one of the main reasons why women stay in situations of chronic intimate partner violence (Anderson 2007). As conceptualized by early urban and gender scholars, while violence against women occurs across cultural contexts and income levels, it has been shown that it lessens with women's economic independence, particularly within the domestic realm (Hayden 1982). Also, and particularly in informal neighborhoods in cities of the global south, clear physical barriers between public and private spaces are often porous or undefined. Therefore, all kinds of violent acts in public spaces, such as those associated with alcohol consumption, often led to increasing violence against women in domestic places as well (Libertun de Duren 2020). In addition, and with great significance, there is also plenty of evidence that exposes to violence in the home during childhood raises the risk of participating in other forms of violence later in life (Bott et al. 2012).

Gender inequalities in Latin American cities intertwine with elevated levels of social and spatial inequality in the access to urban opportunities and services (Falú 2014). Therefore, planners must be aware and initiative taking about the intersectionality of gender, with class, race, and other marks of exclusion in urban spaces (Cho et al. 2013). In that regard, multiple and complex factors are behind the persistence of violence against women. Some of these factors manifest at the individual and the family level, while others are connected to cultural and social dynamics. Among the latter, cities play a central role because unequal access to and distribution of resources—such as the layout and distribution of public transportation networks—will impact the likelihood of violent acts taking place (Sauerborn et al. 2021). For this reason, in the next sections, we explore how the urban planning practice sets the layout for violence against women, both at the symbolic and the physical levels.

## 31.3 Violence Against Women in the City: The Private and the Public Realms

Violence against women, it is critical to understand how women are conceptualized in urban planning practice. To do so, it is important to identify and deconstruct narratives of urban womanhood, how they inform the ways women are seen or expected to behave in the city, and ultimately, how these assumptions impact urban planning decisions. Stated somewhat differently, by identifying discourses about women's roles in the city, built on assumptions about expected behavior, we will understand how and why they are so vulnerable to violence.

One link between narratives and urban spaces is seen in the adoption of two mutually exclusive categories: the domestic realm, which is the site of reproduction and is associated with womanhood, and the public realm, where work and other productive activities occur and are thus associated with manhood. To the extent that the urban planning practice focuses on responding to the needs of production, often in ways that undermine or overlook other ways that women also inhabit public spaces, their needs are disadvantaged. Moreover, public and private spaces do not receive the same care, attention, or programming, precisely because "*urban spaces have been designed to value production and undervalue reproduction*" (Buckingham 2010). As a result, the city is an aggressive and non-inviting place for women who venture to be active both inside and outside the home.[1] One way in which urbanization undermines women before exploring how the built environment of the city creates the conditions for development which is by making it more difficult to combine domestic and non-domestic responsibilities. *Since the city is not designed for the domestic caregiver but for the employed producer, anyone who performs caregiving activities, or who aims to combine these with reproductive work, will not find that the city responds to their needs* (Peters 2013).

**The Private Realm**. The first and original locus of discourses concerning urban womanhood is the home, which is considered a women's territory. This is particularly true in Latin American cities, where cultural norms are deeply impacted by conservative patriarchal traditions (Ramm 2016). Even today, after there is a wide

---

[1] Buckingham (2010) argues that "the everyday experiences of women in cities directly stem from the social constructions of gender and space", thus making it crucial to "examine the inequalities which exist, to identify and satisfy human needs and human rights". Although the role as a caregiver afforded women an actual role in society, finally (re)gaining societal value, it constructed a role that belonged in almost opposite urban spaces to men. Because the public space is where the productive work is carried out, it is male, being planned and designed as a response to men's needs.

acceptance of women being employed outside the home, women are still expected to be responsible for most of the burden of household care; taking care of children, elderly, or disabled family members, and conducting the lion's share of the domestic chores. This expectation implies that women stay-at-home for longer hours than men do (Sanchez de Maradiaga 2013). There is plenty of evidence that caregiving responsibilities are highly correlated with female poverty (Roy 2002). In as much as women remain culturally relegated to fulfilling domestic duties within the house, their freedom to explore other means for personal development, economic growth, or contributing to their community is limited or constrained. It is unclear whether this is because society tends to undervalue duties traditionally associated with women, which happen to take place within the home or whether it is the other way around, namely, that society does not value as much what is not traded in the market, which happens to be most of the output of domestic work. Either way, women's needs and priorities rarely fail to land on the agenda of urban planners.

The ways that rigid boundaries between the public and private realms will limit the opportunities for women to have agency outside (as well as inside) their homes are evident in a variety of ways. For one, there is evidence that as women expand their social network outside the domestic sphere, and they are less likely to remain trapped in abusive relations (Klein 2012). As women can navigate the city independently, their potential for earning money increases, which in turn raises their negotiation power within the household (Fajardo-Gonzales 2020). To be sure, it is worth underscoring that employment has not always empowered women. Low-income women who were already fully enrolled in the workforce outside their own homes were often underpaid and their work undervalued and was still highly exposed to gender violence. Likewise, as the COVID-19 pandemic revealed, the uneven distribution of the burden of care is not fully alleviated by the economic independence or level of education of women, narratives and cultural practices impose a gendered division of labor in the household (Power 2020). But overall, freedom to maneuver in the city without constraints gives women more economic and social power.

When cities' lack of responsiveness to the struggles of women to balance a workload both inside and outside the home, they unleash a vicious circle of exclusion and self-exclusion (Rosenthal and Strange 2012). Evidence of these dynamics is to be found in the persistence of gender pay gaps, even when controlling for education and experience levels of men and women (Vaccaro et al. 2021). This pay gap is often explained as a rational decision by employers, who factor in their compensation the likeliness of higher absenteeism among women employees, as they would be responsible for taking care of family needs (Anker and Hein 1985). Whatever its origins, the gender pay gap reinforces the economic justification for women to take most of the burden of the household's care since the loss of women's wages would impact less on the family budget than the loss of the men's wages. Thus, there is a vicious circle between urbanization and the gendered distribution of labor that ends up undermining women's financial and social independence.

However, one should stress that facilitating women to fully participate in the workforce does not address the core problem of abusive domestic relationships within the domestic realm, where women are more likely to be victims of interpersonal violence.

It is telling that domestic violence increased in response to stay-at-home and lockdown orders adopted after the COVID-19 pandemic, as shown in several studies from different cities, states, and several countries around the world (Piquero et al. 2021). For example, domestic violence calls for service increased by 10% in the United States from March 9 to March 31 during social distance events (Leslie and Wilson 2020). Likewise, in Mexico, there was a sharp rise in requests for psychological services at women's call centers during lockdown periods (Silverio-Murillo et al. 2020). In March 2020, with the beginning of the pandemic, Mexico's government reported that *"the country's emergency call centers were flooded with more than 26,000 reports of violence against women, the highest since the hotline was created"* (Kitroeff 2021). In this context, more Mexican women were murdered than died due to COVID-19 in the first three months of the pandemic (OCHA 2020).

**The Public Realm**. While the domestic space is the prime loci for discourses on womanhood, the discourse on the public realm tends to cast it as a place that endangers women. Violence against women in the public realm becomes a cross-cutting concern that reveals both de facto limitations to women's capabilities in the city as well as gender biases in society. Sexual harassment of women in public spaces has been documented in cities in very diverse contexts (UN Women 2017a, b), although local cultures and the design, maintenance, and governance arrangements of the public realm are important determinants of women's higher risk of being harassed. Even if the data is scant and likely to suffer from underreporting, especially in countries with weak police and legal support, it clearly shows that women are the most common victims of sexual harassment in public spaces (Vilalta et al. 2016). In Mexico, victims in 92% of sexual harassment cases and 83% of rape cases identified themselves as women. Significantly, almost half of these crimes take place in public places (INMUJERES 2020). Recent surveys of women found that 60% of respondents in Lima, Peru, and 62% of respondents in San Jose, Costa Rica, had experienced sexual harassment in public streets (Stop Street Harassment 2021).

Within the public sphere, mobility typically presents special safety challenges for women. In general, women make more multipurpose trips, combining their daily work commute with trips to school, childcare facilities, healthcare centers, and shopping purposes. Despite women being the majority of commuters in Latin American cities, a lack of gender-responsive mobility planning makes traveling more expensive, and more dangerous for women than for men. Tellingly, violence against women is most evident in the public transportation system, where women's safety remains a central concern. Women are twice as likely as men to be victims of gender-based violence while on transit (Libertun et al. 2020). To avoid this violence, many women make difficult decisions about the trade-off between economic opportunity and personal security (Dominguez Gonzalez et al. 2020). It is estimated that lack of access to safe transportation will reduce by 15.5% points women's participation (ILO 2017), further undermining women's economic independence. According to one study on public transport in Mexico City, three out of four women are *"not confident about using the transport system without the risk of sexual harassment and abuse or sexual violence"* (Munro and Moloney 2018). Moreover, victimization surveys indicate that violence against women is acknowledged by both men and women alike. When asked to describe the most harmful types of violence, 65% of Latin American respondents claim that violence against women is the most harmful type of violence, higher than violence organized crime, and gang-related violence (both 51%) (Muggah and Aguirre 2018a, b).

Against this backdrop, it is not surprising that fear plays a significant role in determining the way women experience the city (Pain 2014). In the majority of the cities of Latin America and the Caribbean, fear is one of the concepts that structure the experience of women in the city, furthering the notion of segregation and exclusion in her urban experience (Falú 2014). "The most that women in public spaces can wish for is that no one will notice, address, or whistle at them" observes geographer Leslie Kern. In this way, she presents women's self-restrained behavior in the city as a strategy to avoid unwanted attention and control over their bodies (Learn 2021).

Women most often than not read the urban space concerning the degree of danger they might encounter. As one scholar puts it, the *"relationship between people and places is expressed in the construction of the 'other'"* (Soto Villagrán 2012), such that *"fear of violence (is seen) not only as a result of crime but also as an indicator of the power relations in which women are embedded"* (Koskela 1997). It is worth noting that the most relevant factor in the likelihood of being a victim of a crime is not gender but income level and neighborhood location (Salahub 2019). Even so, the saliency of fear when depicting women in public spaces is also deeply problematic, because it naturalizes the notion of women as an inherently vulnerable and lacking agency. In addition, it reinforces the patriarchal view that women (as opposed to men) should behave as risk-avoiders rather than as risk-seekers. And, significantly, such assumptions prevent serious questioning of the social behaviors of men that led women to feel fearful in the first place.

## 31.4 Urban Policy and Violence Against Women

From the perspective of human development, an urbanization pattern that supports women's development would increase society's capabilities holistically and integrally (Sen 1992; Nussbaum 2000). This would require reducing violence against women by changing gender biases, rather than by solely focusing on treating women as victims who must be protected from threats in productive spaces. Unfortunately, it has been common in the practice of urban planning to promote policies that reinforce gender stereotypes, even when their manifest aim is to reduce violence against women and support women's development. This is not to say that urban scholars have not tried to transcend the conceptual divisions between the private and public realms. According to the feminist historian and urban planner Dolores Hayden (1980, p. S176), a *"program to achieve economic and environmental justice for women requires, by definition, a solution which overcomes the traditional divisions between the household and the market economy, the private dwelling and the workplace."* Still, it is important to underscore that the stark division of productive and reproductive realms remains a feature of affluent societies that have access to multiple buildings and can enforce zoning regulations. In the urban survival economies of the poor, the household itself is a vital source of production, which moves seamlessly from the house to the streets. The boundaries between livelihood and household are extremely porous, both spatially and economically (Jarvis et al. 2009). In these contexts, women are even more exposed to suffering the consequences of street violence within their households (Libertun de Duren 2021).

Let us consider the case of housing policies that target female-headed households. Consistently, the data shows that women are overrepresented among those living in inadequate housing conditions as well as among households located in informal or underserved neighborhoods. On top of this, most land titles are still registered under men's names, and it is estimated that women account for less than a quarter of landowners in Latin America (Chant and McIlwaine 2015). At the same time, women are severely underrepresented in the mortgage markets, even if they are more than three times less likely to default on their payments (Libertun and Hernandez 2020). This high rate of housing deficit serves to increase low-income women's exposure to violence, through several diverse mechanisms. Women living in substandard housing, for example, face increased risks of sexual assault at night when using sanitation facilities located outside their homes. Also, because poor neighborhoods have limited access to efficient means of public transportation, women in these areas are required to make long journeys at early or late hours, further increasing their exposure to sexual violence during the daily commute (Libertun de Duren et al. 2018). In contrast, there is evidence that women's ownership of title correlates with a decline in gender-based violence, either because of a change in men's attitudes (Amaral 2017) or because women are more inclined to leave abusive relationships when they have secure housing (Moser 2017).

Despite the urgency of empowering women (particularly low-income women) outside of the domestic realm to increase their independence and autonomy, most development policy has reinforced the gendered assumptions that women are defined by their domestic roles, something which further constrains them. Take for example the Conditional Cash Transfer programs. In these programs, to qualify as beneficiaries of a monthly payment, women are required to perform motherly and housewife-like duties' (Chant 2008). Tellingly, this approach has been called a 'modernization' of policies (Molyneux 2007). While these programs emphasize the valuable role women play in society, they do so at the expense of limiting and rigidly prescribing these roles to gendered, female activities within the home. Therefore, the position of women as public citizens with responsibilities in the work world continues to be ignored by policy, leading to further undervaluation, such that exposure to domestic violence remains high (Bradshaw et al. 2019).

Violence and the fear of violence that women experience in public spaces are other areas where urbanists have been proactive. Dark areas, isolated parks, empty, and poorly lit streets, underground parking lots, and pedestrian underpasses provide opportunities for criminal activities and cause fear and insecurity for women. Because of that, many planners have advocated for layouts that make it harder for violence to take place and that increase women's sense of security, an approach commonly known as Crime Prevention Through Environmental Design or CPTED (Cozens and Love

2015). Eventually, this approach evolved into one which proactively included women in the design of a safer city. "Women Safety Walks," a design strategy developed in Toronto, Canada, in the late 1980s, allow participants to identify safe and unsafe spaces and provide recommendations on how to improve the safety of the area. It also promotes a partnership between users and experts, residents, and governments (Whitzman et al. 2009). More recently, this concept has incorporated the use of digital technologies and Geographic Information Systems to improve the mapping of the areas in which women feel unsafe, integrating a quantitative and a qualitative approach (Gargiulo et al. 2020). This method has been replicated with success in various urban settings. Significantly, it has allowed women to provide concrete feedback based on their own lived experiences, as well as empowered them to make positive changes in their neighborhoods. But while these are important gains, it would be misguided to assume that violence and fear can be simply "designed out" because it is not the urban space per se that causes violence (Kosekela and Pain 2000). That is, violence is not the outcome of poor design and maintenance, but of a society that condones the behavior and that does not treat women as equals.

A similarly limited approach informs most urban policies that center on the issue of the burden of household care. Instead of aiming for creating equality among men and women's workloads, most policy advocates seek to facilitate the capacities of women to take on this extra load. Urban transportation systems provide an instructive lens to understand how this gender bias works in our cities. As it is well known, the layout of the transportation matrix supports the needs of those who work in paid activities outside of the home. Therefore, those who perform caregiving activities are faced with poor service and less efficient transportation choices (Libertun de Duren et al. 2018). Complicated routes must be carried out in an urban infrastructure that has not been designed to respond to, or designed for, the accommodation of kids, strollers, or groceries. To challenge this concept, urban planners have pushed for the incorporation of the category *"mobility of care to consider the invisibilized, underqualified, unpaid, and undervalued work of taking care of others and the upkeep of the home"* (Sánchez de Maradiaga and Zucchini 2019). But even after this clear call for action, there is a risk that the current gendered experience of the urban transportation systems becomes the blueprint for the urban transportation policies of the future, thus reinforcing the assumption—or the expectation—that women disproportionately carry both work and home burdens while commuting. This approach compounds the *"feminization of obligation and responsibility"* in which women are the service of policy, rather than served by it (Chant 2008).

Transportation systems have also been the focus of policies targeting violence against women. Quite often, their approach has been to prevent gender violence by providing segregated spaces for men and women. These initiatives aim to increase women's safety by compartmentalizing the urban experience into fragments of isolated and disconnected spaces, which may serve to undermine women's freedom and access to opportunities in the city. That is, these measures put *"the burden on women to protect themselves by withdrawing from the male gaze into second-class services…(even though) the onus should be on men to stop harassing women, not on women to escape them"* (Whitzman 2012). The City of Mexico, whose public transportation system ranked as the least safe for women in Latin America and the second-least safe among the 15 largest cities in the world (Granada et al. 2016), adopted a policy based on gender segregation. As a measure to prevent gender violence, the city launched the program "Let's travel safely" (*viajemos seguras*), which provides separate and exclusive cars for the use of female commuters in the Metro, Metrobus, Trolleybus, Light Train, and a women-only bus system (Soto Villagrán et al. 2017). The program has led to mixed outcomes. One positive and significant result of this policy is a reduction of about 11% of the occurrence of sexual violence among women riding the women's only bus. However, at the same time, the program has led to an increase in violence in other spaces.

Indeed, evaluations of the program show that women who do not use segregated buses—which are available less frequently than the regular buses—have been even more exposed to gender violence than before. This is because many men interpret that if a woman do not travel in segregated women-only cars, they do not mind being harassed. This is the view captured in a qualitative study on gender violence and public transportation in Mexico City: *"If you don't like being touched, then go to the all-women metro cars. I don't know why some women like to go with the men during rush hour and sometimes it is inevitable and other times they like it."* (Dunckel Graglia 2016. p. 631). Another unexpected, and deeply disturbing, outcome of the program is a 30% increase in events of non-sexual, physical violence among male commuters (Aguilar et al. 2021). It is unclear the reason behind this increase, but it certainly suggests that violence against women and others could be seen as the emergent expression of a violent society. Accordingly, without addressing the roots of that violence and solely targeting women's behavior, there would be a displacement of the violence and the social exclusion of different targets in society.

The City of Quito, Ecuador, took a different approach to preventing gender violence in public transportation. Rather than isolating men from women, it provided all passengers with the agency to prevent it. The program "Get rid of harassment" (Bájale al acoso) was implemented in 2017 on the Trolleybus system. It allows any passenger who feels threatened with harassment while traveling on a bus to send an SMS message directly to a control center, identifying the bus route. The driver of the bus is immediately informed that there is an incident taking place onboard and plays a recorded alarm and voice recording to warn passengers that inappropriate behavior is occurring and asks passengers if they see someone in distress. At the next stop, specially trained police or other security staff would board the bus so to allow the person affected to identify the perpetrator (Allen et al. 2017). Also, Quito implemented a streamlined and centralized legal process, so the victims do not have to be burdened with a lengthy trial to bring to justice the harasser. Significantly, a continuous impact evaluation showed that the program contributed to reducing situations of sexual harassment in the bus system. Further, by using SMS instead of smartphone applications, the campaign is more inclusive of lower socioeconomic levels. This is an example of how effective reporting can be implemented, helping the victims when they most need it without questioning the legitimacy of what they experienced while identifying and punishing the perpetrator (Granada et al. 2018). More importantly, it empowers the victim to speak up and all passengers to contribute to stopping sexual harassment.

## 31.5 Conclusion

Some of the contemporary patterns of urbanization endanger women in both direct and indirect ways. Concerning the former, women are more exposed to being victims of sexual harassment in public spaces and public transportation systems, as well as being victims of violence within their homes. These occurrences make fear of violence central in many women's lives and lead them to change their behavior in a manner that can further undermine their personal development, such as limiting their participation in labor markets and civic life. Concerning the latter, urbanization patterns—as mediated by infrastructure—also weaken women's status in society and thus make them more likely to be victims of violence. For example, the layout of the public transportation matrix makes it onerous for women with household responsibilities to work in full-time jobs, at the same time as housing titling and financing practices make it more difficult for women than for men to own property. But urbanization has dented women's development in yet another way, which is by promoting urban policies that, under the manifest purpose of helping women, have naturalized gendered roles in society without questioning the inequality in which these assumptions are embedded. Indeed, by reinforcing traditional patterns of female behaviors, these and other urban policies have ignored or failed to engage in conversations about the cultural and institutional practices that have contributed to constraining women's development in the first place (Bradshaw and Linneker 2015). In this way, failing to realize women's right to the city.

An example of this is the frequency in which policy focuses on women solely address the responsibilities of motherhood. When considering programs for women, policymakers tend to prioritize women's access to economic support or have focused on easing the burden of care on women rather than on constructing a society in which men and women equally share the burden of care alike. As Martha Nussbaum has stated in her reading of gender from a "capabilities" framework: *"One might sum all this up by saying that all too often women are not treated as ends in their own right, persons with a dignity that deserves respect from laws and institutions. Instead, they are treated as mere instruments of the ends of others—reproducers, caregivers, sexual outlets, agents of a family's general prosperity. Sometimes this instrumental value is strongly positive; sometimes it may be negative"* (Nussbaum 2001).

A similar focus on the outcomes rather than on the causes of gender inequality is present in many policies aimed at stopping gender violence, whether in the household or the city. They revolve around seeing women as victims, without addressing the problem of abusive men and violence in society. Segregating women and men, or proscribing women's behavior without listening to their concerns, cannot be a long-term approach to curbing violence against women, or any type of violence. In the short term, these strategies displace violence from one place to another or shift the likelihood of victimization from one group of people to another one. Urban planning practitioners and policymakers need to internalize the notion that preventing violence against women is the first step in producing equitable cities. They also must understand that any kind of violence is a manifestation of an unequal society. This demand that urban planners and policymakers think about society and not just (urban) space, and find programs focused on the latter that will positively impact the former. While gender-responsive regulations and interventions are required, particularly in the design and management of housing, transport mobility systems, and public spaces, they should be part of a holistic and integral approach to addressing violence in cities and society simultaneously.

Even when stopping violence against women requires a systemic approach, urban planners can be vigilant about including three specific actions in all projects they tackle, to contribute to gender equality in cities. First, one must always give voice to those whom the project intends to benefit. Otherwise, the risk of misinterpretation, lack of ownership, or continuation of oppressive practice remains. Second, one can never solve social problems by furthering spatial, social, or gender segregation. It will likely create a tiered system of benefits, and it will be unlikely to improve social conviviality. Third, one must consider the whole of the city, in which the public and private realms, the cultural norms, and the forms of the built spaces, are deeply interconnected and intertwined when conceiving of all actions. Boundaries between the two are fabrications that do not reflect the complexity of the lived experiences of urban residents. All in all, this is not to say that urban planning needs to revolve only around the challenges of women, violence, or violence against women. Rather, it needs to integrate gender into its realm, incorporate human relationships as an intrinsic part of the way in engaging in the production of our cities, and focus on the incessant tasks of building cities for all.

## References

Aguilar A, Gutierrez E, Villagran PS (2021) Benefits and unintended consequences of gender segregation in public transportation: evidence from Mexico City's subway system. Econ Dev Cult Change 69(4):1379–1410

Allen H, Leda P, Lake S, Galo C (2017) She moves safely: a study on women's personal security and public transport in three Latin American Cities. In: FIA foundation research series, vol 10. FIA Foundation, London, United Kingdom

Amaral S (2017) Do improved property rights decrease violence against women in India? Institute for Social and Economic Research, University of Essex, Essex, England

Anderson KL (2007) Who gets out? Gender as structure and the dissolution of violent heterosexual relationships. Gend Soc 21(2):173–201

Anker R, Hein C (1985) Why third world urban employers usually prefer men. Int Labor Rev 24(1):73–90

Bott S, Guedes A, Goodwin MM, Mendoza JA (2012) Violence against women in Latin America and the Caribbean: a comparative analysis of population-based data from 12 countries. Pan American Health Organization, Washington, D.C.

Bradshaw S, Linneker B (2015) The gendered destruction and reconstruction of assets and the transformative potential of 'disasters'. In: Gender, asset accumulation and just cities. Routledge, pp 176–192

Bradshaw S, Chant S, Linneker B (2019) Challenges and changes in gendered poverty: the feminization, de-feminization, and re-feminization of poverty in Latin America. Feminist Econ 25(1):119–144

Buckingham S (2010) Examining the right to the city from a gender perspective. In: Sugranyes A, Mathivet C (eds) Cities for all: proposals and experiences towards the right to the city. Habitat International Coalition (HIC), Santiago de Chile, Chile

Caldeira TPR (2000) City of walls: crime, segregation, and citizenship in São Paulo. University of California Press, Berkeley. Print

Castells M (2008) The new public sphere: global civil society, communication networks, and global governance. Annals Am Acad Politi Soc Sci 616(1):78–93. Web

CEPAL (Economic Commission for Latin America and the Caribbean) (2021) Observatorio Igualdad de Género de America Latina y el Caribe. Available at https://oig.cepal.org/es/leyes/leyes-de-violencia

Chant S (2008) The 'feminisation of poverty' and the 'feminizations' anti-poverty programmes: room for revision? J Dev Stud 44(2):165–197

Chant S, McIlwaine C (2015) Cities, slums and gender in the global south: towards a feminized urban future. Routledge, London

Chant S (2020) Geography and gender, hindsight and foresight: a feminist development geographer's reflections on: 'how the other half lives: the geographical study of women. Area (London 1969) 52(4):778–785. Web

Cho S, Crenshaw KW, McCall L (2013) Toward a field of intersectionality studies: theory, applications, and praxis. Signs: J Women Cult Soc 38(4):785–810

Cozens P, Love T (2015) A review and current status of crime prevention through environmental design (CPTED). J Plan Lit 30(4):393–412

Davis DE (2015) Socio-spatial inequality and violence in cities of the global south: evidence from Latin America. In: Salo K, Wilson D, Miraftab F (eds) Cities and inequalities in a global and neoliberal world. Routledge, New York, pp 89–105

Davis D (2014) Modernist Planning and the foundations of urban violence in Latin America. Built Environ 40(3):376–393

Dominguez Gonzalez K, Machado AL, Bianchi Alves B, Raffo VI, Guerrero Gamez S, Portabales Gonzalez I (2020) Why does she move?: a study of women's mobility in Latin American cities. World Bank Group, Washington, D.C.

Dunckel Graglia A (2016) Finding mobility: women negotiating fear and violence in Mexico City's public transit system. Gend Place Cult 23(5):624–640

Fajardo-Gonzalez J (2020) Domestic violence, decision-making power, and female employment in Colombia. Rev Econ Household 1–22. https://doi.org/10.1007/s11150-020-09491-1

Falú AM (2014) El derecho de las mujeres a la ciudad: espacios públicos sin discriminaciones y violencias

García-Moreno C, Zimmerman C, Morris-Gehring A, Heise L, Amin A, Abrahams N, Montoya O, Bhate-Deosthali P, Kilonzo N, Watts C (2015) Addressing violence against women: a call to action. The Lancet 385(9978):1685–1695

Gargiulo I, Garcia X, Benages-Albert M, Martinez J, Pfeffer K, Vall-Casas P (2020) Women's safety perception assessment in an urban stream corridor: developing a safety map based on qualitative GIS. Landsc Urban Plan 198:103779

Granada I, Leaño JM, Crotte Alvarado A, Cortés R, Ortiz P (2018) Género y Transporte: Quito. Banco Inter Americano de Desarrollo

Granada I et al (2016) The relationship between gender and transport. Interamerican Development Bank

Harvey D (2003) The right to the city. Int J Urban Reg Res 27(4):939–941. Web

Hayden D (1980) Supplement. Women and the American City (Spring, 1980). Signs 5(3):S170–S187
Hayden D (1982) The grand domestic revolution. MIT Press, Yale
https://www.unodc.org/unodc/en/data-andanalysis/global-study-on-homicide.html
ILO_International Labor Organization (2017) World employment and social outlook: trends for women 2017. International Labor Office, Geneva
Instituto Nacional de las Mujeres (2020) Las mujeres y la violencia en el espacio público. http://cedoc.inmujeres.gob.mx/documentos_download/BA6N04_VoBo_250620_Final.pdf
Jaitman L (2019) Frontiers in the economics of crime: lessons for Latin America and the Caribbean. Latin Am Econ Rev 28(1):1–36
Jarvis H, Paula K, Jonathan C (2009) Homes, jobs, communities and networks. In: Cities and gender. Routledge, New York, NY, pp 186–215. ISBN 9780415415705
Kitroeff (2021) New York Times. https://www.nytimes.com/2020/05/31/world/americas/violence-women-mexico-president.html
Klein R (2012) Responding to intimate violence against women: the role of informal networks. Cambridge University Press
Koskela H (1997) 'Bold walk and breakings': wo men's spatial confidence versus fear of violence. Gender, Place Cult J Feminist Geogr 4(3):301–320
Koskela H, Pain R (2000) Revisiting fear and place: women's fear of attack and the built environment. Geoforum 31(2):269–280. https://doi.org/10.1016/S0016-7185(99)00033-0
Lefebvre H (1968) Le droit à la ville. Anthopos, Paris
Leslie E, Wilson R (2020) Sheltering in place and domestic violence: evidence from calls for service during COVID-19. J Public Econ. https://doi.org/10.2139/ssrn.3600646
Libertun, Hernandez A (2020) Credit markets in Ecuador. Inter-American Development Bank
Libertun de Duren N (2020) Effects of neighborhood upgrading programs on domestic violence in La Paz, Bolivia. World Dev Perspect 19:100231
Libertun de Duren N et al (2018) Inclusive cities: urban productivity through gender equality. IDB, Washington, D.C. https://doi.org/10.18235/0001320
Libertun de Duren N (2021) Slum upgrading programs in Bolivia
Madariaga ISD, Zucchini E (2019) Measuring mobilities of care, a challenge for transport agendas. In: Integrating gender into transport plannin. Palgrave Macmillan, Cham, pp 145–173
Marcuse P (2014) Reading the right to the city. City 18(1):4–9
Molyneux M (2007) Change and continuity in social protection in Latin America: mothers at the service of the state, gender and development paper 1. United Nations Research Institute for Social Development, Geneva. www.unrisd.org
Moser CON (2017) Gender transformation in a new global urban agenda: challenges for habitat III and beyond. Environ Urban 29(1):221–236. https://doi.org/10.1177/0956247816662573
Muggah R, Aguirre K (2018a) Citizen security in Latin America: facts and figures. Strategic Paper 33, Igarapé Inst., Rio de Janeiro
Muggah R, Aguirre K (2018b) Citizen security in Latin America: the hard facts. Irapagué Instit Strateg Paper 33:1–63
Munro N, Moloney A (2018) Exclusive: Mexico city's transport ranked as most dangerous for women. Reuters. https://www.reuters.com/article/us-transport-women-poll-mexico/exclusive-mexico-citys-transport-ranked-as-most-dangerous-for-women-global-poll-idUSKCN1NK059
Nussbaum MC (2001) Women and human development: the capabilities approach, vol 3. Cambridge University Press
Nussbaum MC (2000) Women and human development: the capabilities approach, vol 3. Cambridge University Press
OCHA_United Nations Office for the Coordination of Human Affairs (2020) Pandemic of violence: protecting women during COVID-19. Accesed 15 Jun 2022. Pandemic of Violence: Protecting Women during COVID-19—Mexico|ReliefWeb
Pain R (2001) Gender, race, age and fear in the city. Urban Stud (Edinburgh, Scotland) 38(5/6):899. ISSN: 0042-0980

Pain R (2014) Seismologies of emotion: fear and activism during domestic violence. Soc Cult Geogr 15(2):127–150

Parkes J (2015) Gender violence in poverty contexts. Taylor & Francis

Peters D (2013) Gender and sustainable urban mobility. In: Global report on human settlements 2013. Nairobi, Kenya

Piquero AR, Jennings WG, Jemison E, Kaukinen C, Knaul FM (2021) Domestic violence during the COVID-19 pandemic-evidence from a systematic review and meta-analysis. J Crim Just 74:101806

Power K (2020) The COVID-19 pandemic has increased the care burden on women and families. Sustain Sci Pract Policy 16(1):67–73

Ramm A (2016) Changing patterns of kinship: cohabitation, patriarchy and social policy in Chile. J Lat Am Stud 48(4):769–796

Rosenthal SS, Strange WC (2012) Female entrepreneurship, agglomeration, and a new spatial mismatch. Rev Econ Stat 94(3):764–788

Roy A (2002) Against the feminization of policy, comparative urban studies project policy brief. Woodrow Wilson International Center for Scholars, Washington DC

Sánchez de Maradiaga I (2013) "From women in transport to gender in transport. Challenging conceptual frameworks for improved policymaking", en The gender issue: beyond exclusion, special issue, Journal of International Affairs, vol 67, no 1. Columbia University, NY, pp 43–66

Salahub JE (2019) Reducing urban violence in the global south: towards safe and inclusive cities

Sassen S (2005) The global city: introducing a concept. Brown J World Aff 11(2):27–43

Sauerborn E, Eisenhut K, Ganguli-Mitra A, Wild V (2021) Digitally supported public health interventions through the lens of structural injustice: the case of mobile apps responding to violence against women and girls. Bioethics 36(1):71–76

Sen A (1992) Missing women. BMJ: Br Med J 304(6827):587

Sen P, Borges E, Guallar E, Cochran J (2018) Towards an end to sexual harassment: the urgency and nature of change in the era of# MeToo. UN Women

Silverio-Murillo A, Balmori de la Miyar JR, Hoehn-Velasco L (2020) Families under confinement: COVID-19, domestic violence, and alcohol consumption. In: Andrew young school of policy studies research paper series forthcoming. Available at SSRN https://ssrn.com/abstract=3688384 or https://doi.org/10.2139/ssrn.3688384

Soto Villagrán P, Esteva AA, Fernández EG, Reséndiz CC (2017) Evaluación de Impacto Del Programa 'Viajemos Seguras En El Transporte Público En La Ciudad de México': Aportes al Diseño e Implementación de Políticas de Prevención de La Violencia de Género En Espacios Públicos. Publications. IDB-TN-1305. IDB

Soto Villagrán P (2012) El miedo de las mujeres a la violencia en la ciudad de México: Una cuestión de justicia espacial. Revista INVI 27(75):145–169. https://doi.org/10.4067/S0718-83582012000200005

Stop Street Harassment (2021) Statistics—the prevalence of street harassment. https://stopstreetharassment.org/resources/statistics/statistics-academic-studies/

UN Women (2017a) Mexico City hosts global forum on safe cities for women and girls|UN women—headquarters

UN Women (2017b) Safe cities and safe public spaces. UN Women Headquarters, New York

UN Women (2020) The shadow pandemic: violence against women during COVID-19. [UN Women.website]. https://www.unwomen.org/en/news/in-focus/in-focus-gender-equality-in-covid-19-response/violence-against-women-during-covid-19

UN Women (2021a) Global fact sheet violence against women prevalence estimates, 2018. Available at https://apps.who.int/iris/handle/10665/341604

UN Women (2021b) Measuring the pandemic: violence against women during COVID-19. https://Measuring-shadow-pandemic.pdf (unwomen.org)

UNODC (United Nations Office on Drugs and Crime) (2019) Global Study on Homicide 2019. UN

Vaccaro G, Basurto MP, Beltrán A, Montoya M (2021) The gender wage gap in Peru: drivers, evolution, and heterogeneities. Soc Incl 10(1):19–34

Vilalta CJ, Castillo JG, Torres JA (2016) Violent crime in Latin American cities. Inter-American Development Bank, Washington, DC

Whitzman C, Shaw M, Andrew C, Travers K (2009) The effectiveness of women's safety audits. Secur J 22(3):205–218

Whitzman C (2012) Women's safety and everyday mobility. In: Building inclusive cities: women's safety and the right to the city. Taylor & Francis Group, London, United Kingdom. http://ebookcentral.proquest.com/lib/mit/detail.action?docID=1143752

Young IM (1994) Gender as seriality: thinking about women as a social collective. Signs: J Women Cult Soc 19(3):713–738


**Nora Libertun de Duren** is a Lecturer-in-Urban-Planning at Harvard_University and a Lead Specialist in the Climate Change and Sustainability Sector at the Inter-American-Development-Bank. She was the Director-of-Planning-and-Natural-Resources for New York City. She has managed $2000 million in sovereign loans for urban projects in Latin America, $1000 million for NYC's parks, and $2.8 million in research grants urbanization and gender and diversity. She has published more than 20 peer-reviewed papers and 30 chapters and monographs. Nora holds a Ph.D. from MIT, a master's from Harvard University, and a Master's from the University-of-Buenos-Aires.

**Diane E. Davis** is the Charles Dyer Norton Professor of Regional Development and Urbanism and former Chair of the Department of Urban Planning and Design at Harvard's Graduate School of Design (GSD). Before moving to the GSD in 2011, Davis served as the head of the International Development Group in the Department of Urban Studies and Planning at MIT, where she also was Associate Dean of the School of Architecture and Planning. Trained as a sociologist (BA in Sociology & Geography, Northwestern University; Ph.D. in Sociology, UCLA) Davis teaches courses on the spatial structure, social composition, and governance of cities.

**Maria Lucia Morelli** is an Urban Designer and Project Manager at Practice for Architecture and Urbanism (PAU) in New York. She obtained a Master of City Planning from the Massachusetts Institute of Technology, where her research focused on how to plan for safer cities for women in Latin America. She holds a Bachelor of Architecture from the Pontificia Universidad Católica del Perú, and worked as an architect in Lima for four years.

# Chapter 32
# Border Cities Between Life and Death: Ciudad Juárez and El Paso

**Mauricio Vera-Sánchez and Luis Alfonso Herrera-Robles**

**Abstract** This text contains a reflection on the border cities of El Paso, Texas, and Ciudad Juárez, Chihuahua, from an unconventional reading within the Social Sciences. It is about capturing those physical and imaginary elements such as the wall, so that through notions such as border, emotional, degradation, decomposition, abandonment, and other categories, an interpretation of the urban, citizen, and border frameworks that are put into play is developed these types of geographic regions. In addition to going beyond the academic tradition of geopolitics and culture as a starting point for any explanation of border phenomena. This time, the bet is on an interpretation that speaks of the life and death of both cities based on the aesthetic practices of geopoetics and urban imaginaries that constitute an "other way" of reading the reality of two cities that to the south and to the north of each of their countries (the United States and México) make up the current phenomenon of the gradual death of cities. The urbicide.

**Keywords** Cities · Borders · Geopolitics · Art · Aesthetics · Image of the city

## 32.1 Introduction

The urban border between the cities of El Paso, located in the United States, and Ciudad Juárez, located in México, reveals to us as an object of study with a double dimension: the first one from a geopolitical perspective of military, economic, and migratory control; a geometrical and measurable border. A border that is visible in the media that shows–like a death liturgy–violence, narcotraffic, destruction, and urban degradation: illegal bodies that jump over the wall or go through the desert or

---

M. Vera-Sánchez (✉)
National Open and Distance University (UNAD), Bogotá, Colombia
e-mail: mauricio.vera@unad.edu.co

L. A. Herrera-Robles
Autonomous University of Ciudad Juárez (UACJ), Ciudad Juárez, México
e-mail: lherrera@uacj.mx

© The Author(s), under exclusive license to Springer Nature Switzerland AG 2023
F. Carrión Mena and P. Cepeda Pico (eds.), *Urbicide*, The Urban Book Series,
https://doi.org/10.1007/978-3-031-25304-1_32

cross the river looking for their relatives, trying to reach the so wanted but not always achieved American dream.

The second dimension allows us to understand these border cities as sensible possibilities and realities, as well as anchored aesthetics that are constructed in the citizens' imaginaries and in their artistic expressions that answer–mostly from *geopoetics* rather than geopolitics–to the inevitable human need of getting together with others, of being with others. To achieve this goal, numerous strategies will be designed, showing that any border is invariably porous and that will always have holes in which the unstoppable desire of survival and transcendence will find its way in.

In this order, this chapter approaches some topics related to the urban border between these two cities that also relates in many aspects, to the border between the two countries. First, this paper approaches the reference about the border wall as the most extreme materialization of geopolitics and the exclusion of the south from North. The research *Geopoeticas, memoria e imaginarios en la frontera Mexico-Estados Unidos* (Vera 2020) (*Geopoetics, memory and imaginaries in the Mexican American Border*), retakes the reading of the wall from a geopolitical and *geopoetic* dimension, as well as the concept of urban imaginaries–in Armando Silva's perspective (2014)– of El Paso and Ciudad Juárez citizens. Elements like the differences and similarities between both cities, the otherness perception, the emotions, the media, where the difference becomes sensible, are also studied. Under Silva's methodologic route, which emphasis is hermeneutical, and it is centered in the search of meaning, the perceptual or aesthetical experience as an inherent factor of human knowledge, as well as the construction of the otherness, a survey was created to facilitate the research about the citizens' imaginaries around these aspects.

Thereby, the applied instrument that simultaneously operates as an interview, was structured from axes of a historical order, identity, visualities, and memories from the border. This instrument was applied to a non-probable sample and stratified by levels of age and geographical location in both cities. Thus, the study includes 60 participants between artists, urbanists, architects, and academics from Ciudad Juárez and El Paso, within an age range of 18–60 years. This sample facilitates a broad view of perceptions. Likewise, three deep interviews were completed with people linked to art museums in both cities.

In the same way, an analysis of some artistic processes and artistic work to which the central focus is the border wall was made. These illustrate not only a critical and hopeless reading of the interstitial emptiness that sometimes the border can be, but simultaneously, show the possibility of life. Methodologically, the selection of the analyzed pieces has as a central focus, the relation between art, identity, territory, and the border, which are seen from their conceptual, formal, and aesthetic proposals.

Also, this research works specifically about the abandonment processes and the urban degradation that has taken place in Ciudad Juárez. Some of these because of the city's condition as a border city with the United States. In the same way that the border wall makes itself sensitive to the art's *geopoetic* and in the citizens' imaginaries as sensitive and aesthetic manifestations, it is important to analyze some artistic productions related to the violence experiences, deterioration, and urban death of the city.

## 32.1.1 The Border Wall

The wall as a *border*, meaning the territorial limit in a strict sense, talks about the extreme, radical, and complex geopolitical condition of the border between México and the United States. From the theory of the imaginaries (Silva 2014) it is possible to point at the perceptual relation between the word border and the image of the wall, to the point that its images–inhabited by migrant bodies in transit trying to cross it–mark the feeling about the border condition: the wall imposes with forcefulness, the concept of the border between these nations in the most rational, geometrical, of military power and economical control of a country above the other. Therefore, to think today about the border, more than in any other moment in history, is to think about the wall.

Historically, the discourses and justifications about the wall–specially the Anglo-Saxon ones- have spun around the reaffirmation and maintenance of the asymmetrical relations of surveillance and migratory control that have characterized the position of the United States in its relationship with Mexico and Latin America in general. So, since the late nineteenth century and the beginning of twentieth century it is possible to track expressions like the one used by the immigration inspector Marcus Braun: "*I knew that eternal vigilante on the Mexican border line is the prize of liberty*", as well as actions like the installation of disinfection stations in El Paso, utilized by border authorities to implement surveillance strategies centered in the control of the migrant body that tried to cross the north from south.

In this sense, the Anglo-American border imaginary about Mexicans starts to take shape, and it is the same still today. This imaginary lays out a reality in which Mexico and mexicaness are described and placed as the other, the one that is different, and it needs to be restricted, except for when low-cost workforce is needed or when American products and culture consumers are needed. Under this logic, the wall shows itself as an effective strategy to the urban space exclusion that occurs between both cities and that allows the perpetuation, precisely, of the imaginary of separation in the geopolitical order between both countries.

The border wall reveals then, the complexity in the understanding of the border as the urban object from which international politics and territorial relations revolve around. The same that allows the implementation of the difference between ones and the others. This human exercise of dividing what in terms of space seems like a whole, it also puts into play the cultural, economic, and historical relations between those two parts that are being separated. Simultaneously, affective relationships are also being modified, configuring the territory, the emotions, and collective imaginaries where it stands the understanding of an own and different place.

Like this, in the urban imaginaries, Ciudad Juárez and El Paso citizens identify the economy as one of the biggest historical differences, which is linked to their

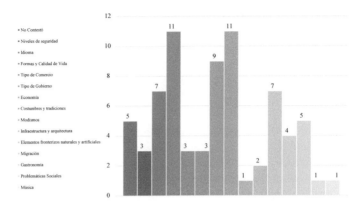

**Fig. 32.1** Differences and similarities between Ciudad Juárez and El Paso. *Source* Own elaboration

ways of life and life quality. In the same way, it is also recognized that customs and traditions are shared and those are represented in everyday life through language idioms derived from the fusion between English and Spanish, as well as gastronomy and music (Fig. 32.1).

From the theory proposed by the urban imaginaries of Silva (2014), we could say that is there, in the border imaginaries–of an emotional base, or better, aesthetic–where border citizens construct, from the perception of their own social reality, what the dividing wall between México and the United States is. In this order, the imaginaries, being emotional, thus aesthetic, allow us to draw, rather than a geopolitically rational map, a sketch of the emotions that pound in the people that inhabits the border. To the questions "What kind of emotions do you believe become sensitive to the differences in the border between the Unites States and México?", an emotional state is revealed. This sentiment goes from urban feelings such as racism, fear, powerlessness, but also unity and affection (Fig. 32.2).

These emotions allow us to understand the historical tensions that frames the relationship between these cities–also reflecting the relationship between México and the United States–and that shows no other feelings more positive or opposite to fear, anger, hopelessness, or racism. The wall then, as an object of urban reference of this separation, operates simultaneously in a material, symbolic, and cultural notion of one nation over the other one.

It is meaningful that fear, as a mutual sentiment between citizens from Ciudad Juárez and El Paso sets the border imaginaries. As described by Silva (2014) in his text *Imaginarios, el asombro social*, fear is one of the urban topics with the most contagious capability:

> The fear in the cities serves as a reference to understand the emotional distribution modes of the imaginaries, in its attachment to the group aesthetic expressions. [...] Perhaps it is fear the feeling with the most consistency in the contemporaneous urbanism: the fear in the city, as a residue of other sentiments (vengeance, illusions, hate) that lead to react (p. 265).

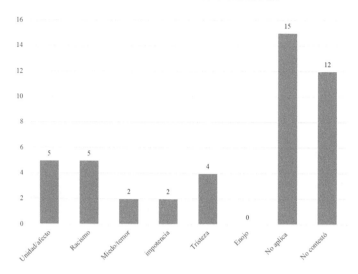

**Fig. 32.2** Emotions that evidence the differences between Ciudad Juárez and El Paso. *Source* Own elaboration

Thus, the border becomes a strategic category to be understood as that social, cultural, and symbolic space. A space where to rethink about the current crisis of a globalized world and an unstoppable capitalist economy.

Now, about the fear of *"the other"*, as a characteristic of the mutual citizen's urbanism of El Paso and Juárez, it is related indeed to the construction of otherness. To the question about how would you defined *"the other"* in the border context, the urban imaginaries give account of answers that range between the *"el pocho"* figures, to the chicano or el gringo. These are cultural identities movements that are historically rooted in the 40 s and 50 s, as a result of the evident discrimination processes against Mexicans, especially in the southwest of the United States (Fig. 32.3).

The figure of the so called *"pachucos"* shows the initial marks of some resistance aesthetics expressed by fashion and language, configuring an identity that is still being resignified, precisely, by the *"cholo"*, chicano en "pocho" culture. This can be seen, not only through the urban imaginaries but also in artistic productions like Paola Tascón Tello's work, who has shown through her photographical project "CHOLOS, Identidad Fronteriza Mexicana" (*CHOLOS, Mexican Border Identity*) the "chola" subculture. This subculture also known as "latino gangs", as a way of self-definition and differentiation of other gangs uses fashion, language, and tattoos as a resistance to an American dominant culture (Biennial 2015, Ciudad Juárez + El Paso). This can also be seen through the graphic work of El Paso artist Gaspar Eriquez, who shows the political tension caused by the *chicano* and how this lifestyle passes from one generation to the next one, surviving war, prison, and conflict. More importantly, it shows that this situation will not change until poverty and the destruction of the neighborhoods in both cities does not stop: "one is born Mexican, but decides to be a chicano", as the artist states (III Biennale Ciudad Juárez-El Paso 2013, p. 39).

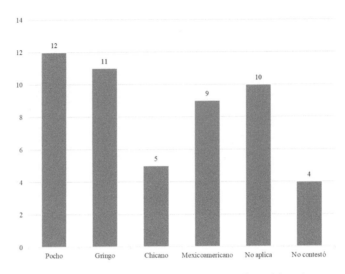

**Fig. 32.3** How do you define the "Other" in the border. *Source* Own elaboration

### 32.1.2 The Geopoetics of the Wall

As it has been proposed previously, it is necessary to think about the border, meaning the wall, rather than as a geopolitical map and barrier, as a sketch of a *geopoetic* order. In this way, both, the imaginaries and art, do not in fact map the border but draw a sketch of feelings and sensitivities from which the border acquires a more complex character that tries precisely to understand its closeness and porosity. The wall is then sort of body in between the divided bodies. Divided by geopolitics, but brought together by *geopoetics* through art. Art is an object that exerts a day to day tension to those who are in both sides, giving them different options like legally cross it though the international bridges, illegally jump it in those far away from the city places or "mock it" through aesthetic strategies, like those made by artist Gracia Chávez, who in her creative process for the 2018 Biennial in Texas, took the place of an immigrant female worker who works legally as a domestic employee in the United States. She crossed from Ciudad Juárez to El Paso instead of her, looking to show how absurd immigration requirements are, as well as to symbolically deactivate the harassment against immigrants.

In a similar way of trying to expose the porous condition of the border wall, artist Martha Palau build an installation titled, precisely, Muros (Walls). In such work, she places in dialog and tension two walls made of rough wood. These vertical walls represent the military power and the migratory rigor of the border Patrol, between these she places a human figure made from jute material. This figure emulates the body of an immigrant that has been killed after trying to cross the border illegally, establishing like this the indissoluble and inevitable relation between life and death in the border. The aesthetic condition of life, meaning, the capacity of feeling and being felt, as José Luis Pardo (1996) refers, the permanent search of emotional insertion with *"the other"* that brings along the explicit possibility of death while trying. Palau shows us a wall not only by its solid structure made of geopolitical engineering, but also in a symbolic and aesthetic dimension.

Even though it is not specifically inspired by the border wall between Ciudad Juárez and El Paso, the performance by artist Jesus Plastic in 2016 puts in evidence the impertinence and foolishness of trying to build the totality of the wall throughout the 3200 km of border, promised by former president Donald Trump. The artist built a small wall surrounding the hall of fame star located in Hollywood of who became the president of the United States but was just a candidate then. Spectators walking around can jump, play, and photograph the small wall, showing that life at the end is an entertainment–not in the Hollywood sense–that recognizes the citizen's dependence on *"the other"*, on the other side of the wall and without whom one cannot live without.

Finally, the wall is a metaphor to understand the universal sense of what the border is, both in the rational sense as in an emotional one. A wall that as a piece of clothes, it is a surface of contact, an in between bodies. The border, incarnated in this case in the wall, is a body, is the way of appearing for the contact between nations, and by being a human construction it is inevitably permeable.

## 32.2 Ciudad Juárez, Abandonment, and Urban Degradation

Currently, cities can be read through various perspectives, most of all from different traditions and theoretical constructions, in this case from urban imaginaries. However, it can also be done from different inputs of urban sociology that allow a fine reading about urban, political, and social frameworks. These incarnate in processes and phenomena produced by a border city full of tensions and multilevel structural contradictions, articulated in a micro and macro social level. These tensions affect the ways the citizens live the city and the way in which they are governed by local and state authorities.

It is common to notice processes that started decades ago and today are part of the city's corporeity. The urban degradation that can be seen in many founding neighborhoods of the old city, and the social decomposition of community and family structures can also be added to the emergence of precariousness as a way of urban life. All the above mentioned, shows a model of urban and economic development that bet everything on creating jobs for the manufacturer industry, as well as the construction of houses, commercial plazas, and big stores, but was careless about implementing a social development model centered on people and not objects. A local development that comes from below, but without the decomposition or degradation of life.

In this sense, the precarious situation of thousands of industrial workers who live on their wages is evident. They waste time of their lives traveling to work centers far away of their homes, and faraway of those spaces where citizenship is constructed.

It is easy to see the configuration of an emergent class within the so call precariat, explained by Guy Standing in the following manner:

> The precariat consists of people who makes a living from unstable jobs, intermingled with periods of unemployment, or forced retirement, living a life of insecurities with an uncertain access to public resources. Experiencing a constant feeling of transience. The proletariat lacks the seven forms of job security that the old working class fought for and that were internationally pursued by the International Labor Organization (ILO). There have always been workers in unsecure conditions. But that, by itself, does not define the current precariat. (Standing 2014, pp. 27–28)

From this, it is important to highlight the current border city, the existent and built city that articulates itself in a *bedroom city* form, in which the organizing logic prioritizes the daily commuting and traveling of hundreds of thousands of industrial workers to the factories where they spend most of their time. To guarantee the arrival of millions of workers to the industrial areas, it was necessary to develop series of road infrastructure and public and private transportation to facilitate an ongoing and never stopping productive apparatus in the city. This kind of city was planned to create large areas of public homes equipped with school, recreative and commercial zones so the tenants could buy groceries, have access to certain areas of consumption, and in a lesser level, access to places where to relax and enjoy their free time with family.

In that sense, the main objective of the city is to make sure that industrial workers can go to their jobs every day, go back home in the afternoon or night, sleep and go back to their jobs the next day for long periods of time. In this, the utilization of free time is limited and is usually spent buying groceries in the big stores, strategically located near to their homes. In this regard, the *bedroom city* was designed to optimize the utilization of the workforce and workers by facilitating a series of accessories for their daily lives.

Unfortunately, cities like Ciudad Juárez, lack of services, equipment, and basic and complementary infrastructure, which ends up making of these urban areas just failed housing developments where there is an urban component, but there is no city. For a city to really exist, there should be certain equipment of schools, health centers, sport, cultural, and recreative spaces. Based on this premise, understanding that an urban area is not the same as a city, it is possible to affirm that what is imposed here is a no city logic, meaning, there is a denial of the city. This anti-city version finally becomes an urban suicide, understanding this concept as the gradual and slow death of the city.

On the other side of the social decomposition and urban degradation as a denial of the city, we find the "*possible city*", the city of those that are far away, the ones that are excluded. This city is formed by millions of poor urban citizens who live the evil consequences of the economical development model that privileged a manufacturer productive system and neglected social and urban policies. Such neglect brought disastrous consequences for the population that will be difficult and very expensive for the local and federal governments to reverse. The creation of an extended, scattered, distant, expensive, and exclusive city finds its foundational bases with the arrival of the Industria Maquiladora de Exportación (IME) (Export Maquiladora Industry) in 1965 within the Programa Nacional Fronterizo (PRONAF) (Border

National program). This program created infrastructure, equipment, and other federal government programs like the one known as *"artículos ganchos"*, that was used to attract tourism from the United States. However, it also encouraged the social decay faced by the border city.

This reading of the city has been repeated in previous research and recognized by the academic and organized civil society, in which unfortunately the economic and social dimensions tend to be juxtaposed or faced against each other. This can be seen in authors like Borja, Carrión and Corti:

> Many social groups and urban thinkers advocate for urban justice as a right, characterized as the right to the city. These two options are not necessarily contradicting each other but could end up doing it in a world in which the economic and social dimensions tend to be dissociated. (Borja et al. 2017, p. 13)

The hopeful part of the material history of the city is the emergence of the rights to the city and the set of urban rights (to lighting, centrality, security, connectivity, universal accessibility, urban mobility, and monumentality) that articulate within alternative approaches to urban models that privileged an extended city. A city that generated an urban periphery full of poor people that could merely survive the day to day. Despite local and federal administration's interventions with community and assistantship projects, these are not enough to eradicate the deep urban problems experienced by the city, like unsafety and violence. There is an interesting characterization given by Professor Alicia Ziccardi from UNAM in one of her visits to the border, where she stated: "In Juárez, lots of houses, but not enough city".

Ziccardi truthfully represented the city from its urban reality. An overwhelming and devastating reality in which public housing projects multiplied since 1998 at the end of the twentieth century. During this period constructors and developers took a big bite of the cake, building hundreds of parcels for the workers in the maquila factories located in the outskirts of the city. Soon, problems linked to the lack of equipment and services appeared. It was the reproduction of a peripherical expansion model, a centrifugal model that expelled thousands to the margins of the city, most of these families were migrants that had just arrived to work in the maquilas factories.

> The housing developments model experiences a deep and direct process of transformation, while in the 40s, the urbanistic logic was the peripherical expansion of the cities, currently that logic goes toward an existent city, causing a mutation in the traditional tendency of urban development, exogenous, and centrifugal toward one that is endogenous and centripetal. (Carrión 2010, p. 15)

The discussion established within the academia and decentralized planning organizations from the municipal and state government, along with the universities and research centers, agrees that the economic projects of the previous decades, that moved forward without a social and community model in which social and health programs are important, ended up creating the reigning city. The city experienced a systemic policy of abandonment based on the concealment of social issues and urban unrest that were visible in popular sectors of the city since the 1980s and 1990s. Youth gangs appeared, disputing territories withing the city and neighborhoods of the city. The same gangs that were later directly linked to criminal structures of narcotraffic.

This abandonment policy was institutionalized for decades and privileged the welfare and political clientelism. At a certain point, welfare was no longer enough to satisfy the population's needs that found in other social and criminal structures a source of financing their economy of poverty. Precariousness as a way of urban life has become a solid way of living as the new urban precariat of the city. Today, the urban degradation has become part of the city's image.

On the other side, and from a different reading, the perspective of abandonment from the surveillance concept of the so called "gothic-Calvinistic" world, oppose to a humanism centered in tolerance and coexistence has been imposed, which fractured humanisms. The city understood as a territory and border geography has been part of a policy of population control in charge of the United States. Historically, the border meant a risk for the United States safety, not only military threat, but also a health and biology one. That is why Texan governments created a *cordon sanitaire* to avoid Mexican and immigrant populations of Syrian, Lebanese, Turkish, Greek, Italian, Chinese, and Japanese descent, could cross the border since the end of the nineteenth century.

The surveillance society established and founded in the United States influenced its own politicians and leaders allowing the establishment of the eugenics as a science to stop and control these foreign populations. Immigrants were needed to work in the construction of the railway infrastructure in the south of the American Union, and later to cultivate the fields. They were seen as necessary evil and as a response to a policy of recruitment for cheap labor, quasi-slave labor. Thus, Mexicans accepted these policies and were subjected for decades to sanitary controls and to the hygienism that permeated a good portion of the American society.

As a result of this, the border region became a city that was besieged at the middle of its own territory and by its relationship with the neighbor up north. A city surrounded by the criminality of delinquency groups that started a process of accelerated corrosion of the institutions and authorities. The northern border society modified its social, political, and economic structures when the biggest transformations of a structural order occurred. The accelerated growth detonated the old issues and aggravated the current ones like narcotraffic and immigration. Therefore, is convenient to revise not only the material history of the city as geopolitical and geocultural, but also from a *geopoetic* view, where aesthetic practices and the city imaginaries negotiate with cultural and social practices.

### 32.2.1 Geopoetics of Abandonment and Urban Degradation

The city does not only extinguish in a material perspective, but the deterioration also emerges as well in the urban imaginaries of its citizens, who from their own perception–and therefore sensible path of the city–identify ugliness, abandonment, and degradation concentrated in zones like Anapra or Riveras del Río Bravo on the northwest and southwest, respectively. Such can be perceived from the answers to the question for the urban space they like the least (Fig. 32.4).

# 32 Border Cities Between Life and Death: Ciudad Juárez and El Paso

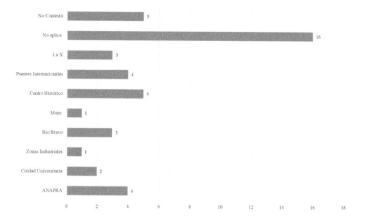

**Fig. 32.4** What is the space you like the least from the border? *Source* Own elaboration

In his work, Máquina del Tiempo (Time Machine) (2012) exhibited in the Ciudad Juárez-El Paso art Biennial III, artist Alfredo Espinoza Gutierrez presented a photographic display recreating a mason's work who is digging–it is not clear if he is building or destroying a determined urban space–from the present's shallowness and the pretension of finding the roots and traces of that liminal phenomenon that is Ciudad Juárez, oscillating between transformation and demolition:

> I constantly try to distance myself a little bit, and taking a stand in history, I try to find references that result pertinent to generate a broader view and to make a complete reading of the city's present. For example: the conditions of the city's foundation, the multiple migratory displacements, the prohibition of alcohol in the United States and the transnational industry are punctual references that turn out to be helpful to understand that the word city is a concept that can easily dilute, but that at the same time entails the opportunity to ask a series of questions of how to rethink it (III Bienal Ciudad Juárez-El Paso 2013, p. 41).

During the presidency of Felipe Calderón, who in his declaration of war against narcotraffic militarized the city instituting a long period of violence, stigmatizing the city in the international environment as the most violent city in the world, the artist Francisco Javier Rosales shows through his work the death liturgy and the abandonment in those "*bedroom areas*" that became obsolete and transformed into ghosts' zones. In his photographic series *Sillón con casa para perro #12* y *Sillón rojo #8*, we see degraded urban spaces, stripped off their life and waiting for their inhabitants to come back and take over those colorful armchairs that were used to rest.

These chairs, deteriorated and rotten, lay peacefully in the middle of what used to be homes, and have now become a testimony of the pause and transit toward an uncertain future. They are evidence of how the city dies because of violence and all the changes that came with the deindustrialization that took over that shadowed time. So, in the same way that the imaginaries allow the consolidation of an aesthetic and ethical scope of thought, the artistic production of Rosales draws attention over the political orientation and the lack of government that characterizes not only Ciudad

Juárez, but most of Latin-American cities. The artist, as the citizen, works from its own perceptions and emotions, materiality, and memory of the city as a social, political, economic, and of course, aesthetic fact. In other words, as a place of the human:

> The chairs in the exteriors, in the sidewalks or outside the houses are an allegory, a fictitious and imaginary *loft* within the abandoned spaces of the city. These chairs offer comfort and security in an ephemeral and timed way, silently inviting people to rest for a moment. They are like a manifestation about the moment in which the streets will be part of the citizens, the ones who build and give life to the cities. Citizens are the ones who are supposed to walk through and retake the public space and declare it their own. A sofa in the street represents the possibility to be able to walk around the city again, feel it, smell it, and rest it. (III Biennial Ciudad Juárez-El Paso 2013, p. 95)

As in the psychic inscription of Silva's (2014) imaginaries–that simultaneously operate with the social and technological orders–we can affirm that in the imaginaries there are new theories that come into play. These new emotion theories do not pretend the creation of a science of feelings, but its projection as a motor for social processes. In other words, to recognize that emotions are not separated from reason, as it is assumed by various classic approaches of sociology, and instead that these are conditions that interact. It could be said that we reason from emotions available in the environment and that are shaping our thoughts. The armchairs then, like an intersection between death and life in the city, as objects that concentrate emotions and awake rationalities, allow–notates Francisco Javier Rosales (2013)–to generate a consideration about the wait and reencounter of the citizens with their city, a necessary rest claimed by the city of the suffered chaotic processes.

Now, the city its equally build in the images broadcasted by communication media outlets, nurturing and enhancing the consolidation of urban imaginaries, and it is reflected in the answers provided by citizens when asked about what the media says about Ciudad Juárez. Television and Internet, mostly made by images, stand out from other media outlets. In regards the predominant topics we find narcotraffic (and the violence generated by it), immigration (illegal), and politics (corrupt) are the topics dominating the agenda (Figs. 32.5 and 32.6).

From the dark explosion of information produce by the media, another explosion follows, one of a creative order that gives origin to a series of artistic productions. These productions anchor themselves to the destruction imaginaries of the city. Imaginaries of war against human and drug trafficking, of death and fear, of the feeling of living in a city that is considered one of the most violent cities in the world. Within these productions, it is worth mentioning the work of Ciudad Juárez artist Alejandro Morales, called *Hallan destazado* (2013) (Found butchered).

It is interesting–according to Morales–the day-to-day aesthetic that we experience through the media and how it can change the identity of a society affected by violence (III Biennial Ciudad Juárez-El Paso 2013, p. 79). In this way, taking as an example the front page of a local newspaper that displays the image of a dismembered body found in the city's downtown by the police, along with the title in capital letters: "THIS MORNING A BUTCHERED BODY WAS FOUND", the artist uses his technique to disappear the body with an eraser, to make him appear again. The body appears, not

# 32 Border Cities Between Life and Death: Ciudad Juárez and El Paso

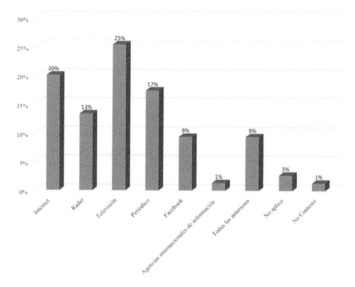

**Fig. 32.5** What media outlets talk about the border? *Source* Own elaboration

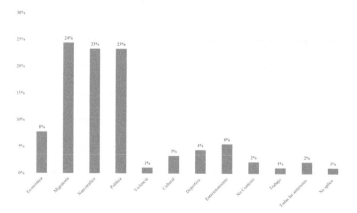

**Fig. 32.6** What kind of information is dominant? *Source* Own elaboration

from the physical or pornographic presence of the journalist exercise but from the absence: the body does not lay in the paper anymore, instead it rests in the feelings of the viewer. A viewer that without seeing the body, still knows that it is there, along with all the other bodies that have accumulated and are still accumulating throughout the multiple violence that has shaken this society generation after generation:

> I place myself in the generation that emerged in the wake of death and pain brought by the war against narcotraffic, a war that does not seem to have an end. Living in the most dangerous city in the world helped me to create my own visual and conceptual vocabulary, full of endemic violent constructions (…) I am from that place where the border between

the legal and illegal, the real and unreal is blurry, where the sinister goes around within stories (…) My work goes through hypothetical situations where I try to make (in)visible the (un)conscious processes from collective imaginaries, where abandoned lines, empty lots and forensic paraphernalia coincide. (III Biennial Ciudad Juárez-El Paso 2013, p. 79)

Urban imaginaries, incarnated in the citizens bodies and engraved in the memories and collective senses, as well as in the forms of art or communication acquire by these, confirm the major symbolic concentration and social perception that the separation has in the inhabitants of the border cities. Following the theory from Armando Silva (2014), the urban imaginaries in Ciudad Juárez allow an artistic and mediatic production that gets together with those border imaginaries that have been identified in the border wall. Imaginaries such as the abandoned parcels, the chairs, the pocho figures, the chicano or cholo allow to understand this territory that we call the border as an aesthetic symbol. The border as a symbol that has become a living part of the body in each person that lives in the community, an assimilated truth that constitutes the existence of the city itself, like an identity certainty: The border is Juárez, Juárez is the border.

Finally, whether the city is approached as a field and group of aesthetic practices, *geopoetics*, from geopolitics and culture or from the abandonment policies, it is important to try not to fall in fatalist discourses or sociological pessimism. On the contrary, it is necessary to show the political, economic, social, cultural, and aesthetic reality where two cities articulate and are knitted together in never ending interstices that open an opportunity for a city that is possible. A city that can be inclusive, integrative, and with enough sensibility for the newcomers and justice for those who are already established. A land that avoids the gradual or accelerated death of the city.

## References

Biennial (2015) Ciudad Juárez+El Paso. 2015. El Paso Museum of Art, Museo de Arte de Ciudad Juárez, El Paso, Ciudad Juárez
Borja J, Carrión F, Corti M (2017) Ciudades resistentes, ciudades posibles. Editorial UOC, Barcelona
Carrión F (2010) Ciudad, memoria y proyecto, OLACHI-Municipio Metropolitano de Quito, Ecuador
III Bienal Ciudad Juárez-El Paso (2013) Catálogo. El Paso Museum of Art, Museo de Arte de Ciudad Juárez, El Paso, Ciudad Juárez.
Pardo JL (1996) La intimidad, Pre-textos, Valencia
Silva A (2014) Imaginarios, el asombro social. Universidad Externado de Colombia, Quito, Bogotá, Ciespal
Standing G (2014) Precariado. Una carta de derechos, Capitán Swing, Madrid
Vera M (2020) Geopoéticas, memoria e imaginarios en la frontera México-Estados Unidos. Universidad Externado de Colombia, Bogotá

**Mauricio Vera-Sánchez** Professor at *Universidad Nacional Abierta y a Distancia*, on Colombia. He is Professor Master in Communication and undergraduate in Social Communication, School of Social Sciences, Arts and Humanities. Fisura research group leader.

**Luis Alfonso Herrera-Robles** is a professor of Sociology at Department of Social Sciences at Autonomous University of Ciudad Juárez, México. Expert on urban studies and social violence.

# Part VIII
# Contraction of Public Management: Privatization

# Chapter 33
# The Metamorphosis of Infrastructure in Latin American Urbanization: From Insufficiency to Presence as Fictitious Capital

**Beatriz Rufino**

**Abstract** This article aims to discuss the significant transformations in the production and operation of infrastructure, recognizing this process as the metamorphosis of infrastructure valorization. Taking into account the socioeconomic differences among Latin American countries, we intend to consider some general reflections about the particular impacts of this global process on the urbanization of the region. Renewing urban spoliation as a historical feature of Latin American urbanization, the metamorphosis of infrastructure will provide important insights into the updating of critical debates on Latin American urbanization and in the understanding of privatization as a relevant dimension of Urbicide.

**Keywords** Urban infrastructure · Process of urbanization · Town planning · Latin America

## 33.1 Introduction

The implementation of neoliberal policies has led to a series of changes in the manner in which infrastructure is produced, financed, and operated (O'Neill 2017; Christophers 2020). This process is perceived globally. However, it also displays specificities tied to different productive sectors, countries, and regions.

Particularly since the 1990s, neoliberal ideas have had a profound effect on the legitimacy of privatization policies in infrastructure, running parallel to the expansion of the role of the private sector in the Global South. The privatization of several spheres of infrastructure became a recurrent directive of the international financial institutions under the Structural Adjustment Programs (SAPs) in the 1980s and 1990s

---

This paper was developed under the support of the research Project funded by FAPESP "Real Estate and Infrastructure under the control of Large Corporations: Financialization and metropolization of space in São Paulo of 21st Century" (2020–2022—Process 18881-0) and the international fellowship award of Urban Studies Foundation.

---

B. Rufino (✉)
University of São Paulo (USP), São Paulo, Brazil
e-mail: beatrizrufino@usp.br

(Loftus et al. 2019). Through various agreements of privatization and concession (Bakker 2003; Christophers 2020), in the three decades that followed the Washington Consensus, a process is evinced in which activities that had previously been under state management and operation are transferred to the private sector (O'Neill 2017).

Privatization processes became more sophisticated with the rise of financial markets in the course of the twenty-first century (Lorrain 2011; O'Neill 2017), with large financial institutions seizing increasingly significant roles in the provision of infrastructure. The expanding complexity of credit markets began to organize new possibilities for capital mobility, especially given the centrality of public and private debt mechanisms, their ability to transact within secondary markets, and their participation in the financing of fixed capital, including urban infrastructure (Harvey 2018).

These changes result in a major shift of the function of infrastructure in the expanded reproduction of capital. Infrastructures, understood as general conditions of capitalist valorization and social reproduction, are projected in contemporary times as an important business sector (Rufino 2021). Thus, we note the constitution of a metamorphosis. The metamorphosis of infrastructure is characterized by the transformation of its dominant form of valorization worldwide, in such a way that infrastructure now functions both as a precondition and as a result of the movement of capital (Lencioni 2021).

Infrastructures, which historically functioned as fixed capital, essential to industrial capitalist accumulation (Harvey 2006), and which became determinant conditions of social reproduction in urban contexts (Lencioni 2007), are redesigned in contemporaneity as important gears of fictitious accumulation, based on the capitalization of rents under the increasing control of finance. Even though infrastructures remain functioning as fixed capital in the process of expanded capital reproduction, with this metamorphosis, their immediate determination became subordinated to the production of fictitious capital.

Certain specificities in the social formation and urbanization processes of Latin American countries demand a particular critical analysis of this global phenomenon, in order to grasp the specific relations of accumulation at play within this relatively new dynamic context, the contradictory centrality of the state, and the reproduction of older and newer forms of inequality in the production of urban spaces across the region.

In line with this new rationality, we argue that the growing investments in Latin American metropolises have given rise to new forms of spoliation, particularly new forms of spoliation in finance and real estate (Pereira and Petrella 2018).

Examining the conditions of urbanization in Brazilian metropolises in the 1970s, Kowarick (1979) defined urban spoliation as the systematic exclusion of the poorest strata of population from access to services of collective consumption. The process of urban spoliation is materialized when this excluded populational stratum takes matters into their own hands by constructing their own housing in precarious territories (Kowarick 1979). This is what has become more widely known as popular urbanization (Connolly 2013). Still, Kowarick recognizes this process as a magic formula of dependent capitalism, enforced to lower the costs in the reproduction of the labor force, balancing a high rate of accumulation with increasingly deteriorated wages (Kowarick 1979).

This article has the purpose of discussing the significant transformations in the production and operation of infrastructure, as these transformations are observed in the metamorphosis of infrastructure valorization. The discussion focuses on the particular impacts of this global process on the phenomenon of Latin American urbanization, bringing to light an analysis of the so-called public–private partnerships (in Portuguese, "Parcerias Público-Privadas" or abbreviated as PPPs), the primary means of promoting this metamorphosis in the region.

Despite recognizing the need for more specific studies that address the various kinds of infrastructures and that consider the particularities of their provision and operation as well as the different intensity and nature of infrastructure privatization and financialization among the countries in the region, in the present work we have opted for a more general approach, to give an overview of the production of urban infrastructures while also attempting to scrutinize and question the urbanization processes of the region.

This article seeks to organize two main contributions. First, it seeks to outline a theoretical reformulation of a critical debate on Latin American urbanization as a general theme and the production of infrastructure, in particular, organizing these themes in association with the intense discussions on global financialization. Additionally, it seeks to bring forth some elements regarding the debate of Urbicide. In line with the book's proposal, we understand Urbicide as a multidimensional process that points to the erosion of cities. Among the various multidimensional components of Urbicide developed in this publication, the metamorphosis of infrastructure helps us to understand how, in a context of global financialization, privatization processes change the social relations in the production of urban spaces and the role of both politics and state in the cities. As we will try to demonstrate, the increased relevance of infrastructures as a private business places it as a central political object for the dominant classes, driving the commodification of its use and a consequent depoliticization of the access to infrastructure as a right.

The article is structured into three main sections. Initially, we will discuss the global process of infrastructure transformation, articulating the political economy concepts of fixed capital and fictitious capital, as well as the growing debate over the financialization of infrastructure, set forth mainly from the Global North. In the second part, based on the categories of urban spoliation and public funding, once formulated for the interpretation of Latin American urbanization processes, we will turn to the discussion of the historical conditions of infrastructure production and the particular repercussions of the accelerated urbanization of the region in the context of Import Substitution Industrialization (ISI) between the 1950s and 1980s. Building on the debate of the two previous sections, we will seek to reflect on the primacy of fictitious capital in the design and operation of infrastructure in the urbanization of Latin American metropolises, discussing the transformations of the public funds and the new forms of spoliation that radicalize the inequalities of Latin American urbanization.

## 33.2 Metamorphosis of Infrastructure as a Global Process: From Fixed to Fictitious Capital

Attentive to Marx's contribution to the Critique of Political Economy in which he designates infrastructure as a specific form of fixed capital, Harvey (2006) emphasizes that fixed capital represents, like all capital, a value in motion; therefore, it is not a thing "[…] but a process of circulation of capital through the use of material objects, such as machines" (Harvey 2006, p. 205). In this sense, fixed capital can be defined as a working instrument employed to facilitate the production of surplus value. A machine "becomes fixed capital as soon as it is bought and incorporated into a production process by a capitalist" (Harvey 2006, p. 208–209).

Due to this specific function in the production processes, the value of fixed capital presents itself in a state of eternal flux, being simultaneously determined: "by initial purchase price, by the surplus value it helps to produce through productive consumption, or by replacement cost" (Harvey 2006, p. 210). Thus, technological innovations and broader social change can lead to abrupt variations in the value of fixed capital, calling into question its use-value and exchange-value. As a general rule, by intensifying the productivity of labor, fixed capital enhances exploitation.

Infrastructure, in the form of railways, waterways, roads, and aqueducts, does not appear as a simple instrument of production within the production process but as an autonomous form of capital which becomes particularized in the valorization process by functioning as a general precondition of production. Constituting themselves as necessary premises for the reproduction of capital, these (pre)conditions are termed general because they are comprehensive conditions and not restricted or exclusive. Their usufruct is always collective and not individual or private, hence the idea that their use or consumption is socialized (Lencioni 2021).

As Lencioni (2007) argues, in the urbanization process, these general conditions of production many times are not detached from the provision of means of collective consumption favoring the reproduction of the labor force. As Pírez (2013) further argues, infrastructure and, more specifically, urban services should be understood as essential components of urbanization that define the urban landscape and allow the daily functioning of the entire material support system.

In the course of capitalist development, the periodic processes of overaccumulation, distinctive to capitalism, determine new rounds of investment. This recurrent transfer of capital from a primary circuit (of production) to a secondary circuit (of consolidation of the built environment) is mediated and accelerated by the consolidation of a credit system.

Conventionally, in their function as fixed capital and general (pre)condition of production, infrastructures stand out for their large scale, durability, and fixity in space. These specificities are reflected in the great amount of resources required for construction. These conditions, associated with their established feature as means of collective consumption, determine specific difficulties and risks in their production and circulation while, at the same time, placing them as a central strategy in the context of the capitalist crises of overaccumulation. By capturing a large volume of excess capital, investments in infrastructure become an important engine for the creation of new cycles of economic expansion, oriented toward an increase in the social productivity of labor. At the same time, the mobilization of large sums of capital for extended periods of time turns infrastructure into an obstacle for the broader circulation of capital (Harvey 2006).

These characteristics of infrastructure highlight a basic contradiction of fixed capital. This contradiction is due to the fact that, from the point of view of production, infrastructure is the pinnacle of capital success but, from the perspective of circulation, it becomes "a simple barrier to further accumulation" (Harvey 2006, p. 238). The following passages from the work of Harvey explain both sides of this matter. First, from the perspective of production:

> Fixed capital raises the productive powers of labour to new heights at the same time as it ensures the domination of past 'dead' labour (embodied capital) over living labour in the work process. From the standpoint of the production of surplus value, fixed capital appears as 'the most adequate form of capital' (Harvey 2006, p. 237).

However, from the standpoint of circulation, "Fixed capital limits the trajectory of future capitalist development, inhibits further technological change and coerces capital precisely because it is tied to its existence as a determined use value" (Harvey 2006, p. 320).

In a different text, Harvey (1978) points out that the aforementioned tendency to transfer capital from the first circuit to the second circuit in the course of an overaccumulation crisis can take the form of over-investments in the infrastructure sector. Harvey argues that this over-investment occurs only in accordance with the needs of capital, having no relation to the needs of the population. These needs, as a rule, remain unfulfilled (Harvey 1978, p. 112). The over-investment may put pressure on the devaluation of these very infrastructural assets in subsequent periods.

If, in the short term, infrastructure emerges as an outlet that mitigates economic fluctuations, in the long run, these investments can lead to new crises through the devaluation of the fixed capital itself (Harvey 2006). In capitalist production, periodic devaluations of fixed capital, which, in the case of infrastructure, can also be understood as the devaluation of the built environment itself, provide one of the inherent means of containing the deflation of the rate of profit and accelerating the accumulation of capital value through the formation of new capital (Harvey 1978, p. 116).

Due to the fact that most infrastructures have the distinctive feature of being fixed in space, a set of other contradictions, inherent to the production of capitalist space under the domain of private property, emerge, reinforcing infrastructure as a fundamental mechanism in the unequal development of capitalism. Harvey asserts that promoting investments in a given physical space also generates a greater flow of capital into that space, delimited by fixed capital, with risk of devaluation and significant consequences for the interest-bearing capital that funded it (Harvey 2018, pp. 149–150).

Although they are used as a social symbol of inclusive growth, infrastructures have historically responded primarily to the interests of the ruling classes, providing them not only with conditions for expanding labor exploitation, but also ensuring new levels of income acquisition.

Infrastructure as capital incorporated into the land in the form of improvements (Harvey 2006, p. 307) tends to lead to transformations both direct and indirect in the possibilities of land rent capitalization through ownership. Developing the idea that the city in its entirety was constituted as a socialized form of fixed capital, Folin (1977) argued, in the late 1970s, that infrastructure was realized in a particular way through land rent. In Folin's interpretation, this income was earned by landowners through the processes of property fragmentation. If we take into consideration that infrastructures can also be constituted under private property regimes (Arboleda and Purcell 2021), control over them earns the owner a revenue, as well as conditions for regulating labor in providing the service.

Not only do we perceive, from these particularities, an important function of infrastructures within capitalism, their contradictory features, and their pivotal role in moments of crises, but we are also able to discern significant variations of this phenomena in the course of the development of capitalism and among its distinct spaces of accumulation.

Historically, the development of infrastructure has been a precondition for the advancement of industrial capitalism (O'Neill 2017). Initially, what we have come to know as the general conditions were produced by processes of appropriation, conversion, and primitive accumulation. "The early industrialists acquired much of their fixed capital by putting old structures (mills, barns, houses, transport systems, etc.) to new productive uses" (Harvey 2006, p. 218). Starting in the nineteenth century, infrastructures were developed as local networks, structured by private enterprises (Graham and Marvin 2001). However, when private sector activity proved to be inefficient, with several cases of competition and bankruptcy among investors (O'Neill 2017), the consolidation and integration of infrastructures became a defining task of the modern nation-state, essential in the definition of its territory, ensuring a certain sense of national cohesion (Graham and Marvin 2001).

A broad social consensus over the centrality of public intervention in providing infrastructure was consolidated during the Keynesian period, consecrating infrastructure as the main area of investment by the state in the core capitalist countries (O'Neil 2017). State-developed planning, financing, and operation of infrastructure were proven to be successful, evincing the pinnacle of a mode of production conducted by the state. This centrality of the state in providing infrastructure, based on the fiscal strategies of Keynesianism, was reinforced in the post-World War II period and in the three decades of economic growth that followed the conflict, as the state supported the necessary volume of investments.

By organizing the possibilities of direct and indirect profitability of capital, the effective participation of the state as a public fund (Oliveira 1998) in the production of fixed capital was fundamental in concealing the previously discussed contradictions of infrastructure. It also provided an important decommodification of the social reproduction of the workforce in the post-war context of the core capitalist countries (Pírez 2012).

Even though this form of infrastructure provision was decisive in sustaining capitalist expansion, it was progressively dismantled. The shift from what was then the effective paradigm took place during the 1980s and was linked to radical transformations in capitalism, such as "declining sovereignty in the face of globalization, slowing economic growth rates, constraints on states' fiscal capacities, and a loss of confidence in the effectiveness of the state apparatus in the supply of essential public services" (O'Neill 2017, p. 176).

The promotion of neoliberal ideals and their prescriptive method of privatization paved the way for the stature of infrastructure to surge as an economic sector of global capitalist accumulation (O'Neil 2017; Lorrain 2011). Now undertaking a predominantly regulatory role, governments began to ensure extraordinary dividends to private agents, through the privatization of assets under market value, while also guaranteeing the constitution of monopolies and oligopolies and instituting several guarantees for the value of tariffs (Pírez 1999).

Studies on the progress of infrastructure privatization have indicated significant changes in this process at the turn of the twenty-first century. Above all, the changes came about due to an increased coordination between the infrastructure sector and the stock market, repositioning the debate on the matter of financialization at the core of discussions over infrastructure production and operation (Loftus et al. 2019). Backer (2003) calls attention to the fact that the widespread use of the term "privatization" can lead to a series of misunderstandings. The term is used broadly to refer, in many cases, to distinct or overlapping strategies, ranging from the total alienation of the asset to various forms of public–private partnerships. Each of these strategies results from specific political and economic contexts, with different implications for the term as it relates to asset ownership, prospects of accumulation, and transformations to the urban space.

Although providing infrastructure is a recurring solution to the contradictions of capitalist accumulation in the course of a crisis, the growing dominance of finance has imposed both the intensification of this strategy and the creation of new relations of production. Currently, it is possible to identify a broad global agenda of investments in infrastructure. For Dodson (2017), the emergence of a coordinated effort to stimulate the development of infrastructure, at national and global levels, through a series of international apparatuses, is responsible for the "global infrastructure turn". At the basis of this movement is the excess of capital over productive ends and the consequent decrease in expectations about future economic growth, in a context of persistently low interest rates.

Infrastructure products are recognized, in contexts of major instability, as an asset class capable of generating stable and long-term returns (Dodson 2017). The international financial crisis of 2008, while revealing the dimension and power of global finance in the production and operation of infrastructure (Lorrain 2011), also consolidated changes in the way financial agents and corporations acted. They began to prioritize direct ownership of land and urban infrastructure for rent extraction to the detriment of participation as credit agents (O'Neil 2019; Purcell et al. 2020).

Land rent, as the historic form of realizing infrastructure, is renewed with the extension of the forms of private monopoly and becomes increasingly sought after by fictitious capital (Purcell et al. 2020). According to Chesnais (2005), the current regime of accumulation, characterized by the dominance of finance, is sustained by the ascendancy of a mind-set of financial valorization of capital and by the cardinal presence of fictitious capital, which contributes to the centrality of private property of a rentier ethos. In dialog with the work of Marx, Chesnais (2005) postulates that fictitious capital is all financial assets whose value rests on the capitalization of a stream of future revenue, which have no counterpart in effective industrial capital.

The processes of valorization occurring through infrastructure are transformed in this context. Purcell et al. (2020) point out that understanding these new accumulation strategies requires an interpretation in which value, rent, and finance are seen as strongly intertwined, revealing the political, economic, and spatial implications of these processes, as well as the underlying class interests.

With the increasing transfer of infrastructure ownership to financial institutions and corporations, infrastructures have come to function as a form of fictitious capital, determined by the possibilities of capitalizing future streams of rent, which is ensured by the continuous flow of tariffs and enhanced by sophisticated financial engineering (Pryke and Allen 2019; Ashton et al. 2012). This is not new, since the insertion of infrastructures in the general process of valorization has always propounded the possibility of pricing by future determinations. What is new is the intensity and celerity of these capitalization processes which transform the role of infrastructures as fictitious capital, giving this new function a dominant stance over the historical and permanent functioning of infrastructure as fixed capital.

Thus, we could say that, under this rationale, infrastructures are increasingly designed as a function of capitalization possibilities. They are becoming disconnected, at least in part, from the valorization of capital and the needs of social reproduction, leading to a form of "inverted capitalism" (Blank 2018). According to Blank (2018), this inverted capitalism would be characterized by the fact that the long-term expansion of fictitious capital no longer reflects the development of working capital. Rather, the growth of productive capital has been "inverted" into a dependent variable of the growth of fictitious capital.

The transformation of infrastructure into financial assets is the subject of a broad debate that has suggested strong impacts on the development of urban policy and on the production of space itself (Graham and Marvin 2001; O'Neill 2017; Dodson 2017; Purcell et al. 2020).

## 33.3 The Insufficiency of Infrastructure in Latin American Urbanization: The Limits of the Public Funds and Urban Spoliation as an Accumulation Strategy

In Latin American countries, as a result of varying sociopolitical situations, the pace of emergence and development of infrastructure was uneven in terms of quality and quantity (Pradilla 1994). As a general tendency, the historical limitation of investments in infrastructure is related to the particular conditions of Latin American configuration, marked by relative shortcomings in the accumulation of capital at a global level, which is derived precisely from the region's subordinate position in global capitalism (Jaramillo 1988).

Even so, it was through infrastructural networks that colonial and imperialist relations were configured in the region, fomenting economic and technological dependence (Davies 2021). In articulation with the colonial dynamics of extractivism, a set of infrastructures—railroads, telegraphs, and port improvements—were produced, in most cases, with foreign technology and capital (Campos 2021). As Seabra (1987) argues based on observations concerning the city of São Paulo, the predominance of concessions that were granted to foreign companies (especially to British companies) comes both from the dynamics generated by agricultural-exportation activities and from the fact that the capitalist system had already developed an international apparatus for infrastructure operations at the end of the nineteenth century, with elaborate financing arrangements and concentration of capital in large companies. When private investments started being directed toward the cities, infrastructure also played an instrumental role in the structural transition from a dominant rural oligarchy to an urban economic elite (Davies 2021).

With the crisis of the primary economic regime of exportation, the national-developmentalist governments of the region initiated processes of nationalization directed toward infrastructure companies of foreign capital, aiming at the interruption of harmful practices and the need to boost national industrialization (Pradilla 1994). As Pradilla (1994) states, given the low profitability of many infrastructures and the inability of private interests to centralize capital for the development of general conditions, especially in periods of recession, the states took on a fundamental role in infrastructure provision, by socializing the costs.

Within the developmentalist project, infrastructures are taken as essential general conditions of production for the emerging industrialization of the region, with the states taking upon themselves the task of assembling the economic infrastructure to promote industrial development, thus organizing the conditions of accumulation for multinational companies to thrive in the environment of Import Substitution Industrialization (ISI), an environment observed in the most dynamic economies of the region between the 1950s and the 1980s.

In prioritizing the immediate interests of dominant sectors of society, governments played a limited role as unifying figures in the global agenda of the exploiting classes and in promoting long-term strategies for developing interests. The result is that the provision of infrastructure gave privilege to use-values directly linked to capital accumulation, particularly to the production of goods, to the detriment of those linked to the needs of the working classes in consumption and reproduction (Jaramillo 1988, pp. 29–31). Although state actions led to significant development in some countries of the region between the 1940s and the 1980s (Argentina, Brazil, and Mexico in particular), as stated by Pírez (2013), these measures did not succeed in diminishing the weight of the working-class population that was out of mercantile conditions, nor did they succeed in institutionalizing a Welfare State as in core capitalist countries (Pírez 2013).

If, in those core countries, the infrastructure that was executed by the use of public funds became a precondition for the accumulation of capital and the reproduction of the labor force, being dispensed globally for the entire population by social spending (Oliveira 1998, p. 8), in Latin American the accumulation of capital is constituted as a primary objective, resulting in limited and partial decommodification of the reproduction of the labor force (Pírez 2013).

In this sense, the role of public funding in the constitution of infrastructure in Latin America generally serves to heighten accumulation beyond the limits imposed by the generation of profit, using public wealth that is not capital and, therefore, is not retributed in the general equation. This distorted mobilization of public funds would result in the constitution of an "Un-welfare State" (Oliveira 1998), with very significant repercussions on the urbanization process of the region.

As a result of the urgency of peripheral capital to create urban centers with a certain degree of consolidation for industrial development, there was a concentration of the limited amount of resources in a few primary centers, with very unequal distribution of infrastructure within these cities and in the larger regional scale as well (Pradilla 1994).

In the majority of great Latin American metropolises, whose accelerated growth began in the 1950s, the spatial result is the reinforcement of inequality in the distribution of infrastructure and facilities, with the state exonerating itself from the provision of some of these services in the more peripheral sectors. In turn, this leads to what has come to be recognized as popular urbanization (Jaramillo 1988). As Emilio (1998) has shown, popular urbanization represented half of the land area and population of the large Latin American metropolises at the end of the twentieth century.

The lack of infrastructure became a structural element of urban spoliation (Kowarick 1979), that is, the systematic exclusion of working classes from access to services of collective consumption resulting from the urbanization processes. By allowing a significant reduction in the reproduction costs of labor power, this form of spoliation ensured the expansion of industrial accumulation. The formation of huge peripheral areas in Latin American metropolises is, in this sense, the product of urban spoliation, and they constitute a margin of accumulation without value production (Pereira and Petrella 2018).

The peripheral areas are characterized as those most distant and of lowest differential ground rent, occupied by the section of the population with lowest income and who are inserted into the working environment in the most precarious manner (Kowarick 1979). Up until the 1980s, they represent territories which are lacking in state presence and influence, almost totally untouched by public policies, except for the mass housing developments that started to be implemented from the late 1960s onward in the continent's most industrialized countries (Rufino 2015).

Given the relative exhaustion of the ISI model in the region in the 1980s, countries have tried to restructure their position in world capitalism (Jaramillo 1988). This process, in conjunction with the exhaustion of the Fordist model in the core capitalist countries, had immediate implications for the provision of infrastructure, with lasting impacts on urbanization.

As a general rule, the economic asphyxiation Latin American countries experienced during the 1980s favored a shift in the conception of what the role of the state is, with a particular emphasis on the crescent push for privatizations (Pírez 2013). According to Pradilla (1994), the breadth, depth, and speed of these processes in Latin America can be explained, at least in part, by the deep relations of dependence toward the core capitalist countries. They are also exacerbated by the large amount of external debt, which grants great power and influence to international banks and agencies in establishing privatization policies.

The region experienced accelerated privatization processes in the 1980s and 1990s. Those were initially marked by the integral sales of public companies to national and foreign private investors, but with significant differences among the various countries in the region. Chile, for instance, was a pioneer in the privatization of infrastructure, given that the global tendency was reinforced locally by an economically liberal dictatorship that took power in 1973. Other countries, such as Mexico, Venezuela, Argentina, and Peru, followed suit in the 1980s, once the local governments abandoned their interventionist alignments (Pradilla 1994). Under varied political circumstances, Brazil, Colombia, and Paraguay, only timidly and partially, began promoting privatizations in the 1980s, with processes accelerating during the 1990s. In the remainder of the countries, the liberal agenda would only move forward in the twenty-first century, with a new round of privatization processes more largely tied to finance.

For the infrastructure companies, which, in most cases, were established by the developmentalist efforts of the previous decades, the privatization processes were marked, on the part of the private sector agents, by an emphasis on financial self-sufficiency and on the application of market mechanisms. These processes transformed the social relations that permeate the provision of infrastructure by introducing into the field powerful companies, usually foreign, with great ability to influence policy and decision-making (Pírez 2013). For the most part, private capital (especially transnational capital) was only interested in infrastructures that presented possibilities of immediate or short-term profit. This led to strangle holds and several areas of infrastructure being voided (Pradilla 1994; Rocha 2013).

According to Pradilla (1994), the most profitable businesses and territorial fragments were concentrated mainly in energy, telecommunication, highway, and railroad sectors. These were benevolently accepted by national and international investors. In addition to choosing the most profitable sectors, foreign capital expanded mainly in the control of companies and sectors whose tendencies were to produce oligopolies at regional or global levels, such as telecommunications and energy (Rocha 2013). In general, private investment tended to focus on the acquisition of existing facilities, adding little to the capacity of provision in sectors and territories with significant historical shortcomings (Bayliss and Fine 2008).

In the 1990s, Latin America also began to stand out for the implementation of the PPPs, concentrating the largest number of contracts worldwide (Michelitsch and Szwedzki 2017). In theory, PPPs differ from company privatizations insofar as they are established as a contract between public and private agents in which risks and management responsibilities are transferred to the private sector through control of the infrastructure production and operation (Bayliss and Van Waeyenberge 2018). In practice, the instrument expanded coordination between private investors and the state, altering the rationale for mobilizing state subsidies. Instead of favoring some level of decommodification of services, subsidies were now subject to competition by private agents as means of ensuring higher levels of profitability.

In general, the processes described above resulted in a growing difference in access to infrastructure, reinforcing the logic of investment in the most profitable projects, with the energy and highway sectors accounting for 60% of investments in PPPs throughout the 1990s. The permanent restriction of investments in urban infrastructure through these new instruments of privatization, while reinforcing the extreme polarization of income distribution in the territories of Latin American metropolises, indicated structural limits to the advancement of PPPs given the very high poverty rates of the region.

By the end of the decade, there was a severe drop in investments through PPPs, and several of the projects implemented under this modality faced financial difficulties (Michelitsch and Szwedzki 2017). The case of the Mexican highway concessions gained notoriety as one of the most symbolic cases of failure to implement PPPs in the region, resulting in massive expenditures for the local government (Pradilla and Lopéz 2017).

The justifications for the various attempts at privatizing infrastructure were usually drawn from an ideological standpoint and are based on the supposed greater efficiency of private agents. On the other hand, they failed to take into account the material determinations and contradictions which had once led Latin American capitalist states to control infrastructure (Pradilla 1994, p. 57).

## 33.4 The Presence of Infrastructure as Fictitious Capital in Latin America Cities: The Transformation of the Role of Public Funds and The Emerge of New Forms of Spoliation

During the first decade of the twenty-first century, as part of the global tendency toward the metamorphosis of infrastructure into fictitious capital, Latin America has experienced a new momentum in the expansion of PPPs. Between 2006 and 2015, $361.3 billion dollars were invested in the region, in more than a thousand infrastructure PPP projects. These projects were heavily concentrated in Brazil, with Mexico and Colombia in distant second and third places (Michelitsch and Szwedzki 2017). As the World Bank data shows, the volume of investments in the region became more significant after 2009, reflecting the global expansion of financial liquidity and the economic growth in the region.

Bayliss and Van Waeyenberge (2018) named the general process the "PPP revival". The name speaks to the expanding mobilization of the instrument, evidenced since 2005, particularly in developing countries. According to the authors, although this resurgence is based on previous privatization initiatives, it is differentiated, during the "revival", by the central role played by global finance. While in the 1990s, the main argument for the advancement of PPPs was the potential gains derived from greater efficiency, thanks to control being handed to private agents, in this new stage, the large availability of financial capital on a global scale becomes the main justification for the reemergence of the instrument (Bayliss and Van Waeyenberge 2018).

After the 2008 crisis, traditional PPP financing models were strongly affected by the substantial increase in the cost of credit, with negative implications for the viability of PPPs, given their dependence on long-term financing (O'Neill 2019; Bayliss and Van Waeyenberge 2018). The answer to this question was to reorganize the financing of PPPs in order to tap into the possibilities of a great demand by institutional investors for stable, long-term returns. The massive injection of liquidity by Central Banks of the Global North (Fernandez and Aalbers 2020), through policies of "quantitative easing" that compress the profitability of debt securities, has made infrastructure investments more attractive by offering higher returns with relative stability (Bayliss and Van Waeyenberge 2018). And, once the logistics of finance were at the forefront, infrastructure provision became increasingly shaped by investor interest.

Overall, the ways in which governmental and intergovernmental agencies provide support for the expansion of institutional investors are shaped by the configurations emerging from financial markets (Bayliss et al. 2021). Strengthening the institutional regulatory capacity of states is a measure recognized by multilateral agencies as a key aspect in developing well-structured projects that ensure leveraging infrastructure as an asset class to channel private savings into the sector (Serebrisky et al. 2015, p. 17).

Thus, the potential of infrastructure to generate attractive financial revenue is a central element in the structuring of the projects. On the other hand, the appeal to financial investors is a central strategy for solving the huge demand for investments in the region (Vassalo 2017). The significant economic growth experienced in several Latin American countries from 2004 onward, which derived in large part from the commodities boom and from an association with a state of high international liquidity, was an essential element in the resurgence of PPPs. The expansion of national reserves was fundamental in the decision by local governments to develop a set of countercyclical policies as a response to the 2008 crisis, boosting investments in infrastructure. These investments were made possible, in many cases, by the implementation of PPPs, with the states taking key roles in financing and structuring projects, as well as defining new financial instruments (Rufino 2021).

Given a scenario in which the development of the capital market was relatively limited, bank financing functioned as a vital catalyst for the operation of infrastructures, one that was more closely tied to the logistics of finance. Although multilateral agencies such as the World Bank had played a relevant role in financing PPPs in Latin America since the 1990s, in this new cycle the presence of national and regional development banks gained prominence. In the Brazilian case, for instance, the National

Bank for Economic and Social Development (Banco Nacional de Desenvolvimento Social e Econômico—BNDES) became the primary agent of PPP promotion and a privileged source of subsidized loans for the provision of infrastructure in Brazil. Moreover, the bank also played a significant role in the development of other countries in the region (Rufino 2021). Through its credit-lines for the exportation of engineering services, BNDES made it possible to expand the activities of large Brazilian contractors, as they began to acquire control over PPPs in other countries, by associating with local groups and international investors, thus, consolidating a complex network of interests in the expansion of PPPs.

Bayliss et al. (2021) argue in their review of the critical literature on PPPs that the attraction of the private sector to infrastructure provision in developing countries has not necessarily reduced the demand for public funds. Substantial public investments are usually needed to attract financial investments in sectors with limited commercial returns, both to offset the risks of long-term uncertainties and to ensure that the benefits reach the entire population, not just those who can afford them. According to World Bank data concerning Latin America, about one-third of PPP financing comes from public funds, with half of all PPPs receiving some form of governmental guarantee (Bayliss et al. 2021).

As a general rule, selectivity in the expansion of investments continues, with PPPs being concentrated in more profitable segments and in the most dynamic regions within countries (Rufino 2016). According to the PPI-World Bank data, of the 1056 PPPs contracted in the region between 2005 and 2015, 588 were within the energy sector. These accounted for over 43% of total investments. On the other hand, general expansion of investments and improvements in project structuring has ensured an expansion in the mobilization of PPPs in the case of urban infrastructures, with significant advances in sectors such as urban mobility and sanitation (Vassalo 2017).

Still, the provision of urban infrastructure via PPPs faces a number of other challenges related to the historical characteristics of Latin American urbanization, the persistent socioeconomic inequalities, and the financial deficits experienced by local governments. As Pradilla and Lopéz (2017) demonstrated, the historical lack of interest displayed by private investors in areas such as transportation, water supply, and solid waste treatment derives from the high level of investments required for such infrastructure and the high risks associated with their urban locations, as well as from the uncertainties regarding the tariff prices charged to users.

In the perception of investors who design PPPs, allocation and management of risks are extremely complex variables in the context of projects developed in Latin America (Vassalo 2017). More generally, the provision of new infrastructures in an urban context involves additional risks related to land expropriation and construction, increasing the need for credit and extending the wait on investment returns. Other difficulties are in predicting the demand for services and the recovery of investments through tariffs. The very ability, on the part of the population, to pay for the services is a structural limit to the extent of this rationality in the region. A history of demonstrations against the increase of public service tariffs makes the dependence on these revenues a social risk for PPP investors. The differentiation between the "political tariff" (or "social tariff"), paid by the user for the service, and the "contractual tariff" (or "compensation tariff"), to which the concessionaire is entitled for the provision of the service (Magalhaes 2021), became the norm in PPP contracts in the region. Subsidies, then, start to be mobilized not as means of decommodification of services, but as a way of ensuring an expected profitability.

In this sense, while in countries of the Global North, the predictability of tariffs is the main foundation for sophisticated securitization processes (Purcell et al. 2020; Pryke and Allen 2019), in developing countries, and particularly in Latin America, it is the guarantees set forth in contracts that ensure the viability of PPPs. Closer analyses of various contracts have demonstrated how PPPs, rather than taping private resources to fill the financing gap, have tied governments to long-term contracts while absorbing a substantial portion of governmental budget (Bayliss and Wanguerbele 2018). Additionally, the recurrent incidence of contract renegotiations has resulted in the transferring of complementary resources to PPP controllers. As Sánchez and Lardé (2020) demonstrated, renegotiations have been a common feature of PPP contracts in the region, occurring in more than 55% of cases between 2004 and 2015. Between 1985 and 2000, renegotiations occurred in 30% of contracts.

In order to ensure higher returns and greater predictability in revenues, PPP investors receive different forms of government guarantees and various forms of subsidies (Bayliss and Wanguerbele 2018). As a result, PPPs have been enshrined as the dominant form of accumulation in infrastructure provision. In this context, the public funds emerge as essential components in enhancing mechanisms for the capitalization of revenue streams in infrastructural projects, serving as the basis for expanding the instrument of PPPs to new geographies and sectors that had hitherto been perceived as unprofitable.

Although the use of public funding in the provision of infrastructure is not something new, as we have already discussed, the form and direction in which the funds are now mobilized impose a particularly new movement to the metamorphosis of infrastructure in the region. It is based on very close ties with public funding that infrastructures come to be determined by fictitious capital.

Through infrastructure, the use of public funds is imposed as a condition for the fruition of fictitious capital in the region (Blank 2018). The detailed analysis of the structuring of PPPs shows that the state itself starts to produce fictitious capital through the creation of new financial instruments. As the financing for long-term projects has generally not been sufficient, governments in the region have increasingly explored financing mechanisms such as structured credits or debt markets bonds (Valenzuela 2021), thus resorting to the production of fictitious capital in the very establishment of public funds. In this sense, these funds have not only become a fundamental place of rapine and plunder of the wealth appropriated by the state, but a locus of production for new fictitious wealth (Blank 2018).

The particularities that the metamorphosis of infrastructure assumes in Latin America allow for the renewal of expedients of spoliation as a determining aspect in urban transformations through accumulation in finance and real estate, illuminating a particular dimension of Urbicide in the Latin American cities directly connected with privatization processes.

The state, by subordinating public funds to capitalization mechanisms, becomes the central agent in the development of financial spoliation, ensuring to companies and financial institutions an accumulation that is independent from the production of value. Additionally, in the production of contemporary Latin American cities, the expedients of real estate spoliation overlap with those of urban spoliation. These are also greatly coordinated with the processes of infrastructure renewal. Thus, it can be affirmed that the growing investments in infrastructure have not disrupted the historical processes of unequal urbanization in the region. On the contrary, despite the fact that the major metropolises of the region have received large investments in infrastructure (especially in transportation) and have continued to grow in a dispersed manner, an acceleration in the densification of popular territories has been observed. According to Jaramillo (2018), although the proportion of self-built houses in relation to the total number of dwellings has been decreasing since the 1980s, from the 1990s onward this trend has lost strength, and, in some of the most populated cities of the region, it has been reversed and currently reaches figures similar to forty years ago.

Even while the traces of urban spoliation remain evident in several precarious territories of the large metropolises of the region, the possibility of better access to services and infrastructures in some of these territories is capitalized by the high costs of housing, leading to the densification and commodification of popular urbanization.

On the other hand, investments in infrastructure have become increasingly associated with processes of urban renewal and urban expansion, which take on increasingly innovative profiles. Supported by discourses and labels such as sustainability, compact cities, and more recently smart cities, these processes have intensified the centralization of capital in the emergence of sophisticated real estate complexes. These processes originated from changes observed simultaneously in the financial markets, the real estate industry, the provision of infrastructure, and the design of public policies, as this rapprochement between real estate and infrastructure becomes a fundamental aspect in the acceleration and expansion of accumulation in the production of built space, constituting a key process for the expansion of inequalities in these metropolises (Rufino 2022).

## 33.5 Final Considerations

In this chapter, we have argued that a metamorphosis of valorization occurs in the production of contemporary infrastructure. This metamorphosis is apparent in the dominant function of infrastructure as fictitious capital to the detriment of its traditional role as fixed capital. The process is unfurled in a generalized global scale but takes on particularities in distinct social contexts.

Despite the fact that, in Latin America, the insufficiency of urban infrastructure is a historical trace mobilized by politicians, public officials, and financial investors as a central aspect to promote an agenda of privatizations, the growing investment in infrastructure provision results in an overall presence of infrastructure that is still timid, inconstant, and selective. With the mobilization of more sophisticated financial instruments in the modeling of infrastructure projects (Bayliss et al. 2021), Latin America has seen priority given to more profitable projects with lower risks. Meanwhile, there have been limited responses to the historical infrastructural shortcomings of Latin American urbanization.

Public–private partnerships (PPPs), by enabling significant allocation of public subsidies to private investors and letting the state absorb a large part of the risks involved (Bayliss et al. 2021), have been established as the primary means of driving infrastructure metamorphosis in the region.

The general analysis of the process in Latin America reveals how, through the implementation of this instrument, infrastructure provision establishes itself as a powerful mechanism for the transfer of wealth. This is supported by the structuring of capitalization processes established through property control by powerful private agents. These processes, which begin to form and are designed by contracts established in times of economic growth, are prolonged in contexts of crises and seize past, present, and future social wealth in the form of fictitious capital.

The state, by means of the public funds, becomes a central figure in this rise of infrastructures as fictitious capital. Of course, this constitutes a considerable transformation of its historical function. From a previously omissive position concerning the decommodification of social conditions of reproduction in the industrialization processes of some of the most advanced economies in the region during the 1960s and 1970s, the state emerges as a presence at the turn of the century, operating sophisticated expedients of spoliation in real estate and finance.

As we have tried to argue, these new forms of spoliation overlap with the more historically observed forms of urban spoliation to result in direr living conditions in these Latin American metropolises. At the same time, they provide forms of capital accumulation without any direct relation to value production.

The provision of new infrastructures has instrumentalized significant appropriation of public funds in such a way that it enables the engineering of financial spoliation, leading those very funds to operate as fictitious capital, by mobilizing future wealth through the use of public debt securities as collateral for PPPs. Therefore, even if the infrastructures remain as a fixed capital essential to the reproduction of the urban, the way in which they will be produced and operated is conditioned by the production of fictitious capital.

In turn, the real estate spoliation operates through transformations and increased coordination between infrastructure provision and real estate production. On the one hand, the slow advance of infrastructure in the most precarious territories consolidates its limited presence as a catalyst for the increase in the price of popular housing. On the other hand, the provision of infrastructure, in conjunction with other urban policies, has consolidated real estate capitalization and opened new fronts for its valorization, supporting unprecedented levels of capital centralization and construction densification in the large metropolises of the region, despite the continuity of the processes of dispersed urban growth.

These processes illuminate a particular dimension of the Urbicide in the Latin American cities directly connected with the specific nature of the infrastructure privatization process and its impact in the urbanization. At the base of these processes, there is a significant movement toward a depoliticization of the urban. As we have already noted, the transformation of urban infrastructures into instruments of capitalization employs the elimination of social risks as a fundamental strategy for its expansion. What investors perceive as "social risk" is precisely the possibility that the despoiled population might organize themselves politically to reclaim their rights. This poses an objective limit to planned capitalization. By bearing most of the risks and ensuring greater predictability of profits, the state conceals underlying or imminent conflicts and transforms the rights of the population into a lucrative business, narrowing the horizons for transformative political actions in the cities.

# References

Arboleda M, Purcell TF (2021) The turbulent circulation of rent: towards a political economy of property and ownership in supply chain capitalism. Antipode 53(6):1599–1618

Ashton P, Doussard M, Weber R (2012) The financial engineering of infrastructure privatization: what are public assets worth to private investors? J Am Plann Assoc 78(3):300–312

Bakker K (2003) Archipelagos and networks: urbanization and water privatization in the south. Geogr J 169:328–341. https://doi.org/10.1111/j.0016-7398.2003.00097.x

Bayliss K, Fine B (2008) Privatization in practice. In: Privatization and alternative public sector reform in sub-Saharan Africa. Palgrave Macmillan, London, pp 31–54

Bayliss K, Van Waeyenberge E (2018) Unpacking the public private partnership revival. J Dev Stud 54(4):577–593

Bayliss K, Romero MJ, Waeyenberge EV (2021) Uneven outcomes from private infrastructure finance: evidence from two case studies. Dev Prac 31(7):934–945. https://doi.org/10.1080/09614524.2021.1938513

Blank J (2018) Um museu de grandes novidades: capital fictício, fundo público e a economia política da catástrofe. Revista Maracanan 18:181–197

Campos P (2021) As empreiteiras brasileiras na transformação dos padrões de acumulação: estratégias hegemônicas e conflitos contemporâneos. In: Rufino B, Wehba C, Faustino R (eds) Infraestrutura na reestruturação do capital e do espaço: Análises em uma perspectiva crítica. Letra Capital, São Paulo

Chesnais F (ed) (2005) Finança Mundializada. Boitempo, São Paulo

Christophers B (2020) Rentier capitalism: who owns the economy, and who pays for it? Verso

Connolly P (2013) La ciudad y el hábitat popular: Paradigma latinoamericano. Teorías sobre la ciudad en América Latina 2:505–562

Davies A (2021) The coloniality of infrastructure: engineering, landscape and modernity in Recife. Environ Plann D Soc Space. 02637758211018706

De Lima Seabra OC (1987) Os meandros dos rios nos meandros do poder: Tietê e Pinheiros: valorização dos rios e das várzeas na cidade de São Paulo. Tese (Doutorado em Geografia)—Faculdade de Filosofia, Letras e Ciências Humanas, Universidade de São Paulo, São Paulo

Dodson J (2017) The global infrastructure turn and urban practice. Urban Policy Res 35(1):87–92

Emilio D (1998) Hábitat popular y política urbana. México. Grupo Editorial Miguel Ángel Porrúa. Universidad Autónoma Metropolitana Unidad Azcapotzalco, UAM-A

Fay M, Andres AL, Fox C, Narloch U, Staub S, Slawson M (2017) Rethinking infrastructure in Latin America and the Caribbean: spending better to achieve more. World Bank, Washington, DC. https://openknowledge.worldbank.org/handle/10986/26390

Fernandez R, Aalbers MB (2020) Housing financialization in the global south: in search of a comparative framework. Hous Policy Debate 30(4):680–701

Folin M (1977) La ciudad del capital y otros escritos. GG, México

Graham S, Marvin S (2001) Splintering urbanism—networked infrastructures, technological mobilities and the urban condition. Routledge, London/New York

Harvey D (1978) The urban process under capitalism: a framework for analysis. Int J Urban Reg Res 2(1–3):101–131

Harvey D (2006) The limits to capital. Verso Books

Harvey D (2018) A loucura da razão econômica. Boitemp, São Paulo

Jaramillo S (1988) Crisis de los medios de consumo colectivo urbano y capitalismo periférico. In: Cuervo LM et al (eds) Economía política de los servicios públicos. Una visión alternativa. CIDEP, Bogotá, 303, pp 15–37

Jaramillo S (2018) Producción no mercantil en la economía capitalista. Universidad de los Andes-CEDE

Kowarick L (1979) A espoliação urbana. Editora Paz e Terra, Rio de Janeiro

Lencioni S (2007) Condições gerais de produção: um conceito a ser recuperado para a compreensão das desigualdades de desenvolvimento regional. Scripta Nova: revista electrónica de geografía y ciencias sociales 11:6

Lencioni S (2021) Condições gerais de produção e espaço-tempo nos processos de valorização e capitalização. In: Rufino B, Wehba C, Faustino R (eds) Infraestrutura na reestruturação do capital e do espaço: Análises em uma perspectiva crítica. Letra Capital, São Paulo

Loftus A, March H, Purcell TF (2019) The political economy of water infrastructure: an introduction to financialization. Wiley Interdiscip Rev Water 6(1):e1326

López LM, Cobos EP (2017) La privatización y mercantilización de lo urbano. In: La ciudad latinoamericana a debate, p 17

Lorrain D (2011) La main discrète: la finance globale dans la ville. Revue française de science politique 61(6):1097–1122. https://doi.org/10.3917/rfsp.616.1097

Magalhaes ALC (2021) Financeirização da infraestrutura: as grandes empreiteiras nacionais e o metrô de São Paulo em transformação. Master's Dissertation, Faculdade de Arquitetura e Urbanismo, University of São Paulo, São Paulo. https://doi.org/10.11606/D.16.2021.tde-24112021-122705. Retrieved 2023-01-31, from www.teses.usp.br

Michelitsch R, Szwedzki R (2017) A decade of PPPs in Latin America and the Caribbean: what have we learned? World Bank Blog. Available at https://blogs.worldbank.org/ppps/decade-ppps-latin-america-and-caribbean-what-have-we-learned

Oliveira FD (1998) Os direitos do antivalor: a economia política da hegemonia imperfeita. Vozes, Petrópolis

O'Neill P (2017) Infrastructure's contradictions: how private finance is reshaping cities. In: Money and finance after the crisis: critical thinking for uncertain times. Wiley. 9781119051428

O'Neill P (2019) The financialisation of urban infrastructure: a framework of analysis. Urban Stud 56(7):1304–1325. https://doi.org/10.1177/0042098017751983

Pereira PCX (ed) (2018) Imediato, global e total na produção do espaço: a financeirização da cidade de São Paulo no século XXI. FAU-USP, São Paulo

Pereira PCX, Petrella G (2018) Introdução. In: Pereira PCX (ed) Imediato, global e total na produção do espaço: a financeirização da cidade de São Paulo no século XXI. FAU-USP, São Paulo

Pírez P (1999) Gestión de servicios y calidad urbana en la ciudad de Buenos Aires. EURE 25(76), dez. 1999, Santiago, Chile

Pírez P (2012) Servicios Urbanos y Urbanización en América Latina: su orientación entre el bienestar y la reestructuración. Geo UERJ—Ano 14 23(2):793–824

Pírez P (2013) La urbanización y la política de los servicios urbanos en América Latina. Andamios 10(22). Universidad Autónoma de la Ciudad de México

Pradilla E (1994) Privatización de la infraestructura y los servicios públicos. Argumentos. Estudios críticos de la sociedad 21:57–79

Pradilla Cobos E, López LM (2017) La privatización y mercantilización de lo urbano. Daniel Hiernaux-Nicolas y Carmen Imelda González-Gómez (coords.), La ciudad latinoamericana a debate: perspectivas teóricas, Querétaro, Universidad Autónoma de Querétaro, Editorial Universitaria (Col. Academia, Serie Nodos)

Pryke M, Allen J (2019) Financialising urban water infrastructure: extracting local value, distributing value globally. Urban Stud 56(7):1326–1346

Purcell TF, Loftus A, March H (2020) Value–rent–finance. Prog Human Geog 44(3):437–456. https://doi.org/10.1177/0309132519838064

Rocha MAMD (2013) Grupos econômicos e capital financeiro: uma história recente do grande capital brasileiro. Instituto de Economia da UNICAMP, Tese de Doutorado, Campinas

Rufino B (2015) Transformação da periferia e novas formas de desigualdades nas metrópoles brasileiras: um olhar sobre as mudanças na produção habitacional. Cadernos Metrópole 18(35):217–236

Rufino B (2016) Public-private partnerships and their implications for inclusive urbanisation in Brazil. Reg Mag 303(1):14–15

Rufino B (2021) Privatização e financeirização de infraestruturas no Brasil: agentes e estratégias rentistas no pós-crise mundial de 2008. urbe. Revista Brasileira de Gestão Urbana [online], v. 13

Rufino B (2022) Imobiliário e infraestrutura de mãos dadas na cidade neoliberal latino-americana: reestruturação financeira na ampliação das desigualdades do espaço na metrópole de São Paulo. In: Pradilla E (ed) Rede Latino Americana de Teoria Urbana. Cidade do México

Sánchez RJ, Lardé J (2020) Public-private partnerships under the "people-first" approach

Serebrisky T, Suárez-Alemán A, Margot D, Ramirez MC (2015) Financing infrastructure in Latin America and the Caribbean: how, how much and by whom. Inter-American Development Bank, Washington, DC

Valenzuela A (2021) El financiamiento de infraestructuras públicas con instrumentos financieros en México. In: Seminario Latino Americano de Teoria Urbana Latino America. Buenos Aires

Vassallo JM (2017) Public-private partnership in Latin America. Facing the challenge of connecting and improving cities. CAF, Caracas. Retrieved from http://scioteca.caf.com/handle/123456789/1549

**Beatriz Rufino** is a Professor and Researcher at the College of Architecture and Urbanism of the University of São Paulo (FAU-USP). She carried out Postdoctoral research at Kings College London through the Urban Studies Foundation's International Fellowship Program (2021–2022). She participates in and coordinates research in the area of Urban Studies, focusing on the themes of Real Estate Production and Infrastructure and Urban and Housing Policies.

# Chapter 34
# Public Policies (Or Their Absence) as Part of Urban Destruction

Marcelo Corti

**Abstract** In our times, cities are constantly threatened: authoritarian or technocratic attempts at demographic deconcentration, based on moralistic, economic or political arguments; on the dispersion induced by mobility based on the private car or by erroneous policies of public housing in the peripheries; on the various privatopias of closed neighbourhoods, on urbanizations by "Communities of Special Interest", etc. So in this article, I am going to address those threats or attacks that are directly and specifically related to urban public policies. There are three basic issues that I would like to comment on in relation to these; it is about the application of certain erroneous urban policies or the conscious application of urban policies that, whatever their objective, ended up being disastrous for the city. There are three issues, at least from my point of view: the conception of the "facilitating" State of private markets, an equivocal notion of "subsidiarity" of local governments and the abuse of "tactical" conceptions of urban planning. All of this basically confronts us with the need to recover a public role for urban development. It is about recovering the role of the public sector as the great city builder, the actor that configures the city and the territory.

**Keywords** Disasters · Public policies · Urban planning

## 34.1 Introduction

First of all, I confess that the very concept of "urbicide" arouses caution in me, as it refers to an instance of intentionally caused death, a murder of the city. Surely, there are actors and agents directly interested in that death, and many more who do it out of negligence or incompetence. And certainly, we are going to have to work hard to defend that urban way of life that defines the very concept of life in society. Nancy (2013) argues that the art of the city is to live together, "for that purpose it was founded, built, organized". It is not defined by the functions of protection, government

M. Corti (✉)
National University of Córdoba, Córdoba, Argentina
e-mail: marcelo.corti@unc.edu.ar

and exchange; "Through the domus or the villa one can reach the village, but not the city, whose constitution requires that living together is not given in advance". For Nancy, the city is the "other one" of the countryside, "from the beginning it is a new country, the 'country' of uprooting". But for that reason, because it is a logical consequence of the way in which human beings organize our lives, I am convinced that the city will not disappear.

Let us never forget that civilization is ultimately a word that shares its origin with that of city. To destroy the city, or the concept of the city, is practically to end civilization and that is why I think we must to work, we must seek and find the ways to maintain the city in any of its scales, in its forms. Basically, the concept of the city is that of a meeting place for different people, a place of interaction and specialization but, at the same time, a place for interaction between diverse human beings. And that is the reason why I believe the city will prevail. I remember in this regard a very funny phrase by Fran Lebowitz. You may have seen—and if you have not seen it, I highly recommend that you do—the series "Pretends is a City", about New York, directed by Martin Scorsese, who also conducts the interviews with the writer (on Netflix, 2021). In one episode, Lebowitz recalls the famous phrase of then US President Gerald Ford when the New York City Council was declared bankrupt in 1975: "New York is dead". Nearly half a century later, Fran Lebowitz wonders: "Who is dead now?".

Rumours about the death of the city seem, therefore, exaggerated... But it is true that our civilizing heroine is constantly threatened: authoritarian or technocratic attempts at demographic deconcentration (of which the genocide of Pol Pot and the Khmer Rouge in Cambodia was the most extreme and abject expression), based on moralistic (the city as "great harlot of Babylon"), economic or political arguments; on the dispersion induced by mobility based on the private car or by erroneous policies of public housing in the peripheries; on the various privatopias of closed neighbourhoods, on urbanizations by "Communities of Special Interest", etc. So in this article, I am going to address those threats or attacks that are directly and specifically related to urban public policies.

There are three basic issues that I would like to comment on in relation to these; it is about the application of certain erroneous urban policies or the conscious application of urban policies that, whatever their objective, ended up being disastrous for the city. There are three issues, at least from my point of view: the conception of the "facilitating" State of private markets, an equivocal notion of "subsidiarity" of local governments and the abuse of "tactical" conceptions of urban planning. Surely, there will be others.

## 34.2 The State as Facilitator of Private Action

The first of these is the idea that began to prevail from the seventies and especially in the eighties of the last century, the one that places the State as a facilitator of private actions in the city, rather than as a promoter or an instance higher control and regulation regarding urban management. This has to do with the flaws and cracks in the Welfare State, as it was conceived in the period roughly between the Great Depression of 1929, World War II, the Cold War and the oil crisis of 1975. And especially the so-called Thirty Glorious Years from 1945 to 1975, characterized by policies of strong state intervention and regulation in the economy (both in production in numerous sectors of activity and in the provision of social welfare services, health, education, recreation, etc.). This model of production and distribution within capitalist societies with different degrees of development entered into crisis in the 1970s, together with many elements of the so-called Fordist economy, characterized by mass and serial production in large locally established factories and with a large and highly unionized staff. Not many years later became the collapse of the socialist economy, which had also shown its limits and shortcomings in the previous period—the socialism that really existed, as it was said at that time.

Based on these facts, a deregulatory tendency against state interventionism in the economy grew up. The idea spread worldwide that the State should, ultimately, give up the control of many issues—among them, the management of the city—to hand it over to the private sector. At most, the role of the public sector would be to facilitate the task of the private sector. This is expressed in many instruments, in many processes and procedures of that time that continue to this day. In general, this voluntary or forced withdrawal of the State is manifested in the idea of public–private partnerships or commitments, where the public sector is often the secondary guest of private business. Or for example, in operations such as the British "Right to buy" the houses built by the municipalities at the time before Margaret Thatcher. It provided the possibility of purchase by tenants, and later this led to unfortunate processes, especially in the city of London. That is how that heritage of public housing built for much of the twentieth century was wasted, and this gave rise to other forms of speculation on the city. A tour of London today gives us several immediate visual impressions: luxury, high technology and a scale of outrageous intervention, close to the grotesque. Anna Minton, a journalist for The Guardian, with postgraduate studies in architecture, reveals in her book Big Capital (2017) the political-economic structure that explains the visible city and its social consequence: the displacement of thousands of Londoners from their city and the enormous difficulty of working sectors, middle classes and professionals to access housing in the British capital and in other cities of the UK.

The book combines testimonies of people displaced from neighbourhoods where they spent their entire lives, through an impeccable description of the device with which the expulsion was implemented and the formidable increase in land value that sustains that process. How is it possible, for example, to demolish an entire social housing neighbourhood like Heygate, in Southwark Council, to make way for a luxury complex like Elephant Park, whose apartments none of the previous residents can even dream of accessing? For Minton, this process cannot even be called "gentrification"; it is a completely different phenomenon, a new politics of space in which global capital reconfigures the entire territory based on rates of return on the value of property that far exceed those that productive activity can support.

In the beginning, it was the anti-modernist rhetoric of the Charles Jencks versus Pruitt-Igoe matrix that fed the bad reputation of the post-war social housing complexes, (produced by the councils), where at the early eighties a third of the London population lived. This interpretation of architectural and urban criticism was combined with the Thatcherian discourse of individual effort and entrepreneurism to get out of the Welfare State. The Right to Buy programme, designed to "turn proletarians into homeowners", allowed for the scrapping of the housing structure by giving resident tenants access to the property. This operation implied a trap: the housing units were sold but not the land on which they are located, which continued to be municipal. What has skyrocketed in this century is the forced purchase of homes for demolition to make room for extreme luxury interventions, dedicated to a global elite (the "alpha elite") and closely linked to money laundering operations—according to Transparency International, UK is a "premium" destination for the cream of international corruption to launder illicit wealth and to access a luxurious lifestyle. The "owners" received a payment that does not consider the value of the land and whose amount, even combined with life savings, is not enough to buy an equivalent property, not just in the new luxury development but even in the same district. Families are thus forced to move to peripheral neighbourhoods, losing social capital, relationship networks, access to work and urban services that they had achieved in their original neighbourhood. In their host boroughs, the same mechanism is reproduced and the economic filter displaces families and households out of London to other cities where the process starts all over again, in a devastating domino effect. The councils that thus recover the land from the previous social housing do not even capitalize to take advantage of the process by developing an urban regeneration that allows them to maintain their residents, because the rent paid by developer corporations barely covers the cost of the sum of the forced purchases.

The elimination of the higher land value tax, another legacy of the Thatcher era, deprives the public sector of the possibility to act in land management. Minton recalls that this tax (historically promoted by figures as far removed from left-wing thought as Adam Smith, Loyd George and Winston Churchill) and the Greater London Green Belt were key in urban planning policy after the Second World War, which in addition, through the Housing Act of 1949, promoted integrated and socially mixed neighbourhoods. The device of expulsion and social "cleansing" that currently replaces it generates personal and family situations almost similar to those portrayed by Charles Dickens in the nineteenth century. This system is made up of

- the so-called demand subsidy (we will return to this topic) through the granting of housing vouchers to the supposed beneficiaries,
- the deregulation of prices and rental terms,
- the decoupling between property prices and the subsidies that can be granted by councils,
- the facilities that Housing and Planning Act of 2016 grants to the demolition and "regeneration" of built-up areas,

- and the legal traps that allow developers to exempt themselves from complying with the affordable housing quotas that they should incorporate in their projects, based on the distortion of their financial evaluations...

This process cannot be separated from the disenchantment that gave rise to the Brexit victory in the 2016 plebiscite. Minton quotes Terry and Brenda, former Heygate residents confined to Sidcup, an hour and a half from central London by public transport, who voted leave even though it was not very clear whether or not it was the best option. "But then I thought that the Southwark Labour municipal government kicked us out of our house and did nothing to protect us. They left us with no options...", says Terry; voting to leave the European Union thus became an option.

In order to overcome this scenario, Minton deems necessary the consideration of the Right to the City, a new democratic and participatory conception of urban planning processes, the encouragement of cooperative housing associations and policies for the completion of fabrics in consolidated areas—in addition, of course, to establishing the tax on the speculative land valuation. She presents European examples that could serve as a reference to a progressive urban policy for London and the UK, such as deFlat in Amsterdam, Sluseholmen in Copenhagen and, above all, the extensive public housing stock of Vienna or Berlin, cities that base their urban policy "in not looking like London".

In our Latin American continent, housing demand subsidy policies, with the aim or under the pretext of favouring those who really need housing or help from the State to gain access to housing, have often given rise to strong speculation and very profitable operations for the companies, but very unfavourable for the supposed beneficiaries, especially in countries like Mexico and Chile. Ana Sugranyes and Alfredo Rodríguez made it clear that the problem in Chile used to be the homeless, and now it is... those "with home". In an interview we had in 2004, Rodríguez told me about it:

> What we find is that there is neither an urban policy nor a housing policy, but rather a housing financing system. [...] Particularly, because as this financing system, which dates back to the time of the dictatorship, has worked very well (the State guarantees real estate companies an annual income) and a lot of houses have been built, the scheme is repeated or copy in Latin America. That worries us, and a lot, because we are seeing and demonstrating what its results are: in the 1980s, the housing problem in Chile was that of the "homeless"; today, on the other hand, the problem belongs to those "with a roof". The public authorities consider the problem solved, but we think that it is not. We've done a very large survey, tracking the 500 or so developments that have been built in the last few years, and we found that there is actually a very serious problem in the housing stock. That is why it is necessary to give another turn to the policies, and start a policy of improving what has been built. Which is quite similar to what happened in other parts of the world, like what the French did with the HLM (Habitation à Loyer Moderé), or the case of New York itself. A massive response to the housing problem is fine, but then you have to go back to review the maintenance of those neighbourhoods, and sometimes this is not so easy.
>
> MC: What is the essence of this model (which apparently worked so well) managed between the State and real estate companies?
>
> AR: It's what they call the "housing subsidy": government annually places some 300 million dollars in subsidies, people opt for them, and with that subsidy (plus a small

saving in some cases) a demand is generated which construction companies take advantage of. Initially, it was called a demand subsidy model, but in practice it is actually a supply subsidy. In the mid-1980s, construction companies bought large extensions where was then the periphery of the most important cities. Every year, each company places 2,000 or 3,000 units, in turn the Ministry of Housing and Urbanism grants these 20, 30 or 40 thousand annual subsidies, and then there is an almost stable operation of the system.

MC: Are the same companies that choose the land and, therefore, define where the city will grow?

AR: Yes, the only thing the government does is give this financial advance.

MC: That seems consistent with Pinochet's policy of deregulating land use, assuming that in this way its cost would be lowered.

AR: Yes, but this is only part of it. Another fundamental issue is the operation of financial capital within the real estate sector, and its impact on the city: nobody studies on this subject. Considering social housing and the rest of the residential developments, only in Santiago the financial capital moves between 3 thousand and 6 thousand million dollars per year, but there are no studies. And after 20 years there are many millions accumulated. In this matter of land management, the growth of the city and its limits, there is a very strong capitalist accumulation. The Italians had studied it well in the 1970s (among other studies, there is that of Francesco Indovina and his book "Real estate waste"), and this is the same thing, the alliance between real estate companies and the State, in which a side incorporates rural land into urban land and because of that single administrative act the price of land rises in an extraordinary way.

In the Chilean case, a student wrote an excellent article called "The Virtual Urbanization", because in the year '94 around 60 or 70 thousand hectares were incorporated in the northern area of Santiago for housing of middle and middle low-income sectors. He had estimated that it would take around 70 years for the land to be fully occupied, and he wondered "why is it then said that the turnover of capital in the construction sector is so fast?" And he realized that we were really chasing ghosts, because the deal had already been done! From the moment that, due to an administrative act of the Ministry (someone who says "it has to be here"), those lands had gone from a zoning that allowed a density of 10 people per hectare, to another of 100 people per hectare. Owners went so to the banks and obtained funds, while they already had them invested in another sector of the economy. But we worried about the houses...!

[...] And furthermore, according to the survey we did, 65% of the people want to leave there, what means their situation has not improved at all. There are high rates of violence too; the reason that people generally give to explain why they want to leave is not the small size of the house, but the problems of living together: daily violence, aggressiveness, drugs, etc. In short: it is poverty (Rodríguez 2004).

In short, a big number of procedures and strategies have had the objective of withdrawing the State from the management of the city and placing it rather as a facilitator of private tasks. I think that a good part of the problems that we are seeing in this book and in the seminar that originated it are related to those policies (Corti 2019).

## 34.3 A Misunderstood Subsidiarity

A second front of the attack on the city was generated by a misunderstood subsidiarity of the State, which tried to delegate responsibility for the development of cities and territories to bodies that were closer—or supposedly closer—to the citizenry, such as the municipality. Based on this principle, the entire weight of urban policies tended to fall on local governments, withdrawing the national State and even subnational governments from their obligations and competencies.

Many times in the twentieth century—let us not even talk about earlier times—the national or subnational States had a very strong importance in the production of the city. It was the state powers which founded cities, connected them, provided them with infrastructure, equipment and services, expanded them and directed their growth (especially in capital cities). With the advance of deregulatory rhetoric, this role was left to the cities, a delegation policy often accompanied even by that catchphrase so typical of the 1980s and 1990s: the cities competition.

This idea was complemented by an implicit promise: cities would surpass their countries, since they were more relevant and had a greater impact on the new world order and on globalization and could therefore, compete between themselves to access the benefits of this new phase of the capitalism and, consequently, of economic growth. But looking back, what this hypothesis actually covered up was an abandonment by national and—to a lesser extent—subnational States in their commitments to cities. Especially, in large metropolitan areas, where the most peripheral and youngest municipalities often have very little management capacity and/or are very weak economically, have little financing capacity and are very susceptible to co-optation by large landowners and private developers.

In my opinion, this "Greek present" received by the cities is at the origin of a good part of the problems that we are analysing at this book. It is true, in contrast, that as a consequence of this new role many cities (especially those of great size and prosperity) positioned themselves as relevant actors in international politics and developed important management capacities. Among the circumstances that accompanied and explained this rise, two positive developments can even be pointed, particularly in Latin America: the recognition of municipal autonomy in many laws and constitutions, and the possibility of citizens to elect their municipal authorities, a right that even after the nineties was denied too many cities, especially the national capitals. Buenos Aires (which gallantly assumed the character of Autonomous City in its own institutional denomination), Bogotá, Mexico City (which also went from being a Federal District to a city recognized in that legal specificity), among others, benefited from this expansion of rights. But, in general, subsidiarity and its perverse sister, competition with other cities, created agendas that cities were neither in a position to fulfil, and were not relevant to them. It is overwhelming to think of the valuable hours wasted by technical teams, officials and civil society participating in meetings and strategic planning studies to determine "which cities ours competes with". Currently, this fetish of competition between cities has been highly discredited and it is preferred to think and work for territorial collaboration and, in any case, for the competitiveness of cities and regions to insert themselves into the global economy.

It is almost embarrassing to refer to a document as justly criticized as the New Urban Agenda of UN-Habitat, approved at the III International Conference held in Quito in 2016. It is a compromise agenda, which to be approved by nearly two hundred countries had to agree on contents and lead to a compendium of truisms with little utility. But I think that one of its very few salvageable points is the one that returns—at least in discourse—the need for national and subnational States to resume their responsibilities in the urban and territorial issue. Thus, its point 15 states that

> We commit ourselves to working towards an urban paradigm shift for a New Urban Agenda that will: […]
>
> (b) Recognize the leading role of national Governments, as appropriate, in the definition and implementation of inclusive and effective urban policies and legislation for sustainable urban development, and the equally important contributions of subnational and local governments, as well as civil society and other relevant stakeholders, in a transparent and accountable manner;

And at point 29:

> We commit ourselves to strengthening the coordination role of national, subnational and local governments, as appropriate, and their collaboration with other public entities and non-governmental organizations in the provision of social and basic services for all, including generating investments in communities that are most vulnerable to disasters and those affected by recurrent and protracted humanitarian crises.

## 34.4 The Abuse of "Tactical Urbanism"

A third type of public policies that compromise the virtuous development of cities is linked to the application of our disciplines (such as urban planning or urban design); it is about the hasty, trivialized and simplified application of the concepts that define the so-called tactical urbanism. Beyond conceptual differences between tactics and strategy, the practice of "tactical urbanism" suffers something similar to which years ago was identified in some uses of "strategic planning": misunderstood, it is limited to the idea that our discipline and its convergent disciplinary fields must resign any ability for structural planning. In this way, a kind of resignation is established; we are no longer going to try to act on the large structures or components of the city, but rather, we are will limit ourselves to installing some flower pots, changing some direction of circulation, removing some lane from some street, putting together some element that proposes or simulates a more pleasant use of the city.

Of course, in its good manifestations, this tactical urbanism is put into practice with excellent intentions and very good results. There are, for example, the implementation of Salvador Rueda's superblocks in Barcelona or the proposal to plant trees in Paris by Mayor Anne Hidalgo, just to name a few. There are a large number of experiences that I separate from this criticism, which is not related to tactical urbanism itself.

What I want to question is this idea that in any case what we can do in cities are just a few small things that, in many cases, become a kind of "urban franchises". In them, each city has a superblock that is actually located in a place where change is no longer necessary, or where it is easier to make a difference and so we forget about the most vulnerable neighbourhoods, we forget about the outskirts and about all those issues that refer to the more structural mandate of urban planning.

Of course, it is not my intention to discourage the production of superblocks in the consolidated city or the provision of urban attributes that transform each urban sector into a "15-min city" or the provision of bike lanes for healthier and more sustainable urban mobility. They are good ideas, they are fine, it is just a matter of making sure they are implemented well and in harmony with the rest of the urban strategies that make up an urban plan.

The city is not a franchise of a company that forces us to respect signage, service standards, offers and staff uniforms (Corti 2020). The city is a social and territorial configuration that in a few points replicates all the cities that exist and, in the rest, is divided between those to which it responds in a similar way to some of its same region, scale or function and those which must be fixed with their own solutions, difficult to repeat. It is so disruptive that whoever plans and manages it feels obliged to imitate a platonic idea of the city (sustainable, intelligent, competitive, functional, organic, slow, safe, inclusive or any of the etceteras that adjective it at a given historical moment) as if they fantasized about being in charge of a unique and unprecedented entity that escapes any general or particular law of the formation of cities.

It is good to look at the superblocks proposed by Salvador Rueda in Barcelona or José María Ezquiaga in Madrid (Rueda 2016). They are manifestations of an idea that is always effective in the modern city: a macro-plot that channels mass mobility and a calmer internal plot that guarantees cordiality in neighbourhood life. It is very good that we think of the city as a federation of neighbourhoods where 99% of the things we need are available within a 15-min walking or biking distance. It is fine that we think of low-cost urban solutions and quick results.

The problem is not that good proposals are disseminated, but that we see them as mandates, recipes and magical solutions. Some time ago, in an Argentine city, a developer of peripheral urbanizations finished off an intervention in a journalistic debate by pointing out that their projects "had been ahead of the pandemic" (due to the social distance of low density…) and that now they are planning to adapt them to the superblock scheme (in a suburbanization of 20 people per hectare!). What happened at the end of the last century with strategic planning should not happen with tactical urbanism, which hopefully will not become an excuse for not thinking and transforming our cities.

## 34.5 The Reconfiguration of the Public

Those three trends are ways in which public policies have generated obstacles and impediments to the virtuous development of cities and fair and inclusive urbanity.

Surely, there are other problems or other ways in which States have been backtrackings, and have been contracting their policies regarding the city. But independently how partial or complete the analysis may be, all of this basically confronts us with the need to recover a public role for urban development. It is about recovering the role of the public sector as the great city builder, the actor that configures the city and the territory.

With colleagues who are part of the alliance that gave rise to this book and its supporting seminar, we are currently finalizing the preparation of a book generated from some ideas, especially those expressed in Pedro Abramo's intervention at the III Permanent Seminar Rethinking the City, organized by Civitic and Polistic and dedicated to the Planning and Management of the Urban Land Market, on 19 November 2020. In his conference, titled *Land market, real estate dynamics and capitalist production*, Abramo formulated some concepts that we considered interesting in order to expand and discuss regarding the need to recover this role of the State. In particular, four strong ideas:

- the possibility of a new urban pact, which we could also call the *urban social pact or urban contract*;
- the purpose of moving from the mere vindication of the right to the city to the production of a city of rights;
- the verification that it is no longer enough to recover capital gains, which was one of the great themes or leitmotifs of urban discussions at the end of the twentieth century and the beginning of the twenty-first century. Now, we have to move to a State which must strongly intervene in the production of the city. Not just a State that recovers the capital gains that society produces with its actions, but also one that intervenes in a very strong way in producing the city;
- and that these ideas would not be incompatible with a political-economic reform that occurs within capitalism. That means that unlike other positions that require a total change of system to produce improvements in living conditions and the quality of our cities, some very interesting reforms with many possibilities of generating a better quality of life in cities are not incompatible with a reformist vision of capitalism.

A case that responds to this model is the city-State of Singapore, whose first objective was to lift out of poverty the high percentage of its population that was in these conditions when it obtained its independence in 1965 (Heng and Yeo 2016). For this purpose, a state agency, the Housing and Development Council (HDB), manages almost all of the urban land and develops in an integral manner and with criteria of sustainable development the construction and management of housing neighbourhoods, which conceived as "new cities". The State,—with particular characteristics, since it is a city-State of only 718 $km^2$ in area—is the largest individual owner of land, administered by various official government agencies as a means to facilitate the planning and management of transportation ways, dwelling, schools and parks. The Land Acquisition Act (LAA) of 1967 gives the State control and coordination in the timing and planning of land development for public projects. The Government Land Sales (GLS) programme allows the government to free up state land for sale to the private sector, thereby enabling the government to engage the private sector in meeting market demands as well as the materialization of development plans for housing, offices, shops and industries. State land tenure operates on a 99-year lease basis for public housing and 30 years for industrial sites. This land tenure system allows the government to repossess land nearing the end of its lease and reallocate it for future development needs. Currently, more than 80% of Singapore's resident population (5.5 million people) live in housing built and managed by the HDB, most of them in some of the 23 high-density, mixed use (including industry) self-sufficient new cities. In 1974, the Urban Redevelopment Authority (URA) was established, with primary responsibility for the management and control of land development in Singapore, in accordance with the intentions and strategies articulated by the planning instruments.

The Singapore experience deserves two fundamental clarifications: on the one hand, this "socialist" management of the city's land and production coexists with a liberal capitalist management of the rest of the economy, in fact, one of the most competitive in the world (could we say that this grassroots socialism sustains and not only coexists with the liberalism of other economic activities?). On the other hand, it hurts to say that it is a deeply authoritarian political society, with a democracy based on a single party in government since its foundation and with a large number of restricted human and civil rights.

Fortunately, other experiences such as those of the Scandinavian countries or the Netherlands show that, just as public land management can coexist with a capitalist market economy, it can also coexist with political democracy. All land in entire regions, like Amsterdam, is public. Public or private constructions are authorized under a surface right regime through the payment of a canon, renewable over time.

Returning to the Asian continent, South Korea is another example of a very advanced capitalist development country in which the State is extremely interventionist in urban matters: it is the public sector that defines where the city is going to expand, the model of new cities—new towns, a land use plan, its subdivision, the architectural plan and marketing. South Korean public companies even carry out operations in other countries and continents. For example, Nueva Santa Cruz de la Sierra, a Bolivian project whose name itself makes explicit its urban concept and location is proposed as a whole new city, just a few minutes from the original Santa Cruz, very close to its airport and with all the attributes and functions of urbanity. It is expected to house 370,000 inhabitants that at the end of its construction; Santa Cruz has more than a million inhabitants (1.8 in its metropolitan area) scattered in an urban sprawl that the new city would expand by 50%. The project is led by a consortium made up of a Bolivian company, Grupo Empresarial Lafuente (the most important real estate holding company in Bolivia) and a Korean State urbanization agency LH—Korea Land and Housing Corporation (Corti 2022).

In addition to these experiences, there are other very diverse and successful ones, and not only in highly developed or affluent societies. To name just a few:

- The Medellín Urban Development Company, which finances and manages the Inclusive Urban Projects of "social urbanism" promoted by the municipality of Antioquia;
- the French Mixed Economy Companies, key in carrying out large urban projects through the modality known as Zone d'Aménagement Concerté (ZAC), public–private concerted action zones (Garay 2011);
- Corporación Antiguo Puerto Madero de Buenos Aires, a public limited company in charge of the Government of the Autonomous City of Buenos Aires and the National Government of Argentina in equal parts; the company was in charge of the integral management of the urbanization of the old port area of Buenos Aires. Beyond the criticism of the segregating and elitist characteristics that the project acquired in its development, its opening generated an expansion of the historical centrality and a significant contribution of public space for the Argentine capital, within the framework of a public policy that resisted the political fluctuations at both jurisdictions.

These and other experiences demonstrate the benefits of a public urbanization company. These organizations can be implemented with the objective of producing urbanized land provided with the required services at an affordable cost for the generation of social housing and, in general, housing for low, medium–low or medium income sectors with difficulties to access that land in the particular conditions of the real estate market and the different types of facilities and services that make up city and territory in general. For this purpose, they may buy or sell land and real estate, carry out urbanization works on their own account or on behalf of third parties, carry out parcel subdivisions and subdivisions, enter into agreements with State, private or community organizations and, in general, develop any operation necessary for the fulfilment of their social mission, including the construction and adjudication and/or sale of housing units and other buildings. Its creation represents, for the municipality or another level of the State concerned, to assume a new challenge of a political, economic and management nature that entails a new form of organization and implies the decentralization of resources. In this way, the constitution of a society with majority state participation is one of the forms of state intervention in local development, in which the administration pursues public goals, becoming an active subject of development, resulting in this society in a management tool that allows obtaining financing from the private sector.

It will be necessary to discuss whether all of the above is aimed at reconstructing the sense of the public of the nineteenth century (and in particular the Welfare State) or if it is about building a new sense of the public for the twenty-first century; a movement for the creation of cities that, at all scales, would be much more humane, more reasonable, more fair, more appropriate to what we really want for our societies.

Obviously, this also leads us to the issue of financing: how are these public policies paid for? Who finances this role of the State? And it carries us to some discussions that are also present in other disciplinary fields closely related to the construction of the city. In this sense, I especially rescue the idea of a great economist, Mariana Mazzucato (2021), who postulates the concept of mission.

The proposal of her recent book, *Mission Economy*, is to address the response to the environmental crisis and climate change with mission criteria, in a similar way to the approach that allowed the arrival of a man to the Moon and, no less important, his return to the planet that today we want to preserve (or something less photogenic, the way in which the great world wars were faced). This mission approach implies a radically different role of the State that the one promoted by the deregulatory and privatizing consensus that was born with the crises of the seventies and reached hegemonic status with Reagan, Thatcher and the fall of the Berlin Wall and of the really existing socialisms in Eastern Europe. And as far as we are concerned as urban planners, this transformation of the economy and politics also involves cities, beyond the clamour about "unrealistic technological panaceas such as artificial intelligence or smart cities" that Mazzucato cites (Corti 2021).

This idea of mission implies putting the State, not only the public sector but also society as a whole, on a mission that aims to recover and build the city of the twenty-first century and overcome that feeling of "urbicide" that we are facing in this book and in the seminary that originated it.

## References

Corti M (2019) Diez principios para ciudades que funcionan. Café de las ciudades, Buenos Aires
Corti M (2020) Contra el urbanismo de franquicias. El problema de las recetas. Revista digital Café de las ciudades, Buenos Aires
Corti M (2021) Una guía para cambiar el capitalismo. Misión Economía, la propuesta de Mariana Mazzucato. Revista digital Café de las ciudades, Buenos Aires
Corti M (2022) Nueva Santa Cruz de la Sierra. Un ambicioso emprendimiento urbanístico e inmobiliario en Bolivia. Revista digital Café de las ciudades, Buenos Aires
Garay A (2011) Modalidades de gestión de grandes proyectos. Revista digital Carajilllo de la ciudad N° 10, Barcelona.
Heng K, Yeo S (2016) Singapur: planificación y desarrollo físicos. In: Del conocimiento al desarrollo, Amette, Corti, Jaimes, Janches (coord) Eudeba, Buenos Aires
Mazzucato M (2021) Misión economía. Una guía para cambiar el capitalismo (1ª edn). Ciudad Autónoma de Buenos Aires, Taurus
Minton A (2017) Big capital: who is London for? Penguin Books, UK
Nancy J (2013) La ciudad a lo lejos. Manantial, Buenos Aires
Rodríguez A (entrevista) (2004) El problema de los "con techo"... Alfredo Rodríguez describe las paradojas del subsidio habitacional en Chile. Revista digital Café de las ciudades N° 19
Rueda S (2016) La supermanzana, nueva célula urbana para la construcción de un nuevo modelo funcional y urbanístico de Barcelona. BCN Ecología, Barcelona. http://www.bcnecologia.net/sites/default/files/proyectos/la_supermanzana_nueva_celula_poblenou_salvador_rueda.pdf

**Marcelo Corti** Director of *Café de las ciudades* (digital magazine and editorial). Director of *Maestría en Urbanismo FAUD-UNC*. Associated in *Estudio Estrategias* and *La Ciudad Posible*. He is author of many books like La ciudad posible (2015) and Diez principios para ciudades que funcionen (2019).

# Chapter 35
# Metropolitanicide? *Urbs*, *Polis* and *Civitas* Revisited

**Mariona Tomàs**

**Abstract** In this chapter, we challenge the concepts of *urbs*, *polis* and *civitas* as they have been traditionally applied to the city. Our exploratory analysis is an invitation to a discussion on the different forms that "urbicide" takes place. Indeed, we argue that we need to analyse this phenomenon with a metropolitan perspective, since cities cannot not be conceived as isolated units but as part of large metropolitan systems. This is why we discuss the idea of "metropolitanicide", or the death of metropolises, through the analysis of two main dimensions (*polis* and *civitas*). First, the idea of *polis* relates to the way metropolitan regions are politically organized: do they have the adequate powers and financial means? We explore the different models of metropolitan governance, from metropolitan governments to informal cooperation, showing the limits of their powers. Second, the notion of *civitas* refers to the existence of a political community at the metropolitan scale. Do citizens of metropolitan regions have the possibility to choose their metropolitan representatives and are called to participate in political decisions? Are citizens engaged in political struggles and mobilizations at a metropolitan scale? Do they have a sense of belonging to this larger territory? In other words, we wonder if metropolises have a political meaning for citizens, which has been an under researched topic. We argue that, although the physical expansion of cities (*urbs*) has certainly a metropolitan dimension, both institutions and political practices are mainly conceived at a city scale.

**Keywords** Urban zone · Urbicide · Urban policies

## 35.1 Introduction

Capel (2003) defines the city as the combination of *urbs* (related to the physical and territorial form), *polis* (the political and institutional organization of cities) and *civitas* (the political community constituted by the citizens who live in the city). However,

M. Tomàs (✉)
Department of Political Science, University of Barcelona, Barcelona, Spain
e-mail: marionatomas@ub.edu

© The Author(s), under exclusive license to Springer Nature Switzerland AG 2023
F. Carrión Mena and P. Cepeda Pico (eds.), *Urbicide*, The Urban Book Series,
https://doi.org/10.1007/978-3-031-25304-1_35

contemporary cities are not isolated entities: they have become part of a larger territory and a set of interrelations. There is not a single word to define this reality that expands the classic notion of city: metropolises, megacities, urban agglomerations, functional urban areas, metropolitan areas, metropolitan zones, metropolitan regions or city-regions are some examples (Tomàs 2015). The different names vary according to legal, administrative, political, economic or cultural criteria, but they all share a common idea: these are territories characterized by strong economic, social and environmental interdependencies which need to be managed in an integrative way. The United Nations Statistical Commission defines metropolitan areas as a city and its commuting zone, consisting of suburban, periurban and rural areas economically and socially linked to the city (United Nations 2020a: 4). In this article, we share this definition of the metropolitan phenomenon and we use it as a general framework, but we refer indistinctively to metropolises, metropolitan areas and regions.

The goal of this chapter is to conceptualize the notion of "urbicide" (Carrión 2018) from a metropolitan perspective. We are aware of the heterogeneity of metropolitan spaces in the world in terms of size, population, economic development, inequalities and governance capacity: metropolitan areas differ from country to country and within countries. However, in this text we do not intend to make a systematic and exhaustive analysis of the metropolitan areas in the world. The article has an exploratory purpose, and the case studies serve as examples that support the main argument of the article. Indeed, this an invitation to a discussion on the different forms that "urbicide" takes place. This is why we discuss the idea of "metropolitanicide", or the death of metropolises.

The idea of an existing "metropolitanicide" paradoxically cohabits with the fact that 55% of the world's population lives in urban areas (United Nations 2018). Moreover, it is estimated that the amount of people living in urban areas will still be growing in future, specially in the regions of Asia–Pacific and Africa. According to the data of UN Habitat, almost 1 billion people will become metropolitan inhabitants in the next fifteen years. In 2020, 34 metropolises have surpassed 10 million inhabitants; while 51 have a population of 5–10 million; 494 of 1–5 million and 1355 of 300,000 to 1 million (United Nations 2020a).

The physical extent of urban areas is growing much faster than their population, thereby consuming more land for urban development. Indeed, many cities have been growing beyond the boundaries of their core municipality, challenging the idea of *urbs*. Moreover, the continuous expansion of cities and the configuration of metropolitan regions questions also the notions of *polis* and *civitas*. In this chapter, we are going to focus on these two last dimensions. The first section is devoted to the analysis of *polis* with a metropolitan perspective, that is, the capacity of metropolitan institutions (their powers, their financial needs and their position in the political system). The second section explores the notion of *civitas* at a metropolitan scale, by considering the different elements that build a political community (mechanisms of political participation, political orientations, sense of belonging). We end the chapter with some reflections on the articulation between *urbs*, *polis* and *civitas* at the metropolitan scale and on the implications in the research in urban politics and urban studies.

## 35.2 Metropolitan Government or Misgovernment?

The concept of *polis* with a metropolitan perspective refers to the powers, financing and political influence of metropolises. Indeed, one of the classic debates in the literature on metropolitan governance is about the type of institutions that should be created to face metropolitan challenges (Heinelt and Kübler 2005). There are opposite views about the political and institutional recognition of metropolitan areas, ranging from high institutionalized models to low institutionalized models (for a synthesis of debates, see Savitch and Vogel 2009; Tomàs 2012). Four metropolitan governance models are generally recognized based on the type of institutional arrangements in place, ranging from models of hard to soft governance (OECD 2015; United Council of Local Governments 2016; Tomàs 2017).

In one extreme, metropolitan governments are structures created explicitly to face metropolitan challenges. There are two types of metro governments. They can be constituted as a one-tier government, after the merger of municipalities—like in Toronto—or after being defined as a "metropolitan city"—like in Tokyo. The more common type is the two-tier government, where municipalities co-exist with a supramunicipal institution, as is the case of many European metropolis like London, Stuttgart, Hannover, Barcelona, Lyon or Milan. Some cases are found in America, too: Portland, Montreal, Quito, Lima, Medellín (Valle de Aburrá), San Salvador or Santiago de Chile. According to the ideal type (Sharpe 1995), these structures rely on a directly elected metropolitan council, exclusive competencies and funding established by law, meaning the maximum expression of a metropolitan political recognition. However, as we will see below most metropolitan governments do not meet these conditions.

Metropolitan agencies are the second type of arrangements. In contrast to metropolitan governments, they have the competence for managing or planning one sole service (public transport, environment, police, etc.). In most of the countries, we find agencies for the planning or/and delivery of one service (the most common, transport, waste management, water, fire and emergency services). This approach is frequently used in the USA (called special districts) and is also a formula used after removal of existing metropolitan authorities (like in the UK and Barcelona in the 1980s).

The third model is that of vertical coordination, in which metropolitan policies are not made by a specific metropolitan institution but de facto by other already existing levels of government (a region, a province, a county, etc.), for example, intermediate governments like the Stockholm County, the Region Hovedstaden (in Copenhagen) or the regional government of Madrid (Comunidad de Madrid). These institutions were not created to make this metropolitan function but they exercise it in practice.

Finally, the less institutionalized models of horizontal collaboration are based on voluntary cooperation between municipalities. In this case, the initiative is local and its existence does not mean reforms in existing political structures. The most traditional form is that of a Union or Association of Municipalities, like in Wroclaw or Vancouver. Its capacity for action in terms of competencies and funding will depend on the legal framework in which the partnership is located as well as the capacity of local representatives to agree on collective decisions. Another form is through public–private cooperation like Montréal International or Barcelona Strategic Metropolitan Plan Association, which are non-profit organizations financed both by the private sector and regional and local governments.

Institutional fragmentation exists in most metropolitan areas in the world and models with an average degree of institutionalization prevail: robust metropolitan governments and voluntary associations of municipalities are in the minority. Indeed, 51% of the metropolitan areas of OECD countries have some sort of metropolitan body, but without the ability to regulate: only 18% have metropolitan authorities with powers (OECD 2015). In practice, models of metropolitan governance vary according to the tradition of cooperation, political alliances, relations between levels of government and the local configuration of public and private stakeholders. These balances modulate the kind of governance that evolves over time. There are many examples of metropolises that have been through different stages of institutionalization of metropolitan governance, moving from a metropolitan government to sectoral agencies, from a strategic plan to cooperation between municipalities, to cooperation between municipalities to a metropolitan government, etc.

In any case, to determine the real power of metropolises, we need to know if they have the appropriate competencies and funding, and if they are a key institution in the political system.

## *35.2.1 Competencies, Funding and Multilevel Governance*

Many comparative studies have highlighted that metropolitan areas enjoy no political recognition (Feiock 2004; Heinelt and Kübler 2005; Geróházi and Tosics 2018; Zimmermann et al. 2020, Nieto and Niño Amézquita 2021). The lack of political power at a metropolitan scale has also been claimed by mayors over the world through the Montreal Declaration on Metropolitan Areas approved in October 2015 for Habitat III (United Nations 2015). In most cases, when metro institutions exist, their competences are related to hard policies (urban planning, public transport, infrastructure and the environment), while they lack competences related to soft policies (education, health, social services and economic development). In the case of Europe, in the last decade new metro institutions have been created with the inclusion of competences related to social and economic development, although these are secondary compared to the traditional hard policies (Tomàs 2020). Furthermore, most of the competences are shared with other levels of government (local, regional or state-related). The binding or non-binding nature of the decisions made at a metropolitan scale is another key point, that is, whether the actions set by a metropolitan plan are mandatory or not for municipalities. Without this exclusive and binding nature, the capacity of planning and implementing metro policies is very limited.

To a large extent, funding determines a metropolis' degree of autonomy. This is not only true with regard to material resources (the amount), but to the source of this funding (own or by other means). As shown in several comparative researches (Bahl and Linn 2014; Slack 2017, 2018) and also expressed in the Barcelona Statement in March 2015 by European metropolitan mayors, metropolitan areas lack the financial resources to meet urban challenges. Metropolitan governments have little fiscal autonomy, since most of the resources are transfers from municipalities or regional/state governments (European Metropolitan Authorities 2015). In most cases, funding from metropolitan institutions comes from a mixture of sources, mainly transfers from other levels of governments and taxes. Relying heavily on own source revenues (taxes and user fees) and having the freedom to levy taxes creates more fiscal autonomy than reliance on intergovernmental transfers, which can be unpredictable and restrict the ability of metro institutions to control their own destiny.

At the same time, metropolitan areas are situated in an environment of multilevel governance, where they have to deal with multiple public institutions, agencies, private sector, non-profit associations, etc. First of all, metropolitan authorities relate to the private sector. Various public, private or joint venture entities of different territorial scopes, diverse compositions and varied functions operate in urban agglomerations. As such, another challenge of metropolitan governance consists of coordinating them all (John 2001). In addition, there is the issue of guaranteeing the economic efficiency and viability of the management of the services in areas with major profits like water and those related to the sustainability of the territory. In this regard, major multinational corporations' growing interest in smart cities puts the capacity of governance to the test, since public–private partnership is inescapable in this area. In fact, neither city councils nor metropolitan governments possess the technology or the knowledge to deploy the smart city: the connection between public and private interests is at the heart of metropolitan governance (Clark and Moonen 2015).

In the second place, the governance of metropolitan areas is affected by relations with other levels of government (municipal, regional and national). In this case, the political and legal consideration of the municipality and of the metropolitan area is crucial: if it is an important level of government (with competencies and funding), if it plays a prominent political role in the country (high turnout in the elections), etc. Moreover, to understand these vertical relations, the importance of the agglomeration in the region or country as a whole is essential (according to its relatively decentralized political structure). In fact, metropolitan institutions are not created in a vacuum, but in an already existing political structure. The main reason why powerful metropolitan governments are not created is the political resistance generated by this type of intervention from municipalities and from other levels of government that already exist, like provinces, regions or the central government itself. In fact, few governments dare to create new metropolitan governments that group together most of the population of the country and/or capital city. When that has happened, they were given limited powers (of management, implementation and planning) in very specific fields (especially transport and the environment, and to a lesser extent spatial planning and economic development). Political recognition of metropolitan areas therefore requires acceptance by higher levels of government, which are those that legislate and determine their capacities (Andersson 2015).

In addition to the relations with the private sector and public administrations, metropolises have been developing an urban diplomacy, including a variety of initiatives and activities such as the creation of networks of cities, both at the national and international scale, study visits and partnerships and exchanges. We find several examples of consolidated international networks of cities and metro regions, such as Metropolis, United Cities and Local Governments (UCLG), the Network of European Metropolitan Regions and Areas (METREX), the European Metropolitan Authorities (EMA), the Euro-Latin American Alliance for Cooperation between Cities (ALLAs) or the Global Taskforce of Local and Regional Governments, just to cite a few. These international activities are driven with many objectives, mainly to mobilize resources for urban projects and to develop long-term systems of stakeholders (Beal and Pinson 2014). At the same time, metropolises try to become global stakeholders in order to influence the global agenda. Enabling the exchange of experiences and best practices may serve to stimulate policies in other cities and make metropolitan problems visible on a global scale (Kosovac and Pejic 2021), although the real impact of these activities is difficult to quantify.

### 35.2.2 A Practical Example: The Implementation of SDGs[1]

An example of the limits of current metro institutions is the implementation of Sustainable Development Goals (SDGs). Indeed, metropolitan areas have become a key battleground for reducing inequalities, addressing climate change challenges and protecting human rights and, as specifically highlighted by the New Urban Agenda, establishing the "right to the city" (right to gender equality, housing, mobility, safety, basic services and culture). The 2030 Agenda needs to be translated into actions and policies at different scales, including the metropolitan scale (United Nations 2020b).

As the work of de Fernández de Losada and Tomàs (2019) shows, metropolitan governance is key to the success of an integrated approach to sustainable development, as required in the implementation of the SDGs. According to the authors, the first model, that of metropolitan governments, does not in itself guarantee effective implementation of the SDGs, especially when there is a lack of binding mechanisms (often the case in the two-tier model). In other words, the absence of exclusive powers for these institutions in key areas such as metropolitan infrastructures is a weakness

---

[1] This part is based on the research done for the Metropolitan Chapter of the Fifth Global Report on Decentralization and Local Democracy financed by the United Cities and Local Governments (see de Fernández de Losada and Tomàs 2019). The research was based on documentary analysis (academic literature, international reports and databases, official documents) of the main metropolitan areas with more than 1 million inhabitants. All the primary and secondary sources are cited in the published chapter. In this article, we synthetize the main elements because of space limitations.

in metropolitan arrangements. Competence for key infrastructures such as highways, railways, ports and airports is typically in the hands of national or sub-national governments (federated states and regions). Another obstacle is the lack of fiscal autonomy of metropolitan institutions, which is especially problematic in the light of the fact that municipal expenditures per capita tend to be higher in metropolitan areas because of the nature of services (e.g. public transportation and waste collection). The absence of powerful metropolitan governments means that, in practice, the actions of metropolitan governments are often bypassed by the municipalities (for example in Barcelona and Montreal) or central government (for example, Bangkok). That said, it is also true that having a metropolitan government at least provides the institutional framework to legitimize the development of urban agendas, as the example of Seoul shows.

In contrast, the second model of metropolitan governance based on sectorial metropolitan agencies (and utilities) that manage or plan a single task or service (public transport, environment, police, etc.) can be useful for the implementation of one of the Goals (e.g. mobility, water and sanitation, etc.), but the main weakness of this model is that it lacks an integrated vision. To compensate for this single-issue focus, coordination with other agencies and levels of government is essential, which is frequently one of the difficulties that metropolitan regions face due to the high degree of institutional fragmentation.

As for the third model of vertical coordination, where metropolitan policies are not carried out by a metropolitan body but by other levels of government that already exist (a region, a province, a county, etc.), the development of SDGs depends mainly on the competences and financing of this layer of government (and how it is coordinated with other layers). Berlin stands out as a successful example.

Finally, the fourth less institutionalized models are based on municipalities' voluntary cooperation, whether through an association of municipalities or by means of strategic planning. These are soft forms of metropolitan governance, where other actors can participate in the development of the SDGs. This model is often used as a mechanism to gather all actors together where there is high institutional fragmentation, like in the case of New York City.

To sum up, weak metropolitan governance undermines the potential of metropolitan areas to function as cornerstones of national sustainable development. At the same time, the extent to which multilevel governance works is a key factor in creating an enabling institutional environment for the implementation of the SDGs. Indeed, lack of coordination between the different institutions involved in metropolitan management with competences related to the development of the SDGs clearly affects their implementation. In the case of metropolitan areas, this coordination is in the hands of cities when no metropolitan institutions exist or when metropolitan institutions have limited powers (de Fernández de Losada and Tomàs 2019).

## 35.3 The Difficulties of Building a Metropolitan *Civitas*

The previous section has focused on the institutional dimension of metropolises (*polis*). Most metropolitan areas—and specially in developing countries—do not have well established governance arrangements or instruments for planning, coordination and financing at that scale (Andersson 2015: 11). In this context of institutional fragmentation, we explore in this section the construction of a political community at a metropolitan scale, which has been less analysed in the literature on metropolitan governance (Zimmermann 2020). An analysis of a metropolitan *civitas* includes a variety of elements. First of all, building a political community needs mechanisms of political participation: the most classic one, the elections. Secondly, *civitas* can be understood as the citizen's expression of political attitudes (ideology, opinions on metropolitan reforms). Thirdly, it also comprises an awareness or consciousness of the issues affecting the territory, mainly shaped by commuting and spatial mobility. Metropolitan areas can be also be places for the implementation of deliberative practices and the space for citizens' struggles for "the right to the city" and a new practice of citizenship. Finally, a metropolitan *civitas* is linked to the construction of a sense of belonging, a sense of place.

### *35.3.1 Metropolitan Elections*

Elections are the classic mechanism of political participation. The academic literature on the direct election of local executives, and especially the mayor, is abundant in relation to the municipal level, both from a comparative perspective and in terms of case studies (for a review of the literature, see Copus 2004; Magre and Bertrana 2007; Sweeting 2017). According to the studies, direct election makes political leadership more visible: citizens recognize their leader and this facilitates access and, a priori, interest. Direct elections establish a closer link between citizens and elected officials, increasing their legitimacy and accountability. At the same time, the mayor gains political strength, not only in the council but also in relation to other levels of government. Direct election also makes it easier for alternative candidates to emerge from established political parties and thus renew elites. It also stands out as an advantage that, by being directly elected, the mandate can be fulfilled and therefore there is a stability that does not happen in cases of indirect election in minority governments. However, various studies also warn of the disadvantages of direct election, especially the concentration of power. Indeed, direct election means greater individualization of power, leaving less room for different voices and can facilitate the emergence of populist politicians. In addition, local government media treatment often focuses on the figure of the mayor and less on his or her policies (Sweeting 2017: 5).

The debate over direct election has important nuances in the metropolitan case. As in the municipality, one of the arguments for defending direct election lies in the legitimacy that emanates from the institutions and the confrontation with the electorate (what is called input legitimacy) (Scharpf 1999). Directly choosing metropolitan representatives can help renew local political elites and find people committed to metropolitan challenges. In addition, having direct representation at the metropolitan level means that there is an election campaign and programme on which to debate and make commitments. Direct election gives visibility to the institution and the metropolitan fact, strengthening the metropolitan leadership. It also allows for greater transparency and accountability to the public.

Instead, in a model of indirect election, the mayors and councillors must defend a common metropolitan interest, while being elected at the base of their municipality. They are accountable to the voters of their municipality, not to the citizens of the whole metropolitan area. In addition, the mandate within the metropolitan government is linked to the municipal electoral calendar: if there are changes in the local political majorities, the composition of the metropolitan government changes accordingly. Therefore, the continuity of metropolitan councillors does not depend on their performance on a metropolitan scale but on a municipal level.

However, direct metropolitan election is more expensive (new elections need to be held) and can lead to political resistance due to the magnitude of the election (metropolitan areas have a large population and the chosen person becomes a prominent political figure). Resistance is especially strong in the case of capitals. In any case, the possibility of creating political rivalry depends largely on the power of metropolitan governments. If these authorities have strategic and management competencies and weak funding, the possibility of them emerging as counter powers is minimal (Tomàs 2020).

In OECD, countries and Latin America indirect election models predominate (OECD 2015). When direct election occurs, it can take three different forms: (a) the presidential model (election of the metro mayor); (b) the parliamentary model (election of the metro assembly) and c) the mixed model (election of both the metro mayor and assembly). Figure 35.1 shows the last election's results in seven metro regions with direct election of representatives. Figure 35.1 compares the turnout at the local level (average in turnout of all municipalities of the metro region) and at the metropolitan scale (following the presidential, parliamentary and mixed model). Data on presidential models comes from the cases of Liverpool and Manchester; on parliamentary model through the cases of Lyon and Stuttgart and on mixed models there are the cases of Hannover, London and Santiago the Chile.

Figure 35.1. Comparative turnout between local and metropolitan elections (2018–2021). The selection of cases has been based on two criteria: (a) the availability of data and (b) the inclusion of the three models (presidential, parliamentary and mixed). Data has been collected from official sources (local or regional official governmental pages depending on the case). In the case of double-turn electoral system (Hannover, Lyon and Santiago the Chile), we have just considered the turnout at the first turn

Different ideas can be highlighted from the analysis of turnout. First, voter turnout is quite low in the majority of cases. The German metro regions (Stuttgart and Hannover) stand out as the metro regions with greatest turnout, exceeding 50%. In Stuttgart, after a first vote in 1994 in which turnout bordered on 70%, it began

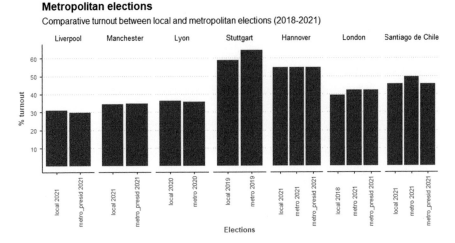

**Fig. 35.1** Metropolitan elections (2018–2021). *Source* Own elaboration

to stabilize at over 50% (between 52 and 54%). In the last elections, the turnout has increased up to 64%. In Hannover, participation in choosing the metropolitan assembly has also oscillated around 50%, in some elections, a little above (2001 and 2014) and in the rest, below (2006 and 2011). In contrast, the elections for a metropolitan mayor, which are held separately every eight years, have shown slightly lower turnouts, between 44 and 46%, except in these last elections (more than 50%).

On the opposite, in Manchester and Liverpool the turnout is around 30% (confirming the low participation in the first metro elections in 2017, which did not reach 30%). In London, turnout has been stable at under 40%, except in the elections in 2008 (45%), 2016 (46%) and 2021 (42%). In fact, Blair's government held a referendum prior to the creation of the Greater London Authority in 2000: 72% of the people voted in favour, but only 35% of the citizens with the right to vote actually exercised that right. In Lyon, the first elections of a metro assembly were held in the middle of the COVID-19 pandemic, which may explain the low turnout, less than 40% (also in local elections, being held simultaneously). In Chile, the election of metro mayor (*Gobernador*) was held for the first time in 2021, while the metro assembly has been directly elected since 2013. Turnout for the metro assembly has been slightly higher than for the metro mayor and local elections.

The turnout in metropolitan elections is an indicator of the democratic legitimacy of directly elected metropolitan governments and assemblies. The data shows that there is no great difference with local elections. Citizens vote as much (or as less) in metro elections than in local elections, which makes us think on the purpose of direct elections at a metropolitan scale: the emergence of a clear leadership appears as the key reason. However, the question of legitimacy is connected to the political power of urban agglomerations. If the choice is to have strong metropolitan governments with exclusive and binding competences and fiscal autonomy, the direct election of metropolitan representatives is inevitable. Contrarily, if output legitimacy is privileged (legitimacy coming from the provision of services and public policies) (Scharpf 1999), metropolitan institutions remain technical and unaccountable to citizenry. As we have seen in Sect. 35.2, current metropolises have limited powers and little fiscal autonomy. The debate on direct election of metropolitan representatives is therefore necessarily connected to the political importance of metropolises.

### 35.3.2 Political Orientations and Practices

*Civitas* is made of a political community constituted by the citizens who live in the city. How this political community is built at a metropolitan scale? What citizens think of metropolitan governance? While mayors' opinions on metropolitan governance have been deeply analysed (Dlabac et al. 2018), the academic debate has not paid much attention to the question of what citizens in metropolitan regions think of institutional reforms. It has been broadly assumed that citizens think locally, and that support for metropolitan authorities is rather low. In recent years, a group of scholars has been working on public opinions on metropolitan governance (see the special issue in the *Journal of Urban Affairs* edited by Lidström and Schaap 2018; also Strebel 2022). These studies, based mainly on European countries, analyse if citizens support stronger metropolitan institutions and what are the factors that influence their views on metropolitan governance. According to Strebel and Kübler (2021), ideology (voter of left-wing or right-wing parties) and place attachment influence citizen's support for institutionalized forms of metropolitan governance. Other studies have focused on vote and ideology considering the place of residence of citizens living in a metropolitan region: do people from suburbs vote more or less than people from central cities? Do their political orientations differ (more conservative, more progressive)? Both comparative studies (Sellers et al. 2013) and specific case studies (Walks 2004) have shown disparities on the political behaviour of citizens within metropolitan regions.

Cognitive factors may be important to understand citizens' views on metropolitan governance and their attachment to a metropolitan territory. Different authors (Kübler 2005; Lidström 2013; Wicki et al. 2019) argue that the awareness of being part of a metro region is related to commuting and spatial mobility. According to their research based on Swedish and Swiss cities, cross-jurisdictional mobility for multiple reasons (daily commuting, shopping purposes and residential moves) increases the awareness of problems in other municipalities in the region. Citizens in metropolitan regions may also have common needs in relation to transportation; or they may own property in other neighbouring municipalities. According to the authors, the consciousness of belonging to a metropolitan region affects their support for more institutionalized forms of metropolitan governance. The research of Walter-Rogg (2018), based on the case of Germany, shows that having more knowledge about metropolitan politics is associated to a bigger attachment to their metropolitan area.

Another strand of literature sustains the existence of an urban citizenship (Purcell 2003), urban and regional forms of citizenship (García 2006) or even a metropolitan citizenship or "metrozenship" (Yiftachel 2015). According to the authors, the traditional notion of citizenship associated to the nation-state is in crisis, since new forms and practices of citizenship are arising. According to Purcell (2003), citizenship is being rescaled, territorialized and reoriented. Is the metropolitan region a scale where a political community is being created? Purcell (2003: 573) states that new citizenship forms are being pursued by social movements mostly at the neighbourhood, the city and the urban region. Indeed, following Lefebvre's work, those who fight for the right to the city are those who live in the city and create urban space. According to Yiftachel (2015: 734): "Metropolitan regions are, gradually, forming a living and political space, where material (rather than formal or legal) 'citizenship' is being attained through residence, investment, work, invasions and struggles". We agree that metropolitan regions are more than territorial or statistical units for planning; they are living territories where political struggles take place around the issues of economic development, social cohesion, sustainability, etc. However, little empirical research has been done to analyse whether social mobilization occurs at the metropolitan scale (meaning that social mobilizations are scaled up), or whether that occurs at municipal and neighbourhood level *within* a metropolitan region. In this sense, we share Yiftachel's concern (2015: 736) about the need to adopt new methods for studying urban society and struggles.

Another way to build a political community is through citizen participation (Heinelt 2012). Traditionally, democratic innovations such as participatory budgets or deliberative councils have been rooted at a local level (city or neighbourhood level): the proximity with citizens and their day-to-day practices is one of the key issues on citizen participation (Martins 1995). Can this participation be scaled up to metropolitan regions? Are the metropolises a suitable space for democratic innovations? There is little empirical research available on this topic. The experiences linked to citizen participation at a metropolitan scale are limited and mainly related to urbanism and urban planning (Kahila-Tani et al. 2016; Roy-Baillargeon 2017). The more institutionalized systems of citizen participation are to be found in the French *métropoles* (through a permanent consultative council), while there are some specific experiences in Helsinki and Barcelona (for the design of the master plan), London (a youth council) and Montréal (a metropolitan agora) (Medir et al. 2022). The scarcity of mechanisms of citizen metropolitan participation is not a surprise. Firstly, because citizen participation is still an emerging practice at the local level, specially in developing countries. Secondly, because of the absence of strong metro governments, as we have seen in the previous section. Last but not least, because metropolises are still spaces where a strong sense of belonging is not developed.

Indeed, a metropolitan *civitas* is also related to the sense of belonging, the sense of place. As Nelson et al. (2020) show, this is a large concept that includes different terms such as place attachment, place meaning, place identity, and territorial identification. *Civitas* is then a collective social construction based on the day-to-day experience of citizens and their perception of the experience. One key element of this social construction is built narratives: do they exist at a metropolitan scale?

Metropolitan narratives are part of the social process of metropolitan building, where different actors have their own definition of the territory: ideas on nature, landscape, the built environment, culture/ethnicity, dialects, economic success/recession, periphery/centre relations, stereotypic images of a people/community, etc. (Paasi 2003: 477). Metropolitan narratives will be more difficult to build if citizens do not share symbols or imaginaries associated to metropolitan regions. And this is often the case because of three reasons: (a) there is no simple or single definition of what a metropolitan region is; (b) metropolitan regions are weakly institutionalized and (c) different actors have their own imaginations of what metropolitan regions are and are for. As Harrison et al. (2020: 135) explain, metropolitan regions are imagined and constructed across space and time: metropolitan spatial imaginaries are dynamic, constantly evolving and always contested. Moreover, the imaginaries are not necessarily shared by all the actors.

To sum up, a metropolitan *civitas* is associated to the political meaning that citizens acknowledge to this territory. The scarcity of empirical research done in this field makes it difficult to prove that metropolises have (or have not) a political meaning for citizens. However, having political institutions at a metropolitan scale with elected representatives should, a priori, enhance the political mobilization of citizens. At the same time, if metropolises constitute a political community, there would be citizen's claims for direct election of representatives and citizen participation. Zimmermann's quotation illustrates this idea (2020: 70):

> As metropolitan regions are growing and functional interdependencies between municipalities are getting denser, many policy controversies also have a regional dimension. Congestion, protection of green belts, airport extensions or other infrastructure decisions clearly are more than just local issues but only a few metro regions have strong governance institutions with directly elected assemblies that are capable of procuring legitimate decisions. Why should citizens engage in arrangements that are comparatively weak (compared to local government where they are entitled to vote and pay taxes), abstract (in terms of size), and closed in terms of access points for participation?

## 35.4 Conclusion

As Jouve (2005:6) states: What distinguishes a metropolis from a city? Is the metropolitan reality different in essence from the urban? Is the legal and institutional analysis relevant to define it? In this chapter, we have argued that while the metropolitan *urbs* is a reality, the creation of metropolitan institutions and a metropolitan political community (*polis* and *civitas*) is at stake.

Is this the end of metropolises? Can we talk about a "metropolitanicide"? Are we assisting a phase of major changes in urban and metropolitan development? As statistics are showing, we are shifting towards a metropolitan society, where many adjustments are taking place. These adjustments are quicker when it comes to the territorial expansion (*urbs*), but slower when it affects the creation of institutions in metropolitan regions (*polis*) and the construction of a metropolitan political community (*civitas*).

We think that the analysis of Beal and Pinson (2014) on urban mayors can be applied to the metropolitan scale. Using Scharpf's differentiation of output and input legitimacy (1999), and based on the French case, the authors argue that urban mayors are involved in a process of transforming forms of legitimacy, giving more importance to legitimacy through "outputs" as a result of public policies. Indeed, their main political activity would have shifted to the creation of urban policies, the construction of coalitions and the articulation of resources that enable the elaboration and implementation of these policies. In the case of metropolitan regions, legitimacy would come in a similar way: through "outputs" rather than through "inputs", not without difficulties. Indeed, most metropolitan regions lack of resources to meet urban challenges.

While comparative analysis on metropolitan institutions are numerous, *civitas* in metropolitan regions has only been researched to a limited extent. As Lidstrom and Schaap (2018: 2) affirm in their introduction to the special issue devoted to this topic: "This is surprising as the potential conflict between traditional municipal forms of citizenship and the everyday practice of being a citizen in a city-region is becoming increasingly significant all over the world". Indeed, one of the main problems of analysing a metropolitan *civitas* is that we are referring to a territory that evolves constantly and that has not a single definition. Maybe *civitas* in metropolitan regions cannot be associated to a specific *polis*: they are several to cohabit in the same territory. Moreover, metropolitan regions are heterogenous in relation to their size, population, economic development, inequalities and governance capacity, just to cite a few. In this sense, more empirical research in urban studies is necessary to show both the diversity of forms of "urbicide" and the different elements that characterize an existing "metropolitanicide".

We can conclude that the weakness of a metropolitan political community is a common characteristic in metropolitan governance, which could foresee the death of the metropolises. However, we prefer to understand it in the opposite sense: that means that there is still room for the creation of a metropolitan political community. In other words, the metropolitan region is potentially a territory where mobilization and creation of a political meaning can still take place. The key question is which are the conditions to enable that this promising opportunity occurs in the short or mid-term.

# References

Andersson M (2015) Unpacking Metropolitan governance for sustainable development. UN Habitat-GIZ. Retrieved 4 Apr 2022, from https://unhabitat.org/sites/default/files/download-manager-files/Unpacking%20Metropolitan%20Governance.pdf

Bahl RW, Linn JF (2014) Governing and financing cities in the developing world. Columbia University Press

Beal V, Pinson G (2014) When mayors go global: international strategies, urban governance and leadership. Int J Urban Reg Res 38(1):302–317. https://doi.org/10.1111/1468-2427.12018

Capel H (2003) A modo de introducción. Los problemas de las ciudades. Urbs, civitas y polis. In Capel H (ed) Ciudades, arquitectura y espacio urbano, vol 3, pp 9–12. Cajamar Caja Rural

Carrión F (2018) Urbicidio o la muerte litúrgica de la ciudad. Oculum Ensaios 15(1):5–12

Clark G, Moonen T (2015) The role of metropolitan areas in the global agenda of local and regional governments for the 21st century. Retrieved 29 Apr 2022, from https://www.gold.uclg.org/sites/default/files/BoC_Report.pdf

Copus C (2004) Directly elected mayors: a tonic for local governance or old wine in new bottles? Local Gov Stud 30(4):576–588. https://doi.org/10.1080/0300393042000318003

Dlabac O, Medir L, Tomàs M, Lackowska M (2018) Metropolitan challenges and reform pressures across Europe—the perspectives of city mayors. Local Gov Stud 44(2):229–254. https://doi.org/10.1080/03003930.2017.1411811

European Metropolitan Authorities (2015) Statement. Retrieved 20 July 2022 https://www.amb.cat/en/web/amb/area-internacional/ema/ema-barcelona/declaracio

Feiock RC (2004) Metropolitan governance conflict, competition, and cooperation. Georgetown University Press

Fernández de Losada A, Tomàs M (2019) Metropolitan Areas. Thematic chapter. In: The localization of the global agendas. How local action is transforming territories and communities. Fifth global report on decentralization and local democracy, pp 324–355. UCLG. Retrieved 28 Apr 2022, from https://www.uclg.org/sites/default/files/goldv_en.pdf

García M (2006) Citizenship practices and urban governance in European cities. Urban Stud 43(4):745–765. https://doi.org/10.1080/00420980600597491

Gerőházi É, Tosics I (2018) Addressing the metropolitan challenge in Barcelona metropolitan area. Lessons from five European metropolitan areas. Amsterdam, Copenhagen, Greater Manchester, Stuttgart and Zürich. Retrieved 29 Apr 2022, from https://docs.amb.cat/alfresco/api/-default-/public/alfresco/versions/1/nodes/31fdd71d-a1cf-46de-a736-10c36b5084b3/content/20180615_Addressing+the+Metropolitan+Challenge+in+AMB.pdf?attachment=false&mimeType=application/pdf&sizeInBytes=821252

Harrison J, Fedeli V, Feiertag P (2020) Imagining the evolving spatiality of metropolitan regions. In: Zimmermann K, Galland D, Harrison J (eds) Metropolitan regions, planning and governance. Springer, pp 135–154

Heinelt H (2012) Local democracy and citizenship. In: John P, Mossberger K, Clarke SE (eds) The oxford handbook of urban politics. Oxford University Press, pp 231–253

Heinelt H, Kübler D (eds) (2005) Metropolitan governance: capacity, democracy and the dynamics of place. Routledge

John P (2001) Local governance in western Europe. Sage Publications

Jouve B (2005) La démocratie en métropoles: gouvernance, participation et citoyenneté. Revue Française De Science Politique 55(2):317. https://doi.org/10.3917/rfsp.552.0317

Kahila-Tani M, Broberg A, Kyttä M, Tyger T (2016) Let the citizens map—public participation GIS as a planning support system in the Helsinki master plan process. Plan Pract Res 31(2):195–214. https://doi.org/10.1080/02697459.2015.1104203

Kosovac A, Pejic D (2021) What's next? New forms of city diplomacy and emerging global urban governance. CIDOB. Retrieved 29 Apr 2022, from https://www.cidob.org/en/articulos/monografias/global_governance/what_s_next_new_forms_of_city_diplomacy_and_emerging_global_urban_governance

Kübler D (2005) La métropole et le citoyen. Presses polytechniques et universitaires romandes

Lidström A (2013) Citizens in the city-regions: political orientations across municipal borders. Urban Aff Rev 49(2):282–306

Lidström A, Schaap L (2018) The citizen in city-regions: patterns and variations. J Urban Aff 40(1):1–12. https://doi.org/10.1080/07352166.2017.1355668

Magre J, Bertrana X (2007) Exploring the limits of institutional change: the direct election of mayors in Western Europe. Local Gov Stud 33(2):181–194. https://doi.org/10.1080/03003930701198557

Martins MR (1995) Size of municipalities, efficiency, and citizen participation: a cross-European perspective. Eviron Plann C Gov Policy 13(4):441–458. https://doi.org/10.1068/c130441

Medir Ll, Tomàs M, Sokolovska S (2022) Citizen participation at the metropolitan level: new experiences for old results? In: ECPR General Conference 2022, University of Innsbruck

Nelson J, Ahn JJ, Corley EA (2020) Sense of place: trends from the literature. J Urbanism: Int Res Placemaking Urban Sustain 13(2):236–261. https://doi.org/10.1080/17549175.2020.1726799

Nieto AT, Niño Amézquita JL (2021) Metropolitan governance in Latin America. Routledge

Organisation for Economic Co-operation and Development (OECD) (2015) Governing the city: Policy highlights. OECD. Retrieved 29 Apr 2022, from https://www.oecd.org/regional/regional-policy/Governing-the-City-Policy-Highlights%20.pdf

Paasi A (2003) Region and place: regional identity in question. Prog Hum Geogr 27(4):475–485. https://doi.org/10.1191/0309132503ph439pr

Purcell M (2003) Citizenship and the right to the global city: reimagining the capitalist world order. Int J Urban Reg Res 27(3):564–590. https://doi.org/10.1111/1468-2427.00467

Roy-Baillargeon O (2017) La symbiose de la planification et de la gouvernance territoriales: Le cas du Grand Montréal. Canadian J Urban Res 26(1):52–63

Savitch H, Vogel RK (2009) Regionalism and urban politics. In: Davies J, Imbroscio D (eds) Theories of urban politics. SAGE Publications Ltd, pp 106–124

Sharpe LJ (1995) the government of world cities: the future of the metro model. John Wiley

Scharpf F (1999) Governing in Europe: effective and democratic? Oxford University Press

Sellers JM, Kübler K, Walter-Rogg M, Walks A (2013) The political ecology of the metropolis. ECPR Press

Slack E (2017) How much local fiscal autonomy do cities have? A comparison of eight cities around the world. University of Toronto

Slack E (2018) Financing metropolitan public policies and services. Issue Paper, 6. Metropolis Observatory. Retrieved 29 Apr 2022, from https://www.metropolis.org/sites/default/files/2018-11/MetObsIP6_EN.pdf

Strebel MA (2022) Who supports metropolitan integration? Citizens' perceptions of city-regional governance in Western Europe. West Eur Polit 45(5):1081–1106. https://doi.org/10.1080/01402382.2021.1929688

Strebel MA, Kübler D (2021) Citizens' attitudes towards local autonomy and inter-local cooperation: evidence from Western Europe. Comparative Eur Politics 19(2):188–207. https://doi.org/10.1057/s41295-020-00232-3

Sweeting D (ed) (2017) Directly elected mayors in urban governance. Bristol University Press. https://doi.org/10.2307/j.ctt1t89hzf

Tomàs M (2012) Exploring the metropolitan trap: the case of Montreal. Int J Urban Reg Res 36(3):554–567. https://doi.org/10.1111/j.1468-2427.2011.01066.x

Tomàs M (2015) If urban regions are the answer, what is the question? Thoughts on the European experience. Int J Urban Reg Res 39(2):382–389. https://doi.org/10.1111/1468-2427.12177

Tomàs M (2017) Explaining Metropolitan Governance. The case of Spain. Raumforschung Und Raumordnung I Spatial Res Plann 75(3):243–252. https://doi.org/10.1007/s13147-016-0445-0

Tomàs M (2020) Metropolitan revolution or metropolitan evolution? The (dis)continuities in metropolitan institutional reforms. In: Zimmermann K, Galland D, Harrison J (eds) Metropolitan regions, planning and governance. Springer, pp 25–39

United Council of Local Governments (2016) GOLD IV report: co-creating the urban future. Edition 2016. Retrieved 29 Apr 2022, from http://www.gold.uclg.org/reports/other/gold-report-iv

United Nations (2015) Montreal declaration. Outcome document of the HABITAT III Thematic meeting on metropolitan areas. Retrieved 20 July 2022 https://habitat3.org/wp-content/uploads/Montreal-Declaration.pdf

United Nations. Department of Economic and Social Affairs, Population Division (2018) World urbanization prospects 2018. Highlights. Retrieved 29 Apr 2022, from https://population.un.org/wup/publications/Files/WUP2018-Highlights.pdf

United Nations. Human Settlements Programme (2020a) Global state of metropolis 2020—population data booklet. Retrieved 20 July 2022, from https://unhabitat.org/sites/default/files/2020/09/gsm-population-data-booklet-2020_3.pdf

United Nations. Human Settlements Programme (2020b) New Urban agenda. Retrieved 29 Apr 2022, from The New Urban Agenda Illustrated | UN-Habitat (unhabitat.org)

Walks A (2004) Place of residence, party preferences, and political attitudes in Canadian cities and suburbs. J Urban Aff 26(3):269–295

Walter-Rogg M (2018) What about Metropolitan citizenship? Attitudinal attachment of residents to their city-region. J Urban Aff 40(1):130–148

Wicki M, Guidon S, Bernauer T, Axhausen K (2019) Does variation in residents' spatial mobility affect their preferences concerning local governance? Polit Geogr 73:138–157. https://doi.org/10.1016/j.polgeo.2019.05.002

Yiftachel O (2015) Epilogue-from "Gray Space" to equal "Metrozenship"? Reflections on urban citizenship. Int J Urban Reg Res 39(4):726–737. https://doi.org/10.1111/1468-2427.12263

Zimmermann K (2020) What is at stake for metropolitan regions and their governance institutions? In: Zimmermann K, Galland D, Harrison J (eds) Metropolitan regions, planning and governance. Springer, pp 59–75

Zimmermann K, Galland D, Harrison J (eds) (2020) Metropolitan regions. Springer, Planning and Governance

**Mariona Tomàs** Professor of Political Science at the University of Barcelona, her research focuses on metropolitan governance, urban policies and local democracy. Dr. Tomàs has participated in numerous international, European and national research projects and has published several articles in the most prestigious journals. She received the most important awards for her Ph.D. in Urban Studies at the University of Quebec and the First Prize of the Political Book of the Quebec Parliament. Professor Tomàs is a member of different networks and advisory councils related to urban policies.

# Chapter 36
# International Tourism, Urban Rehabilitation and the Destruction of Informal Income-Earning Opportunities

**Alan Middleton**

**Abstract** As enablers of accumulation through the promotion of international tourism and the refurbishment of buildings and public spaces, city governments become tools for the displacement, exclusion and destruction of income-earning opportunities for the urban poor. Through the expulsion of street traders in an attempt to cleanse the city and bring the artefacts of elite history into view, the neoliberal city denies free trade to its poorest inhabitants. In addition, its policies for modernisation and rehabilitation have an unforeseen indirect impact on informal producers, who are irrecoverably displaced from their workshops as improvements are made to city transport systems and central places are made safe for a new international elite. Crime is displaced into surrounding poorer neighbourhoods, catalysing the decline of vibrant communities of artisans and shopkeepers. These processes are exemplified in the city of Quito, Ecuador.

**Keywords** Urbicide · Public space · Urban rehabilitation

## 36.1 Introduction

In this chapter, Urbicide is treated as a process in which certain social, economic and cultural sectors of the city are losing a struggle for survival. There are destructive processes, some of them located in the advance of globalisation and urbanisation, that are having a negative impact on the largest sector of the urban workforce in the cities of developing countries, the so-called 'informal sector'. Not all the destructive processes, however, are global. Global forces are experienced on the streets through the policies and practices of national and local governments. Local authorities respond to international pressures and opportunities, increasing or relieving the negative impact on the urban workforce, through the generation of policies that have intended and unintended consequences. This chapter will focus on a selection of urban policies that are generating destructive outcomes for income generation in the

A. Middleton (✉)
Birmingham City University, Birmingham, England
e-mail: alan.middleton@bcu.ac.uk

© The Author(s), under exclusive license to Springer Nature Switzerland AG 2023
F. Carrión Mena and P. Cepeda Pico (eds.), *Urbicide*, The Urban Book Series,
https://doi.org/10.1007/978-3-031-25304-1_36

informal sector. First, it is important to be clear about what we mean by 'the informal sector'.

It is thought that the literature on informality over the 50 years since the ILO Kenya Report extends to tens of thousands of books, articles and reports, covering Asia, Africa and Latin America, as well as the urban areas of Europe and the USA. Numerous review articles and PhD introductory chapters have highlighted the evolving definitions in the trajectory of the concept. An excellent summary and discussion of this evolution can be found in Bromley and Wilson (2018). It is not our intention to repeat this work, but it is important to set out our use of the concept.

Defining informality has always been a contentious issue. For the purposes of this chapter, we understand the informal sector to be a heterogeneous mix of social and economic relations that underpin production and exchange in the cities of the South. It includes small enterprises with up to seven employees, where the owner of the enterprise continues to be involved in manual labour. In the 1970s, the concept referred to small-scale income-earning opportunities that were thought to exist outside the formal capitalist system, free from the regulations that applied to the 'formal sector' (Hart 1973; ILO 1972). As research progressed it became clear that even the most marginal activities were linked to the capitalist system of production and distribution (Bromley 1978; Birkbeck 1978), and that many of the firms that were considered to be formal also evaded the regulations that should have applied to them (ILO 2002; ILO and WIEGO 2013).

Informal sector enterprises are intimately linked to formal capitalism through, for example, the purchase of raw materials—by tailors who buy textiles and shoemakers who buy leather from national or international sources. Capitalist production lies behind artisan repair workshops involved in the repair of televisions, computers, and German or Japanese cars. These workshops rely on the supply of spare parts from all over the world and their existence supports the development of local markets for international manufacturing. There are street traders involved in the sale of cigarettes, chewing gum, sunglasses, watches, CDs, DVDs, clothes and other manufactured goods, made by national or international capitalist firms whose operations are either legal or are based on the theft of intellectual property. There are also waste pickers, whose labour gives value to discarded glass, plastic, metals and other materials that can be recycled and find new uses within capitalist production processes. There are small shops and bars selling the products of national and international capitalism.

Defining and measuring informality has been made even more difficult by international agencies and national governments agreeing to expand the coverage of the 'informal economy' to include the capitalist firms that abuse local laws covering the wages and conditions of their workers. As the ILO and others have expanded the scope of the definition of 'informality' to cover up to 70% of the urban labour force in developing countries, the concept of the informal has not only become meaningless, it has also become unmeasurable with any consistency across the nations of Africa, Asia and Latin America (Middleton 2020). Despite the efforts of the international community to standardise the meaning, different governments have different measures for an economic concept whose analytical value has been reduced by every extension of its definition.

The heterogeneity of the 'informal sector', even if there is agreement that it refers to small firms employing less than 5 or 10 people outside an identified range of state legislation, also makes this a generic concept with little analytic value. Its definition varies from country to country and, in any single country, it covers such a variety of economic activity (different types of producers, traders and a range of other services) that it needs to be broken down into finer constituent parts. However, the variation of the different types of economic activity from country to country makes international comparisons an exercise in mythmaking.

Nevertheless, for around 50 years there has been general consensus that high rates of urbanisation, combined with the capitalist sector's inability to absorb the growing labour force, has made understanding informality vitally important for academics and policymakers. A supply of labour that is far more than the structural demand for waged labour has 'flooded the streets' with vendors and has encouraged what Bremen calls 'the myth of the infinite absorption capacity of the informal sector' (Bremen 2010, 121). However, this apparently 'free and flexible' labour market creates a dilemma for city administrators who wish to create 'world-class' cities that will compete internationally for footloose dollars in tourism and global investment in industry, finance and property. In this context, international organisations, urban elites and city planners collude to eliminate street trading from the urban landscape. The neo-liberal promotion of the free market and deregulation appear to apply to everyone except street traders, who are often repressed with state violence.

It should be no surprise that this contradiction between the international aims of the urban elites and the reality of the daily struggle to survive of most of the population often leads to conflict that can turn violent. These conflicting pressures exist right across the developing South, occasionally finding a peaceful resolution through some form of participatory planning, but more often resulting in the suppression of those working in the public spaces and buildings of modernising city centres.

The influences on the fortunes of artisans and traders in the informal sector are the product of global, national and local social relations of production and exchange. Understanding how macro-level changes impact on street level activities involves combinations of macro and micro analyses that overlap and interact to connect the informal sector to the wider social, economic and political systems in which they operate (Middleton 2020, 2022). The study of single cities cannot discover all the complexities of these forces. Case studies can, however, raise questions about the global trends and their local consequences.

In this chapter, we have chosen to look at both traders and artisans in the context of global pressures. The chapter provides examples of how certain micro-economic urban changes that are related to global forces have had an impact on informal activities at a local level. It focuses on the ramifications of the promotion of international tourism for street traders and utilitarian artisans.[1] The growth of global tourism is, of course, only one of many international developments that have had an impact on the informal sector. The micro-economic urban changes we have chosen to highlight

---

[1] Utilitarian artisans produce goods and services for everyday consumption, as distinct from the small number of craft artisans in urban areas who produce specifically for the tourist market.

are concerned with the refurbishment of urban infrastructure and the reduction of crime, both of which serve to attract international tourists. Obviously, infrastructure upgrading and crime reduction do not always depend on such international pressures, but they are typical responses where local policies aim to attract tourists or other forms of international investment.

The chapter will therefore examine the consequences of tourism attraction on informal traders and artisans, before going on to discuss the unintended repercussions of urban refurbishment and crime displacement on artisan production. The later discussion of refurbishment and crime will rely on information gathered in Quito to illustrate the main points. The case of Quito, particularly in relation to artisans, is used to exemplify how global phenomena reach into urban neighbourhoods in unintended ways and become destructive forces in developing cities. First, let us deal with the competition for space between the promoters of international tourism and informal street traders.

## 36.2 International Tourists and Street Traders

Creating a world-class city or developing tourism policies that will attract international visitors generally demands the removal of street traders. Planners often have an authoritarian perspective, a culture of control that is an essential part of their professional training in the theory and methods of planning practice. This culture is rooted in the planning experience of the North and the historical values of colonising nations. Citizen participation has been an important aspect of planning education in the UK and USA since the 1960s, but the culture of control has been the dominant feature of planning practice since 1947 (Government of the UK 1947). When this culture of control confronts the basic needs of a large section of the urban labour force, the result is a devastatingly negative outcome for the city and its labour force.

The attempted urbicide of this section of the labour force in the interests of world status, however, is based on another myth: that the economic activities of some of the world's poorest people can be destroyed by repression and violence. These economic activities are embedded in social relations of the family, the community and the workplace. They are overlapping relations that exist between individuals, within and between family units, within and between small units of non-capitalist[2] manufacturing and commerce employing wage labour, and between informal and capitalist firms. The precise nature and structure of the social and economic relations varies from city to city. They are resilient and they lead displaced street traders to continue their activity nearby, but there are certain common processes that restrict the opportunities for the urban poor to earn incomes and accumulate wealth.

The restrictive processes are seldom linear and direct, by which I mean there are sometimes small gains in the defence of declining sectors that are later wiped out by

---

[2] They are non-capitalist in the sense that the owner of the enterprises also works 'on the tools' as a manual worker. They are, however, fully integrated into the world capitalist system.

more powerful forces. These forces may be systemic, generated by socio-economic or cultural relations beyond the control of the individuals or urban communities affected. Other forces may be more local, operating through the governance of the city. The relations are embedded in the city itself, but the sectors persist or decline in the context of national and global forces that influence urban change, as well as urban policy responses to these forces. Sometimes, the city can be prospering according to international development indicators, while some of its essential historical components are dying.

From the perspective of first world economists and international organisations, the informal sector is mainly seen as a backward and marginal part of the economy. Despite the decades of evidence that it is intimately connected to national and international capitalism, it is treated as pathological, and it apparently needs to be 'formalised'. The perspective of the workers in the informal economy depends on their position within this heterogeneous socio-economic network of income-earning opportunities but being 'formalised' is the least of their concerns. The survival and development of the family unit is their primary consideration.

There are, however, other processes that lead to the social and economic exclusion of street traders and artisans. These involve the destruction of the social and economic world of small family enterprises which, despite their precarious nature, have nevertheless offered income-earning opportunities where such opportunities have been in short supply. The pursuit of international tourists through city promotion is at the heart of this decline.

The earliest studies of city promotion for the attraction of inward investment in the UK go back to the 1960s and 1970s (Cameron and Clark 1966; Northcott 1977). In response to the growth of 'world class-itis' in the 1980s, authorities developed 'prestige projects', such as International Convention Centres, and promoted cities through 'civic boosterism'[3] (Hambleton 1990; Loftman 1990). In the 1990s, the fascination with world-class cities spread to the South and the relationship between street traders and planners was influenced by competition between cities for an increased share of the global tourist market.

According to the WTO, tourism is 'the world's largest industry and one of the fastest-growing, accounting for over one-third of the value of global services trade' (WTO 2009, 1). Although the statistics are notoriously difficult to gather and compare, before Covid and over the previous six decades, tourism was one of the largest and fastest growing sectors in the world economy, supporting a very large number of small and micro-businesses in a wide range of industries (WTTC 2022; OECD 2021).

However, part of the untold story about tourism is how the world's largest economic activity has an impact on the non-tourist activities of precarious microfirms in the developing world. Its acknowledged global position as a foreign earner makes it an important target of economic policy and it has the potential for generating large-scale employment in small firms, but policies aiming to stimulate international

---

[3] The aggressive marketing and promotion of the city in the context of increasing global competition between city administrations.

tourism also have the potential for the destruction of existing informal economic activity.

As cities compete for global tourism across Latin America, Africa and Asia, their history and heritage, as expressed in the architecture of their churches and cathedrals, the art and artifacts of their museums and the design of their squares and streets, are presented by city authorities as unique selling points. However, these buildings and spaces, along with their cultural contents, are also expressions of the wealth and power of a colonial history whose values reach into the present. They tell the story of an historic elite for the entertainment of a new elite. Particular representations of history become an integral part of the world's political economy, producing a continuation of elite politics and influencing social and economic relations in urban communities across the South.

These elite interpretations of history and heritage are reinforced by planners and other city administrators when they view the people who depend on the economic vitality of the streets as the main barriers to economic regeneration and the modernisation of city centres. Street traders, for example, are seen to create the congestion, rubbish and insecurity that send tourists elsewhere, while their enterprise and their contribution to the local and national economies are ignored.

Based on elite perceptions of class and race, the removal of street traders from the streets of Latin American cities has been an issue since the eighteenth century, long before there was a tourism industry. Since the growth of international tourism in the final decades of the twentieth century, however, a flurry of activity across the global south has seen the violent application of intervention against the traders in Asia and Africa as well as Latin America. From the 1970s, street traders have been increasingly harassed, displaced, and had their goods confiscated without compensation (Skinner 2008; Jimu 2010; Middleton 2003; Brown and Lyons 2010; Lindell 2010a).

Heritage tourism ignores the fact that the colonial economy was based on forced labour and debt bondage, that the church's wealth depended on the super-exploitation of the indigenous population, and that these relations extended into the twentieth century. Heritage tourism, through which middle-class tourists from Europe and North America demand access to historic city centres, is based on interpretations of history that can be contested by social groups whose ancestors have suffered at the hands of colonial masters (Baud and Ypeij 2009; Middleton 2009). This is the historical socio-economic context for tourism policy and the values that drove the history continue to have an impact. The promotion of heritage tourism illuminates the political nature of international tourism. For some, it highlights the continuity of inequalities from the past to the present.

When planners and street traders are in conflict over the characteristics of an area that ought to be preserved, they are expressing different histories of meanings and interpretations. When planners emphasise the importance of historic monuments and seek to remove informal traders from city centres, their rational-technocratic approach can obscure a violent cultural attachment to ideas that have racist roots and deny indigenous interpretations of history. When different social, economic and cultural interests come into conflict in these contested spaces, planners, who believe they are acting in the public interest, tend to identify with local elite interpretations.

The enforcement of these interpretations, when they are funded by international organisations and they ignore the needs of local populations, usually results in the destruction of urban livelihoods.

This is true across much of the global south, although the precise details will vary from city to city; and over time there may be variations in the levels of collaboration and repression. The people who are most directly affected are traders, who have become the direct targets for urban policy. However, these policies for attracting international tourism also have deleterious consequences for pre-existing small-scale urban production and services. Every city will have a different story to tell about the fine detail, but there are common characteristics that can be identified. There are direct and intentional consequences for traders and indirect and unintended consequences for artisans.

## *36.2.1 The Direct Impact on Traders*

The removal of street traders from the streets and pavements of the cities of the developing world existed as a policy of urban governments long before international tourism offered its substantial riches. Across Africa, Asia and Latin America, local and national governments have opted for violent measures to restrict the movement of street traders (Lindell 2010b; Brown 2006a; Shrestha 2006; Middleton 2003). In Africa and Asia, violence against indigenous traders was an integral part of the violent history of colonialism, an expression of apartheid and the inequalities of ethnicity and race that continued into the post-colonial cities such as Durban (Skinner 2008), Cape Town (Jordhus-Lier 2010), Harare (UK Government 2005), Lusaka (Potts 2008) and Karachi (Zaidi 2019). In the neoliberal world city or megacity, there is an intensification of social cleansing, economic inequalities, spatial displacement and social exclusion. In cities such as Lagos, for example, the already marginalised conditions of informal workers deteriorate further (Lawanson 2014; Ojalade et al. 2018; Ojalade and Lawanson 2022).

However, some city authorities may have a complex mix of policies that involve not only persecution, harassment and eviction, but also tolerance, regulation and promotion (Brown 2006b, 12); and this policy mix can change over time. Different departments within a local authority can have different policies (Middleton 2003), but harassment, theft of merchandise and violence becomes the dominant behaviour of the state in almost every case at some time.

At any one time, there can be differences in the approaches of central and city governments, each of which can be transitioning in different directions over time. These alternative approaches are linked to the perspectives on world cities and traders that planners bring to the discussion. The ultimate aim of the local authority is to remove the traders from the streets, but the approach taken by any authority, local or national, can fall somewhere along the continuum between trader participation and outright violent eviction. In the 'world city' or the 'African model Megacity' this global economic and financial hub is described as clean, safe, secure, functioning, modern, orderly, productive, progressive, prosperous, and business-friendly (Olajide et al. 2018; Crentsill and Owusu 2018; de Gramont 2015). On the other hand, the justification for the removal of traders from public spaces is based on language that dehumanises the traders.

Traders are characterised as unsanitary, unhygienic, filthy, illegal, and immoral. They are seen as the causes of traffic congestion, overcrowding, obstruction, poor sanitation, crime, extortion, noise, hazardous products, uncleanliness, litter, health hazards, and the pollution of the environment. According to the local state and local elites, they are involved in dysfunctional activities that need cleansing and eradication (de Gramont 2015; Crentsil and Owusu 2018; Lawanson and Omoegun 2018; Olajide et al. 2018; Njaya 2014). The practical responses of local government cover a range of inhumane treatment, involving raids, forced evictions, confiscation of goods, fines, imprisonment, beatings, harassment, abuse, cruelty, repression, bribery and exploitation (Njaya 2014, 77; Crentsil and Owusu 2018; Roever 2005; Lawanson 2014; Lawanson and Omoegun 2018; de Gramont 2015).

In a climate of bribery, corruption and repression, encouraged by the low-income levels of metropolitan police officers, street traders are harassed, beaten and robbed. Women food sellers in particular are relieved not only of their profits but also their capital.[4] The state vilifies and punishes the poor, making sustainable urban development unattainable. Its action contradicts the essence of sustainable development and poverty alleviation when it deliberately targets the informal activities of the urban poor. The repression of informal traders in the interests of international capital in world-class megacities destroys livelihood opportunities and exacerbates inequalities.

As indicated above, not all city authorities consistently behave in this way. In the cases of Quito, Accra, Durban and across the major cities of India, there have been periods of collaboration between the authorities and traders' organisations (Middleton 2003; Oteng-Ababio and Arthur 2015; Skinner 2009; Madhav 2022). In several cities, the authorities have been working with informal organisations, building relations of trust, responding to the informal groups' expectations of local government through providing training programmes, health clinics and improved roads, all of which have reduced the propensity of the state to resort to violence. The traders' associations offer political leaders a bridge to informal workers that they otherwise would not be able to reach. Nevertheless, the outcome remains a reduction in the income-earning possibilities for the urban poor. In addition, there are unintended consequences for other parts of the informal sector, such as artisan producers.

---

[4] Many borrow money on a daily basis from moneylenders, who charged exorbitant rates of interest, and then they have their goods stolen by the state's enforcers (Middleton 2003).

## 36.3 The Indirect Impact on Artisans

Some of the processes that have an impact on artisans are based on market forces where, for example, urban consumption expands and it becomes viable for capitalist enterprises to serve the growing market. There is increasing profit to be made from an expanding middle class, which is first exploited by national firms, causing the hand-made production of artisan tailors and shoemakers to be replaced by industrially produced cheap products from an expansion of local capitalist production. The experience of recent decades shows that this is eventually supplanted by international capital, usually based in China or South Korea. Some destructive processes are based on changing technology. This occurs when, for example, printing workshops are displaced by photocopiers in the small shops in every neighbourhood or by computers and printers in almost every home.

Paradoxically, these destructive processes can also depend on national economic policies that intend to help, such as those for the promotion of small firms. This happens when the needs of the smallest manufacturing firms are ignored, either because they are difficult to identify, count and reach, or where the owners of these micro-firms are squeezed by the interests of larger enterprises when they come into competition for scarce development resources. When government finance is made available, for example, it inevitably fails to reach the smallest informal firms (Middleton 2020). If these firms do not have the collateral or banking history that will allow them to access the funds, larger businesses will step into the space and sweep up the benefits.

Just as important as these national and international forces, however, are the consequences, intended or unintended, of local responses to the wider relations and processes. The degradation of informal income-earning opportunities for the urban poor is amplified by local governance policies, such as the promotion of international tourism, the destruction and creation of buildings and public spaces, and the failure to deal with escalating violence in the poorest urban neighbourhoods. Within each of these elements, there are contradictions between public policy and public interest. When public policy promotes tourism, it also promotes urban regeneration and the elimination of crime. These are worthy aims, but their promotion can have unfortunate side effects that ought to be considered, in the public interest. In the case of Quito, the first city in the world to be recognised by UNESCO as a Cultural Heritage Site of Humanity in 1978, there were negative repercussions for the small-scale urban productive enterprises in and around the historic city centre. The perspective of the artisans of Quito is worth looking at more closely.

In Quito, artisans have been in decline and they are being restructured by global forces (Middleton 2020). The restructuring is based on the accelerated decline of artisans producing the means of subsistence (such as tailors, shoemakers and carpenters) and a slower demise or, in some cases increase, of artisans who can repair the products of twentieth century capitalism (such as the repair of cars, cookers, refrigerators and computers). The main reasons for the demise of subsistence artisans have been competition with industrial capitalism, with its higher levels of capital investment in new labour-saving technologies. In the case of Ecuador, the catalysts for periods of rapid decline include a series of neoliberal-induced crises that have sucked demand from the economy, such as the (temporary) importing of second-hand clothes, the national banking collapse, the dollarisation of the economy, unfair competition from Chinese manufacturers, the international financial crisis, and the instability of the price of oil. These macro-influences are not directly related to international tourism. However, local urban policies related to this aspect of globalisation have also had an impact in Ecuador. When the transformation of the Historic Centre of Quito was underway, the displacement of street traders was a priority for making the city centre more attractive as a destination for international tourists, but artisans were also affected by the regeneration.

### 36.3.1 Artisan Decline and Displacement

Between 1975 and 2015, censuses and surveys of artisans were carried out in three areas of Quito, covering part of the Historic Centre of Quito (HCQ), neighbourhoods immediately around the HCQ (the peri-centre), and what was initially known as the South of the city (Fig. 36.1). Stratified random samples of artisans in these areas were carried out in 1975, 1982, 1995, 2005 and 2015. In each year, the first stage of the fieldwork was to carry out a mapping of all the artisans in the study area—referred to as 'the census' of artisans. In 1982, the census showed that most of the artisans were no longer in their 1975 workshop locations. From 1982 onwards, the second stage of the fieldwork was to find out what had happened to those who had disappeared, using a survey of the neighbours and new occupiers for a randomly selected number of missing artisans. From 1982 to 2015, the third stage was to repeat the 1975 survey for a stratified random sample of new and old artisans. In 1995 and 2015, there was a fourth stage, when a number of in-depth life-history interviews were also carried out with elderly artisans. In the remainder of the paper, we will call on information that derives from stages 1, 2, 3 and 4. The types of artisans included tailors, shoemakers, carpenters, mechanics, painters, jewellers, printers and others. A full explanation of the methodology can be found in Middleton 2020 and 2022, along with a discussion of the heterogeneity of this section of the urban workforce.[5]

As detailed in the key ('Leyenda'), the solid line represents the 'Edge of the study area' and the dotted line represents the edges of three 'Neighbourhood groupings'. These three grouping were identified as the Colonial Centre ('Centro Colonial', now HCQ), the Peri-centre ('Peri-centro') and the South ('Sur').

In Tables 36.1 and 36.2, we can see the changing distribution of artisans between these three sectors of the city between 1975 and 2015. In 1975, the census identified 2308 workshops but by 2015, the numbers had declined by 40%. In the historic centre and the pericentre around it, the decline in the numbers was much greater, over

---

[5] For an online description of the methodology and categories of artisans, see Birmingham City University: http://www.open-access.bcu.ac.uk/12428/.

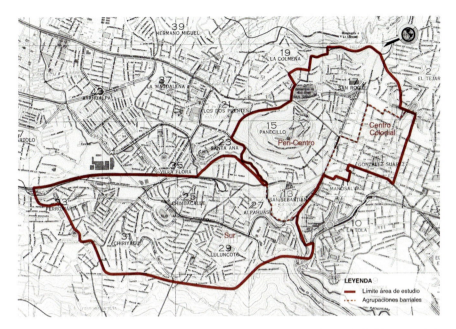

**Fig. 36.1** Map of the study area. *Source* Own elaboration updated by Valeria Vegara and the Instituto de la Cuidad, Quito

60%. However, the decline in numbers was accompanied by a redistribution of the artisans in space. When the second census was carried out in 1982, the total numbers were slightly up and there was a highly significant redistribution in space towards the Historic Centre.[6] These were the years of the oil boom and before structural adjustment was introduced in 1982. Between 1982 and 1995, the total numbers declined significantly and there was a significant redistribution away from the centre towards the south.[7]

In 1982, however, the remapping of the artisans across the whole area showed that almost two thirds of the workshops we had identified in 1975 no longer existed. When we surveyed the new occupiers of the properties or their neighbours, asking what had happened to the firms that had disappeared, we found that almost 70% had not gone out of business but had relocated, mainly within the same neighbourhood. That is, they had been displaced, rather than destroyed.

When we asked why people had moved, we expected many to have done so because of expansion, as free market theorists would have predicted. In fact, very few of them had experienced growth (between 2 and 7%) (Table 36.3). In 1982, only

---

[6] A Chi-square test showed a *p*-value of 0.0013. The number of firms in the historic centre grew by 28%.

[7] *P*-value = 2.93665E−10.

**Table 36.1** Artisans in each sector, 1975–2015

|  | 1975 |  | 1982 |  | 1995 |  | 2005 |  | 2015 |  | Decline 1975–2015 |
|---|---|---|---|---|---|---|---|---|---|---|---|
|  | No | % | No | % | No | % | No | % | No | % | % |
| Peri-centre | 1031 | 44.7 | 980 | 41.2 | 598 | 33.5 | 427 | 32.3 | 388 | 28.1 | −62.4 |
| South | 899 | 39.0 | 917 | 38.5 | 870 | 48.7 | 656 | 49.7 | 853 | 61.7 | −5.1 |
| Historic centre | 378 | 16.4 | 483 | 20.3 | 318 | 17.8 | 238 | 18.0 | 142 | 10.3 | −62.4 |
| Total | 2308 | 100.0 | 2380 | 100.0 | 1786 | 100.0 | 1321 | 100.0 | 1383 | 100.0 | −40.1 |

*Source* Middleton (2020, 40)

**Table 36.2** Average annual growth and decline of artisan workshops, 1975–2015

|  | Average annual Growth 1975–1982 | Average annual Growth 1982–1995 | Average annual Growth 1995–2005 | Average annual Growth 2005–2015 |
|---|---|---|---|---|
|  | % | % | % | % |
| Peri-centre | −0.7 | −3.7 | −3.3 | −1.0 |
| South | 0.3 | −0.4 | −2.8 | 2.7 |
| Historic centre | 3.6 | −3.2 | −2.9 | −5.0 |
| Total | 0.4 | −2.2 | −3.0 | 0.5 |

*Source* Table 36.1

14% had moved because of business growth or because the workshop was too small.[8] In 2005, only 10% had moved for these reasons, and in 2015, only 6%. By far the greatest reasons for people moving their businesses were not related to the internal dynamics of the firms nor their relationship to the market. Almost 70% were property related. The main reasons that re-locations took place were because the owner of the property demanded the return of the workshop, because the artisan moved home, or because the rents were too high. With rehabilitation and gentrification in the HCQ, fragile firms were pushed over the edge, never to be replaced.[9]

---

[8] Saying the workshop was too small did not mean that they moved for reasons of business expansion. It usually meant overcrowding could be eased without employing more workers or investing in new machinery.

[9] A components of change analysis showed that the producers of the means of consumption that were closing were not being replaced; while the producers whose labour or goods were linked to capitalist production, who had similar closure rates, were being re-placed by new firms (Middleton 2007).

**Table 36.3** Reason for change of location

|  | 1982 % | 1995 % | 2005 % | 2015 % | 1982–2015 % |
|---|---|---|---|---|---|
| Owner demanded workshop | 37.3 | 37.2 | 12.9 | 24.1 | 33.5 |
| Change of house | 26.4 | 31.9 | 29.0 | 29.6 | 28.6 |
| High rent | 4.1 | 5.3 | 16.1 | 16.7 | 7.2 |
| Workshop too small (without growth) | 10.9 | 2.7 | 3.2 | 3.7 | 6.9 |
| Badly situated | 7.8 | 3.5 | 9.7 | 9.3 | 6.9 |
| Business growth | 3.1 | 7.1 | 6.5 | 1.9 | 4.3 |
| Lack of services | 1.0 | 0.9 | 3.2 | 0.0 | 1.0 |
| Other | 9.3 | 11.5 | 19.4 | 14.8 | 11.5 |
| Total % | 100.0 | 100.0 | 100.0 | 100.0 | 100.0 |
| Total workshops | 193 | 113 | 31 | 54 | 391 |

*Source* Stage 2 fieldwork survey, 1982–2015, to find out what had happened to those who had disappeared

### 36.3.2 Artisans and the Refurbishment of Public Spaces

Contrary to what neoliberalism predicts, over the four surveys there were no reported cases of businesses closing or moving because of government bureaucracy, regulations or red tape. When we asked about reasons for closure in 1982, around 30% had merely 'failed', another 30% had either retired or died, and around 40% gave a wide variety of other reasons, including bad debt, bad workmanship, robbery, having no clients, being ejected by the owner of the building, alcohol problems and simply owning 'unprofitable businesses'. In 2005, however, after death and retirement, the next most important reasons were related to issues concerning the property: artisans moving home and giving up the workshops, owners asking for the property to be returned for their own use or so that they could raise rents, the Municipio buying the buildings through compulsory purchase for the regeneration of the HCQ, and so on.

From the 1940s, the upper and middle classes of Quito had been steadily moving out of the old colonial-style houses of the city centre, leaving large multi-roomed buildings for the lower classes. As the housing deteriorated, the 20–40 rooms that were occupied by a single family and servants became the homes for 20–40 families. The internal patios of these large buildings became the locations for artisan workshops and other small enterprises. Some of the rooms of the houses opened directly onto the streets, which contained the passing trade of customers for artisans, shops, bars, and cafes, mainly people who worked in what was to become the Historic Centre of Quito. The process of intensification of artisan activity in the centre continued through 1978, when Quito was declared by UNESCO as a World Cultural Heritage site, to 1982 (Table 36.1). As the buildings deteriorated, ground rents declined relative to the north and south of the city, but landlords' income was increasing at a faster rate than ground rents through the increasing concentration of families in single rooms in the multiroom buildings. Many of these rooms had a dual function as living spaces and places of work, particularly the rooms facing directly onto the streets.

When the Company for the Development of the Historic Centre of Quito (ECH) was set up in 1995, its programme of work consisted of the rehabilitation of public spaces and buildings of particular historical and architectural value, improvement of the urban infrastructure and the provision of services that would rescue the historic area from its declining functional importance. It was thought that this would improve the quality of life for those living in the HCQ, but the main challenge was to rescue, rehabilitate and conserve this part of the national and international patrimony and generate a dynamic and secure environment that would attract investment from the private sector (Empresa de Desarrollo del Centro Histórico de Quito, n.d.). The ECH, supported by a $41 million loan from the Inter-American Development Bank (BID 1994),[10] sought to consolidate and enhance the many development activities that were already under way in the HCQ. At the time of the company's inauguration, the first phase of a new transport infrastructure was coming into operation, which was to have a devastating effect on a number of small firms.

## 36.4 Artisans and Transport Infrastructure Improvements

Upgrading the transport infrastructure is an important element in the development of a city, for economic, social and environmental reasons. A new tram line connects people's homes with workplaces, shops, bars and restaurants while reducing car journeys and pollution. Creating new roads, tunnels and bridges can also reduce journey times between home and work and speed up the journey times for the delivery of goods and services. Leaving the complexity of environmental considerations aside, they can be the stimulus for increased productivity and economic growth. In Quito, the opening of the trolly service in 1995 was an essential ingredient the city's investment in infrastructure. However, the unintended consequences of such investments can include the cutting off street access to small businesses, often resulting in business failure.

The displacement of small enterprises can occur during both the construction phase and the operational phase. During construction, land is bought up by the public authorities, buildings are demolished and businesses are destroyed. If the existing roads survive, buildings can be made inaccessible over long periods and businesses can be cut off from their customers.

In 1995, when Quito's trolley-bus service was put in place to connect the centre of the city to the expanding residential areas in the north and south, the route ran along Avenida Maldonado in the southern section. In 1975, this section of Maldonado, between the Machángara River at La Recoleta and Villa Flora, was the location

---

[10] The BID/IDB argued that the informal traders were holding back private sector investment in the HCQ.

**Table 36.4** Decline of Avenida Maldonado, 1975–2015, southern section

|  | 1975 | 1975 | 1975 | 1982 | 1995 | 2005 | 2015 |
|---|---|---|---|---|---|---|---|
|  | Total micro-firms | Shops | Artisans | Artisans | Artisans | Artisans | Artisans |
| Total | 144 | 62 | 82 | 81 | 54 | 28 | 13 |
| % Artisans remaining |  |  | 100 | 98.8 | 65.9 | 34.1 | 15.9 |
| % Artisans lost |  |  |  | 1.2 | 34.1 | 65.9 | 84.1 |

*Source* Stage 1 Census of informal firms in Quito, 1975–2015

for 82 artisan workshops. These included tailors, hatmakers, shoemakers, leather bag makers, carpenters, mechanics, guitar-makers, jewellers, upholsterers, printers, painters, glaziers, electricians, bakers, hairdressers, and workshops repairing electric motors, bicycles, washing machines, refrigerators and automobiles (including body repair and tyre repair). In Table 36.4, we can see that the numbers held up in 1982 and declined thereafter. By the time the construction phase was completed in 1995, the number of workshops had declined by one third. Across the South as a whole, the decline between 1982 and 1995 was only 0.4% (Compare Table 36.4 with Table 36.2). After 1995, the displacement effect of the operation of the trolly on the workshops on Maldonado was considerable. By 2005, there were only 28 out of 82 firms left and in 2015 the figure was just 13.

Before the trolly, Avenida Maldonado was a vibrant shopping road that not only served the residents of the surrounding barrios of Mexico, Chimbacalle and Los Andes, but also attracted the custom of those who walked to their work in the city centre from the expanding barrios to the south of the city centre. The artisans' customers contributed to the vibrancy and were served by 62 small shops of every description as well as bars and restaurants. By 1995, the decline had set in but after the trolley began to operate, the decline due to construction was compounded by the loss of passing pedestrian trade. By 2015, the street had lost 84% of its artisan workshops, compared to a 5% decline across the whole of the South (Compare Table 36.4 with Table 36.1).

The loss of passing customers also became an issue in other parts of the city as the rehabilitation of the HCQ progressed. Once again, the issue was one of social cleansing, rather than the rehabilitation of infrastructure, but it was relating to petty criminals and prostitutes. It was about making improvements to the security of the HCQ for tourists and businesses. Unfortunately, when city authorities crack down on crime in areas that attract tourists, they displace the activities towards the surrounding areas. In the case of Quito, these were dynamic urban streets that were the centre of artisan production. Before we consider this detail, however, let us look at the nature of the crime and violence associated with informal enterprises and their customers.

## 36.5 Crime, Violence and The Informal Sector

In their paper on the impact of urban–rural change as an underlying cause or trigger point for increased violence and insecurity, Moser and Rodgers (2005) acknowledge that globalisation has brought unprecedented benefits for some, but they also argue that globalisation has a dark side, whereby there are reduced employment opportunities for some individuals and communities who are excluded from the central processes of production and exchange. It is thought that this has exacerbated livelihood insecurity and increased different forms of violence linked to alienated, frustrated or excluded populations. They distinguish between political, institutional, economic and social violence, conceiving of these categories as 'an overlapping and interrelated continuum with important reinforcing interconnections' (Moser and Rodgers 2005, 4). They also provide some examples of the types of violence in each category and identify a few examples of the diversity and complexity of the violence endured in the daily lives of the poor. Most of the types of violence endured by the section of the poor who depend on the informal sector, either as workers or consumers, fall under the category of economic violence. Within this category, they include the broad headings of delinquency, robbery, business interests and organised crime (Moser and Rodgers 2005, 5).

These types of violence can be elaborated further. The most common types of crime and violence suffered by the owners and customers of the informal sector can be seen in Table 36.5. The range includes state violence against street traders, various forms of in-market violence, street attacks on customers and businesses, armed robbery, and participation in riots against the government.[11] In almost every case, the victims are informal traders. This does not mean that they are passive in the face of attack. They resist in a variety of ways and this resistance often involves collaboration with allies to minimise the potential for violence. We have already discussed the actions of the police against street traders. Let us deal with the impact of in-market violence, which includes trader-on-trader confrontations, and insecurity on the streets.

### 36.5.1 In-Market Violence

In-market violence can be perpetrated by the traders, their private security, the customers and thieves. Traders sometimes also have links to smugglers and organised crime. However, the increasing demands to secure the tourist areas of the city and the reduced levels of public investment by the neo-liberal state in policing, mean that the police do not have the capacity to secure the markets against crime. The traders therefore mainly rely on private security firms who make the markets more secure for both the traders and customers. Because of ethnic, political and other divisions within the traders, however, security teams can be also used against each

---

[11] I have left out prostitution and the robbery of their clients, to which we will return later.

**Table 36.5** Types of informal sector violence

| Types | Perpetrators | Victims |
|---|---|---|
| State violence against traders | State | Traders |
| *In-market violence* | | |
| Political violence between groups of traders | Traders | Traders |
| Ethnic violence between groups of traders | Traders | Traders |
| Mafia-type turf violence | Traders | Traders |
| Violence by private security officers | Security | Thieves |
| Customer violence | Customers | Traders |
| Violence of thieves | Thieves | Traders/customers/security |
| *Street violence* | | |
| Against customers | Thieves | Customers |
| Against businesses | Thieves | Traders/artisans |
| Armed robbery | Thieves | Traders/artisans/customers |
| Riots against the government | Traders | State |

*Source* Elaborated by the author for this chapter

other. Private security officers get involved if clashes over space break out. They get involved if violence based on politics or ethnicity takes place within the market. The private security enforces the use of space when this happens, resorting more readily to violence and showing little respect for human rights (Moser and Rodgers 2005).

Ethnic violence is most likely to occur when rival groups of migrants are in competition for scarce resources, including space and customers. Moser and Rodgers argue that migration can be both a cause and a consequence of violence and insecurity. Cultural identity reinforces competition for space and underpins demands for the resolution of perceived inequalities in the distribution of physical capital. Market traders are often networked actors in a power struggle, within the boundaries of the markets as well as within the surrounding community, city, region and nation. In the struggle, informal trading spaces are produced and defended, in extreme cases through violent confrontation. The networks are based on social capital, and collaboration between individuals is often cemented by ethnic and/or political affiliation. In this overlap of power, culture and social capital, the capacity of the participants to flourish economically is diminished.

In Karachi's markets, for example, violence is based on overlapping ethnic and political tensions between different migrant groups (Zaidi 2019). These tensions are city-wide, but they find their expression in conflict over space in the markets. While Karachi may be an extreme example, it is a case that demonstrates how the market can become a place where communal violence is exacerbated because of the competition for scarce resources. Zaidi shows how identity-based networks result in internal subdivisions in markets, how this leads to turf wars, how ethnic leaders provide access to financial and other support, how they provide links from the trader to the municipal authorities, and how political networks overlay ethic divisions.

She explains how local politics and ethnicity reinforce each other. However, she also explains how criminal gangs, working with local politicians of the same ethnic groups, take over markets, run the associations of vendors, control the access of political parties to the market traders, exclude the police and other authorities, offer 'protection' at a price, and control the access to the market of potential new traders.

Corruption, extortion and bribery are widespread, as police, politicians and gangsters collaborate in their own interests. Ethnic violence can therefore be carried out with the collusion of the local and regional state, and it is also related to the insertion of different ethnic groups in the urban economy. In Karachi's case, The Muhajirs dominate the city's business and industry sectors; the Pathans control the urban transport system; and Sindhi power is exercised through their dominance in provincial politics and education (Moser and Rodgers 2005). Their long-standing rivalries are expressed in sporadic ethnic riots and in the competition for space in the city's markets, where the police defer to the gangsters and local politicians are controlled by the party's head office. Through the elite capture of the power structures of market places, the social capital of market traders can become subverted to the interests of state predation and violence.

## 36.5.2 Street Violence

Fear of violence in the streets, even if there is little evidence of the existence of violence, can lead to feelings of insecurity and vulnerability in small firm owners and their customers, which contribute to the destruction of informal enterprises. This process and outcome can be compounded by prostitution, which is often perceived as a form of anti-social behaviour where it is not criminalised, and non-violent petty crime.

Petty crime, violence and the fear of violence threaten the livelihood security of artisans and shopkeepers by scaring away potential customers, dissuading them from travelling into the neighbourhoods where informal workshops and shops are located. In city centres, crime is likely to involve shoplifting, people who snatch articles from others in the street and run off, stealing from handbags by snatching or slitting them open, and pickpocketing. Armed robbery, of an individual or a business, is less likely to happen in busy city centres, which are normally heavily policed, than it is in the immediate surrounding areas. Street crime can also involve robbing vehicles or parts of vehicles, which is also more likely outside of heavily policed central areas. Armed robbery and vehicle theft, however, are more likely to occur in affluent areas that in poor neighbourhoods.

Sometimes, as cities promote themselves as world cities or as locations for international tourism, the cleansing of crime and violence from the inner city can lead to the displacement of street crime into surrounding poorer neighbourhoods, with devastating effects for small firms. This happens when thieves target businesses and their customers in the peri-centre, while police are focusing on the safety of tourists. In this situation, the artisans and the owners of small stores, along with their customers, become the victims of a surge in neighbourhood crime that is rooted in the gentrification of the historic centres.

### 36.5.3 Catching Criminals Versus the Management of Tourist Space

In Quito, before the emergence of tourism promotion as a key objective of the Municipio, policing was driven by the need to catch criminals wherever they were operating. When the administrative priority became attracting tourists to the HCQ, urban policy became more concerned with the management of tourist space and the main aim of municipal policing became the removal of the criminals from the HCQ. This displacement resulted in criminal and anti-social behaviour moving into near-by residential barrios where artisans lived and worked. The new spatial form of governance that promoted the HCQ, created not only new spatial patterns of economic organisation but also a redistribution of violence and crime.

A similar process occurred with prostitution. Across the world, in developed and developing countries, women are not as heavily involved in crime as men. If they become involved in prostitution, whether this is designated as a criminal activity or not, they are more likely to become involved in criminal activity through robbing clients, but not necessarily in other illegal behaviour (Birkbeck 1979).

If a woman rejects prostitution and crime, the income-earning opportunities for someone with no capital and few skills are severely restricted in many cultures. In Latin America a woman can work in a bar, café or shop but this is not possible in many parts of the world. Entry into informal trade is difficult without capital or the assistance of someone who will vouch for a loan. Domestic service may be an option, as is begging, but for a young woman with children to feed and clothe, the situation can be desperate.

Prostitution is not illegal in most of the developing world. It is a low-status and heavily regulated occupation, but different from petty crime, which is an illegal income-earning opportunity that brings the threat of imprisonment. For a single mother of small children, working as a prostitute on the street on a part-time basis may support her family, but petty crime is a step too far. Arrest and imprisonment put the very existence of the family at risk. Prostitutes working on the streets may only want part-time work because of family responsibilities, and a lengthy period in jail is a risk that could result in losing their children.

Street prostitutes operate in the same urban space as petty criminals and may be part of the same low-status community. They are also likely to be displaced, along with the criminals, into the areas surrounding the managed tourist spaces. The social cleansing that takes place to create safe tourist spaces creates crime and fear of crime in neighbourhoods where artisan customers are increasingly unwilling to visit.

**Table 36.6** Workshops and Shops on Loja Street and Ambato Street, 1975–1982

|  | 1975 Shops No | 1975 Artisans No | 1982 Artisans No | Percentage decline 1995 Artisans No | 2005 Artisans No | 2015 Artisans No | 1975–2015 Artisans % |
|---|---|---|---|---|---|---|---|
| Calle Loja | 103 | 93 | 76 | 27 | 30 | 21 | 77.4 |
| Calle Ambato | 150 | 159 | 139 | 64 | 30 | 28 | 82.4 |
| Total Loja and Ambato | 253 | 252 | 215 | 91 | 60 | 49 | 80.6 |
| % Artisans remaining |  | 100.0 | 85.3 | 36.1 | 23.8 | 19.4 |  |
| % Artisans lost |  | 0.0 | 14.7 | 63.9 | 76.2 | 80.6 |  |

*Source* Stage 1 Census of informal firms in Quito, 1975–2015

### 36.5.4 Street Crime in Artisan Neighbourhoods

In Quito, the impact of the displacement of street crime and street prostitution was to help undermine the viability of the main artisan neighbourhood in the city. In 1975, Loja and Ambato streets, located immediately outside the HCQ, were the beating heart of fixed place informal activity in Quito. On these two streets, there were over 500 small shops and artisan workshops, including 252 artisan businesses making and repairing all types of consumer goods (Table 36.6). Over the next twenty years, the workshops on these streets declined by 64% and by 2015 the number had fallen by 80%. This is double the 40% decline in the study area as a whole (Compare Table 36.6 with Table 36.1).

Most of this decline was due to market forces exacerbated by neoliberal policies of austerity and structural adjustment, described elsewhere (Middleton 2020, 2022). However, the regeneration of the HCQ had a role to play, particularly the security policies that sought to make the HCQ safe for tourists. The management of security in the central area gave added local impetus to a general increase in crime and delinquency in Quito.

Crime and delinquency first appeared as the most important problem for artisans in 1995, when jewellers complained that, because of street robberies, 'people do not buy gold now. They prefer to buy cheaper imported jewellery'. In 1975 and 1982, no one identified delinquency or crime as their most important problem (Table 36.7). The problem has grown over time. In 1995 it was the main problem for less than 5% of artisans but by 2015 crime and delinquency had become the main concern of 12% of artisans. It was also mentioned by many others as the second or third most important issue facing them.

As the transformation of the HCQ got underway in the 1990s, while the street traders were facing forced eviction, the council's priority of making the city centre more attractive for tourists also had an impact on artisans. Some were displaced or cut

**Table 36.7** Most important problems

|  | 1975 % | 1982 % | 1995 % | 2005 % | 2015 % |
|---|---|---|---|---|---|
| None | 10.6 | 15.1 | 15.3 | 7.5 | 5.6 |
| Demand | 12.9 | 8.1 | 17.4 | 26.2 | 17.3 |
| Competition | 4.2 | 0.8 | 13.2 | 17.7 | 13.4 |
| Delinquency and crime | 0.0 | 0.0 | 4.5 | 10.1 | 11.9 |
| Materials (cost, supply, quality) | 19.8 | 23.1 | 6.4 | 6.7 | 11.8 |
| Govt./economy | 4.3 | 2.2 | 6.5 | 8.4 | 10.2 |
| Problems with labour (incl. shortage) | 7.4 | 15.8 | 3.6 | 4.7 | 6.7 |
| Workplace (size, cost, location) | 9.6 | 7.7 | 6.4 | 6.0 | 5.5 |
| Not enough time | 0.0 | 3.6 | 0.6 | 0.0 | 2.3 |
| Clients (too poor/don't pay) | 7.7 | 3.1 | 4.1 | 1.3 | 2.1 |
| Lack of capital | 17.9 | 11.0 | 5.3 | 5.6 | 0.9 |
| Finance/debt | 0.0 | 1.4 | 1.4 | 0.2 | 0.5 |
| Electricity | 0.0 | 0.0 | 6.9 | 0.6 | 0.0 |
| Other | 5.8 | 8.0 | 8.4 | 5.6 | 12.2 |
| Total | 100.0 | 100.0 | 100.0 | 100.0 | 100.0 |
| Total workshops | 188 | 296 | 321 | 313 | 255 |

*Source* Middleton (2020, 237)

off from their markets by the building of bridges and tunnels. Others suffered from the displacement of the red-light district across the Avenue of 24 de May, into the heart of the artisan economy. As thieves and other 'delinquents' were moved out of the HCQ and the city centre became more secure for the tourists, displaced criminals and prostitutes created insecurity for the artisans in their traditional heartlands. They were more prone to robbery in their businesses and their customers were scared off. As their businesses declined, some elderly artisans continued to exercise their skills in the face of increasing adversity.

Jewellers and watchmakers[12] are the artisans who are most likely to suffer theft from their businesses, and their customers wearing expensive jewellery and watches are most likely to be robbed in the streets. A watchmaker in the HCQ whose workshop was robbed before the security situation improved, and whose house was robbed a year later, lost everything on both occasions. He reflected on the time the area was full of prostitutes, 'which attracted delinquency'.[13] In 2015, he pointed out that the situation had improved and that robbery was no longer a problem. His main problem now was that people were buying cheap Chinese watches, which are not worth either repairing or stealing. Demand for luxury jewellery and watches was down, and mobile phones, which also have timepieces, were more likely to be the target of thieves in

---

[12] 'Relojero' in Spanish, which also applies to watch repairers.

[13] Life history interview with Jeweller in the Historic Centre of Quito, September, 2015.

the streets. However, the robbery of clients was more likely to happen in the barrios around the historic centre, which have lower levels of policing.

When a jeweller-watchmaker in the peri-centre lost everything and was psychologically devasted by the experience, he sold what was left of his business to one of his workers. After the new owner had rebuilt the business, he too was robbed. Because of street robbery of his clients, his business further declined, to the extent that it had become no more than a hobby.[14] However, the problem of rising crime in the artisan neighbourhood is not restricted to jewellers and their customers.

As security declined in the peri-centre, other types of businesses in the same area suffered similar experiences, such as the seamstress who invested in a new German-made industrial sewing machine, only to have it stolen. Or the shoemaker who has been robbed several times and whose clients were constantly telling him that the barrio was dangerous and that he ought to move to another neighbourhood. Or the hairdresser who works alone close to the San Roque market whose family is constantly trying to convince her that she should move to a more secure area. To some extent they are trapped in these barrios. They have local clientele and many have customers who travel from other parts of the city for their services. Their clients know where they are, and rents in more secure areas or in shopping malls are out of their reach.

For the shoemaker, his income is low and his rent is cheap. He would have to pay more if he moved to a better barrio, and he does not have the clientele to justify moving. He concedes that it is dangerous, but he says he is content where he is. He has been attacked inside his workshop on more than one occasion by people he knows. He makes friends with them, but they are drug addicts and when they are high, he says, they are no one's friend. They come back the next day and apologise. He shrugs his shoulders: 'That's the way it is'.[15] He has lost clients, who apologise, saying 'I'm sorry but I cannot continue coming here'. Women clients have been assaulted on the street corner and they are afraid, asking him why he does not move. His reply is that the people around here know him.

The hairdresser, who works alone in her salon next to the San Roque market, has been working in the neighbourhood for 30 years. She is highly skilled, attends courses to learn about the latest styles, but charges her clients very little 'because of the barrio' where she is located.[16] Her daughters and friends have been asking her why she continues to work in that neighbourhood, why does she not move to a mall in the Valle de los Chillos, where she lives; or to the north where she could get have a wealthier clientele. No, she says, I have clients here. Some of them have been customers for 30 years, and now she cuts and styles the hair of their children and grandchildren. She thought about moving to the Valle but she did not want to leave her clients. She knows everyone in San Roque and would have to leave them behind. Like the shoemaker, she has no guarantee that she would be able to build up a new clientele in a new more expensive location.

---

[14] Life history interview with jeweller in the peri-centre, Quito, September, 2015.

[15] Life history interview with shoemaker in the peri-centre, Quito, September, 2015.

[16] Life history interview with hairdresser in the peri-centre, Quito, September, 2015.

These older artisans are seeing out their final years, clinging to what they know. They are one final robbery away from closure.

## 36.6 Conclusions

In this chapter, we have considered some aspects of urbicide in relation to the informal sector. City governments, as enablers of accumulation through the promotion of tourism, the refurbishment of public spaces and the removal of 'delinquents' from city centres, become tools of the displacement and socio-economic exclusion of the urban working poor. The dispossession of buildings and the social cleansing of informal sector workers from the streets and public spaces promote a speculative urbanism that increases the precarity of all sections of the urban poor. Well-intended development efforts that are meant to promote economic growth and reduce poverty result in displacement, instability and insecurity, leading to a systematic denial of opportunities and resources. Assumed trickle-down does not happen, socio-spatial inequalities increase, and the already marginal conditions of low-income workers deteriorate further.

Neoliberalism promotes increased openness to market mechanisms, but this is not extended to the informal sector. Despite the positive development aspirations, the supremacy of the interests of capital produces an urban transformation in which the economic structures of poorer citizens are destroyed. In the creation of urban modernity there are negative outcomes for informal firms that transform intended beneficiaries into victims. The development strategy damages the livelihoods of the urban poor and, instead of economic recovery and job creation, much of the city's social and economic fabric is destroyed. In the entrepreneurial world city, there is a disconnect between the development aims of the local state and the livelihood aspirations of local residents and workers.

The workers and families that rely on income from the informal sector, however, are not just passive victims. An outcome we have not dealt with in this paper is how urban informal workers, in Quito and elsewhere in the South, have joined forces with indigenous social movements against the power of the state. A consequence of ignoring the needs of the informal sector is their attachment to wider social movements in times of crisis. This is a matter for future research.

## References

Banco Interamericano de Desarrollo (BID) (1994) Ecuador: Programa de Rehabilitación del Centro Histórico, Quito, BID

Baud M, Ypeij A (eds) (2009) Cultural tourism in Latin America: the politics of space and imagery. CEDLA Latin American Studies, Amsterdam

Birkbeck C (1978) Self-employed proletarians in an informal factory: the case of Cali's garbage dump. World Dev 6(9/10):1173–1185

Birkbeck C (1979) Women, crime and prostitution in Cali, Colombia. Paper presented to the conference on marginalized women workers. Organised by the British Sociological Association's Development Study Group, London

Bremen J (2010) Outcast labour in Asia: circulation and informalization of the workforce at the bottom of the economy. Oxford University Press, New Delhi

Bromley R (1978) The informal sector: why is it worth discussing? Introduction to a special edition of World Dev 6(9/10):1033–1039

Bromley R, Wilson TD (2018) Introduction: the urban informal economy revisited. Latin Am Perspect 45(1):4–23. Issue 218

Brown A (ed) (2006a) Contested space: street trading, public space, and livelihoods in developing cities. ITDG Publishing, Rugby

Brown A (2006a) Challenging street livelihoods. In: Brown A (ed), pp 3–16

Brown A, Lyons M (2010) Seen but not heard: urban voice and citizenship for street traders. In: Lindell I (ed), pp 33–45

Cameron GC, Clark BD (1966) Industrial movement and the regional problem. Soc Econ Stud. University of Glasgow. Occasional Paper, No. 5, Edinburgh and London, Oliver and Boyd

Crentsil AO, Owusu G (2018) Accra's decongestion policy: another face of urban clearance or bulldozing approach? In: International development policy, vol 10. African cities and the development conundrum, pp 213–228. https://journals.openedition.org/poldev/2719

De Gramont D (2015) Governing Lagos: unlocking the politics of reform. Carnegie Endowment for International Peace, Washington

Empresa de Desarrollo del Centro Histórico de Quito (EMD, later ECH) (n.d.) Empresa de Desarrollo del Centro Histórico de Quito: Economía Mixta. ECH, Quito. Undated leaflet produced by the EMD/ECH

Government of the United Kingdom (1947) Town and country planning act, 1947. UK Government, London

Hambleton R (1990) Urban government in the 1990s: lessons from the USA. School of Advanced Urban Studies, University of Bristol

Hart K (1973) Informal income opportunities and urban employment in Ghana. J Mod Afr Stud 2(1):61–89

ILO (1972) Employment, incomes and equality: a strategy for increasing productive employment in Kenya. ILO (The Kenya Report), Geneva

ILO (2002) Decent work and the informal economy. Report IV, International labour conference, 90th Session. Geneva, ILO

ILO, WIEGO (2013) Women and men in the informal economy: a statistical picture, 2nd edn. ILO, Geneva

Jimu IM (2010) Self-organised informal workers and trade union initiatives in Malawi: organising the informal economy. In: Lindell I (ed), pp 99–114

Jordhus-Lier DC (2010) Moments of resistance: the struggle against informalisation in Cape Town. In: Lindell I (ed), pp 115–129

Kosny M, Loftman P (1991) Birmingham's unitary development plan and Toronto's city plan 1991: two responses to a similar problem of world class-itis. Paper presented to ACSP-AESOP joint international congress, Oxford

Lawanson T (2014) Illegal urban entrepreneurship? The case of street vendors in Lagos, Nigeria. Arch Environ 13(1):45–60

Lawanson T, Omoegun A (2018) In transitioning to Africa's model megacity—where is Lagos for everyday people? Heinrich Boll Stiftung, Abuja. https://ng.boell.org/en/2018/03/05/transiting-africa%E2%80%99s-model-megacity-where-lagos-everyday-people

Lindell I (ed) (2010a) Africa's informal workers: collective agency, alliances and transnational organising in urban Africa. Zed Books, London

Lindell I (2010b) Introduction: the changing politics of informality—collective organising, alliances and scales of engagement. In: Lindell I (ed), pp 1–30

Loftman P (1990) A tale of two cities: Birmingham the convention and unequal city. Research Paper No. 6, Built Environment Development Centre, Birmingham Polytechnic

Madhav R (2022) India's street vendor protection act: good on paper but is it working? WIEGO Blog. https://www.wiego.org/blog/indias-street-vendor-protection-act-good-paper-it-working

Middleton A (1989) The changing structure of petty production in Ecuador. World Dev 17(1):139–155

Middleton A (2003) Informal traders and planners in the regeneration of historic city centres: the case of Quito, Ecuador. Prog Plann 59(2):71–123

Middleton A (2007) Globalization, free trade, and the social impact of the decline of informal production: the case of artisans in Quito, Ecuador. World Dev 35(11):1904–1928

Middleton A (2009) Trivializing culture, social conflict and heritage tourism in Quito. In: Baud M, Ypeij A (eds) Cultural tourism in Latin America: the politics of space and imagery. CEDLA Latin American Studies, Amsterdam, pp 199–216

Middleton A (2020) The informal sector in Ecuador: artisans, entrepreneurs and precarious family firms. Routledge, Abingdon and New York

Middleton A (2022) El Sector Informal en el Ecuador: Artesanos, Empresarios, y Empresas Familiares Precarias. FLACSO, Quito

Moser CON, Rodgers D (2005) Change, violence and insecurity in non-conflict situations. ODI Working Paper 245, London, Overseas Development Institute

Njaya T (2014) Challenges of negotiating sectoral governance of street vending sector in Harare Metropolitan, Zimbabwe. Asian J Econ Model 2(2):69–84

Northcott J (1977) Industry in the development areas: the experience of firms opening new factories. PEP, Vol. XLIII, Broadsheet No. 573, London, PEP

Olajide OA, Agunbiade ME, Bishi HB (2018) The realities of Lagos Urban Development Vision on livelihoods of the urban poor. J Urban Manage 7:21–31

Olajide O, Lawanson T (2022) Urban paradox and the rise of the neoliberal city: case study of Lagos, Nigeria. Urban Stud 59(9):1763–1781

OECD (2021) Managing tourism development for sustainable and inclusive recovery. OECD Tourism Papers 2020/21, OECD Publishing, Paris. https://www.oecd.org/cfe/managing-tourism-development-for-sustainable-and-inclusive-recovery-b062f603-en.htm

Oteng-Ababio M, Arthur IK (2015) Discontinuities in scale, scope and complexities of the space economy: the shopping mall experience in Ghana. Urban Forum 26:151–169

Potts D (2008) The urban informal sector in sub-Saharan Africa: from bad to good (and back again?). Dev South Africa 25(2):151–167

Roever, SC (2005) Negotiating Formality: Informal Sector, Market, and State in Peru. PhD dissertation. University of California, Berkeley. https://www.wiego.org/publications/negotiating-formality-informal-sector-market-and-state-peru

Shrestha S (2006) The new urban economy: governance and street livelihoods in the Kathmandu Valley, Nepal. In: Brown A (ed), pp 153–172

Skinner C (2008) The struggle for the streets: processes of exclusion and inclusion of street traders in Durban, South Africa. Dev South Afr 25(2):227–242

Skinner C (2009) Challenging city imaginaries: street traders' struggles in Warwick Junction. Agenda Empowering Women Gender Equity 81:101–109

UK Government (2005) Freedom of Information request: eGrams from the British Embassy at the time. http://www.un.org/News/dh/infocus/zimbabwe/zimbabwe_rpt.pdf

WTO (2009) Tourism services: background note by the secretariat, S/C/W/298

World Travel and Tourism Council (WTTC) (2022). https://wttc.org/About/About-Us

Zaidi N (2019) The production of informal trading spaces. Ph.D. thesis, Cardiff University

**Alan Middleton** Emeritus Professor of Urban Studies at Birmingham City University, UK. Before retiring he was Director of Research at BCU and Associate Dean of the Faculty of the Built Environment. His Book on *The Informal Sector in Ecuador* (Routledge) has been published in Spanish by Flacso Ecuador.

# Chapter 37
# De-urbanization: From the Shock to the Revolution of a New Urban Logic

Paulina Cepeda Pico

> *When few spend too much, others spend little to survive.*
> *Martínez de Mata*

**Abstract** Cities and urban territories in their accelerated urbanization processes go through logics adjusted to contemporary demographic, spatial, economic, social and political trends that cannot be avoided. In this way, certain disasters are presented as opportunities to transform urban planning from the power and interest of certain sectors (urban shock) or from the social sense of urbanism (urban revolution). This chapter analyzes three global disasters that become urban shocks and raise the question of whether they produce massive urbanization or incomprehensible de-urbanization. For this, it is analyzed under five dimensions or parameters: demographic or migration trends, urban gaps, settlement logics, residence-work relation and outsourcing of urban fixed capital. To conclude that, the incoherent assimilation of de-urbanization processes leads to voracious urbanization and therefore to the production of ghost cities, smart cities and abandoned and degraded territories.

**Keywords** Population decline · Urban planification · Capitalism · Urbanization

## 37.1 Introduction

Cities around the world have gone through an accelerated, voracious and even chaotic and unequal urbanization process, mainly during the last century; now this trend mutates to a transformation of an inverse vector, where citizens reinhabited or move to different territories, accompanied by suburbanization and counter-urbanization logics, framed in de-urbanization processes and breaking the traditional development logic. Where the classic linear cycle of the continuous process of urbanization, with a centripetal tendency (center-periphery) in which cities are the epicenters of development, enter new logics of centrifugal, opposite and even inverse spirals, due to their loss of value or obsolescence.

P. Cepeda Pico (✉)
Facultad Latinoamericana de Ciencias Sociales, Flacso Ecuador, Quito, Ecuador
e-mail: pccepedafl@flacso.edu.ec

Initially, two meanings should be highlighted for the development of this theme. First, urbanization has several connotations: as a massive, exclusive, unnatural and industrial *production process*; as the materialization of capitalist life and its support, or as a creative *development process*. Hence, it can be understood that capitalism and urbanization are the product of each other and generate externalities in terms of the destruction of nature, accumulation of power, increased social and economic vulnerability, among others. Second, in this sense, the difference with cities is that they are centers of communal life, built not materially but from the need for citizen diversity and were formed with historical collective praxis, according to Sadri and Zeybekoglu (2018).

The city, therefore, is unique, spontaneous and irreplaceable, contrary to urbanization that is conceived as artificial. Thus, according to Bookchin in his book *Urbanization without cities* (1992), he questions urbanization as a process that puts the sense of the city at risk. Along this path, he highlights that pre-capitalist society did not endanger natural and social ecosystems, and the author even calls them *eco-communities* in which social and biological nature were integrated but also included a sense of autonomy and resilience. The impact of industrialization and massive urbanization transformed cities into a sense of agglomeration centers for both buildings and people, which respond to the needs of the capitalist system and, also, to the sense of consumption.

Consequentially, Bookchin (1992) defines urbanization as a process of *social cannibalism* that destroys the city and also its relationship with nature and human interactions, that means the transformation of the city from "ecological societies" into "urban agglomerations of capitalist societies." In that sense, the debate postulates that social values are replaced by material values based on interests of economic gain and consumption. Therefore, citizens become taxpayers, customers or users, opposed to being considered as entities with responsibility and social power in decision-making, which leads to the collapse of the social sphere that is the basis of the city. Then it is imperative to pose this question: if we witness the death of the city or of certain urban territories and if they are caused by the current processes of apparent *de-urbanization* or rather *decitizenization*, what does the loss of the eco-community (city) mean? Added to this is the introduction of the debate on the new spiral and integral logics that are articulated within the current processes of de-urbanization and that drive the transformation of traditional city planning.

In this context, the city, being understood as a commodity, a product of the urbanization process, responds to the logic of capital and thus to factors such as immobility, circulation and speed. According to Harvey (2014) just as light is considered a particle and a wave, capital is a process and a thing, and in the same way, the city is submerged in parallel processes of movement and immobility. From this approach, cities are structured with a fixed capital that is the entire material urban landscape and a circulating capital related to value and its expectation (use and change), leaving aside the sense of community.

The process of capital begins with an object (money) that, in the interest of the capitalist, becomes: workers, means of production, technology and others, which allows to obtain a commodity, which is later transported and distributed, to return to its original state (money) but with a remainder that guarantees the continuity and repetition of the process. The continuity of the flow is a fundamental condition of capital, it must circulate, and the faster it is, the more convenient it is; therefore, speed is essential, which produces certain externalities (competition, illegal markets, power and inequity).

In this sense, to ensure the life cycle of the commodity, two main actions are presented: (i) technological innovations and revolution of production systems in order to speed up the process; and (ii) strategies and shortcuts, such as including intermediaries that speed up circulation. According to Harvey (2014), for this, the capitalist must control the process and the barriers that may arise, since the delay in circulation means that commodity capital becomes dead capital.

In the paradox of movement and immobility of capital, both are complementary. Fixed capital supports the process of circulating capital, and the process of circulation generates the means to recover the value invested in fixed capital. The value of fixed capital is related to its use, when the commodities cease to be useful, they lose value, they stop flowing and they are abandoned, putting their cycle and mainly their income at risk. Then, another variable must be considered, which is the rentier class, which has the interest of ensuring the flow of capital through the surpluses produced from immobility and expectation (Harvey 2014).

In urban territories, the material base for the development of fixed capital is land (fictitious capital), which, being scarce, is monopolized by certain sectors that gain control over the immobility of capital through rent. Therefore, capital manages to influence space and time and produce the right environments for its reproduction. According to Marx (2000), one of the greatest aspirations of capital is the annihilation of space through time. Hence, urbanization processes have been framed in the production of mechanisms that reduce time and put the use and value of space at risk. Such is the case that, the agglomeration of individuals and infrastructures (reduce time), with the migration from the countryside to the city, begins in the first industrial revolution, later the expansion of the territories occurs as a result of the production of infrastructure and connectivity goods. This cycle shows that the increase in the costs of a place due to its success causes the capitalist to seek new geographical spaces, which is called the *creative destruction of capital*.

Starting in the 1970s, the industrialized territories began post-industrialization processes, accompanied by the decline of the central nucleus, managed in a specialized manner, which became obsolete. In addition, due to the stagflation of the 1970s, the Keynesian system was abandoned, and liberalization policies were preferred, which meant making trade and investment more flexible. This was accompanied by events such as the fall of the Berlin Wall in 1989, market policies in China in 1978 and the obvious and progressive arrival of globalization. According to Marcela Darinka, this urbanization process gave rise to complex systems of cities, which break with the sense of hierarchy with unique conditions. The structuring of diffuse territories and urban expansion generated a polycentric and fragmented system, producing diseconomies of congestion, agglomeration and high connectivity costs.

In this sense, capital, space and time are a fundamental equation in the urban production[1] processes of cities from a market logic. But this cycle or process cannot be understood in a linear way of continuous urbanization and also unique for all realities, but certain disturbances in the system affect its rhythm and dynamics in a heterogeneous way. Based on what has been mentioned to support this thesis, it is imperative to highlight three moments. First, prior to the turn of the century, the expectation for technology, the decentralization processes, the continuous post-industrialization processes and certain political crises, generated a loss of confidence in circulating capital (asset at excessive risk), and this led to the investment of surpluses in fixed capital (safe haven assets), expanding and abandoning certain urban territories, mainly in the US and Latin America. Second, in 2008, the pressure for the circulation of surplus capital generated the bursting of the real estate bubble and the production of ghost towns, with a stock lower than demand, mainly in cities in Europe and Asia. Third, in 2019, the multidimensional crisis caused by COVID-19 caused the migration of the urban population to other territories and the acceleration of urban depopulation processes with the rise of digital, platform or even smart cities, in most of the cities globally.

Therefore, the processes of apparent de-urbanization are not a phase that can be predicted in the urban life cycle but can occur within sudden global shocks that rapidly transform cities. But by not responding quickly to a disaster, a structure and its systems can collapse. Therefore, catastrophic scenarios are generated that are not planned and are taken advantage of by the speed of capital. But, in addition, they are not presented as a process only of depopulation of the consolidated city and a logic of new or old territories with the production of greater physical infrastructure in relation to the amount of population (urban gaps, abandonment and degradation), but rather, insufficient and incoherent planning collapses the community and its condition as polis and civitas.

In this sense, five factors stand out to be analyzed in this dynamic: demographic and migration trends (rural, dispersed, decrease); urban gaps (territorial, economic and social); residence-work relation (delocation, allocation, relocation); settlement location trends (com-fused to multiple territories); and outsourcing, commodification and financialization of fixed capital for its flow. As a result, we have new urban landscapes[2] such as: obsolete and abandoned territories and centralities; ghost towns and; digital and multifunctional cities. Scenarios where the opportunity can mean establishing justice and law, or, on the other hand, producing deregulation with a game of interest and power.

So if urbanization is capitalist, de-urbanization can be the loss of capitalist production. That is why it is necessary to separate the processes of urbanization from the processes of de-urbanization, but also to contrast the loss of citizenship rights or the loss of the city, which means both the loss of the reproductive sense, social interaction and with the natural environment. Therefore, do urban shocks deepen urbanization

---

[1] It concentrates on the productive and abandons the reproductive sense of cities.

[2] Chaotic (urban shock) or collective opportunity (urban revolution).

processes that mainly generate decitizenization? So, do we lose the city more than the urbanization? And, is de-urbanization an opportunity for an urban revolution?

## 37.2 Shock and De-urbanization. Cycle or Rupture?

Urbanization processes have been directly related to production systems, we can highlight three main phases: (i) 1st. Industrial Revolution, with Taylorism a process of migration from the countryside to the city was generated, late in certain regions, but present in all urban areas; (ii) 2nd. Industrial Revolution, with Fordism the nuclear city and its primacy and hierarchy are produced; (iii) 3rd. Industrial Revolution, with Post-Fordism, new market and urbanization logics began, with unprecedented expansion and the formation of global urban systems (Carrión and Cepeda, 2020). Thus, urbanization processes are defined as the phases that urban territories (cities) go through, which are initially related to concentration and later to dispersion.

Accordingly, urban territories are entities in constant transformation, changing and articulated according to economic, territorial, political and social conditions. In this sense, it is questioned whether urbanization processes are linear as they have traditionally been established. Since the industrial transformation, post-industrialization, the rise of technology, the logic of accumulation, the migration of financial sectors and finally the pandemic, the changes have been profound and accelerated. Therefore, urbanization processes have triggered logics of dispersion and consolidation of territorial units, different from those foreseen or experienced.

The limited assimilation of the meaning of urban systems anchored in globalization puts at risk the planning of cities, understood as unitary and unifunctional organisms and not as a network within a system. In this sense, Jacobs (1992) analyzed the social consequences of the mono-functionality of territories and how this trend can lead to their failure. Condition that was evidenced in the cities of the US and Europe since the 1970s with the crisis of industrialization, but that has also been part of the processes of Latin America, demonstrated in the emptying of its founding centers.

Thus, a new process within the development of cities begins with the collapse of industrial society and the impulse of the neoliberal model. Explained in three main theories: (i) population movement (intra-urban, inter-urban, regional); (ii) evolution of metropolitan areas (urbanization, suburbanization, de-urbanization and re-urbanization); and (iii) urbanization cycle (absolute centralization, population and relative loss, absolute decentralization, population and relative loss) (Ferras 2007; Van den Berg et al. 1987). The third is associated to the relation between the center and the periphery; the second, to the expansion of territorial crowns; and the first, to the sense of rank-size hierarchy within an urban system. Three conditions that in the current urban reality do not become determinant.

The transition from the growth trend of the most urbanized cities to other territorial units of the same system is called *reverse polarization* or *de-urbanization*. According to the three theories, these transformations respond to linear cycles related to social, economic and political conditions. But these patterns can be broken due to sudden shocks, where opportunities are created for certain sectors or actors to implement what Friedman called *ideas that have been around already (shock)*. The urban shock breaks and/or accelerates the urbanization trend (all the same, everyone differently), where what remains is not re-urbanization but urban decline by not proposing adequate de-urbanization policies and plans. Therefore, the urban shock does not generate a new logic to plan, but the massive expansion of urbanization, and with it comes the loss of the city.

## 37.2.1 The Urban Shock Doctrine. Blank Slate or Ruins?

According to Naomi Klein (2007), in the 1950s, psychiatry, in order to treat patients not from a traditional therapy, that is, the cause of their psychological damage, but from their fears, experimented with shock therapy. The goal was to enter the minds of individuals through external mechanisms and break their old patterns of behavior to finally mold them. To do this, the first step was based on eradicating patterns (through torture) producing a first shock, this meant the disturbance in everything that was apparently normal, to reach what Aristotle calls a *blank writing table*, that is, the loss of memory. The objective of reaching a blank slate meant erasing the history of someone's life (memory), but that was not easy to achieve and, on the contrary, what was obtained were *minds in ruins*. Subsequently, the second moment was to implement a behavior, but by not obtaining a blank slate, the reconstruction did not obtain the expected results, and this generated a new and final shock.

In the 1960s, the Central Intelligence Agency (CIA) saw in shock therapy a mechanism to torture and obtain information, which is why it financed the investigation (Klein 2007). What they were looking for was to generate a *mental hurricane* and cause the individual to collapse, by producing a psychological shock. In this process, the experts found the precise moment between the rupture prior to the shock to obtain their objective. Evidencing the search for the loss of two factors that allow us to recognize who we are and where we are in a temporal and spatial way: our memory and our source of sensory information. This is what shock therapy aims to annihilate.

In the same way that the shock transformed an individual, Milton Friedman sought to generate a collective shock to transform the system. According to Naomi Klein (2007), Friedman sought to establish pure capitalism by removing all existing patterns from society. This capitalism must not have tariff barriers or state regulations. To do this, the first moment is to take advantage of a catastrophe (produce a blank slate), to later generate an economic shock (rewrite), but finally, the consequences were accompanied by a final shock that is the force of resistance (metamorphosis). This is the basis for disaster capitalism, for which the moment after a (social, economic, natural) disaster is an opportunity, mainly for the market and capital to find new ways of flowing.

In the development of this doctrine, in 1947, Friedman and Hayek established the Mont Pelerin Society (MPS), which defended the free market based and argued that it would be the solution to the Great Depression of 1929. At that time, Keynes, on the contrary, managed to establish a system based on state intervention to ensure the economic cycle. The depression generated a surplus evidenced in unemployment, so according to Keynes the state should increase its spending to motivate investment and thereby reduce the surplus. But in the 1970s, a new crisis generated a change in the system to a neoliberal trend, where Friedman was trying to return to the freedom of the market with the following actions: eliminate regulations (deregulate); sell public assets (privatize); cut social spending (cuts); and ignore borders (globalization).

The evolution of this doctrine was structured with the training of economists in the Chicago School (Chicago Boys). So, to be able to introduce these economic recipes, the most desired thing was a crisis or a disaster, which was the moment of opportunity based on fear, where ideas that were already around were launched. For this, the first experimental laboratory was Chile. In 1970, Salvador Allende takes the presidency of Chile, which worries foreign capital and large capital that invested in the country. For this reason, in 1973, with a coup d'état, Pinochet became dictator, and, with the advice of the Chicago Boys, he launched a project that lasted almost a decade. Initially, the shock meant taking advantage of a sudden contraction, with a death squad and generating as much collective fear as possible, to later establish measures of privatization and market freedom. But in the 1980s, the crisis was even more intense, and Pinochet had to turn back, nationalizing companies and taking public money to pay even private debt (corporatist state). The result was: the widening of the gap between classes, with the enrichment of the elite (inequality) (Klein 2007).

Furthermore, on September 11, 2001, terrorists hijacked four planes that hit the Twin Towers, the Pentagon and a common field in Pennsylvania. The event meant the death of more than 3000 people and sowed deep fear in the US and the world (collective shock). This resulted in President Bush launching a war on terror with the aim of exterminating external evil and preserving the protection of the territory (nationalism). Then an industry of weapons, security, armies, technology and reconstruction (disaster capitalism) was developed, giving way to the accumulation of power and control of this sector with public investment. This shock in the US spread and mainly generated a series of different impacts in Iraq and Afghanistan, managing to implant fear and take advantage of certain resources for the enrichment of the elites in any part of the world (globalization).

The shock doctrine has not only become relevant in the study of economic transformations of a neoliberal type from strong impacts in certain territories, but it is also part of political and social research, as is the case of community psychology studies (Rojas et al. 2021), culture shock (Newsinger 2015), future shock (Toffler 1972), climate shock (Wagner and Weitzman 2016), among others. In these studies, it is understood that the shock allows to deepen neoliberal recipes impacting not only economically but also on the society, nature and politics. This builds a sense and expectation of fear and danger in the face of sudden impacts or changes, seeking transformations with previously established patterns.

Then, there are three holy grails that according to the shock doctrine Friedman was looking for: globalization, privatization and deregulation; accompanied by a reduced state and free enterprise (Klein 2007). With the argument of producing an apparent development but that results in: an increase in the gap between rich and poor, an increase in defensive nationalism and a panoptic society. Therefore, paradoxically, more reactionary and conscious societies are produced, where a resistance to shock generates a metamorphosis in the systems, since the blank slate and behavioral change are not obtained and therefore the systems collapse.

In this sense, the urban shock is to take advantage of a sudden impact (economic, social, natural, etc.) in urban territories that rapidly transform their current conditions, within others that were already preconfigured (Carrión and Cepeda 2020). This sudden impact is not premeditated and is presented as an opportunity that is taken advantage of by certain sectors and powers that transform cities. The first moment is a disaster (natural, economic or social), the second moment is to establish actions that produce the urban shock (transformation of the urban landscape and the search to implement a behavior, from certain interests and powers) and finally, the resistance arrives (New logics). Sudden impact management fails to establish a blank slate in cities, in fact, it generates ruins, decay and abandonment. Memory or history cannot be erased, on the contrary, what is produced, without proper management, are vestiges that continue to be part of the city. Traditional planning, by regulating in an unnatural way from a utopian society and in a market logic, causes capital to abandon and expand its territories but also causes society to collapse. The territories resulting from the shock (abandonment and decline) cannot be erased and produce externalities in the urban configuration (social, political and territorial).

To understand this phenomenon, we have three disasters and their shocks to analyze. At the turn of the century, the rise of technology generated an unbridled interest in investing and generating companies in this sector. In 1993, the first browser called Mosaic appears and Netscape in 1994, which a year later becomes public, that is, its shares are sold on the stock market. With this context, a war of browsers and websites is created, their usefulness was not of interest, and the only thing that mattered was that they could grow rapidly. According to business logic, the merchandise depends on the process to produce and distribute, but in the stock market what matters is not the production but the expected value of the merchandise. Therefore, technology companies became Public Offer for Sale (IPOs) and from 1995–2000 the NASDAQ index tripled as shown in Fig. 37.1. But in that year, a generalized fear about the turn of the century began to haunt, in which at the beginning of 2000 meant measures that generated the inflation of dotcom shares, accompanied by the recession in Japan and the monopoly lawsuits against Microsoft, which caused the fall of shares up to $-39\%$. The dotcom companies that survived the bust found advantages to grow: cheap real estate, a growing and cheap workforce and less competition. This generated a monopoly, the stability of technology and that it is currently the best valued.

# 37 De-urbanization: From the Shock to the Revolution of a New Urban Logic

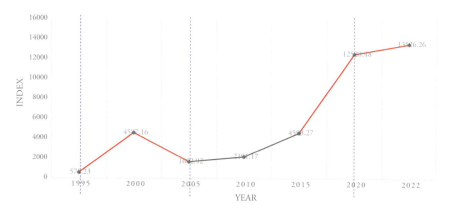

**Fig. 37.1** NASDAQ index. *Source* Own elaboration based on data from Nasdaq Stock Market

In 2008, the real estate bubble that began to develop in the 1990s burst, mainly impacting the most developed countries. After the 9/11 tragedy in 2001, Bush established market deregulation measures, lower interest rates and taxes and credit expansion. This generated a real estate boom with a global effect, added to the financial model that was put into operation, since the banks found a mechanism to deliver a large number of loans despite the credit capacity of the debtors, which, according to experts, meant large private gains at the cost of public loss. At the beginning of the 90s, a financial innovation was produced based on the explosion of market derivatives, which were not regulated. Therefore, there was a transformation in the lending system where there was only a borrower and a lender, introducing securitization, which meant that in the lending process there was a home buyer (borrower), a lender, a bank of investments and an investor. So, when a person was looking to buy a home, they went to the bank, which took the money from the lenders and created a combination of loans called Collateralized Debt Obligations (CDOs), which were rated by rating agencies and sold to investors within the stock market. In this way, the payment of a mortgage ended up in investors from anywhere in the world and the bank was an intermediary. The bubble process took time; according to Domínguez (2009), from 1998–2005, it was created and evolved, and from 2007–2009, it burst throughout the world, with certain moments of transition and apparent calm as can be seen in the Graph 2. What happened at first was that a large number of mortgages were granted and later the price of housing began to rise drastically, but simultaneously, the banks also borrowed money to buy CDOs, a process called leverage. At the same time, American International Group (AIG), the insurance company, was selling not only CDO insurance policies, but also default swaps.

Then another innovation was generated, while the traditional insurance industry could generate an insurance for a good, and the universe of derivatives allowed a good to have several insurances and several people who insure it. What, in reality, meant betting on the payment or non-payment of the mortgage, where AIG had no other option but to pay and they went bankrupt, which many organizations began to anticipate since 2005. Since the banks owned CDOs and swaps, they sold and lent, but also, they bet on their failure, earning an AAA rating from credit rating agencies. In 2008, the bubble collapsed, and lenders and investors went bankrupt, and Lehnman Brothers, the bank with the most CDOs, ran out of cash and went bankrupt at the instruction of the US Federal Reserve, seeking some calm within the collapse of the system. Here, the governments had to enter public money to save the banks. This crisis produced: evictions, increased unemployment and reduction of social spending, this meant the collapse of other sectors and the expansion of the crisis throughout the world (Fig. 37.2).

At the end of 2019, a viral contagion began in a market in China (Wuhan), and in less than four months, it spread throughout the world, causing the closure of land boarders and confinement of the population. In November 2019, the first COVID-19 patient appeared, accompanied by a first phase of denial, where all governments saw it as something oblivious to their realities and no measures were taken in this regard. Gradually, the virus spread to Thailand and Japan, and in January 2020, it reached the US, Europe, later Mexico and finally South America. Then, by March 2020, the World Health Organization (WHO) declared COVID-19 a pandemic, establishing protocols and recommendations. It was until then that countries began to close their territories and confine the population, canceling public and massive events and restricting people from leaving their homes, plunging life into the virtual world. The contagion and lethality rates began to rise drastically and the capacities of the health system in several territories collapsed, while other territories installed the necessary infrastructure in record time. With a strong debate on health measures or economic measures in the face of the crisis, high percentages of public funds were allocated to its management; at the same time, the world's stock markets began to contract, oil fell to -$37.63 a barrel and began an inflationary spiral. At the same time, with an accelerated pace by December 2020, the first people were vaccinated, and plans were launched in all countries, but with initiatives such as Global Access Funds for Vaccines (COVAX) that guarantee the equitable distribution of vaccines among poor and rich societies. Finally, the economies have not recovered, and the confinement meant the transformation in the lifestyle of everything.

In all cases, the disaster generated a shock that meant the collapse and loss of utility of certain fixed capitals and with it the movement of flow capitals and also an opportunity for power sectors. For Marxism, the balance is achieved from the revolution and for the Chicago Boys from shock. So the use of these crises on an urban scale resulted in an urban shock that did not generate a blank slate, but ruins. On the way to rewrite, the capitals find no interest and abandon the ruins, while the local government deregulates and tries to implement an inconsistent policy and planning with the new scenarios as a corporatist state. Finally, societies in shock renounce the values they defended and adapt among the ruins. In the last shock of resistance, the city looks for solutions to the collapse from a logic of restructuring its memory and senses.

So, while the economic shock seeks deregulation, privatization, globalization and cuts, the urban shock is heading for commodification and financialization (flexibility), primacy (contradiction and local incoherence/hierarchy—global/urban

37 De-urbanization: From the Shock to the Revolution of a New Urban Logic

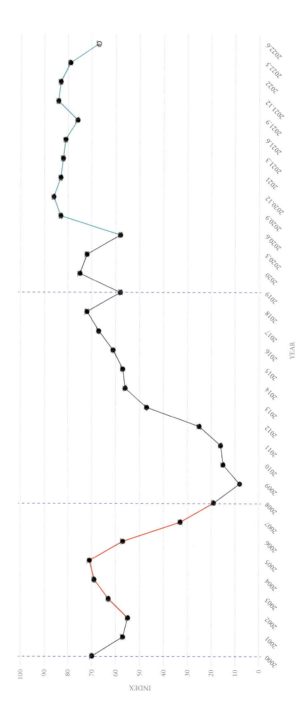

**Fig. 37.2** Nahb index. *Source* Own elaboration based on data from Nahb

system), centralization and monopolization (statization and privatization). With a result of unequal and degraded, ghostly and highly competitive territories (winners and losers), the shock does not interrupt the urban cycle, but rather accelerates and alters it, generating greater uncertainty and confusion. The urban shock transforms processes that had already been preconfigured, but the impacts are not the absolute death of cities, because from their sense of citizenship rather than urbanization, they seek to resist.

## 37.2.2 City and Urbanization. Historical Construction or Human Caprices?

Urbanization is, on the one hand, the mass production of private, fragmented, homogeneous and powerful spaces, which organizes the daily life of citizens in an exploitative, exclusive and hierarchical way. Lefebvre (2020) calls a process of space abstraction, which transforms collective and natural life with lower use values, into high exchange values. Then the urbanization processes are the representation of capitalism. But from other points of view, urbanization is a process of economic and social development that allows structuring centers of innovation and progress (Florida 2005).

In this contrast of value of use and value of exchange, cities are born as the container of exchanges and relations. So much so that Lefebvre (2020) calls the city an *oeuvre* (work of art), to which citizens have the right to use and appropriate, which is inclusive and collective. But, in addition, cities are a historical construction and a human creation of community, which becomes the scene of social transformation that is why according to Bookchin (1992) cities are not less important than their citizens, for which a balance between the city, its environment (nature) and its citizens is essential, through an institutional control that supports the rights and protection as a whole and its elements.

When capitalism and urbanization transform the city, according to Topalov (1979), it becomes a set of production units and the result of the division of labor, and then its use value is contained in its productive and creative force, where the need for movement of capital, according to Marx (2000), generates a constant transformation of productive forces (systems, resources, work, space and time). This transformation results in a process of adjustment of the set of production units, both in their configuration and in their cooperation and interaction. This has been a historical construction process of grouping and separating of production units. So, in the production and urbanization relation, the stages were articulated in: simple cooperation (grouping in the city); manufacturing (separation of tasks and expansion of territory); large industry (monopoly and territorial hierarchy); automation and technology (division of labor and platformization).

As from 1900, society transitions to a world of cities, which are submerged in processes and transformations, are as shown in Fig. 37.3. In 1970s, in North

37 De-urbanization: From the Shock to the Revolution of a New Urban Logic

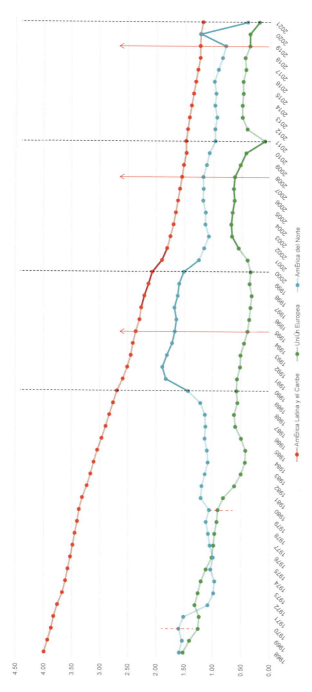

**Fig. 37.3** Urban population growth. *Source* Own elaboration based on World Bank data

America, with the cessation of industrialization, an urban decline process begins related to suburbanization; while in Europe at the beginning of the 1980s, a process of urban deconcentration took place, and finally, in Latin America from the turn of the century a different process of urban development began toward metropolization. These processes are heterogeneous, temporarily distant but directly connected, since they begin with the development of the neoliberal model and the end of industrial society.

So, urban territories enter the logic of urban systems as a result of complex and decentralized processes of urbanization, which, in the post-industrialization, evolved into new compact and dispersed, polycentric, dense and empty structures. Three theories manage to summarize the way of analyzing the urbanization process from its cycles. The first theory, according to Van der Berg et al. (1987), mentions that there are four states or phases in the evolution of metropolitan areas: urbanization, suburbanization, de-urbanization and re-urbanization. The latter being the desire and argued utopia in avoiding urban ruin. In that moment, urban centers expand into new crowns that have now become multiple and imperceptible.

The second theory, according to Hall (1981), states that the five stages of the urbanization cycle are based on a relationship of rank and size, within the urban system, where, despite the hierarchies, they are all related. Therefore, each unit is in different phases, and they are part of the same process. The main city begins with an absolute centrality and the other territorial units lose population. Subsequently, the centrality becomes relative and starts a decentralization process where the other units gain centrality.

The third theory expresses that these processes, according to Ferras (2007) and Camacho Ramírez et al. (2015), are explained from the population movement and the new logics of population migration, where the center loses its use value. The first crown establishes the suburbanization that grows linked to the loss of population in the center and integrates new territories with conurbation. In a subsequent moment, a deconcentration and an absolute decentralization reverses the development trend toward rural territories within the urban system, understood as de-urbanization or counter-urbanization.

These phases can be understood as the uncontrolled and massive expansion of urbanization, resulting in discontinuous territories and what happens is the loss of the city, due to unregulated and incoherent urban legislation, where the movement of capital is prioritized over its use value. According to Pedro Abramo (2009), framed in a game of market speculation that is not perfect, there are urban noises that Keynes (1987) calls *human caprices*. That is, there is no rational decision, as established by the orthodox economy in urban production, but uncertainties and disorder are presented.

The movement of capital in the city generates a game of speculation that is the same as the expectation in the stock market, where variables such as caprice, sentimentality and chance (Keynes 1987) generate a kind of uncertainty. This produces a fragile system that can collapse, but to reduce uncertainty and reduce the damage of these variables, economic policies and institutions must be managed so they can generate control over this game, or as Keynes calls *common sense rules* to alleviate irrational nature (Montalvo 2003). According to Moore's philosophy, common sense is based on accepting as certain the existence of material things (time and space) and acts of consciousness. In that sense, this breaks the idea of rational individuals and Keynes replaces it with a real irrationality.

Therefore, the city creates two fundamental conditions in its configuration: a space for interaction, sociability and collectivity; and time of experience and development, being as old as humanity itself. The city gives rise to citizens and according to Patricia Ramírez (2007) one of the problems that arise in social participation is the relation between citizens and the market. This opens a debate, if the citizens contradict themselves, is a tension or support the market and capitalism. But, in addition, the citizens, being subjected to different historical processes, structure a heterogeneity in their configuration. In an urban shock, citizenship is transformed without erasing its history, but at the same time they respond to powers and interests.

According to Weber, when the focus was placed on the cities, the era of the cities ended, then the new territorial conditions that generated regions, areas and urban systems did not become clear, and according to the author, this lack of distinction generates the crisis of the cities. The urban life initially destined for a specific space is now diffused in urban and rural areas, the condition of the city and its citizens is transformed, thus Weber (1997) mentions that the cosmopolitan appears, which is the citizen of the world integrated from any part around the globe, but at the same time, a pre-industrial native also appears who is the left behind citizen resulting from the new logics of migration and spatial configuration.

Returning the use value over the exchange value to the city, distances it from the capitalist urbanization (caprices) that currently monopolizes urban territories and restores citizenship (memory). Therefore, it is under debate to recognize the temporary (historical) and spatial (de-urbanization) condition that occurs and plan with these variables, contrary to taking advantage of and generating a shock that deepens pre-existing problems. But that does not ignore the new logic to which we must respond, transforming traditional urban planning and policies (zoning and consolidation), to be better prepared for crises and shocks. So it should not be ruled out that there is obviously a losing group in this dynamic and also an opportunity to rebuild the urban system, from the cities. In practice, recognizing the city from its temporality and its social and spatial role means avoiding the current results of urbanization related to unregulated capitalism, product of uncertainty that generates social consequences and a metamorphosis of the urban area.

## 37.3 De-urbanization and Decitizenization. Production or Waste? Process or Flow?

The sudden transformations of global impact have generated new urban territories with highly diverse social tendencies. A specialized group with the ability to locate in multiple territories and a highly excluded group suffers constant expulsion. This produces attractive territories and others that are in decline, which can be a temporary or permanent phase, while the territory begins to merge into one (rural and urban). In this sense, the flow of capital generates new urban centers, and it expands the existing ones and abandons those that become obsolete.

It should be noted that the demographic growth trend of the world population is undergoing new dynamics and is on decline. This is due to the better condition of certain services, but in a segregated and mostly unequal manner, to the aging of the population and to the decrease in the birth rate. Thus, urban contraction is presented as a massive loss of population due to a social, economic, political or natural situation that generates a change in the system as a product, and consequently, of uneven development. As a result, there is an urbanized overstock that generates an increase in disuse and abandonment of land and urban infrastructure. But in addition, the de-urbanization processes include not only the consolidated urban territories but also the new ones that were planned in a boom, and when there are failures, they go into a contraction.

There have been several theories of de-urbanization or counter-urbanization, the most prominent are: (i) clean break or rupture of the past, which implies the abandonment of metropolitan areas to improve the quality of life and also, with post-industrialization, new service economies are produced; (ii) rural–urban continuum or urbanization of the countryside (related to spillover or urban spill), where there is a relation between the countryside and the city in which rural activities change to urban activities in rural territories; (iii) Cloke's rural perspective or rural attractors, where quality, price and other community and environmental factors attract the population; and (iv) urban expansion, a product of urban polarization and population reduction in central areas (Ferras 2007; Berry 1976; Mitchell 2004).

In this sense, these theories allowed us to describe de-urbanization processes such as: in 1937, in relation to the dynamics of North American cities, urban decline appeared; in 1970, in relation to industrial decentralization, shrinking cities appeared (Schlappa and Neill 2013); in 1987 Schrumpfende Städte appears in Germany, accounting for the demographic, spatial and economic decline of large cities related to industrial location. These dynamics are related to monofunctional linear growth as a result of consumerism and the market that subsequently seek to outsource their surpluses to the peripheries.

Thus, there are two trends: (i) A hyper-urbanization or massive urban expansion occurs throughout the territory, without limits, accompanied by decitizenization; while from a legal and ecological point of view, (ii) deconcentration makes it possible to generate coexistence with the rural and natural environment and can mean incorporating collective rights. In both cases, understanding and planning cities from their de-urbanization processes means going with reality and also the opportunity to introduce the social function of urbanism.

The impact of disasters and subsequent urban shocks can be observed on different scales, here we are interested in two. On a first scale, the growth of the urban population of regions and countries has gone through times of contraction, as can be seen in Graphs 3 and 4, suffering a disturbance in the analysis periods. While on a city scale, it is the oldest centralities and the consolidated areas that are the most affected, experiencing new logics that expel or attract, but directly articulated to the production of new under occupied cities and dispersed urban territories.

From 1990 to 2000, the technological trend generated growth in both the urban population and the Nahb index, mainly in US cities, accompanied by a period of contraction, as shown in Graph 4. In addition, decentralization and political and economic crises in Latin America generated the progressive decrease of centralities accompanied by the rise of new business centers, once again specialized and mono-functional. On the other hand, the bubble of 2008 transformed the cities mainly in Europe, having an important moment of construction boom and population growth, to later suffer a decline. Finally, COVID-19 produces a global urban shock with a sudden decrease in urban population accompanied by a multidimensional crisis (Fig. 37.4).

Although certain countries may experience growth and development, several of their urban territories are in decline, and then disasters become shocks where societies unexpectedly produce not only the movement of the population, but also the disuse of infrastructure and urban land abandonment. So, this can be understood as a process where the loss of population generates a reduction in the residential market, the use of services and public infrastructure and produces underutilized territories with the perspective and expectation for local governments to act and decide whether to demolish, redensify or take advantage of and generate a balance with the ecosystem and society.

From this, the following dimensions can be synthesized to analyze the processes of de-urbanization and decitizenization:

- Demographic trend and migration (rural, dispersed, decreasing)

The current demographic trend is the result of migration to more prosperous territories with a better quality of life, accompanied by the aging of the native population and the decrease in the birth rate. Thus, the developed and apparently successful territories attract what Richard Florida (2005) calls the *creative class*; this class causes the development of large cities and for this they need, according to the author: technology, talent and tolerance. This phenomenon produces an increase in inequality, by becoming a knowledge elite. For this, it must be taken into account that the territories compete to be attractive for this development, and as a result, we have the expansion without borders of the power of large corporations and financial capital, in addition to the endless extraction of resources.

Subsequently to urban shocks, the result was the bankruptcy of certain groups, and this caused a decrease in competitiveness, so the sectors that survived became stronger and with more opportunities to expand. Thus, after the dotcom real estate bubble and during the pandemic, the least affected economies, with their creative classes, had the fastest capacity to adapt. So, technology quickly managed to position itself as a monopoly and currently as a new public space of relevance in cities. In addition to this, territorial and international conflicts have increased migration trends and the movement of the population to areas with a better quality of life. Finally, the world population is aging rapidly, accompanied by downward trends in births that are presented as a new urban challenge. It should be noted that despite the fact

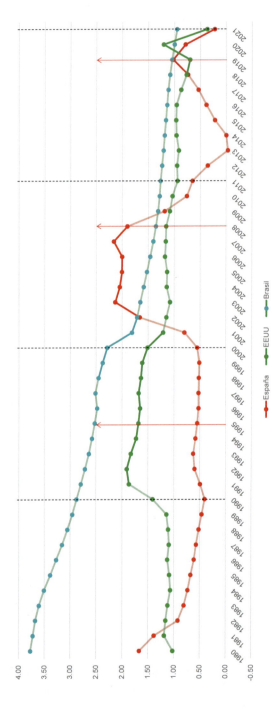

**Fig. 37.4** Population growth Spain—USA—Brazil. *Source* Own elaboration based on World Bank data

37 De-urbanization: From the Shock to the Revolution of a New Urban Logic

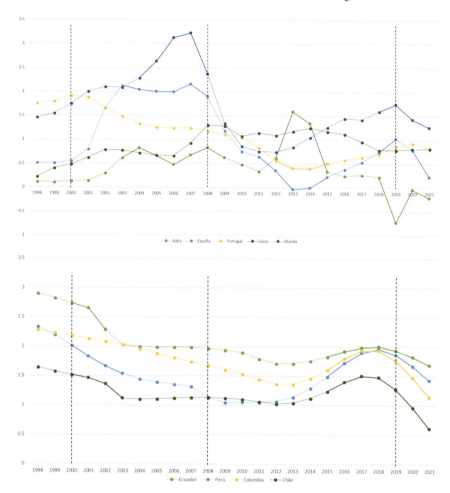

**Fig. 37.5** Population growth countries Europe and Latin America. *Source* Own elaboration based on information from Work Bank data

that the urban population is increasing, its growth is declining, and three moments demonstrate these sudden changes: 1990–2000, 2001–2008 and 2019–2022.

Thus, as a result, we have hyper-urbanization, with an unfolding of capitalist urbanization and the decrease in urban population that means the increase in the loss of community, where citizens become aliens and foreign. While de-urbanization means new logics of proximity, superimposing the reproductive before the productive, where the new demographic trends transform traditional planning and go from zoning to functional diversity (Fig. 37.5).

- Urban gaps (urban, economic and social)

The moment when cities and urban territories stop growing demographically, unoccupied spaces increase, evidenced in vacant lots, underused or abandoned buildings. This produces certain negative externalities to the city. First, it increases the cost, without citizens paying taxes and the maintenance of services that are in disuse; in addition, the emptying is accompanied by insecurity, decadence, etc. According to Berruete (2017), there are three types of gaps: urban, which are part of the urban fabric; economic, related to production sectors such as industrial zones; social, as enclaves for temporary use; and gradients. Without proper planning and policies that assimilate these processes, the result is a long wait to put them back into use.

Urban shocks generate these three types of gaps, and it is not even the product of a historical process of the city, but of a new creation. In the real estate bubble, large residence enclaves were created, in addition to new cities that, after its bursting, were left abandoned as a result of evictions or that were never acquired by anyone. In Spain, one of the cities with the highest number of empty spaces is Yebes, as can be seen in Fig. 37.6, which was a new city between the community of Castilla-La Mancha and Madrid, planned for 10,000 inhabitants and currently has approximately only 5300 residents. But this is a phenomenon that is observed in several cities, where they maintain a tendency to contain buildings and obsolete areas where traditional planning leaves ruins, while from a logic of de-urbanization, new public spaces are considered and used to improve the quality of life of cities (Cañizares and Rodríguez 2020).

- Work-residence relation (delocation, allocation, relocation)

The urban, demographic and physical transformation is accompanied by new social behaviors that cannot be controlled. During the COVID-19 pandemic, the city became the center of contagion; agglomeration, public space and transportation directed and forced a trend of life into the virtual world. The urban population was mobilized to smaller cities and rural territories according to their needs. The expansion of technological services in this sense opened up a privileged space, and this deepened an inequality gap for the population that had the conditions to adapt to that reality. Thus, the constant relation between work and residence, which is the basis of the urban configuration, mutated to new logics.

According to Carrión and Cepeda (2020), these trends are summarized in three: relocation, allocation and delocation. Relocation of consumption and production activities from physical to virtual space, thus the productive world enters the domestic (reproductive) one. Delocation of work and services to other cities or other urban peripheries due to the ability to locate remotely to carry out an activity anywhere in the world. Allocation of work and services, due to the digital nature of global integration, allowing access to goods and services beyond national borders.

This creates a new group within the creative class, called *digital nomads* who provide services through digital jobs, with high quality and low cost. This way, the urban shock gives way to globalization and the submission of all aspects of life into the virtual world. This type of work became specialized and globalized, overcoming

37 De-urbanization: From the Shock to the Revolution of a New Urban Logic

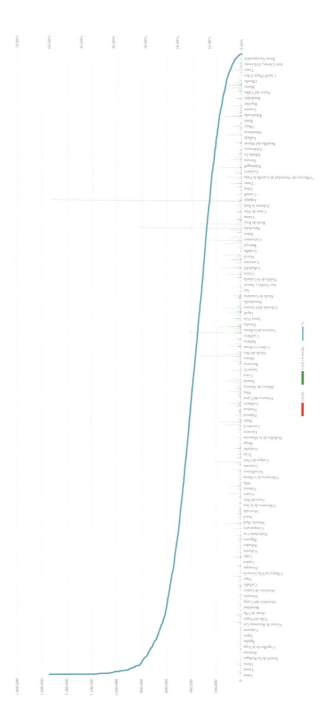

**Fig. 37.6** Empty homes by municipality in Spain. *Source* Own elaboration based on data from the National Statistics Institute (INE) of Spain

its initial dynamics of a strictly local and zoned nature. Therefore, from urbanization, we continue in a logic of planning by zoning, which goes against reality, while from de-urbanization, new logics of spatial production are introduced to which the variable of the virtual world is added (Fig. 37.7).

- Settlement trends (confusing to multiple territories)

The new conditions of development produce new trends in the location of citizens, urban flexibility allows freedom of location in small cities, large cities and rural areas with urban conditions. The territories expand beyond their metropolitan areas and their traditional urban–rural separation; this way, the configuration of urban regions raises the importance of planning from their urban systems. The big cities from their great success expel a large amount of population, this means a large workforce and here people manages to find better conditions in smaller areas located near development centers.

During COVID-19, for example, US cities have shown changes in terms of population location trends, such as the case of New York with a decrease in population in its State and city, unlike smaller and closer cities as is the case of Sleepy Hollow Village, which has had a progressive increase, as can be seen in Fig. 37.8. The urban location trend is not only in new urban rings (peripheries), but also in new and old territorial units that are attractive for the urban population, but that are articulated to the big cities due to their agglomeration of power. This means that large urban centers contain offices and shops derived from industrial centers in other cities and with a creative class that lives in other territories. This dispersion, without being channeled into a kind of urban system, generates significant economic and social losses for cities. From the processes of urbanization, this focuses on observing urban territories from their hierarchical condition, while de-urbanization analyzes the local and its interaction with the entire global urban system.

- Outsourcing, commodification and financialization (fixed capital for capital flow)

The outsourcing of the economy transfers what is productive from the industrial sector to the service sector. Thus, a strategy for local governments after the industrial slowdown and their limited ability to modernize their economy is to generate apparent economic growth from real estate development. For this, local governments become owners of the land but also monopolistic providers of urban land, so they expropriate and sell to real estate developers.

During an urban shock like the one in 2008, the real estate sector could grow beyond real demand, causing municipal governments to go into debt to adapt to that growth. With the bursting of the bubble, the cities were abandoned and were up for sale, despite being a business of local governments, with the lack of tax revenue enters a spiral of attrition and crisis. This way, the production of new cities and urban expansion is justified in the process of constant growth that means nothing more than the source of power conglomerates. So urbanization continues beyond the demand as well as that real dynamic of depopulation that cities go through. As can be seen

37 De-urbanization: From the Shock to the Revolution of a New Urban Logic

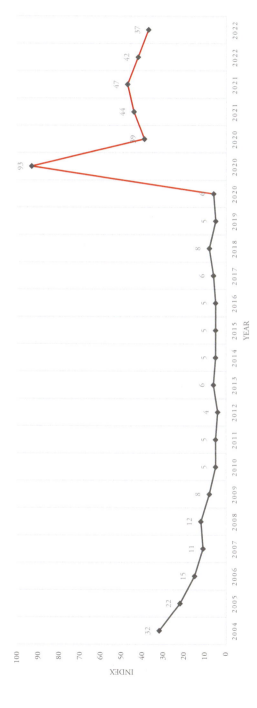

**Fig. 37.7** Remote work worldwide. *Source* Own elaboration based on Google Trends

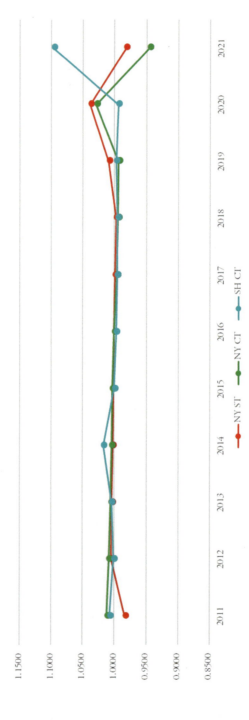

**Fig. 37.8** Population growth New York State—New York City—Sleepy Hollow Village. *Source* Own elaboration based on the United States Census Bureau

in Graph 2, the expectation of this sector has grown more than in the evolution stage of the 2008 bubble.

These dynamics result in new highly competitive and unequal urban landscapes. On the one hand, platform cities or digital cities stand in contrast to smart cities that focus on technological power. On the other hand, the ghost cities that are territories that work with less capacity in relation to those that were planned. Finally, urban territories in degradation and abandonment that have unfortunately become ruins.

## 37.4 Conclusions

The fragility of urban systems in the face of natural disasters, social inequities and their dependence on external resources is increasing and has always been present. Urbanization is a process questioned by the way it destroys nature and reduces the habitability of certain citizens. Therefore, certain sudden ruptures are presented as opportunities where urbanization breaks its traditional life cycle and can deepen (shock) or reverse (revolution) all the social and natural impact it produces. According to Abramo (2009) and Ostrom (2014), the market is not the only entity that can regulate society, because it has a historical construction, and despite the idea of a free and rational market, institutions play a fundamental role in regulating them.

So, how can we plan cities from their de-urbanization processes? Can cities be prepared for sudden transformations? De-urbanization can be an opportunity for urban balance and institutional reconstruction that establishes the right to the city. Several scenarios may arise along this path. A first path in which the negligent public action does not recognize the de-urbanization processes and in a shock attempts a transformation of behavior from a blank slate, obtaining more unequal and degraded territories. A second path in which understanding the reasons for the contraction and the new logics of adaptation allow policies and planning to be established that transform the city of the market into a city that has rights.

The urban shock can become an urban revolution that transforms power response tendencies into adaptive tendencies of society, with a sense of social and natural cohesion. Return *the right to the city*, where it is recognized as a living entity, to which a disaster does not turn it into a capitalist opportunity, but rather a citizen reconstruction. Taking action and using urban gaps and planning with a sense of an urban system beyond the urban–rural boundary are actions that must be controlled by the city government. Return the use value and restrict the power of flow capital to adapt to economies of sustainable production and consumption, with a spiral and nonlinear system of use and disuse. Finally, the disaster channeled as an urban shock generates expansive urbanization or hyper-urbanization, while as a revolution, it enters the logic of de-urbanization processes that allow justice and law.

# References

Abramo P (2009) La ciudad calidoscópica. Apuntes del CENES, pp 125–196

Berruete Martínez FJB (2017) Los vacíos urbanos: una nueva definición. Urbano 20(35):114–122

Bernardos Domínguez G (2009) Creación y destrucción de la burbuja inmobiliaria en España. Información Comercial Española. Revista de Economía ICE, 2009, num. 850, pp 23–40

Berry BJL (1976) Urbanization and counter-urbanization. Sage Pub, Londres

Bookchin M (1992) Urbanization without cities: the rise and decline of citizenship. vol 171. Black Rose Books Limited

Camacho Ramírez MD, Hoyos Castillo G, Sánchez-Nájera RM, Tapia Quevedo J, Franco Ayala R (2015) Estructuración metropolitana de Toluca. Los impactos de la desurbanización en el territorio en el periodo 1990 a 2010.

Cañizares MC, Rodríguez-Domémech MÁ (2020) "Ciudades fantasma" en el entorno del Área Metropolitana de Madrid (España). Un análisis de la Región de Castilla-La Mancha. EURE (Santiago) 46(139):209–231

Carrión Mena F, Cepeda Pico P (2020) Ciudades de plataforma: La Uberización. Revista FORO Colombia, pp 80–90

Ferrás C (2007) El enigma de la contraurbanización: Fenómeno empírico y concepto caótico. Eure (santiago) 33(98):5–25

Florida R (2005) Cities and the creative class. Routledge

Keynes JM (1987) The general theory of unemployment, interest and money. Mcmillan, Londres

Hall P (1981) Urban change in Europe. En: Pred A, Geographical Essays for Torsten Hägerstrand. Gleerup, Lund

Harvey D (2014) Diecisiete contradicciones y el fin del capitalismo

Jacobs J (1992) The death and life of great American cities. Vintage Books edition. New York

Klein N (2007) The shock doctrine: the rise of disaster capitalism. Macmillan

Lefebvre H (2020) El derecho a la ciudad. Capitán Swing Libros

Marx K (2000) El capital: Libro 1. T. 1. vol. 1. Ediciones AKAL

Mitchell CJ (2004) Making sense of counterurbanization. J Rural Stud 20(1):15–34

Montalvo M (2003) La Filosofía de Keynes o The common sense view of economic world. Isegoría 29:69–89

Newsinger J (2015) A cultural shock doctrine? Austerity, the neoliberal state and the creative industries discourse. Media Cult Soc 37(2):302–313

Ramírez P (2007) La ciudad, espacio de construcción de ciudadanía. Revista Enfoques 5(7):85–108

Sadri H, Zeybekoglu S (2018) Deurbanization and the Right to the Deurbanized City. ANDULI, Revista Andaluza De Ciencias Sociales 17:205–219

Schlappa H, Neill W (2013) From crisis to choice: re-imagining the future in shrinking cities. URBACT Secretariat. Paris. Saint-Denis

Toffler A (1972) El "shock" del futuro. Plaza & Janés. Barcelona

Topalov C (1979) La urbanización capitalista: algunos elementos para su análisis. Edicol, México

Van den Berg L, Klaassen LH (1987) The contagiousness of urban decline. En: Van Den Berg L, Burns LS, Kaassen LH (eds) Spatial cycles, Gower, Aldershot, pp 84–99

Wagner G, Weitzman ML (2016) Climate shock. In: Climate shock. Princeton University Press

Weber M (1997) Economía y sociedad: esbozo de sociología comprensiva. (2ª edición). Fondo de Cultura Económica

**Paulina Cepeda Pico** Architect and Master of Research in Urban Studies with a research scholarship at Facultad Latinoamericana de Ciencias Sociales, Flacso Ecuador. Senior architect of developing heritage housing restoration and rehabilitation projects, design and construction of architect projects and making of urban consulting. Her main areas of research are: housing policies, urban planning, urbanization process, governance, feminism urbanism, digital city, market and land policies. She is currently part of the team of researchers at Flacso Ecuador.

# Part IX
# Urbicide: Cities Cases

# Chapter 38
# Grassroots Spaces Make London Exciting: The Relationship Between the *Civitas* and the *Urbs*

**Pablo Sendra**

**Abstract** This chapter explores the contradictory nature of London's processes of urban regeneration and proposes alternative approaches that can nurture the emergence of social infrastructure. While London has become a world-famous city because of its cultural spaces and social infrastructure—many of which have emerged from the grassroots—many of these spaces are at risk of disappearing due to the neoliberal and financialised logic of urban planning and development. Using Klinenberg's (Palaces for the people. Penguin, London, 2018) social infrastructure framework, and through a multiple case study approach, the chapter explores alternative approaches. First, it explores the support that public institutions need to give to these spaces, which imply a radical change in the planning system that departs from prioritising financial viability. Second, based on Sennett's (Building and Dwelling. Allen Lane, 2018) framework of the *civitas* and the *urbs*, it proposes an approach for planners for better understanding the relationship between people and place, as well as the networks of solidarity and care linked to spaces. Third, it proposes re-assembling this relationship between the *civitas* and the *urbs* in a way that creates conditions for the emergence of social infrastructure.

**Keywords** Urban regeneration · Social infrastructure · Cultural infrastructure · Grassroots spaces · London

## 38.1 Introduction

What makes London an exciting city? As a South European person living in London, I get asked many times what is it that I like about London. It is always difficult where to start, but I usually say that it is its diversity and how this diversity is experienced in the streets. The plurality of cultures that live in London's neighbourhoods and how these cultures are manifested. London's diversity responds to Vertovec's concept of super-diversity, who argues that diversity in Britain can no longer be measured

P. Sendra (✉)
The Bartlett School of Planning, University College London (UCL), London, UK
e-mail: pablo.sendra@ucl.ac.uk

just in terms of ethnicity. It is much more complex and has many variables such as "differential immigration statuses and their concomitant entitlements and restrictions of rights, divergent labour market experiences, discrete gender and age profiles, patterns of spatial distribution, and mixed local area responses by service providers and residents" (Vertovec 2007, p. 1025). London's diversity has not just been a phenomenon discussed by academics. It is something that has been celebrated by (some of) its politicians, from Ken Livingstone championing diversity in both his periods as Leader of the Greater London Council (GLC) (1981–1986) (see Hatherley 2020) and in his period as Mayor of London (2000–2008) when the metropolitan authority in London was reinstated[1] (2000–2008), to Sadiq Khan's "London is Open" or "You are all Londoners", which came in reaction to the United Kingdom's decision to leave the European Union. Both Livingstone and Khan have been conscious of the value that migration and diversity bring to London, and how they contribute to London's success.

London is an exciting city also for its street markets, which range from some that have become very touristic, such as Camden or Portobello Road, to locally based markets that provide affordable access to foods and goods, and/or which cater for diverse cultures, bringing to London specific kinds or food or goods from all over the world. These markets also reflect London's diversity and how this is manifested in the streets and in everyday life.

Diversity is also manifested in London's night life, cultural and music scene, from venues where you can listen to live music from all over the World—such as the now closed Passing Clouds, where currently is The Jago—to the Sunday informal jam sessions that take place in many small bars and venues across the capital. This ecosystem of small cultural and music venues across London has a great importance to London's city life. They are not limited to nightlife, but these cultural spaces also include community-based theatres—such as the Arcola Theatre in Dalston—as well as neighbourhood-based community centres where people organise cultural activities.

Some of these neighbourhood-based community spaces are also places where social movements and networks of solidarity and care flourish, such as the Granville Community Kitchen (see location in Fig. 38.1), which organises free community dinners most Wednesdays and Fridays in the Granville community centre in South Kilburn (London Borough of Brent), where people from diverse socioeconomic and cultural backgrounds come to gather and have dinner. During the COVID-19 pandemic, the Granville Community Kitchen became an essential infrastructure of solidarity and care for the neighbourhood, delivering food to those at higher risk that

---

[1] Since 1965, London is composed of 33 municipalities, each of which have their own elections, representatives and provide a series of local services. Since 1965, there has been two periods of metropolitan administration, which provide strategic services: the Greater London Council (GLC) between 1965 and 1986, and the Greater London Authority (GLA) from 2000 onwards. The GLC had a "Leader of the GLC" and the GLA has a "Mayor of London". London did not have a metropolitan administration between 1986 and the year 2000. In 1986, Margaret Thatcher abolished the Greater London Council. The Greater London Authority was created in 2000, during the government of Tony Blair.

were shielding at home and to hundreds of families in the neighbourhood (Sendra et al. 2022). Another example of grassroots space of solidarity is The Village under the Westway flyover, in Bay 56 Acklam Road (see location in Fig. 38.1). The space was taken over by a group of activists and musicians during the aftermath of the Grenfell Tower fire in June 2017, a fire that killed at least 72 people and which devasted the community in North Kensington. The Village became a space for storing donations for those affected by the fire and then a space for healing and overcoming trauma. These spaces where people gather and interact respond to Klinenberg's definition of social infrastructure: "the physical places and organisations that shape the way people interact" (Klinenberg 2018, p. 5). The Mayor of London produced in 2020 a report on "Connective Social Infrastructure" (Mayor of London 2020), where he highlights the importance of these spaces for London. He also connects London's diversity to these spaces, since social infrastructure facilitates that interaction between different cultures.

A lot of the places and activities that make London special and exciting have emerged from the grassroots, from social movements and from struggles related to class, race, gender, or sexual identity, among others. This is not limited to the small spaces and locally based activities that I have described above. Even if we think about some of the major events in London, such as the Notting Hill Carnival, it emerged from racial struggles in the 1950s and then grew to become a large event.

The fact that London is a global financial capital is not what makes it attractive and exciting, although there are links between the flows of capital and everything that happens in London. London has become a world city because of its cultural richness, and much of it comes from the grassroots and from various struggles. As García Vázquez (2022) explains in *Cities After Crisis*, urban places that have emerged from the ground-up have turned some neighbourhoods in attractive places to live.

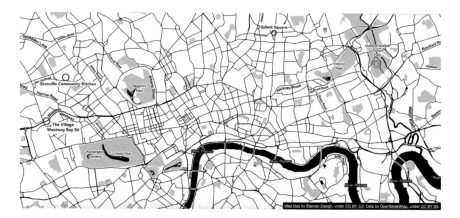

**Fig. 38.1** Map locating the case studies mentioned in this chapter: Granville Community Kitchen, The Village at the Westway Bay 56 and Gillett Square. *Source* Produced by the author from map tiles by Stamen Design, under CC BY 3.0, and data by OpenStreetMap, under CC BY SA

These grassroots spaces—their creation and their maintenance—do not respond to the logic of financial viability that defines urban planning in London nowadays. They are not spaces that provide a short-term economic profit. However, they provide a social value that in the mid and long term can alleviate the pressures on the care systems, and which at the same time make London an attractive city to live in, which is a key driver of London's economy and success as a global city. Since these spaces and activities do not respond to the logic of financial viability and do not provide short-term economic profit, they are becoming spaces at risk of disappearing due to real estate speculation and a financialised planning system.

In the last decade, we have witnessed how many of the cultural venues—including music venues such as Passing Clouds or LGBTQ+ venues such as The Joiners Arms (Campkin and Marshall 2017)—have had to close because of rent increases, issues with their landlords, or because of redevelopment. In addition to this, community spaces and social infrastructure are facing redevelopment and a significant loss of space (Penny 2019; Robinson and Sheldon 2019) to leave more space for other uses that are more profitable such as housing or workspaces.

As Lathan and Layton (2019) explain with social infrastructure, it is important to study the value of these spaces in order to protect them from the financialised planning system. In this chapter, I focus on grassroot spaces, places generated from struggles and/or which provide support for minorities and communities at risk, which add value to London and make it an exciting city to live in. If these spaces disappear, London can lose its essence and become a homogenised urban landscape, dominated by the currently trending architectural style of the "New London Vernacular". As the petition to Save Portobello Road Market from a shopping-mall-style redevelopment in 2015 puts it:

> "It is important because such revolting developments are chipping away at our London and sucking it dry of its lifeblood and individuality. This is a marvellous piece of London where independent traders can sell their wares which allow W11 a truly authentic edge; and now they are proposing a sanitized shopping experience akin to the Westfield that does not enhance or reflect our beloved Portobello." (Sullivan 2015): Petition to Save Portobello Road from the Portobello Village/Westway Space.

Given the value of these grassroots spaces, places generated from struggles and/or which support minorities and communities at risk, the aim of this chapter is to explore how to better support them and avoid jeopardising them in regeneration schemes that can affect them.

For doing so, I first explore which are the existing structures of support that exist for these spaces. While doing so, I also explore which are the struggles these spaces go through, the kind of help they need, and the potential relationship between grassroot spaces and institutions. After that, as a first step for approaching regenerations schemes that may affect such spaces, I propose an approach for understanding the relationship between the "civitas" and the "urbs". This approach departs from Richard Sennett's framework on differentiating the "civitas" and the "urbs", developed in *Building and Dwelling* (Sennett 2018), as well as from some of the discussions in our book *Designing Disorder* (Sendra and Sennett 2020). From understanding the relationship between the "civitas" and the "urbs", I continue explaining how these relationships can be re-assembled (see Sendra 2015 and McFarlane 2011) to explore creative ways of supporting existing initiatives and encouraging others to take place. During the chapter, I refer to various case studies of place in London, in some of which I have been directly involved, while in others I have been an "participant observant" while living in London. For each of the case studies, I outline a series of methods used in my action research work, which include co-design and co-production workshops where participants develop collective reflections, meetings with stakeholders, interviews and participant observation.

## 38.2 From Institutional Support to Struggles on the Ground

Owen Hatherley's (2020) book *Red Metropolis* narrates how throughout the history of metropolitan London institutions in the past 130 years, the capital has experimented with left wing politics–from the early days of the London County Council[2] (LCC) and its social housing programme to the last days of the Greater London Council (GLC) in 1981–1986 before Margaret Thatcher abolished it. In this 1980s period, the GLC provided support to migrant and migrant-descent communities, LGBTQ+ communities, organisations related to women's rights, and the intersection of the above. It supported the creation of various groups, community centres, as well as various community-led initiatives. Its Popular Planning Unit illustrated well a constant interaction between city institutions and grassroots initiatives.

In this chapter, I focus mainly on the current situation. Since the year 2000, a new metropolitan institution was created, the Greater London Authority (GLA) (also discussed at the end of Hatherley's book). London is composed of 33 municipalities, each of which have their own elections, representatives and provide a series of local services. The GLA is a strategic authority, which develop metropolitan strategies that then local authorities have to follow. It is also responsible for some metropolitan services such as Transport for London (for more information see London Councils, n.d.). However, the GLA has limited planning powers since most of planning power lies in local authorities (municipal rather than metropolitan institutions). The Greater London Authority produces the spatial strategy for London—the London Plan—as well as other strategies and guidance; it can refer, stop and require amendments to planning applications of large schemes and is responsible for various sources of funding, including the funding for social housing, as well as other funding schemes.

Soon after the global financial crisis of the 2008 hit London, general elections took place, and a Coalition Government of Conservatives and Liberal Democrats was formed. They implemented substantial cuts on welfare provision and restrictions on the capacity of local authorities in borrowing money. This post-2010 period of

---

[2] The LCC was the metropolitan administration between 1889 and 1965, when the GLC was created. The LCC only included the inner-city boroughs and had a smaller boundary than the GLC and the current GLA.

austerity had a strong effect on urban development in London. Local authorities suffered strong cuts of funding from central government and had to rely on partnering with private developers and put together regeneration schemes that put profit and financial viability at the centre. This resulted in the loss of many community spaces, places where diverse communities gathered, cultural spaces, night venues and many grassroots spaces. London's built environment started to become homogenised with large housing schemes on the New London Vernacular style.

In 2016, Sadiq Khan was elected Mayor of London, with pledges that included more affordable housing, an equal and diverse city, and a support for arts and culture, including a cultural infrastructure plan (Khan 2016). Soon after he was elected Mayor, the United Kingdom decided to leave the European Union (although in London the majority of votes were for remaining in the EU). Sadiq Khan's administration took "London is Open" as a moto that countered xenophobic discourses emerging from some Brexiteers and has been a champion of diversity, emphasising how important migrant and migrant-descent communities are for London.

A lot of his policies, strategies and guidance reflect this importance of London diversity, as well as the importance of community spaces, places for gathering and social interaction, and cultural and night venues. The recent Good Growth by Design guide on "Connective Social Infrastructure" (Mayor of London 2020) starts with "London's diversity is a strength, not a weakness" in the foreword by the mayor. When defining social infrastructure in the first chapter, it says:

> "London's social infrastructure is one of its great assets. From bumping into friends and neighbours in the park café, to visiting a local nail salon, recycling unwanted furniture on a Facebook group, using the library to find information, or getting help from a community support network, social infrastructure plays an important role in supporting and enriching the lives of Londoners." (Mayor of London 2020)

The report recognises a lack of protection of social infrastructure and asks local authorities to implement mechanisms that defend these spaces against other commercial and economically profitable uses. This is also reflected in the London Plan, which says that the social infrastructure loss should be replaced. The problem of this statement in the London Plan, as it happens with many of its policies, is the lack of mechanisms for implementing it. Firstly, it says "where possible", which opens many possibilities for local authorities and developers to justify that something is not possible (e.g. due to financial viability). Secondly, in many cases, the replacement does not necessarily respond to the community needs as much as the lost space did.

> "The **loss of social infrastructure** can have a detrimental effect on a community. Where possible, boroughs should protect such facilities and uses, and where a development proposal leads to the loss of a facility, require a replacement." (Mayor of London 2021, p. 217)

The London Plan also includes the replacement of small cultural venues, but we face the same problems with the lack of mechanisms for implementation and the phrase "where possible".

> "The loss of cultural venues, facilities or spaces can have a detrimental effect on an area, particularly when they serve a local community function. Where possible, boroughs should

protect such cultural facilities and uses, and support alternative cultural uses, particularly those with an evening or night-time use, and consider nominations to designate them as **Assets of Community Value**. Where a development proposal leads to the loss of a venue or facility, boroughs should consider requiring the replacement of that facility or use." (Mayor of London 2021, pp. 300–301)

On cultural infrastructure, indeed, Sadiq Khan has done a lot of work on supporting venues at risk, as this was an important part of its original pledge. Soon after coming to the office, he appointed the Night Czar, Amy Lame, to protect night venues as well as LGBTQ+ spaces in the capital. The mayor and his team put together a Cultural Infrastructure Plan (Mayor of London 2019), as well as a toolkit to protect cultural venues in London. This toolbox includes an interactive map with a wide diversity of cultural venues in London, as well as many other resources. As part of this, they created the Culture and Communities at Risk Unit, which supports both cultural and community organisations that are struggling and/or may be at risk of losing their space or closing down, and/or which may disappear because of a redevelopment scheme. They offer tailored support to these grassroots organisations.

The spirit of the Communities and Culture at Risk Unit has some of the characteristics of the "networks and municipalism" (Sendra and Sennett 2020) that Richard Sennett and I advocate for in our book *Designing Disorder*—open institutions that learn from and support grassroots organisations to thrive. However, a lot of the organisations that the Communities and Culture at Risk Unit are supporting are victims of a planning system dominated by financial viability, of which the Mayor of London (and its spatial strategy, the London Plan) is also responsible for. The London Plan acts as a guide for developers to get their planning application forward. Rather than being pro-active in the support and provision of cultural and social infrastructure, it sets a series of conditions that developers have to go through, where they may need to replace it "where possible". Even when this is achieved, when a much-loved space where people gathered is replaced in square metres elsewhere or as part of a multifunctional building that concentrates various services, the "social infrastructure ecosystem" (as named by the Mayor of London 2020) that exists around it may disappear–i.e. the attachment between people and buildings is broken, and the momentum around community gathering, social interaction and community solidarity dissipates.

Back in the 1980s, the Popular Planning Unit and the GLC had a pro-active approach to "networks and municipalism". The Popular Planning Unit instigated the Royal Docks People's Plan (see Sendra and Fitzpatrick 2023) and the GLC facilitated the community spaces such as the London Lesbian and Gay Centre, as well as funding many community groups and activities from ethnic minorities, LGBTQ+ rights and women's rights (see Hatherley 2020).

In our book *Designing Disorder*, Richard Sennett and I propose open city institutions, which learn from and actively support community organisations and grassroots spaces such as those described in this chapter—locally based social and cultural infrastructure, where different people gather and where diverse forms of culture are manifested and shared. Such spaces exist in London but are at risk partly because of the current neoliberal planning system. Therefore, one of the first steps to implement this municipalist open institution approach is to change the planning system from one that prioritises financial viability and profit to one that understands and prioritises the social and cultural value generated. Social infrastructure and community spaces should not be seen as a token for developers to include in a little corner of their scheme, but as the driver of London's urban planning, since it is what make our city more exciting.

## 38.3 Understanding the Relationship Between the *Civitas* and the *Urbs*

When we learn about urban planning at university, we learn about what a "good city" is supposed to be like. Of course, measuring what constitutes "good city" has evolved in history in response to the challenges of each time (Amin 2006). Currently, given the environmental challenges brought by the climate emergency, urban planning schools teach that cities should be sustainable, safe and provide a good quality of life to people living there. For achieving this "good city", cities should be compact, pedestrian friendly, clean, with greenery and with various types of activities in proximity. However, some planning courses fail to teach something very important, which relates to understanding the relationship between the *civitas* and the *urbs*—this is people's relationship to their built environment, how they feel attached to building and places, what these represent to them, where they gather, what activities take place in them, and what kind of social relations and interactions take place in them. The preconceived ideas about what a good city is do not grasp the importance for people of an old library, or a community centre that may not fulfil the canons of a "beautiful" built environment. Ongoing changes in the planning system in England are making more emphasis on "Building Beautiful" (Building Better, Building Beautiful Commission 2020), but this in occasions does not consider the meaning some buildings might have for some communities and which might not fulfil these cannons of beauty.

At The Bartlett School of Planning, UCL, I teach a course titled Civic Design,[3] where each year we collaborate with a community group based in London. Community members from the group we are collaborating with take the course along with the students, since they get a free registration to the short course. This guarantees that there is a peer-to-peer collaboration between communities and students, and a knowledge exchange that works both ways. Through this collaboration, students learn and appreciate the importance of people's attachment to the buildings, the memories associated to them, the activities they do there and the social relationships they have in them. One of the key things we teach in the course is co-producing evidence with communities. Building on Fals Borda's (1987) framework for Participatory Action Research and on Domenico Di Siena's (2019) Civic Design Method, we teach them

---

[3] The course started first as a summer school in 2018. Then, from 2019, it became a short course for continuing professional development (CPD). From 2020/2021, it became a master's module, and it continues being taught as a short course.

methods for co-producing evidence with communities in order to understand what is people's attachment with places. As muf architecture/art proposed in the project "Making Space in Dalston" (Gibbons and LLP; muf architecture/art 2009): "value what is there, nurture the possible, define what is missing".

I have experimented with this approach in the co-design processes that I have facilitated—both in the action research projects that I have done at UCL in partnership with community groups in London and in consultancy projects that I have done with my practice Lugadero. With UCL, I have completed three projects that engage in a Participatory Action Research methodology in which researchers co-produce evidence with community members. These are William Dunbar and William Saville Community Plan (see Sendra et al. 2020; Colombo et al. 2021), Alton Estate People's Plan (Sendra et al. 2021), and a report on the importance of the community buildings Granville and Carlton during COVID-19 (Sendra et al. 2022). The first two are counter proposals for two council estates that are going to be demolished. With residents, we co-produced a Social Impact Assessment (see Colombo et al. 2021) that evaluate the impact of the potential redevelopment scheme. After that, we continued working with residents on co-designing an alternative scheme, which look at retrofitting the existing homes and building new ones through infill development, instead of demolition. In both cases, we used methods where residents could collectively discuss their attachment to the buildings, their relationships of care and solidarity with their neighbours, and where those relationships took place. We did this through workshops where participants could collectively reflect on these topics, using a diversity of co-production methods that were tailored for these residents and this place (see Colombo et al. 2021). This depth of the co-produced evidence cannot be achieved with other types of analysis of the built environment. It is necessary to generate that empathy with those living there and using the spaces. We took a similar approach with the research report on the Granville and Carlton, where we analysed the importance of these buildings as a social infrastructure, particularly during COVID-19, and where we also used co-production workshops, stakeholders meeting and semi-structured interviews with key actors as methods. Granville is where the previously mentioned Granville Community Kitchen is based, which is an essential social infrastructure for care and solidarity in the area. For this report, we also organised workshops to collectively reflect with residents and users of the buildings on the importance of them.

I have used a similar approach in my consultancy work. In addition to my academic work, I do some consultancy work facilitating co-design processes. In the previously mentioned Westway, I worked with various local activists on facilitating a co-design process to improve the public spaces. The Westway Trust got match funding from the Mayor of London to improve the public spaces through a project named "Community Street". For this work, the Westway Trust called for proposals from teams to facilitate a co-design process. For tendering to the work, I put together a team with local activists and local people with extensive experience on community engagement. We used a diversity of methods, which started from understanding which are the diverse local cultures in the area and how they use the spaces. This included workshops, drop-in sessions, a continuous presence on the streets by having a pop-up office in a pod on the public realm during the four months of the co-design process, and also by having in-depth meetings with many community organisations in the area. The methods we used, which are explained in our final report (see Lugadero and The Grove Think Tank for the Westway Trust 2021), consisted of collectively producing the evidence with the 876 participants that took part, as well as co-designing a series of recommendations/proposals for the appointed architects (DK-CM) to take on.

In my previous work, I have explored the importance of understanding the relationship between the *civitas* and the *urbs* through the framework of assemblage theory, which is a body of work that looks at the agency generated from the connections between human and non-human actors (see McFarlane 2011). In my previous work, I have looked at the importance of understanding which are the existing sociomaterial connections that exist in a place (such as the attachment between people, the built environment, existing policies, social relationships, forms of behaviour, the space, objects and material things) in order to propose re-assembling some of these connections, enhance some of the activities taking place, and propose new ones (see Sendra 2015, 2018; McFarlane 2011). A lot of the methods outlined above aim to understand which are these "assemblages"—these connections between people, place and the activities and social interaction that occur. From understanding these relationships, we can re-imagine how they can be re-assembled.

## 38.4  Re-assembling the *Civitas* and the *Urbs*

When praising the value of grassroots and informal spaces, it is important to clarify that this praise does not mean that we should leave everything unplanned and let activities to emerge. It does not mean that urban designers and planners are not necessary, and our knowledge is meaningless. As Richard Sennett and I argue in *Designing Disorder* (Sendra and Sennett 2020), there are places where these grassroots initiatives need some support in order to emerge, and/or where the conditions need to be created. Urban designers can indeed use their skills creatively to propose situations and alternative arrangements that people that live in an area and/or use a space cannot imagine. However, for doing so, they need to understand the existing relationships between human and non-human actors, material and immaterial things, in order to re-assemble them and introduce new elements that support and create conditions for the emergence of grassroots spaces.

I have explained how this can take place in my previous work using the case study of Gillett Square in Dalston, East London (see Sendra 2015; Sendra and Sennett 2020) (see location in Fig. 38.1 and image in Fig. 38.2). For this piece of research on Gillett Square, I carried out participant observation between 2011 and 2013. As an East London resident, it is also a space that I still frequent for various cultural activities. This is a useful case study because urban design has actively supported the emergence of unplanned activities and has transformed the space into a flexible public realm that is continuously evolving. Where currently is Gillett Square, there used to be a carpark. It was the design intervention of placing a series of kiosks/pods

**Fig. 38.2** Kiosks/pods with local businesses in Gillett Square. The table tennis is stored in one of the shipping containers in the square. *Source* Photography by Estrella Sendra, April 2012

in the edge of the carpark to lease at affordable rents to local businesses what made people gather around the kiosks and turn the car park into a gathering space. The fact that people were gathering on the edge of the carpark made evident the need to create a public square where the carpark was.[4] Years later, it was another urban design intervention which transformed the square into a flexible public space that is continuously changing: two ship containers were placed in the square, which serve as storage for structures for market stalls, table tennis, AV equipment, cinema screens, soft play games for children and many other props and elements that allow a continuous re-definition of the square.[5]

All the subsequent intentions that took place in Gillett Square were urban design decisions. These urban design interventions go beyond material construction and distribution of functions. They understood the existing (human and non-human) actors, the relationship between these actors and the material and non-material elements of the built environment and re-arranged these elements and introduced new components to release the potential space and create the conditions for activities to emerge.

Currently, at the time of writing in 2022, Gillett Square is going through another phase of development, which will expand the workspace building next to the square, remove the kiosks and replace them with some additional shopfront space in both the new building and the other side of the square. The last few times I have visited the

---

[4] The kiosks and the transformation of the carpark into Gillett Square was designed by Hawkins/Brown for Hackney Co-operative Developments.

[5] This intervention was led by muf architecture/art and it is included in the Making Space in Dalston study.

square, I have seen that the kiosks have move temporarily into ship containers in the carpark that there is next to the square, where they are much more hidden. Some local campaigners–including the former CEO of Hackney Co-operative Developments that led the creation of the square and the previously mentioned phases—have raised concerns on the potential impact that the development could have on the traders, on affordable workspaces, as well as on the life of the square (Barltholomew 2018). There have been also concerns about the potential gentrification effect that the scheme could have (Open Dalston 2018), particularly given the gentrification dynamics of Dalston. While it is not clear what effect the scheme will have on traders and on the openness and flexibility of the square, we can see that even the most successful grassroots spaces are continuously at risk due to the current planning system based on economic viability.

## 38.5 Conclusions

The example of the earlier transformations that led to the creation of an open and flexible space in Gillett Square responds to the idea of pro-actively supporting the creation of grassroot spaces, rather than having to react to "save" those that are at risk. Places like Gillett Square, where people from different cultures and socioeconomic background convive in the square, where informal gatherings take place, with a public space that can support different activities, and which bring together various cultural venues, make the neighbourhood of Dalston—and London—an exciting place to live. Gillett Square form the kind of social infrastructure described by Klinenberg (2018), which both spaces and organisations that facilitate social relationships. Some of the community-led schemes that I have been involved with through my action research at UCL emerge as a counter proposal to development schemes that put community spaces and neighbourhoods at risk. Rather than having to support communities to "save" community spaces, planners and institutions should work with communities since the very beginning and collectively put together proposals that create conditions for the emergence of grassroots spaces, as the first phases of Gillett Square did. These are the pro-active municipalist institutions that I propose, which work directly on the ground and collaborate with the networks of grassroots groups to come out with creative proposals to make our city much more exciting and inclusive.

## References

Amin A (2006) The Good City. Urban Studies 43 (5–6):1009–1023
Barltholomew E (2018) Tensions run high in Dalston's Gillett Square over £2m 'gentrification' fear. Hackney Gazette. Available: https://www.hackneygazette.co.uk/news/tensions-run-high-in-dalston-s-gillett-square-over-2m-3593210
Building Better, Building Beautiful Commission (2020) Living with beauty: promoting health, well-being and sustainable growth. Ministry of Housing, Communities and Local Government

Campkin B, Marshall L (2017). LGBTQ+ cultural infrastructure in London: night venues, 2006–present. UCL, research report. Available at: https://www.ucl.ac.uk/urban-lab/sites/urban-lab/files/LGBTQ_cultural_infrastructure_in_London_nightlife_venues_2006_to_the_present.pdf

Colombo C, Devenyns A, Manzini Ceinar I, Sendra P (2021) Co-producing a social impact assessment with affected communities: evaluating the social sustainability of redevelopment schemes. Sustainability 13(23):1–22

Di Siena D (2019) Civic design method whitepaper. Available at: https://civicdesignmethod.com

Fals-Borda O (1987) The application of participatory action-research in Latin America. Int Soc 2(4):329–347

García Vázquez C (2022) Cities after crisis: re-inventing neigbourhood design from the ground-up. Routledge, Abingdon, Oxon

Hatherley O (2020) Red metropolis. Repeater, London

Gibbons J, Llp L, muf Architecture/art (2009) Making space in Dalston. London: London Borough of Hackney. Design for London/LDA. Available at: http://issuu.com/mufarchitectureartllp/docs/making_space_big

Klingenberg E (2018) Palaces for the people. Penguin, London

Khan S (2016) Sadiq Khan for London: a manifesto for all Londoners. Sadiq Khan and London's Labour Manifesto

London Councils (n.d.) The essential guide to London local government. Available at https://www.londoncouncils.gov.uk/who-runs-london/essential-guide-london-local-government (Accessed 13 July 2022)

Latham A, Layton J (2019) Social infrastructure and the public life of cities: studying urban sociality and public spaces. Geogr Compass 13(7):1–15

Lugadero & The Grove Think Tank for the Westway Trust (2021) Westway community street co-design report. Available at: https://www.communitystreet.org/_files/ugd/a7dbe5_51564a04f8144ae981ebc6cb22a6e5cf.pdf

Mayor of London (2019) Cultural infrastructure plan: a call to action. Available at: https://www.london.gov.uk/sites/default/files/cultural_infrastructure_plan_online.pdf

Mayor of London (2020) Good growth by design: connective social infrastructure. How London's social spaces and networks helps us live well together. Available at: https://www.london.gov.uk/sites/default/files/connective_social_infrastructure_0_0.pdf

Mayor of London (2021) The London plan. Available at: https://www.london.gov.uk/sites/default/files/the_london_plan_2021.pdf

McFarlane C (2011) Assemblage and critical Urbanism. City 15(2):204–224

Open Dalston (2018) Gillett square plans—co-operative development or gentrification? Open Dalston. Available at: http://opendalston.blogspot.com/2018/04/gillett-square-plans-co-operative.html

Penny J (2019) 'Defend the ten': everyday dissensus against the slow spoiling of Lambeth's libraries. Environ Plann D: Soc Space 0(0):1–19

Robinson K, Sheldon R (2019) Witnessing loss in the everyday: community buildings in Austerity Britain. Sociol Rev 67(1):111–125

Sendra P (2015) Rethinking urban public space: assemblage thinking and the uses of disorder. City 19(6):820–836

Sendra P (2018) Assemblages for community-led social housing regeneration: activism, big society and localism. City 22(5–6):738–762

Sendra P, Fitzpatrick D (2023-forthcoming) People's plan: the political role of architecture and urban design for alternative community-led futures. In: Bobic N, Haghighi F (eds) The Routledge handbook of architecture, urban space and politics, Vol II: ecology, social participation and marginalities. Routledge, London

Sendra P, Colombo C, Devenyns A, Manzini Ceinar I (2020) Civic design exchange: co-designing neighbourhoods with communities. UCL, research report. Available at https://reflect.ucl.ac.uk/community-led-regeneration/civic-design-exchange/projects-and-collaborations/community-plan-north-west-london-estate/

Sendra P, Short M, Fitzpatrick D, Livingstone N, Manzini Ceinar I, Navabakhsh S, Turner W, Goldzweig S (2021) People's plan of Alton estate. UCL, research report. Available at: https://reflect.ucl.ac.uk/community-led-regeneration/civic-design-exchange/projects-and-collaborations/peoples-plan-of-alton-estate/

Sendra P, Manzini Ceinar I, Pandolfi L (2022) Co-designing resilient social infrastructure in post-Covid cities: Granville and Carlton. UCL, research report. Available online: https://reflect.ucl.ac.uk/community-led-regeneration/civic-design-exchange/projects-and-collaborations/granville-and-carlton/

Sendra P, Sennett R (2020) Designing disorder: experiments and disruptions in the City. Verso, London

Sennett R (2018) Building and dwelling: ethics for the city. Allen Lane, London

Sullivan C (2015) Save Portobello Road from Portobello Village. 38 degrees. Retrieved from https://you.38degrees.org.uk/petitions/save-portobello-raod-from-the-portobello-village-westway-space-1

Vertovec S (2007) Super-diversity and its implications. Ethn Racial Stud 30(6):1024–1054

**Pablo Sendra** Associate Professor at The Bartlett School of Planning, University College London (UCL). He is the programme director of the MSc in Urban Design and City Planning programme. He combines his academic career with professional work through his urban design practice, Lugadero Ltd, which focuses on facilitating co-design processes with communities. He graduated as an Architect at the Universidad de Sevilla, MArch Urban Design at The Bartlett School of Architecture, University College London (UCL), and PhD in Architecture at the Universidad de Sevilla.

# Chapter 39
# Rio de Janeiro: The Trajectory of the Wonderful City, Violence, and Urban Disenchantment

**Mauro Osorio, Maria Helena Versiani, and Henrique Rabelo**

**Abstract** The city of Rio de Janeiro is a national and international reference perceived through the concept of "capitality," inspired by Argan in L'Europe des capitales (1964). In line with Argan's conceptions, the attribute of capitality is applied to cities distinguished in their respective countries as political, cultural, and intellectual centers, axes of modernity, places of affective memory, and spatial landmarks of nationality representation. This article intends to reflect on the aspects of the trajectory of the city of Rio de Janeiro that give it the seal of capitality, as well as the specificities of its economic development in markedly unequal socio-territorial bases. Different series of data and socioeconomic indicators, organized by statistical monitoring institutions, will be presented and analyzed, revealing particularly dramatic portraits of the population's living conditions in the city of Rio de Janeiro and in its Metropolitan Region, as well as the specificities of its insertion on a state and national scale. The proposal is to incorporate these data and indicators into the analysis of Rio de Janeiro's specificities, structuring public planning subsidies, and formulating coordinated policies covering the various spheres of government.

**Keywords** Violence · Modern cities · Urban culture

## 39.1 Building Capitality

For a good understanding of the signs of capitality of Rio de Janeiro, it is convenient to briefly review its forms of political-administrative insertion from the context of the colonization of Brazil by the Portuguese empire. Such a retreat in time brings us to the first decades of the sixteenth century when an administrative system was implemented in the Brazilian colony that divided the territory into extensive lots called "captaincies." Each captaincy was an administrative unit organized by a donee appointed by the King of Portugal. With the ostensible use of slave labor, it was up to the donees to populate, defend, and exploit the captaincies, which ultimately belonged

M. Osorio (✉) · M. H. Versiani · H. Rabelo
Federal University of Rio de Janeiro (UFRJ), Rio de Janeiro, Brazil
e-mail: mauroosorio@uol.com.br

to the King of Portugal. In 1548, the donees became subordinate to a governor-general, and the headquarters of the general government was then established in the city of Salvador, within the territory of the captaincy of Baía de Todos os Santos. This region included the "ground zero" of the colonialist project in Brazil.

It was in 1565 that the foundation of the city of São Sebastião do Rio de Janeiro takes place. The city integrates the newly created captaincy of Rio de Janeiro following the episode known as "Antarctic France," which marked the French attempt of occupation of the Guanabara Bay—which failed after the Portuguese military reaction led by Estácio de Sá.[1] The dispute triggers the Lusitanian concern to ensure strict control of the Bay, with the strategic creation of a city on its coast, establishing a position to face the corsair activity and consolidate the Portuguese domain. The enterprise involved the establishment of settlers, enabling displacements to the continent's interior.

"This is one of the safest and best bars in these parts, through which any ships may enter and leave at all times without fear of any danger" (*apud* Bicalho 2003, p. 32). Testimonies of travelers, such as this one from the Portuguese Pero de Magalhães Gandavo (1540–1579), reported on the privileged geographical position of Guanabara Bay and, in fact, the city of Rio de Janeiro soon matured to port activity, open to the tremendous maritime flow of ships, goods, and information. Fortifications and churches emerged. In the words of Paul Knauss: "The fort protected the harbor militarily, and the church guarded it spiritually" (Knauss 2005, p. 9).

In 1763, the headquarters of the general government of Brazil was transferred from Salvador to the city of Rio de Janeiro, corroborating "the central character that this city and its governors had been assuming since the middle of the seventeenth century, head or articulating *locus* of vast territories, interests, businesses and policies in America and the South Atlantic." The proximity to Minas Gerais was also strategic, ensuring good surveillance of the exploration and circulation of gold discovered in abundance in the region (Bicalho 2011, p. 53).

With the move of the Portuguese Royal Family to Brazil in 1808, the city of Rio de Janeiro became the seat of the Kingdom of Portugal and the Algarves (from 1815, Kingdom of Portugal, Brazil, and the Algarves). Rio was then targeted with investments of extraordinary proportions, being equipped for the function of the headquarters of the Portuguese Court. Thus, a new and capital ambiance was consolidated in the city of Rio de Janeiro, reaffirmed over the years and until today as a cultural way of experiencing the urban space molded in the seal of capitality.[2]

Still in 1821, the captaincies were renamed "provinces" and, in 1822, the year of Independence of Brazil, the city of Rio, without losing its administrative link with the province of Rio de Janeiro, became the Capital of the Empire of Brazil. In other words, unlike the other provinces, that of Rio de Janeiro remained linked to the imperial administration: the Minister and Secretary of State for Imperial Affairs was also the governor of the province of Rio de Janeiro.

---

[1] On the episode called Antarctic France, see Mendonça (1991).

[2] On the subject, see Lessa (2000) and Osorio (2005).

Subsequently, in 1834 the city of Rio was converted into a Neutral Municipality, established as an administrative unit of direct intervention by the imperial government, independent of the province of Rio de Janeiro. However, it would continue to be the first electoral district of the province of Rio de Janeiro during the elective processes for the General Assembly of the Empire.

In the course of the nineteenth century, coffee production began in the city of Rio, which expanded to the province of Rio de Janeiro, on a slaveholding basis. This process was financed by the city's mercantile capital. At that time, coffee became the main product of exploitation of the Brazilian colony, and its high profits ended up also providing investments in other economic activities, such as industry.

Given the logic of the maximum reduction in production costs, while there were new and fertile lands for coffee in Rio de Janeiro, the appropriation of these lands occurred without any investments in implementing productive techniques that could minimize their exhaustion process. This movement ended up leading, from the last quarter of the nineteenth century, to the progressive displacement of coffee production to the region of the current state of São Paulo.

The prohibition of the slave trade in Brazil in 1850 generated an increase in prices and the prospect of a growing scarcity of enslaved, creating conditions more favorable to the wage-labor regime. The new coffee entrepreneurs invested in encouraging European immigration, especially the Italian one, which became massive from 1880, with most immigrants heading to São Paulo.

At that time, the incentive to the immigration of white people was echoed in racial theories anchored in the notion of eugenics—translated as "good generation," a term proposed by the British scientist Francis Galton, whose ideas had worldwide influence. Galton (1979) sought to scientifically prove that human capacity was regulated by hereditary physical characteristics and not by education, proposing that it would be quite desirable to reproduce a race of highly gifted men from marriages committed to interbreeding between whites for several consecutive generations. Such thinking found impressive strength in the Brazil of the 1800s, discouraging miscegenation said to be harmful to society and, in its most hateful version, advocating the elimination of inferior races. It was also argued about the need to import "qualified" labor from Europe, optimizing the territory's occupation and its population's characteristics, with a view of a broad "civilizing" process. The white and European immigrant was perceived as a bearer of civilization. Thus, the emergence of a scientific racism corroborated the affirmation of the superiority of the European worker and the backwardness of the Africans.[3]

In the last decades of the nineteenth century, Brazilian coffee production grew significantly, and São Paulo became its driving force with the emergence of the so-called São Paulo coffee complex. With the adoption of salaried labor and investments in infrastructure, such as the implementation of railways, the São Paulo consumer market and the demand for industrial goods have undergone a substantial expansion, which initiated the industrialization process in São Paulo.[4]

---

[3] On the subject, see Schwarcz (1993).

[4] On the subject, see Mello (1982) and Mariani (2015) (documentary).

With the Proclamation of the Republic in 1889 and the promulgation of the first Brazilian Republican Constitution in 1891, the Neutral Municipality of Rio de Janeiro became the Federal District of Rio de Janeiro, the Capital of Brazil. The term "province" was replaced by "state."

In this new political-administrative configuration, the city/Federal District of Rio de Janeiro continued as the nucleus of the nation's power and cultural center, a hub for sophisticated services, the *locus* of the national financial system, the main headquarters of large companies located in the national territory, and a privileged gateway to tourism in the country. The Federal District continued to receive injections of public spending and investments in creating new companies, hosting the prominent multinationals that settled in Brazil and a series of public agencies and state-owned companies that emerged in the 1930s, 1940s, 1950s, and 1960s. Examples include the creation of Petróleo Brasileiro S/A-Petrobras, the National Economic Development Bank-BNDE, Centrais Elétricas Brasileiras S/A-Eletrobras, Vale do Rio Doce Mining Company, and the Brazilian Institute of Geography and Statistics-IBGE.[5]

Thus, the Federal District of Rio de Janeiro benefited from the high growth of the Brazilian economy between the 1930s and 1960s. Until the 1960s, it presented a growth rate of gross domestic product-GDP close to that of Brazil and other states in the country, except for the state of São Paulo, the leading region of the national economic development process in the course of the twentieth century.[6]

The leverage of industrialization in the state of São Paulo caused that, in 1919, São Paulo's industrial production had already surpassed the sum of industrial output in the Federal District and the state of Rio de Janeiro. According to the 1907 Industrial Census, the gross value of industrial production in the Federal District represented 30.2% of the country's total industrial production. In the state of Rio de Janeiro, it represented 7.55%, and in the state of São Paulo, 15.92%. The 1919 Brazilian Industrial Census showed that the participation of the Federal District had dropped to 22.29%, that of the state of Rio de Janeiro to 6.16%, while that of the state of São Paulo had more than doubled to 32.99%, even exceeding the sum of the gross value of industrial production in the Federal District and the state of Rio de Janeiro, of 28.64%.

However, if the São Paulo economy gained strength and centrality, the entire economy of the city and state of Rio de Janeiro still maintained its economic dynamism for many decades.

---

[5] These organizations were created during the governments of President Getúlio Vargas (1930–1945 and 1951–1954). Specifically, BNDE was created to generate medium- and long-term financing for Brazilian industrialization. In the second half of the 1980s, this bank incorporated an area to support public and social policies, changing its name to Banco Nacional de Desenvolvimento Econômico e Social-BNDES. So far, few countries in the world have banks like the BNDES, which is the case of Germany, Japan, China, and Mexico.

[6] On the dynamism of the Federal District of Rio de Janeiro by keeping itself close to the Brazilian dynamism until 1960, see Osorio (2005) and Lessa (2000).

## 39.2 The Capital Leaves Rio

The first Republican Constitution in Brazil promulgated in 1891, provided for the transfer of the Federal Capital to a sparsely populated area of the central plateau located in the Brazilian Midwest. However, this proposal was only forwarded in 1956 by the then President of the Republic Juscelino Kubitschek (1956–1960), popularly known as JK, being approved by the National Congress.

JK assumed the Presidency of Brazil, continuing the country's modernization and industrialization project initiated by Getúlio Vargas. Based on an economic program known as Plano de Metas (Goals Plan), the prognosis was that Brazil would grow "fifty years in five." The euphoria of development tended to become widespread, and, at the end of the JK government, national economic growth reflected a significant boost given to the consumer and essential goods industry, a new scale of production of the country's energy matrices, as well as investments in the construction of highways and railways, in shipbuilding and automobile industry—in a scenario where the struggle for agrarian reform was demoted and with little priority in combating social inequalities.

Juscelino was nicknamed "president bossa nova," an allusion to his proposal to build a new Brazil, a project also animated by the moment then lived of significant innovations in the world of the arts, reiterating the idea of a modern and renewed country. Among other examples, a new musical style was gaining strength in Brazilian urban centers—bossa nova—which proposed an original way of playing and singing samba; Cinema Novo (New Cinema) adopted new aesthetic standards for national cinema; proposals emerged in the direction of nationalizing and popularizing theater; the Concrete Movement innovated within the scope of Brazilian literary standards and also in the Visual Arts. Finally, the country seemed to confirm its trajectory toward complete modernization through the escalation of economic growth and in different movements that operated in the field of aesthetic renewal of the arts.[7]

A meta-synthesis of this process, in 1960, a new Federal Capital was created for Brazil, with the construction of Brasília, inaugurated on April 21, 1960. In line with JK's modernizing discourse, planting a new capital city in the central plateau was part of Brazil's march toward expansion and progress.

Thus, in 1960, with the transfer of the Federal Capital to Brasília, the Federal District of Rio de Janeiro became the city-state of Guanabara, without fair and necessary financial compensation from the Federal Government in order to mitigate the fact that Rio de Janeiro had lost the *status* of capital and the historical specificity of its participation in federal public spending arising from this function.[8]

---

[7] On the convergence between politics and culture that marked this period, revolutionizing the arts field, see Ridenti (2007).

[8] As an example, at the time of the unification of Germany, the transfer of the capital from Bonn to Berlin was conducted with compensation for the city of Bonn, helping to prevent it from going into decline. Two-fifths of the ministries were kept in Bonn, nearly two billion euros were invested by the central government, and the establishment of international agencies, such as the UN, and the headquarters of privatized companies, such as the post office and telecommunications, were

## 39.3 Rio's Structural Crisis in the Post-1960s

From the transfer of the capital to Brasília and especially when the transfer was consolidated in the 1970s, the economy of the city of Rio lost dynamism and detaches itself from the Brazilian trajectory. For example, between 1970 and 2019, the city of Rio went from participation in the national GDP of 12.84% to only 4.80%, which meant a loss of participation in the national gross domestic product of an expressive 62.60% (data from the Brazilian Institute of Geography and Statistics-IBGE). In this scenario, the city of Rio de Janeiro, despite the loss that occurred, continued to be the second-largest GDP among the country's state capitals. However, the city, which in 1970 was the third-largest GDP per capita among the capitals (behind São Paulo and Vitória), became, in 2019, the fifth-largest GDP per capita.

In the same direction, the historical series of available data on formal employment from 1985 onward indicate that, between 1985 and 2020, the growth of formal employment in the city of Rio de Janeiro was only 6.3%, against growth in the total of the country of 125.6% (Federal Government/Annual List of Social Information-RAIS). In other words, for 35 years, formal employment has practically not grown in the city of Rio.[9]

The dramatic loss of economic dynamism in Rio de Janeiro was due to a combination of factors. First, the transfer of the Federal Capital to Brasília took place without compensation, as already pointed out. It was also due to the fact that it basically occurred without public debate on what should be the new directions for the city of Rio, which obviously made it difficult to define regional strategies and policies.

The lack of debate, in turn, resulted from several aspects. First, there was a disbelief in the country that Juscelino Kubitschek would indeed build a new Federal Capital, which had already been promised since the inscription of this determination in the distant first Constitution of the Republic in 1891.

In addition, a strong tradition of reflection on national and international issues is developed in the city of Rio de Janeiro, to the detriment of local issues. This resulted from the combination of two factors: first, the fact that the city of Rio de Janeiro was the seat of the Portuguese Court and the capital of the country for almost two hundred years and second, due to Rio's capital culture, which, on the one hand, stimulates a concentration of public universities and centers of excellence in the city, such as the Pontifical Catholic University of Rio de Janeiro, the Oswaldo Cruz Foundation, the Getúlio Vargas Foundation, and the Brazilian Academy of Sciences and, on the other hand, favors these institutions to focus their lines of research on national and international themes to the detriment of local ones.

The relegation of regional reflection in Rio was also due to the fact that, between the Proclamation of the Republic in 1889 and the transfer of the Federal Capital to

---

encouraged. Bonn managed to expand the number of local jobs. (See https://veja.abril.com.br/mundo/berlim-e-bonn-20-anos-depois/. Access: 2/3/2022).

[9] The issue can be understood from the formulation of Frank Moulaert (2000) when the author proposes that a region is going through a process of stagnation when in its economic dynamics, there is a significant detachment concerning the national economic dynamics.

Brasilia in 1960, there was never a direct election for the city manager, when we know that electoral moments help to broaden the look and reflection on local issues.

A commonly cited example of Rio's tradition of focusing on national rather than regional themes is the fact that the two leading newspapers in São Paulo are Estado de São Paulo and Folha de São Paulo; in Minas Gerais, it is Estado de Minas; while in Rio de Janeiro, two historically influential newspapers are O Globo (The Globo) and Jornal do Brasil (Brazilian Newspaper). In the same direction, in the academic world, the postgraduate programs in Economics in Rio de Janeiro have a much lower amount of reflection on Regional Economics than the master's and doctoral programs offered in the vast majority of other Brazilian regions.

The difficulty of developing the habit of regional reflection in Rio de Janeiro refers to the idea of economist Douglass North (1993) that the consolidation of specific cultural patterns of behavior and habits can make social life an experience marked by routines in which questions of choice present themselves as something regular, repetitive, "so that about 90% of our actions in life would not require much reflection" (North 1993, p. 37). Breaking this logic in Rio de Janeiro remains a challenge 62 years after the transfer of the capital.

The lack of regional reflection in Rio de Janeiro, which is not limited to academic environments, but instead constitutes a cultural mode of carioca society,[10] makes it difficult or even impossible to design adequate diagnoses of the severity of its economic crisis. As a result, it makes it difficult or unfeasible to propose a strategy that allows Rio to get out of its vicious circle.[11]

## 39.4 City and State of Rio de Janeiro: Intertwinings

With the consolidation of the transfer of the capital to Brasília, the economy of the state of Rio de Janeiro was also affected, and the state now has the lowest economic dynamism among all Brazilian states.[12] This is because, from the moment that Rio's coffee production began to decline, the economic dynamism in the state of Rio until the 1960s came mainly from federal investments, with the creation of state-owned companies in the state, such as National Steel Company, National Engine Factory, Duque de Caxias Refinery, and National Company of Alkalis (dedicated to the production of sodium carbonate and salt). The main factors that made these investments possible were the proximity and historical synergies of the state with the Capital of the Republic.

The dynamism of the state's economy also came from the production of industrial and agricultural consumer goods produced in the state of Rio to meet the demands of the Capital of the Republic. And finally, it benefited from the tourism of the carioca elite in mountain, sea, and historical regions of the state of Rio, who stayed in hotels

---

[10] "Carioca" is a terminology that refers to one who is a native of the city of Rio de Janeiro.

[11] On the concept of the vicious circle, see Myrdall (1968) and Krugman (1999, 2002).

[12] On the subject, see Osorio et al. (2017) and Osorio (2005).

and inns or constituted their own homes for summer vacations and enjoyed second homes on weekends.

In 1975, the merger of the city-state of Guanabara with the then state of Rio de Janeiro was established. The city of Rio de Janeiro—now its official name—became the capital of the new state of Rio de Janeiro.[13] The merger took place at the beginning of the presidential term of General Ernesto Geisel (1974–1979), amid the military dictatorship imposed on Brazil between 1964 and 1985. It was in line with the purposes of the National Development Plan-II PND, implemented by Geisel. Among its objectives, the II PND aimed to reduce economic concentration in the state of São Paulo. In this line, the proposal was to expand the economic importance of the states of Minas Gerais and the new state of Rio de Janeiro (created from the merger of Guanabara with the old state of Rio de Janeiro). Thus, a second metal-mechanic complex was installed in Minas Gerais, anchored by the Fiat automobile industry. The project also involved transforming the former Guanabara and the former state of Rio de Janeiro into a single state, based on the thesis that Guanabara was "a head without a body" and the former state of Rio "a body without a head." In other words, the former state of Rio de Janeiro would become a retro-area for the installation of industries and other economic activities, with the city of Rio (former Guanabara) as a *hub*.[14]

The II PND also planned to support, in the new state of Rio de Janeiro, activities with some technological density from the implementation of a nuclear power plant and the support to research institutions, such as the Oswaldo Cruz Foundation, with research and production activities in health, and the area of Postgraduate Studies and Research in Engineering at the Federal University of Rio de Janeiro—COPPE/UFRJ.[15]

On the one hand, the II PND fulfilled its functions in order to make investments to expand the production of machinery and equipment in the country and to stimulate the creation of industries producing industrial inputs to meet the demands of the industrial and agricultural sectors.[16] On the other hand, the II PND did not fulfill the function of reviving the Rio de Janeiro economy, expanding its participation in the national economy, and reducing the economic power of the state of São Paulo.

According to Carlos Lessa (2000), this was mainly because the investments of the II PND in the state of Rio de Janeiro were linked to investments with greater technological density, which require longer maturation and which were primarily swallowed up by the economic crisis that settled in Brazil in the 1980s, generating what became known as the "lost decade".[17] This crisis represented a severe drop in the fiscal capacity of the Brazilian public power, thus limiting the state's investment capacity. In addition, it decreased its ability to cover expenditures, particularly

---

[13] On the merger, see Brasileiro (1979).

[14] On the subject, see Lessa (2000) and Osorio (2004).

[15] On the subject, see Lessa (2000).

[16] On the subject, see Castro and Souza (2004).

[17] On the Brazilian economic crisis of the 1980s, also known as the "lost decade," see Lacerda et al. (2000).

affecting the economy of the new state of Rio de Janeiro—considering that, in the 1980s, the presence of federal agencies in the state of Rio de Janeiro still had important significance.[18]

Thus, the territory that makes up the new state of Rio de Janeiro, from the merger, maintains, from the 1980s to the present day, a loss of participation in the national economy within a vicious circle.

This can be seen from a number of economic indicators. For example, between 1980 and 2019, the state of Rio de Janeiro, according to the IBGE, showed a loss of participation in the national GDP of 24.49%, the biggest loss among all Brazilian states. In the same direction, data from the Ministry of Labor and Social Security show that, between 1985 and 2020 (the most extended series available with the same methodology), the state of Rio de Janeiro showed a growth in formal employment of only 40.9%, against an increase in the country's total of 125.6%. The lowest growth is among all Brazilian states.

During this period, the state of Rio de Janeiro, the second federative unit in terms of formal employment in the country, behind only the state of São Paulo, was surpassed by the state of Minas Gerais, falling to the third position.

With regard to total formal employment in the manufacturing industry, the performance of the state of Rio de Janeiro in relation to other Brazilian regions was even worse. Between 1985 and 2020, the state of Rio de Janeiro experienced a drop in employment in the manufacturing industry of -34.4%, against growth in the country's total of 37.9%. Throughout this period, the state of Rio de Janeiro went from second to sixth position with regard to the number of formal jobs in the manufacturing industry. It was surpassed by Minas Gerais, Santa Catarina, Paraná, and Rio Grande do Sul.

The lack of economic dynamism and the low growth of the productive base for tax collection affected the fiscal revenue of the Government of the State of Rio de Janeiro. In 2004, according to data from the Ministry of Economy, the state of Minas Gerais surpassed the state of Rio de Janeiro with regard to the revenue generated by the leading state tax, the Tax on Circulation of Goods and Services—ICMS.

## 39.5 Impacts of the 1964 Civil-Military Coup on the Politics of Rio de Janeiro[19]

Alongside the transfer of the capital without fair compensation and the lack of integrated and systemic regional reflection on the city of Rio, its metropolis, and the state of Rio de Janeiro as a whole, a third factor also decisive for the vicious circle in Rio

---

[18] Although by the end of the 1970s, the new Capital of the Brazilian Republic was already consolidated in Brasília, with the transfer of all ministries, most public organizations, and foreign embassies, including federal hospitals and a significant part of the Armed Forces, remained in Rio. In the 1980s, the entire national directive structure of Social Security was transferred to Brasília.

[19] This article topic benefited from reflections developed in Versiani (2016).

de Janeiro since the 1960s was the civil-military coup that took place in 1964, which prevailed in Brazil for 21 years.[20] The coup was followed by a repressive cycle triggered by the publication of Institutional Acts: decrees with constitutional force issued by military personnel who then assumed the Presidency of the Republic. Against the resistance movements, there were innumerable arrests of the regime opponents and orders to revoke parliamentary mandates, in which Guanabara stood out as the most affected federative unit.[21]

In 1965, the direct elections to the Presidency of the Republic were suspended and all the political parties then existing were extinguished, and then the bipartisan system was established in the country, with the creation of the MDB and the ARENA, respectively, the opposition parties and the situation. The MDB was created then as an opposition party to operate in an authoritarian political system, with all exception legislation granted to allow for the revocation of mandates of opponents of the regime, suspension of political rights, and arrests, in addition to the cover-up of the practice of torture.

Formed by members who were far from being a group of shared ideas and ideals, the MDB emerged as a parliamentary front. At that time, it was the only formal channel for the opposition in the country and, within the party, politicians of different partisan origins and ideological tendencies that were roughly classified as "authentic," (parliamentarians with more combative positions against the dictatorship), "moderates," and "adherents."[22] Supposedly gathered around the common goal of opposing the dictatorship and defending the re-establishment of the democratic regime in Brazil, in practice, however, some members of the MDB joined the party evaluating the possibility of extracting electoral advantages by joining a legend that, they supposed, had the preference of the electorate they wanted to conquer.

Precisely in this direction moved Antônio de Pádua Chagas Freitas, then federal deputy for the city-state of Guanabara, who joined the opposition party and became part of the wing of the MDB then recognized as adherent: an opposition that in practice did not oppose. Chagas had understood that the existence of an opposition party was admitted by the military government within the limits of an authoritarian regime. Thus, he assessed the behavior of the Rio electorate, concluding that, with a well-established network of local support and the propaganda machine of the newspapers owned by him—*A Notícia* and *O Dia*—and also taking extreme caution in relation to any type of radicalization, his electoral possibilities, and political strength would be greater within the opposition party.

Chagas used his newspapers as the primarily vehicle for an intense campaign of affiliation to the MDB, trying to minimize the weight of the opposition to his leadership within the Rio de Janeiro MDB—already favored by the removal from the political scene of the most active opposition sectors of the state, from the cassations.

---

[20] On the Coup of 1964, see Fico (2004).

[21] See Santos (1990). The work, entitled "Que Brasil é esse?" (Which Brazil is this?), presents a series of electoral indicators, showing, among other points, that Guanabara was the federative unit that suffered the most from cassations in Brazil.

[22] On the different ideological currents of the MDB, see Versiani (2007).

This campaign gained control over all 25 zonal directories and the party's regional directory in Guanabara, which allowed the *Chaguista* current to facilitate/hinder the processes of party affiliation to the taste of their interests and still decisively influence the indication and veto of names for the formation of the party's electoral slates.[23] All these control mechanisms achieved by Chagas Freitas ended up inhibiting the strength of non-*Chaguistas* within the Rio de Janeiro MDB.

In 1968, the military regime hardened with the decree of the so-called Institutional Act $n^r$ 5, with a new wave of cassations. Despite the Act, in Guanabara, the MDB was victorious in the elections of October 15, 1970, for the State Legislative Assembly and the National Congress, being the only regional scope in which the MDB emerged victorious.[24]

The MDB's electoral strength in Guanabara, however, at that time, expressed less the victory of an active opposition to the military regime and more the influence that Chagas Freitas had achieved in Rio's politics, including his indirect election as the state governor, on October 3, 1970. Chagas Freitas was the only governor elected by the MDB in 1970, in an election held under the terms of Institutional Act $N^r$ 3, enacted on February 5, 1966, which determined that state governors and vice-governors would be indirectly elected by the State Legislative Assemblies. On that occasion, President of the Republic, Emílio Garrastazu Médici, warned that he would not accept opposition candidacies "that could represent a challenge to the revolutionary process" (Dias 2001, p. 3,683).

Therefore, within this authoritarian and coercive framework, Chagas Freitas' hegemony was constituted in Rio's politics. His indirect election to the government of Guanabara in 1970 was supported by the then commander of the First Army, General Adalberto Pereira dos Santos, and the minister of the Army, General Orlando Geisel—who awarded Chagas Freitas, in December 1970, the Peacekeeper's Medal.

Although the elected governor of Guanabara by the opposition party, Chagas Freitas adopted a line of conciliation with the regime, including the disclosure on *O Dia* of a good image of the military government and its achievements.[25] Célio Borja, elected, in 1970, as a federal deputy for ARENA, summarized: "Governor Chagas Freitas (…) did the mime of the opposition when it was necessary, especially at the time of the election, and then he understood and lived with the military government. He did not face the regime" (Motta 1999, pp. 155–156).

The election results of 1970 guaranteed the *Chaguista* group nothing less than occupying more than 50% of the Rio de Janeiro MDB bench for the Federal Chamber, which expressed a new force of *Chaguism* in Rio's politics. The speeches of these parliamentarians, in plenary, during the 1971/1975 legislature, were one of the homogenizing aspects of the group, with no criticism of the current military order.

---

[23] On the subject, see, for example, an article by Flávio Pinheiro published in *Veja magazine*, 06/28/1978, p. 28–31, entitled: "The miracle of Chagas—where the chief of Rio takes the votes, he has distributed among his disciples for 25 years".

[24] On the subject, see Motta (2000).

[25] See, for example, editorials from *O Dia*, 9/17/1970, 10/22/1970, 10/24/1970, 11/1-2/1970, and 11/7/1970.

In their pronouncements, the subliminal understanding was that the outbreak of the 1964 coup was a wise move and that the opposition had to unite with the government to share the responsibilities of power. In addition, the performance of parliamentary functions, for the members of the Chagas Freitas' political group, consisted basically in abstaining from issues related to the construction of broad social policies and dealing with the mediation of the restricted interests of their electorate.[26]

Thus, it was with the MDB hindered within an authoritarian political reality, in which violence was institutionalized, that the phenomenon of *chaguismo* was consolidated in the Rio de Janeiro political field. The cycle of cassations introduced by the military regime from 1964 onward, by removing from the electoral dispute the names of the combative opposition with political force in the state of Guanabara, ended up inducing a renewal of the opposition bench, with the strengthening of what became known as the *chaguista* style of performing, representing the consecration of the physiological and clientelist political bias.

The end of Chagas Freitas' first term in office in the Rio de Janeiro government coincided with the merger between Guanabara and the former state of Rio de Janeiro on March 15, 1975. At that moment, control of the new MDB in Rio de Janeiro was disputed between Chagas Freitas himself and Ernani do Amaral Peixoto, the party's main leader in the former state of Rio de Janeiro. Chagas Freitas won the contest, extending his dominance of the party in Rio de Janeiro to the entire Rio de Janeiro area. Then, in 1978, he was an elected governor again, now from the new state of Rio de Janeiro.

The clientelistic logics and practices anchored in the political hegemony of Chagas Freitas produced legacies and vicissitudes that survive in time until today, influencing behaviors in the universe of power relations in the city and state of Rio de Janeiro, and demanding concrete actions in the direction of overcoming them.

## 39.6 Socio-territorial Inequalities in the City of Rio de Janeiro

Inequality in the city of Rio de Janeiro is not just social. It is also territorial. The city is divided into five Planning Areas and 33 Administrative Regions.

Planning Area-1 (AP-1) covers the historic center and surrounding neighborhoods. It contains about 37% of formal jobs and 5% of the city's residents (data for 2020, from the Pereira Passos Institute/Rio de Janeiro City Hall and RAIS/Ministry of Labor and Welfare).

AP-2 is where a middle- and upper-class circle hegemonically exists. In it you will find the South Zone of Rio, where famous beaches such as Copacabana and Ipanema are located, and Administrative Regions-ARs in the North Zone of Rio with hegemony of middle-class residents: Tijuca and Vila Isabel.

---

[26] See Diniz (1982).

AP-3, where about 37% of the population of Rio de Janeiro lives, is a predominantly popular housing area, in which there are large favela complexes. AP-3 borders three large municipalities on the outskirts of the Metropolitan Region of Rio de Janeiro-RMRJ: Nilópolis, São João de Meriti, and Duque de Caxias.[27]

AP-4 covers two ARs, one for the middle and upper classes (Barra da Tijuca) and another with a strong presence of popular housing and favelas, such as Rio das Pedras and Cidade de Deus.

AP-5 is hegemonically a territory for popular housing, with several clandestine subdivisions organized by militias. It is located in a peripheral area of the city of Rio, where about 27% of the population live, presenting only about 8% of the formal jobs in the city of Rio.

The Social Progress Index-IPS,[28] organized by the Pereira Passos Institute-IPP for the 33 Administrative Regions-RAs of the city of Rio de Janeiro, taking into account indicators in the areas of health, sanitation, education, access to information technology, urban mobility, violence, and access to cultural equipment, corroborates the thesis that inequality in the city of Rio de Janeiro is also territorial.

Among the 33 RAs, the six that appear with the best IPS are exactly those where middle- and upper-class people reside hegemonically. They are Botafogo (IPS 85.0), Copacabana (80.2), Lagoa (79.0), Vila Isabel (73.0), Tijuca (71.6), and Barra da Tijuca (69.7). The first five are on AP-2 and Barra da Tijuca is on AP-4.

The six RAs that present the worst IPS are Portuária (42.0), Pavuna (43.0), Guaratiba (43.5), Complexo do Alemão (43.7), Jacarezinho (45.1), and Maré (47.2). Pavuna, Complexo do Alemão, Jacarezinho, and Maré are on the AP-3. Guaratiba, on the AP-5, and the Port Region, on the AP-1, are in a degraded area of downtown Rio.

Another indicator, the Social Development Index-IDS, created by the IPP, also analyzes the issue of inequality in the territory of Rio by the Administrative Region of the city. This index, however, considers not only indicators related to the supply and quality of social policies and existing infrastructure in each region but also indicators related to the per capita income of households.[29]

As pointed out in the IPS, the IDS shows that the six RAs in the best social situation are Botafogo (0.764), Copacabana (0.735), Lagoa (0.730), Vila Isabel (0.697), Tijuca (0.679), and Barra da Tijuca (0.676).

---

[27] The Metropolitan Region of Rio de Janeiro is made up of 22 municipalities: Belford Roxo, Cachoeiras de Macacu, Duque de Caxias, Guapimirim, Itaboraí, Itaguaí, Japeri, Magé, Maricá, Mesquita, Nilópolis, Niterói, Nova Iguaçu, Paracambi, Petrópolis, Queimados, Rio Bonito, Rio de Janeiro, São Gonçalo, São João of Meriti, Seropédica, and Tanguá. The indication "metropolitan periphery" refers to all municipalities except the capital of the state of Rio de Janeiro.

[28] The latest edition of IPS is for the year 2020. For detailed knowledge of the IPS parameters and methodology, see https://www.data.rio/documents/2b5e629dbe304a17a4fe6a5b58dd1d05/explore (accessed on 2/15/2022).

[29] The IDS was organized in 2013, based on data from the Demographic Census conducted by the IBGE. The last one available is from 2010. The closer the index is to one, the better the situation of the site. For detailed knowledge of the IDS parameters and methodology, see https://www.data.rio/maps/PCRJ::%C3%ADndice-de-desenvolvimento-social-da-regi%C3%A3o-metropolitana-do-rio-de-janeiro-2010/about (access in 15/2/2022).

The six RAs that present the worst IDS are Portuária (0.493), Pavuna (0.528), Complexo do Guaratiba (0.532), Alemão (0.533), Jacarezinho (0.534), and Maré (0.547).

Again, the AP-3 and AP-5 ARs in Rio appear in negative prominence. An exception is Rocinha, the largest favela in the South Zone of the city of Rio, located on AP-2, which appears among the six worst IDS. That must be related to the fact that the IDS includes indicators related to income per household. Many people live in Rocinha because it is in the South Zone of Rio—an affluent area with an offer of jobs in commerce, services, and upper- and middle-class residences, jobs that do not require high qualifications of workers and low-paid wages.

Within the vicious circle that was established in Rio, one of the central aspects, as mentioned, was, from the 1964 Coup, the emergence of a clientelist and destructive political logic. This has generated significant corruption in the police[30] and facilitated, especially in the city and its Metropolitan Region, the existence of "parallel states" controlled by drug trafficking or militias, progressively expanding their presence.

An article from the online news channel globo.com, dated 10/19/20, based on a study entitled "Maps of armed groups in Rio de Janeiro," points out that 3.7 million of the 6.7 million inhabitants of the city of Rio de Janeiro live in territories under the yoke of militia or drug trafficking, which control 57% of the city's area.[31] Again, most of these parallel territories are located in the Zona Suburbana and Zona Oeste of the city of Rio.

According to the same article: "In the Metropolitan Region, researchers have found that the tendency is for militiamen to gain wide dominance. According to the study, in 2019, there were 199 neighborhoods (21.8%) controlled by these criminal groups.

In the scenario in which the city of Rio and the territory that forms the current state of Rio de Janeiro are the regions with the lowest economic dynamism in the country and in which the public machine suffers particular degradation, the most affected part is the periphery of the Metropolitan Region of Rio de Janeiro-RMRJ.

Several studies that we have conducted, with indicators of health, education, formal employment, and violence, show that the metropolitan periphery of Rio de Janeiro is the most precarious compared to the metropolitan peripheries of the capitals of the states of São Paulo and Minas Gerais in Southeast Brazil.[32]

---

[30] On corruption in the police in Rio de Janeiro, see article by journalist Xico Vargas, entitled "The golden path," in 2006. http://ozeas.blogspot.com/2006/05/o-caminho-do-ouro.html (acesso em 15/2/2022).

[31] https://g1.globo.com/rj/rio-de-janeiro/noticia/2020/10/19/rio-tem-37-milhoes-de-habitantes-em-areas-dominadas-pelo-crime-organizado-milicia-controla-57percent-da-area-da-cidade-diz-estudo.ghtml (Access: 15/02/2022).

[32] The Southeast Region of Brazil has four states: Rio de Janeiro, São Paulo, Minas Gerais, and Espírito Santo. The first three have an estimated population in 2021 of, respectively, 17,463,349, 46,649,132, and 21,411,923. The state of Espírito Santo has an estimated population of only 4,108,508. For this reason, when comparing the situation of the metropolitan peripheries of the capitals of the states of the Southeast Region, we only compare those of the three largest states. For a more detailed analysis of the situation of the metropolitan periphery of Rio de Janeiro compared to the metropolitan peripheries of the capitals of the Southeast states, see Osorio (Coord) (2017).

For example, in a ranking for 2019 of the homicide rate per hundred thousand inhabitants of the municipalities with more than 50 thousand inhabitants of the metropolitan peripheries of the capitals of the states of Rio de Janeiro, São Paulo, and Minas Gerais, it can be seen that, for the 63-ranked municipalities, nine of the ten municipalities in the worst situation are on the periphery of the RMRJ. The municipality of Itaboraí is the one with the worst situation, with a homicide rate of 56.94. None of the ten municipalities with the lowest homicide rate are on the outskirts of the RMRJ. It should be noted that all these ten municipalities have a homicide rate per 100,000 inhabitants of less than 6.00.

In the same direction, in a ranking of the Basic Education Development Index-IDEB/Ministry of Education,[33] for teaching from the first to the fifth year of public schools, in 2019, among the 63 municipalities with more than 50 thousand inhabitants of the metropolitan peripheries of the capitals of the states of Rio de Janeiro, São Paulo, and Minas Gerais, it appears that all 19 municipalities of the metropolitan periphery of Rio de Janeiro are in the last positions.

In the health area, the Federation of Industries of the State of Rio de Janeiro created an indicator for all Brazilian municipalities called the FIRJAN Municipal Development Index-IFDM/Saúde.[34] In 2016 (last year with available data), among the ten towns with the best results, none were on the periphery of the RMRJ. Among the ten worst results, six municipalities were on the periphery of Rio de Janeiro.

Finally, the particular lack of formal private employment in the periphery of the RMRJ is evident in the ranking of the relationship between formal private jobs and population. For the year 2020, among the sixty-three municipalities with more than fifty thousand inhabitants on the outskirts of the Metropolitan Regions of the states of Rio de Janeiro, São Paulo, and Minas Gerais, no municipality in the periphery of the RMRJ is among the ten municipalities with the best results. Among the ten worst results, five municipalities were on the periphery of Rio de Janeiro.

## 39.7 Facing Social and Territorial Inequalities in Rio de Janeiro

As presented, the city of Rio, its Metropolitan Region, and the state of Rio de Janeiro, since the 1960s and mainly from the 1970s, have gone through a long vicious circle. The challenge, therefore, is to place these regions in a virtuous circle.

Albert Hirschman, in the classic "Strategy of economic development" (1958), points out that, for the socioeconomic development of a region, public coordination is, in the first place, necessary. The author proposes that it is not possible to say, a

---

[33] For a more detailed analysis of this indicator, see https://www.gov.br/inep/pt-br/areas-de-atuacao/pesquisas-estatisticas-e-indicadores/ideb.

[34] The variables used in the IFDM Saúde are number of prenatal consultations, deaths from ill-defined causes, and infant deaths from preventable causes. For a more detailed analysis of this indicator, see https://www.firjan.com.br/ifdm/.

priori, what starts the virtuous circle of a region because it is always a conjunction of factors that depends on each historical moment and territory's characteristics. In addition, the most challenging thing is to start defining the factors that would allow reversing the situation from a vicious circle into a virtuous one. Hirschman argues that, to the extent that one succeeds in building the *start* of a virtuous circle in a given region, its population begins to believe in the strategy, and the self-esteem generated helps in its consolidation and continuity.

The city of Rio de Janeiro is the capital of a territorially small state in the Brazilian scenario and is quite urban. In the city of Rio and its Metropolitan Region live 38.86% and 75.62% of the total population of the state, respectively.

In this sense, it is necessary to think of a development strategy for the city of Rio integrated with a policy for the whole of its metropolis, which does not yet exist. In addition, it is necessary to design and implement a policy that generates win–win games between the eight Government Regions of the state.[35]

It is urgent to design and implement a policy for Rio de Janeiro that addresses the severe inequality in the city, with impacts on the improvement of social indicators in those most vulnerable areas, located mainly in AP-3 and AP-5. In these two Planning Areas reside 64% of the population of Rio de Janeiro. For example, the Municipality of Rio de Janeiro, in partnership with the State Government, could organize a territorialized budget in order to establish priorities, with transparency, for the expenses incurred in the different regions of the city.

In the Metropolitan Region of Rio de Janeiro, a Metropolitan Chamber was created, based on a law called the Statute of Metropolises created in the country in 2015.[36] The proposal is to build public policies that affect two or more municipalities by that body and in partnership with city halls. It is also necessary to establish a metropolitan budget within the scope of the Metropolitan Chamber.

The precarious infrastructure concerning the provision of telecommunications, electricity, and logistical access services affects not only the population's quality of life on the periphery of the RMRJ but also makes it difficult to attract companies and consequently generate local jobs. Thus, there is a need for some *New Deal* for the metropolitan periphery of Rio de Janeiro, with the design and implementation of an integrated infrastructure policy. To that end, municipal, state, federal, and international funding institutions can be articulated, such as the New Development Bank—BRICS Bank, created in 2014 by the governments of Brazil, Russia, India, China, and South Africa.

Likewise, for the seven Regions of Government of the interior of Rio de Janeiro, it is necessary to design and implement social and infrastructure policies through partnerships between the municipalities and the State Government. This is urgent in light of rankings organized by the Observatory of Studies on Rio de Janeiro,[37] which

---

[35] The State of Rio de Janeiro comprises the following Regions of Government: Metropolitan, North, Northwest, Mountains, Coastal Lowlands, Center-South, Middle Paraíba, and Costa Verde.
[36] Law Nʳ 13,089/2015.
[37] Research Group linked to the National Law School of the Federal University of Rio de Janeiro, registered with the CNPq.

uses indicators related to employment density, health, education, and public safety. In these rankings, one can see a particularly terrible situation of the municipalities of Rio de Janeiro in the scenario of the cities of the Southeast Region of Brazil.

In addition to social and infrastructure policies, an economic strategy is needed. Take into account that Rio de Janeiro is the state in Brazil with the lowest rate of economic growth and an exceptionally high unemployment rate.[38]

It should be noted that implementing an economic strategy is also decisive for expanding public revenue, which allows for improving social and infrastructure policies. That is because financial promotion with a diversification of the productive structure broadens the basis for tax collection.

In the current technological scenario, it is increasingly challenging to define economic activities as activities of the primary, secondary, and tertiary sectors. Besides, it is more and more difficult to determine where industry ends and services begin or where services end and industry begins. Accordingly, a strategy to promote economic development must be based on the analysis of production systems, that is, a set of economic activities that present synergy, regardless of whether they cover more than one productive sector.

For example, in the Brazilian case, one of the proposals put forward for debate aimed at systematically promoting economic activities is the so-called Industrial Economic Complex of Health-CEIS. This complex brings together service activities in the areas of public and private health and the production of goods and services linked to health and technological innovation.[39]

Brazil, one of the world's largest countries in terms of population and territory and having a universal public health system, has an immense demand for goods and services related to this productive complex. According to data from the Ministry of Economy, in 2021, Brazil imported goods and services linked to the CEIS worth US$ 26.9 billion.

The scenario of the COVID-19 pandemic has exposed the fact that it is dramatic to depend on imports of goods and services related to health. Furthermore, importing these goods and services does not generate income and employment in the country.

The CEIS proposal contains, as one of its main points, the conduction of an import substitution program, seeking to expand production in Brazil and reduce imports in the health area—also developing technological autonomy in this area.

The proposal presents essential synergies with the characteristics of the city and state of Rio de Janeiro economy. It can be an impact factor in an inclusive development strategy.[40] In the city of Rio and its Metropolitan Region, there are respectful public production and research health institutions, such as the Oswaldo Cruz Foundation, universities with research centers, and significant pharmaceutical industry.

---

[38] According to the IBGE Continuous National Household Sampling Survey (PNAD Contínua) for the fourth quarter of 2021, the unemployment rate for the state of Rio de Janeiro was 14.2%, against an unemployment rate of 11,1% for the country.

[39] On the subject, see Gadelha (2021).

[40] For a discussion of the productive systems with the most significant potential for boosting the economy and generating employment in the city of Rio, its metropolis, and the state of Rio de Janeiro as a whole, see Osorio (Coord) (2014) and Osorio et al. (2021).

In the state of Rio de Janeiro as a whole, CEIS also stands out among the activities of production of goods and services, provision of public and private services, and health research. According to the Ministry of Economy, in 2021, there were 479,179 direct formal jobs linked to CEIS in the state of Rio de Janeiro.

Thus, the confrontation of socio-territorial inequalities in the city, its metropolis, and the state of Rio de Janeiro requires a coordinated policy involving, simultaneously, expansion and improvement of social and infrastructure policies, with priority for the areas most in need of these services, and an integrated economic approach, with preference given to the production systems with more significant potential in the city and state territory.

## References

Argan G (1964) L'Europe des capitales. Albert Skira
Bicalho MF (2011) A cidade do Rio de Janeiro e o sonho de uma capital americana: da visão de D. Luís da Cunha à sede do vice-reinado (1736–1763). Revista História 30(1):37–55
Bicalho MF (2003) A cidade e o Império: o Rio de Janeiro no século XVIII. Civilização Brasileira
Brasileiro AM (1979) A fusão: análise de uma política pública. Série Estudos para Planejamento -21. IPEA
Castro AB, Souza FEP (2004) A economia brasileira em marcha forçada. Paz e Terra
Dias S (2001) Emílio Garrastazu Médici. In: Abreu AA et al (Coord) Dicionário Histórico-Biográfico Brasileiro pós-1930. FGV/CPDOC, III, p 3.678–3.692
Diniz E (1982) Voto e máquina política: patronagem e clientelismo no Rio de Janeiro. Paz e Terra
Fico C (2004) Além do golpe: versões e controvérsias sobre 1964 e a ditadura militar. Record
Gadelha CAG (2021) O Complexo Econômico-Industrial da Saúde 4.0: por uma visão integrada do desenvolvimento econômico, social e ambiental. Cadernos do Desenvolvimento 16(28):25–49
Galton F (1979) Hereditary genius. Julyan Friedman (Original work published 1869)
Hirschman A (1958) Estratégia do desenvolvimento econômico. Fundo de Cultura
Knauss P (2005) Apresentação. In: Figueiredo C O Porto e a cidade: o Rio de Janeiro entre 1565 e 1910. Casa da Palavra, pp 8–12
Krugman P (1999) The role of geography in development. Int Reg Sci Rev 22(2):142–161
Krugman P, Fujita M, Venables AJ (2002) Economia espacial: urbanização, prosperidade econômica e desenvolvimento humano no mundo. Futura
Lacerda AC et al (2000) Economia Brasileira. Saraiva
Lessa C (2000) O Rio de todos os Brasis: uma reflexão em busca de auto-estima. Record
Mariani J (2015) Um sonho intense. Documentário
Mello JMC (1982) O capitalismo tardio. Brasiliense
Mendonça PK (1991) O Rio de Janeiro da Pacificação. Secretaria Municipal de Cultura, Turismo e Esportes
Motta MS (1999) Celio Borja. Conversando sobre Política. FGV
Motta MS (2000) Saudades da Guanabara. FGV
Moulaert F (2000) Globalization and integrated area development in European cities. Oxford University Press
Myrdall G (1968) Teoria econômica e regiões subdesenvolvidas. Saga
North DC (1993) Instituciones, cambio institucional y desempeño económico. Fondo de Cultura Económica
Osorio M, Rego HR, Versiani MH (2017) Rio de Janeiro: trajetória institucional e especificidades do marco de poder. In: Marafon GJ, Ribeiro MA (Orgs) Revisitando o território fluminense VI. Eduerj, pp 3–24

Osorio M (ed) (2017) Trajetória e evolução recente da RMRJ. In: Petraglia, CL, Leite VF (Coords) Caderno metropolitano 2. Centralidades: territórios de perspectivas para políticas públicas. Câmara Metropolitana de Integração Governamental, pp 8–21

Osorio M (2005) Rio local, Rio Nacional: mitos e visões da crise carioca e fluminense. Senac

Osorio M (2004) A fusão: equívocos e memória. Jornal Dos Economistas 181:3–4

Osorio M (Coord) (2014) A capacidade indutora dos serviços no estado do Rio de Janeiro. Sebrae/RJ

Osorio M, Marcellino I, Veiga L, Rego H, Falcão M (2021) O potencial representado pelo Sistema Produtivo de Petróleo e Gás no Rio de Janeiro e implicações para o desenvolvimento regional. Cadernos do Desenvolvimento, 16(29):165–196

Ridenti M (2007) Cultura e Política: Os anos 1960–1970 e sua herança. In: Ferreira J, Delegado LA (Orgs) O Brasil republicano: O tempo da ditadura - regime militar e movimentos sociais em fins do século XX. Civilização Brasileira, pp 133–166

Santos WG (ed) (1990) Que Brasil é este? Manual de indicadores políticos e sociais. Vértice/Iuperj

Schwarcz LM (1993) O espetáculo das raças. Cientistas, instituições e questão racial no Brasil - 1870–1930. Companhia das Letras.

Versiani MH (2007) Padrões e práticas na política carioca: os deputados federais eleitos pela Guanabara em 1962 e 1970. Universidade Federal do Rio de Janeiro, Dissertação de Mestrado

Versiani MH (2016) O Rio de Janeiro na República da Ditadura. In: Osorio M, Magalhães AF, Versiani MH (Coords) Rio de Janeiro: reflexões e práticas. Fórum, pp 128–147.

**Mauro Osorio** PhD in Urban and Regional Planning (IPPUR/Federal University of Rio de Janeiro-UFRJ). Professor at the National Faculty of Law at UFRJ. Coordinator of the Observatory of Studies on Río de Janeiro (FND/UFRJ). Director-President of the Fiscal Advisory Board of the Legislative Assembly of the State of Rio de Janeiro. Founding member of the Institute of Studies on Rio de Janeiro-IERJ. Author of the book "Rio Nacional, Rio Local: myths and views on the crisis in the state and city of Rio de Janeiro".

**Maria Helena Versiani** PhD in History, Politics and Cultural Goods (CPDOC/Fundação Getúlio Vargas), with Post-Doctorate in History (Universidade Federal Fluminense-UFF). Researcher at the Museum of the Republic. Guest professor at the Professional Masters in History at UFF. Researcher at the Observatory of Studies on Rio de Janeiro (FND/UFRJ) and at the Institute for Studies on Rio de Janeiro. She publishes and develops studies especially in the field of relations between history, memory, political culture and cultural heritage.

**Henrique Rabelo** Economist and Master in Urban and Regional Planning (IPPUR/Federal University of Rio de Janeiro-UFRJ). Researcher at the Observatory of Studies on Rio de Janeiro (FND/UFRJ). Director of the Fiscal Advisory Board of the Legislative Assembly of the State of Rio de Janeiro. Founding member of the Institute of Studies on Rio de Janeiro-IERJ.

# Chapter 40
# The Implosion of Memory. City and Drug Trafficking in Medellín and the Aburrá Valley

Luis Fernando González Escobar

**Abstract** In Medellín (Colombia), the implosion of a building by local authorities, as it was considered a negative symbol inasmuch as it caused a bad image for the city, is the leitmotiv to question the reasons for its demolition and to propose a more complex insight with regard to the origins and permanence of drug trafficking in urban planning and urban culture. This is what this text proposes, a look not at an episodic or circumstantial event but as part of a complex framework that involved a strong urban conflict, with triggering effects not only on the architectures built by illegally injected besieged capital, but on recognized urban works, while, in the configuration of the urban landscape, and in the social and cultural practices that are projected up to the present and that in some way have already been naturalized. The implosion of a building, in turn an "uncomfortable heritage", is the metaphor of a city and a society besieged, that although at the time were almost in "ruins", did not succumb and were reconfigured, even incorporating new non-recognized practices, but that are latent there as part of their cultural language and landscape.

**Keywords** Medellín · City · Drug trafficking · History · Cultural landscape

## 40.1 Introduction

The 1980s and 1990s were decades marked by urban violence in the city of Medellín (Colombia). A time of terror in which "homicide was the leading cause of overall mortality since 1986 and its share of total deaths increased from 3.5% in 1976, to 8.0% in 1980, to 17.0% in 1985 and peaked at 42.0% in 1991" (Cardona et al. 2005:

---

Translated by Maria Clara Echeverria Ramirez.

L. F. González Escobar (✉)
National University of Colombia, Medellín, Colombia
e-mail: lfgonzal@unal.edu.co

© The Author(s), under exclusive license to Springer Nature Switzerland AG 2023
F. Carrión Mena and P. Cepeda Pico (eds.), *Urbicide*, The Urban Book Series,
https://doi.org/10.1007/978-3-031-25304-1_40

840), the year in which it was considered the city with the highest homicide rate in the world.[1]

The Latin American economic crisis of the 1980s expressed at the local scale in the decline of industry, in what was considered the "industrial city" of Colombia. With unemployment and informality on the rise, drug trafficking was the way in which this crisis was largely mitigated. But after a few years of certain collusion and social acceptance of that new emerging class, represented by drug traffickers, the violent actions and terrorism of the so-called Medellín Cartel marked one of the critical and cruelest periods of the city,[2] with selective assassinations, bombs and car-bombs, massacres, among many other cruelties. By having former allies against them in collusion with sectors of the state, a confrontation of such magnitude was generated that not only murders but also bomb attacks and massacres were the way they reacted. The city experienced moments of great uncertainty. Fear gripped its streets. The use of public space and the revels was lost. The drama of its urban landscape was intensified with the paralysis of the main construction work: the viaduct of the metro system, which turned the city into an apocalyptic scenario. It became known as *Medellín No Futuro* (Non-future Medellín).

This can be considered as a true *urbicide* since it typifies, not as much the absolute death of the city, but rather that it was introduced into an anomie, due to the fear that violence produced. It was not just about the violent destruction and urban chaos, but also the loss of the meeting and socialization place for many social sectors. At the same time, during these years, the prejudice and discrimination and socio-spatial segregation of low income barrios were accentuated, being seen only as places where death emerged. Although, since the last decade of the twentieth century, the city, the ruling class, and society responded to such dynamic by means of a process of urban transformation and resilience that has been recognized on a global scale, it is also true that those *urbicidal* considerations were also transformed in something less visible, but no less dramatic through territorial controls, the management of illicit income and drug trafficking. Other expressions, such as fear and silence, illegal and legal controls of spaces and social relations, intra-urban displacement, direct or symbolic violence mark the territorial current situation not only in Medellín but on a metropolitan scale.[3]

---

[1] The Medellin Report *¡Basta Ya!* (Stop Now!) recognizes 6819, but underreporting has been pointed out in these data. Due to it, the figures even rise to 7237 murders, with a rate of 266 homicides per 100,000.

[2] The aforementioned Medellin Report ¡Basta Ya! establishes four periods: a first period, 1965–1985, that of the antecedents that shaped the factors that later triggered the armed conflict; the second, from 1982 to 1994, properly associated with the Medellín Cartel, but also associated with political violence, and a "dirty war" due to the complicity of state institutions; the third, 1995–2005, when the paramilitary and rural guerrilla expansion turned Medellín into the geographical and logistical center of the conflict; and the fourth period, 2006–2014, marked by the decline of the guerrilla and paramilitary forces, and the containment of violence by the state (National Center for Historical Memory 2014; 23–25). Although, as will be seen at the end of this text, it is a reconfiguration of the conflict.

[3] The *Área Metropolitana del Valle de Aburrá* (Valle de Aburrá Metropolitan Area) incorporates 10 municipalities: Barbosa, Girardota, Copacabana, Bello, Medellín, Envigado, Itagüí, Sabaneta, La Estrella, and Caldas.

This *urbicidal* mutation is seen in this work, viewed from a historical perspective. Perhaps, it is necessary to state that *urbicide* can also be typified and recognized in urban historical denialism, insofar as the causes that limit or obstruct the validity of the city with its multiple problems are—intentionally or not—disregarded.

This work sets as its starting point the discussion raised by the demolition of the Mónaco building,[4] which has become emblematic and a space of dispute related to how to deal with the memory of drug trafficking in the city of Medellín. Its demolition does not bury the intricate causalities that are woven between the illegal and the legal, the material and the immaterial, of many of those factors that determined the serious and complex urban problems. Throughout the narrative developed in this work, this near past and some of its causalities are explored, in order to arrive again at the present moment, where determining elements are perceived as belonging to a new urban culture, while others continue to have a negative impact on the desired transformations and urban vitality or, at least, in those which are promoted as such.

## 40.2 Demolishing the Ignominious Past

On February 22th, 2019, in the city of Medellín (Colombia), a building was imploded. This action was not just any act, as it involved demolishing one of the buildings considered emblematic, one of the several that the infamous drug trafficker Pablo Escobar Gaviria ordered to be built. He collected works of art and cars on his eight floors, where his family lived, and for several years, it was his place of residence and refuge from persecution by the authorities and his own illegal enemies. It was precisely there that on January 13th, 1988, the drug rivals attacked him, with a "car bomb" loaded with 700 k of dynamite. That only affected a little the reinforced structure of the building, although it affected a lot the elite neighborhood of the Barrio Santa María de los Ángeles. It was the beginning of the bloody war between rival groups of drug traffickers from the cities of Cali and Medellin, known by the DEA and the US government as the "drug cartels".

The demolition was accompanied by a ceremony presided over by the mayor of the city, Federico Gutiérrez (2016–2019), with the presence of hundreds of guests who were located in the parking lot of the *Club Campestre* (Campestre social club), just two blocks away from the building. Among those guests were a representative group of relatives of drug trafficking victims, such as the widow and a son of the murdered presidential candidate Luis Carlos Galán, among others. Perfectly aligned chairs looking toward the building to implode were located, offer of drinks, a stage with renowned artists, souvenirs designed and delivered at the entrance of the commemorative event… In short, it was a true media event, not only at the demolition site and its surroundings, but also due to the live broadcast on the official local and regional

---

[4] Monaco was one of the real estate properties owned by Colombian drug trafficker Pablo Escobar, located in the barrio Santa María de los Ángeles, in the exclusive sector of El Poblado.

television channels, as well as on social networks, in addition to the presence of national and foreign journalists.

Those who were not invited to the Club Campestre parking lot found themselves in the immediate vicinity, on nearby streets, or watched the event from the balconies and terraces of apartments in neighboring buildings. The buildings, the demolished one and the one for the event, were located in the Barrio El Poblado, the site that was by excellence a territory of economic and social power, to the point that not only there was the elite club and in its vicinity the building of the most persecuted drug trafficker, but also mansions like Montecasino. This neighborhood was also an enclave of paramilitarism, where the clan of the Castaño brothers, meeting with political and business allies, determined actions against Pablo Escobar, executions of urban and rural social leaders, scorched earth actions in different parts of the Colombian territory or metropolitan territorial control.

After the sirens alert, the explosion of the charges sounded in a sequence separated by microseconds. The culminating moment lasted just three seconds. The structure collapsed, succumbing in a cloud of dust that rose with force, despite the powerful streams of water. Dragged by the wind, it began to cover the place of the guests. While the salsa music singer, Yuri Buenaventura, made an effort to continue his song, he at the same time was devoured by that grayish veil that transformed him into a phantasmagoria. While the people applauded, someone shouted "Pablo son of a bitch, there you left us…"!

## 40.3 Medellín Embraces Its History

The media spectacle of the implosion was part of a project with which the municipal administration tried to give context to this event, that is, the proposal called *Medellín abraza su historia* (Medellín embraces its history).

The annoyance of the municipal mayor with regard to the bad image generated by the visit of tourists and, above all, urban music singers who made tours of places referenced as significant milestones in the memory of the drug lord, prompted the idea of demolishing the Mónaco building. The so-called *narco tour*, that tourists did and continue to do, includes stops at different places considered emblematic in the criminal life of that sinister character, such as the building he constructed for his family, the prison that he also made to his measure where he confined himself for a time while he continued committing crimes, the house where he was murdered, and even his tomb, converted into a place of pilgrimage and sanctuary of people's religiosity. It was not only such visits, but also the images that circulated on social networks of those visitors and which, in some cases, included the use of drugs on said sites. Those replicating images were considered apologetic, in a narrative that followed the line of famous television series and streaming platforms. The criticism was hardly logical, being considered as an imposition of the narrative seen from the point of view of the perpetrator and an apology for evil, ignoring the victims and the efforts of local administrations to change the state of affairs.

In this sense, demolishing the Monaco building and instead designing and building a memorial for the victims was considered a change of perspective. Given the initial simplicity of the proposal and with the aim of presenting it as an articulated project, the administration hastily included the strategy called Medellín Embraces its History, outside the municipal Development Plan, approved during its administration. This strategy included other components, besides the implosion and the memorial: a *Tour por la memoria* (tour through memory) as a reply to the *narco tours*, in this case that cover sculptures, plaques, and places where violence was exerted and where the victims were highlighted; a chair of memory: *En colegio, Medellín abraza su historia* (At school, Medellín embraces its history), a pedagogical activity in educational institutions, both private and public, in order to bring young people closer and sensitize them to that painful past and ethical changes in urban culture; the construction of the second phase of the Casa de la Memoria Museum, whose first phase had been inaugurated and put into operation between 2011 and 2012, but the project was not completed, among other projects that were mentioned but not developed, such as the continuity of the program *parceros* (partners, as bro), with the intention of inclusion and employment for young people.

The final result of the program was really languid. The implosion of the Monaco building was fulfilled, and in its place, the memorial was built, apart from commissioning the realization of some works of art from local artists—some of them deplorable—and the publication of the memoir *Historias para no olvidar* (Histories not to be forgotten), which circulated almost clandestinely, as is the case with much of the official production.

The *Parque Memorial Inflexión* (Inflection Memorial Park), as it was called, was the result of an *Concurso público internacional de anteproyecto arquitectónico para el diseño de un espacio de memoria y reflexión* (Public International Contest for an Architectural Blueprint for the Design of a Space for Memory and Reflection), convened by the Mayor's Office and organized by the *Empresa de Desarrollo Urbano—EDU* (Urban Development Enterprise—EDU) and the Sociedad Colombiana de Arquitectos—SCA (Colombian Society of Architects—SCA) in 2018. Among the 45 proposals, in December of that year, the one proposal presented by the firm *Montajes de Marca S. A.*, from Medellín, was chosen as the winner, whose designers were actually a group of young architects from the *Universidad Nacional de Colombia, Sede Medellín* (National University of Colombia, Medellin campus)—Carolina Henao Salazar, Felipe Zapata, Tomás del Gallego and Germán Tamayo—gathered in the office: *Pequeña Escala Arquitectura* (Small Scale Architecture office). After the implosion, the removal of debris, and the transition from the preliminary project to the architectural project, the work was carried out in a hasty manner, hurrying in order to be inaugurated on December 20, 2019, while the cement was still fresh, and the construction was unfinished. Again, another media event was organized with great pomp, presided over not only by the municipal mayor but also by the president of the republic and, again, with the presence of relatives of drug trafficking victims.

The memorial park is developed, in accordance with the designers themselves, with the victims as the central axis of the urban proposal. In an area of 5300 m$^2$, three fundamental spaces are included: (1) *The Camino de los Héroes* (The Path of the Heroes) as a pedestrian entrance path, flanked by monoliths in homage to personalities immolated by drug traffickers; (2) The *Memorial Inflexión* (Inflection Memorial), a wall 70 m long by 5 m high, but broken into four bodies or types of panels, with 46,612 holes, which are illuminated at night, to commemorate the same number of people who were calculated as the victims in the fateful period of the city of Medellín between 1981 and 1994; in addition to a chronology carved in stone of 208 violent events that occurred in the city in those years; (3) The *Bosque de la Resiliencia* (Resilience Forest) which, as its name indicates, the planted tree species symbolize the resistance and survival capacity of society, to stabilize and rise above uncertainty. This forest is separated from the neighboring buildings by means of gabions filled with the debris from the demolition of the Monaco building, as a historical trace of its pre-existing. Additionally, sculptural elements and grandstands were implanted for the events to be held.

Consistente with the narrative that the urban project was attempting to implement, the book *Historias Para No Olvidar* raises the ideal of recognizing the victims over the perpetrators, defining new heroes that would overcome those established by the mass media, the series and narco-novels. In addition, it talks about the social groups and the projects that opposed the overwhelming action of drug trafficking, valuing citizen resistance, and a memory that exalts the achievements and the overcoming of those hard, violent, and ignominious moments that we face before the world. To a certain extent, the attempt is representative because it brings cruelly sacrificed personalities out of oblivion and exalts their valor, but it is still an official memory, as are the historical contextual elements that were used, without deep causal explanations of that past that seems to have been overcome and have been left behind.

A certain simplification and an interested narrative clipping are evident, as when it is pointed out in the book:

> To understand that reality you have to remember. In the sixties, this violence emerges with the appearance of the first traffickers. The urban and dynamics of Medellín were changing in a swifting way, due to the arrival of thousands of people from the countryside in search of opportunities. The smuggling of liquor, cigarettes, and electrical appliances took a back seat with the appearance of cocaine trafficking. (Alcaldía de Medellín 2019: 103)

Thus, in an almost elementary and naive way, it is described that drug traffickers and drug trafficking "appeared" out of nowhere, that they emerged without apparent reasons of causality, outside of all the conditions of possibility in the historical, economic, demographic, social, cultural, and urban context.

Facts that are so obvious, derived, and interdependent, such as the history of drug trafficking illegality in relation to the different forms of smuggling—a long history that establishes relationships and continuities with markets, ports, routes, organizations, and practices rooted in society—are not looked at as determinants but only as something surpassed by a new illegal phenomenon. It even seems that the figure of Pablo Escobar was an anomaly within society. An exceptional evil: It arises unexpectedly, shocks society, and affects it with its illegal and criminal practices, but such society overcomes it, not without drama, as a result of the action of men and women who managed to eradicate the greater evil even at the cost of their own lives. Society, then thus, managed to return to its normal channels.

Memory in such narrative is seen as the necessary remembrance of that fatal historical moment with its terror and its horrors that, with few exceptions, was already something overcome, with which the city had been transformed in the "physical" and in the social; as it was stated in the texts of the book and proclaimed by the mayor in the media events with regard to the implosion of the building and the inauguration of the memorial: "Medellín is a resilient city, Medellín has a society that knew how to get up when it had its worst difficulties, and today, we are a great symbol of resilience that honors the victims".[5]

And in a certain way, it is true. The city did not entirely succumb to terror. It overcame many difficulties—not all of them, nor most of them—and projected a renewed international image, but this was done from partial constructions of recent history and intentional forgetfulness. As such: without establishing in-depth causal relationships with that immediate past and, above all, without understanding which dynamics were maintained, how others changed and which, even today, are part of its own social, cultural, economic, and territorial reality. In one sense or another, there are always unconcealable traces.

## 40.4 A Benefactor of the Low Income Barrios

The Pablo Escobar-Medellín relationship, no matter how much urban marketing is done, will not dissolve as easily as it is intended. It is and will be in an indissoluble relationship like Al Capone's Chicago. Rooted both in the imagination of tourists who visit the city, fascinated by the narratives of the series that are uncritically consumed in the world, as well as in many low income sectors that saw him and still see him as a true hero, both by those who had the direct experience with him or as an imaginary product built in the barrios. To a large extent, for these sectors, he represents that character who built himself from below, going over all opposition—legally or illegally—until reaching the top, disputing capital and power with the powerful and bending the state despite its criminal and murderous forms, but that are taken as obvious, they are ignored in an interested way and, even, considered as venial; from that ethical relativism; insofar as the image of a generous being who shared wealth with the poor had been configured, in such a way that at a certain moment, he was considered the *Robin Hood Paisa*[6] (The Paisa Robin Hood), as he was titled in 1983 by the magazine Semana, an important printed medium of the capital of the republic.

---

[5] This quote, taken from official press information, appears on several virtual pages which gave the news of the inauguration of the Inflection Memorial Park, such as: https://www.eje21.com.co/2019/12/medellin-inaugura-park-in-memory-of-the-thousands-of-victims-of-drug-trafficking/, among others that repeat the phrase.

[6] *Paisa* is the way in which those who belong to the region of Antioquia and the Coffee Region of Colombia are commonly called.

Escobar created a social and political base which led him to occupy a seat in the lower house of the Congress of the Republic and even be part of the official Colombian delegation in the possession of the Spanish socialist president, Felipe González, in 1982. He achieved all this by working strategically on the projects that he promoted from the *Medellín Cívico* (Civic Medellín) newspaper. This written medium belonged to his uncle, Hernando Gaviria Berrío, since 1958, was promoted as "the first ecological newspaper in Colombia", which even had a discreet national recognition among those who promoted ideas that were so little widespread at that time.[7] Leaning on his uncle, on other well-known journalists in the city, on poets who became his ghost writers, signing some written columns, Escobar promoted two great campaigns from there: *Civismo en Marcha* (Civism on the Move) and *Medellín sin Tugurios* (Medellin without Slums).

The *Civismo en Marcha* project was promoted as something unprecedented in Latin America and in the world. A weekly civic feast took place, where the problems and concerns of the city's neighborhoods were aired. It involved the prior formation of a Civic Committee, in charge of coordinating activities, receiving materials and improving its spaces, schools, health centers, highways, and sports facilities. The inaugurations were held on Sundays, with live broadcasts by radio stations in the city. It included a stage for festive, sporting, and recreational activities, such as prize pools, representative artists from the city, and the inclusion of neighborhood artists. It was an inclusive dynamic, where the communities were benefited and felt represented, including the most excluded, such as the black communities, for whom he promoted *la casa de las negritudes* (the house of black people). In this project, he developed one of the most effective interventions that he called the *estadios populares* (people's stadiums), through which he intervened, adapted, and illuminated some 120 soccer fields in the neighborhoods of Medellín, Envigado, Caldas, among other municipalities of the Valle de Aburrá. The Saturday night openings, with massive attendance, in some cases included the participation of well-known local soccer players, born in these *barriadas*, but turned into idols of the professional teams and of the Colombian national team that were summoned by Escobar, who took a dip in popularity doing the honorary kick-off and even playing.

The other project, *Medellín Sin Tugurios*, arose after the fires of more than one hundred "slums" between February 1982 and 1983, in a sector of the Barrio Moravia invasion (Moravia quarter), that is, in the Medellín city's garbage disposal mountain. The proposal to move the affected families to a new barrio built in the central eastern part of the city, financed by Escobar, was promoted by the newspaper in a laudatory manner: "They will come out of the pestilent hell of the garbage dump to start a life

---

[7] Television presenter Gloria Valencia de Castaño wrote in 1983: I belong to the group of lucky people who receive the first ecological newspaper in Colombia on a monthly basis. 25 years ago "Medellín Cívico" appeared, and since then, the director has put its pages at the service of a cause that at that time was quite strange, and today is everyone's concern: the defense of the environment. Hernando Gaviria goes so far in his goals that in the delivery for the month of January he asks for the creation of an "ecologist party" and a "ministry of the environment" (Valencia de Castaño 1983).

of noble dignity", to finish off in a rhetorical and anti-state way: "…there is a background of rebellion against misery, against social inequality, against abandonment and oblivion. But it is a creative rebellion. That is why we say: It is a man and his work. A work that will make history in the American continent"! (Anónymous 1984: 1). The slower response of the local government was used for the electoral purposes that he promoted at that time from the *Movimiento de Renovación Liberal* (Liberal Renovation Movement), with which he would reach the House of Representatives.

He built from scratch, a neighborhood of narrow streets with 200 houses. The new inhabitants baptized it Barrio Pablo Escobar, contrary to the official attempts to name it in a different way, either by extending the toponym of the adjoining barrio of Loreto or that of *Medellín sin Tugurios*. But the tribute to the benefactor was maintained, which is projected until the present. The inhabitants remember him as someone who, although "he did many bad things… he also did good deeds", as demonstrated by the barrio itself through its houses with simple but significant architecture for them, where they could live near the city center and as Christians. Thus, from people's religiosity, they took up and enthrone the saint to whom the drug trafficker was devoted, to whom they thanks, as a plaque says: "Holy Infant Jesus of Atocha. You have come to the Barrio Pablo Escobar Gaviria to help us. You will be our king and our savior. Thank you for taking care of our benefactor. Inhabitants of the Barrio Pablo Escobar Gaviria. November 29, 1992".

A sanctuary whose center is the image of the Holy Infant Jesus of Atocha in a large niche, with a dome with a mural painting of the Holy Spirit radiating a beam of light, with thanks plaques around it. It also has mural paintings by graffiti artists from the barrio: the Paso Fino horse, *Terremoto* (Earthquake), owned by Pablo's brother; a landscape with the façade of the Hacienda Nápoles that belonged to Escobar and above that façade the iconic HK-677 airplane, with which he transported the first shipments to the United States; a helicopter arriving at the plantation swimming pool, portraits of the other drug lords such as his brother Roberto and his cousin Gustavo Gaviria; Gonzalo Rodríguez, alias the Mexican; the Ochoa brothers—Juan David, Jorge Luis and Fabio—the most recognized and feared assassins of the Medellín Cartel, with their aliases, such as Otto, Arete, Negro Pabón, Cacho Chino… Being it as a museum with the statue and various photographs of Pablo Escobar. Other plates of thanks for the favors received appear at the foot of the sculpture of the holy child, which they say was brought from Spain, is a photo of the eternally smiling Pablo. And in the immediate vicinity, there are micro-businesses selling t-shirts with prints of Pablo's face, including one with a black and white photo taken by the judicial authorities with the plaque *Cárcel DTTO (distrito) Judicial Medellín 128482* (DTTO Medellín Judicial District Jail). That prisoner number, which was supposed to be a moral sanction, ended up being a people's iconic element, enhanced by the cynical smile of that man who looks at us from the past.

## 40.5 It Was not the Rule, but It Was not the Exception Either…

The so-called *Patrón del Mal* (Boss of evil) is far from being an exception. It was largely the result of complex dynamics, in a time of transition of the city and society in various aspects.

It is necessary to observe the very history of drug trafficking. Colombians occupied a secondary place in the 1960s in an illegal market dominated by Cubans and Chileans, but in which Italians, French, Americans, Argentines, Bolivians, and Ecuadorians also participated; but by the early 1970s, they were already recognized as the main exporters of cocaine. In these years, there was a systematic and profound incorporation in such traffic, in which different social sectors from Medellín played an important role. Between the 1950s and 1960s, researcher Eduardo Sáenz Rovner states that:

> drug trafficking "across borders was not a business for the poor; it required a certain know-how, capital, and international connections. As Colombia became more integrated into the North American economy, it improved its communications with the rest of the world, it rapidly urbanized and incorporated broader social layers into drug trafficking, and crime became a perverse means of economic and social mobility". (Sáenz Rovner 2021: 45)

The records show that Colombia did not produce the cocaine. Instead, the cocaine entering the United States arrived legally and in a controlled manner from the Merck laboratories in Germany, and then, it was illegally re-exported to the United States. Some of the pioneers were the twin brothers Rafael and Tomás Herrán Olózaga. Rafael was a pharmacist, owner of the *Farmacia Unión* in the city of Medellín, who, since 1944, was on the radar of the police for suspected illegal activities.[8] Apparently since 1952, they stopped being re-exporters to produce and export directly, when they installed their own laboratory in a villa in the exclusive Barrio El Poblado, on the outskirts of Medellín. This same year, the Medellin Pharmacy Inspection discovered several cases of illicit cocaine trafficking in the city. It became known that the production was not only for local consumption but also for the international market, when in 1956, the brothers were captured in *La Habana* with a shipment of heroin, for which Tomás, who was a pilot and apparently the leader of the gang, was sentenced to jail. The following year, that is, in 1957, the laboratory was raided by Colombian and North American authorities (Sáenz Rovner 2021: 45). This particular case raises two aspects for the investigator Sáenz Rovner: (1) The Herrán Olózaga brothers were pioneers of a criminal business; (2) These events somehow marked the transition from the "prehistory" of drug trafficking in Colombia to the "modern" history of the phenomenon (Sáenz Rovner 2021: 46). In addition, according to his own ideas, drug trafficking "becomes the dark side of globalization and of the integration of economies" (Sáenz Rovner 2021: 69).

---

[8] In that year, "he asked a German factory for the prices of cocaine and heroin in quantities greater than one kilogram. Both the Colombian authorities and the German police suspected that it was illicit trafficking" (Sáenz Rovner 2021: 40).

The Herrán Olózaga brothers, descendants of former presidents of the republic, were not the only people from the elite who joined drug trafficking, as other descendants of former presidents such as the Ospina Baraya brothers[9] also did so, as well as members of prestigious families of society, such as the Bravo brothers,[10] the Londoño White brothers, among other characters from well-known families[11] who participated in and controlled politics not only in Medellín but also in the urban centers of the north and south of the Valle de Aburrá.

The emergence and consolidation of drug trafficking in the city of Medellín are coupled, on the one hand, with the economic recession and the irreversible decline of the productive sector, "which testifies the beginning of the decline of hegemony and the reorganization of the Antioquia's industrial fraction",[12] and on the other hand, the loss of the political control that it held in an absolute manner, and the social and political ascent of subordinate regional sectors. The researcher Vilma Liliana Franco summarizes the 1970s as the transition from the local plutocracy—traditional and dominant until then—known as the "notables", and the consolidation of the "emergents", in the spheres of local and regional power. As such, she makes reference to the arrival of politicians of popular origin, of peasant parents, coming from the towns of Antioquia, "who were an integral part of the subordinate sectors of the socioeconomic structure and the territorial periphery of the region" (Franco Restrepo 2005: 207), not without resistance, and strong class and racist reactions.

For the same reason, it does not seem to be a coincidence either that in the consolidation of drug trafficking, there were also characters from those same subordinate sectors, those who allied themselves with members of the elite or directly assumed their own risks. Thus, personalities such as Benjamín Herrera Zuleta, from the town of San Roque (Antioquia), called the *Papa Negro* (Black Pope), who is credited with being the first great *narco* that Colombia had; Griselda Blanco, who arrived with her mother from the Caribbean Coast in the Barrio Lovaina (Louvain), where she worked as a prostitute, before settling in Barrio Antioquia, which she turned into the great center of her activities; Santiago Ocampo Zuluaga from the town of Santuario in eastern Antioquia and considered the *Papá de los Pollitos* (The Chickens' Father), as he was considered to be the true initiator of this new stage of drug trafficking in Antioquia; or Alfredo Gómez, from Caramanta, in the southwest of Antioquia, who

---

[9] The brothers, Javier, Rodolfo, and Mariano Ospina Baraya, were grandchildren of former President Mariano Ospina Pérez (Castle 1987: 66 and 67).

[10] Alberto Bravo "high-class paisa gentleman, well-spoken, and graduated from *Colegio San Ignacio*". His brother, Carlos Bruno, was a former soldier who was in the Korean War (Soto 2013: 22).

[11] Like the case of Alberto Prieto Escobar, "a powerful individual who managed the strings of power in Antioquia for years" (Soto 2013: 61).

[12] "The fundamental change is in the transition from the large corporations and the economic guilds of those fractions of the ruling class to the formation of an economic group. This constitutes both a new form of property organization and a new center of power, which also integrates different fractions of the ruling class" (Franco Restrepo 2005: 163). That economic group, called *Sindicato Antioqueño* (Antioquenian Union), was formed in 1978 and would be renamed the *GEA—Grupo Empresarial Antioqueño*—as it is known today.

was known by the alias of *El Padrino* (The Godfather), among many others, who preceded and were in different ways referents and sponsors of Pablo Escobar himself. And, with them were other urban subaltern groups, already from the same middle class that also played a crucial role in seeking spaces of social and economic power that they did not have or were quite limited in the crisis of those years,[13] or from low income neighborhoods such as barrios Antioquia, Castilla, Villahermosa, Manrique, or Aranjuez, which emerged to dispute power, either as hitmen, lieutenants, middle managers and even at the top, albeit in an ephemeral manner.

The lineage of smugglers turned into drug traffickers was key in the transition to drug trafficking. Smuggling was a long-standing practice, taking advantage of the geostrategic location of Antioquia, in the northwest corner of South America, which allowed it to go out to the Pacific and, especially, to the Caribbean, since the end of the eighteenth century when it was trading with Jamaica, without informing to the Spanish Crown. Some and others, notable and emergent, used the routes from Panama to Turbo or to the different ports of the Gulf of Morrosquillo—especially Tolú and Coveñas—and from those ports, by road to Medellín, cigarettes, liquors, electrical appliances, and merchandise entered surreptitiously. Those were sold in the so-called *Sanandresitos*[14] and in stores in the center of the city. The routes, networks, and practices of this smuggling supported a good part of the beginning of drug trafficking, and it was also backed by the social acquiescence and venality of the authorities, who did not see these practices as problematic at all.

With the substantial increase in money from drug trafficking, this activity served as a front to "launder" the assets and legalize them, although the torrent was so great that it was not enough. Other possibilities were necessary, such as the purchase of real estate and land, the construction of houses and buildings. Suburban farms or large extensions of land served to establish airports where planes stopped with coca paste from Bolivia, Ecuador, and Peru; and substantial investments in legal industries were attended by bankrupt entrepreneurs and industrialists who received resources to get their companies afloat. Some underhand and others openly took advantage of that bonanza. Many preceded Pablo Escobar, others used him or were partners of his partners, but almost all of them conveniently forget and feed the idea of exceptionality.

---

[13] Among many family groups, the *Clan de los Ochoa* (Ochoa clan) is referents of this various social class, several of whose members were leaders of the Medellin Cartel, in whose family background there were prominent and recognized professionals, and before being part of the mafia, they already stood out as horsemen and managed the Las Margaritas restaurant an stables, which would later become a symbol of the new emerging power; or the so-called *Clan de los Tomates* (Tomato clan), headed by the brothers Diego and Carlos Arcila, also partners of Pablo Escobar who "came from a middle-class family from Antioquia, with a certain degree of education and social renown, typical of their social status" (Salazar Pineda 2005: 54).

[14] This term derived from the declaration by the government of General Gustavo Rojas Pinilla in 1953 of the Island of San Andrés, located in the Caribbean Sea, as a Free Port. Colombian tourists who entered the island were allowed to return, bringing a quota of merchandise without paying tariffs. This initiative to encourage the island economy led to a practice in which the quotas were bought by merchants, and, incidentally, they managed to get a larger quota. The places where these merchandise were sold were known as the "sanandresitos" throughout the country.

## 40.6 The Metro System: A Symbol of the Times in the Metropolitan City

On November 30, 1995, line A of the new metro system in the city of Medellín was inaugurated. After 10 years of the beginning of its construction, the commissioning of this main section of the system was celebrated and associated with the urban resurgence of a city. Its construction had begun in the midst of one of the most dramatic economic crises in the city, produced by deindustrialization, high unemployment and homicide rates, rising of drug trafficking, unleashed crime, and urban fear and insecurity. Hence why, after overcoming the difficulties of that construction, its start-up was felt as a way to demonstrate the urban resurgence of that city overwhelmed and stigmatized as the capital of drug trafficking on a global scale. This was the opportunity to show a new face and look at the future more positively.

The metro did not escape the circumstances of the time. While its benefits and contributions to the city were strategically shown, its negative elements, in technical and urban terms, and the opaque part that it implied were also minimized or hidden since then, in a careful and systematic manner, related to: the award of the construction contract, financing and cost overruns of the project, in addition to its relations with other dynamics, which are generally excluded from any narrative. So the metro, between its achievements and problems, was the great symbol of its time, for better or for worse.

When the *Empresa de Transporte Masivo del Valle de Aburrá*, ETMVA (Massive Transport Company of the Aburrá Valley) was created on May 31, 1979, ETMVA, as the one in charge of leading the process of planning, construction, operation, and administration of the system,[15] its promoters, among politicians, guilds and local businessmen, promulgated their works as a lifeline, and closed ranks to pressure the national government, given the doubts it had not only regarding the technical aspects, but also the financial and sustainability aspects of the project. After overcoming the doubts and obtaining approval, on February 17, 1983, the bidding for the design and construction was opened, and it was awarded in November of that same year to the German Hispanic Metromed Consortium. An award was not exempt from

---

[15] In the immediate term, since 1977, the *Plan Metropolitano de Medellín* (Medellín Metropolitan Plan) contemplated a project in this sense, by including a study called *Proyecto de Transporte Rápido Masivo en el Valle de Aburrá* (Massive Rapid Transportation Project in the Aburrá Valley). But the Pilot Plan drawn up by Paul Wiener and José Luis Sert, between 1948 and 1951, raised the need to create a road system parallel to the river that would connect the urban settlements of the Aburrá Valley. After that, the Regulatory Plan and its derivatives of road plans were formulated. Among them the one of 1968, with the so-called *Espina dorsal del Valle de Aburrá* (Backbone of the Aburrá Valley), with which the strips on both sides of the river were reserved for a multimodal corridor that include the metro layout.

controversy regarding the indications of favoring the award[16] and corruption through commissions given to officials to favor the interests of the winning consortium.[17]

Facing the denounces from the national control bodies, the guilds, the institutions, and the local press closed ranks to defend the technical rigor of the action, the honorability of the Board of Directors, and the social and economic importance of the project; in addition, they resorted to the feeling of regional identity, against the centralism, in case of any adverse decision. Despite the fact that since then, it has become clear that it was more a decision of interested political will than a clear and coherent technical and planning vision, the project was imposed, for which the contract was signed on July 19, 1984: "for the design and execution of works, supply and transportation of rolling stock and fixed equipment; the training of personnel and delivery of the metro for the Aburrá Valley, between Diego Londoño White on behalf of the company and each of the legal representatives of the companies that make up the German-Hispanic Consortium" (General Comptroller 1994: 282).[18] April 30, 1985 was the official day for the start of the works, the same ones that were suspended on October 31, 1989, which was maintained until December 31, 1992, due to the problems and differences that surfaced between the contractor company and the ETMVA, caused by unresolved claims, on construction costs, lack of capitalization and financial problems, that the central government did not help to resolve, in a context of devaluation of the Colombian peso and inflation. A series of actions and determinations that were taken during the 38 months of paralysis[19] enable to resume

---

[16] In statements given on April 28, 2001 in the local newspaper El Colombiano, in the note entitled *El poder que había era superior a la justicia* (The power that existed was superior to justice), the lawyer Ignacio Mejía Velásquez, who had the power of attorney for one of the companies that participated in the bidding for the works: the French company S.G.T.—Societé Generale de Techniques et d'Etudes—pointed out: "It was always said that this was not a contract for a metro, but a metro for a contract, because what mattered was awarding it, not so much the work. Through this, it became known that there were posh characters who were behind the benefits of the contract" (Restrepo 2005: 368).

[17] Denounces made at the time were either not investigated or died slowly in the courts. Although a posteriori, in 2001, the Attorney General's Office "confirmed the payment of at least 20 million dollars in commissions... nothing could be done about it, since the criminal action had already prescribed" (Restrepo 2005: 369).

[18] This consortium was made up of three Spanish companies—Entrecanales y Tavora S.A., Construcciones y Contratas S.A., Ateinsa—and three German companies—Man AG, Dickerhoff & Widmann, Siemens AG.

[19] For example, the enactment of Law 86 of December 29, 1989, "by which rules are issued on urban public service systems for mass passenger transportation, and resources are provided for their financing", known as the Metro Law. It included in Chap. II, "Of the Aburrá Valley mass transportation system", that in order to "deal with the expenses caused by the construction of the Aburrá Valley mass transportation system, and primarily the debt service, a surtax will be charged on the consumption of motor gasoline of 10% of its price to the public on the sales of Ecopetrol at the supply plant located in the Aburrá Valley as of January 1, 1990". In addition, "as a condition for the granting of the guarantee of the Nation, rents must be pledged in sufficient amounts that, added to the resources generated by the gasoline surcharge referred to in the previous article, cover in present value the entire initial cost of the project, equivalent to US$650 million in 1984". Check at: https://www.funcionpublica.gov.co/eva/gestorrmativo/norma.php?i=3426. With the approved law, the Medellín protocol was signed in December 1990, in which the commitments between the

the works in December 1992 until the inauguration of line A in 1995, line B in February 1996, and the fulfillment of all the system on September 30, 1996.

Corruption problems, to a lesser extent, and financial difficulties, for the most part, have captured the interests of denunciations, texts, and investigations since then. But very little has been highlighted about the effects of changing the route, with the consequent constructive, technical and cost overruns. From the pre-feasibility study of 1979, carried out by the English company Mott, Hay & Anderson, it was clear that it was a cross layout, with a line A of 26 km from Machado—at north of the Aburrá Valley—to Sabaneta—at to the south, with the use of the strip of the river reserved for that purpose; and a line B of 3.6 km, from *Parque de Berrío* (Berrío Park)—at the center of the city—to Barrio La Floresta—at west—in which a tunnel was contemplated between the Parque de Berrío and the Cisneros station, as well as a viaduct to the station located in the old La Macarena bullring, which was to be the interchange station, to continue along the La Hueso stream, to the final station.

This layout and its characteristics meant a low-cost and *self-paying* metro, as was proposed from the beginning and as was argued when the national government tried to prevent its construction. Álvaro Pachón and Gustavo Esguerra, engineers from the Colombian team in charge of reviewing the technical studies in 1981, in parallel to a French group that presented its report separately (Bigey and Hunel 1991), all considered that the costs were reasonable.

Despite this, there war a change in the route moving the line from the river edge to the center of the city. This change of layout began to take shape on July 10, 1984, when the study *Alternativa Centro de la Ciudad* (City Center Alternative) was presented to the Board of Directors of the Company, with which the benefits were shown, and the change of the layout was justified, since it was considered that the river route was not advisable and, on the contrary, the entrance to the center allowed the "optimization of the system". In addition, the manager himself considered that the change in alignment did not affect the overall value of the project, "because compensation of items (sic) was present, that is, several activities were stopped, and, in their place, new ones were carried out" (General Comptroller 1994: 71). However, the reality was not like that since the additional works and cost overruns were even greater than expected, hence a work contracted for 650 million dollars ended up being worth 2174 million dollars, a debt that will only be paid in the year 2085.

Moreover, it was extremely strange that a week after the construction of the works had begun, the change in the layout was made official, which in part demonstrated the

---

parties were established and to comply with the provisions of the law, specifically the collection of the gasoline surcharge and the pledging of tobacco rents. It is also necessary to point out the expiration of the contract in December 1991, that forced agreements between the parties, such as taking the differences to an international court, the payment of an advance of 50 million dollars and the revocation of the expiration in August 1992. The settlement between the parties was only achieved 14 years later, that is, on September 12, 2009, when the metro agreed to pay only 3.5 million dollars, something insignificant given the initial claim of the construction consortium.

improvisation and lack of technical rigor[20] of a work that was tendered and started without defined designs, so that the construction consortium would develop them with a form of contracting called "Turnkey", considered harmful to the interests of the company itself. This new route, which left aside the river strip, implied entering the city center through a 6.6-km-long viaduct that had not been contemplated, with its enormous square columns and large cantilevered lintels cast in situ, prefabricated multi-support beams—weighing between 100 and 200 tons—stations with their roof vaults, among other concrete works, which particularly favored the steel and cement interests of entrepreneurs from the regional economic and political elite.

The cost overruns of the work were not only related to the construction of the viaduct, but also included those for the purchase of the real estate necessary for the passage of the viaduct along the route of the Carrera Bolivar (Bolívar Street), including its demolition. This is something that had and it continues to have an unclear side or whose purchase process is murky, to the point that cases of figurehead have been established, payments at prices much higher than the cadastral and commercial values or unjustified payments according to the few investigations carried out.

All these changes on the fly and purchases were made during the management of ETMVA by Diego Londoño White, Business Administrator of the Eafit University, who had been director of the Administrative Planning Department of Medellín between 1978 and 1979, when he was promoter and manager of the metro project. He was in the management of the Project Company until 1987, being considered not only its great defender but also an interlocutor of the Colombian Government itself before the Spanish authorities, industrialists, and businessmen who, in the end, were the great responsible and beneficiaries of the work. But, while the changes in the subway layout and the purchase of real state goods—lots and buildings—were taking place, with their urban and financial implications, at the same time, this character was the project manager for the Monaco building, as well as for the Dallas building, built between 1984 and 1986 for Pablo Escobar, by the company Londoño Vayda Limitada, formed by his brother Gabriel Londoño White and his partner, Isaias Vayda. The Londoño White brothers, Diego, Luis Guillermo, and Santiago, were part of the city's social, economic, and political elite; entrepreneurs, builders of many projects over the years, and owners of a real estate office. But, as the journalist Edgar Torres Arias points out, it was "a situation that is difficult to believe or even to suspect"[21]

---

[20] A posteriori, the Comptroller General pointed out that "it was the consequence of poor planning and an administrative structure made up of technical personnel with little experience in managing urban transport planning" (General Comptroller 1994: 66).

[21] "The importance of his last names and his image as an enterprising and thriving executive had made him the first manager of the most ambitious company started by any city in the country: the Medellín metro. Due to this position, to say the least, Diego Londoño became for several years, before the Spanish government of Felipe González, the official spokesman for Metromed, a consortium that operated with binational funds and was in charge of the construction of the first metro nationwide.

In defense of the project, Diego Londoño White had represented President César Gaviria and shared, side by side, with the Colombian ambassador, Ernesto Samper Pizano, before industrialists and high-ranking officials from Madrid. Perhaps for this reason, it was difficult to even imagine the magnitude of his unofficial activities. Due to this *good will* and his permanent contact with the that they were part of the gear of the so-called "Medellín Cartel": Luis Guillermo[22] being Pablo Escobar's personal adviser, and Diego himself would end up convicted of figurehead, conspiracy to commit a crime and collaboration in kidnapping.[23]

Diego Londoño White's relationship with Pablo Escobar had apparently begun since 1980, due to the former's real estate and property activities, through which he sold him "a piece of land on the upper transversal, in the San Lucas sector, in Medellín, where the mafia boss built *La Cascada* (The Waterfall), a luxurious nightlife center" (Torres Arias 1995: 410). But later, there were more lots, houses, apartments, entire condominiums, construction of buildings of various types, among other businesses for Pablo Escobar and Gustavo de Jesús Gaviria, Pablo's cousin and main partner, for other subordinate bosses and an undetermined number of hitmen at the service of the cartel, who benefited from his intermediation. Even the engineer Juan Fernando Toro Arango, who became one of the most dangerous hitmen and criminals, known by the alias of *La Monja Voladora* (The Flying Nun), was a partner in the real estate company.[24] A lucrative real estate business, not only carried out by the Londoño Whites, which completely changed the urban landscape of Medellín, and in a very special way, the southwestern slopes of the Barrio El Poblado, an elite neighborhood, where the Dallas and Mónaco buildings were located; whose large country mansions built between the end of the nineteenth century and the first half of the twentieth century were swept away by the building fever with the capitals of drug trafficking.

The strange case of Mr. Londoño White raises the issue of that ambiguity of the society of the moment, in the midst of the collapse and the economic crisis, who did business for his benefit and looked to the other side without being amazed by that dark side. Apart from being managed by a partner of the drug cartel, the subway was not affected in anything else? The hasty changes of the layout—weakly supported in technical, financial, and urban planning terms—directly benefited some business sectors that, regardless of the devastation of the landscape and urban memory, managed to get their companies in crisis afloat. In addition, the use of privileged information allowed the purchase and speculation on real estate along the route of the viaduct,

---

cream of Colombian society and industry, Diego Londoño White ended up pointing out, indirectly for Pablo Escobar and the bandits, some of the potential victims of plagiarism, at the request of La Monja Voladora (The Flying Nun) [alias of engineer Juan Fernando Toro] after he joined and became a partner of Londoño White, a prestigious real estate brokerage firm in Medellín. Millionaire sums thus entered the saddlebags of Pablo Escobar Gaviria himself and the bandits, meanwhile The Flying Nun received juicy commissions for each kidnapping. Diego and Guillermo Londoño White even turned out, on some occasions, to be intermediaries and good offices managers" (Torres Arias 1995: 407–408).

[22] … "another notable member of Medellín society, club member, businessman, horseman, and golf player, who in the late 1980s turned out to be one of the key links of the drug lord Pablo Escobar Gaviria". https://www.elespectador.com/judicial/el-emulo-de-londono-white-article-34492/.

[23] The judicial persecution of Diego Londoño White began in 1990. After some time in freedom, he surrendered on March 1, 1993. He was sentenced in 1999–14 years in prison, of which he paid nine. After being released from prison, he was assassinated in November 2002.

[24] This sinister character, curiously and probably due to the established closeness, changed his name to Fernando Londoño White, as he was known before he was himself assassinated.

and there is a strong suspicion about beneficiary sectors that were never investigated, or if the investigations were initiated, they were buried in the judicial archives.

The 1980s were difficult and traumatic times for the city. In the midst of the economic crisis and the increase in unemployment and informality, drug trafficking, hired assassins, and bombs took over the city. In the second half of that decade, when the subway works were paralyzed, the urban landscape was gray and dilapidated. It was literally a city between undone, half done, and bombed. It was an apocalyptic landscape. That is why the resumption first, and then the commissioning of the system, was considered an urban rebirth. If at the beginning the regionalist, spirit of Antioquia was a fundamental reason put forward before the "centralist" organizations for it to be built and restarted, with its implementation the calls for identity pride and civility multiplied, added to the construction of the imaginary of the *Cultura Metro* (metro culture) as a new determining element of society. The gray and chaotic city remained at ground level and outside the system, within it, was education, order, and asepsis. Everything negative was left behind and buried.

It was evident that there was a before and after in the city. It was proposed as the resurgence of a new city model, but it was not as it was thought at that time, in that transition from the industrial city of Colombia to the city of services that it is now. Medellín was and continues to be the only city in Colombia to have a metro system for mass transportation, which improved the quality of life in many aspects, generated unforeseen environmental externalities—reduction of $CO_2$ emissions, so fundamental today—created a "culture" that is sold and exported, somehow managed to alleviate unemployment and reactivate the economy and allowed connecting spaces and social sectors that hardly had links. But that model of a linear city along the metro, which would stop the settlement of the hillsides, was not implemented, and on the contrary, the criticism made at the time was fulfilled: "the valorization process produced by the metro, instead of allowing the densification of the corridor by people with limited resources, will definitely drive them away".

But, also, the metro stands as a symbol of silencing and concealment of that dark side in which it arises but wants to be buried; of its actors who acted between legality and illegality; but, above all, of the construction of a new city imaginary that, supposedly, left behind its dynamics of exclusion/inclusion, which simply varied, and the idea of a city that overcomes its crisis and conflicts, including those of the drug trafficking; which is not true either because it simply mutated, with other urban social and territorial expressions.

## 40.7 The Dramatic Change of the Metropolitan Landscape

The Monaco building, sober on the outside, without outstanding architecture, almost anodyne, but dazzling inside, as it was described at the time, with its walls covered in marble, collector cars and works of art, between originals and forgeries, was just one among many properties bought or ordered to be built as a residence, investment or to be used as a strategic route for permanence, mobility or escape. The then-exclusive Barrio El Poblado was one of the favorite places where several of the most representative and bloodthirsty drug traffickers and paramilitaries settled, following the example of Godfather Alfredo Gómez himself, who since the early 1970s had established his fortress, with complete impunity, in the sector of Santa María de los Ángeles, in the midst of "notable" members of society, the same sector where Escobar would later build that emblematic demolished building.

Since the end of the nineteenth century, El Poblado was the place where the social, economic, and political elites built their recreational houses and then in different decades of the twentieth century their residences, in luxurious villas and mansions, but with restrained esthetic and the architectural severity of the different fashionable languages, stamped by the most outstanding architects of the city. On the other hand, between 1937 and 1938, middle-class barrios were built in the vicinity of the central park that encompassed this part of the city, such as the Barrio Lleras, made up of two-story single-family houses, promoted by the BCH Banco Central Hipotecario (Central Mortgage Bank).[25] Between the serial houses of the middle class and the villas of the notables, there were some bastions of peasants and workers who lived in small houses on the side of the roads that crossed up from the Aburrá Valley to the high plains of the Near East.

Few houses survived the real estate speculation and the construction fever unleashed since the 1980s, as most of them were demolished and along with them the gardens and the rich arborization were razed, while the creek beds of the rivulets, tributaries, and streams that crossed through these slopes of the southeast of the valley were constricted. Such was the topography between El Poblado and the adjoining municipality of Envigado that was known by the name of its hills—Los González, Los Parra, Los Mango, Las Brujas, El Esmeraldal—where the buildings of the new rich and emerging people were implanted and reproduced. From the bucolic landscape, we passed to the vertical constructive density. A mostly anodyne architectural saturation like that already mentioned in Monaco was the almost general norm, with some honorable exceptions. El Poblado is the outstanding example of transformation due to real estate speculation, which brought with it saturation and urban densification. A place where not only the mansions of the big bosses was located, but also the aspirational place to live in apartment towers by the middle and subordinate sectors of the drug traffickers who came from the low income popular barrios. Additionally, illegal capitals were injected contributing to this real estate speculation, both in apartment towers, shopping centers, and commercial premises that, since the end of the twentieth century, were displacing the residential activity, until almost absolute colonization in the first decades of the twenty-first century. While most of the upper and upper middle-class inhabitants, without links to drug trafficking and illegality, found themselves invaded, besieged, and even expelled from their homes and apartments, being this a type of displacement that has been scarcely studied.

---

[25] Later, other neighborhoods would be built such as Barrios Lalinde, Manila, Alejandría, Astorga, Provenza, (this last one consolidated in the 1960s).

It remains contradictory, paradoxical, and of great urban contrasts that separated by the Río Medellín (Medellín River) but connected by the emblematic 10th street which crosses the south of the Aburrá Valley from east to west, the Barrio El Poblado and the Barrio Antioquia neighborhood are located. This was and is the opposite of El Poblado. One, the elite neighborhood, above, on the slopes, and the other, the popular neighborhood, below, on the flat area. This Barrio was designed in the 1920s by the architect Félix Mejía Arango, far from the center, on the outskirts of the city, the same one that by 1932 already appears in urban cartography divided into two sectors: *Fundadores*, later renamed *Santísima Trinidad* (Holy Trinity), but recognized as simply as Trinidad, and the Barrio Antioquia itself. Its character would change when the mayor of Medellín in 1951 decreed the transfer of the Lovaina (Louvain) prostitution zone, located next to the Barrio Prado—an elite neighborhood at the time in the northern part of the center of the urban area—to that distant barrio to the southwest of the city, despite to the protest of its then inhabitants. It soon became a barrio of bars and cantinas, of procuress and whores, of kidnappers and thieves who, taking advantage of its strategic location next to the Olaya Herrera airport, derived from the 1960s in pioneers of drug trafficking as recipients of coca paste loads and transporters of marijuana and cocaine to the North American market. Precisely here, Griselda Blanco established part of her command. She was the one who in the vicinity of the barrio theater began to establish her economic emporium and her power that made her the godmother and a myth of the underworld, to the point of being recognized as *La Viuda Negra* (The black widow).[26] As the journalist and former mayor of the city, Alonso Salazar, points out, "... as if it were some kind of social retaliation, the sons of the bitches, who the city had thrown to the peripheral barrio, in their condition of new rich moved to the exclusive Barrio El Poblado, which for a time earned the name of Altos de la Santísima Trinidad" (Highs of the Holy Trinity) (Salazar 2002: 51). That revenge was led by Griselda herself who, without leaving her neighborhood environment, also had her mansion in El Poblado.

These opposite but complementary barrios, El Poblado and Barrio Antioquia, were and continue to be the urban mark of the rapid and profound transformations that were suffered, in which nothing was left unscathed. It was such a frenetic and vertiginous process that when a lethargic society—participant or silent accomplice or hypocritical—woke up, it was no longer the same: lax ethics and the power of money. Shameless exhibitionism and unbridled consumption were expressions of social ascent. The esthetics of architectural forms and spaces led to delirium far exceeding the limits of kitsch; where the popular, the bad taste, the "cajamanesque" barrio inheritance were amalgamated, with what is called the "*mañé*" (ordinary or ugly style) and the "*lobo*" (wolf), the esoteric, the pop and the neoclassical factorized series, to configure something like the *Narc-decó*. A narco-esthetic also expressed in the shapes of the body, especially of women, which changed to external exuberance with the introduction of silicone, while men flowed their natural obesity, framed

---

[26] The story of her starting point in Medellín, her vertiginous rise and immense and cruel power in the United States, his imprisonment and decline, her return and her murder in 2012, in a meat shop in the Barrio Belén of Medellín, is narrated by Soto (2013).

by chains, rings, and golden watches. The explosions between the frightening and the festive; the bombs, interspersed with shots fired into the air—or not—and the explosion of gunpowder to celebrate the *coronados* (crowned) shipments—when their successful arrival of drug in foreign markets was reported. Pyrotechnics taken to its maximum fury. Noise, in a broad sense, is as a symbol of the new society. The silence itself was different.

The house, the street, the barrio, the commune, the city and the metropolis were transformed. Nothing was left without feeling the drama of the transformation. Just as the styles of furniture, vases, paintings and interior decorations were also changing, as well as the houses with their architectures that shout to the public space the power acquired by their owner. The streets change youthful routines and rituals, from lovemaking practices to corner sociability. Musics, speeches, and languages created to the point of configuring an own way to communicate that was then named *El Parlache*.[27] All exercised from the barrio's power and hitman violence, as the writer Gilmer Mesa narrates in a dramatic way in his novel *La Cuadra*.

The bunker buildings with their coves (as hiding places), as a new urban architectural typology, as well as the parking buildings and warehouses converted into strategic places to store, hide in coves, distribute or, even, torture and murder. The dance and party venues, discos and *estaderos* (typical places where the restaurant is combined with horse stables) with unprecedented architectural forms and decorations, increasingly challenging that, likewise, were festive, sociable, and business spaces; where the female body could be exhibited like the trot of a thoroughbred horse, settle disputes with bullets, closing deals or even conducting political academic forums to propose laws that would prevent the extradition of drug traffickers.

*Estaderos*, stables, horse tracks, and bullrings in suburban areas of Medellín or other urban centers of the metropolitan area, either to the south between Caldas, La Estrella, Envigado and Sabaneta, or to the north between Copacabana, Girardota and Barbosa, where rural property significantly increased its value to prohibitive extremes. Once again, power, ostentation, and strategy came together as determining factors so that each capo would have his large, emblematic farm, close to the metropolitan area, but with its coves and exit routes through strategic corridors to other regional environments, such as Las Lomas, El Bizcocho, La Cascada, among many more that were recognized. Some architectures between the chalet and the mansion, ostentatious and heterodox: with their balusters, big vases and rosettes, with entrances flanked by rigorously aligned palm trees; its central focus was the swimming pool, the new fetish and exhibitionist place for parties and orgies, surrounded by exquisite gardens with their sculptures, lights, elves, giant mushrooms, and cement or fiberglass animals; fences to delimit paddocks that also became part of the ostentation of power. All was conveniently delimited by electric fences, and hidden behind the *suiglia* (swinglea or lemon swingla)—shrub of Asian origin, of a beautiful intense green color, which was introduced to Colombia by cattle ranchers

---

[27] A street language, cryptic, as a way to communicate and evade control. After its emergence and consolidation in the 1980s, the RAE—Real Academia Española since 2001 recognized and incorporated some of its words. A language that has its own dictionary as of 2006.

as a natural alive fence, although it was also used as a windbreak barrier, probably in the Magdalena Medio region, a place of drug trafficking and paramilitarism. It is not strange, since exotic species of all kinds arrived there and spread throughout the region, from bushes to fauna, such as hippopotamuses. Its accelerated, bushy growth, and thorny foliage was the reason why it was also quickly used not only to delimit properties but also to hide intimacies from farms to motels. Veritable green walls that grow uncontrollably until they hide not only that private inner world, but even cut off the landscape, which has been denied, both in its near and distant views. In many places, the landscape was lost as a common good and the right to enjoy it was impeded, in order to build a landscape of concealment, denial, impediment, and fear.

It was not only the profane but also the sacred that was appropriated, transformed, and placed at the service of drug trafficking. Although the saints included the masculine with the aforementioned Jesus Child of Atocha, Saint Judas Tadeo, or the Miraculous Lord of Buga, the reason for devotion was mostly virgins. The sacralization of the mother—called *La Cucha* (The Oldie Mom), in the particular language of the *Parlache*—is related to the Marian centrality that the narco-fetishist religiosity configured. They put at their service nuns, priests, bishops, and the virgins themselves who were the patron saints of urban centers where they were in power, such as the *Virgin of Chiquinquirá* from La Estrella or *María Auxiliadora* (Mary Help pf Chistians) from Sabaneta[28]; or failing that they built other sanctuaries, such as that of the *Rosa Mística de la Aguacatala* (Mystic Rose of La Aguacatala), in the Barrio El Poblado. They sought her protective mantle, with rituals, offerings, and festivals that guaranteed their own life, not that of their enemies, and wealth. Some devotions were the most recognized, but the metropolitan landscape in all neighborhoods or rural areas is dotted with caves, altars, and sanctuaries, public and private, where the cult derived from these particular ways of understanding the sacred was rendered and continues to be rendered and claim twisted profits.

Emblematic and enigmatic places elevated to myths and urban legend. Symbolic referents of drug waste and horror. Unintentionally converted into places of memory and uncomfortable heritage, like the same drug tour route, which is promoted on the Web by *Pablo Escobar's route*. Between short and long tours, they take their clients to places that some of the promoters elevate to the rank of "historic places closely related to his life and death",[29] whether they were the places where he lived, was imprisoned, discharged, or his own grave; as well as to visit the works that he promoted—the Barrio Pablo Escobar or the soccer fields—the museum that his brother Roberto

---

[28] It is precisely this cult that gives its name to the novel *La Virgen de los sicarios* (The Hitmen Virgen) (1993), by the writer Fernando Vallejo, made into a film by the director Barbet Schroeder in 1999. The writer Héctor Abad Faciolince speaks precisely from this novel of the genre of the *sicaresca* (*hitmanesque*).

[29] Recorrido sobre Pablo Escobar en Medellín 2022—Viator. Consulted: 07.30.2022 https://www.viator.com/es-ES/tours/Medellin/Pablo-Escobar-Historical-Tour-of-Medellin/d4563-5549ESCOBAR?m=26374&supag=67067036863&supsc=dsa-694098304004&supai=420430240007&supap=&supdv=c&supnt=nt:g&suplp=1003654&supli=&supti=dsa-694098304004&tsem=true&supci=dsa-694098304004&supap1=&supap2=&gclid=Cj0KCQjw54iXBhCXARIsADWpsG8F_WG9sIuu3q3JDuXDxC_AHqPqxMexIJEGkkFpWA9xipaWYk4mCrgaAuenEALw_wcB.

established, or, to the dismay of the mayor who promoted its demolition, to the "ruins of the Monaco Building, today a memorial park". Even in some cases, "you can take a souvenir (included) of the great capo of capos".[30] All sprinkled with anecdotes, exaggerations, and inventions to feed the desire, the feverish mentality, and the morbidity of *narcoturismo* (drug tourism). But beyond that, there is a cartography of ruins, traces, and presences that not only account for that past that many want to let go, forget or bury, but that still shape or determine the present. They emerge not as ghosts but as true presences that, on the one hand, face or determine the way of conceiving the world and the metropolitan landscape.

## 40.8 The *Criminalis* Metropolis

Historically, Medellín configured its regional centrality in the nineteenth century, since then it has been consolidating itself at different times as the center of political, social, economic, educational, religious, demographic, or functional power. A dominant hegemony that has been maintained to the present but has been reconfigured in recent decades, with the transition from the industrial city to the city of services. The dynamics of drug trafficking played a prominent role in this. The smuggling land routes converged toward Medellín and on it the new drug trafficking routes were raised, combined with the new insertions into international markets, either by maritime routes or by air routes that were multiplied. All of them established the course to or from the Caribbean Coast, especially the *Gulf of Urabá*, with its epicenter in the port of Turbo.

In the new phase of drug trafficking, the coca base brought from Peru or Bolivia arrived directly at the Olaya Herrera airport in Medellín, where the strategic contiguity of the Barrio Antioquia allowed it to be collected in order to be taken to the laboratories in the territories of neighboring towns. After it was processed, it was taken by land to Turbo and exported on ships, especially by the then famous *Flota Mercante Grancolombiana* (Grancolombiana Merchant Fleet). But given the growth in demand and production, both the laboratories and the tracks were relocated to regions where drug traffickers bought large tracts of land in *Magdalena Medio* (Middle Magdalena), *Bajo Cauca* (lower Cauca), Urabá, and the Colombian-Panamanian Darién, and even in the Pacific, in this case, in ports such as Cupica, where tracks and large laboratories were installed. The planes landed and took off on runways that were even authorized by the authorities in charge of aviation control. Huge farms such as *Hacienda Nápoles* (Nápoles Plantation), with more than three thousand hectares or the *Hacienda Virgen del Cobre* (Copper Virgin Plantation), in the port of Necoclí, were recognized both for their extensions and eccentricities as well as for their strategic location and their illegal activities.

---

[30] https://pabloescobartour.co/landing-2/?gclid=Cj0KCQjw54iXBhCXARIsADWpsG_JzFziN KmNY24rhIDa9oQ9ZlLk-SeS8sOByF-1zhZMME3Y5BuEimMaAu81EALw_wcB.

Despite the displacement of certain activities to other territories, Medellín maintained its centrality as the financial, money laundering, logistical support, or decision-making center for those territories and their inhabitants. In this way, a wide geography was configured that exceeded the limits of the city and, even, of the metropolitan area due to the communicating vessels that were established; hence, the strategic corridors configured in the four cardinal points of the Valle de Aburrá, for the movement of narcotics, weapons, people, and resources. As the journalist and researcher Alonso Salazar points out, this was clear to Pablo Escobar, who "moved in the area that included the Valle de Aburrá—Medellín and seven neighboring municipalities, a metropolitan area of some three million inhabitants—and a rural strip that from this valley, through the eastern mountains, it passed through the municipalities of El Retiro, Guarne, Marinilla and El Peñol—in the eastern plateau—and Granada, San Carlos, San Luis—in the mountainous slope that declines over the extensive and thick jungle toward Magdalena Medio, where the Hacienda Nápoles was located" (Salazar 2002: 223).

This expansion of activities toward these territories coincided with the emergence of armed paramilitary, anti-guerrilla, and extreme right-wing groups on these outskirts, which were financed by landowners and drug traffickers themselves. Those initiated as self-defense groups—as in the case of the *Autodefensas de Puerto Boyacá* (Self-Defense Forces of Puerto Boyacá), in Magdalena Medio, created in 1982—and ended up being veritable paramilitary confederations at the service of drug trafficking, as in the case of the *Autodefensas Unidas de Colombia* (United Self-Defense Forces of Colombia) which, between 1997 and 2006, grouped together various blocs or regional armed associations. They were expressions of the struggles for territorial control, but supported operationally, logistically, economically, and politically from Medellín. It was not a simple coincidence that the Montecasinos mansion, also located in the Barrio El Poblado, was the residence of the top commanders of the self-defense groups, the Castaño brothers—Carlos and Fidel—where they met with businessmen, politicians, and leaders to define support, counterinsurgent actions, assassinations, and massacres; or that from the Padilla parking lot, in the center of the city, a few blocks from the Administrative Center of La Alpujarra—where the main regional political and judicial authorities are located—worked the center of logistical, financial, and accounting operations of these *autodefensas* (self-defense organization).[31] Even in the persecution of Pablo Escobar, the Castaño brothers and the self-defense groups participated actively, in tacit alliances with the authorities.[32] After Escobar's

---

[31] On April 30, 1998, a raid was carried out, where "many pieces of evidence were found to prosecute dozens of merchants, businessmen, industrialists, cattle ranchers, soldiers, and policemen who in one way or another were related to the structure of the ACCU"—*Autodefensas Campesinas de Córdoba y Urabá* (Peasant Self-Defense Groups of Córdoba and Urabá) (Memoria de la impunidad en Antioquia 2010: 55) The macabre thing was that instead of continuing the investigations and prosecuting those responsible, and the investigators in charge of the raid were either dismissed or killed. This fact was conveniently hidden in the byways of justice, and much of the evidence was lost.

[32] A group of drug traffickers, including those from the Cali Cartel and former members of the Medellín Cartel, victims of Pablo Escobar's abuses, formed Los *PEPES—Perseguidos por*

death at the end of 1993, the power gap was filled by the paramilitaries, first under the control of the *Bloque Metro* (Metro Bloc),[33] after 2001 dividing up the territory with the *Bloque Cacique Nutibara* (Cacique Nutibara Bloc), and, since the end of 2003, after a bloody armed struggle throughout the urban and rural metropolitan territory, it remained under the control of the latter, commanded by Diego Fernando Murillo, alias Don Berna.

In his youth, Don Berna had been a member of a guerrilla group, then a car washer and driver for a drug trafficker from one of the Medellín Cartel clans. After saving himself from a massacre ordered by Pablo Escobar against his bosses, over time he allied himself with the Castaño brothers, controlling fearsome criminal gangs, first in the municipality of Itagüí, to the south of de Metropolitan Area and later in the rest of the valley.[34] As such, he began to impose his power and hierarchy, until consolidating it through the self-defense groups *Cacique Nutibara* (Nutibara Cacique), *Héroes de Granada* (Granada Heroes), and *Héroes de Tolová* (Tolová Heroes). His power was imposed in a bloodless manner, product of the armed action against those other criminal and paramilitary sectors, resorting to murder, bomb attacks, forced disappearance or people´s displacement. He even participated in the largest urban military operation carried out in Colombia, known as Operation Orion, between October 16 and 19, 2002, in which with the participation of combined forces of the Army, Police, *DAS—Departamento Administrativo de Seguridad* (Security Administrative Department), *CTI—Cuerpo Técnico de Investigación* (Technical Research Corps) of the Prosecutor's Office, and Colombian Air Force. With this intervention, they sought to expel the *Milicias* (Militians) of the ELN—*Ejército de Liberación Nacional* (National Liberatior Army) and of the FARC—*Fuerzas Armadas Revolucionarias de Colombia* (Revolutionary Armed Forces of Colombia) and the so-called CAP—*Comandos Armados Populares* (Popular Armed Commands), who had turned

---

*Pablo Escobar* (Persecuted by Pablo Escobar), as an organizations who attacked him, conducted intelligence and passed on information to the authorities to find him until his death.

[33] It operated in Medellín and in part of the Valle de Aburrá between 1997 and 2004 and also had territorial control in the Northeast and East of Antioquia. It was commanded by the lawyer and ex-military Carlos Mauricio García, known as Rodrigo or Doble Cero. Before commanding this bloc, he was an ideologue for the Autodefensas Campesinas de Córdoba y Urabá-ACCU. He refused the entry of pure drug traffickers to the self-defense groups. After being defeated in Medellín, he settled in the city of Santa Marta, where he was assassinated in May 2004.

[34] He was a member of the EPL—Ejercito Popular de Liberación. "When Escobar betrayed the Galeanos and murdered them in the La Catedral prison, where he was being held, Murillo almost fell into the trap. That is why he joined the PEPES—Perseguidos Por Pablo Escobar and collaborated with the authorities to finish off the head of the Medellín Cartel. With the Galeanos dead, Murillo took over his illegal emporium in Itagüí. Like many members of the PEPES, once Escobar died in December 1993, Murillo eventually ended up allied with the paramilitary group created by the brothers Fidel and Vicente Castaño. With the capo dead, Murillo emerged as the new head of Medellín's criminal gangs—the most fearsome of them, La Terraza (The Terrace), he prospered in the business of armed robbery (the most remembered, in which they took 13 billion pesos of a valuables truck in Bucaramanga)", see: https://verdadabierta.com/perfil-diego-fernando-murillo-bejarano-alias-don-berna/. Consultation: 07.31.2022.

that vast area of the city into a criminal stronghold that was difficult to penetrate.[35] Nothing was moved without Don Berna's consent. Absolute dominance was reflected in the decrease in homicide rates for the year 2004. That *calma chicha* (false stillness) was known as *donbernabilidad* (donbernability), to speak about the silent connivance of this state of affairs, from which the official authorities drew political benefits and international image, more than about the control of those authorities that took pride in their security policies.

In the first government of Álvaro Uribe Vélez, the *Acuerdo de Santa Fe de Ralito* (Santa Fe de Ralito Agreement) was signed with the AUC—*Autodefensas Unidas de Colombia* (United Self-Defense Forces of Colombia) in July 2003, which led to negotiations for the demobilization and disarmament of 34 paramilitary blocks, including the Cacique Nutibara, who was the first to do so in November of that same year. Don Berna demobilized, but was charged with murder, captured in 2005, and in May 2008 extradited to the United States. Since then, different organizations have competed and succeeded in power and control. Although there are hierarchies and superior control, in some cases regarded as a great urban myth, as is the case of the *Oficina de Envigado* (Envigado's Office) started by Pablo Escobar and inherited by Don Berna, for the collection of rents, murder and control from the gangs, they all learned from paramilitarism the importance and value of the territorial issue, hence they distributed its control from which they derive their income.

Today, studies show the existence of a criminal government that, like many cities in the world, controls criminal activity and governs a good part of the communities. In the particular case of Medellín, researchers from *Innovations for Poverty Action— IPA*, the University of Chicago and the EAFIT University, point out how "a hierarchical structure of highly organized criminal organizations persists", whose pyramidal shape has a base composed of around 350 groups called *combos*. The *combos* are usually made up of young men, and their criminal activities are usually confined to small and relatively well-defined territories. The middle part of the pyramid is made up of between 15 and 20 groups called bandas o *razones* (gangs or *reasons*). The gangs are usually made up of adults with long careers in organized crime and their activities are not always limited to specific territories (in fact, some extend to other areas of the country). While gangs are relatively autonomous, most *combos* are subservient to some gangs. At the top level of the pyramid are different instances of collective decision-making in which the gangs coordinate some of their operations (Blattman et al. 2022: 4), but which has been known as *La Oficina* (The Office).[36]

---

[35] This operation was carried out in the barrios of Comuna 13, in the eastern center of the city of Medellín. "Don Berna explained in a free version rendered from his place of detention in the United States, in March 2009, how paramilitaries from the Cacique Nutibara Bloc participated in the planning of this operation, which was under the command of then-general Mario Montoya, from the Colombian army; and Leonardo Gallego, from the Valle de Aburrá Metropolitan Police". https://verdadabierta.com/la-tenebrosa-maquina-de-guerra-que-dirigio-don-berna/. Consultation: 07.31.2022.

[36] *La Oficina* (The Office) has its origin in the Medellín ¡Basta Ya! (Medellín Stop Now!) campaign, which calls the complex armed structure that Pablo Escobar formed and that allowed him to consolidate himself as head of the Cartel, which was "possible thanks to a division of labor between

Although the main sources of income for both combos and gangs are extortion and the sale of drugs, they differ in that in the latter case the *combos* are wholesalers and the gangs sell them in the different *plazas de vicio* (vice plazas) that they control, in the streets and public spaces of the city. In addition to providing contract killing services, known as *sicariato* (hitmenship), debt collection and security for the transportation of drugs, they derive other income from theft, high-interest loans—called *gota a gota* (drop by drop)—sales of exclusive products in the barrios, games of chance, procurers, among other activities. But, in the same way, they have also ventured into activities of selling lots in peripheral pirate urbanizations, in other cases, they demand participation in the works[37] of legal builders or, even, they have already ventured into building their own high-rise building projects, through which have laundered their assets. Although the levels of criminal government vary, in some cases, the combos exercise "more government functions than the state" (Blattman et al. 2022: 3), in other cases, the same residents see them as a complement to the State itself.

## 40.9 Between Permanence and Mutation…as a Colophon

There is no doubt that the different local governments have made efforts to overcome that past with important housing projects and educational and cultural infrastructure, with symbolic architectures; costly public space and infrastructure projects, sometimes megalomaniac and unnecessary; the realization of massive events, media and marketing projects to project the city, until obtaining prizes and international recognition of the city. But apparently, it has not been done enough and in a sustained manner, nor often in the direction that it should be. Perhaps for this reason, the dark past continuous casting its shadow, resurfaces, transforms and even seems to be incorporated naturally into the ways of living, being, and inhabiting the metropolitan city; and, even, to have metastasized into a globally projected urban culture. It is not enough to demolish a building or make a faded and naive project of historical memory, because reality is there lurking in the urban space, in the architectures,

---

gangs and the "offices" to which one could go to settle disputes, from the loss of some shipments to revenge, or go to coordinate criminal activities" (Centro Nacional de Memoria Histórica 2014; 133). After Escobar's death, the organization has oscillated between the hierarchical—in the times of Don Berna—to a kind of federation, as they present themselves today, by sending a letter on July 28, 2022 to the current President of the Republic, Gustavo Petro, as a reaction to his proposal for Paz Total (Total Peace), heading the letter as follows: "We, the organization called "La Oficina", which brings together armed groups with presence in territories of the Medellin metropolitan area and other regions of the country, we are the oldest in the urban armed conflict in Colombia, due to our participation since the 1980s in the different transitional phenomena of the conflict". Members of that office continue to control the legal and illegal dynamics of many metropolitan territories, surviving the purges, conflicts and even the action of the state, since they lead their actions from prisons.

[37] It is pointed out that to allow the construction of buildings or urbanizations, they claim for themselves a certain number of apartments and control of the stock of materials, although the construction guilds have denied this.

in the cultural practices… It can be seen in the center of the city in the *El Hueco* (The Hollow) sector, in the south in Parque Lleras and its surroundings, including Provenza, and, again, on the other side of the river, in the permanence of the Barrio Antioquia as a significant enclave.

El Hueco is an important sector of the city center known by that name. Sometimes, it is confused with the barrio Guayaquil, since it became the territorial, historical heir, and of some of its economic practices, when the latter was the great dry port of the city and the popular economic center, since the terminal stations of the railways were located there—of Amagá and Antioquia—and the covered market square was also there. This used to be a dynamic, popular environment that subverts social and cultural orders, erected into a myth by the urban novel *Aire de tango* written by Manuel Mejía Vallejo, which portrayed the environment at the limit moment between splendor and the beginning of the end. With the burning of the market and the transfer of its activities, with the end of the railways in the midst of an urban economic crisis, Guayaquil was one of the most affected. In this decadent environment and half-abandoned buildings, smuggling, money laundering, and the beginnings of drug trafficking began to operate. Habitat of The Godfather, Alfredo Gómez, as evidenced by Jairo Gómez in the novel *Familia*.

The insertion in the global economy to the city through the smuggling route and the new and emerging capitals, first set the course from the free port of Colon in Panama and, starting in the 1980s, from the markets of Chinese cities, before the national government itself raised it and achieved it. Containers loaded with merchandise and trinkets were scattered through the informal markets of the center and the commercial premises of El Hueco. But the merchants transmuted into entrepreneurs. From contraband to the maquilas there, in the Far East, and here in the new premises that began to expand and grow densely and vertically. Buildings with glazed facades, perforated, or superimposed sheets were erected, which already spoke of other esthetics that combined Chinese shapes and colors with elements of North American neighborhoods, mixed with the popular local and the emerging *narc-decó*. Showcases with voluptuous mannequins fitting the new esthetic aspirations, along with butt *levantacola* (lift-tail) blue jeans or the baggy ones of the new reggaeton singers, with t-shirts, caps, chains, rings, and a long list of bright and colorful beads. From these dense and compact blocks, crossed by dozens of commercial passageways, not only merchandise but also the new esthetics of Medellín's urban culture are exported on a global scale.

In other ways, this globalizing insertion affected the transformation of the old barrio Lleras and its usual residential world into an environment of services for consumption and tourism. Since the last decade of the twentieth century, the offer of sex, drugs, and music was the triad that transformed this environment, with Parque Lleras as its epicenter. Around it, hotels and hostels of all conditions and costs, bars, restaurants, art galleries, supermarkets, small markets, liquor stores were established, turning the peaceful homes of the barrio into trembling, exotic, and colorful places that live outside the urban routine of work and rest schedules, features of the *rumba nocturna* (night partying), and the daytime offers. An offer promoted by social networks and Web pages with tourist packages that include parties, women, and drugs, but that is palpable in the streets, where the illegal control of the combos and the offer are evident. An offer that, far from being controlled or regulated, has been strengthened, either directly or indirectly, with the implementation of the so-called economic use of public space; a policy proposed by the municipal administration of Medellín, through the Agencia APP—*Agencia para la Gestión del Paisaje, el Patrimonio y las Alianzas Público-Privadas* (Agency for the Management of Landscape, Heritage and Public–Private Partnerships) that privileged private interests more than the management for recreation, meeting, landscape, and heritage, as argued at the time of the Agency's creation in 2015. Hence, since 2017, when the Pilot Plan for *Parque Lleras-Provenza* was implemented, privatization, economic use, and the predominance of supply places for this growing tourism have been promoted; which has even been denounced in 2022 as an "open-air brothel" by the writer Carolina Sanín. Nothing new, since sex tourism has been denounced for years both in Medellín and in other tourist cities in Colombia, such as the case of Cartagena. It is simply a reissue of old denunciations and a new scandal in a growing and uncontrolled dynamic, between the legal and the illegal.

This environment, as well as others in the city, expresses all that narco-esthetic heritage, expressed in decorations, bodies, costumes, and music, where reggaeton is the new musical band in the city. A music that began in Puerto Rico but that found a niche in Medellín and, then, it was re-exported to the world, with fashion singers, from Maluma and J. Balvin to the new phenomenon of Karol G. By the way, it collected and recycled sounds, languages and esthetics, and imaginaries and urban myths projected globally. As Daniel Rivera Marín points out: When a "city becomes a song, it already has a place in the collective imagination". To exemplify what has been pointed out, he reminds us, a fragment of Provenza, the fashionable song by Karol G.: "I do not know if it convinces you, we give each other a little graze through Provenza (…) we go through the Barrio, for grass".[38] It is the city of contrasts, of the new official urban fashion site with its international restaurants, in a sector of El Poblado, the old place of the notables, where the drug traffickers and their emerging capitals established their mansions, which is connected in the song as in daily life with that other side of the river, designated in code as "The Barrio", to refer only to Barrio Antioquia, also the old enclave of drug supply. That world of *chiaroscuro* re-signifies itself but remains presently valid.

# References

Alcaldía de Medellín (2019) Historias para no olvidar. Panamericana Formas e Impresos S. A., Medellín
Anónymous (1984) Aquí Viven. Aquí Vivirán. Medellín Cívico, 1627
Bigey M, Hunel A (1991) Revisión de estudios del Metro del Valle de Aburra, 1981

---

[38] Daniel Rivera Marín, *Esta eterna canción de reguetón* (This eternal reggaeton song), in: Generación, Medellín, El Colombiano newspaper, 07.18.2022, p. 20.

Blattman C, Duncan G, Lessing B, Tobón S, Mesa Mejía JO (2022) Gobierno criminal en Medellín: panorama general del fenómeno y evidencia empírica sobre cómo enfrentarlo. IPA-Proantioquia-The University of Chicago-Universidad EAFIT, Medellín

Cardona M et al (2005) mai–jun. Homicidios en Medellín, Colombia, entre 1990 y 2002: actores, móviles y circunstancias. Cad. Saúde Pública, Rio de Janeiro 21(3)

"Don Berna". Diego Fernando Murillo Bejarano (2009). Consulted 07.31.2022. https://verdadabierta.com/perfil-diego-fernando-murillo-bejarano-alias-don-berna/

El emulo de Londoño White (2008). Consulted: 07.28.2022 de https://www.elespectador.com/judicial/el-emulo-de-londono-white-article-34492/

Franco Restrepo VL (2005) Poder regional y proyecto hegemónico: el caso de la ciudad metropolitana de Medellín y su entorno regional 1970–2000. IPC Instituto de Capacitación Popular, Medellín

La tenebrosa máquina de guerra que dirigió "Don Berna" (2014). Consulted: 07.31.2022 de https://verdadabierta.com/la-tenebrosa-maquina-de-guerra-que-dirigio-don-berna/

Ley 86/1989, de 29 de diciembre de 1989, sobre sistema de servicios públicos urbanos masivo de pasajeros. Consulted: 07.29.2022 de https://www.funcionpublica.gov.co/eva/gestornormativo/norma.php?i=3426. With the approved law

Medellín. Contraloría General (1994) El Metro una decisión no planificada. Medellín

Memoria de la impunidad en Antioquia (2010) IPC-Instituto Popular de Capacitación-Corporación Jurídica Libertad, Medellín

Pablo Escobar Tour (2022) Consulted: 07.28.2022 de https://pabloescobartour.co/landing-2/?gclid=Cj0KCQjw54iXBhCXARIsADWpsG_JzFziNKmNY24rhIDa9oQ9ZlLk-SeS8sOByF-1zhZMME3Y5BuEimMaAu81EALw_wcB

Recorrido sobre Pablo Escobar en Medellín (2022). Consulted: 07.30.2022 de https://www.viator.com/es-ES/tours/Medellin/Pablo-Escobar-Historical-Tour-of-Medellin/d4563-5549ESCOBAR?m=26374&supag=67067036863&supsc=dsa-694098304004&supai=420430240007&supap=&supdv=c&supnt=nt:g&suplp=1003654&supli=&supti=dsa-694098304004&tsem=true&supci=dsa-694098304004&supap1=&supap2=&gclid=Cj0KCQjw54iXBhCXARIsADWpsG8F_WG9sIuu3q3JDuXDxC_AHqPqxMexIJEGkkFpWA9xipaWYk4mCrgaAuenEALw_wcB

Restrepo JD (2005) El Metro de Medellín, obra del mito antioqueño que sacrificó la verdad. In: Isaza O et al (eds) Estrategias de la corrupción en Colombia. Discursos y realidades. Instituto Popular de Capacitación, Medellín, pp 325–395

Sáenz Rovner E (2021) Conexión Colombia. Una historia del narcotráfico entre los años 30 y los años 90. Editorial Planeta Colombia S. A., Bogotá

Salazar JA (2002) La parábola de Pablo. Auge y caída de un gran capo del narcotráfico. Editorial Planeta, Bogotá

Salazar Pineda G (2005) El confidente de la mafia se confiesa. El narcotráfico colombiano al descubierto. Madrid, Nombrelatino, Madrid

Soto M (2013) La viuda negra. La verdadera historia de Griselda Blanco, la narcotraficante más poderosa y sanguinaria que ha tenido Colombia. Intermedio Editores, Bogotá

Torres Arias E (1995) Mercaderes de la muerte. Intermedio Editores-Círculo de Lectores, Santafé de Bogotá

Valencia de Castaño G (1983) 22 de febrero. El primer periódico ecológico de Colombia. Cromos, 3397

**Luis Fernando González Escobar** Builder architect, Magister in Urban and Regional Studies and Doctorate in History. Titular Professor of the Escuela del Hábitat (School of Habitat) and Academic Coordinator of the Doctorate in Urban and Territorial Studies—DEUT—of the Faculty of Architecture, National University of Colombia, Medellin.

# Chapter 41
# Caracas. Urbicide and Precariousness of Urban Life at the Beginning of the Venezuelan Twenty-First Century. The Worst of Capitalism and Savage Populism

**Alberto Lovera**

> *Puede que nuestras ciudades no mueran, pero su renacimiento va a ser difícil, costoso y prolongado*
> Our cities may not die, but their rebirth is going to be difficult, costly and prolonged
> Marco Negrón, ¿Pueden morir nuestras ciudades? Caracas, 2018

**Abstract** This text presents two issues that seem pertinent to us; the exploration of the concept of urbicide as a line of reflection and analysis seems promising to identify the processes of destruction of the essence of the city as a place for diversity and quality of life. On the other hand, to show how the Venezuelan case and its capital city have been an example of the destruction of the city during the two decades of the twenty-first century, not because of processes derived from neoliberal logics, but mainly from an authoritarian and voluntarist populism orientation.

**Keywords** Caracas · Urbicide · Urban ruin · Public services · Urban living

## 41.1 Introduction

Cities undergo processes of construction but also of destruction. The acute analyst Jane Jacobs reminds us that although decades have passed since she stated it, her warnings are still valid: "Societies and civilizations whose cities stagnate do not develop or flourish again. They deteriorate" (Jacobs 1961).

In the Venezuelan case, different views have shown how the deterioration of urban conditions is perceived, with the scrutiny of a diary that records how this multidimensional debacle is operating. "Sometimes, I have the impression of witnessing

---
A. Lovera (✉)
Central University of Venezuela (UCV), Caracas, Venezuela
e-mail: alberto.lovera@gmail.com

the death of a city," Ana Teresa Torres confesses (Torres 2018). Or from a socio-anthropological point of view that tells us: "Daily life in this country out of service, whose infrastructure is destroyed, is like a science fiction scenario, a dystopia" (Vásquez 2020).

As we write this text, we are in the presence of a live urbicide operation: the destruction of Ukrainian cities and their people. The most radical form, the warlike onslaught with all the resources of the military apparatus, to erase together with its inhabitants a slowly built heritage, full of history, of relationships, of urban and social fabrics, by means of physical violence, although there are other modalities that achieve the same through more sophisticated procedures. We must also pay attention to them. All of them threaten the conditions that make urban life possible, its diversity and coexistence, endangered by actions of different kinds.

In the following, we have a dual purpose. First, to contribute to the exploration and debate on the concept of urbicide and its possible contribution to clarify the processes of destruction of cities in the different variants that would fit in it for urban analysis. And second, to show the destructive processes suffered by the city of Caracas, Venezuela, which have led to the precariousness of urban life at the beginning of the twenty-first century, although with antecedents, which in turn show that urbicide can operate in socio-political and economic contexts different from those shown by many of the studies on this phenomenon.

Due to the space available to show our arguments and evidence, we refer the reader to a series of bibliographical references where the issues addressed in this text can be studied in greater depth.

## 41.2 Urbicide: A Useful Concept for Urban Analysis

Although the concept of urbicide has antecedents in urban interpretation since the 1960s and the following decades of the twentieth century, both in its warlike variant (destruction of cities as a result of conventional and unconventional warfare), as well as in the actions that have produced the degradation of the basic functions of the city (severe deterioration of public services and infrastructure, gentrification, demolition of large complexes, mutation of urban centers that have changed their vocation due to socio-technological change, processes that dynamize the functioning of the city, as well as the impacts of climate change on the urban fabric and the urban fabric, demolition of large complexes, mutation of urban centers that have changed their vocation due to socio-technological change, processes that dynamize the functioning of the city, as well as the impacts of climate change on the urban fabric and daily life in cities), it is worth highlighting the effort that for years put Fernando Carrión to revalue the power of this concept (Carrión 2014, 2018), which shows us that without abandoning the analysis of the production of the city, it is necessary to investigate the mechanisms that operate for its destruction, not to consider them an inevitable outcome but to seek ways to preserve the diversity of the city, which gives it vitality, as shown by the study of Jane Jacobs that relates the struggle between life and death of cities, in this case of the USA, who more than six decades ago bequeathed us one of the most accurate writings on the socio-anthropological reality of the city, its vital sources and what threatens them (Jacobs 1961), which does not fail to give us relevant clues, despite the need to add new data of contemporary reality. Something that today challenges us to look for ways to revitalize the city in the service of its inhabitants (Delgadillo 2021). As Jacobs showed us, there are many things that die in urban life, but others are born again, with new physiognomies, reviving urban diversity.

Finally, from our point of view, it is as important to identify the processes of construction and reproduction of the city as those of its destruction. In this dialectic, we will be able to gage both the perverse effects that hinder the democratic, inclusive, and sustainable city, as well as the virtuous effects that help it to make headway.

For the analysis of both the production and reproduction of the city and its destruction, it is necessary to refine the theoretical and methodological tools to reflect the different edges of a complex reality, where different logics coexist at the same time: that of the market, that of the state, and that of necessity (Abramo 2011), which act for the structuration and deconstruct of the city and its functioning. Latin American social sciences have made important contributions to highlight the modalities of urban development in our continent, among the concepts that stand out are those of structural heterogeneity and unequal and combined development, which show a dynamic coexistence of diverse forms of production, distribution, and consumption of the built environment in the evolution of our cities and their diverse manifestations (Pradilla 2013; Jaramillo 2021).

The so-called critical approach to urban research, of which we feel a part, has made and continues to make important contributions to the interpretation of urban reality, but it has unfinished business. In its origins, it was inspired by the Marxist and neo-Marxist interpretation of urban reality. Today, it incorporates other optics, trying to reflect new phenomena that avoid the "lazy security" of which Pierre Bourdieu spoke to us, when consecrated formulas are repeated that do not show the novel configurations. That is, the methodological resource of appealing to eclectic visions, capable of capturing the new data of reality, without losing its critical focus. It is to make use of eclecticism, not as a philosophical system but as a method of "open reflections, that is to say, dialogical (polemic and controversial) and interdisciplinary" (Álvarez 2002: 278). Something that critical thought has been doing for some time, as the cited author shows, although some do not recognize it, that its theoretical reflections are open to add contributions fed from other sources, but pertinent and reinterpreted from an integrating vision that reflects the complexity of multiple relations and new phenomena.

This critical thinking in urban research and in the historical-structural interpretation of socioeconomic phenomena has shed much light on the functioning of our societies and their tensions and contradictions. It has focused on the negative consequences of the capitalist accumulation model, now in its neoliberal variant, which has marked the economic and urban transformation in recent times, but, we insist, it has a pending matter: a critical analysis of the anti-neoliberal experiences, which have, like the experiences of really existing socialism old and new, perverse, and unacceptable consequences for good living and sustainable development, which have led to despotic regimes (Del Bufalo 2011), absolutely incompatible with an emancipatory thinking. Extractivism has not been absent from the most recent anti-neoliberal attempts, which makes it not to move away from the political economy of industrialism, common to existing capitalism and socialism (Naredo 1996), which is not a viable foundation to seek the paths to a society of free and equal men, which continues to be the fundamental force that motivates social conflict in the world, which manifests itself in different scenarios and places (Del Bufalo 2011). To avoid many of the deviations of critical thinking, it is necessary, following the orientation recommended by Kornai (2016), to avoid the versions of the apologists of an imaginary capitalism or socialism, of the devotees of each of these systems, but of the existing one (so far), which does not exclude that, like any historical construction, it may be replaced at some point.

A second aspect: The urbicide, the destruction of the city, is not only a product of the capitalist model but also the result of other logics that claim to be anti-capitalist and anti-neoliberal. They also produce destruction of the city. For different reasons, not necessarily mutually exclusive: Because the political economy of both models of accumulation is associated with a predatory pattern of nature; because it is believed that the city can be organized at will, ignoring the rules and inertias of urban reality; because an anti-urban conception prevails; because, especially in certain military sectors, the slums are considered a danger to be eradicated. Additionally, as we will see below, in regimes such as the one established in Venezuela since 1999, which has become an authoritarian populism (Arenas and Gómez Calcaño 2006; López-Maya 2016), due to its attempt to destroy the previous regime and its purpose of initiating a "new era," it tries to physically and symbolically demolish the emblems where liberal democracy is based, trying to rewrite a story of both the past and the future, as well as material supports that represent them, destroying the city. Although urbicide has occurred in different times and cultures, our emphasis is to show that it is wrong to assert that this phenomenon is only associated with capitalism; other anti-capitalist regimes (socialist or populist) can lead to the destruction of cities.

Third: Urbicide is also associated with Manichean visions of market and state. Several issues must be considered to manage the democratic city in a way that does not assassinate its diversity and complexity. This is clear in the processes of gentrification, widely documented by critical urban research, which transform certain areas of the city into a place for elites, expelling other social sectors, and impoverishing its heterogeneity. But this is not exclusive to a capitalist orientation. Polanyi (2001) convincingly argued that the self-regulating market is a utopia. It turns out that the historical experience of the twentieth century also showed the unfeasibility of centralized administration to manage the economy and society. The analysis of Kornai (1992) and Prezeworski (1991) on how centrally planned economies work and their inevitable distortions, and the reading of these issues on the Venezuelan reality (Silva Michelena 2014), are conclusive in the same sense: the need for the coexistence of market mechanisms with regulatory actions of the state. There is enough evidence on the unfeasibility of central planning, as well as of a management exclusively managed with market criteria free of regulations. The prominent Venezuelan economist Héctor Silva Michelena puts it very acutely: "The most important legacy that social scientists left to their colleagues of the 21st century is the concept of mixed economy. They have overthrown the idea that the state and the market are contradictory and have established that, rather, they create a synergy that enhances growth with equity and sustainability" (Silva Michelena 2013).

In terms of Karl Polanyi's most emblematic work, the market is a great servant, but a disastrous master (Bienefeld 1990). That is precisely why it must be regulated to protect society from its excesses. This leads us to the orientation proposed by Naredo (1996): to move from the market as a panacea to the market as an instrument, subject to regulation.

Just as there are no good economic, socio-political, urban, or environmental sustainability results from the application of the so-called "market fundamentalism," neither are there good results when it is substituted by a "statist fundamentalism" (Todorov 2010). The results tend to be unsatisfactory in both cases. State and market must work in synergy. When this is ignored, as in the case of Venezuela in the twenty-first century, the results are a deterioration of the living conditions of society and cities, because sooner or later it becomes economically, socially, and environmentally unviable, as we will show below. The result is that the destructive forces of urbicide are unleashed.

## 41.3 Urbicide in a Petro-state

Although our analysis focuses on one city in particular, Caracas, the capital of Venezuela, to understand its processes, it is necessary to provide some keys to interpret the Venezuelan reality. Additionally, much of the deterioration of urban life in Caracas is not exclusive to that city but extends throughout the national geography, particularly in other urban centers, where precariousness is even greater. The origins of the problems are the same with nuances according to each urban agglomeration.

Since the first third of the twentieth century, Venezuela became an oil and rentier country. It was able to evade the classic rigors that accompany the original accumulation to establish capitalism, thanks to the oil rent, which allowed a long period of growth without radical restrictions on the income of wage earners and allowing profitability to private agents, thanks to a state with broad intervention capabilities, both in the economic and socio-political fields (Coronil 1977; Baptista 2004, 2007, 2010; Urbaneja 2013).

A long period of political stability, economic growth, and social cohesion created an "illusion of harmony" (Naim and Piñango 1985) due to the good performance of the first decades of the democratic regime that began in 1958 but ignored that problems were brewing, both for economic (Baptista 2004, 2010) and socio-political reasons (Álvarez 1996; López-Maya 2006). That Venezuelan exceptionality had many weaknesses. And they began to manifest themselves. To paraphrase Bienefeld

(1990) on how the framework of the international economic system that was designed in the post-war period could get into trouble with satisfactory results, showing that this was possible because it was a victim of its own good performance, when it began to neglect the elements on which its success depended. Something similar happened in the Venezuelan economic and socio-political system, which did not heed the warning signs of the crisis of rentier capitalism and neglected the mechanisms of social cohesion and political representation, which were losing vitality and began to show alarming signs of poverty, social exclusion, and institutional deterioration.

The social outburst of the "Caracazo" (1989), like other similar ones in Latin America at the end of the twentieth century, showed that certain sectors could not find spaces in the current institutional framework to channel their demands. Although there were already indicators of economic and social imbalances in Venezuelan society prior to that date (the end of a long period of fixed dollar exchange rate in February 1983 with the so-called "black Friday" is emblematic), that outbreak marked a turning point for the generalization of situations of anomie and growth of urban violence, until then of very low occurrence in comparison with Venezuela's Latin American neighbors. Analyzes on this social outburst and its subsequent consequences (López-Maya 2000, 2006; Vázquez 2020; Moleiro 2021) can be read as the beginning and development of a process of urbicide in the city of Caracas, as well as in other urban centers.

A multidimensional structural crisis had been installed in Venezuelan society, following a process that we cannot detail here, but that has been analyzed and followed by many scholars (Baptista 2010; Coronil 1977; López-Maya 2006, 2016; Arenas 2018; Corrales and Penforld 2010, among others).

## 41.4 The Urbicide in Caracas

Although our analysis takes as the period of inquiry the beginning of the twenty-first century, this does not mean that only since then, there have been manifestations of deterioration in the conditions of urban life in Venezuela and Caracas in particular. Since the last two decades of the twentieth century, Venezuela was undergoing a process of structural crisis with manifestations at different levels due to the exhaustion of the accumulation pattern of rentier capitalism (Baptista 2004, 2007, 2010), as well as of its socio-political arrangements that had begun to show cracks (Álvarez 1996). Many things were changing, but for the worse, and urban violence was one of them (Briceño-León 2016). Also, the growth of poverty (España 2009). All this accompanied by a growing anomie (Durkheim), as sociologists call the deterioration of the rules of the social game, which ended up overflowing from the social outburst of the so-called "Caracazo" of February 1989 and subsequent events, such as the failed attempts of military coups in 1992. A cocktail with many economic, socio-political, and cultural components showed unsatisfied demands and the search for new paths (López-Maya 2006).

The social outbreak of the "Caracazo" of 1989 was accompanied by looting of stores and later by an intense repression that left thousands of victims. It was not only a rupture of the social pact but also the presence of looting, which in Venezuelan, history has been in some circumstances a violent form of the exercise of power, but also a form of resistance and protest, as will be manifested again during the so-called Tragedy of 1999, the socio-natural phenomenon of floods and landslides, which produced thousands of deaths and hundreds of thousands of displaced people due to the destruction, it caused in the northern coastal area near Caracas and in some other areas of the country (Vázquez 2010, 2012).

Beyond the spurious re-reading and reinterpretation of these events by the new cast in power since 1999, there is no doubt that these events meant a significant rupture in the Venezuelan future ("nothing will ever be as before"), both for the dominant and dominated sectors.

Many of these factors explain the emergence of new actors and proposals for political change that attracted significant support. This is not the place to dwell on this process and its characteristics, which, although important, would take us away from our focus of analysis. There is a profuse bibliography in this regard, which with varied criteria analyze the scenario that led a new political cast to power and the characterization of the resulting regime and its changes (López-Maya 2006, 2016, 2021; Arenas and Gómez Calcaño 2006; Arenas 2018; Petkoff 2010; Corrales and Penfort 2010; Casanova 2016, among many others).

Although there was a period of improvement in socioeconomic indicators over a decade, between 1997 and 2007, in which there was a reduction of the population living in poverty, this was not the result of structural changes but of the increase in oil income and its distribution through public spending (España 2009). After more than a decade of deterioration in the living and working conditions of the popular sectors at the end of the twentieth century, the new government's orientation led to a (fully justified) revenge that improved social welfare indicators but went no further than that. When oil income was reduced, it could no longer be sustained because its source was a dwindling oil income.

Many of the misunderstandings in the interpretation of the Venezuelan reality of the new regime come from confusing the evident improvements that took place in the reduction of poverty and inequality in the first decade of the twenty-first century, but since there was no change in the bases of sustenance and a socioeconomic model with viability, when the sources that fed that improvement (oil income) declined, the weaknesses of public policies that had not addressed the unresolved problems of the Venezuelan economy and society began to manifest themselves. Or they had done so with formulas that aggravated the problems: nationalization of the economy with unsatisfactory performance, unresolved macroeconomic imbalances, which led to an increase in inflation (including a long hyperinflation), the unraveling of the economy, the dramatic fall of the GDP, an unprecedented increase in poverty (in 2021: 94.5% of the population in poverty, 76.6% in extreme poverty), according to the Survey of Living Conditions (ENCOVI) (UCAB, ENCOVI Project, UCAB.org.edu.ve), which since 2014, three universities (UCAB, UCV and USB) carry out annually, to remedy the statistical silence of the state in many areas (social, economic, epidemiological, etc.), in this case of social indicators.

It must be borne in mind that the change of political course that began with the presidential election of Hugo Chávez in 1999 until his death in 2013, and which was prolonged by his successor Nicolás Maduro, who has ruled Venezuela since then, has gone through various economic and socio-political orientations, although all of them can be sheltered under a populism of multiple faces, depending on the circumstances, but all of them oriented to achieve the consolidation of political power and the hegemony of the new ruling cast, with an anti-capitalist orientation, but antimodern for its (unfeasible) pretension of restoring in republican ideology of the nineteenth century, which has derived in patrimonialism of the ruling leadership, losing strength the socialist ideology, which is not abandoned, but fades away (Guerra 2013; López-Maya 2016, 2021; Spiritto 2017; Casanova 2016). We use the metaphor of wild populism to show that, even with its variants over time, the resulting authoritarian centralism has adopted different clothes according to changing situations.

Although it may not seem obvious, these elements are associated with the core of our analysis: the urbicide and the deterioration and destruction of living conditions in Venezuela and its capital city. The orientation of public affairs has led to the precariousness of urban life. It has done so with a mixture of capitalism and populism, although its verbiage is tinged with expressions of authoritarian socialist thinking.

Like other processes of anti-neoliberal socio-political change, the Venezuelan case shows that it was, at least in its early stages, as has been argued for similar processes, defensive rather than offensive movements (Gilly 2006). Measures were taken to stop the excesses of the neoliberal variant of capitalism, but they were not capable of founding a new political economy. It seems that what Walter Benjamin said: "For Marx, revolutions are the locomotives of history, but perhaps, things are different. Perhaps, revolutions are the way in which humanity, traveling on that train, pulls the emergency brake" (quoted by Gilly 2006). The proposal of a new model remained an unfulfilled promise. There are many analyzes that reflect the persistence of the rentier model, and the little oxygen that social economy initiatives have had from the central power. The alarm signals came from analysts of various orientations, but it is striking that over the years, those coming from those who originally supported the process of change have multiplied, and now disqualify it for its authoritarianism, its rentierism and extractivism, not only oil but now also mining, and for the destruction of the productive forces and levels of poverty and inequality, which had been temporarily reversed, and now show unprecedented figures (ENCOVI Project: https://www.proyectoencovi.com; López Caldera 2015; Lander et al. 2013; Lander 2016).

Venezuela has been in the first decades of the twenty-first century an emblematic case of destruction of productive forces, without the mediation of war conflagrations or natural disasters (Guerra 2013, 2017; Vera 2018). The destruction of cities and their social, political, economic, and cultural fabrics is a manifestation of this. The case of Caracas appears as an emblematic case, although not unique. And it shows that as destructive of urban life and coexistence can be savage capitalism as the different variants of savage populism that have been tested in the Venezuelan territory and society in this beginning of the century, which, although in the name of reversing the previous neoliberalism has been marked by diverse orientations and emphases. Always prioritizing the political control of the upper echelons of power, but according to time and place, trying moderate reforms, exacerbated statist socialism and economic pragmatism, which has been very hostile to the previous business world, but not to the new economic elites that have prospered in the shadow of the state, the so-called "bolibourgeoisie," a sort of capitalism of the regime's friends, who have made enormous fortunes, together with allied countries, as has been documented by investigative journalism (Armando.info and Chavismo Inc.) and the investigations of Transparencia Venezuela.

## 41.5 Information Opacity

As in the initial part of this text, we refer to various bibliographic and documentary sources that allow us to expand and verify the statements and facts shown of deterioration of urban living conditions, including a set of observatories in different areas (violence, economy, finance, public services, social indicators, etc.) that have been developing in Venezuela in the face of the silence or the spasmodic and biased way in which official information is presented. Since 2018, the Central Bank of Venezuela (BCV) stopped publishing information on GDP disaggregated by sectors and balance of payments; since 2019, there is also no information on industrial production and trade sales; since 2009, there is no information on fiscal management and public debt; since 2016, there is no information on oil production, and in 2015, it was the last information on poverty. There is no information on the national budget and official information on other key areas.

## 41.6 The Urbicide Demonstrations in Caracas

The urbicide and precariousness of urban life are manifested in both symbolic and material forms.

With regard to what has happened in the first two decades of the twenty-first century in Venezuela, and in Caracas in particular, it should be noted that both in terms of political design and in terms of its economic and urban policy, the prevailing regime has been marked by diverse orientations and emphases, all under a populist atmosphere, which has ranged from rather reformist moments to proposals of radical socialism and more recently of a pragmatism that supports the continuity of those in power at any price.

The discourse and purpose of the new cast in power were to refound the republic, this meant dismantling the material and symbolic structures.

### 41.6.1 Symbolic Urbicide

Since the new political leadership was installed in Venezuela in the twilight of the twentieth century, a deliberate attempt has been made to modify the symbolic references of the old regime.

It was not only the change of the name of the Republic in the new Constitution, but also in that of emblematic sites: streets and avenues, squares and other sites that are part of the historical memory of the inhabitants of Caracas, as well as the demolition of monuments. The most recent, that of the coat of arms of the Capital City, trying to erase a controversial but inescapable colonial past. In the name of a justified vindication of the legacy of the indigenous and black population, the aim is to erase those derived from the Spanish and the mixture of all of them.

As Ana Teresa Torres tells us in her novel Diorama (Torres 2018, 2021), it is an operation, the *damatio remorie*, like that of the Romans, of the condemnation of memory.

### 41.6.2 Urban Policies

In terms of urban and territorial policies, the "catastrophic vision of the urban phenomenon" (Negrón 2011) has prevailed, which has been predominant in the political leadership of different ideological signs throughout time, in the case of the new post-1999 political cast, it has been exacerbated, dreaming of the possibility of reversing the settlement pattern of the population. It is possible to speak of an anti-urban vision, which in the first years was not only in discourse, but resources were also dedicated to the Orinoco-Apure Axis, thinking of a reversal of the settlement patterns of the population and activities. Although they did not materialize due to their unfeasibility, time and resources were invested that delayed the recognition that the population had to be attended to where it was settled, urban inertias prevailed over territorial voluntarism.

### 41.6.3 Territorialization of the Political Conflict

Socio-political polarization has accompanied the "populist rupture in Venezuela" (López-Maya 2016, 2021). The case of Caracas has been emblematic. Throughout more than two decades of political conflict, this has manifested itself over the territory. Friend/enemy, us/them, are keys to polarization and intolerance. Spaces where only supporters of a single side (whether pro-government or opposition) can be present. These phenomena have been identified by social research and shown their perversion for citizen and political coexistence (García Guadilla 2012; Lozada 2008, 2016, 2021; Isidoro Losada 2015). That different groups of citizens cannot share a space because they have diverse views, and sympathies is a tragedy for urban and democratic coexistence. It is to prevent diversity and pluralism, and it is urbicide that blocks paths for a collective construction of urban life where respect for the other can prevail.

Although there have been different intensities of the conflict and its territoriality, even in the stages where it has been reduced, this pattern of segregation persists and also presents in popular sectors where there have been changes of adhesion to political options, but the territorial control of political or social organizations affected by the regime persists. Plurality is intimidated. In other areas of the city, similar situations are taking place, but of the opposite sign, in this case, censorship is against those who are not opponents. Polarization is not very different in any of its poles. Regardless of who is responsible for unleashing the mechanism, the effects are there.

We will see that the destructive nature of social and political polarization for social coexistence and to address the problems of daily life in the absence of an outcome in favor of the contending political sectors (a catastrophic tie) will lead to the emergence of social arrangements to address the problems of daily life and the overflow of urban violence, beyond political preferences, seeking agreements that address issues whose manifestations do not discriminate sectors by their political preferences (deficit of public services, violence, access to food, etc.).

### *41.6.4 Urban Violence and the Proliferation of Organized Crime*

The Caracas Metropolitan Area, like Venezuela, was for a long time a territory where violence rates were much lower than its Latin American neighbors that changed in the 1990s. The indicators not only rose dramatically, but they also surpassed the cities that until then were considered the most violent in the continent. The statistical record skyrocketed, and the analyzes made year after year show that not only urban violence grows, but organized crime is enthroned. This is documented by the Venezuelan Violence Observatory (Briceño-León et al. 2009, 2016; Briceño-León and Camardiel 2011, 2016; Briceño-León 2016).

It is clearly established in urban research that violence in cities is a factor in the deterioration of urban life and coexistence. It is an indicator of urbicide. It also shows a paradox that less inequality does not necessarily imply less violence. In the early years of the new political regime, distributive policies, underpinned by a favorable oil scenario, produced a significant reduction in inequality, but not a reduction in violence (Zubillaga 2013; Briceño-León 2016). Years later, inequality increased, and intense levels of urban violence remain, as shown in the reports of the Venezuelan Violence Observatory.

The conclusion reached by these studies is that a latent (or hidden) variable has been underestimated: institutionality. Its absence or deterioration explains a lot.

When, as in the case of Venezuela and Caracas in the elapsed time of the twenty-first century, a systematic deterioration of the rules of the social game and of the rule of law, of discretion in the exercise of power, of impunity, can be verified, only rescuing a pact can stop the escalation of violence, which deteriorates the daily life of citizens, which includes manifestations of different types of urban violence, as well as systematic violations of human rights by repressive state agencies, as has been widely documented by both NGOs defending human rights in Venezuela (PROVEA, among others) and on the international scene (Amnesty International), as well as the reports of the UN Commissioner for Human Rights, Michelle Bachelet.

### *41.6.5 The Deterioration of the City's Government and Infrastructure*

Like other Latin American capital cities, the governance of Caracas has been problematic. Its territorial scope extends to two political-administrative entities (Capital District and Miranda State). The opportunity to provide a practical and viable solution seemed to have arisen with the approval of the new 1999 Constitution. Different options had been outlined previously, but for different reasons, this was not possible. Both the constitutional design and the laws derived from it established an oxymoron that left the problem unresolved.

The coexistence of the municipal mayor's offices and a Metropolitan Mayor's Office has always been problematic, both when the latter was governed by officials affiliated with the central government and when it was governed by the opposition forces. The centralist vision was imposed and was finally eliminated, ignoring the provisions of the Constitution.

This journey, which once again leaves the issue of governance of the Metropolitan Area of Caracas unresolved, is well documented and analyzed in a text by one of the keenest observers of the Venezuelan urban reality, and which shows that recentralization does not help to address the basic conditions of urban life in Caracas; on the contrary, it has perpetuated the deterioration of urban life (Negrón 2021).

It is not only a deterioration of the governance of the city, but the basic conditions of its functioning have been deteriorating. Basic and infrastructural services show precariousness indexes. Electricity, water, domestic gas, telecommunications, waste collection, etc., do not meet the needs of citizens. The same can be said of public transportation, both subway (Caracas Metro) and above ground, as well as the deterioration of streets, avenues, and highways. To which he later added the access to gasoline. All this has been documented by a set of observatories that allow us to identify the magnitude of the problems in these areas.

There is a very significant fact about Caracas and the deterioration of its public services and infrastructure. Traditionally, political decisions have sought to keep the capital city as less affected as possible by deficits in this area, due to its role as the nerve center of the country and its impact on social conflict. This was the case both in the government of Hugo Chávez in 2009, when electricity rationing was imposed, or the more dramatic crisis of the great national blackout of 2019 during the government of Nicolás Maduro. The same criterion has been maintained for the problems of water supply, gasoline, or at the time of food shortages. What must be highlighted is that this protection to Caracas ended up overflowing and did not prevent keeping it exempt from the crisis of public services. Although less than in the interior of the country, there are failures in the supply of electricity, water, domestic gas (particularly gas cylinders, which is accessed by low-income sectors), not to mention telecommunications services where there are serious problems with Internet connection and fixed telephony, and to a lesser extent, cellular telephony. What we mean to say is that Caracas could not be kept unscathed from a generalized deterioration of public services. Electricity blackouts are frequent, but not as prolonged as in the interior of the country; there are difficulties in the public transportation service, sometimes associated to electrical problems, others to maintenance problems; the water supply is irregular, only for a short time, it is what in colloquial language is called "la hora loca," you have to suspend any other activity to collect water, wash clothes and belongings, take a bath, and all this in sixty, thirty minutes or less. What the rest of the country was already living and sometimes with more precariousness, many hours without electricity, many days without water, without telephone connection was extended to Caracas. And it is still like that when we write this text in 2022.

### *41.6.6 The Popular Neighborhoods and the Gran Misión Vivienda Venezuela (Great Mission Housing Venezuela)*

Caracas has been impacted by the urban policies undertaken by the new government administration that began almost at the turn of the century. The housing and habitat policy that was formulated had enormous strengths: It was the result of the synergy between state and university institutions that had worked together to identify the key factors of a housing and habitat policy that would address all its aspects (existing housing, both in popular neighborhoods and in the housing complexes promoted by the state over many decades; construction of new housing, rehabilitation of popular housing and neighborhoods for their recognition and provision of services and infrastructure to guarantee the full exercise of citizenship; attention to the homeless population) (Baldó and Villanueva 1998; CONAVI 1999; Baldó 2004; Lovera 2013).

This orientation was predominant during the first year and a half of Hugo Chávez's administration, but soon, the antagonistic visions coming from the civilian and military sectors (Camacho 2004) were confronted, where the latter prevailed, underestimating, and demonizing the popular neighborhoods and favoring the construction of new housing, next to anti-urban visions with their unfeasible pretension of reverting the urban settlement of the population in the territory.

After many failed trials of new housing production over the following years and many changes of the protagonists in the state institutions in this field, the Gran Misión Vivienda Venezuela (GMVV) was conceived in 2011. It was a new housing production program of enormous proportions, which was possible to finance due to the good years that were being experienced in the oil market. The idea of attending to existing housing in unison with the construction of new housing complexes was abandoned. Now, the priority was new housing units of different sizes. Without underestimating the effort to multiply the number of new housing units, the approach of simultaneously attending to existing housing issues was abandoned, particularly those of the popular neighborhoods, which due to their precarious conditions are a source of vulnerability (Lovera 2013; Cilento and Trocelli 2020).

There are varied analyzes of Gran Misión Vivienda in Venezuela, some highlighting the importance of this initiative (Carriola 2015), others showing its strengths and weaknesses (Valente 2018), still others on its scarce transparency, pockets of corruption, and the inconsistency of the figures of its results (Transparencia Venezuela 2012, 2017a; PROVEA 2022).

From our focus of analysis, the great weaknesses are in its "viviendist" vision that produces housing, but not a city. Buildings located in different locations, without consideration of the urban context, with huge deficits of public services and collective facilities, as shown in a study conducted for the case of Caracas (IMUTC 2013). The result is that a lot of new housing may be being produced, while the urban fabric deteriorates. To these should be added the urban and social problems that large housing estates have generated in distinct parts of the world, and which have led to the need to demolish them due to their unviability. It is not only attending to the homeless and their precariousness, in both cases with unsatisfactory results (Rodríguez and Sugranyes 2004; Cilento 2018).

In short, if all the factors at play are not considered, residential construction may be taking place while the city is being destroyed, and that part of the city that has been produced after many hardships by the inhabitants of the poor neighborhoods is ignored (Bolívar 2011). And this has been happening in Caracas.

### *41.6.7 The City Center: Deterioration and Spurious Recovery*

A series of circumstances coincided to cause a systematic deterioration of the historic center of Caracas for years. The growth of urban violence, the proliferation of anomie, the weakening of planning and urban control mechanisms, the proliferation of building takeovers, the unjustified expropriation of premises whose use was uncertain, plus the growing ungovernability of the city due to the prolongation of the fragmentation of the city government (Negrón 2021), and the prolongation of the territorialization of the political conflict, turned the central zone of Caracas into a no man's land, without order or agreement, where there were no shared rules of the game, but rather segments precariously led by formal and informal authorities trying to establish their rules with different results.

The generalized deterioration of the central area of the city later led to a rehabilitation action. Public spaces (squares and boulevards) were recovered, and the creation of commercial services for the public (restaurants and cafeterias, among others) was encouraged. The result is an improvement of public spaces, but access to most of the available businesses is only possible for sectors with a high payment capacity; for the rest, only small businesses and the informal market, which coexists in this part of the city, are available. The manifestations of segregation persist. Not everyone has access to what is offered, although the urban fabric is somewhat better. There is much scenographic recovery, but many issues of the democratic and inclusive city are still pending.

In other areas of the central city, there are very few initiatives to create new public spaces or restore deteriorated ones. Paradoxically, much more has been done by the municipal governments of the fringe of the Metropolitan Area of Caracas (those located in the State of Miranda), governed by the opposition, in terms of the creation of public spaces, but with the absence of urban development plans to avoid the perverse effects of a real estate production that does not adequately articulate the private and collective interests of the city.

### *41.6.8 Utility Problems Persist*

The crises of public services: electricity, water, gas, gasoline, transportation, urban cleaning, Internet, and fixed and cellular telephony are still present in Caracas. The most catastrophic events, such as those that have occurred in the supply of electricity, water, and gasoline at various times, are less prolonged over the years, but they are still present. Even in different areas of the city, they continue to suffer, as shown by complaints and protests. The conditions of a precarious urban life are not being overcome. The destruction of these conditions has not been reversed.

The figures are eloquent. A June 2021 report tells us the following for the Capital District, the central area of Caracas, about the percentages of the population suffering from deficits in public services: electricity interruptions: 41.5%; lack of public transportation: 43.7%; food insecurity: 43.4%; water access restrictions: 29.0%, multifunctional poverty (households or people with multiple deficiencies in health, education, and standard of living): 48.9% (Hum Venezuela 2021). Although they are lower than in other regions of the country, they are of a significant dimension, which shows the vulnerability of urban life in Caracas. In the same sense, it shows us another observation of the public services of the capital city, much better than in other localities, but with precariousness (CEDICE 2022).

## 41.6.9 Adding Up Problems: Urbicide, Complex Humanitarian Emergency, Pandemic, and Climate Change

It is well known that humanitarian crises are defined by international organizations as those resulting from natural disasters or armed conflicts. But another modality has been identified, which is defined as a Complex Humanitarian Emergency (CHE) resulting from a combination of political instability, conflicts and violence, social inequalities, and poverty. This is how the UN has characterized the case of Venezuela. It is the only country in the Americas under this denomination.

If the situation of urban precariousness was delicate in Caracas, and in Venezuela, due to the condition of complex humanitarian crisis, in the terms that the UN defines such an event, it worsened with the arrival of the COVID-19 pandemic, as witnessed by the analyzes and reports in this regard (Cilento and Trocelli 2020; Hum Venezuela 2021).

Many hardships were prolonged because the arrival of international humanitarian aid was prevented, given that the executive power continued to explain the crisis by the economic sanctions, when it is sufficiently documented that the collapse of the Venezuelan economy is prior to them, since 2013 at least, although obviously, they aggravated the situation (Sutherland 2020; Cartaya et al. 2020) and made the situation of daily life more complicated that of Caracas among them.

## 41.7 Urban Management in the Wild Populism

The problems of urban destruction in Caracas illustrate that their origin does not necessarily derive from a capitalist orientation of the city as a source of profit. This modality is well documented in many cases, but the scrutiny does not end there, as certain explorations tend to affirm. Other forms of urban management destroy urban conditions when the production of goods and services and the operation of the city are inadequately managed.

The problems that have been experienced in the city of Caracas, and Venezuela, at the beginning of the twenty-first century, do not stem from the fact that the state has been given prominence. Far from the neoliberal narrative that demonizes all state initiatives as pernicious and all private initiatives as virtuous, what explains a good part of the failures is that the basic rules of any good management, whether public or private, have been ignored. The period of the so-called civil republic (1958–1998) shows emblematic examples of successful performance of public enterprises, which were models of good management (not only PDVSA at that time, but also the electric company EDELCA, the Corporación Venezolana de Guayana CVG, among others).

This has not been the case in the cycle of nationalization of companies undertaken from 2002 onwards by the Venezuelan government, which meant an exponential growth of public companies, which totaled more than 900 by mid-2021, a good part of them by processes of expropriation and nationalization of production activities of goods and services that were privately owned. The analysis of the results of such state management is largely negative and is widely documented (Obuchi 2011; Transparencia Venezuela 2017b, 2018, 2020; Lovera 2017). Goods and services companies paralyzed, with reduced production or with serious deficits in attention to the target population. There are ways for an efficient operation of state enterprises, and guidelines have been proposed for their good management and for their performance to be transparent (Transparencia Venezuela 2021).

This is closely related to the deterioration of the city's operating conditions and quality of life. Caracas and its inhabitants suffer from this, and it is the product of a management that subjects the population to enormous precariousness. The case of Venezuela and Caracas shows us, once again, that the road to hell is full of good intentions, but it is imperative to build a route that can combine equity, economic, social, and environmental viability, which is not possible without a variable mix of market forces, state regulation, and citizen leadership.

The case of Caracas must be analyzed in the context of one of the most dramatic events of the destruction of the productive forces. In the last few years, the Venezuelan economy has reduced its GDP by 80%. Poverty is at 95% and 70% of extreme poverty. We had 4 years of hyperinflation, and we still have the highest inflation in the world. This has been accompanied by an international migration of enormous proportions that has reduced Venezuela's population from the 32 and a half million that official sources say, to 28 million as established by estimates that speak of between 4 million and more than 5 million Venezuelans in the diaspora spread in different parts of the world. These figures also speak of urbicide. People fleeing their cities and their country because their living conditions have deteriorated in their place of origin.

## 41.8 Weaving the Threads of the Analysis of the Urbicide and Its Manifestations in Caracas

There are several conclusions that can be drawn from the analysis that we have been proposing throughout this text, both in its theoretical-methodological dimension, as well as in what the examination of the case of Caracas teaches us.

I begin with a reflection made to me by a veteran researcher of the CENDES of the UCV, Alberto Urdaneta, with whom I worked when I started my career as an urban researcher. When the Cuban revolution triumphed, they decided to "punish" Havana, and the priority was the interior of the island, trying to modify the distribution of the population in the territory. Twenty years later, the Cuban capital had reduced its population by only 1%, but the city was in ruins. Anti-urban voluntarism would only have produced deterioration. It was a costly experiment in social engineering.

The Venezuelan experience of the first decades of the twenty-first century tells us nothing different. After attempting the chimerical development of the Orinoco-Apure Axis and dedicating many efforts and resources to it, seeking to modify the distribution of the population in the territory, reality imposed itself, there are socio-territorial inertias that are difficult to revert. It is necessary to attend to the population and its problems where it is settled, but by then, there had been a great destruction of urban life, an urbicide, in the most important urban centers of the country, Caracas among them, and its recovery has been laborious, also because the great resources of the oil boom, which had been very useful to attend to the enormous deficits to make cities progress more equitable and sustainable economically, socially, and environmentally, had vanished.

We should not underestimate the importance of an institutional framework that addresses urban problems of various kinds. It is possible, and often necessary and imperative, to change the rules of the social game, but they must be replaced by others to guarantee a social cohesion that helps coexistence in a diverse conglomerate with interests that preserve democratic coexistence. The growth of urban violence while poverty and inequality were attenuated in Caracas and other urban centers is a warning that a latent and ignored variable was being left aside, the need for new rules of the game. When this does not occur, the people themselves generate an alternative institutionality, not only to avoid the most perverse effects of violence on daily life, but also in other areas to address other issues that the destruction of living conditions in the city demands of its inhabitants, particularly those most affected in their living and working conditions. Depending on the situation, how to respond to the deterioration of access to public services, food, fuel (gasoline or domestic gas), the long period of hyperinflation, the restrictions imposed by the pandemic on low-income sectors and other situations that conspire against urban life, in addition to the restrictions already present and unresolved.

What we have documented throughout this text is that the processes of precariousness of urban life and destruction of cities, the urbicide, are not only a product of what is called savage capitalism, the predominant neoliberalism, it can also be generated by savage populism, even in its leftist and allegedly anti-neoliberal version. Also in its right-wing version, equally predatory, as Bolsonaro's Brazil has shown us.

The destruction of the city generated both by the dynamics of neoliberal capitalism and those inspired by wild populism are equally pernicious and do not bode well for the daily life of the inhabitants of our cities. Even with their strong inertias, they can be reversed, it will not be easy, but it is possible. Overcoming urbicide is an arduous task, but it will require restraining the most perverse aspects of capitalism and savage populism. That is the importance of identifying the sources of urbicide, to eradicate them and give life again to diversity and coexistence, nourishing sources of the democratic city, where we can live with the common and the different.

**Acknowledgements** We are very grateful to the editors of this book: Fernando Carrión and Paulina Cepeda, who invited us to participate in this important collective inquiry in the field of urban research, as well as to the other colleagues who share this editorial effort, who in the seminar we held as part of the preparation of this publication, provided us with insights to enrich our contribution. We would also like to thank the two anonymous referees, whose comments helped us to perfect the final version of our text and to our friend Diana Rodríguez for her assistance in the translation of this text.

# References

Abramo P (2011) Mercado inmobiliario y la producción de las ciudades en América Latina. OLACCHI, Quito

Álvarez Á (ed) (1996) El Sistema político venezolano: Crisis y transformaciones. UCV, Caracas

Álvarez F (2002) La respuesta imposible. Eclecticismo, marxismo y transmodernidad. Siglo XXI editores, México

Arenas N (2018) Chávez: Populismo con uniforme. Diálogos. Revista electrónica de Historia, vol 19 especial

Arenas N, Gómez Calcaño L (2006) Populismo autoritario en Venezuela. 1999–2005. Ediciones CENDES, Caracas

Baldó J (2004) La política de vivienda para Venezuela. In: Venezuela en perspectiva Carlos Genatios (comp). Fondo Editorial Question, Caracas

Baldó J, Villanueva F (1998) Un plan para los barrios de Caracas. MINDUR-CONAVI, Caracas

Baptista A (2004) El relevo del capitalismo rentístico: hacia un nuevo balance de poder. Fundación Polar, Caracas

Baptista A (2007) La sociedad capitalista: ¿Hacia su estadio final? Academia Nacional de Ciencias Económicas, Caracas

Baptista A (2010) Teoría económica del capitalismo rentístico. Banco Central de Venezuela, Caracas

Bienelfeld M (1990) Las lecciones de la historia y el mundo en desarrollo. In: Bienelfeld M, Valecillos H (eds) El reajuste estructural de la economía: Desafíos y oportunidades para el movimiento obrero, Caracas

Bolívar T (2011) Desde adentro. Viviendo la construcción de las ciudades con su gente. OLACCHI/Municipio Metropolitano de Quito, Quito

Briceño-León R (ed) (2016) Ciudades de vida y muerte. Editorial Alfa, Caracas

Briceño-León R, Camardiel A (2011) Violencia e Institucionalidad. Informe del Observatorio Venezolano de Violencia 2011. Editorial Alfa, Caracas

Briceño-León R, Camardiel A (2016) Delito Organizado, mercados ilegales y democracia en Venezuela. Editorial Alfa, Caracas

Briceño-León R et al (2009) Inseguridad y violencia en Venezuela. Informe 2008. Editorial Alfa, Caracas

Briceño-León R et al (2016) Los nuevos rostros de la violencia. Empobrecimiento y letalidad policial. Editorial Alfa, Caracas

Camacho OO (2004) Actores universitarios y militares en la instrumentación de los programas habitacionales del gobierno de Hugo Chávez: 1999–2005. Encuentro: Yo también quiero mi techo., Organización Conciencia Activa/UCV/Fundación de la Vivienda Popular en Caracas, Venezuela

Carriola C (ed) (2015) La Gran Misión Vivienda Venezuela. Hacia una política socioterritorial de vivienda. Una mirada desde Caracas Metropolitana. Fondo Editorial Mendez Castellanos/FUNDACREDESA/CENDES, Caracas

Carrión F (2014) Urbicidio o la producción del olvido. In: Durán L et al (eds) Habitar el patrimonio. Nuevos aportes al debate en América Latina. IMP/FLACSO/UBA, Quito

Carrión F (2018) Urbicidio o la muerte litúrgica de la ciudad. Oculum Ensaios, v.15, N° 1

Cartaya V et al (2020) Venezuela's complex humanitarian crisis. Responses and challenges for civil society. WOLA, Washington (Spanish version: Cartaya V et al (2020) Venezuela Emergencia Humanitaria Compleja. Desafíos para la Sociedad Civil. WOLA/Acción Solidaria)

Casanova R (2016) La gramática del chavismo. Entre la pulsión socialista y el redentorismo popular. Notas de investigación. Cuadernos del CENDES, Año 33, N° 91

CEDICE (2022) Monitoreo de Servicios. Marzo. 2022, Caracas

Cilento A (2018) Sobre el fracaso de la construcción masiva de viviendas completas. Red Nacional de Investigación Urbana. Puebla, México. http://www.rniu.buap.mx/infoRNIU/ago18/2/art_sobre-fracaso-construccion%20masivaviviendas-completas.pdf

Cilento A, Trocelli M (2020) Coronavirus, emergencia humanitaria, crisis ambiental, teletrabajo, hospitales, escuelas y vecindarios y barrios sostenibles. ACADING, Boletín N° 48, Caracas

Consejo Nacional de la Vivienda (CONAVI) (1999) Política de vivienda 1999–2004 (qué hacer y cómo hacerlo en relación con el programa de vivienda). Tecnología y Construcción, N° 15-I

Coronil F (1977) The magical state. Nature, money and modernity in Venezuela University of Chicago Press, Chicago (Spanish version: (2013) El Estado Mágico. Naturaleza, dinero y modernidad en Venezuela. Editorial Alfa, Caracas)

Corrales J, Penfold M (2010) Dragon in the tropics. Brookings Institution (Spanish version: (2012) Un dragón en el trópico. Cyngular, Caracas)

Del Bufalo E (2011) Adiós al socialism. Bid&co. Editor, Caracas

Delgadillo V (2021) Muerte simbólica y material de la ciudad, Una aproximación al urbicidio. Revistarquis, vol 10, N° 1

España LP (2009) Más allá de la pobreza. Diez años después UCAB, Caracas

García Guadilla MP (2012) Caracas de la colonia al socialismo del siglo XXI. Espacio, clase social y movimientos ciudadanos. In: Almandóz A (ed) Caracas, de la metrópolis súbita a la meca roja. OLACCHI, Quito

Gilly A (2006) Historia a contrapelo. Una constelación. Ediciones Era, México

Guerra J (2013) El legado del Chávez. Editorial Libros Marcados, Caracas

Guerra J (2017) Ante el fin del socialismo rentista, una propuesta de reforma económica. In: Spiritto F (coord) La nueva economía venezolana. Propuestas ante el colapso del socialismo rentista. Editorial Alfa, Caracas

Hum Venezuela (2021) Impactos de la Emergencia Humanitaria Compleja en Venezuela con la Pandemia COVID, Caracas

Instituto Municipal de Urbanismo. Taller Caracas (IMUTC) (2013) Construir vivienda y hacer ciudad. Un binomio inseparable. In: XXX Jornadas de Investigación del IDEC, FAU, UCV, Caracas

Isidoro Losada AM (2015) Estrategias territoriales de control político en Venezuela. Iberoamericana. Nueva Época, Año 15, N°59

Jacobs J (1961) The death and life of great American cities. Random House Inc., New York (Spanish version: (2013) Muerte y vida de las grandes ciudades. Capital Swing, Madrid)

Jaramillo S (2021) Heterogeneidad estructural en la ciudad latinoamericana. Más allá del dualismo. Ediciones Uniandes, Bogotá

Kornai J (1992) The socialist system. The political economy of communim. Princenton University Press

Kornai J (2016) The system paradigm revised. Classification and additions in the light of experiences in the Post Socialist Región. Acta Oeconomica 66(4) (Spanish version: (2017) Revisión del paradigma de los sistemas. Clasificación y agregados a la luz de la transición post-socialista. Documentos de Trabajo N° 125, FIEL, Buenos Aires)

Lander E (2016) La implosión de la Venezuela rentista. Cuadernos de la Nueva Política, Ámsterdam

Lander E et al (2013) Promesas en su laberinto. Cambios y continuidades en los gobiernos progresistas en América Latina. IEE/CEDLA/CIM, La Paz

López Caldera A (ed) (2015) Transición, transformación y rupturas en la Venezuela Bolivariana. CELARG/Fundación Rosa Luxemburgo, Caracas

López-Maya M (2000) ¡Se rompieron las fuentes! La política está en la calle. In: Baptista A (coord/ed) Venezuela Siglo XX. Visiones y testimonies. Fundación Polar, Caracas

López-Maya M (2006) Del viernes negro al referéndum revocatorio. Editorial Alfa, Caracas

López-Maya M (2016) El Ocaso del chavismo. Venezuela 2005–2015. Editorial Alfa, Caracas

López-Maya M (2021) Democracia para Venezuela ¿representativa, participativa o populista? Editorial Alfa, Caracas

Lovera A (2013) De espaldas a las barriadas populares. Las paradojas de la política habitacional de Hugo Chávez. In: Bolívar T et al (coord) Ciudades en construcción permanente, vol II. Ediciones Abya-Yala/CLACSO/UCV/USP, Quito

Lovera A (2017) La maraña de las empresas públicas de cemento. Opacidad y desorden. Transparencia Venezuela, Empresas de Propiedad Pública, Caracas

Lozada M (2008) Nosotros o ellos. Representaciones sociales, polarización y espacio público en Venezuela. Cuadernos del CENDES, Año 25, N° 60

Lozada M (2016) Conflicto y polarización en tiempos de revolución: Representaciones e imaginarios del otro en Venezuela. SOMEPSO vol 1, N° 1

Lozada M (2021) Reparación integral y reconciliación nacional. Desafíos de la transición democrática en Venezuela

Moleiro A (2021) La nación incivil. El Caracazo, sus consecuencias y el fin de la democracia. Editorial Dahbar, Caracas

Naim, Piñango R (ed) (1985) El caso Venezuela: Una ilusión de armonía, IESA, Caracas

Naredo JM (1996) La economía en evolución. Historia y perspectivas de las categorías básicas del pensamiento económico, 2ª. Edición actualizada, Siglo XXI editores de España Editores, Madrid

Negrón M (2011) La democracia y el miedo a la ciudad. Cuarenta años de contradicciones urbanas y territoriales en Venezuela. In: Otamendi F, Straka T (eds) Venezuela: República Democrática. Grupo Jirahara, Barquisimeto

Negrón M (2018) ¿Pueden morir nuestras ciudades? Talcualdigital, Caracas

Negrón M (2021) La accidentada travesía de la ciudad de Caracas y su área metropolitan. In: Carrión F, Cepeda P (comp) Ciudades capitales en América Latina. Capitalidad y autonomía. FLACSO Ecuador, Quito

Obuchi R (ed) (2011) Gestión en rojo. Evaluación de desempeño de 16 empresas estatales y resultados generales del modelo productivo socialista. Ediciones IESA, Caracas

Petkoff T (2010) El Chavismo como problema. Editorial Libros Marcados, Caracas

Polanyi K (2001) The great transformation. The political and economicc origins of our time. Beacon Press, Boston, Massachusetts (Spanish version: (2003) La Gran Transformación. Los orígenes políticos y económicos de nuestro tiempo. FCE, México)

Pradilla E (2013) La economía y las formas urbanas en América Latina. In: Ramírez B, Pradilla E (comp) Teorías sobre la ciudad en América Latina, vol I. UAM, México

Prezeworski A (1991) ¿Podríamos alimentar a todos? En torno a la irracionalidad del capitalismo y la inviabilidad del socialism. In: Lechner N (ed) Capitalismo, democracia y reformas. FLACSO, Santiago de Chile

PROVEA (2022) Informe Anual. Situación de los Derechos Humanos en Venezuela. Enero-Diciembre 2021. PROVEA, Caracas

Rodríguez A, Sugranyes A (2004) Los con techo. Un desafío para la política de vivienda social, Ediciones SUR, Santiago de Chile

Silva Michelena H (2013) Maduro, talud de la economía. Talcualdigital. Opinión, 18.03.2013

Silva Michelena H (2014) Estado de siervos. Desnudando el Estado communal. Ediciones del Rectorado UCV/ bid&co. Editor, Caracas

Spiritto F (ed) (2017) La nueva economía venezolana. Propuestas ante el colapso del socialismo rentista. Editorial Alfa, Caracas

Sutherland M (2020) Las sanciones económicas, contra Venezuela: consecuencias, crisis humanitaria, alternativas y acuerdo humanitario. ILDIS, Caracas

Todorov T (2010) La experiencia totalitaria. Galaxia Gutemberg. Círculo de Lectores, Barcelona

Torres AT (2018) Diario en Ruinas (1998–2017). Editorial Alfa, Caracas

Torres AT (2021) Diorama. Monroy Editores, Caracas

Transparencia Venezuela (2012) Riesgos a la integridad de la Gran Misión Vivienda Venezuela, Caracas. https://transparencia.org.ve

Transparencia Venezuela (2017a) La Gran Misión Vivienda Venezuela. Entre el riesgo y la corrupción. https://transparencia.org.ve

Transparencia Venezuela (2017b) Empresas propiedad del Estado en Venezuela. Un modelo de control de Estado, Caracas. https://transparencia.org.ve

Transparencia Venezuela (2018) Empresas propiedad del estado II. 10 años de opacidad, decadencia y destrucción, Caracas. https://transparencia.org.ve

Transparencia Venezuela (2020) Empresas propiedad del estado III. Un conglomerado de marcado por la ineficiencia y la opacidad, Caracas. https://transparencia.org.ve

Transparencia Venezuela (2021) Gobernanza para las empresas propiedad del Estado venezolano. Una propuesta para institucionalizar la transparencia y la rendición de cuentas, Caracas. https://transparencia.org.ve

Urbaneja DB (2013) La renta y el reclamo. Ensayo sobre petróleo y economía política en Venezuela. Editorial Alfa, Caracas

Valente X (2018) Análisis de la formulación e implementación de la Gran Misión Vivienda Venezuela (2011–2017) desde el enfoque de los Derechos Humanos. CENDES-UCV, Caracas

Vázquez P (2010) Poder y catástrofe. Venezuela bajo la tragedia de 1999. Editorial Santillana, Caracas

Vázquez P (2012) El Caracazo (1989) y la Tragedia (1999). Economía moral e instrumentalización política del saqueo en Venezuela. Cuadernos Unimetanos, N° 30

Vázquez P (2020) País fuera de servicio. Venezuela de Chávez a Maduro. Siglo XXI editores, México

Vera L (2018) ¿Cómo explicar la catástrofe económica venezolana? Nueva Sociedad, N° 274

Zubillaga V (2013) Menos desigualdad, más violencia: la paradoja de Caracas. Nueva Sociedad, N° 243

## *Websites*

Academia Nacional de Ingeniería y Hábitat (ANIH). http://www.acading.org.ve
Anova Policy Research. https://thinkanova.org
Armando.info. Plataforma de periodistas de investigación. https://armando.info
Chavismo Inc. Los engranajes del capitalismo bolivariano, investigación desarrollada por Transparencia Venezuela, Alianza Rebelde investiga (ARI) y la plataforma latinoamericana de periodismo CONNECTAS. https://chavismoinc.com
En inglés: Chavismo Inc. The gears of bolivariam capitalism. https://chavismoinc.com/en
HumVenezuela. Independent platform for the monitoring, documentation and follow-up of the Complex Humanitarian Emergency in Venezuela, developed by Venezuelan civil society organizations. https://humvenezuela.com
Observatorio de Ecología Política de Venezuela. https://www.ecopoliticavenezuela.org
Observatorio Venezolano de Conflictividad Social. https://www.observatoriodeconflictos.org.ve
Observatorio Venezolano de Derechos Humanos Ambientales. https://clima21.net
Observatorio Venezolano de Finanzas. https://observatoriodefinanzas.com
Observatorio Venezolano de Servicios Públicos. http://observatoriovsp.org
Observatorio Venezolano de Violencia. https://observatoriodeviolencia.org.ve
Programa Venezolano de Educación Acción en Derechos Humanos (PROVEA). https://provea.org
Proyecto ENCOVI (Encuesta Condiciones de Vida). https://www.proyectoencovi.com
Transparencia Venezuela. https://transparencia.org.ve
WOLA (The Washington Office on Latin America), ONG. https://wola.org

**Alberto Lovera** Professor and Researcher at the Institute of Experimental Development of Construction (IDEC), Faculty of Architecture and Urbanism (FAU), Central University of Venezuela (UCV). Sociologist (UCAB), M.Sc. in Development Planning: Science and Technology (CENDES-UCV), Specialist in Human Settlements (University of Chile), Doctor in Architecture (UCV). Former Director of IDEC-UCV. Author of numerous essays and refereed books on urban research, public policies, socioeconomics of construction and R&D in the field of construction and habitat.

# Chapter 42
# Santiago, the Non-city? Destruction, Creation, and Precariousness of Verticalized Space

Loreto Rojas Symmes, Alejandro Cortés Salinas, and Daniel Moreno

**Abstract** This chapter situates urbicide as an intentional process that requires the devastation of the public, the destruction of the private and domestic, to decompose the urban structure and the city, generating a (re)production of urban space, in some cases precarious. This interpretation logic situates verticalization processes and their "precarious" expression as a spatialization of urbicide. Beyond its identification and documentation, it provides patterns of localization, diversification, and even intensification in contemporary urban production that transcend the city of Santiago de Chile, expressing a process of recursive urbicide in the logic of production of urban space. This methodology is structured in two main parts using quantitative methods. The first focuses on the analysis of the evolution of verticalization in the Santiago Metropolitan Area (AMS), explained through three hypotheses. The second develops the argument of how verticalization has transcended from precarious areas and communes to precarious silos, a new way of understanding the precariousness of affordable high-rise housing. We use the building permits for new construction and regularization of new construction of high-rise buildings of more than nine floors (National Institute of Statistics, INE) to construct the evolution of verticalization and its transition towards precariousness with an index in the AMS from 1990 to 2019.

**Keywords** Urban development · Housing · Urban renewal

---

L. Rojas Symmes (✉) · A. Cortés Salinas
Universidad Alberto Hurtado, Santiago de Chile, Chile
e-mail: lorojas@uahurtado.cl

A. Cortés Salinas
e-mail: alcortes@uahurtado.cl

D. Moreno
Pontifical Catholic University of Chile, Santiago de Chile, Chile
e-mail: dlmoreno@uc.cl

© The Author(s), under exclusive license to Springer Nature Switzerland AG 2023
F. Carrión Mena and P. Cepeda Pico (eds.), *Urbicide*, The Urban Book Series,
https://doi.org/10.1007/978-3-031-25304-1_42

## 42.1 Introduction

Talking about the city or its negation as a non-city implies referring to a systemic process of spatial production. This spatial production of urban space is generated by the agglomeration of infrastructure, population, and activities, thus understanding the city as the physical base for locating citizens, services, and public space (Carrión 2018).

The buildings are its most significant material expression, denoting diverse, complex, and specific forms of relationship of its city dwellers with the urban space. These historic qualities of the buildings are constantly and permanently threatened by the action of violent acts exercised in and against the cities, which can even lead to the city's death (Berman 1987).

The production model based on destruction and violent action in the urban space exacerbates the already distant relationship between construction and living in the contemporary city (Sennet 2019). Under the theoretical-methodological approach of urbicide, understanding these processes allows revealing one of the most violent expressions under which urban space is produced today in the geographies we inhabit.

The challenge that opens this chapter situates urbicide as an intentional process that requires the devastation of the public (systemic understanding of the city), the destruction of the private and domestic (pre-existing buildings), to decompose the urban structure and the city, generating a (re)production of urban space, in some cases precarious.

This case study on the city of Santiago de Chile accounts for a process of urbicide initially expressed in the materialization of an intensive verticalization process and its consequent impacts on the resulting urban space. It begins for this metropolis in the 1990s, under an evident pattern of concentration in central and pericentral areas and through real estate projects that introduce structural transformations in the residential morphological landscape (Contreras 2011; López-Morales et al. 2012; Pumarino 2014; Vergara Vidal 2017).

This extreme intensification materializes a precariousness of housing and residential habitat from a physical-spatial perspective since the height of the building increases, the number of units per tower and floors, and the size of the new house decreases. Various investigations have reported the negative aspects of this type of verticalization, highlighting the destruction of the existing urban fabric, the increase in land prices, the destruction of architectural heritage, the collapse of road infrastructure and services, the poor quality of construction, the creation of residual spaces and the changes in the forms and patterns of social interaction (Carrasco 2007; Contreras 2011; Contrucci 2011; Herrmann and Van Klaveren 2013; Vergara 2013; Innocenti et al. 2014; Inzulza and Galleguillos 2014; López-Morales et al. 2015).

However, this precariousness is not uniformly distributed in the urban space. It tends to be located where urban regulations are permissive (to attract investment to the commune) or outdated, generating regulatory gaps, giving space to the development of large verticalized real estate projects, and sometimes precarious.

The materialization of "precarious verticalization" as a residential typology is gaining strength in Chilean cities. This concept analyzed under the theoretical-methodological structure of urbicide allows us to account for the impossibility of reading this type of process in a linear or closed way. On the contrary, this process accounts for a recursive character in the (re)production of urban space, whose genesis in the case of Santiago de Chile was the process of precarious verticalization of the Estación Central commune (Rojas 2017).

The chapter onwards reports how the literature addresses the identification and interpretation of urbicidal acts on the city, to then place one of the interpretations that account for the need to reinterpret the destruction and subsequent construction on the rubble that the violent action leaves. This interpretation logic situates verticalization processes and their "precarious" expression as a spatial expression of urbicide. Beyond its identification and documentation, it provides patterns of localization, diversification, and even intensification in contemporary urban production that transcend the city of Santiago de Chile, expressing a process of recursive urbicide in the logic of production of urban space.

## 42.2 The City as an Object of Destruction in the Production of Urban Space

The city can be defined as the intervened space where communities live, build and prepare a territory to coexist collectively in buildings, roads, homes, etc. (Llorente 2015). It's most significant material expression, the buildings, in their diversity generate volume, dimensions, perspectives, and other spatial qualities with which their inhabitants coexist.

The application of violence to the city and, particularly on its buildings, is not something recent or new. Whether due to the arrival of migrants from the countryside or other countries, the development of infrastructure and industrialization processes, urban gentrification, or the result of armed conflicts, cities build their urban space under a permanent threat of attack.

In this time of destruction, the systematic attack on urban spaces requires expanding the perspective of the impact or damage to the city. That is to say, the urbicidal act as an agent of material destruction in the city that impacts human life and the infrastructures that make up their ways of life. In other words, in the ways of being an inhabitant of a built and shared space, where to live in common is no longer possible (Ilich 1984 in Aguirre and Perea 2020).

The conceptual discussion around urbicide is based on the systematic processes of destruction of the city from the use of violence. Although it is an open concept in terms of its conceptual limits, today, exist three lines of interpretation of the concept that are not mutually exclusive.

**Urbicide as anti-urban, anti-city, non-city**: This approach understands urbicide as an attack on the objective of what cities represent: cosmopolitan and tolerant living spaces, which are expressed in their buildings, assets, institutions, industries, and infrastructures that extend towards their symbolic meanings, both in acts of organized war as through bureaucratic actions and urban planning policies.

**Urbicide as a policy of exclusion, anti-heterogeneity**: Urbicide destroys plurality and heterogeneity, manifested in the spatial configuration of the urban system. It is the destruction of urbanity as a physical form and as a way of life (Coward 2004).

**Urbicide as a war of (or against) terror**: This proposal is based on the inseparability of war, terror, the destruction of places, and modern urban planning as a mechanism of reparation in a context of very initiative—taking and fruitful colonialism. In this context, cities are perceived in the imaginary of the political elite as places of conflict that need to be regularized and reorganized, either due to war or urban planning policies (Graham 2004; Gregory 2004).

## *42.2.1 The Expressions of Destruction: Debris and (Re)construction*

The production of space from violent destruction de-produces space. It destroys and disperses the parts of the city. It generates a destructive production (Gordillo 2014), making the parts of the city indifferent to each other; without a sense of relationship, where what was once a city, in its destruction, evidences the intention to take away the sense of place.

The destroyed spaces often leave behind ruins or rubble. Both residues generate a new urban landscape that acquires a forensic connotation due to the violent destruction and the victim dimension that the destroyed spaces achieve (Aguirre and Perea 2020).

Faced with an increasingly systematic debris generation process, it is necessary to understand what is done with the debris and what processes could be related to it. Making a comparison with David Harvey's (2014) proposal on the fundamental contradictions of capitalism, crises—which are often expressed in acts of destruction—can be something essential in the reproduction of capital since its imbalances are confronted, remodeled, and rearranged to create a new version of its dynamic core.

Capitalism takes advantage of the violence in the face of a continuous process of capital accumulation and dispossession, transfers a use-value (conceived under the logic of city, urbs, cité, ville) to one of exchange value, speculative and isolated, where the land and housing become their main assets.

De Mattos (2016) argues that as the availability of building land was reduced in the most consolidated (and profitable) parts of each urban space, real estate investors intensified their "creative destruction" operations, by which they replaced parts of the low-density built environment for new buildings capable of housing a more significant number of families. With this, when they managed to generate new real estate businesses, they also influenced a persistent re-densification and verticalization of important parts of the respective urban spaces. In this way, through the verticalization and re-densification of these areas, it was possible to promote a new type of real estate business, which has been having a significant impact on the transformation and commodification of urban development.

On the other hand, also due to the scarcity of urban land, reconversion and revaluation operations have been promoted in certain central parts of the consolidated urban areas, which has been associated with the expulsion of lower-income sectors to be replaced by new sectors of higher socioeconomic status, reinforcing socio-spatial segregation and urban fragmentation (Sabatini 2006). However, and as it develops from now on, the verticalization process today exposes new edges and expressions of an urban production of non-city, which accounts for how today the destruction of pre-existing buildings is not precisely carried out to generate a replacement with better conditions both in the physical characteristics of the building and in the socio-economic conditions of those who reside in them.

## 42.3 Verticalization and Transit Towards Its "Precarious" Expression

Verticalization is a process of production and transformation of the urban space in which the building is the concrete object of its representation (Brito and Clair 2016). It causes changes in a city's landscape and economic and social relations (Hipólito de Oliveira et al. 2015). The authors converge that for Chilean cities, the verticalization is represented by primarily residential buildings located in central and inner-city areas of the metropolitan and intermediate cities, explosively densifying the area (Insulza and Galleguillos 2014; Vergara Vidal 2017; Rojas 2017).

Verticalization is usually associated with compact city models or renewal of central and inner-city spaces, encouraged under the argument of the benefits of having a greater density in the occupation of the land, propitiating mixed uses, being accessible to the main transport networks, many of the common principles of New Urbanism (Lin 2018).

Unfortunately, this misses some crucial points considering a planning model that, when it intensively promotes this urbanization mechanism, all the components of urban design—including circulation, land uses, public spaces, ecologies, and human activities—are articulated under a pattern of congestion, overload, which is not only expressed in relationships on an urban scale but instead, have a high impact within the same verticalized buildings.

For Rolnik (2013), the State plays a fundamental role in the financialization of housing and built space. It creates the legal instruments that allow the global real estate finance business. Also, it reduces the forms of access to housing to conduct them to a single route that favors and delegates specific financial and real estate actors to access to homeownership.

For their part, local governments strengthen the construction of this type of vertical residential production through the relaxation and corruption of urban regulations, fiscal and administrative facilities, but also through the construction of physical works (roads, parks, infrastructure) and social repression, when there is strong opposition to these real estate businesses (Delgadillo 2021).

Studies that address the impacts of verticalization in cities show how they reinforce a process of selective investment under widely studied real estate financialization mechanisms (Theurillat 2009; López et al. 2012; Rossi 2013), which impact not only the physical attributes of the residential offer but also in the way in which this type of housing is inhabited. These elements configure a new typology of verticalization, whose central attribute is the precariousness.

### 42.3.1 The Context of Chilean Cities

In the case of Chilean cities and specifically from the analysis of the central commune of Santiago, three historical moments linked to typologies of urban renewal are identified: the first focused on the need to restructure the existing city, order it and beautify it, under hygienists principles (1872–1939); the second, manifested in the construction of housing projects with an urban vision, influenced by the Modern Movement (1965–1976); and a third moment (1985 to the present) that begins with the Repopulation Plan of the Santiago commune, implemented through a housing subsidy in defined areas of urban renewal that are still materializing (Arizaga 2019).

In general, the urban renewal processes in Chile are in the last phases described, under a business approach that the high mobility of assets has driven in the previous decades. This has overshadowed the concerns and actions toward an urban redevelopment process, rather than towards a renewal, understood as the first phases described (Rojas 2019).

Under this scenario, in the 1990s, Chilean state regulation assumed a facilitating role in introducing real estate capital in the country's central areas of the most important cities (Pérez et al. 2019). Local governments, through land-use regulatory adjustments, sought to attract real estate projects, supported by a subsidiary logic proposed by the State, and applied through the Urban Renewal Subsidy, stimulating the land market and subsidizing access to a property in central areas (Pérez et al. 2019).

In this context, verticalized residential projects become products with industrial characteristics, from the design to the construction. Real estate developers have such a weight in urban development that they already have typological models evaluated countless times, which are profitable as a business. They occupy the property to the maximum to make the most of the land, which translates into a more significant number of apartments for sale or administration (Fedele and Martínez 2015; Pérez et al. 2019).

The verticalized buildings are built on plots between dividing walls, with reduced fronts, which implies a strong visual presence as large blind walls that do not have a search from the form, language, or materiality (Fedele and Martínez 2015). Inside, a homogeneous product is generated, on which units of limited dimensions are offered, even producing 90% of one- and two-bedroom apartment typologies (López et al. 2014 in Pérez 2019).

Although what has been described accounts for the general characteristics of the verticalization process in an important area of the Santiago Metropolitan Area (center and inner-city, excluding the high-income cone), there are territories in which the verticalization processes have materialized with a greater degree of violence in the physical attributes of the building. The emblematic case is the inner-city commune of Estación Central, a territory that, since 2012, begins to build apartment blocks that exceed thirty floors, a thousand apartment units, and an average of 30 m$^2$ apartment area, mega blocks that become actual walls in the receiving space and that give rise to an unprecedented building process for Chilean cities, known as the new form of housing precariousness (Rojas 2017, 2020).

The discussion around housing precariousness materializes in the construction of an indicator, which allows analyzing the physical attributes (infrastructure) of the new vertical building (floors, apartments, apartments per floor, elevators, parking lots, and apartment area), relocating the discussion from the point of view of the housing typology subject to analysis, the location of the phenomenon and the producer actor. It questions the standards of what is today considered precarious considering the newly emerging urban processes, such as verticalization (Rojas 2020).

This edge of verticalization has been scarcely studied, even though the case of Estación Central is only the beginning of a process that is spreading to other districts of the Santiago Metropolitan Area (AMS), even to peripheral areas of the city. Therefore, the city look is entirely relevant, identifying patterns, spatial forms, and chronologies of a process already part of Chilean cities' urban dynamics.

This vision or observation lens seeks to complement the gaze of critical urbanism that reports the impacts of neoliberal policies to build and inhabit the city, under the emphasis about deregulation of planning and capture of surplus value at the urban level by the private sector (Hidalgo and Janoscka 2014 in Pérez 2019). Today's high-rise building is far from being a punctual element praised as a sign of progress and an indicator of progress as it was initially (Pérez et al. 2019).

High-rise buildings are units with their own functioning and the leading actor within a phenomenon associated with the urban and massive scale of living, where it is not only necessary to consider the effect of their shape on the urban scale (Brenner 2004) but also the effect that the fabric of actors involved in its operation has on the urban scale and its surroundings (Vergara Vidal 2017).

## 42.4 Evolution of Verticalization in the City of Santiago de Chile: Transition Towards New Forms of Precariousness

The dynamics of verticalization in Chile have not been the same since urban renewal was used as a policy to promote greater housing and population density, taking advantage of the centrality.

Analyzing these dynamics allows us to understand the logic of verticalization and how it generates precariousness in high-rise housing from two integrated perspectives: temporal and territorial. This section is structured in two main parts using quantitative methods. The first focuses on the analysis of the evolution of verticalization in the AMS, explained through three hypotheses: verticalization transits from a phenomenon that is initially concentrated to one that is dispersed and, at the same time, atomized in the territory; the increase and adjustment of the building height materialize as a systemic process; and the supply of verticalization implies a critical increase in the magnitude of the number of dwellings per building. The second part develops the argument of how verticalization has transcended from precarious areas and communes to precarious silos, a new way of understanding the precariousness of affordable high-rise housing through the change in the relationship between the building and the urban context where it is located. We use the building permits for new construction and regularization of new construction of high-rise buildings of more than nine floors (National Institute of Statistics, INE) to construct the evolution of verticalization and its transition towards precariousness in the AMS from 1990 to 2019.

For the development of this exercise, the AMS is divided into four territories with a certain socioeconomic and territorial homogeneity due to their location with respect to the center: center, inner city, high-income cone, and periphery (Table 42.1).

The variables to analyze in the evolution of verticalization are the number of buildings, the height, and the number of apartments. Subsequently, a precariousness index based on Rojas (2020) is estimated, which considers the weighting of the variables number of floors in the building, number of apartments, number of apartments per floor within the building, number of elevators per apartment, number of parking lots by apartment and area of the apartment in square meters. The same variables are considered for this study, except for the number of elevators and parking spaces per department, since this information is not available in the INE databases. The step to follow is to standardize into five categories, where 1 represents the absence of precariousness and 5 represents greater precariousness (Table 42.2), using natural breaks of the values of each variable, except for the department surface variable, assuming the minimum surface plus high required by Chilean regulations for social housing (55 $m^2$) as the maximum value of high precariousness. Finally, the index's weighting is similar to that of Rojas (2020) but adjusted to the available variables.

**Table 42.1** Zones and communes of the AMS

| Zone | Commune |
|---|---|
| Center | Santiago |
| Inner city | Cerrillos |
| | Conchalí |
| | Estación Central |
| | Independencia |
| | La Cisterna |
| | La Granja |
| | Lo Espejo |
| | Lo Prado |
| | Macul |
| | Ñuñoa |
| | Pedro Aguirre Cerda |
| | Quinta Normal |
| | Recoleta |
| | Renca |
| | San Joaquín |
| | San Miguel |
| | San Ramón |
| High-income cone | La Reina |
| | Las Condes |
| | Lo Barnechea |
| | Providencia |
| | Vitacura |
| Periphery | Cerro Navia |
| | El Bosque |
| | Huechuraba |
| | La Florida |
| | La Pintana |
| | Maipú |
| | Peñalolén |
| | Pudahuel |
| | Quilicura |
| | Puente Alto |
| | San Bernardo |

*Source* Own elaboration

**Table 42.2** Levels of precariousness

| Level | Range |
|---|---|
| Not precarious | 1 |
| Low precariousness | 1.1–1.9 |
| Medium precariousness | 2.0–2.9 |
| High precariousness | 3.0–3.9 |
| Very high precariousness | 4.0–5.0 |

*Source* Rojas (2020: 27)

## 42.4.1 Verticalization as a Temporal and Territorial Process: Between Dispersion and Atomization, Systemic Height Adjustment and Critical Magnitude of the Residential Building

### 42.4.1.1 Transit of the Verticalization Process: From Concentration to Dispersion and Territorial Atomization

When dividing the analysis period into three decades, fewer buildings above nine floors are observed in the first (more than 800) and figures close to each other in the second and third (approximately 1200). Considering the four defined territories, the high-income cone has decreased the construction of new buildings over the three decades, going from 629 to 376 and 189. The center and inner-city have been increasing it, starting with 183, increasing to 744 and 882. This shows a transition from the phenomenon concentrated in certain areas to one that is dispersed towards new territories but also atomized by the property-to-property operation that this form of urban development implies. This relocation was intentionally directed towards the center and then to the inner-city, becoming a large territory with substantial real estate interest (Figs. 42.1, 42.2 and 42.3).

Las Condes, Santiago, Ñuñoa, San Miguel and Providencia concentrated 68% of buildings with more than nine floors in the thirty years analyzed, which are mostly

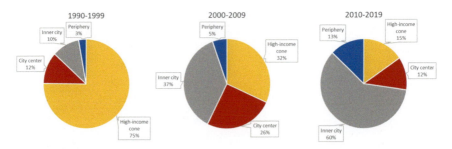

**Fig. 42.1** Percentage of buildings over nine stories by zone and decade, 1990–2019. *Source* Own elaboration based on INE

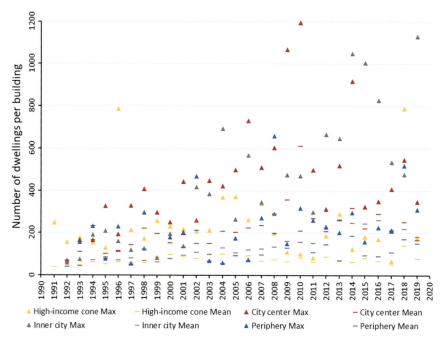

**Fig. 42.2** Number of dwellings per building, annual averages, and maximum by zone, 1990–2019. *Source* Own elaboration based on INE

high and medium–high-income communes. Urban development in these areas is linked to more exclusive residential markets associated with the city's central financial hub. However, most of the permits for Las Condes were granted in the first ten years, unlike several peripheral communes that have had more prominence in the second and third decades. In fact, in the AMS, it is observed that as of 2005, the number of buildings was higher in the inner-city than in the high-income cone. Likewise, the buildings were more numerous in the center from 2004 to 2009. In more recent years, since 2015, buildings with more than nine floors have been increasing in the peripheral zone, going from 6 average permits per year to more than 20 and even 40 and 30 in 2018 and 2019, respectively. The preceding signifies a new phase of verticalization that, in its expansive process, is reaching the periphery (particularly the commune of La Florida), an area that was not usually intricately linked to this housing typology.

What is described in this section reveals how verticalization has gone from a concentration to a territorial dispersion, evidenced initially in the cone of high income in the first decade and then in the center and inner-city in the second and third. This dispersion has even reached peripheral communes in recent years, showing a possible new phase of the redevelopment process. However, this dispersion is linked to an atomized development, looking for properties with the appropriate conditions to build a residential tower and thus take advantage of the profitability of urban land.

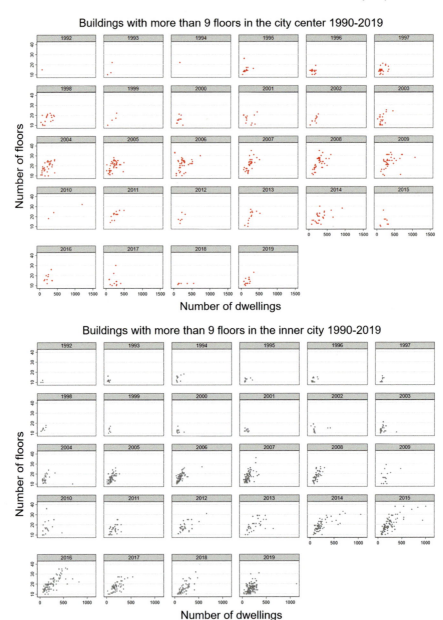

**Fig. 42.3** Evolution of verticalization by zone, 1990–2019. *Source* Own elaboration based on INE

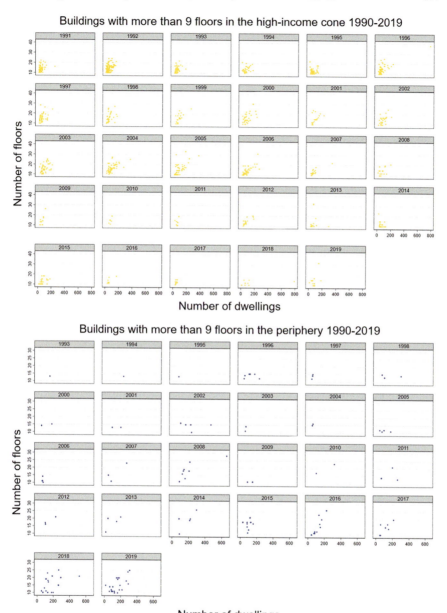

**Fig. 42.3** (continued)

As the spaces to renovate are exhausted due to the exact replication of the process, developers are looking for new business niches due to the high costs involved in demolishing a building with more than nine floors to build another higher, even if it has an attractive location. Therefore, the spaces that remain and approach optimal conditions for housing sales are those located on the outskirts with sub centralities.

### 42.4.1.2 Increase and Adjustment of Building Height: Verticalization as a Systemic Process

The temporal and territorial patterns of verticalization are verified concerning its crucial variable: the height expressed in the number of floors. Considering the average of this variable, it is evident that the height has increased since 1990 throughout the AMS, between 2004 and 2010 and between 2012 and 2016 (Fig. 42.3).

The average annual height of the high-income cone is 15 stories, with a variation from 13 to 20 throughout the period. The center anticipated higher heights since the mid-1990s, with averages exceeding 20 and 18 floors in some years, a height that intensified between 2004 and 2014, exceeding 25 and 30 stories in some cases. The highest heights began to appear in the inner-city in 2005, from 16 annual average floors and maximums that exceeded 25 floors. This average does not decrease in the following years until today. The periphery has experienced increasing heights in the last decade, with an average of 15, 18, and 20 floors in some years, evidencing a possible new phase of expansion of verticalization, which has materialized with greater force in the last two decades in several inner-city communes. In fact, in the previous ten years, the inner-city has surpassed downtown Santiago in historical maximums, mainly since 2012, reaching towers of 36 and 38 floors, compared to the maximum of the center, which was 35 floors.

Only 18 communes of the AMS have had an annual average of 16 stories or more in the analyzed period. Estación Central registered an average close to 23 floors, a more recent figure than other historical communal averages. They are followed by Santiago, Macul, San Miguel, Independencia, and Recoleta, with between 18 and 19 average floors per year.

Verticalization has had reactive processes at its height despite higher average heights, which explains the decrease in the number of floors in the last six years. However, these adjustment processes have not yet been developed in all verticalized territories with verticalization potential. These height adjustments are attributed to the impact that some projects have generated, the population's reaction, and the authorities to regulate urban norms. This situation leaves a couple of facts to discuss. The first is that once the tower is built and inhabited, it is impossible to reverse this so that the impacts may last over time despite the change in the urban norm. The second fact is that once the maximum height is adjusted and decreased, the territory ceases to be attractive and gives space to other developers with projects that have fewer floors. However, the real estate companies that maintain their project scale with larger floors look for other areas with favorable land prices, height, and residential density conditions. This is evidenced in the transit that the center-inner-city-periphery verticalization has had, a systemic transit, understanding the multiple actors that urban development has, as well as the interactions that they exert between them that change the course of the verticalization, causing the implementation of higher-rise projects in other territories due to impacts on the quality of life and subsequent regulatory adjustments in areas where they were previously allowed.

### 42.4.1.3 Supply Scale of Verticalization: Critical Increase in the Magnitude of Housing per Building

Verticalization has also implied an increase in housing supply in the last 30 years (Fig. 42.3). The inner-city is the territory in which the most apartments have been built, with more than 220,000 units, followed by the center with 123,000 and the high-income cone with 81,000. In the first decade, it was the high-income cone where this type of housing was most developed, with 33,000. In the second it was the center with more than 70,000, mainly between 2005 and 2009 (54,000). Finally, the inner-city led the housing supply in the third, reaching 161,000 units, widely surpassing other territories, particularly since 2012.

Santiago is the commune that has produced the most high-rise housing supply (123,000), followed by Ñuñoa (54,000), Las Condes (53,000), Estación Central (40,000), San Miguel (37,000), Independencia (23,000), Macul (19,000) and Florida (18,000). However, the number of departments generated by verticalization has had different temporalities. The first decade registered many departments for the Las Condes commune (24,000). Likewise, the second decade reported large numbers for Santiago (70,000). In contrast, in the third decade, large numbers were registered for this commune (38,000), Estación Central (37,000), Ñuñoa (30,000), San Miguel (26,000), and other inner-city communes.

In short, it is a verticalization that has materialized initially, on the one hand, thanks to urban renewal policies to revitalize the center and, on the other, by real estate development aimed at population-territorial sectors with higher incomes (cone of high-income). This verticalization has been expanding to other territories searching for business niches with income potential. The inner-city has been the primary objective since the 2000s and recently in more peripheral communes, but with a particular functional centrality. However, verticalization is not a complete synonym for residential precariousness. Still, it has led to this condition in some territories, urging it as a new typology to analyze.

A first look at this precariousness is to observe the magnitude of housing supply per building (Fig. 42.2). In the first decade, most of the buildings had between 39 and 234 units on average, even though maximums were recorded that reached 400 and almost 800 units, the center, and the high-income cone, respectively. In the second decade, the averages and the maximums increase, mainly in the center and inner-city. The annual averages varied between 79 and 357 units per building in these territories, while the maximums reached between 139 and 1069 units per building. In this decade, the periphery also registers a couple of extreme cases that reach 468 and 660 units per building. More cases have exceeded 800 units in the last decade, all located in the center and inner-city. The averages in these two zones now hover between 153 and 612 units per building. The periphery, meanwhile, has averages with ranges from 70 to 210 and maximums of up to 520.

The previous reflects a relevant growth in the averages and even more critical in the maximum values of units per building. This means a considerable increase in residential densification due to verticalization, which is an excessive agglomeration of people in the same building and the complications that this implies for their quality of life. In other words, the verticalization transition leads to the precariousness of affordable high-rise housing in various territories. This transition is consolidating in the center and inner-city and extending to some communes on the periphery.

The transition mentioned above can be seen in Fig. 42.4, which shows the location of building permits with more than 9 floors in the last decade of analysis (2010–2019). Currently, this is the only information available since there is no access to georeferenced information from the previous two decades.

### 42.4.2 Transcendence of Verticalization: From Zones to Silos of Precariousness

The index of the precariousness of verticalized housing comprises a smaller housing area, a greater number of apartments in the building and per floor, and a higher height expressed in floors. This index has a different manifestation depending on the territory. Considering the same division by zones of the AMS, it is possible to observe the evolution of this index by communes for each building.

In the center of the city, represented by the commune of Santiago, an average precariousness is observed that prevailed in the first decade (Fig. 42.4). However, in the second and third decades, high and very high precariousness is kept to a greater extent, mainly explained by smaller housing surfaces.

The inner-city, for its part, has a precariousness that grows over time (Fig. 42.4). In the first decade, the communes with the most permits, Ñuñoa and San Miguel, presented low and medium precariousness. In the second, the real estate market manifested itself in these and other communes, concentrating a medium precariousness followed by a high precariousness in buildings present in Recoleta, Ñuñoa, Macul, San Joaquín, Independencia, and Estación Central. In the third decade, specifically from 2012, buildings with very high precariousness began to be observed, mainly in the communes of Estación Central and Independencia, followed by similar cases in Ñuñoa, Macul, and Quinta Normal. In this decade, many buildings with medium and high precariousness have been built in these communes and others, such as San Miguel and La Cisterna. The precariousness in the second and third decades is mainly explained by the surface area of the dwelling, which also tends to be smaller.

The high-income cone presents primarily low and medium precariousness throughout the period (Fig. 42.4). These even decrease in the third decade. In the

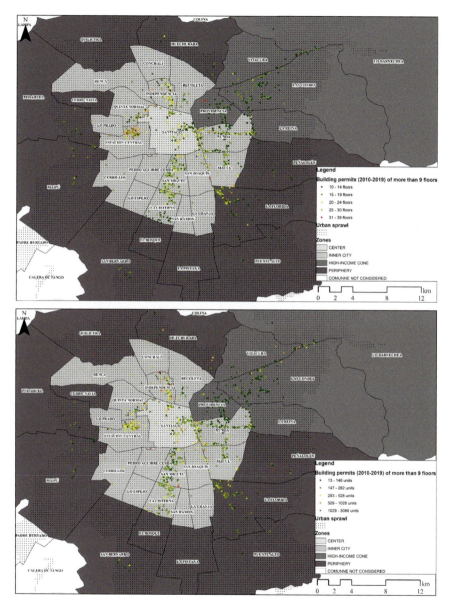

**Fig. 42.4** Location of building permits with more than 9 floors and units, 1990–2019. *Source* Own elaboration based on INE

second decade, there are few buildings with high and very high precariousness, mainly in the districts of Las Condes and Providencia. This is explained again by smaller housing sizes followed by the height of the buildings.

Finally, the periphery does not have many buildings in the first and second decade, and these are distributed in low, medium, and high precariousness. However, in the third decade, specifically from 2015, there is a more significant presence of high-rise residential buildings with medium and high precariousness. Of the communes present in this territory, La Florida stands out for registering a significantly higher number than the other communes and because most of its new buildings are highly precarious, a factor that has been maintained since the second decade. The precariousness in these cases is explained by the surface of the dwelling, followed by the number of apartments per floor.

When observing the 25 most extreme cases, those with the highest precariousness (with an index greater than 4.5), specific trends can be observed (Table 42.3). For this group of buildings, the housing surface has an average of 37.35 m$^2$, which is in a range that goes from 28.11 to 49.71 m$^2$. Likewise, these cases have an average of 778 units per building, with a minimum of 604 and a maximum of 1196. Related to this, they present from 19 to 57 units per floor, with an average of 28. Finally, they have a height of 19–38 floors, with an average close to 29. Of the 25, 15 are in the inner-city, mainly in Estación Central with 11 cases, Independencia with 2, and Macul and Quinta Normal with 1 each. Santiago, in turn, concentrates on 10 cases, while Maipú has only one, being the only one from the periphery. 7 of the 25 buildings were built between 2006 and 2009, almost all in Santiago. The remaining 18 were built between 2010 and 2019, mainly in 2014 at Estación Central and Santiago.

This analysis allows us to understand that the precariousness of high-rise housing has tended to be concentrated in certain areas, which implies that it does not have a homogeneous distribution in the territory. There are significant differences between zones and even between communes in the same zone, as is the case of the inner-city and the periphery. There is also evidence that the verticalization that has triggered precariousness has increased in the center and the inner-city for 30 years, contrary to what has happened in the high-income cone. The periphery, and specifically La Florida, for its part, tends to follow the process of the precariousness of the central areas but with a different temporality.

In short, high-rise real estate development structures areas and communes with precarious housing in high-rise buildings. However, some communes present dissimilar precariousness among the buildings presents in their territory. This occurs in districts such as Santiago, Ñuñoa, San Miguel, and La Florida, which have buildings with very high and low precariousness in short periods. The preceding implies a dual real estate market, focused on the first residence and on a second (or third, fourth, etc.) home to invest in, manifesting another scale of precariousness in spatial terms. In other words, it is no longer only possible to speak of areas and communes with vertical housing precariousness but also islands and even silos of precariousness. This is also due to the scale factor of real estate development and urban planning, which has found in the market and land-to-land operations the optimal way to multiply the profitability of land through housing.

**Table 42.3** Cases with the highest rate of precariousness

| Year | Commune | Zone | Dwelling area (m²) | Units per building | Units per floor | Floors | Precariousness index |
|---|---|---|---|---|---|---|---|
| 2006 | Santiago | City center | 41.00 | 731 | 24 | 30 | 4.76 |
| 2008 | Maipú | Periphery | 45.39 | 660 | 24 | 28 | 4.76 |
| 2008 | Santiago | City center | 43.39 | 604 | 19 | 31 | 4.76 |
| 2009 | Santiago | City center | 28.11 | 1069 | 38 | 28 | 5.00 |
| 2009 | Santiago | City center | 32.04 | 708 | 23 | 31 | 4.76 |
| 2009 | Santiago | City center | 42.27 | 689 | 22 | 31 | 4.76 |
| 2009 | Santiago | City center | 30.95 | 641 | 24 | 27 | 4.56 |
| 2010 | Santiago | City center | 35.51 | 1196 | 37 | 32 | 5.00 |
| 2012 | Estación Central | Inner-city | 28.43 | 667 | 22 | 31 | 4.76 |
| 2014 | Estación Central | Inner-city | 32.21 | 1050 | 35 | 30 | 5.00 |
| 2014 | Estación Central | Inner-city | 29.06 | 680 | 21 | 33 | 5.00 |
| 2014 | Estación Central | Inner-city | 32.83 | 616 | 21 | 30 | 4.76 |
| 2014 | Estación Central | Inner-city | 32.83 | 616 | 21 | 30 | 4.76 |
| 2014 | Estación Central | Inner-city | 32.83 | 616 | 21 | 30 | 4.76 |
| 2014 | Estación Central | Inner-city | 32.83 | 616 | 21 | 30 | 4.76 |
| 2014 | Estación Central | Inner-city | 36.07 | 725 | 23 | 31 | 4.76 |
| 2014 | Santiago | City center | 39.15 | 919 | 33 | 28 | 4.76 |
| 2014 | Santiago | City center | 49.71 | 663 | 35 | 19 | 4.59 |
| 2014 | Santiago | City center | 48.75 | 612 | 24 | 26 | 4.56 |
| 2015 | Estación Central | Inner-city | 34.35 | 714 | 20 | 35 | 4.80 |
| 2015 | Estación Central | Inner-city | 34.71 | 840 | 23 | 37 | 4.76 |
| 2015 | Estación Central | Inner-city | 35.85 | 1006 | 26 | 38 | 4.76 |
| 2015 | Quinta Normal | Inner-city | 45.97 | 940 | 39 | 24 | 4.76 |
| 2016 | Independencia | Inner-city | 41.97 | 827 | 38 | 22 | 4.80 |
| 2016 | Independencia | Inner-city | 35.45 | 695 | 28 | 25 | 4.56 |
| 2019 | Macul | Inner-city | 49.47 | 1130 | 57 | 20 | 4.59 |

*Source* Own elaboration based on INE

The foregoing opens the challenge of thinking of this type of building as a set of multiple layers and multidimensional organization. It makes up a complex system that can be analyzed as a silo, whose internal dynamics and relationship with the outside account for an urban holistic functioning, under a scale of analysis that is even micro-neighborhood, which hopes to be developed in greater depth in future research.

## 42.5 Conclusions: Precarious Verticalization as a Recursive Process and Conformation of an Urbicide

This work documents a dynamic verticalization process, which has acquired not only a new spatiality (from precarious areas to silos) but also specific construction characteristics during the three decades under analysis. This work also allows us to affirm that the urbanization of the precariousness expands to the different zones of the Metropolitan Area of Santiago under the figure of precarious towers (Fig. 42.5).

The urbicide process, therefore, begins with planning—a deliberate act—that allows the destruction of previous homes or urbanizations, where their rubble is then rebuilt under vertical residential projects whose characteristics in some of them reveal a structure of "precarious silo," which is attentive to the systemic understanding of what we have historically understood as a city.

Complementarily, urbicide exceeds impacts on the building and its site and location conditions. They affect all city dwellers as far as the absence of a vision of production in a shared space: "the city" today is more of a "non-city." This systematic process of de-production of urban space on rubble today converts these new verticalized and precarious residential projects into a consequence but also into the antecedent of a process of creating places without urbanity.

From the chronological perspective, the process of precarious vertical residential production turns out to be recursive, revealing a series of relevant edges that invite us to a systemic reading of the city.

It is considered that, although the concept of recursion comes from cybernetics and systems theory, it offers elements that allow us to discuss urban phenomena from new perspectives that are widely developed in cities, including verticalization. The concept of recursion from its generic conception, alludes to replication or reproduction. What is repeated is reintroduced in the constitution process (Innerarity 2011), recursion being only possible under an absolute unit (Hui 2019: 67).

The above definition is key to this research to the extent that in the expansion of the verticalization process of the AMS, a series of repetitions are visualized—in various territories—in the construction patterns that constitute the variables of the precariousness index, under a scenario that requires constant contingencies (Hui 2019), which in this specific case is linked to the regulatory processes of each territory. However, as Derrida (1994) points out, "every case is another" and needs an entirely different interpretation, to the extent that "repetition is never pure (…) there must be

# 42 Santiago, the Non-city? Destruction, Creation, and Precariousness ...

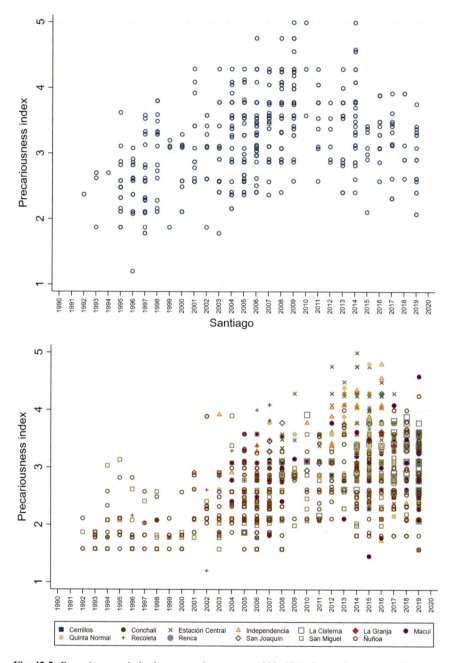

**Fig. 42.5** Precariousness index by zone and commune, 1990–2019. *Source* Own elaboration based on INE

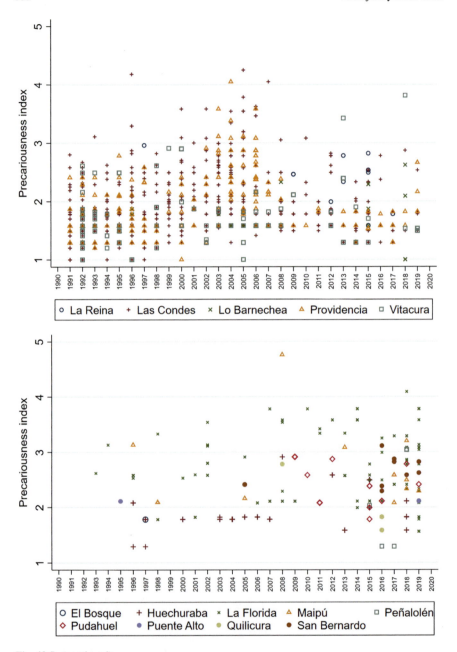

**Fig. 42.5** (continued)

a supplement, something additional, subsequent, given that the application of rules is never repetition in the sense of assured replication, mere reproduction" (Innerarity 2011: 29).

Based on what has been pointed out, under the idea of recursion lies the notion that space production does not end with the materialization of a verticalization project (construction of buildings) but rather that the same materialization reinforces new "decision-taking," this time in other territories, with new actors, different regulations, but under the same logic. Thus, a process that would have a linear character, historically linked to urban renewal policies of central and inner-city areas, today expands to new locations, for example, to peripheral communes of the AMS, even becoming -by way of hypothesis- a new mechanism of urban expansion, a kind of suburb, but vertically, questioning the classic models of residential production (Hall 1996; Janoschka 2002).

In this way, the key is the understanding of the whole and its parts, where each of these is explanatory of the process, nourishes, and provides elements for systemic learning. But, in addition, where the character of recursion in the (re)production of urban space in Chilean cities is realized, whose genesis is the process of precarious verticalization of the Estación Central commune.

Now, the precarious verticalization as a recursive process has a precise spatial and temporal expression, showing how, throughout the three decades of study, the process is not only dispersed and atomized but also increases the levels of housing precariousness.

The preceding must be understood from a systemic point of view to the extent that the regulatory adjustment measures explain the expansion in the communes that preceded the process, which leads real estate companies to seek other spaces with favorable conditions to replicate the pattern of precarious vertical production. What is interesting in this regard is how, despite this adjustment in building conditions, the new territories welcome processes with even greater degrees of precariousness, expressed in the height of the building, the number of apartment units, and, above all, in the surface of the dwelling.

Based on the above, we face a new expression of verticalization, whose central character is the precariousness of the physical dimension. A process of expansion towards non-traditional areas, with a heterogeneous distribution in spatial terms, allows us to point out that more than areas or communes with vertical precariousness, we are facing islands or silos of precariousness.

These silos of precariousness, whose spatial manifestation today is a recursive process, are also the expression of an urbicide whose conception is neat and planned but which, in its construction and subsequent use, is situated in a context of apparent crisis Harvey (2014) proposes. Not only reinforce a process of capital accumulation but, in their actions, consider the dispossession of a city vision—to a non-city vision—as well as the right to reside in a home with minimum attributes that ensure the quality of urban life, from precarious buildings. Today, the precarious verticalization is the faithful reflection of an urbicide where the building transfers a value of use (conceived under the logic of city, urbs, cité, villé) to one of exchange value, speculative, isolated, and precarious, where the land and housing become their main assets.

## References

Aguirre A, Perea G (2020) Urbicidio: violencia bélica contra las urbes. Bajo Palabra. II Época 24(1):319–336. https://doi.org/10.15366/bp.2020.24.016

Arizaga X (2019) Propuesta de caracterización de la renovación urbana en Chile. El caso de la comuna de Santiago Centro. EURE (Santiago) 45(134):169–191

Berman P (1987) Among the ruins. New Internationalist 178. https://newint.org/features/1987/12/05/among

Brenner N (2004) New states spaces. Urban Governance and the rescaling of statehood. Oxford, New York

Brito J, Clair M (2016) A verticalização urbana em cidades de porte médio: o caso da cidade de campo mourão - Paraná, Brasil. Revista de Geografia (Recife) 33(1):48–67

Carrión F (2018) Urbicidio o muerte litúrgica de las ciudades. Oculum Ensaios: Revista de arquitetura e urbanismo 15(1):5–12

Carrasco G (2007) Santiago de Chile: propuesta para la recuperación y revitalización del centro urbano. Final report, sirchal 2.2, 22/10/2007. http://repositorio.uchile.cl/bitstream/handle/2250/118203/Carrasco_santiago.pdf?sequence=1

Contreras Y (2011) La recuperación urbana y residencial del centro de Santiago: Nuevos habitantes, cambios socioespaciales significativos. EURE 112(37):89–113. http://dx.doi.org/10.4067/S0250-71612011000300005

Contrucci P (2011) Vivienda en altura en zonas de renovación urbana: Desafíos para mantener su vigencia. EURE 37(111):185–189. http://dx.doi.org/10.4067/S0250-71612011000200010

Coward M (2004) Urbicide: the politics of urban destruction. Routledge, London

Delgadillo V (2021) Financiarización de la vivienda y de la (re)producción del espacio urbano. Revista INVI 36(103):1–18. https://doi.org/10.4067/S0718-83582021000300001

Derrida J (1994) Force de loi, París, Galilée

De Mattos C (2016) Financiarizacion, valorización inmobiliaria y mercantilización de la metamorfosis. Sociologías 18(42):24–52. https://doi.org/10.1590/15174522-018004202

Fedele J, Martínez I (2015) Verticalización y desarrollo inmobiliario del núcleo urbano central de Santa Fe: Cambios morfológicos, conflictos urbanos y regulaciones edilicias en la recuperación postcrisis 2001. Cuaderno Urbano. Espacio, Cultura, Sociedad 18(18):65–88

Graham S (2004) Cities, war, and terrorism. Towards urban geopolitics. Blackwell Publishing Ltd.

Gregory D (2004) Palestine and the "War on Terror". Comp Stud South Asia Africa Middle East 24(1):183–195. https://www.muse.jhu.edu/article/181207

Gordillo G (2014) Rubble. The afterlife of destruction. Duke University Press, Durham

Hall P (1996) Ciudades del mañana: Historia del urbanismo en el siglo XX. Ediciones del Serbal, Colección La Estrella Polar, Barcelona

Herrmann G, Van Klaveren F (2013). ¿Cómo densificar? Problemas y desafíos de las tipologías de densificación en la ciudad de Santiago. Revista 180(31). http://dx.doi.org/10.32995/rev180.Num-31.(2013).art-73

Hipólito de Oliveira PV, Hipólito de Oliveira PH, de Faria Mendes W, Batista de Oliveira M, Soraggi M (2015) Verticalização consciente: edificar integrando ao meio urbano. Revista Interdisciplinar do Pensamento Científico 1(1). https://doi.org/10.20951/2446-6778/v1n1a2

Hui Y (2019) Recursivity and contingency. Rowman & Littlefield, London

Innerarity D (2011) La democracia del conocimiento. Paidós Ibérica, Madrid

Innocenti D, Mora P, Fulgueiras M (2014) ¿Densificación como vía para conciliar negocio inmobiliario e integración social? El caso de la comuna de Santiago de Chile. Final Report, Centro

UC Políticas Públicas. https://politicaspublicas.uc.cl/wp-content/uploads/2014/07/D2_Innocenti MoraFulgueiras.pdf

Inzulza J, Galleguillos X (2014) Latino gentrificación y polarización: transformaciones socioespaciales en barrios pericentrales y periféricos de Santiago, Chile. Revista Geografía Norte Grande 58:135–159. https://doi.org/10.4067/s0718-34022014000200008

Janoschka M (2002) El nuevo modelo de la ciudad latinoamericana: fragmentación y privatización. EURE (Santiago) 28(85):11–20. https://doi.org/10.4067/S0250-71612002008500002

Lin Z (2018) Vertical urbanism: re-conceptualizing the compact city. Routledge, London

Llorente M (2015) La ciudad: Huellas en el espacio habitado. Acantilado, Barcelona

López-Morales E, Gasic I, Meza D (2012) Urbanismo pro-empresarial en Chile: políticas y planificación de la producción residencial en altura en el pericentro del Gran Santiago. Revista INVI 27(76):75–114. https://doi.org/10.4067/s0718-83582012000300003

López-Morales E, Arriagada-Luco C, Gasic-Klett I, Meza-Corvalán D (2015) Efectos de la renovación urbana sobre la calidad de vida y perspectivas de relocalización residencial de habitantes centrales y pericentrales del Área Metropolitana del Gran Santiago. EURE 41(124):45–67. http://dx.doi.org/10.4067/S0250-71612015000400003

Pérez L, González G, Villouta D, Pagola L, Ávila C (2019) Procesos de reestructuración y verticalización en el centro de Concepción: Barrio Condell. Revista de Urbanismo 41:1–17. https://doi.org/10.5354/0717-5051.2019.53926

Pumarino N (2014) Edificio residencial: un Gigante Egoísta. AUS 15:46–51. https://doi.org/10.4206/aus.2014.n15-09

Rojas L (2017) Ciudad vertical: la nueva forma de la precariedad habitacional. Comuna de Estación Central, Santiago de Chile. Revista 180 39(1):1–17. https://doi.org/10.32995/rev180.Num-39.(2017).art-365

Rojas L (2019) La precariedad habitacional en el contexto del neoliberalismo urbano chileno: reflexiones en torno al proceso de verticalización de la comuna de Estación Central, Santiago de Chile. Quid 16: Revista del Área de Estudios Urbanos (12):96–113

Rojas L (2020) Ciudad vertical: la nueva forma de la precariedad habitacional, comuna de Estación Central (2008–2018). Doctoral dissertation, Pontificia Universidad Católica de Chile. https://repositorio.uc.cl/handle/11534/52759

Rolnik R (2013) Late neoliberalism: the financialization of homeownership and housing rights. Int J Urban Reg Res 37(3):1058–1066

Rossi J (2013) Hedging, selective hedging, or speculation? evidence of the use of derivatives by Brazilian firms during the financial crisis. J Multinat Financ Manage 23(5):415–433. Recuperado de: https://doi.org/10.1016/j.mulfin.2013.08.004

Sabatini F (2006) La segregación social del espacio en las ciudades de América Latina. Banco Interamericano de Desarrollo, Washington DC

Sennet R (2019) Construir y habitar. Ética para la ciudad. Anagrama, Barcelona

Theurillat T (2009) The negotiated city: between financialisation and sustainability. University of Neuchâtel, Neuchâtel

Vergara C (2013) Gentrificación y renovación urbana. Abordajes conceptuales y expresiones en América Latina. Anales de Geografía de la Universidad Complutense. 33(2):219–234. https://doi.org/10.5209/rev_AGUC.2013.v33.n2.43006

Vergara Vidal J (2017) Verticalización. La edificación en altura en la Región Metropolitana de Santiago. INVI 32(90):9–49. https://doi.org/10.4067/s0718-83582017000200009

**Loreto Rojas Symmes** Bachelor of Environmental Management, Master in Urban Development and Ph.D. in Architecture and Urban Studies from the Pontifical Catholic University of Chile. Director and professor at the Geography Department of the Alberto Hurtado University, Santiago de Chile.

**Alejandro Cortés Salinas** Geographer, Master in Urban Development, and Ph.D. Candidate in Architecture and Urban Studies from the Pontifical Catholic University of Chile. Professor at the Geography Department at Alberto Hurtado University, Santiago de Chile.

**Daniel Moreno** Economist from the Externado University of Colombia, Master in Urban Development, and Ph.D. Candidate in Architecture and Urban Studies from the Pontifical Catholic University of Chile. He is a researcher at the Research Center on Governance and Territorial Planning (NuGOT) and a Consultant of the same university.

# Chapter 43
# Neoliberal Urbicide in Barcelona. The Case of Ciutat Vella

**Pedro Jiménez-Pacheco**

**Abstract** Cities of global leisure face a historical movement of urban neoliberalization that instrumentalizes urban space, subjecting society to the domination of the market and the very institutions that regulate it. In this sense, the political economy of space and critical urban theory support the application of the Lefebvrian method, consisting of spatial analysis in the historical centrality of Barcelona, to reveal an organized system of neoliberal urbicide. This critical category allows the study of the range of urbanistic operations over time through concrete devices and protocols accredited by urbanism consensus that reproduce real estate production relations, promoting spaces of social destruction, injustice, and spatial inequality. The aim is to unveil how urbicide has been produced in Barcelona centrality by the action of capitalist urbanization and urban neoliberalism. The phase of real estate financialized destruction allows us to articulate the critical theoretical elements of urbicide with the idea of the end of the city and the theoretical projection of a scenario of hope for the city through the construction of transformative urban demands around housing and the practice of the de facto right to the city for the production of a radical social space.

**Keywords** Urbicide · Neoliberalism · Urban theory

## 43.1 Introduction

> It is here, in the debris of city and life, that here has to be poetic creation, or rather *it is not here*. It is elsewhere, in other forms, which remain to be invented by using the immense resources that are wasted today in monstrous pseudocreations: 'conurbations', clusters, 'estates' of whatever size. What is currently called 'urbanism' is nothing more an ideology (that of technocratic groups), designed, like any ideology, to mask real problems, propose false solutions, dissimulate the 'real', although not without revealing it involuntarily. (Lefebvre 2016, p. 112)

---

P. Jiménez-Pacheco (✉)
University of Cuenca, Cuenca, Ecuador
e-mail: pedro.jimenezp@ucuenca.edu.ec

In the context of understanding the causes of the organized destruction of cities, the category of urbicide is appropriate to study the process of city extermination or some of its essential systems. This radical critique of urban development models is also relevant because it seeks to give back to the city its utopian sense (Carrión Mena 2018). It could be said, however, that a city can survive only to the extent that its inhabitants resist and fight back, but do the rest of the cities die?

Henri Lefebvre, in *Métaphilosophie* (1965), first translated into English in 2016, approached the idea of the end of the city. He posits it as a much more productive and creative task than thinking about its continuity or modernization, as it does not lend to useful prospective studies. Thus, it is up to metaphilosophical thinking to imagine and propose new forms that can be built practically and perform the philosophical project, transforming people's daily lives.

Since the early 1960s, Lefebvre announced that the rural population would disappear and cities and agro-cities would replace villages reduced to an antediluvian, folkloric, or touristic existence (Lefebvre 2016). He knew that the phenomenon of urban growth was tearing the city apart in a disappearance process. This seductive and fruitful hypothesis would ultimately be strategic. For the author, demographic pressure, industrialization, and the influx of people dedicated to services (tertiary sector, professionals, bureaucracy, and commerce) transformed cities into enormous human conglomerates that no longer had any form. Territories surrounded by suburbs, and beyond these suburbs, by bungalows and state residential complexes that no longer had anything in common with the city, while the heart of the city decayed.

A neoliberal urbicide has annihilated the social space of the old city of Barcelona and has instrumentalized it for the devastation of shared resources, the dissolution of social cohesion, the detachment of institutional covenants of public service, and the privilege of the consumer over the daily life of the population.

> It is in this practical context that the everyday establishes itself, and it is here that everydayness can and must change. The end of the city, the dislocation of what was the finest work and dwelling place of man, gives us notice to create new works [*oeuvres*]. (Lefebvre 2016, p. 112)

This article seeks to understand how the phenomenon of urbicide has appeared in Ciutat Vella (district of Barcelona). A global leisure center due to the action of capitalist urbanization and urban neoliberalism. To this end, reflections coming from the political economy of the city (Morton 2013) are linked with those that move toward a critical urban theory (Brenner 2009), articulating the theoretical elements of urbicide and the end of the city with the notion of radical social space, a possible counterproject at the service of the transformation of urban life and resistance. The research hypothesis proposes that it is possible to apply the Lefebvrian method of spatial analysis in the centrality of Barcelona to define the category of a neoliberal urbicide in global leisure cities and, in turn, to postulate certain antagonistic practices to the phenomenon of urbicide that weighs on society as a whole.

The understanding of the *transduction method* adopted by Lefebvre in the 1960s[1] is useful for the theoretical and critical analysis of the current urban space with

---

[1] This mental operation considers the projection of virtual objects from the exploration of the possible, in the process of taking hypotheses to the limit, leaving behind the classical operations of

its formal and material reality, as well as its attributes of social reality. A tripartite analysis technique (*formal, functional, and structural*), supported by other Lefebvrian classifications, is required for the historical and current understanding of a complex urban reality (Jiménez-Pacheco 2018). This critical theoretical exercise allows postulating the concept of *radical social space* as a strategic theoretical object.

The levels of social space announced by Lefebvre and considered in the study of Ciutat Vella are *inhabiting* (habitual residence and tourist accommodation), *urban* (street, boulevard, neighborhood, district, city), and *global* (Spain, Europe, world). The hierarchy of these levels consists of Lefebvre's radical consideration that gives priority to inhabiting and primacy to the urban (Jiménez-Pacheco 2017). Using this method allows us to cross Lefebvrian concepts with the existing factors involved in the mode of spatial production in global Barcelona (dispossession of housing, mass tourism, migration, corruption, housing mafias linked to drug trafficking, terrorism, etc.). In this way, the analysis grid guides the process of the concatenation of the concepts applied in Barcelona centrality, allowing us to define a general characterization of the neoliberal urbicide in the case of Ciutat Vella and to outline a counterproject to *the end of the city*.

### *43.1.1 The Instrumental Space of Urban Neoliberalism in Barcelona*

Who Killed Los Angeles? A political autopsy. (Mike Davis 2007)

At the beginning of the 1970s, from a planetary perspective, Lefebvre foresaw the advance of the process of production of specific space, which is formed based on differentiation in the modes of production. Thus, the edges of the Mediterranean were becoming the leisure space of industrial Europe. Neocolonialism settling on these new spaces of pleasure and non-work manifests itself in the social and economic fields but also in architectural and urban planning (Lefebvre 2013).

During the 1970s, in countries of advanced capitalism, neoliberalism was launched on an instrumental space (Mediterranean coast) produced in the previous decade as a force of reproduction of the relations of production in Europe. Similarly, in 2008, it initiated the production of an organized space to allow capitalism to operate in the postneoliberal era. On the one hand, the region-state space and the fixed income of debt lost their attractiveness for financial capital; on the other hand, cities gained prominence in the recovery of the economy. In this scenario, a competitive and global city will not be enough; cities will be organized in the name of innovation sciences, tourism, collaboration, citizenship, etc.

---

induction and deduction, as well as the development of models or the simulation of statements. It consists of elaborating and constructing a virtual theoretical object from information on reality, as well as an issue posed by this reality, assuming a dialectical relationship between the conceptual framework used and empirical observations.

**Fig. 43.1** Battle for housing and the right to the city in Barcelona. *Source* Own elaboration

> We have seen this space take shape and perform before our eyes. It enables current neoliberalism, which allows the device set in place in the previous period to function. Capitalism allows the luxury of neo-capitalism, which survives thanks to what it has now, to the impulse and inertia of the previous period. (Lefebvre 1976a, p. 233)

In cities such as Barcelona, after three decades of urban conciliation between regulatory and predatory agents, until 2008, the former space of control was pacified in the service of directed technology and consumption (Jiménez-Pacheco 2018). Humans filled with insecurities, infallible cybernantropos (Lefebvre 1972), *citizenized* by local authorities (Alain 2001; Delgado 2017), and declared resilient by supranational frameworks and arrangements (Chandler and Reid 2016) consume this space. We affirm that over almost half a century, social relations of capitalist production, fueled by economic profit, have not only been reproduced but intensified in leisure centralities such as Ciutat Vella due to the revolution of the financial agents that control them (Fig. 43.1).

The production of an instrumental space, following Lefebvre (2013), is not only a space of police control but is also an economically organized space in which all kinds of flows are regulated (energy, raw materials, labor, products, people, cars, etc.). Technocrats succeed in coordinating these heterogeneous flows, both in terms of their points of origin and their terminals in space. This instrumental space aims to regularize flows and control the population, as opposed to the chaotic space of capitalist interests, which seeks to establish where surplus value is easy to perform. While technocrats conceive of a homogeneous regulated space, the predatory space of capitalism is fragmented. The result is a contradictory space trapped in the neoliberal solution.

Over the decades, the Barcelona Council has configured an official narrative to explain its recent transformation from an urbanistic vision. To explain this intention, three documents were revised. The catalog by architect Bohigas (1983) contains the presentation of Mayor Pasqual Maragall and focuses on the period 1981–1982, detailing 50 projects of various kinds. However, a second edition (1984) describes in annexes the situation of 19 projects in progress, 27 works in progress, 72 works executed by the *Serveis de Projectes Urbans* and by the *Servei de Parcs i Jardins*, and 26 plans in progress (including 8 Special Interior Reform Plans) in different phases of approval.

After the Olympics, Roig's catalog (1996) is published under Maragall's administration. This report frames the actions in the period 1991–1995 within the so-called *Barcelona Model* as a general definition of a *renewed and ordered* city based on spatial integration and *consensus* among various agents to improve the quality of life and citizen welfare. According to architect Ricard Fayos i Molet, who was director of the *Serveis de Planejament Urbanistic*, urban planning is configured in seven lines of action: continuity of previous projects, the complementarity of the operations of 1992, new proposals on roads, major projects of the second renovation, actions on neighborhoods, strategic actions, and the orientation from planning to housing.

Montaner's work (1999) is headed by the figure of Mayor Joan Clos and represents two main events: the commemoration of the renewed city and the projection of the Forum 2004. On the theoretical level, Montaner tries to show that within the general model of Barcelona, different models coexist, always directed toward *sustainability*. Thus, the model is disaggregated into three stages:

- 1st Stage (1976–1986): Democratic municipal management and the reconstruction of Barcelona (Bohigas 1985).
- 2nd Stage (1986–1992): The Olympic race was characterized by infrastructure operations, interventions in four Olympic areas, and areas of *new centrality* (Busquets 1999).
- 3rd Stage (1992–1997): Maragall's replacement by Clos as mayor, the postOlympic *logical recession*, the hypothesis of a *second renovation*, and the importance of the *metropolitan scale* in the ideology of urban development (Montaner 1999).

This last stage, which comes almost at the end of the century, clearly shows the private immersion in the logic of urban management, which far from causing any alarm in Professor Montaner's interpretation, is a matter of ostentation. What might be seen as neoliberal fermentation in this study:

> …in these two years, one feature has been characteristic of city management and logistics. Within its complexity, the management mechanism of the city has been based on a conciliation model between public management and private initiative, peculiar to the Barcelona model. This alliance consisted of reformulating and updating the typical alliance that occurred in the liberal city of the second half of the 19th century, when the emblematic *Pla Cerdà* was applied in Barcelona. (Montaner 1999, pp. 25–26)

Some of the clearest expressions of advanced urban neoliberalism are evident in Barcelona at the beginning of the twenty-first century. This can be explained by a planned move from the urbanistic model to the trademark (Charnock et al. 2014; Aricó et al. 2016). However, such a movement does not mean an impersonation of the *citizen urbanism* model by an *urbanism of developers and business*, as suggested by Borja (2009); rather, the opening to free real estate exploitation of land previously prepared concerted institutional urbanism (Federació d'Associacions Veïnals de Barcelona 1993). All this, with the help of privatizing and controlling management devices of an instrumental space for the reproduction of capitalist production relations.

Toward the first decade of this century, technocrats and intellectuals such as Borja minimized the consequences of gentrification and segregation, warned by critics and neighborhood movements for not having enough interest from the masses to become *more real* problems.

> In Ciutat Vella, a very ideological and minority critique has prevailed, denouncing a relatively modest *gentrification* and some projects considered *speculative*, while the population is concerned with more immediate problems (if I may say so, much more real issues) related to housing, poverty, cleanliness, and safety in public space and coexistence between different populations. (Borja 2009, p. 192)

From these considerations, the fundamental hypothesis emerges, under which we believe that the neoliberal urbanism of the late twentieth century and the beginning of the new millennium has played a decisive conciliatory role between regulators and predators. Therefore, we propose that the neoliberal urbicide in Barcelona constitutes conciliatory or concerted urbanism, of particular interest for the process that the Ciutat Vella district has undergone.

## 43.2 Neoliberal Urbicide in Ciutat Vella (1970–2018)

For the last forty years, we have witnessed a revolutionary process of capitalist urbanization in Barcelona, organized in two stages in its historical centrality. The first stage covers the period from the definitive renovation of Ciutat Vella as a result of a phase of spatial neoliberalization deployed from the second half of the 1980s and extended until the global city of the real estate and financial crisis of 2008. Here begins a period of preparation of an urban and legal field for the recomposition of real estate capital, confining in the current space the bases of an urban postneoliberalism tied to the emergence of housing in the city, not only of the popular classes but also of the middle classes. The second stage was analyzed with data up to 2016, in which territorial policies started to reverse, and a significant shift in planning occurred due to the first government of Ada Colau (June 2015), without prejudice to the fact that we have early warning of a potential postneoliberal urbicide.

This section presents the analysis of the social space in Ciutat Vella, always trying to connect the previous ideas with the possibilities of interpretation of a counterproject to the planned neoliberal urbicide. The analysis starts from the previous understanding of the most relevant characteristics of the urban phenomenon in Barcelona. The projection of social relations in a capitalist city, namely property relations, suggests the analysis of a multiplicity of agents and markets operating. However, this research is limited to the financialized real estate market and its specialization in the study area, which systematically transformed into a space of leisure and tourist consumption (Cócola 2016).

After a century of restoration operations in the old town of Barcelona, accompanied by a budget and specific policies for the Ciutat Vella district (Jiménez-Pacheco 2012), we are witnessing the progressive dismantling of social relations, carefully filtered by the depredation of capitalism over the years. This social continent of public and private life translated into the daily life of the *center* has exploded, allowing the ultimate domination of the world of commodities by imposition in the space of exchange values over use values (Jiménez-Pacheco 2017).

Ciutat Vella is formed by four neighborhoods (Raval, Gótico, Barceloneta, and Sant Pere-Santa Caterina-La Ribera), with an approximate population of 100,000 inhabitants, representing 6.2% of the total number of registered inhabitants in the city. This population in 1991 was 90,612 inhabitants, representing 5.5% of the city's population, divided into 96.2% of inhabitants born in Catalonia, including the rest of Spain, and 3.8% of inhabitants born abroad. This ratio changed in 2016, so the local (Spanish) population dropped to 56.8%, and foreign-born inhabitants reached 43.2% (Ajuntament de Barcelona 1900–2016). Although the income index of the district's population has been increasing considerably since 2000—reaching the total figure of 85.5 in 2015—when observing the index by neighborhoods, a sharp and sustained inequality is observed between the Raval (75.8) and Gothic (108.5) neighborhoods until 2016. Knowingly, Raval remains on the threshold for the poorest neighborhoods, and Gótico remains among the ones with the highest incomes in the city (Ajuntament de Barcelona 2000–2016).

The predominance of activities in the tertiary sector of the economy, mainly hotel and restaurant services in the Ciutat Vella district, is also observed, reflecting that tourism numbers have an exceptional increase. During the second quarter of 2016, for example, hotels located in this district welcomed 3.35 million people, generating 7.2 million overnight stays, 6.1% and 4.2% more, respectively, over the same period of 2015. Such a growth rate (which maintains the upward trajectory of recent years) highlights the exceptionality of the results of the first four months of 2016. In this direction, during the period June 2015–June 2016, overnight stays increased by 6.7%, as did the number of tourists by 8.2% (LABturisme 2016), significant growth rates in their projection for 2016 and the following years until before the pandemic.

### 43.2.1 The Capitalist Production of Space in the Old Quarter of Barcelona During the Late Francoism Period and the Spanish Transition

The work of Pere López entitled *El centro histórico: un lugar para el conflicto*, carried out in the first half of the 1980s, is part of the study of the transformations of the old quarter of Barcelona under capitalist urbanization in the period between the 1950s and 1970s. From the beginning of the twentieth century until 1930, López Sánchez (1986) shows with empirical data that capital strategies to recover a revaluable space in Barcelona's historic center were developed.

> The hygienist ideology disguises these hidden interests by promoting renovation and the destruction of the space infected by workers, who are left on the way, in capital's struggles between land owners and the other more dynamic fractions. Meanwhile, an attempt to make the center profitable, that is, to use it temporarily as a place destined for the reproduction of the labor force at the minimum cost and the noninvestment of capital causes its degradation. (López Sánchez 1986, p. 21)

According to the geographer, around the mid-1950s, the construction of the periphery began in Spain (a moment of implosion-explosion of the center). Thus, the new metropolitan city project required a central space, which rekindled the renovating and demolishing impulses. However, faced with the dilemma of easy speculation and evidence of resistance to expulsion, they opted exclusively for the occupation of the periphery. However, under planning protection, local administration carried out expropriations with a double effect: expelling uses and residents and concentrating the property as preparation for reuse. During this period, a residual space was consolidated as a ghetto of marginalization. Its physical deterioration accelerated, and its social disintegration was encouraged.

With the crisis of the growth model observed in the second half of the 1970s, urban space production techniques were reconsidered. There was a shift from the construction of new housing without rehabilitation demand and from the expansive city to the existing city. At this juncture, the historic center appeared, once again, as a territory for speculation. However, destruction would be displaced by new forms of *culturalist revitalization*, laying the foundations for the *spectacle city* (López Sánchez 1986). To absorb processes of resistance, the expulsion of residents combined rehabilitation with tactics of prior disintegration of the community. It was compounded by economic restrictions, as there was only a choice between high rents or ownership of apartments. There was also an increase in the price of basic needs due to the reactivation or replacement of small business networks, which were unable to support the social component due to their low income. Against this, some residents' voices demanded the right to stay, to enjoy centrality, understood in its complexity of uses and as a place for unscheduled meetings, budgets far removed from the *spectacle city*.

According to López Sánchez (1986), the housing situation responded to specific policies and budgets that unleashed the first wave of social expulsion in the historic center of Barcelona. Two different stages can be distinguished in the large housing complex construction up to 1970. A first stage (1950–1962) of protagonism of the state-promoting bodies. At the end of this period, an intense private intervention of *devalued capital* began, thanks to the promotion received from the municipal administration. The second period (1960–1970) consisted of a progressive delegation of public powers to benefit private investment promotion. State intervention was exclusively in charge of building large industrial estates located on the periphery of the Barcelona area. The decongestion of the old city facilitated its subsequent renovation intending to recover the land of a revalued centrality, which at the social level meant replacing the former proletarian population with social strata of higher status.

In line with the urban policy developed by the Spanish state applied in Barcelona city, López demonstrates the main factors affecting the enormous population decline in neighborhoods, such as Santa Caterina-Portal Nou, where the construction of new social housing is accessible to low-income economies through the intervention of state agencies. On the other hand, the awarding of contracts enables deferred access to the property, an offer combined with the systematic degradation of the housing stock in these neighborhoods. On another level, the policy of not providing public facilities was also influenced. Last, Lopez decodes the threats of social expulsion arising from local urban planning.

In the mid-1980s, López concludes that the administration contributed directly to the expulsion of the social component in conditions that were not at all helpful to the tenants. He also accuses the City Council of using rehabilitation as an expulsion tactic while prioritizing compensation in contrast to the request for housing by those affected. In the same sense, he verifies that the installation of some residents in rehabilitated apartments was temporary or precarious, without guaranteeing their permanence and foreseeing the possibility of a new eviction. Finally, he anticipates the possible extension of the property regime in rehabilitated housing as a new displacement device. Working-class neighborhoods were especially characterized by a cheap rental system. In short, Lopez argues that there was a shift from the aggressive expulsion tactic of urban renewal to underhanded expulsion of an economic nature. His detailed study allowed him to show that this was not a process of natural depopulation due to voluntary mobility but of provoked processes of the expulsion of social groups from the old city, with the premise of reproducing this central space for other uses and other social strata (López Sánchez 1986).

## 43.2.2 Ciutat Vella SA: Urban Entrepreneurialism for a Middle-Class Center

A note published in La Vanguardia (1987) about a meeting held on March 6 between Mayor Pasqual Maragall and representatives of the residents of the Ciutat Vella district, attended by technicians from the City Council, gives insight into the political atmosphere of the time. This note gathers the mayor's words on the eve of his re-election and summarizes the philosophy of municipal urban management after five years of Maragall's consensus. It explains how Maragall plays a definitive political role in the neoliberal project implemented in Barcelona and the old city in particular. Such a project requires new planning devices and a new management model, accompanied by some *values* to govern such changes.

Maragall asked the neighbors of Ciutat Vella to *break* with certain *ideological dogmas* (Fig. 43.2), to immediately start up the mixed company *Promoció Ciutat Vella SA* (PROCIVESA), and to save the degraded center with private favor, forging in the popular imagination, in short, the idea that the Casc Antic would have no future without wealthy classes. During those years, Bohigas promptly came to a well-known rationale for the monumentalization of the periphery and neo-hygienism through operations such as *urban sponging* (Delgado 2008; Aricó et al. 2016).

The mayor's speech at that meeting was perfectly structured. First, he accepted that the district degradation had outpaced the improvements that had been made since the transition. Therefore, while recognizing the limitations of public action, he was contributing to the stigmatization of the *degraded district*, two words that allowed justifyment of any intervention. Second, he asked the neighbors to reflect on overcoming three *ideological dogmas*: The interests of the business sector are contrary to those of the neighbors, the rehabilitation of an area displaces poor people, and the action of the private sector does not serve to improve a neighborhood.

Getting rid of these dogmas did not mean much at this moment beyond installing a common sense about the importance of being helped and its consequences, but Maragall had a concerted plan with sectors of economic influence, and he had to stir

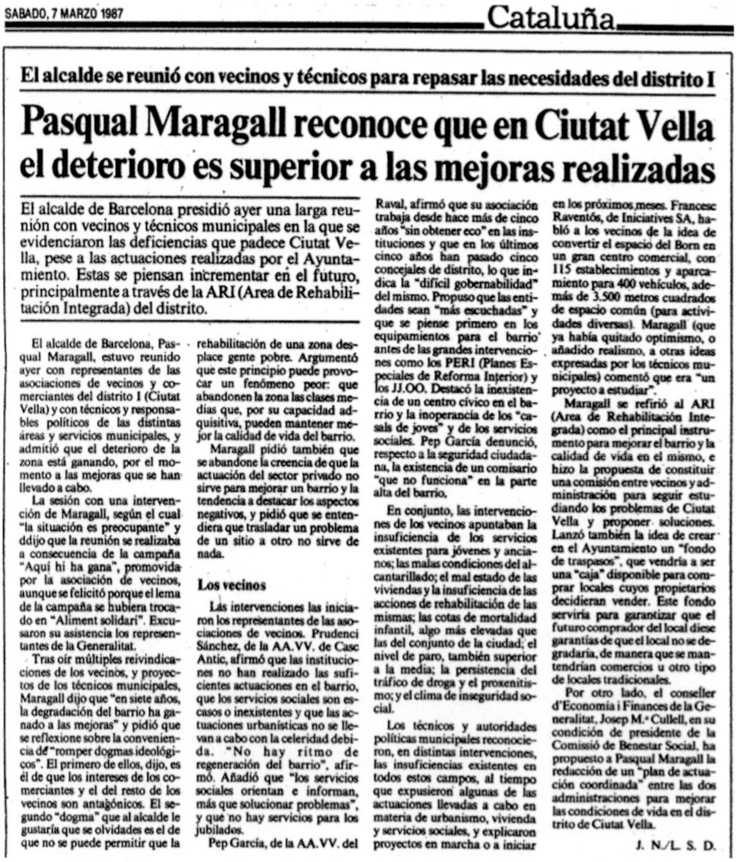
SABADO, 7 MARZO 1987 — Cataluña

**El alcalde se reunió con vecinos y técnicos para repasar las necesidades del distrito I**

**Pasqual Maragall reconoce que en Ciutat Vella el deterioro es superior a las mejoras realizadas**

El alcalde de Barcelona presidió ayer una larga reunión con vecinos y técnicos municipales en la que se evidenciaron las deficiencias que padece Ciutat Vella, pese a las actuaciones realizadas por el Ayuntamiento. Estas se piensan incrementar en el futuro, principalmente a través de la ARI (Area de Rehabilitación Integrada) del distrito.

El alcalde de Barcelona, Pasqual Maragall, estuvo reunido ayer con representantes de las asociaciones de vecinos y comerciantes del distrito I (Ciutat Vella) y con técnicos y responsables políticos de las distintas áreas y servicios municipales, y admitió que el deterioro de la zona está ganando, por el momento a las mejoras que se han llevado a cabo.

La sesión con una intervención de Maragall, según el cual "la situación es preocupante" y ddijo que la reunión se realizaba a consecuencia de la campaña "Aquí hi ha gana", promovida por la asociación de vecinos, aunque se felicitó porque el lema de la campaña se hubiera trocado en "Aliment solidari". Excusaron su asistencia los representantes de la Generalitat.

Tras oír múltiples reivindicaciones de los vecinos, y proyectos de los técnicos municipales, Maragall dijo que "en siete años, la degradación del barrio ha ganado a las mejoras" y pidió que se reflexione sobre la conveniencia de "romper dogmas ideológicos". El primero de ellos, dijo, es el de que los intereses de los comerciantes y el del resto de los vecinos son antagónicos. El segundo "dogma" que al alcalde le gustaría que se olvidades es el de que no se puede permitir que la rehabilitación de una zona desplace gente pobre. Argumentó que este principio puede provocar un fenómeno peor: que abandonen la zona las clases medias que, por su capacidad adquisitiva, pueden mantener mejor la calidad de vida del barrio.

Maragall pidió también que se abandone la creencia de que la actuación del sector privado no sirve para mejorar un barrio y la tendencia a destacar los aspectos negativos, y pidió que se entendiera que trasladar un problema de un sitio a otro no sirve de nada.

**Los vecinos**

Las intervenciones las iniciaron los representantes de las asociaciones de vecinos. Prudenci Sánchez, de la AA.VV. de Casc Antic, afirmó que las instituciones no han realizado las suficientes actuaciones en el barrio, que los servicios sociales son escasos o inexistentes y que las actuaciones urbanísticas no se llevan a cabo con la celeridad debida. "No hay ritmo de regeneración del barrio", afirmó. Añadió que "los servicios sociales orientan e informan, más que solucionar problemas", y que no hay servicios para los jubilados.

Pep García, de la AA.VV. del Raval, afirmó que su asociación trabaja desde hace más de cinco años "sin obtener eco" en las instituciones y que en los últimos cinco años han pasado cinco concejales de distrito, lo que indica la "difícil gobernabilidad" del mismo. Propuso que las entidades sean "más escuchadas" y que se piense primero en los equipamientos para el barrio antes de las grandes intervenciones como los PERI (Planes Especiales de Reforma Interior) y los JJ.OO. Destacó la inexistencia de un centro cívico en el barrio y la inoperancia de los "casals de joves" y de los servicios sociales. Pep García denunció, respecto a la seguridad ciudadana, la existencia de un comisario "que no funciona" en la parte alta del barrio.

En conjunto, las intervenciones de los vecinos apuntaban la insuficiencia de los servicios existentes para jóvenes y ancianos; las malas condiciones del alcantarillado; el mal estado de las viviendas y la insuficiencia de las acciones de rehabilitación de las mismas; las cotas de mortalidad infantil, algo más elevadas que las del conjunto de la ciudad; el nivel de paro, también superior a la media; la persistencia del tráfico de droga y el proxenitismo; y el clima de inseguridad social.

Los técnicos y autoridades políticas municipales reconocieron, en distintas intervenciones, las insuficiencias existentes en todos estos campos, al tiempo que expusieron algunas de las actuaciones llevadas a cabo en materia de urbanismo, vivienda y servicios sociales, y explicaron proyectos en marcha o a iniciar en los próximos meses. Francesc Raventós, de Iniciatives SA, habló a los vecinos de la idea de convertir el espacio del Born en un gran centro comercial, con 115 establecimientos y aparcamiento para 400 vehículos, además de 3.500 metros cuadrados de espacio común (para actividades diversas). Maragall (que ya había quitado optimismo, o añadido realismo, a otras ideas expresadas por los técnicos municipales) comentó que era "un proyecto a estudiar".

Maragall se refirió al ARI (Area de Rehabilitación Integrada) como el principal instrumento para mejorar el barrio y la calidad de vida en el mismo, e hizo la propuesta de constituir una comisión entre vecinos y administración para seguir estudiando los problemas de Ciutat Vella y proponer soluciones. Lanzó también la idea de crear en el Ayuntamiento un "fondo de traspasos", que vendría a ser una "caja" disponible para comprar locales cuyos propietarios decidieran vender. Este fondo serviría para garantizar que el futuro comprador del local diese garantías de que el local no se degradaría, de manera que se mantendrían comercios u otro tipo de locales tradicionales.

Por otro lado, el conseller d'Economia i Finances de la Generalitat, Josep M.ª Cullell, en su condición de presidente de la Comissió de Benestar Social, ha propuesto a Pasqual Maragall la redacción de un "plan de actuación coordinada" entre las dos administraciones para mejorar las condiciones de vida en el distrito de Ciutat Vella.

J. N./L. S. D.

**Fig. 43.2** Journalistic note of the meeting of Mayor Maragall with neighbors of Ciutat Vella. *Source* La Vanguardia Ediciones S.L (1987), public domain all rights reserved the consciences of the traditional inhabitants. Therefore, at the heart of his discourse, he sought to take care of the wealthy middle classes of Ciutat Vella since their income and level of consumption could better maintain the district's quality of life. We can interpret this moment before Maragall's second term as a starting point of urban entrepreneurialism in Barcelona (Harvey 2001). What it represents is a new model of planning, project execution, design, and implementation of urban policy in the city, specifically to intervene in Barcelona's centrality.

In his research, Ruiz de Somocurcio (2005) studies the management structures in the urban area of Ciutat Vella from the late 1980s to the beginning of the new century. From his work, we show what the mixed economy companies (SEM[2]) consisted of, as well as the creation and development of the public–private machinery to fulfill the objectives declared by Maragall, through PROCIVESA and its successor Foment de Ciutat Vella SA (FOCIVESA). It will shed light on the business scaffolding set up for the revaluation of the land and the emergence of a new stage in surplus value extraction from Ciutat Vella, with arrangements for the private sector.

The first step was the designation in 1986 of the Ciutat Vella District as an *Área de Rehabilitación Integrada* (ARI), which enabled the arrival of PROCIVESA as an advanced implementation device for urban management (Fig. 43.3). In July 1988, the City Council approved the creation of this mixed capital company, and two months later, the first SEM was constituted. The main objective of PROCIVESA was to concentrate investment in Ciutat Vella for a fourteen-year period. To better understand the scope of the activities foreseen in its statutes, the company's main lines of action are development of urban planning studies, preparation of reparcelling, expropriation, or compensation projects, acquisition of land and buildings, execution of infrastructure, construction, remodeling and rehabilitation works, sale of plots and buildings, demolitions, all types of real estate operations, signing of agreements with individuals and organizations, and management of aid and subsidies.

According to Santos (2013), investment throughout the entire life of PROCIVESA was concentrated mainly in public spaces, housing, and equipment.[3] The system operated when the Barcelona City Council assigned the benefit of the expropriations to this SEM, which assumed the role of land manager for public use. Simultaneously, the *Generalitat de Catalunya* built public housing through *Instituto Catalán de Suelo* (INCASOL), housing constructed on the land expropriated by PROCIVESA. In 1999, the *Sociedad Anónima Foment Ciutat Vella* was created, whose activity revolved around municipal needs and public–private integration. The public–private action had as its driving force the acquisition of land and the dynamization of its management, all promoted by private initiative.

---

[2] The SEMs were corporations with public capital, capital of the citizens represented in the municipalities, regional, autonomous, sectorial, or state governments, as well as companies with entirely public capital that may also be participating in their formation.

[3] PROCIVESA's basic actions in areas of public space and housing by 2002. *Public space*: 37 new squares and public spaces, 30 existing parks redeveloped, and 12 new roads. *Housing*: Construction of 2700 new public housing units for the rehousing of families living in buildings affected by the urban development program, release of 24,000 m$^2$ of land, 275 housing units for young people and the elderly, and 286 privately developed affordable housing units.

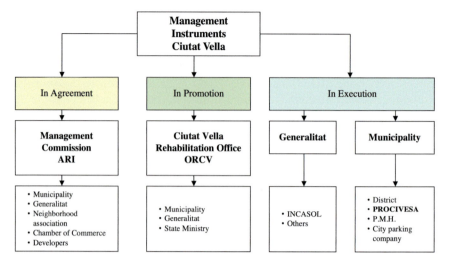

**Fig. 43.3** Organizational management structure under which PROCIVESA used to operate. *Source* Own elaboration

This societal apparatus mobilized 2.8 billion euros up to 2002, of which 60% corresponded to the private sector. The private initiative contributed approximately 300 million euros to rehabilitation and new construction, while the remaining 1.4 billion euros came from investment in commerce, professional activities, hotels and restaurants, and renovation promotions carried out by service companies. There were almost 13,000 economic activities installed in the district by 2000, of which 11,000 were business activities. Services and commerce represented 76% of the business activities in Ciutat Vella at the beginning of 2000. Finally, according to Ruiz de Somocurcio (2005), in these years, there was an increase in the number of homes renovated by owners (55,872 houses rehabilitated), fulfilling the rescue of the middle class, which had been undergoing a previous trend of mass exodus.

After the 2008 crisis, the conditions for the operation of these joint ventures changed. The City Council, presided by X. Trias, decided to reassemble a dispersed business fabric, concentrating its tasks in a single municipal firm. Thus, in 2012, *Barcelona d'Infraestructures Municipals* (BIMSA) was created. With this maneuver, the City Council considered saving two million euros, placing under the sole command of the third deputy mayor, Antoni Vives, the investment programs in the districts that until now were managed by several companies.

## 43.2.3 Ciutat Vella Premium: Violence of the Financialized Real Estate Circuit

First, it is necessary to establish an account that puts the *Barcelona Model* in perspective until the crisis hit (2008) and reflects on its effects. The work of Jordi Borja, who has an authorized and concerned voice,[4] helps explain the equidistance between the defenders and critics of the model. His ambiguity allows for a new evaluation since our analysis does not criticize the model per se or its evolutionary discourse but a unitary process of capitalist urbanization, which has ended up placing Ciutat Vella in the *premium zone*, in the jargon of a real estate agency, for its final absorption in the real estate-financialized circuit. In this way, we extract the ideas of the urban planner and manager about how the center was thought and intervened before the crisis. In addition, what was made to face it?

> ... the old historic center transformed its urban functions and social uses without changing its structural morphology. The operation initiated at the end of the 1980s is an intervention that seeks to make city over city, to make center in the center and centrality over the marginality of a degraded area, while maintaining the popular and patrimonial content of a historical center. A middle way was sought between the Lecorbuserian radicalism of the project of the thirties (almost total demolition, only a few isolated monumental elements were allowed to survive) and the absolute conservatism that could avoid decadence, degradation, and a kitsch tourist specialization. (Borja 2009, p. 108)

At the time of the crisis, Borja thought that public action in the old town had to take place in three dimensions:

- For the 100,000 people living in the neighborhood, it would be small-scale intervention, housing rehabilitation, and the recovery of public space in the form of small squares and pedestrian streets.
- For the city at large, or a good part of its inhabitants, who work, consume, or use the old centrality daily.
- The scale of metropolitan populations and everywhere, tourists or visitors from Europe and the world, who were increasingly present every day.

> Tourists are very present (too much?) in the public space during an increasing period of the year almost always. The immigrant population is part of the landscape and has colonized specific neighborhoods (Pakistanis, North Africans, and South Americans in the Raval and the area between Santa Caterina and the Born). All of this may have provoked a certain sense of *dispossession* on the part of traditional citizens, but the previous negative dynamics were much more dispossessing. Tourism has a significant impact on the physical image of the city. (Borja 2009, p. 110)

In his book, Borja explains some of the urban planning operations carried out to justify demolition actions, rehousing, and other problems that were beginning to emerge concerning the role of tourism, one of the main industries of the city and which

---

[4] Borja held the position of Deputy Mayor (second on board) in the Barcelona City Council for the PSUC from 1983 to 1995 and Executive Vice President of the Metropolitan Area of Barcelona between 1987 and 1991.

is mainly concentrated in Ciutat Vella (Borja 2009). Years later, the Masala journal, a serial publication for information, denunciation, and localized social criticism of the Ciutat Vella district, in its dossier entitled *Especulación, turismo y naufragio habitacional* (December 2016), subscribes to another tone that the officialism explained: both institutionally and in the media, tourism figures as the salvation board of the Barcelona model and, in the case of Ciutat Vella, it is the unquestionable dogma (Masala 2016).

Researchers consider that after the effects of the crisis, *touristization* has substituted gentrification as the: "hitching flag in the heads... of developmental Barcelona, legitimizing the vertiginous substitution of housing for tourist accommodations" (Masala 2016, para. 14). They explain that such a level of tourist overcrowding could only be reached under coincidence between the slogans of the institutions and the tourist lobby: *construction is dead, long live tourism*. The journal issue details the strategy applied in Ciutat Vella since 2008. The deregulation of everything related to the tourist market for the attraction of visitors and the impulse to promote new public and private spaces for the deployment of the leisure industry. For example, in November 2012, in a meeting with the hotel sector, Mayor Trias stated that *Barcelona has in the tourism sector a great condenser of many of the efforts we are making to get out of the crisis* (Masala 2016). This resulted in the *Modificación del Plan especial de establecimientos de concurrencia pública, hostelería y otras actividades del Distrito de Ciutat Vella* (Ajuntament de Barcelona 2013), which in practice deregulated residential uses for the installation of larger hotel space in buildings listed as architectural heritage and the adequacy of new service spaces.

Since the opening of the national regulations for the arrival of *Sociedades Anónimas Cotizadas de Inversión Inmobiliaria*, there has been a new wave of investment, which would be aggravating the already deeply problematic conditions of life in the district (Jiménez-Pacheco 2018). According to data from the city council, between 2013 and 2015, there were 3133 sales and purchases, of which 2705 were for used housing and the rest were for new housing. This indicates an aggressive intervention of the real estate development system in the existing residential infrastructure, accompanied by the evolution of sales prices. The price per square meter built in Ciutat Vella in 2013 was 2598.9 €, and in 2015, it reached 3075.7 €. In the second quarter of 2016, it amounted to 3536.2 €. In 2013, Ciutat Vella remained below the Barcelona average (2719 €/m$^2$); currently, the district is above the city average (3208 €/m$^2$). It places Ciutat Vella among the areas of the city that have experienced one of the highest price increases, such as Sarrià-Sant Gervasi, Les Corts, or the Eixample.

The rental market follows a similar trend. Between 2008 and 2014, there was a slight downward trend, from 12.6 to 11.3 €/m$^2$, with an increase to 12.1 €/m$^2$ in 2015. In any case, in the first quarter of 2017, the rental price in the district reached its historical maximum at 20.6 €/m$^2$, the most expensive rent in Barcelona and, on average, the most expensive rent in Spain. As Masala (2016) expresses, the centrality of still working-class neighborhoods, with entrenched situations of structural poverty, opens the door to a new process of mass population expulsion. Not in vain, the Oficina d'Habitatge de Ciutat Vella intervened, only between January and September 2016, in 260 cases of eviction, in which 720 people were affected, among them, 258 minors.

Masala's research (June 2017) demonstrates how real estate pressure is organized through companies that execute different layers of structural harassment and have made the expulsion of neighbors with alternative methods of real estate violence in Ciutat Vella a daily occurrence. Within the financialized real estate circuit, a *mafia-like* structure has grown to launch companies specializing in *intermediation* in illegally occupied real estate.[5] The existence of an *opaque network* is documented in which relationships between insurance corporations, private security, and *anti-squatting companies* are identified. It denounces *police and judicial passivity* in the face of the cases of real estate violence, as well as *political insufficiency* to stop or halt harassment.

Jiménez-Pacheco (2018) reveals that municipal policies have played a determining role as an accelerator of this process and that its impact and shockwave have not stopped with the measures of the Barcelona en Comú government in the urban planning field. Having put on the free offer a decade ago, an entire district for the tourist business facilitated the sprouting of more than "the old oil stain of the architects of urbanicide… the blood smelled for the umpteenth time by the brick sharks in the historical center" (Masala 2017, para. 9). As of 2018, the Household Disposable Income per capita (HDIpc) of Barcelona was 21,484 €. Between 2010 and 2018, Barcelona's HDIpc increased by 4.9%. However, with the inflation variable in the same period, the city's HDIpc decreased by − 7.2%. The district of Barcelona with the highest HDIpc level is Sarrià-Sant Gervasi (33,113 €), while Ciutat Vella presents the lowest income per capita with 14,505 € (Ajuntament de Barcelona 2021).

In short, the urbicide guided by the capitalist production of space has perpetrated a long moment of implosion-explosion in Barcelona's centrality, now introduced in the financialized real estate circuit, which is carried out within the process of urban neoliberalization under the mantle of concerted urbanism. Since its inception, regulatory and predatory agents have set the machinery of urban entrepreneurialism for the salvation of a *degraded center* with the favor of private investment and the fetish of the middle classes. It is no coincidence how articulation with the temporary neo-hygienist urbanistic vision translated into the *progressive operations* of urban sponging since it was necessary to forge in the social feeling the idea that Ciutat Vella could be transformed but that it would have no future without the wealthy classes. After 2008, it was made clear that there would be no future without tourism, which implied the deregulation of land use and occupation in Ciutat Vella. This organized urbicide (with the support of national legislation) has unleashed a wave of real estate violence that drags down both the lower and middle strata, presaging that the city death could not be avoided (El Periódico de Catalunya 2018).

---

[5] See: https://www.desokupa.com.

## 43.3 The Production of Radical Social Space Against Neoliberal Urbicide

From this point, we ask how to oppose the urbicide organized by real estate-financialized capital. How to reconnect with the social pulses, since the battle against capitalist urbanization, is not only theoretical. Here, it is necessary to put to work one of the conclusions of Stanek's study (2011), where he argues that Lefebvre's project is not limited to analyzing the interstices between the dominant social practices in charge of producing space but manifests itself in its universal dimension, allowing us to recognize what particular struggles and singular events that give meaning to universal claims share. Thus, it is worth asking ourselves, in the face of the crushing global urban reality that weighs on Barcelona, what are the possibilities of building an effective popular resistance against real estate violence? What can we do from the sociospatial demands to actualize the right to the city in real battles such as the one in Barcelona? How can the real estate-financial circuit be sabotaged to try to go beyond the limits of the reproduction of the social relations of neoliberal production of the city?

The Barcelona Neighborhood Plan (2021–2024), launched in the second term of Mayor Ada Colau, proposes a framework of multiple objectives that go beyond the urban planning issue and focus on the populations most affected by the health, economic, social, and housing crises. The instrument under implementation aims to empower neighbors through their self-organization and a shared proposal of challenges for improving collective life in their neighborhoods, promoting social innovation practices and citizen action. This neighborhood capacity thought from below should propose the key in areas such as education and public health, social rights, gender equity, community action, housing, economic boost and social economy, public space, accessibility, environmental sustainability, and climate emergency.[6]

### 43.3.1 Transformative Urban Demands

Marcuse and Madden (2016) open up the discussion on housing, pointing out that residential housing is political, meaning that the shape of the housing system is always the result of struggles between different groups and classes. Thus, the ways in which social antagonisms shape housing are hidden and need to be revealed. His approaches assume that housing is caught in a series of simultaneous social conflicts. There is the conflict between housing as a lived social space and housing as an instrument for profit, namely the conflict between housing as a home and real estate. There is also the root problem stemming from social inequalities and antagonisms. In this sense, a *true right* to housing needs to take the form of an ongoing effort to democratize

---

[6] See: https://www.pladebarris.barcelona.

**Fig. 43.4** Social organization in the streets of Barcelona. *Source* Own elaboration

and decommodify housing, ending the alienation engendered by the existing housing system (Fig. 43.4).

For the authors, transformative demands seek to address the systemic causes of inequalities and injustices, comprehensively examining the sources of a particular problem and the systemic and institutional factors that feed it. At the same time, they challenge the system because the aim is not to make the current system more resilient. However, it proposes actions that improve present conditions and progressively enable the construction of a different world (Marcuse and Madden 2016). Transformative demands (at a specific level of development) would bring to maturity existing potentialities, which are often blocked by conditions of domination. Thus, its radicality consists of the search to solve problems not on the surface but at the root.

In his conference in Barcelona, *Culture and cities: The Challenge of Tourism*, David Harvey (November 2016) expressed three fundamental elements (Fig. 43.5). First, *touristification* is not the problem that we must eliminate precisely in the city, disappointing certain activists and some associations (especially merchants). The real cause to fight is capitalism, he explained. It operates in *the urban area*, mainly through investment flows within a real estate-financial circuit, entering the city mobilized by corporations and developers. Second, the question is not what city we want but what kind of society we want to be. This implies considering that the reproduction of social relations takes place in everyday life in the city, which is planned for those who seek maximum profitability. It reinforces Lefebvre's proposal, under which a global anti-capitalist project is required to advocate for another society in another space. Third, there are alternatives to private property, deindustrialization, predatory economics, and urban neoliberalism. These depend strongly on the maximum differences that sprout and resprout in each place. Such differences have been eliminated under

**Fig. 43.5** David Harvey in Barcelona talking about the city and tourism. *Source* Own elaboration

a homogenizing project during the last decades. There, policies and budgets can contribute as facilitators of the means, without the institutions being the means and much less the end. In this way, Harvey argues that all the alternative paths that we can take under an anti-capitalist program will be forced by the struggle between those who want to return us to Keynesianism and those who think that ending neoliberal urbicide is the key to beginning to solve the needs and desires of society.

The *Sindicat de Llogaters*, created in May 2017, has been a beacon to confront what has been called a "rental price bubble" in Barcelona (Jiménez-Pacheco 2018). The collective emerged with the conviction that only through tenant organizing can the speculative surge be confronted. Their experience over the years has allowed them to consolidate a space for mutual aid and self-defense. During this time, it has managed to stop thousands of evictions, confront the financial and real estate power, and give a respite to thousands of tenants who lived harassed by the real estate power. The Barcelona Tenants Union has grown and extended its strategies of resistance and transformation to 13 consolidated sections in Catalonia.[7] They have also achieved legislative changes, such as the regulation of rental prices, which has shaken landlords and speculators. Currently, the organization of the *Sindicat* is a mature tool to manufacture transformative demands.

---

[7] See: https://sindicatdellogateres.org.

## 43.3.2 (De Facto) Righ to the City

> The necessary *rights*, from the habeas corpus to the right to the city, are no longer enough. It is also essential for the urban to become threatening. (Lefebvre 1976b, p. 8)

The Diputació and the Barcelona City Council organized a broad colloquium called *Del Civisme al Dret a la Ciutat* (June 2017) with the involvement of several social organizations, academics, and administrative dependencies, which proposed as a general framework not precisely the vindication of the struggles against urban violence or real estate violence but rather the institutional ineffectiveness through the codes of civility to manage conflicts in the city in terms of coexistence in urban space. Its objective would not be the search for mechanisms to empower collectives or to deepen the analysis of the understanding of conflicts related to social life in urban space. Thus, a right to the city appeared stripped of the revolutionary sense proposed by Lefebvre in 1968.

This institutional phenomenon became clearer during another international colloquium, this time in Paris, entitled *The Right to the City in the South, Urban Experiences and Rationalities of Government* (November 2017), organized by the University of Paris Diderot with the municipality and other institutions. This meeting sought to understand the relationships between the daily practices of city dwellers and governmental rationalities. The discussions recalled that inhabitants, through their spatial activities in the city, experience the process of social exclusion, relegation, and marginalization. However, political and social inclusion and the affirmation of forms of local citizenship whose contents are difficult to interpret also exist.

However, this right to the city would imply considering urban practices in their predictable dimensions to capture them at the institutional level. Thus, urban practices seem to constitute a privileged object of study to deepen reflection on this right, understood not as the result of open political conflicts (mobilizations, urban struggles) but as a process of adjustment between urban experiences and the normative production of governmental rationalities. These approaches, which seek to understand the right to the city through possible normative arrangements, are harmonious with postneoliberal and citizen theories of urban pacification and civic adaptation in Barcelona, with the epilog of greater control, privatization, hygienism, and real estate violence, to the detriment of the possibility of exploring a right to de facto action, in which institutionality would be obliged to understand and adapt to certain insurgent situations found outside legality.

It is therefore necessary to review community practices that offer glimpses of the potential for social change without the constraints imposed by the rules of the game in an alienated society. Despite their minority status, the creation of autonomous environments where gaps open up to develop activities involving cooperation, mutual support, and the stimulation of associative and neighborhood fabrics implies questioning the foundations of the capitalist production of space and the pragmatic advice of the reformists. This opening of the possible has achieved, for example, in the heart of Madrid, that the human collective La Ingobernable (Fig. 43.6), thanks to

**Fig. 43.6** First building squatted by La Ingobernable in Gobernador street, Madrid. *Source* Own elaboration

their drives and desires, rage and needs, dared to occupy an empty building.[8] Thus, the production of the body materializes, of the spaces of representation for a current reality (not of representation of space, nor commodified symbols), the building of the sensible practical, the possible-impossible, the maximum differences. Above all, the production of new social relations of appropriation, not ownership. It is the beginning of the end of urbicide, the production of radical social space through true right to the city.

> Philosophical utopia, like technocratic utopia, must be surmounted in the name of everyday life… This equally implies a different way of living, extending to the creation of a new social space, a different social time; the creation of a different mode of existence of social relations and different situations, liberated from models that reproduce the existing order. (Lefebvre 2014)

## 43.4 Conclusions

We value that the theoretical dimension of urbicide has repercussions in the study of a case with a capacity for replication in cities of global leisure. When we asked ourselves how urbicide has been produced in the centrality of Barcelona by the

---

[8] See: https://ingobernable.net.

action of urban neoliberalism, we faced the theoretical and methodological need to apprehend a new category of critical analysis in a concrete historical process. This meant modifying, not without risk, previous official and scientific accounts. The spatial analysis indicates a panorama of daily drama at the base of the society that inhabits the Ciutat Vella district, concerning the structural system of property relations and access to housing. An urbicide with its specific, based on urbanism concerted with real estate power groups and the tourist industry, becomes a producer of chronic social inequalities and urban violence.

Argumentative connections have been established between the different moments of a unitary process of capitalist urbanization, supported by localized and global literature and with previous foundations emerging from our studies on Barcelona. Thus, this chapter proposes four fundamental contributions: (a) the application of our method of spatial analysis supported by in-depth studies of the work and thought of Henri Lefebvre; (b) an original critique of the standardized periodization of stages of urban development in Barcelona and a counterproposal through a continuous process of reproduction of the relations of production in capitalist urbanization in the financial real estate system; (c) the approach of the thesis of urbanism of conciliation for a contradictory space and its integration with a theory of neoliberal urbicide; and (d) facing the distrust of the urbanistic fact, we show the design of a conceptual framework of alternatives of change to the urban reality, based on the search for transformative urban demands and de facto social practices.

The structural problem in the historic center of Barcelona, which expands to the metropolitan city and the towns of the province of Catalonia and is spreading throughout the cities of advanced capitalism, becomes a current and future problem for Latin American cities, especially those convinced of a tourist vocation. The question of access to housing in its different modes of tenure is dissolved in the interests of the real estate and financial powers that, when converging with the tourist industry, turn urban development processes into allies that move from the government to government, facilitating the programmed urbicide on cities of daily violence.

## References

Alain C (2001) L'impasse citoyenniste: contribution à une critique du citoyennisme. In: Collectif L'anti citoyennisme (coord) Critique du démocratisme radical. http://www.theyliewedie.org/ressources/biblio/fr/Collectif_-_L%27anti_citoyennisme.html

Ajuntament de Barcelona (1900–2016) Població i Demografia. In: Anuari Estadístic de la Ciutat de Barcelona. Ajuntament de Barcelona, Barcelona

Ajuntament de Barcelona (2000, 2007, 2011, 2013, 2015, 2016) Distribució territorial de la Renda Familiar per càpita a Barcelona. http://ajuntament.barcelona.cat/barcelonaeconomia/

Ajuntament de Barcelona (2013) Plan especial de establecimientos de concurrencia pública, hostelería y otras actividades del distrito de Ciutat Vella. https://ajuntament.barcelona.cat/ciutatvella/sites/default/files/informacio/09-022013021378.pdf

Ajuntament de Barcelona (2021) La renda de les llars a Barcelona. Distribució per districtes, barris i seccions censals 2018. https://ajuntament.barcelona.cat/premsa/wp-content/uploads/2021/07/LA-RENDA-DE-LES-LLARS_2018_def.pdf

Aricó G, Mansilla JA, Stanchieri ML (2016) El legado Porciolista: Extracción de rentas, dinastías de poder y desplazamiento de clases populares en las políticas urbanísticas de la Barcelona contemporánea. Working paper series contested_cities. Universidad Autónoma de Madrid, Madrid, España, pp 1–22

Bohigas O (1983) Plans i projectes per a Barcelona. Ajuntament de Barcelona

Bohigas O (1985) Reconstrucció de Barcelona. Edicions 62

Borja J (2009) Luces y sombras del urbanismo de Barcelona. UOC

Brenner N (2009) What is critical urban theory? City 13(2):198–207

Busquets J (1999) La ciudad como resultado de planes y proyectos: Desde los tejidos suburbanos a las nuevas centralidades. In: Montaner JM (coord) Barcelona 1979–2004: del desarrollo a la ciudad de calidad. Ajuntament de Barcelona, pp 157–164

Carrión Mena F (2018) Urbicidio o la muerte litúrgica de la ciudad. Oculum Ensaios 15(1):5–12. https://doi.org/10.24220/2318-0919v15n1a4103

Chandler D, Reid J (2016) The neoliberal subject. Resilience, adaptation and vulnerability. Rowman & Littlefield International, Ltd.

Charnock G, Purcell T, Ribera-Fumaz R (2014) City of rents. The limits to the Barcelona model of urban competitiveness. IJURR 38(1):198–217

Cócola A (2016) La producción de Barcelona como espacio de consumo. Gentrificación, turismo y lucha de clases. In: Grupo de Estudios Antropológicos La Corrala (coords) Cartografía de la ciudad capitalista. Transformación urbana y conflicto social en el Estado español, pp 31–56

Davis M (2007) Ciudades muertas. Ecología, catástrofe y revuelta. Traficante de Sueños

Delgado M (2008) Barcelona: urbanisme versus urbà. In: Gerència de Serveis d'Habitatge (eds) Ciutats en (re)construcció-necessitats socials, transformació i millora de barris. Diputació de Barcelona, pp 151–156

Delgado M (2017) Els moviments socials ara no hi són, tornem a la catàstrofe de la Transició (interview). Carrer 13:145–146

El Periódico de Catalunya (2018) Colau: 'Los fondos buitre acosan nuestra ciudad'. https://www.elperiodico.com/es/barcelona/20180305/colau-los-fondos-buitre-acosan-nuestra-ciudad-6668371?fbclid=IwAR3R9teLxTnX1XTDZCeZnjItyTPupGbTkbFEjmRnk8BQoE6Yjr5t67kkw6U

Federació d'Associacions Veïnals de Barcelona (1993) El futur urbanístic de Barcelona (editorial). Carrer (17):9

Harvey D (2001) Spaces of capital: towards a critical geography. Routledge.

Harvey D (2016, Noviembre 14) Culture and city: the challenge of tourism (conference). Centre de Cultura Contemporània de Barcelona

Jiménez-Pacheco P (2012) Aproximación a la labor y el pensamiento del arquitecto restaurador-funcionario Adolfo Florensa Ferrer y su presencia en la ciudad de Barcelona. Master thesis, Universidad Politécnica de Cataluña

Jiménez-Pacheco P (2017) Re-imaginar el centro para cambiar la vida. Ciutat Vella como obra del habitar. Quaderns de Recerca en Urbanisme 8(2017):80–97. http://hdl.handle.net/2117/109156

Jiménez-Pacheco P (2018) After planning, the production of radical social space in Barcelona: real-estate financial circuit and (de facto) right to the city. Urban Plann 3(3):83–104. https://doi.org/10.17645/up.v3i3.1360

LABturisme (2016) Activitat turística de la Destinació Barcelona. Informe anual de la província 2016. Diputació de Barcelona, Barcelona

Lefebvre H (1968) Humanisme et Urbanisme. Archit Forme Fonction (14):22–26 (original work published 1962)

Lefebvre H (1972) Contra los tecnócratas. Hacia el Cibernantropo. Granica (original work published 1967–1971)

Lefebvre H (1976a) Tiempos equívocos. Kairós (original work published 1975)

Lefebvre H (1976b) Espacio y política. Península (original work published 1972)

Lefebvre H (2013) La producción del espacio. Capitán Swing (original work published 1974)

Lefebvre H (2014) Critique of everyday life. Volume III: From modernity to modernism (towards a metaphilosophy of daily life). Verso e-book (original work published 1981)
Lefebvre H (2016) Metaphilosophy. Verso Books (original work published 1965)
López Sánchez P (1986) El Centro histórico: un lugar para el conflicto: estrategias del capital para la expulsión del proletario del centro de Barcelona. Edicions Universitat Barcelona
Marcuse P, Madden D (2016) In defense of housing. The politics of crisis. Verso Books
Masala (2016) Especulación, turismo y naufragio habitacional (dossier). Masala (73). Revista d'informació, denúncia i crítica social a Ciutat Vella. https://masala.cat/especulacion-turismo-y-naufragio-habitacional/
Masala (2017) Inversionisme i violència immobiliària organitzada. Masala (74). Revista d'informació, denúncia i crítica social a Ciutat Vella. https://masala.cat/inversionisme-i-violencia-immobiliaria-organitzada/
Montaner JM (ed) (1999) Barcelona 1979–2004: del desarrollo a la ciudad de calidad. Ajuntament de Barcelona
Morton AD (2013) Spatial political economy. J Aust Polit Econ 79:21–38
Redacción La Vanguardia (1987, Marzo 7) Pasqual Maragall reconoce que en Ciutat Vella el deterioro es superior a las mejoras realizadas. La Vanguardia
Roig M (ed) (1996) Barcelona: La segona renovació. Ajuntament de Barcelona
Ruiz de Somocurcio I (2005) Sociedades de economia mixta: una herramienta de gestion urbanistica. Master thesis, Universidad Politécnica de Cataluña
Santos MA (2013) La transformation de Ciutat Vella à Barcelone. Le centre historique revitalisé, un processus continu. In: Casanovas X (ed) Réhabilitation et revitalisation urbaine à Oran, pp 37–49. Rehabimed, from http://openarchive.icomos.org/id/eprint/1401/5/Aureli%20Santos.pdf
Stanek L (2011) Henri Lefebvre on space. Architecture, urban, research and the production of theory. University of Minnesota Press

**Pedro Jiménez-Pacheco** Architect and Ph.D. in Urban Theory (Barcelona School of Architecture). Professor of Urban Planning at the Faculty of Architecture and Urbanism of the University of Cuenca. Research Coordinator of this faculty and Director of the Estoa scientific journal.

# Part X
# Epilogue

## Chapter 44
# Epilogue. Remake Us from Ruins, Collective Memories and Dreams

**Víctor Delgadillo**

**Abstract** This epilogue reviews how social groups have imagined better urban futures and other cities in the face of catastrophic situations. The city is a social and historical product built by generations of people so that human beings can live better in society. This is why most of humanity lives in "cities". However, various fundamentally economic and political processes are leading to the material and symbolic death of the city. Therefore, the city no longer unites, integrates and relates people. Recognizing and understanding the causes of this destruction is essential to build and defend the city and to dispute the future of the public city against market forces. The epilogue recognizes people's capacity for resistance and recovery in adverse conditions. For this reason, utopias as well as futuristic visions of science fiction, architecture and urban planning are reviewed, and the history of urban destruction and the reconstruction of cities is learned, since this is part of the recovery of identities and collective memories, mourning and social reconciliation. If the end of the city is drawing near, as various pieces of evidence support, many social groups and scholars recognize that another end of the city is possible.

**Keywords** Reconstruction · Cities · Urban ruins · Urban development · Social history

## 44.1 The Life of People and Life of Cities

The life of human beings is short, while the life of cities is much longer, so we tend to conceive of them as everlasting. However, the ruins of ancient cities and civilizations remind us that cities and civilizations also perish. The cities are also made of reconstructions (partial or total), which show the persistence of human beings in rebuilding from ruins and the obstinacy to take root in a territory. Many cities in the world have been rebuilt after earthquakes, floods, tsunamis, fires, bombardments, etc. Here, people, together with the reconstruction of their habitable spaces, have

V. Delgadillo (✉)
National Autonomous University of Mexico (UNAM), Mexico City, Mexico
e-mail: victor.delgadillo@uacm.edu.mx

privileged the rebuilding of their symbolic spaces. Urban reconstruction is part of the reconstruction of identities and collective memories, mourning and social reconciliation. The material and symbolic reconstruction of cities is and has been a way of aspiring to better futures.

In recent decades, a new type of urban ruins and constructions with a very short useful life has emerged. It is about the production of buildings and urbanizations that are not finished (due to various crises) or are finished but not inhabited; and of urbanizations with a programmed obsolescence that accelerates the capitalist cycle of construction, deterioration and new construction. Both types of urbanizations are multiplying in several cities of the world, in distant peripheries or in select inner-city areas, and contribute to the production of an enormous wastefulness.

In the twenty-first century, the death of the city is not only material, but also and above all symbolic. Cities disappear (even if the physical structures remain or are newly built) when they do not fulfill or cease to fulfill the functions for which they have been socially constructed by generations of people. This death occurs progressively or abruptly due to direct social causes of a political and economic nature (deindustrialization, urban renewal, modernization, ideology, creative destruction, speculation) or due to indirect social causes, such as natural phenomena (earthquakes, tsunamis, floods) or viruses produced by the interaction of human beings with living organisms, which affect socially constructed conditions of vulnerability. Disasters are not natural.

The COVID-19 pandemic came to deepen the processes that in recent decades have been eroding the attributes of the city and the materiality that allows proximity relationships and multidimensional physical contacts (social, political, economic, etc.) among different people, as well as the social heterogeneity in its multiple dimensions. The pandemic did not bring new things. Long before there was already ample evidence that cities were losing sociocultural diversity, socioeconomic heterogeneity and land-use zones.

Indeed, the urbanism that has taken place in the twentieth and twenty-first centuries has gradually led to the annihilation of what we call the city. The new human settlements in ever more distant urban peripheries are single-functional and homogeneous bedroom communities. The historic centers and neighborhoods that until recently constituted vestiges of a "city" are depopulated, touristized, park-themed, gentrified and become exclusive for higher-income consumers. Walled neighborhoods and ultra-guarded intelligent buildings constitute an insular urbanism, while the financialization of real estate produces a ghost urbanism with new buildings that are not used. Thus, the city (symbol of emancipation and social integration) is confronted with an expansive dynamic, a neoliberal globalization, a financialized real estate market and the virtualization of social contacts, which divide, disperse, fragment, privatize, decentralize and separate the social tissue and urban fabrics; and create new and diverse urban and territorial barriers. The city no longer brings people together, does not integrate and does not relate. In addition, during the first weeks of the COVID-19 pandemic, the city practically emptied itself due to the confinement of its citizens to private space. Thus, the measures of the governments in the face of the pandemic have not only suppressed, but repressed social relations in the public spaces of the cities, as if it were also a question of suffocating the city and its citizens.

In this epilogue, which aims to close this book in an optimistic way, we do not intend to list what "should be done" to confront the urban crisis. That would be an unhelpful exercise. Instead, we find it productive to review how social actors and groups have imagined another society and another possible city and have set out to confront the causes that are killing our cities. In this sense, Schávelzon (1990) maintains that in order to preserve and defend urban heritage, in this case the city, it is necessary to understand the causes that lead to its destruction.

The title of this epilogue alludes to people's capacity for resistance and recovery in adverse conditions. Therefore, it reviews how past, recent and remote generations imagined the city of the future in the face of disaster: what utopias did they build, what visions did they imagine? The purpose is to recover and learn from the history of utopias, science fiction, the destruction and reconstruction of cities after disasters caused by social, natural, economic and political causes. Reviewing these experiences and imaginations from a historical and critical perspective may produce contributions to be better prepared to dispute, in the globalized neoliberal capitalist world in which we live, a future that places people at the center of the life of cities and the planet, not private profit. The chapter has the following sections: Urbicide, recent trends; learning from history, utopias, futuristic visions and the destruction and reconstruction of cities; and colophon: the dispute for the future.

## 44.2 Urbicide, Recent Trends

In very recent years, a new literature has symptomatically emerged that addresses the material and symbolic death of the city from different perspectives (Agulles 2017; Aquilé 2021; Delgadillo 2021; Glaeser and Cutler 2021; Indovina 2017; Koolhaas 2021; Moskowitz 2021; Settis 2019). These publications transcend the contributions that two decades ago already warned us about "the kingdom of the urban and the death of the city" (Choay 2009) and investigate the various causes that led to the destruction of the city, which we briefly review.

### 44.2.1 Physical Destruction Due to Economic, Political and Social Causes

The physical causes of the death of (parts of) the city are economic, political and social and act in a disruptive or progressive way. The destruction of buildings, neighborhoods and cities is often legitimized by the discourse of decay and the "natural" exhaustion of structures with its different speeds and multiple causes (physical, functional or economic obsolescence).

For Jacobs (1967), getting rid of places that are still useful in the name of "natural" deterioration responded to real estate speculation. Lynch (2005) states in *Wasting Away* how the capitalist logic of making everything new destroys buildings, neighborhoods and cities. Byles (2005) in *Rubble, Unearthing the History of Demolition* analyzes the moral and urban renewal policy in the USA, which between 1949 and 1973 destroyed 97,000 social housing units in 2500 neighborhoods and 992 cities and dispossessed one million people of their homes as a "curative" measure. Berman (1993), a native of the Bronx, exemplifies the tragedy of modernity in the construction of the Cross Bronx highway that divided his neighborhood and brutally destroyed local ways of life.

The deindustrialization produced by the transition from Fordist Keynesian capitalism to post-Fordist neoliberal capitalism led to the international fragmentation and dispersion of industrial production and the loss of industries and jobs in many cities around the world. Perhaps the worst example is Detroit, a city that housed two million inhabitants, but the exodus of factories, jobs and people led to the loss of residents (in 2005 there were less than 700 thousand inhabitants) and the abandonment of 78 thousand houses (Byles 2005).

The strategies and mechanisms of creative destruction are very diverse. They include the physical destruction of goods with various methods, technological innovation, the imposition of fashions, the creation of new needs and the programming of an artificial obsolescence, using perishable materials in short term. The latter is the case of the millions of "social housing" units built in distant urban peripheries of countries such as Chile, Mexico and Brazil. Thus, many new constructions no longer survive more than a human generation or less, and many real estate megaprojects quickly turn into ruins (Agulles 2017; Consejo Nocturno 2018; Koolhaas 2021).

Among the economic causes that kill the city is also tourism: this activity transforms cities and historic centers into open-air museums and theme parks for the homogenized consumption of visitors eager for "authentic" places (Augé 2008). Many cities and historic centers are gradually left without residents and become ultra-guarded shells, uninhabited and destined for the consumption of visitors (Consejo Nocturno 2018). Venice, for example, attracts millions of tourists, at the cost of displacing its residents (Settis 2019). Brossat (2019), responsible for housing in Paris between 2014 and 2018, denounces the cannibalization of neighborhoods produced by Airbnb and other digital platforms that profit from homes and neighborhoods that are not theirs, increase rents, expel residents and replace shops and local services with others, at the service of visitors.

Among the political and social causes, wars and revolutions stand out, as well as urban modernization and beautification projects, which often respond to dominant ideologies and real estate deals. Wikipedia (2020) has an entry on the Urbicide concept, which refers to the destruction of cities by wars to destroy the enemy, undermine the material foundations of the lives of its inhabitants, expel people and destroy memory and collective identity. In this sense, Jouannais (2017) recognizes that the destruction of cities by wars is very old and very recent. The annihilation is often concentrated in certain buildings and places, not because they are strategic militarily, but because of the symbolic importance associated with the buildings and places. Bevan (2006) argues that in wars certain enemy buildings and structures are destroyed to terrorize, divide or eradicate their cultural symbols, destroy their identities and collective memories. It is not a collateral damage, but a systematic destruction of architecture to which collective, religious, profane, civic, spiritual meanings and values are attributed. Here can be mentioned Beirut, Dresden, Guernica, Hiroshima, Mexico Tenochtitlán, Palmira, Rome, Sarajevo and very recently Kiev, Mariupol and other Ukrainian cities. Among recent cases, the Balkan war stands out, where the urban heritage of Bosnia Herzegovina and Croatia was a specific target of the conflict, as mosques, temples, libraries and other iconic buildings perished. Coward (2009) argues that in the Balkans war it was explicitly intended to annihilate the sociocultural diversity and ethnic heterogeneity of those cities. Thus, the city, synonymous with multiculturalism, respect and tolerance of diversity, was the object of ethnic cleansing.

Regime changes also lead to the partial destruction of the city. During the French Revolution monuments and buildings of the aristocracy were demolished, the insurgents took revenge with the symbols of the oppression of the overthrown regime that constituted an insult to poverty and the morality of the Revolution. In this sense, revolutionary vandalism has political reasons; it is not a barbaric act (Gamboni 2014). Stalin's regime attacked the religious architecture and vestiges of Czarist Russia. Mussolini destroyed Roman and medieval vestiges to monumentalize fascist Rome with new avenues. In Berlin, after German reunification, buildings of the socialist past have been destroyed in the name of the ugliness of the architecture and that the buildings were built with asbestos, to rebuild a baroque castle destroyed for ideological reasons around 1957. In this same sense, in the 2020s, in Canada, the USA, Chile, and Colombia, we have witnessed the collapse of commemorative statues of historical figures associated with slavery and colonialism, within the framework of various social protests.

Finally, also as social caused, we locate the destruction of (parts of) cities derived from the incidence of natural phenomena in conditions of socially constructed vulnerability. Floods, hurricanes, tornadoes, cyclones, earthquakes, volcanic eruptions or tsunamis are natural phenomena, some of them intensified, in their dimension and frequency, by climate change, which is unevenly produced by the capitalist depredation of nature. Natural phenomena act in socially constructed environments, but if a city is well-built (with good materials and construction systems, and well-located places), it should not be (so) devastated by a natural event. People cannot predict natural phenomena, but they can undertake certain measures in order to confront risks. In turn, the population can reduce its conditions of vulnerability to diminish the effects of these natural phenomena and avoid a social disaster.

### *44.2.2 Symbolic Destruction*

The city is a multidimensional concept and reality that has its roots in the Greco-Latin world and refers to a space built by generations of people to live well and collectively.

The city is a physical space for collective use (*urbs* in Latin), a community of citizens with rights and obligations over what is public (*civitas* in Latin) and a political community that governs itself (*polis* in Greek) (Borja 2011; Carrión 2016; Choay 2009; Delgadillo 2021). Thus, the city is part of the civilizing process and the conquest of freedoms and human rights.

A city is also a high density of diverse and heterogeneous people (in socioeconomic, cultural, political terms, etc.) with different interests that cohesively live and coexist in a limited physical space, thanks to tolerance and respect for the social pact. For this reason, the city, our cultural heritage, has been defined as a public space par excellence, of common and general interest for the population that lives or visits it (Borja 2011; Carrión 2016).

The city is a historical, social and political product, the site of the reproduction of the social relations of production. It is a space that is socially disputed in unequal conditions of power by various social, political and economic actors and agents, who appropriate that space to use it, inhabit it, exploit it, dominate it and/or control it (Lefebvre 2013). For Choay (2009) and Mongin (2006) a city had physical limits and responded to the culture of integrating and relating different people. The city favored social mixing, confluence and encounter, while the intrinsic social conflict was mediated by civility and tolerance. However, in the twenty-first century, "city" is no longer built. For decades, the city, a symbol of emancipation and social integration, has been confronted with a metropolitan dynamic and globalization that divide, disperse, fragment, privatize, decentralize, separate and create new and diverse urban and territorial hierarchies. Meanwhile, the spread of new communication technologies has de-territorialized the interaction that people used to have in a physically delimited space and time. Thus, belonging to communities is no longer based on territorial proximity. COVID-19 intensified work at home, distance education, the consumption of communication and information technologies and strengthened the capitalism of digital entertainment platforms and home delivery services. In recent decades, we have witnessed processes of urban sprawl made up of bedroom communities and shopping malls in various regions of the world, which blur the notion of limits and deny the city as a historical and social way of inhabiting space. Thus, there is no longer a "city" but a dispersed planetary urbanization, without form or order, unlimited and indefinite on a global continuum (Agulles 2017; Consejo Nocturno 2018; Lefebvre 2022).

Carrión (2016) adds violence and urban crime as causes of Urbicide in Latin America, since it provokes fear of public space, the self-confinement of social groups in closed settlements, the closure of streets and the legitimization of various punitive public security policies. During the COVID-19 confinement, crimes on public roads and home robberies decreased, but domestic violence increased, particularly against women.

The new real estate businesses contribute to killing the attributes of the city because they produce exclusive and exclusionary urban islands and make select neighborhoods more expensive and directly and indirectly displace the resident population, through gentrification processes that deepen the historical urban segregation (Carrión

2016; Delgadillo 2021). This urbanism of isolated fragments fears the city, diversity and socioeconomic mix and is often accompanied by codes of good behavior in public spaces, which expel informal vendors and the homeless population (Borja 2011). In short, this neoliberal urbanism erodes collective and political life, places urban management at the service of profit, commodifies the city and its parts and conceives of citizens as simple consumers who must reside and consume where they can pay.

The death of the city also occurs due to social prophylaxis, derived from health, social or political problems. Indeed, in recent history there have been authoritarian regimes that kill the city, in symbolic and material terms, to remove what they consider to be social cancer. Rodríguez and Rodríguez (2009) says that, just as Machiavelli affirmed that to control a city one has to destroy it, the Chilean military dictatorship in Santiago de Chile dedicated itself to killing several neighborhoods sympathetic to the ousted president Salvador Allende, dispersing the people to distant outskirts, under the discourse of combating the housing deficit. We have a different example in the USA, where in the middle of the twentieth century some neighborhoods inhabited by poor people, African Americans and the unemployed were destroyed because they were considered to be the cause of antisocial behavior: the urban and moral renewal removed the social cancer (Byles 2005).

### 44.2.3 Obsolete City

Choay (2007) warned us that social practices change faster than the lexicon we use, so we tend to continue calling certain things for what they are no longer: we say "center" and "city" to refer to territories that have lost those functions and qualities. Perhaps for this reason, Koolhaas (2021) called one of his latest publications *Studies on (what once was called at the time) the city*. Here he affirms that: in Europe and other regions "city" is no longer built, in many cities the urban center has practically died and the periphery tends to evaporate; and architectures are no longer built with minimum economic expenses to obtain maximum social benefits, but are designed as financial packages that last 25–30 years. Not only is it not intended to be built for eternity, but also for a short time in an accelerated cycle of construction–destruction–reconstruction, which adjusts to the cycles of return of capital. Here, the planned obsolescence has replaced the tabula rasa.

Thus, there are cities (such as Detroit) that end up in ruins, because capital no longer needs their existence. The city was the seat of the reproduction of capital, through the production of material goods. However, since the band of industrial production was fragmented and dispersed in the world, the workers disappeared. Now, in many cities a tertiary economy focused on consumption, services and financial flows predominates. The city of services is the one that serves the capital of the twenty-first century and conflicts no longer confront capital with work, but civil society with the government (Indovina 2017).

However, not only the cities have been reconfigured, but the countryside as well. Agricultural production in Europe has been digitized, the countryside has been urbanized with second homes, while the picturesque rural towns, many of them listed as heritage sites for their historical, artistic and/or landscape attributes, have become touristic. Modern farmers work with computers and sensors, while tractors are driven by foreign engineers. There are many people who live in the countryside, but do not work in agriculture. The countryside has become a theme park, because rural buildings are used for tourist accommodation and second homes. German companies have acquired entire Tuscan villages (Koolhaas 2021), while Pienza and its natural surroundings, once declared a UNESCO World Heritage Site, for the beauty and harmony of the city with the landscape, began to be devastated with an urbanization that affects the landscape and profits from this World Heritage status (Settis 2013).

The increasingly gigantic digitization and virtualization of functions, activities and economic, social, cultural, entertainment relationships, etc., tend to disconnect more and more from built forms and replace face-to-face relationships. In this sense, Batty (2018) recognizes that built forms or containers distance themselves more and more from the functions (contents) for which they were built, that is, the buildings become obsolete for what they were designed for, which does not mean that cannot be used and adapted for other uses and activities. Indeed, many economic transactions and social relations happen independent of the geographical location and the original function of the buildings, so that people—even before the pandemic—remained increasingly physically distant, but electronically connected. Thus, information networks are already replacing various traditional functions of cities as sites of exchange and innovation.

### 44.2.4 War and Disaster as Business

In an investigation that analyzes Urbicide for reasons of public security, wars and reconstructions, Aquilé (2021) says that the city is a palimpsest in which various actors over time write, erase and rewrite. The city is often confronted with conflictive, adverse and uncertain situations, which we should understand in order to enhance its ability to overcome and be resilient. She studies three cases of Urbicide and adheres to Coward's (2009) definition, for whom Urbicide is the deliberate destruction of the built environment that enables diverse social relations among a heterogeneous population.

- The almost complete destruction of the Bijlmermeer neighborhood in Amsterdam, built in 1962 under the paradigm of the modern movement, was carried out to combat social problems of insecurity and drug addiction.
- The destruction of Sarajevo, besieged for more than four years during the Balkan war, was a plan of ethnic cleansing. Here, the city divided between Bosnia and Herzegovina was rebuilt according to each political power: the tourist parts were

rebuilt as they were before the war, while foreign capitals imposed out-of-scale buildings in select parts of the city.
- Beirut is a city that suffered a double Urbicide: one during the long war and another in the reconstruction of the central district that created a center with new skyscrapers and exclusive and exclusionary shopping malls, many of which remain empty. Here, a single company adopted all the land as a single property, where before there were 30,000 owners and 50,000 residents.

War and disasters have become an opportunity for lucrative private business, a kind of "natural" creative destruction. Meskell (2015) and Rolnik (2017) report how on some Indonesian islands, after the tsunami, the affected population was relocated and the beaches were rebuilt as international tourist paradises, while Hurricane Catrina was used by authorities to rebuild downtown New Orleans, particularly the Bywaters neighborhood, for a higher-income, white population. That is, a gentrification that according to Moskowitz (2021) occurs in other forms in most American cities, including the famous West Village neighborhood, from where Jane Jacobs defended living, inhabited neighborhoods, characterized by the diversity of people, residents and transients, activities and land (urban) soil uses.

## 44.2.5 Ghost Urbanism

Recently, speculative urbanization has reached a colossal scale in many cities around the world. It is about the production of large buildings, skyscrapers, megaprojects, urbanizations and cities that are built and not inhabited. If architecture and urbanism have historically served to provide shelter, manifest culture and store wealth, the rise of finance capitalism has dramatically accelerated the "store wealth" function on a colossal scale.

Traditionally, land and buildings have been difficult to buy and sell quickly, because they are immobile, expensive, require maintenance and are located in sites with a complex social and political context. However, the revolution in financial instruments has made it possible to make the immobile mobile. Through securitization real estate tied to local territories is converted into mobile financial assets that are sold on international stock markets. Thus, local real estate markets are connected to huge flows of global financial capital (De Mattos 2016; Marcinkoski 2018; Rolnik 2017; Soules 2021).

This logic of real estate production has given rise to various phenomena: an enormous number of new buildings and developments that remain unoccupied for a long time; the production of new urban ruins in countries and regions where the housing bubble burst and many buildings and developments were not completed or were completed and were not sold. In Spain Schulz-Dornburg (2012) and Marcinkoski (2018) have given an account of the *Modern Ruins* and *the City that Never Was*. There is already a literature that analyzes this phenomenon in China, Dubai, Ireland, Spain, the USA, Panama, Turkey and in several African countries. Marcinkoski

(2018) reports how the real estate financial crisis in the global north led many capitals to find new destinations and new megaprojects emerged from the metropolitan peripheries of African countries. One of the most conspicuous examples is New Kilamba City, 30 km from Luanda, Angola. In this project that would house half a million inhabitants, only 20,000 apartments have been built and only 20% have been sold, because these houses are unattainable for the majority of Angolans who live on 2 dollars a day and are located far from the city. This shows that these urbanizations are not born of social demands or local market needs, but of the profit ambitions of international and local investors and of politicians who seek to project economic stability to attract more investment through the production of urban structures.

Soules (2021) qualifies this as zombie and ghost urbanism based on the degree of abandonment or under-occupation presented by the new developments. Vacancy can be due to three causes that occur alone or in combination: storing wealth, secondary housing and speculation. In many cities (London, Toronto, Venice), the scale of vacancy has increased colossally because many investors buy houses and buildings as investments and because real estate prices continue to rise.

## 44.3 Learning from History

Historically, people dream of better futures, cities and societies and some actors represent such aspirations in words, paintings, plans and digital media. Such dreams are neither neutral nor depoliticized. The utopian and fictitious expressions of society and the city are made by subjects with a social and political position in the world in which they live, although this condition is usually hidden under a Universalist discourse of a better future. Many *Ideal City* projects have been created by architects, urban planners, painters and writers who produce or work for the elite. Thus, some futuristic visions not only ignore the causes of problems and class inequalities, but also show the rejection of other social groups.

An in-depth analysis of utopias and ideal visions of the cities of the future has already been carried out by historians and architects such as Eaton (2001), Ramírez (1988) and very recently by Beanland (2021) and Dobraszczyk (2019). The purpose of this brief review is to recall that, in other crises, derived from industrialization, wars, epidemics, overpopulation or climate change, visions of better human and urban futures have been created.

### *44.3.1 Utopias and Ideal Cities*

Since ancient times, humanity has recorded its aspirations for better, ideal and utopian cities and societies. In *The Republic*, Plato described his idea of a democratic city, which, unlike Athens with its elevated Acropolis, would be a horizontal city in the shape of a circle and the agora, with market and temples, would occupy the center of the circle accessible to all citizens (Bobbio et al. 1991). At that time, neither slaves nor women were citizens and in his ideal city there were always be people who lived closer to the center than those on the periphery. Campanella (2017) proposed in 1602 an ideal republic founded on love and harmony, in a city that replicates the solar system with seven wall and a center with a circular temple, where private property does not exist.

For Eaton (2001) and Heffes (2013) the *Ideal* is reactionary and reformist because it refers to the perfection of the functioning of the existing. While *Utopia*, a neologism introduced by Thomas More, designates a non-place, a place that does not exist, but works better than the world in which one lives. Thus, the *Ideal City* accepts the status quo, while the utopias constitute radical and alternative proposals to build a better world different from the existing and chaotic inhabited order.

Some utopias and ideal city projects seek the lost unity with nature, the cosmos or God; they seek to overcome human misery and poverty; abolish inequality; and so on. In this sense, human efforts to imagine a better world are not only built on dreams, but are also populated by nightmares. Thus, for example, at a time when the recently invented perspective was spreading, the Italian Quattrocento paintings by Fran Carnevale and Pier della Francesca, called *Ideal City*, represent a city with orthogonal streets and squares… without people, as if they were portraying cities during the twenty-first century pandemic.

Most of the utopias and proposals for ideal cities of the nineteenth century arose to confront the problems of the industrial city: traffic congestion, overcrowding and substandard housing for workers, pollution, noise, etc. It is no coincidence that many *Ideal City* projects consist on creating new towns in the countryside on the outskirts of cities. Some factory owners carried out philanthropic, moralistic or humanitarian actions, such as the *Familistères* and *Falanstérios* developed or projected by Charles Fourier, Victor Considerant and André Godins in France and Robert Owen's projects in Scotland and England. Some of them not only want better cities, but real social, economic and even land ownership changes. Ebenezer Howard's Garden Cities movement is anchored in these proposals to join the best of the countryside with the best of the city, creating small satellite neighborhoods with airy houses and agricultural cultivation areas, linked to the cities through railways and highways.

In the nineteenth century, some literary utopias also emerged, such as Etienne Cabet's *Travels in Icaria* (1840). John Minter Morgan's *The Christian commonwealth* (1845) inspired the founding of a town in the USA. A doctor, Benjamin Ward Richardson, wrote *Hygeia, a City of Health* (1876), where a city of one hundred thousand people who live and work in healthy and dignified conditions appears, thanks to advances in industry and science.

In the first decades of the twentieth century, visions of ideal cities emerged based on advances in industrialized production and new materials and construction systems, whose validity lasted well into the 1960s. We are referring to the architects and urban planners of the Bauhaus and of the International Congresses of Modern Architecture (CIAM) that conceived the city and housing as machines to inhabit and consequently designed cities to function with such perfection. These visions of urban planning based on the separation of functions, high-rise buildings integrated into superblocks with large green areas and the replacement of the old city, though a tabula rasa, became the paradigm that guided the reconstruction of European cities devastated by World War II and guided the modernization of several Latin American and US cities in the 1960s and 1970s. These visions of the *Ideal City* never could be realized on the planned scale in Latin America, with the great exception of Brasilia, the Brazilian capital, which by its scale and completion time constitute a world milestone.

In a very different sense, Frank Lloyd Wright proposed *Broad Acre City*: a form of de-urbanism, where very low-density human settlements were evenly distributed across the USA in orthogonal parcels. This position is similar to the Russian de-urbanists and the Nazi anti-urbanists: the first conceived the city as capitalist product and the second tried to disperse the cities in the territory, to make it more difficult for them to be destroyed by air attacks. Interestingly, during the COVID-19 pandemic, there were some people who pointed out how opportune it would be to disperse the population of the cities in the countryside to prevent the spread of the virus.

### 44.3.2 Latin American Ideal Urban Visions

The foundation of cities, villages and reductions[1] in Latin America in the sixteenth century was part of the strategy of the Spanish military and cultural conquest. The city, both as an expression of civilization and as a form of military control, is a legacy of the Roman Empire, which the Spanish repeat in America. Heffes (2013) reports on the production of utopias in Latin America since colonial times, most of them are urban. In the sixteenth century, Bartolomé de las Casas proposed creating communities of free Indians and Vasco de Quiroga, who had read Thomas More and it seems, translated *Utopia* into Spanish, built the *Hospital Cities* in Mexico City and Michoacán with six thousand families, six-hour workdays and the limitation of the use of money only toward foreign trade relations. We can also mention the Jesuit missions in South America in the seventeenth and eighteenth centuries, based on community work.

In the mid-nineteenth century, Faustino Sarmiento proposed *Argirópolis*, where new cities as a vehicle for civilization would put an end to uncultivated fields. By the end of the nineteenth century and in the first decades of the 20th, concerns were already about urban hygiene: Emilio Coni, in 1919, proposed *The Ideal Argentine City* as a healthy and perfect space. Heffes (2013) also reviews alternative futuristic visions, always urban, produced by anarchists, avant-gardes, cooperatives and various community projects, some promoted by liberation theologians, carried out outside the state and the market. Ainsa (2013) also reviews anti-utopian and catastrophic visions produced from the discredit of progress and development based on blind faith in science and technology. If in Europe, Nietzsche and Spengler expressed negative and decadent visions of the future, in Latin America academics and writers

---

[1] Reduction: Colonial policy, which concentrates dispersed population in a place in order to evangelize and control them.

denounced pessimistic scenarios: cities ceased to be places that integrate and where people progress, improve and ascend socioeconomically, to become monstrous sites, condemned to ruin and precariousness. Here, novels parade that present the ruination of emblematic buildings and entire neighborhoods, with suicides who survive in decadent and precarious neighborhoods that "cannot be called a city". Some nostalgic visions for the lost past also appear.

### 44.3.3 Science Fiction and Futuristic Cities

In *Illusory Constructions*, Ramírez recognizes that, as the urbanization process progressed in the twentieth century, "overpopulation" was considered one of the great problems that would lead to the end of the world. It was believed that there would be so many billions of humans that they would not fit on the earth and that even air would be lacking: "suffocation will occur until the year 2560" (Ramírez 1988, p. 218). These concerns gave rise to the creation of futuristic visions that sought to create space to house the unstoppable new inhabitants. Thus, architects and urban planners from different geographical basins produced an endless number of ideal cities and megastructures that range from underground, marine and underwater cities to artifacts suspended in the air, spaceships and even the colonization of the moon and other planets.

From a broader and updated vision, Dobraszczyk (2019) reviews the visions of the cities of the future and imagined cities, through literature, art, architecture, cinema and video games. To the concerns of overpopulation are added more contemporary fears: the urbanization of Asia and Africa, climate change, the increase in poverty and inequalities. Dobraszczyk recognizes that no matter how so fantastic they may seem, the images of the future of the city are always based on the present, and that the digital age we live in has intertwined the real and the imagined more closely, since cities always are sites physically and mentally rooted. Here, fictional and real images of flooded cities appear that repeat apocalyptic visions in the face of climate change, such as the universal flood: cities submerged by hydroelectric dams or drowned by hurricanes, while the ideal city projects practically consist of contemporary versions of an exclusive and excluding Noah's Ark; and in city projects, such as those carried out by the Seastanding Institute, to attract investors and save rich people. In his book, projects that colonize the subsoil and the air continue to be reproduced as in the 2013 *BioShock Infinite* video game with skyscrapers, suspended buildings with atomic energy, balloon dwellings and flexible inflatable structures.

### 44.3.4 Over Ruins and Devastations

There is a long tradition of imagined scenarios of catastrophes that expel the inhabitants of cities and paradises and leave cities in ruins. The Christian and Greco-Roman traditions are rich in it: Atlantis, Troy, Sodom, Gomorrah, etc. (Preti and Settis 2015). However, beyond the myths, the history of many cities is marked by traumatic events and destruction, and the ruins are testimonies of it.

The ruins have different destinations: sometimes they are recognized, relegated, valued, destroyed or replaced, but they never stop marking the material and immaterial culture, they never stop awakening dreams and fantasies. Ruins evoke different things and events and are valued differently at different times. Sometimes they are understood as signs of decline and death, other times they are signs of hope or of the great past. These changing attitudes are related to the origin of the production of the ruins: destruction by enemies, economic decline, abandonment, disaster. Thus, the modern ruins and abandoned buildings of Detroit are the product of deindustrialization, the abandonment of Pripyat is the product of the nuclear power plant accident of Chernobyl, and Pompeii was the product of a volcanic eruption. However, ruins are increasingly converted into touristic sites.

For Settis (2022) human beings are attracted to ruins by a feeling linked to the fragility of human nature. Ruins evoke absences and presences. They are intersections between the visible and the invisible, between the death of antiquity and the announcement of rebirth. In addition, he recognizes that *heritage-ization* leads to killing living and inhabited neighborhoods and cities and turning them into sterile tourist sites: Venice is practically a dead island. This position is very similar with respect to the production of other forms of modern ruins and uninhabited urbanizations. For O'Callaghan and Di Feliciantonio (2021), ruin is the most visible form of vacancy, an intrinsic function of the capitalist city that reproduces capital, speculates with the moment of the best investment and is a central component of devaluation processes prior to reinvestment and gentrification. Here, the modern ruins and uninhabited spaces are not inert, and they are active and disputed, especially since the real estate financial crisis of 2008 with its aftermath of evictions due to non-payment of debts and mortgages, dispossession, rent increases, etc. Since then, ruins and empty new buildings have become visible and politicized and have been the object of imaginative futures such as squatting that give a social use to abandoned private properties. Thus, urban vacancy plays an increasingly important role in public policies and in the city of the future, since empty spaces have grown on a colossal scale in several cities.

### *44.3.5 Reconstructions of Cities*

The history of humanity is full of city reconstructions carried out after disasters caused by social or natural causes. Although it is not always possible, the idea of reconstructing "how it was and where it was" is legitimate and natural reactions that tend to remove the trauma of the discontinuity produced by the catastrophe. Rebuilding "as it was" is complex for many reasons: building safety, physical stability, disuse of construction systems and materials, etc. "Rebuilding where it was" is not always possible due to political power or physical and economic issues.

In urban reconstructions, there are always debates and disputes: the new versus the old, conservation of the pre-existing morphology versus its replacement.

Destructions and reconstructions historically occur in countries like Chile, Italy, Guatemala, Japan or Mexico, where frequent earthquakes and other calamities happen. However, there is but a very scarce literature that analyzes the post-disaster reconstruction experiences of cities from a historical and comparative perspective.

The German literature that delves into the history of reconstruction and the reconstruction of history stands out (Nerdinger 2010), to justify the reconstructions of "historic" buildings and urban centers that have recently been carried out there. Over centuries, many original buildings have been reconstructed, and there was not much debate about whether those reconstructions were forgeries. However, this is not the case with controversial reconstructions such as the "new medieval city" of Frankfurt, inaugurated in 2018. Several German colleagues describe this reconstruction as the *Disneyification of History*: in eight thousand square meters and under a single building license, they rebuilt 35 buildings that pretend to appear as isolated ones. A single company manages this megaproject with shops, restaurants and a museum, as if it were a shopping center. Against the detractors, Langer (2019) quotes German representatives from various political parties who see such reconstruction as "a form of sentimental urbanism" and "nostalgia for history".

There is a vast literature on the reconstruction of cities in the two German countries, but there is almost no comparative literature between countries. One of the exceptions is the book by Ferlenga and Bassoli (2018), which deals with post-World War II reconstruction and recent years in Germany, France and England, particularly in Italy. This book accounts for diverse reconstructions: in some cases dominated by the tabula rasa and CIAM-type zoning; in other cases, the urban layout and morphology were respected, with the reconstruction of shapes and volumes, but not of ornaments; and in many others only a few of the great religious buildings were faithfully rebuilt as they once were.

Italy, due to its history and geography, has historically had to confront the destruction of its rich cultural heritage, unfortunately without continuity. In different times, including the fascist stage, its cities have restored and overcome what fire, water or earth destroyed. In some reconstructions, the center of attention has been the affected population, followed by the building heritage. However, in other cases, economic and political interests have prevailed, and the population has been displaced to remote territories. Giglia (2000) studied the population affected by an earthquake that was displaced from the historic center of Pozzuoli to a homogeneous, harsh and sullen periphery, designed according to abstract rules that do not respond to the needs of the people. Here, some colleagues speak of a criminal management of the territory and the reconstruction of open-air museums, to attract tourism. In this sense, Settis (2013) denounces that the historic center of L'Aquila was abandoned after the earthquake, while in Emilia Romagna bell towers were dynamited rather than repaired.

A recent book by Arefian et al. (2021), in 37 chapters, addresses the physical reconstruction of historic centers and neighborhoods in the world, particularly in the Middle East and Asia. There, various armed conflicts and natural phenomena have led to disasters that have affected the population and their built heritage. The shared position is that urban heritage is not reduced to the physical, but encompasses the social relationships that signify and value it as such, socially constructed knowledge and ways to prevent risks and confront disasters. Here, the slogan "the future will be better" is proposed to confront destruction and reconstruction, but involving citizens on different scales: housing, public space and infrastructure, including self-help. This book recognizes that reconstruction is political and politicized, and multiple actors (economic, political and social) participate in reconstruction processes in unequal conditions of power on a local and international scale. In the case of World Heritage institutions with a clear international agenda (donors, UNESCO, ICOMOS, development banks) participate, as do investors and governments with their own interests. Thus, the reconstruction of urban heritage navigates between the state and the market, tourism, profitability, privatization and, in a very limited way, citizen participation.

Latin America is a subcontinent that, since its origin, has historically been a prey to disasters and where the construction and reconstruction of housing, neighborhoods and cities have been a constant. McGuirk (2015) records a large number of heroic and exemplary experiences of social production of housing and the city, in a region where historically the poor are relegated to the urban peripheries. He reviews housing and urban milestones, such as the contributions of the anarchist architect John FC Turner and the PREVI project, of progressive housing, developed by the Peruvian president and architect Fernando Belaúnde in the 1960s, with the participation of the most famous architects in the world of that time. McGuirk (2015) and Díaz and Ortiz (2017) highlight various housing and urban experiences conquered in the context of the neoliberal urbanism, some of which have been exported to European countries such as the improvement of neighborhoods and housing, participatory budgets, the Metrobús, the cable cars and electric stairs as collective transportation systems, and very diverse self-management and cooperative experiences from Buenos Aires to Tijuana. Here are specific experiences of confronting disasters, where resilience is built from below.

## 44.4 Colophon

### 44.4.1 Between Urban Justice and the Right to the City

In many cities of the world, social movements of resistance and opposition to the plundering of cities and people, the economic crisis, socioeconomic inequalities have multiplied, with different timing, intensity and geographical location (central areas and urban peripheries). In some cities and countries, social forces have coalesced with political forces and have won municipal and national elections, such as the "governments of change" in Spain, "Socialism of the 21st century" (South America) (with alternation of openly neoliberal governments in some periods) and the recent self-styled "non-neoliberal governments" (Mexico, Argentina, Peru, Honduras). These are municipal, intermediate and national governments in a hegemonic neoliberal world, in which local and global economies are interconnected, and economic and often political decisions transcend borders and sovereignty.

They are experiences with diverse and contradictory results. Both (the so-called turn to the left in South America, Carrión 2015, and the current "non-neoliberal" national governments) are characterized—beyond the duration in government with continuities and ruptures—by ambivalent social and economic policies, which meet some social needs of the majority with various direct supports and subsidies. However, they simultaneously facilitate and promote lucrative transnational private businesses, including the financialized reproduction of urban space and housing. For example, there are the millions of new social housing units built in distant peripheries of Mexico, Brazil, and Ecuador that, in the name of combating the housing deficit, allowed lucrative real estate deals to a few and deported the "benefited" population to distant urban peripheries.

The underlying question is, in a hegemonic neoliberal world with interconnected economies and where fundamental decisions transcend borders and sovereignty, how far can progressive governments advance in social policies? What public policies can be deployed to achieve social justice that goes beyond the simple recognition of the various social rights embodied in legislation? Is it possible to achieve social justice in its multiple dimensions (spatial, economic, etc.) in the capitalist system?

In this context, the debates on the Just City and the Right to the City emerge. One current maintains that urban justice is possible in the capitalist world, when people conquer rights in laws, because they become enforceable rights and social movements can pressure to improve their housing and neighborhoods. Another current maintains that spatial justice is incompatible with capitalism and that the only possibility is to build another city and another possible world.

## 44.4.2  *Another End of the City is Possible*

Faced with those who warn about the end of the city, there are alternative visions that hold "another end of the city is possible." In fact, in recent months it has been repeated that "Destiny has already reached us", paraphrasing the Spanish title "When destiny reaches us" of Richard Fleischer's 1973 futuristic film *Soylent Green*.

The future cannot be predicted, but it can be built and is disputed daily between economic, political and social actors, in unequal power conditions. That is, between the market, the state and civil society, but taking into account the enormous diversity of economic actors (large and small capitalists), politicians (civil servants, representatives, executive and legislative power; neoliberals, social democrats) and social actors (tenants and owners, workers and unemployed, etc.).

Imagining possible futures refers to the imagination, which is an important human faculty, although it is often associated with fantasy, opposed to scientific rationalism and separated from reality. However, imagination is not an escape from reality, but a transformative and political human faculty. As we have seen in this text, there is no imagination without a past, even the most futuristic and innovative buildings include references to the past and the present. Cities are made of dreams and nightmares, let us confront the nightmares and aspire to the dreams of vibrant, inclusive and optimistic cities.

### 44.4.3 Reinvent Urbanism

Modern urban planning and urbanism were born in the nineteenth century to confront urban problems derived from the impacts of the industrial revolution: health, hygiene and urban disorder. Thus, urban congestion and malaria were confronted through the elaboration of urban expansion and ordering (zoning) plans, and the introduction of sewers and drinking water systems, which made the cities healthier. However, normative urban planning has been overtaken by the market for decades. Thus, in the twenty-first century, post-COVID-19, urbanism and town planning should overcome the social and moral reformism with which they were born and serve the needs of the majority of people.

### 44.4.4 Public Policies for the People, not for the Market

In recent decades, many people have demanded urban policies that put people at the center of attention and not private profit. In the years of the pandemic, voices have multiplied that demand a moratorium on urban megaprojects that are carried out based on global capitalist interests; and a stop to the increase in urban rents and evictions. During the pandemic, some governments adopted unprecedented measures and public policies, which is why several researchers and social actors point out that these extraordinary measures taken in an emergency situation should be extended as ordinary measures after COVID-19 (O'Callaghan and Di Feliciantonio 2021; Ávila and Shrecko 2021).

In this sense, one of the greatest demands is the promotion of affordable housing alternatives for citizens and an increase in measures to defend housing and housing land use for the benefit of the people, limiting digital platforms and second homes that make housing more expensive. In some cities, in pre-pandemic times, this has already been done: Paris introduced a 20% tax on second homes and in 2017 increased it to 60%; Vancouver enacted a vacant housing tax in 2019; while Washington DC and Oakland have their own version of taxes on vacant real estate (Brossat 2019).

In Latin America there is a demand to stop thinking about the Parisian "city of fifteen minutes" (a luxury that only tourists and the middle classes pay for) and to rethink the future of our cities from their neighborhoods: autonomous, self-sufficient and sustainable. In this book, the women of Antofagasta express the dream of "our ideal city", which has little to do with the Airbnbized and Uberized centers of Paris and Barcelona.

Likewise, in the face of the *Smart City*, many actors demand to recover the urban intelligence accumulated over centuries and to democratize access to broadband and the collective control of socially produced data.

### 44.4.5 Radical Positions

Due to the measures exposed, in the face of temporary housing digital platforms, it is not so utopian to demand the regulation of large capitals (that they remain where they do their business and pay their taxes), put an end to tax paradises and regulate and collect taxes from the digital platforms, which could not be transferable to consumers. These regulations cannot be only local or national, because this type of virus (global financial capital) is pandemic by definition, that is, it is global and international, and the solutions can never be only local.

Perhaps one of the most radical positions in 2022 is no longer to build (almost) anything new in the cities, but to occupy what already exists: the unoccupied skyscrapers, the empty houses in the central areas of the cities left by speculation and capitalist crises. In this sense, local governments should provide themselves with legal instruments that allow the social function of dilapidated, abandoned and underutilized properties. In addition, there are interesting antecedents, because after the real estate financial crisis in Spanish cities, empty buildings were squatted. While in Sao Paulo, vacant buildings are squatted to oppose the commercial logic of owners and investors who produce buildings that are not used. Thus, vacancy is at the center of equity and justice issues in post-pandemic cities.

### 44.4.6 The Future Dispute

The future is disputed on a local and global scale on various fronts and trenches. One of them is the production of knowledge. Here the proposal of Glaeser and Cutler (2021) cannot be overlooked. They recognize that the US health system is one of the worst in the world to prevent deaths. However, they demand more of the same. A better health system requires that people pay more. Housing will become cheaper if governments enlarge the buildable urban surface. This is a neoliberal myth that is constantly refuted by the real estate bubbles of Spain, the USA and Mexico. Finally, faced with the limitations of the World Health Organization versus COVID-19, they demand a NATO-type organization that controls the spread of future pandemics and ostracizes those countries that maintain unhealthy practices. What will those practices be and who decides what they are?

Faced with this neoliberal posture, this Urbicide book in fifty chapters provides scientific knowledge from cities on two continents that breaks these myths and shows the perverse effects of the neoliberal commodification of the city (including the privatization of health). The book offers the reader a broad understanding of the multidimensional processes that lead to the symbolic and material death of the city and allows us be better prepared for the defense of the public city, our most precious cultural heritage.

## References

Agulles J (2017) La destrucción de la ciudad. El mundo urbano en la culminación de los tiempos modernos. Catarata, Madrid

Ainsa F (2013) La ciudad entre la nostalgia del pasado y la visión apocalíptica. In: Heffes G (ed) Utopías urbanas: geopolítica del deseo en América Latina. Iberoamericana Vervuert, Frankfurt–Madrid pp 49–86

Aquilé I (2021) Ciudad e Incertidumbre. Ediciones Asimétricas, España

Arefian F, Reyser J, Hopkins A, Mackee J (eds) (2021) Historic cities in the face of disasters: reconstruction, recovery and resilience society. Springer, Switzerland

Augé M (2008) El tiempo en ruinas. Gedisa, Barcelona

Ávila R, Srecko H (eds) (2021) ¡Todo debe cambiar! El mundo después de Covid 19. Rayo Verde, Barcelona

Batty M (2018) Inventing future cities. MIT Press, Cambridge

Beanland Ch (2021) Unbuilt: radical vision of a future that never arrived. Batsford, United Kingdom

Berman M (1993) Tudo sólido desmancha no ar. Companhia das Letras, Sao Paulo

Bevan R (2006) The destruction of memory. Architecture at war. Reaktion Books, London

Bobbio N, Matteucci N, Pasquino G (1991) Diccionario de Política. Siglo XXI, México

Borja J (2011) Revolución urbana y derecho a la ciudad. OLACCHI – MMQ, Quito

Brossat I (2019) Airbnb la ciudad uberizada. Katakrak, Pamplona

Byles J (2005) Rubble. The unearthing history of demolition. Harmony Books, New York

Campanella T (2017) La imaginaria Ciudad del Sol. Idea de una república filosófica. FCE, México

Carrión F (2015) El giro a la izquierda. Los gobiernos locales de América Latina. Café de las Ciudades, Buenos Aires

Carrión F (2016) El espacio público es una relación, no un espacio. In: Ramírez P (ed) La reinvención del espacio público en la ciudad fragmentada. UNAM, México, pp 13–47

Choay F (2007) Alegoría del patrimonio. Gustavo Gili, Barcelona

Choay F (2009) El reino de lo urbano y la muerte de la ciudad. Andamios, Revista de Investigación Social 6(12):157–187

Consejo Nocturno (2018) Un habitar más fuerte que la metrópoli. Pepitas de Calabaza, Logroño

Coward M (2009) Urbicide. The politics of urban destruction. Routledge, New York

Delgadillo V (2021) La muerte simbólica y material de la ciudad: una aproximación sobre el urbanicidio. Revista ARQUIS 10:14–22

De Mattos C (2016) Financiarización, valorización inmobiliaria del capital y mercantilización de la metamorfosis urbana. Sociologías 18(42):24–52

Díaz J, Ortiz E (2017) Utopías en construcción. Experiencias latinoamericanas de producción social del habitat. Rosa Luxemburg Stiftung – MISEREOR, México

Dobraszczyk P (2019) Future cities, architecture and the imagination. Reaktionbooks, Londres

Eaton R (2001) Die ideale Stadt. Von der Antike bis zum Gegenwart. Nicolai, Berlín

Ferlenga A, Bassoli N (eds) (2018) Ricostruzioni. Architettura, città, paesaggio nell'epoca delle distruzioni. Silvana Editorial, Milán

Gamboni D (2014) La destrucción del arte. Iconoclasia y vandalismo desde la Revolución Francesa. Cátedra, Madrid

Giglia A (2000) Terremoto y reconstrucción. Un estudio antropológico en Pozzuoli, Italia. FLACSO – Plaza y Valdés, México

Glaeser E, Cutler D (2021) Survival of the city. Living and thriving in an age of isolation. Penguin Press, New York

Heffes G (ed) (2013) Utopías urbanas: geopolítica del deseo en América Latina. Iberoamericana Vervuert, Frankfurt–Madrid

Indovina F (2017) Vieja y nueva cuestión urbana. In: Becchi A et al (eds) La ciudad del siglo XXI. Conversando con Bernardo Secchi. Catarata, Madrid, pp 122–157

Jacobs J (1967) Muerte y vida de las grandes ciudades. Península, Barcelona

Jouannais J (2017) El uso de las ruinas. Retratos obsidionales. Acantilado, Barcelona

Koolhaas R (2021) Estudios sobre (lo que en su momento se llamó) la ciudad. Gustavo Gili, Barcelona

Langer F (2019) Frankfurts neue Altstadt. Insel Verlag, Ulm

Lefebvre H (2013) La producción del espacio. Capitán Swing, Madrid

Lefebvre H (2022) La Revolución Urbana. Alianza, Madrid

Lynch K (2005) Echar a perder, análisis del deterioro. Gustavo Gili, Barcelona

Marcinkoski Ch (2018) The city that never was. Princeton Architectural Press, New York

McGuirk J (2015) Ciudades radicales: un viaje a la nueva arquitectura latinoamericana. Turner, Madrid

Meskell L (ed) (2015) Global heritage. A reader. Willey Blackwell, UK

Mongin O (2006) La condición urbana, la ciudad a la hora de la mundialización. Paidos, Buenos Aires

Moskowitz PE (2021) How to kill a city. Gentrification, inequality and the fight for the neighboorhod. Bold Type Books, New York

Nerdinger W (2010) Geschichte der Rekonstruktion, Rekonstruktion der Geschichte. Prestel, München–New York

O'Callaghan C, Di Feliciantonio C (2021) The new urban ruins. Urban politics and international experiments in the post-crisis city. Policy Press, Bristol

Preti M, Settis S (2015) Villes en Ruine. Museo del Louvre, Paris

Ramírez JA (1988) Construcciones ilusorias. Arquitecturas descritas, arquitecturas pintadas. Alianza, Madrid

Rodríguez A, Rodríguez P (eds) (2009) Santiago, una ciudad neoliberal. OLACCHI, Quito

Rolnik R (2017) La guerra de los lugares. La colonización de la tierra y la vivienda en la era de las finanzas. Santiago de Chile, LOM

Schávelzon D (1990) La conservación del patrimonio cultural en América Latina: restauración de edificios prehispánicos en Mesoamérica 1750–1980. FADU UBA - The Getty Grant Program, Buenos Aires

Schulz-Dornburg J (2012) Ruinas modernas. Una topografía del lucro. Ambit serveis Barcelona

Settis S (2013) Paisaje, patrimonio cultural, tutela: Una historia italiana. Universidad de Valparaíso

Settis S (2019) Wenn Venedig stirbt. Streitschrift gegen den Ausverkauf der Städte. Wagenbach, Berlin

Settis S (2022) Eclipse y resurrección de la ciudad histórica. In: Delgadillo V, Niglio O (ed) El asedio inmobiliario y turístico al patrimonio. Tabedizioni, Roma

Soules M (2021) Icebergs, zombis and the ultra-thiny. Architecture and capitalism in XXI century. Princeton Architectural Press, New York

Wikipedia (2020) Urbicide. Wikipedia, the free encyclopedia. https://en.wikipedia.org/wiki/Urbicide

**Víctor Delgadillo** Ph.D. on Urbanism (National Autonomous University of Mexico), Urban Planner (Stuttgart University, Germany), Architect (Autonomous University Puebla, Mexico). National Researcher level two (SNI CONACYT). Professor of the College of Humanities and Social Sciences—National Autonomous University of Mexico. He has carried out research, consultancies and plans on heritage and urban development for Mexican and Latin American cities. He has published 10 books and more than 120 articles in magazines and books in various countries.

# Index

**A**

Aesthetics, 15, 63, 79, 81, 86, 128, 129, 133, 137, 174, 336, 337, 339, 340, 347, 349, 351, 467, 469, 470, 472, 486, 512, 521, 523, 527, 528, 542, 545, 627, 655, 656, 658–661, 664–666, 668, 797

Archetypes, 94–98, 363, 457

Architecture, 16, 79, 84, 139, 165, 173, 182, 183, 185, 194, 308, 325, 333, 334, 336–340, 343–352, 382, 387, 421, 425, 471, 472, 474, 477, 481, 490, 491, 495, 497, 507, 518, 519, 525, 537, 544, 546, 550, 551, 556, 565–567, 575, 581, 607, 614, 695, 730, 787, 789, 813, 817, 821, 830, 833, 839, 917, 921, 923, 925, 927, 929

Art, 57, 78, 133, 162, 164, 324, 338, 349, 401, 402, 404, 405, 411, 472, 474, 478, 491, 496, 497, 546, 547, 572, 598, 656, 660, 663, 665, 668, 693, 730, 762, 784, 787, 789, 797, 815, 817, 830, 840, 929

**B**

Borders, 16, 18, 39, 90, 109, 145, 148, 150, 271, 318, 334, 361–365, 377, 378, 387, 395, 397, 398, 402–404, 406, 407, 410–415, 419, 420, 423, 435, 438, 440, 448–454, 456–458, 460, 518, 531, 536, 543, 580, 588, 591–593, 595–599, 605, 609, 611, 618, 620, 622, 623, 626, 627, 630, 655–665, 667, 668, 757, 767, 770, 805, 822, 933

**C**

Capitalism, 65, 78, 81, 89, 93, 101–103, 128, 129, 131, 147, 148, 153, 166, 227, 268, 294, 336–339, 363, 379, 381, 391, 399, 404, 420, 430, 433, 440, 441, 470, 532, 568, 595, 619, 674, 676–682, 699, 702, 726, 729, 733, 752, 756, 757, 762, 765, 846–848, 850, 851, 860, 868, 893, 894, 897, 907, 911, 920, 922, 925, 933

Caracas, 20, 36, 208, 214, 215, 237, 844, 847–860

Catastrophe, 59, 60, 69–71, 79, 84, 96, 99, 134, 136, 235, 239, 240, 336, 497, 756, 929, 930

Cities, 3–20, 25–43, 48–50, 52–54, 56, 57, 59–63, 65–72, 77–90, 93–103, 107, 108, 111–113, 119–122, 127–141, 145–154, 157–168, 172–187, 189–194, 198–200, 205–209, 211–219, 221, 222, 224, 226–228, 235, 237–245, 248–260, 263–274, 276, 278–285, 293–298, 300–302, 304, 308, 310–312, 315–321, 323–330, 333–338, 340–352, 359–374, 377–392, 395–415, 419–423, 425–427, 429, 430, 433–441, 447–460, 467–475, 478–484, 486, 489–503, 505–508, 511–528, 531–547, 550, 551, 555–558, 560–563, 565–581,

© The Editor(s) (if applicable) and The Author(s), under exclusive license to Springer Nature Switzerland AG 2023
F. Carrión Mena and P. Cepeda Pico (eds.), *Urbicide*, The Urban Book Series, https://doi.org/10.1007/978-3-031-25304-1

585–600, 603–632, 637–645,
647–650, 655–668, 675, 678, 681,
682, 684, 687, 688, 690, 693–705,
707–709, 711–714, 717–720,
725–734, 737–742, 744–747,
751–756, 758, 760, 762, 764–767,
770, 772, 774, 775, 779–786, 790,
793–802, 804–810, 813–815,
818–825, 827, 828, 830–833,
835–841, 843–848, 850, 852–854,
856–860, 865–872, 874, 875,
878–880, 882, 884, 887, 891–899,
901–911, 917–936

Climate change, 5, 10, 11, 25, 26, 29,
31–33, 35, 47, 49, 77, 78, 89, 99,
127, 140, 141, 148, 189, 194, 199,
235, 236, 238, 239, 241, 259, 274,
300, 303, 380, 540, 546, 611, 705,
712, 844, 858, 921, 926, 929

Colonialism, 16, 395, 397, 415, 522, 524,
731, 868, 921

Colonization, 10, 15, 26, 29, 79, 82, 89,
179, 181, 378, 467, 471, 482, 483,
486, 618, 622, 793, 831, 929

Community, 5, 6, 8–11, 17, 30–32, 37, 41,
48, 53, 60, 62, 66, 69, 79, 87, 88, 97,
102, 113, 114, 116, 119, 121, 132,
151, 154, 167, 168, 171, 175, 182,
186, 191, 198, 236, 237, 242, 244,
309, 310, 330, 335, 336, 341, 343,
348, 349, 351, 361, 363, 366, 367,
378–382, 388, 389, 391, 396, 408,
409, 413, 414, 422, 425, 435, 437,
439–441, 447–450, 456–460, 467,
469–473, 476, 482–486, 496, 499,
504, 506, 513, 516, 521, 524, 533,
536, 541–543, 546–548, 551, 562,
568, 570, 574, 576, 577, 580, 589,
590, 592–594, 598, 605–609, 613,
615, 616, 625, 628–630, 643, 661,
663, 668, 693, 694, 700, 704, 707,
708, 714, 717–720, 725, 726,
728–730, 740, 741, 743, 752, 754,
757, 762, 766, 769, 770, 780–788,
790, 820, 838, 867, 898, 906, 909,
918, 922, 928

Corruption, 245, 253, 315, 316, 318–320,
326, 329, 330, 380, 383, 384, 450,
588, 592, 603, 604, 611, 628, 629,
696, 732, 742, 806, 826, 827, 856,
870, 893

Crime, 10, 18, 33, 38, 61, 62, 209, 218,
248, 379, 380, 385, 414, 433, 460,
533, 561, 585–593, 595, 596,
598–600, 611, 612, 615, 624, 625,
629–631, 640, 644–646, 725, 728,
732, 733, 739, 740, 742–744, 746,
816, 822, 825, 829, 922

Cultural identity, 55, 473, 481–484, 659,
741

Cultural landscape, 472, 473

Cultural tourism, 504, 534

Culture, 6, 15, 17, 25, 32, 40, 63, 71, 81,
82, 87, 88, 93, 95, 97, 140, 161, 162,
173, 175, 183, 185, 237, 336, 338,
343, 344, 349, 372, 378, 391, 402,
404, 420, 421, 469, 471, 472, 477,
478, 481–486, 494, 499, 506, 515,
516, 526, 528, 532, 534, 535,
567–569, 571, 572, 575, 576, 579,
589, 641, 644, 655, 657, 659, 668,
712, 719, 728, 741, 743, 757,
779–781, 784, 785, 787, 790, 797,
798, 830, 846, 907, 922, 925, 930

**D**

Degradation, 3, 12, 13, 15–17, 31, 37, 42,
102, 235, 236, 245, 253, 259, 293,
303, 308, 309, 378, 408, 422, 489,
501, 502, 508, 539, 544, 567, 572,
573, 575, 655, 656, 661, 662, 664,
733, 754, 775, 806, 844, 898–900,
903

Democracy, 107, 116, 117, 119, 120, 128,
129, 146, 147, 149, 151, 346, 512,
517, 528, 532, 535, 537, 538, 558,
630, 703, 712, 846

Demolition, 79, 102, 139, 168, 319, 321,
324, 337, 338, 350, 468, 565, 566,
622, 665, 696, 787, 813, 815, 818,
828, 835, 844, 852, 901, 903, 920

Digital networks, 15, 145–149, 151, 154

Digital space, 145–147, 149–154, 489

Disasters, 4, 14, 19, 35, 69, 131, 141, 235,
236, 238–240, 243–245, 248,
250–252, 258, 270, 271, 285, 303,
378, 559, 593, 609, 610, 621, 700,
751, 754, 756–758, 760, 766, 767,
775, 918, 919, 921, 924, 925,
930–932

Diversity, 4, 20, 25, 30, 38, 60, 63, 112,
113, 119, 184, 197, 198, 227, 239,
309, 321, 333, 335, 336, 345, 348,
360, 378, 382, 386, 392, 426, 435,
455, 456, 473, 484–486, 489, 493,
502, 503, 514, 525, 526, 546, 551,

# Index

566, 571, 572, 581, 607, 720, 740, 752, 769, 779–781, 784, 785, 787, 843–846, 853, 860, 867, 918, 921, 923, 925, 933

Drug trafficking, 18, 20, 36, 61, 72, 83, 90, 592, 594, 596, 598, 599, 603, 604, 611, 621, 622, 626, 627, 630–632, 666, 806, 813–815, 817, 818, 822–825, 829–832, 834–836, 840, 893

## E

Education, 6, 38, 54, 55, 71, 81, 89, 114, 138, 178, 206, 249, 310, 381, 383, 404, 484, 486, 490, 517, 532, 538, 548, 580, 581, 627, 643, 695, 710, 728, 742, 795, 805–807, 809, 824, 830, 857, 906, 922

Execution of works, 826

## G

Gender, 18, 33, 36, 62, 114, 189, 215, 227, 365, 383, 391, 460, 502, 516, 541, 588, 623, 628, 632, 637–639, 641–650, 712, 780, 781, 906

Genealogy, 77, 589

Geopolitics, 595, 655, 656, 660, 668

Governance, 14, 77, 78, 80–83, 88, 97, 107–109, 117, 127–130, 145–154, 188, 253, 301, 303, 346, 347, 403, 470, 535, 557, 568, 625, 627, 629, 630, 632, 644, 707–714, 717, 719, 720, 729, 733, 743, 854

## H

Heritage, 7, 17, 31, 38, 39, 43, 94, 100, 101, 194, 293, 294, 309, 341, 344, 346, 348, 349, 377, 378, 380, 391, 399, 400, 419, 420, 425–429, 440, 452, 467, 472, 473, 479, 480, 482, 486, 489, 491–498, 500–508, 512, 514, 518, 519, 521, 522, 527, 568, 572, 575, 695, 730, 733, 737, 813, 834, 841, 844, 866, 904, 922, 924, 930–932, 936

Historic city, 340, 428, 469, 471, 491, 492, 495, 500, 503, 581, 730, 733

History, 3, 4, 6–8, 11, 13, 17, 26, 27, 29–31, 39–43, 56, 64, 66, 79, 95–97, 99, 103, 109, 131, 138, 148, 158–161, 165, 167, 186, 259, 307, 324, 335, 336, 339, 342–346, 363, 373, 388, 399, 402, 410, 419, 437, 440, 453, 459, 460, 467, 469–474, 478, 481, 483, 484, 491, 494, 496–498, 500, 503, 506, 507, 519, 523, 524, 526–528, 532, 537, 544, 546, 550, 565, 567, 568, 571–575, 580, 581, 610, 611, 657, 663–665, 686, 725, 730, 731, 733, 734, 756, 758, 783, 786, 816–819, 821, 822, 844, 849, 850, 917, 919, 920, 923, 926, 930, 931

## I

Image of the city, 360, 425, 903

Immigration, 86, 396, 397, 399, 435, 436, 440, 657, 660, 664, 666, 780, 795

Incontinent urbanization, 498

Inequality, 6, 13, 16, 20, 26, 31, 32, 37, 47, 48, 54, 55, 98, 108, 109, 113, 115–117, 119, 121, 145–147, 152, 189, 200, 205–208, 210, 211, 213, 218, 220, 221, 224, 226–228, 244, 245, 248, 253, 256, 257, 294, 295, 300, 345, 362, 363, 365, 366, 371, 373, 374, 378, 382–384, 386, 391, 395, 398, 404, 407, 412, 415, 419, 420, 423, 424, 435–437, 439–441, 448, 455, 458, 489, 490, 493, 515, 516, 518, 525, 532, 624, 625, 637, 638, 640–642, 649, 674, 675, 682, 686, 688, 708, 712, 720, 730–732, 741, 747, 757, 767, 770, 804, 805, 807, 808, 810, 849, 850, 853, 860, 891, 897, 907, 926, 927, 929, 932

Inhabiting, 93, 94, 99–103, 336, 341, 374, 378, 382, 414, 491, 558, 839, 893, 922

## L

Latin America, 7, 8, 12, 40–42, 52, 53, 59, 60, 65, 79, 81, 129, 205, 207, 209, 213, 216, 221, 224222, 225, 227, 228, 235, 249250, 268, 273, 336, 342, 364, 379, 381, 386–388, 395–397, 399, 401, 402, 404, 407, 408, 411, 415, 426, 437, 448, 450, 451, 490, 507, 515–517, 521, 531, 533, 545, 556, 558, 565, 568, 625, 639, 640, 645, 646, 648, 657, 682–689, 697, 699, 715, 726, 730,

731, 743, 754, 755, 764, 767, 769, 820, 848, 922, 928, 932, 934
Lithuania, 467, 473–475, 477, 478, 480, 483–485
Local project, 506, 508

## M

Man-made disaster, 263
Medellín, 10, 20, 224, 309, 310, 312, 579, 598, 606, 621, 704, 709, 813–826, 828–830, 832–838, 840, 841
Methodology, 13, 27, 269, 341, 347, 348, 469, 473, 734, 787, 801, 805, 865
Metropolis, 101, 172–174, 177, 179, 180, 183, 184, 205–208, 215, 217, 227, 239, 240, 363, 470, 500, 501, 532, 606, 630, 709, 711, 712, 719, 783, 801, 808–810, 833, 835, 866
Mexico, 9, 35–37, 40, 59–61, 68, 69, 162, 209, 211, 214, 227, 239, 242, 315–321, 323–329, 342, 362, 364, 379, 381–391, 395, 396, 399–402, 404, 406, 407, 410–412, 415, 422, 426, 489, 492, 493, 495–498, 501–503, 505–507, 516, 591–593, 595, 598, 599, 605, 606, 610, 611, 621–623, 626, 627, 644, 648, 656, 657, 681, 683, 684, 697, 699, 739, 760, 796, 920, 921, 928, 931–933, 935
Mobility, 6, 15, 18, 83, 84, 88, 96, 101, 102, 109, 112, 119, 131–133, 135–137, 139, 148, 150, 151, 171–173, 175, 180, 181, 185, 186, 188, 189, 191, 192, 194, 198–200, 205–215, 218–222, 224–228, 268, 308, 363, 365, 366, 384, 385, 387, 388, 395–397, 402, 403, 406, 415, 434–436, 441, 455, 495, 515, 534, 573, 644, 647, 650, 674, 693, 694, 701, 712–714, 717, 822, 831, 870, 899
Modern cities, 11, 34, 43, 63, 84, 138, 293, 335, 343, 470, 471, 497, 518, 567, 568, 701
Monuments, 17, 18, 31, 40, 78, 161, 163, 388, 400, 469, 473–475, 478–483, 494, 495, 498, 501–504, 506, 512–514, 516, 519–524, 526, 528, 533, 567, 609, 610, 613, 619, 730, 852, 921

## N

Natural disaster, 70, 79, 339, 572, 775, 850, 858
Natural hazard, 236, 241, 254, 255, 263, 264, 283–285
Neoliberalism, 16, 108, 117, 129, 133, 377, 382, 384, 387, 498, 515, 537, 547, 607, 737, 747, 851, 860, 891–895, 907, 911

## O

Organized crime, 78, 585, 588–591, 594–596, 598, 599, 624, 645, 838, 853

## P

Pandemic, 10, 14, 30, 32, 33, 36, 48, 49, 51, 54–56, 78, 90, 94, 107–120109, 111–113, 115–121, 133–135, 138–141, 189, 212, 219–221, 224225, 228, 239, 360, 361, 371, 377, 378, 380, 439, 470, 483, 528, 531–533, 540, 541, 543, 545, 547, 562, 568, 576–579, 630, 643, 644, 701, 716, 755, 760, 767, 770, 780, 809, 858, 860, 897, 918, 924, 927, 928, 934, 935
Paradigms, 5, 8, 13, 14, 32, 38, 80, 129, 141, 182, 184, 365, 391, 392, 457, 485, 513, 537, 590, 679, 700, 924, 928
Politics, 6, 25, 26, 42, 63, 78, 81, 90, 146, 150, 151, 346, 380, 452, 492, 506, 528, 569, 617, 657, 666, 675, 695, 699, 705, 708, 717, 730, 741, 742, 757, 797, 801, 803, 823
Population decline, 899
Poverty, 48, 54, 62, 71, 102, 114, 119, 151, 182, 235, 244, 245, 248, 249, 253, 256, 298, 364, 378–380, 382, 383, 390, 455, 485, 490, 505, 516, 535, 537, 538, 550, 558, 562, 573, 576, 624, 641, 643, 659, 664, 684, 698, 702, 732, 747, 838, 848–851, 857–860, 896, 904, 921, 927, 929
Process of urbanization, 7, 8, 26, 301, 302, 455, 459, 489, 751
Protest, 53, 152, 168, 210, 228, 511–514, 516–522, 524, 525, 527, 528, 533, 610, 621, 628, 832, 849, 857, 921
Public policies, 5, 6, 8, 15, 19, 69, 135, 151, 206, 250, 367, 395, 403, 447, 449,

Index 943

451, 455, 459, 535, 547, 548, 607,
608, 610, 613, 614, 626, 629, 682,
688, 693, 694, 700, 701, 704, 717,
720, 733, 808, 849, 930, 933, 934
Public services, 33, 51, 54, 71, 117, 151,
192, 319, 321, 390, 435, 490, 514,
561, 679, 686, 826, 844, 851,
853–857, 860, 892
Public space, 4, 8, 9, 13–18, 31, 32, 37–41,
66, 67, 86, 87, 103, 109, 120, 127,
129–131, 133, 135, 136, 140, 141,
145–147, 149, 152–154, 171, 172,
179, 180, 182, 186, 188, 191, 193,
194, 198, 199, 213–215, 225, 226,
294, 308, 309, 339, 343, 363, 368,
372, 373, 377–381, 383–387,
390–392, 426, 433–435, 441, 447,
448, 450, 453, 456–460, 482, 499,
502, 503, 511–514, 519–528,
531–533, 536, 538–540, 547, 550,
562, 566, 574, 575, 579, 585, 589,
597, 598, 623, 628, 629, 638,
640–642, 644–646, 649, 650, 666,
704, 725, 727, 732, 733, 737, 738,
747, 767, 770, 787, 789, 790, 814,
833, 839, 841, 857, 866, 869, 896,
901, 903, 906, 919, 922, 923, 932

## Q
Quito, 4, 11, 31, 36, 37, 40, 48, 55, 79, 84,
87, 219, 242, 263, 265–269,
271–274, 278, 279, 282–285, 342,
426, 503, 517–520, 565, 570–579,
581, 626, 648, 700, 709, 725, 728,
732–735, 737–739, 743, 744, 747

## R
Reconstruction, 39, 62, 63, 72, 100, 243,
284, 377, 378, 383, 391, 398, 458,
461, 473–483, 506, 558, 606, 610,
620, 629, 756, 757, 775, 895,
917–919, 923–925, 928, 930–932

## S
Sightseeing, 430
Social cohesion, 13, 20, 31, 107–109, 113,
114, 120, 186, 198, 437, 450, 460,
498, 718, 847, 848, 860, 892
Social conflict, 152, 208, 846, 854, 906, 922
Social inequality, 99, 111, 112, 149, 206,
207, 228, 253, 318, 364, 377, 412,
419, 423, 439, 457, 640, 797, 821,
858, 906, 911
Social process, 69, 150, 154, 335, 359, 360,
451, 666, 719
Sustainable development, 47, 48, 206, 211,
236, 457, 485, 702, 712, 713, 732,
846

## T
Territory, 3, 4, 6, 8, 9, 14–20, 26, 31, 32,
39, 40, 65, 71, 78, 79, 81, 82, 94–98,
101, 102, 128, 130, 141, 150, 168,
172, 173, 175, 176, 178, 179,
181–184, 199, 200, 213, 218, 220,
221, 224, 238, 245, 252, 268, 273,
282, 284, 293–296, 299–304,
309–312, 315, 317, 334, 335, 337,
345, 347, 349, 421, 422, 424, 430,
431, 435, 440, 447–460, 467, 470,
472, 483–485, 491–493, 499, 500,
503, 505, 506, 512, 528, 531, 533,
534, 541, 547, 557, 568–571,
585–600, 604–606, 611, 618, 620,
623–627, 630, 631, 642, 656, 657,
663, 664, 668, 674, 678, 682–684,
688, 689, 693, 695, 699, 702, 704,
707, 708, 711, 714, 717–720,
751–755, 757, 758, 760, 762,
764–767, 770, 772, 775, 793–796,
801, 805, 806, 808–810, 816,
835–839, 850, 852, 853, 855, 859,
860, 867, 871, 872, 874, 878–880,
882, 884, 887, 892, 898, 917, 923,
925, 928, 931
Town planning, 317, 326, 934

## U
Urban culture, 15, 95, 140, 172, 186, 620,
813, 815, 817, 839, 840
Urban deterioration, 16, 61
Urban development, 3, 5, 6, 11, 25, 26, 32,
35, 36, 67, 68, 79, 129, 132, 146,
214, 235, 237, 238, 244, 245, 252,
257, 258, 265, 284, 305, 307,
315–317, 321, 328, 329, 378, 380,
381, 383, 386, 422, 429, 431, 437,
450, 501, 502, 514, 531, 534, 537,
557, 559, 606, 607, 663, 693, 700,
702, 704, 708, 732, 764, 784, 817,
845, 857, 869, 870, 874, 875, 879,
892, 895, 901, 911

Urban environment, 15, 55, 61, 63, 72, 93, 94, 101, 103, 108, 111, 112, 114, 121, 172, 199, 263, 282, 308, 346, 365, 380, 390, 489, 561, 585, 603, 604, 606, 615, 617, 620, 627
Urban heritage, 108, 489, 491–493, 498, 500, 502, 503, 507, 508, 537, 919, 921, 932
Urban history, 168, 345, 435, 512, 528
Urban infrastructure, 18, 19, 70, 245, 249, 347, 369, 449, 457, 497, 647, 674, 675, 679, 684, 686, 688, 690, 728, 738, 766
Urbanization, 3, 10, 12, 15, 16, 18, 19, 26, 29, 30, 41, 42, 61, 78, 81, 83, 85–88, 90, 94, 95, 100–102, 179, 183, 184, 199, 206, 293–300, 302, 304, 305, 307, 309–311, 336, 351, 360, 361, 363, 364, 366, 367, 369, 370, 373, 374, 380, 382, 390, 419, 421–423, 425, 433, 447–450, 452, 454, 456–458, 460, 489, 491, 492, 494, 534, 541, 542, 544–546, 566, 607, 609, 638, 639, 642, 643, 645, 649, 673–676, 680, 682, 683, 686, 688–690, 693, 694, 698, 701, 703, 704, 751–756, 762, 764–766, 769, 772, 775, 839, 869, 884, 891, 892, 896, 897, 903, 906, 911, 918, 922, 924–926, 929, 930
Urban legislation, 500, 764
Urban living, 55, 851
Urban mobility, 112, 131, 171, 172, 185, 188, 199, 200, 211, 663, 686, 701, 805
Urban norms, 878
Urban planning, 17–19, 36, 37, 61, 62, 66, 67, 80, 85, 90, 108, 109, 111, 113, 120, 132–135, 137–139, 141, 151, 171, 180, 193, 194, 198, 199, 235, 245, 253, 257, 259, 269, 284, 296, 317, 318, 320, 321, 325, 333, 334, 336, 337, 343, 344, 347–349, 352, 363, 382–384, 391, 392, 422, 447, 468, 471, 489, 492, 500–503, 507, 514, 524, 544, 556, 557, 573, 589, 637, 642, 645, 650, 693, 694, 696, 697, 700, 701, 710, 718, 751, 765, 779, 782, 786, 813, 829, 868, 882, 893, 895, 899, 901, 903, 905, 906, 917, 927, 934
Urban policies, 3, 19, 27, 28, 30–32, 49, 52, 60, 62, 66, 127, 141, 189, 207, 214, 228, 367, 372, 382–385, 423, 426, 427, 448, 468, 500, 506, 513, 531, 535–538, 562, 575, 604, 606–608, 615, 637, 639, 645, 647, 649, 662, 680, 689, 693, 694, 697, 699, 700, 720, 725, 729, 731, 734, 743, 851, 852, 855, 899, 901, 934
Urban rehabilitation, 422, 574
Urban renewal, 61, 63, 72, 84, 102, 337, 338, 482, 534, 607, 609, 613, 614, 627, 688, 870, 872, 879, 887, 899, 918, 920
Urban ruins, 764, 918, 925
Urban sociology, 89, 661
Urban space, 16, 31, 93, 112, 130, 132, 133, 135, 137, 140, 146–154, 176, 185, 200, 214, 226, 237, 238, 253, 258, 294, 296, 298, 301, 309, 311, 337, 338, 340, 351, 364, 365, 367, 370, 372, 374, 379, 383, 384, 395–398, 402, 407–410, 412, 420, 421, 423–425, 433, 436, 485, 490, 492, 498, 508, 528, 558, 585, 586, 590, 594, 600, 603–606, 612, 626, 628, 632, 637, 639, 641, 642, 645, 647, 657, 664, 665, 674, 675, 679, 718, 743, 794, 839, 865–869, 884, 887, 891, 892, 898, 909, 933
Urban structure, 9, 15, 29, 32, 34, 207, 211, 238, 350, 384, 385, 436, 439, 468, 629, 865, 866, 926
Urban studies, 3, 8, 32, 49, 359, 360, 363, 367, 448, 450, 708, 720
Urban theory, 14, 363, 364, 891, 892
Urban transformation, 18, 49, 61, 99, 101, 107, 108, 120, 121, 207, 226, 228, 298, 360, 363, 408, 426, 436, 511, 513, 566, 687, 747, 814, 845
Urbicide, 3, 5–14, 19, 20, 25, 27–39, 41–43, 59–66, 71, 72, 77–80, 82, 84, 87, 89, 90, 93–96, 98–103, 107–112, 114, 116–118, 121, 122, 127, 157, 165, 205, 207, 210, 213, 218, 227, 238–240, 258, 260, 263–265, 285, 293–295, 300–302, 307, 308, 311, 315–317, 329, 333, 334, 336, 337, 340, 351, 359, 360, 362, 363, 373, 374, 377–380, 392, 397, 399, 412, 447–450, 452–456, 458–461, 491, 500, 507, 508, 511, 513, 524, 527, 531–534, 546, 550, 551, 555, 559, 563, 565–570, 572, 573, 575, 576, 580, 581, 585, 603, 604, 606, 607,

Index
945

609–624, 627, 628, 630, 632, 639,
655, 673, 675, 687, 690, 693, 705,
707, 708, 720, 725, 728, 747, 814,
815, 843, 844, 846–848, 850–853,
858–860, 865–868, 884, 887,
891–893, 896, 905, 906, 908, 910,
911, 919, 920, 922, 924, 925, 935

**V**
Vilnius, 473–476, 478–482
Violence, 4, 5, 7, 10, 14, 18, 20, 26, 28, 29,
31, 32, 36, 48, 59–66, 71, 72, 81,
103, 114, 145, 221, 227, 242, 263,
264, 284, 334, 336–338, 366, 372,
377, 378, 380, 385, 389, 390, 392,
413, 414, 447–451, 453, 454, 456,
458–461, 511, 513, 518–520, 525,
526, 533, 536, 548, 551, 558, 561,
569, 578, 585, 586, 589–591, 593,
595, 600, 603, 604, 607–618,
620–632, 637–650, 655, 657, 663,
665–667, 698, 727, 728, 731–733,
739–743, 804–806, 813, 814, 817,
818, 833, 844, 848, 851, 853, 854,
856, 858, 860, 867, 868, 871, 903,
905, 906, 909, 911, 922

Printed in the United States
by Baker & Taylor Publisher Services